# Lecture Notes in Computer Science 4485

*Commenced Publication in 1973*
Founding and Former Series Editors:
Gerhard Goos, Juris Hartmanis, and Jan van Leeuwen

Fiorella Sgallari Almerico Murli
Nikos Paragios (Eds.)

# Scale Space and Variational Methods in Computer Vision

First International Conference, SSVM 2007
Ischia, Italy, May 30 - June 2, 2007
Proceedings

Springer

Volume Editors

Fiorella Sgallari
University of Bologna, Department of Mathematics - CIRAM
via Saragozza, 8, 40123 Bologna, Italy
E-mail: sgallari@dm.unibo.it

Almerico Murli
University of Naples Federico II, Department of Mathematics and Applications
Complesso Universitario Monte Sant'Angelo, Via Cintia, 80126 Naples, Italy
E-mail: almerico.murli@dma.unina.it

Nikos Paragios
MAS, Ecole Centrale Paris
Grande Voie des Vignes, 92295 Chatenay-Malabry, France
E-mail: nikos.paragios@ecp.fr

Library of Congress Control Number: 2007927099

CR Subject Classification (1998): I.4, I.5, I.3.5, I.2.10, I.2.6, G.1.2, F.2.2

LNCS Sublibrary: SL 6 – Image Processing, Computer Vision, Pattern Recognition, and Graphics

ISSN 0302-9743
ISBN-10 3-540-72822-8 Springer Berlin Heidelberg New York
ISBN-13 978-3-540-72822-1 Springer Berlin Heidelberg New York



Springer is a part of Springer Science+Business Media

springer.com


Printed in Germany

Typesetting: Camera-ready by author, data conversion by Scientific Publishing Services, Chennai, India
Printed on acid-free paper SPIN: 12070893 06/3180 5 4 3 2 1 0

# Preface

Image processing, computational vision, robot and machine vision are terms that refer to automatic visual perception through intelligent processing of image content. Such a demand requires the development of appropriate mathematical models which reformulate the answer to the perception problem as the lowest potential of a specifically designed objective function. The development of such models capable of reproducing human vision is a long-shot objective in the domain. Variational methods are a very popular selection for addressing a number of components of visual perception, while scale space methods introduce the notion of hierarchical representation of image content or property often present in biological autonomous perception organisms. This development has been made possible by two factors: first, the advent of computers powerful enough to cope with the large dimensionality of the image data and the complexity of the algorithms that operate on them (Teraflop, Terabyte); second, the availability of new models, methods and algorithms, thanks to many excellent mathematicians, computing scientists and engineers from all over the world.

The 1st International Conference on Scale Space and Variational Methods in Computer Vision (SSVM 2007) was an attempt to bring together two different communities with adjacent research interests, the one of scale-space analysis and the one of variational, geometric and level sets (VLSM). This conference was the joint edition of the 4th VLSM and 6th Scale Space with the aim of bringing together various disciplines working in the area of visual perception (mathematicians, physicists, computing scientists, etc.). It gathered the attention of an important international scientific crowd with submissions and presentations from approximately 26 countries (Austria, Australia, Belgium, Canada, Switzerland, China, Germany, Denmark, Spain, France, Greece, Honk Kong, Israel, India, Ireland, Italy, Japan, Korea, Mexico, The Netherlands, Norway, Poland, Sweden, Turkey, England, USA) from leading scientists in the field

We received 133 high-quality full-paper double-blind submissions. Each paper was reviewed by at least three members of the Program Committee. These reviews were considered from the Area Chairs, who finally proposed 79 to be accepted. We selected 24 manuscripts for oral presentation and 55 for poster presentation. Both oral and poster papers were attributed the same length of pages in the conference proceedings.

Furthermore, we invited keynote speakers who could provide valuable additional inspiration beyond the mainstream topics in scale-space analysis and variational methods. It was our pleasure to welcome Franco Brezzi of the University of Pavia, Institute for Advanced Study and IMATI-CNR (Italy), Emmanuel Candes of California Institute of Technology (USA), and Peter Schröder of California Institute of Technology (USA), as keynote speakers.

We would like to thank the authors for their contributions and the members of the Program Committee for their time and valuable comments during the review process. We would also like to acknowledge the support of Christian Trocchi for his help with the Website and Daniela Casaburi and Livia Marcellino for their help with the organization. Last but not least, special thanks to Francesca Incensi for handling the submission/review/decisions and proceedings aspects of the conference. Finally, we are grateful to the University of Bologna, the University of Naples Federico II, GNCS-INDAM, CINECA Bologna and CIRAM (Research Centre of Applied Mathematics) Bologna for their sponsorship.

It is our belief that this conference will become a reference in the domain, and will contribute on the development of new ideas in the area of visual perception through processing images with mathematical models.

May-June 2007

Fiorella Sgallari
Almerico Murli
Nikos Paragios

# Organization

## General Co-chairs and Organizers

| | |
|---|---|
| Fiorella Sgallari | (University of Bologna, Italy) |
| Almerico Murli | (University of Naples Federico II, Italy) |
| Nikos Paragios | (MAS, Ecole Centrale de Paris, France) |

## Conference Chairs

| | |
|---|---|
| Alfred Bruckstein | (Technion IIT, Israel) |
| Bart ter Haar Romeny | (Eindhoven University of Technology, The Netherlands) |
| Guillermo Sapiro | (University of Minnesota, Minneapolis, MN, USA) |
| Joachim Weickert | (Saarland University, Germany) |

## Program Committee

Luis Alvarez
Jonas August
Benedicte Bascle
Bernhard Burgeth
Vicent Caselles
Tony F. Chan
Yunmei Chen
Laurent Cohen
Daniel Cremers
Françoise Dibos
Remco Duits
Maurizio Falcone
Michael Felsberg
Luc Florack
Nicola Fusco
Lewis D. Griffin
Anders Heyden
Atsushi Imiya
Marie-Pierre Jolly
Renaud Keriven
Ron Kimmel
Arjan Kuijper
Petros Maragos
Étienne Mémin
Fernand Meyer
Karol Mikula
Farzin Mokhtarian
Mads Nielsen
Mila Nikolova
Ole Fogh Olsen
Stanley Osher
Emmanuel Prados
Martin Rumpf
Otmar Scherzer
Christoph Schnörr
Stefano Soatto
Nir Sochen
Xue-Cheng Tai
Hugues Talbot
Demetri Terzopoulos
Jean-Philippe Thiran
David Tschumperlé
Michael Unser
Baba C. Vemuri
Martin Welk
James Williams
Anthony Yezzi
Hong-Kai Zhao
Lilla Zöllei
Steven W. Zucker

## Other Reviewers

Luc Brun
Andrés Bruhn
Lorina Dascal
Stephan Didas
Irena Galic
Jalal Fadili
Gabriel Peyré
Luis Pizarro
Christian Schmaltz

## Invited Speakers

| | |
|---|---|
| Franco Brezzi | (University of Pavia, Italy) |
| Emmanuel Candes | (California Institute of Technology, USA) |
| Peter Schröder | (California Institute of Technology, USA) |

## Sponsoring Institutions

University of Bologna, Italy
University of Naples Federico II, Italy
GNCS-INDAM, Italy
CINECA, Bologna, Italy
CIRAM-Research Centre in Applied Mathematics, Bologna, Italy

# Table of Contents

## Oral Presentations

### 1. Scale Space and Features Extraction



### 2. Image Enhancement and Reconstruction

## 3. Image Segmentation and Visual Grouping



## 4. Motion Analysis, Optical Flow, Registration and Tracking

## 5. 3D from Images



# Poster Presentations

## 1. Scale Space and Feature Extraction

## 2. Image Enhancement, Reconstruction and Texture Synthesis

## 3. Image Segmentation and Visual Grouping

## 4. Motion Analysis, Optical Flow, Registration and Tracking

## 5. 3D from Images



## 6. Biological Relevance

# Full Affine Wavelets Are Scale-Space with a Twist

Yossi Ferdman[1], Chen Sagiv[2], and Nir Sochen[1]

[1] Department of Applied Mathematics
University of Tel Aviv
Ramat-Aviv, Tel-Aviv 69978, Israel
yosseffe@post.tau.ac.il, sochen@post.tau.ac.il
[2] Eliezer Yaffe 37
Ra'anana, 43451, Israel
ChenSagivRon@gmail.com

**Abstract.** In this work we study the relation between the Gabor-Morlet wavelet transform and scale-space theory. It is shown that the usual wavelet transform is a projection of scale-space on a specific frequency component. This result is then generalized to the full two-dimensional affine group. A close relation between this generalized wavelet transform and a family of scale-spaces of images that are related by $SL(2)$ is established. Using frame theory we show that sampling from different images in this family, and from different scales enables a complete reconstruction of the image.

## 1 Introduction

Images, as other signals, contain information at different scales. This is a well-known attribute that is expressed in various domains of research such as scale-space [5,15,10], multi-grid [4], wavelets [2,12] and more. This study revolves around the relation between scale-space and wavelets. This relation was studied before [9] on the numerical level, and the connection between the Haar wavelet shrinkage and inhomogeneous diffusion was established.

In this paper we take a different viewpoint and discuss the relation of the Morlet-Gabor wavelet transform to the linear scale-space. Moreover, the relation between a generalized Gabor-Morlet wavelet and a scale-space of images is established. We show that the Affine-Gabor-Morlet wavelet is related to a *family* of scale-spaces. These scale spaces are obtained by transforming the original image by an affine transform, and then applying a (projected) scale-space on the transformed image. The projection is done for one frequency component of the scale-space.

Then, using frame analysis, we show that an image can be reconstructed by a *discrete* combination of scale-space coefficients at different scales and for different affine transformations. We generalize to two dimensions the frame criterion for one-dimensional wavelets of [2] and use this generalization to assess the tightness of the frames obtained using our scale-orientation-shear and frequency analysis.

F. Sgallari, A. Murli, and N. Paragios (Eds.): SSVM 2007, LNCS 4485, pp. 1–12, 2007.

The rest of this paper is organized as follows: In §2 we provide the background for this study, present previous work, and fix the notations .We introduce the continuous wavelet transform and show the equivalence to linear scale-space. Then in §3 and §4 we generalize the Gabor-Morlet transform to the full two-dimensional affine group and show the equivalence of this generalized transform to a family of scale-spaces that are related by the $SL(2)$ group. In §5 the discretization of these transforms via sampling of the affine group is discussed . In §6 we present a formula for reconstruction of the image in terms of its corresponding scale-spaces. In §7 we present some examples of filter banks and calculate few frame bounds using Daubechies's frame criterion to the full Affine group in two-dimensions. We conclude by providing preliminary reconstruction results.

## 2 Multi-scale Analysis

Two of the common methods in multi-scale analysis are the linear diffusion filtering a.k.a Scale-space, and wavelet transform. Here we recall the basic ideas of linear diffusion filtering (see [9] and references therein) and the wavelet transform [12,2]. We show the equivalence between these methods by adding a constraint to the solution of the diffusion equation and express it in terms of the wavelet transform.

### 2.1 Linear Diffusion Filtering

Our model for gray level images is of a continuous bounded function $f : \mathbb{R}^2 \rightarrow \mathbb{R}$. A common approach for representing an image $f$ at different scales is to regard the image as the initial value of the homogeneous linear diffusion equation:

$$\begin{aligned} \partial_t u &= \triangle u, \\ u(\boldsymbol{x}, 0) &= f(\boldsymbol{x}). \end{aligned} \tag{1}$$

Then, scaled versions of the original image are given in terms of the solution of this equation for $t > 0$:

$$u(\boldsymbol{x}, t) = (K_{\sqrt{2t}} * f)(\boldsymbol{x}) = \int_{\mathbb{R}^2} f(\boldsymbol{y}) K_{\sqrt{2t}}(\boldsymbol{x} - \boldsymbol{y}) d\boldsymbol{y}, \tag{2}$$

where $K_\sigma$ denotes a 2D Gaussian with standard deviation $\sigma$:

$$K_\sigma(x_1, x_2) \doteq \frac{1}{2\pi\sigma^2} e^{-\frac{x_1^2 + x_2^2}{2\sigma^2}}. \tag{3}$$

Suppose that we are interested only in the part of the diffused image $u(x, t)$ with frequency components that are close to some fixed spatial frequency $\boldsymbol{\kappa_0}$, for each value of $t$. Thus, we look at the image at a certain scale and a certain frequency. Then, the integral kernel becomes a two-dimensional Gabor function :

$$\mathcal{G}_\sigma(\boldsymbol{x}) = K_\sigma(\boldsymbol{x}) e^{i\boldsymbol{\kappa_0} \cdot \frac{\boldsymbol{x}}{\sigma}} \tag{4}$$

and therefore the resultant image ,which we denote by $\mathcal{U}(x,t)$ is,

$$\mathcal{U}(\boldsymbol{x},t) =(\mathcal{G}_{\sqrt{2t}} * f)(\boldsymbol{x}) = \int_{\mathbb{R}^2} f(\boldsymbol{y})\mathcal{G}_{\sqrt{2t}}(\boldsymbol{x}-\boldsymbol{y})d^2\boldsymbol{y} = \tag{5}$$
$$= \int_{\mathbb{R}^2} f(\boldsymbol{y})\overline{\mathcal{G}_{\sqrt{2t}}(\boldsymbol{y}-\boldsymbol{x})}d^2\boldsymbol{y}$$

### 2.2 The Continuous Wavelet Transform

The one-dimensional wavelet transform is a useful tool for analyzing signals. It enables the extraction of local information about the signal $f$.
The continuous wavelet transform is explicitly given by

$$\mathcal{W}_{\psi}f(a,b) = |a|^{\frac{1}{2}} \int_{\mathbb{R}} f(t)\overline{\psi\Big(\frac{t-b}{a}\Big)}dt, \tag{6}$$

where at each location $b$ we view details of the signal in different resolutions that are determined by the scale parameter $a$. An important property of the wavelet transform is the time-frequency localization that is not an attribute of the usual Fourier transform. The function $\psi$ is called the *analyzing wavelet* or *mother wavelet.*

In our analysis, we choose the mother wavelet $\psi_{Gab} \in L^2(\mathbb{R}^2)$ to be the two-dimensional Gabor wavelet [7,8,13] that is widely used in computer vision for its scale and orientation sensitivity

$$\psi_{Gab}(\boldsymbol{x}) = \frac{1}{\sqrt{2\pi}}e^{-\frac{\|\boldsymbol{x}\|^2}{2}}e^{i\boldsymbol{\kappa_0}\cdot\boldsymbol{x}} \tag{7}$$

where $x$ is the spatial vector and $\kappa_0$ denotes the wave vector.
Using our selection for the mother wavelet, the continuous wavelet transform is given by:

$$\mathcal{W}_{\psi_{Gab}}(a,b) = \langle f, \psi^{a,b}_{Gab}\rangle = \int_{\mathbb{R}^2} f(\boldsymbol{y})\overline{\frac{1}{a}\psi_{Gab}\Big(\frac{\boldsymbol{y}-\boldsymbol{b}}{a}\Big)}d^2\boldsymbol{y} \tag{8}$$

Now we can establish the connection between linear diffusion filtering as in §2.1 and the Wavelet transform :

$$\mathcal{U}(\boldsymbol{x},t) = \frac{1}{\sqrt{2\pi}a}\mathcal{W}_{\psi_{Gab}}(a,\boldsymbol{b}), \quad a=\sqrt{2t}, \boldsymbol{b}=\boldsymbol{x}, \tag{9}$$

where the factor $\frac{1}{2\pi a}$ appears due to different normalization requirements : $\|\psi^{a,b}_{Gab}\|_2 = 1$ in usual wavelet analysis whereas $\|\mathcal{G}_\sigma\|_1 = 1$ in the usual scale-space formalism. In the one-dimensional case more can be said about the relation between wavelet-based processing and differential equations [1].

The wavelets $\psi^{a,b}_{Gab}(\boldsymbol{x}) = \frac{1}{a}\psi_{Gab}\Big(\frac{\boldsymbol{x}-\boldsymbol{b}}{a}\Big)$ are associated with a group in $\mathbb{R}^2$: $G_{aff} = \mathbb{R}^2 \ltimes \mathbb{R}_+$ with group law

$$(a,\boldsymbol{b}) \circ (a',\boldsymbol{b}') = (aa', \boldsymbol{b}+a\boldsymbol{b}') \tag{10}$$

which is a two-dimensional generalization of the one-dimensional Affine group, the "$ax + b$" group. From computer vision point of view this wavelet transform is very poor since it gives no attention to the directionality in two dimensions.

One possible extension to two dimensions considers the $SIM(2) = \mathbb{R}^2 \ltimes (\mathbb{R}_*^+ \times SO(2)))$ group [7]. In this paper we explore a more general approach that seems like a natural extension of the 1D case: The full two dimensional Affine group.

## 3 The 2D Affine Group and the Continuous Wavelet Transform

Before going into the details of wavelet theory and the associated generalized scale-space, we want to mention a matrix decomposition that is similar to the Iwasawa decomposition [6].

### 3.1 Matrix Decomposition

A general, non-singular, two-by-two matrix with real coefficients (G), can be represented uniquely as a product of three two-by-two matrices ,$G = OUD$ where $O$ is an orthogonal matrix, $U$ is an upper triangular matrix with 1's on the diagonal and $D$ is a diagonal matrix.

$$G = \begin{bmatrix} a & b \\ c & d \end{bmatrix} = \begin{bmatrix} \cos\theta & -\sin\theta \\ \sin\theta & \cos\theta \end{bmatrix} \begin{bmatrix} 1 & s \\ 0 & 1 \end{bmatrix} \begin{bmatrix} a_1 & 0 \\ 0 & a_2 \end{bmatrix} \tag{11}$$

where $\theta = \angle(a, c)$, $s = \frac{\langle (a,c),(b,d) \rangle}{\det G}$, $a_1 = \|(a, c)\|$ and $a_2 = \frac{\det G}{a_1}$.
Thus, any $G \in GL(2, R)$ can be represented using four parameters: rotation ($\theta \in [0, 2\pi]$), shear ($s \in \mathbb{R}$) and dilation in x and y ($a_1, a_2 \in \mathbb{R}$).

Let $G \in GL(2, \mathbb{R})$. We denote its inverse by $H \in GL(2, \mathbb{R})$. Therefore, if we write $H = OUD$ then $G = D^{-1}U^{-1}O^{-1}$.
Therefore, every $G \in GL(2, \mathbb{R})$ can be also decomposed in the following way: $G = DUO$ (but obviously not with the same parameters used for $H$).

### 3.2 Representation of the Full Affine Group

Let $\mathcal{A}$ be the affine group ,

$$\mathcal{A} := \{(G, \boldsymbol{b}) | G \in GL(\mathbb{R}^2), \boldsymbol{b} \in \mathbb{R}^2\} \tag{12}$$

with group law

$$(G, \boldsymbol{b}) \circ (G', \boldsymbol{b}') = (GG', \boldsymbol{b} + G\boldsymbol{b}') \tag{13}$$

A unitary representation of this group can be obtained by the action of $\mathcal{A}$ on $\psi(x)$

$$[\Omega(G, \boldsymbol{b})\psi](\boldsymbol{x}) = |\det G|^{-\frac{1}{2}} \psi\Big(G^{-1}(\boldsymbol{x} - \boldsymbol{b})\Big). \tag{14}$$

Using the decomposition introduced in §3.1 we may write $G = DUO$, and the representation of the Affine group can be viewed as a combination of four operators in $L^2(\mathbb{R}^2, d^2\boldsymbol{x})$:

$$\text{Translation operator}: \quad (T^{\boldsymbol{b}} f)(\boldsymbol{x}) = f(\boldsymbol{x} - \boldsymbol{b})$$

$$\text{Dilation operator}: \quad (D^{\boldsymbol{a}} f)(\boldsymbol{x}) = \frac{1}{|a_1 a_2|^{\frac{1}{2}}} f\Big(\frac{x_1}{a_1}, \frac{x_2}{a_2}\Big)$$

$$\text{Shear operator}: \quad (S^s f)(\boldsymbol{x}) = f(x_1 - s x_2, x_2)$$

$$\text{Rotation operator}: \quad (R^\theta f)(\boldsymbol{x}) = f(x_1 \cos\theta + x_2 \sin\theta, -x_1 \sin\theta + x_2 \cos\theta).$$

and is given by

$$\Omega(G, \boldsymbol{b}) = \Omega(\theta, s, \boldsymbol{a}, \boldsymbol{b}) = T^{\boldsymbol{b}} D^{\boldsymbol{a}} S^s R^\theta. \tag{15}$$

### 3.3 The Continuous Wavelet Transform Revisited

We may use the notation introduced in the previous section to reformulate the continuous wavelet transform.

The coefficients generated by $\langle f, \Omega(G,t)\psi \rangle$ provide the continuous wavelet transform of a function $f$, where $\psi$ is the mother wavelet.
The continuous wavelet transform can now be reformulated as follows:

$$\big(\mathcal{W}_\psi f\big)(G, \boldsymbol{b}) = |\det G|^{-\frac{1}{2}} \int_{\mathbb{R}^2} f(x) \overline{\psi\Big(G^{-1}(\boldsymbol{x} - \boldsymbol{b})\Big)} d\boldsymbol{x}. \tag{16}$$

We use the decomposition introduced in the previous section, and write $G = DUO$ to obtain a parameterized continuous Affine system as follows:

$$\Big\{\psi_{\theta s a b} = T^{\boldsymbol{b}} D^{\boldsymbol{a}} S^s R^\theta \psi(\boldsymbol{x}) \ : \ \theta \in [0, 2\pi], a \in \mathbb{R}^2_+, s \in \mathbb{R}, \boldsymbol{b} \in \mathbb{R}^2\Big\} \tag{17}$$

This continuous affine system can be described by specifying the action of these operators on $\psi$ in the frequency domain:

$$\begin{aligned}
\hat{\psi}(\boldsymbol{\xi}) &= (\mathcal{F}\psi)(\boldsymbol{\xi}) = \int_{\mathbb{R}^2} \psi(\boldsymbol{x}) e^{-i\boldsymbol{\xi}\cdot\boldsymbol{x}} d^2\boldsymbol{x} \\
(\widehat{T}^{\boldsymbol{b}}\psi)(\boldsymbol{\xi}) &= (\mathcal{F}T^{\boldsymbol{b}}\psi)(\boldsymbol{\xi}) = e^{-i\boldsymbol{\xi}\cdot\boldsymbol{b}}\hat{\psi}(\boldsymbol{\xi}) \\
(\widehat{D}^{\boldsymbol{a}}\psi)(\boldsymbol{\xi}) &= (\mathcal{F}D^{\boldsymbol{a}}\psi)(\boldsymbol{\xi}) = (D^{\frac{1}{\boldsymbol{a}}}\hat{\psi})(\boldsymbol{\xi}) \\
(\widehat{S}^s\psi)(\boldsymbol{\xi}) &= (\mathcal{F}S^s\psi)(\boldsymbol{\xi}) = \hat{\psi}(\xi_1, s\xi_1 + \xi_2) \\
(\widehat{R}^\theta\psi)(\boldsymbol{\xi}) &= (\mathcal{F}R^\theta\psi)(\boldsymbol{\xi}) = (R^\theta\hat{\psi})(\boldsymbol{\xi})
\end{aligned}$$

$$\hat{\psi}_{\theta s a b}(\xi) = (\widehat{D}^{\boldsymbol{a}}\widehat{S}^s\widehat{R}^\theta\widehat{T}^{\boldsymbol{b}}\psi)(\boldsymbol{\xi}) \tag{18}$$

We can now write the parameterized **continuous Wavelet transform** of $f \in L^2(\mathbb{R})$ as

$$\big(\mathcal{W}_\psi f\big)(\theta, s, a, b) = \langle f, \psi_{\theta s a t}\rangle, \theta \in [0, 2\pi], \boldsymbol{a} \in \mathbb{R}^2_+, s \in \mathbb{R}, \boldsymbol{b} \in \mathbb{R}^2. \tag{19}$$

# 4 From the Continuous Wavelet Transform to Linear Diffusion Formalism

In this section we present a one-to-one relationship between wavelet transformed versions of an image and some twisted linear diffusion filtering that accounts for scale (naturally), orientation, shear and frequency.

In [11], [7] the following mother wavelet is considered

$$\psi(x_1,x_2) = \frac{1}{\sqrt{2\pi}} e^{-\frac{1}{8}(4x_1^2+x_2^2)} \left[ e^{i\kappa x_1} - e^{-\frac{\kappa^2}{2}} \right] \tag{20}$$

where the term $e^{-\frac{\kappa^2}{2}}$ can be neglected for $\kappa \geq 2.5$. In the examples we show in the results section we used $\kappa = 2.5$. Our mother wavelet, therefore, is the following two-dimensional Morlet wavelet that can be written as a two dimensional Gabor function (7):

$$\psi_{Mor}(\boldsymbol{x}) = \psi_{Gab}\big(A\boldsymbol{x}\big) = \frac{1}{\sqrt{2\pi}} e^{-\frac{1}{8}(4x_1^2+x_2^2)} e^{i\kappa x_1}, \tag{21}$$

where $\boldsymbol{\kappa_0} = (\kappa, 0)$ and $A = \mathrm{diag}\{1, \frac{1}{2}\}$ .
Assuming that $\det G > 0$ we may write the following equality:

$$\begin{aligned} \psi_{Mor}(G^{-1}\boldsymbol{x}) =& \psi_{Gab}\big(AG^{-1}\boldsymbol{x}\big) = \psi_{Gab}\big(\sqrt{\det G}AG^{-1}\frac{\boldsymbol{x}}{\sqrt{\det G}}\big) \\ &= (\det G)\sqrt{2\pi}\ \mathcal{G}_{\sqrt{2t}}(\tilde{G}^{-1}\boldsymbol{x}) \end{aligned}$$

where $\tilde{G} = \frac{1}{\sqrt{\det G}} G A^{-1}$, $\sqrt{2t} = \sqrt{\det G}$ and $\mathcal{G}$ was defined in (4).

We use this result to show that:

$$\begin{aligned} \big(\mathcal{W}_{\psi_{Mor}} f\big)(G,\boldsymbol{b}) &= |\det G|^{-\frac{1}{2}} \int_{\mathbb{R}^2} f(\boldsymbol{y}) \overline{\psi_{Mor}\Big(G^{-1}(\boldsymbol{y}-\boldsymbol{b})\Big)} d^2\boldsymbol{y} = \\ &= \sqrt{2\pi} |\det G|^{\frac{1}{2}} \int_{\mathbb{R}^2} f(\boldsymbol{y}) \overline{\mathcal{G}_{\sqrt{2t}}(\tilde{G}^{-1}(\boldsymbol{y}-\boldsymbol{b}))} d^2\boldsymbol{y} \\ &= \frac{\sqrt{2\pi}}{2} |\det G|^{\frac{1}{2}} \int_{\mathbb{R}^2} f(\tilde{G}\tilde{\boldsymbol{y}}) \overline{\mathcal{G}_{\sqrt{2t}}(\tilde{\boldsymbol{y}}-\boldsymbol{x})} d^2\tilde{\boldsymbol{y}} \\ &(\text{ by change of variables: } \tilde{\boldsymbol{y}} = \tilde{G}^{-1}\boldsymbol{y} \text{ and replacing } \boldsymbol{x} = \tilde{G}^{-1}\boldsymbol{b}) \\ &= \frac{\sqrt{2\pi}}{2} \sqrt{2t}\,\mathcal{U}(\boldsymbol{x},t) = \sqrt{\pi t}\ \mathcal{U}(\boldsymbol{x},t). \end{aligned} \tag{22}$$

Now, $\mathcal{U}(\boldsymbol{x},t) = (\mathcal{G}_{\sqrt{2t}} * f_{\tilde{G}})(\boldsymbol{x})$ is actually the solution of the following diffusion equation:

$$\partial_t u = \triangle u, \tag{23}$$

$$u(\boldsymbol{x},0) = f(\tilde{G}\boldsymbol{x}), \tag{24}$$

with the constraint that this solution is projected on the vicinity of the frequency $\kappa$ in the $x_1$ direction. We should note that the initial condition must satisfy: $\tilde{G}A \in SL(2)$

So, we have shown here that the wavelet decomposition of an image can be considered as the solution of the diffusion equation for certain scale, orientation, shear and frequency. An obvious question that arises now is whether we may use this decomposition to obtain an adequate reconstruction of our image from these projections. In the next section we show that by an appropriate discretization of the matrix $\tilde{G}$ and the scales $t$ we obtain a rather simple approximation formula for this reconstruction. As the family of Gabor-Morlet wavelets does not constitute an orthogonal basis, we consider the ability to provide an almost tight frame using this decomposition. In the next sections we offer a discretization of the wavelet transform and provide the constraints for a tight frame reconstruction.

As we have demonstrated the connection between wavelet analysis and linear diffusion formalism implies that, reconstructing an image using wavelets is equivalent to reconstructing using the scale-space information.

## 5 Discrete Wavelets-Synthesis

In this section we provide a sampling of the continuous transform using an appropriate discrete set of rotation, shear, scaling and translation parameters $(\theta, s, a, t) \in [0, 2\pi] \times \mathbb{R} \times \mathbb{R}^2_+ \times \mathbb{R}^2$, to obtain a frame for $L^2(\mathbb{R}^2)$.
As a first step an arbitrary set of rotations $\{\theta_l\}_{l\in\mathbb{Q}} = \{l\theta_0\}_{l\in\mathbb{Q}} \subset [0, 2\pi]$ and scales $\{\boldsymbol{a}_j\}_{j\in\mathbb{Z}} = \{(a^j_{0_1}, a^j_{0_2})\}_{j\in\mathbb{Z}} \subset \mathbb{R}^2_+, a_0 > 1$ can be chosen. Next, the shear parameters $\{s_{jk}\}_{k\in\mathbb{Z}} = \{s_{0_j}k\}_{k\in\mathbb{Z}} \subset \mathbb{R}$ depend on $j$, so the directionality of the representation is allowed to change with scale. In this way we ensure the self-similarity in each scale. Finally, to obtain a "uniform covering", we set the translation parameter $\boldsymbol{b}_{jm} = \boldsymbol{a}_j \times \boldsymbol{b}_0 \times \boldsymbol{m}, \boldsymbol{m} \in \mathbb{Z}^2, \boldsymbol{b}_0 = const$ in $\mathbb{R}^2_+$. This leads to the discrete set

$$\{\psi_{ljkm} = T^b_{jm} D^{\boldsymbol{a}}_j S^s_{jk} R^\theta_l \psi(\boldsymbol{x}) \quad : \; l, j, k \in \mathbb{Z}, m \in \mathbb{Z}^2\}. \tag{25}$$

The Fourier transform of the two dimensional wavelet is given by,

$$\hat{\psi}_{lkjm}(\xi) = (\widehat{D}^{\boldsymbol{a}}_j \widehat{S}^s_{jk} \widehat{R}^\theta_l \widehat{T}^{\boldsymbol{b}}_{jm} \psi)(\boldsymbol{\xi}) \tag{26}$$

where we choose $s_{0_j} = (\frac{a_1}{a_2})^j$.
The Fourier transform of the Gabor-Morlet wavelet we chose in section §4 is:

$$\hat{\psi}_{Mor}(\xi_1, \xi_2) = \sqrt{8\pi} \left\{ e^{-\frac{1}{2}\left[(\xi_1 - \kappa)^2 + 4\xi_2^2\right]} \right\} \tag{27}$$

In Fig.1 we show different filters, that are generated by the action of different group elements, on mother wavelet in the frequency domain.
The discrete wavelet transform is:

$$(\mathcal{W}_{lkjm} f) = \iint f(\boldsymbol{x}) (\overline{S^s R^\theta \psi})(a^{-j}_{0_1} x_1 - m_1 b_{0_1}, a^{-j}_{0_2} x_2 - m_2 b_{0_2}) d^2\boldsymbol{x} \tag{28}$$

Note that we take a discrete set of parameterized wavelets but the integral is taken over the continuous image domain. In applications the image domain should be discrete as well.

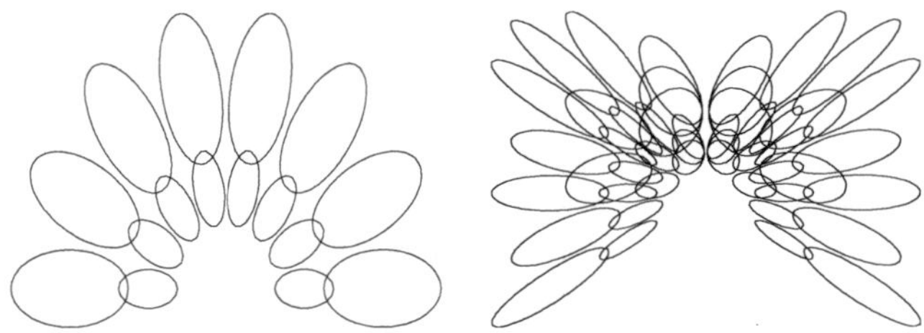

**Fig. 1.** The contours indicate the half-peak magnitude of Morlet wavelets filter bank in the frequency domain generated by: (a) SIM(2) group (b) full affine group

## 6 Tight Frame and a Reconstruction Formula in Terms of Scale-Space and Wavelets

In the previous section we have provided a sampling of the continuous wavelet transform using an appropriate discrete set of rotation, shear, scaling and translation parameters. In this section, we recall the definition of frame and the reconstruction formula for tight frames, and apply it to the framework we developed. We start with the definition of a frame given in [3],[17].

**Definition:** A family of functions $\{\varphi_j\}_{\in J}$ in a Hilbert space $H$ is called a *frame* if there exist $0 \leq A \leq B < \infty$ so that for all $f \in H$ the following holds,

$$A\|f\|^2 < \sum_{j\in J} |\langle f, \varphi_j \rangle|^2 \leq B\|f\|^2, \tag{29}$$

where $A$ and $B$ are the frame bounds. A frame is called a *tight frame* if the frame bounds are equal, $A = B$. Then, for every $f \in H$ , $\sum_{j\in J} |\langle f, \varphi_j\rangle|^2 = A\|f\|^2$ and we have [2] the following reconstruction formula :

$$f = \frac{1}{A}\sum_{j\in J}\langle f, \varphi_j\rangle\varphi_j, \tag{30}$$

which means that $f$ can be reconstructed by linear superposition of wavelets with the discrete wavelet transform coefficients obtained for $f$.
Using our previous analysis in §4 we may write the reconstruction formula (30) in terms of the scale space framework. First, we sample the scale, $t$, by writing it as a product of any two positive numbers $a_{0_1}^j, a_{0_2}^j$ to obtain $t_j = \frac{a_{0_1}^j a_{0_2}^j}{2}$. Next, we have to choose an appropriate set of initial values by sampling $\tilde{G}$. Using (11) we may write $\tilde{G}$ in the following way:

$$\tilde{G} = \begin{bmatrix} (\frac{a_{0_1}}{a_{0_2}})^{j/2} & 0 \\ 0 & (\frac{a_{0_2}}{a_{0_1}})^{j/2} \end{bmatrix} \begin{bmatrix} 1 & -s_{jk} \\ 0 & 1 \end{bmatrix} \begin{bmatrix} \cos\theta_l & \sin\theta_l \\ -\sin\theta_l & \cos\theta_l \end{bmatrix} \begin{bmatrix} 1 & 0 \\ 0 & 2 \end{bmatrix}$$

where $s_{jk} = s_{0_j}k = \frac{a_1^j}{a_2^j}k, \quad \theta_l = l\theta_0$ for any $\theta_0 \in [0, 2\pi]$.

Thus, we can reconstruct the original image $f(\boldsymbol{y})$ by "collecting the information" in each scale space that is obtained from a diffusion equation with different initial conditions that reflect the orientation and shear selected. We may write it in the following explicit form:

$$\begin{aligned} f(\boldsymbol{y}) &= \frac{1}{A}\sum_{ljkm} 2\pi t_j \mathcal{U}(\boldsymbol{x}_{ljkm}, t_j)\mathcal{G}_{\sqrt{2t_j}}(\tilde{G}^{-1}\boldsymbol{y} - \boldsymbol{x}_{ljkm}) \\ &= \frac{2\pi}{A}\sum_{j} t_j\Big(\mathcal{U}(\cdot, t_j) * \mathcal{G}_{\sqrt{2t_j}}\Big)(\tilde{G}^{-1}\boldsymbol{y}), \end{aligned} \tag{31}$$

where $x_{ljkm} = \tilde{G}^{-1}\boldsymbol{b}_{jm}$. ($\boldsymbol{b}_{jm}$ is as in the previous section)

## 7 Results

In this work we present the relation between the full Affine Gabor-Morlet Wavelet transform and linear diffusion with a certain initial value projected on a certain frequency component.

First, we present several filters that have different scale, orientation and shear attributes. We also present the effect of applying these filters to an image, which is actually a visualization of a certain slice in scale space. In Fig. 2 top we have an image that is convolved with several filters. we present different filters along with the responses of the image to these filters. The first column consists of the different filters, the second column provides the absolute value of the response of the image to this filter and the third column the phase. In the first row we reduced the scale parameter, and therefore the filter detects finer elements in the image. In the last row, we have obtained a filter that has the same attributes as the one in the first row, but without the shear parameter. Thus, rather than the full two-dimensional Affine group, this last filter is generated only with the $SIM(2)$ group. As can be seen, a weaker response was obtained. We may assume that the shear parameter allows for a better extraction of scale-space information from our image.

Our next task is to assess the goodness of our decomposition. Using frame theory, we calculated the tightness of the frame for the selection of the following parameters: $\theta_0 = \frac{\pi}{7}$ [1]. The ratio between the frame bounds was $\frac{B}{A} = 1.22$ which means that it is nearly a tight frame. Therefore, we can use the following reconstruction formula:

$$f \approx \frac{2}{A+B}\sum_{ljkm} \langle \psi_{lkjm}, f\rangle \psi_{lkjm}. \tag{32}$$

It is interesting to note that in [11] the same parameters were used, but for the $SIM(2)$. Then, a less tight frame is obtained, $\frac{B}{A} = 1.5$ and this is also

[1] The formulas and their derivation is long and tedious and they are omitted here because of lack of space. The calculation can be found in [16].

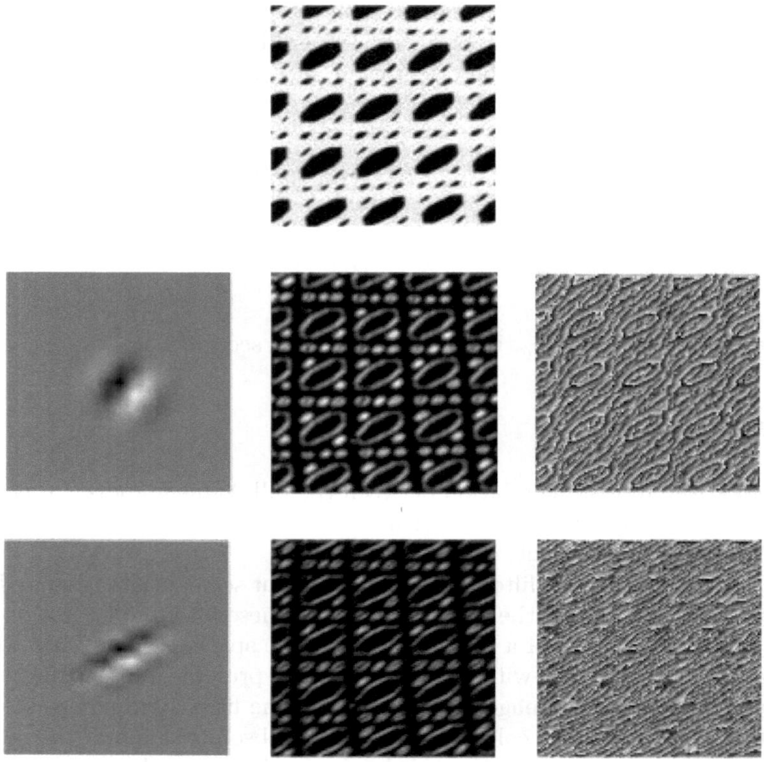

**Fig. 2.** The response of the test image to different filters generated by the full affine group. The filters are: $\psi_{lkj}$ with parameters:$l = [1,1], k = [1,0], j = [0,0], (\theta_0 = \pi/3, s_0 = 1, a_{0_{1|2}} = 2)$ (left in each row), $|\langle f, \psi\rangle|$ (middle in each row) , $\tan^{-1}(\frac{Im\langle f,\psi\rangle}{Re\langle f,\psi\rangle})$ (right in each row).

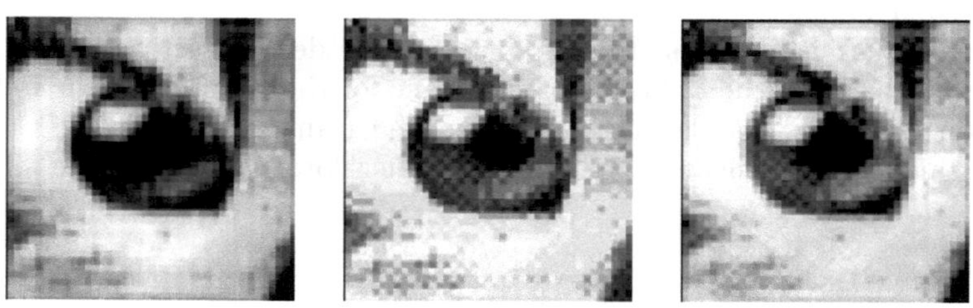

**Fig. 3.** Reconstruction of an image (center) by wavelets generated by the SIM(2) group with parameters : ,$\theta_0 = \pi/25, l = 0,..,24$ and $a_{0_{1|2}} = 2, j = 0,..,29$(left) and by the full affine group with parameters :$\theta_0 = \pi/7, l = 0,..,6, s_0 = 1, k = -7,..7$ and $a_{0_{1|2}} = 2, j = 0,..,6$(right)

reflected on the actual reconstruction results. In Fig. 3 we show an image that was reconstructed by using two families of filters: The ones that are generated by the $SIM(2)$ group and the filter bank related to the full affine group. It can be seen that using the full Affine group leads to a better detection of image features. Fine details that are recovered by the full affine group are missing from the more limited $SIM(2)$ reconstruction.

**Table 1.** The ratio $\frac{B}{A}$ for different orientations , $\theta_0 = \frac{\pi}{L}$, where $a_{0_1} = a_{0_2} = 2, s_0 = 1, b_{0_1} = b_{0_2} = 0.25$ are constant

| $L$ | $\frac{B}{A}$ |
|---|---|
| 7 | 1.22 |
| 9 | 1.13 |
| 11 | 1.12 |

## 8 Discussion and Conclusions

Analysis of signals and images in a multi-scale approach yields better insight of their contents. The wavelet transform and scale-space analysis are closely related in the sense that both aim at providing multi-resolution/scale information for signals and images. The mechanism offered in this manuscript provides a mathematical analysis of this relation, in addition to the obvious conceptual similarity. It relates the wavelet transform of an image to the (projection of) the solution of diffusion equations of the image and of those images that relate to it by group transformations. The image is thus represented in terms of its scaled-rotated and sheared versions.

After providing this twisted Gaussian approach, we have established the goodness of such reconstruction by the generalization to two-dimensions the Daubechie's frame criterion. The generalization of this criterion to two dimensions was considered in the past [11,14], but here we account for the full two-dimensional Affine group.

We show that using the full Affine group rather than the $SIM(2)$ group leads to a tighter frame, and not surprisingly , also to better reconstruction results.

To conclude, we have presented here a generalized scale-space, using the full two-dimensional Affine group, and have shown the relation to wavelet and frame analysis. We have also demonstrated that the twisted scale space, that accounts for scale, shear, orientation and frequency, may serve as a tool for representing and analyzing images.

## Acknowledgment

This research was supported by MUSCLE Multimedia Understanding through Semantic, Computation and Learning, an European Network of Excellence funded by the EC 6th Framework IST Programme, and by the Adams super-center for Brain Research."

## References


1. K. Bredies, D. A. Lorenz, and P. Maass, *Mathematical concepts of multiscale smoothing*, Applied and Computational Harmonic Analysis **19** (2005), no. 2, 141–161.
2. I. Daubechies, *Ten Lectures on Wavelets*, SIAM, Philadelphia, 1992.
3. J. Duffin and A.C. Scheffer, *A Class of Nonharmonic Fourier Series*, Trans. Amer. Math. Soc. **72** (1952), 341–366.
4. W. Hackbusch, *Multi-Grid Methods and Applications*, Springer-Verlag, Berlin, 1985.
5. T. Ijima, *Theory of Pattern Recognition*, Electronics and Communications in Japan (Nov. 1963).
6. K. Iwasawa, *On some types of topological groups*, Ann. of Math. **50** (1949), 507–558.
7. R.Murenzi J.-P.Antoine, *Two-dimentional directional wavelets and the scale-angle representation*, Signal Processing **52** (1996), 259–281.
8. R.Murenzi J.-P.Antoine, P.Vanderheynst, *Two-dimentional directional wavelets in image processing*, International Journal of Imaging Systems and Technology **7** (1996), 152–165.
9. J.Weickert, *A Review of Nonlinear Diffusion Filtering*, Scale-Space Theory in Computer Vision (B.M.ter HaarRomeny, et al., Eds.) (1997), 3–28.
10. J. J. Koenderink, *The structure of images*, Biol. Cybern.
11. T.S. Lee, *Image Representation Using 2D Gabor Wavelets*, IEEE Transactions on PAMI **18(10)** (1996), 959–971.
12. S. G. Mallat, *A wavelet tour of signal processing*, second ed., Academic Press, San Diego, 1998.
13. J. Morlet and D. Giard. G. Arens, I. Fourgeau, *Wave propagation and sampling theory*, Geophysics **47** (1982), 203–236.
14. C. Sagiv, *Uncertaty principles and Gabor framework in segmentation analysis and synthesis*, PhD Thesis, Tel-Aviv University, 2006.
15. A.P. Witkin, *Scale-space filtering*, Proc. Eighth Int. Joint Conf. on Artificial Intelligence **2** (1983), 1019–1022.
16. N. Sochen Y. Ferdman, C. Sagiv, *Two dimentional generalization of daubechies's frame criterion for the full affine wavelets*, http://www.math.tau.ac.il/ sochen/.
17. M. Young, *An Introducton to Nonharmonic Fourier Series*, Academic Press, New York, 1980.

# Iterated Nonlocal Means for Texture Restoration

Thomas Brox and Daniel Cremers

CVPR Group, University of Bonn
Römerstr. 164, 53117 Bonn, Germany
{brox,dcremers}@cs.uni-bonn.de

**Abstract.** The recent nonlocal means filter is a very successful technique for denoising textured images. In this paper, we formulate a variational technique that leads to an adaptive version of this filter. In particular, in an iterative manner, the filtering result is employed to redefine the similarity of patches in the next iteration. We further introduce the idea to replace the neighborhood weighting by a sorting criterion. This addresses the parameter selection problem of the original nonlocal means filter and leads to favorable denoising results of textured images, particularly in case of large noise levels.

## 1 From Neighborhood Filters to the Nonlocal Means Filter

In recent years, increasingly sophisticated filtering techniques have been developed in order to remove noise from a given input image $f : (\Omega \subset \mathbb{R}^2) \to \mathbb{R}$. While linear Gaussian filtering

$$u(x) = G_\rho * f(x) = \int G_\rho(x') f(x - x')\, dx' \tag{1}$$

with a Gaussian $G_\rho$ of width $\rho > 0$ is known to blur relevant image structures, more sophisticated nonlinear filtering techniques were developed, such as the total variation filtering [10], also known as the ROF model, which minimizes the cost functional:

$$E(u) = \int (f - u)^2\, dx + \lambda \int |\nabla u|\, dx. \tag{2}$$

The ROF model is closely related to nonlinear diffusion filters [9], in particular to the total variation flow [1]

$$\begin{aligned} u(x,0) &= f(x) \\ \partial_t u(x,t) &= \operatorname{div}\left(\frac{\nabla u(x,t)}{|\nabla u(x,t)|}\right). \end{aligned} \tag{3}$$

In the space-discrete, one-dimensional setting, it was shown that the solution of this diffusion equation at a time $t$ is equivalent to the solution of the ROF model with $\lambda = t$ as well as a certain implementation of wavelet soft shrinkage [12].

F. Sgallari, A. Murli, and N. Paragios (Eds.): SSVM 2007, LNCS 4485, pp. 13–24, 2007.

**Fig. 1.** Nonlocal means filter. **From left to right:** (a) Reference image of size 119×121 pixels. (b) Gaussian noise with $\sigma = 20$ added. (c) Denoising result of the nonlocal means filter ($h = 0.95\sigma$). (d) Denoising result of the changed nonlocal means filter using a sorting criterion ($n = 8$).

Despite an enormous success in image enhancement and noise removal applications, approaches like the ROF filtering remain spatially local, in the sense that at each location $x \in \Omega$ the update of $u$ is determined only by derivatives of $u$ at that same location $x$ – see equation (3).

A class of image filters which adaptively takes into account intensity information from more distant locations are the Yaroslavsky neighborhood filters [16]:

$$u(x) = \frac{1}{C(x)} \int K(x, y, f(x), f(y)) \quad C(x) = \int K(x, y, f(x), f(y))\, dy. \tag{4}$$

$K$ is a nonnegative kernel function which decays with the distances $|x - y|$ and $|f(x) - f(y)|$. Thus the application of this filter amounts to assigning to $x$ a weighted average over the intensities $f(y)$ of all pixels $y$ which are similar in the sense that they are close to $\{x, f(x)\}$ in space and in intensity. These filters are also known as local M-smoothers [4,15]. A similar, but iterative, filter is the bilateral filter [11,13]. Relations between such neighborhood filters and nonlinear diffusion filters have been investigated in [8].

A drastic improvement of these neighborhood filters is the nonlocal means filter which was recently proposed by Buades et al. [3]. Its application to video processing and surface smoothing has been demonstrated in [7,17] and a very related statistical filter was presented in [2]. The nonlocal means filter can be written as:

$$u(x) = \int w_f(x, y)\, f(y)\, dy, \tag{5}$$

with normalized weights of the form

$$w_f(x, y) = \frac{g_f(x, y)}{\int g_f(x, y)\, dy}, \tag{6}$$

where

$$g_f(x, y) = \exp\left(-\frac{d_f^2(x, y)}{h^2}\right) \tag{7}$$

and

$$d_f^2(x, y) = \int G_\rho(x')\big(f(x - x') - f(y - x')\big)^2\, dx'. \tag{8}$$

In contrast to the neighborhood filters, the nonlocal means filter quantifies the similarity of pixels $x$ and $y$ by taking into account the similarity of whole patches around $x$ and $y$. The similarity is expressed by a dissimilarity measure $d_f^2(x,y)$, which contains the size $\rho$ of the compared patches as a parameter, and a weighting function $g$ with parameter $h$, which quantifies how fast the weights decay with increasing dissimilarity of respective patches. Since the above similarity measure takes into account complete patches instead of single pixel intensities, the nonlocal means filter is able to remove noise from textured images without destroying the fine structures of the texture itself. This amazing property is demonstrated in Fig. 1.

The key idea of nonlocal means filtering is that the restoration of a destroyed texture patch is improved with support from similar texture patches in other areas of the image. The filter is, hence, based on a similar concept as the texture synthesis work of Efros and Leung [5]. In this paper, we will show that this property is further enhanced by iterated nonlocal means which shall be developed in the following.

## 2 Iterated Nonlocal Means

The nonlocal means filter assigns to each pixel $x$ a weighted average over all intensities $f(y)$ of pixels $y$ which share a similar intensity neighborhood as the point $x$. A trivial variational principle for this filter can be written as:

$$E(u) = \int \left( u(x) - \int w_f(x,y) f(y)\, dy \right)^2 dx. \tag{9}$$

An alternative variational formulation of nonlocal means was proposed in [6]. In this paper, we propose an iterated form of the nonlocal means filter which arises when extending the above functional by replacing $w_f$ by $w_u$ in the following manner:

$$E(u) = \int \left( u(x) - \int w_u(x,y) f(y)\, dy \right)^2 dx. \tag{10}$$

Thus, rather than imposing similarity of $u(x)$ to $f(y)$ for locations $y$ where the input image $f(y)$ is similar to $f(x)$, we impose similarity to $f(y)$ for locations $y$ where the filtered image $u(y)$ is similar to $u(x)$. This induces an additional feedback and further decouples the resulting image $u$ from the input image $f$. The idea is that the similarity of patches can be judged more accurately from the already denoised signal than from the noisy input image.

### 2.1 Fixed Point Iteration

Due to the introduced dependence of $w$ on $u$, the minimizer of (10) is no longer the result of a weighted convolution, but the solution of a nonlinear optimization problem. A straightforward way to approximate a solution of (10) is by an iterative scheme with iteration index $k$, where we start with the initialization $u^0 = f$.

For fixed $u = u^k$ we are in a similar situation as with the conventional nonlocal means filter. In particular, we can compute the similarity measure $w_{u^k}(x,y)$ for the current image $u^k$ and, as a consequence, we obtain an update on $u$

$$u^{k+1}(x) = \int w_{u^k}(x,y)\, f(y)\, dy. \tag{11}$$

Whether this iterative process converges to a stationary solution, is subject of future investigation.

## 2.2 Euler-Lagrange Equation and Gradient Descent

An alternative way to find a solution of (10) is by computing its Euler-Lagrange equation, which states a necessary condition for a (local) minimum. We are seeking for the gradient

$$\frac{\partial E(u)}{\partial u} = \left.\frac{\partial E(u+\epsilon h)}{\partial \epsilon}\right|_{\epsilon \to 0}. \tag{12}$$

After evaluation and substitution of integration variables, we end up with the following Euler-Lagrange equation:

$$\begin{aligned}
\frac{\partial E(u)}{\partial u} = 0 = &\left(u(x) - \int f(y) w_u(x,y) dy\right) \\
+2 \iint &\left(u(z) - \left(\int f(y') w_u(z,y') dy'\right) \frac{f(y) g'(z,y)}{\int g(z,y') dy'} G_\rho(y-x)(u(z-y-x) - u(x))\right) dy dz \\
+2 \iiint &\left(u(z) - \left(\int f(y') w_u(z,y') dy'\right) \frac{f(y) g(z,y) g'(z,y'')}{\left(\int g(z,y') dy'\right)^2} G_\rho(z-x)(u(x) - u(y''-z-x))\right) dy'' dy dz
\end{aligned}$$

After initializing $u^0 = f$, the gradient descent equation

$$u^{k+1} = u^k - \tau \frac{\partial E(u)}{\partial u} \tag{13}$$

yields the next local minimum for some sufficiently small step size $\tau$ and $t \to \infty$. Obviously, gradient descent with the gradient being reduced to the first term in (2.2) leads for $\tau = 1$ to the same iterative scheme as in Section 2.1. The additional two terms take the variation of $w_u$ into account and ensure convergence for sufficiently small step sizes $\tau$. However, these terms induce a very large computational load in each iteration. In particular, the time complexity of the third term in each iteration is $O(MN^4)$, where $N$ is the number of pixels in the image and $M$ the number of pixels in the compared patch. For comparison, the first term only has a time complexity of $O(MN^2)$ and a nonlinear diffusion filter like TV flow has a time complexity of $O(N)$ in each iteration. Hence, in our experiments, we took only the first term into account.

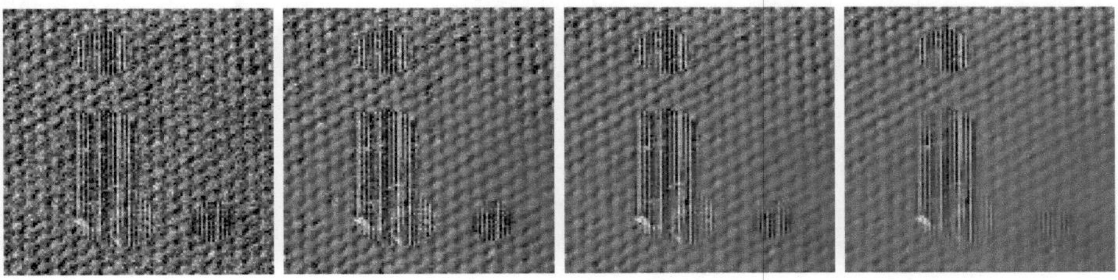

**Fig. 2.** Parameter sensitivity of nonlocal means filter in case of high noise levels. **From left to right:** (a) Gaussian noise with $\sigma = 30$ added to reference image in Fig.1a. (b,c,d) Denoising result of nonlocal means filter for $h = 0.65\sigma$, $h = 0.7\sigma$, and $h = 0.75\sigma$. The fact that the restored image contains both unfiltered, noisy areas, as well as over-smoothed areas shows that there exists no ideal choice of the parameter $h$.

## 3 A Robust Threshold Criterion

While the nonlocal means filter can yield astonishing denoising results, a deeper experimental investigation reveals a large sensitivity to the parameter $h$ in

$$g_f(x,y) = \exp\left(-\frac{d_f^2(x,y)}{h^2}\right), \tag{14}$$

which is responsible for steering the decay of weights for decreasing similarity of patches. This parameter sensitivity increases with the noise variance $\sigma$ in the image. Moreover, if the noise level exceeds a certain value, it is no longer possible to choose a global $h$ such that the noise is removed everywhere without destroying repetitive structure somewhere else in the image. This is demonstrated in Fig. 2.

The reason for this effect is the weighting function $g$. By definition we have $g_f(x,x) = 1$. Suppose there is a highly repetitive patch and the noise level is rather small. In this case, there will be many similar patches with $g(x,y) \approx 1$ and the smoothing between these patches works well. Now suppose a patch that is hardly similar to other patches in the image, or only very few of them. Consequently, there will be almost no change at $x$ since $g(x,x) = 1$ and $g(x,y) \approx 0$ almost everywhere. In this case, one has to increase $h$ such that there are enough $y$ with $g(x,y) > \epsilon$ in order to see a smoothing effect.

Buades et al. have been aware of this problem and suggested to set $g(x,x)$ to $\max_{y \neq x} g(x,y)$. Although this attenuates the problem, it does not resolve it, as it only ensures the averaging of at least two values. The results shown in Fig. 2, where we implemented this idea, reveal that the averaging of at least two values is in many cases not sufficient.

Here, we suggest to approach the problem from a different direction. Instead of defining a function $g$ that assigns weights to positions $y$, we choose the number $n$ of positions that is appropriate to remove a certain noise level. We then simply take those $n$ patches with the smallest dissimilarity $d^2(x,y)$.

**Fig. 3. From left to right:** (a) Gaussian noise with $\sigma = 50$ added to the reference image in Fig.1a. (b,c,d) Denoising result of nonlocal means filter with the new sorting criterion for $n = 10$, $n = 20$, and $n = 40$. Although the noise level is significantly higher than in Fig. 2, the exact parameter choice is less critical and the results look favorable.

**Fig. 4.** Iterated nonlocal means. **From left to right:** (a) Gaussian noise with $\sigma = 70$ added to reference image in Fig.1a. (b,c,d) Denoising result of the iterated nonlocal means filter with the sorting criterion after 1, 2, and 5 iterations ($n = 20$). Iterations improve the regularity of the texture.

By considering for any pixel $x$ the $n$ most similar pixels rather than all those pixels of similarity above a fixed threshold, we allow for denoising which does not depend on how repetitive the respective structure at $x$ is in the given image.

## 4 Experimental Results

Figure 3 shows the effect of the new sorting criterion. Although the noise level in the input image has been chosen much higher than in Figure 2, the result looks more appealing. The noise has been removed while the repetitive texture patterns have been preserved. Even the contrast did not suffer severely. Moreover, the sensitivity with respect to the parameter choice has been attenuated considerably: All three results depicted in Figure 3 are satisfactory, although the parameter $n$ has been varied by a factor 4. For all experiments, also in the previous sections, we fixed $\rho = \sqrt{8}$ and used a $9 \times 9$ window for implementation.

Figure 4 demonstrates the impact of iterating the nonlocal means filter. Again we increase the noise level. Note that due to the high amount of noise and clipping intensities that exceed the range of $[0, 255]$, the noise is not fully Gaussian anymore. Nevertheless, the results that can be obtained with the modified

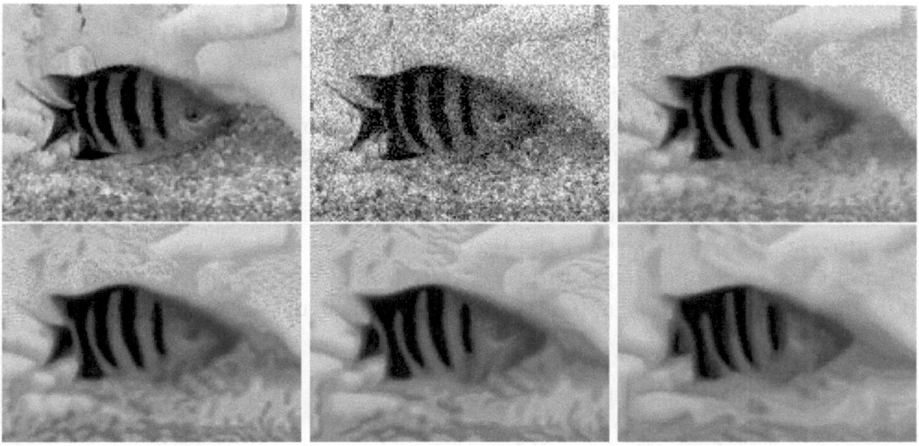

**Fig. 5.** Denoising of a natural, non-regular image. **From left to right, top to bottom:** (a) Reference image of size 162 × 120 pixels. (b) Gaussian noise with $\sigma = 30$ added. (c,d,e,f) Denoising result of the iterated nonlocal means filter with the sorting criterion after 1, 2, 4, and 10 iterations ($n = 100$).

nonlocal means filter, in particular with its iterated version, look quite satisfactory. Clearly, iterating the filter improves the regularity of the texture pattern.

This can lead to interesting effects, if the filter is applied to an image that is mainly non-regular. Such a case is shown in Fig. 5. For an increasing number of iterations, the filter acts more and more coherence enhancing reminiscent of curvature motion or coherence enhancing anisotropic diffusion [14].

While the denoising in Fig. 5 has been achieved with a quite large number of partners, namely $n = 100$, Fig. 6 shows what happens if one decreases $n$ and instead increases the number of iterations. Due to the small number of neighbors, the filter is not able to fully remove the noise. Interestingly, with further iterations, the filter detects structures in the noise that have not been present in the image and starts enhancing these.

**Fig. 6.** Hallucination of regular patterns in noise for many iterations and small mask size. **From left to right:** (a) Gaussian noise with $\sigma = 30$ added to reference image in Fig.5a. (b,c) Denoising result of the iterated nonlocal means filter with the sorting criterion after 1 and 300 iterations ($n = 10$). For small $n$, the iterated filter creates structures from the noise.

In Fig. 7 we show a comparison of the modified, iterated nonlocal means filter to the ROF model. In order to demonstrate the robustness of the parameter settings, the image contains various different textures of different scales. The results, in particular the two closeups in Fig. 8, reveal a very precise reconstruction of all textures despite the fixed parameter setting. Even very fine texture details are preserved. This is in contrast to the ROF model, which preferably removes small scale structures. In most cases such structures are noise pixels, but they may also be important parts of the texture.

Finally, we performed a quantitative evaluation of various filters, including the ROF model, conventional nonlocal means, nonlocal means with the suggested

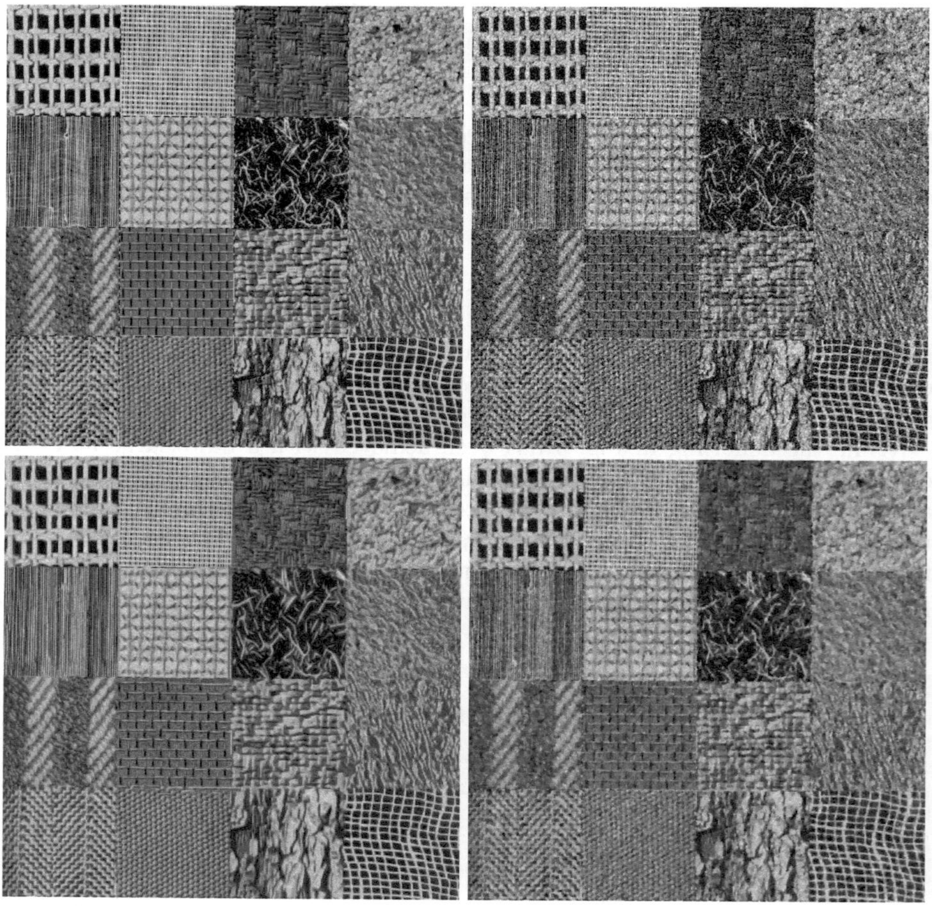

**Fig. 7.** Comparison to ROF denoising model. **Top left:** (a) Reference image of size $512 \times 512$ pixels. **Top right:** (b) Gaussian noise with $\sigma = 40$ added. **Bottom left:** (c) Denoising result with iterated nonlocal means and the sorting criterion ($n = 20$) after 2 iterations. **Bottom right:** (d) ROF model for $\alpha = 20$.

**Fig. 8.** Zoom into two regions in Fig. 7. **Each row from left to right:** (a) Noisy input image. (b) ROF model. (c) Iterated nonlocal means.

**Fig. 9.** Undisturbed reference images for the quantitative comparison in Table 1

**Table 1.** Root mean square error for the input images in Fig. 9 and different denoising techniques. See Fig. 10 for the resulting images. The value in brackets is the optimized setting of the free parameter(s).

| | texture ($\sigma = 70$) | | owl ($\sigma = 40$) | | Lena ($\sigma = 30$) | |
|---|---|---|---|---|---|---|
| Input image | 62.28 | | 38.06 | | 29.16 | |
| ROF model | 34.28 | (28) | 22.78 | (16) | 11.33 | (18) |
| Nonlocal means | 38.73 | ($0.45\sigma$) | 27.29 | ($0.6\sigma$) | 13.69 | ($0.6625\sigma$) |
| Sorting criterion | 32.02 | (175) | 23.95 | (40) | 13.38 | (100) |
| Iterated sorting criterion | 28.16 | (75/3) | 23.95 | (40/1) | 12.57 | (65/2) |

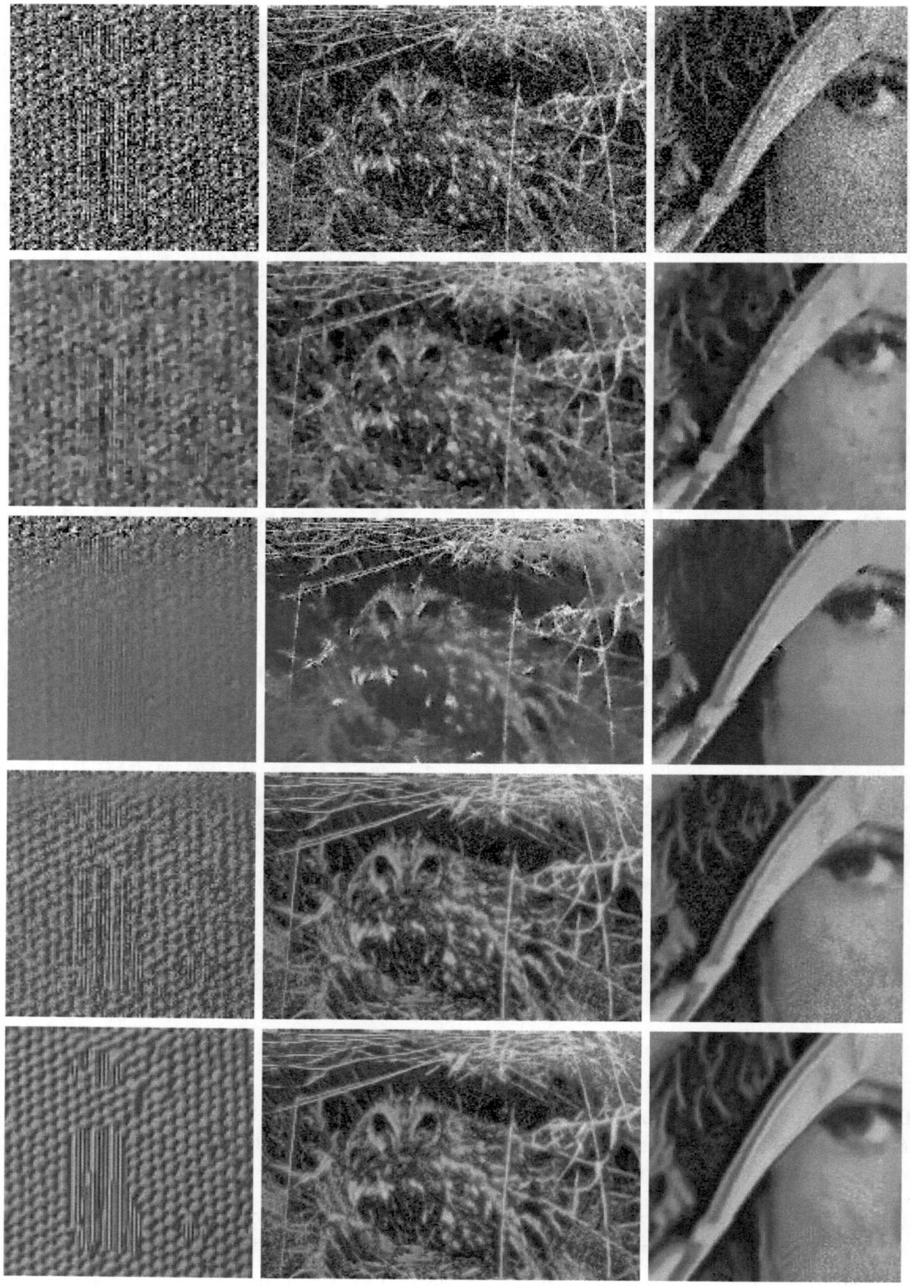

**Fig. 10.** Images corresponding to the result in table 1. **Each column from top to bottom:** Noisy input image, TV flow, nonlocal means filter, nonlocal means with sorting criterion, iterated nonlocal means. In case of the owl image, the iterated and non-iterated results are identical.

sorting criterion, as well as its iterated version. Table 1 lists the root mean square (RMS) error

$$e_{RMS} := \sqrt{\frac{1}{N}\sum_{i=1}^{N}(r(i) - u(i))^2} \tag{15}$$

between the outcome $u$ of the filter and the undisturbed reference image $r$. The test images and the filtering results are shown in Fig. 9 and Fig. 10, respectively. For a fair comparison, we ensured that exactly the same noise was added in all test runs and optimized the free parameters.

The numbers only partially support the visual impression, as the results of the ROF model reveal more unpleasant artifacts than its good RMS errors indicate. However, the numbers are in line with the impression that the nonlocal means filter using the sorting criterion, in particular its iterated version, performs best for regular texture patterns. In case of images that are dominated by piecewise homogeneous areas, the conventional nonlocal means filter still yields the most appealing results.

## 5 Conclusions

We proposed a variational formulation of the recently developed nonlocal means filter and introduced an additional feedback mechanism at the variational level. We showed that the solution by a fixed point iteration gives rise to an iterated version of the nonlocal means filter. Moreover, we proposed to replace the neighborhood weighting in the original formulation by a sorting criterion which assures that the amount of filtering no longer depends on how repetitive respective image structures are in the given image. Experimental results demonstrate that the iterated nonlocal means filter outperforms both nonlocal means and total variation filtering when applied to the restoration of regular textures. At the same time, our experiments indicate that the increased feedback may lead to a hallucination of regular patterns in noise for large iteration numbers.

## References


1. F. Andreu, C. Ballester, V. Caselles, and J. M. Mazón. Minimizing total variation flow. *Differential and Integral Equations*, 14(3):321–360, Mar. 2001.
2. S. Awate and R. Whitaker. Unsupervised, information-theoretic, adaptive image filtering for image restoration. *IEEE Transactions on Pattern Analysis and Machine Intelligence*, 28(3):364–376, Mar. 2006.
3. A. Buades, B. Coll, and J. M. Morel. A review of image denoising algorithms, with a new one. *SIAM Interdisciplanary Journal*, 4(2):490–530, 2005.
4. C. K. Chu, I. Glad, F. Godtliebsen, and J. S. Marron. Edge-preserving smoothers for image processing. *Journal of the American Statistical Association*, 93(442): 526–556, 1998.
5. A. Efros and T. Leung. Texture synthesis by non-parametric sampling. In *Proc. International Conference on Computer Vision*, pages 1033–1038, Corfu, Greece, Sept. 1999.

6. S. Kindermann, S. Osher, and P. W. Jones. Deblurring and denoising of images by nonlocal functionals. *SIAM Interdisciplinary Journal*, 4(4):1091–1115, 2005.
7. M. Mahmoudi and G. Sapiro. Fast image and video denoising via nonlocal means of similar neighborhoods. *Signal Processing Letters*, 12(12):839–842, 2005.
8. P. Mrázek, J. Weickert, and A. Bruhn. On robust estimation and smoothing with spatial and tonal kernels. In R. Klette, R. Kozera, L. Noakes, and J. Weickert, editors, *Geometric Properties from Incomplete Data*, pages 335–352. Springer, Dordrecht, 2006.
9. P. Perona and J. Malik. Scale space and edge detection using anisotropic diffusion. *IEEE Transactions on Pattern Analysis and Machine Intelligence*, 12: 629–639, 1990.
10. L. I. Rudin, S. Osher, and E. Fatemi. Nonlinear total variation based noise removal algorithms. *Physica D*, 60:259–268, 1992.
11. S. M. Smith and J. M. Brady. SUSAN: A new approach to low-level image processing. *International Journal of Computer Vision*, 23(1):45–78, May 1997.
12. G. Steidl, J. Weickert, T. Brox, P. Mrázek, and M. Welk. On the equivalence of soft wavelet shrinkage, total variation diffusion, total variation regularization, and SIDEs. *SIAM Journal on Numerical Analysis*, 42(2):686–713, May 2004.
13. C. Tomasi and R. Manduchi. Bilateral filtering for gray and color images. In *Proc. Sixth International Conference on Computer Vision*, pages 839–846, Bombay, India, Jan. 1998. Narosa Publishing House.
14. J. Weickert. Coherence-enhancing diffusion of colour images. *Image and Vision Computing*, 17(3/4):199–210, Mar. 1999.
15. G. Winkler, V. Aurich, K. Hahn, and A. Martin. Noise reduction in images: some recent edge-preserving methods. *Pattern Recognition and Image Analysis*, 9(4):749–766, 1999.
16. L. P. Yaroslavsky. *Digital Picture Processing - An Introduction.* Springer, 1985.
17. S. Yoshizawa, A. Belyaev, and H.-P. Seidel. Smoothing by example: mesh denoising by averaging with similarity-based weights. In *Proc. International Conference on Shape Modeling and Applications*, pages 38–44, June 2006.

# The Jet Metric

Marco Loog

Department of Computer Science, Nordic Bioscience A/S
University of Copenhagen, Herlev
Copenhagen, Denmark
loog@diku.dk

**Abstract.** In order to define a metric on jet space, linear scale space is considered from a statistical standpoint. Given a scale $\sigma$, the scale space solution can be interpreted as maximizing a certain Gaussian posterior probability, related to a particular Tikhonov regularization. The Gaussian prior, which governs this solution, in fact induces a Mahalanobis distance on the space of functions. This metric on the function space gives, in a rather precise way, rise to a metric on $n$-jets. The latter, in turn, can be employed to define a norm on jet space, as the metric is translation invariant and homogeneous. Recently, [1] derived a metric on jet space and our results reinforce his findings, while providing a totally different approach to defining a scale space jet metric.

## 1 Introduction

Feature vectors made up of $n$-jets or other Gaussian derivative features are used in many image analysis and processing applications. Examples can be found in image retrieval and matching, representation, and image labelling and segmentation tasks (see for instance [2,3,4,5,6,7,8,9]). In many of the previous applications, the need arises to define a distance between such measurement vectors for which one should define a metric in this multidimensional space of measurements. The decision of which metric to actually use can be based on a study of the problem at hand, its optimality with respect to the given task, the right prior knowledge, educated guessing, etc. It is however also of interest to consider the possibility of defining such a metric in 'empty' space in a principled way, *i.e.*, provide a metric given the only thing we know is that the derivative measurements are obtained by means of, say, Gaussian apertures.

### 1.1 Outline

Section 2 provides a sketch of a recently proposed approach to deriving a metric on jet space. Subsequently, Section 3 provides a new approach to it based on statistical considerations, starting from a known scale space regularization framework, and relating this to a statistical interpretation. In four steps, this section identifies the Gaussian prior governing scale space regularization, discusses the Mahalanobis distance induced by it, suggests a marginalization of the prior to define distances in subspaces, and finally defines the general $n$-jet metric using the latter suggestion. Section 4 concludes the paper.

F. Sgallari, A. Murli, and N. Paragios (Eds.): SSVM 2007, LNCS 4485, pp. 25–31, 2007.

Finally, as a note, this work should not be confused with the work presented in [10,11,12], which deal with an a priori metric in scale space itself, or rather its absence. In addition, we should note that this work is not concerned with distances between distribution, which may dealt with using information geometry.

## 2 [1]'s Jet Norm

Recently, as part of a considerably larger effort, an attempt to define the proper distance on scale space $n$-jets—with a focus on 2-jets—has been undertaken in [1] (*cf.* [13]). The author defines a norm on jet space, which induces the desired metric. In order to find the appropriate jet space norm, several requirements that should characterize it are specified, e.g., it is supposed to be invariant to translation, rotation, and reflection of the image domain from which the jet is derived and also an invariance to constant image offsets is assumed. The latter implies, for instance, that the zeroth order jet should be of no influence on the norm and therefore we are in fact dealing with a semi-norm.

However, these requirements are, indeed, requirements on images and not on jets. In order to relate jets to images, the authors use the language of metamerism, which relates a collection of image measurements to the class of images that give rise to precisely these observations [14,15,16]. Another point to relating jets to images is that in image space there is a 'natural' choice for the so-called scale space norm based on which the actual scale space jet norm can then be constructed. Given an image $I : \mathbb{R}^n \to \mathbb{R}$, the scale space $||\cdot||_\sigma$ norm takes on something like the form of a weighted $L^2$ norm

$$||I||_\sigma = \left( \int\limits_{\mathbb{R}^d} g_\sigma(x) I^2(x) - \left( \int\limits_{\mathbb{R}^d} g_\sigma(x) I(x) dx \right)^2 dx \right)^{\frac{1}{2}} , \qquad (1)$$

where $g_\sigma$ is the Gaussian kernel at scale $\sigma$. Note that the latter part in the definition of the norm makes sure that it is invariant to constant image offsets and therefore, in fact, we are dealing with a semi-norm again.

The idea is now to take a unique representative from the metameric class of images related to the jet and define its norm to be equal to the norm of this unique representative. The choice made in [1] is the function from the class that minimizes the scale space norm. This turns out to be a polynomial of the same degree as the jet is. Without going into any additional details concerning the exact derivation, we merely provide the end result, in which only the norm for the 2-jet of two-dimensional images is fully worked out.

Let $c_{uv}$ be the two-dimensional image derivative at scale $\sigma$ at the origin, which is given by

$$c_{uv} = \int\limits_{\mathbb{R}^2} \frac{\partial^{u+v}}{\partial x^u \partial y^v} g_\sigma(x) I(x) dx \qquad (2)$$

and let the corresponding 2-jet be denoted by $J_2$:

$$J_2 = (c_{00}, c_{01}, c_{10}, c_{11}, c_{02}, c_{20})^T . \qquad (3)$$

The 2-jet norm is now given by

$$||J_2|| = \sigma(c_{10}^2 + c_{01}^2 + \tfrac{1}{2}\sigma^2(c_{20}^2 + c_{11}^2 + c_{02}^2))^{\frac{1}{2}} . \tag{4}$$

It is straightforward to check that this norm fulfills the requirements mentioned earlier. As always, the norm also induces a metric on the space. Given two jets $J_2$ and $Y_2$, $||J_2 - Y_2||$ gives the distance between the two.

## 3 Prior Induced Jet Metric

The derivation of the norm in [1] and its induced metric, may involve several nontrivial steps to some. An important one, for instance, being the choice of representative function from the metameric class defined by the jet. As [1] remarks, the function taking on the maximum scale space norm is not uniquely determined and is therefore excluded. However, this does not necessarily imply that the minimum should be considered. Moreover, it may be questioned whether it should actually be the scale space norm to base the right choice of representative on, given that the definition of a jet norm should anyway be based on such metameric representative.

In this section, a different approach to defining a jet norm is proposed. Linear scale space is considered from a statistical standpoint from which it is clear that a certain Gaussian prior in the space of functions actually governs its behavior. This prior induces a Mahalanobis distance on this space, which in turn can be used to define a metric on $n$-jets. Finally, as the metric is translation invariant and homogeneous, it can also be used to define a norm on scale space jets. For two-dimensional images, as it turns out, the jet metric defined is the same as the one given in [1].

### 3.1 Gaussian Prior

In [17], scale space is related to a specific instance of Tikhonov regularization [18]. The regularized scale space image $I_\sigma$ at scale $\sigma$ associated to the initial image $I$ on $\mathbb{R}^d$ minimizes the functional $E$ defined as

$$E[\Upsilon] := \tfrac{1}{2} \int_{\mathbb{R}^d} (\Upsilon(x) - I(x))^2 + \sum_{i=1}^{\infty} \sum_{|N|=i} \frac{\sigma^{2i}}{2N!} \left( \frac{\partial^{|N|}\Upsilon(x)}{\partial x^N} \right)^2 dx , \tag{5}$$

where

$$N = N_1, \dots, N_d \tag{6}$$

is a multi-index used to denote derivatives of order

$$|N| = \sum_{i=1}^{d} N_i \tag{7}$$

and $N!$ equals the product over all factorials of all $d$ indices, *i.e.*,

$$N! = \prod_{i=1}^{d} N_i! , \tag{8}$$

where $d$ is the dimensionality of the images in the space. The first term on the right hand side penalizes deviations of the function $\Upsilon$ from the given image $I$, while the second part is the regularization term for $\Upsilon$, not involving $I$. It is this latter part that can be readily interpreted as a prior $P$ on the space of image through a multiplication by $-1$ and subsequently exponentiating it (see for example [19,20]):

$$P(I) = \frac{1}{Z} \exp \int_{\mathbb{R}^d} - \sum_{i=1}^{\infty} \sum_{|N|=i} \frac{\sigma^{2i}}{2N!} \left( \frac{\partial^{|N|} I(x)}{\partial x^N} \right)^2 dx \,. \tag{9}$$

This is a well-known *Gaussian* prior used, for instance, in areas that are concerned with kernel methods and machine learning [21] (*cf.* [22]).

### 3.2 Scale Space Mahalanobis Distance

Given that $P$ defines a Gaussian prior on image space, a straightforward choice of metric on this space is the Mahalanobis distance induced by the covariance structure of the Gaussian distribution, which is in analogy with the finite dimensional case [23]. That is, given a covariance matrix $C$ on the space, the Mahalanobis distance $d_M$ between two vectors $x$ and $y$ is given by

$$d_M(x, y) = \sqrt{(x - y)^t C^{-1} (x - y)} \,. \tag{10}$$

Initially, the fact that one needs the inverse of the covariance matrix, may seem troublesome as the Gaussian density specified in Equation (9) is given on an infinite dimensional space for which it may not be readily clear how to define such inverses. Luckily Gaussian densities are actually defined by means of their inverted covariance matrices and Equation (9) directly provides us with the appropriate scale space Mahalanobis distance $d_S$, which is defined for two images $I$ and $\Upsilon$ as

$$d_S(I, \Upsilon) = \sqrt{\int_{\mathbb{R}^d} \sum_{i=1}^{\infty} \sum_{|N|=i} \frac{\sigma^{2i}}{2N!} \left( \frac{\partial^{|N|} I(x)}{\partial x^N} - \frac{\partial^{|N|} \Upsilon(x)}{\partial x^N} \right)^2 dx} \,. \tag{11}$$

Note that in the previous equation, the summation and integration go over all the image derivatives and all spatial locations, however there is no covariance between any of these. In other words, cross terms between image $I$ and image $\Upsilon$ involving different locations or different multi-indices for the respective images do not occur, only squares of differences between corresponding components of $I$ and $\Upsilon$ do, *i.e.*, $\frac{\partial^{|N|} I(x)}{\partial x^N}$ and $\frac{\partial^{|N|} \Upsilon(x)}{\partial x^N}$ for the same choice of $x$ and $N$. In a sense, the covariance matrix and its inverse are diagonal.

### 3.3 Marginalization of the Scale Space Metric

Given that none of the components in the prior model governing scale space are correlated, if one is interested in a metric only involving a subset of the components, it may be reasonable to simply restricting the distance calculation to this subset in order to provide a metric in this lower-dimensional space. A similar situation arises in three-dimensional Euclidean space, when one is merely interested in the first two dimensions. In that case, the distance between points would be given by considering their distance in the two-dimensional Euclidean space corresponding to the first two coordinates, the third dimension is simply discarded in the calculation of the metric.

Clearly, Euclidean space corresponds to a covariance that equals the identity matrix and a deviation from the identity as covariance may render the approach suggested invalid. However, realizing that restricting the distance calculations to a subspace of the original space actually means that a *marginalization* of the distribution has taken place, we can further substantiate the approach sketched above. A result on the marginalization of Gaussian processes [23] states that if $X$ is normally distributed, any set of components of $X$ is distributed in a multivariate normal way for which the means, variances, and covariances are obtained by taking the corresponding components of the original mean vector and covariance matrix. Therefore, one can simply pick out the locations $x$ and multi-indices $N$ from Equation (11) in which one is interested and only integrate or sum over these.

## 3.4 The *n*-Jet Metric

In order to define any $n$-jet metric coming from any $d$-dimensional images, the spatial location has to be restricted to the origin and the derivative order has to bounded by $n$, which would directly lead to the following metric

$$d'_J(J_n, Y_n) = \sqrt{\sum_{i=1}^{n} \sum_{|N|=i} \frac{\sigma^{2i}}{2N!} \left( \frac{\partial^{|N|} I(0)}{\partial x^N} - \frac{\partial^{|N|} \Upsilon(0)}{\partial x^N} \right)^2}, \tag{12}$$

in which the jets $J_n$ and $Y_n$ are obtained from the images $I$ and $\Upsilon$, respectively.

However, we are not necessarily interested in a metric on image derivatives at scale zero. The scale specific prior was used because, the images were assumed to be scale space regularized at that scale. Therefore, the interest is not with Equation (12), but rather with its adaptation into which $\sigma$ blurred images are substituted, *i.e.*

$$\begin{aligned} d_J(J_n, Y_n) &= \sqrt{\sum_{i=1}^{n} \sum_{|N|=i} \frac{\sigma^{2i}}{2N!} \left( \frac{\partial^{|N|} I_\sigma(0)}{\partial x^N} - \frac{\partial^{|N|} \Upsilon_\sigma(0)}{\partial x^N} \right)^2} \\ &= \sqrt{\sum_{i=1}^{n} \sum_{|N|=i} \frac{\sigma^{2i}}{2N!} (c_N - \gamma_N)^2}, \end{aligned} \tag{13}$$

in which the coefficients $c_N$ and $\gamma_N$ are the jet coefficients corresponding to derivative multi-index $N$, which is the general form of the coefficients given in Equation (2). Equation (13) defines our general jet metric for which it is easy to demonstrate that it equals [1]'s metric on 2-jet space for 2-dimensional images—or generally any higher-order jets on two-dimensional images.

Note that $d_J$ is both translation invariant and homogeneous, *i.e.*,

$$d_J(J_n, Y_n) = d_J(J_n + Q_n, Y_n + Q_n) \tag{14}$$

and

$$d_J(cJ_n, cY_n) = |c| d_J(J_n, Y_n) \tag{15}$$

for all $n$-jets $J_n$, $Y_n$, and $Q_n$, and constant $c \in \mathbb{R}$, which implies that $d_J(J_n, 0)$ is a norm on jet space. We also should remark that the requirements imposed in [1] on the norm and the metric are 'automatically' fulfilled by our approach and they do not have to be enforced explicitly.

Besides, note also that when dealing with *scale-normalized* derivatives—*i.e.*, dimensionless measurements, the expression for metric becomes scale independent:

$$d_J(J_n, Y_n) = \sqrt{\sum_{i=1}^{n} \sum_{|N|=i} \frac{1}{2N!} \left(c'_N - \gamma'_N\right)^2 \, dx}\,, \tag{16}$$

where $c'_N$ and $\gamma'_N$ are the scale-normalized jet coefficients. In this way, jets from different scales may actually be compared to each other using the jet metric.

## 4 Conclusion and End Remarks

Through an approach that substantially differs from the initial one suggested in [1], a definition of a scale space jet metric in 'empty' space was proposed. Our findings support the original definition of this metric, strengthening the validity of it. The statistical approach employed to come to our derivation of the metric may also be of interest in itself, as it provides a general tool to formulate more exotic metrics, norm, distances, or even similarity measures.

One remark that should be made is that in the Gaussian case, the choice of the Mahalanobis distance may seem obvious. However, one should realize that other choices may be possible as well. One of these is to consider image space endowed with a Gaussian density as a realization of Gauss space [24]. Such an approach would, however result in highly nonlinear behavior, but we have no argument yet to discard this option beforehand. Another option might be to consider the so-called canonical distance measure proposed in [25]. Even though the latter is normally used in supervised learning algorithms, it might provide additional insight with respect to our current approach.

## Acknowledgement

We thank Lewis Griffin, University College London, for providing us with a preprint of the work presented in [1].

## References

1. Griffin, L.D.: The 2nd order local-image-structure solid. IEEE Transactions on Pattern Analysis and Machine Intelligence – (in press) –
2. Cao, G., Chen, J., Jiang, J.: A novel local invariant descriptor adapted to mobile robot vision. In: Proceedings of the 2004 American Control Conference. Volume 3. (2004) 2196–2201

3. Gevers, T., Smeulders, A.W.M.: Image retrieval by multi-scale illumination invariant indexing. In: Proceedings of the IAPR International Workshop on Multimedia Information Analysis and Retrieval, Springer-Verlag (1998) 96–108
4. van Ginneken, B., Stegmann, M.B., Loog, M.: Segmentation of anatomical structures in chest radiographs using supervised methods: a comparative study on a public database. Medical Image Analysis **10** (2006) 19–40
5. Kanters, F., Platel, B., Florack, L.M.J., ter Haar Romeny, B.M.: Content based image retrieval using multiscale top points. In: Proceedings of the 4th international conference on Scale Space Methods in Computer Vision, Isle of Skye, UK, Springer (2003) 33–43
6. Lillholm, M., Pedersen, K.S.: Jet based feature classification. In: Proceedings of the 17th International Conference on Pattern Recognition. Volume 2. (2004)
7. Pham, T.V., Smeulders, A.W.M.: Sparse representation for coarse and fine object recognition. IEEE Transactions on Pattern Analysis and Machine Intelligence **28** (2006) 555–567
8. Ravela, S., Manmatha, R.: Retrieving images by appearance. In: Proceedings of the 6th International Conference on Computer Vision. (1998) 608–613
9. Schmid, C., Mohr, R.: Local grayvalue invariants for image retrieval. IEEE Transactions on Pattern Analysis and Machine Intelligence **19** (1997) 530–535
10. Eberly, D.: A differential geometric approach to anisotropic diffusion. In: Geometry-Driven Diffusion in Computer Vision. Kluwer Academic Publishers (1994) 371–392
11. Eberly, D.: Ridges in Image and Data Analysis. Kluwer Academic Publishers (1996)
12. Florack, L.M.J.: Deep structure from a geometric point of view. In: Proceedings of the Workshop on Deep Structure, Singularities, and Computer Vision. Volume 3753., Springer-Verlag (2005) 135–145
13. Griffin, L.D., Lillholm, M.: Hypotheses for image features, icons and textons. International Journal of Computer Vision **70** (2006) 213–230
14. Koenderink, J.J., van Doorn, A.J.: Receptive field assembly specificity. Journal of Visual Communication and Image Representation **3** (1992) 1–12
15. Koenderink, J.J., van Doorn, A.J.: Metamerism in complete sets of image operators. In: Advances in Image Understading '96. (1996) 113–129
16. Lillholm, M., Nielsen, M., Griffin, L.D.: Feature-based image analysis. International Journal of Computer Vision **52** (2003) 73–95
17. Nielsen, M., Florack, L.M.J., Deriche, R.: Regularization, scale-space, and edge detection filters. Journal of Mathematical Imaging and Vision **7** (1997) 291–307
18. Tikhonov, A.N., Arseninn, V.Y.: Solution of Ill-Posed Problems. Winston and Sons, Washington D.C. (1977)
19. Marroquin, J., Mitter, S., Poggio, T.: Probabilistic solution of ill-posed problems in computational vision. Journal of the American Statistical Association **82** (1987) 76–89
20. Wahba, G.: Spline models for observational data. Society for Industrial and Applied Mathematics, Philadelphia (1990)
21. Rasmussen, C.E., Williams, C.K.I.: Gaussian processes for machine learning. The MIT Press (2005)
22. Loog, M.: Support blob machines. the sparsification of linear scale space. In: Proceedings of the ECCV. Volume 3024 of Lecture Notes in Computer Science., Springer-Verlag (2004) 14–24
23. Anderson, T.W.: An Introduction to Multivariate Statistical Analysis. John Wiley & Sons (1984)
24. Ledoux, M., Talagrand, M.: Probability in Banach Spaces: Isoperimetry and Processes. Springer (1991)
25. Baxter, J.: Learning internal representations. In: Proceedings of the eighth annual conference on computational learning theory, New York, USA, ACM Press New York (1995) 311–320

# Scale Selection for Compact Scale-Space Representation of Vector-Valued Images

Cosmin Mihai[1], Iris Vanhamel[1], Hichem Sahli[1], Antonis Katartzis[2], and Ioannis Pratikakis[3]

[1] Vrije Universiteit Brussel, ETRO-IRIS, Pleinlaan 2, 1050 Brussels, Belgium
cmihai@etro.vub.ac.be, iuvanham@etro.vub.ac.be, hsahli@etro.vub.ac.be
[2] Imperial College, EEE Dep., CSP Group, Exhibition Road, SW7 2AZ London, UK
a.katartzis@imperial.ac.uk
[3] NCSR "Demokritos", IIT, Computational Intelligence Laboratory
GR-153 10 Agia Paraskevi, Athens, Greece
ipratika@iit.demokritos.gr

**Abstract.** This paper investigates the scale selection problem for vector-valued nonlinear diffusion scale-spaces. We present a new approach for the localization scale selection, which aims at maximizing the image content's presence by finding the scale having a maximum correlation with the noise-free image. For scale-space discretization, we propose to address an adaptation of the optimal diffusion stopping time criterion introduced by Mrázek and Navara [1], in such a way that it identifies multiple scales of importance.

## 1 Statement of the Problem

Scale-space theory provides the computer vision and image processing communities with a powerful tool for multiscale signal representation. Yet, scale-spaces contain a very large amount of information, most of it being redundant. Considering the increasing size of the actual image data, a reduced variant of this kind of signal representation is required for e.g. segmentation approaches employing the scale-space framework [2, 3, 4, 5, 6]. Deriving a compact version of the scale-space is desirable also for computational accuracy reasons: by processing an image only at certain scales, known to contain the most important information concerning image embedded features, we eliminate the additional spurious data involved in the case that a larger number of scales is processed.

In scale-space theory, the original image is embedded in a family of gradually smoother versions. Scale-space filters using nonlinear PDE-based diffusion processes are particularly interesting since they avoid blurring and dislocating important features. In this work, we focus on edge-affected diffusion processes in which the diffusion varies spatially aiming to favor intra-region instead of inter-region smoothing, thus overcoming the dislocation of region boundaries. The mathematical form of this type of processes is given by [7, 8]:

$$\partial_t \mathbf{u}^{(i)}(t) = \operatorname{div}\left[g(|\nabla_{\sigma^r}\mathbf{u}(t)|)\nabla \mathbf{u}^{(i)}(t)\right] \ \forall i = 1, 2, \ldots, M \text{ and } \forall t \in \mathbb{R}_+. \quad (1)$$

F. Sgallari, A. Murli, and N. Paragios (Eds.): SSVM 2007, LNCS 4485, pp. 32–42, 2007.

The scale-space image $\mathbf{u}$ is obtained by evolving the above PDE using the original $M$-banded image $\mathbf{f}(\mathbf{x}) = \{f^{(1)}(\mathbf{x}), \ldots, f^{(M)}(\mathbf{x})\}$, $\forall \mathbf{x} \in \Omega \subset \mathbb{Z}^2$, as the initial condition for the time/scale parameter $t = 0$, and homogeneous von Neumann boundary conditions, with $\mathbf{x}$ denoting a 2D position vector on the image plane $\Omega$. Note that $|\nabla_{\sigma^r}\mathbf{u}(t)|$ denotes the regularized vector-valued gradient magnitude (obtained by convolving the image with a Gaussian kernel of size $\sigma^r$ [7]), and $g$ denotes the diffusivity function, which is a bounded, positive, decreasing function that discriminates the different diffusion models. The above formulation includes the homogeneous heat equation, filters related to the Perona and Malik's edge preserving diffusion [9, 10], and filters that originally were expressed as energy minimization problems in [11, 12].
In this paper we discuss the following two issues:

- **Localization Scale Selection.** The selection of an appropriate stopping time for the diffusion, denoted as "localization scale selection". The localization scale marks the minimum size/contrast an image feature should posses to be discriminated from noise. Hence, it can be considered as the scale at which the diffusion filter has removed the noise from homogeneous areas without destroying or dislocating the edges of the important features in the scene.
- **Scale-space Discretization.** The identification of the scales in which higher scale image features are meaningful. Since generally, objects persist over a range of scales [13], compactness requires that a single representative scale for each range is included. Hence, scale-space discretization deals with the detection of all important scales in the scale-space where the image objects are seen cleared of noise in the form of smaller internal structures.

Formally, starting from a naturally sampling of the scale-space image [13]

$$t_j = \begin{cases} 0 & \text{if } j = 0 \\ \exp\left[2(j-1)\tau\right] & \text{if } j > 0 \end{cases} \tag{2}$$

with $\tau$ being the considered time step for time discretization, the *localization scale* will be $s_0 = t_j$ for which $\mathbf{u}(t_j)$ contains a minimum amount of noise whilst retaining all important image features at their exact location. The set of discrete scales $\mathcal{S} = \{s_0, s_1, \ldots s_{\text{end}}\}$ is an ordered subset of the sampled scales obtained by (2) and is denoted here as *scale-space discretization.*

This paper provides an overview of localization scale selection methods under the framework of PDE-based nonlinear diffusion scale-spaces (Sec.2.1), with emphasis on regularization kernel methods [7] (Sec.2.2), and the Mrázek-Navara decorrelation criterion [1] (Sec.2.3). A novel criterion, for localization scale selection, is proposed based on the idea that the "best" diffused image is obtained when the correlation between the noise-free and the enhanced image is maximized (Sec.2.4). For the scale-space discretization, we examine the possibility whether a reported drawback of the Mrázek-Navara decorrelation criterion, for optimal stopping time selection in diffusion-based denoising, may be exploited to identify important coarser scales (Sec.3). Finally, experimental results and discussion are given in Section 4.

# 2 Localization Scale Selection

## 2.1 General Overview

Localization scale selection denotes the process that aims at selecting an unique scale in the scale-space, at which the diffusion filter has removed the noise from homogeneous areas without affecting or dislocating the edges of the important features in the image. Therefore, it is closely related to an optimal diffusion time selection in image denoising and restoration problems. It aims at stopping the diffusion process when the best approximation of the original noise free image is obtained. The literature reports on several approaches that can be categorized according to noise modeling assumptions [14, 15], filtering characteristics [8, 16], and the portion of the scale-space stack that needs to be available before the optimal localization scale can be estimated. Among the latter methods, one can identify (i) methods for which the original noisy image suffices [7, 8], (ii) methods that require the entire scale-space for which a global extremum (maximum or minimum) of some scale measure function is estimated [1, 14, 17], and (iii) methods which require only part of the scale-space image. Approaches in the latter category include Lin and Shi's [16] method which is based on the second order derivative of a functional that models the homogeneity (the underlying idea is that the derivative marks the time at which the flat areas have been "homogenized").

## 2.2 Regularization Kernel Method

Several methods use the notion that the optimal selection corresponds to a scale which should contain a minimum level of noise. An early suggestion by Catté *et.al.* [7] relates the ideal localization scale to the size of the regularization kernel. If we denote by $\sigma^r_{\mathbf{f}}$ the estimated regularization kernel for the image $\mathbf{f}$, then the localization scale is given by:

$$s_0 = t_j \text{ with } j = \begin{cases} 0 & \text{if } \sigma^r_{\mathbf{f}} \leq round\left[\sqrt{2}e^{-\tau}\right] \\ round\left[\frac{1}{2\tau}\ln\frac{(\sigma^r_{\mathbf{f}})^2}{2}+1\right] & \text{otherwise} \end{cases} \tag{3}$$

where $round\left[\cdot\right]$ denotes the rounding to the nearest integer operator. This approach is very fast and relies solely on the noisy input from which the regularization kernel of the diffusion process is estimated. However, in [7], no estimation method was provided for $\sigma^r_{\mathbf{f}}$. Moreover, as reported in [1], this approach has the tendency to produce under-estimates and on occasion an over-estimate.

## 2.3 Decorrelation Method

Some approaches start from the notion that the ideal localization scale can be defined as the scale that corresponds to the noise-free image. In [14], considering an additive uncorrelated noise model with zero mean value, Weickert proposed a method where the optimal localization scale is determined in terms of an estimate of the signal-to-noise ratio. The method tends to under-estimate the localization scale. Analyzing Weickert's method, Mrázek et al. [1] demonstrated that overestimates are more likely to occur when the image is contaminated with high levels of noise. Additionally, they stated that the diffusion process may introduce a correlation between noise and signal, and proposed a *decorrelation criterion.* In this case, the underlying idea is to minimize the correlation between the noise ($\widetilde{\mathbf{n}}(t) = \mathbf{u}(0) - \mathbf{u}(t)$) and the signal $\mathbf{u}(t)$. Formally, this implies the minimization of the following expression:

$$\mathcal{C}_{MN}(t) = |\rho\left[\mathbf{u}(0) - \mathbf{u}(t), \mathbf{u}(t)\right]| \tag{4}$$

where $\rho$ denotes the correlation operator. The localization scale will then be

$$s_0 = \arg\min_{t_j} \left(|\rho\left[\mathbf{u}(0) - \mathbf{u}(t_j), \mathbf{u}(t_j)\right]|\right) \tag{5}$$

In this way, they attempt to choose a scale where the noise and its filtered substitution are as uncorrelated as possible. The advantage against the method in [14] is that the decorrelation method does not require any a priori knowledge concerning the noise variance. However, without extra assumptions on the noise, the signal or the filter, one can not guarantee that the criterion possesses a global minimum [1].

In this paper, the application of criterion (4) to vector-valued images is achieved by estimating the correlation between the "noise" and the diffused image at each band separately, and then combining them as follows:

$$\rho\left[\widetilde{\mathbf{n}}(t), \mathbf{u}(t)\right] = \sum_{i=1}^{i=M} w_i \left|\rho\left[\widetilde{\mathbf{n}}^{(i)}(t), \mathbf{u}^{(i)}(t)\right]\right| \tag{6}$$

where the weights $w_i$ stress the structural importance of the image bands and verify $\sum_{i=1}^{M} w_i = 1$.

## 2.4 Maximum Correlation Method - Proposed Approach

Usually localization scale selection methods assume that the PDE-based diffusion is able to eliminate the noise without affecting the noise-free image. This assumption is far from reality. In this work, we aim at selecting the scale which mostly resembles the noise-free image, or, in other words, the scale where the correlation between the noise-free and the diffused image is maximal. In our model, we assume that at any scale we are able to split the diffused image $\mathbf{u}(t)$, into the noise-free image $\widehat{\mathbf{f}}$, and the noise $\mathbf{n}(t)$:

$$\mathbf{u}(t, \cdot) = \widehat{\mathbf{f}}(\cdot) + \mathbf{n}(t, \cdot) \tag{7}$$

Note that, $\mathbf{n}(t, \cdot)$ is *not* considered here as the diffused version of $\mathbf{n}(0, \cdot)$.
Using basic statistics we get:

$$\sigma^2\left[\mathbf{n}(t)\right] = \sigma^2\left[\widehat{\mathbf{f}}\right] + \sigma^2\left[\mathbf{u}(t)\right] - 2\rho\left[\widehat{\mathbf{f}}, \mathbf{u}(t)\right]\sigma\left[\widehat{\mathbf{f}}\right]\sigma\left[\mathbf{u}(t)\right] \tag{8}$$

$$\sigma^2\left[\widehat{\mathbf{f}}\right] = \sigma^2\left[\mathbf{u}(t)\right] + \sigma^2\left[\mathbf{n}(t)\right] - 2\rho\left[\mathbf{u}(t),\mathbf{n}(t)\right]\sigma\left[\mathbf{u}(t)\right]\sigma\left[\mathbf{n}(t)\right] \tag{9}$$

where $\sigma^2[\mathbf{a}]$ denotes the variance of $\mathbf{a}$, and $\rho[\mathbf{a},\mathbf{b}]$ is the (cross-)correlation between $\mathbf{a}$ and $\mathbf{b}$. Summing (8) and (9) yields the expression of the correlation between the noise-free image $\widehat{\mathbf{f}}$, and the diffused image at scale $t$:

$$\rho\left[\widehat{\mathbf{f}},\mathbf{u}(t)\right] = \frac{\sigma\left[\mathbf{u}(t)\right] - \rho\left[\mathbf{u}(t),\mathbf{n}(t)\right]\sigma\left[\mathbf{n}(t)\right]}{\sigma\left[\widehat{\mathbf{f}}\right]} \tag{10}$$

Since $\widehat{\mathbf{f}}$ is independent of $t$, maximizing (10) with respect to $t$ is equivalent to maximizing the following criterion:

$$\mathcal{C}_{m\rho}(t) \;=\; \sigma\left[\mathbf{u}(t)\right] - \rho\left[\mathbf{u}(t),\mathbf{n}(t)\right]\sigma\left[\mathbf{n}(t)\right] \tag{11}$$

Next we develop upon an approximation of (11).

The noise $\mathbf{n}(t)$, at any scale is defined as the difference between the diffused and the noise-free image. It consists of outlier-noise ($\mathbf{n}_{out}$) and inlier-noise ($\mathbf{n}_{in}$) created when the diffusion process smooths away image structures:

$$\mathbf{n}(t) = \mathbf{n}_{in}(t) + \mathbf{n}_{out}(t) \tag{12}$$

Incorporating (12) in the correlation criterion (11), we get:

$$\mathcal{C}_{m\rho}(t) \;=\; \sigma\left[\mathbf{u}(t)\right] - \mathcal{D}(t)\sigma\left[\mathbf{n}_{out}(t)\right] \tag{13}$$

with

$$\mathcal{D}(t) = \rho\left[\mathbf{u}(t),\mathbf{n}_{out}(t)\right] + \rho\left[\mathbf{u}(t),\mathbf{n}_{in}(t)\right]\frac{\sigma\left[\mathbf{n}_{in}(t)\right]}{\sigma\left[\mathbf{n}_{out}(t)\right]} \tag{14}$$

and $\sigma\left[\mathbf{n}_{out}(t)\right]$ estimated as in [18]:

$$\sigma\left[\mathbf{n}_{out}(t)\right] = 1.4826 * \mathrm{MAD}\left[|\nabla\mathbf{u}(t)|\right] \tag{15}$$

Assuming that the outlier-noise and the noise-free image are independent, i.e. $\rho\left[\widehat{\mathbf{f}},\mathbf{n}_{out}(t)\right] = 0$, and that the inlier-noise and the outlier-noise are uncorrelated, i.e. $\rho\left[\mathbf{n}_{out}(t),\mathbf{n}_{in}(t)\right] = 0$, $\mathcal{D}(t)$ simplifies to:

$$\mathcal{D}(t) \;=\; \frac{\sigma\left[\mathbf{n}_{out}(t)\right]}{\sigma\left[\mathbf{u}(t)\right]} + \frac{\sigma^2\left[\mathbf{n}_{in}(t)\right]}{\sigma\left[\mathbf{u}(t)\right]\;\sigma\left[\mathbf{n}_{out}(t)\right]}\left(1 + \rho\left[\widehat{\mathbf{f}},\mathbf{n}_{in}(t)\right]\frac{\sigma\left[\widehat{\mathbf{f}}\right]}{\sigma\left[\mathbf{n}_{in}(t)\right]}\right) \tag{16}$$

Since the inlier-noise is created by the diffusion process $\sigma\left[\mathbf{n}_{in}(0)\right] = 0$ and $\mathcal{D}(0) = \frac{\sigma[\mathbf{n}_{out}(0)]}{\sigma[\mathbf{u}(0)]}$. Furthermore, as for the diffusion equation given in (1) it is known that $\lim\limits_{t\to\infty}\mathbf{u}^{(i)}(\cdot,t) = \frac{\sum_{\mathbf{x}\in\Omega} f^{(i)}(\mathbf{x})}{\sum_{\mathbf{x}\in\Omega} 1}$, one can demonstrate that $\lim\limits_{t\to\infty}\sigma^2\left[\mathbf{n}_{in}(t)\right] = \sigma^2\left[\widehat{\mathbf{f}}\right]$ and $\lim\limits_{t\to\infty}\rho\left[\widehat{\mathbf{f}},\mathbf{n}_{in}(t)\right] = -1$. Moreover, we can be reasonably sure that the latter is negative for all scales, hence $\lim\limits_{t\to\infty}\mathcal{D}(t)$ approaches $-\infty$.

From our experiments, given the adopted method for estimating the standard deviation of the outlier-noise, we ascertained that $\mathcal{D}$ remains approximately constant at the finer scales. Therefore, we shall assume that it remains constant until the localization scale is selected. Although $\mathcal{D}(t)$ should depend on the initial level of outlier noise and the diffusion model, further investigation of its behavior is needed to obtain an efficient modeling.

In this work, we assume $\mathcal{D}(t) = \mathcal{D}(0) = \frac{\sigma[\mathbf{n}_{out}(0)]}{\sigma[\mathbf{u}(0)]}$, hence, the localization scale corresponds to the scale that maximizes the following criterion:

$$\widetilde{\mathcal{C}}_{m\rho}(t) \;=\; \sigma\left[\mathbf{u}(t)\right] - \frac{\sigma\left[\mathbf{n}_{out}(0)\right]}{\sigma\left[\mathbf{u}(0)\right]}\sigma\left[\mathbf{n}_{out}(t)\right] \tag{17}$$

Note that, for noise-free signals $s_0 = t_0$ since for diffusion processes, based on (1), the variance of $\mathbf{u}$ decreases with scale.

For vector-valued images, we estimate the correlation coefficients, the variances, and noise variance on each channel separately and combine the results in a weighted sum as in (6).

## 3 Scale-Space Discretization

The objective of scale-space discretization (**SSD**) is to create a compact version of the scale-space stack composed solely of meaningful scales. To achieve this, common practice suggests the definition of a *scale information measure* to describe the relevance of the image structure at a certain scale. Depending on the nature of this measure we can classify the existing SSD approaches in two main categories: (i) Local SSD and (ii) Global SSD. The first category consists of methods which define the scale information measure on a portion of the image domain, considering that image structures differ in size locally from region to region. Usually, these methods integrate scale in the feature detector and adopt a methodology derived from the one proposed by Lindeberg [2]. These concepts are generalized for nonlinear scale-spaces using local Lyaponov functionals [19]. The second category includes methods for which the scale information measure is defined on the entire image domain, representing the importance of each scale for the original image content [20, 21]. These global measures present many local extrema which correspond to objects of different contrast/size in the image.

One of the remarks made by Mrázek *et al.* in [1] is that the decorrelation criterion can not guarantee the uniqueness of the minimum in the curve given by (4). However, the appearance of multiple local minima seems to be an indication that a significant change in the image structure occurs at these scales. In this work, starting from the localization scale $s_0 = t_j$, as estimated in the previous section, we iteratively locate significant scales by considering the first local minimum of

$$\mathcal{C}_{MN}(t)_{[t_k,t_N]} = \left|\rho\left[\mathbf{u}(t_k) - \mathbf{u}(t), \mathbf{u}(t)\right]\right|, t \in [t_{k+1}, t_M]; k = j_1, \ldots, j_L \tag{18}$$

with $t_{j_1}, \ldots t_{j_L}$ being the iteratively selected scales. In this way, at each iteration, we select a coarser scale $t_{j_{i+1}}$ which is uncorrelated with the finer scale $t_{j_i}$.

## 4 Experimental Results and Discussion

The proposed localization scale selection and scale-space discretization methods are evaluated using the following nonlinear diffusion model [6, 22]:

$$\partial_t \mathbf{u}^{(i)}(t) = \text{div}\left[g(|\nabla_{\sigma_r}\mathbf{u}(t)|)\,\frac{\nabla \mathbf{u}^{(i)}(t)}{|\mathbf{u}^{(i)}(t)|}\right], \quad \forall i = 1..M \tag{19}$$

where $g(y) = \left[1 + k^2/y^2\right]^{-1}$.

Two noise-free artificial RGB images (Figure 1.(a)&(b)) to which we added increasing levels of Gaussian noise (Figure 1.(c)-(f)) are used to provide ground-truth for the localization scale methods of Section 2.3 and Section 2.4. The weights used for the estimation of the correlation coefficients, variances, and noise variances for equations (6) and (17) are set to $w = \frac{[1,1,1]}{3}$. We employ four criteria to measure the closeness of the selected localization scale with the noise-free image: (i) *mean absolute difference* between the pixel-values of the selected scale and the noise-free image (**MMAD**), (ii) the PSNR, (ii) the *structural similarity index* (**SSIM**)[23], and (iv) the correlation between the selected scale and the noise-free. The fulfillment of the first criterion implies small values of the corresponding measure, while for the last three criteria large values are desired.Table 1 shows the results in the case that channel-uncorrelated noise is added. Both selection methods perform rather well for low amounts of noise. For severe noise contamination, the proposed method is clearly the better in terms of correlation.

Visual inspection of the selected scales (Fig.1), indicated that the decorrelation criterion $\mathcal{C}_{MN}$ often underestimates the localization scale. The latter is due to the presence of negative correlation between the noise estimate and the diffused image. In [1], it was mentioned that this deteriorates the decorrelation method. We also did notice that there exists a local minima in $\mathcal{C}_{MN}(t)$ at a coarser scale which is a better match for the localization scale.

**Table 1.** Ground-truth evaluation measures for the localization scale selection

| Criterion | *MMAD* | | *PSNR* | | *SSIM* | | $\rho\left[\tilde{\mathbf{f}}, \mathbf{u}(s_0)\right]$ | |
|---|---|---|---|---|---|---|---|---|
| $\sigma\left[\mathbf{n}(0)\right]$ | $\mathcal{C}_{MN}$ | $\tilde{\mathcal{C}}_{mp}$ | $\mathcal{C}_{MN}$ | $\tilde{\mathcal{C}}_{mp}$ | $\mathcal{C}_{MN}$ | $\tilde{\mathcal{C}}_{mp}$ | $\mathcal{C}_{MN}$ | $\tilde{\mathcal{C}}_{mp}$ |
| 5.772 | 0.022 | 0.002 | 92.016 | 177.697 | 0.961 | 0.963 | 0.927 | 1.000 |
| 27.918 | 0.083 | 0.052 | 19.099 | 23.360 | 0.961 | 0.961 | 0.921 | 0.973 |
| 37.125 | 0.115 | 0.069 | 15.917 | 20.991 | 0.960 | 0.960 | 0.842 | 0.955 |
| 43.912 | 0.138 | 0.083 | 14.338 | 19.618 | 0.959 | 0.959 | 0.785 | 0.940 |
| 49.343 | 0.155 | 0.094 | 13.258 | 18.657 | 0.958 | 0.958 | 0.737 | 0.927 |
| 53.974 | 0.171 | 0.104 | 12.439 | 17.840 | 0.957 | 0.957 | 0.695 | 0.912 |
| 58.250 | 0.183 | 0.112 | 11.800 | 17.263 | 0.956 | 0.956 | 0.659 | 0.901 |
| 62.104 | 0.195 | 0.120 | 11.254 | 16.788 | 0.955 | 0.956 | 0.628 | 0.892 |
| 65.513 | 0.205 | 0.127 | 10.805 | 16.337 | 0.954 | 0.955 | 0.598 | 0.882 |
| 68.601 | 0.214 | 0.134 | 10.422 | 15.877 | 0.953 | 0.954 | 0.573 | 0.867 |
| 71.293 | 0.222 | 0.139 | 10.111 | 15.648 | 0.953 | 0.954 | 0.551 | 0.866 |
| Average | 0.155 | 0.094 | 20.132 | 32.734 | 0.957 | 0.958 | 0.720 | 0.919 |

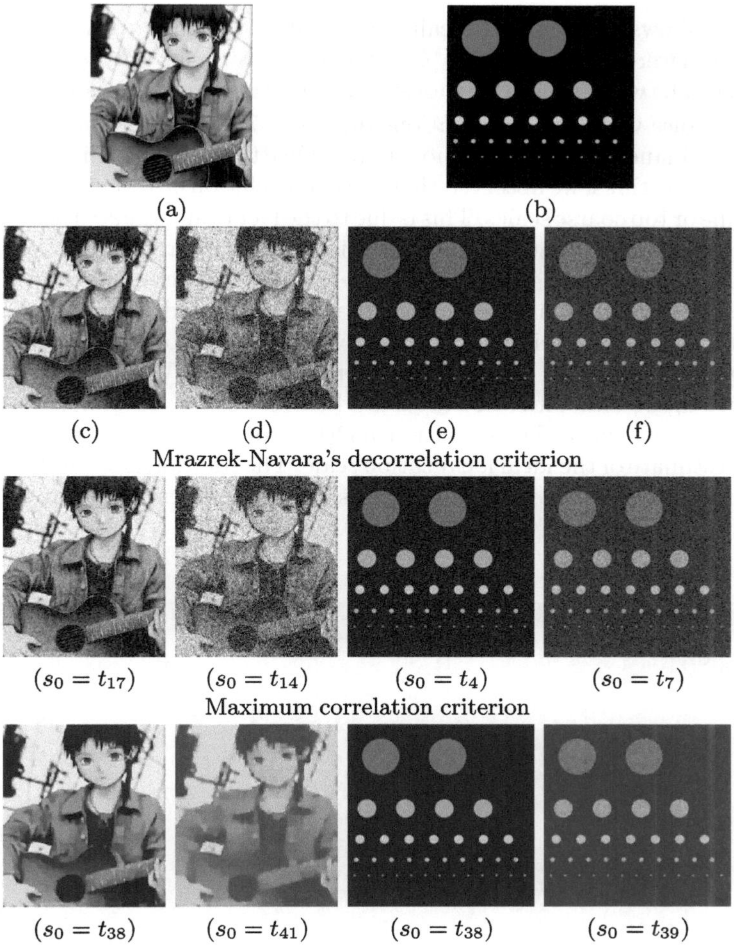


**Fig. 1.** Localization scale selections. Noise-free images: (a)&(b). Noise corrupted images: (c)&(e) with $\sigma\left[\mathbf{n}_{out}(0)\right] \simeq 5.772$, and (d)&(f) with $\sigma\left[\mathbf{n}_{out}(0)\right] \simeq 65.513$.

**Fig. 2.** Localization scale selections in natural scene images. (a): Input image, (b) Mrazrek-Navara's decorrelation criterion: $s_0 = t_{56}, t_1$, (c) excluding overestimates at $u(t) \approx u(\infty)$: $t_1, t_1$, and (d) Maximum correlation criterion: $s_0 = t_0, t_{13}$.

Figure 2 shows the selected localization scales for two natural scene images. These images were processed in $La^*b^*$ color space, using the fully generated scale-space stack. The weights used for the estimation of the correlation coefficients, variances, and noise variances for equations (6) and (17) are set to $w = \frac{[6,1,1]}{8}$, which gives the luminance $L$ three times more importance than the chrominance information. In these settings, we observed that the decorrelation criterion $\mathcal{C}_{MN}$ selects either too fine or too coarse scales. This is due to the fact that large structural changes occur when a large amount of outlier-noise is removed or large objects are merged into the background inducing very low correlation values between removed and remaining scale content. This problem can be partially avoided by selecting a scale corresponding to a local minimum which is closer, in terms of noise level, to a noise-free image. The proposed criterion $\widetilde{\mathcal{C}}_{m\rho}$ performs better. We did however notice that it encounters difficulties in the cases when very low levels of noise are added. In these situations the initial estimation of $\mathcal{D}(t)$ is too small and the selection yields an under-estimate of the ideal localization scale. The solution is to model $\mathcal{D}(t)$ taking into account diffusion characteristics as well as noise estimates at each scale.

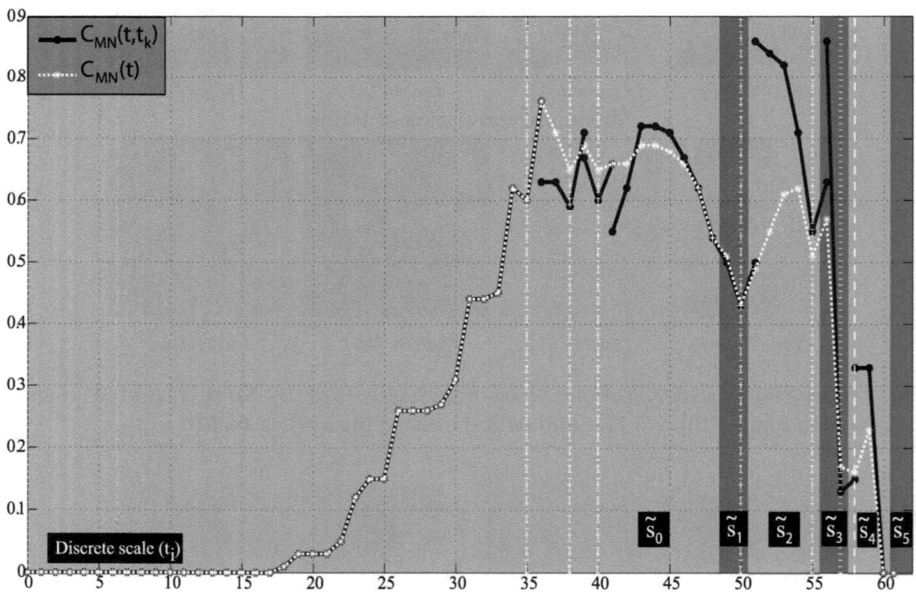


**Fig. 3.** Scale-space discretization of the image 1.b using as localization scale $t_0$: Evolution of the decorrelation value obtained via (4) and (18). The selected scales correspond to the local minima of the curves.

Figure 3 illustrates the scale-space discretization results using the local minima of $\mathcal{C}_{MN}(t)$ and the successive minima $\mathcal{C}_{MN}(t)_{[t_k,t_N]}$, using the generated scale-space diffusion stack of the noise-free dots image of Figure 1.b. Dark and light grey backgrounds correspond to the number of rows of dots visually perceived for each scale. Since all scales in each of these intervals have similar appearance, they are considered as equal candidates for the ideal scale $\tilde{s}$ containing the corresponding number of rows of dots. The white, dotted plot represents $\mathcal{C}_{MN}(t)$ and the continuous black curve corresponds to $\mathcal{C}_{MN}(t)_{[t_k,t_N]}$. The gaps in the latter curve indicate scales where new iterations are started. The proposed method identifies most of the scales of importance in the scale-space stack. However, many local minima are detected at finer scales which do not indicate any obvious structural change in the image. Furthermore, we experienced that coarser scale structural changes are not always detected due to a too fine sampling, which implies smoother transitions. The main difference between the proposed method and the successive local minima of the decorrelation criterion is that the local minima are better discriminated at the coarser scales, which results in selecting more relevant coarser scales.

## 5 Conclusions

In this paper we propose a new method for the selection of a localization scale and explore the scale-space discretization capabilities of the Mrázek and Navara's decorrelation criterion [1]. We apply the proposed methods for vector-valued images and use them to produce a compact scale-space representation of images. The conducted experiments show that the proposed maximum correlation method for localization scale selection produce encouraging results. Further investigation of its potential is necessary for the inclusion of diffusion specific information, hence modeling $\mathcal{D}(t)$ in function of noise. The decorrelation criterion in [1] provides good performance, for scale-space discretization, in cooperation with some scale noise indicator, such as the proposed estimate for the regularization kernel.

## Acknowledgements

This work has been partially funded by the EU-IST project STREAM "Technology to Support Sustainable Humanitarian Crisis Management" (contract EU-IST-FP6-2-511 705).

## References

[1] Mrázek, P., Navara, M.: Selection of optimal stopping time for nonlinear diffusion filtering. Int. J. of Comp. Vis. **52**(2/3) (2003) 189–203
[2] Lindeberg, T.: Feature detection with automatic scale selection. Int. J. of Comp. Vis. **30**(2) (1998) 77–116
[3] Pratikakis, I., Sahli, H., Cornelis, J.: Low level image partitioning guided by the gradient watershed hierarchy. Signal Processing **75**(2) (1998) 173–195
[4] Vanhamel, I., Pratikakis, I., Sahli, H.: Multi-scale gradient watersheds of color images. IEEE Trans. on IP **12**(6) (2003) 617–626
[5] Petrovic, A., Divorra Escoda, O., Vandergheynst, P.: Multiresolution segmentation of natural images: From linear to non-linear scale-space representations. IEEE Trans. on IP **13**(8) (2004) 1104–1114

[6] Katartzis, A., Vanhamel, I., Sahli, H.: A hierarchical markovian model for multiscale region-based classification of vector-valued images. IEEE Trans. on Geoscience and Remote Sensing **43**(3) (2005) 548–558
[7] Catté, F., Lions, P.L., Morel, J.M., Coll, T.: Image selective smoothing and edge detection by nonlinear diffusion. SIAM J. on Numerical Analysis **29**(1) (1992) 182–193
[8] Whitaker, R., Gerig, G.: Vector-valued diffusion. In: Geometry-Driven Diffusion in Computer Vision. Volume 1 of Computational Imaging and Vision. Kluwer Academic Publishers (1994) 93–134
[9] Perona, P., Malik, J.: Scale-space and edge detection using anisotropic diffusion. IEEE Trans. on PAMI **12**(7) (1990) 629–639
[10] Black, M., Sapiro, G., Marimont, D., Heeger, D.: Robust anisotropic diffusion. IEEE Trans. on IP **7**(3) (1998) 421–432
[11] You, Y.L., Xu, W., Tannenbaum, A., Kaveh, M.: Behavioral analysis of anisotropic diffusion in image processing. IEEE Trans. on IP **5**(11) (1996) 1539–1553
[12] Geman, D., Reynolds, G.: Constrained restoration and the recovery of discontinuities. IEEE Trans. on PAMI **14**(3) (1992) 367–383
[13] Koenderink, J.: The structure of images. Biological Cybernetics **50** (1984) 363–370
[14] Weickert, J.: Coherence-enhancing diffusion of colour images. Image and Vision Computing **17**(3-4) (1999) 201–212
[15] Mrázek, P.: Selection of optimal stopping time for nonlinear diffusion filtering. In: Lec. Notes in Computer Science. Volume 2106., Springer (2001) 290–298
[16] Lin, Z., Shi, Q.: An anisotropic diffusion PDE for noise reduction and thin edge preservation. In: Int. Conf. on Image Analysis and Processing. (1999) 102–107
[17] Gilboa, G., Sochen, N., Zeevi, Y.: Estimation of optimal pde-based denoising in the snr sense. CCIT report 499, Technion-Israel (2004)
[18] Hampel, F.: The influence curve and its role in robust estimation. J. Amer. Statist. Assoc. **69** (1974) 383–393
[19] Sporring, J., Colios, C., Trahanias, P.: Generalized scale-selection. In: IEEE Int. Conf. on Image Processing. Volume 1. (2000) 920 – 923
[20] Sporring, J., Weickert, J.: Information measures in scale-spaces. IEEE Transactions on Information Theory **45** (1999) 1051–1058
[21] Hadjidemetriou, E., Grossberg, M., Nayar, S.: Resolution selection using generalized entropies of multiresolution histograms. In: European Conf. on Computer Vision. (2002)
[22] Vanhamel, I.: Vector valued nonlinear diffusion and its application to image segmentation. PhD thesis, ETRO/IRIS: Vrije Universiteit Brussel, Brussels-Belgium (2006)
[23] Wang, Z., Bovik, A.C., Sheikh, H.R., Simoncelli, E.P.: Image quality assessment: From error visibility to structural similarity. IEEE Trans. on Image Processing **13**(4) (2004) 600–612

# An High Order Finite Co-volume Scheme for Denoising Using Radial Basis Functions

Serena Morigi and Fiorella Sgallari

Department of Mathematics-CIRAM, University of Bologna, Bologna, Italy
{morigi,sgallari}@dm.unibo.it

**Abstract.** In this work we investigate finite co-volume methods for solving Partial Differential Equation (PDE) based diffusion models for noise removal in functional surfaces. We generalized the model proposed by Tai et al. [1][2] based on the reconstruction of a noise-reduced surface from the smoothed normal field, considering a curvature preserving term. The discretization of the PDE model by basic finite co-volume schemes on unstructured grids is investigated. The accuracy of the numerical model is then improved by using an higher order optimal recovery based on Radial Basis Functions (RBF). Preliminary numerical results demonstrate the effectiveness of the new numerical approach.

## 1 Introduction

Removal of noise is a necessary pre-processing step for several surface processing tasks such as surface reconstructions, curvature detections, structure recognition, and so on. This problem arises for example in 3D scanner acquisition of object surfaces, when the equipment provides an unstructured cloud of points corrupted by measurement errors. In this work we will consider the case of reconstruction of functional surfaces that are typical of range image 3D scanners, where the range image represents an unstructured set of scalar values corresponding to a set of points on a 3D acquisition plane.

This framework can be considered an extension of the classical image denoising approach since a gray-scale image can be interpreted as a functional surface by assigning the image intensity to the elevation along the $z$ direction of the $xy$ image plane. In this case the problem is simplified because the data domain is a rectangular structured grid.

In this work we generalized the ideas proposed in [1] and [2] based on a TV-norm approach. This results into two coupled non-linear second order PDEs. The first equation smoothes the normal field of the corrupted surface, while the second equation reconstructs a noise-reduced surface from the smoothed normal field. This model has been proposed for noise removal in digital images where the purpose is to preserve the edges. In the functional surface case the structures that characterize the data are the creases, that is areas of high curvature. The original model will be changed according to this purpose.

The contribution of this paper is twofold. First, we provide a discretization of the surface denoising PDE model by means of a basic finite co-volume scheme

F. Sgallari, A. Murli, and N. Paragios (Eds.): SSVM 2007, LNCS 4485, pp. 43–54, 2007.

on unstructured grid. Then, we improve the accuracy of the basic computational scheme using an higher order optimal recovery based on Radial Basis Functions.

High order schemes currently form a major research direction in the theory of finite volume methods [3],[4]. In order to increase the spatial order of accuracy of the basic finite volume schemes, the reconstruction of a piecewise smooth function from the cell average is required. The first higher order reconstruction schemes employed were based on polynomials and suffered from the typical behavior of multivariate polynomials, such as oscillation and ill-conditioning. To obtain stable schemes a strategy from Essentially Non-Oscillatory (ENO) algorithms can be used as trivial recovery operators, thus involving a significative computational effort. Recovery algorithms of ENO-type based on RBFs were developed in [3] and theoretical results have been discussed in [4]. The optimal recovery with RBF that we propose employs a particular stencil which avoids the ENO-construction.

The paper is organized as follows. The surface denoising PDE model is introduced in Section 2, while its discretization based on a basic finite volume scheme is discussed in Section 3. The radial recovery strategy is introduced in Section 4 in order to derive an higher order finite volume scheme. A few selected numerical examples will be presented in Section 5 to demonstrate the effectiveness of the proposed computational scheme. Section 6 contains concluding remarks.

## 2 A Model for Denoising by Smoothing and Surface Fitting

We consider an intensity function $u : \Omega \subset \mathbb{R}^2 \to \mathbb{R}$ as a parameterized surface

$$S = (x, y, u(x, y))$$

along the $xy$-plane. Let us denote the graph $S$ by *functional surface*. This turns out to be a common way to represent bivariate data such as height fields and grey-scale images. In fact an image can be interpreted as a discretization of a continuous function defined on $\Omega \subset \mathbb{R}^2$ by assigning the image intensity to the elevation along the $z$ direction.

By introducing the function $d(x, y, z) = z - u(x, y)$ then the surface $S$ is implicitly defined by $d(x, y, z) = 0$. We will denote by $\mathbf{n}$ the unit normal vector of the surface $S$ given by

$$\mathbf{n} = \frac{\nabla d}{|\nabla d|} = \frac{(-u_x, -u_y, 1)}{\sqrt{1 + u_x^2 + u_y^2}},$$

where $u_x$ denotes the first partial derivative of $u$ with respect to $x$, and analogously for $u_y$.

We are interested in restoring a functional surface $d(x, y, z)$ which is corrupted by noise in such a way that the process should recover the main structures of the surface. Let us denote by $d_0$ the observed surface and by $d$ the noisy-free surface. The model of degradation we assume is $d_0 = d + \eta^\sigma$, where $\eta^\sigma$ represents some additive noise in the given data, due for example to measurement errors. Throughout this paper, we assume that a fairly accurate bound $\sigma$ of the norm of the noise is available, i.e.,

$$\int_\Omega (\eta^\sigma(x))^2 dx \leq \sigma^2. \tag{1}$$

In the image processing field, the Total Variation (TV) models proposed, see for example [5] and the references herein, have been shown to be quite effective in removing noise without causing excessive smoothing of the edges. The original formulation of this filter is to obtain $d(x, y, z)$ as a solution of the constrained optimization problem:

$$\inf_d \int_\Omega |\nabla d| dx \quad subject \quad to \quad \int_\Omega |d - d_0|^2 dx = \sigma^2. \tag{2}$$

However it is well known that the TV norm filter has the disadvantage of staircase effect: smooth functions get transformed into piecewise constant functions. Following [1] we propose to modify equation (2) to get a smoothed normal vector field, that is to minimize the TV norm of $\nabla d/|\nabla d|$, which gives the following equation:

$$\inf_d \int_\Omega |\nabla \frac{\nabla d}{|\nabla d|}| dx \quad subject \quad to \quad \int_\Omega |d - d_0|^2 dx = \sigma^2. \tag{3}$$

The fourth order PDE model that results from directly minimizing (3) is difficult to solve numerically in a stable manner. Therefore, equation (3) is split into two steps. The first step involves the smoothing of the normal vectors $\mathbf{n}_0 = \nabla d_0/|\nabla d_0|$ minimizing the functional:

$$\inf_{|\mathbf{n}|=1} \left\{ \int_\Omega |\nabla \mathbf{n}| dx + \frac{\alpha}{2} \int_\Omega |\mathbf{n} - \mathbf{n}_0|^2 dx \right\}, \tag{4}$$

where $\alpha > 0$ is a parameter that balances smoothing and fidelity to the original vector field. The second step recovers the functional surface from the smoothed normal field that results from solving (4), by minimizing the functional

$$\inf_d \left\{ \int_\Omega (|\nabla d| - \nabla d \cdot \mathbf{n}) dx + \int_\Omega g(curv_S) |\nabla d| dx + \frac{\beta}{2} \int_\Omega (|d - d_0|^2 - \sigma^2) dx \right\} \tag{5}$$

where $\beta > 0$, is a given parameter, and

$$curv_S = div(\mathbf{n}) = div\left(\frac{(-u_x, -u_y, 1)}{\sqrt{1 + u_x^2 + u_y^2}}\right) = \frac{(1 + u_y^2)u_{xx} - 2u_x u_y u_{xy} + (1 + u_x^2)u_{yy}}{(1 + u_x^2 + u_y^2)} \tag{6}$$

is the mean curvature of the functional surface $S$. The central functional term in (5) has been added to model the curvature driven diffusion. The diffusivity function $g(\cdot)$ considered in this work is the well-known Perona-Malik diffusivity $g(s) = \frac{1}{(1+s^2/K^2)}$, $\quad K \geq 0$.

For a fixed parameterization along the flow, the evolution of $S$ is given by

$$S_t = \frac{\partial}{\partial t}(x, y, u(x, y)) = (0, 0, \frac{\partial u}{\partial t}(x, y)).$$

Thus, for tracking the evolution of the flow field along the direction $z = (0, 0, 1)$, we let $u$ evolve in time. In order to simplify the computation we reduce the problem (4) considering $\mathbf{n} = (\cos(\theta), \sin(\theta))$ analogously to what done in [1] using polar coordinates to represent the unit normal vector $\mathbf{n}$, thus obtaining $|\nabla \mathbf{n}| = |\nabla \theta|$. This observation leads us to reduce the optimality conditions for (4) as follows

$$\frac{\partial \theta}{\partial t} = \nabla \cdot (\frac{\nabla \theta}{|\nabla \theta|}) - \alpha(\sin(\theta - \theta_0)) \tag{7}$$

where an artificial time variable is added, and homogeneous Neumann boundary conditions can be considered.

A minimizer for (5) can be computed by

$$\frac{\partial d}{\partial t} = \nabla \cdot (\frac{\nabla d}{|\nabla d|}(1 + g(curv_S)) - \mathbf{n}) - \beta(d - d_0) \tag{8}$$

where an artificial time variable has been added and homogeneous Neumann boundary conditions can be considered. In [1] a formula for an exact computation of $\beta$ is provided. Alternatively, when an estimate of the norm of the error is available as in (1), the evolution driven by the PDE model

$$\frac{\partial d}{\partial t} = \nabla \cdot (\frac{\nabla d}{|\nabla d|}(1 + g(curv_S)) - \mathbf{n}), \tag{9}$$

can be terminated as soon as the condition

$$|d - d_0| \leq c\sigma, \tag{10}$$

with $c > 1$ be a fixed constant, is satisfied. This stopping criterium is a generalization of the well-known discrepancy principle used in the solution of ill-posed problems [6].

In the next section we discuss a discretization of the PDE model given by (7) and (9) in terms of basic finite co-volume schemes.

## 3 Finite Co-volume Discretization

In the finite volume spatial discretization the computational domain $\Omega \subset \mathbb{R}^2$, that we assume for simplicity bounded by a piecewise polygonal curve, is first tessellated into a collection $\mathcal{T}$ of non-overlapping triangles $T_i$, $i = 1, \ldots, \mathcal{N}$, such that

$$\Omega = \bigcup_{i=1}^{\mathcal{N}} T_i \tag{11}$$

forms an unstructured mesh characterized by a length scale $h$. We follow the cell-centered finite volume scheme where the triangle themselves serve as control volumes with unknown solutions stored on each triangle. Other finite volume schemes have been successfully applied to the image processing field [7].

Fundamental to finite volume schemes is the introduction of the operator called *cell average* for each $T_i \in \mathcal{T}$:

$$L_i\, u = \overline{u_i}(t) := \frac{1}{|T_i|} \int_{T_i} u(t,x)dx, \tag{12}$$

where $|T_i|$ denotes the area of the $i$th triangle.

Let us describe the spatial discretization by a basic finite volume scheme of the following PDE model which generalizes both (7) and (9)

$$\frac{\partial u}{\partial t} = \nabla \cdot \left( \frac{\nabla u}{|\nabla u|} - \mathbf{a} \right) - f(u - u_0), \tag{13}$$

where $\mathbf{a}$ represents a generic vector and $f$ a generic function. Applying in the usual way the Gauss-Green theorem to (13), leads to:

$$\frac{d\overline{u}(t)}{dt} = \frac{1}{|T_i|} \sum_{j \in N(i)} \int_{e_{ij}} (\frac{\nabla u}{|\nabla u|} - \mathbf{a}) \cdot \mathbf{n} ds - \frac{1}{|T_i|} \int_{T_i} f(u - u_0) ds \tag{14}$$

where $e_{ij} = \partial T_i \bigcap \partial T_j$ is the common edge between triangles $T_i$ and $T_j$, $\mathbf{n} = (nx, ny)^T$ is the outer unit normal vector on the edge $e_{ij}$, and $N(i) = \{j \in \mathbb{N} | e_{ij} \ \ is \ \ edge \ \ of \ \ T_i\}$. In order to numerically compute the line integral in (14) on the edge $e_{ij}$ we use Gaussian quadrature of the form

$$\int_e G(\phi(s))ds = \sum_{k=1}^{ng} w_k G(\phi(s_k)) + \mathcal{O}(h^{2ng}) \tag{15}$$

where $ng$ denotes the number of integration nodes, $w_k$ are certain quadrature weights, and $h$ is in this case the length of the control volume edge. In particular, on the edge $e_{ij}$ defined by the vertices $\mathbf{x}_i$, $\mathbf{x}_j$, we chose a three-point Gaussian quadrature in the interval $[-1, 1]$ and the integral nodes were given by the parametrization

$$\phi_{ij}(s) = \frac{1}{2}(\mathbf{x}_i - \mathbf{x}_j) + \frac{s}{2}(\mathbf{x}_j - \mathbf{x}_i).$$

Thus the first integral term on the right-hand side of equation (14) becomes:

$$\frac{1}{|T_i|} \sum_{j \in N(i)} \int_{-1}^{1} \left( \frac{u_x(\phi_{ij}(s))}{|\nabla u|} nx + \frac{u_y(\phi_{ij}(s))}{|\nabla u|} ny - \mathbf{a} \cdot \mathbf{n} \right) ds$$

and applying the quadrature rule (15) we get

$$\frac{1}{|T_i|} \sum_{j \in N(i)} \frac{|e_{ij}|}{2} \left\{ \sum_{k=1}^{ng} w_k \frac{u_x(\phi_{ij}(s_k), t)nx + u_y(\phi_{ij}(s_k), t)ny}{\sqrt{u_x(\phi_{ij}(s_k), t)^2 + u_y(\phi_{ij}(s_k), t)^2}} + \mathcal{O}(h^{2ng}) \right\} + |e_{ij}|(\mathbf{a} \cdot \mathbf{n}).$$

The second integral term on the right-hand side of equation (14) can be accurately computed in a similar way by a two-dimensional Gaussian quadrature formula ensuring that the errors in computing (14) are due to inaccuracies in the first integral term and not in the second one. In this work we used a seven points Gaussian quadrature rule.

According to the classical conservation theory literature [8], a numerical approximation of the flux function in (14) can be obtained by basic finite volume method replacing the unknown values $u(\phi(s_k), t)$ simply by the cell averages, that is $F(\overline{u}, \overline{u}; \mathbf{n})$, thus obtaining the following semi-discrete finite volume basic scheme.

**Definition 1.** ***Basic FV.*** *The semi-discrete finite volume approximation of (14) utilizing continuous in time solution representation, $t \in [0, +\infty)$, and piecewise constant solution representation in space $u_h$, such that*

$$\overline{u}_j(t) = \frac{1}{|T_j|} \int_{T_j} u_h(x, t) dx,$$

*with initial data $\overline{u}_j(0) = \frac{1}{|T_j|} \int_{T_j} u_0(x) dx$, and suitable boundary conditions, is given by the following system of ordinary differential equations*

$$\frac{d\overline{u_i}}{dt} = \frac{1}{|T_i|} \sum_{j \in N(i)} \frac{|e_{ij}|}{2} \sum_{k=1}^{ng} w_k F(\overline{u_i}, \overline{u_j}; \mathbf{n}_{ij}) - \frac{1}{|T_i|} \sum_{j \in N(i)} w_k f(\overline{u}_k - \overline{u}_{0,k}), \quad \forall T_i \in \mathcal{T}. \tag{16}$$

If the weak solution $u$ and the numerical flux function $F$ are continuously differentiable, then the basic finite volume method is of first order in space [8].

The system of ordinary differential equations (16) can be solved in time using a variety of explicit and implicit time integration schemes. A particularly simple time integration scheme we applied is the well-known forward Euler scheme.

Using the basic finite volume scheme and the Euler method, we end up with a first order scheme in space and in time. To improve the time order of convergence we have applied third order strongly stability preserving Runge Kutta method. Moreover, since recovery has to be done on each time level, we suppress dependency on time in the remaining sections.

## 4 Optimal Recovery Using Radial Basis Function

Unfortunately, first order accurate schemes are generally considered too inaccurate for most quantitative calculation unless the mesh spacing is made excessively small thus rendering the scheme inefficient. For a good reconstruction in regions where the solution of (14) is known or expected to be smooth, a higher order reconstruction scheme is desirable. Such high order schemes currently form a major research direction in the theory of finite volume [4].

To increase the spatial order of accuracy of the basic finite volume method, we need to reconstruct a piecewise smooth function from the cell averages. This reconstruction is known as *recovery*. Trivially it is not reasonable to build a reconstruction using all the cell averages. Instead, for each cell (triangle $T_i$) a local reconstruction is computed using all the cell averages of cells in a neighborhood of $T_i$, that we will denote by *stencil* $S_i = \{T_1, \cdots, T_M\}, \quad T_1 = T_i$.

**Definition 2.** *On each stencil, a function $T_i \ni x \mapsto R_i(x)$ is called an $r$th-order recovery function if the following conditions hold*

*1)* $R_i := \frac{1}{|T_i|} \int_{T_i} R(t,x)dx = \overline{u_i}(t) \quad i = 1, \cdots, M,$
*2)* $\lim_{h\to 0} \|R_i - u\|_{L_\infty(\Omega)} = O(h^r)$.

According to [4], the next result states the accuracy of the finite volume approximation.

**Theorem 1.** *Assume $R_i : \mathbb{R}^2 \to \mathbb{R}$ is an $r$th-order recovery function on the triangle $T_i$ for all $T_i \in \mathcal{T}$. Let the weak solution $u$ as well as the numerical flux $F(u, u, \mathbf{n})$ be differentiable up to order* $\min\{r, ng\}$. *Then $\forall T_i \in \mathcal{T}$, the finite volume method*

$$\frac{d\overline{u_i}}{dt} = \frac{1}{|T_i|} \sum_{j\in N(i)} \frac{|e_{ij}|}{2} \sum_{k=1}^{ng} w_k F(R_i(t, \phi_{i,j}(s_k)), R_i(t, \phi_{i,j}(s_k)); \mathbf{n}_{ij}), \tag{17}$$

*has spatial order $O(h^{\min\{r,2ng\}})$.*

However, to avoid unwanted oscillations typical from polynomial interpolation, the ENO approach chooses for each cell a set of different stencils, it computes for each stencil a local reconstruction and then selects the reconstruction where the oscillation of the solution is least. In the next paragraph we consider the possibility to find local functions with good approximation properties in order to avoid the oscillatory behavior. Supported by numerical experiments, summarized in Section 5, we claim that the use of a suitable large stencil in the radial recovery step, avoids the computational cost of the ENO construction.

### 4.1 Radial Recovery Step

For a given triangle $T_i \in \mathcal{T}$ we construct the stencil $S_i$ which contains the triangles that share an edge with $T_i$, as well as the triangles which share a vertex with $T_i$, together with $T_i$ itself. Instead of using polynomial reconstruction we will use a radial recovery function $R(x)$ on a triangle $T_i$ given by

$$R_i(x) = \sum_{j=1}^{M} \lambda_j L_j^{\mathbf{x}_j} \varphi(\|\mathbf{x} - \mathbf{x}_j\|_2), \tag{18}$$

where $\varphi : \mathbb{R}_{\geq 0} \to \mathbb{R}$ is a radial function and $M$ is the number of cells in the stencil $S_i$. Here $L^{\mathbf{x}_j}$ means the application of the cell average operator to the variable $\mathbf{x}_j$. In order to satisfy condition 1) in Definition (2), the coefficients

$\lambda_1, \ldots, \lambda_M$ in (18) are determined by the following recovery conditions on the centers $\{\mathbf{x}_{c_j}\}_{j=1,\cdots,M}$, that are the centroids of the cells in the stencil

$$L_j R_i = L_j u, \quad j = 1, \cdots, M,$$

where $L_j u$ are the cell averages of the triangles in the stencil. These conditions can be conveniently written in a matrix-vector form

$$A\Lambda = U, \qquad A \in \mathbb{R}^{M \times M}, \qquad U \in \mathbb{R}^M, \qquad \Lambda \in \mathbb{R}^M, \tag{19}$$

where $A = [L_i^{\mathbf{x}} L_j^{\mathbf{x}_j} \varphi(\|\mathbf{x} - \mathbf{x}_j\|_2)]_{i,j=1,\cdots,M}$, while the right-hand side is given by $U = \{\overline{u}_j\}_{j=1}^M$. In order to compute the elements of $A$ we need to approximate the term

$$L_j^{\mathbf{y}} \varphi(\|\mathbf{x} - \mathbf{y}\|_2) = \frac{1}{|T_i|} \int_{T_i} \varphi(\|\mathbf{x} - \mathbf{y}\|_2) dy \tag{20}$$

by a midpoint quadrature rule that is $L_j^{\mathbf{y}} \varphi(\|\mathbf{x} - \mathbf{y}\|_2) \approx \varphi(\|\mathbf{x} - \mathbf{x}_{\mathbf{c_j}}\|_2)$.

The application of the operator $L_i^{\mathbf{x}}$ is accounted by means of a seven points Gaussian quadrature rule within the triangle $T_i$.

In what follows we will consider special classes of radial functions $\varphi$ which allow generalized interpolants in the form (19), thus we restrict ourselves to positive definite functions, such as inverse multiquadrics ($\varphi(r) = 1/\sqrt{r^2 + \gamma^2}, \gamma > 0$), Gaussians ($\varphi(r) = e^{-\delta r^2}, \delta > 0$), and compactly supported radial basis function of continuity $C^{2\ell}$, see [4], ($\varphi(r) = (1 - r)_+^{2+2\ell} p(r)$, $p$ polynomial $\partial p = \ell$). In these cases, $A$ is a symmetric positive definite matrix, and the interpolation problem (19) is theoretically uniquely solvable [4]. Thus these RBFs do not require to be augmented with the polynomial part which is instead necessary in a general RBF form [3].

It is well known that the accuracy of a RBF interpolant strongly depends on the condition number of the linear system in (19), see [9]. In the case of optimal recovery this depends on the separation distance of the set of centroids in the stencil. In the present work, the linear system (19) has been regularized by means of Truncated Singular Value Decomposition.

Given the radial recovery function (18), and considering (20), an approximation of $\nabla R$ is given by

$$\nabla R_i(x) = \begin{bmatrix} \sum_{j=1}^M \lambda_j |T_j| (\frac{\mathbf{x}^{(1)} - \mathbf{x}_{c_j}^{(1)}}{\|\mathbf{x} - \mathbf{x}_{c_j}\|}) \varphi'(\|\mathbf{x} - \mathbf{x}_{c_j}\|) \\ \sum_{j=1}^M \lambda_j |T_j| (\frac{\mathbf{x}^{(2)} - \mathbf{x}_{c_j}^{(2)}}{\|\mathbf{x} - \mathbf{x}_{c_j}\|}) \varphi'(\|\mathbf{x} - \mathbf{x}_{c_j}\|) \end{bmatrix}. \tag{21}$$

The radial recovery step takes the cell averages $\overline{u}_i, i = 1, \cdots, \mathcal{N}$, associated to the triangles $T_i$ as input and compute the unknowns $\lambda_j, j = 1, \cdots, M$ in (19) for each control volume $T_i$. The complete semi-discrete finite volume higher order scheme is defined as follows.

**Definition 3.** ***Higher Order FV.*** *The semi-discrete finite volume approximation of (14) utilizing continuous in time solution representation, $t \in [0, +\infty)$,*

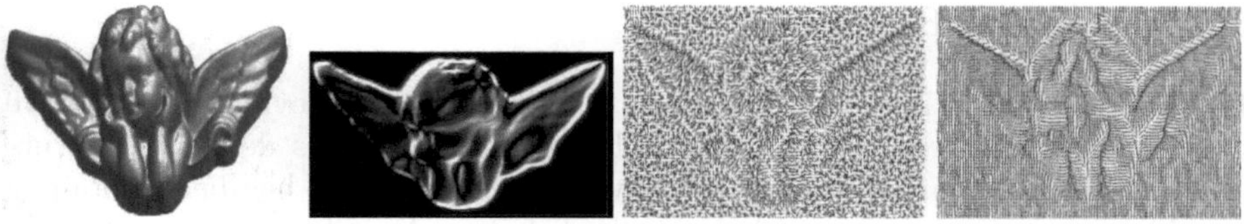

**Fig. 1.** Example 1: (from left to right) original surface on structured grid; curvature map; noisy normal field; smoothed normal field after 10 time steps

**Fig. 2.** Example 1: (from left to right) a detail from Fig.1(left); corrupted surface with noise level $5 \cdot 10^{-3}$; reconstructed surface after 4 time steps

*and higher order radial recovery in space* $u_h$*, such that*

$$\overline{u}_j(t) = \frac{1}{|T_i|}\int_{T_j} u_h(x,t)dx,$$

*with initial data* $\overline{u}_j(0) = \frac{1}{|T_i|}\int_{T_j} u_0(x)dx$*, and suitable boundary conditions, is given by the following two steps for each* $T_i \in \mathcal{T}$*:*

**STEP 1: radial recovery** *solve (19) for* $\Lambda$
**STEP 2: cell average computation** *solve the system of ODEs:*

$$\frac{d\overline{u}_i}{dt} = \frac{1}{|T_i|}\sum_{j\in N(i)} \frac{|e_{ij}|}{2}\sum_{k=1}^{ng} w_k F(R_i(t,\phi_{i,j}(s_k)), R_i(t,\phi_{i,j}(s_k)); \mathbf{n}_{ij}) \\ -\frac{1}{|T_i|}\sum_{j\in N(i)} w_k f(\overline{u}_k - \overline{u}_{0,k}).$$

A final remark concerns the approximation error between the solution of (14) in a Sobolev space $W_2^k(\Omega)$ of all $u$ with distributional derivatives $D^\alpha u \in L_2(\Omega), |\alpha| \leq k$, and the optimal recovery $u_h$ given by the Higher Order FV scheme. At this aim, the weak solution $u$ is required to be more regular than $u \in W_2^1(\Omega)$, more precisely, $u \in W_2^k(\Omega)$, with $k > D/2$, if $D$ is the current space dimension. Following [4], under the assumption of $u \in W_2^k(\Omega)$, the reconstruction error for $u_h$ in the finite dimensional subspace $V_h$ of $W_2^1(\Omega)$, can be bounded by

$$\|u - u_h\|_{L_\infty(\Omega)} \leq Ch^{k-1}\|u\|_{W_2^k(\Omega)}. \tag{22}$$

This result is applied to the $C^{2\ell}$ compactly supported RBF with $\ell \geq k - \frac{D+1}{2}$, and to the Gaussian RBF, see [4]; for the class of inverse multiquadrics RBF this bound is still an open problem.

## 5 Numerical Results

To illustrate and validate the methods described in the previous sections, we show experimental results obtained with functional surfaces defined on structured and unstructured meshes. A noisy functional surface has been obtained by adding an error vector $\eta$ with normally distributed random entries with zero mean to the functional values $z$. The vector $\eta$ is scaled to correspond to a specified noise level $\mu = \frac{\|\eta^\sigma\|}{\|z\|}$. Applying the stopping rule (10) we set $c = 1.1$ for all the numerical experiments. According to the notation in Section 4, for the compactly supported RBFs we set $\ell = 4$, for the Gaussian RBFs we set $\delta = 0.5$, while for the inverse multiquadrics we set $\gamma = 0.5$. All computations are carried out in Matlab with machine epsilon $\epsilon \approx 2 \cdot 10^{-16}$.

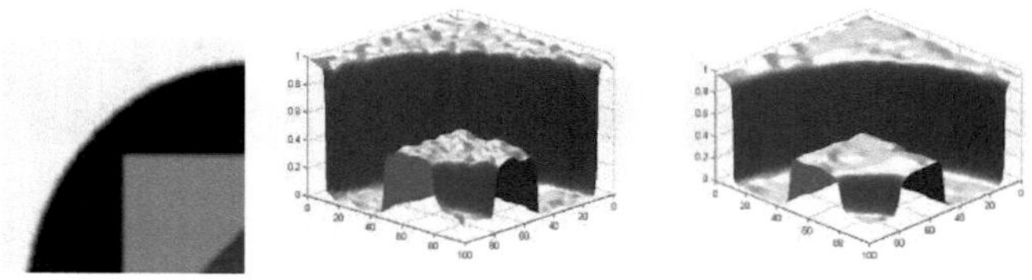

**Fig. 3.** Example 2: (from left to right) noise-free image; reconstructed surface using model (9) without curvature contribution; reconstructed surface with curvature term $g(curv_S)$

Example 1. In the first test we consider the original surface shown in Fig.1(left) and defined on a structured grid of 5000 triangles 2601 vertices. Additive white noise with noise level $\mu = 5 \cdot 10^{-3}$ was added in the $z$ direction. The noisy normal-field and its smoothed version obtained applying model (7) after 10 time steps, are shown in Fig.1. In Fig.2 (left) a detail of the original and noisy surface is shown together with the corresponding damaged version (middle). The reconstructed surface applying model (9) with curvature map depicted in Fig.1 (second from left) and discretized with Higher Order FV scheme and Inverse Multiquadrics RBF after 4 time steps is shown in Fig.2(right). Visual inspection shows a very good reconstruction after only few time steps. □

Example 2. A test image has been used as a functional surface to show the effectiveness of the curvature term introduced in the PDE model (9). The noise-free image is shown in Fig.3(left). The reconstructed surface obtained applying model (9) with $g(curv_S)$ is shown in Fig.3(right), while the reconstruction shown in Fig.3(middle) has been obtained without curvature contribution. An Higher Order FV scheme has been used for the discretization. □

Example 3. To demonstrate the effectiveness of the Higher Order FV scheme we compare the results obtained by the sophisticated models (7) and (9), with the results obtained using the standard Total Variation (TV) model $\frac{\partial u}{\partial t} = \nabla \cdot (\frac{\nabla u}{|\nabla u|})$, see [10] for reference. The TV model is applied to the damaged surface shown in Fig.4(second from left), defined on a unstructured grid of 13208 triangles

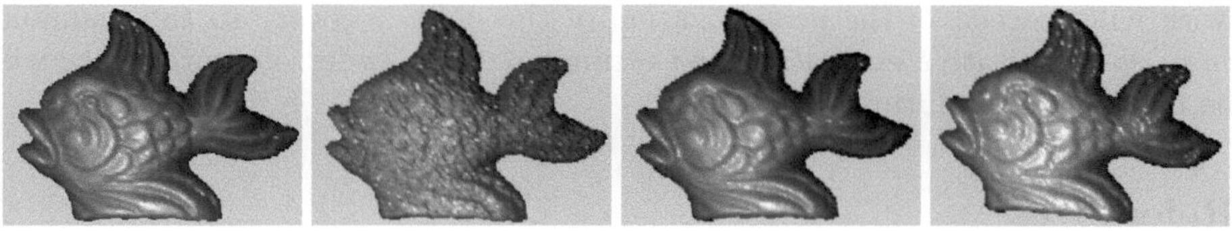

**Fig. 4.** Example 3: (from left to right) noise-free surface on unstructured grid; corrupted surface with noise level $1 \cdot 10^{-2}$; reconstructed surface applying model (7) and (9); reconstructed surface using TV model

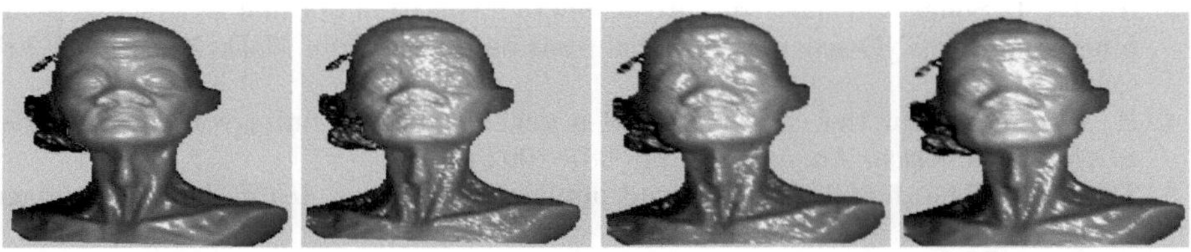

**Fig. 5.** Example 4: (from left to right) noise-free surface; corrupted surface with noise level $1 \cdot 10^{-2}$; reconstructed surface with TV model using Basic FV scheme; reconstructed surface with TV model using Higher Order FV scheme

and 6833 vertices, noise level $\mu = 1 \cdot 10^{-2}$, with the aim to recover the noisy-free surface illustrated in Fig.4(left). In Fig.4(third from left) the reconstruction using models (7) and (9) after 4 and 10 time steps are respectively illustrated, while in Fig.4(right) the surface obtained after 10 time steps of the TV model is shown. The satisfactory reconstruction we get after only a few steps can be motivated by the use of Higher Order FV scheme, using Gaussian RBF recovery strategy. □

Example 4. For a comparison between the accuracy provided by the Basic and the Higher Order FV schemes, we considered the reconstruction obtained by the TV model of a noisy-free surface shown in Fig.5(left) defined on an unstructured grid of 9471 vertices and 18359 triangles. The damaged surface shown in Fig.5(second from left), has been obtained with a noise level $\mu = 0.01$. The reconstruction carried out by TV model using a Basic FV scheme is illustrated in Fig.5 (third from the left) after 100 time steps, while the reconstruction after only 8 time steps using Higher Order FV scheme for the discretization of the TV model, with Inverse Multiquadrics RBFs is shown in Fig.5(right). The restoration determined by the RBF recovery yields the highest quality. □

## 6 Conclusions and Ongoing Work

In this paper we have investigated an higher order co-volume schemes for diffusion using RBF optimal recovery which provides high quality results for the recovery of functional surfaces corrupted by additive noise. A natural ongoing work will proceed in the generalization of this RBF recovery to the manifold case. A quantitative evaluation and estimates of the order of convergence will be investigated in future works.

## References


1. D.Krishnan, P.Lin, X.C.Tai, An efficient operator splitting method for noise removal in images, *Commun. Comput. Phys.* **1** pp. 847–858 (2006).
2. M.Lysaker, S.Osher, X.C.Tai, Noise removal using smoothed normals and surface fitting, *IEEE Transactions on Image Processing* **13/10** pp. 1345–1357 (2004).
3. A.Iske, T. Sonar, On the Structure of function spaces in optimal recovery of point functionals for ENO-schemes by radial basis functions, *Num.Math.* **74** pp. 177–201 (1996).
4. H. Wendland, On the convergence of a general class of finite volume methods, *SIAM J. Numerical Analysis* **43**, pp. 987–1002 (2005).
5. T.F.Chan, The digital TV filter and nonlinear denoising, *IEEE Trans. on Image Processing,* **10/2** pp. 231–241 (2001).
6. H. Engl, M.Hanke and A. Neubauer, *Regularization of inverse problems*, Kluwer, Dordrecht, 1996.
7. S.Corsaro, K.Mikula, A.Sarti, F.Sgallari, Semi-implicit co-volume method in 3D image segmentation, *SIAM Journal on Scientific Computing* **28/6** pp. 2248–2265 (2006).
8. T. Sonar, Optimal Recovery Using Thin Plate Splines in Finite Volume Methods for the Numerical Solution of Hyperbolic Conservation Laws, *IMA Journal of Numer. Analysis* **16** pp. 549–581 (1996).
9. G.Casciola, D.Lazzaro, L.B.Montefusco, S.Morigi, Fast surface reconstruction and hole filling using Radial Basis Functions, *Numerical Algorithms* **39/1-3** pp. 1017–1398 (2005).
10. L.Rudin, S. Osher, E. Fatemi, Nonlinear total variation based noise removal algorithms, *Physica D* **60** pp. 259–268 (1992).

# Linear Image Reconstruction by Sobolev Norms on the Bounded Domain*

Bart Janssen, Remco Duits, and Bart ter Haar Romeny

Eindhoven University of Technology, Dept. of Biomedical Engineering
{B.J.Janssen,R.Duits,B.M.terHaarRomeny}@tue.nl

**Abstract.** The reconstruction problem is usually formulated as a variational problem in which one searches for that image that minimizes a so called prior (image model) while insisting on certain image features to be preserved. When the prior can be described by a norm induced by some inner product on a Hilbert space the exact solution to the variational problem can be found by orthogonal projection. In previous work we considered the image as compactly supported in $\mathbb{L}_2(\mathbb{R}^2)$ and we used Sobolev norms on the unbounded domain including a smoothing parameter $\gamma > 0$ to tune the smoothness of the reconstruction image. Due to the assumption of compact support of the original image components of the reconstruction image near the image boundary are too much penalized. Therefore we minimize Sobolev norms only on the actual image domain, yielding much better reconstructions (especially for $\gamma \gg 0$). As an example we apply our method to the reconstruction of singular points that are present in the scale space representation of an image.

## 1 Introduction

One of the fundamental problems in signal processing is the reconstruction of a signal from its samples. In 1949 Shannon published his work on signal reconstruction from its equispaced ideal samples [17]. Many generalizations [16,18] and applications [3,13] followed thereafter.

Reconstruction from differential structure of scale space interest points, first introduced by Nielsen and Lillholm [15], is an interesting instance of the reconstruction problem since the samples are non-uniformly distributed over the image they were obtained from and the filter responses of the filters do not necessarily coincide. Several linear and non-linear methods [10,12,14,15] appeared in literature which all search for an image that (1) is indistinguishable from its original when observed through the filters the features were extracted with and (2) simultaneously minimizes a certain prior. We showed in earlier work [10] that if such a prior is a norm of Sobolev type on the *unbounded* domain one can obtain visually attractive reconstructions while retaining linearity. However, boundary problems degrade the reconstruction quality.

---

* The Netherlands Organisation for Scientific Research (NWO) is gratefully acknowledged for financial support.

F. Sgallari, A. Murli, and N. Paragios (Eds.): SSVM 2007, LNCS 4485, pp. 55–67, 2007.

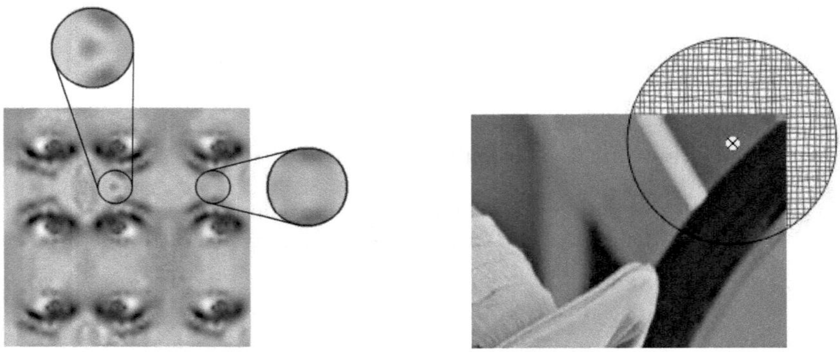

**Fig. 1.** An illustration of the bounded domain problem: Features that are present in the center of the image are lost near its border, since they, in contrast to the original image $f$ are not compactly supported on $\Omega$

The problem that appears in the unbouded domain reconstruction method is best illustrated by analyzing Figure 1 . The left image is a reconstruction from differential structure obtained from an image that is a concatenation of (mirrored) versions of Lena's eye. One can clearly observe that structure present in the center of the image does not appear on the border. This can be attributed to the fact that, when features are measured close to the image boundary, they partly lay outside the image and are "penalized" by the energy minimization methods that are formulated on the unbounded domain. This is illustrated by the right image in Figure 1 . We solve this problem by considering bounded domain Sobolev norms instead. An additional advantage of our method is that we can enforce a much higher degree of regularity than the unbounded domain counter part (in fact we can minimize semi-norms on the bounded domain). Furthermore we give an interpretation of the 2 parameters that appear in the reconstruction framework in terms of filtering by a low-pass Butterworth filter. This allows for a good intuition on how to choose these parameters.

## 2 Theory

In order to prevent the above illustrated problem from happening we restrict the reconstruction problem to the domain $\Omega \subset \mathbb{R}^2$ that is defined as the support of the image $f \in \mathbb{L}_2\left(\mathbb{R}^2\right)$ from which the features $\{c_p(f)\}_1^P, c_p(f) \in \mathbb{R}$ are extracted. Recall that the $\mathbb{L}_2\left(\Omega\right)$-inner product on the domain $\Omega \subset \mathbb{R}^2$ for $f, g \in \mathbb{L}_2\left(\Omega\right)$ is given by

$$(f,g)_{\mathbb{L}_2(\Omega)} = \int_\Omega \overline{f(x)} g(x) dx \ . \tag{1}$$

A feature $c_p(f)$ is obtained by taking the inner product of the $p^{\text{th}}$ filter $\psi_p \in \mathbb{L}_2\left(\Omega\right)$ with the image $f \in \mathbb{L}_2\left(\Omega\right)$,

$$c_p(f) = (\psi_p, f)_{\mathbb{L}_2(\Omega)} \ . \tag{2}$$

An image $g \in \mathbb{L}_2(\Omega)$ is equivalent to the image $f$ if they share the same features, $\{c_p(f)\}_1^P = \{c_p(g)\}_1^P$, which is expressed in the following equivalence relation for $f, g \in \mathbb{L}_2(\Omega)$.

$$f \sim g \Leftrightarrow (\psi_p, f)_{\mathbb{L}_2(\Omega)} = (\psi_p, g)_{\mathbb{L}_2(\Omega)} \text{ for all } p = 1, \ldots, P. \tag{3}$$

Next we introduce the Sobolev space of order $2k$ on the domain $\Omega$,

$$\mathbb{H}^{2k}(\Omega) = \{f \in \mathbb{L}_2(\Omega) \mid |\Delta|^k f \in \mathbb{L}_2(\Omega)\}\ ,\ k > 0\ . \tag{4}$$

The completion of the space of $2k$-differentiable functions on the domain $\Omega$ that vanish on the boundary of its domain $\partial\Omega$ is given by

$$\mathbb{H}_0^{2k}(\Omega) = \{f \in \mathbb{H}^{2k}(\Omega) \mid f|_{\partial\Omega} = 0\}\ ,\ k > \frac{1}{2}\ . \tag{5}$$

Now $\mathbb{H}_0^{2k,\gamma}(\Omega)$ denotes the normed space obtained by endowing $\mathbb{H}_0^{2k}(\Omega)$ with the following inner product,

$$(f,g)_{\mathbb{H}_0^{2k,\gamma}(\Omega)} = (f,g)_{\mathbb{L}_2(\Omega)} + \gamma^{2k}\left(|\Delta|^{\frac{k}{2}} f, |\Delta|^{\frac{k}{2}} g\right)_{\mathbb{L}_2(\Omega)} = (f,g)_{\mathbb{L}_2(\Omega)} + \gamma^{2k}\left(|\Delta|^k f, g\right)_{\mathbb{L}_2(\Omega)}, \tag{6}$$

for all $f, g \in \mathbb{H}_0^{2k}(\Omega)$ and $\gamma \in \mathbb{R}$.

The solution to the reconstuction problem is the image $g$ of minimal $\mathbb{H}_0^{2k,\gamma}$-norm that shares the same features with the image $f \in \mathbb{H}_0^{2k,\gamma}(\Omega)$ from which the features $\{c_p(f)\}_1^N$ were extracted. The reconstruction image $g$ is found by an orthogonal projection, within the space $\mathbb{H}_0^{2k,\gamma}(\Omega)$, of $f$ onto the subspace $V$ spanned by the filters $\kappa_p$ that correspond to the $\psi_p$ filters,

$$\arg\min_{f \sim g} ||g||^2_{\mathbb{H}_0^{2k,\gamma}(\Omega)} = \mathbb{P}_V f\ , \tag{7}$$

as shown in previous work [10]. The filters $\kappa_p \in \mathbb{H}_0^{2k,\gamma}(\Omega)$ are given by

$$\kappa_p = \left(I + \gamma^{2k}|\Delta|^k\right)^{-1} \psi_p\ . \tag{8}$$

As a consequence $(\kappa_p, f)_{\mathbb{H}_0^{2k,\gamma}(\Omega)} = (\psi_p, f)_{\mathbb{L}_2(\Omega)}$ for $(p = 1 \ldots P)$ for all $f$. Here we assumed that $f \in \mathbb{H}^{2k}(\Omega)$ however, one can get away with $f \in \mathbb{L}_2(\Omega)$ if $\psi$ satisfies certain regularity conditions. The interested reader can find the precise conditions and further details in [7].

The two parameters, $\gamma$ and $k$, that appear in the reconstruction problem allow for an interesting interpretation. If $\Omega = \mathbb{R}$ the fractional operator $\left(I + \gamma^{2k}|\Delta|^k\right)^{-1}$ is equivalent to filtering by the classical low-pass Butterworth filter [2] of order $2k$ and cut-off frequency $\omega_0 = \frac{1}{\gamma}$. This filter is defined as

$$B_{2k}\left(\frac{\omega}{\omega_0}\right) = \frac{1}{1 + |\frac{\omega}{\omega_0}|^{2k}}\ . \tag{9}$$

A similar phenomena was recently observed by Unser and Blu [20] when studying the connection between splines and fractals. Using this observation we can

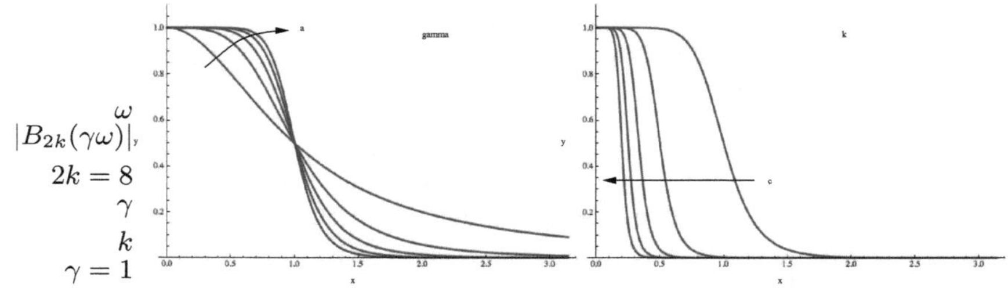


**Fig. 2.** The filter response of a Butterworth filter. On the left $\gamma$ is kept constant and the filter responses for different $k > 0$ are shown. On the right the order of the filter,$2k$, is kept constant and the filter responses for different $\gamma > 0$ are shown.

interpret equation (7) as finding an image, constructed by the $\psi_p$ basis functions that, after filtering with the Butterworth filter of order $2k$ and with a cut-off frequency determined by $\gamma$, is equivalent (cf. equation (3)) to the image $f$. The filter response of the Butterworth filter is shown in Figure 3. One can observe the order of the filter controls how well the ideal low-pass filter is approximated and the effect of $\gamma$ on the cut-off frequency.

### 2.1 Spectral Decomposition

For now we set $k = 1$ and investigate the Laplace operator on the bounded domain: $\Delta : \mathbb{H}_0^2(\Omega) \mapsto \mathbb{L}_2(\Omega)$ which is bounded and whose right inverse is given by the minus Dirichlet operator, which is defined as follows.

**Definition 1 (Dirichlet Operator).** *The Dirichlet operator $\mathcal{D}$ is given by*

$$g = \mathcal{D}f \Leftrightarrow \begin{cases} \Delta g = -f \\ g|_{\partial\Omega} = 0 \end{cases} \tag{10}$$

*with $f \in \mathbb{L}_2(\Omega)$ and $g \in \mathbb{H}_0^2(\Omega)$.*

The Green's function $G : \Omega \times \Omega \mapsto \mathbb{R}^2$ of the Dirichlet operator is given by

$$\begin{cases} \Delta G(\mathbf{x}, \cdot) = -\delta_{\mathbf{x}} \\ G(\mathbf{x}, \cdot)|_{\partial\Omega} = 0 \end{cases} \tag{11}$$

for fixed $\mathbf{x} \in \Omega$. Its closed form solution reads

$$G(\mathbf{x}, \mathbf{y}) = -\frac{1}{2\pi}\log\left|\frac{\mathrm{sn}(x_1 + ix_2, \tilde{k}) - \mathrm{sn}(y_1 + iy_2, \tilde{k})}{\mathrm{sn}(x_1 + ix_2, \tilde{k}) - \overline{\mathrm{sn}(y_1 + iy_2, \tilde{k})}}\right| . \tag{12}$$

Here $\mathbf{x} = (x_1, x_2), \mathbf{y} = (y_1, y_2) \in \Omega$ and $\tilde{k} \in \mathbb{R}$ is determined by the aspect ratio of the rectangular domain $\Omega$ , sn denotes the well know Jacobi-elliptic function [9]. In Appendix A we derive equality (12), and show how to obtain $\tilde{k}$. Figure 3 shows a graphical representation of this non-isotropic Green's function for a square domain ($\tilde{k} \approx 0.1716$). Notice this function vanishes at its boundaries and is, in the center of the domain, very similar to the isotropic fundamental solution on the unbounded domain [5]. In terms of regularisation this means the Dirichlet operator smoothens inwards the image but never "spills" over the border of the domain $\Omega$.

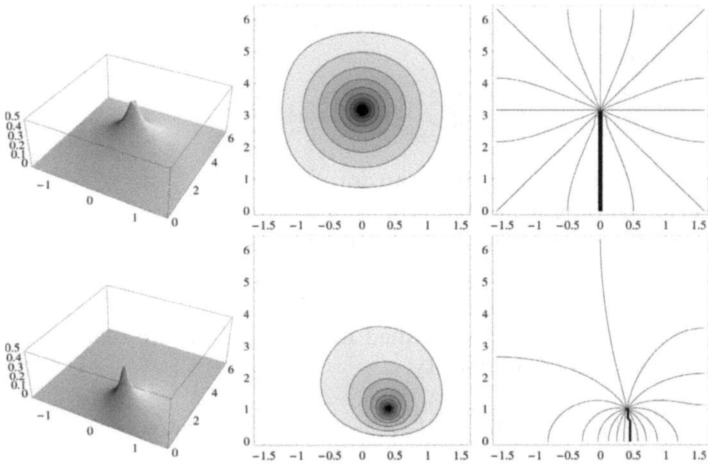

**Fig. 3.** From left to right plots of the graph of $\mathbf{x} \mapsto G_{\mathbf{x},\mathbf{y}}$, isocontours $G(\mathbf{x},\mathbf{y}) = c$ and isocontours of its Harmonic conjugate $H(\mathbf{x},\mathbf{y}) = \frac{-1}{2\pi}\arg\left(\frac{\mathrm{sn}(x_1+ix_2,\tilde{k})-\mathrm{sn}(y_1+iy_2,\tilde{k})}{\mathrm{sn}(x_1+ix_2,\tilde{k})-\overline{\mathrm{sn}(y_1+iy_2,\tilde{k})}}\right)$

When the Dirichlet operator as defined in Definition 1 is expressed by means of its Green's function, which is presented in equation (12),

$$(\mathcal{D}f)(\mathbf{x}) = \int_{\Omega} G(\mathbf{x},\mathbf{y})f(\mathbf{y})d\mathbf{y}, \quad f \in \mathbb{L}_2(\Omega) \ , \mathcal{D}f \in \mathbb{H}_0^2(\Omega) \tag{13}$$

one can verify that it extends to a compact, self-adjoint operator on $\mathbb{L}_2(\Omega)$. As a consequence, by the spectral decomposition theorem of compact self-adjoint operators [21], we can express the Dirichlet operator in an orthonormal basis of eigen functions. The normalized eigen functions $f_{mn}$ with corresponding eigen values $\lambda_{mn}$ of the Laplace operator $\Delta : \mathbb{H}_0^2(\Omega) \mapsto \mathbb{L}_2(\Omega)$ are given by

$$f_{mn}(x,y) = \sqrt{\frac{1}{ab}}\sin(\frac{n\pi x}{a})\sin(\frac{m\pi y}{b}) \qquad \lambda_{mn} = -\left(\left(\frac{n\pi}{a}\right)^2 + \left(\frac{m\pi}{b}\right)^2\right) \tag{14}$$

with $\Omega = [0,a] \times [0,b]$. Since $\Delta\mathcal{D} = -I$, the eigen functions of the Dirichlet operator coincide with those of the Laplace operator (14) and its corresponding eigen values are the inverse of the eigen values of the Laplace operator.

## 2.2 Scale Space on the Bounded Domain

The spectral decomposition presented in the previous subsection, by (14), will now be applied to the construction of a scale space on the bounded domain [6] .

Before we show how to obtain a Gaussian scale space representation of an image on the bounded domain we find, as suggested by Koenderink [11], the image $h \in \mathbb{H}^2(\Omega)$ that is the solution to $\Delta h = 0$ with as boundary condition that $h = f$ restricted to $\partial\Omega$. Now $\tilde{f} = f - h$ is zero at the boundary $\partial\Omega$ and can serve as an initial condition for the heat equation (on the bounded domain). A practical method for obtaining $h$ is suggested by Georgiev [8] . Now $f_{mn}(x, y)$ is obained by expansion of

$$\tilde{f} = \sum_{m,n \in \mathbb{N}} (f_{mn}, \tilde{f})_{\mathbb{L}_2(\Omega)} f_{mn} \ , \tag{15}$$

which effectively exploits the sine transform.

The (fractional) operators that will appear in the construction of a Gaussian scale space on the bounded domain can be expressed as

$$|\Delta|^{2k} f_{mn} = (\lambda_{mn})^k f_{mn} \quad , \quad e^{-s|\Delta|} f_{mn} = e^{-s\lambda_{mn}} f_{mn} \ . \tag{16}$$

We also note that the $\kappa_p$ filters, defined in equality (8), are readily obtained by application of the following identity

$$\left(I + \gamma^{2k}|\Delta|^k\right)^{-1} f_{mn} = \frac{1}{1 + \gamma^{2k}\left(\lambda_{mn}\right)^k} f_{mn} \ . \tag{17}$$

Consider the Gaussian scale space representation[1] on bounded domain $\Omega$

$$u_{\tilde{f}}^{\Omega}(x, y, s) = \sum_{m,n \in \mathbb{N}} e^{-(\lambda_{mn})s} (f_{mn}, \tilde{f})_{\mathbb{L}_2(\Omega)} f_{mn}(x, y) \tag{18}$$

where the scale parameter $s \in \mathbb{R}^+$. It is the unique solution to

$$\begin{cases} \frac{\partial u}{\partial s} = \Delta u \\ u(\cdot, s)|_{\partial\Omega} = 0 \quad \text{for all } s > 0 \\ u(\cdot, 0) = \tilde{f} \end{cases} . \tag{19}$$

The filter $\phi_p$ that measures differential structure present in the scale space representation $u_{\tilde{f}}^{\Omega}$ of $\tilde{f}$ at a point $p$ with coordinates $(x_p, y_p, s_p)$, such that

$$\left(D^{\mathbf{n}_p} u_{\tilde{f}}^{\Omega}\right)(x_p, y_p, s_p) = \left(\phi_p, \tilde{f}\right)_{\mathbb{L}_2(\Omega)} \ , \tag{20}$$

is given by (writing multi-index $\mathbf{n}_p = (n_p^1, n_p^2)$)

$$\phi_p(x, y) = \sum_{m,n \in \mathbb{N}} e^{-(\lambda_{mn})s_p} \, (D^{\mathbf{n}_p} f_{mn})(x_p, y_p) \, f_{mn}(x, y) \ , \tag{21}$$

where we note that

$$(D^{\mathbf{n}_p} f_{mn})\,(x_p, y_p) = \sqrt{\frac{1}{ab}} \left(\frac{m\pi}{b}\right)^{n_p^2} \left(\frac{n\pi}{a}\right)^{n_p^1} \sin\left(\frac{m\pi y_p}{b} + \frac{\pi}{2}\, n_p^2\right) \sin\left(\frac{n\pi x_p}{a} + \frac{\pi}{2}\, n_p^1\right) \ ,$$

$\mathbf{x} = (x, y) \in \Omega$, $\mathbf{x}_p = (x_p, y_p) \in \Omega$ and $\mathbf{n}_p = (n_p^1, n_p^2) \in \mathbb{N} \times \mathbb{N}$.

[1] The framework in this paper is readily generalized to $\alpha$-scale spaces in general (see e.g. [6]) by replacing $(-\lambda_{mn})$ by $(-\lambda_{mn})^{2\alpha}$.

### 2.3 The Solution to the Reconstruction Problem

Now we have established how one can construct a scale space on the bounded domain and shown how to measure its differential structure we can proceed to express the solution to the reconstruction problem (recall equations (7) and (8)) in terms of eigen functions and eigen values of the Laplace operator:

$$\mathbb{P}_V \tilde{f} = \sum_{p,q=1}^{P} G^{pq} (\phi_p, \tilde{f})_{\mathbb{L}_2(\Omega)} \kappa_q = \sum_{p,q=1}^{P} G^{pq} c_p(\tilde{f}) \kappa_q \ , \tag{22}$$

where $G^{pq}$ is the inverse of the Gram matrix $G_{pq} = (\kappa_p, \kappa_q)_{\mathbb{H}_0^{2k,\gamma}(\Omega)}$ and the filters $\kappa_p$ satisfy

$$\kappa_p(x,y) = \sum_{m,n \in \mathbb{N}} \frac{e^{-(\lambda_{mn}) s_p}}{1 + \gamma^{2k} (\lambda_{mn})^k} (D^{\mathbf{n}_p} f_{mn})(x_p, y_p)\, f_{mn}(x,y) \ . \tag{23}$$

This is the unique solution to the optimization problem

$$\arg \min_{\tilde{f} \sim g} ||g||^2_{\mathbb{H}_0^{2k,\gamma}(\Omega)} \ , \tag{24}$$

which was introduced in equation (7). Instead of Dirichlet boundary conditions one can impose Neumann-boundary conditions. In this case the eigenvalues $\lambda_{mn}$ are maintained, the eigen functions are given by

$$f_{mn}(x,y) = \sqrt{\frac{1}{ab(1+\delta_{m0})(1+\delta_{n0})}} \cos(\frac{n\pi x}{a}) \cos(\frac{m\pi y}{b}) \ . \tag{25}$$

## 3 Implementation

The implementation of the reconstruction method that was presented in a continuous Hilbert space framework is completely performed in a discrete framework in order to avoid approximation errors due to sampling, following the advice: *"Think analog, act discrete!"* [19].

First we introduce the discrete sine transform $\mathcal{F}_S : l_2(I_N^{\mathcal{D}}) \mapsto l_2(I_N^{\mathcal{D}})$ on a rectangular domain $I_N^{\mathcal{D}} = \{1, \ldots, N-1\} \times \{1, \ldots, M-1\}$

$$(\mathcal{F}_S f)(u,v) = -\frac{2}{\sqrt{MN}} \sum_{i=1}^{M-1} \sum_{j=1}^{N-1} \underbrace{\sin\left(\frac{iu\pi}{M}\right) \sin\left(\frac{jv\pi}{N}\right)}_{(\varphi_i \otimes \varphi_j)(u,v)} f(i,j) \ , \tag{26}$$

with $(u,v) \in I_N^{\mathcal{D}}$. Notice that this unitary transform is its own inverse and that

$$(\varphi_i, \varphi_j)_{l_2(I_N^{\mathcal{D}})} = \delta_{ij} \ , \tag{27}$$

so $\{\varphi_i \otimes \varphi_i \mid \begin{smallmatrix} i=1,\ldots,M-1 \\ j=1,\ldots,N-1 \end{smallmatrix}\}$ forms an orthonormal basis in $l_2(I_N^{\mathcal{D}})$.

The Gaussian scale space representation $u_f^{I_N^{\mathcal{D}}}(i,j,s)$ of an image $f \in l_2(I_N^{\mathcal{D}})$ introduced in the continuous domain in equality (18) now reads

$$u_f^{I_N^{\mathcal{D}}}(i,j,s)=\left(e^{s\Delta}f\right)(i,j)=-\frac{2}{\sqrt{MN}}\sum_{u=1}^{M-1}\sum_{v=1}^{N-1}\hat{f}(u,v)e^{-s\left(\frac{u^2}{M^2}+\frac{v^2}{N^2}\right)\pi^2}\left(\varphi_u\otimes\varphi_v\right)(i,j)$$

where $\hat{f}(u,v) = (\mathcal{F}_S f)(u,v)$. Differential structure of order $\mathbf{n}_p = (n_p^1, n_p^2) \in \mathbb{N}\times\mathbb{N}$ at a certain position $(i_p, j_p) \in I_N^{\mathcal{D}}$ and at scale $s_p \in \mathbb{R}^+$ is measured by

$$\left(D^{\mathbf{n}_p}u_f^{I_N^{\mathcal{D}}}\right)(i_p,j_p,s_p) = -\frac{2}{\sqrt{MN}}\sum_{u=1}^{M-1}\sum_{v=1}^{N-1}\hat{f}(u,v)e^{-s_p\left(\frac{u^2}{M^2}+\frac{v^2}{N^2}\right)\pi^2}$$
$$\left(\frac{u\pi}{M}\right)^{n_p^1}\left(\frac{v\pi}{N}\right)^{n_p^2}\sin\left(\frac{i_p u\pi}{M}+\frac{\pi}{2}n_p^1\right)\sin\left(\frac{j_p v\pi}{N}+\frac{\pi}{2}n_p^2\right).$$

The filters $\phi_p$, with $p = (i_p, j_p, s_p, \mathbf{n}_p)$ a multi-index, are given by

$$\begin{aligned}\phi_p(i,j,s) = -\frac{2}{\sqrt{MN}}\sum_{u=1}^{M-1}\sum_{v=1}^{N-1}e^{-s_p\left(\frac{u^2}{M^2}+\frac{v^2}{N^2}\right)\pi^2}\left(\varphi_u\otimes\varphi_v\right)(i,j)\\ \left(\frac{u\pi}{M}\right)^{n_p^1}\left(\frac{v\pi}{N}\right)^{n_p^2}\sin\left(\frac{i_p u\pi}{M}+\frac{\pi}{2}n_p^1\right)\sin\left(\frac{j_p v\pi}{N}+\frac{\pi}{2}n_p^2\right)\end{aligned} \tag{28}$$

and the filters $\kappa_p$ corresponding to $\phi_p$ read

$$\begin{aligned}\kappa_p(i,j,s) = -\frac{2}{\sqrt{MN}}\sum_{u=1}^{M-1}\sum_{v=1}^{N-1}\frac{e^{-s_p\left(\frac{u^2}{M^2}+\frac{v^2}{N^2}\right)\pi^2}}{1+\gamma^{2k}\left(\frac{u^2}{M^2}+\frac{v^2}{N^2}\right)^k}\left(\varphi_u\otimes\varphi_v\right)(i,j)\\ \left(\frac{u\pi}{M}\right)^{n_p^1}\left(\frac{v\pi}{N}\right)^{n_p^2}\sin\left(\frac{i_p u\pi}{M}+\frac{\pi}{2}n_p^1\right)\sin\left(\frac{j_p v\pi}{N}+\frac{\pi}{2}n_p^2\right).\end{aligned} \tag{29}$$

An element $G_{pq} = (\phi_p, \phi_q)_{l_2(I_N^{\mathcal{D}})}$ of the Gram matrix can, because of the orthonormality of the transform, be expressed in just a double sum,

$$G_{pq} = -\frac{2}{\sqrt{MN}}\sum_{u=1}^{M-1}\sum_{v=1}^{N-1}\frac{e^{-(s_p+s_q)\left(\frac{u^2}{M^2}+\frac{v^2}{N^2}\right)\pi^2}}{1+\gamma^{2k}\left(\frac{u^2}{M^2}+\frac{v^2}{N^2}\right)^k}\left(\frac{u\pi}{M}\right)^{n_p^1}\left(\frac{v\pi}{N}\right)^{n_p^2}\sin\left(\frac{i_p u\pi}{M}+\frac{\pi}{2}n_p^1\right)$$
$$\sin\left(\frac{j_p v\pi}{N}+\frac{\pi}{2}n_p^2\right)\left(\frac{u\pi}{M}\right)^{n_q^1}\left(\frac{v\pi}{N}\right)^{n_q^2}\sin\left(\frac{i_q u\pi}{M}+\frac{\pi}{2}n_q^1\right)\sin\left(\frac{j_q v\pi}{N}+\frac{\pi}{2}n_q^2\right).$$

In order to gain accuracy we implement equality (3) by summing in the reverse direction and multiplying by $\gamma^{2k}$. Then we compute

$$\tilde{g} = \sum_{p,q=1}^{P} G^{pq}\gamma^{2k}c_p(f)\phi_q \tag{30}$$

and find the reconstruction image $g$ by filtering $\tilde{g}$ by a discrete version of the 2D Butterworth filter of order $2k$ and with cut-off frequency $\omega_0 = \frac{1}{\gamma}$.

The implementation was written using the sine transform as defined in equality (26) where we already explicitly mentioned the transform can be written as

$$(\mathcal{F}_S f)(u,v) = -\frac{2}{\sqrt{MN}} \sum_{i=1}^{M-1} \sum_{j=1}^{N-1} (\varphi_i \otimes \varphi_j)(u,v) f(i,j) \; . \tag{31}$$

Now we define the cosine transform $\mathcal{F}_S : l_2(I_N^{\mathcal{N}}) \mapsto l_2(I_N^{\mathcal{N}})$ on a rectangular domain $I_N^{\mathcal{N}} = \{0,\ldots,N+1\} \times \{0,\ldots,M+1\}$ in a similar manner

$$(\mathcal{F}_C f)(u,v) = \sum_{i=0}^{M+1} \sum_{j=0}^{N+1} (\tilde{\varphi}_i \otimes \tilde{\varphi}_j)(u,v) f(i,j) \; . \tag{32}$$

where$(\tilde{\varphi}_i \otimes \tilde{\varphi}_j)(u,v) = \cos\left(\frac{\pi(i+\frac{1}{2})u}{M}\right)\sqrt{\frac{2-\delta_{u0}}{M}} \cos\left(\frac{\pi(j+\frac{1}{2})v}{M}\right)\sqrt{\frac{2-\delta_{v0}}{N}}$. These cosine basis functions $\{\tilde{\varphi}_i \otimes \tilde{\varphi}_j \mid \begin{smallmatrix} i=0,\ldots,M+1 \\ j=0,\ldots,N+1 \end{smallmatrix}\}$ form an orthogonal basis in $l_2(I_N^{\mathcal{N}})$ and can thus be used to transform the reconstruction method that was explicitly presented for the Dirichlet case into a reconstruction method based on Neumann boundary conditions.

## 4 Experiments

We evaluate the reconstruction by applying our reconstruction method to the problem that was presented in the introduction. The upper row of Figure 4 shows, from left to right, the image from which the features were extracted, a reconstruction by the unbounded domain method [10] (parameters: $\gamma = 50, k = 1$) and a reconstruction by the newly introduced bounded domain method using Dirichlet boundary conditions (parameters: $\gamma = 1000, k = 1$). Features that were used are up to second order derivatives measured at the singular points [4] of the scale space representation of $f$. One can clearly see that the structure that is missing in the middle image does appear when the bounded domain method is used. The bottom row of Figure 4 shows reconstructions from second order differential structure obtained from the singular points of the scale space of the laplacian of $f$. On the left the unbounded domain method was used with $\gamma = 100$ and $k = 1$, this leads to a reconstructed signal that has "spilled" too much over the border of the image and therefore is not as crisp as the reconstruction obtained by our newly proposed method using Dirichlet boundary conditions (parameters: $\gamma = 1000$ and $k = 1$). Due to this spilling the Gram matrix of the bounded domain reconstruction method is harder to invert since basis functions start to become more and more dependent, this problem gets worse when $\gamma$ increases. Our bounded domain method is immune to this problem.

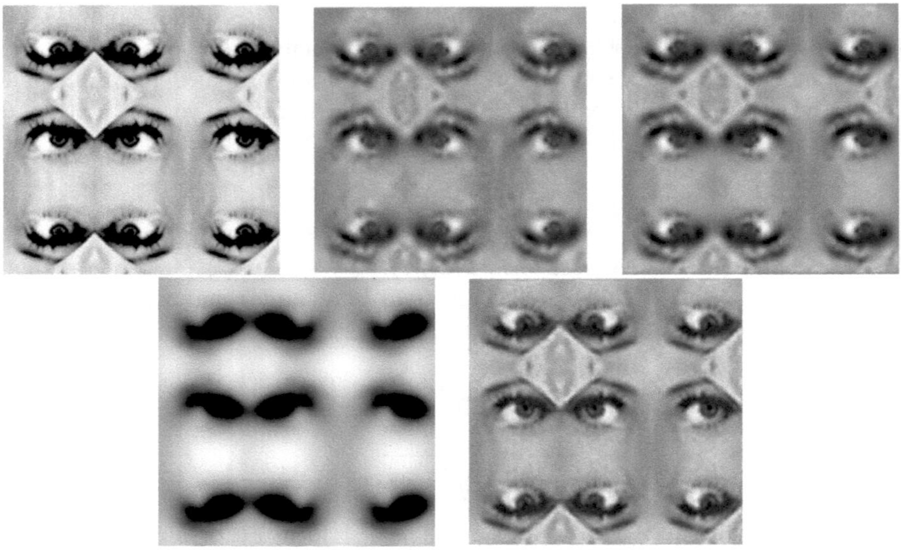

**Fig. 4.** Top left: The image $f$ from which the features were extracted. Top center and right: reconstruction from second order structure of the singular points of $f$ using the unbounded domain method[10] (parameters: $\gamma = 50, k = 1$) and the bounded domain method (parameters: $\gamma = 1000, k = 1$). Bottom row: unbounded domain (left) and bounded domain (right) reconstruction from singular points of the laplacian of $f$ with $k = 1$ and $\gamma$ set to 100 and 1000 respectively.

## 5 Conclusion

In previous work we considered the image as compactly supported in $\mathbb{L}_2(\mathbb{R}^2)$ and we used Sobolev norms on the unbounded domain including a smoothing parameter $\gamma > 0$ to tune the smoothness of the reconstruction image. Due to the assumption of compact support of the original image components of the reconstruction image near the image boundary are too much penalized. Therefore we proposed to minimize Sobolev norms only on the actual image domain, yielding much better reconstructions (especially for $\gamma \gg 0$). We also showed an interpretation for the parameter $\gamma$ and the order of the Sobolev space $k$ in terms of filtering by the classical Butterworth filter. In future work we plan to exploit this interpretation by automatically selecting the order of the Sobolev space.

## References

1. J. Boersma, J.K.M. Jansen, F.H. Simons, and F.W. Sleutel. The siam 100-dollar, 100-digit challenge - problem 10. SIAM News, January 2002.
2. S. Butterworth. On the theory of filter amplifiers. *Wireless Engineer*, 7:536–541, 1930.

3. J.E. Candes, J. Romberg, and T. Tao. Robust uncertainty principles: Exact signal reconstruction from highly incomplete frequency information. *IEEE Transactions on Information Theory*, 52(2):489–509, February 2006.
4. J. Damon. Local Morse theory for solutions to the heat equation and Gaussian blurring. *Journal of Differential Equations*, 115(2):368–401, January 1995.
5. J. Duchon. Splines minimizing rotation-invariant semi-norms in Sobolev spaces. Springer Verlag, 1977.
6. R. Duits, M. Felsberg, L.M.J. Florack, and B. Platel. $\alpha$ scale spaces on a bounded domain. In Lewis Griffin and Martin Lillholm, editors, *Scale Space Methods in Computer Vision, 4th International Conference, Scale Space 2003*, pages 494–510, Isle of Skye, UK, June 2003. Springer.
7. Remco Duits. *Perceptual Organization in Image Analysis.* PhD thesis, Technische Universiteit Eindhoven, 2005.
8. T. Georgiev. Relighting, retinex theory, and perceived gradients. In *Proceedings of Mirage 2005*, March 2005.
9. I. S. Gradshteyn and I. M. Ryzhik. *Table of Integrals, Series, and Products.* Boston, fifth edition, 1994. (Edited by A. Jeffrey.).
10. B.J. Janssen, F.M.W. Kanters, R. Duits, L.M.J. Florack, and B.M. ter Haar Romeny. A linear image reconstruction framework based on sobolev type inner products. *International Journal of Computer Vision*, 70(3):231–240, 2006.
11. J. J. Koenderink. The structure of images. *Biological Cybernetics*, 50:363–370, 1984.
12. J. Kybic, T. Blu, and M. Unser. Generalized sampling: a variational approach – part I: Theory. *IEEE Transactions on Signal Processing*, 50:1965–1976, 2002.
13. J. Kybic, T. Blu, and M. Unser. Generalized sampling: a variational approach – part II: Applications. *IEEE Transactions on Signal Processing*, 50:1965–1976, August 2002.
14. M. Lillholm, M. Nielsen, and L. D. Griffin. Feature-based image analysis. *International Journal of Computer Vision*, 52(2/3):73–95, 2003.
15. M. Nielsen and M. Lillholm. What do features tell about images? In *Scale-Space and Morphology in Computer Vision: Proceedings of the Third International Conference*, pages 39–50. Springer Verlag, 2001.
16. A. Papoulis. Generalized sampling expansion. *IEEE Transactions on Circuits and Systems*, 24:652–654, 1977.
17. C.E. Shannon. Communication in the presence of noise. In *Proc. IRE*, volume 37, pages 10–21, January 1949.
18. M. Unser. Sampling—50 Years after Shannon. *Proceedings of the IEEE*, 88(4):569–587, April 2000.
19. M. Unser. Splines: On scale, differential operators, and fast algorithms. In *Fifth International Conference on Scale Space and PDE Methods in Computer Vision (ICSSPDEMCV'05)*, Hofgeismar, Germany, April 6-10, 2005. Plenary talk.
20. M. Unser and T. Blu. Self-similarity: Part I-splines and operators. *IEEE Transactions on Signal Processing, to appear*, 2006.
21. K. Yosida. *Functional Analysis.* Springer Verlag, Berlin, sixth edition, 1980.

# A Closed from Expression of the Green's Function of the Dirichlet Operator

The Green's function $G : \Omega \times \Omega \mapsto \mathbb{R}^2$ of the dirichlet operator $\mathcal{D}$ (recall Definition 1) can be obtained by means of conformal mapping[2]. To this end we first map the rectangle to the upper half space in the complex plane. By the Schwarz-Christoffel formula the derivative of the inverse of such a mapping is given by

$$\frac{dz}{dw} = -\frac{1}{\tilde{k}}(w-1)^{-\frac{1}{2}}(w+1)^{-\frac{1}{2}}(w-\frac{1}{\tilde{k}})^{-\frac{1}{2}}(w+\frac{1}{\tilde{k}})^{-\frac{1}{2}} = \frac{1}{\sqrt{1-w^2}}\frac{1}{\sqrt{1-\tilde{k}^2w^2}},$$

where $w(\pm 1/\tilde{k}) = \pm a + ib$. As a result

$$z(w,\tilde{k}) = \int_0^w \frac{dt}{\sqrt{1-t^2}\sqrt{1-\tilde{k}^2t^2}} \Leftrightarrow w(z) = sn(z,\tilde{k}),$$

where $sn$ denotes the well-known Jacobi-elliptic function. We have $sn(0,\tilde{k}) = 0$, $sn(\pm a,\tilde{k}) = \pm 1$, $sn(\pm a + ib,\tilde{k}) = \pm(1/\tilde{k})$ and $sn(i(b/2),\tilde{k}) = i/\sqrt{\tilde{k}}$, where the elliptic modulus $\tilde{k}$ is given by

$$(b/a)z(1,\tilde{k}) = z(1,\sqrt{1-\tilde{k}^2}).$$

For example in case of a square $b/a = 2$ we have $\tilde{k} \approx 0.1715728752$.

The next step is to map the half plane onto the unit disk $\overline{B_{\mathbf{0},1}}$. This is easily done by means of a linear fractional transform

$$\chi(z) = \frac{z - sn(y_1 + i\,y_2,\tilde{k})}{z - \overline{sn(y_1 + i\,y_2,\tilde{k})}}.$$

To this end we notice that $|\chi(0)| = 1$ and that the mirrored points $sn(y_1 + i\,y_2,\tilde{k})$ and $\overline{sn(y_1 + i\,y_2,\tilde{k})}$ are mapped to the mirrored points $\chi(sn(y_1 + i\,y_2,\tilde{k})) = 0$ and $\chi(\overline{sn(y_1 + i\,y_2,\tilde{k})}) = \infty$.

Now define $F : \mathbb{C} \to \mathbb{C}$ and $\mathbf{F} : \overline{\Omega} \to \overline{B_{\mathbf{0},1}}$ by

$$\begin{aligned} &F = \chi \circ sn(\cdot,\tilde{k}), \text{ i.e. } F(x_1 + i\,x_2) = \tfrac{sn(x_1+i\,x_2,\tilde{k}) - sn(y_1+i\,y_2,\tilde{k})}{sn(x_1+i\,x_2,\tilde{k}) - \overline{sn(y_1+i\,y_2,\tilde{k})}} \\ &\mathbf{F}(x_1,x_2) = (\,\mathrm{Re}(F(x_1 + i\,x_2)),\, \mathrm{Im}(F(x_1 + i\,x_2)))^T, \end{aligned} \tag{33}$$

then $\mathbf{F}$ is a conformal mapping of $\overline{\Omega}$ onto $\overline{B_{\mathbf{0},1}}$ with $\mathbf{F}(\mathbf{y}) = \mathbf{0}$. As a result we have by the Cauchy-Riemann equations

$$\Delta_{\mathbf{F}(\mathbf{x})} = |\mathbf{F}'(\mathbf{x})|^{-1}\Delta_{\mathbf{x}}, \tag{34}$$

[2] Our solution is a generalization of the solution derived by Boersma in [1]

where the scalar factor in front of the right Laplacian is the inverse Jacobian:

$$|\mathbf{F}'(\mathbf{x})|^{-1} = (\det \mathbf{F}'(\mathbf{x}))^{-1} = \left( \left( \frac{\partial F_1}{\partial x}(\mathbf{x}) \right)^2 + \left( \frac{\partial F_2}{\partial x}(\mathbf{x}) \right)^2 \right)^{-1} = |F'(x_1 + ix_2)|,$$

for all $\mathbf{x} = (x_1, x_2) \in \Omega$.

Now $\tilde{G}(\mathbf{u}, \mathbf{0}) = \frac{-1}{2\pi} \log \|\mathbf{u}\|$ is the unique Greens function with Dirichlet boundary conditions on the disk $\overline{B_{\mathbf{0},1}} = \{\mathbf{x} \in \mathbb{R}^2 \mid \|\mathbf{x}\| \leq 1\}$ with singularity at $\mathbf{0}$ and our Green's function is given by $G = \tilde{G} \circ \mathbf{F}$, i.e.

$$G(\mathbf{x}, \mathbf{y}) = -\frac{1}{2\pi} \log |(\chi \circ sn(\cdot, \tilde{k}))(x_1 + i\, x_2)| = -\frac{1}{2\pi} \log \left| \frac{sn(x_1 + i\, x_2, \tilde{k}) - sn(y_1 + i\, y_2, \tilde{k})}{sn(x_1 + i\, x_2, \tilde{k}) - \overline{sn(y_1 + i\, y_2, \tilde{k})}} \right| . \tag{35}$$

# A Nonconvex Model to Remove Multiplicative Noise

Gilles Aubert[1] and Jean-François Aujol[2]

[1] Laboratoire J.A. Dieudonné , UMR CNRS 6621
gaubert@math.unice.fr
[2] CMLA, ENS Cachan, CNRS, PRES UniverSud
Jean-Francois.Aujol@cmla.ens-cachan.fr

**Abstract.** This paper deals with the denoising of SAR images. We draw our inspiration from the modeling of multiplicative speckle noise. By using a MAP estimator, we propose a functional whose minimizer corresponds to the denoised image we want to recover. Although the functional is not convex, we prove the existence of a minimizer. Then we study a semi-discrete version of the associated evolution problem, for which we derive existence and uniqueness results for the solution. We prove the convergence of this semi-discrete scheme. We conclude with some numerical results.

## 1 Introduction

Image denoising is a recurrent problem in image processing. The problem consists in recovering an image $u$ from a noisy image $f$ knowing some statistical information about the noise. Of course the difficulty is not the same depending on the noise under consideration. For example it is easier to remove white Gaussian additive noise than speckle noise. Many approaches in the literature have been proposed to tackle this problem. Among them stochastic approaches [9] , wavelet approaches [8], pdes and variational approaches [17]. In this paper we are interested in the denoising of SAR images, i.e. on models of the form $f = uv$ where $f$ is the observed image, $v$ the speckle noise and $u$ the true image to be recovered. Our approach is mainly variational: we search for $u$ as a minimizer of a functional $E(u) = J(u) + H(f,u)$, where $J(u)$ is the Total Variation of $u$ and $H(f,u)$ a data fidelity term. We obtain the expression of $H(f,u)$ in the discrete setting by considering explicitly the probability density function (pdf) of the speckle noise and by using a maximum likelihood principle. This computation leads us to propose, in the continuous setting, the following data fidelity term: $H(f,u) = \int_\Omega (\log u + \frac{f}{u})$, so that our global variational problem is

$$\inf_{u \in S(\Omega)} J(u) + \lambda \int_\Omega \left( \log u + \frac{f}{u} \right) \tag{1}$$

Where $S(\Omega) = \{u \in BV(\Omega), u > 0\}$.

As far as we know, the only variational approach devoted to multiplicative noise is the one by Rudin et al [16], as used for instance in [18].

F. Sgallari, A. Murli, and N. Paragios (Eds.): SSVM 2007, LNCS 4485, pp. 68–79, 2007.

The paper is organized as follows. We draw our inspiration from the modeling of active imaging systems, which we remind to the reader in Section 2. We use the classical MAP estimator to derive a new model to denoise non textured SAR images in Section 3. We consider this model from a variational point of view in Section 4. We carry out the mathematical analysis of the functional in a continuous setting. We then study a semi-implicit time discretization scheme in Section 5. We prove the convergence of the algorithm to a solution of the stationary problem. In these last two sections we mainly state the mathematical results skipping their proofs when technicalities are beyond the scope of this paper. We refer the reader to [2] for more details. We give some numerical examples in Section 6. We conclude the paper in Section 7.

## 2 Speckle Noise Modeling

Synthetic Aperture Radar (SAR) images are strongly corrupted by a noise called speckle. A radar sends a coherent wave which is reflected on the ground, and then registered by the radar sensor [12]. If the coherent wave is reflected on a coarse surface (compared to the radar wavelength), then the image processed by the radar is degraded by a noise with large amplitude: this gives a speckled aspect to the image, and this is the reason why such a noise is called speckle [10]. To illustrate the difficulty of speckle noise removal, Figure 1 shows a slice of noise free image, and the corresponding slice in the speckled image: one can see that almost all the information has disappeared. If we denote by $I$ the image intensity

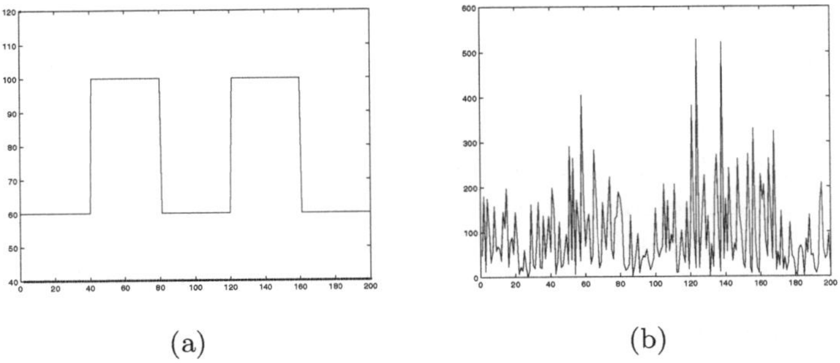


**Fig. 1.** Speckle noise in 1D (notice that the vertical scale is not the same on both images): (a) slice in a noise free synthetic image; (b) corresponding slice in the speckled image

considered as a random variable, then $I$ follows a negative exponential law. The density function is: $g_I(x) = \frac{1}{\mu_I} e^{-\frac{x}{\mu_I}} \mathbf{1}_{\{x \geq 0\}}$, where $\mu_I$ is both the mean and the standard deviation of $I$. In general the image is obtained as the summation of $L$ different images (this is very classical with satellite images). If we assume that the variables $I_k$, $1 \leq k \leq L$ are independent, and have the same mean $\mu_I$, then the intensity $J = \frac{1}{L} \sum_{k=1}^{L} I_k$ follows a gamma law, with density function:

$g_J(x) = \left(\frac{L}{\mu_I}\right)^L \frac{1}{\Gamma(L)} x^{L-1} \exp\left(-\frac{Lx}{\mu_I}\right) \mathbf{1}_{\{x \geq 0\}}$, where $\Gamma(L) = (L-1)!$. Moreover, $\mu_I$ is the mean of $J$, and $\frac{\mu_I}{\sqrt{L}}$ its standard deviation.

The classical modeling [20] for SAR image is: $I = RS$, where $I$ is the intensity of the observed image, $R$ the reflectance of the scene (which is to be recovered), and $S$ the speckle noise. $S$ is assumed to follow a gamma law with mean equal to 1: $g_S(s) = \frac{L^L}{\Gamma(L)} s^{L-1} \exp(-Ls) \mathbf{1}_{\{s \geq 0\}}$.

Speckle removal methods have been proposed in the literature. There are geometric filters, such as Crimmins filter [6] based on the application of convex hull algorithms. There are adaptive filters, such as Lee filter, Kuan filter, or its improvement proposed by Wu and Maitre [21]: first and second order statistic computed in local windows are incoporated in the filtering process. There are adaptive filters with some modelization of the scene, such as Frost filter: the criterion is based on a MAP estimator, and Markov random fields can be used such as in [19, 7]. Another class of filters are multi-temporal ones, such as Bruniquel filter [4]: by computing barycentric means, the standard deviation of the noise can be reduced (provided that several different images of the same scene are available). A last class of methods are variational ones, such as [17, 16, 3], where the solution is computed with PDEs.

## 3 A Variational Multiplicative Denoising Model

The goal of this section is to propose a new variational model for denoising images corrupted by multiplicative noise and in particular for SAR images. We start from the following multiplicative model: $f = uv$, where $f$ is the observed image, $u > 0$ the image to recover, and $v$ the noise. We consider that $f$, $u$, and $v$ are instances of some random variables $F$, $U$ and $V$. In the following, if $X$ is a random variable, we denote by $g_X$ its density function. We refer the interested reader to [11] for further details about random variables. In this section, we consider discretized images. We denote by $\mathcal{S}$ the set of the pixels of the image. Moreover, we assume that the samples of the noise on each pixel $s \in S$ are mutually independent and identically distributed (i.i.d) with density function $g_V$.

### 3.1 Density Laws with a Multiplicative Model

We aim at using the MAP estimator. We therefore need to compute $g_{F|U}$.

**Proposition 1.** *Assume $U$ and $V$ are independent random variables, with continuous density functions $g_U$ and $g_V$. Then we have for $u > 0$:*
$g_V\left(\frac{f}{u}\right) \frac{1}{u} = g_{F|U}(f|u)$.

*Proof:* Let $\mathcal{A}$ an open subset in $\mathbb{R}$. We have:
$\int_{\mathbb{R}} g_{F|U}(f|u) \mathbf{1}_{\{f \in \mathcal{A}\}} = P(F \in \mathcal{A}|U) = \frac{P(F \in \mathcal{A}, U)}{P(U)} = \frac{P\left(\left(V = \frac{F}{U}\right) \in \frac{\mathcal{A}}{U}, U\right)}{P(U)}$. Using the fact that $U$ and $V$ are independent, this is equal to:
$P\left(\left(V = \frac{F}{U}\right) \in \frac{\mathcal{A}}{U}\right) = \int_{\mathbb{R}} g_V(v) \mathbf{1}_{\left\{v \in \frac{\mathcal{A}}{u}\right\}} dv = \int_{\mathbb{R}} g_V(f/u) \mathbf{1}_{\{f \in \mathcal{A}\}} \frac{df}{u}$.

Since this last equality holds for any $\mathcal{A}$ open subset in $\mathbb{R}^*$, this concludes the proof. ∎

### 3.2 Our Model Via the MAP Estimator

We assume the following multiplicative model: $f = uv$, where $f$ is the observed image, $u$ the image to recover, and $v$ the noise. We assume that $v$ follows a gamma law with mean 1, and with density function:

$$g_V(v) = \frac{L^L}{\Gamma(L)} v^{L-1} e^{-Lv} \mathbf{1}_{\{v \geq 0\}} \tag{2}$$

Using Proposition 1, we therefore get: $g_{F|U}(f|u) = \dfrac{L^L}{u^L \Gamma(L)} f^{L-1} e^{-\frac{Lf}{u}}$. We also assume that $u$ follows a Gibbs prior: $g_U(u) = \frac{1}{Z} \exp\left(-\gamma \phi(u)\right)$, where $Z$ is a normalizing constant, and $\phi$ a non negative given function. We aim at maximizing $P(U|F)$. This will lead us to the classical Maximum a Posteriori estimator. From Bayes rule, we have: $P(U|F) = \frac{P(F|U)\,P(U)}{P(F)}$. Maximizing $P(U|F)$ amounts to minimizing the log-likelihood: $-\log(P(U|F) = -\log(P(F|U)) - \log(P(U)) + \log(P(F))$.

Since $\log(P(F))$ is a constant, we just need to minimize: $-\log\left(P(F|U)\right) - \log(P(U)) = -\sum_{s \in S} \left(\log\left(P(F(s)|U(s))\right) - \log(P(U(s)))\right)$. Using $g_{F|U}(f|u)$, and since $Z$ is a constant, we eventually see that minimizing $-\log\left(P(F|U)\right)$ amounts to minimizing: $\sum_{s \in S} \left( L \left( \log U(s) + \frac{F(s)}{U(s)} \right) + \gamma \phi(U(s)) \right)$. The previous computation leads us to propose the following functional for restoring images corrupted with gamma noise: $\int \left( \log u + \frac{f}{u} \right) dx + \frac{\gamma}{L} \int \phi(u)\, dx$.

Remarks: *1)* It is easy to check that the function $u \to \log u + \frac{f}{u}$ reaches its minimum value $1 + \log f$ over $\mathbb{R}_*^+$ for $u = f$.

*2)* In the additive noise case, the most classical assumption is to assume that the noise is a white Gaussian noise. However, this can no longer be the case when dealing with multiplicative noise, except in the case of tiny noise. Indeed, if the model is $f = uv$ where $v$ is a Gaussian noise with mean 1, then some instances of $v$ are negative. Since the data $f$ is assumed positive, this implies that the restored image $u$ has some negative values which is of course impossible. Nevertheless, numerically, if the standard deviation of the noise is smaller than 0.2 (i.e. in the case of tiny noise), then it is very unlikely that $v$ takes some negative values.

## 4 Mathematical Study of the Variational Model

This section is devoted to the mathematical analysis of the variational model $\inf_u \int_\Omega \left( \log u + \frac{f}{u} \right) + \lambda \int_\Omega \phi(u)$. We only state the mathematical results without giving always their complete proofs and we refer the reader to [2] for more details.

### 4.1 Preliminaries

Throughout our study, we will use the following classical distributional spaces. $\Omega \subset R^2$ will denote an open bounded set of $R^2$ with Lipschitz boundary.

- $C_0^\infty(\Omega)$ is the set of functions in $C^\infty(\Omega)$ with compact support in $\Omega$.
- $BV(\Omega)$ is the subspace of functions $u \in L^1(\Omega)$ such that the following quantity is finite:

$$J(u) = \sup\left\{\int_\Omega u(x)\mathrm{div}\,(\xi(x))dx/\xi \in C_0^\infty(\Omega, \mathbb{R}^2), \|\xi\|_{L^\infty(\Omega,\mathbb{R}^N)} \leq 1\right\} \tag{3}$$

If $u \in BV(\Omega)$, the distributional derivative $Du$ is a bounded Radon measure and (3) corresponds to the total variation, i.e. $J(u) = \int_\Omega |Du|$. For $\Omega \subset R^2$, if $1 \leq p \leq 2$, we have: $BV(\Omega) \subset L^p(\Omega)$. Moreover, for $1 \leq p < 2$, this embedding is compact. For further details on $BV(\Omega)$, we refer the reader to [1].
- Since $BV(\Omega) \subset L^2(\Omega)$, we can extend the application $J$ (which we still denote by $J$) over $L^2(\Omega)$:

$$J(u) = \begin{cases} \int_\Omega |Du| \text{ if } u \in BV(\Omega) \\ +\infty \text{ if } u \in L^2(\Omega)\backslash BV(\Omega) \end{cases} \tag{4}$$

We can then define the subdifferential $\partial J$ of $J$ [15]: $v \in \partial J(u)$ iff for all $w \in L^2(\Omega)$, we have $J(u+w) \geq J(u) + \langle v, w\rangle_{L^2(\Omega)}$ where $\langle .,. \rangle_{L^2(\Omega)}$ denotes the usual inner product in $L^2(\Omega)$.
- *Decomposability of $BV(\Omega)$:* If $u$ in $BV(\Omega)$, then $Du = \nabla u\, dx + D_s u$, where $\nabla u \in L^1(\Omega)$ and $D_s u \perp dx$. $\nabla u$ is called the regular part of $Du$.
- If a function $f$ belongs to $L^\infty(\Omega)$, we denote by $\sup_\Omega f$ (resp. $\inf_\Omega f$) the supess of $f$ (resp. the infess of $f$). We recall that supess $f = \inf\{C \in \mathbb{R};\ f(x) \leq C\ a.e.\}$ and infess $f = \sup\{C \in \mathbb{R};\ f(x) \geq C\ a.e.\}$.

### 4.2 The Variational Model

The application we have in mind is the denoising of non textured SAR images. Inspired by the works of Rudin et al [17, 16], we decide to choose $\phi(u) = |Du|$. We thus propose the following restoration model ($\lambda$ being a regularization parameter), where $S(\Omega) = \{u \in BV(\Omega), u > 0\}$:

$$\inf_{u \in S(\Omega)} J(u) + \lambda \int_\Omega \left(\log u + \frac{f}{u}\right) \tag{5}$$

From now on, without loss of generality, we assume that $\lambda = 1$. We denote by $h(u) = \log u + \frac{f}{u}$.

### 4.3 Existence of a Minimizer

Problem (5) has at least one solution.

**Theorem 1.** *Let $f$ be in $L^\infty(\Omega)$ such that $\inf_\Omega f(x) > 0$, then problem (5) has at least one solution $u$ in $BV(\Omega)$ satisfying:*

$$0 < \inf_\Omega f \leq u \leq \sup_\Omega f \tag{6}$$

*Proof:* See [2]. ■

### 4.4 Euler-Lagrange Equation Associated to Problem (5)

Let us now write an "Euler-Lagrange" equation for any solution of problem (5), the difficulty being that the ambient space is $BV(\Omega)$.

**Proposition 2.** *Let $f$ be in $L^\infty(\Omega)$ such that $\inf_\Omega f(x) > 0$. If $u$ in $BV(\Omega)$ is a solution of Problem (5) with $\inf_\Omega u(x) > 0$, then:*

$$-\text{div}\left(\frac{\nabla u}{|\nabla u|}\right) + h^{'}(u) = 0 \tag{7}$$

*holds in the distributional sense, and $u$ satisfies Neumann conditions $\frac{\partial u}{\partial N} = 0$ on the boundary $\partial\Omega$ ($\nabla u$ denotes the regular part of $Du$).*

*Proof:* Let us denote by: $E(u) = \int_\Omega |Du| + \int_\Omega \left(\frac{f}{u} + \log u\right) = J(u) + \int_\Omega h(u)$, and $S(\Omega) = \{u \in BV(\Omega),\, u > 0\}$. Remark that $S(\Omega)$ is an open convex set. Let $w \geq 0$ in $C_0^\infty(\Omega)$ be a test function, and $Du = \nabla u\, dx + D_s u$ the Lebesgue decomposition of $Du$ (with $\nabla u \in L^1(\Omega)$). For $\rho \in (0,1)$, we have $u + \rho(w-u) > 0$ and $|Du + \rho D(w-u)| = |(1-\rho)Du + \rho Dw| = |((1-\rho)\nabla u + \rho \nabla w)|\, dx + (1-\rho)\,|D_s u|$. We thus have for all $w \geq 0$ in $C_0^\infty(\Omega)$, and $\rho \in (0,1)$: $0 \leq \frac{E(u+\rho(w-u))-E(u)}{\rho} = \int_\Omega \frac{|((1-\rho)\nabla u+\rho\nabla w)|-(1-\rho)|\nabla u|}{\rho}\, dx + \int_\Omega \frac{h(u+\rho(w-u))-h(u)}{\rho}\, dx$. Letting $\rho \to 0^+$, we get (in the distributional sense): $-\text{div}\left(\frac{\nabla u}{|\nabla u|}\right) + h^{'}(u) \geq 0$. Let us now assume that $u$ is a solution of Problem (5) and that there exists $\gamma > 0$ such that $u \geq \gamma > 0$. Then, for any $w$ in $C_0^\infty(\Omega)$, $u + \rho w$ belongs to $S(\Omega)$ for $\rho$ sufficiently small. And then is is easy to show as above that the following equality holds in the distributional sense:

$$-\text{div}\left(\frac{\nabla u}{|\nabla u|}\right) + h^{'}(u) = 0 \tag{8}$$

But we have: $h^{'}(u) = \frac{-f}{u^2} + \frac{1}{u} = \frac{u-f}{u^2}$ and we easily see that if $u \geq \gamma > 0$: $\left|h^{'}(u)\right| \leq \frac{f}{\gamma} + \frac{1}{\gamma}$. This implies that equality (8) holds in $L^\infty(\Omega)$. Then, by choosing the test function $w \in C^\infty(\overline{\Omega})$, we get Neumann conditions $\frac{\partial u}{\partial N} = 0$ on the boundary $\partial\Omega$. ■

## 5 Evolution Equation: Discrete Setting

In this section, we consider a semi-discrete version of the problem: the space $\Omega$ is still included in $R^2$, but we discretize the time variable. We consider the case of a regular time discretization, $(t_n)$, with $t_0$ given, and $t_{n+1} - t_n = \delta t$ in $\mathbb{R}_+^*$ (in this section, $\delta t$ is fixed). We define $u_n = u(., t_n)$, and we consider the following implicit scheme.

$$0 \in \frac{u_{n+1} - u_n}{\delta t} + \partial J(u_{n+1}) + h'(u_{n+1}) \tag{9}$$

where $J$ is the extended total variation as defined in (4). We first need to check that indeed (9) defines a sequence $(u_n)$. To this end, we intend to study the following functional: $\inf_{u \in BV(\Omega), u>0} F(u, u_n)$, with: $F(u, u_n) = \int_\Omega \frac{u^2}{2}\, dx - \int_\Omega u_n u\, dx + \delta t \left(J(u) + \int_\Omega h(u)\, dx\right)$. We want to define $u_{n+1}$ as:

$$u_{n+1} = \operatorname*{argmin}_{\{u \in BV(\Omega), u>0\}} F(u, u_n) \tag{10}$$

### 5.1 Existence and Uniqueness of a Sequence $u_n$

**Theorem 2.** *Let $f$ be in $L^\infty(\Omega)$ such that $\inf_\Omega f(x) > 0$, and $u_0$ in $L^\infty(\Omega) \bigcap BV(\Omega)$ with $\inf_\Omega u_0(x) > 0$ be given. There exists a sequence $(u_n)$ in $BV(\Omega)$ satisfying (10). If $\delta t < 27(\inf_\Omega f)^2$, then $(u_n)$ is unique. Moreover, the following estimates hold:*

$$\inf \left(\inf_\Omega f\, ,\, \inf_\Omega u_0\right) \leq u_n \leq \sup\left(\sup_\Omega f\, ,\, \sup_\Omega u_0\right) \tag{11}$$

*and*

$$J(u_n) \leq J(u_0) + \int_\Omega h(u_0)\, dx - \int_\Omega h(f) \tag{12}$$

*Proof:* For the existence of the sequence $(u_n)$, we refer the reader to [2]. Here we just show the uniqueness. We consider: $g(u) = \delta t h(u) + u^2/2 - u_n u$. A sufficient condition ensuring the uniqueness of a minimizer for problem (10) is that $g$ be strictly convex on $\mathbb{R}_+^*$. We have: $g''(u) = 1 + \delta t \frac{f-u}{u^2} = \frac{u^3 - \delta t u + 2\delta t f}{u^3}$. A simple computation shows that if $\delta t < 27(\inf_\Omega f)^2$, then $g''(u) > 0$ for all $u > 0$, i.e. $g$ strictly convex on $\mathbb{R}_+^*$. ■

### 5.2 Euler-Lagrange Equation

We have the following "Euler-Lagrange" equation :

**Proposition 3.** *The sequence $(u_n)$ defined by (10) satisfies (9). Moreover, if we denote by $\nabla u_{n+1}$ the regular part of $Du_{n+1}$, then*

$$0 = \frac{u_{n+1} - u_n}{\delta t} + \left(-\operatorname{div}\left(\frac{\nabla u_{n+1}}{|\nabla u_{n+1}|}\right) + h'(u_{n+1})\right) \tag{13}$$

*holds in the distributional sense, and $u_{n+1}$ satisfies Neumann boundary conditions $\frac{\partial u_{n+1}}{\partial N} = 0$ on the boundary of $\Omega$.*

*Proof:* The proof is similar to the one of Proposition 2. ■

### 5.3 Convergence of the Sequence $u_n$

The following convergence result holds:

**Proposition 4.** *Let $f$ be in $L^\infty(\Omega)$ such that $\inf_\Omega f(x) > 0$, and $u_0$ in $L^\infty(\Omega) \bigcap BV(\Omega)$ with $\inf_\Omega u_0(x) > 0$ be fixed. Let $\delta t < 27(\inf_\Omega f)^2$. The sequence $(u_n)$ defined by equation (9) is such that there exists $u$ in $BV(\Omega)$ with $u_n \rightharpoonup u$ (up to a subsequence) for the $BV(\Omega)$ weak * topology, and $u$ is solution of $0 \in \partial J(u) + h'(u)$ in the distributional sense.*

*Proof:* Since $F(u_{n+1}, u_n) \leq F(u_n, u_n)$, we have:
$\frac{1}{2}\int_\Omega (u_{n+1} - u_n)^2 \leq \delta t \left(J(u_n) - J(u_{n+1}) + \int_\Omega h(u_n) - \int_\Omega h(u_{n+1})\right)$. By summation, we obtain:

$$\frac{1}{2}\sum_{n=0}^{N-1}\int_\Omega (u_{n+1} - u_n)^2 \leq \delta t \left(J(u_0) - J(u_N) + \int_\Omega h(u_0) - \int_\Omega h(u_N)\right)$$
$$\leq \delta t \left(J(u_0) + \int_\Omega h(u_0) - \int_\Omega h(f)\right) < +\infty$$

(since $\int_\Omega h(u_N) \geq \int_\Omega h(f)$). In particular, this implies that $u_{n+1} - u_n \to 0$ in $L^2(\Omega)$ strong. From estimate (12), we know that there exists $u$ in $BV(\Omega)$ such that up to a subsequence $u_n \rightharpoonup u$ for the $BV(\Omega)$ weak * topology. Moreover, $u_n \to u$ in $L^1(\Omega)$ strong. Let $v \in L^2(\Omega)$. From (9), we have: $J(v) \geq J(u_{n+1}) + \left\langle v - u_{n+1}, -\frac{u_{n+1} - u_n}{\delta t} - h'(u_{n+1})\right\rangle_{L^2(\Omega)}$. Using estimate(11) and the fact that $u_n \to u$ in $L^1(\Omega)$ strong, we deduce from Lebesgues dominated convergence theorem that (up to a subsequence) $u_n \to u$ in $L^2(\Omega)$ strong. Moreover, since $u_{n+1} - u_n \to 0$ in $L^2(\Omega)$ strong, and thanks to the lower semi-continuity of the total variation, we get: $J(v) \geq J(u) + \langle u - v, -h'(u)\rangle_{L^2(\Omega)}$. ■

# 6 Numerical Results

## 6.1 Algorithm

To numerically compute a solution to Problem (5), we classically use the Euler-Lagrange equation. We embed it into the following dynamical process which we drive to a steady state: $\frac{\partial u}{\partial t} = \text{div}\left(\frac{\nabla u}{|\nabla u|}\right) + \lambda \frac{f-u}{u^2}$ with initial data $u(x,0) = \frac{1}{|\Omega|}\int_\Omega f$. We denote this model as the AA model.

## 6.2 Other Models

We compare our results with some other classical variational denoising models.

*ROF model:* The first one is the Rudin-Osher-Fatemi (ROF) model:
$\inf_{(u) \in BV}\left(J(u) + \frac{1}{2\lambda}\|f - u\|^2_{L^2}\right)$. To compute the solution, we use Chambolle's projection algorithm [5].

*RLO model:* The second model we compare is the multiplicative version of the ROF model: it has been proposed by Rudin, Lions, and Osher in [16, 14], and

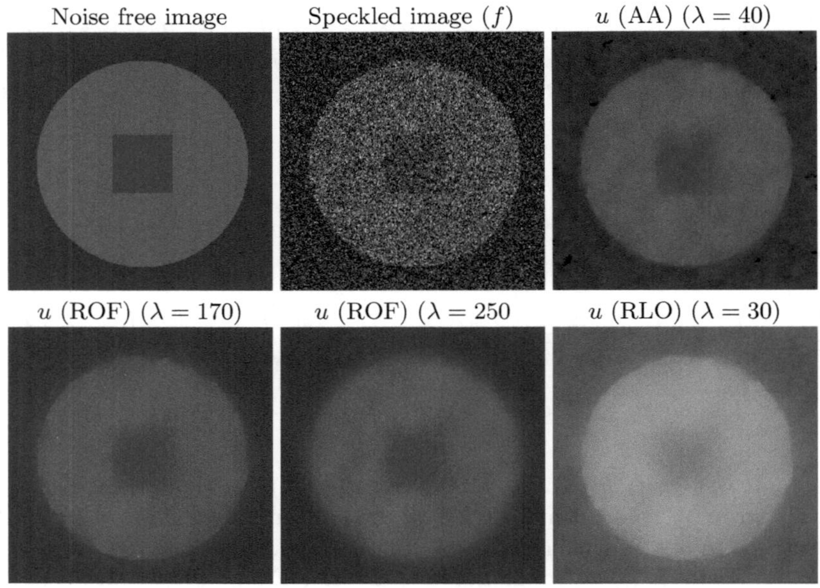


**Fig. 2.** Denoising of a synthetic image with gamma noise. $f$ has been corrupted by some mutliplicative noise with gamma law of mean one. $u$ is the denoised image.

we will call it the RLO model: $\inf_{(u) \in BV} \left( J(u) + \lambda \left\| \frac{f}{u} - 1 \right\|_{L^2}^2 \right)$. To numerically compute a solution, we classically use the Euler-Lagrange equation.

### 6.3 Results

On Figure 2, we show a first example. The original synthetic image is corrupted by some mutliplicative noise with gamma law of mean one (see (2)). We display the denoising results obtained with our approach (AA), as well as with the ROF and the RLO models. Due to the very strong noise, the ROF model has some difficulties to bring back in the range of the image some isolated points (white points on the denoised image): to remove these artefacts, one needs to regularize more the solution, and therefore some part of the edges are lost. The RLO model gives very good results when used to remove small multiplicative noise [16], but it does not perform well with strong multiplicative noise. On Figure 3, we show how our model behaves with a complicated geometrical image. We compare with the ROF model (which has the same drawback as on Figure 2). On Figure 4, we show the result we get on a SAR image provided by the CNES (French space agency: http://www.cnes.fr/index_v3.htm). The reference image (also furnished by the CNES) has been obtained by amplitude summation.

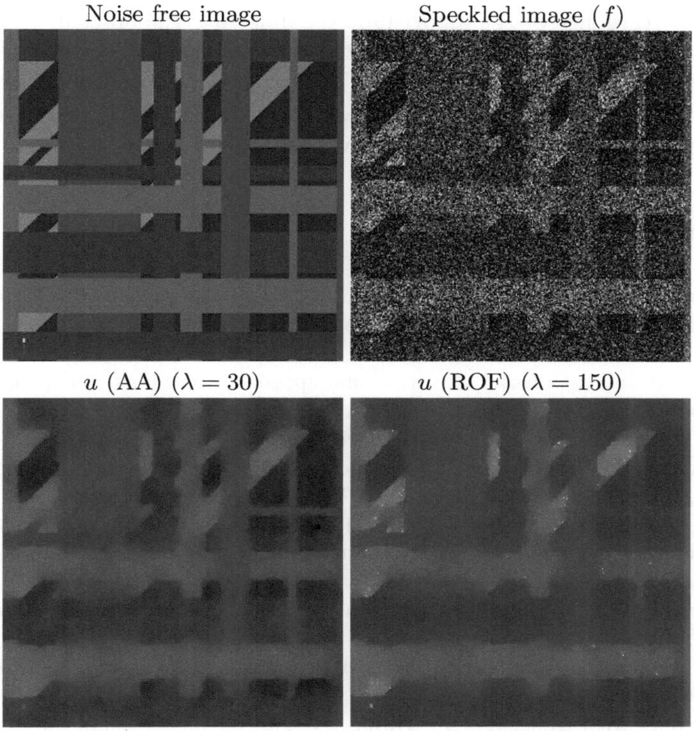


**Fig. 3.** Denoising of a synthetic image with gamma noise. $f$ has been corrupted by some mutliplicative noise with gamma law of mean one.

**Fig. 4.** Denoising of a SAR image provided by the CNES (French space agency)

# 7 Conclusion

In this paper, we propose a new algorithm to remove multiplicative noise. The approach is based on the modeling of speckle noise and the use of variational methods. We introduce a new functional, and we carry out a complete mathematical anlysis of the model. We show existence and uniqueness results, and we prove the convergence of the algorithm. We illustrate the efficiency of the algorithm with some numerical examples, and we show that it compares well with other variational denoising approaches. The proposed approach works well with non textured images. In a future work, we intend to denoise textured images, by using image decomposition methods as introduced in [13].

## References


[1] L. Ambrosio, N. Fusco, and D. Pallara. *Functions of bounded variations and free discontinuity problems*. Oxford mathematical monographs. Oxford University Press, 2000.

[2] G. Aubert and J-F. Aujol. A variational approach to remove multiplicative noise, October 2006. CMLA Preprint 2006-11, http://www.cmla.ens-cachan.fr/Utilisateurs/aujol/publications.html.

[3] J.F. Aujol, G. Aubert, L. Blanc-Féraud, and A. Chambolle. Image decomposition into a bounded variation component and an oscillating component. *Journal of Mathematical Imaging and Vision*, 22(1):71–88, January 2005.

[4] J. Bruniquel and A. Lopes. Analysis and enhancement of multi-temporal sar data. In *SPIE*, volume 2315, pages 342–353, Septembre 1994.

[5] A. Chambolle. An algorithm for total variation minimization and applications. *JMIV*, 20:89–97, 2004.

[6] T.R. Crimmins. Geometric filter for reducing speckle. *Optical Engineering*, 25(5):651–654, May 1986.

[7] J. Darbon, M. Sigelle, and F. Tupin. A note on nice-levelable MRFs for SAR image denoising with contrast preservation, September 2006. Preprint.

[8] D.L. Donoho and M. Johnstone. Adapting to unknown smoothness via wavelet shrinkage. *Journal of the American Statistical Association*, 90(432):1200–1224, December 1995.

[9] D. Geman and S. Geman. Stochastic relaxation, Gibbs distributions, and the Bayesian restoration of images. *IEEE PAMI*, 6:721–741, 1984.

[10] J.W. Goodman. *Statistical Properties of Laser Speckle Patterns*, volume 11 of *Topics in Applied Physics*. Springer-Verlag, second edition, 1984.

[11] G. Grimmett and D. Welsh. *Probability: an introduction*. Manual of Remote Sensing. Oxford Science Publications, 1986.

[12] Henderson Lewis. *Principle and applications of imaging radar*, volume 2 of *Manual of Remote Sensing*. J.Wiley and Sons, third edition, 1998.

[13] Yves Meyer. Oscillating patterns in image processing and in some nonlinear evolution equations, March 2001. The Fifteenth Dean Jacquelines B. Lewis Memorial Lectures.

[14] S. Osher and N. Paragios. *Geometric Level Set Methods in Imaging, Vision, and Graphic*. Springer, 2003.

[15] T. Rockafellar. *Convex Analysis*, volume 224 of *Grundlehren der mathematischen Wissenschaften*. Princeton University Press, second edition, 1983.

[16] L. Rudin, P-L. Lions, and S. Osher. Multiplicative denoising and deblurring: Theory and algorithms. In S. Osher and N. Paragios, editors, *Geometric Level Sets in Imaging, Vision, and Graphics*, pages 103–119. Springer, 2003.

[17] L. Rudin, S. Osher, and E. Fatemi. Nonlinear total variation based noise removal algorithms. *Physica D*, 60:259–268, 1992.

[18] E. Tadmor, S. Nezzar, and L. Vese. A multiscale image representation using hierarchical $(BV, L^2)$ decompositions. *SIAM MMS*, 2(4):554–579, 2004.

[19] F. Tupin, M. Sigelle, A. Chkeif, and J-P. Veran. Restoration of SAR images using recovery of discontinuities and non-linear optimization. In J. Van Leuween G. Goos, J. Hartemis, editor, *EMMCVPR'97*, Lecture Notes in Computer Science. Springer, 1997.

[20] M. Tur, C. Chin, and J.W. Goodman. When is speckle noise multiplicative? *Applied Optics*, 21(7):1157–1159, April 1982.

[21] Y. Wu and H. Maitre. Smoothing speckled synthetic aperture radar images by using maximum homogeneous region filters. *Optical Engineering*, 31(8):1785–1792, August 1992.

# Best Basis Compressed Sensing

Gabriel Peyré

Ceremade, Université Paris Dauphine,
Place du Marchal De Lattre De Tassigny,
75775 Paris Cedex 16 France
gabriel.peyre@ceremade.dauphine.fr,
http://www.ceremade.dauphine.fr/~peyre/

**Abstract.** This paper proposes an extension of compressed sensing that allows to express the sparsity prior in a dictionary of bases. This enables the use of the random sampling strategy of compressed sensing together with an adaptive recovery process that adapts the basis to the structure of the sensed signal. A fast greedy scheme is used during reconstruction to estimate the best basis using an iterative refinement. Numerical experiments on sounds and geometrical images show that adaptivity is indeed crucial to capture the structures of complex natural signals.

## 1 Introduction

### 1.1 Classical Sampling vs. Compressed Sensing

The classical sampling theory of Shannon is based on uniform smoothness assumptions (low frequency spectral content). Under band limited condition, finely enough sampled functions can be recovered from a set of $n$ pointwise measurements.

However, most natural signals $f$ are characterized by very different prior assumptions, such as a decomposition with few elements on some fixed orthogonal basis $\mathcal{B}$. This is the case for a sparse expansion of a sound in a 1D local Fourier basis or the compression of a natural image using a wavelet expansion. Under such sparseness assumption, one can hope to use a much smaller number $n < N$ of measurements, which are linear projections $\Phi f = \{\langle f, \varphi_i \rangle\}_{i=1}^{n}$ on a set of fixed vectors $\varphi_i \in \mathbb{R}^N$. The price to pay for this compressed sampling strategy is a non-linear reconstruction procedure to recover $f$ from the compressed representation $\Phi f$. This theory of compressed acquisition of data has been pushed forward during last few years conjointly by Candès and Tao [1] and Donoho [2].

In order for this recovery to be effective, one needs sensing vectors $\varphi_i$ that are incoherent with the vectors of $\mathcal{B}$. A convenient way to achieve this property is to use random vectors $\varphi_i$, which cannot be sparsely represented in basis $\mathcal{B}$.

*Application in imaging.* Compressed sensing acquisition of data could have an important impact for the design of imaging devices where data acquisition is expensive. For instance in seismic or magnetic imaging one could hope to use few random projections of the object to acquire together with a high precision reconstruction.

F. Sgallari, A. Murli, and N. Paragios (Eds.): SSVM 2007, LNCS 4485, pp. 80–91, 2007.

*Analogies in physiology.* This compressed sampling strategy could potentially lead to interesting models for various sensing operations performed biologically. Skarda and Freeman [3] have proposed a non-linear chaotic dynamic to explain the analysis of sensory inputs. This chaotic state of the brain ensures robustness toward unknown events and unreliable measurements, without using too many computing resources. While the theory of compressed sensing is presented here as a random acquisition process, its extension to deterministic or dynamic settings is a fascinating area for future research in signal processing.

### 1.2 The Best Basis Approach

*Frames vs. dictionary of bases.* Fixed orthogonal bases are not flexible enough to capture the complex redundancy of sounds or natural images. For instance the orthogonal wavelet transform [4] lacks of translation and rotation invariance and is not efficient to compress geometric images [5,6]. It is thus useful to consider families of vectors that are redundant but offer a stable decomposition. For instance, frames of translation invariant wavelets have been used for image denoising and frames of rotation-invariant Gabor functions are useful to characterize textures [4].

However, to capture the complex structure of sounds or the geometry of natural images, one needs a very large number of such elementary atoms. Frame theory suffers from both theoretical difficulties (lack of stability) and technical problems (computational complexity) when the number of basis vectors increases too much. To cope with these problems, one can consider a dictionary $\mathcal{D} = \{\mathcal{B}^\lambda\}_{\lambda \in \Lambda}$ of orthogonal bases $\mathcal{B}^\lambda$. Choosing an optimal basis in such a dictionary allows to adapt the approximation to the complex content of a specific signal.

*Sound processing.* Atoms with a broad range of localizations in space and frequency are needed to represent the transient features that exist in sounds such as the one depicted in figure 3, (a). Local cosine bases [4] divide the time axis in segments that are adapted to the local frequency content of the sound. Other kinds of dictionaries of 1D bases have been proposed, such as the wavelet packets dictionary [7].

*Images and geometry.* The set of cartoon images is a simple model that captures the sketch content of natural images. Figure 4, (a), shows such a geometrically regular image, which contains smooth areas surrounded by regular curves. The curvelet frame of Candès and Donoho [8] can deal with such a regularity and enjoys a better approximation rate than traditional isotropic wavelets. This result can be enhanced using a dictionary of locally elongated functions that follow the image geometry. Bandelets bases of Le Pennec and Mallat [6] provide such a geometric dictionary together with a fast optimization procedure to compute a basis adapted to a given image.

*Adaptive biological computation.* Hubel and Wiesel have shown that low level computation done in area V1 of the visual cortex are well approximated by multiscale oriented linear projections [9]. Olshausen and Field proposed in [10] that redundancy is important to account for sparse representation of natural inputs. However further non-linear processings are done by the cortex to remove high order geometrical correlations present in natural images. Such computations are thought to perform long range groupings over the first layer of linear responses [11] and thus correspond to an adaptive modification of the overall neuronal response. The best-basis coding strategy could thus offer a signal-processing counterpart to this neuronal adaptivity. This paper only deals with orthogonal best basis search but extensions of this approach to dictionaries of redundant transforms are likely to improve numerical results and better cope with biological computations.

### 1.3 Best Basis Computation

*Notations.* The $\ell^p$ norms are defined by

$$\|f\|_{\ell^p}^p = \sum_i |f[i]|^p \qquad \text{and} \qquad \|f\|_{\ell^0} = \#\left\{i \setminus f[i] \neq 0\right\}.$$

A dictionary $\mathcal{D}_\Lambda = \{\mathcal{B}^\lambda\}_{\lambda\in\Lambda}$ is a set of orthogonal bases $\mathcal{B}^\lambda = \{\psi_m^\lambda\}_m$ of $\mathbb{R}^N$. A cost $\text{pen}(\lambda)$ is defined as a prior complexity measure associated to each basis $\mathcal{B}^\lambda$ and satisfies $\sum_{\lambda\in\Lambda} 2^{-\text{pen}(\lambda)} = 1$. A fixed weight $\text{pen}(\lambda) = \log_2(M)$ can be used if the size $M$ of $\mathcal{D}$ is finite. The parameter $\text{pen}(\lambda)$ can be interpreted as the number of bits needed to specify a basis $\mathcal{B}^\lambda$. Following the construction of Coifman and Wickerhauser [7], a best basis $\mathcal{B}^{\lambda^*}$ adapted to a signal minimizes a Lagrangian $\mathcal{L}$

$$\lambda^* = \underset{\lambda\in\Lambda}{\text{argmin}}\, \mathcal{L}(f,\lambda,t) \qquad \text{where} \qquad \mathcal{L}(f,\lambda,t) = \|\Psi^\lambda f\|_{\ell^1} + C_0\, t\, \text{pen}(\lambda), \quad (1)$$

where $\Psi^\lambda = [\psi_0^\lambda, \ldots, \psi_{N-1}^\lambda]^{\mathrm{T}}$ is the transform matrix defined by $\mathcal{B}^\lambda$. This best basis $\mathcal{B}^{\lambda^*}$ is thus the one that gives the sparsest description of $f$ as measured by the $\ell^1$ norm. The Lagrange multiplier $C_0\, t$ weights the penalization $\text{pen}(\lambda)$ associated to the complexity of a basis $\mathcal{B}^\lambda$. The parameter $t$ is the noise level due to acquisition errors or approximate sparsity, whereas $C_0$ is a scaling constant that can be tuned for a specific dictionary $\mathcal{D}_\Lambda$. Note that $t$ is estimated iteratively during the recovery algorithms presented in subsections 2.2 and 3.2.

*Practical best basis computation.* For a dictionary $\mathcal{D}$ that enjoys a multiscale structure, the optimization of $\mathcal{L}$ is carried out with a fast procedure. This algorithm uses the fact that if a basis splits as $\mathcal{B}^\lambda = \mathcal{B}^{\lambda_0} \cup \mathcal{B}^{\lambda_1}$, the Lagrangian satisfies $\mathcal{L}(f,\lambda,t) = \mathcal{L}(f,\lambda_0,t) + \mathcal{L}(f,\lambda_1,t)$. It is thus only needed to compute the value of $\mathcal{L}$ on elementary bases from which each basis $\mathcal{B}^\lambda$ can be decomposed. The number of such elementary bases is usually much smaller than the total number of bases. A bottom-up regression algorithm is then used to search for $\lambda^*$, see [7,6] for practical examples of this process.

# 2 Compressed Sensing Reconstruction

In this paper, the sampling matrix $\Phi = [\varphi_0, \ldots, \varphi_{n-1}]^{\mathrm{T}}$ is defined by random points $\{\varphi_i\}_i$ of unit length although other random sensing schemes can be used, see [1].

## 2.1 Basis Pursuit Formulation

Searching for the sparsest signal $f^*$ in some basis $\mathcal{B} = \{\psi_m\}_m$ that matches the sensed values $y = \Phi f$ leads to consider

$$f^* = \underset{g \in \mathbb{R}^N}{\operatorname{argmin}} \|\Psi g\|_{\ell^0} \quad \text{subject to} \quad \Phi g = y. \tag{2}$$

where $\Psi = [\psi_0, \ldots, \psi_{N-1}]^{\mathrm{T}}$ is the transform matrix defined by $\mathcal{B}$. The combinatorial optimization (2) is NP-hard to solve and convexification of the objective function is introduced in the following basis pursuit [12] formulation

$$f^* = \underset{g \in \mathbb{R}^N}{\operatorname{argmin}} \|\Psi g\|_{\ell^1} \quad \text{subject to} \quad \Phi g = y. \tag{3}$$

Candès and Tao in [1] and Donoho in [2] have shown that, if $f$ is sparse enough in some basis $\mathcal{B}$, $f$ can be recovered from the sensed data $y = \Phi f$. More precisely, they show that it exists a constant $C$ such that if $\|\Psi f\|_{\ell^0} \leqslant k$ then if $n \geqslant C \log(N)\, k$ one recovers $f^* = f$.

To deal with noisy measurements $y = \Phi f + w$, where $w$ is a Gaussian noise of variance $t^2$, one can turn the constrained formulations (3) into a penalized variational problem

$$f^* = \underset{g \in \mathbb{R}^N}{\operatorname{argmin}} \left(\frac{1}{2}\|\Phi g - y\|_{\ell^2}^2 + t\|\Psi g\|_{\ell^1}\right). \tag{4}$$

The Lagrange multiplier $t$ accounts both for stabilisation against noise and approximate sparsity, which is common in practical applications.

This approximate sparsity is characterized by the decays of the best $M$-term approximation $f_M$ of $f$ in basis $\mathcal{B}$ defined by

$$f_M = \sum_{m \in I_t} \langle f,\, \psi_m \rangle \psi_m \qquad \text{where} \qquad I_t = \{m \setminus |\langle f,\, \psi_m \rangle| > t\}$$

and $M = \mathrm{Card}(I_t)$. This can also be understood as an approximation using a thresholding of the coefficients of $f$ in $\mathcal{B}$ that are below $t$. The approximation power of $\mathcal{B}$ is measured by the decay of the approximation error $\|f - f_M\| \leqslant C\, M^{-\alpha}$. The parameter $\alpha$ is a measure of the smoothness of $f$ with respect to the basis $\mathcal{B}$.

The function $f$ with an approximation exponent $\alpha$ can be seen as a $k$-sparse function $f$ corrupted by a deterministic noise of amplitude $k^{-\alpha}$. One thus needs to choose a threshold $t \sim p^{-\alpha}$ for a compressed sensing scenario where

$p \geqslant C\log(N)k$. Candès et al. [1] shows that the $\ell^1$ minimization (4) leads to a recovery error of order $n^{-\alpha}$ up to some logarithmic factors.

The relationship $t \sim p^{-\alpha}$ (up to logarithmic factors) is not precise enough to be useful in real applications, but reveals the insight behind the regularization formulation (4). When the number $n$ of acquired samples decreases, the noise in the function reconstructed with basis pursuit (3) increases and the regularization imposed by the sparsity $\|\Psi f\|_{\ell^1}$ needs to be increased.

## 2.2 Iterative Thresholding for Sparsity Minimization

The recovery procedure suggested by equation (4) corresponds to the inversion of the operator $\Phi$ under sparsity constraints on the observed signal $f$. In this paper we follow Daubechies et al. that propose in [13] an algorithm based on iterated thresholding to perform such regularized inversion. This algorithm had been previously used for image restauration by Figueiredo and Nowak [14] and is derived in [15] as the iteration of two projections on convex sets.

Other greedy algorithms such as orthogonal matching pursuit (OMP) [16] and stagewise orthogonal matching pursuit [17] (StOMP) have been used to solve the compressed sensing reconstruction in a fast way. These algorithms do not fit very well into the best-basis extension exposed in section 3, mainly because OMP is too computationally intensive for imaging applications and because StOMP performed a very agressive iterations (which speed up computation but bias the choice of best-basis toward the initially chosen basis).

We make repeated use of the soft thresholding operator defined in an orthogonal basis $\mathcal{B} = \{\psi_m\}_m$

$$S_t(g) = \sum_m s_t(\langle g,\, \psi_m\rangle)\, \psi_m \qquad \text{where} \qquad s_t(x) = \operatorname{sign}(x)(|x| - t)_+.$$

*The algorithm.* The steps of the algorithm are

- **Initialization.** Set $s = 0$, $f_0 = 0$.
- **Step 1: Updating the estimate.** Set $f_{s+1} = f_s + \Phi^{\mathrm{T}}(y - \Phi f_s)$. This update is an orthogonal projection on the convex set $\{g \setminus \Phi g = y\}$.
- **Step 2: Denoising the estimate.** Estimate the noise level $\sigma_s$ using a median estimator in the transformed domain $\sigma_s = \operatorname{median}(|\Psi f_{s+1}|)/0.6745$ and set the threshold to $t = 3\sigma_s$. Compute $f_{s+1} = S_t(f_{s+1})$ the thresholding of $f$ in basis $\mathcal{B}$. This step corresponds to a projection on a $\ell^1$ ball $\{g \setminus \|\Psi g\|_{\ell^1} \leqslant c\}$ for some $c$ that depends on $t$.
- **Stopping criterion.** If $s < s_{\max}$, go to step 1, otherwise stop iterations.

In all our experiments, the number of iterations is set to $s_{\max} = 20$. The strategy of updating $t$ through iterations has been adopted by Donoho et al. in their extension of orthogonal matching pursuit [17]. The median estimator of the noise is based on the assumption, accurate in practice, that the current estimate $f_{s+1}$ is corrupted by a Gaussian noise which diminishes during iterations.

# 3 Best Basis Compressed Sensing

## 3.1 Variational Formulation

To enhance the quality of the recovery in real life compressed sensing applications, the use of redundant frames has been proposed by several authors [18,15]. Redundancy can impact compressed sensing efficiency, because random sensing vectors can be correlated once inverted through the frame operator $\Psi$, since it is not orthogonal. This paper uses a different approach relying on orthogonal transforms which does not lower the recovery property of the sampling matrix $\Phi$. It also allows a more efficient approximation through the use of a best basis in a large dictionary.

The compressed sensing machinery is extended to a dictionary of bases $\mathcal{D}_\Lambda$ by imposing that the recovered signal is sparse in at least one basis of $\mathcal{D}_\Lambda$. To avoid using too complex basis the recovery process from noisy measurements takes into account a complexity $\text{pen}(\lambda)$ of the optimal basis $\mathcal{B}^\lambda$. This is coherent with the best-basis approximation scheme of [5,6], although penalization is not strictly required in classical statistical estimation. The original recovery procedure (4) is replaced by the following minimization

$$f^* = \underset{g\in\mathbb{R}^N}{\text{argmin}} \; \min_{\lambda\in\Lambda} \Big(\frac{1}{2}\|\Phi g - y\|_{\ell^2}^2 + t\,\|\Psi^\lambda g\|_{\ell^1} + C_0\, t^2 \,\text{pen}(\lambda)\Big), \tag{5}$$

where the penalization $C_0\, t^2 \,\text{pen}(\lambda)$ is the same as in equation (1).

## 3.2 Best Basis Recovery Algorithm

Searching in the whole dictionary $\mathcal{D}$ for the best basis that minimizes formulation (5) is not feasible for large dictionaries, which typically contain of the order of $2^N$ bases. Instead we propose a greedy search for the best basis during the recovery process. This leads to the following algorithm.

- **Initialization.** Set $s = 0$, $f_0 = 0$ and choose $\lambda_0 \in \Lambda$ at random or using some default choice (such as a DCT basis in 1D or a wavelet basis in 2D).
- **Step 1: Updating the estimate.** Set $f_{s+1} = f_s + \Phi^{\mathrm{T}}(y - \Phi f_s)$.
- **Step 2: Denoising the estimate.** Compute the noise level

  $$\sigma_s = \text{median}(|\Psi f_{s+1}|)/0.6745$$

  and set the threshold to $t = 3\sigma_s$. Compute $f_{s+1} = S_t(f_{s+1})$ where $S_t$ is the threshold operator at $t$ in the basis $\mathcal{B}^{\lambda_s}$.
- **Step 3: Update best basis.** Compute $\lambda_{s+1} = \text{argmin}_\lambda\ \mathcal{L}(f_{s+1}, \lambda, t)$. For typical dictionaries such as the ones considered in this paper, this minimization is carried out with a fast procedure, as seen in subsection 1.3.
- **Stopping criterion.** If $s < s_{\max}$, go to step 1, otherwise stop iterations.

# 4 Best Local Cosine Basis Compressed Sensing

## 4.1 Adapted Local Cosine Transform

For each scale $j > 0$, the set of locations $\{0, \ldots, N-1\}$ is subdivided using $N/2^j$ intervals $[x_p^j, x_{p+1}^j]$, where the endpoints are given by $x_p^j = 2^{-j}Np - 1/2$. A local cosine basis $\mathcal{B}_j^p = \{\psi_k^{jp}\}_k$ is defined for each of these intervals using

$$\forall\, k = 0, \ldots, N/2^j - 1, \quad \forall\, \ell,$$
$$\psi_k^{jp}[\ell] = b\big(2^j(\ell - x_p^j)\big)\sqrt{\frac{2}{2^{-j}N}}\cos\left[\pi\big(k + \frac{1}{2}\big)\frac{\ell - x_p^j}{2^{-j}N}\right],$$

where $b$ is a smooth windowing function that satisfies some compatibility conditions [4]. A local cosine basis $\mathcal{B}^\lambda$ of $\mathbb{R}^N$ is parameterized by a binary tree segmentation $\lambda$ of $\{0, \ldots, N-1\}$, see [4]. The set of leaves $L = \{L_j^p\}$ of $\lambda$ are indexed by their depth $j > 0$ and position $p$ in the binary tree, see figure 1, left. The basis $\mathcal{B}^\lambda$ is the union of the various elementary bases $\mathcal{B}^\lambda = \bigcup \{\mathcal{B}_j^p \setminus L_j^p \in L\}$, see figure 1, right. For this local cosine basis dictionary, the penalization $\mathrm{pen}(\lambda)$ is defined as the number of leaves in the binary tree $\lambda$.

The decomposition of a signal $f$ in each of the basis $\mathcal{B}_j^p$ is computed in $O(N\log(N)^2)$ time using FFT. From these atomic decompositions, a best basis $\mathcal{B}^{\lambda^*}$ that minimizes (1) can be extracted by a tree pruning procedure in $O(N)$ time, see [7,4].

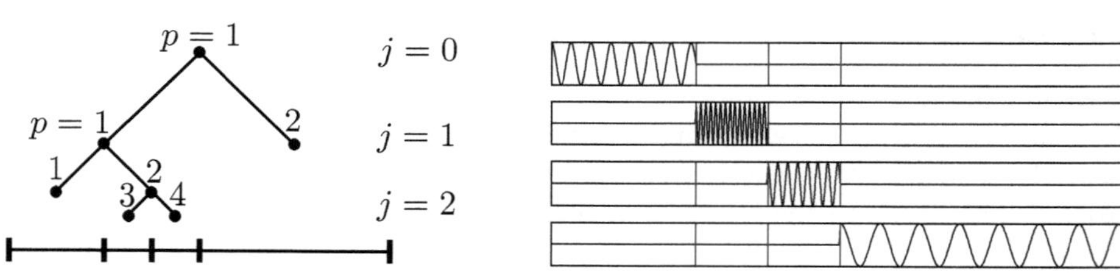


**Fig. 1.** A dyadic tree $\lambda$ defining a spatial segmentation (left) ; some local cosine basis functions $\psi_k^{jp}$ of the basis $\mathcal{B}^\lambda$ (right)

## 4.2 Numerical Results

A synthetic sparse signal $f = (\Psi^\lambda)^{-1}h$ is generated using a random local cosine basis $\mathcal{B}^\lambda$ and a random signal of spikes $h$ with $\|h\|_{\ell^0} = 30$, see figure 2, (a). The signal recovered by the non-adaptative algorithm of subsection 2.2 in an uniform cosine basis $\mathcal{B}^{\lambda_0}$ is significantly different from the original, figure 2, (b). This is due to the fact that $f$ is less sparse in $\mathcal{B}^{\lambda_0}$, since $\|\Psi^{\lambda_0} f\|_{\ell^0} = 512$ and $\|\Psi^{\lambda_0} f\|_{\ell^1} \approx 2.8\|\Psi^\lambda f\|_{\ell^1}$. During the iterations of the algorithm presented in subsection 3.2, the estimated best basis $\mathcal{B}^{\lambda_s}$ evolves in order to match the best basis $\mathcal{B}^\lambda$, see figure 2, (c1–c3). The recovered signal (c3) is nearly identical to $f$.

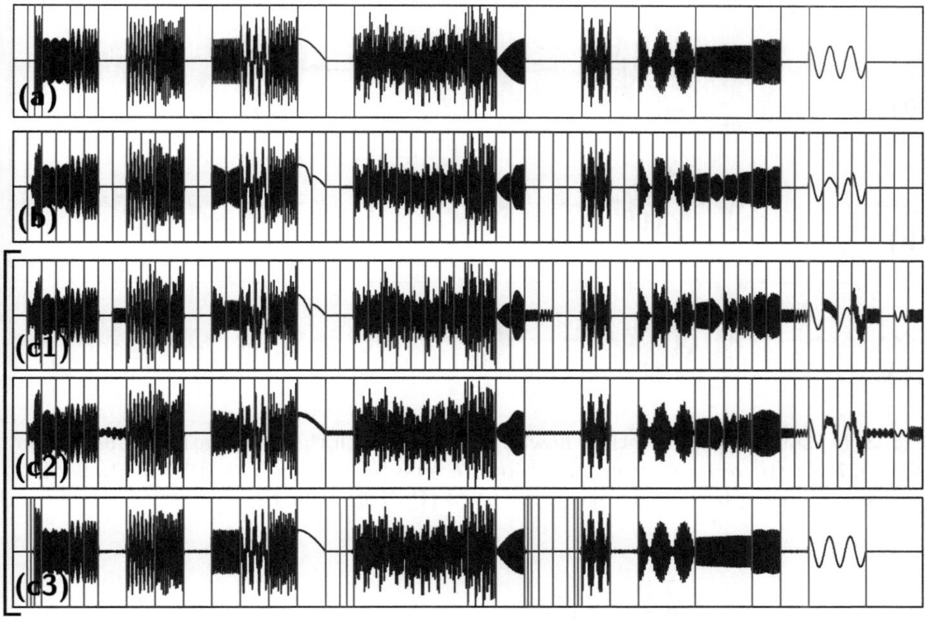


**Fig. 2.** (a) synthetic sound signal with 30 random cosine atoms $N = 4096$ ; (b) recovery using a fixed cosine basis ; (c1) first iteration of the best basis recovery algorithm, $n = N/3$ ; (c2) iteration $s = 5$ ; (c3) iteration $s = 20$

On figure 3 one can see a real sound signal of a tiger howling, together with the signals recovered from fixed basis and adapted basis iterations. Although the final adapted basis is not the same as the one of the original signal, it still provides an improvement of 2dB with respect to a fixed spatial subdivision.

## 5 Best Bandelet Basis Compressed Sensing

### 5.1 Adapted Bandelet Transform

The bandelet bases dictionary was introduced by Le Pennec and Mallat [6,19] to perform adaptive approximation of images with geometric singularities, such as the cartoon image in figure 4, (a). We present a simple implementation of the bandelet transform inspired from [20].

A bandelet basis $\mathcal{B}^\lambda$ is parameterized by $\lambda = (Q, \{\theta_S\}_{S \in Q})$, where $Q$ is a quadtree segmentation of the pixels locations and $\theta_S \in [0, \pi[ \cup \Xi$ is an orientation (or the special token $\Xi$) defined over each square $S$ of the segmentation, see figure 4, (a). The bandelet transform corresponding to this basis applies independently over each square $S$ of the image either

- if $\theta_S = \Xi$: a 2D isotropic wavelet transform,
- if $\theta_S \neq \Xi$: a 1D wavelet transform along the direction defined by the angle $\theta_S$.

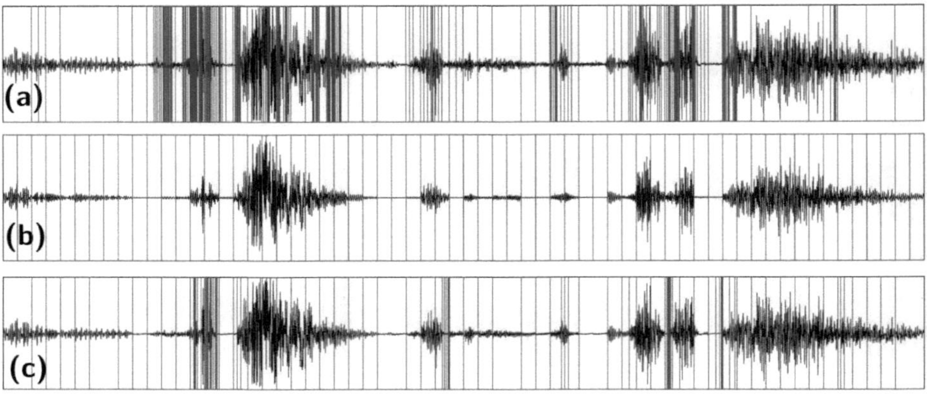


**Fig. 3.** (a) sound signal of a tiger howling, together with the best spatial segmentation, $N = 32768$ ; (b) recovery using fixed local cosine basis, $n = N/3$ (PSNR=19.24dB) ; (c) recovery using best cosine basis, $n = N/3$ (PSNR=21.32dB)

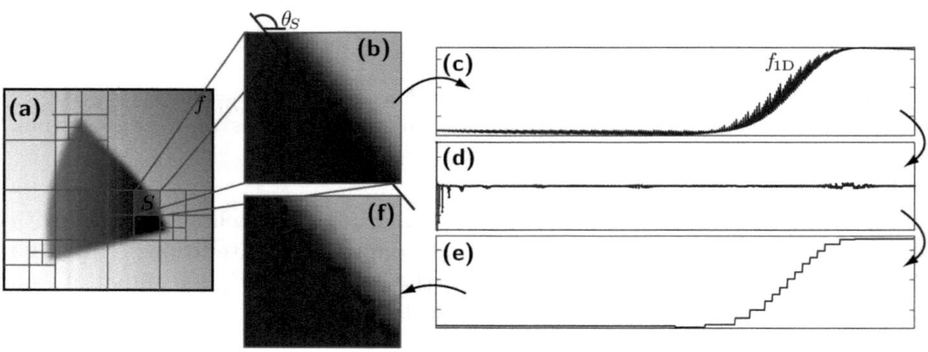


**Fig. 4.** (a) a geometric image together with some adapted dyadic segmentation $Q$ ; (b) a square $S$ together with some adapted direction $\theta_S$ ; (c) the 1D signal $f_{1\text{D}}$ obtained by mapping the pixels values $f(x^{(i)})$ on a 1D axis ; (d) the 1D Haar coefficients of $f_{1\text{D}}$ ; (e) the 1D approximation obtained by reconstruction from the 20 largest Haar coefficients ; (f) the corresponding square approximated in bandelet

We now detail the latter transform. The position of a pixel $x = (x_1, x_2) \in S$ with respect to the direction $\theta_S$ is $p_x = \sin(\theta_S)x_1 - \cos(\theta_S)x_2$. The $m$ pixels $\{x^{(i)}\}$ in $S$ are ranked according to the 1D ordering $p_{x^{(0)}} \leqslant p_{x^{(1)}} \leqslant \ldots \leqslant p_{x^{(m-1)}}$. This ordering allows to turn the image $\{f[x]\}_{x \in S}$ defined over $S$ into a 1D signal $f_{1\text{D}}[i] = f[x^{(i)}]$, see figure 4, (c). The bandelet transform of the image $f$ inside $S$ is defined as the 1D Haar transform of the signal $f_{1\text{D}}$, see figure 4, (d). This process is both orthogonal and easily invertible, since one only needs to compute the inverse Haar transform and pack the retrieved coefficients at the

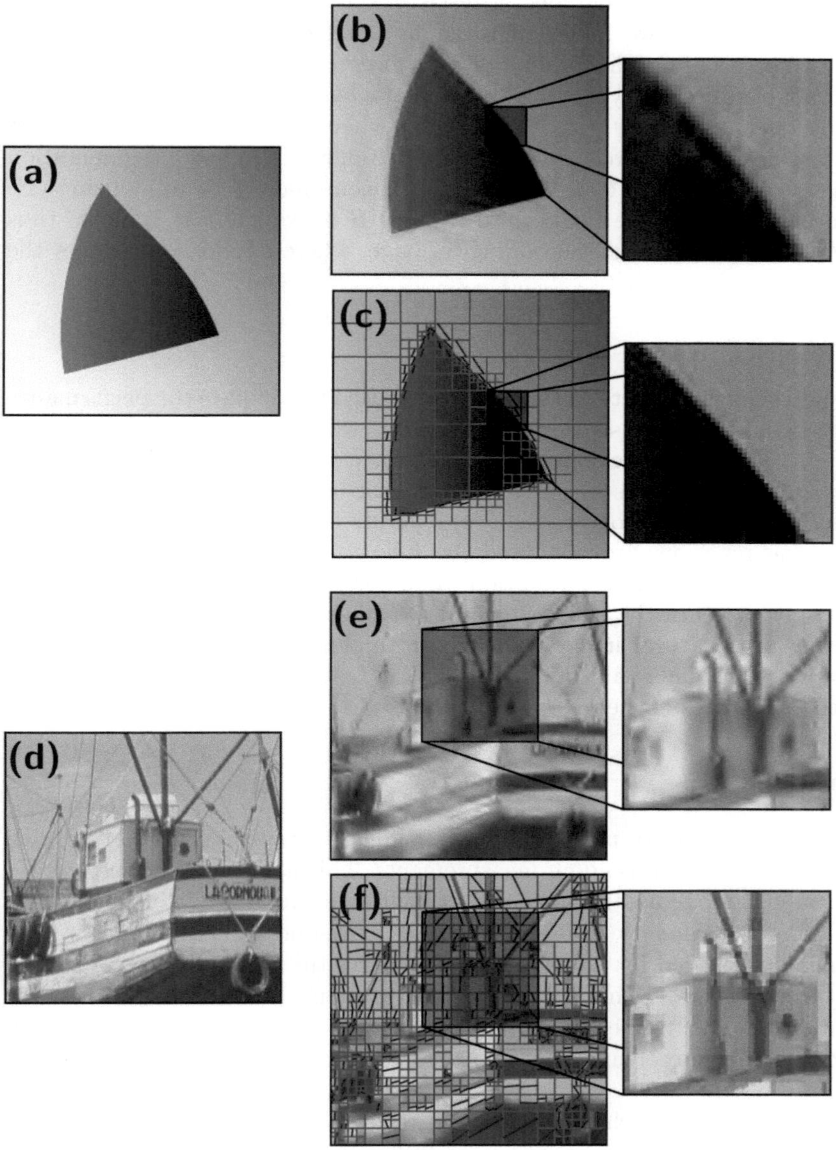


**Fig. 5.** (a/d) original image ; (b/e) compressed sensing reconstruction using the wavelet basis, $n = N/6$ (b:PSNR=22.1dB, e:PSNR=23.2dB) ; (c/f) reconstruction using iteration in a best bandelet basis (c: PSNR=24.3dB, f:PSNR=25.1dB)

original pixels locations. Keeping only a few bandelet coefficients and setting the others to zero performs an approximation of the original image that follows the local direction $\theta_S$, see figure 4 (f).

In order to restrict the number of tested geometries $\theta$ for a square $S \in Q$ containing $\#S$ pixels, we follow [20] and use the set of directions that pass through two pixels of $S$. The number of such directions is of the order of $(\#S)^2$. For this bandelet dictionary, the penalization of a basis $\mathcal{B}^\lambda$ where $\lambda = (Q, \{\theta_S\}_S)$ is defined as $\text{pen}(\lambda) = \#Q + \sum_{S \in Q} 2\log_2(\#S)$, where $\#Q$ is the number of leaves in $Q$. A fast best basis search, described in [20], allows to define a segmentation $Q$ and a set of directions $\{\theta_S\}_S$ adapted to a given image $f$ by minimizing (1). This process segments the image into squares $S$ on which $f$ is smooth, thus setting $\theta_S = \Xi$ and squares containing an edge, where $\theta_S$ closely matches the direction of this singularity.

### 5.2 Numerical Results

The geometric image depicted in figure 5, (a) is used to compare the performance of the original compressed sensing algorithm in a wavelet basis to the adaptative algorithm in a best bandelet basis. Since the wavelet basis is not adapted to the geometric singularities of such an image, reconstruction (b) has strong ringing artifacts. The adapted reconstruction (c) exhibits fewer such artifacts since the bandelet basis functions are elongated and follow the geometry. The segmentation is depicted after the last iteration, together with the chosen direction $\theta_S$ which closely matches the real geometry. On figure 5, (d/e/f), one can see a comparison for a natural image containing complex geometric structures such as edges, junctions and sharp line features. The best bandelet process is able to resolve these features efficiently.

## 6 Conclusion

The best basis framework presented in this paper allows to recover signals with complex structures from random measurements. This approach is successful for natural sounds and geometric images that contain a broad range of sharp transitions. Using a dictionary of bases decouples the approximation process from the redundancy needed for adaptivity and requires the design of a penalization cost on the set of bases. This lowers the computational burden and the numerical instabilities. This framework is not restricted to orthogonal bases, although it is a convenient mathematical way to ensure the compressed sensing recovery condition.

This best basis approach to sensing and recovery is also a promising avenue for interactions between biological processing, where a deterministic or chaotic process is highly probable and signal processing, where randomization has proven useful to provide universal coding strategies.

## References

1. Candès, E., Romberg, J., Tao, T.: Robust uncertainty principles: Exact signal reconstruction from highly incomplete frequency information. IEEE Trans. Inform. Theory (2004) Submitted.

2. Donoho, D.: Compressed sensing. IEEE Transactions on Information Theory **52**(4) (2006) 1289–1306
3. Skarda, C., Freeman, W.: Does the brain make chaos in order to make sense of the world? Behavioral and Brain Sciences **10** (1987) 161–165
4. Mallat, S.: A Wavelet Tour of Signal Processing. Academic Press, San Diego (1998)
5. Donoho, D.: Wedgelets: Nearly minimax estimation of edges. Annals of Statistics **27**(3) (1999) 859–897
6. Le Pennec, E., Mallat, S.: Bandelet Image Approximation and Compression. SIAM Multiscale Modeling and Simulation **4**(3) (2005) 992–1039
7. Coifman, R., Wickerhauser, V.: Entropy-based algorithms for best basis selection. IEEE Trans. Inform. Theory **IT–38**(2) (1992) 713–718
8. Candès, E., Donoho, D.: New tight frames of curvelets and optimal representations of objects with piecewise $C^2$ singularities. Comm. Pure Appl. Math. **57**(2) (2004) 219–266
9. Hubel, D., Wiesel, T.: Receptive fields and functional architecture of monkey striate cortex. Journal of Physiology (London) **195** (1968) 215–243
10. Olshausen, B.A., Field, D.J.: Emergence of simple-cell receptive-field properties by learning a sparse code for natural images. Nature **381**(6583) (1996) 607–609
11. Lee, T.S.: Computations in the early visual cortex. J Physiol Paris **97**(2-3) (2003) 121–139
12. Chen, S., Donoho, D., Saunders, M.: Atomic decomposition by basis pursuit. SIAM J. Sci. Comp. **20**(1) (1998) 33–61
13. Daubechies, I., Defrise, M., Mol, C.D.: An iterative thresholding algorithm for linear inverse problems with a sparsity constraint. Comm. Pure Appl. Math **57** (2004) 1413–1541
14. Figueiredo, M., Nowak, R.: An em algorithm for wavelet-based image restoration. IEEE Transactions on Image Processing **12**(8) (2003) 906–916
15. Candès, E., Romberg, J.: Practical signal recovery from random projections. IEEE Trans. Signal Processing (2005) Submitted.
16. Tropp, J., Gilbert, A.C.: Signal recovery from partial information via orthogonal matching pursuit. Preprint (2005)
17. Donoho, D.L., Tsaig, Y., Drori, I., Starck, J.L.: Sparse solution of underdetermined linear equations by stagewise orthogonal matching pursuit. Preprint (2006)
18. Donoho, D., Tsaig, Y.: Extensions of compressed sensing. Preprint (2004)
19. Le Pennec, E., Mallat, S.: Sparse geometric image representations with bandelets. IEEE Transactions on Image Processing **14**(4) (2005) 423–438
20. Peyré, G., Mallat, S.: Surface compression with geometric bandelets. ACM Transactions on Graphics, (SIGGRAPH'05) **24**(3) (2005)

# Efficient Beltrami Filtering of Color Images Via Vector Extrapolation

Lorina Dascal, Guy Rosman, and Ron Kimmel

Computer Science Department, Technion,
Institute of Technology, Haifa 32000, Israel

**Abstract.** The Beltrami image flow is an effective non-linear filter, often used in color image processing. It was shown to be closely related to the median, total variation, and bilateral filters. It treats the image as a 2D manifold embedded in a hybrid spatial-feature space. Minimization of the image area surface yields the Beltrami flow. The corresponding diffusion operator is anisotropic and strongly couples the spectral components. Thus, there is so far no implicit nor operator splitting based numerical scheme for the PDE that describes Beltrami flow in color. Usually, this flow is implemented by explicit schemes, which are stable only for very small time steps and therefore require many iterations. At the other end, vector extrapolation techniques accelerate the convergence of vector sequences, without explicit knowledge of the sequence generator. In this paper, we propose to use the minimum polynomial extrapolation (MPE) and reduced rank extrapolation (RRE) vector extrapolation methods for accelerating the convergence of the explicit schemes for the Beltrami flow. Experiments demonstrate their stability and efficiency compared to explicit schemes.

**Keywords:** *Color processing, geometric heat flow, Laplace-Beltrami operator.*

## 1 Introduction

Nonlinear partial differential diffusion equations are used extensively for various image processing applications. The Beltrami framework was first introduced in [20], then in [21], followed by [26]. This filter was applied for edge preserving denoising and deblurring for signals and especially multi-channel images, see for example [1]. In this paper we demonstrate an efficient scheme for computing the Beltrami color flow. This flow is usually implemented by a discrete approximation of a partial differential equation (PDE). Standard explicit finite difference schemes for heat equations require small time-steps for stability, that lead to a large number of iterations required for convergence to the desired solution.

The additive operator splitting (AOS) scheme was first developed for solving the Navier-Stokes equations [10,11]. It was later used in [24] for implementing the regularized Perona-Malik filter [4]. The AOS scheme is first order in time, semi-implicit, and unconditionally stable with respect to its time step. It can

F. Sgallari, A. Murli, and N. Paragios (Eds.): SSVM 2007, LNCS 4485, pp. 92–103, 2007.

be used for single channel diffusion filters, like the Perona-Malik [12], the Total variation (TV) [14], or the anisotropic diffusion filter for grey-level images [25]. Unfortunately, due to the strong coupling and its anisotropic nature, splitting is difficult for the Beltrami color operator. In fact, so far, there is no PDE based implicit scheme for Beltrami flow in color.

In [22] the short time kernel for the Beltrami operator was computed in order to approximate the Beltrami flow. This method is still computationally demanding, as computing the kernel operation involves geodesic distance computation around each pixel. The bilateral operator was studied in different contexts (see for example [18], [23], [19], [7], [2]), and can be shown to be an approximation of the Beltrami kernel.

In this paper we propose to apply vector extrapolation methods to accelerate the convergence rate of standard explicit schemes for the Beltrami flow in color. The minimum polynomial extrapolation algorithm (MPE) of Cabay and Jackson [3] and the reduced rank extrapolation algorithm (RRE) of Eddy [6] are two vector extrapolation methods derived in order to accelerate the convergence of vector sequences. They are obtained from an iterative solution of linear and nonlinear systems of equations. Both MPE and RRE algorithms are detailed in Section 4 with their respective definitions as solutions for least squares problems. An efficient solution is achieved by the modified Gram-Schmidt algorithm [16]. These vector extrapolation methods can be computed directly from the elements of the sequence. Furthermore, unlike alternative acceleration techniques, they can be applied not only to linearly generated sequences, but also to nonlinear ones. This allows us to apply the RRE/MPE methods for accelerating the convergence of the vector sequence generated in the explicit scheme for the Beltrami geometric flow.

We demonstrate the efficiency and accuracy of the vector extrapolation methods in color image processing applications such as: scale-space analysis, denoising, and deblurring.

This paper is organized as follows: Section 2 gives a brief summary of the Beltrami framework. In Section 3 we review approximations based on standard explicit finite difference schemes. Section 4 describes the minimal polynomial extrapolation (MPE) and the reduced rank extrapolation (RRE) algorithms, the two methods we use for accelerating the convergence of the standard explicit scheme. In Section 5 we apply the RRE algorithm to our Beltrami color flow and demonstrate the resulting speed-up. Section 6 concludes the paper.

## 2 The Beltrami Framework

Let us briefly review the Beltrami framework for non-linear diffusion in computer vision [9,20,21,26]. We represent images as embedding maps of a Riemannian manifold in a higher dimensional space. We denote the map by $U : \Sigma \to M$, where $\Sigma$ is a two-dimensional surface, with $(\sigma^1, \sigma^2)$ denoting coordinates on it. $M$ is the spatial-feature manifold, embedded in $\mathbb{R}^{d+2}$, where $d$ is the number of image channels. For example, a gray-level image can be represented as a 2D surface embedded in $\mathbb{R}^3$. The map $U$ in this case is $U(\sigma^1,\sigma^2)=(\sigma^1,\sigma^2,I(\sigma^1,\sigma^2))$, where $I$ is the image intensity. For color images, $U$ is given by $U(\sigma^1,\sigma^2) = (\sigma^1,\sigma^2,I^1(\sigma^1,\sigma^2),I^2(\sigma^1,\sigma^2),I^3(\sigma^1,\sigma^2))$, where $I^1,I^2,I^3$ are the three components of the color vector.

Next, we choose a Riemannian metric on this surface. The canonical choice of coordinates in image processing is Cartesian (we denote them here by $x^1$ and $x^2$). For such a choice, which we follow in the rest of the paper, we identify $\sigma^1 = x^1$ and $\sigma^2 = x^2$. In this case, $\sigma^1$ and $\sigma^2$ are the image coordinates. We denote the elements of the inverse of the metric by superscripts $g^{ij}$, and the determinant by $g = \det(g_{ij})$. Once images are defined as embedding of Riemannian manifolds, it is natural to look for a measure on this space of embedding maps.

Denote by $(\Sigma, g)$ the image manifold and its metric, and by $(M, h)$ the space-feature manifold and its metric. Then, the functional $S[U]$ attaches a real number to a map $U : \Sigma \to M$,

$$S[U, g_{ij}, h_{ab}] = \int d^m\sigma \sqrt{g}||dU||^2_{g,h}, \tag{1}$$

where $m$ is the dimension of $\Sigma$, $g$ is the determinant of the image metric, and the range of indices is $i,j = 1,2,\dots\dim(\Sigma)$ and $a,b = 1,2,\dots\dim(M)$. The integrand $||dU||^2_{g,h}$ is expressed in a local coordinate system by $||dU||^2_{g,h} = (\partial_{x_i}U^a)g^{ij}(\partial_{x_j}U^b)h_{ab}$. This functional, for $\dim(\Sigma) = 2$ and $h_{ab} = \delta_{ab}$, was first proposed by Polyakov [13] in the context of high energy physics, in the theory known as *string theory*. The elements of the induced metric for color images with Cartesian color coordinates are

$$G = (g_{ij}) = \begin{pmatrix} 1+\beta^2\sum_{a=1}^3 (U^a_{x_1})^2 & \beta^2\sum_{a=1}^3 U^a_{x_1}U^a_{x_2} \\ \beta^2\sum_{a=1}^3 U^a_{x_1}U^a_{x_2} & 1+\beta^2\sum_{a=1}^3 (U^a_{x_2})^2 \end{pmatrix},$$

where a subscript of $U$ denotes partial derivation and the parameter $\beta > 0$ determines the ratio between the spatial and spectral (color) distances. Using standard methods in the calculus of variations, the Euler-Lagrange equations with respect to the embedding (assuming Euclidean embedding space) are

$$0 = -\frac{1}{\sqrt{g}}h^{ab}\frac{\delta S}{\delta U^b} = \underbrace{\frac{1}{\sqrt{g}}\mathrm{div}\,(D\nabla U^a)}_{\Delta_g U^a}, \tag{2}$$

where the matrix $D = \sqrt{g}G^{-1}$. See [20] for explicit derivation. The operator that acts on $U^a$ is the natural generalization of the Laplacian from flat spaces to manifolds, it is called the Laplace-Beltrami operator, and is denoted by $\Delta_g$.

The parameter $\beta$, in the elements of the metric $g_{ij}$, determines the nature of the flow. At the limits, where $\beta \to 0$ and $\beta \to \infty$, we obtain respectively a linear diffusion flow and a nonlinear flow, akin to the TV flow for the case of grey-level images (see [21] for details).

The Beltrami scale-space emerges as a gradient descent minimization process

$$U^a_t = -\frac{1}{\sqrt{g}}\frac{\delta S}{\delta U^a} = \Delta_g U^a. \tag{3}$$

For Euclidean embedding, the functional in Eq. (1) reduces to

$$S(U) = \int \sqrt{g}\, dx^1\, dx^2$$
$$= \int \sqrt{1 + \beta^2 \sum_{a=1}^{3} |\nabla U^a|^2 + \frac{1}{2}\beta^4 \sum_{a,b=1}^{3} |\nabla U^a \times \nabla U^b|^2}\, dx^1\, dx^2.$$

This geometric measure can be used as a regularization term for color image processing. In the variational framework, the reconstructed image is the minimizer of a cost-functional. This functional can be written in the following general form,

$$\Psi = \frac{\alpha}{2} \sum_{a=1}^{3} ||KU^a - U_0^a||^2 + S(U),$$

where $K$ is a bounded linear operator. In the denoising case, $K$ is the identity operator and in the deblurring case, $K$ is a convolution operator of $U^a$ with a given filter. The parameter $\alpha$ controls the smoothness of the solution.

The modified Euler-Lagrange equations as a gradient descent process for each case are

$$U_t^a = -\frac{1}{\sqrt{g}} \frac{\delta \Psi}{\delta U^a} = -\frac{\alpha}{\sqrt{g}}(U^a - U_0^a) + \Delta_g U^a \quad \text{(denoising)}, \tag{4}$$

$$U_t^a = -\frac{1}{\sqrt{g}} \frac{\delta \Psi}{\delta U^a} = -\frac{\alpha}{\sqrt{g}} k(-x,-y) * (k * U^a - U_0^a) + \Delta_g U^a \quad \text{(deblurring)}, \tag{5}$$

where $Ku = k * u$ and $k$ is often approximated by the Gaussian blurring kernel. The above equations provide an adaptive smoothing mechanism. In areas with large gradients (edges), the fidelity term is suppressed and the regularizing term becomes dominant. At homogenous regions with low-gradient magnitude, the fidelity term takes over and controls the flow.

## 3 Standard Explicit Finite Difference Scheme

Our goal is to speed-up the convergence of the explicit scheme in Beltrami color processing. In this section, we detail the common explicit scheme. The applications we address are the Beltrami based smoothing, Beltrami based denoising, and Beltrami based deblurring.

We work on a rectangular grid with step sizes $\Delta t$ in time and $h$ in space. The spatial units are normalized such that $h = 1$. For each channel $U^a$, $a \in \{1, 2, 3\}$, we define the discrete approximation $(U^a)_{ij}^n$ by

$$(U^a)_{ij}^n \approx U^a(ih, jh, n\Delta t).$$

On the boundary of the image we impose the Neumann boundary condition.

The explicit finite difference scheme is written in a general form as

$$(U^a)_{ij}^{n+1} = (U^a)_{ij}^n + \Delta t O_{ij}^n(U^a), \tag{6}$$

where $O_{ij}^n$ is the discretization of the right hand side of the relevant continuous equation (3), (4), or (5). Below, we give the exact form of the operator $O_{ij}^n$ for each of the above cases.

- *Beltrami-based smoothing.*
  The explicit scheme (6) for discretizing Eq. (3) takes the form

$$(U^a)_{ij}^{n+1} = (U^a)_{ij}^n + \Delta t L_{ij}^n(U^a), \tag{7}$$

  where $L_{ij}^n(U^a)$ denotes a discretization of the Laplace-Beltrami operator $\Delta_g U^a$, for example, using a backward-forward approximation.
- *Beltrami-based denoising.*
  The explicit scheme (6) is given in this case by

$$(U^a)_{ij}^{n+1} = (U^a)_{ij}^n + \Delta t\left[L_{ij}^n(U^a) + \frac{\alpha}{\sqrt{g}}((U_0^a)_{ij}^n - (U^a)_{ij}^n)\right]. \tag{8}$$

- *Beltrami-based deblurring.*
  Similarly, in the deblurring case, the explicit scheme (6) reads

$$(U^a)_{ij}^{n+1} = (U^a)_{ij}^n + \Delta t\left[L_{ij}^n(U^a) + \frac{\alpha}{\sqrt{g}}\bar{k}_{ij}^n * \left((U_0^a)_{ij}^n - k_{ij}^n * (U^a)_{ij}^n\right)\right], \tag{9}$$

  where $\bar{k} = k(-x, -y)$.

Due to stability requirements (see [5], [8]), explicit schemes limit the time-step $\Delta t$ and usually require a large number of iterations in order to converge. We propose to use vector extrapolation techniques in order to accelerate the convergence of these explicit schemes.

## 4 MPE/RRE Acceleration Techniques

Let us start by describing extrapolation methods for accelerating convergence of vector sequences. These techniques do not require information on the sequence generator, but are computed directly from the elements of the sequence. Following [16], we review the Minimal Polynomial Extrapolation (MPE) and Reduced Rank Extrapolation (RRE) methods.

These methods were first introduced for the case of linearly generated vector sequences (see [3], [6]) and were analyzed from the point of view of convergence, rate of convergence, and stability in [15]. In [17] these methods' convergence behavior was analyzed in the case of nonlinear problems . It is important to note that various related methods such as Krylov subspace methods and generalized conjugate residuals can be applied *only to linear systems*. Unlike these methods, the MPE and RRE techniques are applicable to nonlinearly generated sequences.

Let $\mathbf{x}_n, ..., \mathbf{x}_{n+k}$ be a given sequence of $N$ dimensional column vectors, and denote by $\mathbf{s}_{n,k}$ its limit. These vectors are usually taken to be the last $k+1$ vectors of an iterative process. The vector $\mathbf{x}_n$ is not necessarily the initial solution, but rather can be taken at an arbitrary position in the sequence. In practice, after applying the acceleration technique and obtaining an approximate solution vector, this vector can be considered as the new $\mathbf{x}_n$.

The extrapolation methods have been derived based on differences, and below we use the abbreviated notation for first and second differences of the sequence of vectors. Denote by $\mathbf{u}_j$ and $\mathbf{w}_j$ the first and the second differences of the vectors $\mathbf{x}_i$.

$$\mathbf{u}_j = \mathbf{x}_{j+1} - \mathbf{x}_j, \quad \mathbf{w}_j = \mathbf{u}_{j+1} - \mathbf{u}_j, \quad j = 0, 1, 2, .... \tag{10}$$

Define the $N \times (j+1)$ matrices $\mathbf{U}_j^{(n)}$ and $\mathbf{W}_j^{(n)}$ by

$$\mathbf{U}_j^{(n)} = [\mathbf{u}_n|\mathbf{u}_{n+1}|\cdots|\mathbf{u}_{n+j}] \tag{11}$$

and

$$\mathbf{W}_j^{(n)} = [\mathbf{w}_n|\mathbf{w}_{n+1}|\cdots|\mathbf{w}_{n+j}], \tag{12}$$

respectively.

### 4.1 MPE Definition

Let $k$ be a positive integer number. The approximation $\mathbf{s}_{n,k}$ to the desired limit $\mathbf{s}$ is given by

$$\mathbf{s}_{n,k} = \sum_{j=0}^{k} \gamma_j \mathbf{x}_{n+j}, \tag{13}$$

where the coefficients $\gamma_j$ are determined as follows:

1. Obtain the least squares solution $\mathbf{c}$ for the overdetermined linear system
$$\mathbf{U}_{k-1}^{(n)} \mathbf{c} = -\mathbf{u}_{n+k},$$
using the modified Gram-Schmidt algorithm [16].
2. Denote $\mathbf{c} = (c_0, c_1, ..., c_{k-1})^T$. Set $c_k = 1$ and compute the coefficients $\gamma_j$
$$\gamma_j = \frac{c_j}{\sum_{i=0}^{k} c_i}, 0 \le j \le k, \tag{14}$$
assuming that $\sum_{i=0}^{k} c_i \neq 0$. When this condition is violated, $\mathbf{s}_{n,k}$ does not exist.

### 4.2 RRE Definition

For the RRE method, the approximation $\mathbf{s}_{n,k}$ of $\mathbf{s}$ is defined as

$$\mathbf{s}_{n,k} = \mathbf{x}_n + \sum_{i=0}^{k-1} \xi_i \mathbf{u}_{n+i}. \tag{15}$$

The coefficients $\xi_i$ are determined by solving the overdetermined linear system

$$\mathbf{W}_{k-1}^{(n)}\xi = -\mathbf{u}_n, \tag{16}$$

with coefficients $\xi = (\xi_0, \xi_1, ..., \xi_{k-1})^T$.

Since there always exists a solution to the least square problem (16), $\mathbf{s}_{n,k}$ always exists. In particular, $\mathbf{s}_{n,k}$ exists uniquely when the matrix $\mathbf{W}_{k-1}^{(n)}$, has a full rank.

The essential difference between MPE and RRE is the way of determining the coefficients of the extrapolation. In our experiments, the MPE algorithm leads to visual results similar to the RRE, but converges more slowly than the RRE. Thus for the experiments displayed in the next section we chose the RRE as the extrapolation method.

Next, we describe the way of applying the vector extrapolation method in the context of the Beltrami framework. Assume the vector sequence generated by an explicit scheme is given by

$$\mathbf{x}_{n+1} = \mathbf{F}(\mathbf{x}_n), \tag{17}$$

where the nonlinear operator $\mathbf{F}$ is

$$\mathbf{F} = I + \Delta t O, \tag{18}$$

and $O$ is given in Section 3 by either Eq. (7),(8), or (9).

The *vector extrapolation* method is then applied as follows:

1. We start with $n = 0$ and an initial guess $\mathbf{x}_0$, which is, in fact, the starting vector for the sequence to be generated. A few iterations of the explicit scheme (17) can be used as a preconditioner.
2. We use the explicit scheme (17) and $\mathbf{x}_n$ to generate a sequence of vectors $\mathbf{x}_n$,..., $\mathbf{x}_{n+k}$.
3. A vector extrapolation method (MPE/RRE) is used in order to extrapolate the approximation $\mathbf{s}_{n,k}$ of $\mathbf{s}$ from the last $k+1$ vectors, $\mathbf{x}_n$,..., $\mathbf{x}_{n+k}$.
4. The approximation $\mathbf{s}_{n,k}$ is used as a new starting point $\mathbf{x}_{n'}$, $n' = n + k + 1$ for a new sequence.
5. The procedure is repeated from Step 2 until convergence.

Sidi [16] used the term *cycling* to refer to this kind of scheme.

Note that the nonlinear eqs. (17), (18) describe the discretization of the Beltrami flow in the same form used by Smith et al. in the convergence analysis of nonlinearly generated sequences ([17], Section 6).

## 5 Experimental Results

We proceed to demonstrate experimental results of the Beltrami scale-space and restoration of color images processed by the explicit and the RRE accelerated schemes, specifying the CPU runtime and the resulting speed-up. Although in each application we display results with respect to a single image, the behavior exhibited is similar to other input data.

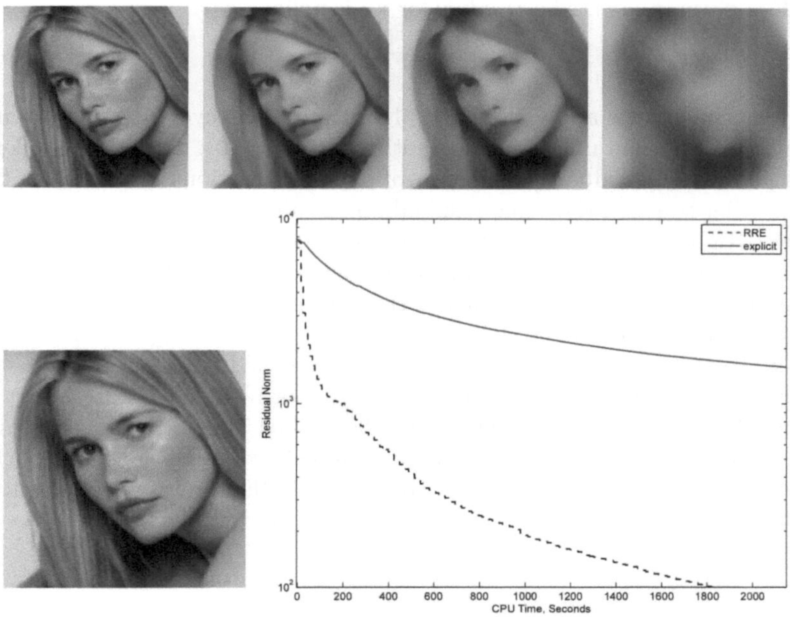


**Fig. 1.** Top (left to right): RRE iterations: 50, 150, 300, 450. Bottom: left: Original picture. right: Comparison of the residual norms versus CPU time. Parameters: $\beta = \sqrt{1/0.0005} \simeq 44$, $\Delta t = 0.0042/\beta^2$.

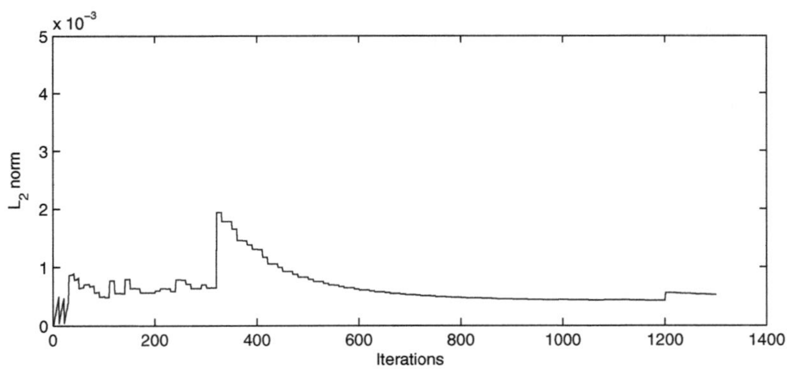


**Fig. 2.** $l^2$-norm error between the RRE iterations sequence and the corresponding explicit iterations images. A sequence using $k = 15$ is shown, which gave the worst $l^2$-norm values in practice. Parameters: $\beta = \sqrt{1/0.001} \simeq 31$, $\Delta t = 0.0042/\beta^2$.

Let the sequence of vectors $\mathbf{x}_1, \mathbf{x}_2, ...$ be generated according to

$$\mathbf{x}_{n+1} = \mathbf{F}(\mathbf{x}_n), \quad n = 0, 1, ...$$

where $\mathbf{F}$ is given in (18). The *residual* for $\mathbf{x}_n$ is defined as

$$Res(\mathbf{x}_n) = \mathbf{F}(\mathbf{x}_n) - \mathbf{x}_n. \tag{19}$$

In our experiments we apply the RRE algorithm in the cycling mode (20 initial explicit iterations, and unless specified otherwise, $k = 10$). The RRE accelerated scheme allows us to reduce the number of explicit iterations by at least a factor of 10, in order to reach the same residual norm value. Experiments demonstrate that the RRE scheme remains stable as the number of iterations increases.

Figure 1 top row depicts the scale-space behavior of the Beltrami color flow, obtained using Eq. (7). At the bottom right it shows the speed-up obtained by using the RRE scheme for the Beltrami-based scale-space analysis. The speed-up gain is up to 40, as can be seen in the graph of the residual norm.

We measure the approximation error using $l^2$-norm values. Figure 2 shows the $l^2$-norm values of the images generated by the explicit and the RRE schemes during the scale-space evolution. Comparison is done by running the explicit scheme and the RRE scheme simultaneously. After each RRE iteration, we advance the explicit sequence, starting from the previous result until it diverges from the RRE

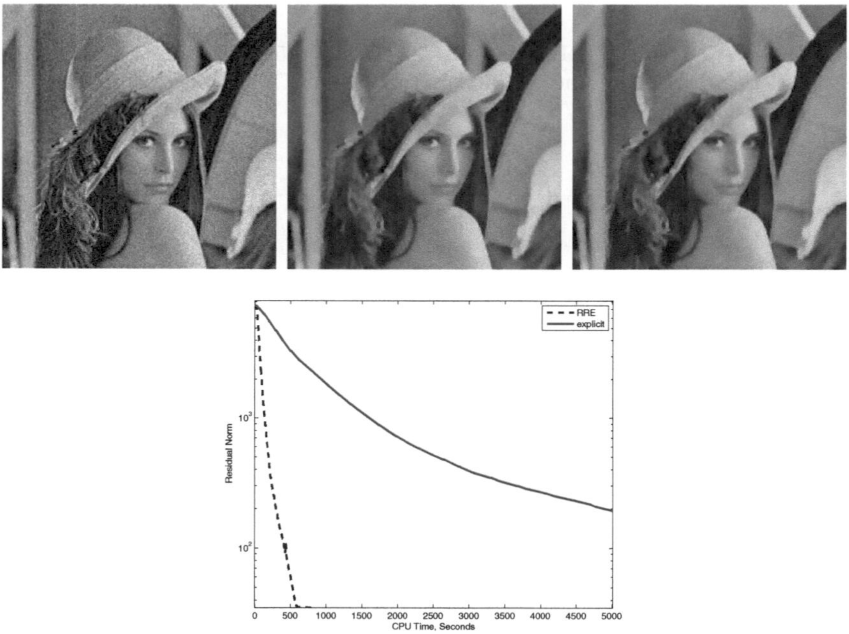


**Fig. 3.** Beltrami based denoising. Top left: Noisy image. Middle: Denoised image obtained by the RRE (901 iterations). Right: Denoised image obtained by the explicit scheme (11541 iterations). Bottom: Comparison of the residual norms versus CPU time. Parameters: $\beta = \sqrt{1000}, \lambda = 0.02, \Delta t = 0.0021/\beta^2$.

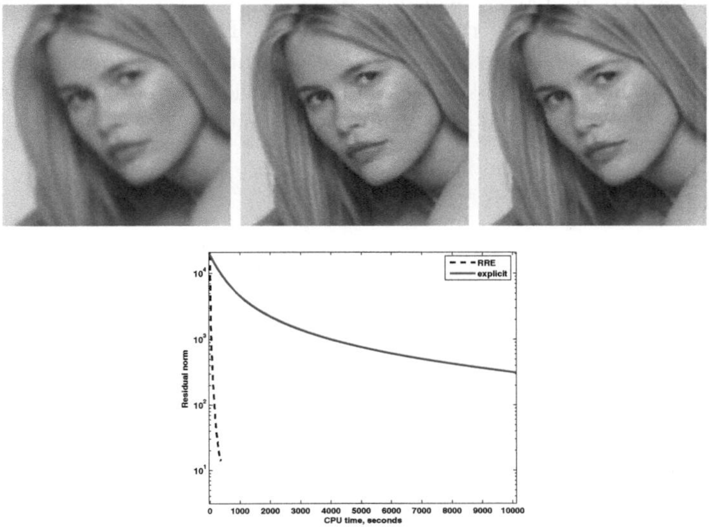


**Fig. 4.** Beltrami-based deblurring. Top left: Blurred image. Middle: Deblurred image obtained by the RRE scheme (1301 iterations). Right: Deblurred image obtained by the explicit scheme (196608 iterations). Bottom: Comparison of the residual norms versus CPU time. Parameters: $\beta = \sqrt{1/0.0005} \simeq 44$, $\Delta t = 0.0021/\beta^2, \lambda = 0.03$.

result. $l^2$-norm values indicate that the images obtained by the explicit and RRE techniques are "numerically" the same. The maximum $l^2$-norm value observed during scale-space evolution was 0.194%. This validates numerically the convergence of the scheme.

### 5.1 Beltrami-Based Denoising

Figure 3 displays the restoration of an image from its noisy version by applying Eq. (8). The speed-up in this case is about 10.

### 5.2 Beltrami-Based Deblurring

In the next example the original image was blurred by a Gaussian kernel, as shown in Figure 4 top-left. The image was restored using Eq. (9). A significant speed-up is obtained in this case, as seen in Figure 4 bottom.

## 6 Concluding Remarks

Due to its anisotropic nature and non-separability, Beltrami color flow discretizations are usually performed with explicit schemes. Low computational efficiency limits their use in practical applications. We accelerated the convergence of the explicit scheme using vector extrapolation methods. Experiments of denoising and deblurring color images based on the RRE algorithm have demonstrated the efficiency of the method. This makes vector extrapolation methods useful and attractive to the Beltrami filter and potentially other image processing applications.

## Acknowledgment

We would like to thank Prof. Avram Sidi for his advice regarding vector extrapolation algorithms.

This research was supported in part by the European 6th framework program NOE-MUSCLE, in part by the Israeli Science Foundation (ISF) grant No. 738/04, and in part by the Horowitz fund.

## References

1. L. Bar, A. Brook, N. Sochen, and N. Kiryati. Color image deblurring with impulsive noise. *Proceedings of the International Conference on Variational, Geometry and Level Sets Methods in Computer Vision (VLSM 05)*, 3752:49–60, 2005.
2. D. Barash. A fundamental relationship between bilateral filtering, adaptive smoothing and the nonlinear diffusion equation. *IEEE Transactions on Pattern Analysis and Machine Intelligence*, 24(6):844–847, 2002.
3. S. Cabay and L. Jackson. Polynomial extrapolation method for finding limits and antilimits of vector sequences. *SIAM Journal on Numerical Analysis*, 13(5):734–752, 1976.
4. F. Catte, P. L. Lions, J. M. Morel, and T. Coll. Image selective smoothing and edge detection by nonlinear diffusion. *SIAM Journal on Numerical Analysis*, 29(1): 182–193, 1992.
5. R. Courant, K. Friedrichs, and H. Lewy. Uber die partiellen differenzengleichungen der mathematischen physik. *Mathematische Annalen*, 100(1):32–74, 1928.
6. R. P. Eddy. Extrapolationg to the limit of a vector sequence. pages 387–396, 1979.
7. M. Elad. On the bilateral filter and ways to improve it. *IEEE Transactions On Image Processing*, 11(10):1141–1151, 2002.
8. B. Gustafsson, H. Kreiss, and J. Oliger. *Time dependent problems and difference methods.* Wiley, 1995.
9. R. Kimmel, R. Malladi, and N. Sochen. Images as embedding maps and minimal surfaces: Movies, color, texture, and volumetric medical images. *International Journal of Computer Vision*, 39(2):111–129, 2000.
10. T. Lu, P. Neittaanmaki, and X.-C. Tai. A parallel splitting up method and its application to Navier-Stokes equations. *Applied Mathematics Letters*, 4(2):25–29, 1991.
11. T. Lu, P. Neittaanmaki, and X.-C. Tai. A parallel splitting up method for partial diferential equations and its application to Navier-Stokes equations. *RAIRO Mathematical Modelling and Numerical Analysis*, 26(6):673–708, 1992.
12. P. Perona and J. Malik. Scale-space and edge detection using anisotropic diffusion. *IEEE Transactions on pattern Analysis and machine intelligence*, 12:629–639, 1990.
13. A. M. Polyakov. Quantum geometry of bosonic strings. *Physics Letters*, 103 B: 207–210, 1981.

14. L. Rudin, S. Osher, and E. Fatemi. Non linear total variation based noise removal algorithms. *Physica D Letters*, 60:259–268, 1992.
15. A. Sidi. Convergence and stability properties of minimal polynomial and reduced rank extrapolation algorithms. *SIAM Journal on Numerical Analysis*, 23(1):197–209, 1986.
16. A. Sidi. Efficient implementation of minimal polynomial and reduced rank extrapolation methods. *J. Comput. Appl. Math.*, 36(3):305–337, 1991.
17. D. A. Smith, W. F. Ford, and A. Sidi. Extrapolation methods for vector sequences. *SIAM Review*, 29(2):199–233, 1987.
18. S. M. Smith and J. Brady. Susan - a new approach to low level image processing. *International Journal of Computer Vision*, 23:45–78, 1997.
19. N. Sochen, R. Kimmel, and A. M. Bruckstein. Diffusions and confusions in signal and image processing. *Journal of Mathematical Imaging and Vision*, 14(3):195–209, 2001.
20. N. Sochen, R. Kimmel, and R. Maladi. From high energy physics to low level vision. *Proceedings of the First International Conference on Scale Space Theory in Computer Vision*, 1252:236–247, 1997.
21. N. Sochen, R. Kimmel, and R. Maladi. A general framework for low level vision. *IEEE Trans. on Image Processing*, 7:310–318, 1998.
22. A. Spira, N. Sochen, and R. Kimmel. Efficient Beltrami flow using a short time kernel. *Proceedings of Scale Space 2003, Lecture Notes in Computer Science*, 2695: 511–522, 2003.
23. C. Tomasi and R. Manduchi. Bilateral filtering for gray and color images. *Proceedings of IEEE International Conference on Computer Vision*, pages 836–846, 1998.
24. J. Weickert. Efficient and relible schemes for nonlinear diffusion filtering. *IEEE Trans. on Image Processing*, 7:398–410, 1998.
25. J. Weickert. Coherence-enhancing diffusion filtering. *International Journal of Computer Vision*, 31:111–127, 1999.
26. A. J. Yezzi. Modified curvature motion for image smoothing and enhancement. *IEEE Transactions on Image Processing*, 7(3):345–352, 1998.

# Vector-Valued Image Interpolation by an Anisotropic Diffusion-Projection PDE*

Anastasios Roussos and Petros Maragos

School of ECE, National Technical University of Athens, Greece
{troussos,maragos}@cs.ntua.gr

**Abstract.** We propose a nonlinear image interpolation method, based on an anisotropic diffusion PDE and designed for the general case of vector-valued images. The interpolation solution is restricted to the subspace of functions that can recover the discrete input image, after an appropriate smoothing and sampling. The proposed nonlinear diffusion flow lies on this subspace and its strength and anisotropy effectively adapt to the local variations and geometry of image structures. The derived model efficiently reconstructs the real image structures, leading to a natural interpolation, with reduced blurring, staircase and ringing artifacts of classic methods. This method also outperforms other existing PDE-based interpolation methods. We present experimental results that prove the potential and efficacy of the method as applied to graylevel and color images.

## 1 Introduction

Image interpolation is among the fundamental image processing problems and is often required for various image analysis operations. It is therefore of interest for many applications such as biomedical image processing, aerial and satellite imaging, text recognition and high quality image printing. In this paper, the term *image interpolation* is used in the sense of the operation that takes as input a discrete image and recovers a continuous image or a discrete one with higher resolution. The case where the output image is discrete appears in the literature with several other names: *digital zooming*, *image magnification*, *upsampling*, *resolution enhancement*.

There exists a large variety of image interpolation methods, which can be classified to two main classes, linear and nonlinear methods (see [1] for a detailed review). The *linear* methods (e.g. bicubic, quadratic and spline interpolations) perform convolution of the image samples with a single kernel, equivalent to a lowpass filtering. These methods yield relatively efficient and fast algorithms, but they cannot effectively reconstruct the high-frequency part of images and inevitably introduce artifacts. *Nonlinear* methods perform a processing adapted to the local geometric structure of the image, with main goal to efficiently reconstruct image edges. This class includes variational (e.g. [2,3,4]) and PDE-based

* Work supported by the European research program ASPI (IST-FP6-021324).

F. Sgallari, A. Murli, and N. Paragios (Eds.): SSVM 2007, LNCS 4485, pp. 104–115, 2007.

(e.g. [5,6]) methods, some of which will be presented in the following sections. Such methods (e.g. [7,8,9]) have also been developed for two closely related problems, image inpainting [10] and scattered data interpolation.

In this paper a novel nonlinear method for the interpolation of vector-valued images is proposed. We pose a constraint, which effectively exploits the available information of input image. Then, we design an anisotropic diffusion PDE, which performs adaptive smoothing but also complies with this constraint, thanks to an appropriate projection operation. The diffusion strength and anisotropy adapt to the local variations and geometry of image structures. This method yields a plausible result even when the resolution of input image is relatively low and reduces the artifacts that usually appear in image interpolation. The paper is organized as follows: In Sect. 2, some interpolation models related to the proposed method are discussed. Sect. 3 presents our novel interpolation PDE model. In Sect. 4, we demonstrate results from interpolation experiments, that show the potential and efficacy of the new method.

## 2 Preliminaries and Background

### 2.1 Reversibility Condition Approach to Interpolation

The problem of image interpolation is viewed here in a way similar to [2,3]. The continuous solution of interpolation $u(x,y)$ should yield the known, low resolution discrete image $z[i,j]$, after a lowpass filtering followed by sampling. To pose this *reversibility condition* formally, let us consider that $z[i,j]$ is defined on an orthogonal grid of $N_x \times N_y$ points with vertical and horizontal steps $h_x$ and $h_y$ respectively. Also let $u(x,y)$ be defined in the domain $\Omega = [\frac{h_x}{2}, N_x + \frac{h_x}{2}] \times [\frac{h_y}{2}, N_y + \frac{h_y}{2}]$, which contains the grid points. Then, the reversibility condition for the solution $u(x,y)$ can be written as follows:

$$(S * u)(ih_x, jh_y) = z[i,j]\,, \text{ for all } (i,j) \in \{1,..,N_x\} \times \{1,..,N_y\}\ , \tag{1}$$

where $h_x$, $h_y$ are hereafter considered unitary ($h_x = h_y = 1$), "$*$" denotes convolution and $S(x,y)$ is a smoothing kernel that performs the lowpass filtering and has Fourier transform with nonzero values for all the baseband frequencies $(\omega_1, \omega_2) \in [-\pi, \pi]^2$. For example, $S(x,y)$ could be the *mean kernel*, i.e. $S(x,y) = \mathbb{1}_{[-\frac{1}{2},\frac{1}{2}]^2}(x,y)$, where $\mathbb{1}_B$ denotes the indicator function for any set $B \subset \mathbb{R}^2$. Note that (1) degenerates to the exact interpolation condition when $S = \delta(x,y)$ (2D unit impulse). However, condition (1) with an appropriate lowpass filtering can be more realistic, as it can better model the digitization process, which is the final step of image acquisition systems [4]. In addition, this lowpass filtering is desirable, as it reduces the aliasing effects at the acquired image. The problem of finding $u(x,y)$ in (1) is ill-posed, as (1) is satisfied by infinitely many functions. Let $\mathcal{U}_{z,S}$ be the set of these functions. It is easy to see that $\mathcal{U}_{z,S}$ is an affine subspace of the functions defined in $\Omega$. Therefore, some extra criterion must be posed to choose among the functions of $\mathcal{U}_{z,S}$.

A simple linear interpolation method arises by imposing the additional constraint that $u(x, y)$ is a bandpass 2D signal, similarly to Shannon's theory. Then, the solution of (1), which we refer to as *(frequency) zero-padding interpolation*, is unique and can be easily derived using the Sampling theorem (note that it depends on the kernel $S(x, y)$). This method reconstructs image edges without blurring or distorting them, but usually introduces strong oscillations around edges [3]. The cutoff of high frequencies is thus undesirable, as the bandlimited assumption is not true for most real-world images. Therefore, a more appropriate method of selection among the functions of $\mathcal{U}_{z,S}$ is needed. Such methods will be presented in the following sections.

**Total Variation Based Interpolation.** Guichard and Malgouyres [2,3] proposed to choose as solution of the interpolation the image that minimizes the Total Variation (TV), $E[u] = \iint_\Omega \|\nabla u\| \, dxdy$ , under the constraint that $u \in \mathcal{U}_{z,S}$. This minimization problem is solved in [2] by applying a constrained gradient descent flow, described by the following PDE:

$$\partial u(x, y, t)/\partial t = P_{\mathcal{U}_{0,S}} \{\operatorname{div} (\nabla u / \|\nabla u\|)\} \ , \tag{2}$$

supplemented with the initial condition that $u(x, y, 0)$ is the zero-padding interpolation of $z[i, j]$. $P_{\mathcal{U}_{0,S}}\{\cdot\}$ denotes the operator of orthogonal projection on the subspace $\mathcal{U}_{0,S}$, which corresponds to the condition (1) with $z[i, j] = 0$ for all $(i, j)$. This projection ensures that $u(x, y, t) \in \mathcal{U}_{z,S}, \forall t > 0$, since $u(x, y, 0) \in \mathcal{U}_{z,S}$. The authors propose two options for the smoothing kernel of condition (1): the mean kernel and the *sinc kernel*, which provides an ideal lowpass filter as its Fourier transform is $\mathbb{1}_{[-\pi,\pi]^2}(\omega_1, \omega_2)$.

This method leads to reconstructed images without blurring effects, as it allows discontinuities and preserves 1D image structures. However, TV minimization is based on the assumption that the desirable image is almost piecewise constant, which yields a result with over-smoothed homogeneous regions. In addition, the diffusion in (2) is controlled by the simple coefficient $1/\|\nabla u\|$, therefore it cannot remove block effects, especially in the regions with big image variations. Further, the mean kernel vanishes too sharply, so the projection $P_{\mathcal{U}_{0,S}}\{\cdot\}$ reintroduces block effects and the sinc kernel is badly localized in space and oscillates, so $P_{\mathcal{U}_{0,S}}\{\cdot\}$ causes formation of oscillations in reconstructed edges.

**Belahmidi-Guichard (BG) Method.** Belahmidi and Guichard [5] have improved the TV-based interpolation by developing a nonlinear anisotropic PDE, hereafter referred as *BG interpolation method*. In order to enhance edge preservation, this PDE performs a diffusion with strength and orientation adapted to image structures. The reversibility condition (1) is taken into account (with the choice of mean kernel for $S(x, y)$) by adding to the PDE an appropriate fidelity term, so that the flow $u(x, y, t)$ stays close to the subspace $\mathcal{U}_{z,S}$ (see [5] for details). This method balances linear zooming on homogeneous regions and anisotropic diffusion near edges, trying to combine the advantages of these two processes. Nevertheless, the diffusion is not always desirably adapted to real image structures and the fact that the PDE flow is not constrained to lie inside $\mathcal{U}_{z,S}$ may decrease the accuracy of the result.

## 2.2 PDE Model of Tschumperlé and Deriche (TD)

Tschumperlé and Deriche [11,6] proposed an effective PDE method for vector-valued image regularization. This PDE scheme, which we refer to as *TD PDE*, is mainly designed for image restoration applications, but it is presented here because we utilize it to the design of the new interpolation PDE (Sect. 3). Their model is an anisotropic diffusion flow, which uses tensors to adapt the diffusion to the image structure. Let $\boldsymbol{u}(x, y, t) = [u_1,..,u_M]^{\mathrm{T}}$ be the output vector-valued image at time $t$ and $M$ be the number of vector components. Then, the TD PDE model can be described by the following set of coupled PDEs:

$$\frac{\partial u_m(x,y,t)}{\partial t} = \mathrm{trace}\Big( T\big(J_\rho(\nabla \boldsymbol{u}_\sigma)\big) \cdot D^2 u_m \Big) \, , \; m{=}1,..,M \; , \qquad (3)$$

with initial condition that $\boldsymbol{u}(x, y, 0)$ is the input vector-valued image. $D^2 u_m$ denotes the spatial Hessian matrix of the component $u_m(x, y, t)$ and $T$ is the $2{\times}2$ *diffusion tensor*:

$$T\big(J_\rho(\nabla \boldsymbol{u}_\sigma)\big) = \big[1 + (\mathcal{N}/K)^2\big]^{-\frac{1}{2}} \cdot \boldsymbol{w}_- \boldsymbol{w}_-^{\mathrm{T}} + \big[1 + (\mathcal{N}/K)^2\big]^{-1} \cdot \boldsymbol{w}_+ \boldsymbol{w}_+^{\mathrm{T}} \; , \qquad (4)$$

where $\mathcal{N} = \sqrt{\lambda_+ + \lambda_-}$ and $K$ is a threshold constant similar to the diffusivity of [12].[1] Also, $\lambda_- \leq \lambda_+$ and $\boldsymbol{w}_-$,$\boldsymbol{w}_+$ are the eigenvalues and unit eigenvectors of the $2{\times}2$ *structure tensor*:

$$J_\rho(\nabla \boldsymbol{u}_\sigma) = G_\rho * \sum_{m=1}^{M} \nabla(G_\sigma * u_m) \; (\nabla(G_\sigma * u_m))^{\mathrm{T}} \; . \qquad (5)$$

The 2D isotropic Gaussian kernels $G_\sigma$ and $G_\rho$ are of standard deviation $\sigma$ and $\rho$ respectively.[2] The structure tensor $J_\rho(\nabla \boldsymbol{u}_\sigma)$ measures the local geometry of image structures (convolutions with $G_\sigma$,$G_\rho$ make this measure more coherent [13]). The eigenvectors $\boldsymbol{w}_-$ and $\boldsymbol{w}_+$ describe the orientation of minimum and maximum vectorial variation of $\boldsymbol{u}$ and the eigenvalues $\lambda_-$ and $\lambda_+$ describe measures of these variations (the term $\mathcal{N}$ is an edge-strength predictor which effectively generalizes the norm $\|\nabla u\|$). Thus, the diffusion is strong and isotropic in homogenous regions (small $\mathcal{N}$), but weak and mainly oriented by image structures near the edges (big $\mathcal{N}$). Consequently, this method offers a flexible and effective control on the diffusion process (see [6] for more details).

**Application to the Interpolation.** Among various applications, the generic PDE model (3) of [11,6] is applied to image interpolation (we refer to the derived

[1] This is a slightly more general version of the original model [11,6], where $K = 1$.

[2] The original model corresponds to $\sigma{=}0$ but we use the more general version of [13].

method as *TD interpolation method*). This method casts image interpolation as a special case of the image inpainting problem [10]. It imposes the constraint that the solution must coincide with the input at the appropriate pixels in the new coarser grid (exact interpolation condition). Thus, the inpainting domain (i.e. the domain where the image values are unknown) consists of the remaining pixels. The image values in this domain are processed according to PDE (3), with a modified diffusion tensor [11]:

$$T\big(J_\rho(\nabla \boldsymbol{u}_\sigma)\big) = \big[1 + (\mathcal{N}/K)^2\big]^{-\frac{1}{2}} \cdot \boldsymbol{w}_- \boldsymbol{w}_-^{\mathrm{T}} \ . \tag{6}$$

The bilinear interpolation of the input image is chosen as initial condition $\boldsymbol{u}(x, y, 0)$ and the interpolation solution is derived from the equilibrium state. Contrary to the effectiveness of the TD PDE model for image restoration, the derived interpolation method suffers from some inefficiencies. The initialization by the bilinear interpolation contains edges with significant blurring. Also, the information of each input value $z[i, j]$ is not spread to all the corresponding pixels of the coarser grid, as some pixels stay anchored whereas the rest pixels change without constraint. Furthermore, the diffusion tensor (6) is fully anisotropic even in regions with small image variations, therefore it may distort image structures and create false edges.

## 3 The Proposed Anisotropic Diffusion-Projection PDE

The aforementioned PDE interpolation methods outperform classic linear methods, as they reconstruct the edges without blurring them. In some cases though, they yield artifacts such as over-smoothing of homogeneous regions, block effects or edge distortion. In order to improve the effectiveness of these methods, we propose a novel PDE model, which performs a nonlinear interpolation. It is based on an efficient combination of the reversibility condition approach and TD PDE (3). The model is designed to deal with vector-valued images in general and processes the different channels in a coupled manner.

More precisely, the design of our model has been based on the observation that the TV-based interpolation PDE (2) can be derived from a non-minimization point of view: it is in fact a modification of the zero-fidelity ($\lambda$=0) TV PDE [14]:

$$\partial u(x, y, t)/\partial t = \operatorname{div}\left(\nabla u / \left\|\nabla u\right\|\right) \ , \tag{7}$$

which can be viewed as a special case of the general nonlinear diffusion of [12]. This modification is done by replacing the right hand side (RHS) of the PDE with its projection to $\mathcal{U}_{0,S}$. Thanks to this projection, the whole flow remains into the subspace $\mathcal{U}_{z,S}$, provided that $u(x, y, 0) \in \mathcal{U}_{z,S}$. We followed a similar approach to design the proposed PDE model, but instead of TV PDE (7), we chose to modify the TD PDE (3), as it is an effective and robust diffusion PDE model for image regularization (see Sect. 2.2).

Before we proceed to the description of our model, let us mention that we straightforwardly generalize the condition (1) for vector-valued images: it should be satisfied independently by every channel. This generalized reversibility condition can be equivalently written as:

$$\langle S_{ij}, u_m \rangle_{L^2(\Omega)} = z_m[i,j] \ , \tag{8}$$

where $(m,i,j) \in \{1,..,M\} \times \{1,..,N_x\} \times \{1,..,N_y\}$ and $z_m[i,j]$, $u_m(x,y)$ are the m-th of M components of the discrete input and interpolated image respectively. Also, $S_{ij}(x,y) = S(i-x, j-y)$ and $\langle \cdot, \cdot \rangle_{L^2(\Omega)}$ denotes the inner product of $L^2(\Omega)$. Let $\boldsymbol{\mathcal{U}}_{z,S}$ be the set of vector-valued images $\boldsymbol{u}(x,y)$ that satisfy the generalized reversibility condition (8).

**Description of the Model.** We derive the interpolated image from the equilibrium solution of the following system of coupled PDEs:

$$\frac{\partial u_m(x,y,t)}{\partial t} = P_{\mathcal{U}_{0,S}} \left\{ \text{trace}\Big( T\big(J_\rho(\nabla \boldsymbol{u}_\sigma)\big) \cdot D^2 u_m \Big) \right\} \ , \ m{=}1,..,M \ , \tag{9}$$

where $P_{\mathcal{U}_{0,S}}\{\cdot\}$ denotes the operator of orthogonal projection on the subspace $\mathcal{U}_{0,S}$ and the tensors $T(J_\rho(\nabla \boldsymbol{u}_\sigma))$ and $J_\rho(\nabla \boldsymbol{u}_\sigma)$ are again given by (4) and (5) respectively. We have chosen the following initial conditions for (9): every $u_m(x,y,0)$ is derived from the zero-padding interpolation (see Sect.2.1) of the symmetrically extended $z_m[i,j]$. This initialization, which is similar to the one of PDE (2) proposed in [2], can be easily computed and contains efficient reconstructions of image edges (see also the following discussion of the model's properties). The reflection that we added before zero-padding offers a slight improvement of the initial estimate, as it eliminates the ringing effects near the image borders. Note that $\boldsymbol{u}(x,y,0) \in \boldsymbol{\mathcal{U}}_{z,S}$, so $P_{\mathcal{U}_{0,S}}\{\cdot\}$ ensures that $\boldsymbol{u}(x,y,t) \in \boldsymbol{\mathcal{U}}_{z,S}$, $\forall t > 0$.

Let us now derive an expression for the projection $P_{\mathcal{U}_{0,S}}\{\cdot\}$. First of all, the subspace $\mathcal{U}_{0,S}$ can be defined as the set of functions $v(x,y)$ that satisfy $\langle S_{ij}, v \rangle_{L^2(\Omega)} = 0$, for all $(i,j) \in \{1,..,N_x\} \times \{1,..,N_y\}$. If we assume for the chosen smoothing kernel that:

$$S(x,y) = 0 \, , \text{ for all } (x,y) \notin [-1/2, 1/2]^2 \ , \tag{10}$$

each $S_{ij}$ takes nonzero values only inside a different square of $\Omega$. Therefore $\langle S_{ij}, S_{i'j'} \rangle_{L^2(\Omega)} = \|S\|^2_{L^2(\mathbb{R}^2)} \, \delta_{i-i', j-j'}$ (where $\delta_{i,j}$ is the 2D discrete unit impulse), which means that the set of all $S_{ij}$ is an orthogonal basis of $\mathcal{U}_{0,S}$. Consequently, a relatively simple expression for the projection $P_{\mathcal{U}_{0,S}}\{\cdot\}$ can be derived:

$$P_{\mathcal{U}_{0,S}}\{v\} = v - \|S\|^{-2}_{L^2(\mathbb{R}^2)} \sum_{i=1}^{N_x} \sum_{j=1}^{N_y} \langle S_{ij}, v \rangle_{L^2(\Omega)} \cdot S_{ij} \ , \tag{11}$$

The assumption (10), apart from simplifying the expression for $P_{\mathcal{U}_{0,S}}\{\cdot\}$, it is also realistic for most image acquisition systems: during the digitization process, the measured value at any pixel $(i,j)$ depends mainly on the intensities of points that lie in the interior of this pixel's area, i.e. the domain $\Omega_{ij} = [i-\frac{1}{2}, i+\frac{1}{2}]\times[j-\frac{1}{2}, j+\frac{1}{2}]$. Therefore we have chosen the following smoothing kernel:

$$S(x,y) = \mathbb{1}_{[-\frac{1}{2},\frac{1}{2}]^2}(x,y) \cdot \frac{G_{\hat{\sigma}}(x,y)}{\iint_{[-\frac{1}{2},\frac{1}{2}]^2} G_{\hat{\sigma}}(x',y')\mathrm{d}x'\mathrm{d}y'} \ , \tag{12}$$

where $G_{\hat{\sigma}}(x,y)$ is the 2D isotropic Gaussian of standard deviation $\hat{\sigma}$. Multiplication with $\mathbb{1}_{[-\frac{1}{2},\frac{1}{2}]^2}(x,y)$ is done to satisfy the assumption (10) and the denominator of (12) normalizes the kernel to have unitary mean value. Note that $\hat{\sigma}$ must be neither too small nor too big. If $\hat{\sigma}$ is too small, $S(x,y)$ is too localized in space and the information of each input value $z[i,j]$ is not spread properly to all the corresponding pixel area $\Omega_{ij}$. In addition, if $\hat{\sigma}$ is too big, $S(x,y)$ reduces to the mean kernel. This kernel though is undesirable, because we want to relax the constraints near the border of each $\Omega_{ij}$ and thus prevent $P_{\mathcal{U}_{0,S}}\{\cdot\}$ from producing block effects.

**Properties of the Model.** As already mentioned, the zero-padding interpolation, which we use as initial condition of (9), efficiently reconstructs image edges without blurring or distorting them, but also introduces strong oscillations around edges (see [3]). It can thus be viewed as a desirable interpolation result degraded by a significant amount of noise. The scope of the proposed PDE (9) is to effectively regularize the image $\boldsymbol{u}(x,y,0)$ by removing these oscillations. Note also that (11) shows that the projection $P_{\mathcal{U}_{0,S}}\{v\}$ subtracts the component of $v$ that does not comply with the reversibility condition. This subtraction does not affect the basic characteristics of the regularization that the velocity $v_m = \mathrm{trace}(T{\cdot}D^2u_m)$ tends to apply to the image. Therefore, PDE (9) performs an anisotropic smoothing with properties very similar to (3). This fact, in combination with the analysis of Sect. 2.2, shows that the proposed PDE efficiently removes the undesirable oscillations and simultaneously preserves the important image structures. Namely, the proposed PDE can be considered as a diffusion flow towards elements of $\boldsymbol{\mathcal{U}}_{z,S}$ with "better" visual quality. Additionally, the projection $P_{\mathcal{U}_{0,S}}\{v\}$ offers the advantage that there is no need to specify the stopping time as an additional parameter. The best regularized image is derived at $t{\rightarrow}\infty$, where the flow equilibrates thanks to the term that $P_{\mathcal{U}_{0,S}}\{v\}$ subtracts from the velocity $v$.

**Numerical Implementation.** The continuous result $\boldsymbol{u}(x,y)$ of the proposed model is approximated by a discrete image $\boldsymbol{u}[i',j']$, defined to a coarser grid than the input image $\boldsymbol{z}[i,j]$, i.e. a discrete interpolation is performed. We consider only the case where the grid step of $\boldsymbol{z}[i,j]$ is a multiple of the grid step of $\boldsymbol{u}[i',j']$ by an integer factor $d$, which we call *zoom factor*. Namely, the input image is magnified $d{\times}d$ times. For the sake of simplicity, we hereafter assume that the coarser grid of $\boldsymbol{u}[i',j']$ has unit step, hence the grid of input $\boldsymbol{z}[i,j]$ has step $d$. In the discretization of PDE (9), we used an explicit numerical scheme with finite differences, similar to [11]. The discrete time step $\delta t$ was chosen sufficiently small for stability purposes (the typical value of $\delta t$=0.2 was used). Due to the fact that the output image is given at the equilibrium, we stop the iterative process when $\boldsymbol{u}^{n+1}$ differs from $\boldsymbol{u}^n$ by a small constant, with respect to an appropriate norm.

## 4 Experimental Results and Comparisons

In order to compare the interpolation methods and extract performance measures, we use the following protocol: We choose a reference image with a relatively good resolution and negligible noise. We reduce the dimensions of this image by an integer factor $d$ (i.e. the image is reduced to $\frac{1}{d}\times\frac{1}{d}$ of its size), using a decimation process, i.e. (anti-aliasing) lowpass filtering followed by subsampling. We implement the lowpass filtering by a convolution with a bicubic spline, which results to a reliable and commonly used decimation process. Finally, we apply the interpolation methods to enlarge the decimated image by the zoom factor $d$, so that the output images have the same size as the reference image. Note that we implemented the other PDE-based interpolation methods with a way similar to the implementation of the proposed method, as briefly described in Sect. 3. Also, we used the range $[0, 255]$ for image values and in the case of color, we applied the corresponding PDE methods representing the images in the RGB color space. The reference image can be considered as the ideal output of the interpolation, as it is noiseless. Therefore the difference between the reference image $\boldsymbol{r}[i,j]$ and the output of a method $\boldsymbol{u}[i,j]$ can be viewed as reconstruction error and is quite representable of the method performance.

We use two measures for this error, the classic peak signal-to-noise-ratio (PSNR)[3] and the mean structural similarity (MSSIM) index [15], which seems to approximate the perceived visual quality of an image better than PSNR or various other measures. MSSIM index takes values in [0,1] and increases as the quality increases. We calculate it based on the code available at: `http://www.cns.nyu.edu/~lcv/ssim/`, using the default parameters. In the case of color images, we extend MSSIM with the simplest way: we calculate the MSSIM index of each RGB channel and then take the average.

We repeat the above procedure for different reference images from a dataset and for zoom factors $d$=2,3 and 4. For every zoom factor and interpolation method, we compute the averages of PSNR and MSSIM for all the images in the set, which we consider as final measures of performance.

We followed the above experimental protocol using a dataset of 23 natural images of size 768×512 pixels.[4] We run two series of experiments, the first for the graylevel versions (where we applied bicubic, TV-based, BG and the proposed method) and the second for their color counterparts (where we applied bicubic, TD and our method.[5]) For the methods that needed specification of parameter(s), we utilized fixed values in all the dataset, which we empirically derived based on a visual plausibility criterion. We have hence chosen the parameters

[3] We use the definition $\text{PSNR}=10\log_{10}\left(255^2 M/\text{var}\left\{\|\boldsymbol{u}[i,j]-\boldsymbol{r}[i,j]\|\right\}\right)$, where $\|\cdot\|$ denotes here the Euclidean norm of vectors with $M$ components.

[4] The kodak collection, available from `http://www.cipr.rpi.edu/resource/stills/`.

[5] TV-based and BG methods are applicable to graylevel images only.

$\sigma$=0.3$d$, $\rho$=0.4$d$, $\hat{\sigma}$=0.6$d$ and $K$=1 for the proposed method. Also, in TD method we used the same values $\rho$=0.4$d$, $K$=1 and in BG method we used $K$=3 for the corresponding threshold constant. An extensive demonstration of these results can be found at `http://cvsp.cs.ntua.gr/~tassos/PDEinterp/ssvm07res`.

Figure 1 is a snapshot of the results for graylevel image interpolation (for the sake of demonstration, the input image has been enlarged by the simple zero order hold (ZOH)). It can be observed that the bicubic interpolation significantly blurs the edges (e.g. note the flower boundary in Fig. 1(b)). The TV-based interpolation over-smooths some homogeneous areas (e.g. the interior of the flower in Fig. 1(c)), creates block effects (e.g. the thin black branch at the upper right of Fig. 1(d)) and oscillations in reconstructed edges (e.g. the shutter behind the flower in Fig. 1(c)). BG interpolation shows an improved performance but it maintains the block effects in some regions (e.g. the flower boundary in Fig. 1(e)). Figure 1(f) shows that the proposed method yields the most effective reconstruction of image structures and the most plausible result. Observe finally how the shutter is desirably reconstructed only by the proposed method.

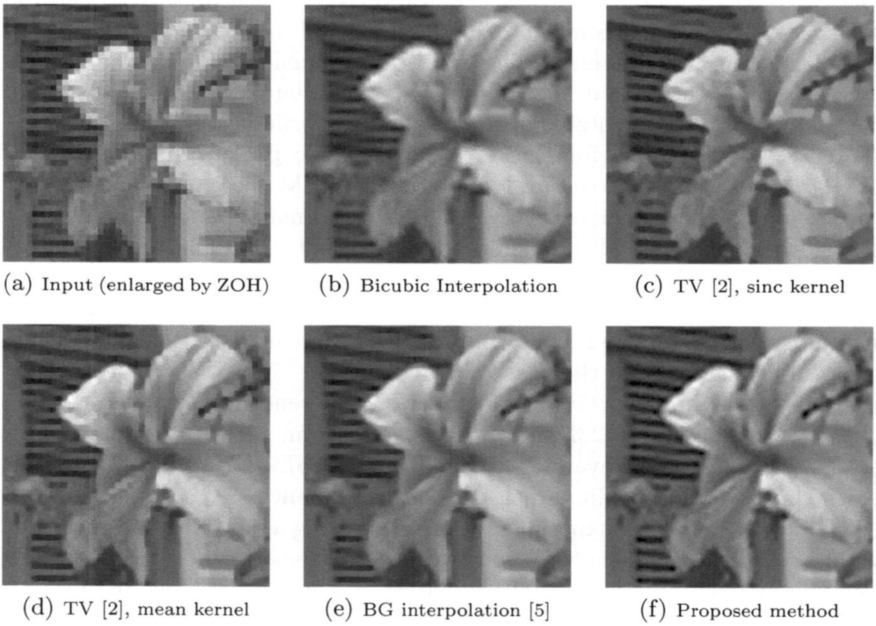

(a) Input (enlarged by ZOH) (b) Bicubic Interpolation (c) TV [2], sinc kernel

(d) TV [2], mean kernel (e) BG interpolation [5] (f) Proposed method

**Fig. 1.** Details of 4×4 graylevel interpolation using the 7$^{\text{th}}$ image of dataset

Figure 2 demonstrates a detail of the results for interpolation in color images. We observe that bicubic interpolation gives again a result with blurring but also significant staircase effects (e.g. note the edges of motorbikes in Fig. 2(b)). Fig. 2(c) shows that TD interpolation yields an excessively synthetic aspect to the result, as it has distorted image edges and created false thin edges around the real ones. The proposed method yields again a result (Fig. 2(d)) which has not any notable artifact and seems the most aesthetically satisfying. This result contains sharper and better localized edges than the bicubic interpolation (e.g. note the more effective reconstruction of motorbikes' edges) and looks much more natural than the result of TD interpolation.

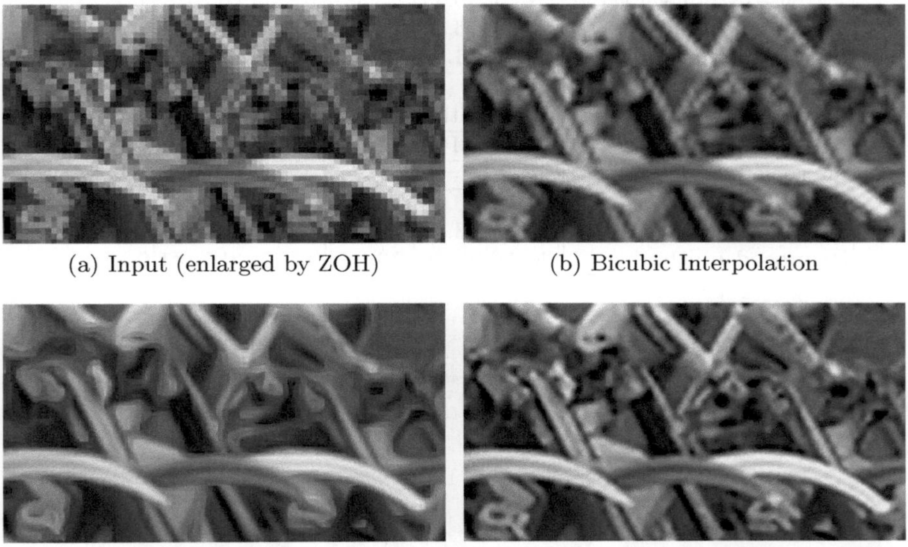

(a) Input (enlarged by ZOH) (b) Bicubic Interpolation

(c) TD interpolation [11,6] (d) Proposed method

**Fig. 2.** Details of 4×4 color interpolation using the $5^{\text{th}}$ image of dataset

Table 1 contains the overall performance measures of the interpolation methods, for the two series of experiments in the dataset of the 23 images. We see that the proposed method yields improved PSNR and MSSIM results in all the cases of zoom factors and series of experiments. This improvement may be attributed to the fact that the proposed method performs a more flexible adaptive smoothing and reliably exploits the input image data to increase the accuracy of the result.

*Interpolation of Biomedical Vocal Tract images.* In this experiment, we have used an MRI midsagittal image of a speaker's vocal tract from: `http://www.speech.kth.se/~olov/`. Fig. 3(a) is a close-up of a denoised (using anisotropic diffusion) version of this image. Image data of this type are important for the analysis and modeling of the human speech production system. Similarly to the above experiments, we used this image as reference and we reduced its dimensions by a factor $d$=3 (see Fig. 3(b)). Finally, we applied the proposed method to (3×3) interpolate the decimated image (Fig. 3(c)). We see that the

**Table 1.** Average error measures in all results using the 23 images, for different zoom factors $d$

Experiments with graylevel images

| Interpolation Method | Average PSNR (dB) | | | Average MSSIM | | |
|---|---|---|---|---|---|---|
| | $d$=2 | $d$=3 | $d$=4 | $d$=2 | $d$=3 | $d$=4 |
| Bicubic interpolation | 29.14 | 26.68 | 25.55 | 0.8561 | 0.7464 | 0.6953 |
| TV based [2], sinc kernel | 29.75 | 26.87 | 25.94 | 0.8739 | 0.7567 | 0.7105 |
| TV based [2], mean kernel | 29.53 | 26.83 | 25.82 | 0.8714 | 0.7578 | 0.7114 |
| BG interpolation [5] | 28.36 | 26.58 | 25.60 | 0.8253 | 0.7402 | 0.7004 |
| **Proposed method** | **30.22** | **26.96** | **26.05** | **0.8816** | **0.7671** | **0.7194** |

Experiments with color images

| Interpolation Method | Average PSNR (dB) | | | Average MSSIM | | |
|---|---|---|---|---|---|---|
| | $d$=2 | $d$=3 | $d$=4 | $d$=2 | $d$=3 | $d$=4 |
| Bicubic interpolation | 29.11 | 26.66 | 25.56 | 0.8524 | 0.7425 | 0.6921 |
| TD interpolation [11] | 26.77 | 23.89 | 23.37 | 0.7925 | 0.6330 | 0.6147 |
| **Proposed method** | **30.16** | **26.96** | **26.06** | **0.8779** | **0.7631** | **0.7157** |

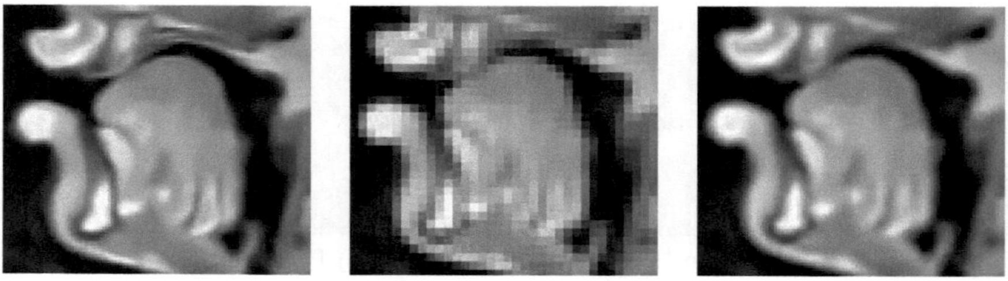

(a) Reference image (b) Input(enlarged by ZOH) (c) Proposed method

**Fig. 3.** Interpolation (3×3) of a vocal tract image using the proposed method

proposed method yields a very satisfactory reconstruction of vocal tract shape, even though the decimated input image has notably low resolution. This simple example reveals that the proposed model can be also used to effectively enhance the resolution of medical image data of the vocal tract.

## 5 Conclusions

In this paper, we have proposed a model for the interpolation of vector-valued images, based on an anisotropic diffusion PDE. Our main contribution is an efficient combination of the reversibility condition approach [2] with the Tschumperlé-Deriche PDE model [6]. The proposed mopdel reduces the undesirable effects of classic linear and similar PDE based interpolation methods. Extensive experimental results have demonstrated the potential and efficacy of the method as applied to graylevel and color images.

Finally, we remark that the proposed PDE (9) is only one possible choice, as it is derived using the RHS of TD PDE (3), and a similar approach can be obtained using the RHS of other effective regularization PDEs. For example, based on the general anisotropic diffusion model of [13], one can use the RHS $\mathrm{div}(T{\cdot}\nabla u_m)$, with the tensor $T$ given by (4). This method performs very similarly to the proposed method, yielding a slight improvement, as revealed by some preliminary experiments that we performed. In addition, note that the proposed model assumes that the input image is noise free. It can be modified to handle noisy inputs, if the projection operator is replaced by an appropriate fidelity term. These issues are part of our ongoing research and we plan to present them in a following paper.

## References


1. Meijering, E.: A chronology of interpolation: from ancient astronomy to modern signal and image processing. Proc. IEEE **90**(3) (2002) 319–342
2. Guichard, F., Malgouyres, F.: Total variation based interpolation. In Proc. EUSIPCO **3** (1998) 1741–1744
3. Malgouyres, F., Guichard, F.: Edge direction preserving image zooming: a mathematical and numerical analysis. SIAM J. Num. Anal. **39**(1) (2001) 1–37
4. Aly, H.A., Dubois, E.: Image up-sampling using total-variation regularization with a new observation model. IEEE Tr. Im. Pr. **14**(10) (2005) 1647–1659
5. Belahmidi, A., Guichard, F.: A partial differential equation approach to image zoom. ICIP (2004) 649–652
6. Tschumperlé, D., Deriche, R.: Vector-valued image regularization with PDE's : A common framework for different applications. IEEE -PAMI **27**(4) (2005) 506–517
7. Caselles, V., Morel, J.M., Sbert, C.: An axiomatic approach to image interpolation. IEEE Tr. Im. Pr. **7**(3) (1998) 376–386
8. Chan, T.F., Shen, J.: Mathematical models for local nontexture inpaintings. SIAM J. Appl. Math. **62**(3) (2002) 1019–1043
9. Weickert, J., Welk, M.: Tensor field interpolation with PDEs. In: Visualization and Processing of Tensor Fields. Springer, Berlin (2006) 315–325
10. Bertalmio, M., Sapiro, G., Caselles, V., Ballester, C.: Image inpainting. Proc. SIGGRAPH 2000 (2000) 417–424
11. Tschumperlé, D.: PDE's Based Regularization of Multivalued Images and Applications. PhD thesis, Univ. of Nice-Sophia Antipolis (2002)
12. Perona, P., Malik, J.: Scale space and edge detection using anisotropic diffusion. IEEE -PAMI **12**(7) (1990) 629–639
13. Weickert, J.: Anisotropic Diffusion in Image Processing. Teubner, Stuttgart (1998)
14. Rudin, L., Osher, S., Fatemi, E.: Nonlinear total variation based noise removal algorithms. Physica D **60** (1992) 259–268
15. Wang, Z., Bovik, A., Sheikh, H., Simoncelli, E.: Image quality assessment: from error visibility to structural similarity. IEEE Tr. Im. Pr. **13**(4) (2004) 600–612

# Faithful Recovery of Vector Valued Functions from Incomplete Data

## Recolorization and Art Restoration

Massimo Fornasier*

Program in Applied and Computational Mathematics
Princeton University, Princeton NJ 08544, USA
mfornasi@math.princeton.edu

**Abstract.** On March 11, 1944, the famous Eremitani Church in Padua (Italy) was destroyed in an Allied bombing along with the inestimable frescoes by Andrea Mantegna *et al.* contained in the Ovetari Chapel. In the last 60 years, several attempts have been made to restore the fresco fragments by traditional methods, but without much success. We have developed an efficient pattern recognition algorithm to map the original position and orientation of the fragments, based on comparisons with an old gray level image of the fresco prior to the damage. This innovative technique allowed for the partial reconstruction of the frescoes. Unfortunately, the surface covered by the fragments is only 77 $m^2$, while the original area was of several hundreds. This means that we can reconstruct only a fraction (less than 8%) of this inestimable artwork. In particular the original color of the blanks is not known. This begs the question of whether it is possible to estimate *mathematically* the original colors of the frescoes by making use of the potential information given by the available fragments and the gray level of the pictures taken before the damage. Can one estimate how *faithful* such restoration is? In this paper we retrace the development of the recovery of the frescoes as an inspiring and challenging real-life problem for the development of new mathematical methods. We introduce two models for the recovery of vector valued functions from incomplete data, with applications to the fresco recolorization problem. The models are based on the minimization of a functional which is formed by the discrepancy with respect to the data and additional regularization constraints. The latter refer to joint sparsity measures with respect to frame expansions for the first functional and functional total variation for the second. We establish the relations between these two models. As a byproduct we develop the basis of a theory of fidelity in color recovery, which is a crucial issue in art restoration and compression.

* The author acknowledges the financial support provided by the European Union's Human Potential Programme under contract MOIF-CT-2006-039438. The paper also contributes to the project WWTF Five senses-Call 2006, Mathematical Methods for Image Analysis and Processing in the Visual Arts.

F. Sgallari, A. Murli, and N. Paragios (Eds.): SSVM 2007, LNCS 4485, pp. 116–127, 2007.

# 1 Introduction

**Mathematical Imaging in Art Restoration.** We address the problem of the faithful reconstruction of vector valued functions from incomplete data, with special emphasis in color image recovery. We are inspired by a real-life problem, *i.e.* the rebirth of one of the most important masterpieces of the Italian Renaissance, by making use of mathematical imaging techniques. We refer to the decorative cycle in the Ovetari Chapel in the Eremitani Church in Padua. The chapel was seriously damaged by an air strike in 1944 and a large section of the contained frescoes were *sparsely* fragmented. A digital cataloging of pictures of the remaining fragments made it possible to count the number (78.561) of those with an area larger than 1 $cm^2$. The distribution of the areas shows that most of them are relatively small (5-6 $cm^2$). There is no information on the possible location of the pieces on the huge original surface and also unknown is the angle of rotation with respect to the original orientation. These *a priori* data demonstrated the lack of contiguous fragments for any given fragment.These difficulties explain the unsuccessful attempts of recomposition by traditional methods. In simple words, it is an incomplete *puzzle* which is too big to be solved by human eyes only. There exist some fairly good quality black and white photographs of the frescoes dated from between 1900 and 1920.This heritage gave rise to the hope that a computer-based comparison between the fresco digital images and those of the fragments could help to recognize their original location. We have developed an efficient pattern recognition algorithm based on *circular harmonic* expansions [20]. This method has been implemented for the solution of the fragment recollocation problem and we illustrate some of the final results in Fig. 1. On the basis of the map produced by this *computer assisted anastylosis*, part of the frescoes has already been physically restored. We refer to the book chapter [8] for more details. Even though the collocation of one single fragment is of historical and cultural importance, the success of the computer assisted anastylosis has been partially spoiled by the limited surface that the fragments can cover. This begs the question of whether it is possible to estimate *mathematically* the original colors of the missing parts of the frescoes by making use of the potential information given by the available fragments and the gray level of the pictures taken before the damage. Can one estimate how *faithful* such restoration is?

**Mathematical Inpainting and Recolorization.** *Mathematical inpainting*, an artistic synonym for image interpolation, has been introduced by Sapiro et al. [3] with the specific purpose of imitating the basic approaches used by professional restorers when filling blanks in paintings. Their algorithm amounts to the solution of an evolutionary differential equation whose steady-state is the prolongation of the incomplete image in the inpainting region to make constant the information along isophotes. See also further recent developments [1]. Closely related to inpainting is the contribution by Masnou and Morel [25,26,27] who addressed the so-called *disocclusion problem.* Essentially it amounts to an application of the principle of *good continuation, i.e.*, without forming undesired T-junctions and abrupt direction changes, of the image level curves into the

**Fig. 1.** Fragmented A. Mantegna's frescoes (1452) by a bombing in the Second World War. Computer based reconstruction by using efficient pattern matching techniques [20].

region where an occlusion occurred in order to restore the essential morphology. This work can be seen as a development of the theory of Euler's elastica curves by Mumford [28]. Chan and Shen contributed to inpainting with other models similar or related to the ones previously cited, see [9,10,11,12,13]. In simple words, mathematical inpainting is the attempt to guess the morphology of the image in a relatively small missing part from the level curves of the relevant known part.

The recolorization problem, like that of the frescoes, can be viewed as a particular case of inpainting. Nevertheless, in this case two significant differences occur with respect to the classical problem: 1) The region of missing color is usually much larger than the one with known colors 2) the morphology of the image in the missing part can be determined by the known gray level, see also [7]. Several approaches to the recovery of colors in gray level images have been recently proposed based on different intuitions. Neighboring pixels with similar intensities should have similar color. By using non-local fitting term an algorithm based on optimization has been proposed in [23] to match the colors. Similarly, a fast algorithm using a weighted distance image blending technique is studied in [31]. From the assumption that the color morphology is essentially determined by the gradient of the gray level, Sapiro proposed in [29] a recolorization method based on minimizing the difference between the gradient of gray level and the gradient of color. The problem reduces to the solution of a (nonlinear) boundary value problem. Based on similar assumptions two variational approaches are proposed in [22] where the authors minimize the discrepancy with respect to the color datum and impose a smoothness constraint on the solution out of the gray level discontinuity set. All the proposed solutions show that a very limited amount of color is sufficient information to recover a pleasant result. However none of these contributions seem to emphasize the problem of the *fidelity* of the recovered color.

**The Fidelity of Restoration.** Recolorization can be indeed a controversial practice, especially if we claim to "estimate" the color of an art masterpiece. It is therefore crucial to investigate the relations between *amount of color data,* *model of reconstruction, and fidelity of the solution.* Clearly this issue can also have a relevant role for color image compression.

In this paper we review two different approaches to recolorization previously proposed by the author *et al.* in [18,19]. Since color images are modeled as multichannel signals, the problem is reformulated as the recovery of vector valued functions from incomplete data. The vector components are assumed to be coupled. The difference between the proposed methods is the way we couple the information. For both the approaches, the recovery is realized as the minimization of a functional which is formed by the discrepancy with respect to the data and additional regularization constraints. The latter refer to joint sparsity measures with respect to frame expansions for the first functional (Section 2) and functional total variation for the second (Section 3). We establish the relations between these two models. As a byproduct we develop the basis of a corresponding theory of fidelity in color recovery.

## 2 Recovery of Vector Valued Data with Joint Sparsity Constraints

### 2.1 Sparse Frame Expansions

A *sparse representation* of an element of a Hilbert space is a series expansion with respect to an orthonormal basis or a frame that has only a small number of large coefficients. Several types of signals appearing in nature admit sparse frame expansions and thus, sparsity is a realistic assumption for a very large class of problems [24]. The recent observation that it is possible to reconstruct sparse signals from vastly incomplete information [6,5,15] stimulated a new fruitful line of research which is called *sparse recovery* or *compressed sensing.* This section is devoted to reveal the relations between faithful sparse recovery, vector valued functions, and the application to color images. Indeed multi-channel signals (i.e., vector valued functions) may not only possess sparse frame expansions for each channel individually, but additionally the different channels can also exhibit common sparsity patterns. Color images are multi-channel signals, exhibiting a very rich morphology. In particular, discontinuities may appear in all the channels at the same locations. This will be reflected, *e.g.*, in sparse curvelet expansions [4,16] with relevant coefficients appearing at the same labels, or in turn in sparse gradients with supports at the same locations.

### 2.2 Inverse Problems with Joint Sparsity Constraints

Let $\mathcal{K}$ and $\mathcal{H}_j$, $j = 1, \dots, N$, be (separable) Hilbert spaces and $A_{\ell,j} : \mathcal{K} \to \mathcal{H}_j$, $j = 1, \dots, M$, $\ell = 1, \dots, N$, some bounded linear operators. Assume we are given data $g_j \in \mathcal{H}_j$, $g_j = \sum_{\ell=1}^{M} A_{\ell,j} f_\ell$, $j = 1, \dots, N$. Then our basic task consists in reconstructing the (unknown) elements $f_\ell \in \mathcal{K}$, $\ell = 1, \dots, M$. In practice, it happens that the corresponding mapping from the vector $(f_\ell)$ to the vector $(g_j)$ is not invertible or ill-conditioned, as for the recolorization problem. In order to exploit sparsity ideas we assume that we have given a suitable frame

$\{\psi_\lambda : \lambda \in \Lambda\} \subset \mathcal{K}$ indexed by a countable set $\Lambda$. This means that there exist constants $C_1, C_2 > 0$ such that $C_1\|f\|_{\mathcal{K}}^2 \leq \sum_{\lambda\in\Lambda} |\langle f, \psi_\lambda\rangle|^2 \leq C_2\|f\|_{\mathcal{K}}^2$, for all $f \in \mathcal{K}$. Orthonormal bases are particular examples of frames. Frames allow for a (stable) series expansion of any $f \in \mathcal{K}$ of the form $f = Fu := \sum_{\lambda\in\Lambda} u_\lambda \psi_\lambda$, where $u = (u_\lambda)_{\lambda\in\Lambda} \in \ell_2(\Lambda)$. Introduce the operators $T_{\ell,j} = A_{\ell,j}F : \ell_2(\Lambda) \to \mathcal{H}_\ell$ and

$$T : \ell_2(\Lambda, \mathbb{R}^M) \to \mathcal{H}, \quad Tu = \left(\sum_{\ell=1}^M T_{\ell,j} u^\ell\right)_{j=1}^N = \left(\sum_{\ell=1}^M A_{\ell,j} F u^\ell\right)_{j=1}^N,$$

$u^\ell := (u_\lambda^\ell)_{\lambda\in\Lambda}$. We denote also $u_\lambda = (u_\lambda^\ell)^{\ell=1,\dots,M}$, and $u = (u_\lambda^\ell)_{\lambda\in\Lambda}^{\ell=1,\dots,M}$. Our recovery model [19] is based on the minimization of the functional

$$J(u,v) = J_{\theta,\rho,\omega}^{(q)}(u,v) := \|Tu - g|\mathcal{H}\|^2 + \sum_{\lambda\in\Lambda} v_\lambda \|u_\lambda\|_q + \sum_{\lambda\in\Lambda} \omega_\lambda \|u_\lambda\|_2^2 + \sum_{\lambda\in\Lambda} \theta_\lambda (\rho_\lambda - v_\lambda)^2. \tag{1}$$

restricted to $v_\lambda \geq 0$. Here, $(\theta_\lambda)_\lambda$, $(\omega_\lambda)_\lambda$, and $(\rho_\lambda)_\lambda$ are some suitable positive sequences. We can show that for the minimizer $(u^*, v^*)$ of $J$ we have $u^* = \mathbb{S}_{\theta,\rho,\omega}(u^* + T^*(g - Tu^*))$, where $\mathbb{S}_{\theta,\rho,\omega}$ is a vector valued *firm shrinkage* operator, see, e.g., [21], depending on the parameters $\theta, \rho, \omega$. The functional depends on two variables. The first belongs to the space of signal coefficients to be reconstructed, the second belongs to the space of sparsity indicator weights. We minimize $J(u,v)$ jointly with respect to both $u, v$. Analyzing $J(u,v)$ we realize that for the minimizer $(u^*, v^*)$ we will have $v_\lambda^* = 0$ (or close to 0) if $\|u_\lambda^*\|_q = \left(\sum_{\ell=1}^M |u_\lambda^\ell|^q\right)^{1/q}$ is large so that $v_\lambda^* \|u_\lambda^*\|_q$ gets small. On the other hand, if $\|u_\lambda^*\|_q$ is small then the term $\theta_\lambda(\rho_\lambda - v_\lambda^*)^2$ dominates and forces $v_\lambda^*$ to be close to $\rho_\lambda$. Moreover, for the parameter $q > 1$, the model imposes a further coupling of the sparsity pattern through different channels, see [19,30]

The recovery algorithm consists in alternating a minimization with respect to $u$ and a minimization with respect to $v$. More formally, for some initial choice $v^{(0)}$, for example $v^{(0)} = (\rho_\lambda)_{\lambda\in\Lambda}$, we define

$$\begin{aligned} u^{(n)} &:= \arg\min_{u \in \ell_2(\Lambda, \mathbb{R}^M)} J(u, v^{(n-1)}), \\ v^{(n)} &:= \arg\min_{v \in \ell_{\infty,\rho^{-1}}(\Lambda)_+} J(u^{(n)}, v). \end{aligned} \tag{2}$$

The minimization of $J(u, v^{(n-1)})$ with respect to $u$ can be done by means of an iterative thresholding algorithm [14]. The minimizer $v^{(n)}$ of $J(u^{(n)}, v)$ for fixed $u^{(n)}$ can be computed explicitly. Indeed, it follows from elementary calculus that

$$v_\lambda^{(n)} = \begin{cases} \rho_\lambda - \frac{1}{2\theta_\lambda}\|u^{(n)}{}_\lambda\|_q & \text{if } \|u^{(n)}{}_\lambda\|_q < 2\theta_\lambda\rho_\lambda \\ 0 & \text{otherwise}. \end{cases} \tag{3}$$

We have the following result about the convergence of the above algorithm [19].

**Theorem 1.** *Assume $q \in \{1, 2, \infty\}$ and $\theta_\lambda \omega_\lambda \geq \sigma > \phi_q/4$ for all $\lambda \in \Lambda$, where $\phi_1 = M$, $\phi_2 = 1$, $\phi_\infty = 1$. Moreover, we assume that $\omega_\lambda \geq \gamma > 0$ for all $\lambda \in \Lambda$. Then the sequence $(u^{(n)}, v^{(n)})_{n \in \mathbb{N}}$ converges to the unique minimizer $(u^*, v^*) \in \ell_2(\Lambda, \mathbb{R}^M) \times \ell_{\infty, \rho^{-1}}(\Lambda)_+$ of $J$. The convergence of $u^{(n)}$ to $u^*$ is strong in $\ell_2(\Lambda, \mathbb{R}^M)$ and $v^{(n)} - v^*$ converges to 0 strongly in $\ell_{2,\theta}(\Lambda)$.*

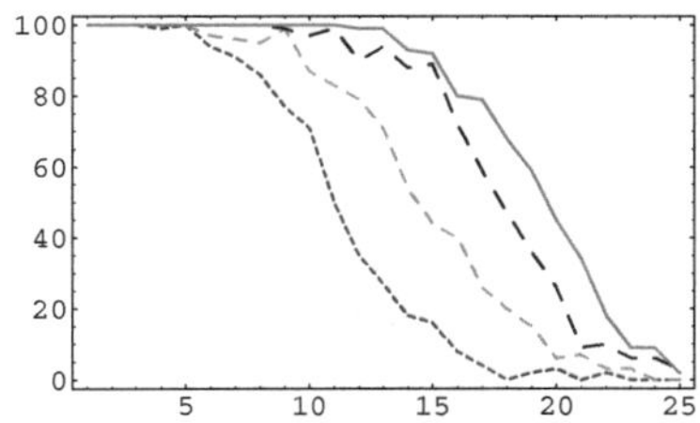


**Fig. 2.** In ordinate we illustrate the probability of *exact* support recovery from a fixed number of random linear measurements. In abscissa we illustrate the size of the support to be reconstructed. The probability is assessed as the number of successful trials over the number of total trials. The different curves represent the probability of reconstruction for increasing number $M = 2, 4, 8, 16$ of channels under joint sparsity constraints, from the most dashed to the solid one. The fidelity of the reconstruction increases with the number of channels.

### 2.3 Fidelity in Joint Sparse Recovery

We want to discuss the effect of the coupling due to joint sparsity for the fidelity of the reconstruction, assessed by the probability of *exact reconstruction* of sparse model signals. Let us assume that $M = N$, the matrices $T_{\ell,j} = 0$ for $j \neq \ell$, and $T_{\ell,\ell}$ are generic matrices, *i.e.*, random matrices with zero mean i.i.d. Gaussian entries. This means that we do not consider at the moment either a specific color model or a particular color datum. We only assume that the color channels are coupled in terms of sparsity. This situation appears previously in the literature under the name of *distributed compressed sensing*, see [2]. Since random matrices have, "with overwhelming probability", the so-called *Restricted Isometry Property* (see [5,6] for details), it is essentially sufficient to recover the location of the non-zero coefficients $u^\ell_\lambda$ in order to recover their value too. Applications of our joint sparsity algorithm (2) show that the probability of perfect reconstruction of the support of $u$ (hence of $u$ itself) increases with the number of channels $M$, see Fig. 2, see also [2]. *The moral is that it is more probable the faithful reconstruction of a color image encoded into multiple channels as soon as we couple the channels in terms of their sparsity, which, in turn, means coupling derivatives.* This further explains the positive results obtained in the papers [22,29].

## 3 Restoration of Vector Valued BV Functions from Projections

The gray level can be also interpreted as a combination of the color (*e.g.*, RGB) intensities to impose, besides derivatives, an additional constraint to fidelity. In this section we want to recall the model proposed by the author in [17,18] and to discuss its fidelity. In this recolorization model, the color image is encoded into RGB channels and no coupling of derivatives is explicitly imposed. This situation is opposite to the one previously discussed, where no coupling was claimed with respect to data instead.

A digital image can be modeled as a function $u : \Omega \subset \mathbb{R}^2 \to \mathbb{R}^3_+$, so that, to each "point" $x$ of the image, one associates the vector $u(x) = (r(x), g(x), b(x)) \in \mathbb{R}^3_+$ of the color represented by the different channels red, green, and blue. The gray level of an image can be described as $\mathcal{L}(r, g, b) := L(\alpha r + \beta g + \gamma b), (r, g, b) \in \mathbb{R}^3_+$, where $\alpha, \beta, \gamma > 0$, $\alpha + \beta + \gamma = 1$, $L : \mathbb{R} \to \mathbb{R}$ is a non-negative increasing function. The map $\mathcal{L}$ is learned a priori by fitting the known color and gray levels. Then, it is always possible to re-equalize the gray level to make $\mathcal{L}(r, g, b) = \frac{1}{3}(r, g, b)$. The recolorization is modeled as the minimum solution of the functional

$$F(u) = \mu \underbrace{\int_{\Omega \setminus D} |u(x) - \bar{u}(x)|^p dx}_{=G_1(u)} + \lambda \underbrace{\int_D |\mathcal{L}(u(x)) - \bar{v}(x)|^p dx}_{=G_2(u)} + \int_\Omega \sum_{\ell=1}^M |\nabla u^\ell(x)| dx, \tag{4}$$

where we want to reconstruct the vector valued function $u := (u^1, \ldots, u^M) : \Omega \subset \mathbb{R}^2 \to \mathbb{R}^M$ ($M = 3$ for RGB images) from a given observed couple of color/gray level functions $(\bar{u}, \bar{v})$. Without loss of generality let us assume $\lambda = \mu = 1$. For the computation of minimizers, we use a similar approach as in (2). For simplicity we assume $d = p = 2$. Let us introduce a new functional given by

$$\mathcal{E}_h(u, v) := 2\left(G_1(u) + G_2(u)\right) + \int_\Omega \sum_{\ell=1}^M \left( v^\ell |\nabla u^\ell(x)|^2 + \frac{1}{v^\ell} \right) dx, \tag{5}$$

where $u \in W^{1,2}(\Omega; \mathbb{R}^M)$, and $v \in L^2(\Omega; \mathbb{R}^M)$ is such that $\varepsilon_h \leq v^\ell \leq \frac{1}{\varepsilon_h}$, $\ell = 1, \ldots, M$, $\varepsilon_h \to 0$ for $h \to \infty$. While the variable $u$ is again the function to be reconstructed, we call the variable $v$ the *gradient weight*. For any given $v^{(0)} \in L^2(\Omega; \mathbb{R}^M)$ (for example $v^{(0)} := 1$), we define the following iterative double-minimization algorithm:

$$\begin{cases} u^{(n+1)} = \arg \min\limits_{u \in W^{1,2}(\Omega; \mathbb{R}^M)} \mathcal{E}_h(u, v^{(n)}) \\ v^{(n+1)} = \arg\min_{\varepsilon_h \leq v \leq \frac{1}{\varepsilon_h}} \mathcal{E}_h(u^{(n+1)}, v). \end{cases} \tag{6}$$

We have the following convergence result [19].

**Theorem 2.** *The sequence $\{u^{(n)}\}_{n \in \mathbb{N}}$ has subsequences that converge strongly in $L^2(\Omega; \mathbb{R}^M)$ and weakly in $W^{1,2}(\Omega; \mathbb{R}^M)$ to a point $u_h^{(\infty)}$. We have that $(u_h^{(\infty)})_h$*

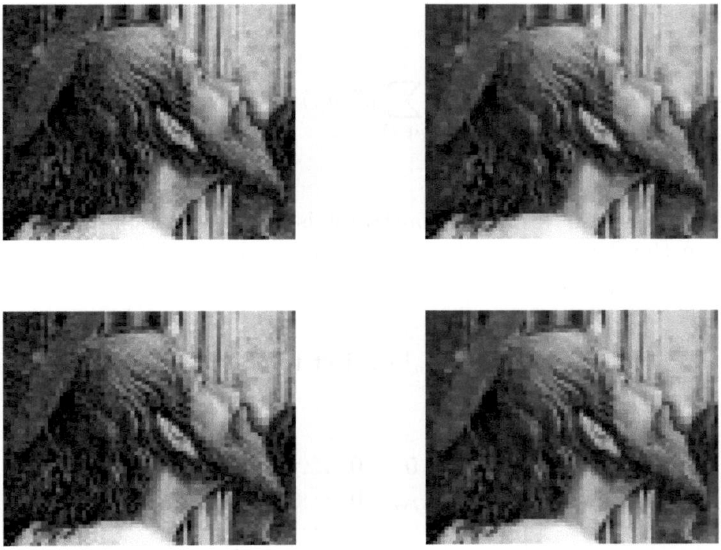

**Fig. 3.** The first column illustrates two different data for the recolorization problem. The second column illustrates the corresponding $10^{\text{th}}$ iteration algorithm (6). In the bottom-left position we illustrate a datum with only the 3% of original color information, randomly distributed.

*converges for $h \to \infty$ in $BV(\Omega; \mathbb{R}^M)$ to a solution of the Euler-Lagrange equations of $F$.*

### 3.1 Fidelity in the Linear Projection Method

We want to highlight the relations between the models (4) and (1), with particular emphasis on the fidelity of reconstruction. We assess the performances by the probability of *exact reconstruction* of piecewise constant model signals. For the sake of simplicity we consider 1D *discrete* signals, *i.e.* $d = 1$, and a linear projection $\mathcal{L}(x_1, \ldots, x_M) = \frac{1}{M}(x_1 + \cdots + x_M)$. The set $\Omega = \{0, 1, \ldots, \omega - 1\}$, for $\omega := |\Omega| > 0$. For a discrete signal of length $\omega$, we define the total variation as follows. Let us denote with $\mathcal{D}$ the $(\omega - 1) \times \omega$ *derivative matrix*

$$\mathcal{D} := \begin{pmatrix} -1 & 1 & 0 & \ldots & 0 & 0 \\ 0 & -1 & 1 & \cdots & 0 & 0 \\ \ldots & \ldots & \ldots & \ldots & \ldots & \ldots \\ 0 & 0 & 0 & \cdots & -1 & 1 \end{pmatrix}. \tag{7}$$

The discrete total variation of $v = (v_0, \ldots, v_{\omega-1})^T$ is given by

$$TV(v) := \|\mathcal{D}v\|_1 = \sum_{m=1}^{\omega-1} |(\mathcal{D}v)_m|.$$

The discrete version of (4) reads

$$F_d(u) = \mu \sum_{n \in \Omega \setminus D} |u_n - \bar{u}_n|^2 + \lambda \sum_{n \in D} |\mathcal{L}(u_n) - \bar{v}_n|^2 + \underbrace{\sum_{\ell=1}^{M} \sum_{m=1}^{\omega-1} |(\mathcal{D}u^\ell)_m|}_{\text{TV-constraint}}. \qquad (8)$$

Observe that the total variation constraint is now re-formulated as a *sparsity constraint* on the derivative. Let us first analyze the model with no noise, *i.e.*, for $\lambda, \mu \to \infty$, the minimization problem becomes

$$\min \sum_{\ell=1}^{M} \sum_{m=1}^{\omega-1} |(\mathcal{D}u^\ell)_m| \text{ subject to } Gu = (\bar{u} \quad \bar{v})^T, \qquad (9)$$

where

$$G := \begin{pmatrix} I_{\Omega\setminus D} & 0 & 0 & \dots & 0 & 0 \\ 0 & I_{\Omega\setminus D} & 0 & \cdots & 0 & 0 \\ \cdots & \cdots & \cdots & \cdots & \cdots & \cdots \\ 0 & 0 & 0 & \cdots & 0 & I_{\Omega\setminus D} \\ \frac{1}{M} I_D & \frac{1}{M} I_D & \cdots & \cdots & \cdots & \frac{1}{M} I_D \end{pmatrix}.$$

Let us denote $f := (\bar{u} \quad \bar{v})^T$.

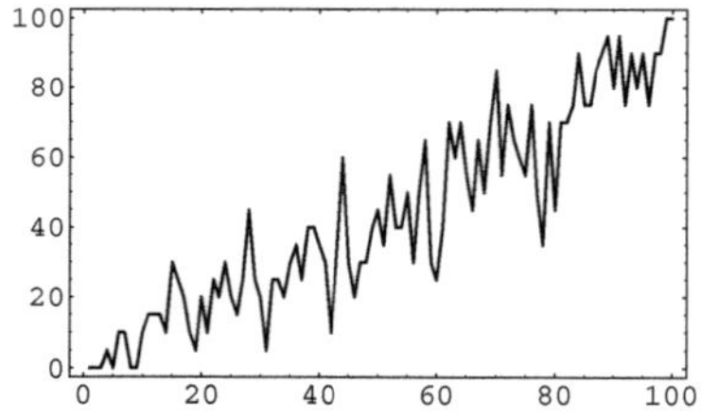


**Fig. 4.** In ordinate we illustrate the probability of *exact recovery* for piecewise constant random signals with two channels (M=2). The probability is assessed as the number of successful trials over the number of total trials. In abscissa we illustrate the percentage of color data which is assumed randomly distributed on $\Omega$. Exact recovery is indeed equivalent to the correct detection of the discontinuities (see Proposition 1). The fidelity of the reconstruction increases with the percentage of the given color.

Numerical examples on piecewise constant signals confirm that by coupling the color channels to match the gray level (here reproduced as a linear combination of the colors) and by minimizing at the same time the sum of the total variations of each individual channel, we obtain a coupling of the derivatives of the solution. Of course, this is ensured with high probability only when sufficient color information is provided compared to the number of discontinuities of the signal, see Fig. 4. Therefore, although it is not explicitly required by the formulation of the constraints, this second model produces the same effect of coupling derivatives as the first one. This explains why also this model is very effective in recolorization, see examples in [18]. In Fig. 3 we show two examples of applications of algorithm (6) depending on two different initial configurations of the color datum. It is clear that few uniformly distributed samples of color are more representative than lots of samples only locally distributed. We formalize this observation in the following proposition.

**Proposition 1.** *We assume that the signal $u$ to be reconstructed is consistent, i.e. $Gu = f$, and with joint sparse derivative,*

$$\mathrm{supp}(\mathcal{D}u^\ell) \subset \mathrm{supp}(\mathcal{D}\mathcal{L}(u)), \quad \ell = 1, \dots, M, \text{ where } |\mathrm{supp}(\mathcal{D}\mathcal{L}(u))| = K \leq \omega - 1.$$

*This means that the signal $u$ is piecewise constant. For*

$$\mathcal{J} := \{I \subset \Omega : I \text{ is an interval and } u|_I \text{ is constant }\},$$

*we also assume that $(\Omega \setminus D) \cap I \neq \emptyset$ for all $I \in \mathcal{J}$ and, without loss of generality, $0 \in (\Omega \setminus D)$. If the solution $u^*$ to (9) has the property*

$$\mathrm{supp}(\mathcal{D}(u^*)^\ell) \subset \mathrm{supp}(\mathcal{D}\mathcal{L}(u)), \quad \ell = 1, \dots, M,$$

*then necessarily $u^* = u$.*

*Proof.* We sketch the proof of the proposition. Let us consider the $\omega \times (\omega - 1)$ *integration matrix*

$$\mathcal{I} := \begin{pmatrix} 0 & 0 & 0 & \dots & 0 & 0 \\ 1 & 0 & 0 & \cdots & 0 & 0 \\ 1 & 1 & 0 & \cdots & 0 & 0 \\ \dots & \dots & \dots & \dots & \dots & \dots \\ 1 & 1 & 1 & \cdots & 1 & 1 \end{pmatrix}.$$

We have $\mathcal{D}\mathcal{I} = I_{(\omega-1)\times(\omega-1)}$ and $v = \mathcal{I}\mathcal{D}v + C_v$ for all $v \in \mathbb{R}^\omega$, where $C_v = (c_v, \dots, c_v)$ is a constant. Then for $z = (\mathcal{D}u^\ell)_{\ell=1}^M$, (9) can be equivalently reformulated as

$$\min \sum_{\ell=1}^{M} \sum_{m=1}^{\omega-1} |z_m^\ell| \text{ subject to } \tilde{G}z = f - \tilde{f}_z, \tag{10}$$

where $\tilde{f}_z := (I_{\Omega \setminus D} C_{z^1}, \dots, I_{\Omega \setminus D} C_{z^M}, \frac{1}{M} \sum_{\ell=1}^M I_D C_{z^\ell}, \dots, \frac{1}{M} \sum_{\ell=1}^M I_D C_{z^\ell})^T$. The constants $c_{z^\ell} = u_0^\ell$ for all $\ell = 1, \dots, M$. Moreover, by assumption we know already that $\mathrm{supp}(z^\ell) \subset \mathrm{supp}(\mathcal{D}\mathcal{L}(u)) := T, \quad \ell = 1, \dots, M$. It is sufficient now to observe that $(I_{\Omega \setminus D}\mathcal{I})|_T$ is necessarily a full rank matrix and $\tilde{G}z = f - \tilde{f}_z$ has a unique solution $z^*$. Therefore, we have $u^* = \mathcal{I}z^* + C_{z^*} = u$.

Of course, to model images as piecewise constant functions is quite unrealistic, especially in presence of noise. We may better model $u$ as belonging to the class of signals with $(\varepsilon, K)$-sparse derivatives defined by $S_{\varepsilon,K} := \{u \in \mathbb{R}^{\omega \times M} : \#\{m : |\mathcal{D}u_m^\ell| > \varepsilon\} \leq K\}$, for $\varepsilon > 0$ and $K \leq \omega - 1$. If an *oracle* could tell us that $\{m : |\mathcal{D}u_m^\ell| > \varepsilon\} \subset T$, for a fixed $T \subset \Omega$, for all $\ell = 1, \dots, M$ and

$(\Omega \setminus D) \cap I \neq \emptyset$ for each interval $I = [a,b] \subset \Omega$ such that $\partial I = \{a,b\} \subset T$, but $(I \setminus \partial I) \cap T = \emptyset$, then again $(I_{\Omega \setminus D}\mathcal{I})|_T$ would be a full rank matrix and we could easily compute $u_\varepsilon^*$ such that $\|u_\varepsilon^* - u\|_2 \leq C\varepsilon$, for a constant $C = C(\omega; K)$ independent of $u$ and $\varepsilon$. One may argue that this oracle can be furnished directly by the gray level, *e.g.*, by segmentation. Nevertheless, although the matrices $\tilde{G}$ do not have the Restricted Isometry Property, numerical experiments indicate that such an oracle can be indeed directly provided by the support of the derivative of the minimizer $u_\varepsilon^*$ of (8), for suitable constants $\lambda(\varepsilon), \mu(\varepsilon) > 0$. Moreover, such minimizers do already satisfy the property $\|u_\varepsilon^* - u\|_2 \leq C\varepsilon$.

## 4 Conclusion

We discussed the relationship between amount of color data, model of reconstruction, and the faithful recolorization of images. With the first model, we have shown that, regardless of the given color model, the probability of exact reconstruction increases with the number of channels as soon as the derivatives of the channels are mutually coupled. With the second model, we have shown that the coupling of the channels, but not of their derivatives, yields anyway to faithful reconstructions whenever sufficient color information is provided. The combination of these two models justify the results obtained in [17,18,22] where faithful recolorizations are obtained by variational principles even with limited amount of color information.

## References


1. Ballester, C., Bertalmio, M., Caselles, V., Sapiro, G., and Verdera, J.: *Filling-in by joint interpolation of vector fields and gray levels.*, IEEE Trans. Image Process. **10** (2001), no. 8, 1200–1211.
2. Baron, D., Wakin, M. B., Duarte, M. F., Sarvotham S., and Baraniuk R G.: *Distributed compressed sensing*, preprint (2005).
3. Beltramio, M., Sapiro, G., Caselles, V., and Ballester, B.: *Image inpainting*, SIGGRAPH 2000, 2001.
4. Candès, E. J. and Donoho, D. L.: *New tight frames of curvelets and optimal representations of objects with piecewise $C^2$ singularities.*, Commun. Pure Appl. Math. **57** (2004), no. 2, 219–266.
5. Candès, E. J., Romberg, E.J., and Tao, T.: *Exact signal reconstruction from highly incomplete frequency information*, IEEE Trans. Inf. Theory **52** (2006), no. 2, 489–509.
6. Candès, E. J, Romberg, E.J., and Tao, T.: *Stable signal recovery from incomplete and inaccurate measurements*, preprint (2005).
7. Caselles, V.,Coll, V., and Morel, J.-M.: *Geometry and color in natural images*, Journal of Mathematical Imaging and Vision, **16** (2002) no. 2 89–105.
8. Cazzato R., Costa G., Dal Farra A., Fornasier M., Toniolo D., Tosato D., and Zanuso C.: *Il Progetto Mantegna: storia e risultati* (Italian), in "Andrea Mantegna e i Maestri della Cappella Ovetari: La Ricomposizione Virtuale e il Restauro" (Eds. Anna Maria Spiazzi, Alberta De Nicolò Salmazo, Domenico Toniolo), Skira, 2006.

9. Chan, T. F. and Kang, S. H.: *Error analysis for image inpainting*, UCLA CAM 04-72, (2004).
10. Chan, T. F. and Kang, S. H. and Shen, J.: *Euler's elastica and curvaure-based inpainting*, SIAM J. Appl. Math. **63** (2002), no. 2, 564–592.
11. Chan, T. F. and Shen, J.: *Inpainting based on nonlinear transport and diffusion*, Contemp. Math. **313** (2002), 53–65.
12. Chan, T. F. and Shen, J.: *Mathematical models for local nontexture inpaintings*, SIAM J. Appl. Math. **62** (2002), no. 3, 1019–1043.
13. Chan, T. F. and Shen, J.: *Variational image inpainting*, Commun. Pure Appl. Math. **58** (2005), no. 5, 579–619.
14. Daubechies, I., Defrise M. and DeMol, C.: *An iterative thresholding algorithm for linear inverse problems*, Comm. Pure Appl. Math. **57** (2004), no. 11, 1413–1457.
15. Donoho, D. L.:*Compressed Sensing*, IEEE Trans. Inf. Theory **52** (2006), no. 4, 1289–1306.
16. Elad, M.,Starck, J.-L.,Querre, P., and Donoho, D.L.,: *Simultaneous cartoon and texture image inpainting using morphological component analysis (MCA)*, Appl. Comput. Harmon. Anal. **19** (2005), 340–358.
17. Fornasier M.: *Nonlinear projection recovery in digital inpainting for color image restoration*, J. Math. Imaging Vis. **24** (2006), no. 3, 359–373.
18. Fornasier M. and March R.: *Restoration of color images by vector valued BV functions and variational calculus*, Johann Radon Institute for Computational and Applied Mathematics (RICAM) preprint 2006-30 (2006). http://www.ricam.oeaw.ac.at/publications/reports/06/rep06-30.pdf
19. Fornasier M. and Rauhut H.: *Recovery algorithms for vector valued data with joint sparsity constraints*, Johann Radon Institute for Computational and Applied Mathematics (RICAM) preprint 2006-27 (2006). http://www.ricam.oeaw.ac.at/publications/reports/06/rep06-27.pdf
20. Fornasier M. and Toniolo D.: *Fast, robust, and efficient 2D pattern recognition for re-assembling fragmented digital images*, Pattern Recognition **38** (2005), 2074–2087.
21. Gao H.-Y. and Bruce A. G.: *WaveShrink with firm shrinkage*, Statist. Sinica **7** (1997), no. 4, 855–874.
22. Kang, S. H. and March, R.: *Variational models for image colorization via Chromaticity and Brightness decomposition*, preprint (2006).
23. Levin, A., Lischinski, A. and Weiss Y,: *Colorization using optimization*, Proceedings of the 2004 SIGGRAPH Conference, **23** (2004) no. 3 689–694.
24. Mallat, S.: *A Wavelet Tour of Signal Processing. 2nd Ed.*, San Diego, CA: Academic Press., 1999.
25. Masnou, S.: *Filtrage et Désocclusion d'Images par Méthodes d'Ensembles de Niveau*, PhD Thesis, Université Paris-Dauphine (1998).
26. Masnou, S.: *Disocclusion: a variational approach using level lines*, IEEE Trans. on Image Processing, **11** 68–76, no. 2 68–76
27. Masnou S. and Morel J.-M.:, *Level lines based disocclusion*. Proceedings of 5th IEEE IEEE Intl Conf. on Image Process.,Chicago, **3** (1998) 259–263.
28. Mumford, D.: *Elastica and computer vision*, Algebraic geometry and applications, ed. C. Bajaj, Springer-Verlag, Heidelberg, (1994) 491–506.
29. Sapiro G.: *Inpainting the colors*, ICIP 2005. IEEE International Conference on Image Processing, **2** (2005), 698–701.
30. Tropp, J.: *Algorithms for simultaneous sparse approximation. Part II: Convex relaxation*, Signal Processing **86** (2006), 589–602.
31. Yatziv, L. and Sapiro, G.: *Fast image and video colorization using chrominance blending*, IEEE Transactions on Image Processing **15** (2006), no. 5, 1120–1129.

# Discrete Regularization on Weighted Graphs for Image and Mesh Filtering

Sébastien Bougleux[1], Abderrahim Elmoataz[2], and Mahmoud Melkemi[3]

[1] GREYC CNRS UMR 6072 - Image,
ENSICAEN, 6 BD du Maréchal Juin, 14050 Caen Cedex France
sebastien.bougleux@greyc.ensicaen.fr
[2] LUSAC - VAI
Site Universitaire, BP 78, 50130 Cherbourg-Octeville, France
elmoataz@chbg.unicaen.fr
[3] LMIA - MAGE
4 rue des Frères Lumière, 68093 Mulhouse Cedex France
mahmoud.melkemi@uha.fr

**Abstract.** We propose a discrete regularization framework on weighted graphs of arbitrary topology, which unifies image and mesh filtering. The approach considers the problem as a variational one, which consists in minimizing a weighted sum of two energy terms: a regularization one that uses the discrete $p$-Laplace operator, and an approximation one. This formulation leads to a family of simple nonlinear filters, parameterized by the degree $p$ of smoothness and by the graph weight function. Some of these filters provide a graph-based version of well-known filters used in image and mesh processing, such as the bilateral filter, the TV digital filter or the nonlocal mean filter.

## 1 Introduction

In many computer vision applications, it is necessary to filter and to simplify images or meshes. In the context of image processing, smoothing and denoising constitute important steps of filtering processes. Among the existing methods, the variational ones, based on regularization, provide a general framework to design efficient filters. Solutions of variational models can be obtained by minimizing appropriate energy functions. The minimization is usually performed by designing continuous partial differential equations (PDEs), whose solutions are discretized in order to fit with the image domain. A complete overview of these methods can be found in [1][2][3][4] and references therein.

Another important problem of computer vision is mesh smoothing or denoising. This process is dedicated to noise removal, causing minimal damage to geometric features. Most of mesh smoothing methods are based on the discrete Laplace-Beltrami regularization or on the discrete curvature regularization [5][6]. Variational Beltrami flows have also been used to denoise and regularize data defined on manifolds [7]. Other mesh smoothing methods, based on feature preserving, were mostly inspired by anisotropic diffusion in image processing [8][9][10].

F. Sgallari, A. Murli, and N. Paragios (Eds.): SSVM 2007, LNCS 4485, pp. 128–139, 2007.

Inspired by continuous regularization of images, we propose a general framework based on a discrete regularization on weighted graphs of arbitrary topology. This framework unifies the regularization of images and meshes. Let $G_w = (V, E)$ be a weighted graph which consists of a set of vertices $V$, a set of edges $E \subset V \times V$, and a similarity weight function $w$ defined on edges. Let $\mathcal{H}(V)$ be a Hilbert space defined on the vertices of $G_w$. We formalize the discrete regularization of a function $f^0 \in \mathcal{H}(V)$ by the following minimization problem:

$$\min_{f \in \mathcal{H}(V)} \left\{ E_p(f, f^0, \lambda) = \sum_{v \in V} \| \nabla_v f \|^p + \lambda \| f - f^0 \|^2_{\mathcal{H}(V)} \right\}, \qquad (1)$$

where $p \in [1, +\infty)$ is the smoothness degree, $\lambda$ is the fidelity parameter, and $\nabla f$ represents the weighted gradient of the function $f$ over the graph. The solution of problem (1) leads to a family of nonlinear filters, parameterized by the weight function, the degree of smoothness and the fidelity parameter. There exist two main advantages of using this framework, which can be considered as a discrete analogue of continuous regularization on weighted graphs. Firstly, the regularization is expressed directly in discrete settings. Secondly, filters are computed by simple and efficient iterative algorithms, without solving any PDEs. Second, the topology of graphs can be arbitrary. Since the proposed approach is general, any discrete data set can be transformed into a weighted graph, by using a similarity measure between data. Thus, we can consider any function defined on these data as a function defined on the vertices of the graph.

The family of filters we propose includes graph-based versions of well-known filters used in image and mesh processing. If $w = 1$, they correspond exactly to the digitized PDE filters, introduced in the context of image restoration on grid graphs [11] ($L_2$ digital filter for $p = 2$ and TV digital filter for $p = 1$). If $w \neq 1$, our the filters behave like weighted TV regularization or weighted $L_2$ regularization. In particular, if $p = 2$ and $\lambda = 0$, the choice of the weight function $w$ allows to find the bilateral filter [12] and the nonlocal mean filters [1]. In this particular case, we also show that the discrete regularization is linked to spectral graph theory [13] and to Markov matrix filtering [14]. We can quote other existing methods, developed in the context of image filtering, that can be considered as discrete regularizations on unweighted graphs [15] [16]. These regularizations yield to Markov random fields where only binary variables are involved in the minimization.

The solution of problem (1) is obtained by defining local differential geometry operators on weighted graphs. The idea of using differential geometry on graphs, in a regularization process, has also been proposed in other context, such as semi-supervised data learning [17] and image segmentation [18].

In this paper, we firstly define differential geometry operators on weighted graphs in Section 2. Section 3 presents the regularization problem (1) and the associated family of filters. Section 4 analyzes the obtained filters and gives relations to existing methods. In Section 5, we show some regularization examples, in the context of image denoising, image segmentation and mesh denoising.

## 2 Differential Geometry on Weighted Graphs

In this section, we recall some basic definitions on graphs, and we define local differential geometry operators which can be considered as discrete versions of continuous differential operators. Analogue definitions and properties have also been used in the context of semi-supervised learning [17], and differential calculus on graphs [19][20].

### 2.1 Preliminary Definitions

A *graph* $G_w = (V, E)$ consists of a finite set $V$ of $N$ vertices and a finite set $E \subseteq V \times V$ of edges. We assume $G_w$ to be undirected, with no self-loops and no multiple edges. Let $uv$ be the edge that connects the vertices $u$ and $v$. An undirected graph is *weighted* if it is associated with a weight function $w : E \rightarrow \mathbb{R}_+$ satisfying $w_{uv} = w_{vu}$, for all $uv \in E$, and $w_{uv} = 0$ if $uv \notin E$. The weight function represents a similarity measure between two vertices of the graph.

Let $\mathcal{H}(V)$ denotes the Hilbert space of real-valued functions on vertices. A function $f : V \rightarrow \mathbb{R}$ in $\mathcal{H}(V)$ assigns a vector $f_v$ to each vertex $v \in V$. Clearly, $f$ can be represented by a column vector of $\mathbb{R}^N$, $f = [f_1, \ldots, f_N]^T$. The function space $\mathcal{H}(V)$ is endowed with the usual inner product $\langle f, h \rangle_{\mathcal{H}(V)} := \sum_{v \in V} f_v h_v$, where $f, h \in \mathcal{H}(V)$. Similarly, one can define $\mathcal{H}(E)$, the space of real-valued functions on edges, endowed with the inner product $\langle f, h \rangle_{\mathcal{H}(E)} := \sum_{v \in V} \sum_{u \sim v} f_{uv} h_{uv}$, where $f$ and $h$ are two functions in $\mathcal{H}(E)$.

### 2.2 Weighted Gradient and Divergence Operators

Let $G_w = (V, E)$ denotes a weighted graph. The *difference operator* $d : \mathcal{H}(V) \rightarrow \mathcal{H}(E)$ of a function $f \in \mathcal{H}(V)$ on an edge $uv \in E$, is defined by:

$$(df)_{uv} := \sqrt{w_{uv}}(f_v - f_u), \ \forall uv \in E. \tag{2}$$

The *directional derivative* (or *edge derivative*) of a function $f \in \mathcal{H}(V)$ at a vertex $v$ along an edge $e = uv$, is defined as $\partial_v f_u := (df)_{uv}$. This definition is consistent with the continuous definition of the derivative of a function, e.g., if $f_v = f_u$ then $\partial_v f_u = 0$. Moreover, note that $\partial_v f_u = -\partial_u f_v$, and $\partial_v f_v = 0$.

The *weighted gradient operator* $\bigtriangledown$ of a function $f \in \mathcal{H}(V)$ at a vertex $v$ is the vector operator defined by $\bigtriangledown_v f = (\partial_v f_u : u \sim v)^T$. The *local variation* of $f$ at $v$, is defined to be:

$$\|\bigtriangledown_v f\| := \sqrt{\sum_{u \sim v} (\partial_v f_u)^2} = \sqrt{\sum_{u \sim v} w_{uv}(f_v - f_u)^2}. \tag{3}$$

It can be viewed as a measure of the regularity of a function around a vertex. The amplitude of the graph gradient is defined by $\| \bigtriangledown f\| := \langle \bigtriangledown f, \bigtriangledown f \rangle_{\mathcal{H}(E)}^{1/2}$.

The *adjoint operator* of the difference operator, denoted by $d^* : \mathcal{H}(E) \rightarrow \mathcal{H}(V)$, is defined by $\langle df, h \rangle_{\mathcal{H}(E)} := \langle f, d^* h \rangle_{\mathcal{H}(V)}$, with $f \in \mathcal{H}(V)$ and $h \in \mathcal{H}(E)$.

Using the definitions of the inner products in $\mathcal{H}(V)$ and $\mathcal{H}(E)$ (see Section 2.1), and definition (2), we obtain the expression of $d^*$ at a vertex $v$:

$$(d^* h)_v = \sum_{u \sim v} \sqrt{w_{uv}}(h_{uv} - h_{vu}). \tag{4}$$

The *divergence operator*, defined by $-d^*$, measures the net outflow of a function in $\mathcal{H}(E)$ at each vertex of $V$.

### 2.3 A Family of Weighted $p$-Laplace Operators

The *weighted $p$-Laplace operator*, $\Delta_p : \mathcal{H}(V) \rightarrow \mathcal{H}(V)$ with $1 \leq p < +\infty$, is defined by $\Delta_p f := d^*(\| \bigtriangledown f \|^{p-2} df)$. Substituting (2) and (4) into the definition of $\Delta_p f$, we obtain the expression of $\Delta_p$ at a vertex $v$:

$$(\Delta_p f)_v = \sum_{u \sim v} \gamma_{uv} \left(f_v - f_u\right), \tag{5}$$

where $\gamma$ is the function defined by $\gamma_{uv} := w_{uv} \left(\| \bigtriangledown_v f\|^{p-2} + \| \bigtriangledown_u f\|^{p-2}\right)$.

The operator $\Delta_p$ is nonlinear, with the exception of $p = 2$. Furthermore, $\Delta_p$ is positive semi-definite:

$$\begin{aligned}\langle f, \Delta_p f\rangle_{\mathcal{H}(V)} &= \langle f, d^*(\| \bigtriangledown f\|^{p-2} df)\rangle_{\mathcal{H}(V)} = \langle df, \| \bigtriangledown f\|^{p-2} df\rangle_{\mathcal{H}(E)} \\ &= \sum_{v \in V} \| \bigtriangledown_v f\|^{p-2} \sum_{u \sim v} (df)^2_{uv} = \sum_{v \in V} \| \bigtriangledown_v f\|^{p} \geq 0. \end{aligned} \tag{6}$$

The definition of $\Delta_p$ can be considered as the discrete analogue of the $p$-Laplacian in the continuous case. When $p = 2$, $\Delta_2$ represents the *weighted Laplace operator* on $G_w$, and (5) reduces to:

$$(\Delta f)_v := (\Delta_2 f)_v = 2 \sum_{u \sim v} w_{uv}(f_v - f_u). \tag{7}$$

When $p = 1$, $\Delta_1$ represents the *weighted curvature operator* on $G_w$, and expression (5) reduces to:

$$(\kappa f)_v := (\Delta_1 f)_v = \sum_{u \sim v} w_{uv} \left( \frac{1}{\| \bigtriangledown_v f\|} + \frac{1}{\| \bigtriangledown_u f\|} \right) (f_v - f_u). \tag{8}$$

In practice, to avoid zero denominator in (8), the local variation (3) is replaced by its regularized version: $\| \bigtriangledown_v f\|_\epsilon := \sqrt{\| \bigtriangledown_v f\|^2 + \epsilon^2}$, with $\epsilon \rightarrow 0$ fixed.

## 3 $p$-Laplace Regularization on Weighted Graphs

In this section, we present the discrete regularization problem (1) and associated filters. Let $G_w = (V, E)$ be a weighted graph. The regularization of a given function $f^0 \in \mathcal{H}(V)$, using the weighted $p$-Laplace operator, consists in seeking for a function $f^* \in \mathcal{H}(V)$ which is not only smooth enough on $G_w$, but also close enough to $f^0$. This optimization problem can be formalized by the minimization of a weighted sum of two energy terms:

$$f^* = \min_{f \in \mathcal{H}(V)} \left\{ E_p(f, f^0, \lambda) := \sum_{v \in V} \| \nabla_v f \|^p + \lambda \| f - f^0 \|^2_{\mathcal{H}(V)} \right\}. \tag{9}$$

The first energy in (9) is the smoothness term or regularizer, meanwhile the second is the fitting term. The parameter $\lambda \geq 0$ is a fidelity parameter, called the Lagrange multiplier, which specifies the trade-off between the two competing terms. Both energy functions in $E_p$ are strictly convex functions of $f$. In particular, by standard arguments in convex analysis, the problem (9) has a unique solution, for $p = 1$ and $p = 2$, which satisfies:

$$\left. \frac{\partial E_p(f, f^0, \lambda)}{\partial f} \right|_v = \frac{\partial}{\partial f} \| \nabla_v f \|^p + 2\lambda (f_v - f_v^0) = 0, \;\; \forall v \in V. \tag{10}$$

Using equation (6) to compute the derivative of the first term in $E_p$, the system of equations (10) is rewritten as:

$$(\Delta_p f)_v + 2\lambda (f_v - f_v^0) = 0, \, \forall v \in V. \tag{11}$$

The solution of problem (9) is also the solution of the system of equations (11). This is a nonlinear system, with the exception of $p = 2$ (see Section 2.3). Substituting the expression of the $p$-Laplace operator into (11), we obtain:

$$\left( 2\lambda + \sum_{u \sim v} \gamma_{uv} \right) f_v - \sum_{u \sim v} \gamma_{uv} f_u = 2\lambda f_v^0, \;\; \forall v \in V. \tag{12}$$

Among the exiting methods to solve the system of equations (12), we use the Gauss-Jacobi iterative algorithm. Let $t$ be an iteration step, and let $f^{(t)}$ be the solution of equation (12) at the step $t$. The corresponding linearized Gauss-Jacobi algorithm is given by:

$$\begin{cases} f^{(0)} = f^0 \\ \gamma_{uv}^{(t)} = w_{uv} \left( \| \nabla_v f^{(t)} \|^{p-2} + \| \nabla_u f^{(t)} \|^{p-2} \right), \;\; \forall uv \in E \\ f_v^{(t+1)} = \left( 2\lambda + \sum_{u \sim v} \gamma_{uv}^{(t)} \right)^{-1} \left( 2\lambda f_v^0 + \sum_{u \sim v} \gamma_{uv}^{(t)} f_u^{(t)} \right), \;\; \forall v \in V \end{cases} \tag{13}$$

where $\gamma^{(t)}$ is the function $\gamma$ at the step $t$. The weights $w_{uv}$ are computed from $f^0$, or can be given a priori. We define the function $\varphi$ at an iteration $t$ of algorithm (13) by:

$$\varphi_{vu}^{(t)} = \frac{\gamma_{uv}^{(t)}}{2\lambda + \sum_{u \sim v} \gamma_{uv}^{(t)}} \text{ if } u \neq v, \text{ and } \; \varphi_{vv}^{(t)} = \frac{2\lambda}{2\lambda + \sum_{u \sim v} \gamma_{uv}^{(t)}}$$

Then, an iteration of the regularization algorithm (13) is rewritten as:

$$f_v^{(t+1)} = \varphi_{vv}^{(t)} f_v^0 + \sum_{u \sim v} \varphi_{vu}^{(t)} f_u^{(t)}, \; \forall v \in V. \tag{14}$$

At each iteration, the new value $f^{(t+1)}$, at a vertex $v$, depends on two quantities, the original value $f_v^0$, and a weighted average of the existing values in a neighborhood of $v$. This shows that the proposed filter, obtained by iterating (14), is a low-pass filter which can be adapted to many graph structures and weight functions.

# 4 Filter Analysis and Examples

The case of an arbitrary $p$ is not considered in this paper. In the sequel, we discuss the cases of $p = 2$ and $p = 1$, and we give some existing results related to our filter. When $p = 2$, it follows from equation (11) that the solution of problem (9) is based on the weighted Laplace operator defined by equation (7). Equation (11) reduces to $\Delta f^* + 2\lambda(f^* - f^0) = 0$. In this case, the iterative filter (13) is linear on the graph structure, and the coefficients given by the function $\gamma$ do not have to be updated at each iteration because they depend on the function $w$. When $p = 1$, it follows from equation (11) that the solution of problem (9) is based on the weighted curvature operator defined by equation (8). Equation (11) reduces to $\kappa f^* + 2\lambda(f^* - f^0) = 0$. In this case, the iterative filter (13) is nonlinear, and the coefficients given by the function $\gamma$ are adaptively updated at each iteration in addition of updating the function $f$.

## 4.1 Regularization of Functions on Discrete Data

The family of filters presented in Section 3 can be used to regularize any function defined on discrete data by constructing a weighted graph, and by considering the function to be regularized as a function defined on graph vertices.

Let $V = \{v_1, \ldots, v_N\}$ be a finite data set such that $v_i$ is a vector of $\mathbb{R}^m$. There exist several popular methods that transform the set $V$ with a given pairwise similarity measure $w$ into a graph $G_w = (V, E)$. Constructing similarity graphs consists in modelizing local neighborhood relationships between data points. Among the existing methods, we can quote *the $\epsilon$-neighborhood graph* where two points $u, v \in V$ are connected by an edge if $\|v-u\| < \epsilon$, $\epsilon > 0$. Another important graph is *the $k$-nearest neighbors graph* where two points $u, v \in V$ are connected by an edge if $u$ is among the $k$ nearest neihgbors of $v$. This definition leads to a directed graph because the neighborhood relationship is not symmetric. In order to make this graph symmetric, let $nn_k(u)$ be the set of $k$-nearest neighbors of the point $u$. Then, a point $v$ is connected to $u$ if $u \in nn_k(v)$ or $v \in nn_k(u)$.

Let $f \in \mathcal{H}(V)$ be a function defined on each point of the data set $V$. Similarities between data points are estimated by comparing their features. Features generally depend on the function $f$ and the set $V$. Every point $v \in V$ is assigned with a feature vector denoted by $F_f(v) \in \mathbb{R}^q$. We propose to use one of the two general weight functions:

$$w_{uv} = \exp\left(-\frac{\|F_f(v) - F_f(u)\|^2}{h^2}\right), \ \forall uv \in E, \qquad (15)$$

and,

$$w_{uv} = \exp\left(-\frac{\|u-v\|^2}{2\sigma^2}\right)\exp\left(-\frac{\|F_f(v)-F_f(u)\|^2}{h^2}\right), \ \forall uv \in E. \qquad (16)$$

where $\sigma$ and $h$ are two parameters depending on the variations of $\|u-v\|$ and $\|F_f(v)-F_f(u)\|$ over the graph.

The graph structure, associated with one of the above weight functions, describe a general family of filters. This family is linked to several filters defined in the context of image and mesh processing. When $p = 2$, filter (13), associated with the weight function (16), is equivalent to the bilateral filter, introduced in the context of image denoising [12] [21]. It is a nonlinear filter that combines geometric and range filtering. Bilateral filtering is also used to denoise meshes [22]. It is obtained by using the scalar feature $F_f(v) = f_v$ for all $v \in V$. Using the same parameters, filter (13) can also be considered as a discrete nonlocal mean filter, introduced in the context of images [1]. Indeed, it is obtained by using the weight function (15) with the feature vector $F_f(v) = [f_u : u \in B_{v,s}]^T$ and $B_{v,s}$ a bounding box of size $s$ centered at $v$.

When $\lambda \neq 0$ and $w$ is constant, filter (13) corresponds exactly to the digitized PDE filters proposed in the context of image restoration [11]. If $p = 1$, it is the TV regularization. If $p = 2$, it is the $L_2$ regularization. In general, if the weight function is not constant, filter (13) corresponds to the weighted $L_2$ regularization and the weighted TV regularization on arbitrary graphs.

### 4.2 Relationships with Spectral Graph Filtering

We consider the regularization problem (9) for $p = 2$ and $\lambda = 0$, and we show that it can be expressed in terms of spectral graph theory [13]. From expression (14), the filter reduces to:

$$f_v^{(t+1)} = \sum_{u\sim v} \varphi_{vu} f_u^{(t)}, \ \forall v \in V, \qquad (17)$$

where $\varphi_{vu} = w_{uv}/\sum_{u\sim v} w_{uv}$, $\forall uv \in E$. As we have $\varphi_{vu} \geq 0$ and $\sum_{u\sim v}\varphi_{vu} = 1$, $\varphi_{vu}$ can be interpreted as the probability of a random walker to jump from $v$ to $u$ in a single step [14]. Let $P$ be the $N \times N$ Markov matrix defined by: $P(v,u) = \varphi_{vu}$ if the edge $uv \in E$, and $P(v,u) = 0$ otherwise. Let $F$ be the matrix form of the function $f$. With these notations, expression (17) is rewritten as:

$$F^{(t+1)} = PF^{(t)} = P^t F^{(0)}. \qquad (18)$$

An element $P^t(v,u)$, $vu \in E$, describes the probability of transition in $t$ steps. The matrix $P^t$ encodes local similarities between vertices of the graph and it diffuses or integrates this local information for $t$ steps to larger and larger neighborhoods of each vertex.

The spectral decomposition of the matrix $P$ is given by $P\phi_i = \lambda_i\phi_i$, with $1 \geq \lambda_1 \geq \ldots \geq \lambda_i \geq \ldots \geq \lambda_N \geq 0$ the eigenvalues of $P$, and $\phi_i$ its eigenvectors. The eigenvectors associated with the $k$ first eigenvalues contain the principal

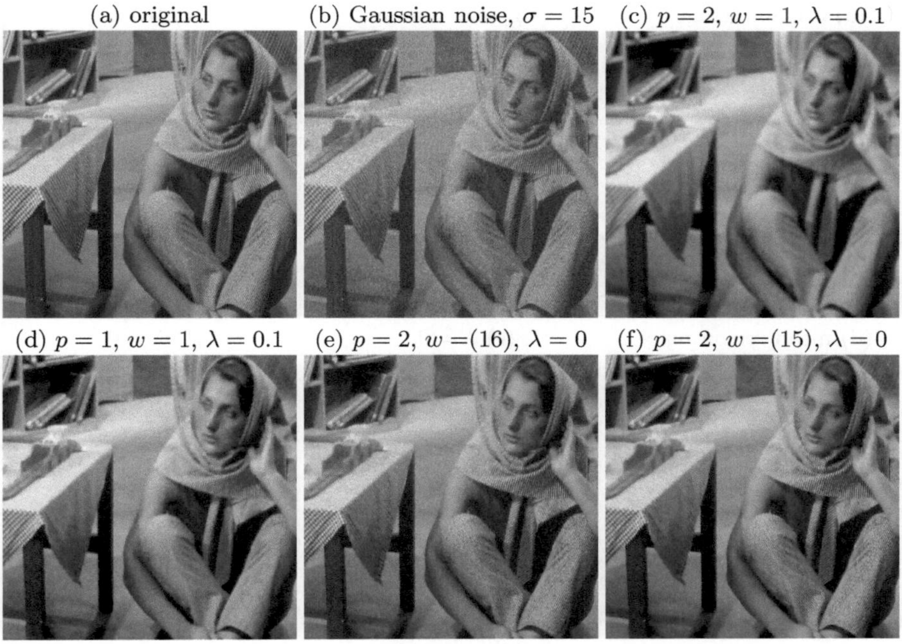


**Fig. 1.** Application to image denoising. Regularizations performed with $t = 5$. (c) Unweighted $L_2$ regularization (13) on an 8-adjacency graph. (d) Unweighted $TV$ regularization (13) on an 8-adjacency graph ($p = 1$). (d) Regularization (13) with $p = 2$ and weight function (16) and scalar feature $F_f = f$ on an 8-adjacency graph. (e) Regularization (13) with weight function (15) and feature vector $F_f(v) = [f_u : u \in B_{v,7}]^T$ on an $\epsilon$-neighborhood graph with $\epsilon = 4$.

information. Thus, an equivalent way to look at the power of $P$ in filter (18) is to decompose each value of $F$ on the first eigenvectors of $P$. Moreover, the eigenvectors of the matrix $P$ can be seen as an extension of the Fourier transform basis functions with eigenvalues representing frequencies [23]. It defines a basis of any function $f$ in $\mathcal{H}(V)$, and the function $f$ can be decomposed on the $k$ first eigenvectors of $P$ as: $f = \sum_{i=1}^{i=k} \langle f, \phi_i \rangle_{\mathcal{H}(V)} \phi_i$. This can be interpreted as a filtering process in the spectral domain.

## 5 Applications

The family of filters proposed in Section 3 can be used to regularize any function defined on the vertices of a graph, or on any discrete data set. Through examples, we show its efficiency in the case of image denoising, image simplification, polygonal curve denoising and surface mesh denoising. We also compare several results obtained for different weight functions and regularization parameters.

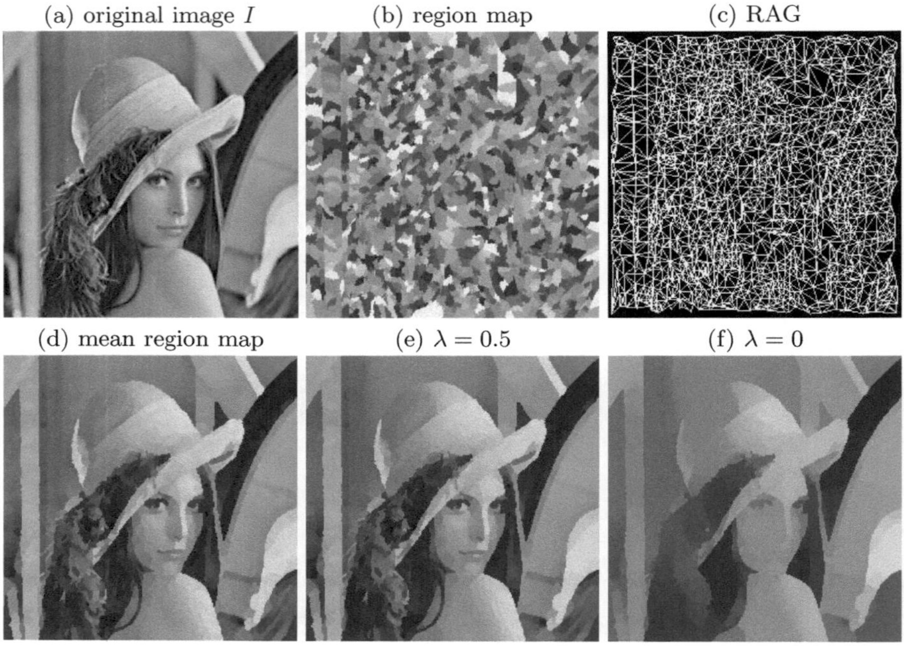


**Fig. 2.** Image simplification. (a) Original image $I$. (b) The region map $R$ obtained by an energy partition of $I$. (c) The RAG $G_w$ associated with $R$, weighted with weight function (15), $h = 8$ and $F_f(v) = f$. The function considered on the vertices corresponds to the mean value of a region. (d) The mean region map associated with $R$. (e,f) Regularization of $G_w$ with filter (13).

**Image Smoothing and Denoising:** Let $f^0$ be a gray level image of $N$ pixels, $f^0 = [f_1^0, \ldots, f_N^0]^T$ with $f^0 : V \subset \mathbb{Z}^2 \to \mathbb{R}$. Figure 1 illustrates the regularization of a noisy image $f^0$ (Fig. 1(b)) on a grid graph of 8-adjacency (Fig. 1(c), (d) and (e)), and on a $\epsilon$-neighborhood graph (Fig. 1(f)). The weight function is chosen such that Fig. 1(c) corresponds to the unweighted $L_2$ regularization, Fig. 1(d) to the unweighted TV regularization, Fig. 1(e) to the bilateral filter, and Fig. 1(f) to the nonlocal mean filter. The use of a non-constant weight function implies an anisotropic diffusion which better preserves image features.

In the case of color images, the function $f^0 : V \subset \mathbb{Z}^2 \to \mathbb{R}^3$ is a mapping from the vertices of the graph to a vector of color channels. The regularization can be applied on each channel leading to an iteration of filter (13) rewritten as:

$$\begin{bmatrix} f_{c1} \\ f_{c2} \\ f_{c3} \end{bmatrix}_v^{(t+1)} = \left( p\lambda + \sum_{u \sim v} \gamma_{uv}^{(t)} \right)^{-1} \left( p\lambda \begin{bmatrix} f_{c1} \\ f_{c2} \\ f_{c3} \end{bmatrix}_v^0 + \sum_{u \sim v} \gamma_{uv}^{(t)} \begin{bmatrix} f_{c1} \\ f_{c2} \\ f_{c3} \end{bmatrix}_u^{(t)} \right), \quad (19)$$

where $\gamma_{uv}^{(t)}$ depends on the norm of the $p$-Laplace operator defined by $\| \nabla_v f \| = \sqrt{\| \nabla_v f_{c1} \|^2 + \| \nabla_v f_{c2} \|^2 + \| \nabla_v f_{c3} \|^2}$. This norm takes into account the inner

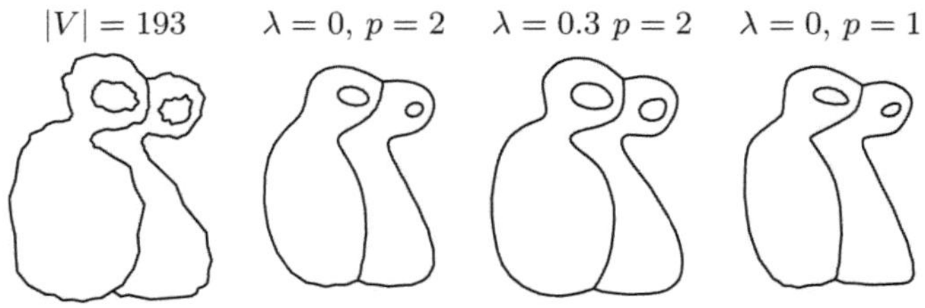


**Fig. 3.** Polygonal curve denoising. Edges of polygons are weighted with equation (15). The regularization of vertices position is performed in 10 steps. When $\lambda = 0$, the regularization introduces shrinkage effects. They are reduced using a value of $\lambda \neq 0$.

correlation aspect of color vector data (in the case of $p \neq 2$). The above iteration can be extended to any vector-valued function $\mathbb{R}^m \to \mathbb{R}^k$.

**Image Simplification:** One can simplify an image by first considering a fine partition of this image (or over-segmentation), where the pixel values of each region of the partition is replaced by the mean or the median pixel value of this region. The partition can be structured by a region adjacency graph (RAG), where each vertex represents a region and where edges are linking adjacent regions. Figure 2(b) and (c) illustrate a fine partition of an image and its associated RAG. Let $G_w = (V, E)$ be a RAG. Let $f^0 : V \subset \mathbb{Z}^2 \to \mathbb{R}^m$ be a mapping from the vertices of $G_w$ to the mean or median value of their regions. Then, the simplification is achieved by regularizing the function $f^0$ on the subgraph of $G_w$ composed of the edges of $E$ for which $w_{uv} > \mu > 0$. Figure 2(e) shows the graph $G'_w$ used to regularize the partition of Fig. 2(d). The filtered partition is depicted in Fig. 2(f).

**Mesh Smoothing and Denoising:** By nature, polygonal curves and surface meshes have a graph structure. Let $V$ be the set of mesh vertices, and let $E$ be the set of mesh edges. If the input mesh is noisy, we can regularize vertex coordinates or any other function $f^0 : V \subset \mathbb{R}^3 \to \mathbb{R}^m$ defined on the graph $G_w = (V, E)$. Results of filter (19) are given in Fig. 3 for polygonal curves, and in Fig. 5 and Fig. 4 for surface meshes.

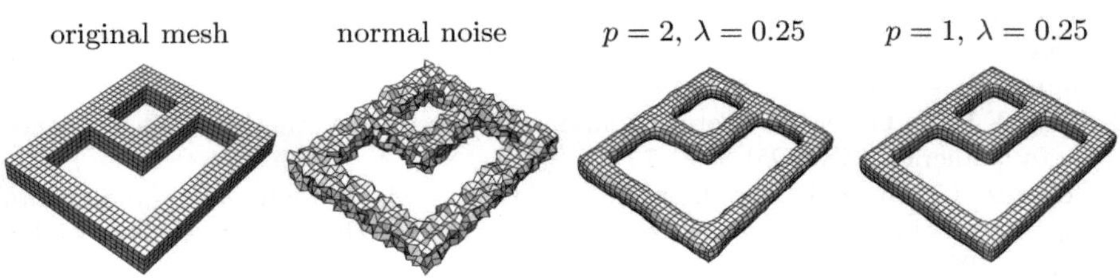


**Fig. 4.** Mesh denoising by regularizing the position of vertices ($t = 10$ and $w =$(15))

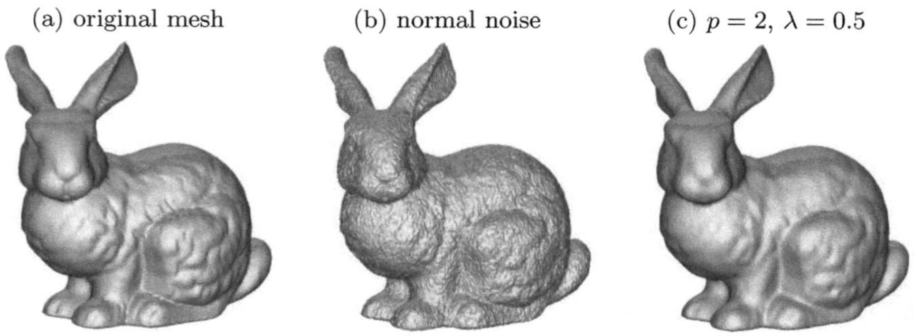


**Fig. 5.** Mesh denoising. (a) Original Stanford Bunny with $|V| = 35949$. (b) Noisy Bunny with normal noise. (c) Regularization of vertex coordinates on mesh edges in 8 iterations. We use the weight function (15) with the scalar feature $F_f(v) = f_v$.

## 6 Conclusion

We propose a general discrete framework for regularizing real-valued or vector-valued functions on weighted graphs of arbitrary topology. The regularization, based on the $p$-Laplace operator, leads to a family of nonlinear iterative filters. This family includes the TV digital filter, the nonlocal mean filter and the bilateral filter, both widely used in image processing. Also, the family is linked to spectral graph filtering.

The choice of the graph topology and the choice of the weight function allow to regularize any discrete data set or any function on a discrete data set. Indeed, the data can be structured by neighborhood graphs weighted by functions depending on data features. This can be applied in the context of image smoothing, denoising or simplification. We also show that mesh smoothing and denoising can be performed by the same filtering process. The main ongoing work is to use the proposed framework in the context of hierarchical mesh segmentation and point cloud clustering.

## References


1. Buades, A., Coll, B., Morel, J.M.: A review of image denoising algorithms, with a new one. Multiscale Modeling and Simulation **4**(2) (2005) 490–530
2. Chan, T., Shen, J.: Image Processing and Analysis - variational, PDE, wavelets, and stochastic methods. SIAM (2005)
3. Tsai, Y.H.R., Osher, S.: Total variation and level set methods in image science. Acta Numerica **14** (2005) 509–573
4. Alvarez, L., Guichard, F., Lions, P.L., Morel, J.M.: Axioms and fundamental equations of image processing. Archive for Rational Mechanics and Analysis **123**(3) (1993) 199–257
5. Taubin, G.: Geometric signal processing on polygonal meshes. In: Eurographics, State of the Art Report. (2000)

6. Ohtake, Y., Belyaev, A., Bogaeski, I.: Mesh regularization and adaptive smoothing. Computer-Aided Design **33** (2001) 789–800
7. Sochen, N., Deriche, R., Lopez-Perez, L.: Variational Beltrami flows over manifolds. In: ICIP'03: Proc of the Inter. Conf. on Image Processing. Volume I., IEEE Computer Society (2003) 861–864
8. Desbrun, M., Meyer, M., Schröder, P., Barr, A.: Anisotropic feature-preserving denoising of height fields and bivariate data. Graphics Interface (2000) 145–152
9. Bajaj, C.L., Xu, G.: Anisotropic diffusion of surfaces and functions on surfaces. ACM Trans. on Graph. **22**(1) (2003) 4–32
10. Hildebrandt, K., Polthier, K.: Anisotropic filtering of non-linear surface features. Eurographics 2004: Comput. Graph. Forum **23**(3) (2004) 391–400
11. Osher, S., Shen, J.: Digitized PDE method for data restoration. In Anastassiou, E.G.A., ed.: In Analytical-Computational methods in Applied Mathematics. Chapman&Hall/CRC (2000) 751–771
12. Tomasi, C., Manduchi, R.: Bilateral filtering for gray and color images. In: ICCV'98: Proc. of the 6th Int. Conf. on Computer Vision, IEEE Computer Society (1998) 839–846
13. Chung, F.: Spectral graph theory. CBMS Regional Conference Series in Mathematics **92** (1997) 1–212
14. Coifman, R., Lafon, S., Maggioni, M., Keller, Y., Szlam, A., Warner, F., Zucker, S.: Geometries of sensor outputs, inference, and information processing. In: Proc. of the SPIE: Intelligent Integrated Microsystems. Volume 6232. (2006)
15. Chambolle, A.: Total variation minimization and a class of binary MRF models. In: LNCS 3757, Proc. of the 5th Int. Work. EMMCVPR, Springer-Verlag (2005) 136–152
16. Darbon, J., Sigelle, M.: Exact optimization of discrete constrained total variation minimization problems. In R.Klette, Zunic, J., eds.: LNCS 3322: Proc. of the 10th Int. Workshop on Combinatorial Image Analysis. (2004) 548–557
17. Zhou, D., Schölkopf, B.: Regularization on discrete spaces. In: LNCS 3663, Proc. of the 27th DAGM Symp., Springer-Verlag (2005) 361–368
18. Bougleux, S., Elmoataz, A.: Image smoothing and segmentation by graph regularization. In: LNCS 3656, Proc. Int. Symp. on Visual Computing, Springer-Verlag (2005) 745–752
19. Bensoussan, A., Menaldi, J.L.: Difference equations on weighted graphs. Journal of Convex Analysis **12**(1) (2005) 13–44
20. Friedman, J., Tillich, J.P.: Wave equations for graphs and the edge-based laplacian. Pacific Journal of Mathematics **216**(2) (2004) 229–266
21. Barash, D.: A fundamental relationship between bilateral filtering, adaptive smoothing, and the nonlinear diffusion equation. IEEE Trans. Pattern Anal. Mach. Intell. **24**(6) (2002) 844–847
22. Fleishman, S., Drori, I., Cohen-Or, D.: Bilateral mesh denoising. ACM Trans. on Graphics **22**(3) (2003) 950–953
23. Coifman, R., Lafon, S., Lee, A., Maggioni, M., Nadler, B., Warner, F., Zucker, S.: Geometric diffusions as a tool for harmonic analysis and structure definition of data. Proc. of the National Academy of Sciences **102**(21) (2005)

# Counter-Examples for Bayesian MAP Restoration

Mila Nikolova

CMLA, ENS Cachan, CNRS, PRES UniverSud
61 Av. President Wilson, F-94230 Cachan, France
nikolova@cmla.ens-cachan.fr

**Abstract.** Bayesian MAP is most widely used to solve various inverse problems such as denoising and deblurring, zooming, reconstruction. The reason is that it provides a coherent statistical framework to combine observed (noisy) data with prior information on the unknown signal or image. However, this paper exhibits a major contradiction since the MAP solutions substantially deviate from both the data-acquisition model and the prior model. This is illustrated using experiments and explained based on some known analytical properties of the MAP solutions.

**Keywords:** MAP estimation, restoration, regularization, modeling.

## 1 MAP Estimators to Combine Noisy Data and Priors

We address inverse problems where an unknown $x \in \mathbb{R}^p$ (an image or a signal) is recovered from a realization[1] of noisy data $Y = y \in \mathbb{R}^q$ using a statistical model for their production as well as a prior model for the original $X = x$. Typical applications are signal and image restoration, segmentation, motion estimation, sequence processing, color reproduction, optical imaging, tomography, seismic and nuclear imaging, and many others. The likelihood function $f_{Y|X}(y|x)$—the distribution for the observed data $Y = y$ given any original $X = x$—is governed by the data-acquisition model. The most common models are of the form

$$Y = AX + N, \tag{1}$$

where $A : \mathbb{R}^p \to \mathbb{R}^q$ is a linear operator (e.g. a blurring kernel, a Fourier or a Radon transform, or a subsampling operator) and $N$ is additive noise which is independent of $X$. If the noise samples $N_i, 1 \leq i \leq q$ are independent and identically distributed (i.i.d.) with marginal distribution $f_N$, then

$$f_{Y|X}(y|x) = \prod_{i=1}^{q} f_N\big(a_i^T x - y_i\big), \tag{2}$$

where $a_i^T$ are the rows of $A$. Usually $f_N$ is a zero-mean Gaussian density on $\mathbb{R}$.

[1] Random variables are in uppercase letters and their values in lowercase letters.

F. Sgallari, A. Murli, and N. Paragios (Eds.): SSVM 2007, LNCS 4485, pp. 140–152, 2007.

A meaningful solution $\hat{x}$ can seldom be recovered from the data-acquisition model $f_{Y|X}$ solely [41,39,12,42], without priors on $X$. Two ways to model statistical priors $f_X$ on $X$ are sketched next. Markov random models [7,16,22,8,19,3] focus on the local characteristics of $X$—the interaction between each pixel $x_i$ and its adjacent neighbors. Assume that the prior has the usual Gibbsian form

$$f_X(x) \propto \exp\{-\lambda \Phi(x)\}, \tag{3}$$

where $\Phi$ is a prior energy and $\lambda > 0$ is a parameter. Usually[2] [8,6,21,10]

$$\Phi(x) = \sum_{i=1}^{r} \varphi(\|G_i x\|), \tag{4}$$

$G_i$, $1 \le i \le r$, are linear operators (e.g. they can yield the differences or discrete derivatives of $x$). Various functions $\varphi : \mathbb{R}_+ \to \mathbb{R}_+$ have been proposed, see [10] for a full list and Table 1 for examples. The fit of $\varphi$ to the empirical distribution of the differences $x_i - x_j$ in images is considered e.g. in [35].

**Table 1.** Commonly used functions $\varphi$ where $\alpha > 0$ is a parameter

| | |
|---|---|
| (f1) $\varphi(t) = t^\alpha,\ 1 < \alpha \le 2$ | (f8) $\varphi(t) = t$ |
| (f2) $\varphi(t) = \sqrt{\alpha + t^2}$ | |
| (f3) $\varphi(t) = \log(\cosh(\alpha t))$ | |
| (f4) $\varphi(t) = \min\{\alpha t^2, 1\}$ | (f9) $\varphi(t) = t^\alpha,\ 0 < \alpha < 1$ |
| (f5) $\varphi(t) = \dfrac{\alpha t^2}{1 + \alpha t^2}$ | (f10) $\varphi(t) = \dfrac{\alpha t}{1 + \alpha t}$ |
| (f6) $\varphi(t) = \log(\alpha t^2 + 1)$ | (f11) $\varphi(t) = \log(\alpha t + 1)$ |
| (f7) $\varphi(t) = 1 - \exp(-\alpha t^2)$ | (f12) $\varphi(0) = 0,\ \varphi(t) = 1$ if $t \ne 0$ |

Another approach is to use wavelet expansions from the outset. Let $\{w_i : 1 \le i \le r\}$ be a family of wavelet functions on $\mathbb{R}^p$. In numerous papers [23,43,25,37, 24,5,40,1,9] the coefficients $u_i = \langle w_i, x\rangle$, $1 \le i \le r$, are assumed i.i.d. and their statistical distribution is described using priors $f_{U_i}$ of the form

$$f_{U_i}(t) = \exp\big(-\lambda_i \varphi(t)\big)/Z, \quad t \in \mathbb{R}, \tag{5}$$

where $\varphi$ is a function as those given in Table 1 and $\lambda_i > 0$. In (5) and in what follows, $Z$ denotes a normalization constant.

The posterior distribution $f_{X|Y}(x|y)$, given by the Bayesian chain rule,

$$f_{X|Y}(x|y) = f_{Y|X}(y|x) f_X(x) \frac{1}{Z}, \quad Z = f_Y(y).$$

Bayesian estimators are based on $x \to f_{X|Y}(x|y)$ and they realize an *optimal compromise* between $f_{Y|X}(y|.)$ and $f_X$ with respect to a loss function. Our focus

[2] For $\Phi$ as in (4), $x \to \exp(-\lambda\Phi(x))$ is non-integrable on $\mathbb{R}^p$, so $f_X$ is an improper prior [8,6]. This impropriety can be easily removed either by restricting $x$ to belong to a bounded domain or by preventing $x$ to shift up and down. As noticed by many authors, this is hardly worthwhile whenever the posterior $f_{X|Y}(x|y)$ is proper.

is on the most popular Bayesian estimator, namely the *Maximum a Posteriori (MAP)*, which selects $\hat{x}$ as *the most likely solution given the recorded data* $Y = y$:

$$\hat{x} = \arg\max_x f_{X|Y}(x|y) = \arg\min_x \big( - \ln f_{Y|X}(y|x) - \ln f_X(x)\big).$$

For $\Psi(x, y) \propto - \ln f_{Y|X}(y|x)$ and using (3), $\hat{x}$ equivalently minimizes a posterior energy $E_y$ of the form

$$E_y(x) = \Psi(x, y) + \lambda\Phi(x). \tag{6}$$

Under the classical model that the noise in (1) is i.i.d. with a zero-mean Gaussian density with variance $\sigma^2$, $N \sim \mathcal{N}(0, \sigma^2 I)$, the MAP estimate $\hat{x}$ minimizes

$$E_y(x) = \|Ax - y\|^2 + \beta\Phi(x) \quad \text{where} \quad \beta = 2\sigma^2\lambda. \tag{7}$$

Since [13], denoising of signals and images is efficiently dealt by restoring the noisy wavelet coefficients $\langle w_i, y\rangle$, $1 \le i \le r$, with the aid of priors of the form (5). Such methods were considered by many authors, e.g. [23,43,25,37,24,5,40,1,9], and they amount to calculating $\hat{u} = \arg\min_u E_y(u)$ for

$$E_y(u) = \sum_i \Big( (u_i - \langle w_i, y\rangle)^2 + \lambda_i\varphi(|u_i|)\Big). \tag{8}$$

The solution is then $\hat{x} = W^\dagger\hat{u}$ where $W^\dagger$ is a left-inverse of $\{w_i, 1 \le i \le r\}$.

Realistic statistical modeling of the physical phenomena in data-acquisition devices on the one hand, and modeling of priors for real-world images and signals on the other hand, focuses more and more efforts in research and applications, and references are abundant, e.g. [3, 6, 20, 35, 26, 38, 36, 18]. This is naturally done with *the expectation to obtain solutions* $\hat{x}$ *that are coherent with all the two models* $f_{Y|X}$ *and* $f_X$. The adequacy of the most popular MAP estimator has essentially been considered in asymptotical conditions when $\beta \searrow 0$, $q \to \infty$, or $\beta \nearrow \infty$, i.e. when either $f_X$ or $f_{Y|X}$ vanishes. However, in regular conditions, we exhibit important contradictions in the MAP approach since *the MAP solutions substantially deviate from the data-acquisition model* $f_{Y|X}$ *on the one hand, and from the prior model* $f_X$ *on the other hand.* The practical consequences of this gap between modeling and solution can be embarrassing... This gap is illustrated in section 2 using examples that consider the ideal case when both the data-acquisition and the prior models are known exactly. The remaining of the paper explains rigorously the reasons for this gap based on the analytical properties characterizing the MAP solutions as a function of the shape of $E_y$, see e.g. [28, 29, 32]. Conversely, the latter kind of results, aimed at characterizing the minimizers of $E_y$, can provide a rigorous alternative for modeling.

## 2 Gaps Between Models and Estimates

Let us consider a measurement model of the form (1)-(2) where the sought-after $X$ and the noise $N$ have distributions $f_X$ and $f_N$, respectively. In full rigor, an estimator $\hat{X}$ for $X$, based on data $Y$, can be said to be coherent with the underlying model if $\hat{X} \sim f_X$ (i.e. $\hat{X}$ has the same distribution as the prior) and if the resultant noise estimator $\hat{N} = Y - A\hat{X}$ satisfies $\hat{N} \sim f_N$. In general, neither $f_{\hat{X}}$ nor $f_{\hat{N}}$ can be calculated. A simple analytical example presented in [33] clearly shows that the distribution of $\hat{X}$ is dissimilar to $f_X$ and that $\hat{N} \not\sim f_N$.

Instead we present an experiment on the MAP restoration of noisy wavelet coefficients that was sketched in (5) and (8). The statistical distribution of the (noise-free) wavelet coefficients in real-world image data has been shown to be fairly described using generalized Gaussian (GG) distribution laws [23, 24, 5]

$$f_X(x) = \frac{1}{Z} e^{-\lambda |x|^\alpha}, \quad x \in \mathbb{R}, \quad \text{for } \lambda > 0, \alpha > 0. \tag{9}$$

Under i.i.d. Gaussian noise, the MAP estimate $\hat{u}_i$ of each noisy coefficient $\langle w_i, y \rangle$ in (8) is done independently, by minimizing a scalar function $E_y$ of the form

$$E_y(x) = (x - y)^2 + \beta |x|^\alpha \quad \text{for} \quad \beta = 2\sigma^2 \lambda. \tag{10}$$

This is a situation where both the prior and the data-acquisition models are pertinent. It is then natural to require that the solution $\hat{x}$ fits these models.

For $(\alpha, \lambda)$ and $\sigma$ fixed, we realize 10 000 independent trials. In each trial, an original $x \in \mathbb{R}$ is sampled from $f_X$, then $y = x + n$ for $n \sim \mathcal{N}(0, \sigma^2)$, and the true MAP solution $\hat{x}$ is calculated using (10).

*(a) Case* $\alpha \geq 1$. Fig. 1 corresponds to $\alpha = 1.2$, $\lambda = 0.5$ and $\sigma = 0.6$. The histogram of $x \sim f_X$ is shown in Fig. 1(a). For every $y \in \mathbb{R}$, $E_y$ is strictly convex and has a unique minimizer $\hat{x}$. Unlike the prior $f_X$, the histogram the MAP solutions $\hat{x}$ plotted in Fig. 1(b) is concentrated near zero. The histogram of the resultant noise estimates $\hat{n} = y - \hat{x}$ in Fig. 1(c) is far from being Gaussian: it is clearly bounded while its value in the vicinity of zero is very small which means that almost all MAP solutions $\hat{x}$ are biased.

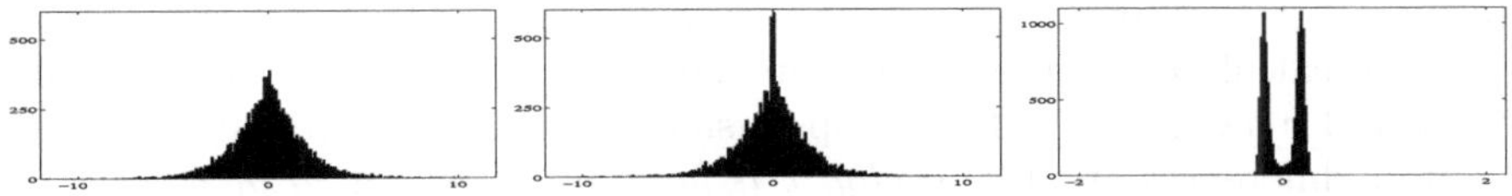


(a) Prior (9), $\alpha$=1.2, $\lambda$=0.5 (b) True MAP $\hat{x}$ by (10) (c) Noise estimate $\hat{n}$=$y$−$\hat{x}$

**Fig. 1.** Histograms for 10 000 independent trials, case $\alpha \geq 0$

*(b) Case* $0 < \alpha < 1$. Fig. 2 corresponds to $\alpha = 0.5$, $\lambda = 2$ and $\sigma = 0.8$. The samples drawn from $f_X$ are sown in Fig. 2(a). For every $y \neq 0$, $E_y$ has two local minimizers [33], $\hat{x}_1 = 0$ and $\hat{x}_2$ such that $|\hat{x}_2| > \theta$ for $\theta = \left(\frac{2}{\alpha(1-\alpha)\beta}\right)^{\frac{1}{\alpha-2}} \approx 0.47$. Hence the distribution $f_{\hat{X}}$ of the true MAP solution $\hat{X}$ contains a Dirac-delta at 0 and is null on a subset containing $(-\theta, 0) \bigcup (0, \theta)$. This is corroborated by the histogram of the MAP estimate $\hat{x}$, Fig. 2(b): we have $\hat{x} = 0$ in 77% of the

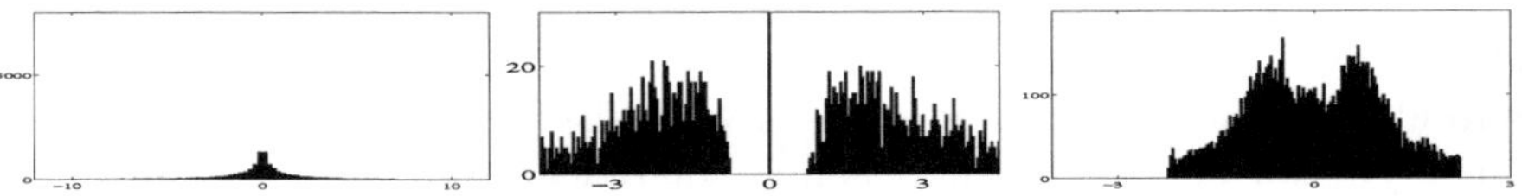


(a) Prior (9), $\alpha$=0.5, $\lambda$=2 (b) True MAP $\hat{x}$ (zoom) (d) Noise estimate $\hat{n}=y-\hat{x}$

**Fig. 2.** Histograms for 10 000 independent trials, case $0 < \alpha < 1$

trials and the smallest non-zero $|\hat{x}|$ is $0.77 > \theta$. It is essentially different from the prior $f_X$. The resultant estimate of the noise $\hat{n}$ shown in (c) is bounded on $\mathbb{R}$.

The MAP estimate does not fit neither the GG prior model nor the data-acquisition model. This is especially unfortunate if these models are faithful.

## 3 Non-smooth at Zero Priors

Let us consider Gibbsian priors (3)-(4) where $\varphi : \mathbb{R}_+ \to \mathbb{R}_+$ is an increasing $\mathcal{C}^m$-function with

$$\varphi'(0) > 0,$$

such as (f8)-(f12) in Table 1. We suppose that the data $Y$ correspond to a likelihood $f_{Y|X}(y|x) \propto \exp\big(-\Psi(x,y)\big)$ where $\Psi$ is $\mathcal{C}^m$, $m \geq 2$. The MAP estimator $\hat{X}$ then minimizes a posterior energy $E_y$ of the form (6) where $\Phi$ is nonsmooth.

**Laplacian Markov chain with Gaussian noise.** Let $x$ correspond to

$$\Phi(x) = \sum_{i=1}^{p-1} |x_i - x_{i+1}|, \quad \lambda > 0. \tag{11}$$

The differences $X_i - X_{i+1}$ are i.i.d. with Laplacian density $\frac{\lambda}{2}\exp(-\lambda|t|)$. We fix $\lambda = 8$ and $p = 500$ and consider data $Y = X + \mathcal{N}(0, \sigma^2 I)$ for $\sigma = 0.5$. Realizations $X = x$ and $Y = y$ are shown in Fig. 3(a). The MAP solution $\hat{x}$ in Fig. 3(b) is calculated for the true parameters $(\lambda, \sigma)$. Compared to the original $x$, the restored $\hat{x}$ has a very different aspect since it is constant on many regions—92% of its differences are 0. *Visually, $\hat{x}$ is far from fitting the prior model.*

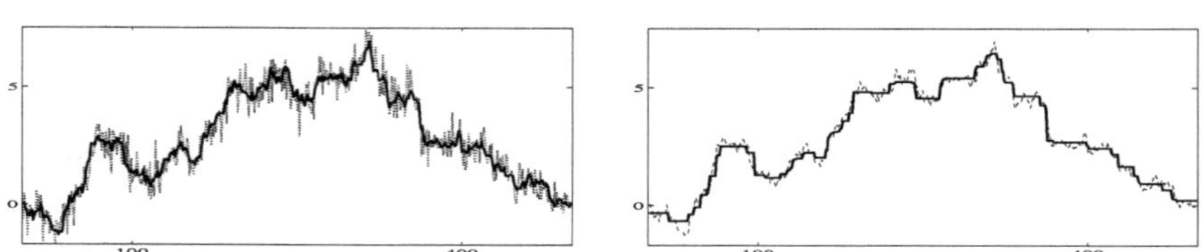


(a) Original $x$ (—), data $y = x + n$ ($\cdots$). (b) True MAP $\hat{x}$ (—), original $x$ (- - -).

**Fig. 3.** True MAP for a Laplacian Markov chain with Gaussian noise

Next we repeat the same experiment 40 times. The histograms of all original differences $x_i - x_{i+1}$, and all differences restored by MAP $\hat{x}_i - \hat{x}_{i+1}$, for all trials, are shown in Fig. 4(a) and (b), respectively. They are completely dissimilar.

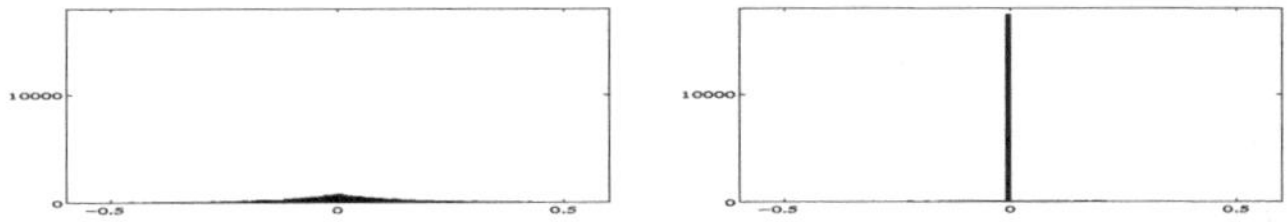


(a) 40×499 differences $x_i - x_{i+1}$ sampled from $f_{\Delta X}$ for $\lambda = 8$. (b) The differences $\hat{x}_i - \hat{x}_{i+1}$ of the true MAP solutions.

**Fig. 4.** Histograms for 40 trials as in Fig. 3

The observed incoherence between the prior and the solution will be explained using some results from [28] and [31] (Theorems 6.1 and 2, resp.) $\delta E_y(x)(u)$ will denote the one-sided derivative of $E_y$ at $x$ in the direction of $u \neq 0$.

**Theorem 1.** *Given $y \in \mathbb{R}^q$, let $\hat{x} \in \mathbb{R}^p$ be such that for $J = \big\{i \in \{1, \ldots, r\} : G_i\hat{x} = 0\big\}$ and $K_J = \big\{u \in \mathbb{R}^p : G_i u = 0,\ \forall i \in J\big\}$, we have*

*(a) $\delta E_y(\hat{x})(u) > 0$ for every $u \in K_J^{\perp} \setminus \{0\}$;*
*(b) $DE_y|_{K_J}(\hat{x})u = 0$ and $D^2E_y|_{K_J}(\hat{x})(u,u) > 0$, for every $u \in K_J \setminus \{0\}$.*

*Then $E_y$ has a strict (local) minimum at $\hat{x}$. Moreover, there are a neighborhood $O_J$ of $y$ and a continuous function $\mathcal{X} : O_J \to \mathbb{R}^p$ such that $\mathcal{X}(y) = \hat{x}$ and that for every $y' \in O_J$, $E_{y'}$ has a (local) minimum at $\hat{x}' = \mathcal{X}(y')$ satisfying*

$$G_i\hat{x}' = 0 \quad \forall i \in J,$$

*or equivalently, that $\hat{x}' \in K_J$ for every $y' \in O_J$.*

Conditions (a)-(b) are very general [31]. Since $O_J$ contains an open subset of $\mathbb{R}^q$

$$y \in O_J \ \text{ and } \ \hat{x} = \arg\max_{x \in \mathbb{R}^p} f_{X|Y}(x|y) \ \Rightarrow \ G_i\hat{x} = 0 \ \ \forall i \in J, \tag{12}$$

or equivalently $\hat{x} \in K_J$. Then the probability to have $\hat{X} \in K_J$ satisfies

$$\Pr(\hat{X} \in K_J) \geq \Pr(Y \in O_J) = \int_{O_J} f_Y(y)dy > 0.$$

The "prior" model which is *effectively* realized by the MAP estimator $\hat{X}$ hence corresponds to images and signals such that $G_i\hat{X} = 0$ for a certain number of indexes $i$. If $\{G_i\}$ are first-order differences or discrete gradients, then we have an *effective prior model for locally constant images and signals.* This is in contradiction with the prior model: since $f_X$ is continuous and $K_J \subset \mathbb{R}^p$ is a subspace of $\mathbb{R}^p$ of dimension $< p$, the probability that $X \in K_J$ is null:

$$\Pr(X \in K_J) = \int_{K_J} f_X(x)dx = 0,$$

**Laplacian Markov chain.** Consider that $E_y$ is defined by (7) and (11) and that $A$ is invertible. The following striking phenomena occur (see [31]):

(a) for every $\hat{x} \in \mathbb{R}^p$, there is a polyhedron $Q_{\hat{x}}$ of dimension $\#\{i : G_i\hat{x} = 0\}$, such that for every $y \in Q_{\hat{x}}$, the same $\hat{x}$ is the unique minimizer of $E(.,y)$;
(b) for every $J \subset \{1,\ldots,p-1\}$, there is a subset $\tilde{O}_J \subset \mathbb{R}^q$, composed of $2^{n-\#J-1}$ unbounded polyhedra of $\mathbb{R}^q$, such that for every $y \in \tilde{O}_J$, the minimizer $\hat{x}$ of $E_y$ satisfies $\hat{x}_i = \hat{x}_{i+1}$ for all $i \in J$ and $\hat{x}_i \neq \hat{x}_{i+1}$ for all $i \in J^c$. Moreover, their closure forms a covering of $\mathbb{R}^q$.

As a consequence, for every $J \subset \{1,\ldots,p-1\}$ we have $\Pr\big(\hat{X}_i = \hat{X}_{i+1}, \forall i \in J\big) \geq \Pr\big(Y \in \tilde{O}_J\big) > 0$. These are solutions composed of constant pieces. However, the prior model (11) yields $\Pr\big(X_i = X_{i+1}\big) = 0$, $\forall i$.

## 4 Non-smooth at Zero Noise Models

Consider a measurement model of the form (1)-(2) where

$$f_N(t) = \frac{1}{Z} e^{-\sigma\psi(|t|)}, \quad \psi(0) > 0, \tag{13}$$

where $\psi : \mathbb{R}_+ \to \mathbb{R}$ is $\mathcal{C}^m$. Using (2), $f_{Y|X}(y|x) \propto \exp(-\Psi(x,y))$ where

$$\Psi(x,y) = \sigma \sum_{i=1}^{q} \psi(|a_i^T x - y_i|). \tag{14}$$

Furthermore, let $X$ correspond to a Gibbsian prior (3) where $\Phi$ is $\mathcal{C}^m$—e.g. of the form (4) with $\varphi$ as (f1)-(f7) in Table 1. Given $y \in \mathbb{R}^q$, the MAP solution $\hat{x}$ minimizes $E_y$ as given in (6). We start with an experiment.

**GG Markov chain under Laplace noise.** Let $X$ be a 100-length Markov chain whose differences $X_i - X_{i+1}$ are i.i.d. with a GG density as given in (9). Let $Y = X + N$ where $N_i$ are i.i.d. with marginal density $\frac{\sigma}{2}e^{-\sigma|t|}$. The posterior distribution $f_{X|Y}(x|y)$ corresponds to $E_y(x) = \sigma\sum_i |x_i - y_i| + \lambda \sum_i |x_i - x_{i-1}|^\alpha$.

We repeat 1000 times the experiment where an original GG Markov chain $X = x$ is generated for $\alpha = 1.2$, $\lambda = 1$, then $Y = y$ is obtained by adding white Laplacian noise with $\sigma = 2.5$ and the MAP solution $\hat{x}$ is calculated using the same parameters $\alpha$, $\lambda$ and $\sigma$. Fig. 5(a) shows the histogram of all original differences $x_i - x_{i+1}$. The histogram of the Laplacian noise samples is shown in Fig. 5(b). The histogram of all differences $\hat{x}_i - \hat{x}_{i+1}$ for all MAP solutions is shown in Fig. 5(c) and the histogram of the resultant noise estimates $\hat{n}_i = y_i - \hat{n}$ in Fig. 5(d). For 87% of the samples in all trials, $\hat{x}_i = y_i$, hence the huge spike at 0 in (d): most of the samples $\hat{x}_i$ of the MAP solution keep the noise intact which explains why the histogram in (b) is flattened near the origin.

The key to explain the observed behavior is the result stated next [29].

**Theorem 2.** *Given $y \in \mathbb{R}^q$, suppose that $\hat{x} \in \mathbb{R}^p$ is such that for $J = \big\{i \in \{1,\ldots,q\} : a_i^T\hat{x} = y_i\big\}$ and $K_J = \{u \in \mathbb{R}^p : \ a_i^T u = 0 \ \forall i \in J\}$, we have:*

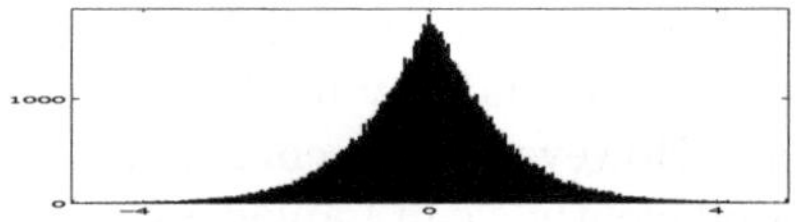

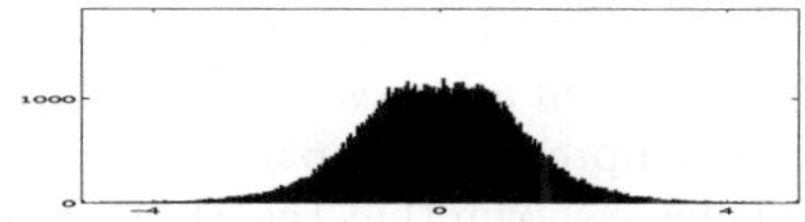

(a) Original differences $x_i - x_{i+1}$ (GG) (c) Differences $\hat{x}_i - \hat{x}_{i+1}$ of the MAP

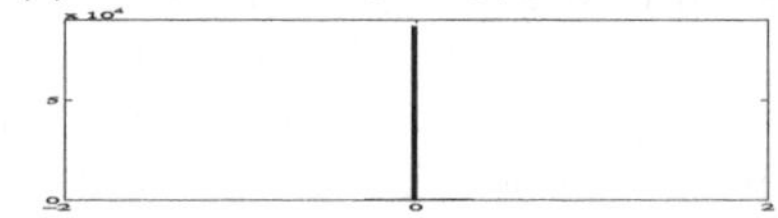

(b) Laplacian i.i.d. noise. (d) All the residuals $y - \hat{x}$.

**Fig. 5.** Histograms for 1000 independent trials with 100-length signals. Left: original models. Right: the models *effectively* realized by the MAP.

*(a) the set* $\{a_i : i \in J\}$ *is linearly independent;*
*(b)* $DE_y|_{\hat{x}+K_J}(\hat{x})u = 0$ *and* $D^2E_y|_{\hat{x}+K_J}(\hat{x})(u,u) > 0$, *for every* $u \in K_J \setminus \{0\}$;
*(c)* $\delta E_y(\hat{x})(u) > 0$, *for every* $u \in K_J^\perp \setminus \{0\}$.

*Then* $E_y$ *has a strict (local) minimum at* $\hat{x}$. *Moreover, there are a neighborhood* $O_J \subset \mathbb{R}^q$ *containing* $y$ *and a* $\mathcal{C}^{m-1}$ *function* $\mathcal{X} : O_J \to \mathbb{R}^p$ *such that for every* $y' \in O_J$, $E_{y'}$ *has a (local) minimum at* $\hat{x}' = \mathcal{X}(y')$ *and that the latter satisfies*

$$\begin{array}{ll} a_i^T \hat{x}' = y_i' & \text{if} \quad i \in J, \\ a_i^T \hat{x}' \neq y_i' & \text{if} \quad i \in J^c. \end{array} \tag{15}$$

*Hence* $\mathcal{X}(y') \in \hat{x} + K_J$ *for every* $y' \in O_J$.

Assumption (a) holds for almost all $y \in \mathbb{R}^q$, and (b)-(c) are weak conditions for a strict local minimum at $\hat{x}$, see [29]. It is crucial that $O_J$ contains an open subset of $\mathbb{R}^q$. By Theorem 2, the distribution of the MAP estimator $\hat{X}$ is such that

$$\Pr\left(a_i^T \hat{X} - Y_i = 0\right) \geq \Pr\left(Y \in O_J\right) = \int_{O_J} f_Y(y)dy > 0, \quad \forall i \in J.$$

This contradicts the model for the noise assumed in (13) since by the latter

$$\Pr\left(a_i^T X - Y_i = 0\right) = \Pr\left(N_i = 0\right) = 0, \quad \forall i.$$

Consider now that $A$ is invertible and that $\Phi$ is of the form (4). Define $O_\infty = \{y \in \mathbb{R}^p : \|D\Phi(A^{-1}y)\| < \frac{\psi'(0^+)}{\beta} \min_{\|u\|=1} \sum_{i=1}^p |a_i^T u|\}$ which clearly contains an open subset of $\mathbb{R}^q$. It is amazing to see that if $y \in O_\infty$, then $a_i^T \hat{x} = y_i$, $\forall i$, which means that the prior has no influence on the solution. In words, $\Pr(A\hat{X} = Y) \geq \Pr(Y \in O_\infty) > 0$. This violates the prior model.

**A Laplace noise model to remove impulse noise**
For any $y \in \mathbb{R}^p$ $(p = q)$, let us consider the minimization of $E_y$ below:

$$E_y(x) = \sum_{i=1}^p |x_i - y_i| + \frac{\beta}{2} \sum_i \sum_{j \in \mathcal{N}_i} \varphi(x_i - x_j) \tag{16}$$

where $\varphi$ is a symmetric $\mathcal{C}^1$ strictly convex edge-preserving function, e.g. (f1)-(f3) in Table 1. From a Bayesian standpoint, the $\ell_1$ data-fidelity in (16) corresponds to data corrupted with Laplacian white noise. However, Theorem 2 and the latter example have shown that the MAP cannot efficiently clean Laplacian noise since all $\hat{x}_i$ such that $\hat{x}_i = y_i$ keep the noise intact. Based on (15), the data samples $y'_i$, $i \in J$ are fitted exactly, hence they must be free of noise. Some results in [30] show that $y'_i$, $i \in J$ can be dissimilar with respect to their neighbors only up to some degree. Otherwise, $i \in J^c$ and the value of $y'_i$ is replaced by an estimate $\hat{x}'_i = \mathcal{X}_i(\{y'_i : i \in J\})$ which depends only on noise-free data samples. The samples $y'_i$ for $i \in J^c$ are hence *outliers* that can take any value on the half-line contained in $O_J$. *In fact, the MAP estimator defined by (16) corresponds to an impulse noise model on the data.*

Fig. 6(a) shows an original GG Markov chain $x$ and data $y$ containing 10% random-valued impulse noise and Fig. 6(b) the minimizer $\hat{x}$ of $E_y$ in (16) for $\beta = 0.4$. The noisy samples are restored well and $\hat{x}_i = y_i$ for 89/90 of the noise-free samples. Similar applications were considered in [30, 4].

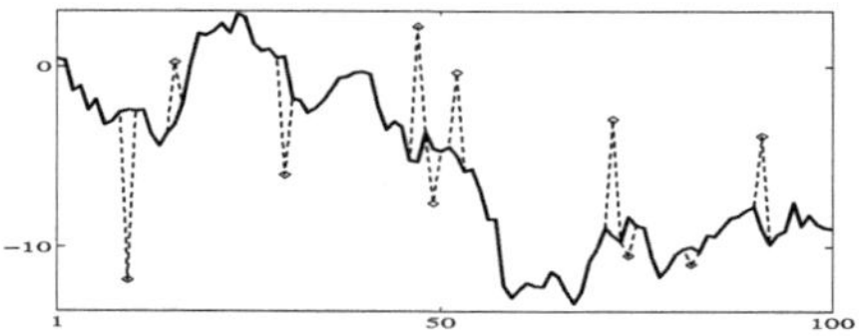


(a) Original $x$ (—), data $y$ (- - -) with 10% impulse noise.

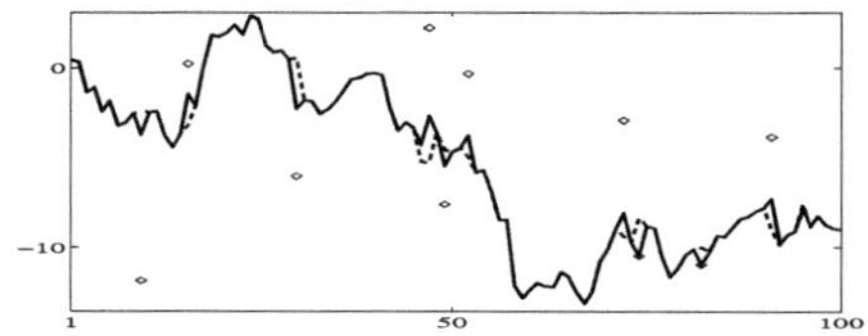


(b) Minimizer $\hat{x}$ of $E_y$ for $\beta = 0.4$ (—), Original $x$ (- - -), Impulses $y_i \neq x_i$ ($\diamond$) .

**Fig. 6.** Restoration of a GG Markov chain corrupted with impulse noise by minimizing an energy $E_y$ with an $\ell_1$ data-fidelity term

# 5 Priors with Non-convex Energies

Let us now consider a linear model for the data (1) with $N \sim \mathcal{N}(0, \sigma^2 I)$ and a Gibbsian prior (3) with a nonconvex prior energy $\Phi$ of the form (4) where $G_i \in \mathbb{R}^{1\times p}$ are difference operators and $\varphi : \mathbb{R} \to \mathbb{R}_+$ is *nonconvex*, $\mathcal{C}^2$ and increasing on $(0, +\infty)$, admitting $\theta > 0$ such that $\varphi''(\theta) < 0$ and $\lim_{t\to\infty} \varphi''(t) = 0$. Given $y \in \mathbb{R}^q$, the MAP solution $\hat{x}$ is the (global) minimizer $\hat{x}$ of a posterior energy $E_y$ of the form (7). Since [16], nonconvex functions $\varphi$ are used to produce solutions $\hat{x}$ comprising well smoothted regions and sharp edges [17,27,8,34,14,15,21,10,2]. Our reasoning is based on some results published in [32].

## 5.1 MAP for Smooth Regularization

Assume also that $\varphi'(0) = 0$ and that $\exists \tau > 0$ and $\exists \mathcal{T} \in (\tau, \infty)$ such that $\varphi''(t) \geq 0$ if $t \in [0, \tau]$ and $\varphi''(t) \leq 0$ if $t \geq \tau$, and $\varphi''$ decreases on $(\tau, \mathcal{T})$ and increases on $(\mathcal{T}, \infty)$. These assumptions are satisfied e.g. by (f4)-(f7) in Table 1. Below we write $G$ for the $r \times p$ matrix with rows $G_i$, $1 \le i \le r$ and $e_i$ for the $i$th vector of the canonical basis of $\mathbb{R}^p$.

**Theorem 3.** *Assume that* $\operatorname{rank} G = r$ *and that* $\beta > \frac{2\mu^2 \|A^T A\|}{|\varphi''(\mathcal{T})|}$ *where* $\mu = \max_{1 \le i \le r} \|G^T(GG^T)^{-1} e_i\|$. *Then there are* $\theta_0 \in (\tau, \mathcal{T})$ *and* $\theta_1 \in (\mathcal{T}, \infty)$ *such that for every* $y \in \mathbb{R}^q$, *every minimizer* $\hat{x}$ *of* $E_y$ *satisfies*

$$\text{either} \quad |G_i \hat{x}| \le \theta_0, \quad \text{or} \quad |G_i \hat{x}| \ge \theta_1, \quad \forall i \in \{1, \ldots, r\}.$$

Hence the distribution of the MAP estimator $\hat{X}$ satisfies

$$\Pr\big(\theta_0 < |G_i \hat{X}| < \theta_1\big) = 0, \quad \forall i \in \{1, \ldots, r\}.$$

*The prior model effectively realized by the MAP estimator corresponds to images and signals whose differences are either smaller than* $\theta_0$ *or larger than* $\theta_1$. Nothing similar holds for the prior model $f_X$ since $\Pr\big(\theta_0 < |G_i X| < \theta_1\big) > 0$, $\forall i$.

**Piecewise Gaussian Markov chain in Gaussian noise.** The constants $\theta_0$ and $\theta_1$ have an explicit form for the global minimizers of $E_y$ for (f4) in Table 1 which is the discrete version of the Mumford-Shah regularization [16, 27, 11]. These constants are derived in [32]. For illustration, we repeat 200 times the following experiment. We generate an original $X = x$ of length $p = 300$ whose differences $x_i - x_{i+1}$, are sampled on $[-\gamma, \gamma]$ from $\frac{1}{Z} \exp\big(-\lambda\varphi(t)\big)$ for $\alpha = 1$, $\lambda = 5$ and $\gamma = 15$. The histogram of all original differences $x_i - x_{i+1}$ is shown in Fig. 7(a). For each original $x$ we generate $y = x + \mathcal{N}(0, \sigma^2 I)$ with $\sigma = 4$ and compute the global minimizer $\hat{x}$ of $E_y$ for the true parameters. The histogram of the differences $\hat{x}_i - \hat{x}_{i+1}$ of the MAP solutions in Fig. 7(b) is very different from the prior and does not contain differences with magnitude in $(\theta_0, \theta_1)$.

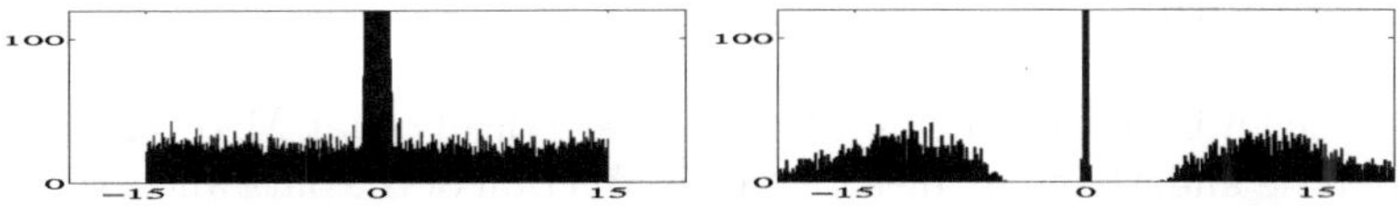


(a) Original differences (zoom) (b) Differences for the MAP (zoom)

**Fig. 7.** Discrete one-dimensional Mumford-Shah model

## 5.2 MAP for Non-smooth Regularization

Beyond the assumptions made in the introduction of section 5, let $\varphi'(0^+) > 0$ and $\varphi''$ be increasing on $(0, \infty)$ with $\varphi''(t) \le 0$, for all $t > 0$. This assumption is satisfied by all nonsmooth at zero nonconvex functions $\varphi$ in Table 1.

**Theorem 4.** *There is a constant* $\mu > 0$ *such that if* $\beta > \frac{2\mu^2 \|A^T A\|}{|\varphi''(0^+)|}$, *then there exists* $\theta_1 > 0$ *such that for every* $y \in \mathbb{R}^q$, *every minimizer* $\hat{x}$ *of* $E_y$ *satisfies*

$$\text{either} \quad |G_i \hat{x}| = 0, \quad \text{or} \quad |G_i \hat{x}| \ge \theta_1, \quad \forall i.$$

It follows that the distribution of the MAP estimator $\hat{X}$ is such that

$$\Pr\Big(|G_i\hat{X}| = 0\Big) > 0 \quad \text{and} \quad \Pr\Big(0 < |G_i\hat{X}| < \theta_1\Big) = 0, \quad \forall i. \tag{17}$$

This was already observed in Fig. 2. If $\{G_i\}$ correspond to the first-order differences between neighboring samples, (17) shows that every minimizer $\hat{x}$ of $E_y$ is composed of constant patches separated by edges higher than $\theta_1$. This is the *effective* prior model realized by the MAP estimator $\hat{X}$. This result is in disagreement with the prior $f_X$ for which $\Pr\big(|G_iX| = 0\big) = 0$ and $\Pr\big(0 < |G_iX| < \theta_1\big) > 0$.

The original $x$ in Fig. 8(a) is a realization of a 100-length Markov chain whose differences $X_i - X_{i+1}$, are i.i.d. on $[-\gamma, \gamma]$ with density $\propto \exp\big(-\lambda\varphi(t)\big)$ where $\varphi$ is (f10) in Table 1 for $\alpha = 10$, $\lambda = 1$ and $\gamma = 4$. The data $Y = X + \mathcal{N}(0, \sigma^2 I)$ for $\sigma = 5$ is shown in Fig. 8(a). The MAP solution $\hat{x}$ in Fig. 8(b), calculated for the true parameters, is constant on many pieces which are separated by large edges. Its visual aspect is totally different from the original $x$.

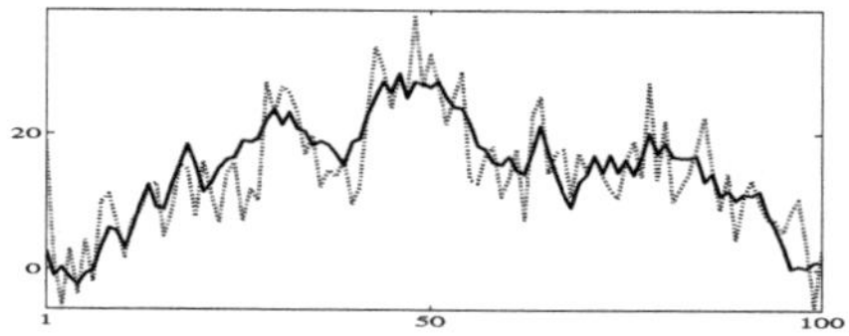


(a) Original $x$ (—), data $y = x + n$ $(\cdots)$

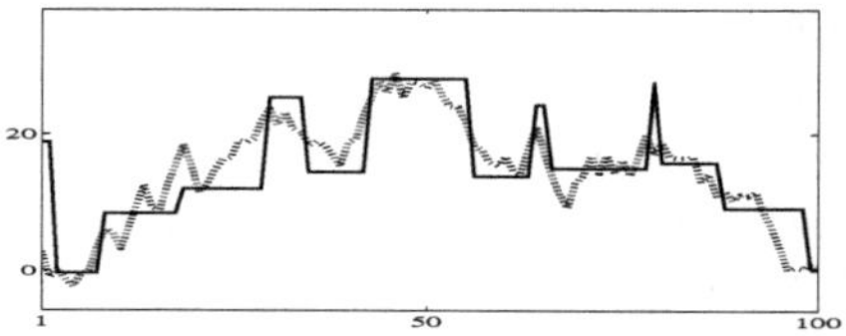


(b) True MAP $\hat{x}$ (—), original $x$ $(\cdots)$

**Fig. 8.** True MAP restoration of a Markov chain with a nonsmooth nonconvex prior energy from data $y = x + n$ corrupted with white Gaussian noise

# 6 Conclusion

We have shown both experimentally and theoretically that MAP estimators do not match the underlying models for the production of the data and for the prior. Instead, based on some analytical properties of the MAP solutions, we partially characterize the models that are *effectively realized by the MAP solutions.*

# References

1. A. Antoniadis and J. Fan, *Regularization of wavelet approximations*, J. of Acoustical Society America, 96 (2001), pp. 939–967.
2. G. Aubert and P. Kornprobst, *Mathematical problems in images processing*, Springer-Verlag, Berlin, 2002.
3. R. G. Aykroyd and P. J. Green, *Global and local priors, and the location of lesions using gamma-camera imagery*, Phil. Trans. R. Soc. Lond. A, 337 (1991).
4. L. Bar, N. Sochen, and N. Kiryati, *Image deblurring in the presence of salt-and-pepper noise*, in Proceeding of 5th International Conference on Scale Space and PDE methods in Computer Vision, ser. LNCS, vol. 3439, 2005, pp. 107–118.

5. M. Belge, M. Kilmer, and E. Miller, *Wavelet domain image restoration with adaptive edge-preserving regularization*, IEEE Trans. on Image Processing, 9 (2000), pp. 597–608.
6. J. Besag, P. Green, D. Higdon, and K. Mengersen, *Bayesian computation and stochastic systems*, Statistical Science, 10 (1995), pp. 3–66.
7. J. E. Besag, *Spatial interaction and the statistical analysis of lattice systems (with discussion)*, J. of the Royal Statistical Society B, 36 (1974), pp. 192–236.
8. ——, *Digital image processing : Towards Bayesian image analysis*, J. of Applied Statistics, 16 (1989), pp. 395–407.
9. J. M. Bioucas-Dias, *Bayesian wavelet-based image deconvolution: A gem algorithm exploiting a class of heavy-tailed priors*, IEEE Trans. on Image Processing, 15 (2006), pp. 937–951.
10. M. Black and A. Rangarajan, *On the unification of line processes, outlier rejection, and robust statistics with applications to early vision*, International J. of Computer Vision, 19 (1996), pp. 57–91.
11. A. Blake and A. Zisserman, *Visual reconstruction*, The MIT Press, Cambridge, 1987.
12. G. Demoment, *Image reconstruction and restoration : Overview of common estimation structure and problems*, IEEE Trans. on Acoustics Speech and Signal Processing, ASSP-37 (1989), pp. 2024–2036.
13. D. L. Donoho and I. M. Johnstone, *Ideal spatial adaptation by wavelet shrinkage*, Biometrika, 81 (1994), pp. 425–455.
14. D. Geman and G. Reynolds, *Constrained restoration and recovery of discontinuities*, IEEE Trans. on Pattern Analysis and Machine Intelligence, PAMI-14 (1992), pp. 367–383.
15. D. Geman and C. Yang, *Nonlinear image recovery with half-quadratic regularization*, IEEE Trans. on Image Processing, IP-4 (1995), pp. 932–946.
16. S. Geman and D. Geman, *Stochastic relaxation, Gibbs distributions, and the Bayesian restoration of images*, IEEE Trans. on Pattern Analysis and Machine Intelligence, PAMI-6 (1984), pp. 721–741.
17. S. Geman and D. E. McClure, *Statistical methods for tomographic image reconstruction*, in Proc. of the 46-th Session of the ISI, Bulletin of the ISI, vol. 52, 1987, pp. 22–26.
18. J. Gutiérrez, F. Ferri, and J. Malo, *Regularization operators for natural images based on nonlinear perception models*, IEEE Trans. on Image Processing, 15 (2006), pp. 189–200.
19. F. Jeng and J. Woods, *Compound Gauss-Markov random fields for image estimation*, IEEE Trans. on Signal Processing, SP-39 (1991), pp. 683–697.
20. R. Kass and L. Wasserman, *The selection of prior distributions by formal rules*, J. of Acoustical Society America, 91 (1996), pp. 1343–1370.
21. S. Li, *Markov Random Field Modeling in Computer Vision*, Springer-Verlag, New York, 1 ed., 1995.
22. J. Marroquin, S. Mitter, and T. Poggio, *Probabilistic solution of ill-posed problems in computational vision*, J. of Acoustical Society America, 82 (1987), pp. 76–89.
23. P. Mathieu, M. Antonini, M. Barlaud, and I. Daubechies, *Image coding using wavelet transform*, IEEE Trans. on Image Processing, 1 (1992), p. 205220.
24. P. Moulin and J. Liu, *Analysis of multiresolution image denoising schemes using generalized gaussian and complexity priors*, IEEE Trans. on Image Processing, 45 (1999), pp. 909–919.
25. P. Muller, B. Vidakovic, and Eds., *Bayesian Inference in Wavelet-Based Models*, Springer-Verlag, New York, 1999.

26. D. Mumford, *The Dawning of the Age of Stochasticity*, AMS, ed. by arnold, atiyah, lax and mazur ed., 2000.
27. D. Mumford and J. Shah, *Boundary detection by minimizing functionals*, in Proc. of the IEEE Int. Conf. on Acoustics, Speech and Signal Processing, 1985, pp. 22–26.
28. M. Nikolova, *Local strong homogeneity of a regularized estimator*, SIAM J. on Applied Mathematics, 61 (2000), pp. 633–658.
29. ——, *Minimizers of cost-functions involving nonsmooth data-fidelity terms. Application to the processing of outliers*, SIAM J. on Numerical Analysis, 40 (2002), pp. 965–994.
30. ——, *A variational approach to remove outliers and impulse noise*, J. of Mathematical Imaging and Vision, 20 (2004).
31. ——, *Weakly constrained minimization. Application to the estimation of images and signals involving constant regions*, J. of Mathematical Imaging and Vision, 21 (2004), pp. 155–175.
32. ——, *Analysis of the recovery of edges in images and signals by minimizing nonconvex regularized least-squares*, SIAM J. on Multiscale Modeling and Simulation, 4 (2005), pp. 960–991.
33. ——, *Model distorsions in Bayesian MAP reconstruction*, Inverse Problems and Imaging, 2 (2007), pp. 399-422.
34. P. Perona and J. Malik, *Scale-space and edge detection using anisotropic diffusion*, IEEE Trans. on Pattern Analysis and Machine Intelligence, PAMI-12 (1990), pp. 629–639.
35. S. Saquib, C. Bouman, and K. Sauer, *ML parameter estimation for Markov random fields, with appications to Bayesian tomography*, IEEE Trans. on Image Processing, 7 (1998), pp. 1029–1044.
36. H. Sidenbladh and M. Black, *Learning the statistics of people in images and video*, J. of Mathematical Imaging and Vision, 54 (2003), pp. 183–209.
37. E. P. Simoncelli, *Bayesian denoising of visual images in the wavelet domain*, Lecture Notes in Statistics, Vol. 41, Springer Verlag: Berlin, 1999.
38. A. Srivastava, A. B. Lee, E. P. Simoncelli, and S.-C. Zhu, *On advances in statistical modeling of natural images*, J. of Mathematical Imaging and Vision, 18 (2003).
39. A. Tarantola, *Inverse problem theory : Methods for data fitting and model parameter estimation*, Elsevier Science Publishers, Amsterdam, 1987.
40. L. Tenorio, *Statistical regularization of inverse problems*, SIAM Review, 43 (2001), pp. 347–366.
41. A. Tikhonov and V. Arsenin, *Solutions of Ill-Posed Problems*, Winston, Washington DC, 1977.
42. C. Vogel, *Computational Methods for Inverse Problems*, Frontiers in Applied Mathematics Series, Number 23, SIAM, 2002.
43. G. Wang, J. Zhang, and G.-W. Pan, *Solution of inverse problems in image processing by wavelet expansion*, IEEE Trans. on Image Processing, 4 (1995), pp. 579–593.

# New Possibilities with Sobolev Active Contours*

Ganesh Sundaramoorthi[1], Anthony Yezzi[1], Andrea C. Mennucci[2], and Guillermo Sapiro[3]

[1] School of Electrical Engineering, Georgia Institute of Technology, Atlanta, USA
[2] Scuola Normale Superiore, Pisa, Italy
[3] Dept. of Electrical & Computer Engineering, University of Minnesota, Minneapolis, USA

**Abstract.** Recently, the Sobolev metric was introduced to define gradient flows of various geometric active contour energies. It was shown that the Sobolev metric out-performs the traditional metric for the same energy in many cases such as for tracking where the coarse scale changes of the contour are important. Some interesting properties of Sobolev gradient flows are that they stabilize certain unstable traditional flows, and the order of the evolution PDEs are reduced when compared with traditional gradient flows of the same energies. In this paper, we explore new possibilities for active contours made possible by Sobolev active contours. The Sobolev method allows one to implement new energy-based active contour models that were not otherwise considered because the traditional minimizing method cannot be used. In particular, we exploit the stabilizing and the order reducing properties of Sobolev gradients to implement the gradient descent of these new energies. We give examples of this class of energies, which include some simple geometric priors and new edge-based energies. We will show that these energies can be quite useful for segmentation and tracking. We will show that the gradient flows using the traditional metric are either ill-posed or numerically difficult to implement, and then show that the flows can be implemented in a stable and numerically feasible manner using the Sobolev gradient.

## 1 Introduction

Active contours [1] is a popular technique for the segmentation problem. Over the years there has been a progression of active contours derived from edge-based energies (*e.g.*, [2,3]), to region-based energies (*e.g.*, [4,5]), to more recently, prior-based energies (*e.g.*, [6,7,8]) and energies incorporating complex geometrical information (*e.g.*, [9,10,11]). The progression from simple to more complicated energies is not only due to a desire to segment more complicated images, but it can also be attributed to the traditional gradient descent technique becoming trapped by (undesirable) local minima of the energy being optimized. Therefore there have been efforts to design optimization schemes that can obtain the global minimum curve of a generic energy. For example, the minimal path technique [12] was designed to find the global minimal solution of the edge-based energy considered in [2,3]. Another technique, called graph cuts [13,14], was designed for minimizing discrete approximations to some active contour energies.

* G. Sundaramoorthi and A. Yezzi were supported by NSF CCR-0133736, NIH/NINDS R01-NS-037747, and Airforce MURI; G. Sapiro was partially supported by NSF, ONR, NGA, DARPA, and the McKnight Foundation.

F. Sgallari, A. Murli, and N. Paragios (Eds.): SSVM 2007, LNCS 4485, pp. 153–164, 2007.

The limitation of these global methods is that they may be applied to only certain types of energies, and therefore gradient descent methods must be used in many cases. Recently, [15,16] have noticed that the gradient of an energy that is used in descent algorithms depends on a metric chosen on the space of curves. This fact has been ignored in previous active contour literature; indeed previous active contours were always derived from the geometric $\mathbb{L}^2$-type ($H^0$) metric. Accordingly, [17,18] have considered new metrics in the space of curves. It was shown that the metric choice affects the path taken to minimize an energy, and that certain local minima of an energy can be avoided by designing an appropriate metric. In particular, Sobolev metrics were considered. It was shown that gradient flows according to Sobolev metrics give smooth global flows, which avoid many local minima of energies that trap the usual $\mathbb{L}^2$ gradient flow. In [19], it was shown that Sobolev active contours move successively from coarse to finer scale motions, and therefore the method is suitable for tracking.

The main purpose of [17,19] was to show advantages of using Sobolev active contours over the traditional active contour based on the same energy. In contrast, in this paper we introduce *new* active contour energies that are quite useful for various segmentation tasks, but *cannot* be minimized with the traditional $\mathbb{L}^2$ active contour (nor other global optimization techniques), and the Sobolev active contour must be used. We show a few examples of these energies, which include simple geometric priors for active contours and new edge-based energies. These new energies fall into two categories: one in which the resulting $\mathbb{L}^2$ flows are not stable, and another in which the traditional gradient flow results in high order PDEs that are numerically difficult to implement using level set or particle based methods. We propose to use Sobolev active contours, which avoid both of these problems.

This paper is meant to illustrate that energies that result in $\mathbb{L}^2$ unstable or high order flows can still be considered for optimization with the Sobolev method (and these energies need not be discarded or adjusted). As such we give a few simple examples of the energies that fall into these categories. Experiments in this paper show the types of behaviors that can be obtained from the simple energies considered, and one can obtain good results on more complex images by combining these results with other energies.

The graph cut method is known to be able to minimize three types of geometric energies: weighted length, flux of a vector field, and weighted area [14]. Some of the energies we consider are non-simple operations (such as division) of the previous energies mentioned, and the technique in [14] does not apply; indeed we are not certain that a graph for the energies we consider can be constructed. Other energies we consider have curvature inside the integral; since edge weights in graph cuts depend on an edge (between a pixel and its neighbor), it is unclear whether a curvature term may be incorporated in the framework since to compute curvature one needs three points. In any case, since graph cut methods do not have sub pixel accuracy, curvature computations would be extremely inaccurate.

In [18], the authors consider various different metrics resulting in 'coherent' gradient flows; indeed they construct flows that favor certain group motions such as affine motions. In the case of the affine group (others are analogous), the flow is formed by re-weighting the affine component of the traditional gradient higher and the component orthogonal (according to the $\mathbb{L}^2$ inner product) lower. For the class of energies that we wish to explore in this paper however, the metrics proposed by [18] based on group motions also suffer from the same problems as the traditional $\mathbb{L}^2$ metric; namely, these flows are either not stable or are high order PDEs and are difficult to implement numerically.

## 2 Some Useful Energies Precluded by $\mathbb{L}^2$

In this section, we introduce three geometric "energies", which can be used as building blocks to produce a variety of other useful energies (to be described in subsequent sections). We then derive the $\mathbb{L}^2$ gradient and show that the gradient descent flow is either ill-posed or very difficult to implement numerically. We then derive the Sobolev gradient flows, and justify that they are well-posed and numerically feasible to implement.

Before we proceed, we introduce the notation used in this paper. A contour will be denoted, $c$; its length is $L$. We define $\overline{\cdot} := L^{-1} \int_c \cdot \, \mathrm{d}s$, where $\mathrm{d}s$ is the arc length measure of $c$. Throughout this paper, we define the Sobolev metric as $\|h\|^2_{\text{Sobolev}} := |\overline{h}|^2 + \lambda \|h'\|^2_{\mathbb{L}^2}$, where $h$ is a perturbation of $c$, $h'$ is the arc parameter derivative, and $\lambda > 0$ is a scaling factor. We note the fact from [17] that the Sobolev gradient can be computed from the $\mathbb{L}^2$ gradient as $\nabla_{\text{Sobolev}} E = K * \nabla_{\mathbb{L}^2} E$, where $*$ is circular convolution and $K$ is a kernel found in [17] whose second derivative exists in a distributional sense.

The first "energy" that we introduce is the following generalization of average weighted length:

$$E(c) = \frac{1}{L} \int_c \phi(c(s)) \, \mathrm{d}s = \overline{\phi}, \tag{1}$$

where $\phi : \mathbb{R}^2 \to \mathbb{R}^k$ where $k \geq 1$. The $\mathbb{L}^2$ gradient of this energy is

$$\nabla_{\mathbb{L}^2} E(c) = \mathcal{N}[\mathcal{N}^T (D\phi)^T - \kappa(\phi - \overline{\phi})^T] \tag{2}$$

where $T$ denotes transpose, and $D$ denotes derivative. Since $\phi - \overline{\phi}$ is not strictly positive, the gradient descent flow has a component that is reverse heat flow on half of the contour, and therefore the $\mathbb{L}^2$ gradient descent is ill-posed. Note that the reverse heat component attempts to increase the length of certain portions of the contour. Since the ill-posedness of the $\mathbb{L}^2$ flow only arises from the length increasing effect, we expect the Sobolev gradient flow to be well-posed. This is because increasing the length of the contour is a well-posed process using the Sobolev gradient; indeed, the Sobolev gradient ascent for length is simply a rescaling of the contour [17]. Computing the Sobolev gradient of (1) we have

$$\nabla_{\text{Sobolev}} E(c) = -\frac{c - \overline{c}}{\lambda L^2} \overline{\phi}^T + K * (D\phi)^T + K' * (c_s \phi^T) \tag{3}$$

Notice that the component, $\mathcal{N} \overline{\phi}^T \kappa$, of the $\mathbb{L}^2$ gradient that caused the ill-posedness has been converted to the first term of the Sobolev gradient (3), which is a stable rescaling of the contour.

Next, we introduce a scaled version of the weighted area, given by the energy

$$E(c) = \frac{1}{L^2} \int_R \phi(x) \, \mathrm{d}A(x) = \frac{A_\phi}{L^2}, \tag{4}$$

where $\phi : \mathbb{R}^2 \to \mathbb{R}$, $R$ is the region enclosed by $c$, and $\mathrm{d}A$ is the area measure in $\mathbb{R}^2$. Similar to the previous energy, the ill-posedness of the $\mathbb{L}^2$ gradient descent flow of (4) is due to the scale factor of $L^{-2}$, which causes a length increasing component in the gradients, and is ill-posed with respect to $\mathbb{L}^2$. Indeed, calculating the gradient, we have

$$\nabla E(c) = \frac{L^2 \nabla A_\phi - 2A_\phi L \nabla L}{L^4} = \frac{A_\phi}{L^2}\left[\frac{\nabla A_\phi}{A_\phi} - 2\frac{\nabla L}{L}\right].$$

Therefore, we see that

$$\nabla_{\text{Sobolev}} E(c) = -\frac{A_\phi}{L^2}\left[L\frac{K * (\phi \mathcal{N})}{A_\phi} + 2\frac{c - \bar{c}}{\lambda L^2}\right], \tag{5}$$

which leads to a well-posed descent (and ascent).

Lastly, we introduce the following generalization of the elastic energy:

$$E(c) = L \int_c \phi(c(s))\kappa^2(s)\,\mathrm{d}s, \tag{6}$$

where $\phi : \mathbb{R}^2 \to \mathbb{R}$, and $\kappa$ is the signed curvature of $c$. The factor of $L$ multiplying the integral makes the energy scale-invariant when $\phi$ is a constant. Note that without the $L$ factor, one can make the elastic energy arbitrarily small by scaling a contour large enough. We will also consider the scale-varying elastic energy without the $L$. These energies have been used in the past for the "curve completion" problem, which is a curve interpolation problem between two points [20,21]. In [21], for the numerical implementation, a discrete version of the energy is minimized with a "shooting" method. One can show that the $\mathbb{L}^2$ gradient of (6) is

$$\nabla_{\mathbb{L}^2} E(c) = -Ec_{ss} + 2L^2\partial_{ss}(\phi c_{ss}) + 3L^2\partial_s(\phi\kappa^2 c_s) + L^2\kappa^2\nabla\phi \tag{7}$$

We note the result of [22], which considers the $\mathbb{L}^2$ gradient descent flow of an energy similar to (6). The author of [22] considers the $\mathbb{L}^2$ gradient descent flow of the energy

$$E(c) = \int_c (\kappa^2(s) + \alpha)\,\mathrm{d}s$$

where $\alpha > 0$. It is proven that an immersed/regular curve evolving under this fourth-order flow stays immersed/regular, and a solution exists for all time. In the case when $\phi$ is a constant, the flow (7) is similar to the flow that is considered in [22], except that $\alpha$ is time varying in (7). For numerical implementation, the fourth order flow (7) is difficult to implement with marker particle methods because of numerical artifacts arising from fourth order differences, and it is even more problematic to implement with level set methods because the flow is not known to have a maximum principle and because of numerical artifacts. These are the reasons for considering the Sobolev gradient:

$$\nabla_{\text{Sobolev}} E = -\frac{E}{\lambda L^2}(\bar{c} - c) + \frac{2}{\lambda}(\overline{(\phi\kappa\mathcal{N})} - \phi\kappa\mathcal{N}) - 3L^2 K' * (\phi\kappa^2 \mathcal{T}) + L^2 K * (\kappa^2 \nabla\phi). \tag{8}$$

The Sobolev flow is second order, although it is a integral PDE. We can bypass the question about a maximum principle for this flow since the local terms have a maximum principle, and we perform extensions in the level set implementation for global terms.

# 3 Geometric Priors for Active Contours

In this section, we introduce some simple geometric shape priors for use in active contour segmentation. As these energies are formed from the energies presented in the previous section, they cannot be minimized with the usual $\mathbb{L}^2$ gradient descent.

## 3.1 Length and Smoothness Priors

In many active contour models, a curvature term, *i.e.*, $\alpha\kappa\mathcal{N}$ (where $\alpha > 0$ is a weight), is added to a data-based curve evolution. The resulting flow will inherit regularizing properties such as smoothing the curve from the addition of this term. If the active contour model is based on minimizing an energy, then adding a curvature term is equivalent to adding a length penalty to the original energy, that is, if $E_{\text{data}}$ is the original energy then the new energy being optimized (w.r.t the traditional $\mathbb{L}^2$ metric) is

$$E(c) = E_{\text{data}}(c) + \alpha L(c). \tag{9}$$

This may be considered as a simple prior in which we assume that the length of the curve is to be shrunk. In general segmentation situations, this assumption may not be applicable. A more general energy incorporating length information, when such prior length information is known, is

$$E(c) = E_{\text{data}}(c) + \alpha(L(c) - L_0)^2, \tag{10}$$

in which it is assumed that length of the target curve is near $L_0$. See [23] (and references within) for related flows where the length of the curve is preserved. Note that this prior allows for increasing or decreasing the length of the curve based on the current length of the curve and the value of $L_0$. The $\mathbb{L}^2$ gradient is

$$\nabla_{\mathbb{L}^2} E(c) = \nabla_{\mathbb{L}^2} E_{\text{data}}(c) - 2\alpha(L - L_0)\kappa\mathcal{N},$$

which leads to an unstable flow if $L - L_0 < 0$. The Sobolev gradient is

$$\nabla_{\text{Sobolev}} E(c) = \nabla_{\text{Sobolev}} E_{\text{data}}(c) + 2\alpha(L - L_0)\frac{c - \overline{c}}{\lambda L},$$

which is stable if the data term is stable.

In active contour works, the goal of adding the length penalty may have been mainly for obtaining the regularizing properties of the resulting flow, even though the energy itself does not favor more regular curves. It is evident that the Sobolev length descent does not regularize the active contour since the flow is a rescaling of the curve. Thus, to introduce smoothness into the Sobolev active contour (and even the $\mathbb{L}^2$ active contour), we introduce the smoothness prior given by the energy,

$$E(c) = E_{\text{data}}(c) + \alpha L(c) \int_c \kappa^2(s)\, \mathrm{d}s. \tag{11}$$

The energy itself favors smoother contours, and we are not relying on the properties of a particular metric for regularity; it is inherent in the energy itself. The factor of $L$ is for scale-invariance (unlike the length descent, this regularizer does not favor shrinking the length of the contour). Using the scale-varying and scale-invariant elastic energies as smoothness measures for active contours is mentioned but not implemented in [24,25].

### 3.2 Centroid and Isoperimetric Priors

We now consider incorporating prior information on the centroid, length, and area of a curve into active contour segmentation. We consider the energy

$$E(c) = E_{\text{data}}(c) + \alpha\|\bar{c} - v\|^2 + \beta(L - L_0)^2 + \gamma(A - A_0)^2 \quad (12)$$

where $\alpha, \beta, \gamma \geq 0$ are weights, $\bar{c}$ is the centroid of the curve $c$, $v \in \mathbb{R}^2$ is the centroid known *a-priori* (see Section 5.2 for an example of how this may be obtained), $L_0$ and $A_0$ are the prior values for the length and area. If detailed information is not known about the length and area, then that part of the energy may be replaced by the energy

$$E(c) = E_{\text{data}}(c) + \alpha\|\bar{c} - v\|^2 + \beta(\rho(c) - \rho_0)^2 \quad (13)$$

where

$$\rho(c) = \frac{A(c)}{L^2(c)} \quad (14)$$

is the isoperimetric ratio, which is a geometric measure of the relative relation between the length and area of a curve. Note that $\rho$ is scale-invariant. It is a well known fact that the isoperimetric ratio is maximized by circles, and the maximum ratio is $1/(4\pi)$. Thus, the prior ratio must be constrained so that $\rho_0 \leq 1/(4\pi)$. Note that a low (near zero) isoperimetric ratio can be obtained by a snake-like shape, and a high ratio implies a shape that looks close to a circle. The isoperimetric ratio is mentioned to be used as a smoothness measure in [24], but this idea is not pursued.

Note that both the $\mathbb{L}^2$ gradient descents for the centroid constraint and the isoperimetric penalties are ill-posed. The isoperimetric ratio is a special case of (4) (when $\phi = 1$), and the constraint gives a gradient of $(\rho - \rho_0)\nabla\rho$, which gives an unstable $\mathbb{L}^2$ gradient descent flow when $\rho > \rho_0$. Note that the centroid is a special case of (1) (when $\phi : \mathbb{R}^2 \to \mathbb{R}^2$ is $\phi(x) = x$). The gradient of the centroid penalty is $\nabla(\bar{c})(\bar{c} - v)$, which gives an $\mathbb{L}^2$ gradient of

$$[(\bar{c} - v) \cdot \mathcal{N} - (c - \bar{c}) \cdot (\bar{c} - v)\kappa]\mathcal{N}$$

using (2). The gradient descent is unstable when $(c - \bar{c}) \cdot (\bar{c} - v) < 0$. The Sobolev gradient using (3) is

$$(\bar{c} - v) + K' * [c_s(c - \bar{c}) \cdot (\bar{c} - v)].$$

One possible use for (12) and (13) is in tracking applications (see Section 5.2).

## 4 New Edge-Based Active Contour Models

The energy for the traditional edge-based technique [2,3] (called geodesic active contours) is

$$E(c) = \int_c \phi(c(s))\, \mathrm{d}s \quad (15)$$

where $\phi : \mathbb{R}^2 \to \mathbb{R}$ is chosen low near edges (a common example is $\phi = 1/(1 + \|\nabla(G * I)\|)$ where $G$ is a Gaussian smoothing filter). There are several undesirable features of this model (even if a perfect edge-map $\phi$ is chosen). The energy has trivial (undesirable) minima and even minima that are not at the edges of the image (see for example [26]). This is in part due to the bias that the model has in preferring shorter length contours, which may not always be beneficial. Therefore, we propose new edge-based models.

### 4.1 Non-shrinking Edge-Based Model

We propose to minimize the following non-length shrinking edge-based energy:

$$E(c) = \int_c \phi(c(s)) \left(L^{-1} + \alpha L \kappa^2(s)\right) \,\mathrm{d}s, \tag{16}$$

where $\alpha \geq 0$, which we claim alleviates some of the undesirable properties of (15). An energy, which is similar to (16) (except for the factor of $L$ on the curvature term), is considered by [27], but a discrete version of the energy is used for implementation. The first term, $\frac{1}{L}\int_c \phi \,\mathrm{d}s$ (*i.e.*, (16) when $\alpha = 0$), is the same as the energy used for the geodesic active contour model, but there is a scale factor of $1/L$. This removes the length shrinking effect of (15) in descent flows; in particular if there are no edges ($\phi$ is constant), then a descent flow will not shrink the contour. The $\mathbb{L}^2$ gradient of the first term (when $\alpha = 0$ in (16)) as noted in (2) is $-L(\phi - \overline{\phi})\kappa\mathcal{N} + L(\nabla\phi \cdot \mathcal{N})\mathcal{N}$, which is zero when the contour is aligned on true edges of the image (note that this may not be the case with the geodesic active contour model). The flow is stable with respect to the Sobolev metric, but not with respect to $\mathbb{L}^2$.

Dividing the energy (15) by $L$, as in the first term of (16), loses regularizing effects of the original flow, and it is possible that the contour can become non-smooth from irrelevant noise. This observation is the reason for the second term of (16). The second term, $L\int_c \phi\kappa^2 \,\mathrm{d}s$, is an image dependent version of the scale-invariant elastic energy. This term favors smooth contours, but smoothness is relaxed in the presence of edges, which are determined by $\phi$. The factor of $L$ makes the energy scale-invariant when $\phi$ is constant; therefore, a descent flow will not increase or decrease the length of the contour unless these behaviors make the curvature smaller or make the contour align along the edges. The reason for not considering this term alone is for the following reason. Suppose we are considering open curves with two endpoints fixed. Regardless of the $\phi$ that is chosen, the minimum of this term is always zero, and it is minimized by a straight line (the curvature is zero). For closed contours, we have observed in the numerical implementation that the contour sticks to isolated points where there is an edge of the image, and the converged contour is a straight line between these points (even if there is no edge along the line). Thus, the contour looks polygon-like. Even though the $\kappa = +\infty$ at vertices of polygons, this is not true numerically where $\kappa$ is finite. Therefore, in a numerical implementation, the second term of (16) is not useful by itself.

### 4.2 Increasing Weighted Length

Instead of a non-shrinking edge-based model, if we have prior information that the length of the curve should increase, *e.g.*, the initial curve is within the object of interest, then one may want to *maximize* the following energy:

$$E(c) = \int_c \phi(c(s))\,\mathrm{d}s - \alpha \int_c \kappa^2(s)\,\mathrm{d}s \tag{17}$$

where $\alpha \geq 0$, and $\phi$, contrary to the geodesic active contour model, is designed to be large near edges (one example is choosing $\phi = \|\nabla I\|$). The first term of the energy is weighted length, and therefore this term favors increasing the length of the curve while stopping near edges. Considering only the first term ((17) when $\alpha = 0$), since the length of the curve is being increased, it is likely that when the curve has converged on a coarse scale, fine details due to noise become detected and the curve becomes rough, thereby further increasing length. Therefore, we add a regularizer, which is the second term of (17), to the weighted length. Note that we propose to use the scale-varying elastic energy, which in addition to regularity, gives an effect of increasing the length of curve, which is beneficial based on the prior assumption.

The $\mathbb{L}^2$ gradient ascent of the weighted length term results in one term that is $-\phi\kappa\mathcal{N}$, which makes the length of the curve increase and is unstable. If $\alpha > 0$, then the $\mathbb{L}^2$ flow of (17) may become well-posed since this results in higher order regularity terms, but the elastic energy has its own problems using the $\mathbb{L}^2$ gradient flow. Therefore, we use the Sobolev flow.

# 5 Experiments

## 5.1 Regularity of Sobolev Active Contour

In this experiment, we show a case when the scale-invariant elastic regularity term (11) is more beneficial than the using the traditional length penalty (9). Note that the elastic regularizer does not generally have a length shrinking effect, but keeps the contour regular. This length shrinking effect may have a detrimental effect as shown in Fig. 1. Note that the length penalty restricts the curve from moving into the groves between the fingers. The elastic regularity term, on the other hand, has no such restriction, and makes the curve more smooth and rounded.

## 5.2 Tracking with Centroid/ Isoperimetric Prior

In this experiment, we illustrate one possible application of the energy (13) in tracking a man through an occlusion. For the data-based term in (13), we use the Mumford-Shah energy [4]. The prior information on the centroid and isoperimetric ratio can be obtained through a filtering process (indeed, we assume a constant acceleration model of both quantities). We use the tracking framework of [28] for both simulations in Fig. 2. The top row shows the result using the framework of [28] without the use of prior centroid and isoperimetric information; the bottom row incorporates this prior information. Notice that the prior information on the centroid keeps the contour moving through the occlusion, while the isoperimetric ratio (and because we are using Sobolev active contours) keeps the shape constrained.

## 5.3 Edge Detection with Non-shrinking Model

In this experiment, we demonstrate that the traditional edge-based geodesic active contour model has an arbitrary length shrinking effect that causes the contour to pass over

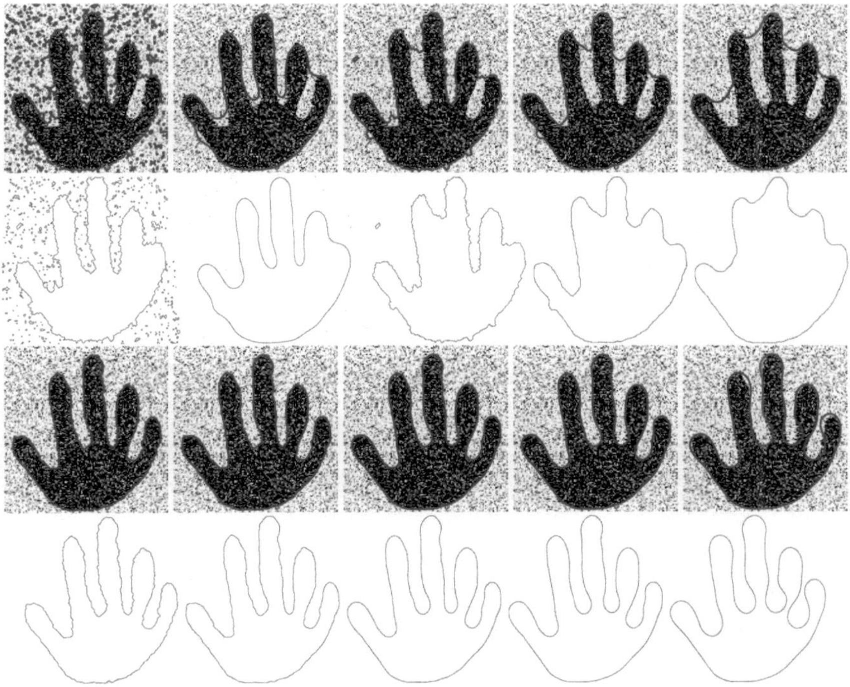

**Fig. 1.** $\mathbb{L}^2$ regularization (top two rows). Left to right: $\alpha = 1000$, $\alpha = 1000$ followed by curvature smoothing to remove the noise (least number of iterations to remove noise), $\alpha = 10000, 50000, 90000$. The image-based term is Chan-Vese. Sobolev elastic regularization (bottom two rows). Left to right: $\alpha = 0, 0.1, 5, 10, 25$. The second and fourth row show the same result as the row above them, but the image is removed for visibility.

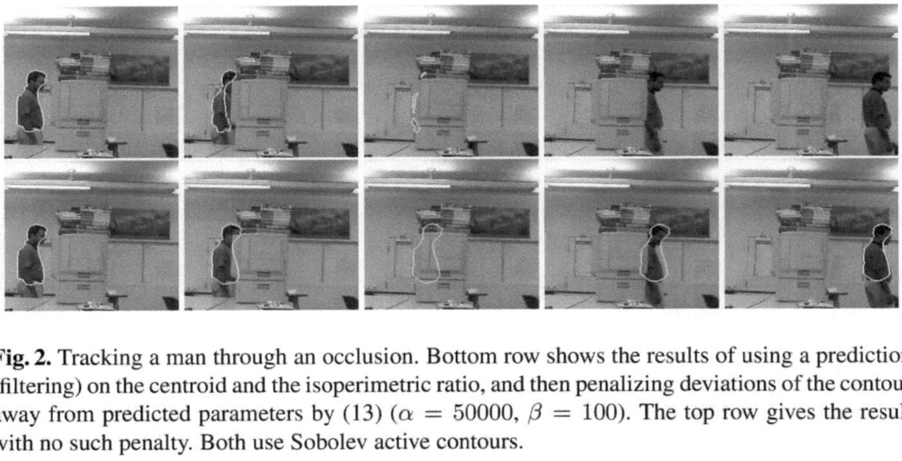

**Fig. 2.** Tracking a man through an occlusion. Bottom row shows the results of using a prediction (filtering) on the centroid and the isoperimetric ratio, and then penalizing deviations of the contour away from predicted parameters by (13) ($\alpha = 50000$, $\beta = 100$). The top row gives the result with no such penalty. Both use Sobolev active contours.

some meaningful edges. We show that the non-shrinking edge-based model (16) can help correct this behavior. We use edge-map, $1/(1+\phi)$, where

$$\phi(x) = \frac{1}{|B_r|}\int_{B_r(x)} (I(y) - \overline{I}_r(x))^2 \,\mathrm{d}A(y), \text{ where } \overline{I}_r(x) = \frac{1}{|B_r|}\int_{B_r(x)} I(y)\,\mathrm{d}A(y), \tag{18}$$

$B_r(x) = \{y \in \mathbb{R}^2 : \|y - x\| \leq r\}$, and $|B_r|$ denotes the area of $B_r$.

In Fig. 3, we segment a cyst image using various initializations. Notice that the contour with the traditional edge-based energy (using the $\mathbb{L}^2$ or the Sobolev descents) consistently passes over the edge on the right side of the cyst. The non-shrinking model consistently captures the correct segmentation.

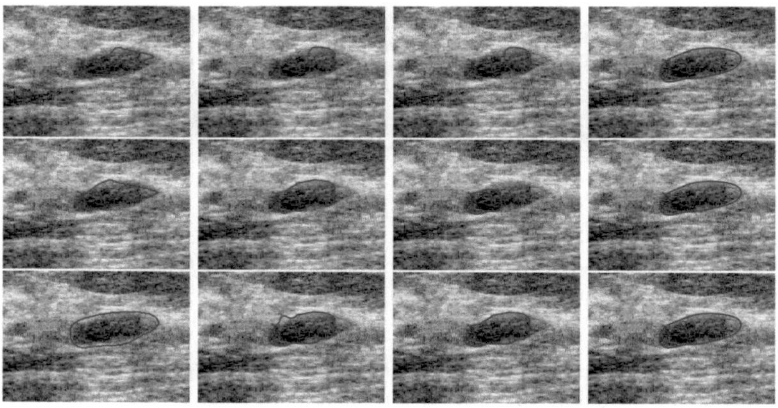

**Fig. 3.** Segmentation of cyst image with three different initializations (first image in each row). Converged results for the (15) and $\mathbb{L}^2$ active contour (second image), (15) with the Sobolev active contour (third image), and the energy (16) (last image).

## 5.4 Edge Detection by Increasing Weighted Length

In this experiment, we apply the weighted length increasing energy (17) to vessel segmentation. We show the results of using the traditional edge-based technique with a balloon; that is, we show results of using the $\mathbb{L}^2$ gradient descent for the energy

$$E(c) = \int_c \phi(c(s))\,\mathrm{d}s - \alpha \int_R \phi\,\mathrm{d}A. \tag{19}$$

We use (18) as the edge-map for the weighted length increasing flow. The edge-map for (19) is $1/(1+\phi)$ where $\phi$ is given in (18).

In the case of vessel segmentation, it beneficial to increase the length of the initial contour more so than area. Since a vessel is characterized as a long, thin structure, a balloon term will fail to capture the global geometry of the vessel. This is demonstrated in Fig. 4: a small weight on the balloon term results in the flow capturing local features close to the initial contour; larger weights on the balloon makes the contour balloon out

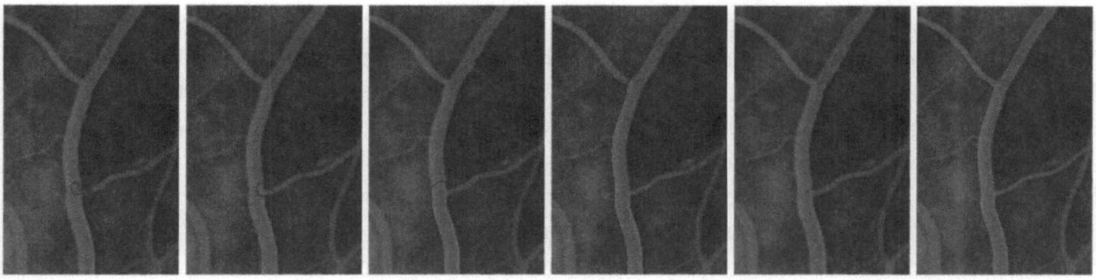

**Fig. 4.** Left to right: initial contour, minimizing (19) $\alpha = 0.2, 0.25, 0.4$ using $\mathbb{L}^2$, and increasing weighted length (17) $\alpha = 0.1$ using Sobolev (all images show converged contour). The contour expands to enclose the entire image (fifth image).

to capture the entire image. Note the weighted length maximizing flow does not pass the walls of the vessel since that does not increase the *length* (although it does increase area) of the contour, and is therefore able to capture the vessel.

## 6 Conclusion

We have demonstrated that the Sobolev gradient method allows one to consider active contour energies that were not considered in the past because the gradient method using the traditional metric cannot be used. In particular, we have given examples of energies that result in $\mathbb{L}^2$ gradients that are ill-posed or high order PDEs (and hence numerically difficult to implement). These energies, as we have shown, result in Sobolev gradient flows that are both well-posed and numerically simple to implement. The experiments have shown potential uses for some energies introduced in segmentation and tracking applications.

## References


1. Kass, M., Witkin, A., Terzopoulos, D.: Snakes: Active contour models. International Journal of Computer Vision **1** (1987) 321–331
2. Caselles, V., Kimmel, R., Sapiro, G.: Geodesic active contours. In: Proceedings of the IEEE Int. Conf. on Computer Vision, Cambridge, MA, USA (1995) 694–699
3. Kichenassamy, S., Kumar, A., Olver, P., Tannenbaum, A., Yezzi, A.: Gradient flows and geometric active contour models. In: Proceedings of the IEEE Int. Conf. on Computer Vision. (1995) 810–815
4. Mumford, D., Shah, J.: Optimal approximations by piecewise smooth functions and associated variational problems. Comm. Pure Appl. Math. **42** (1989) 577–685
5. Chan, T., Vese, L.: Active contours without edges. IEEE Transactions on Image Processing **10**(2) (2001) 266–277
6. Chen, Y., Tagare, H., Thiruvenkadam, S., Huang, F., Wilson, D., Gopinath, K., Briggs, R., Geiser, E.: Using prior shapes in geometric active contours in a variational framework. International Journal of Computer Vision **50**(3) (2002) 315–328
7. Rousson, M., Paragios, N.: Shape priors for level set representations. In: Proc. European Conf. Computer Vision. Volume 2. (2002) 78–93

8. Cremers, D., Soatto, S.: A pseudo distance for shape priors in level set segmentation. In: IEEE Int. Workshop on Variational, Geometric and Level Set Methods. (2003) 169–176
9. Kim, J., Fisher, J., Yezzi, A., Cetin, M., Willsky, A.: Nonparametric methods for image processing using information theory and curve evolution. In: IEEE International Conference on Image Processing. Volume 3. (2002) 797–800
10. Rochery, M., Jermyn, I., Zerubia, J.: Higher order active contours and their application to the detection of line networks in satellite imagery. In: IEEE Workshop on VLSM. (2003)
11. Sundaramoorthi, G., Yezzi, A.J.: More-than-topology-preserving flows for active contours and polygons. In: ICCV. (2005) 1276–1283
12. Cohen, L.D., Kimmel, R.: Global minimum for active contour models: A minimal path approach. In: CVPR. (1996) 666–673
13. Boykov, Y., Jolly, M.P.: Interactive graph cuts for optimal boundary and region segmentation of objects in n-d images. In: ICCV. (2001) 105–112
14. Kolmogorov, V., Boykov, Y.: What metrics can be approximated by geo-cuts, or global optimization of length/area and flux. In: ICCV. (2005) 564–571
15. Michor, P., Mumford, D.: Riemannian geometries on the space of plane curves. ESI Preprint 1425, arXiv:math.DG/0312384 (2003)
16. Yezzi, A.J., Mennucci, A.: Conformal metrics and true "gradient flows" for curves. In: ICCV. (2005) 913–919
17. Sundaramoorthi, G., Yezzi, A., Mennucci, A.: Sobolev active contours. In: VLSM. (2005) 109–120
18. Charpiat, G., Keriven, R., Pons, J., Faugeras, O.: Designing spatially coherent minimizing flows for variational problems based on active contours. In: ICCV. (2005)
19. Sundaramoorthi, G., Jackson, J.D., Yezzi, A.J., Mennucci, A.: Tracking with sobolev active contours. In: CVPR (1). (2006) 674–680
20. Horn, B.K.P.: The curve of least energy. ACM Transactions on Mathematical Software **9**(4) (1983) 441–460
21. Bruckstein, A.M., Netravali, A.N.: On minimal energy trajectories. Comput. Vision Graph. Image Process. **49**(3) (1990) 283–296
22. Polden, A.: Curves and Surfaces of Least Total Curvature and Fourth-Order Flows. PhD thesis, Mathematisches Institut Unversitat Tubingen, Germany (1996)
23. Sapiro, G., Tannenbaum, A.: Area and length preserving geometric invariant scale-spaces. IEEE Trans. Pattern Anal. Mach. Intell. **17**(1) (1995) 67–72
24. Delingette, H.: On smoothness measures of active contours and surfaces. In: VLSM '01: Proceedings of the IEEE Workshop on Variational and Level Set Methods (VLSM'01), Washington, DC, USA, IEEE Computer Society (2001) 43
25. Brook, A., Bruckstein, A.M., Kimmel, R.: On similarity-invariant fairness measures. In: Scale-Space. (2005) 456–467
26. Ma, T., Tagare, H.: Consistency and stability of active contours with euclidean and non-euclidean arc lengths. IEEE Transactions on Image Processing **8**(11) (1999) 1549–1559
27. Fua, P., Leclerc, Y.G.: Model driven edge detection. Mach. Vision Appl. **3**(1) (1990) 45–56
28. Jackson, J., Yezzi, A., Soatto, S.: Tracking deformable moving objects under severe occulsions. In: IEEE Conference on Decision and Control. (2004)

# A Geometric-Functional-Based Image Segmentation and Inpainting

Vladimir Kluzner[1], Gershon Wolansky[2], and Yehoshua Y. Zeevi[3]

[1] Mathematics Department, Technion
mavladim@tx.technion.ac.il
[2] Mathematics Department, Technion, gershonw
@math.technion.ac.il
[3] Department of Electrical Engineering, Technion
zeevi@ee.technion.ac.il

**Abstract.** The Mumford-Shah functional minimization, and related algorithms for image segmentation, involve a tradeoff between a two-dimensional image structure and one-dimensional parametric curves (contours) that surround objects or distinct regions in the image.

We propose an alternative functional that is independent of parameterization; it is a geometric functional which is given in terms of the geometry of surfaces representing the data and image in a feature space. The $\Gamma$-convergence technique is combined with the minimal surface theory in order to yield a global generalization of the Mumford-Shah segmentation functional.

## 1 Introduction

Let $g(x, y)$ be the intensity of the light signal impinging on a planar image domain $B$ at a point $(x, y)$. The image $g(x, y)$ is expected to be discontinuous along the edges of the objects. The definition of the segmentation depends on whether one approaches the problem at the level of the image as a whole or, alternatively, considers the image as a collection of edge fragments. In the first case it is natural to consider the partitioning of the image into smaller structures. In the second case it becomes more natural to consider the problem of grouping of elements into larger structures. In both cases, the following questions arise: (i) What exactly is the goal of the segmentation process? (ii) Is segmentation feasible? These questions are important for understanding of the above process. Without a clear conception of the task and its requirements no satisfactory progress in this area can be made.

The above dichotomy into local vs. global, and related heuristic approaches, were later circumvented by the variational approach to segmentation, adopted by us and further developed in this paper.

Morel and Solimini [6] showed that any heuristic segmentation method may be translated into a variational one. Variational formulations summarize all criteria concerning a set of edges $K$ in a single real-valued functional $F(K)$, i.e. to any set of edges or "segmentation" $K$ is associated a value $F(K)$ which states how

F. Sgallari, A. Murli, and N. Paragios (Eds.): SSVM 2007, LNCS 4485, pp. 165–177, 2007.

"good" the segmentation is. $F$ is defined in such a way that $F(K)$ is better the lower it is. This fact lends itself to an obvious comparison principle: given two segmentations, it must determine which is better. In other words, a variational method implicitly orders all segmentations. Any order can be quantified by $F(K)$ such that $F(K) > F(K^{'})$ if $K^{'}$ is better than $K$.

The present study merges the $\Gamma$-convergence technique (see [9]) and the minimal surface theory (see [3]) to yield a global generalization of the Mumford-Shah segmentation functional. We then apply the functional model to segmentation and inpainting problems. Proofs of theorems, propositions and lemmas are presented elsewhere [5].

## 2 Measure-Based Metric Function

Mumford and Shah [7] proposed the functional

$$F_{MS}(u,K) = \int\limits_{B\setminus K} |\nabla u|^2 + |K| + \alpha \int\limits_{B} |u-g|^2, \qquad (1)$$

where $K \subset B$ is a one dimensional set which represents the edges, and $|K|$ is the length of this set (understood as the one-dimensional Hausdorff measure of $K$). A minimizer $\{K, u\}$ of $F_{MS}$ thus produces the required image $u$ as well as the edges $K$.

We propose an alternative functional which is independent of parameterization, i.e a geometric functional which may be given in terms of the geometry of surfaces representing the data and image in the feature space. Considering the image $u$ as a two-dimensional surface, we shall replace the first two terms of (1) by the area of this surface. This allows sharp discontinuities (edges) of the image in the form of surface folding. This idea is not new (see, e.g, [11]). However, to make the third term of (1) fit the geometrical description, we must replace it by another metric $\mathcal{D}(g,u)$ representing the distance between the two surfaces. Note that the last term connects locally the image $u(b)$ for each $b \in B$ to the data $g$ at *the same* point $b$. It implicitly assumes that there is a one-to-one correspondence between points of the data to points of the image.

The proposed modified functional allows us to replace the deterministic data $g$ by a random one. For this we replace $g(b)$ by a measure $\mu_b(dy)db$, where $y$ is a parameterization of the feature fiber. Further, to different pixels there may correspond different amount of data, so $\int \mu_b(dy)db \leq 1$ for each pixel $b$ where strong inequality and even zero value may not be excluded for some of the pixels. Our objective is, essentially, to define a metric $\mathcal{D}(\mu, U)$.

Let us define first the feature space. Let $\mathbf{Y}$ be a set representing the possible data at a single pixel $b \in B$. It may be considered as a real number (the brightness) or a vector (if several color channels are present or, for example, a Gabor-wavelet filter). We shall assume here that $\mathbf{Y} = \mathbb{R}^m$. The feature space is, then, defined as the cylinder $E := B \times \mathbf{Y} \subset \mathbb{R}^{m+2}$. The data $g$ is represented as a measure $\mu$ supported in $E$.

To define an image $u$, let us consider the two-dimensional unit disc $S : \{|s| < 1\}$ diffeomorphic to $B$. We represent the image as a mapping $U : S \rightarrow E$ such that $U(\partial S) \subset \mathbf{C} := \partial B \times \mathbf{Y}$ is projected onto $\partial B$. Coordinates in the feature space are described by $z := \{b, y\}$.

The feature space $E$ is endowed with a metric

$$dz^2 = \gamma db^2 + dy^2 := \sum_{i,j=1}^{m+2} h_{i,j} dz_i dz_j,$$

where $db^2 = db_1^2 + db_2^2$, $dy^2 = \sum_{i=1}^{m} dy_i^2$ are the Euclidean metric in $\mathbb{R}^2$ and $\mathbb{R}^m$, respectively, while $\gamma > 0$ represents the relation between the geometric (pixel-domain) and feature metrics. With this setting, the embedding $U(S)$ into $E$ is endowed with the induced metric

$$\Gamma_{i,j}(U) = U_{s_i}^T h U_{s_j} \quad \text{for} \quad i, j = 1, 2.$$

Setting $|\Gamma(U)| := \left|\Gamma_{1,1}(U)\Gamma_{2,2}(U) - \Gamma_{1,2}^2(U)\right|^{1/2}$, the surface area of $U(S)$ is

$$A(U) = \int_S |\Gamma(U)(s)| ds. \tag{2}$$

Let us replace the first two terms of (1) by (2), and the last term by a distance between the embedded surface $U(S)$ and the data measure $\mu$:

$$F(U) = A(U) + \alpha \mathcal{D}^2(\mu, U). \tag{3}$$

By our convention, the distance $\mathcal{D}$ should only depend on the image $U(S)$ and not on a particular parameterization. With this assumption, we replace $A(U)$ by the quadratic form

$$\widetilde{A}(U) = \frac{1}{2} \int_S Tr\left(\nabla U(s)^T h \nabla U(s)\right) ds, \tag{4}$$

where $Tr(\cdot)$ is a trace of a given matrix. Note that

$$\widetilde{A}(U) \geq A(U), \tag{5}$$

which reduces to an equality if the parameterization of $U$ is conformal,

$$\Gamma_{1,1}(U) = \Gamma_{2,2}(U), \quad \Gamma_{1,2}(U) = \Gamma_{2,1}(U) = 0.$$

We replace $F(U)$ by

$$\widetilde{F}(U) = \widetilde{A}(U) + \alpha \mathcal{D}^2(\mu, U) \tag{6}$$

and obtain by (5)

$$\widetilde{F}(U) \geq F(U),$$

and equality if the embedded $U$ is a conformal mapping from $S$ to $E$.

For parametric representation of an image used in this section we cannot exclude the possibility that the optimal image is folded over the pixel space $B$. If this is the case, then some pixels of the image over $B$ may have multiple values.

Our first object is to define a distance between two positive measures $\mu$ and $\nu$ supported in $E$, where $\nu$ represents the image and $\mu$ the data. Let

$$d(A,B) = \sqrt{\sup_{v \in B} \inf_{u \in A} |u-v|^2}$$

and

$$D(\mu,\nu) = d(supp(\mu), supp(\nu)). \tag{7}$$

**Lemma 1.** *For fixed, compactly supported $\mu$ in $E$, $D(\mu,\nu)$ is lower semi-continuous with respect to $\nu$ under the topology of weak $C^*$ convergence.*

We now replace the measure $\nu$ by a mapping $U$. Let $l$ be a measure on the parameter space $S$ (say, the uniform Lebesgue measure). We denote the measure $\nu$ associated with an embedding $U \in \mathbb{H}^1(S,E)$ by $\nu_U$ and define it by the pullback $\nu_U(\sigma) = l\left(U^{-1}(\sigma \cap U(S))\right)$ for every Borel measurable set $\sigma \subset E$. In terms of its action on a test function $\phi \in C_0(E)$ the above measure is defined in the following way:

$$\langle \nu_U, \phi \rangle = \int_S \phi(U(s))ds. \tag{8}$$

We conclude that (8) extends to any $U \in \mathbb{H}^1(S,E)$ and define

$$\mathcal{D}(\mu,U) := D(\mu,\nu_U).$$

From Lemma 1 we obtain:

**Lemma 2.** *The metric $\mathcal{D}(\mu,U)$ is weakly lower semi-continuous with respect to $U$ in the $\mathbb{H}^1$ topology.*

We consider now the existence of the minimizer of functional $\widetilde{F}$. Let $U : S \to E$ be written in the form $U := \{U^B, U^{\mathbf{Y}}\}$ where $U^B : S \to B$ and $U^{\mathbf{Y}} : S \to \mathbf{Y}$. Consider the functional (6) written as:

$$\widetilde{F}(U^B,U^{\mathbf{Y}}) = \frac{1}{2}\int_S \left[\gamma Tr\left(\nabla U^B(s)^T \nabla U^B(s)\right)\right.$$
$$\left. + Tr\left(\nabla U^{\mathbf{Y}}(s)^T \nabla U^{\mathbf{Y}}(s)\right)\right] ds + \alpha \mathcal{D}^2(\mu,U).$$

This representation allows us to define the domain of $U^{\mathbf{Y}}$ as $\mathbb{H}^1(S,\mathbf{Y})$ without any boundary condition:

$$\mathcal{DOM}_{\mathbf{Y}} := \{U^{\mathbf{Y}} \in \mathbb{H}^1(S,\mathbf{Y})\}.$$

The mapping $U^B$, on the other hand, must map $S$ onto $B$ such that $U^B|_{C_S} : C_S \to C_B := \partial B$ is a homeomorphism ($C_S := \partial S$ is the "frame" of the image).

This corresponds to treating $U_B$ as in a Plateau problem with *non-free* boundary condition. Following the common wisdom, we need the technical 3-point condition (see [3], p. 235) and set:

$$\begin{aligned}&\mathcal{DOM}_B := \\ &\{U^B \in \mathbb{H}^1(S,B) \cap C^0(C_S, C_B) \,|U^B|_{C_S} : C_S \to C_B \textit{ is a} \\ &\textit{homeomorphism such that } \; U^B(s_i) = \zeta_i \textit{ for } i = 1,2,3, \\ &\textit{where } s_1, s_2, s_3 \textit{ are three distinct points on } C_S, \textit{ while } \zeta_1, \zeta_2, \zeta_3 \\ &\textit{are three distinct points on } C_B \textit{ with the same ordering}\}\end{aligned}$$

**Theorem 1.** *If $\mu$ has a compact support in $E$ then a minimizer of $\widetilde{F}$ is attained in the domain $\mathcal{DOM} := \mathcal{DOM}_B \times \mathcal{DOM}_{\mathbf{Y}}$. Moreover, any minimizer $\overline{U}$ is a minimizer of $F$ (3) as well.*

We use a powerful approach that has appeared in the mathematical theory of approximation of functionals via $\Gamma$-convergence. The idea is to approximate the functional (6) with the lack of regularity (due to the metric term $\mathcal{D}(\mu, U)$) by a series of different, parameter dependent functionals, that are expected to be more regular. In addition, we expect from convergence of functionals to imply the convergence of minimizers.

Given $\beta > 0$, set

$$D_\beta(\mu,\nu) = \sqrt{\beta \ln \left[ \int\limits_E \frac{\nu(dv)}{\int\limits_E e^{-|u-v|^2/\beta} \mu(du)} \right]}.$$

**Lemma 3.** *$D(\mu, \cdot)$ is the $\Gamma-$limit of $D_\beta(\mu,\cdot)$ as $\beta \to 0$ from above, if considered as a functional of the second argument. That is:*

$$\lim_{\beta \to 0; \beta > 0} D_\beta(\mu,\nu) = D(\mu,\nu),$$

$$\liminf_{n \to \infty} D_{\beta_n}(\mu,\nu_n) \geq D(\mu,\nu),$$

*where $\beta_n \to 0$ from above and $\nu_n \rightharpoonup \nu$ in $C^*$.*

We introduce now an implementation of the relaxed distance function $D_\beta$ and define a corresponding version of the metric $\mathcal{D}_\beta(\mu, U)$, $U \in \mathbb{H}^1(S,E)$. This enables us to introduce the relaxation of $\widetilde{F}$ (6) by

$$\widetilde{F}_\beta(U) = \frac{1}{2} \int\limits_S Tr\left(\nabla U(s)^T h \nabla U(s)\right) ds + \alpha \mathcal{D}^2_\beta(\mu, U). \tag{9}$$

We set the $\beta-$distance between a mapping $U : S \to E$ to a measure $\mu$ as

$$\mathcal{D}_\beta(\mu, U) := D_\beta(\mu, \nu_U) = \sqrt{\beta \ln \left[ \int_S \frac{ds}{\int_E e^{-|U(s)-z|^2/\beta} \mu(dz)} \right]}. \tag{10}$$

Thus

$$\mathcal{D}^2_\beta(\mu, U) = \beta \ln \left[ \int_S \Xi_\beta(U)(s) ds \right],$$

where

$$\Xi_\beta(U)(s) = \frac{1}{\int_E e^{-|U(s)-z|^2/\beta} \mu(dz)}.$$

From Lemma 3 we obtain:

**Corollary 1.** *$\mathcal{D}(\mu, \cdot)$ is the $\Gamma-$limit of $\mathcal{D}_\beta(\mu, \cdot)$ as $\beta \to 0$ from above.*

Eventually, we obtain:

**Theorem 2.** *For any $\beta > 0$, the functional (9) attains a minimizer $U_\beta \in \mathcal{DOM}$. The sequence of minimizers $\{U_\beta\}$ is bounded in $\mathbb{H}^1(S, E)$. Any weak limit $U$ in $\mathbb{H}^1$, s.t.*

$$U_\beta \overset{\beta \to 0}{\rightharpoonup} U \in \mathcal{DOM},$$

*is a minimizer of $F$ (3) and $\widetilde{F}$, and is a conformal embedding of $S$ into $E$.*

The functional $\widetilde{F}_\beta$ equipped with the relaxed distance $\mathcal{D}_\beta$ is not parameter independent, and its minimizers are not conformal mappings. However, it converges to parameter independent functional $\widetilde{F}$ as $\beta \to 0$.

## 3 Non-parametric Representation

The parametric model is presented in [5]. We therefore proceed to present the non-parametric representation. Let consider a non-parametric representation of the image $U$. Here we identify $S$ with $B$ and set $U^B$ as identity map. Thus, the image is given in terms of a graph $U^{\mathbf{Y}} := f : B \to \mathbf{Y}$. Considering the image $U$ in terms of a graph of function $f$ does not allow existence of edges in form of the surface folding. In case of non-parametric formulation we cannot use the majorant area functional $\widetilde{A}(U)$ as defined in (4). The area functional is then given as in (2):

$$A(U) = \int_S |\Gamma(U)(s)| ds = \int_B \sqrt{\gamma^2 + \gamma |\nabla f|^2} db.$$

We also separate the scale $\gamma$ into $\gamma_1$ for the area term and $\gamma_2$ for the metric. In order to unify the limits $\gamma_1 \to 0$ and $\gamma_1 \to \infty$ we normalize the area term by $\frac{1+\sqrt{\gamma_1}}{\sqrt{\gamma_1}}$. Thus, the corresponding functional is

$$\mathbf{F}^{\beta}_{\gamma_1,\gamma_2}(f) = (1+\sqrt{\gamma_1}) \int\limits_B \sqrt{\gamma_1 + |\nabla f|^2} db + \alpha \mathcal{D}^2_{\beta,\gamma_2}(\mu, f).$$

We define now the relaxed metric term $\mathcal{D}_{\beta,\gamma_2}(\mu, f)$. For non-parametric formulation we represent the data $\mu(dz)$ as $\mu_b(dy)db$. Under the above assumption, the parameter dependent metric (10) attributes the measure $db$ to the image $f$, and it is written as

$$\mathcal{D}^2_{\beta,\gamma_2}(\mu, f) = \beta \ln \left[ \int\limits_B \Psi^f_{\beta,\gamma_2}(b) db \right],$$

where

$$\Psi^f_{\beta,\gamma_2}(b) = \frac{1}{\int\limits_{\mathbf{Y}} \int\limits_B e^{-[|f(b)-y|^2+\gamma_2|b-b'|^2]/\beta} db' \mu_{b'}(dy)}. \tag{11}$$

In addition, the following analysis is performed for deterministic data, thus $\Psi^f_{\beta,\gamma_2}(b)$ in (11) is given as

$$\Psi^f_{\beta,\gamma_2}(b) = \frac{1}{\int\limits_B e^{-[|f(b)-g(b')|^2+\gamma_2|b-b'|^2]/\beta} db'}. \tag{12}$$

Finally, the corresponding functional is:

$$\mathbf{F}^{\beta}_{\gamma_1,\gamma_2}(f) = (1+\sqrt{\gamma_1}) \int\limits_B \sqrt{\gamma_1 + |\nabla f|^2} db + \alpha\beta \ln \left[ \int\limits_B \Psi^f_{\beta,\gamma_2}(b) db \right], \tag{13}$$

where $\Psi^f_{\beta,\gamma_2}(b)$ is given as in (12). Its minimum is the function $f$ which solves the boundary value problem

$$- \operatorname{div} \left( \frac{(1+\sqrt{\gamma_1})\nabla f}{\sqrt{\gamma_1 + |\nabla f|^2}} \right) + \frac{2\alpha \Psi^f_{\beta,\gamma_2}(b)}{\int\limits_B \Psi^f_{\beta,\gamma_2}(b) db} f$$

$$- \frac{2\alpha \left(\Psi^f_{\beta,\gamma_2}(b)\right)^2}{\int\limits_B \Psi^f_{\beta,\gamma_2}(b) db} \int\limits_B g(b') e^{-[|f(b)-g(b')|^2+\gamma_2|b-b'|^2]/\beta} db' = 0 \tag{14}$$

with Neumann boundary condition

$$\frac{\partial f}{\partial \boldsymbol{n}} = 0 \quad \text{on } \partial B. \tag{15}$$

Note that the function $G(b) = \dfrac{(1+\sqrt{\gamma_1})}{\sqrt{\gamma_1 + |\nabla f|^2}}$ in the first term of the left-hand part of (14) may be considered as edge indicator (penalty) function.

Consider now the $M \times N$ image ($M$ rows and $N$ columns). Let $B \subset \mathbb{R}^2$ be its domain and $g(b)$, $b \in B$ denote its graph. We assume that its desired segmentation is the minimum of relaxed functional (13). We search for a weak solution of the Euler-Lagrange equation (14)-(15).

Due to the robustness and simplicity of implementation of the fixed point algorithm [12], we apply it to solve (14)-(15) in this work. The solution, which is the grey-level image, is the desired segmentation. In order to present the results in a pleasant way, the segmentation contour should be obtained. To achieve this, we threshold the accepted segmentation image by using Otsu's method [8]. Finally, Canny edge detector [2] is applied to the binary image, and achieved segmentation contour is applied to the original image $g$.

In this section we are going to point out the influence of parameters $\beta, \alpha, \gamma_1, \gamma_2$ on the segmentations results. To demonstrate the performance of the proposed method, we used various textured images [5]. Two examples are illustrated here (Fig. 1). We recall that $\beta$ is the convergence parameter and the minimizer of the

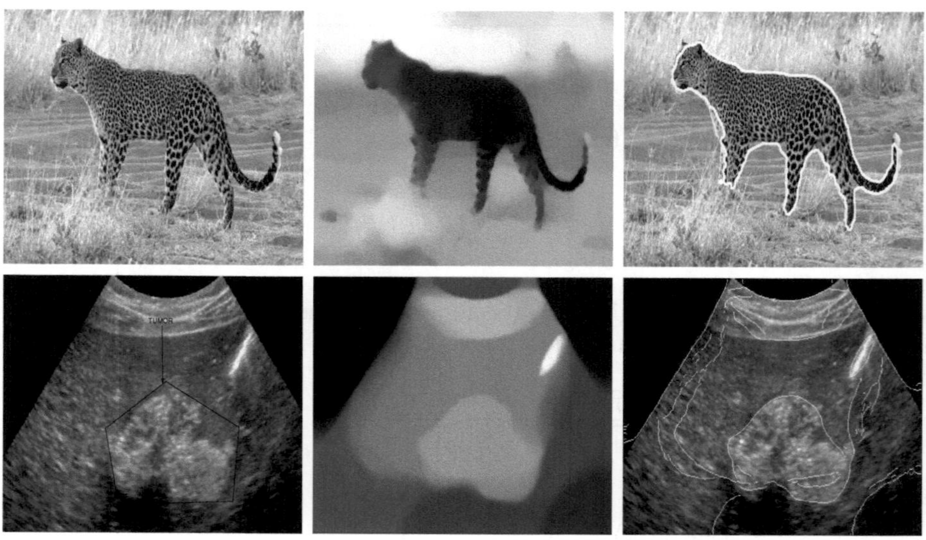

**Fig. 1.** Leopard (upper row) and ultrasound (lower row) image segmentation using the proposed method. Original images (left), algorithm outputs (middle), outlined segmented images (right). Parameters: $\alpha = 185$ (leopard), 50 (ultrasound), $\beta = 0.1$, $\gamma_1 = 0$, $\gamma_2 = \infty$.

original (the limit) functional is achieved when $\beta \to 0$. We also note that the limit $\beta \to 0$ of (13) yields

$$\mathbf{F}^0_{\gamma_1,\gamma_1}(f) = (1+\sqrt{\gamma_1})\int\limits_B \sqrt{\gamma_1 + |\nabla f|^2}db + \alpha \sup_{b\in B}\inf_{b'\in B}\left[|f(b) - g(b')|^2 \right.$$
$$\left. + \gamma_2|b-b'|^2\right],$$

which is the original functional for deterministic images. Thus, for our calculations, $\beta$ is taken sufficiently small. The parameter $\alpha$, as in case of Mumford-Shah functional, measures the trade-off between a good fit of the solution $f$ to data $g$ and the regularity of the solution $f$.

We define now two limiting cases for the functional (13): $\mathbf{F}^\beta_{0,\gamma_2}(f)$ and $\mathbf{F}^\beta_{\infty,\gamma_2}(f)$, when $\gamma_1 \to 0$ and $\gamma_1 \to \infty$, respectively. As $\gamma_1$ tends to 0, one can obtain

$$\mathbf{F}^\beta_{0,\gamma_2}(f) = TV(f) + \alpha\beta \ln\left[\int\limits_B \Psi^f_{\beta,\gamma_2}(b)db\right],$$

where $TV(f) \equiv \int\limits_B |\nabla f|db$ is the *Total Variation* (TV) norm, originally introduced by Rudin et. al. in [10]. On the other hand, $\gamma_1 \to \infty$ yields

$$\mathbf{F}^\beta_{\infty,\gamma_2}(f) \approx H^1(f) + \alpha\beta \ln\left[\int\limits_B \Psi^f_{\beta,\gamma_2}(b)db\right]$$

up to the constant, where $H^1(f) \equiv \int\limits_B |\nabla f|^2 db$ is Sobolev space norm. We note that the $TV$ norm allows discontinuities in $f$, thus making it superior to the $H^1$ regularization in cases where $f$ can have sharp edges. As is shown above, the parameter $\gamma_1$ being defined as representing the scale difference between pixel domain and feature fiber, also determines the kind of regularization for function $f$, defining the various regularization norms.

The role of parameter $\gamma_2$, which similarly to parameter $\gamma_1$ being defined as representing the scale difference between pixel domain and feature fiber, is completely different. Here the case of interest is the limit $\gamma_2 \to \infty$. It is easily verified that, as $\gamma_2 \to \infty$,

$$\lim_{\gamma_2\to\infty} \Psi^f_{\beta,\gamma_2}(b) = C\frac{\gamma_2}{\beta} \cdot e^{|f(b)-g(b)|^2/\beta},$$

where $C$ stands for a generic constant. Thus, the limit $\gamma_2 \to \infty$ yields

$$\mathbf{F}^\beta_{\gamma_1,\infty}(f) = \lim_{\gamma_2\to\infty} \mathbf{F}^\beta_{\gamma_1,\gamma_2}(f) \approx (1+\sqrt{\gamma_1})\int\limits_B \sqrt{\gamma_1 + |\nabla f|^2}db$$
$$+ \alpha\beta \ln\left[\int\limits_B e^{|f(b)-g(b)|^2/\beta}db\right] - \alpha\beta\ln\beta \tag{16}$$

up to the constant multiplied by $\beta$. Finally, we note that the limit $\beta \to 0$ of (16) yields

$$\mathbf{F}^0_{\gamma_1,\infty}(f) = (1+\sqrt{\gamma_1})\int\limits_B \sqrt{\gamma_1 + |\nabla f|^2}db + \alpha \sup_{b\in B}|f(b)-g(b)|^2.$$

The parameter $\gamma_2$ determines the local neighborhood of the pixel $b$, which gives the valuable contribution to the final segmentation.

## 4 Digital Image Inpainting

The same functional approach is now applied to the problem of inpainting. As in [1], it does not require any user intervention, once the region to be inpainted has been selected. Consider the 2D discrete grey level image $g$, being defined on the domain $B \setminus D$, where $D$ is the region to be inpainted. Our functional for the two-dimensional image in the non-parametric case accepts the following form:

$$F_{\gamma_1,\gamma_2}(f) = (1+\sqrt{\gamma_1})\int\limits_B \sqrt{\gamma_1 + |\nabla f|^2}db + \alpha \sup_{b\in B}\inf_{b'\in B}\left[|f(b)-g(b')|^2 + \gamma_2|b-b'|^2\right]. \quad (17)$$

Let $f$ be a desired inpainting of the image $g$ on the region $D$, be defined on the entire domain $B$. We assume that it should be a minimum of the functional (17). Actually, for the inpainting task the above functional is not applicable: let denote

$$Q^g_{\gamma_2}(b) = \inf_{b'\in B}\left[|f(b)-g(b')|^2 + \gamma_2|b-b'|^2\right].$$

We claim that, if $g$ is defined everywhere on $B$, then

$$\lim_{\gamma_2\to\infty} Q^g_{\gamma_2}(b) = |f(b)-g(b)|^2.$$

If $g$ is not defined in an open set $D$, then $Q^g_{\gamma_2}(b)$ is replaced by

$$Q^g_{\gamma_2}(b) = \inf_{b'\in B\setminus D}\left[|f(b)-g(b')|^2 + \gamma_2|b-b'|^2\right]$$

and

$$\lim_{\gamma_2\to\infty} Q^g_{\gamma_2}(b) = \infty \quad \text{for} \quad b \in D.$$

Thus, we replace $Q^g_{\gamma_2}(b)$ by

$$\tilde{Q}^g_{\gamma_2}(b) = Q^g_{\gamma_2}(b) - \gamma_2 Dis^2(b, B\setminus D),$$

where $Dis(b, B\setminus D)$ is the minimal distance between $b$ and $B\setminus D$, and the desired functional is

$$\begin{aligned}\mathbf{F}_{\gamma_1,\gamma_2}(f) &= (1+\sqrt{\gamma_1})\int\limits_B \sqrt{\gamma_1 + |\nabla f|^2}db \\ &+\alpha \operatorname{ess\,sup}_{b\in B}\inf_{b'\in B\setminus D}\left[|f(b)-g(b')|^2 + \gamma_2(|b-b'|^2 - Dis^2(b, B\setminus D))\right].\end{aligned} \quad (18)$$

Let now consider the limit $\gamma_2 \to \infty$. We define

$$\tilde{g}(b) := \begin{cases} g(b), & b \in B \setminus D \\ g(y_0(b)), & b \in D \end{cases},$$

where $y_0(b) \in \partial D$ is defined by

$$|y_0(b) - b| \leq |y - b|, \quad \forall y \in B \setminus D.$$

Note that $y_0(b)$ and thus, $\tilde{g}(b)$ are not defined unambiguously on $D$. According to its definition, the skeleton of $D$ is of zero Lebesgue measure. This fact explains the change from supremum to essential supremum in the definition of metric term of (18). It is also obvious that

$$\lim_{\gamma_2 \to \infty} Q^{\tilde{g}}_{\gamma_2}(b) = |f(b) - \tilde{g}(b)|^2$$

for almost any $b$. Thus, the functional (18) has a limit as $\gamma_2 \to \infty$, that is

$$\lim_{\gamma_2 \to \infty} \mathbf{F}_{\gamma_1,\gamma_2}(f) = (1 + \sqrt{\gamma_1}) \int\limits_B \sqrt{\gamma_1 + |\nabla f|^2} db + \alpha \operatorname*{ess\,sup}_{b \in B} |f(b) - \tilde{g}(b)|^2.$$

As in case of image segmentation model, we approximate the functional (18) by relaxed functional with parameter dependent metric using $\Gamma$-convergence technique. The desired relaxed functional is

$$\mathbf{F}^{\beta}_{\gamma_1,\gamma_2}(f) = (1 + \sqrt{\gamma_1}) \int\limits_B \sqrt{\gamma_1 + |\nabla f|^2} db + \alpha\beta \ln \left[ \int\limits_B \Psi^f_{\beta,\gamma_2}(b) db \right], \qquad (19)$$

where

$$\Psi^f_{\beta,\gamma_2}(b) = \frac{1}{\int\limits_{B \setminus D} e^{-[|f(b) - g(b')|^2 + \gamma_2(|b-b'|^2 - Dis^2(b, B \setminus D))]/\beta} db'}.$$

In the same manner, we set the Euler-Lagrange equation of the functional (19) to be

$$- \operatorname{div} \left( \frac{(1 + \sqrt{\gamma_1}) \nabla f}{\sqrt{\gamma_1 + |\nabla f|^2}} \right) + \frac{2\alpha \Psi^f_{\beta,\gamma_2}(b)}{\int\limits_B \Psi^f_{\beta,\gamma_2}(b) db} f(b)$$

$$- \frac{2\alpha \left( \Psi^f_{\beta,\gamma_2}(b) \right)^2}{\int\limits_B \Psi^f_{\beta,\gamma_2}(b) db} \int\limits_{B \setminus D} g(b') e^{-[|f(b) - g(b')|^2 + \gamma_2(|b-b'|^2 - Dis^2(b, B \setminus D))]/\beta} db' = 0$$

with natural boundary condition

$$\left. \frac{\partial f}{\partial \boldsymbol{n}} \right|_{\partial B} = 0.$$

Fig. 2 shows the results of inpainting a natural image for various $\gamma_2$. The user only supplied the "mask" image, using paintbrush-like program. The basic idea

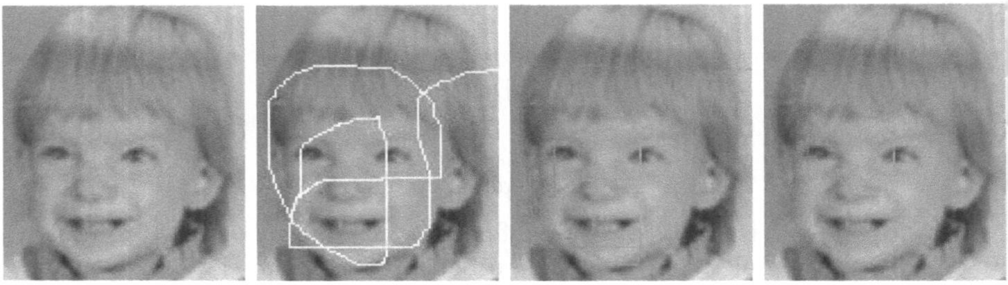

**Fig. 2.** Output results of inpainting algorithm for real-life image: $\alpha = 100000$, $\gamma_1 = 0$; from left to right: original image, "masked" image, reconstructed image with $\beta = 0.07$, $\gamma_2 = 10$, reconstructed image with $\beta = 0.1$, $\gamma_2 = 100$

is to complete the "masked" region according to minimized modified Hausdorff metric between the corrupted image and its reconstruction.

## 5 Discussion and Conclusions

The present study is a step towards the development of a general framework that can deal with segmentation problems in the context of multi-channel images. The main novelty of this study is the replacement of the metric term of Mumford-Shah functional by a metric based on Hausdorff distance function. This may be useful in cases of defocusing and of mapping problems. The proposed change allows us to replace the deterministic data by a random one and include also the case of missing data (inpainting).

Since the new metric term, and thus the functional, suffered from lack of regularity, we utilized an approach adopted from mathematical theory of approximation of functionals via $\Gamma$-convergence to overcome this deficiency. However, we should point out that the developed relaxed functional demands extensive computational effort to obtain its minimum. This is the main drawback of our algorithm. An optional solution to this problem is to apply a multi-resolution analysis, by performing the relative computations on higher levels of a Gaussian pyramid and thereby reduce significantly the amount of required computations.

**Acknowledgement.** This paper is partially supported by the Israel Science Foundation, grant 406/05, founded by the Israel Academy of Sciences and Humanities.

## References

1. M. Bertalmio, G. Sapiro, V. Caselles, and C. Ballester, Image inpainting, *Computer Graphics (SIGGRAPH 2000)*, pp. 417-424, July 2000.
2. J. Canny, A Computational Approach to Edge Detection, *IEEE Transactions on PAMI*, **8**(6): 679-698 (1986).

3. U. Dierkes, S. Hildebrandt, A. Kuster and O. Wohlrab, *Minimal Surfaces I*, Springer-Verlag Ser., **295** (1991).
4. R. Kimmel, R. Malladi and N. Sochen, On the geometry of texture, Report LBNL 39640, LBNL, UC-405 UC Berkeley, November 1996.
5. V. Kluzner, G. Wolansky and Y. Y. Zeevi, Minimal surfaces, measure-based metric and image segmentation, *CCIT Report No. 605, EE Pub No. 1562*, November 2006.
6. J. M. Morel and S. Solimini, *Variational Methods in Image Segmentation*, Birkhauser, Boston, MA, 1995.
7. D. Mumford and J. Shah, Optimal approximations by piecewise smooth functions and associated variational problems, *Comm. Pure Appl. Math.*, **42**: 577-685 (1989).
8. N. Otsu, A Threshold Selection Method from Gray-Level Histograms, *IEEE Transactions on Systems, Man, and Cybernetics*, **9** (1): 62-66 (1979).
9. T. Richardson and S. Mitter, Approximation, computation and distortion in the variational formulation, In *Comm. Geometric-Driven Diffusion in Computer Vision*, Ed. B. M. ter Haar Romeny, Kluwer Academic Publishers, 1994.
10. L. I. Rudin, S. Osher and E. Fatemi, Nonlinear total variation based noise removal algorithms, *Physica D.*, **60**: 259-268 (1992).
11. N. A. Sochen, R. Kimmel and R. Malladi, A general framework for low-level vision, *IEEE Trans. Image Processing*, **7**: 310-318 (1998).
12. C. Vogel and M. Oman, Iterative methods for total variation denoising, *SIAM J. Sci. Statist. Comput.*, **17**(1): 227-238 (1996).

# Level Set Methods for Watershed Image Segmentation

Xue-Cheng Tai[1,⋆], Erlend Hodneland[2], Joachim Weickert[3], Nickolay V. Bukoreshtliev[2], Arvid Lundervold[2], and Hans-Hermann Gerdes[2]

[1] Department of Mathematics, University of Bergen, Johannes Brunsgate 12, 5007, Bergen, Norway
[2] Department of Biomedicine, University of Bergen, Jonas Lies vei 91, 5009 Bergen, Norway
[3] Faculty of Mathematics and Computer Science, Bldg. E11, Saarland University, 66041 Saarbrücken, Germany

**Abstract.** In this work a marker-controlled and regularized watershed segmentation is proposed. Only a few previous studies address the task of regularizing the obtained watershed lines from the traditional marker-controlled watershed segmentation. In the present formulation, the topographical distance function is applied in a level set formulation to perform the segmentation, and the regularization is easily accomplished by regularizing the level set functions. Based on the well-known Four-Color theorem, a mathematical model is developed for the proposed ideas. With this model, it is possible to segment any 2D image with arbitrary number of phases with as few as one or two level set functions. The algorithm has been tested on real 2D fluorescence microscopy images displaying rat cancer cells, and the algorithm has also been compared to a standard watershed segmentation as it is implemented in MATLAB. For a fixed set of markers and a fixed set of challenging images, the comparison of these two methods shows that the present level set formulation performs better than a standard watershed segmentation.

## 1 Introduction

Segmentation is a major challenge in image analysis, referring to the task of detecting boundaries of objects of interest in an image. Several approaches have been proposed and many of them belong to one of the following categories: energy-driven segmentation [1,2,3,4,5,6,7] and watershed-based [8,9,10]. Energy-driven segmentation normally consists of two parts, the data term and the regularizer. The data term assures a solution which is sufficiently close to the desired boundaries and the regularizer controls the smoothness of the boundaries. A smoothing is often required due to noise and artifacts in real images. Watershed segmentation [8,9,10] is a region growing technique belonging to the class of morphological operations. Traditionally, the watershed techniques have been

⋆ Corresponding author. tai@mi.uib.no

F. Sgallari, A. Murli, and N. Paragios (Eds.): SSVM 2007, LNCS 4485, pp. 178–190, 2007.

conducted without a smoothing term, but recent progress has resulted in energy-based watershed segmentations that contain regularizers [11]. In the following the two main approaches for segmentation are treated more carefully.

The energy-driven segmentation methods are mainly divided into two classes, contour-based (snakes) and region-based. The contour based methods rely on strong edges or ridges as a stopping term in a curve evolution which is balanced between a data term and a smoothness term. One of the most well-known region-based method is the Mumford and Shah model[12]. In Chan-Vese [5,13], the Osher-Sethian level set idea [14] was combined with the Mumford-Shal model to solve the region-based segmentation. Recently, some variants of the Osher-Sethian level set idea was proposed by Tai et al. [15,4]. In this work, we shall extend these ideas to watershed segmentation.

The watershed segmentation has proven to be a powerful and fast technique for both contour detection and region-based segmentation. In principal, watershed segmentation depends on ridges to perform a proper segmentation, a property which is often fulfilled in contour detection where the boundaries of the objects are expressed as ridges. For region-based segmentation it is possible to convert the edges of the objects into ridges by calculating an edge map of the image. Watershed is normally implemented by region growing based on a set of markers to avoid severe over-segmentation[10,16]. Different watershed methods use slightly different distance measures, but they all share the property that the watershed lines appear as the points of equidistance between two adjacent minima. Meyer [9] use the topographical distance function for segmenting images using watershed segmentation, while Najman and Schmitt [8] present the watershed differences with classical edge detectors. Felkel et al. [16] use the shortest path cost between two nodes which is defined as the smallest lexiographic cost of all paths between two points, which reflects the flooding process when the water reaches a plateau.

The success of a watershed segmentation relies on a situation where the desired boundaries are ridges. Unfortunately, the standard watershed framework has a very limited flexibility on optimization parameters. As an example, there exists no possibility to smooth the boundaries. However, recent progress allows a regularization of the watershed lines [11] with an energy-based watershed algorithm (watersnakes). In contrast to the standard watershed and the watersnakes, our work is based on partial differential equations which easily allow a regularization of the watersheds. Moreover, the method is flexible with regard to several optimization parameters. As an example, it could allow optimization on the Euler number to avoid internal holes inside the phases.

## 2 Marker-Controlled Watershed Segmentation by Level Set

### 2.1 Creating Markers

The marker-controlled watershed segmentation has been shown to be a robust and flexible method for segmentation of objects with closed contours where the boundaries are expressed as ridges. The marker image used for watershed segmentation is a binary image consisting of either single marker points or larger marker regions where each connected marker is placed inside an object of interest. Each initial marker has a one-to-one relationship to a specific watershed region. The resulting output is very depending on the markers, both for the proposed watershed by level set and for the standard watershed, through the one-to-one relationship and the size and position of the markers. Marker regions generally create results of higher quality than point-markers since they are close to the desired boundaries and therefore there will be a smaller probability of the flooding converging too early. The markers can be manually or automatically selected, but high throughput experiments often require automatically generated markers to save human time and resources. After segmentation, the boundaries of the watershed regions are arranged on the ridges, thus separating each object from its neighbors. In the present work, an adaptive thresholding [17] and filling of closed objects were used to automatically create markers. A similar combination of these operators was used in [18]. First, an adaptive thresholding was performed on the image $f$ to label locally high intensity-valued pixels. In contrast to global thresholding, an adaptive thresholding demonstrates a much higher ability to deal with uneven scene illumination. A binary image $f_b$ was thus constructed, where high intensity pixels are given the value 1 and the others 0. Then, all small objects in $f_b$ were removed since they were considered to be insignificant due to their size. To be able to close minor gaps in the binary structures outlining the approximate boundaries, an iterative morphological closing was conducted. For each iterative closing step, a larger structural element was applied to facilitate the closing of incrementally larger gaps. Directly after each closing step a morphological filling was performed to fill all holes in $f_b$ that were not accessed from the image boundary. All filled regions that had no intersection with earlier filled regions were then assigned to the marker image as a marker. The closing was performed iteratively in order to obtain markers that were as close as possible to the desired boundaries, and it was repeated a predefined number of steps which applied for all cell images in our experiments. Figure 1 demonstrates the process of creating markers from adaptive thresholding and filling, where the image in (a) was used to create a binary image (b) from adaptive thresholding. The smallest objects were removed and iterative closing and filling were applied to (b) to obtain the final binary marker image (c) where the marker regions are labeled white (value 1) and the background is black (value 0).

## 2.2 Topographical Distance Function

The watershed transform relies on the computation of the topographical distance function, closely related to the framework of minima paths [19]. Following [9], we apply the topographical distance function to obtain a watershed segmentation which is formulated within the level set theory. The topographical distance function between two points $\mathbf{x}$ and $\mathbf{y}$ is according to [11] defined as:

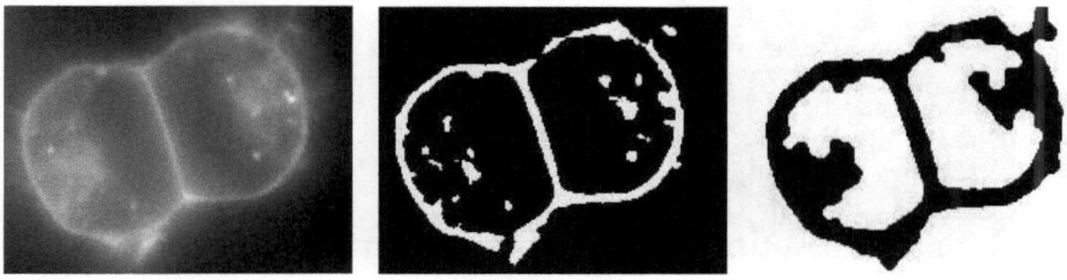

(a) (b) (c)

**Fig. 1.** Creating markers for watershed segmentation with level sets. Adaptive thresholding was applied to the image in (a) to construct a binary image (b). Thereafter, morphological closing and filling was applied to achieve the final marker image (c) which was used in the watershed segmentation.

**Definition 1.** *For a smooth function* $f(\mathbf{x}) : \mathbb{R}^n \to \mathbb{R}$ *the topographical distance between two points* $\mathbf{x}$ *and* $\mathbf{y}$ *is defined as the geodesic distance weighted by the gradient* $|\nabla f|$, *i.e.*

$$L(\mathbf{x}, \mathbf{y}) = \inf_{\gamma \in [\mathbf{x} \to \mathbf{y}]} \int_\gamma |\nabla f(\gamma(s))| ds \tag{1}$$

*where* $[\mathbf{x} \to \mathbf{y}]$ *denotes all possible paths from* $\mathbf{x}$ *to* $\mathbf{y}$ .

In 1D the topographical distance function is straightforward to compute since there is only one possible path between any two points $\mathbf{x}$ and $\mathbf{y}$. For 2D and 3D, the topographical distance function can be calculated using the iterative forest transform (IFT) [16] which computes the shortest path energy between two points. The algorithm has a low time complexity of $O(m + n \log n)$ where $n = n_1 n_2$ is the number of pixels in the image and $m = n_1(n_2 - 1) + (n_1 - 1)n_2$, defined by a 4-connectivity neighborhood. For the present study, we have used the IFT algorithm to calculate the topographical distance function. To exemplify, see the synthetic image in Figure 2 (a) resembling a real cell image. Let $L_1$ and $L_2$ be the topographical distance functions from point markers outside (c, top) and inside (c, bottom) the cell, respectively. For more complex images, the ridge is given as locations where $L_i = L_j$ for the topographical distance functions $L_i$ and $L_j$ associated with adjacent markers. The plot in (b) shows an intensity profile of the image along the dashed line in (a) together with the corresponding intensity profile of the topographical distance functions $L_1$ (dashed line) and $L_2$ (dotted line). Clearly, the ridge in (a) is obtained where $L_1 = L_2$.

### 2.3 Four-Color Theorem

The Four-Color theorem will be applied in the watershed segmentation by level set to distinguish between objects inside the same phase. Using the Four-Color theorem, adjacent objects can be labeled among four colors and they are thus uniquely distinguishable. Chan and Vese [5] note that the Four-Color theorem can be used in image segmentation in the piecewise smooth case to distinguish between any number of objects with as few as four phases. The Four-Color

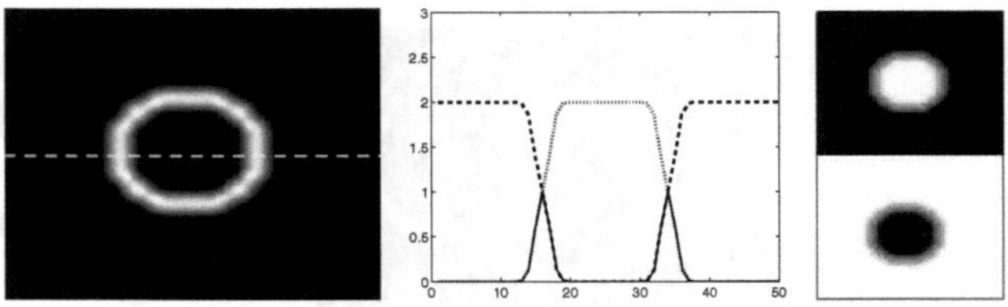


(a) (b) (c)

**Fig. 2.** Topographical distance function. The synthetic cell image (a) was used for calculating the topographical distance functions $L_1$ (c, top) and $L_2$ (c, bottom) from a point marker outside and inside the cell, respectively. The intensity profiles of the topographical distance functions $L_1$ (dashed line) and $L_2$ (dotted line) are plotted in (b). It is evident that the desired ridge from the original image is at the spatial locations where the two topographical distance functions are equal.

theorem was proven first by Appel and Haken in 1976 [20], and it has been validated again by different approaches in recent years [21]. The Four-Color theorem states the following: Define a graph $G$ consisting of a finite set $V(G)$ of vertices and a finite set $E(G)$ of edges. Every edge is connected to two vertices, called the *ends*, whereas an edge with equal ends is called a *loop*. Each vertex is an arbitrary point associated to one region, such that each region has one vertex.

**Theorem 1.** *Every loopless plane graph $G$ can be divided into 4 colors, that is a mapping $c : V(G) \to \{1, 2, 3, 4\}$ such that $c(u) \neq c(v)$ for every edge of $G$ with ends $u$ and $v$[21].*

## 3 The Algorithm

Given a set of $k$ markers, it is possible to divide the image into $k$ regions according to the four-color theorem. This partition must be completed such that no objects belonging to the same color are adjacent after segmentation. If the markers are merely points or small regions this can be demanding since the boundaries of the objects are unknown prior to segmentation. Therefore, markers are preferrably automatically constructed as large marker regions with a certain extent, to allow an approximation of the influence zone of each marker which in turn enables a reliable partitioning of the markers according to the four-color theorem. In principle, this partitioning may fall short of success and turn out to be inconsistent, but in practical applications it is possible to achieve a reliable four-color partitioning prior to segmentation.

### 3.1 Euclidean Influence Zones

Given $k$ markers, an influence zone is calculated around each marker which for all markers will divide the image into $k$ influence zones. Label all markers $\{K_i\}_{i=1}^{k}$, then calculate the Euclidean distance function $d_i(\mathbf{x}) = dist(\mathbf{x}, K_i)$ around each

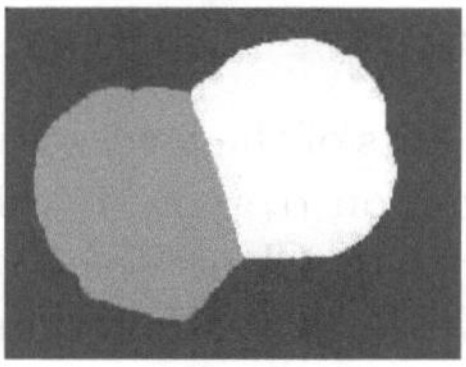

**Fig. 3.** The Euclidean influence zones $f_{IZ}$. The image in Figure 1(a) was used to create the marker image in Figure 1(c) where the three markers are uniquely labeled. Based on the distance transform, the Euclidean influence zones $f_{IZ}$ were calculated to obtain an approximation on the boundaries of the finally segmented image. The influence zones are then later painted with at most four colors.

marker $K_i$. Thus, $k$ distance functions are obtained, $\{d_i(\mathbf{x})\}_{i=1}^k$. The Euclidean influence zone image $f_{IZ}$ is a function such that

$$f_{IZ}(\mathbf{x}) = argmin_i\{d_i(\mathbf{x})\}_{i=1}^k = \{i|\ dist(\mathbf{x}, K_i) \leq dist(\mathbf{x}, K_j),\ \forall j\}. \quad (2)$$

This representation is fast to compute and it divides the image into $k$ disjoint regions suitable for a further labeling within the four-color theorem. See Figure 3 as an example where the given markers were automatically generated by the method described in Section 2.1, and the Euclidean influence zones $f_{IZ}$ were obtained from Equation 2. Thus, the piecewise constant image $f_{IZ}$ is constructed where each region $i$ is uniquely labeled by an integer from $\{1, 2, \cdots, k\}$.

### 3.2 The Four-Color Theorem and the Topographical Distance Functions

The four-color theorem was applied to the Euclidean influence zones $f_{IZ}$ function which contains an approximation to the final boundaries after segmentation. For the images in this paper the painting of the regions were done by hand. Thus, a final partitioning $f_c$ was obtained where adjacent zones in $f_{IZ}$ and their corresponding markers are always assigned different colors. Empty colors will not influence the performance of the algorithm.

Once the markers have been painted with one of the four colors, we can group the marker into four groups, i.e. we define $C_i = \cup_{f_c(K_j)=i} K_j, \quad i = 1, 2, 3, 4$. We then use the method of [16] to compute up to four topographical distance functions from the marker groups $C_i$, $L_i(\mathbf{x}) = \inf_{y \in C_i} L(\mathbf{x}, \mathbf{y}), \quad i = 1, 2, 3, 4$. As was proven in [11], a partition $\{\Omega_i\}_{i=1}^k$ minimizes the functional

$$E(\Omega_1, \ldots, \Omega_k) = \sum_{i=1}^{k} \int_{\Omega_i} \{\alpha_i + L_i(\mathbf{x})\} d\mathbf{x} \quad (3)$$

if and if it is a watershed segmentation. In our calculations, $\alpha_i$ is the minimum value of $f$ on the boundary of marker $i$. In the following, we propose a level set method to solve the above watershed segmentation problem.

### 3.3 Level Set Formulation

We shall use three different variants of the level set idea to perform the watershed segmentation based on the function $\alpha_i + L_i, i = 1, 2, 3, 4$. First, we propose to use the level set idea [22,14] as in Chan-Vese [13] for the segmentation. Let $\phi_1(\mathbf{x}), \phi_2(\mathbf{x}) : \mathbb{R}^2 \to \mathbb{R}$ be two level set functions defined on the domain $\Omega$. These functions will partition the domain into four (possibly disconnected) sub-regions. The characteristic functions for these sub-regions are $\psi_i, i = \{1, 2, 3, 4\}$ given as

$$\psi_1(\phi_1, \phi_2) = H(\phi_1)H(\phi_2), \qquad \psi_2(\phi_1, \phi_2) = (1 - H(\phi_1))H(\phi_2),$$
$$\psi_3(\phi_1, \phi_2) = H(\phi_1)(1 - H(\phi_2)), \quad \psi_4(\phi_1, \phi_2) = (1 - H(\phi_1))(1 - H(\phi_2)).$$

The sub-regions are $\Omega_i = \{\mathbf{x} |\ \psi_i(\mathbf{x}) = 1\}, i = 1, 2, 3, 4$. This partition of the domain has no vacuum and no overlaps. In the above, $H(\cdot)$ denotes the Heaviside function, i.e. $H(x) = 1$ if $x \geq 0$, $H(x) = 0$ if $x < 0$. For the numerical experiments, a regularized Heaviside was used, $H_\epsilon(x) = \frac{1}{2}\left(1 + \frac{2}{\pi}\arctan(\frac{x}{\epsilon})\right)$ where $\epsilon > 0$ is small, see [13]. Using Eq. 3 within the level set formulation, the desired watershed lines can be obtained by minimizing the following functional

$$F = \int_\Omega \sum_{i=1}^{4} \{\alpha_i + L_i(\mathbf{x})\}\psi_i d\mathbf{x} + \beta \int_\Omega \sum_{i=1}^{4} |\nabla \psi_i| d\mathbf{x}. \tag{4}$$

The first term is the data term providing the watershed segmentation and the second term is the regularization.

The second level set method we propose to use is the so-called Binary Level Set Method [4,2]. For this method, we need to find two functions $\phi_1(\mathbf{x}), \phi_2(\mathbf{x}) : \mathbb{R}^2 \to \mathbb{R}$ satisfying $\phi_i(\mathbf{x})^2 = 1,\ i = 1, 2$. The characteristic functions for the sub-regions partitioned by $\phi_i$ are given by

$$\psi_{i+1+2*j} = \frac{1}{4}\left(1 + (-1)^i \frac{\phi_1}{|\phi_1|}\right)\left(1 + (-1)^j \frac{\phi_2}{|\phi_2|}\right), \quad i, j = 0, 1.$$

This will give us the characteristic functions $\psi_k, k = 1, 2, 3, 4$ for the four sub-regions. In our experiments, $\phi/|\phi|$ are replaced by $\phi/\sqrt{|\phi|^2 + \epsilon}$ with a small $\epsilon > 0$. The watershed segmentation for this level set method is obtained by minimizing:

$$F = \int_\Omega \sum_{i=1}^{4} \{\alpha_i + L_i(\mathbf{x})\}\psi_i d\mathbf{x} + \beta \int_\Omega \sum_{i=1}^{4} |\nabla \psi_i| d\mathbf{x} + \frac{1}{\sigma}\sum_{i=1}^{2}\int_\Omega (\phi_i^2 - 1)^2 d\mathbf{x}. \tag{5}$$

The constant $\sigma > 0$ is a penalization constant to enforce $\phi_i^2 = 1$. Due to the special constructions of the characteristic functions $\psi_i$, we can choose any $\sigma > 0$ in the above minimization functional.

The third level set method we propose to use is the "Piecewise Constant Level Set Methods (PCLSM)" [15]. For this method, we just need to use one level set function $\phi : \mathbb{R}^2 \to \mathbb{R}$ satisfying $\kappa(\phi) = (\phi - 1)(\phi - 2)(\phi - 3)(\phi - 4) = 0$ in $\Omega$.

Associated with this $\phi$, the characteristic functions for the sub-regions are given by

$$\psi_i = \frac{1}{\lambda_i}\prod_{\substack{j=1\\ j\neq i}}^{4}(\phi - j) \text{ and } \lambda_i = \prod_{\substack{k=1\\ k\neq i}}^{4}(i-k).$$

In order to use this method for the watershed segmentation, we need to solve the following constrained minimization problem:

$$\min_{\phi,\ \kappa(\phi)=0} \int_\Omega \sum_{i=1}^{4}\{\alpha_i + L_i(\mathbf{x})\}\psi_i d\mathbf{x} + \beta\int_\Omega \sum_{i=1}^{4}|\nabla\psi_i| d\mathbf{x}. \tag{6}$$

As in [15,3], Augmented Lagrangian or penalization methods can be used to solve the above constrained minimization problem.

For minimization problem (4), the Euler-Lagrange equations for $\phi_1$ and $\phi_2$ are:

$$\sum_{i=1}^{4}\{\alpha_i + L_i(\mathbf{x})\}\frac{\partial\psi_i}{\partial\phi_1} + \beta\sum_{i=1}^{4}\nabla\cdot\left(\frac{\nabla\psi_i}{|\nabla\psi_i|}\right)\frac{\partial\psi_i}{\partial\phi_1} = 0,$$
$$\sum_{i=1}^{4}\{\alpha_i + L_i(\mathbf{x})\}\frac{\partial\psi_i}{\partial\phi_2} + \beta\sum_{i=1}^{4}\nabla\cdot\left(\frac{\nabla\psi_i}{|\nabla\psi_i|}\right)\frac{\partial\psi_i}{\partial\phi_2} = 0.$$

As usual, the explicit gradient flow problem must be solved to steady state:

$$\frac{\phi_1^{n+1}-\phi_1^n}{\tau} = \sum_{i=1}^{4}\{\alpha_i + L_i(\mathbf{x})\}\frac{\partial\psi_i^n}{\partial\phi_1^n} + \beta\sum_{i=1}^{4}\nabla\cdot\left(\frac{\nabla\psi_i^n}{|\nabla\psi_i^n|}\right)\frac{\partial\psi_i^n}{\partial\phi_1^n}$$
$$\frac{\phi_2^{n+1}-\phi_2^n}{\tau} = \sum_{i=1}^{4}\{\alpha_i + L_i(\mathbf{x})\}\frac{\partial\psi_i^n}{\partial\phi_2^n} + \beta\sum_{i=1}^{4}\nabla\cdot\left(\frac{\nabla\psi_i^n}{|\nabla\psi_i^n|}\right)\frac{\partial\psi_i^n}{\partial\phi_2^n},$$

where the differentiation of $|\nabla\cdot|$ with respect to $\phi$ is in detail explained in [13], and $\psi^n = \psi(\phi_1^n, \phi_2^n)$. The relation $H'(x) = \delta(x)$ was used for differentiation of $\psi_i$, and a smooth $\delta_\epsilon(x)$ was used in the numerical experiments by calculating the derivative of the smooth Heaviside. For minimization problems (5) and (6), the Euler-Lagrange equations equation can be obtained in a similar way. We shall omit the details.

Here, we have proposed three different level set methods with different advantages and weak points, producing slightly different results which are complementary to each other. The connection between the data term of our three minization methods and the traditional watershed transform can be proven by considering the minimization of

$$F = \int_\Omega \sum_{i=1}^{4}\{\alpha_i + L_i(\mathbf{x})\}\psi_i d\mathbf{x}. \tag{7}$$

In contrast to the traditional watershed, minimizing Eq. 7 provides the watershed lines around each color instead of each marker. As stated in [11], $\alpha_i + L_i(\mathbf{x}) \geq f(\mathbf{x})$ for $1 \leq i \leq K$ where $K$ is the number of markers and $f(\mathbf{x}) = \min_{1 \leq i \leq K} \{\alpha_i + L_i(\mathbf{x})\}$. In our formulation, the markers are grouped together into a set of possible disconnected regions, and $\alpha_i + L_i(\mathbf{x}) \geq f(\mathbf{x})$ for $1 \leq i \leq 4$ and for all $\mathbf{x} \in \Omega$. Therefore, due to the fact that $\psi_i \geq 0$, we have

$$F = \int_\Omega \sum_{i=1}^{4} \{\alpha_i + L_i(\mathbf{x})\} \psi_i d\mathbf{x} \geq \int_\Omega \sum_{i=1}^{4} f(\mathbf{x}) \psi_i d\mathbf{x}.$$

From the definition of the watershed lines, the function $\phi$ corresponding to the watershed lines satisfies $\int_\Omega \sum_{i=1}^{4} \{\alpha_i + L_i(\mathbf{x}) - f(\mathbf{x})\} \psi_i d\mathbf{x} = 0$, c.f. [11]. This equality is correct if we replace $\Omega$ by any $U \in \Omega$. Due to the properties of the characteristic functions, $\psi_i \neq 0$ everywhere in $\Omega$ and therefore $\alpha_i + L_i(\mathbf{x}) - f(\mathbf{x}) = 0$, which is exactly the requirement for a watershed partition around color $i$. The obtained level set segmentation may not be a watershed partition around color $i$ if the regularization term is included in addition to the data term.

## 4 Experimental Results of Segmentation

This section contains experiments involving real cell images taken by fluorescence microscopy showing rat pheochromocytoma PC12 cells. The images are optical planes extracted from 3D stacks, demonstrating the ability of the proposed method. In all experiments values of $\beta = 0.1$ and a time-step of $\Delta t = 0.01$ were chosen for the steepest descent method, and a standard watershed segmentation as implemented in MATLAB [10] was also calculated for comparison of performance between the two methods which used the same automatically generated markers. The images used for experimental testing point out important limitations in the image quality, in the sense that the traditional additive noise is not a serious question of matter, but rather internalized particles that appear as bright spots inside the cells (Figure 4(a), arrows). These internalized particles affect the segmentation negatively by causing more oscillatory watersheds (data not shown).

### 4.1 Two Objects

For this example, the image in Figure 4(a) was chosen for segmentation. It shows one PC12 cell in addition to a background region. Based on the obtained marker image in (b), the watersheds for the watershed by level set (c) and for the standard watershed segmentation (d) were obtained. Note the smoothness of the watersheds in (c) compared to (d).

### 4.2 Four Objects

The image in Figure 5 (a) shows three cells in addition to background. This image was used to automatically obtain a marker image (b). Based on the marker

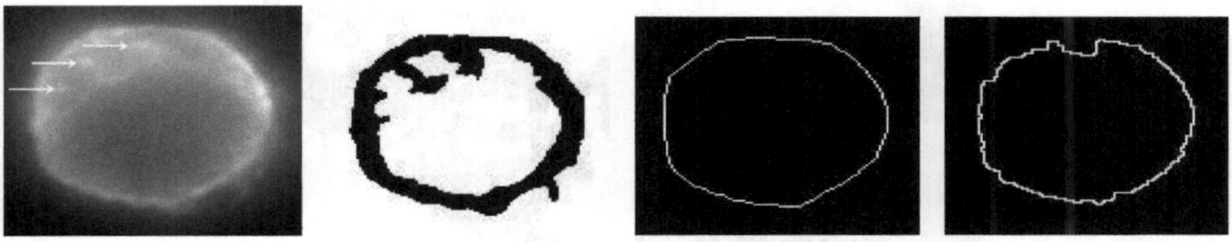

(a) (b) (c) (d)

**Fig. 4.** Watershed segmentation of one cell and a background region. The image in (a) was used to automatically obtain markers (b) which were used for watershed segmentation by the Chan-Vese level set (c) and for a standard watershed segmentation (d). Note how the watershed by level set creates smoother boundaries (c) than the standard watershed (d). The arrows in (a) point to internalized particles appearing as bright spots inside fluorescently labeled cells. These can easily interfere with the standard watershed segmentation such that the watershed lines become oscillatory.

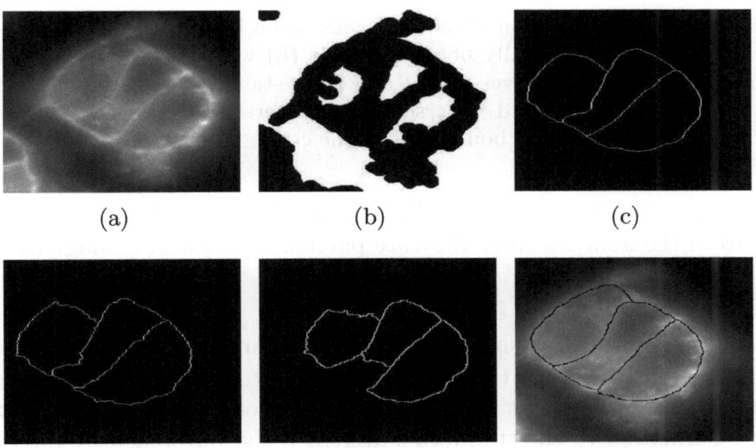

(a) (b) (c)

(d) (e) (f)

**Fig. 5.** Watershed segmentation of three cells and a background region. The image in (a) was used to automatically obtain markers (b) which were used for watershed segmentation by the binary level set, with (c, $\beta = 0.1$) and without (d, $\beta = 0$) regularization. The output from the standard watershed segmentation is displayed in (e) together with a ground-truth created by hand-drawing (black lines) in (f). Compared with the ground-truth, note how the watershed by level set captures more of the cells than the standard watershed segmentation.

image, the watersheds for the watershed by level set with (c) and without regularization (d, $\beta = 0$) and the standard watershed segmentation (e) were obtained. The image in (f) shows a hand-drawn solution. Comparing with the hand-drawn solution, note that the watershed by level set captures more of the cells than the

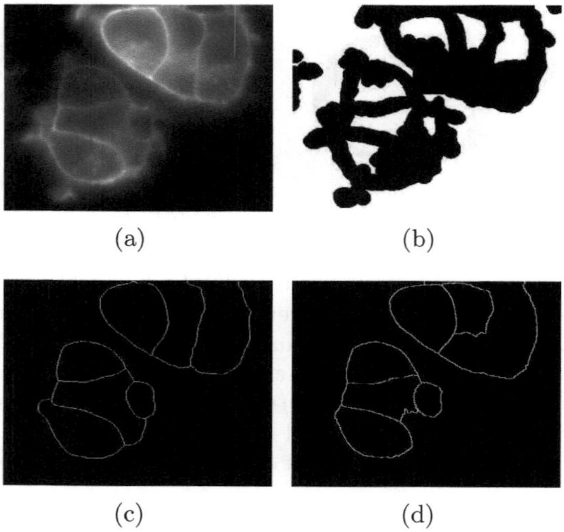

**Fig. 6.** Watershed segmentation of multiple cells and a background region. The image in (a) was used to automatically obtain markers (b) which were used for watershed segmentation by the binary level set (c) and for a standard watershed segmentation (d). Compared to the standard watershed, the watershed by level set has a higher capacity of detecting the weak boundaries of the cell.

standard watershed segmentation. The cells in the lower left corner of (a) were omitted from the analysis since they are partially outside the image frame.

### 4.3 Multiple Objects

The image in Figure 6 (a) shows six PC12 cells and background. Based upon this image, a marker image (b) was automatically created and the watersheds for the watershed by level set (c) and standard watershed segmentation (d) were obtained. Compared to the standard watershed, the watershed by level set has a higher capacity of detecting the weak boundaries of the cell.

From all these studies, we have shown that the proposed watershed by level set is able to segment real cell images containing severe irregularites in a better way than the standard watershed segmentation. Our formulation is based on level set theory which easily allows a regularization of the watersheds, and which is a flexible approach for further optimization parameters.

## References

1. Caselles, V., Catté, F., Coll, T., Dibos, F.: A geometric model for active contours in image processing. Numer. Math. **66**(1) (1993) 1–31
2. Nielsen, L.K., Tai, X.C., Aanonsen, S.I., Espedal, M.: Reservoir Description using a Binary Level Set Model. In: Image Processing Based on Partial Differential Equations. Springer, Heidelberg, Editor, X.-C. Tai, K. A. Lie, T. Chan and S. Osher (2006) 403–426

3. Christiansen, O., Tai, X.C.: Fast Implementation of Piecewise Constant Level Set Methods. In: Image Processing Based on Partial Differential Equations. Springer, Heidelberg, Editor, X.-C. Tai, K. A. Lie, T. Chan and S. Osher (2006) 289–308
4. Lie, J., Lysaker, M., Tai, X.C.: A binary level set model and some applications to Mumford-Shah image segmentation. IEEE Transactions on Image Processing **15**(5) (2006) 1171–1181
5. Vese, L.A., Chan, T.F.: A multiphase level set framework for image segmentation using the Mumford and Shah model. International Journal of Computer Vision **50**(3) (2002) 271–293
6. Cremers, D., Tischhäuser, F., Weickert, J., Schnörr, C.: Diffusion snakes: introducing statistical shape knowledge into the Mumford–Shah functional. International Journal of Computer Vision **50** (2002) 295 – 313
7. Kass, M., Witkin, A., Terzopoulos, D.: Snakes: Active contour models. International Journal of Computer Vision **1**(4) (January 1988) 321–331
8. Najman, L., Schmitt, M.: Watershed of a continuous function. Signal Process. **38**(1) (1994) 99–112
9. Meyer, F.: Topographic distance and watershed lines. Signal Processing **38**(1) (1994) 113–125
10. Vincent, L., Soille, P.: Watersheds in digital spaces: An efficient algorithm based on immersion simulations. IEEE Trans. Pattern Anal. Mach. Intell. **13**(6) (1991) 583–598
11. Nguyen, H., Worring, M., van den Boomgaard, R.: Watersnakes: Energy-driven watershed segmentation. IEEE Trans. on PAMI **25**(3) (2003) 330–342
12. Mumford, D., Shah, J.: Optimal approximation by piecewise smooth functions and associated variational problems. Communications on Pure Applied Mathematics **42** (1989) 577–685
13. Chan, T., Vese, L.: Active contours without edges. IEEE Trans on Image Processing **10** (2001) 266–277
14. Osher, S., Sethian, J.A.: Fronts propagating with curvature-dependent speed: Algorithms based on Hamilton-Jacobi formulations. Journal of Computational Physics **79** (1988) 12–49
15. Lie, J., Lysaker, M., Tai, X.C.: A variant of the level set method and applications to image segmentation. Mathematics of Computation **75**(255) (2006) 1155–1174
16. Felkel, P., Bruckschwaiger, M., Wegenkittl, R.: Implementation and complexity of the watershed-from-markers algorithm computed as a minimal cost forest. Computer Graphics Forum **20**(3) (2001)
17. Chang, S.G., Yu, B., Vetterli, M.: Spatially adaptive wavelet thresholding with context modeling for image denoising. IEEE Transactions on Image Processing **9**(9) (2000) 1522–1531
18. Hodneland, E., Lundervold, A., Gurke, S., Tai, X.C., Rustom, A., Gerdes, H.H.: Automated detection of tunneling nanotubes in 3d images. Cytometry Part A **69A** (2006) 961–972
19. Arbeléz, P.A., Cohen, L.D.: Energy partitions and image segmentation. J. Math. Imaging Vis. **20**(1-2) (2004) 43–57
20. Appel, K.I., Haken, W.: Every planar map is four colorable. Illinois J. Math. **21** (1977) 429–567

21. Robertson, N., Sanders, D., Seymour, P., Thomas, R.: A new proof of the four colour theorem. Electronic Research Announcements of the American Mathematical Society **2**(1) (1996)
22. Dervieux, A., Thomasset, F.: A finite element method for the simulation of Rayleigh–Taylor instability. In Rautman, R., ed.: Approximation Methods for Navier–Stokes Problems. Volume 771 of Lecture Notes in Mathematics. Springer, Berlin (1979) 145–158

# Segmentation Under Occlusions Using Selective Shape Prior*

Sheshadri R. Thiruvenkadam[1], Tony F. Chan[1], and Byung-Woo Hong[2]

[1] Department of Mathematics
[2] Computer Science Department
University of California, Los Angeles, CA-90095, USA
sheshad,chan@math.ucla.edu, hong@cs.ucla.edu

**Abstract.** In this work, we address the problem of segmenting multiple objects, with possible occlusions, in a variational setting. Most segmentation algorithms based on low-level features often fail under uncertainties such as occlusions and subtle boundaries. We introduce a segmentation algorithm incorporating high-level prior knowledge which is the shape of objects of interest. A novelty in our approach is that prior shape is introduced in a selective manner, only to occluded boundaries. Further, a direct application of our framework is that it solves the *segmentation with depth* problem that aims to recover the spatial order of overlapping objects for certain classes of images. We also present segmentation results on synthetic and real images.

## 1 Introduction

Image segmentation is an important step in understanding the composition of the original 3D scene that gave rise to the image. However, it is often considered as a difficult problem due to noise which results in spurious edges and boundary gaps, and occlusions which leads to a overlap of object boundaries. Low-level visual features such as intensity, color and texture are generally not sufficient to overcome such difficulties that would make purely bottom-up segmentation approaches unsuccessful. This naturally leads to a need for integrating low-level features and high-level information in segmentation. Enforcing a prior knowledge on the shape of objects is a common way to facilitate segmentation specially under low contrasts, occlusions and other undesirable noisy conditions.

In this paper, we address the problem of segmenting multiple objects with possible occlusions. Here, a segmentation method incorporating shape prior knowledge is presented in a variational approach using the level set framework [15]. The level set framework is commonly used in segmentation [10,1,8,16,3,17] due to the following favorable properties: it provides an implicit boundary representation that is free of parameterization, easily deals with topological changes of the boundary such as splitting and merging, and can be naturally extended to any dimension.

* Supported by ONR grant N00014-06-1-0345 and NSF grant DMS-0610079.

F. Sgallari, A. Murli, and N. Paragios (Eds.): SSVM 2007, LNCS 4485, pp. 191–202, 2007.

A level set implementation of active contours with region-based image terms using the Mumford-Shah functional [12] has been developed by Chan and Vese [3]. This is shown to be robust with respect to noise due to the intrinsic smoothness terms that regularize the shape of the segmenting contours and also allows the segmentation of multiple objects [18]. However, this method solely relies on image intensity which is not sufficient to overcome occlusions that occur in many practical applications. This consequently leads to the efforts that introduce prior shape information into segmentation schemes based on level sets. The active shape model [5] using principal component analysis from a set of training shapes has been used as prior shape information and incorporated with level set implementation in [17,9,11], where a statistical shape model is employed. In [4,2], the authors have proposed a variational method where the energy function that governs the evolution of the level set function depends on image information and shape information. In contrast to a conventional linear PCA, a nonlinear statistics by means of kernel PCA is considered as a shape model in [6].

Although incorporating a shape prior within our segmentation model was inspired by the above works, our method has a novelty in that, the use of shape prior knowledge is automatically restricted *only* to occluded parts of the object boundaries. This selective use of local prior shape avoids enforcing prior shape on regions where the object boundary is clearly characterized by image intensity.

A direct application of our approach is that it solves the *segmentation with depth* problem that aims to determine the boundaries of the overlapping objects, along with their spatial ordering, based on the intensity distributions in the object regions. The segmentation with depth problem has been addressed before in a variational framework by Nitzberg, Mumford and Shiota (NMS) in [13,14] and numerical methods for minimizing the NMS model have been presented in [7]. The NMS model is closely related to our segmentation model and their differences will be discussed later.

## 2 Occlusion Model

Suppose that $I : \Omega \to \mathbb{R}$, $\Omega \subset \mathbb{R}^2$ is a 2D image of a scene composed of $N$ objects $\{O_p\}_{p=1}^N$. Let $\{A_p\}_{p=1}^N$ be the regions formed on the image domain by the objects. Now suppose that we have the following assumptions:

- The image intensity formed by the object $O_k$ is close to a constant $c_k$, and the background intensity is close to a constant, $\tilde{c}$.
- The objects are not twisted between themselves.

One way to represent such an occlusion scene is as a linear combination of the background, the object regions, and their intersection regions (where occlusions possibly occur). To motivate the form of $I$, we see in the case $N = 3$, suppose that $O_1$, $O_3$, $O_2$ is the spatial order of the objects,

$$I = c_1\chi_{A_1} + c_2\chi_{A_2} + c_3\chi_{A_3} - c_2\chi_{A_1\cap A_2} - c_2\chi_{A_2\cap A_3} - c_3\chi_{A_1\cap A_3}$$
$$+c_2\chi_{A_1\cap A_2\cap A_3} + \tilde{c}(1-\chi_{A_1})(1-\chi_{A_2})(1-\chi_{A_3})$$

where $\chi_S$ is the characteristic function of a set $S$.

In general, for $N$ objects, the image $I$ is of the form,

$$I = \sum_{p=1}^{N} \sum_{k=1}^{\binom{N}{p}} (-1)^{p-1} c_{p,k} \chi_{P_{p,k}} + \tilde{c} \prod_{k=1}^{N} (1 - \chi_{A_k}) \tag{1}$$

Here, $P_{p,k}$, $(p = 1, 2..N, k = 1, 2, ..\binom{N}{p})$ is the $k^{th}$ unordered intersection of $p$ regions from $A_1, A_2, ..., A_N$. $c_{p,k}$ are positive constants, with $c_{1,k} = c_k$, and for $p > 1$, $c_{p,k}$ takes one of the values $c_1, c_2, ...c_N$, depending entirely on the occlusion relationships between the objects.

In fact, there are $N!$ different possible sequences for $c_{p,k}$, $p > 1$, with each sequence corresponding to a particular spatial ordering of the objects. For instance, for the $N = 2$ case, $I = c_1\chi_{A_1} + c_2\chi_{A_2} - c_{2,1}\chi_{A_1 \cap A_2} + \tilde{c}(1 - \chi_{A_1})(1 - \chi_{A_2})$, and we see that, $c_{2,1} = c_2(c_1)$ iff $A_1(A_2)$ occludes $A_2(A_1)$.

In this work, given an image $I_0$, we solve the inverse problem of recovering the object regions $\{A_p\}_{p=1}^{N}$, and constants $c_{p,k}$, $p = 1, 2..N, k = 1, 2, ..\binom{N}{p}$. We formulate the above problem as the following energy minimization. Here, we have also added a term within the energy that captures shape of the objects (to resolve occluded boundaries).

$$E := \int_{\Omega} (I_0 - (\sum_{p=1}^{N} \sum_{k=1}^{\binom{N}{p}} (-1)^{p-1} c_{p,k} \chi_{P_{p,k}} + \tilde{c} \prod_{k=1}^{N} (1 - \chi_{A_k})))^2 dx + \sum_{p=1}^{N} (\lambda \int_{\partial A_p} ds + \beta \int_{\Omega} \hat{S}(A_p) dx) \tag{2}$$

The second term is a length regularization term, and the third term constrains the shape of the boundaries of $A_k$. Note that in the forward model (1), the constants $c_{p,k}$ satisfy at least one of $N!$ constraints corresponding to $N!$ possible spatial ordering for the objects. Here, for computational simplicity, we have assumed the constants $c_{p,k}$ to be independent, and minimize (2) without constraints. Then, we show in section 5. that for images satisfying certain conditions, the recovered constants $c_{p,k}$ can be used to solve occlusion relationships between the objects (hence solving the segmentation from depth problem). Later, in section 6, we show how these relationships can be used to selectively impose shape constraints, only to occluded object boundaries.

## 3 Related Works

As mentioned in the previous section, the inverse problem we are looking at is related to the segmentation with depth problem. Here, we will briefly review the related formulation of NMS [14] for segmentation with depth and present comparisons with our approach.

Using the same terminology used as before, let $A_1, A_2, ..., A_N$ be the regions on the image plane corresponding to the objects $O_1, O_2, .., O_N$, with corresponding constant intensity $c_1, c_2, ..c_N$. Also assume that the objects are spatially ordered, i.e, if $i < j$ then $O_i$ is on top of $O_j$. Then, the non-occluded (visible) part of $O_i$, $\tilde{A}_i$ is $\tilde{A}_i = A_i - \bigcup_{j<i} A_j$, for $i > 1$. Let $\tilde{A}_1 = A_1$. Then the image formed by this ordering of $O_k$ is:

$$I = \sum_{i=1}^{N} c_i \chi_{\tilde{A}_i} \quad (3)$$

To start with, we note that our occlusion model (1) and (3) can be shown to be equivalent. For instance, in the two object case, we verify that if $A_1$ occludes $A_2$, then (3) gives $I = c_1\chi_{A_1} + c_2\chi_{A_2 - A_1}$ which can be rewritten as $I = c_1\chi_{A_1} + c_2\chi_{A_2} - c_2\chi_{A_1 \cap A_2}$, which is our model shown in (1).

In (NMS), the authors look at the inverse problem of solving for $A_k$, $c_k$, and the ordering of the objects $O_k$, from a given image $I_0$, by considering the following minimization:

$$\tilde{E} := \sum_{i=1}^{N} \int_{\tilde{A}_i} (I_0 - c_i)^2 dx + \int_{\partial A_i} (\alpha + \beta\phi(k)) ds \quad (4)$$

where the function $\phi(x)$ is to be a positive, convex, even function. The above energy is minimized for each possible order relation between $O_k$ (a total of $N!$ energy minimizations), to derive the optimal ordering of the objects. In fact, constraining the constants $c_{p,k}$ in the model (2) to adhere to one of $N!$ possible spatial ordering would make (2) equivalent to the (NMS) energy, (4). This approach is currently in progress and will be reported in another publication.

However, in this work we use a segmentation based approach to first solve for the regions $A_k$, and the constants $c_{p,k}$ in the intersection-regions $P_{p,k}$. Then, a comparison of the constants $c_{p,k}, p > 1$ with the object intensities $\{c_i\}_{i=1}^{N}$ would directly give us the ordering of the objects. The main expense in (NMS) and our model (2), is minimizing with respect to the regions $A_k$. Thus, our algorithm seems to be computationally feasible, since unlike (NMS) which has $N!$ shape minimizations involving $A_k, k = 1, 2, ..N$, we have just one minimization problem to deal with.

## 4 Level Set Formulation

In this section, a level set implementation of energy (2) is presented. The regions $A_k$ are represented as the interior of level set functions $\phi_k$, i.e. $H(\phi_k) = \chi_{A_k}, k = 1, 2, ..., N$, where $H(t)$ is the Heaviside function. Thus, we try to recover $\phi_k$ from $I_0$. Let $\Phi = (\phi_1, \phi_2, .., \phi_N)$ and $C = (c_{1,1}, .., c_{1,N}, ...., c_{N,1}, \tilde{c})$. We reformulate the problem (2) as the minimization with respect to $\Phi$ and $C$, of following energy:

$$E[\Phi, C] = \int_{\Omega} (I_0 - \sum_{p=1}^{N} \sum_{k=1}^{\binom{N}{p}} (-1)^{p-1} c_{p,k} S_{p,k} - \tilde{c}\tilde{S})^2 dx + \lambda \int_{\Omega} \sum_{k=1}^{N} |\nabla H(\phi_k)| + \beta \int_{\Omega} \sum_{k=1}^{N} \hat{S}(\phi_k) dx \quad (5)$$

where, $S_{p,k}$ is $k^{th}$ unordered product of $p$ functions from $H(\phi_1), H(\phi_2), ..., H(\phi_N)$ and $\tilde{S} = \prod_{k=1}^{N}(1 - H(\phi_k))$. The second term regularizes $\Phi$, and the last term constrains the shape of the 0-levelset of $\phi_k$. $\lambda$ and $\beta$ balance the three terms.

For simplicity, we will illustrate the $N = 2$ case. The above energy reduces to: $E[\phi_1, \phi_2, c_1, c_2, c_{2,1}, \tilde{c}] =$

$$\int_{\Omega}(I_0 - (c_1 H(\phi_1) + c_2 H(\phi_2) - c_{2,1} H(\phi_1) H(\phi_2) + \tilde{c}(1 - H(\phi_1))(1 - H(\phi_2))))^2 dx$$
$$+\lambda(\int_{\Omega} |\nabla H(\phi_1) + \int_{\Omega} |\nabla H(\phi_2)|) + \beta \int_{\Omega} \hat{S}(\phi_1) + \hat{S}(\phi_2) dx \quad (6)$$

In the above energy (6), it is noted that the first term is equivalent to the multi-phase formulation of [18] for a unique set of constants corresponding to $c_1, c_2, c_{2,1}, \tilde{c}$. However, in our model since occlusions are allowed, segmented objects are represented by $H(\phi_i)$, $i = 1, 2$, whereas, in the multi-phase version of the CV model, objects (i.e phases) are assumed to be disjoint, and hence $H(\phi_1)(1 - H(\phi_2))$, $H(\phi_2)(1 - H(\phi_1))$, $H(\phi_1)H(\phi_2)$ and $(1 - H(\phi_1))(1 - H(\phi_2))$ represent the object regions.

The third term is used to impose shape-based constraints such as curvature, and explicit shape on the objects $H(\phi_i)$. This term is used to avoid local minima of the first term in (6) that occur particularly under occlusions. In this work, we only deal with imposing constraints on length and explicit shape. In the (NMS) model, curvature information is used to segment under occlusions.

## 5 Dis-occlusion

An immediate application of the segmented object-regions $A_p$ and the constants $c_{p,k}$, is finding occlusion relationships between the objects $O_p$. Suppose that the following assumptions are true:

- Objects do not twist between themselves, i.e. if $O_i$ occludes $O_j$ in some region, $O_i$ is never occluded by $O_j$ in any region.
- The intensity in the regions $A_p$ formed by the objects is close to a constant $c_p$, and is *different* for objects that occlude each other.

Then the region covering all the objects, $A = \bigcup_{p=1}^{N} A_p$, can be written down as a disjoint union of $2^N - 1$ regions, where the image intensity is close to a constant in each of those regions. Of these, $N$ regions are the visible parts of the objects $O_k$, with intensity $c_k$. The rest of the $2^N - N - 1$ regions is where occlusions can possibly occur. The intensities in these regions are close to one of the object intensities, $c_k$, i.e. the intensity of the object on top. Hence, in one such nonempty occlusion region (WLOG), say $P = \{\cap_{s=1}^{p} A_s\} \bigcap \{\cap_{s=p+1}^{N} \bar{A}_s\}, p > 1$, we can infer that the topmost object is $O_t$,

$$t = \min_{1 \leq s \leq p} D(\mu_P, c_s),$$

where $\mu_U$ is the average image intensity in a region $U \subseteq \Omega$. In this work, due to the constant intensity assumption in the regions $A_p$, we use mean-intensity to test for occlusions, and $D(x,y) = |x-y|^2$. This can be easily extended to other measures to handle general image distributions. Also, $\mu_P$ can be expressed in terms of the constants $c_{q,k}, q \leq p$, given in (2). We will demonstrate this for $N = 2$ case. From the occlusion model (1), we have:

$$\begin{aligned} I &= c_1\chi_{A_1} + c_2\chi_{A_2} - c_{2,1}\chi_{A_1 \cap A_2} \\ &= c_1\chi_{A_1 - A_2} + c_2\chi_{A_2 - A_1} + (c_1 + c_2 - c_{2,1})\chi_{A_1 \cap A_2}. \end{aligned}$$

Thus, if $P = A_1 \cap A_2 \neq \{\phi\}$, then, $\mu_P = \frac{\int_P I dx}{\int_P dx} = c_1 + c_2 - c_{2,1}$. To test for the occlusion relationship between $A_1$ and $A_2$, we notice that if

$$(c_2 - c_{2,1})^2 = (\mu_P - c_1)^2 < (\mu_P - c_2)^2 = (c_1 - c_{2,1})^2,$$

then $A_1$ occludes $A_2$ and vice versa. Finding occlusion relationships between two objects $O_i$ and $O_j$ gives us relative depth information between them (e.g.$O_i$ is closer than $O_j$). Secondly, assume that:

- if $O_i$ is closer than $O_j$ and $O_j$ is closer than $O_k$, implies that $O_i$ is closer than $O_k$.
- each pair of objects can be compared (either directly or through the above *transitivity* assumption)

Then, we get an ordering of the objects in space, thus solving the segmentation from depth problem for the image.

## 6 Selective Shape Term

In this work, since we are dealing with occlusions in multi-object segmentation, we use prior shape information of the objects to fill in missing boundaries in occluded regions. However, just adding a shape term as in (6) means that the shape term might influence boundary shapes even in unoccluded regions, where the boundary is unambiguous. Hence, we introduce our shape term in a selective manner. That is, the shape term is allowed to take effect only for occluded boundaries. From the previous section, we see that the constants $c_{p,k}$ encode occlusion information. Hence for a nonempty occlusion region $P = \{\cap_{s=1}^{p} A_s\} \bigcap \{\cap_{s=p+1}^{N} \bar{A}_s\}, p > 1$, we look at the shape term,

$$\sum_{s=1}^{p} \int_{P_s} (\mu_P - c_s)^2 \hat{S}(A_s). \tag{7}$$

Here, $P_s = \bigcap_{\substack{t=1 \\ t \neq s}}^{p} A_t$. We see that the above shape term localizes use of shape only to occluded regions. Firstly, the shape term (7) is defined only on $P_s$, the region that occludes the object $A_s$. Secondly, the terms $\hat{S}(A_s)$, that constrain the shape of $A_s$, are weighted by $(\mu_P - c_s)^2$, which is comparably larger for occluded regions and minimal for the region on top. We demonstrate this idea for $N = 2$. Denote $I = c_1 H(\phi_1) + c_2 H(\phi_2) - c_{2,1} H(\phi_1) H(\phi_2) + \tilde{c}(1 - H(\phi_1))(1 - H(\phi_2))$. The energy (6) becomes:

$$E[\phi_1, \phi_2, c_1, c_2, c_{2,1}, \tilde{c}] =$$

$$\int_\Omega (I_0 - I)^2 dx + \lambda(\int_\Omega |\nabla H(\phi_1) + \int_\Omega |\nabla H(\phi_2)|) + \beta \int_\Omega \hat{S}(\phi_1) + \hat{S}(\phi_2) dx$$
$$+\tilde{\beta} \int_\Omega H(\phi_2)(c_{2,1} - c_2)^2 \hat{S}(\phi_1) + H(\phi_1)(c_{2,1} - c_1)^2 \hat{S}(\phi_2) dx \quad (8)$$

Here, the fourth term is the shape term used to nominally influence the shape of the segmented objects to avoid local minima, and the last term weighs the shape term in the regions that occlude $A_1$ and $A_2$, i.e. $H(\phi_2)$ and $H(\phi_1)$ respectively. $\beta$ and $\tilde{\beta}$ balance the shape terms with $\tilde{\beta} \gg \beta$.

# 7 Numerical Implementation

In this paper, given the binary image $S$ of a prior shape, we use the symmetric area measure to compare shapes. Hence, in (5), the shape term is $\hat{S}(\phi_k) = (H(\phi_k) - S \circ T_k)^2$. $T_k$ are rigid transformations, which also needs to be determined during minimization. To minimize (5), we use a finite difference scheme to solve the resulting Euler Lagrange equations. We present the numerical implementation for $N = 2$ case, shown in (8). Denote $T = [T_1, T_2]$, where $T_k = [\mu_k, \theta_k, \mathbf{t}_k]$, $i = 1, 2$ are rigid transformations with scale $\mu_k$, rotation $\theta_k$ and translation $\mathbf{t}_k$. Rewriting (8) using the shape term above,

$$E[\Phi, C, T] =$$

$$\int_\Omega (I - I_0)^2 dx + \lambda(\int_\Omega |\nabla H(\phi_1) + \int_\Omega |\nabla H(\phi_2)|) + \int_\Omega \{\beta + \tilde{\beta} H(\phi_2)(c_{2,1} - c_2)^2)\}$$
$$(H(\phi_1) - S \circ T_1)^2 dx + \int_\Omega \{\beta + \tilde{\beta} H(\phi_1)(c_{2,1} - c_1)^2\}(H(\phi_2) - S \circ T_2)^2 dx \quad (9)$$

Given $(\Phi, T)$, the minimizing constants $C$ of the above energy are easily computed as the solution of a linear system. For an index $k \in \{1, 2\}$, let $\bar{k}$ denote its complement. For $k = 1, 2$, the Euler Lagrange equations for (9) are:

$$\delta(\phi_k)\{(I - I_0)(c_{\bar{k}} - c_{2,1} H(\phi_{\bar{k}}) - \tilde{c}(1 - H(\phi_{\bar{k}}))) - \lambda \nabla \cdot \frac{\nabla \phi_k}{|\nabla \phi_k|} +$$
$$\{\beta + \tilde{\beta} H(\phi_{\bar{k}})(c_{2,1} - c_{\bar{k}})^2)\}(H(\phi_k) - S \circ T_k) + \tilde{\beta}(c_{2,1} - c_k)^2 (H(\phi_{\bar{k}}) - S \circ T_{\bar{k}})^2\} = 0$$

$$\frac{\partial \phi_k}{\partial \boldsymbol{n}} = 0 \text{ on } \partial\Omega. \quad (10)$$

$$\mu_k \int_\Omega \{\beta + \tilde{\beta} H(\phi_{\bar{k}})(c_{2,1} - c_{\bar{k}})^2)\}(S \circ T_k - H(\phi_k)) \nabla S_{T_k x} \cdot R'_{\theta_k} x \, dx = 0, \quad (11)$$

$$\int_\Omega \{\beta + \tilde{\beta} H(\phi_{\bar{k}})(c_{2,1} - c_{\bar{k}})^2)\}(S \circ T_k - H(\phi_k))\nabla S_{T_k x} \cdot R_{\theta_k} x \, dx = 0, \qquad (12)$$

$$\int_\Omega \{\beta + \tilde{\beta} H(\phi_{\bar{k}})(c_{2,1} - c_{\bar{k}})^2)\}(S \circ T_k - H(\phi_k))\nabla S_{T_k x} \, dx = 0 \qquad (13)$$

Given initial values for $\Phi$, $C$, and $T$, we use gradient descent to minimize (9), using the above equations.

## 8 Experimental Results

We present results on synthetic images and a EM (electron microscope) image of erythrocytes, with multiple occluded objects. We demonstrate use of prior shape information (length (Fig.1) and explicit shape (Fig.3,Fig.2) to handle occlusions. Once the mean intensities are obtained for occlusion regions, we show in examples (Fig.3) how occlusion relationships can be deduced for the objects. Finally in (Fig.5 and Fig.6), we see how these relationships have been used to impose shape constraints selectively.

### 8.1 Segmentation with Length and Explicit Shape

In the images in Fig.1, we assume that only linear segments of objects are occluded. Setting $\beta = 0$ in (6), we see that the length term can be used to resolve segmentation in occluded regions. Starting with initial level sets $\{\phi_k\}_{k=1}^N$(N$\leq$ max. number of objects expected) shown in (I), we minimize (6) to obtain segmentation results in (II). The initial level sets have to be overlayed at least

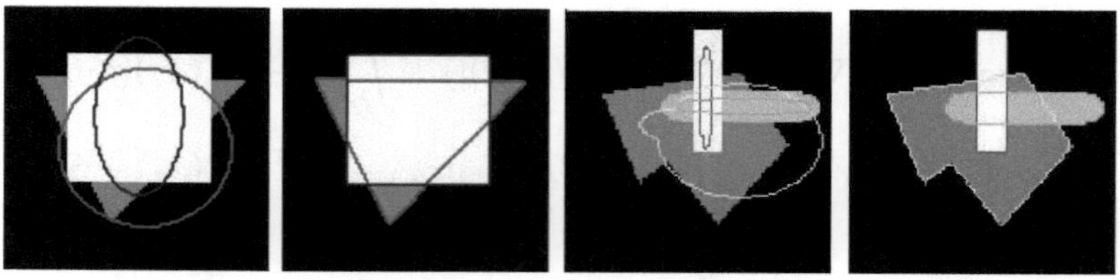

(I) Initialization (II) Result (III) Initialization (IV) Result

**Fig. 1.** Segmentation of occluded objects with length term

Shape prior (I) Initialization (II) Result

**Fig. 2.** Occluded boundaries are detected due to the use of shape

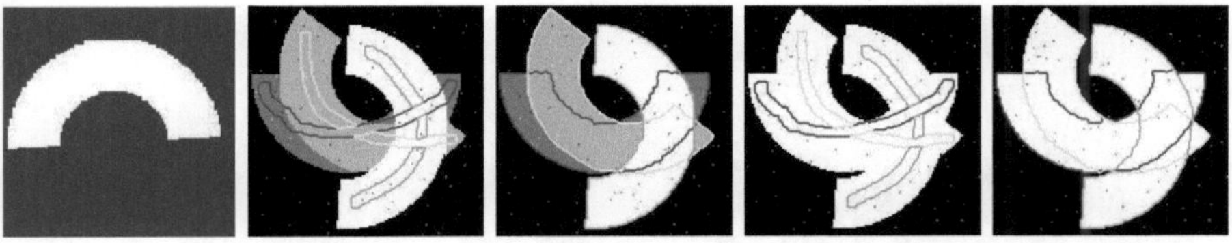


**Fig. 3.** Segmentation of occluded objects with explicit shape

partially on the corresponding objects, to avoid local minima. The mean intensities can be arbitrarily initialized.

In the next set of examples in Fig.2 and Fig.3, we assume that prior shape information on the objects to be segmented is available. In Fig.2, given an image with occlusions (I), in which the objects can be described by a prior shape (binary image shown on top), we minimize (8) using (10-12) to get the result in (II). The only initial values required here are the initial level sets (shown in I). Using this, a few iterations of the equations (11-12), are used to get an initial guess for the rigid transformation $T_k$. Fig.3 shows an example for the 3 objects case. Notice that use of prior shape information has resulted in a good segmentation in spite of lack of any intensity information (e.g. Image II).

In Fig.4 (I), we see a given EM image of erythrocytes (red blood cells), which are generally known to have a circular surface structure. Naturally, we use a prior shape of a circle in this case, to resolve occlusions. Starting with a initial guess for the active contours in (II), we arrive at result (IV), which is not possible without prior shape information (III).

### 8.2 Dis-occlusion

An immediate application of computing the constants $c_{p,k}$ by minimizing (5), is to deduce occlusion relationships between the objects. We demonstrate this procedure for the image shown in Fig.3(I) , with 3 objects. The object regions defined by the characteristic functions $\mathrm{H}(\phi_\mathrm{k}) := \mathrm{H}_\mathrm{k}$, with intensities $c_k$, define 4 occlusion regions, $\mathrm{P}_1 = \mathrm{H}_1\mathrm{H}_2\bar{\mathrm{H}}_3, \mathrm{P}_2 = \mathrm{H}_1\bar{\mathrm{H}}_2\mathrm{H}_3, \mathrm{P}_3 = \bar{\mathrm{H}}_1\mathrm{H}_2\mathrm{H}_3, \mathrm{P}_4 = \mathrm{H}_1\mathrm{H}_2\mathrm{H}_3$. The computed constants $c_{p,k}$ were $c_1 = 128.9, c_2 = 191.9, c_3 = 254.3, c_{2,1} = 129, c_{2,2} = 129.1, c_{2,3} = 192, c_{3,1} = 129.1$. The following table shows a comparison for these intensities in different occlusion regions. The third column shows the mean intensities $\mu_k$ corresponding to the regions $P_k$, that can be computed(shown in the second column) from the constants $c_{p,k}$.

When we compare the mean intensities $\mu_{p,k}$ with $c_k$, we see that,

$$|\mu_1 - c_2| < |\mu_1 - c_1| \text{ gives, } O_2 \text{ top of } O_1$$

$$|\mu_3 - c_3| < |\mu_3 - c_2| \text{ gives, } O_3 \text{ top of } O_2$$

which gives the ordering(with increasing depth) of the objects as $O_3$, $O_2$, $O_1$.

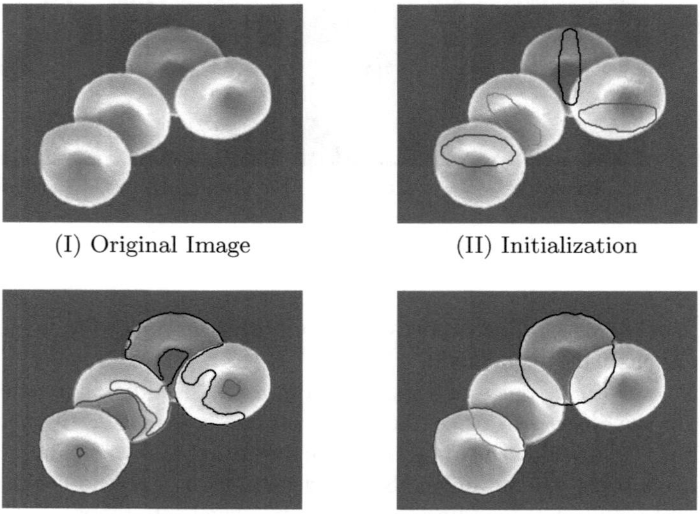

(I) Original Image (II) Initialization

(III) Without shape prior (IV) With a circular shape prior

**Fig. 4.** Segmentation of erythrocytes in EM image

**Table 1.** Comparison of mean intensities in different regions

| Occlusion Region | Mean,($\mu_k$) | Values |
|---|---|---|
| $P_1$ | $c_1 + c_2 - c_{2,1}$ | 191.8 |
| $P_2$ | $c_1 - c_{2,2} + c_3$ | 254.1 |
| $P_3$ | $-c_{2,3} + c_2 + c_3$ | 254.1 |
| $P_4$ | $c_{3,1} + c_{2,1} + c_{2,2} + c_{2,3} - c_1 - c_2 - c_3$ | 254 |

### 8.3 Selective Use of Shape

Finally, we will see examples where we have used occlusion relationships to impose shape constraints selectively. In Fig.5, we see an image of two objects say (A and B), with A occluding B. In addition, the object on top (A) has sharp features in the intersection region, which we want the segmentation to preserve. Assuming occluded boundaries to be linear, we use the length term to resolve occlusions. In (II), we minimized (6) with $\beta = 0$. The resulting segmentation has correctly filled in the missing linear boundaries for B, but has not segmented A correctly, since the use of length term evenly for both the objects has resulted in the loss of sharp features in A. When we use a selective length term as in (8) with $\hat{S}(\phi_k) = |\nabla H(\phi_k)|$ and setting the parameters $\tilde{\beta} = 0, \lambda \ll \beta$, the required boundaries are computed as needed (III). Notice that the length term is effected only for the object that is occluded (i.e. B), hence preserving the features of A.

A similar example is presented in Fig.6, with use of explicit shape. Here (I) shows two *square* shaped objects A & B with A occluding B, and each with one

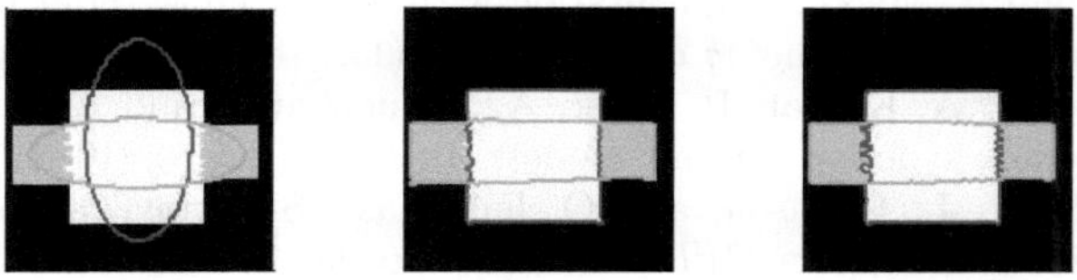


**Fig. 5.** Selective use of length only for occluded boundaries (I) Initialization (II) Without selective length (III) Selective length term

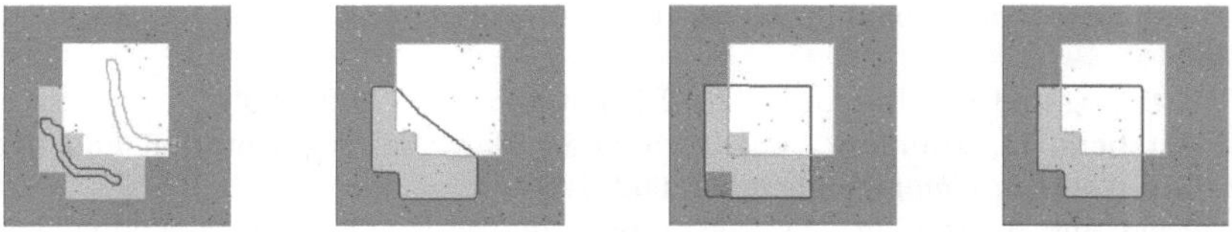


**Fig. 6.** Selective use of Shape (I) Initialization (II) Without shape (III) Without selective shape (IV) Selective shape

of their corners chipped off. We want our segmentation to be able to complete the missing boundary for B in the occlusion region, and also be able to preserve edges that are not occluded. (II) shows the result when a shape term is not used. (III) shows the result with a uniform shape term as in (8). Notice that the corners of both A & B are not segmented properly, due to the influence of the shape term, even in non-occluded regions. Finally, we get the correct segmentation in (IV), using a selective shape term as in (9). Firstly, use of the shape term has filled the missing boundary of B that has been occluded. Secondly, the shape term is applied only to the occluded object B. Hence the corner of A is recovered. Thirdly, the shape term applied to B is effected only within the object A, thus localizing the effect of shape on B. Hence the corner of B is also recovered.

## References

1. V. Caselles, R. Kimmel, and G. Sapiro. Geodesic active contours. In *ICCV*, pages 694–699, 1995.
2. T. Chan and W. Zhu. Level set based shape prior segmentation. In *Proc. CVPR'05*, pages 20–26, 2005.
3. T.F. Chan and L.A Vese. Active contours without edges. *IEEE Trans. Image Processing*, 10(2):266–277, 2001.
4. Yunmei Chen, Hemant Tagare, et al. Using prior shapes in geometric active contours in a variational framework. *IJCV*, 50(3):315–328, 2002.
5. Timothy Cootes, Christopher Taylor, David Cooper, and Jim Graham. Active shape models-their training and application. *CVIU*, 61(1):38–59, 1995.
6. D. Cremers, N. Sochen, and C. Schnörr. Towards recognition-based variational segmentation using shape priors and dynamic labeling. In *Proceedings of International Conference on Scale Space Theories in Computer Vision*, 2003.

7. S. Esedoglu and R. March. Segmentation with depth but without detecting junctions. *J. Mathematical Imaging and Vision*, 18(1):7–15, 2003.
8. S. Kichenassamy, A. Kumar, P. Olver, A. Tannenbaum, and A. Yezzi. Gradient flows and geometric active contour models. In *ICCV*, pages 810–815, 1995.
9. M. Leventon, W. L. Grimson, and O. Faugeras. Statistical shape influence in geodesic active contours. In *CVPR*, volume 1, pages 316–323, 2000.
10. R. Malladi, J.A. Sethian, and B.C. Vemuri. Shape modeling with front propagation: A level set approach. *IEEE Trans. on PAMI*, 17:158–175, 1995.
11. Rousson Mikaël, Paragios Nikos, and Deriche Rachid. Active shape models from a level set perspective. Technical Report 4984, INRIA, October 2003.
12. D. Mumford and J. Shah. Optimal approximations by piecewise smooth functions and associated variational problems. *Comm. on Pure and App. Math.*, 42:577–684, 1989.
13. M. Nitzberg and D. Mumford. The 2.1-d sketch. In *ICCV*, 1990.
14. M. Nitzberg, D. Mumford, and T. Shiota. Filtering, segmentation and depth. *Lecture Notes in Computer Science*, 662, 1993.
15. S. Osher and J. Sethian. Fronts propagating with curvature dependent speed: algorithms based on the Hamilton–Jacobi formulation. *J. of Comp. Phy.*, 79:12–49, 1988.
16. N. Paragios and R. Deriche. Geodesic active regions and level set methods for supervised texture segmentation. *IJCV*, 46(3):223–247, 2002.
17. A. Tsai, A. Yezzi, et al. Model-based curve evolution technique for image segmentation. In *In Proc. CVPR'01*, volume 1, pages 463–468, December 2001.
18. L. Vese and T. Chan. Multiphase level set framework for image segmentation using the mumford and shah model. *IJCV*, 50(3):271–293, 2002.

# On the Statistical Interpretation of the Piecewise Smooth Mumford-Shah Functional

Thomas Brox and Daniel Cremers

CVPR Group, University of Bonn
Römerstr. 164, 53117 Bonn, Germany
{brox,dcremers}@cs.uni-bonn.de

**Abstract.** In region-based image segmentation, two models dominate the field: the Mumford-Shah functional and statistical approaches based on Bayesian inference. Whereas the latter allow for numerous ways to describe the statistics of intensities in regions, the first includes spatially smooth approximations. In this paper, we show that the piecewise smooth Mumford-Shah functional is a first order approximation of Bayesian a-posteriori maximization where region statistics are computed in local windows. This equivalence not only allows for a statistical interpretation of the full Mumford-Shah functional. Inspired by the Bayesian model, it also offers to formulate an extended Mumford-Shah functional that takes the variance of the data into account.

## 1 Introduction

Since the beginning of image analysis research, there has been enormous interest in image segmentation. While the topic was handled in a quite heuristic manner for a long time, a more systematic approach to the problem has been initiated by three seminal works in the 1980s: the Bayesian formulation of Geman and Geman [9], the energy functional of Mumford and Shah [18,19], and the snakes model by Kass, Witkin, and Terzopoulos [13]. In all these works, the formerly purely algorithmic description of a segmentation method has been replaced by its formulation as an optimization problem. This systematic description based on sound mathematical concepts has considerably improved the understanding of image segmentation and, hence, supported the development of new models and better algorithms. The initially large gap between sound energy formulations and efficient ways to find solutions of these energies, in particular in case of the Mumford-Shah functional, was bridged by the works of Ambrosio and Tortorelli [1], Morel and Solimini [16,17], as well as the use of level set representations of contours by Caselles et al. [6], Chan and Vese [7], and Paragios and Deriche [22]. A further type of optimization strategy has emerged in the spatially discrete case with graph cut methods [10,3].

Whereas all three approaches to image segmentation are based on energy minimization, their motivation is quite different. In [26], Zhu and Yuille outlined many relations between the methods and algorithmic implementations such as region merging or region growing. In particular, they established a link between

F. Sgallari, A. Murli, and N. Paragios (Eds.): SSVM 2007, LNCS 4485, pp. 203–213, 2007.

a Bayesian approach to image segmentation and the piecewise constant case of the Mumford-Shah functional, sometimes called the *cartoon limit*. Zhu and Yuille also suggested a more general energy functional that replaces the constant approximation of image regions by arbitrary intensity distributions. This formulation was used particularly in level set based segmentation approaches where full Gaussian distributions [24], Laplace distributions [11], and nonparametric kernel densities [14] have been suggested.

Zhu and Yuille established relations between Bayesian methods and the cartoon limit of the Mumford-Shah functional, yet in their work, they ignored the part of the functional that allows also for piecewise *smooth* approximations. In the present paper, we complete their work by showing that the Mumford-Shah functional can be interpreted as a first-order approximation of a Bayesian model with probability densities estimated in local windows. Such types of densities have been used in [4] in the scope of contour-based pose estimation. Similar to the work of Zhu and Yuille [26], this equivalence allows to generalize the Mumford-Shah functional. We demonstrate this by proposing a functional which allows to approximate the input intensity by a piecewise smooth Gaussian distribution including mean *and* variance.

## 2 The Mumford-Shah Functional

The idea of Mumford and Shah was to find a piecewise smooth approximation $u : (\Omega \subset \mathbb{R}^2) \to \mathbb{R}$ of the image $I : (\Omega \subset \mathbb{R}^2) \to \mathbb{R}$ and an edge set $K$[1] separating the pieces of $u$, such that $u$ is close to $I$ and the total length of the edge set is minimal. This can be expressed as minimizing the functional

$$E(u, K) = \int_{\Omega - K} (u - I)^2 \mathbf{dx} + \lambda \int_{\Omega - K} |\nabla u|^2 \mathbf{dx} + \nu \, |K| \to \min, \qquad (1)$$

where $\lambda \geq 0$ and $\nu \geq 0$ are constant weighting parameters. An interesting special case arises for $\lambda \to \infty$, where $u$ is required to be piecewise constant. This case, already discussed by Mumford and Shah in [19], is also known as the *cartoon limit* and can be written in short form

$$E(u, K) = \sum_i \int_{\Omega_i} (u_i - I)^2 \mathbf{dx} + \nu_0 \, |K| \to \min, \qquad (2)$$

where $\Omega_i$ denotes the piecewise constant regions separated by $K$ and $\nu_0$ is the rescaled version of the parameter $\nu$ in (1). Due to the quadratic error measure, given $\Omega_i$, the solution of $u_i$ is the mean of $I$ within $\Omega_i$. A related approach was independently developed by Blake and Zisserman [2]. In the spatially discrete case, (2) is related to the Potts model [23].

The model in (2) can be simplified further by assuming a fixed number of regions $N$. In particular, the case $N = 2$ and its level set formulation by Chan

[1] Since our focus lies on image segmentation, we will only consider edge sets which are sets of closed curves [17].

and Vese [7] has become very popular. A discrete version of the binary case has been introduced by Lenz and Ising for modeling ferromagnetism already in the 1920s [15,12].

## 3 Bayesian Model and Local Region Statistics

An alternative approach to image segmentation can be derived using Bayes' rule

$$p(K|I) = \frac{p(I|K)p(K)}{p(I)}. \tag{3}$$

Here one seeks for a partitioning by the edge set $K$ that maximizes the a-posteriori probability given the image $I$. The first factor in the nominator is in general approximated by an intensity distribution in the regions $i = 1, ..., N$ separated by $K$. The second factor is the a-priori probability of a certain partitioning $K$. Usually, the total length of the edge set $K$ is assumed to be small,

$$p(K) = \exp(-\nu_B|K|), \tag{4}$$

but other more sophisticated shape priors can be integrated here, as well [8]. Assuming independence of intensities at different locations $\mathbf{x}$, one can write

$$p(I|K) = \prod_{\mathbf{x}\in\Omega} p(I(\mathbf{x})|K,\mathbf{x})^{\mathbf{dx}}, \tag{5}$$

a continuous product with $\mathbf{dx}$ being the infinitesimal bin size. With the partitioning of $\Omega$ by the edge set $K$ into disjoint regions $\Omega = \bigcup_i \Omega_i$, $\Omega_i \cap \Omega_j = \emptyset, \forall i \neq j$, the product over the whole domain $\Omega$ can be separated into products over the regions:

$$\prod_{\mathbf{x}\in\Omega} p(I(\mathbf{x})|K,\mathbf{x})^{\mathbf{dx}} = \prod_i \prod_{\mathbf{x}\in\Omega_i} p(I(\mathbf{x})|\mathbf{x},\mathbf{x}\in\Omega_i)^{\mathbf{dx}}. \tag{6}$$

For convenience we define the conditional probability density to encounter an intensity $s$ at position $\mathbf{x}$ given that $\mathbf{x} \in \Omega_i$ as

$$p_i(s,\mathbf{x}) := p(s|\mathbf{x},\mathbf{x}\in\Omega_i). \tag{7}$$

Note that we have here a family of probability densities $p_i(s,\mathbf{x})$ for all $\mathbf{x} \in \Omega$, i.e.,

$$\begin{aligned} &p_i(s,\mathbf{x}) : \mathbb{R} \to \mathbb{R}_0^+ \\ &p_i(s,\mathbf{x}) \geq 0 \quad \forall s \in \mathbb{R}, \forall \mathbf{x} \in \Omega \int_{\mathbb{R}} p_i(s,\mathbf{x})ds = 1 \forall \mathbf{x} \int \Omega. \end{aligned} \tag{8}$$

In general, it is preferable to express the maximization of (3) by the minimization of its negative logarithm. With the above assumptions, this leads to the energy functional

$$E(K) = \sum_i \int_{\Omega_i} -\log p_i(I(\mathbf{x}),\mathbf{x})\mathbf{dx} + \nu_B\,|K|. \tag{9}$$

It obviously resembles the cartoon limit of the Mumford-Shah functional. We will come back to this issue in the next section.

There are several possibilities how to model the probability densities $p_i$. Typically, one assumes a homogeneous Gaussian distribution in each region $\Omega_i$:

$$p_i(s) = \frac{1}{\sqrt{2\pi}\sigma_i} \exp\left(-\frac{(s-\mu_i)^2}{2\sigma_i^2}\right), \tag{10}$$

where $\mu_i$ and $\sigma_i$ denote the mean and standard deviation of $I$ in region $\Omega_i$. Other choices like a Laplace distribution [11] or a nonparametric density [14] are possible, as well. All these models apply the same probability density to all points in a region. Hence, we will call them spatially *homogeneous* region models.

In contrast, *local* region models take the spatial position into account, i.e., there is in general a different probability density at each point $\mathbf{x}$ in the region. For a Gaussian distribution this yields [4]:

$$p_i(s, \mathbf{x}) = \frac{1}{\sqrt{2\pi}\sigma_i(\mathbf{x})} \exp\left(\frac{(s-\mu_i(\mathbf{x}))^2}{2\sigma_i(\mathbf{x})^2}\right). \tag{11}$$

Estimation of the parameters $\mu_i(\mathbf{x})$ and $\sigma_i(\mathbf{x})$ can be achieved using a window function, e.g. a Gaussian $G_\rho$ with standard deviation $\rho$, and restricting the estimation only to points within this window:

$$\mu_i(\mathbf{x}) = \frac{\int_{\Omega_i} G_\rho(\zeta-\mathbf{x}) I(\zeta)\, d\zeta}{\int_{\Omega_i} G_\rho(\zeta-\mathbf{x})\, d\zeta} \qquad \sigma_i(\mathbf{x}) = \frac{\int_{\Omega_i} G_\rho(\zeta-\mathbf{x}) (I(\zeta)-\mu_i(\mathbf{x}))^2\, d\zeta}{\int_{\Omega_i} G_\rho(\zeta-\mathbf{x})\, d\zeta}. \tag{12}$$

Obviously, the local region model converges to the corresponding homogeneous model for $\rho \to \infty$.

## 4 Bayesian Interpretation of the Mumford-Shah Functional

The Bayesian model from the last section is quite flexible in the choice of the probability density function. It further yields a nice statistical interpretation of the model assumptions and allows for the sound integration of a-priori information. On the other hand, the Mumford-Shah functional combines segmentation and image restoration by a piecewise smooth function. The reader may have already noticed similarities between the models in Section 2 and Section 3. In this section, we will investigate the relation between both segmentation approaches aiming at a statistical interpretation of the *full* Mumford-Shah functional.

We start with the Bayesian model in (9). A comparison to the cartoon model in (2) reveals a large similarity. As shown in [26], for a specific choice of the probability densities, both formulations turn out to be equivalent. Indeed, equivalence of (2) and (9) is established by modeling the probability densities as Gaussian functions with fixed standard deviation

$$p_i(s) = \frac{1}{\sqrt{2\pi}\sigma} \exp\left(-\frac{(s-\mu_i)^2}{2\sigma^2}\right). \tag{13}$$

Applying the logarithm

$$\log p_i(s) = -\frac{1}{2}\log(2\pi\sigma^2) - \frac{(s-\mu_i)^2}{2\sigma^2} \tag{14}$$

and plugging this into (9) yields

$$\begin{aligned} E(K) &= \sum_i \int_{\Omega_i} \frac{1}{2}\log(2\pi\sigma^2) + \frac{(I(\mathbf{x})-\mu_i)^2}{2\sigma^2}\mathbf{dx} + \nu_B\,|K| \\ &= \sum_i \int_{\Omega_i} \frac{(I(\mathbf{x})-\mu_i)^2}{2\sigma^2}\mathbf{dx} + \nu_B\,|K| + \text{const.} \end{aligned} \tag{15}$$

Due to the same fixed standard deviation in all regions, the logarithm term containing $\sigma$ does not depend on $K$ and, hence, is a negligible constant in the energy functional. Also the denominator $2\sigma^2$ is a constant and merely leads to a rescaling of the parameter $\nu_B$. Thus, with $\mu_i \equiv u_i$, $\sigma = \sqrt{0.5}$, and $\nu_B = \nu_0$, (15) states exactly the same energy minimization problem as the cartoon model in (2).

With this equivalence in mind, the question arises, whether there exists a choice of the probability density function that relates the Bayesian model to the full, piecewise smooth Mumford-Shah functional stated in (1). Since (1) explicitly allows the approximation $u$ to vary within a region, a homogeneous region model is obviously not sufficient. Local region statistics, on the other hand, include varying parameters in the region. Hence, having in mind that the equivalence of the Bayesian model and the cartoon model was established for a homogeneous Gaussian region model with fixed standard deviation, we take a closer look at the *local* Gaussian model, again with fixed standard deviation.

Since the standard deviation is fixed, we can focus on the local mean in (12):

$$\mu_i(\mathbf{x}) = \frac{\int_{\Omega_i} G_\rho(\zeta - \mathbf{x}) I(\zeta)\, d\zeta}{\int_{\Omega_i} G_\rho(\zeta - \mathbf{x})\, d\zeta}. \tag{16}$$

The numerator is a convolution of the image $I$ with the Gaussian function $G_\rho$. The denominator is only for normalization in case the window hits the boundary of $\Omega_i$. It ensures the preservation of the average gray value of $\mu_i$ in the domain $\Omega_i$ independent of $\rho$.

In order to bend the bow to the Mumford-Shah functional, we will relate this filtering operation to a regularization framework. Yuille and Grzywacz [25] as well as Nielsen et al. [21] showed that the outcomes of some linear filters are exact minimizers of certain energy functionals with an infinite sum of penalizer terms of arbitrarily high order. More precisely, it was shown in [21] that filtering an image $I$ with the filter

$$\hat{h}(\omega) = \frac{1}{1 + \sum_{k=1}^{\infty} \alpha_k \omega^{2k}} \tag{17}$$

given in the frequency domain, yields the minimizer of the following energy functional:

$$E(u) = \int_{\mathbb{R}} \left( (u-I)^2 + \sum_{k=1}^{\infty} \alpha_k \left( \frac{d^k u}{dx^k} \right)^2 \right) dx. \tag{18}$$

In particular, this includes for $\alpha_k = \frac{\lambda^k}{k!}$, the Gaussian filter

$$\hat{h}(\omega, \lambda) = \frac{1}{1 + \sum_{k=1}^{\infty} \frac{\lambda^k}{k!} \omega^{2k}} = \exp(-\lambda \omega^2). \tag{19}$$

This filter corresponds to the Gaussian $G_\rho$ with standard deviation $\rho = \sqrt{2\lambda}$ in the spatial domain. Nielsen et al. further showed in [20] that for Cartesian invariants, such as the Gaussian, this correspondence can be generalized to higher dimensions. Therefore, the convolution result in (16) is the exact minimizer of

$$E(\mu_i) = \int_{\Omega_i} \left( (\mu_i - I)^2 + \sum_{k=1}^{\infty} \frac{\lambda^k}{k!} \sum_{j_1+j_2=k} \left( \frac{d^k \mu_i}{dx^{j_1} dy^{j_2}} \right)^2 \right) \mathbf{dx} \tag{20}$$

with natural boundary conditions.

Based on these findings, we can proceed to generalize the piecewise constant case in (15). We plug the local Gaussian probability density from (11) with fixed standard deviation $\sigma = \sqrt{0.5}$ into the Bayesian model in (9):

$$\begin{aligned} E(\mu, K) &= \sum_i \int_{\Omega_i} \frac{1}{2} \log(2\pi\sigma^2) + \frac{(I(\mathbf{x}) - \mu_i(\mathbf{x}))^2}{2\sigma^2} \mathbf{dx} + \nu_B \, |K| \\ &= \sum_i \int_{\Omega_i} (I(\mathbf{x}) - \mu_i(\mathbf{x}))^2 \mathbf{dx} + \nu_B \, |K| + \text{const.} \end{aligned} \tag{21}$$

The means $\mu_i$ have in (16) been defined as the results of local convolutions. As we have just found, this convolution result is the minimizer of (20). Hence, we can write the Bayesian energy as:

$$E_B(\mu, K) = \sum_i \int_{\Omega_i} \left( (\mu_i - I)^2 + \sum_{k=1}^{\infty} \frac{\lambda^k}{k!} \sum_{j_1+j_2=k} \left( \frac{d^k \mu_i}{dx^{j_1} dy^{j_2}} \right)^2 \right) \mathbf{dx} + \nu_B \, |K|. \tag{22}$$

Neglecting all penalizer terms of order $k > 1$ yields

$$E_{MS}(\mu, K) = \sum_i \int_{\Omega_i} \left( (\mu_i - I)^2 + \lambda |\nabla \mu_i|^2 \right) \mathbf{dx} + \nu_B \, |K| + \text{const.} \tag{23}$$

which states exactly the Mumford-Shah functional in (1). Consequently, *minimizing the full piecewise smooth Mumford-Shah functional is equivalent to a first-order approximation of a Bayesian a-posteriori maximization based on local region statistics.* In particular, it is the approximation of the Bayesian setting with a Gaussian distribution, fixed standard deviation $\sigma = \sqrt{0.5}$, and a Gaussian windowing function where $\rho = \sqrt{2\lambda}$ and $\nu_B = \nu$.

What is the effect of neglecting the higher order terms, as done by the Mumford-Shah functional? The main effect is that the minimizers $\mu_i$ of the functional in (23) are less smooth than those of the functional in (22). Figure 1 depicts a comparison in case of the whole image domain being a single region. Obviously, the visual difference is almost negligible, and it can be further reduced by choosing $\lambda$ in the first-order approximation slightly larger than in the regularizer containing the infinite sum of penalizers.

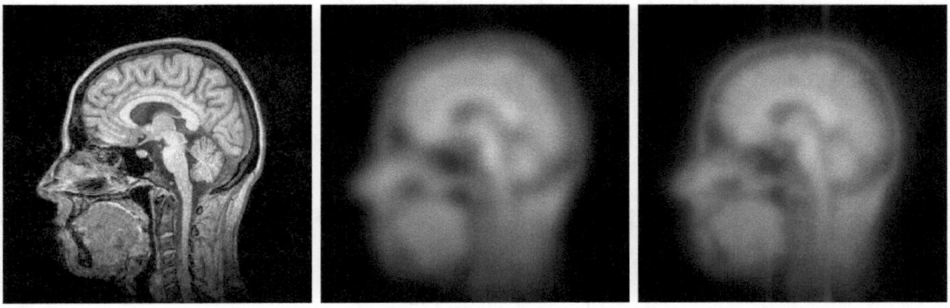

**Fig. 1.** Comparison of regularization with and without higher order penalizers. **Left:** Original image. **Center:** Smoothing result with the regularizer in (22) (Gaussian smoothing) for $\lambda = 20$. **Right:** Smoothing results with the regularizer in (23) for $\lambda = 20$.

## 5 Extending the Mumford-Shah Functional

In the previous section, we have shown that the full, piecewise smooth version of the Mumford-Shah functional is a first-order approximation of a Bayesian segmentation approach assuming local Gaussian distributions with a fixed standard deviation. In this section, we will make use of this relation in order to extend the Mumford-Shah functional in a way that it also takes the variance of the data into account. In the Bayesian formulation, this is easy to achieve, as shown in Section 3. Hence, we can take the Bayesian model and express the convolutions by regularization formulations.

With the full Gaussian model, the probability densities

$$p_i(s, \mathbf{x}) = \frac{1}{\sqrt{2\pi}\sigma_i(\mathbf{x})} \exp\left(\frac{(s - \mu_i(\mathbf{x}))^2}{2\sigma_i(\mathbf{x})^2}\right). \tag{24}$$

depend on two functions $\mu_i(\mathbf{x})$ and $\sigma_i(\mathbf{x})$ given by (12). For $\rho \to \infty$ they are the mean and standard deviation of $I$ in $\Omega_i$, i.e., the minimizers of

$$\int_{\Omega_i} \left(\frac{(\mu_i - I)^2}{2\sigma_i^2} + \frac{1}{2}\log(2\pi\sigma_i^2) + \lambda\left(|\nabla\mu_i|^2 + |\nabla\sigma_i|^2\right)\right) \mathbf{dx} \tag{25}$$

for $\lambda \to \infty$. This yields a generalized cartoon model. For $\rho \ll \infty$ we make use of the relation between Gaussian convolution and regularization stated in the previous section and obtain $\mu_i(\mathbf{x})$ and $\sigma_i(\mathbf{x})$ as the minimizers of

$$\begin{aligned} E(\mu_i, \sigma_i) = &\int_{\Omega_i} \left( \frac{(\mu_i - I)^2}{2\sigma_i^2} + \frac{1}{2} \log(2\pi\sigma_i^2) \right) \mathbf{dx} \\ &+ \int_{\Omega_i} \sum_{k=1}^{\infty} \frac{\lambda^k}{k!} \sum_{j_1+j_2=k} \left( \frac{d^k \mu_i}{dx^{j_1} dy^{j_2}} \right)^2 \mathbf{dx} \\ &+ \int_{\Omega_i} \sum_{k=1}^{\infty} \frac{\lambda^k}{k!} \sum_{j_1+j_2=k} \left( \frac{d^k \sigma_i}{dx^{j_1} dy^{j_2}} \right)^2 \mathbf{dx} \end{aligned} \tag{26}$$

and the Bayesian energy can be written as

$$E_B(\mu, \sigma, K) = \sum_i E(\mu_i, \sigma_i) + \nu |K|. \tag{27}$$

Based on the observation in Section 4, a qualitatively similar approach is obtained by neglecting the penalizer terms with $k > 1$

$$\begin{aligned} E_{MS}(\mu, \sigma, K) = &\int_{\Omega - K} \left( \frac{(\mu - I)^2}{2\sigma^2} + \frac{1}{2} \log(2\pi\sigma^2) \right) \mathbf{dx} \\ &+\lambda \int_{\Omega - K} \left( |\nabla \mu|^2 + |\nabla \sigma|^2 \right) \mathbf{dx} + \nu\, |K|, \end{aligned} \tag{28}$$

which we may call an extended version of the Mumford-Shah functional. Main advantage of this extension over the original Mumford-Shah functional is that the parameter $\nu$ gets invariant with respect to the image contrast. This contrast invariance becomes even more interesting when dealing with vector-valued input images and estimating a separate variance for each vector channel. The influence of each channel on the segmentation then only depends on its discriminative properties and not on the magnitude of the channel values. This allows for the sound integration of different input channels with different contrast and noise levels. For a proof in case of a global Gaussian model we refer to [5]. This proof can be adopted for the local Gaussian model in a straightforward manner.

Another advantage of taking the variance into account is the possibility to distinguish regions that are equal in their mean value, but differ in the variance.

Figure 2 illustrates a result obtained with the extended Mumford-Shah functional. For the experiment we used a level set implementation and expect two regions in the image. Our implementation is based on gradient descent and, hence, can only ensure a local optimum that need not necessarily be the global one. The initial contour is shown in Figure 2a. The background region of the input image has been generated by a constant function at 127 and Gaussian noise with standard deviation 20. The circular foreground region contains a gradient ranging from 0 to 255. Gaussian noise with standard deviation 70 has been added to this region. The resulting contour and the local mean approximation

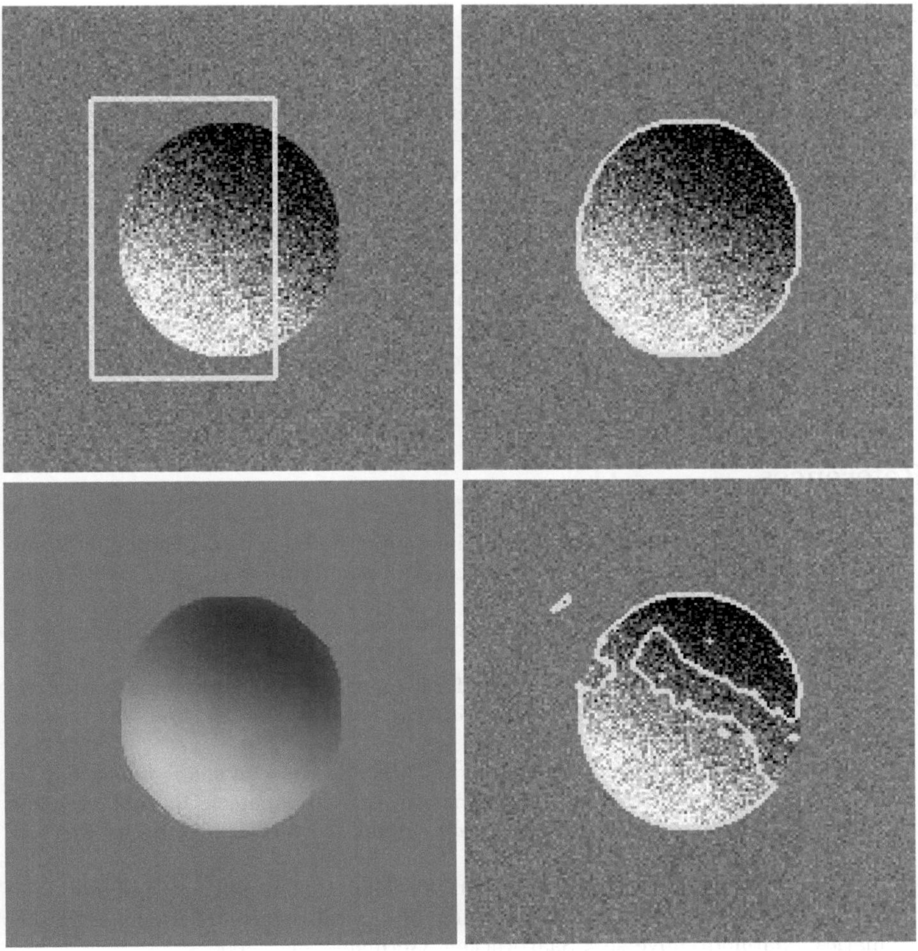

**Fig. 2.** Example for two regions. **Top left:** (a) Original image of size $162 \times 171$ pixels with the initial contour. **Top right:** (b) Contour obtained with the extended Mumford-Shah functional in (28). **Bottom left:** (c) Approximated mean $\mu$. **Bottom right:** (d) Contour obtained with the original Mumford-Shah functional.

for $\lambda = 32$ and $\nu = 32$ are shown in Figure 2b and Figure 2c, respectively. For comparison, we depict in Figure 2d the contour found with the same implementation but the standard deviation set fixed, i.e., the original Mumford-Shah functional. For this case, the parameter $\nu$ had to be increased to $\nu = 1000$ to obtain reasonable results. Since the two regions have different variances, which can only be exploited by the extended Mumford-Shah functional, the extension finds a more attractive solution than the original version. Larger $\nu$ in the original Mumford-Shah functional cannot improve the quality of the result as they lead to an over-smoothed contour not capturing the full circle anymore.

## 6 Summary

We have provided a statistical interpretation of the Mumford-Shah functional with piecewise smooth regions by showing its relations to Bayesian image segmentation with local region statistics. The link has been established by means of a theorem that relates Gaussian convolution to a regularization problem with an infinite sum of penalizers of arbitrarily high order. Based on this relation, we showed that the Mumford-Shah functional is equivalent to a first-order approximation of a Bayesian approach with Gaussian probability densities estimated with a Gaussian windowing function and the standard deviation set fixed. By means of this relation, we derived an extended version of the Mumford-Shah functional from the Bayesian model which includes the standard deviation as a spatially varying, dynamic function.

## References

1. L. Ambrosio and V. Tortorelli. Approximation of functionals depending on jumps by elliptic functionals via $\gamma$-convergence. *Communications on Pure and Applied Mathematics*, XLIII:999–1036, 1990.
2. A. Blake and A. Zisserman. *Visual Reconstruction.* MIT Press, Cambridge, MA, 1987.
3. Y. Boykov, O. Veksler, and R. Zabih. Fast approximate energy minimization via graph cuts. *IEEE Transactions on Pattern Analysis and Machine Intelligence*, 23(11):1222–1239, 2001.
4. T. Brox, B. Rosenhahn, and J. Weickert. Three-dimensional shape knowledge for joint image segmentation and pose estimation. In W. Kropatsch, R. Sablatnig, and A. Hanbury, editors, *Pattern Recognition*, volume 3663 of *LNCS*, pages 109–116. Springer, Aug. 2005.
5. T. Brox, M. Rousson, R. Deriche, and J. Weickert. Colour, texture, and motion in level set based segmentation and tracking. Technical report 147, Dept. of Mathematics, Saarland University, Saarbrücken, Germany, Aug. 2005.
6. V. Caselles, F. Catté, T. Coll, and F. Dibos. A geometric model for active contours in image processing. *Numerische Mathematik*, 66:1–31, 1993.
7. T. Chan and L. Vese. Active contours without edges. *IEEE Transactions on Image Processing*, 10(2):266–277, Feb. 2001.
8. D. Cremers, F. Tischhäuser, J. Weickert, and C. Schnörr. Diffusion snakes: introducing statistical shape knowledge into the mumford-shah functional. *International Journal of Computer Vision*, 50(3):295–313, Dec. 2002.
9. S. Geman and D. Geman. Stochastic relaxation, Gibbs distributions, and the Bayesian restoration of images. *IEEE Transactions on Pattern Analysis and Machine Intelligence*, 6:721–741, 1984.
10. D. Greig, B. Porteous, and A. Seheult. Exact maximum a posteriori estimation for binary images. *Journal of the Royal Statistical Society B*, 51(2):271–279, 1989.
11. M. Heiler and C. Schnörr. Natural image statistics for natural image segmentation. *International Journal of Computer Vision*, 63(1):5–19, 2005.
12. E. Ising. Beitrag zur Theorie des Ferromagnetismus. *Zeitschrift für Physik*, 31:253–258, 1925.

13. M. Kass, A. Witkin, and D. Terzopoulos. Snakes: Active contour models. *International Journal of Computer Vision*, 1:321–331, 1988.
14. J. Kim, J. Fisher, A. Yezzi, M. Cetin, and A. Willsky. A nonparametric statistical method for image segmentation using information theory and curve evolution. *IEEE Transactions on Image Processing*, 14(10):1486–1502, 2005.
15. W. Lenz. Beitrag zum Verständnis der magnetischen Erscheinungen in festen Körpern. *Physikalische Zeitschrift*, 21:613–615, 1920.
16. J.-M. Morel and S. Solimini. Segmentation of images by variational methods: a constructive approach. *Revista Matematica de la Universidad Complutense de Madrid*, 1:169–182, 1988.
17. J.-M. Morel and S. Solimini. *Variational Methods in Image Segmentation.* Birkhäuser, Basel, 1994.
18. D. Mumford and J. Shah. Boundary detection by minimizing functionals, I. In *Proc. IEEE Computer Society Conference on Computer Vision and Pattern Recognition*, pages 22–26, San Francisco, CA, June 1985. IEEE Computer Society Press.
19. D. Mumford and J. Shah. Optimal approximations by piecewise smooth functions and associated variational problems. *Communications on Pure and Applied Mathematics*, 42:577–685, 1989.
20. M. Nielsen, L. Florack, and R. Deriche. Regularization and scale space. Technical Report 2352, INRIA Sophia-Antipolis, France, Sept. 1994.
21. M. Nielsen, L. Florack, and R. Deriche. Regularization, scale-space and edge detection filters. *Journal of Mathematical Imaging and Vision*, 7:291–307, 1997.
22. N. Paragios and R. Deriche. Geodesic active regions: A new paradigm to deal with frame partition problems in computer vision. *Journal of Visual Communication and Image Representation*, 13(1/2):249–268, 2002.
23. R. Potts. Some generalized order-disorder transformation. *Proceedings of the Cambridge Philosophical Society*, 48:106–109, 1952.
24. M. Rousson and R. Deriche. A variational framework for active and adaptive segmentation of vector-valued images. In *Proc. IEEE Workshop on Motion and Video Computing*, pages 56–62, Orlando, Florida, Dec. 2002.
25. A. Yuille and N. M. Grzywacz. A computational theory for the perception of coherent visual motion. *Nature*, 333:71–74, 1988.
26. S.-C. Zhu and A. Yuille. Region competition: unifying snakes, region growing, and Bayes/MDL for multiband image segmentation. *IEEE Transactions on Pattern Analysis and Machine Intelligence*, 18(9):884–900, Sept. 1996.

# Fuzzy Region Competition: A Convex Two-Phase Segmentation Framework

Benoit Mory and Roberto Ardon

Philips Medical Systems Research Paris,
51 rue Carnot, B.P. 301, F-92156 SURESNES Cedex, France
{benoit.mory,roberto.ardon}@philips.com

**Abstract.** We describe a novel framework for two-phase image segmentation, namely the *Fuzzy Region Competition*. The functional involved in several existing models related to the idea of *Region Competition* is extended by the introduction of a fuzzy membership function. The new problem is convex and the set of its global solutions turns out to be stable under thresholding, operation that also provides solutions to the corresponding classical formulations. The advantages are then shown in the piecewise-constant case. Finally, motivated by medical applications such as angiography, we derive a fast algorithm for segmenting images into two non-overlapping smooth regions. Compared to existing piecewise-smooth approaches, this last model has the unique advantage of featuring closed-form solutions for the approximation functions in each region based on normalized convolutions. Results are shown on synthetic 2D images and real 3D volumes.

## 1 Introduction

Several successful variational approaches to region-based image segmentation are based on the minimization of a functional that includes both boundary and region integrals [5,7,11,17]. When a two-phase partition of an image $I$ is considered, a general form of the optimization problem reads :

$$(\mathcal{P}_0) \quad \min_{\substack{\Sigma\subset\Omega \\ \boldsymbol{\alpha}\in A}} \left\{ F_0(\Sigma,\boldsymbol{\alpha}) = \int_\Gamma g(\Gamma(s))ds + \int_\Sigma r_1^{\alpha_1}(x)dx + \int_{\Omega\setminus\Sigma} r_2^{\alpha_2}(x)dx \right\}$$

where $\Omega \subset \mathbb{R}^n$ is the image domain, $\Sigma \subset \Omega$ the foreground region and $\Gamma = \partial\Sigma$ its boundary. The error functions $r_i^{\alpha_i} : \Omega \to \mathbb{R}$ encode the underlying model of each region in terms of intensity properties. They may depend on a set of *region parameters* $\boldsymbol{\alpha} = (\alpha_1, \alpha_2)$, such as a couple of scalars [5], vectors [11,17] or functions [13,14] (see Table 1). $g$ is a positive boundary potential, usually chosen to be a decreasing function of the image gradient. If the optimal $\boldsymbol{\alpha}$ is known *a priori*, $(\mathcal{P}_0)$ is a *supervised* segmentation problem, constrained by the geodesic length of the boundary [11]. Otherwise, the segmentation is *unsupervised* and $\boldsymbol{\alpha}$

F. Sgallari, A. Murli, and N. Paragios (Eds.): SSVM 2007, LNCS 4485, pp. 214–226, 2007.

**Table 1.** Possible error functions $r_i$ and region parameters

| | | |
|---|---|---|
| *Active Contours Without Edges* [5] | $r_i = \lambda\,(I - c_i)^2$ | $A = \mathbb{R}^2$ |
| *Region Competition* [17]<br>*Geodesic Active Regions* [11] | $r_i = -\lambda\,\log P_i(I\|\alpha_i)$ | $A = (\mathbb{R}^k)^2$ |
| *Mumford-Shah* [9,13,14] | $r_i = \lambda\,(I - s_i)^2 + \mu\,\|\nabla s_i\|^2$ | $A = (\mathcal{C}^1)^2$ |

has to be included in the optimization process. A natural approach is then to perform successive minimization steps alternatively on $\Sigma$ (the *partition* variable) and on the components of $\boldsymbol{\alpha}$ (the *region parameters*).

Since the *Region Competition*[1] algorithm [17], a popular way of solving $(\mathcal{P}_0)$ is to rewrite $F_0$ as a functional that only contains boundary integrals, using *Green*'s theorem. Gradient-descent evolution schemes can then be built upon optimality conditions for $\Gamma$ (*Euler-Lagrange* equations). The boundary evolution is most often carried out numerically in the *Level Sets* framework [10] in which $\Gamma$ is represented implicitly by a real-valued function $\phi$ such that $\Gamma = \phi^{-1}(0)$. This boundary-oriented approach has intrinsic convergence limitations, since neither the involved functional nor the optimization space - the set of curves - are convex. In practice, this produces a dependency on initial conditions and prevents the use of powerful tools from convex optimization theory.

A different perspective on $(\mathcal{P}_0)$ is to rewrite $F_0$ as a functional that only contains integrals over the whole domain $\Omega$. To that end, *Level Sets* have also proved useful since the *Active Contours Without Edges* and related models [16,5,14]. The key idea is to replace in $F_0$ the set $\Sigma$ by the Heaviside function $H(\phi)$ and derive *Euler-Lagrange* equations directly for $\phi$, without explicit use of the boundary. This approach inherits some aforementioned drawbacks. It is ill-posed[2] in $\phi$ and the optimization space - the set of characteristic functions - is still clearly non-convex. This recently motivated very promising works [2,4] that consider alternative convex methods based on total variation regularization for the global minimization with respect to $\phi$.

Along the same line, our present contribution is three-fold. In section 2, we first propose a generic formulation that has a number of interesting properties, including convexity with respect to the *partition* variable. This leads to a new framework, namely the *Fuzzy Region Competition*, that can be applied to any problem under the form $(\mathcal{P}_0)$. We illustrate the *supervised* case by using a statistical region term borrowed from the original *Region Competition* [17]. In section 3, we also apply our convex framework to the classical error function of the cartoon *Mumford-Shah* functional [5,9], assuming constant regions. In section 4, motivated by medical applications such as angiography, we finally propose, develop and illustrate a new localized extension of this model that includes an intrinsic notion of scale, namely the *Smooth Region Competition*. Built on windowing and

[1] The name illustrates the *competition* between two repulsive forces applied at the boundary, depending on the *competition function* $r = r_1 - r_2$.

[2] An infinite number of solutions $\phi$ are valid representations of a given optimal $\Sigma$.

convolutions, this last segmentation method is very efficient and gives qualitative results that are similar to existing piecewise-smooth models based on *Level Sets*. Examples are shown in 2D and 3D.

## 2 Two-Phase Fuzzy Region Competition

The crux of our general formulation is that $(\mathcal{P}_0)$ can be solved by considering a closely related problem that is at least convex in its *partition* variable. Indeed, $(\mathcal{P}_0)$ is not convex since the set of sub-domains $\Sigma \subset \Omega$ is not convex. Nonetheless, it can be expressed as an optimization problem in the set of characteristic functions[3] (still non-convex) so that it also reads

$$\min_{\chi,\boldsymbol{\alpha}}\left\{\tilde{F}_0(\chi,\boldsymbol{\alpha}) = \int_\Omega g(x)\,|\nabla\chi(x)|\,dx + \int_\Omega \chi(x) r_1^{\alpha_1}(x)dx + \int_\Omega (1-\chi(x)) r_2^{\alpha_2}(x)dx\right\}$$

Under this form, we propose to extend $(\mathcal{P}_0)$ into a problem that is convex in its *partition* variable, replacing the characteristic function $\chi$ by a fuzzy membership function $u$ belonging to a convex set. A suitable choice for this set is the space of functions of bounded variations taking their values in $[0,1]$, noted hereafter $BV_{[0,1]}(\Omega)$. This extension leads to the new *Fuzzy Region Competition* problem:

$$(\mathcal{P}) \quad \min_{u\in BV_{[0,1]},\boldsymbol{\alpha}}\left\{F(u,\boldsymbol{\alpha}) = \int_\Omega g\,|\nabla u| + \int_\Omega u(x) r_1^{\alpha_1}(x)dx + \int_\Omega (1-u(x)) r_2^{\alpha_2}(x)dx\right\}$$

Being convex in $u$, problem $(\mathcal{P})$ has only solutions globally minimizing $F$ and can be solved (for fixed $\boldsymbol{\alpha}$) with efficient algorithms. Note that a multi-phase formulation with membership functions has recently been used with a different regularization term in [12] for 'soft' segmentation. Our formulation has the advantage of also providing solutions to the 'hard' segmentation problem $(\mathcal{P}_0)$:

**Proposition 1.** *Fixing $\boldsymbol{\alpha}$, if $u^*$ is a global minimizer of $F$ in $BV_{[0,1]}(\Omega)$ then for almost every $t \in [0,1]$, the characteristic function*[4] *$x \mapsto \chi_{u^*}(x,t)$ of the set $\Sigma_t = \{x \in \Omega, u^*(x) > t\}$ is also a global minimizer of $F$. In addition, $\Sigma_t$ is a global minimizer of $F_0$.*

*Proof:* For any functions $u$ in $BV_{[0,1]}(\Omega)$ and $r$ in $L^1(\Omega)$, we have:

*Coarea formula* [6]:
$$\int_\Omega g\,|\nabla u| = \int_0^1 \int_\Omega g(x)\,|\nabla \chi_u(x,t)|\,dxdt \tag{1}$$

$$\int_\Omega ur = \int_\Omega \left(\int_0^{u(x)} dt\right) r(x)dx = \int_\Omega \left(\int_0^1 \chi_u(x,t)dt\right) r(x)dx = \int_0^1\!\!\int_\Omega \chi_u(x,t) r(x)dxdt \tag{2}$$

For sake of simplicity, we shall omit the fixed variable $\boldsymbol{\alpha}$ in the following. Applying (1) and (2) to the minimizer $u^*$, we obtain $F(u^*) = \int_0^1 F(\chi_{u^*}(\cdot,t))dt$, which is equivalent to $\int_0^1 \{F(\chi_{u^*}(\cdot,t)) - F(u^*)\}\,dt = 0$. Since $\forall t,\ F(u^*) \leqslant F(\chi_{u^*}(\cdot,t))$, $F(\chi_{u^*}(\cdot,t)) = F(u^*)$ for a.e. $t \in [0,1]$. This means that the function $x \mapsto \chi_{u^*}(x,t)$

[3] The characteristic function of a set $\Sigma \subset \Omega$ is the function $\chi^\Sigma(x) = 1$ if $x \in \Sigma$, 0 otherwise. The perimeter of its boundary is given by $Per(\partial\Sigma) = \int_\Omega |\nabla\chi|$.

[4] $\chi_u$ is the function defined in $\Omega \times [0,1]$ by $\chi_u(x,t) = 1$ if $u(x) > t$, 0 otherwise.

is also a minimizer of $F(\cdot,\boldsymbol{\alpha})$ for almost every $t \in [0,1]$. In addition, if $\Sigma \subset \Omega$ is such that $F_0(\Sigma) < F(u^*)$ then its characteristic function[3] also satisfies $F_0(\Sigma) = F(\chi^{\Sigma}) < F(u^*)$, which is a contradiction. Since for a.e $t \in [0,1]$ $F_0(\Sigma_t) = F(\chi_{u^*}(\cdot,t)) = F(u^*)$, we finally have $\forall \Sigma \subset \Omega$, $F_0(\Sigma_t) \leqslant F_0(\Sigma)$. □

In other words, we proved that the set of solutions of $(\mathcal{P})$ is stable under thresholding and it is clear that every solution of $\mathcal{P}_0$ corresponds to one of these thresholded functions. We can now develop an efficient optimization strategy capable of solving any two-phase segmentation problem that can be expressed through the general formulation $(\mathcal{P}_0)$, as the classical models listed in Table 1. Indeed, being convex in $u$, problem $(\mathcal{P})$ may be solved through powerful tools from convex optimization theory. In the general *unsupervised* setting, $F$ can be minimized iteratively by alternating the following two steps:

| |
|---|
| **(A)** Considering $u$ fixed, optimize and update the region parameters $\boldsymbol{\alpha}$,<br>**(B)** Considering $\boldsymbol{\alpha}$ fixed, minimize w.r.t the *partition* variable and update $u$.<br>When a steady state $u^*$ is found, simply threshold it. |

Step (A) depends on the specific choice of error functions (see sections 3 and 4). Step (B), however, can be realized by applying a generic minimization scheme for the variable $u$, as in [2]: Fixing $\boldsymbol{\alpha}$, minimizing $F$ with respect to $u$ in $BV_{[0,1]}$ is equivalent to minimizing $\int_\Omega g\,|\nabla u| + \int_\Omega ur$ in $BV$ under the constraint $0 \leqslant u \leqslant 1$, where $r = r_1^{\alpha_1} - r_2^{\alpha_2}$ is the *competition function*. Based on [4], this *constrained* problem has the same set of minimizers than the *unconstrained* problem $(\tilde{\mathcal{P}})$ of minimizing $\tilde{F}(u) = \int_\Omega g\,|\nabla u| + \int_\Omega ur + \beta\nu(u)$, with the exact penalty term $\nu(\xi) = \max(0, |2\xi - 1| - 1)$ and $\beta > \frac{1}{2}|r|_\infty$. Even though $(\tilde{\mathcal{P}})$ can be numerically solved using a gradient-descent scheme based on the *Euler-Lagrange* equation, no advantage would be taken of the convexity of $\tilde{F}$. We thus choose to follow [2] and exploit the fast duality projection algorithm of *Chambolle* [3]. To that end, we add an auxiliary variable $v$ and consider the following weak approximation:

$$(\tilde{\mathcal{P}}_\theta) \qquad \min_{(u,v)\in BV} \left\{ \tilde{F}_\theta(u,v) = \int_\Omega g\,|\nabla u| + \frac{1}{2\theta}\int_\Omega |u-v|^2 + \int_\Omega rv + \beta\nu(v) \right\},$$

where $\theta$ is chosen to be small enough so that the two components of any minimizing couple $(u^*, v^*)$ are almost identical w.r.t. the $L^2$ norm. Note that this approximation is still componentwise convex in $u$ and $v$. Moreover, $u$ being fixed, it is easy to check that the optimal $v$ is directly given by $v = \max(\min(u - \theta\, r, 1), 0))$. Now, $v$ being fixed, the projection algorithm of [3] can be applied for the minimization of $\tilde{F}_\theta$ with respect to $u$. Hence, $u = v - \theta \operatorname{div}(\boldsymbol{p})$ where the vector field $\boldsymbol{p}$ can be computed by using a fixed point algorithm, iterating on $n \geqslant 0$

$$\text{taking } \tau > 0,\ \ \boldsymbol{p}^0 = \mathbf{0},\ \ \boldsymbol{p}^{n+1} = \frac{\boldsymbol{p}^n + \tau\nabla(\operatorname{div}(\boldsymbol{p}^n) - v/\theta)}{1 + \tau\,|\nabla(\operatorname{div}(\boldsymbol{p}^n) - v/\theta)|\,/g}.$$

In practice, this algorithm is numerically very stable. Note that contrary to gradient-descent schemes, it does not rely on the explicit computation of the curvature of $u$. We also observed that decoupling $u$ and $v$ has a relaxation effect on the regularization and a very positive impact on the overall convergence.

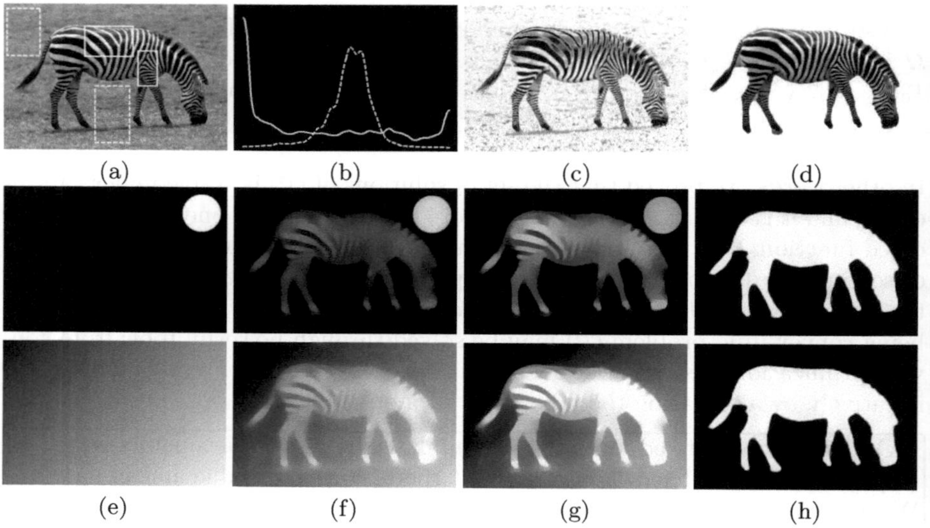

**Fig. 1.** *Fuzzy Region Competition*, supervised segmentation of the zebra image: (a) The original image and patches used to learn the background (dashed) and foreground (solid) histograms given in (b). (c) The corresponding competition function $r = \log(P_2/P_1)$. (d) Final segmentation. Two different (e) initializations, (f)-(g) intermediary states (step 10-20) and (h) final partition function $u$ (step 300).

In figure 1, we show an immediate application of the above generic scheme in a *supervised* segmentation experiment. The *competition function* $r = r_1 - r_2$ has been built *a priori*, following the statistical model given in [11] or [17]. Probability distributions $P_i(I)$ have been estimated from both background and foreground region patches, such that $r = \lambda \log(P_2/P_1)$, where $\lambda$ is a regularization parameter. In practice, we have observed that this supervised method gives always the same binary partition function, regardless of initial conditions. However, there exist 'pathological' cases of $r$ for which several minimizers of $F$, not necessarily binary, could solve the non strictly convex problem $(\mathcal{P})$. Those cases do not contradict proposition 1, neither do they easily show up in practice, but it may be interesting to characterize them in a more in-depth study. In the following sections, we apply the *Fuzzy Region Competition* to some *unsupervised* cases, by first considering the classical two-phase constant model (section 3), then extending it to a new smooth model (section 4).

## 3 Constant Region Competition

In [4], *Chan et al.* propose to solve the minimization involved in the *Active Contours Without Edges* [5] by considering an auxiliary convex problem. They show that if $c_1$ and $c_2$ are fixed and $u^*$ is a solution of

$$\min_{u \in BV_{[0,1]}(\Omega)} \left\{ F_{[4]}(u) = \int_\Omega |\nabla u| + \lambda \int_\Omega u \left( (I - c_1)^2 - (I - c_2)^2 \right) \right\},$$

then the set $\Sigma_t = \{x \in \Omega, u^*(x) > t\}$ is for a.e. $t \in [0,1]$ a minimizer of the following functional, considered in [5]:

$$F_{[5]}(\Sigma) = \int_{\partial\Sigma} ds + \lambda \int_{\Sigma} (I - c_1)^2 + \lambda \int_{\Omega\setminus\Sigma} (I - c_2)^2$$

This result inspired the general formulation of section 2, and also guides its first application. Solving the *Active Contours Without Edges* model using our framework boils down to choosing $r_i^{c_i} = \lambda\,(I - c_i)^2$, $\boldsymbol{\alpha} = (c_1, c_2)$ and $g = 1$ in $(\mathcal{P})^5$, such that the problem becomes

$$(\mathcal{P}_c) \qquad \min_{\substack{u\in BV_{[0,1]}\\ (c_1,c_2)\in\mathbb{R}^2}} \left\{ F_c(u, c_1, c_2) = \int_{\Omega} |\nabla u| + \lambda \int_{\Omega} u(I - c_1)^2 + \lambda \int_{\Omega} (1 - u)(I - c_2)^2 \right\}$$

where $\lambda$ is a parameter balancing the region error terms and the Total Variation regularization. Applied to $(\mathcal{P}_c)$, Proposition 1 gives the result already obtained in [4]. Nonetheless, although closely related by $F_c(u, c_1, c_2) = F_{[4]}(u, c_1, c_2) + \int_\Omega (I - c_2)^2$, the two involved functionals lead to a slightly different perspective. $F_{[4]}$ is not to be considered as a minimization on the triplet $(u, c_1, c_2)$ but as a convex alternative to obtain a minimizer of the original problem [5], when $c_1$ and $c_2$ are fixed. Indeed, no optimization of those constants is involved in [4]. In practice, this implies the choice of an arbitrary level set of $u$ (e.g. 0.5) in step (A) of the alternate minimization scheme. This may at first seem a minor issue since any $\Sigma_t$ should be a minimizer at convergence. However, full convergence of $u$ is rarely obtained in practice at each step (B), in particular in the first iterations, and the levels of $u$ are not equivalent. Hence the arbitrary choice of a level $t$ for the computation of $c_1$ and $c_2$ may introduce a bias in the optimization process. On the contrary, in the proposed approach, the derivatives of $F_c$ with respect to scalars $c_1$ and $c_2$ give directly new optimality conditions that naturally involve all levels of the fuzzy membership function[6] $u$:

$$c_1^* = \frac{\int_\Omega uI}{\int_\Omega u}, \qquad c_2^* = \frac{\int_\Omega (1-u)I}{\int_\Omega (1-u)} \tag{3}$$

Our *Constant Region Competition* algorithm follows the generic scheme presented in section 2: (A) $u$ being fixed, compute the weighted averages $c_1$ and $c_2$ using equation (3). Then, (B) $c_1$ and $c_2$ being fixed, minimize w.r.t. $u$ using the weak formulation $(\widetilde{\mathcal{P}}_\theta)$ and perform a few iterations of *Chambolle's* projection algorithm. At convergence, the final segmentation is obtained by thresholding $u$ at any level in $[0,1]$, as justified by Proposition 1. In figure 2, we give an example of this simple *unsupervised* segmentation algorithm on a synthetic image. As in the *supervised* case, at least above a reasonable value of $\lambda$, we observed that the steady state is again always a binary function that is independent of the initialization. This method can thus be a useful tool for segmenting two-phase images

[5] For sake of simplicity, $g$ will be kept constant in the remainder (euclidian length).

[6] If $u$ is 0 almost everywhere (or 1 almost everywhere), $c_1^* = c_2^* = \int_\Omega I / |\Omega|$.

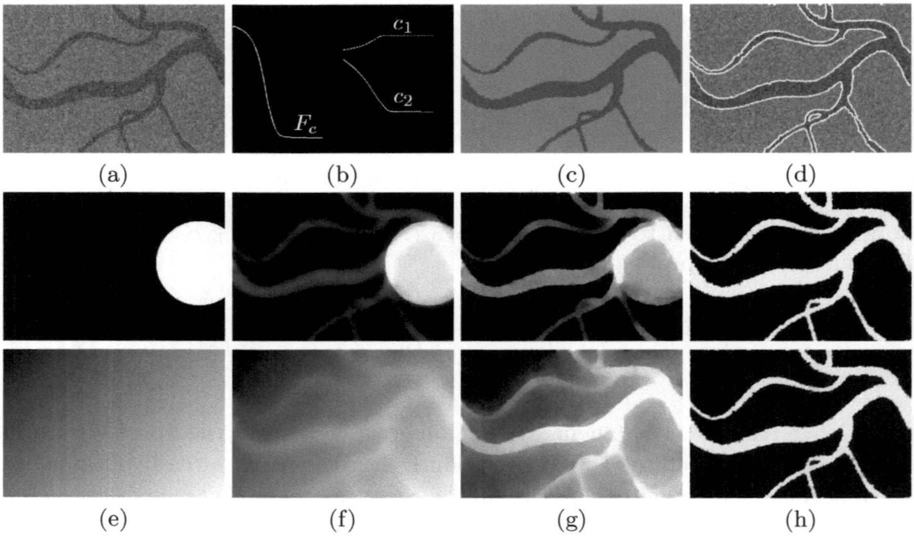


**Fig. 2.** *Constant Region Competition* : (a) A noisy synthetic angiography image where only two intensity means are present. (b) Convergence plots for $F_c$, $c_1$ and $c_2$. (c) Final approximation $uc_1 + (1-u)c_2$. (d) Final segmentation. Two different (e) initial conditions, (f)-(g) intermediary steps and (h) final partition functions $u$.

corrupted by a gaussian noise of constant variance (the underlying assumption). Unfortunately, this is hardly realistic for medical images, in particular in angiography, since vessels usually exhibit a slowly varying contrast across the image (figure 3.a). In addition, extending the model in order to handle more than two phases would not necessarily be of any help if the observed variations are smooth. In the next section, we derive a new segmentation model by adding in the region error a localizing window, which explicitly provides an intrinsic notion of scale to our two-phase separation problem.

## 4 Smooth Region Competition

Even though two-phase images that could be accurately approximated by a single constant in each region are rarely encountered in practice, there are certainly some applications for which the hypothesis may still hold locally, i.e. at a certain *scale*. To that end, our purpose is to formulate a local extension of the previous model that would account for slowly varying intensities and still keep the advantages of the convex general framework ($\mathcal{P}$). A possible way to achieve this localization is to introduce a symmetrical window function $\omega : \Omega \mapsto \mathbb{R}^+$, such that $\omega(x) \to 0$ when $|x| \to +\infty$, and use it to design a continuous sum of local *Constant Region Competitions*. In the previous section, the global contribution of the foreground to the total error was $E = \int_\Omega u(I - c_1)^2$. Now windowing this error with $\omega$ in a neighborhood of $y \in \Omega$ gives :

$$e(y) = \int_{x\in\Omega} u(x)(I(x) - s_1(y))^2\omega(x-y)dx$$

where the previously constant approximation $c_1$ is now an integrable function $s_1(y)$ that is allowed to vary in space. Hence, the new contribution of the foreground to the total error is $\int_{y\in\Omega} e(y)dy$. Adding a similar background contribution and switching the order of integrations finally exhibits our new region error terms, which leads to the *Smooth Region Competition* problem:

$$(\mathcal{P}_s) \qquad \min_{\substack{u\in BV_{[0,1]}\\ (s_1,s_2)\in L^1}} \begin{cases} F_s(u,s_1,s_2) = \int_\Omega |\nabla u| + \lambda \int_{x\in\Omega} u(x) \int_{y\in\Omega} \omega(x-y)(I(x)-s_1(y))^2\, dydx \\ \qquad + \lambda \int_{x\in\Omega} (1-u(x)) \int_{y\in\Omega} \omega(x-y)(I(x)-s_2(y))^2\, dydx \end{cases}$$

where $\lambda$ is a parameter balancing the error terms and the Total Variation regularization. $(\mathcal{P}_s)$ corresponds to the specific choices $\boldsymbol{\alpha} = (s_1, s_2)$, $A = L^1(\Omega)^2$ and $r_i^{s_i} : x \to \lambda \int_{y\in\Omega} \omega(x-y)(I(x) - s_i(y))^2\, dy$ in the general problem $(\mathcal{P})$, such that the generic scheme given in section 2 and Proposition 1 are still valid. Therefore, we only need to specify how to minimize with respect to $s_1$ and $s_2$.

**Proposition 2.** *Fixing the partition variable $u$, the couple of functions $(s_1^*, s_2^*)$ that minimize $F_s(u, s_1, s_2)$ satisfy for almost every $y \in \Omega$ the following equations:*

$$s_1^*(y) \int_\Omega u(x)\omega(x-y)dx - \int_\Omega u(x)I(x)\omega(x-y)dx = 0$$
$$s_2^*(y) \int_\Omega (1-u(x))\omega(x-y)dx - \int_\Omega (1-u(x))I(x)\omega(x-y)dx = 0$$

*Consequently, as soon as there is a measurable set $M_0$ (resp. $M_1$) where $\forall x \in M_0$ (resp. $M_1$), $u(x) \neq 0$ (resp. $u(x) \neq 1$) and if $\omega$ is a positive function, optimal closed-form solutions are given by the following normalized convolutions*[7]*:*

$$s_1^* = \frac{\omega * (uI)}{\omega * u} \quad \text{and} \quad s_2^* = \frac{\omega * ((1-u)I)}{\omega * (1-u)} \tag{4}$$

*Proof:* With $f_\eta(t) = F_s(u, s_1 + t\eta, s_2)$, the functional derivative of $F_s$ with respect to $s_1$ in the direction $\eta$ (a compactly supported $\mathcal{C}^\infty$ function) is given by

$$\left.\frac{df_\eta}{dt}\right|_{t=0} = -2\lambda \int_{x\in\Omega} u(x) \int_{y\in\Omega} (I(x) - s_1(y))\eta(y)\omega(x-y)dydx$$
$$= -2\lambda \int_{y\in\Omega} \eta(y) \int_{x\in\Omega} (I(x) - s_1(y))u(x)\omega(x-y)dxdy$$

A necessary condition for $s_1^*$ to be a minimizer of $F_s$ is : $\forall\eta$, $df_\eta/dt|_{t=0} = 0$. Thus $s_1^*$ must satisfy for almost every $y \in \Omega$: $\int_{x\in\Omega}(I(x) - s_1^*(y))u(x)\omega(x-y)dx = 0$. Similar derivations with respect to $s_2$ complete the proof. □

[7] The convolution between $f$ and $g$ is function $f * g : x \to \int_\Omega f(y)g(x-y)dy$.

The smooth case can be interpreted by analogy to the constant case of section 3, where the weighted averages $c_1$ and $c_2$ have been replaced by the normalized convolutions $s_1$ and $s_2$. They should be considered as smooth approximations of the image within each fuzzy region. Indeed, even though no regularity constraint is explicit in $F_s$, the resulting functions are as regular as the window $\omega$, provided that the latter is positive. We remind that the theory of normalized convolution, introduced in [8], is a simple and useful extension of convolution that accounts for uncertain or missing image samples. Here, normalized convolutions naturally appear to create smooth approximations for each phase, taking into account the fuzzy membership functions $u$ and $(1-u)$ as certainty measures.

Our *Smooth Region Competition* algorithm follows again the generic scheme presented in section 2: (A) $u$ being fixed, compute the normalized convolutions $s_1$ and $s_2$ using equation (4). Then, (B) $s_1$ and $s_2$ being fixed, compute the competition function $r = r_1^{s_1} - r_2^{s_2}$ (this also involves convolutions[8]) and perform a few iterations of the dual projection algorithm . When a steady state is found, the final segmentation is obtained by thresholding $u$ at any level in $[0,1]$. The function $S = us_1 + (1-u)s_2$ gives then a piecewise-smooth approximation of the original image (see figure 3.c). Indeed, a binary $u$ prevents the smoothing of $s_1$ and $s_2$ across the boundary. In this respect, our approach relates to segmentation methods assuming a piecewise-smooth underlying model, such as those based on the *Mumford-Shah* functional [9]. However, it also features an

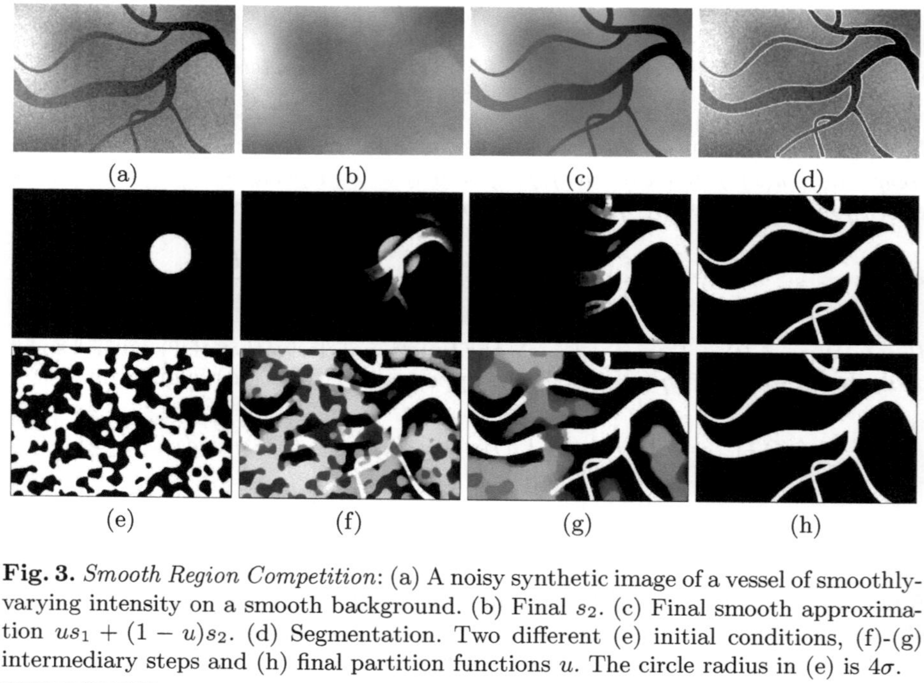

**Fig. 3.** *Smooth Region Competition*: (a) A noisy synthetic image of a vessel of smoothly-varying intensity on a smooth background. (b) Final $s_2$. (c) Final smooth approximation $us_1 + (1-u)s_2$. (d) Segmentation. Two different (e) initial conditions, (f)-(g) intermediary steps and (h) final partition functions $u$. The circle radius in (e) is $4\sigma$.

[8] The competition function here reads $r(x) = [\omega * (s_1^2 - s_2^2)](x) - 2I(x)[\omega * (s_1 - s_2)](x)$.

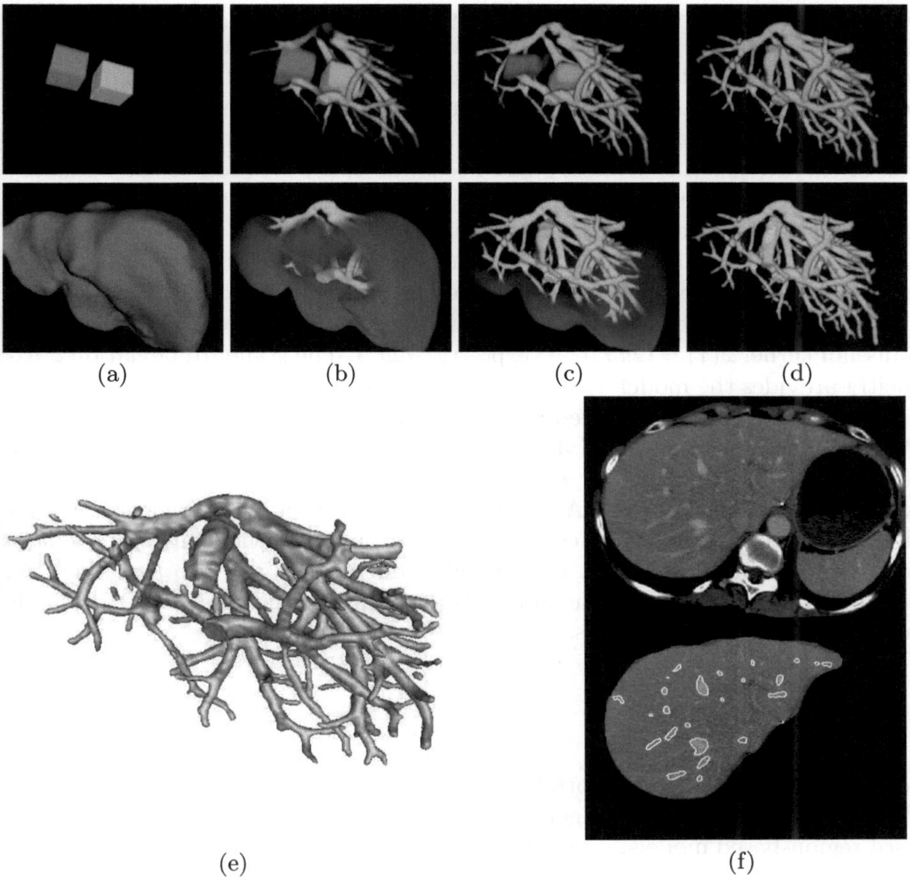

Fig. 4. A typical example of the *Smooth Region Competition* applied to medical images in 3D. We applied our model to the segmentation of a liver vessel tree from a 294x215x101 CT angiography. In this experiment, a segmentation mask of the liver was available, such that the two-phase hypothesis, (vessels + liver tissue), was approximately valid. In (a),(b),(c),(d), two evolutions of the fuzzy partition function $u$ are shown in volume rendering for different initializations. Note that at convergence, the result is almost binary and independent from the initial conditions. Computation time was about 20 sec on a standard PC platform. (e) shows the resulting mesh in surface rendering after the final thresholding. (f) shows a slice of the original volume and the contours of the final vessel segmentation overlaid on the masked liver.

essential practical advantage. Indeed, for instance in [13,14,1], the approximation functions $s_i$ are defined as solutions of second-order differential equations, only defined inside each phase[9]. Although theoretically sound, this has two drawbacks in practice. First, diffusion equations with conditions on the boundary

[9] $\int_\Sigma |\nabla s_1|^2 + \int_{\Omega\setminus\Sigma} |\nabla s_2|^2$ is typically added in the functional.

of an arbitrary-shaped domain (the current contour/surface) must be solved at each step (A) of the alternate minimization scheme. This is tedious since the domain is moving between each iteration. Second, both approximation functions are only defined in their respective domain, but need to be extended in order to compute the competition function, at least in a local vicinity. The construction of the extension is arbitrary and needs the exact location of the boundary separating the phases at all times. For 3D volumes in particular, those steps involve quite heavy computations. In contrast, step (A) of the *Smooth Region Competition* algorithm is much lighter since smooth closed-form solutions for $s_1$ and $s_2$, defined everywhere in $\Omega$, are directly constructed from $u$.

A very suitable choice for the window function is the normalized isotropic Gaussian kernel $\omega(x) = (2\pi\sigma^2)^{-n/2} \exp(-|x|^2/2\sigma^2)$. The standard deviation $\sigma$ explicitly provides the model with an intrinsic notion of scale, related to the intensity variations that are expected to arise in each region. Note that when $\sigma \to \infty$, we recover the constant model. Furthermore, being positive, non-compactly supported and $\mathcal{C}^\infty$, the Gaussian window guarantees the regularity of $s_i$ functions everywhere in the domain $\Omega$. As for the implementation, very efficient recursive filters that approximate the convolution and simulate the non-compact support with a computation cost that is independent on $\sigma$ are also available [15]. In figures 3 and 4, we show the application of this algorithm on a 2D synthetic image and on real 3D volumes.

## 5 Conclusion

We introduced a new framework for two-phase image segmentation, namely the *Fuzzy Region Competition*. It generalizes some existing *supervised* and *unsupervised* region-based models, if only two phases are considered, and can be used in all applications where a distinction foreground / background is meaningful, in 2D or 3D. The functional formulation we proposed is convex in its bounded variation *partition* variable $u$. The convexity guarantees to obtain in the *supervised* case a globally optimal solution. Although this property does not hold in the *unsupervised* setting, for which the *region parameters* have also to be optimized, it still produces very stable algorithms that turn out to be weakly sensitive to initial conditions. Furthermore, those algorithms can be very fast, in particular by using convex optimization tools and recent developments in total variation regularization. We believe that the use of a fuzzy membership function in region-based segmentation models leads to easy implementations that are more efficient and stable than their existing curve evolution *Level Sets* counterparts. This motivates the need for further work, in particular by including some a priori shape knowledge within the model in order to increase robustness.

Based on this framework, a new consistent model for the partition of an image into two smooth components was also derived. It is extensively built around windowing principles, convolutions, and features closed-form approximations in each region. Thus, it can be faster than some of the existing variational methods that give qualitatively similar results. It also provides an intrinsic notion of scale, depending on the intensity variations expected to arise in the objects to be segmented. However, from a segmentation perspective, scale also intuitively relates to other geometrical quantities such as the smoothness of the boundary or the allowed size of foreground objects. Both are presently dependent on the chosen balance between the error terms and the TV regularization. This would probably need also to be better understood and studied in the future.

**Acknowledgments.** We would like to thank Dr X. Bresson, Dr O. Cuisenaire and Prof. J.P. Thiran for many inspiring discussions.

## References

1. X. Bresson, S. Esedoglu, P. Vandergheynst, J.-P. Thiran, and S. Osher. Global minimizers of the active contour/snake model. *No CAM Report 0504, UCLA, Dept. of Mathematics*, 2005.
2. X. Bresson, S. Esedoglu, P. Vandergheynst, J.-P. Thiran, and S. Osher. Fast global minimization of the active contour/snake model. *to appear in Journal of Mathematical Imaging and Vision*, 2006.
3. A. Chambolle. An algorithm for total variation minimization and applications. *Journal of Mathematical Imaging and Vision*, 20(1-2):89–97, 2004.
4. T.F. Chan, S. Esedoglu, and M. Nikolova. Algorithms for finding global minimizers of image segmentation and denoising models. *UCLA CAM Report 04-54*, 2004.
5. T.F. Chan and L.A. Vese. Active contours without edges. *IEEE Trans. on Image Processing*, 10(2):266–277, February 2001.
6. W. Fleming and R. Rishel. An integral formula for total gradient variation. *Archiv Der Mathematik*, 11:218–222, 1960.
7. S. Jehan-Besson, M. Barlaud, and G. Aubert. Dream2s: Deformable regions driven by an eulerian accurate minimization method for image and video segmentation. *Int. Journal of Computer Vision*, 53(1):45–70, 2003.
8. H. Knutsson and C.F. Westin. Normalized and differential convolution: Methods for interpolation and filtering of incomplete and uncertain data. In *Proc. of Computer Vision and Pattern Recognition*, pages 515–523, New York City, USA, June 16-19 1993.
9. D. Mumford and J. Shah. Optimal approximations by piecewise smooth functions and associated variational problems. *Comm. on Pure and Applied Mathematics*, 42(5):577–685, 1989.
10. S. Osher and J.A. Sethian. Fronts propagating with curvature dependent speed: algorithms based on the hamilton-jacobi formulation. *Journal of Computational Physics*, 79:12–49, 1988.
11. N. Paragios and N. Deriche. Geodesic active regions and level set methods for supervised texture segmentation. *Int. Journal of Computer Vision*, 46(3):223–247, 2002.
12. Jianhong (Jackie) Shen. A stochastic-variational model for soft mumford-shah segmentation. *International Journal of Biomedical Imaging*, 2006(92329).
13. A. Tsai, A. Yezzi Jr., and A.S. Willsky. Curve evolution implementation of the mumford-shah functional for image segmentation, denoising, interpolation, and magnification. *IEEE Trans. on Image Processing*, 10(8):1169–1186, 2001.

14. L.A. Vese and T.F. Chan. A multiphase level set framework for image segmentation using the mumford and shah model. *Int. Journal of Computer Vision*, 50(3): 271–293, 2002.
15. I. Young and L. van Vliet. Recursive implementation of the gaussian filter. *Signal Processing*, 44:139–151, 1995.
16. H. K. Zhao, T. Chan, B. Merriman, and S. Osher. A variational level set approach to multiphase motion. *Journal of Computational Physics*, 127:179–195, 1996.
17. S. C. Zhu and A. Yuille. Region competition: Unifying snakes, region growing, and bayes/mdl for multiband image segmentation. *IEEE Trans. On Pattern Analysis and Machine Intelligence*, 18(9):884–900, September 1996.

# A Variational Approach for Multi-valued Velocity Field Estimation in Transparent Sequences

Alonso Ramírez-Manzanares[1], Mariano Rivera[1], Pierre Kornprobst[2], and François Lauze[3]

[1] Centro de Investigacion en Matematicas A.C., AP 402, Guanajuato, Gto., 36000, Mexico
{alram,mrivera}@cimat.mx
[2] INRIA, Odyssée Lab., 2004 route des Lucioles BP 93, 06902 Sophia Antipolis, France
Pierre.Kornprobst@sophia.inria.fr
[3] Nordic Bioscience Imaging A/S, DK-2370 Herlev, Denmark
francois@nordicbioscience.com

**Abstract.** We propose a variational approach for multi-valued velocity field estimation in transparent sequences. Starting from existing local motion estimators, we show a variational model for integrating in space and time these local estimations to obtain a robust estimation of the multi-valued velocity field. With this approach, we can indeed estimate some multi-valued velocity fields which are not necessarily piecewise constant on a layer: Each layer can evolve according to non-parametric optical flow. We show how our approach outperforms some existing approaches, and we illustrate its capabilities on several challenging synthetic/real sequences.

**Keywords:** Variational approaches, transparent motion, multi-valued velocity fields, model competition.

## 1 Introduction

There exists a very wide literature on apparent motion estimation, also called optical flow (OF), due to the number of applications that require motion estimation, and the complexity of the task. Motion estimation methods often rely on an intensity conservation principle and on spatial or spatiotemporal regularity constraints. The simplest conservation principle states that the intensity of a point remains constant along its trajectory. Although widely used, this principle is not satisfied in several real situations which include changing illuminance conditions, specularities, and multiple motions, as it is for instance the case in transparency.

Transparency can be modeled as a linear superposition of moving layers, meaning addition of layer intensities, or a generalized one [1]. A simple superposition model has been introduced by Burt *et al* in [2], from which they derive

F. Sgallari, A. Murli, and N. Paragios (Eds.): SSVM 2007, LNCS 4485, pp. 227–238, 2007.

an iterative *three frame algorithm for estimating two motions.* A more thorough study and extension of this idea is proposed in [3], providing a frequency domain interpretation and explaining a mechanism of "dominant velocity extraction". It is used by Irani and Peleg in [4] (see also [5]). Shizawa and Mase [6,7,8] explore a frequency domain, total least squares formulation of the multiple motion problems, providing a *2-fold OF constraint equation*, a closed-form formula is proved for two motions, but the problem becomes rapidly more complicated for higher orders. Liu *et al.* use it in [9] with Hermite polynomial based differentiation filters, and the authors check for the presence of single or multiple motions. Darell and Simoncelli [10] "dualize" this constraint in order to construct some Fourier "donuts" that respond to one or more velocities. Mota *et al.* have extended these ideas in [11] and Mühlich and Aach have proposed an algebraic framework based on homogeneous components of symmetric algebras in [12]. The nonlinear form of the 2-fold OF constraint provides what one may call the *2-fold displaced frame difference equation*, it can be extended to more than two motions to derive a block-matching approach to the multiple motion problem [13]. In summary, this class of approach is based on a single higher order constraint designed to "react" to multiple motions.

A spatial regularization in the framework of Markov Random Fields of the block matching solution is proposed in [14], in order to promote smooth solutions. However finding the global solution of the energy minimization method results in a *computationally expensive minimization*, because of the use of a field of binary indicator variables. Similarly, in [15] a parametric variant of the above block matching was applied in a layered approach for transparent X-Ray sequences (where the integration of material density produces transparent sequences [16]). The method segments and estimates the OF by alternately apply IRLS and ICM methods in order to compute both layer indicator variables and velocities.

Another important class of approaches is based on multiple low order motion constraints, designed to detect single motions. In the robust statistics approach of Black and Anandan [17], the image plane is assumed to be partitioned into regions, each one corresponding to a parametric motion model. The motion parameters are then assumed to represent the motion of two layers that cover the entire image plane. The layers are recovered by an iterative parameter/region estimation and by a nulling process. Mixture models for multiple motion computation have been introduced by Jepson and Black [18]. One assumes that the motion in layers can be explained by up to $N$ parametric motion fields by computing the best mixture and motion parameters, usually done by using EM-like algorithms. Ju *et al* [19] propose a model, in which, multilayered affine models are defined on small rectangular image patches (bones), and an inter-patch term (skin) introduces a regularization effect in the model parameters estimation. Then layer ownerships and affine model parameters are computed within a robust estimation framework by using an EM-algorithm. Black *et al* [20] compute a set of membership weights in order to link layers with regions. Although the method captures the changes in illumination, it does not allow the computation of the OF of moving transparencies. Weiss and Adelson [21] and Rivera *et al* [22] proposed EM-based approaches for computing different layered motion models. They use as prior knowledge the smooth feature of the velocities. The solution is given by a field of layer probabilities. Both methods produce pixel-wise unimodal solutions (single motions) since they use a distance measure for single motions as well as entropy controls. For more details about the different types of constraints and proposed approaches, we refer the reader to [23].

In this work we propose an approach based on local detectors sensitive to one or more velocities. We observed that the responses of these detectors usually provide a very local, noisy and somewhat too complex description of the velocities (more velocities than actually present may be detected at a given location). So there is a need for integration and regularization of this local information, which can be naturally performed with the framework of variational approaches.

The paper is organized as follows. Section 2 describes the proposed framework based on a finite sampling of the space of velocities and states a discrete variational model to handle multiple motions. Our approach encodes prior knowledge about the OF smoothness and the expected, relatively small, number of motions per pixel (one or two). The method performance is illustrated in Section 3, on synthetic, synthesized realistic and real sequences. We conclude and present future work in Section 4.

## 2 From Local to Global by a Variational Model

Let us introduce the main notations. The function $f : (\mathbf{x}, t) \in \Omega \times \{0, \ldots, T\} \rightarrow R$ denotes the input sequence, defined as a volume over space and time. Let us now define a finite sampling of the velocity space: We consider $N$ vectors

$$\{u_1, \ldots, u_N\}, \text{ with } u_i = (u_{i1}, u_{i2})^T,$$

describing the set of possible velocities. Our goal will be to determine what is the likelihood of having velocity $u_i$ at a given position. To do so, one need an initial **local** estimate of this likelihood. Let us assume that we know a function $d(u_i, r) \in \mathbb{R}^+|_{i=1\ldots N}$ which describe at each position $r = (\mathbf{x}, t)$ whether the velocity $u_i$ can explain locally the apparent motion (characterized by $d(u_i, r) \approx 0$) or not (characterized $d(u_i, r) \gg 0$). In general, $d$ can be implemented as a norm (or quasi-norm) over a similarity operator, and we refer to Section 2.1 for more details.

So, given a sequence $f$ and the local motions estimations $d(u_i, r)|_{i=1\ldots N}$, our goal is to propose an approach which integrates this local information in order to obtain a global and robust velocity estimation. This integration is necessary to deal with noisy sequences but also sequences with complex motions such as transparent motion.

### 2.1 How to Estimate Local Velocity Information?

Two well known similarity operators satisfy these requirements for the single motion case: The non-linear difference (Correlation-based)

$$M_{\mathcal{C}}^{(1)}(u_i) f(\mathbf{x}, t) \stackrel{def}{=} f(\mathbf{x}, t) - f(\mathbf{x} - u_i, t - 1),$$

and its linearized version (Differential-based)

$$M_{\mathcal{D}}^{(1)}(u_i)f(x_1,x_2,t) \stackrel{def}{=} \left(u_{i1}\frac{\partial}{\partial x_1} + u_{i2}\frac{\partial}{\partial x_2} + \frac{\partial}{\partial t}\right) f(x_1,x_2,t),$$

where the superscript indicates the number of displacements that these operators (and the following ones) take into account. Following Shizawa and Mase [8], one can define an operator for two velocities as the composition

$$M_{\mathcal{D}}^{(2)}(u_i,u_j)f(\mathbf{x},t) \stackrel{def}{=} M_{\mathcal{D}}^{(1)}(u_i)M_{\mathcal{D}}^{(1)}(u_j)f(\mathbf{x},t)$$

where products $\frac{\partial}{\partial r}\frac{\partial}{\partial s}$ are naturally expanded as $\frac{\partial^2}{\partial r \partial s}$. Composing instead the nonlinear correlation operators $M_{\mathcal{C}}^{(1)}$ provides the nonlinear operator for two velocities

$$M_{\mathcal{C}}^{(2)}(u_i,u_j)f(\mathbf{x},t) \stackrel{def}{=} f(\mathbf{x},t) - f(\mathbf{x}-u_i,t-1) - f(\mathbf{x}-u_j,t-1) + f(\mathbf{x}-u_i-u_j,t-2),$$

that corresponds to distance reported in [13].

We introduce here the general mechanism we have used in order to select the local velocity descriptors $d(u_i,r)$ from multiple motion operators. Given an integer $k \in [1,N]$, let us assume that we have a family of "$k$ velocities probe" operators

$$\mathbf{M} = \{M^{(k)}(u_{i_1},\dots,u_{i_k}), \quad 1 \le i_1 < \dots < i_k \le N\}$$

where $M^{(k)}(u_{i_1},\dots,u_{i_k})f(r) \approx 0$ if the velocity vectors $u_{i_1},\dots,u_{i_k}$ explain the motion of the image sequence $f$ at the position $r$. We build them by either cascading $k$ correlation based filters, such a probe will be denoted in general $M_{\mathcal{C}}$, or $k$ differential-based ones, these ones will be denoted $M_{\mathcal{D}}$. Then for each vector $u_i$, let us consider the subset $\mathbf{M}_{u_i}$ of all the operators involving $u_i$ and define

$$d_{\mathcal{C}}(u_i,r) = \min_{M_{\mathcal{C}} \in \mathbf{M}_{u_i}} \frac{1}{k} \sum_{s \in W_r} \left(M_{\mathcal{C}} f(s)\right)^2 \tag{1}$$

and

$$d_{\mathcal{D}}(u_i,r) = \min_{M_{\mathcal{D}} \in \mathbf{M}_{u_i}} \frac{1}{k} \sum_{s \in W_r} \left(M_{\mathcal{D}} f(s)\right)^2, \tag{2}$$

where $W_r$ is a $3 \times 3$ spatial window center at $r$. In the present work, we used as input distances (1) and (2) for different experiments, showing the general framework feature of our proposal as is explained in the following. Note that because of the series Taylor's approximation, distance (1) is more suitable than (2) for long displacements.

### 2.2 Global Motion Integration Via a Variational Approach

Let us define the function $r = (\mathbf{x},t) \mapsto \alpha_i(r) \in [0,1]$, which will correspond to the probability that velocity $u_i$ explains the apparent motion at the spatio-temporal position $r$. We define the unknowns of the problem as the vector valued field $\alpha$:

$$\alpha(r) = (\alpha_1(r),\dots,\alpha_N(r)), \quad \alpha_i(r) \in [0,1] \quad \forall r \in \Omega \times \{0,\dots T\}.$$

Note that, although each component of $\alpha(r)$ can be interpreted as a probability, $\alpha(r)$ is not a probability measure (as in [21,22]) in the sense that the sum of its components is not constrained to be one. If two motions $u_i$ and $u_j$ are present at a particular pixel position, then we expect that $\alpha_i(r) \approx \alpha_j(r) \approx 1$. Inversely, the velocity field at a position $r$ can be extracted from $\alpha(r)$ by keeping the velocity $u_i$ with high value of $\alpha_i(r)$.

In order to estimate the **global** multi-valued velocity field from the **local** data $d(u_i, r)$, we propose to minimizes the cost function $\mathcal{E}(\alpha)$ defined by

$$\mathcal{E}(\alpha) = \sum_r \Big\{ \sum_i d(u_i, r)\alpha_i^2(r) \tag{3}$$

$$+ \frac{\lambda_s}{2} \sum_{s:s\in\mathcal{N}_r} \sum_i w_i(r,s)[\alpha_i(r) - \alpha_i(s)]^2 \tag{4}$$

$$+\lambda_c \Big[ c\bar{\alpha}^2(r) - \sum_i \alpha_i^2(r) \Big] \Big\}, \tag{5}$$

subject to the constraints $\alpha_i(r) \in [0,1]$ for all $i$, with $\bar{\alpha}(r) \stackrel{def}{=} \frac{1}{N}\sum_i \alpha_i(r)$, $w_i(r,s)$ are some diffusion weight defined in the sequel, $\mathcal{N}_r \stackrel{def}{=} \{s : r, s \in \Omega \times [0,T], \|r - s\| < 2\}$ is the spatiotemporal neighborhood of the $r$ position, and $c, \lambda_s, \lambda_c$ are some positive constants. Let us now comment the different terms of this energy.

**Term (3): Attach Term.** This term links the input (the functions $d$'s) to the unknown $\alpha$. For computing the presence of the $i$-th model, we use an approach related with the outlier rejection method [24] and with the EM formulation [20,21,22,19]. Minimizing term (3) *w.r.t.* $\alpha_i(r)$ produces $\alpha_i(r)$ close to 0 for high $d(u_i, r)$ values, indicating in this way that such a motion model is not likely at position $r$. Otherwise, the $\alpha_i(r)$ is free and its value is established by the next terms and the bound constraint.

**Term (4): Spatial Regularization.** Local information is integrated through this regularization term. At a given location $r$, we minimize the difference between vector $\alpha(r)$ and all the vectors $\alpha(s)$ in the neighborhood $\mathcal{N}_r$. Because our indicator variables are real valued, we can use differentiable potentials with their well-known algorithmic advantages. We use the approach presented in [25]. The smoothing process is controlled by directional fixed weights, $w_i(r,s) = \left((s-r)^T \bar{\mathbf{I}}_i (s-r)\right) / \|s-r\|^4$, generated from the $i$–th tensor associated to the $i$–th velocity model: $\bar{\mathbf{I}}_i = \gamma I_d + U_i {U_i}^T$, where $I_d$ is the identity matrix, $\gamma = 0.1$ and $U_i = [u_{i1}, u_{i2}, 1]^T / \|[u_{i1}, u_{i2}, 1]\|$ is a homogeneous-coordinate unit vector. For small $\gamma$ values (as the one proposed here) these weights promote a strong smoothness along the $i$–th velocity direction. The arbitrarily fixed 4–th power of the distance restricts the spatial influence of the smoothness term, see also [23]. As a consequence piece-wise smooth OFs are recovered with well-defined boundaries along the velocity model (see Figure 3).

**Term (5): Intra-Model Competition.** Our aim is to detect multiple simultaneous motions (transparent motions), thus we may have problems at sites where multiple spurious matches are locally detected, for example in homogeneous regions, where $d(u_i, r) \approx 0$ for many (maybe all) velocities. For this reason we need a mechanism to eliminate spurious models (i.e. to cancel some of the components of $\alpha(r)$) and to promote the valid ones: We want to recover almost binary solutions (similar to the entropy-control potentials, see for example the Shannon one [21] or the Gini one [22]). In our case, we use potential (5) because $\alpha(r)$ is not a probability measure and also because this potential is well adapted for recovering multimodal solutions. The parameter $c$ is a tuning parameter related to the number of switched-on models. To understand the potential's behavior, we note that the first term penalizes the number of switched-on models while the second term promotes the "switching on" of models and avoids the trivial null solution $\alpha(r) = 0$. Hence for a fixed mean value (controlled by the first term) the second term prefers highly contrasted solutions. Note that (5) can be tuned such that for a given $c$ value, a multimodal solution has lower energy than an unimodal one or conversely. That makes an important distinction with respect to entropy based measures that always have lower energy for unimodal solutions [21,22]. Additionally, our proposed quadratic potential is easily differentiable and therefore simple minimization algorithms can be used as explained below.

**Algorithmic Details.** The cost function $\mathcal{E}(\alpha)$ is quadratic and can be minimized by solving the linear system $\partial\mathcal{E}(\alpha)/\partial\alpha_i(r) = 0, \quad \forall i, \forall r$, with the constraints $\alpha_i(r) \in [0, 1]$. We use a Gauss-Seidel algorithm. Given an estimate $\alpha_i^n$, we iterate until convergence:

$$\alpha_i^{n+1}(r) = \frac{\lambda_s \sum_{s \in \mathcal{N}_r} w_i(r,s)\alpha_i^n(s) - c\lambda_c \bar{\alpha}^n(r)}{d(u_i, r) + \lambda_s \sum_{s \in \mathcal{N}_r} w_i(r,s) - \lambda_c}.$$

The bound constraints on $\alpha_i(r)$ are then enforced by projecting non-feasible values to bounds at each iteration. We note that for obtaining a smooth algorithm convergence, it was important to keep fixed the mean of the previous iteration, $\bar{\alpha}^n(r)$, for updating the current $\alpha(r)$ vector. This can be seen as an over-relaxation strategy. We initially set $\alpha_i^0(r) = 0.5, \forall i, r$.

A Deterministic Annealing strategy in the $\lambda_c$ parameter introduces the *intra-model competition* only until an approximate solution with valid representative models have predominant $\alpha_j(r)$ values: For each iteration $k = 1, 2, \ldots, n$, we set $\lambda_c^{(k)} = \lambda_c a_k$, where $\lambda_c$ is the chosen contrast level and $a_k = 0.1 + 0.9(1 - 0.95^{(100k/n)})$ is a factor that increases to 1 in approximately 80% of the total iterations. Results are sensitive to the annealing speed of $\lambda_c$: Premature increment could lead to an incorrect solution. Nevertheless, we used the same annealing scheduling in all our experiments. The large value $\lambda_s$ eliminates noise but one too large blurred the motion boundaries. We used $\lambda_s \in [50, 100]$ for an adequate noise reduction. Parameter $c = 1$ performs well for most noise-free synthetic sequences. For noise-contaminated, real sequences or when the number of base velocities are increased (then several spurious models may be present) the prominent models are obtained by increasing this parameter, $c \in [1,4]$. Note also that in all cases, we compute our dense optical flow using at most 200 iterations.

## 3 Experiments on Synthetic and Real Sequences

**Local Measurements Are Noise Sensitive.** Figure 1 (a) shows a synthetic sequence (size $54 \times 54 \times 16$) with transparent motion, similar to the one in [14]: There is a moving background (with velocity $\hat{u} = [0,-1]$) and an overlapped moving transparent square (with velocity $\hat{v} = [1,0]$), with additive Gaussian noise. Figure (b) (resp. (c)) show the OFs associated to the minima of distance (1) (resp. (2)). This represents indeed what will be the input of our approach and illustrate the need of velocity integration.

**Regularization of local measurements.** Sequence in Figure 1 (a) is corrupted with Gaussian noise to measure the robustness of our approach. Figures 2 (a)-(c) show the results obtained. The percentages of pixels with a wrong estimation are 2.12%, 2.29% and 2.40% respectively: Our approach can deal with a strong noise corruption, with better results than [14] (see Figures 2 (d) and 2 (e), and results in [14]). The velocity basis was composed by 33 vectors, specified through their magnitudes and orientations, respectively $\{0,1,2,3,4\}$ pixels and $\{0, \frac{\pi}{4}, \frac{\pi}{2}, \frac{3}{4}\pi, \pi, \frac{5}{4}\pi, \frac{3}{2}\pi, \frac{7}{4}\pi\}$ radians. For comparison purposes, Figures 2 (d) and 2 (e) show the computed OF with the computationally expensive Gibbs sampler approach for minimizing the discrete energy function reported in [14]. In [14] a deterministic relaxation ICM algorithm was used, which, differently to Gibbs sampler approach, is prone to converge to local minima. The noise-free case is shown in Figure 2 (d), and the SNR=30 case in Figure 2 (e). The results shown correspond to the computed solution after 150,000 iterations (about 2.5 hours, in a PC Pentium IV, 3.0 GHz) that represent 150 times slower than our approach. For the Gibbs sampler results, the quality decreases for noise corrupted sequences (see Figure 2 (e) and compare it with the one computed with the proposed method in about 1 minute in Figure 2 (f)). The behavior of the spatial regularization and intra-model competition is illustrated in Figure 3. It shows the evolution of the layer associated with velocity [1,0]: Large values appear in the square region whereas small values appear in the background region.

**Realistic Textured Sequences.** High textured sequences are relatively easy to solve using local motion measures. The real performance of the method should be evaluated in realistic textured scenes, where we may find difficulties because several models may locally explain the data. The next experiment is composed of two moving photographs: a face $I_1$ with motion $u = [1,0]$ (limited textured scene) and a rocky Mars landscape $I_2$, with motion $v = [-1,0]$. The sequence was generated with $f = 0.6I_1 + 0.4I_2$, see Figure 4 (a).

Figure 4 (b) shows the OF associated to the minimum distance (2) used in the attach term. The computed velocity field is shown in Figure 4 (c). Note that the right OF is recovered in all the pixels regardless of the high amount of noise.

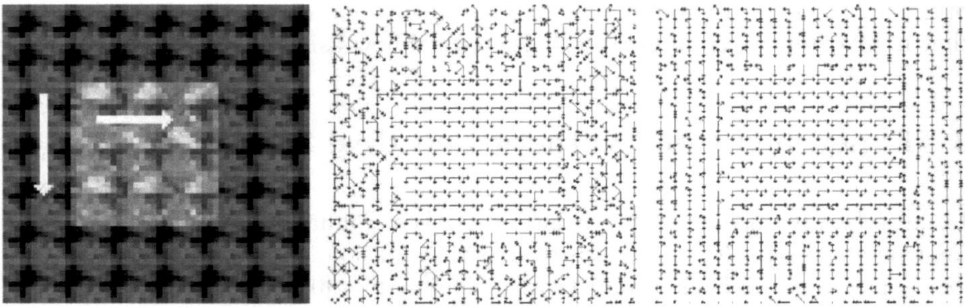

(a) Noise free sequence (b) OFs from $d_{\mathcal{C}}$ (c) OFs from $d_{\mathcal{D}}$

**Fig. 1.** Synthetic noisy sequence (SNR=30) and results obtained with the distances $d_{\mathcal{C}}$ (1) and $d_{\mathcal{D}}$ (2)

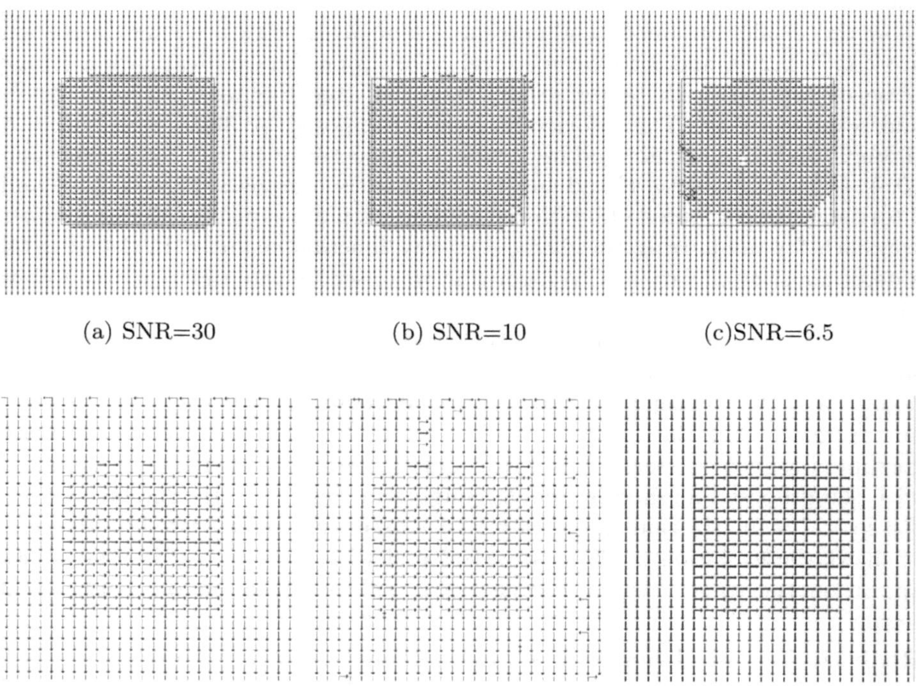

(a) SNR=30 (b) SNR=10 (c)SNR=6.5

(d) Gibbs S., noise-free (e) Gibbs S., SNR=30 (f) Our result, SNR=30

**Fig. 2. First row.** Results obtained with our approach applied to the synthetic sequence presented in Figure 1 (a) with different noise levels (input was $d_{\mathcal{C}}$). **Second row.** Comparison with Gibbs Sampler scheme: (d) noise-free case and (e) noisy case SNR=30. (f) our result for SNR =30, confront with (e).

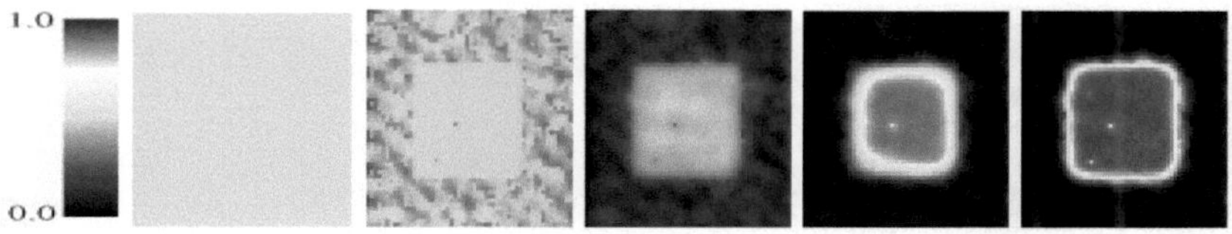


**Fig. 3.** Evolution during minimization (in pseudo–color scale) of $\alpha_i(r)$ for the velocity $u_i = [1, 0]$ for a strongly noise corrupted sequence (SNR=15). Iterates 0, 1, 11, 31 and 200 are shown.

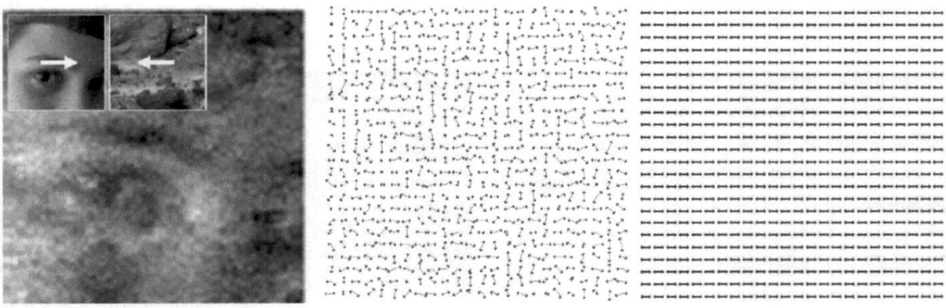

(a) Noisy sequence (SNR=8) (b) OFs from $d_{\mathcal{D}}$ (c) Our result

**Fig. 4.** Two noisy textured patterns in translation. We show the velocities associated to minimum distance (2) and our regularized result.

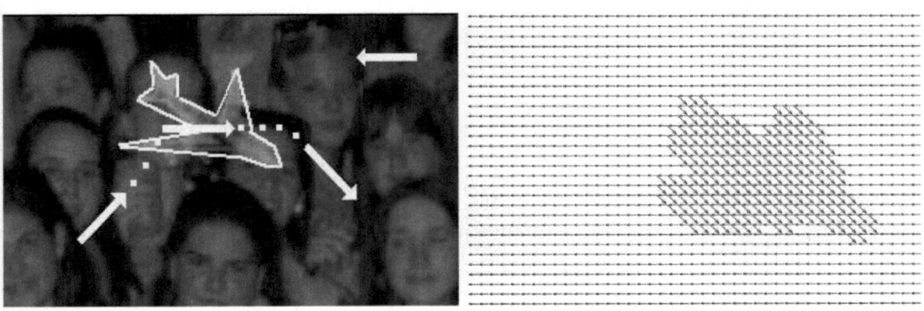

**Fig. 5.** Transparent object moving with changing translational speed over a translating background and one of the recovered motion fields

Figure 5 shows a sequence with a time–varying transparent region and motions. The changing velocities are schemed in Figure 5 (a). An example of the obtained multi-velocity vector field is shown in Figure 5 (b) (results for more frames of this sequence can be found in [23]). For this experiment we used the distance measure in (1).

Figure 6 shows results for a rotating layer sequence. The sequence was made by adding a rotating earth globe (1 degree per frame) and a translating textured

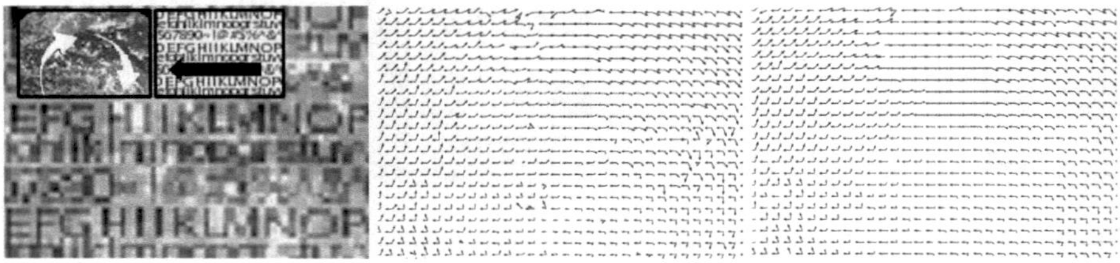


**Fig. 6.** Transparent motion sequence with complex rotating motion, SNR=60. We show the sequence, associated velocities to minimum distances (1) and our method's result.

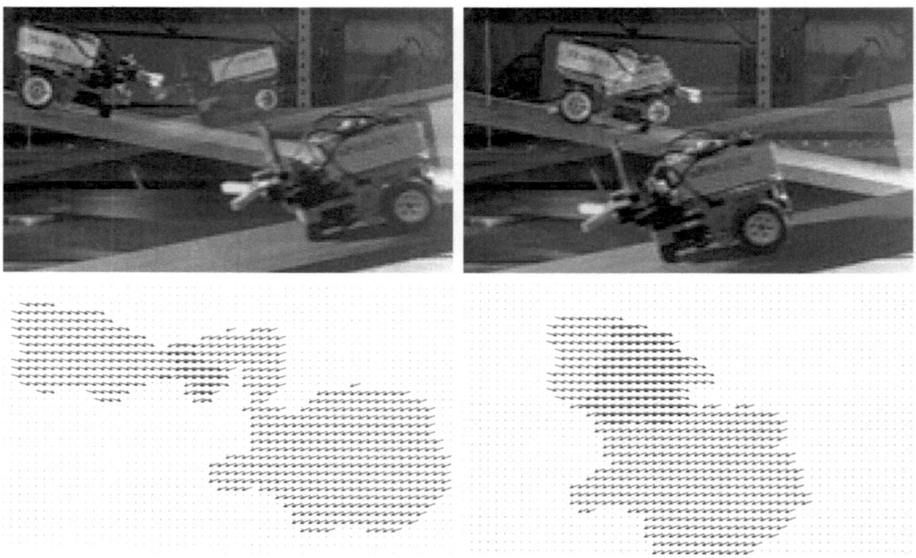

**Fig. 7.** Real transparent sequence and our results in the second row

image with velocity [-1,0], as it is indicated in Figure 6 (a). Note that in such a case a large set of velocities (either different in orientation and magnitude) are computed as the final solution in Figure 6 (c). In this experiment we estimate dense smooth flow which does not rely on any motion assumption or model.

**Transparency and Occlusion in a Real Sequence (Figure 7).** We show the obtained results for a real sequence composed by two robots moving down a slope. The upper-left robot is located behind a glass while the lower-right robot is located in front of the camera. The reflection of the second robot is located in the upper-central part. For this experiment we used as input the distance measure (1). The recovered velocities were [1.5, 0.4] pixels for the upper-left robot and [-1.5,0.5] for both the lower-right robot and its reflection, see second row in

Figure 7. Despite the fact that the lower right-robot is moving a little faster than its reflection (easy to deduct form the projection geometry), both are detected with the same velocity model. That is explained by the low resolution of the discrete velocity basis. For this experiment, we perform a spatiotemporal Gaussian smoothing process ($\sigma = 0.5$) of the sequence and we process only the regions that contain displacements: the static background was removed automatically by thresholding the difference between consecutive frames, and then applying opening-closing morphologic operators.

## 4 Conclusion

We have presented a novel variational formulation for the estimation of multiple motions, and especially transparency. The unknown is a vector valued field that indicates the presence of some given motions at each spatiotemporal location. Our formulation extends previous works based on layered OF computation, by using a distance measure suitable for transparent motions and proposes an intra–model competition mechanism well–suited for multi–valued solutions. In our case, multiple motions, the intra–model competition behaves similar to the mechanisms used for entropy–control for single motion fields. This term is by itself a novel contribution of this work, since we do not need special preprocessing in order to tackle sequences with one or more layers, as was shown in synthetic experiments in textured and non-textured sequences as well as real sequences.

In future extensions of this work, we will study in more depth the diffusion terms and also investigate how the different velocity maps may interact together. We also want to evaluate our approach on test sequences used in psychophysics, which will certainly suggest some improvement of the current model.

## Acknowledgments

A. Ramirez-Manzanares and M. Rivera were partially supported by CONACYT, Mexico, PhD Scholarship and grant 46270, respectivelly.

## References

1. Oppenheim, A.V.: Superpostion in a class of nonlinear systems. In: IEEE International Convention, New York, USA (1964) 171–177
2. Bergen, J., Burt, P., Hingorani, R., Peleg, S.: Computing two motions from three frames. In: ICCV 90, Osaka, Japan (December 1990) 27–32
3. Burt, P., Hingorani, R., Kolczynski, R.: Mechanisms for isolating component patterns in the sequential analysis of multiple motion. In: IEEE Workshop on Visual Motion, Princeton, NJ (October 1991) 187–193
4. Irani, M., Peleg, S.: Motion analysis for image enhancement: resolution, occlusion, and transparency. Journal on Visual Communications and Image Representation **4**(4) (1993) 324–335
5. Irani, M., Rousso, B., Peleg, S.: Computing occluding and transparent motions. IJCV **12**(1) (January 1994) 5–16

6. Shizawa, M., Mase, K.: Simultaneous multiple optical flow estimation. In: ICPR 90. Volume 1. (1990) 274–278
7. Shizawa, M., Mase, K.: Principle of superposition: a common computational framework for analysis of multiple motion. In: IEEE Workshop on Visual Motion. (1991) 164–172
8. Shizawa, M., Mase, K.: A unified computational theory for motion transparency and motion boundaries based on eigenergy analysis. In: CVPR 91. (1991) 289–295
9. Liu, H., Hong, T., Herman, M., Chellappa, R.: Spatio-temporal filters for transparent motion segmentation. In: ICIP 95. (1995) 464–468
10. Darrell, T., Simoncelli, E.: Separation of transparent motion into layers using velocity-tuned mechanisms. In: MIT Media Laboratory Vision and Modeling Group Technical Report. Number 244 (1993)
11. Mota, C., Stuke, I., Aach, T., Barth, E.: Divide-and-Conquer strategies for estimating multiple transparent motions. In: 1st International Workshop on Complex Motion, Schloss Reisensburg, Germany. LNCS Vol. 3417. (2005) 66–78
12. Mühlich, M., Aach, T.: A theory of multiple orientation estimation. In: ECCV. Volume 2. (2006) 69–82
13. Stuke, I., Aach, T., Barth, E., Mota, C.: Estimation of multiple motions by block matching. In: 4th ACIS International Conference on Software Engineering, Artificial Intelligence, Networking and Parallel/Distributed Computing (SNPD 2003). (2003) 358–362
14. Stuke, I., Aach, T., Barth, E., Mota, C.: Multiple-motion-estimation by block matching using MRF. In: ACIS, International Journal of Computer and Information Science. (2004)
15. Auvray, V., Bouthemy, P., Lienard, J.: Motion estimation in x-ray image sequence with bi-distributed transparency. In: ICIP 06, Atlanta, USA (2006)
16. Fitzpatrick, J.: The existence of geometrical density-image transformations corresponding to object motion. CVGIP **44**(2) (November 1988) 155–174
17. Black, M., Anandan, P.: The robust estimation of multiple motions: Parametric and piecewise-smooth flow fields. CVGIP: Image Understanding **63**(1) (1996) 75–104
18. Jepson, A., Black, M.: Mixture models for optical flow computation. In: CVPR 93. (1993) 760–761
19. Ju, S., Black, M., Jepson, A.: Skin and bones: Multi-layer, locally affine, optical flow and regularization with transparency. In: Proceedings of CVPR 96, San Francisco, CA (June 1996) 307–314
20. Black, M., Fleet, D., Yacoob, Y.: Robustly estimating changes in image appearance. Computer Vision and Image Understanding **78** (2000) 8–31
21. Weiss, Y., Adelson, E.: A unified mixture framework for motion segmentation: incorporating spatial coherence and estimating the number of models. In: CVPR 96. (1996) 321–326
22. Rivera, M., Ocegueda, O., Marroquin, J.L.: Entropy controlled gauss-markov random measure field models for early vision. In: VLSM 05, LNCS 3752. (2005) 137–148
23. Ramirez-Manzanares, A., Rivera, M., Kornprobst, P., Lauze, F.: Multi-valued motion fields estimation for transparent sequences with a variational approach. Technical Report RR-5920, INRIA, (Also, Reporte Técnico CIMAT, (CC)I-06-12) (June 2006)
24. Black, M., Rangarajan, P.: On the unification of line processes, outlier rejection, and robust statistics with applications in early vision. IJCV **19**(1) (1996) 57–91
25. Ramirez-Manzanares, A., Rivera, M.: Brain nerve boundless estimation by restoring and filtering intra-voxel information in DT-MRI. In: VLSM 03. (October 2003) 71–80

# Dense Optical Flow Estimation from the Monogenic Curvature Tensor*

Di Zang[1], Lennart Wietzke[1], Christian Schmaltz[2], and Gerald Sommer[1]

[1] Department of Computer Science, Christian-Albrechts-University of Kiel, Germany
{zd,lw,gs}@ks.informatik.uni-kiel.de
[2] Faculty of Mathematics and Computer Science, Saarland University, Germany
schmaltz@mia.uni-saarland.de

**Abstract.** In this paper, we address the topic of estimating two-frame dense optical flow from the monogenic curvature tensor. The monogenic curvature tensor is a novel image model, from which local phases of image structures can be obtained in a multi-scale way. We adapt the combined local and global (CLG) optical flow estimation approach to our framework. In this way, the intensity constraint equation is replaced by the local phase vector information. Optical flow estimation under the illumination change is investigated in detail. Experimental results demonstrate that our approach gives accurate estimation and is robust against noise contamination. Compared with the intensity based approach, the proposed method shows much better performance in estimating flow fields under brightness variations.

**Keywords:** Optical flow estimation, phase, monogenic curvature tensor.

## 1 Introduction

Optical flow estimation is one of the key problems gathering the interest of researchers for decades in the computer vision community. It has a wide application in motion estimation, object recognition, tracking, surveillance and so on.

Various approaches have been proposed to estimate the optical flow. Significant improvements [1] have been obtained since the pioneering work of Horn and Schunck [2] and Lucas and Kanade [3]. In [4], Barron et al. evaluated the performance of optical flow techniques. The local phase-based method [5] was proven to be the best performed due to its subpixel accuracy and its robustness with respect to smooth contrast changes and affine deformations. Differential methods, on the other hand, have become the most frequently used techniques for optical flow estimation because of the simplicity and good performance. Among the differential methods, there exist two classes. They are local methods such as that of Lucas and Kanade and global methods such as that of Horn and

* This work was supported by German Research Association (DFG) Graduiertenkolleg No. 357 (Di Zang), DFG grant So-320/4-2 (Lennart Wietzke), DFG grant We-2602/5-1 (Christian Schmaltz) and DFG grant So-320/2-3 (Gerald Sommer).

F. Sgallari, A. Murli, and N. Paragios (Eds.): SSVM 2007, LNCS 4485, pp. 239–250, 2007.

Schunk. Local methods are known to be more robust under noise, while global approaches yield 100% dense flow fields. Hence, Bruhn et al. [6] proposed the combined local-global (CLG) approach to yield dense optical flow fields which is robust against noise.

In order to have accurate estimation of dense optical flow fields with robustness against noise and brightness variation, we propose a novel approach based on the monogenic curvature tensor, a new image model. In contrast to the classical phase computation in [5], the monogenic curvature tensor can generate multi-scale local phases of intrinsically 1D and 2D image structures in a rotation invariant way. Thus, the proposed approach combines the advantages of the phase-based method and the CLG method. Experiments with synthetic and real image sequences demonstrate the favorable performance of the proposed method when compared with the related work.

## 2 Monogenic Curvature Tensor

The monogenic curvature tensor is a novel 2D image model, from which multi-scale local phases of image structures can be obtained in a rotation invariant way. It is well known that the local phase has the advantage of being invariant to illumination change [7]. In this paper, we will adapt the CLG method to this framework for the dense optical flow estimation. Hence, a brief overview of the novel image model is given in this section.

### 2.1 Basis Functions

The monogenic curvature tensor consists of a curvature tensor and its conjugate part. By employing damped 2D spherical harmonics as basis functions, the monogenic curvature tensor is unified with a scale-space framework. An $n$th order damped 2D spherical harmonic $H_n$ in the spectral domain takes the following form

$$H_n(\rho, \alpha; s) = \exp(in\alpha)\exp(-2\pi\rho s) = [\cos(n\alpha) + i\sin(n\alpha)]\exp(-2\pi\rho s) , \quad (1)$$

where $\rho$ and $\alpha$ denote the polar coordinates in the Fourier domain, $s$ refers to the scale parameter. The damped 2D spherical harmonic is actually a 2D spherical harmonic, $\exp(in\alpha)$, combined with the Poisson kernel, $\exp(-2\pi\rho s)$, [8]. The first order damped 2D spherical harmonic is basically identical to the conjugate Poisson kernel [8]. When the scale parameter is zero, it is exactly the Riesz transform [9].

### 2.2 Curvature Tensor and Its Conjugate Part

In order to evaluate the local phase information, the curvature tensor and its harmonic conjugate part are designed to capture the even and odd information of 2D image structures.

Designing the curvature tensor is motivated by the second fundamental theorem of differential geometry, that is the second order derivatives or Hessian matrix which contains curvature information of the original signal. Let $f$ be a 2D signal, its Hessian matrix is correspondingly given by

$$H = \begin{bmatrix} f_{xx} \; f_{xy} \\ f_{xy} \; f_{yy} \end{bmatrix} , \tag{2}$$

where $x$ and $y$ are the Cartesian coordinates. According to the derivative theorem of the Fourier theory, the Hessian matrix in the spectral domain reads

$$\mathcal{F}\{H\} = \begin{bmatrix} -4\pi^2\rho^2 \frac{1+\cos(2\alpha)}{2} F & -4\pi^2\rho^2 \frac{\sin(2\alpha)}{2} F \\ -4\pi^2\rho^2 \frac{\sin(2\alpha)}{2} F & -4\pi^2\rho^2 \frac{1-\cos(2\alpha)}{2} F \end{bmatrix} , \tag{3}$$

where $F$ is the Fourier transform of the original signal $f$. It is obvious that angular parts of the second order derivatives in the Fourier domain are related to 2D spherical harmonics of even order 0 and 2. Hence, these harmonics represent the even information of 2D structures. Therefore, we are motivated to construct a tensor $T_e$, which is related to the Hessian matrix. This tensor is called a curvature tensor because it is similar to the curvature tensor of the second fundamental form of differential geometry. This curvature tensor $T_e$ can be obtained from a tensor-valued filter $H_e$ in the frequency domain, i.e. $T_e = \mathcal{F}^{-1}\{FH_e\}$, where $\mathcal{F}^{-1}$ means the inverse Fourier transform. Hence, the tensor-valued filter $H_e$, called the even Hessian operator reads

$$\begin{aligned} H_e &= \begin{bmatrix} \frac{H_0+\mathrm{real}(H_2)}{2} & \frac{\mathrm{imag}(H_2)}{2} \\ \frac{\mathrm{imag}(H_2)}{2} & \frac{H_0-\mathrm{real}(H_2)}{2} \end{bmatrix} = \frac{1}{2} \begin{bmatrix} 1+\cos(2\alpha) & \sin(2\alpha) \\ \sin(2\alpha) & 1-\cos(2\alpha) \end{bmatrix} \exp(-2\pi\rho s) \\ &= \begin{bmatrix} \cos^2(\alpha) & \frac{1}{2}\sin(2\alpha) \\ \frac{1}{2}\sin(2\alpha) & \sin^2(\alpha) \end{bmatrix} \exp(-2\pi\rho s) , \end{aligned} \tag{4}$$

where real(·) and imag(·) indicate the real and imaginary parts of the expressions.

In this filter, two components $\cos^2(\alpha)$ and $\sin^2(\alpha)$ can be considered as two angular windowing functions. These angular windowing functions provide a measure of the angular distance. From them, two perpendicular i1D components of the 2D image, oriented along the $x$ and $y$ coordinates, can be obtained. The other component of the filter is also the combination of two angular windowing functions, i.e. $\frac{1}{2}\sin(2\alpha) = \frac{1}{2}(\cos^2(\alpha - \frac{\pi}{4}) - \sin^2(\alpha - \frac{\pi}{4}))$. These two angular windowing functions yield again two i1D components of the 2D image, which are oriented along the diagonals. These four angular windowing functions can also be considered as four differently oriented filters, which are basis functions to steer a filter. They make sure that i1D components of even symmetry along different orientations are extracted. Consequently, the even Hessian operator $H_e$ enables the extraction of differently oriented and even i1D components of the 2D image.

The conjugate Poisson kernel, which evaluates the corresponding odd information of the i1D signal, is in quadrature phase relation with the i1D signal. Therefore, the odd representation of the curvature tensor, called the conjugate curvature tensor $T_o$, is obtained by employing the conjugate Poisson kernel to elements of $T_e$. Besides, the conjugate curvature tensor $T_o$ results also from a tensor-valued odd filter $H_o$, i.e. $T_o = h_1 * T_e = \mathcal{F}^{-1}\{H_1 H_e F\} = \mathcal{F}^{-1}\{H_o F\}$, where $h_1$ denotes the conjugate Poisson kernel in the spatial domain. Hence, the odd Hessian operator $H_o$ in the spectral domain is given by

$$H_o = \frac{1}{2}\begin{bmatrix} H_1(H_0 + \mathrm{real}(H_2)) & H_1(\mathrm{imag}(H_2)) \\ H_1(\mathrm{imag}(H_2)) & H_1(H_0 - \mathrm{real}(H_2)) \end{bmatrix} . \tag{5}$$

### 2.3 Local Amplitudes and Phases

Similar as the Hessian matrix, we are able to compute the trace and determinant of $T_e$ and $T_o$ for detecting the intrinsically one dimensional (i1D) and intrinsically two dimensional (i2D) structures. Note that 2D images can be classified into three categories according to the intrinsic dimensionality [10], a local property of a multidimensional signal. Thus, constant signals are intrinsically zero dimensional (i0D) structures, i1D signals represent straight lines or edges and i2D signals are composed of curved edges and lines, junctions, etc.

Consequently, a novel model for the i1D structures is obtained by combing the traces of $T_e$ and $T_o$. This is exactly the monogenic scale-space as proposed in [8],

$$\mathbf{f}_{i1D}(\mathbf{x}; s) = \mathrm{trace}(T_e(\mathbf{x}; s)) + \mathrm{trace}(T_o(\mathbf{x}; s)) = t_e + \mathbf{t}_o \tag{6}$$

with $\mathbf{t}_o = [\mathrm{real}(\mathrm{trace}(T_o(\mathbf{x}; s))),\ \mathrm{imag}(\mathrm{trace}(T_o(\mathbf{x}; s)))]^T$.

Hence, the multi-scale local amplitude and local phase vector for i1D structures are given by

$$a(\mathbf{x}; s) = \sqrt{t_e^2 + \mathbf{t}_o^2} \tag{7}$$

$$\varphi(\mathbf{x}; s) = \frac{\mathbf{t}_o}{|\mathbf{t}_o|}\mathrm{atan}\left(\frac{|\mathbf{t}_o|}{t_e}\right) , \tag{8}$$

where $\frac{\mathbf{t}_o}{|\mathbf{t}_o|}$ denotes the local orientation of the i1D structure. Correspondingly, combing the determinants of $T_e$ and $T_o$ results in a novel model for the i2D structure, which is called the generalized monogenic curvature scale-space $\mathbf{f}_{i2D}(\mathbf{x}; s)$,

$$\mathbf{f}_{i2D}(\mathbf{x}; s) = \det(T_e(\mathbf{x}; s)) + \det(T_o(\mathbf{x}; s)) = d_e + \mathbf{d}_o \tag{9}$$

with $\mathbf{d}_o = [\mathrm{real}(\det(T_o(\mathbf{x}; s))),\ \mathrm{imag}(\det(T_o(\mathbf{x}; s)))]^T$.

From it, the local amplitude for the i2D structure is obtained as

$$A(\mathbf{x}; s) = \sqrt{d_e^2 + \mathbf{d}_o^2} , \tag{10}$$

and the local phase vector takes the following form

$$\Phi(\mathbf{x}; s) = \frac{\mathbf{d}_o}{|\mathbf{d}_o|}\mathrm{atan}\left(\frac{|\mathbf{d}_o|}{d_e}\right) , \tag{11}$$

where $\frac{\mathbf{d}_o}{|\mathbf{d}_o|}$ decides the local main orientation of the i2D structure.

Since the local phase information of the i1D and i2D structures contains not only phase information but also the local orientation, the evaluation can be done in a rotation invariant way.

# 3 Dense Optical Flow Estimation

Differential methods have become the most widely used techniques for optical flow computation. By combining the advantages of local methods and global methods, Bruhn et al. [6] proposed a new method (CLG), which could yield flow fields with 100% density and are robust against noise. Since the phase-based approach was shown to perform very good with the advantage of being robust against brightness change [5,4], it is very natural to combine the advantages of the phase-based approach and the CLG method. In this section, we will adapt the CLG method into our model framework to estimate two-frame optical flow fields.

## 3.1 2D Combined Local-Global (CLG) Method

Differential methods are based on the assumption that the grey values of image sequences $f(x, y, t)$ in subsequent frames do not change over time

$$f(x+u,\, y+v,\, t+1) = f(x, y, t)\ , \tag{12}$$

where the displacement field $[u,\, v]^T$ denotes the optical flow. Following this, the spatial CLG method aims to minimize an energy function for estimating the flow field

$$E(\mathbf{w}) = \int_{\Omega} (\psi_1(\mathbf{w}^T J_\rho(\nabla_3 f)\mathbf{w}) + \alpha\psi_2(|\nabla \mathbf{w}|^2))dxdy \tag{13}$$

with

$$\nabla = [\partial_x,\, \partial_y]^T \tag{14}$$

$$\nabla_3 := [\partial_x,\, \partial_y,\, \partial_t]^T \tag{15}$$

$$\mathbf{w} := [u,\, v,\, 1]^T \tag{16}$$

$$|\nabla \mathbf{w}|^2 := |\nabla u|^2 + |\nabla v|^2 \tag{17}$$

$$J_\rho(\nabla_3 f) := K_\rho * (\nabla_3 f \nabla_3 f^T)\ , \tag{18}$$

where $\Omega$ denotes the image domain, $\alpha$ serves as regularization parameter, $K_\rho$ means a Gaussian kernel with standard deviation $\rho$, $\psi_1(\cdot)$ and $\psi_2(\cdot)$ indicate two nonquadratic penalisers with the following form

$$\psi_i(z) = 2\beta_i^2 \sqrt{1 + \frac{z}{\beta_i^2}} \quad i \in \{1,\, 2\} \tag{19}$$

with $\beta_1$ and $\beta_2$ as scaling parameters to handle outliers.

## 3.2 New Energy Function with Phase Constraints

In order to combine the phase-based approach with the 2D CLG method, the classical brightness constancy assumption will be replaced by new phase constraints. Two local phase vectors of i1D and i2D structures can be derived from the monogenic curvature tensor. One can assume that local phases of image sequences $f(x, y, t)$ in subsequent frames do not change over time. This results in the following new constancy assumptions

$$\boldsymbol{\varphi}(x+u,\, y+v,\, t+1) = \boldsymbol{\varphi}(x,\, y,\, t) \tag{20}$$
$$\boldsymbol{\Phi}(x+u,\, y+v,\, t+1) = \boldsymbol{\Phi}(x,\, y,\, t)\ . \tag{21}$$

For small displacements, we may perform a first order Taylor expansion yielding the optical flow constraints:

$$\boldsymbol{\varphi}_x u + \boldsymbol{\varphi}_y v + \boldsymbol{\varphi}_t = 0 \tag{22}$$
$$\boldsymbol{\Phi}_x u + \boldsymbol{\Phi}_y v + \boldsymbol{\Phi}_t = 0\ . \tag{23}$$

Let the i1D and i2D local phase vectors be $\boldsymbol{\varphi} = [\varphi_1,\, \varphi_2]^T$ and $\boldsymbol{\Phi} = [\Phi_1,\, \Phi_2]^T$, respectively, we propose to minimize the following energy function

$$E(\mathbf{w}) = \int_\Omega (\psi_1(\mathbf{w}^T J_\rho(\nabla_3\boldsymbol{\varphi} + \gamma\nabla_3\boldsymbol{\Phi})\mathbf{w}) + \alpha\psi_2(|\nabla\mathbf{w}|^2))dxdy \tag{24}$$

with

$$J_\rho(\nabla_3\boldsymbol{\varphi} + \gamma\nabla_3\boldsymbol{\Phi}) = K_\rho * \begin{bmatrix} M_{11} & M_{12} & M_{13} \\ M_{21} & M_{22} & M_{23} \\ M_{31} & M_{32} & M_{33} \end{bmatrix} \tag{25}$$

$$M_{11} = \varphi_{1x}^2 + \varphi_{2x}^2 + \gamma(\Phi_{1x}^2 + \Phi_{2x}^2) \tag{26}$$
$$M_{12} = M_{21} = \varphi_{1x}\varphi_{1y} + \varphi_{2x}\varphi_{2y} + \gamma(\Phi_{1x}\Phi_{1y} + \Phi_{2x}\Phi_{2y}) \tag{27}$$
$$M_{13} = M_{31} = \varphi_{1x}\varphi_{1t} + \varphi_{2x}\varphi_{2t} + \gamma(\Phi_{1x}\Phi_{1t} + \Phi_{2x}\Phi_{2t}) \tag{28}$$
$$M_{22} = \varphi_{1y}^2 + \varphi_{2y}^2 + \gamma(\Phi_{1y}^2 + \Phi_{2y}^2) \tag{29}$$
$$M_{23} = M_{32} = \varphi_{1y}\varphi_{1t} + \varphi_{2y}\varphi_{2t} + \gamma(\Phi_{1y}\Phi_{1t} + \Phi_{2y}\Phi_{2t}) \tag{30}$$
$$M_{33} = \varphi_{1t}^2 + \varphi_{2t}^2 + \gamma(\Phi_{1t}^2 + \Phi_{2t}^2)\ . \tag{31}$$

In this energy function, $\gamma$ is employed to adjust the trade-off between the i1D and i2D structures. According to the new energy function, the minimizing flow field $[u,\, v]^T$ will satisfy the following Euler-Lagrange equations

$$\operatorname{div}(\psi_2'(|\nabla\mathbf{w}|^2)\nabla u) - \frac{1}{\alpha}\psi_1'(\mathbf{w}^T J_\rho(\nabla_3\boldsymbol{\varphi} + \gamma\nabla_3\boldsymbol{\Phi})\mathbf{w})(J_{11}u + J_{12}v + J_{13}) = 0$$
$$\operatorname{div}(\psi_2'(|\nabla\mathbf{w}|^2)\nabla v) - \frac{1}{\alpha}\psi_1'(\mathbf{w}^T J_\rho(\nabla_3\boldsymbol{\varphi} + \gamma\nabla_3\boldsymbol{\Phi})\mathbf{w})(J_{21}v + J_{22}u + J_{23}) = 0$$

with

$$\psi_i'(z) = \frac{1}{\sqrt{1 + \frac{z}{\beta_i^2}}} \qquad i \in \{1,\, 2\}\ . \tag{32}$$

The estimation of optical flow field can thus be obtained iteratively by using an SOR [11] scheme. In our application, we take 200 iterations.

### 3.3 Computation of Phase Derivatives

In order to avoid phase wrapping, phase derivatives are computed from the filter responses in the monogenic scale-space and the generalized monogenic curvature scale-space, respectively. The spatial derivatives of i1D and i2D local phase vectors are given by

$$\nabla \boldsymbol{\varphi} = \frac{t_e \nabla \mathbf{t}_o - \mathbf{t}_o^T \nabla^T t_e}{t_e^2 + |\mathbf{t}_o|^2} \tag{33}$$

$$\nabla \boldsymbol{\Phi} = \frac{d_e \nabla \mathbf{d}_o - \mathbf{d}_o^T \nabla^T d_e}{d_e^2 + |\mathbf{d}_o|^2} . \tag{34}$$

The temporal derivatives of these local phase vectors read

$$\boldsymbol{\varphi}_t = \frac{t_e^t \mathbf{t}_o^{t+1} - t_e^{t+1} \mathbf{t}_o^t}{|t_e^t \mathbf{t}_o^{t+1} - t_e^{t+1} \mathbf{t}_o^t|} \text{atan} \left( \frac{|t_e^t \mathbf{t}_o^{t+1} - t_e^{t+1} \mathbf{t}_o^t|}{t_e^t t_e^{t+1} + \mathbf{t}_o^t \cdot \mathbf{t}_o^{t+1}} \right) \tag{35}$$

$$\boldsymbol{\Phi}_t = \frac{d_e^t \mathbf{d}_o^{t+1} - d_e^{t+1} \mathbf{d}_o^t}{|d_e^t \mathbf{d}_o^{t+1} - d_e^{t+1} \mathbf{d}_o^t|} \text{atan} \left( \frac{|d_e^t \mathbf{d}_o^{t+1} - d_e^{t+1} \mathbf{d}_o^t|}{d_e^t d_e^{t+1} + \mathbf{d}_o^t \cdot \mathbf{d}_o^{t+1}} \right) , \tag{36}$$

where $t_e^t$, $\mathbf{t}_o^t$, $d_e^t$, $\mathbf{d}_o^t$ denote the filter responses of the image frame at time $t$ and $t_e^{t+1}$, $\mathbf{t}_o^{t+1}$, $d_e^{t+1}$, $\mathbf{d}_o^{t+1}$ are the filter responses of the next image frame.

### 3.4 Multi-scale Optical Flow Estimation

The linearized optical flow constraint, stated in section 3.2, is based on the phase constancy assumption. As a consequence, it requires that $u$ and $v$ are relatively small so that the linearization holds. However, this is not always the case for an arbitrary sequence. Hence, multi-scale optical flow estimation technique should be employed to deal with large displacements.

In this paper, we use an incremental coarse to fine strategy. In contrast to the classical multi-scale approach, the estimated flow field at a coarse level is used to warp the image sequence instead of serving as initialization for the next finer scale. This compensation results in a hierarchical modification which requires to compute only small displacement. Once this is done from a coarse to the finest scale, much more accurate estimation will be obtained.

Let $d\mathbf{w}^s$ denote a displacement increment of $\mathbf{w}^s$ at scale $s$, for the coarsest scale ($s = S$), the optical flow field has the initial data $\mathbf{w}^S = [0, 0, 0]^T$. Hence, $d\mathbf{w}^s$ is given by minimizing the following energy function

$$\begin{aligned} E(d\mathbf{w}^s) = & \int_{\Omega} (\psi_1((d\mathbf{w}^s)^T J_\rho(\nabla_3 \boldsymbol{\varphi}(\mathbf{x} + \mathbf{w}^s) + \gamma \nabla_3 \boldsymbol{\Phi}(\mathbf{x} + \mathbf{w}^s)) d\mathbf{w}^s) \\ & + \alpha \psi_2(|\nabla(\mathbf{w} + d\mathbf{w}^s)|^2)) dx dy , \end{aligned} \tag{37}$$

where $\mathbf{x} = [x, y, t]^T$ and $\mathbf{w}^{s+1} = \mathbf{w}^s + d\mathbf{w}^s$. Note that local phase vectors will be warped as $\boldsymbol{\varphi}(\mathbf{x} + \mathbf{w}^s)$ and $\boldsymbol{\Phi}(\mathbf{x} + \mathbf{w}^s)$ via bilinear interpolation. Final result will be obtained when the minimization is done to the finest scale.

## 4 Experimental Results

In order to evaluate the performance of the proposed approach, optical flow estimation on both synthetic and real-world image data is given in this section. We use the so-called average angular error (AAE) [4] as the quantitative quality measure. Given the estimated flow field $[u_e, v_e]^T$ and the ground truth $[u_c, v_c]^T$, the AAE is defined as

$$AAE = \frac{1}{N}\sum_{i=1}^{N} \arccos\left(\frac{u_{ci}u_{ei} + v_{ci}v_{ei} + 1}{\sqrt{(u_{ci}^2 + v_{ci}^2 + 1)(u_{ei}^2 + v_{ei}^2 + 1)}}\right) , \tag{38}$$

where $N$ denotes the total number of pixels.

The *Yosemite* sequence with clouds is employed as the synthetic data for the experiment. This sequence combines divergent and translational motion under varying illumination and hence is usually regarded as the benchmark for optical flow estimations. Fig. 1 demonstrates the ground truth, the estimated magnitudes and optical flow fields from our approach and the 2D CLG model. It is obvious that our approach produces more accurate result than that of the 2D CLG method. Especially in the clouds region, where the illumination varies, the proposed approach shows more stable estimation. Even if we compare with 2D CLG where the intensity is replaced by the gradient, our approach also performs better.

Detail comparisons with other approaches according to the measurement AAE are given in Table 1, where STD indicates the standard deviation. Our approach demonstrates much better performance with lower AAE and STD when compared with most of the methods. When $\gamma = 0$, only i1D phase information is included for the constraint, the AAE now takes $3.37^0$, which is $1.49^0$ lower than that of the 2D CLG method. Interestingly, this result is even lower than that of the 3D CLG method. By adjusting the parameter as $\gamma = 0.1$, i2D phase is also contained to strengthen the constraint. Hence, estimation with even lower error can be obtained. For this experiment, we also extend the two-frame estimation to multi-frame by adding the temporal information. Results also indicate the good performance of the proposed approach.

Even much better results have been reported in [1]. However, they do not perform a first order Taylor expansion of the intensity assumption to yield the optical flow constraint. Thus, it is very promising that our approach can also yield comparably good results by using the non-linearized constancy assumption.

For the following experiments, we simply focus on two-frame flow field estimation. To investigate the robustness of our approach against noise, the 8th frame of the *Yosemite* sequence is degraded with additive Gaussian noise.

The noise contaminated image (signal noise ratio: SNR=10dB) and the estimated flow field are shown in Fig. 2. It is obvious that the original image is seriously degraded, nevertheless, the estimation also shows good performance with AAE=$14.16^0$. More detail information can be found in Fig. 3. When the SNR decreases from 40dB to 10dB, much more noise is added to the original image. However, the estimated result is still not very sensitive to noise.

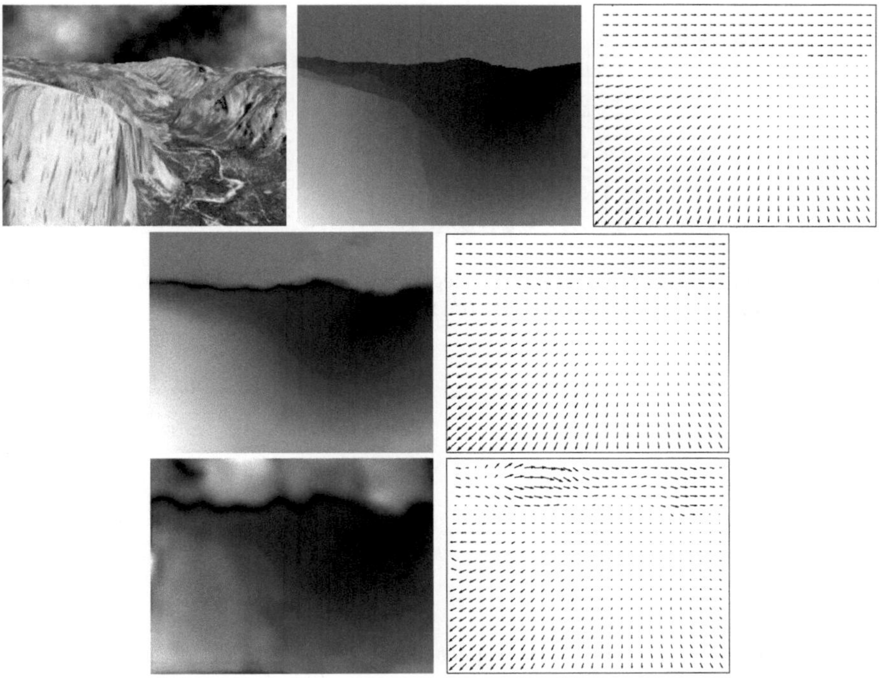

**Fig. 1.** Top row: from left to right are one frame of the Yosemite image sequence, the magnitude and flow field of the ground truth. Middle row: from left to right are the magnitude and flow field estimated from our approach. Bottom row: from left to right are the magnitude and flow field estimated from the 2D CLG method.

**Table 1.** Optical flow estimation comparisons between different approaches (100% density). AAE (average angular error), STD (standard deviation).

| Approach | AAE | STD |
|---|---|---|
| Horn/Schunck (Barron et al., 1994) [4] | $31.69^0$ | $31.18^0$ |
| Nagel (Barron et al., 1994) [4] | $10.22^0$ | $16.51^0$ |
| Uras et al. (Barron et al., 1994) [4] | $8.94^0$ | $15.61^0$ |
| 2D CLG (2005) [6] | $4.86^0$ | $8.48^0$ |
| Mémin and Pérez (1998) [12] | $4.69^0$ | $6.89^0$ |
| 3D CLG (2005) [6] | $4.17^0$ | $7.72^0$ |
| **Our 2D approach** ($\gamma = 0$) | $3.37^0$ | $8.27^0$ |
| **Our 2D approach** ($\gamma = 0.1$) | $3.25^0$ | $8.22^0$ |
| **Our 3D approach** ($\gamma = 0$) | $2.74^0$ | $7.17^0$ |
| **Our 3D approach** ($\gamma = 0.1$) | $2.67^0$ | $7.12^0$ |

This indicates that employing the local method and multi-scale technique into our approach does result in a robust estimation against noise. As mentioned in [5], the phase-based approach has the advantage of being not sensitive to the

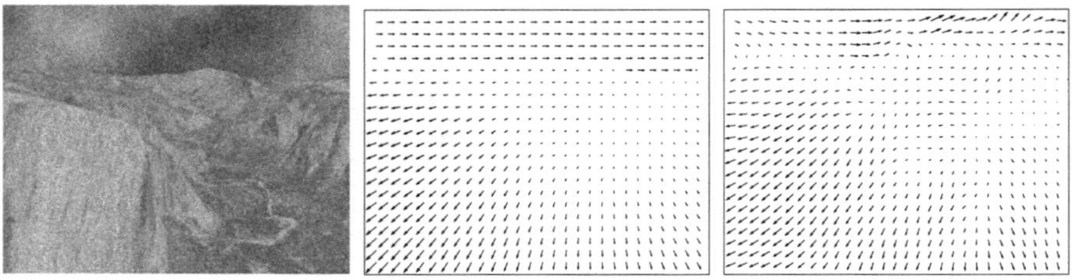

**Fig. 2.** From left to right are the noise degraded image (SNR=10dB), ground truth and the estimated flow field (AAE=14.16$^0$, STD=12.76$^0$)

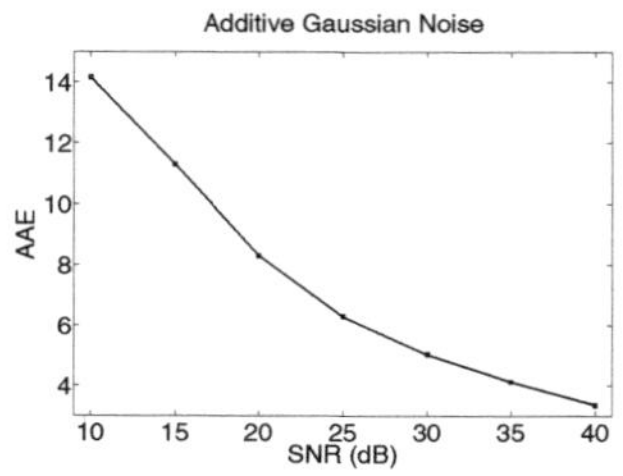


**Fig. 3.** The estimated results with respect to additive Gaussian noise change

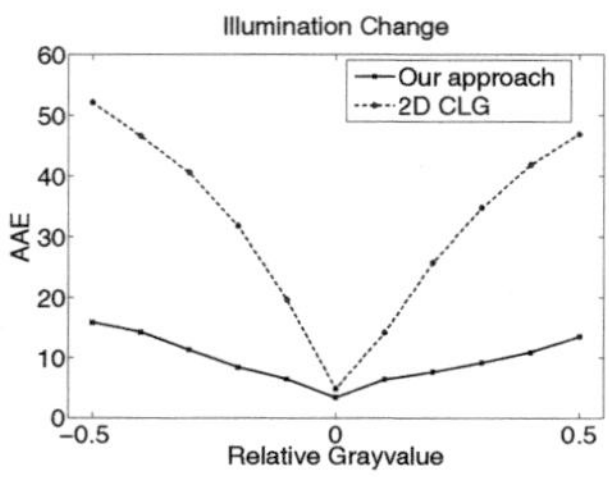


**Fig. 4.** The estimated results with respect to illumination change

illumination variation. Additionally, the proposed approach adapts the CLG method into the framework of the monogenic curvature tensor. As a consequence, this new method combines the advantages of the phase-based approach and of the CLG method. In this way, our approach should also be robust under illumination change within some limits. To this end, another experiment is conducted to test the performance of our approach for the brightness variation. Fig. 5 shows the performance comparison between our approach and the 2D CLG method under brightness change. The 8th frame of the synthetic sequence is degraded with brighter and darker illumination changes of 50%, respectively. Experimental results denote that our approach is much more robust against illumination variation when compared with that of the 2D CLG method.To evaluate the performance of the proposed approach under illumination change in detail, the 8th frame is degraded with different brighter and darker brightness variations. The estimated AAEs with respect to the relative grayvalue changes are shown in Fig.4. The results indicate that our approach is very robust against illumination change. However, the 2D CLG method is very sensitive to it.

The last experiment aims to examine the performance of the proposed approach for real image sequences. In this experiment, two sequences are used. They are the well-known Hamburg taxi sequence and the Ettlinger Tor traffic sequence [13]. Estimated flow fields are illustrated in Fig. 6. It is clear that the proposed approach also yields realistic optical flow for real-world data.

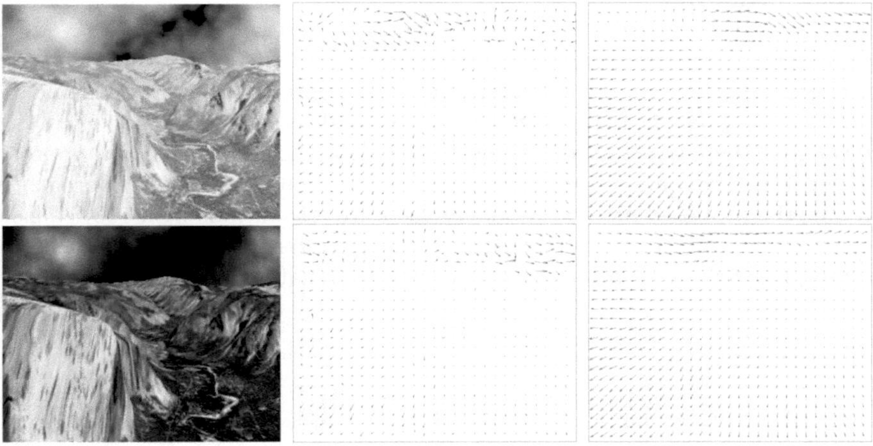

**Fig. 5.** Top row: from left to right are frame 8 degraded with brighter illumination change of 50%, estimated flow fields from the 2D CLG (AAE=46.94$^0$, STD=39.97$^0$) and our approach (AAE=13.50$^0$, STD=17.42$^0$). Bottom row: from left to right are frame 8 degraded with darker illumination variation of 50%, optical flow estimations from the 2D CLG (AAE=52.14$^0$, STD=46.63$^0$) and our method (AAE=15.83$^0$, STD=19.71$^0$).

**Fig. 6.** Top row: one frame of the Hamburg taxi sequence and the estimated optical flow from our approach. Bottom row: one frame of the Ettlinger Tor traffic sequence and the flow fields from our approach.

## 5 Conclusions

We present a novel approach for estimating two-frame dense optical flow field in this paper. This new approach adapts the CLG approach to the monogenic curvature tensor, a new framework which enables multi-scale local phase evaluation of i1D and i2D image structures in a rotation invariant way. Hence, our approach takes both the advantages of phase-based approach and the CLG approach. In this way, the proposed method produces accurate estimations with 100% density and is robust against noise. Compared with the intensity based approach, our method performs much better in the case of illumination variation.

## References


1. Papenberg, N., Bruhn, A., Brox, T., Didas, S., Weickert, J.: Highly accurate optic flow computation with theoretically justified warping. International Journal of Computer Vision **67**(2) (2006) 141–158
2. Horn, B., Schunck, B.: Determining optical flow. Artificial Intelligence **17** (1981) 185–203
3. Lucas, B., Kanade, T.: An iterative image registration technique with an application to stereo vision. In: Proc. 7th International Joint Conference on Artifical Intelligence, Vancouver, Canada (1981) 674–679
4. Barron, J.L., Fleet, D.J., Beauchemin, S.S.: Performance of optical flow techniques. International Journal of Computer Vision **12**(1) (1994) 43–77
5. Fleet, D.J., Jepson, A.D.: Computation of component image velocity from local phase information. International Journal of Computer Vision **5**(1) (1990) 77–104
6. Bruhn, A., Weickert, J., Schnörr, C.: Lucas/Kanada meets Horn/Schunck: Combining local and global optic flow methods. International Journal of Computer Vision **61**(3) (2005) 211–231
7. Oppenheim, A.V., Lim, J.S.: The importance of phase in signals. IEEE Proceedings **69** (May 1981) 529–541
8. Felsberg, M., Sommer, G.: The monogenic scale-space: A unifying approach to phase-based image processing in scale-space. Journal of Mathematical Imaging and Vision **21** (2004) 5–26
9. Felsberg, M., Sommer, G.: The monogenic signal. IEEE Transactions on Signal Processing **49**(12) (December 2001) 3136–3144
10. Zetzsche, C., Barth, E.: Foundamental limits of linear filters in the visual processing of two-dimensional signals. Vision Research **30** (1990) 1111–1117
11. Young, D.M.: Iterative Solution of Large Linear Systems. Academic Press (1971)
12. Mémin, E., Pérez, P.: A multigrid approach for hierarchical motion estimation. In: Proc. 6th International Conference on Computer Vision, Bombay, India, Narosa Publishing House (Jan. 1998) 933–938
13. Nagel, H.H.: Ettlinger tor traffic sequence. Available at `http://i21www.ira.uka.de/image_sequences/`

# A Consistent Spatio-temporal Motion Estimator for Atmospheric Layers

Patrick Héas, Étienne Mémin, and Nicolas Papadakis

INRIA/IRISA, Rennes, Vista Project
Campus universitaire de Beaulieu. 35000 Rennes, France
{Patrick.Heas,Etienne.Memin,Nicolas.Papadakis}@irisa.fr

**Abstract.** In this paper, we address the problem of estimating mesoscale dynamics of atmospheric layers from satellite image sequences. Relying on a physically sound vertical decomposition of the atmosphere into layers, we propose a dense motion estimator dedicated to the extraction of multi-layer horizontal wind fields. This estimator is expressed as the minimization of a global function including a data term and a spatio-temporal smoothness term. A robust data term relying on shallow-water mass conservation model is proposed to fit sparse observations related to each layer. A novel spatio-temporal regularizer derived from shallow-water momentum conservation model is proposed to enforce a temporal consistency of the solution along the sequence time range. These constraints are combined with a robust second-order regularizer preserving divergent and vorticity structures of the flow. In addition, a two-level motion estimation scheme is proposed to overcome the limitations of the multiresolution incremental scheme when capturing the dynamics of fine mesoscale structures. This alternative approach relies on the combination of correlation and optical-flow observations. An exhaustive evaluation of the novel method is first performed on a scalar image sequence generated by Direct Numerical Simulation of a turbulent bi-dimensional flow and then on a Meteosat infrared image sequence.

## 1 Introduction

The analysis of complex fluid flows behaviors is a major scientific issue. In particular, understanding atmospheric dynamics is of great importance for meteorologists interested in weather forecasting, climate prediction, singular system analysis, etc. The use of surface meteorology stations, balloons, and more recently aircraft measurements or first-generation satellite images has improved the estimation of wind fields and has been a subsequent step for a better understanding of meteorological phenomena. However, the measurements provided by the network's temporal and spatial resolutions may be insufficient for the analysis of mesoscale dynamics characterized by horizontal scales in the range of about 10-1000 km. Recently, in an effort to avoid these limitations, an increasing interest has been devoted to motion extraction from images of a new generation of geostationary satellites, with higher acquisition rates and finer spatial resolutions.

F. Sgallari, A. Murli, and N. Paragios (Eds.): SSVM 2007, LNCS 4485, pp. 251–263, 2007.

The analysis of motion in such sequences is particularly challenging due to the great deal of spatial and temporal distortions that luminance patterns exhibit in imaged atmospheric phenomena. Standard techniques from Computer Vision, originally designed for quasi rigid motions and stable salient features along time, are not well adapted in this context. Recently, methods for fluid-dedicated dense estimation have been proposed to characterize atmospheric motion [2,8]. Nevertheless, we will show that due to the underlying three-dimensional nature of the scene, the employed dynamical models remain unadapted to satellite observations. Furthermore, such methods may fail to accurately characterize motion associated to mesoscale structures. Thus, the design of an appropriate approach taking into account the physics of three-dimensional atmosphere dynamics constitutes a widely open domain of research. Our work is a contribution in this direction. Rather than coupling the motion vector estimation process to a complex and complete numerical meteorological circulation model, we propose to incorporate in the motion estimation scheme "image based adequate" dynamics defined as an adaptation of Navier-Stokes equations to infra-red images. The objective being in fine a three-dimensional reconstruction of atmospheric horizontal winds. Alternatively, the challenge also consists in providing accurate estimators able to tackle the motion complexity of sparse and noisy structures.

## 2 Related Work on Optical-Flow Estimation

The problem of wind field recovery in an image sequence $I(x, y, t)$ consists in estimating the real three-dimensional atmospheric motion from observations in the projected image plane. This problem is a complex one, for which we have only access to projected information on clouds position and spectral signatures provided by satellite observation channels. Spatial horizontal coordinates $(x, y)$ are denoted by $\mathbf{s}$. To avoid tackling the three-dimensional wind field $\mathbf{V}(\mathbf{s}, z, t)$ reconstruction problem, up to now all the developed wind field estimation methods rely on the assumption of inexistent vertical winds and consists to estimate an average horizontal wind.

### 2.1 Real Projected Wind Fields and Optical-Flow

The apparent motion $\mathbf{v} = (u, v)$, perceived through image intensity variations, can be computed with the standard Optical Flow Constraint (OFC):

$$I_t(\mathbf{s}, t) + \mathbf{v} \cdot \nabla I(\mathbf{s}, t) = 0. \tag{1}$$

For image sequences showing evolving atmospheric phenomena, the brightness consistency assumption does not allow to model temporal distortions of luminance patterns caused by 3D flow transportation. For transmittance imagery of fluid flows, the so called continuity equation :

$$\frac{1}{\rho}\frac{D\rho}{Dt} + \nabla.\mathbf{V} = 0, \tag{2}$$

may be derived from the 3D mass conservation law, where $\rho$ denotes a three-dimensional density function. In this case, an apparent motion $\mathbf{v}$ is redefined as a density-weighted average of the original three-dimensional horizontal velocity field. For the case of a null motion on the boundary planes, in [3], the author showed that the integration of Eq.2 leads to a 2D Integrated Continuity Equation (ICE):

$$\Big(\int \rho dz\Big)_t + \mathbf{v}.\boldsymbol{\nabla}\Big(\int \rho dz\Big) + \Big(\int \rho dz\Big)\text{div}(\mathbf{v}) = 0, \quad (3)$$

Unlike the OFC, such models can compensates mass departures observed in the image plan by associating two-dimensional divergence to brightness variations. By time integration, an equivalent non-linear formulation can be recovered [2] :

$$\Big(\int \rho dz\Big)(\mathbf{s} + \mathbf{v}, t+1) \exp \text{div}(\mathbf{v}) - \Big(\int \rho dz\Big)(\mathbf{s}, t) = 0. \quad (4)$$

The image formation model for satellite infrared imagery is slightly different. In [2], the authors have directly assumed the unrealistic hypothesis that infrared pixel values $I$ were proportional to density integrals : $I \propto \int \rho dz$. In [16], the authors proposed an inversely proportional approximation for infrared measurements : $I \propto (\int \rho dz)^{-1}$.

## 2.2 Regularization Schemes and Minimization Issues

The previous formulations of Eq. 1, 3 and 4 can not be used alone, as they provide only one equation for two unknowns at each spatio-temporal locations $(\mathbf{s}, t)$. To deal with this problem, the most common assumption consists in enforcing spatial and temporal local coherence.

Disjoint local smoothing methods considers neighborhoods centered at pixel locations. An independent parametric field is locally estimated on each of these supports. In the the work of *Lucas and Kanade* [7], relying on the OFC equation, motion which is assumed to be locally constant is estimated using a standard linear least square approach. In meteorology, classical approaches are Euclidean correlation-based matchings, which corresponds to the OFC constraint associated to a locally constant field and a $\boldsymbol{L}^2$ norm [6,10,12]. On the one hand, these methods are fast and are able to estimate large displacement of fine structures. On the other hand, they present the drawback to be sensitive to noise and inefficient in the case of weak intensity gradients. Moreover, estimation with these approaches is prone to erroneous spatial variability and results in the estimation of sparse and possibly incoherent vector fields.

Globalized smoothing schemes can be used to overcome the previous limitations. These methods model spatio-temporal dependencies on the complete image domain. Thus, dense velocity fields are estimated even in the case of noisy and low contrasted observations. More precisely, the motion estimation problem is defined as the global minimization of a energy function composed of two components :

$$J(\mathbf{v}, I) = J_d(\mathbf{v}, I) + \alpha J_r(\mathbf{v}). \quad (5)$$

The first component $J_d(\mathbf{v}, I)$ called the data term, expresses the constraint linking unknowns to observations while the second component $J_r(\mathbf{v})$, called the regularization term, enforces the solution to follow some smoothness properties. In the previous expression, $\alpha > 0$ denotes a parameter controlling the balance between the smoothness and the global adequacy to the observation model. In this framework, Horn and Schunck [5] first introduced a data term related to the OFC equation and a first-order regularization of the two spatial components $u$ and $v$ of velocity field $\mathbf{v}$. In the case of transmittance imagery of fluid flows, $I = \int \rho dz$, and using the previously defined ICE model (Eq.3) leads to the functional :

$$J_d(\mathbf{v}, I) = \int_{\Omega} \left(I_t(\mathbf{s}) + \mathbf{v}(\mathbf{s}) \cdot \boldsymbol{\nabla} I(\mathbf{s}) + I(\mathbf{s})\mathrm{div}(\mathbf{v}(\mathbf{s}))\right)^2 d\mathbf{s}. \tag{6}$$

Moreover, it can be demonstrated that a first order regularization is not adapted as it favors the estimation of velocity fields with low divergence and low vorticity. A second order regularization on the vorticity and the divergence of the defined motion field can advantageously be consider as proposed in [11][2][15] :

$$J_r(\mathbf{v}) = \int_{\Omega} \parallel \boldsymbol{\nabla} curl \mathbf{v}(\mathbf{s}) \parallel^2 + \parallel \boldsymbol{\nabla} div \mathbf{v}(\mathbf{s}) \parallel^2 d\mathbf{s}. \tag{7}$$

Instead of relying on a $\boldsymbol{L}^2$ norm, robust penalty function $\phi_d$ may be introduced in the data term for attenuating the effect of observations deviating significantly from the ICE constraint [1]. Similarly, a robust penalty function $\phi_r$ can be used if one wants to handle implicitly the spatial discontinuities of the vorticity and divergence maps. In the image plan, these discontinuities are nevertheless difficult to relate to abrupt variations of clouds height . Moreover, layers clouds form unconnected regions which should interact during the motion estimation process.

OFC or ICE model rely on the assumption that the intensity function can be locally efficiently approximated by a linear function. Since the larger the displacement the more narrow the linearity domain, large displacements are difficult to recover directly. The multiresolution approach is a common way to overcome this limitation. However, since the multiresolution schemes estimates principal component displacements only at coarse resolutions where small photometric structures are rubbed out, this approach enables the characterization of large displacements of small structures only in the case when their motion are close enough to the principal component's one. This is often not the case for a multi-layered atmosphere.

## 3 Dense Motion Estimator Dedicated to Atmospheric Layers

### 3.1 Dynamical Model for Layers

The ICE model relies on strong assumptions in the case of satellite infrared imagery. However, it has been demonstrated that this approach is well suited for an image sequence of transmittance measurements.

Since there is a lack of information induced by projection in an image plane, several hypothesis are necessary to tackle the reconstruction problem. We assume that the lower part of the atmosphere is in hydrostatic equilibrium. This assumption provides an excellent approximation for the vertical dependence of the pressure field and enables a layer decomposition of the three-dimensional space [4]. Let us denote pressure and gravity respectively by $p$ and $g$. By the vertical integration of the hydrostatic equation $-\rho g = \frac{dp}{dz}$, density integrals are linked to pressure differences :

$$g \int_{z_b}^{z_t} \rho \, dz = p(z_b) - p(z_t). \tag{8}$$

where the lower and higher boundary functions are denoted by $z_b$ and $z_t$. The problem is thus moved on to the derivation of pressure difference maps from infrared images for each layer.

Let us first explain how the layer are dissociated. The membership of clouds to a layer is determined by a height classification map routinely provided by the EUMETSAT consortium, the European agency which manages the Meteosat satellite data. We denote by $C^k$ a class corresponding to the $k$-th layer in between the altimetric interval function $[z_b^k, z_t^k]$. Note that the top of cloud pressure image denoted by $p_{\bigcup_k}$ is composed of segments of top of clouds pressure functions $p(z_t^k)$ related to the different layers. That is to say :

$$p_{\bigcup_k} = \bigcup_k p(z_t^k, \mathbf{s}); \ \ \mathbf{s} \in C^k. \tag{9}$$

Sparse pressure maps of the layers upper boundaries are computed from infrared images as described in [13]. As in satellite images, clouds lower boundaries are always occluded, we approximate the missing pressure observations $p(z_b^k)$ by an average pressure value observed on top of the underneath layer. Finally, for the $k$-th layer, we define image observations as the pressure difference :

$$p(z_b^k) - p_{\bigcup_k}, = h^k \begin{cases} = g \int_{z_b^k}^{z_t^k} \rho \, dz \ \ if \ \mathbf{s} \in C^k \\ \neq g \int_{z_b^k}^{z_t^k} \rho \, dz \ \ if \ \mathbf{s} \in \bar{C}^k. \end{cases} \tag{10}$$

Thus, if we neglect vertical wind on layer boundaries, the ICE model of Eq.3 holds for observations $h^k$ on regions of $C^k$, and constitute a physically sound model for motion estimation of atmospheric layers evolving independently. $\forall k \in [0, K]$ :

$$\frac{\partial h^k}{\partial t} + \mathbf{v} \cdot \boldsymbol{\nabla} h^k + h^k \text{div}(\mathbf{v}) = 0, \tag{11}$$

where $K$ is the highest layer index and $\mathbf{v}$ corresponds to the density-weighted average horizontal wind related to the $k$-th layer. Note that as vertical wind on the layer boundaries has been neglected, this model assumes independent layer motion. Due to the hydrostatic relation, $h^k$ may be viewed as an atmospheric layer thickness function if we assume shallow layers and thus neglect density variations : $h^k = g\rho(z_b^k - z_t^k)$. The ICE equation then corresponds to shallow-water mass conservation model [4].

### 3.2 Robust Estimator for Sparse Observations

Relatively to the different layers, true pressure differences are sparsely observed only in the presence of clouds. A dense estimator dedicated to layer motion should consider simultaneously all cloudy regions belonging to a given layer while discarding the influence of other clouds. For the $k$-th layer, we previously remarked that outside the class '$C^k$', the so defined pressure difference $h^k$ is not relevant of the $k$-th layer thickness. Thus, we propose to introduce in Eq.4 a masking operator on unreliable observations. We denote by $\mathbf{M}_{\mathbf{s}\in C^k}$ the operator which is identity if pixel belongs to the class, and which returns otherwise a fixed value out of the range taken by $h^k$. Thus, applying this new masking operator in Eq.4, we obtain the robust data term $J_d(\mathbf{v}, h^k)$ :

$$\int_\Omega \phi_d\{exp\text{div}(\tilde{\mathbf{v}}(\mathbf{s}))([\tilde{h}^k(\mathbf{s})\nabla\text{div}(\tilde{\mathbf{v}}(\mathbf{s}))+\nabla\tilde{h}^k(\mathbf{s})]^T\mathbf{v}'(\mathbf{s})+\tilde{h}^k(\mathbf{s}))-\mathbf{M}_{\mathbf{s}\in C^k}(h^k(\mathbf{s}))\}d\mathbf{s}, \quad (12)$$

where $\mathbf{v}$ corresponds to the density-weighted average horizontal wind related to the $k$-th layer. The div-curl regularization term (Eq.7) is conserved. The masking procedure together with the use of robust penalty function on the data term allows to discard implicitly the erroneous observation from the estimation process. It is important to outline that, for the $k$-th layer, the method provides dense motion fields and areas outside class '$C^k$' correspond to an interpolated wind field. Nevertheless, let us point out that in the case of very sparse observations and large displacements, robust estimation becomes unstable and may lead to erroneous minima. Such limitations will be overcome in the following.

## 4 A Two Level Decomposition for Mesoscale Motion Estimation

In order to enhance the estimation accuracy, a collection of correlation-based vectors $\mathbf{v}_c$ is introduced as sparse constraints in a differential estimation scheme for the recovery of a dense displacement field. Contrary to the classical multiresolution approach, this new technique enables to deal with the large displacements of small structures as it relies on a unique representation of the full resolution image. Moreover, in order to preserve spatio-temporal consistency of displacement estimates, we propose to incorporate in the estimation scheme an *a priori* physical knowledge on fluid dynamical evolution. A dense displacement field $\mathbf{v}_p$ is predicted by time integration of a simplified Navier-Stokes dynamical model. The propagated field is then introduced in the estimation process as a spatio-temporal regularizer. Keeping notations of section 2.2, a new functional is defined for the estimation of variable $\mathbf{v}$

$$J(\mathbf{v}) = J_d(\mathbf{v}, h^k) + \gamma J_c(\mathbf{v}, \mathbf{v}_c) + \alpha J_r(\mathbf{v}) + \beta J_p(\mathbf{v}, \mathbf{v}_p), \tag{13}$$

where $J_c(.)$, $J_p(.)$ are energy functions respectively constraining displacements $\mathbf{v}$ to be close to a sparse correlation-based vector field $\mathbf{v}_c$ and to be consistent with a physically sound prediction $\mathbf{v}_p$ relying on Navier-Stokes equations. In the previous expression, $\gamma$ and $\beta$ denote weighting factors. Functionals $J_c(.)$ and $J_p(.)$ will be further detailed in the following.

The displacement field $\mathbf{v}$ is decomposed into a large displacement field $\bar{\mathbf{v}}$ and an additive small displacement field $\tilde{\mathbf{v}}$. The optimization problem is conducted sequentially. Here, an analogous version of the alternate multigrid minimization scheme proposed in [2] has been implemented.

Note that in the case $\alpha, \beta, \gamma \gg 1$, the energy minimization leads to a large displacement field which can be seen as a physically sound spatio-temporal interpolation of the correlation-based vectors.

### 4.1 Variational Approach for a Correlation/Optical-Flow Collaboration

In order to obtain a dense estimation of displacements fitting a sparse correlation-based displacement field, we define a functional where the $i^{th}$ correlation-based vector $\mathbf{v}_c^i = (u^i, v^i)$ located at the point $\mathbf{s}^i = (x^i, y^i)$ influences his neighborhood according to a shifted bi-dimensional Gaussian law $\mathcal{N}^i(\mathbf{s}^i - \mathbf{s})$ of variance $\sigma$ related to the correlation window influence

$$J_c(\mathbf{v}, \mathbf{v}_c) = \int_\Omega \mathbf{M}_{\mathbf{s} \in C^k} \left( \sum_{i=1}^{K} g^i \mathcal{N}^i(\mathbf{s}^i - \mathbf{s}) \phi_c\{\mathbf{v}_c^i - \mathbf{v}(\mathbf{s})\} \right) ds, \tag{14}$$

where $\phi_c$ is a robust penalty function similar to the one attached to the data term. In the previous expression, $g^i$ denote confidence factors. We choose, to define them according to the dissimilarity function. The masking operator $\mathbf{M}_{\mathbf{s} \in C^k}()$ was introduced as the correlation/optical-flow collaboration is not possible in regions with no image observations.

### 4.2 Spatio-temporal Regularization

The functional $J_p(.)$ aims at constraining a motion field to be consistent with a physically predicted wind field. We simply define this functional as a quadratic distance between the estimate field $\mathbf{v}$ and the dense propagated field $\mathbf{v}_p$:

$$J_p(\bar{\mathbf{v}}, \mathbf{v}_p) = \int_\Omega \| \mathbf{v}_p(\mathbf{s}) - \mathbf{v}(\mathbf{s}) \|^2 ds. \tag{15}$$

This approach constitutes an alternative to the spatio-temporal regularizer defined in [14] and is to some extend similar to the temporal constrain introduced in [9]. Our propagation model includes a bi-dimensional divergence component which is equal to zero only for incompressible bi-dimensional flows. As it is detailed below, our approach extends [9] to the spatio-temporal smoothing of the full velocity field in the case of three-dimensional geophysical flows driven by a shallow-water evolution law.

Dynamical models describing wind field evolution are needed here for the prediction at time $t+1$ of a sound field $\mathbf{v}_p$ using the previous motion estimation $\mathbf{v}$ performed for the $k$-th layer between time $t-1$ and $t$. As atmosphere evolution is governed by fluid flows laws, we rely on Navier-Stokes equations in order to derive simplified dynamical models adapted to short time propagation of layer mesoscale motions. A scale analysis of the horizontal momentum equations showed that the Coriolis, the curvature terms and the friction forces are in this case negligible. Denoting by $\nu$ a turbulent viscosity coefficient, imposing incompressibility in the hydrostatic relation, and adding the mass conservation model of Eq.11, we form the complete shallow-water equation system :

$$\begin{cases} \mathbf{v}_t + \mathbf{v} \cdot \boldsymbol{\nabla}(\mathbf{v}) + \frac{1}{\rho_0} \boldsymbol{\nabla} h^k = \nu \Delta(\mathbf{v}), \\ h_t^k + \mathbf{v} \cdot \boldsymbol{\nabla} h^k + h^k \zeta = 0, \end{cases} \tag{16}$$

with the notations $\boldsymbol{\nabla}(\mathbf{v}) = (\boldsymbol{\nabla} u, \boldsymbol{\nabla} v)^T$ and $\Delta(\mathbf{v}) = \boldsymbol{\nabla} \cdot \boldsymbol{\nabla}(\mathbf{v})$. Denoting the vorticity by $\xi = \text{curl}(\mathbf{v})$ and the divergence by $\zeta = \text{div}(\mathbf{v})$, the previous system may be expressed in a vorticity-divergence form :

$$\begin{cases} \xi_t + \mathbf{v} \cdot \boldsymbol{\nabla} \xi + \xi \zeta = \nu \Delta \xi, \\ \zeta_t + \mathbf{v} \cdot \boldsymbol{\nabla} \zeta + \zeta^2 = 2det(J(u,v)) - \frac{\Delta h^k}{\rho_0} + \nu \Delta \zeta, \\ h_t^k + \mathbf{v} \cdot \boldsymbol{\nabla} h^k + h^k \zeta = 0, \end{cases} \tag{17}$$

where $J(.)$ is the Jacobian operator. The dynamical model predict the evolution of 3 variables which may depend on each others. One of the major difficulties is induced by the fact that variable $h^k$ is derived only for cloudy regions corresponding to the $k$-th layer. Therefore, variable $h^k$, and thus all unknowns, can only be propagated on a sparse spatial support. However, in opposition to the classical formulation, the vorticity-divergence equations provide a dynamical model for which the vorticity evolution is independent of variable $h^k$ and for which the divergence evolution depends only weakly on variable $h^k$ (up to a constant). Based on the assumption that divergence is weak almost everywhere and assimilable to noise, we propose to simplify the divergence dynamical model in order to make it independent of variable $h^k$. Divergence $\zeta$ is assumed to be driven by a Gaussian random function with stationary increments (i.e. a standard Brownian motion). As a consequence, divergence expectation asymptotically obeys to a heat equation of diffusion coefficient $\nu$. The simplified vorticity-divergence model reads:

$$\begin{cases} \xi_t + \mathbf{v} \cdot \boldsymbol{\nabla} \xi + \xi \zeta = \nu \Delta \xi, \\ \zeta_t = \nu \Delta \zeta. \end{cases} \tag{18}$$

The curl and divergence completely determine the underlying 2D velocity field and the current velocity estimate can be recovered from these quantities up to a laminar flow. Indeed, the Helmholtz decomposition of the field into a sum of gradients of two potential functions is expressed as $\mathbf{v} = \nabla \times \Psi + \nabla \Phi + \mathbf{v}_{har}$, where $\mathbf{v}_{har}$ is an harmonic transportation part ($div\mathbf{v}_{har} = curl\mathbf{v}_{har} = 0$) of the field $\mathbf{v}$ and where the stream function $\Psi$ and the velocity potential $\Phi$ correspond to the solenoidal and the irrotationnal part of the field. The latter are linked to divergence and vorticity through two Poisson equations. Expressing the solution of both equations as a convolution product with the 2D Green kernel $G$ associated to the Laplacian operator: $\Psi = G * \xi, \quad \Phi = G * \zeta$, the whole velocity field can be recovered with the equation :

$$\mathbf{v} = \nabla \times (G * \xi) + \nabla (G * \zeta) + \mathbf{v}_{har}, \tag{19}$$

which can be efficiently solved in the Fourier domain.

Let us sum up this prediction process. The vorticity and the divergence fields are developed from $t$ to $t+1$ using a discretized form of Eq.18 and time increments $\Delta t$. After each time increment, assuming $\mathbf{v}_{har}$ constant between the same time interval, Eq.19 is used to update the velocity $\mathbf{v}$ needed by Eq.18, with the current vorticity and divergence estimates. Classical centered finite difference schemes are used for the curl and divergence discretization. To avoid instability, a semi-implicit time discretization scheme is used to integrate forward Eq.18. To solve the linear system associated to the semi-implicit discretization scheme, the matrix has been constrained to to be diagonally dominant. Finally, the dynamical model time integration is done independently for each layer. This procedure results in a predicted average horizontal wind field $\mathbf{v}_p$ related to each layer.

## 5 Experimental Evaluation

For an exhaustive evaluation, we first propose to rely on a simulated flow. A Direct Numerical Simulation (DNS) of a 2D, incompressible, and highly turbulent flow has been used to generated an image sequence depicting the motion of a continuous scalar field. The sequence of scalar images of 256 by 256 pixels together with the true vector fields generated by the DNS were provided by the laboratory of fluid mechanics of *Cemagref (center of Rennes, France)*. The thickness conservation model reduces in this 2D case to the classical OFC data model. Note that as divergence vanishes, the spatio-temporal regularization constrains only vorticity to be coherent in time.

In order to experiment our method with correlation-based vectors with different noise level, the correlation-based vectors have been substituted by DNS vectors contaminated by an additive Gaussian noise. As correlation techniques only operate on contrasted regions, vector constraints were attached to regions with sufficient gradient. To be realistic with correlation measurements, DNS vectors have been sub-sampled in those regions. DNS velocity vectors which have been selected as non-noisy correlation measurements are presented in fig.1. They are superimposed to the scalar image. Based on the non-noisy correlation constraints defined previously, we first compare our two-level collaborative scheme to the fluid flow dedicated multiresolution approach described in [2]. In Fig.2, it

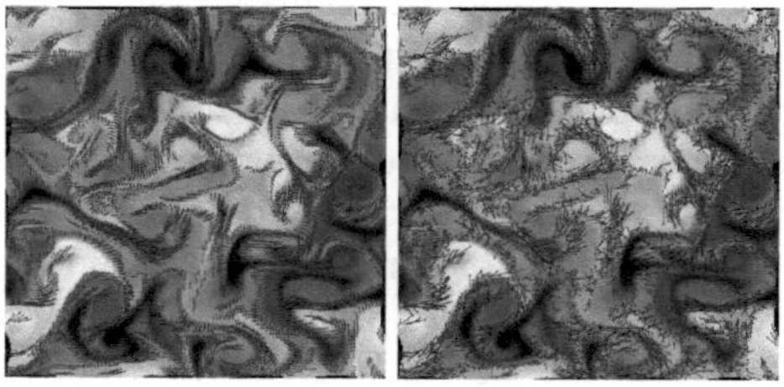

**Fig. 1. Velocity constraints and fluid imagery for a bi-dimensional flow.** Left: velocity vectors provided by the DNS which have been selected as constraints are superimposed on the image. Right: Gaussian noise $\mathcal{N}(0,1)$ has been added to these vectors.

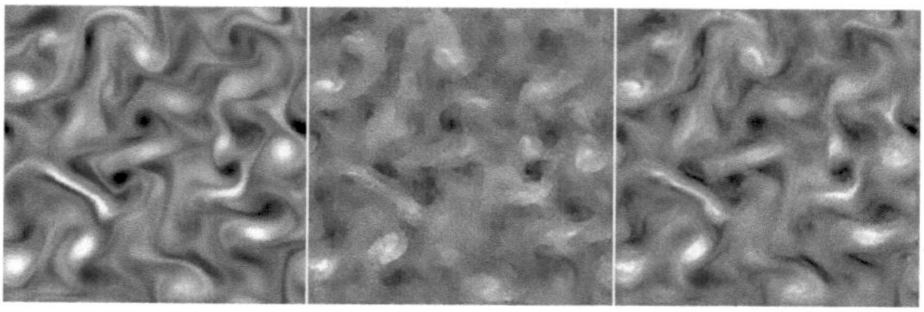

**Fig. 2. Comparison on the image domain of multiresolution and collaborative schemes in the case of a bi-dimensional flow.** Left: vorticity provided by the DNS. Center: vorticity estimation by the fluid flow dedicated multiresolution approach of [2].Right: vorticity estimation after the second level of the collaborative scheme.

clearly appears that the multiresolution approach hardly estimates fine turbulent structure while the collaborative method manages to recover most of the vorticity field structures. Indeed, in scalar imagery, low contrast regions correspond to high vorticity areas. Thus, the multiresolution technique suffers from a lack of information in those crucial regions. And, incorporating motion constraints in contrast areas around vortices reduces the degree of freedom of the solution and thus, considerably enhances the estimated motion field. In order to evaluate the robustness of the collaborative method to inaccurate constraints, Gaussian noise of zero mean and increasing variance has been added to the true velocity vectors provided by the DNS. Constraint examples are displayed in Fig.1. In Fig.3, we can visually inspect the influence of noise on the estimated solution for a particular horizontal slice of the image and for the global image domain by referring to RMS errors on vorticity values. It clearly appears that, even in presence of noise, motion estimation is better achieved by our collaborative scheme than by a classical multiresolution approach. Spatio-temporal regularization benefits which are assessed for both, multiresolution and collaborative methods, are shown in Fig.3.

We then turned to qualitative comparisons on a real meteorological image sequence. The benchmark data was composed by a sequence of 18 Meteosat Second Generation (MSG) images, showing thermal infrared radiation at a wavelength

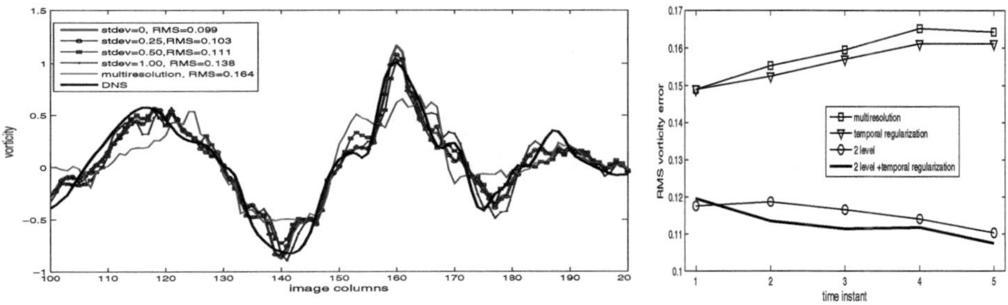


**Fig. 3. Influence of noise and spatio-temporal regularization.** Left: for increasing noise, vorticity estimates on a slice of the image and global RMS vorticity error in comparison to the multiresolution approach. Right: RMS vorticity errors on five consecutive estimations for the multiresolution approach, the collaborative scheme constrained by noisy correlation vectors combined or not with spatio-temporal regularization.

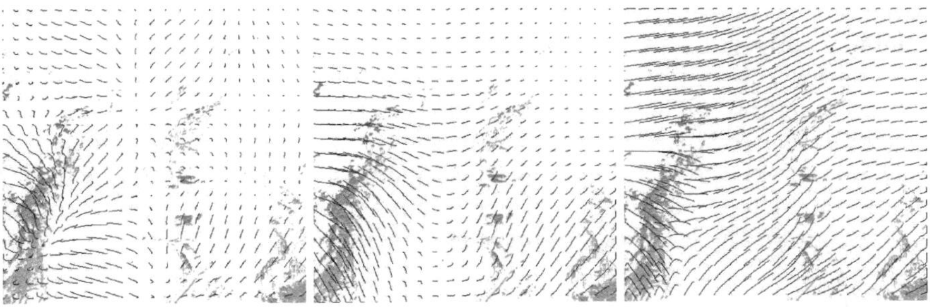

**Fig. 4. Collaborative approach and spatio-temporal regularization influence** on the estimation of wind field for the highest layer. Trajectories reconstruction for an estimation scheme without (left) and with (center) spatio-temporal regularization. Trajectories reconstruction for the two-level collaborative estimation scheme with spatio-temporal regularization (right).

of 10.8 mm. The 512 x 512 pixel images cover an area over the North-Atlantic Ocean, off the Iberian peninsula, during part of one day (5-June-2004), at a rate of one image every 15 minutes. The spatial resolution is 3 kilometers at the center of the whole Earth image disk. Clouds from a cloud-classification product derived from MSG images by the operational centre EUMETSAT, are used to segment images into 3 broad layers, at low, intermediate and high altitude. This 3 layers decomposition is imposed by the EUMETSAT classification. Applying the methodology previously described, pressure difference images were derived for these 3 layers.

Trajectories reconstructed by a Runge-Kutta integration method [2] from the estimated wind fields provide a practical visualization tool to assess the quality of the estimation in time and space.

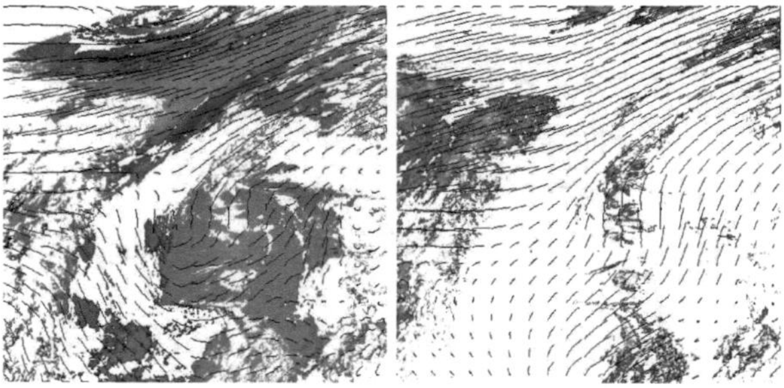

**Fig. 5. Middle layer and lower layer trajectories for a two-level collaborative estimation scheme using spatio-temporal regularization.** The trajectories correspond to the low (right), and to the medium (left) layer motions.

The enhancements brought by the collaborative estimation scheme for the recovery of a wind field related to the highest layer is shown in Fig.4. It can be noticed in this comparative figure that the introduction of spatio-temporal constraints smooths trajectory discontinuities and, together with the introduction of correlation constraints, propagate motion in regions where observations are missing. Using the collaborative framework and the spatio-temporal regularizer, trajectories related to the other layers are presented in Fig.5. In the middle of the image, one can notice the estimation of two perpendicular motions : the upward motion related to sparse clouds of the intermediate layer has been accurately recovered above an underneath stratus moving downward.

## 6 Conclusion

In this paper, we have presented a new method for estimating winds in a stratified atmosphere from satellite image sequences. The proposed motion estimation method is based on the minimization of a functional including a two part globalized regularizer. The data term relies on shallow-water mass conservation model. Indeed, the hydrostatic assumption allows a layer decomposition of the atmosphere. This decomposition is used to derive, relatively to each layer, thickness-based observations from infrared satellite images. Resulting observations verify independent shallow-water mass conservation models. To overcome the problem of sparse observations, a robust estimator is introduced in the data term.

A novel spatio-temporal regularizer is proposed. An approximation of shallow-water momentum equations expressed in a divergence-vorticity form is used to derive temporal coherence constraints. These temporal constraints are combined with a robust second-order regularizer preserving divergent and vorticity structures of the flow. In order to capture mesoscale dynamics, an optic-flow/correlation collaborative estimation scheme is proposed. Relying on two-level of estimation, this approach constitutes an advantageous alternative to the standard multiresolution framework. On both synthetic images and real satellite infrared images, the merit of the novel data-model and of the introduction of correlation-based and temporal constraints have been demonstrated.

## References

1. M. Black and P. Anandan. The robust estimation of multiple motions: Parametric and piecewise-smooth flow fields. *Computer Vision and Image Understanding*, 63(1):75–104, 1996.
2. T. Corpetti, E. Mémin, and P. Pérez. Dense estimation of fluid flows. *IEEE Trans. Pattern Anal. Machine Intell.*, 24(3):365–380, 2002.
3. J. Fitzpatrick. The existence of geometrical density-image transformations corresponding to object motion. *Comput. Vision, Graphics, Image Proc.*, 44(2):155–174, Nov. 1988.
4. J. Holton. *An introduction to dynamic meteorology.* Academic press, 1992.
5. B. Horn and B. Schunck. Determining optical flow. *Artificial Intelligence*, 17:185–203, 1981.
6. J. Leese, C. Novack, and B. Clark. An automated technique for obtained cloud motion from geosynchronous satellite data using cross correlation. *Journal of applied meteorology*, 10:118–132, 1971.
7. B. Lucas and T. Kanade. An iterative image registration technique with an application to stereovision. In *Int. Joint Conf. on Artificial Intel. (IJCAI)*, pages 674–679, 1981.
8. E. Mémin and P. Pérez. Fluid motion recovery by coupling dense and parametric motion fields. In *Int. Conf. on Computer, ICCV'99*, pages 620–625, 1999.
9. P. Ruhnau, A. Stahl, and C. Schnoerr. On-line variational estimation of dynamical fluid flows with physics-based spatio-temporal regularization. In *28th Symposium of the German Association for Pattern Recognition*, Berlin, Sept. 2006.
10. J. Schmetz, K. Holmlund, J. Hoffman, B. Strauss, B. Mason, V. Gaertner, A. Koch, and L. V. D. Berg. Operational cloud-motion winds from meteosat infrared images. *Journal of Applied Meteorology*, 32(7):1206–1225, 1993.
11. D. Suter. Motion estimation and vector splines. In *Proc. Conf. Comp. Vision Pattern Rec.*, pages 939–942, Seattle, USA, June 1994.
12. A. Szantai and F. Desalmand. Using multiple channels from msg to improve atmospheric motion wind selection and quality. In *7th International Winds Workshop, EUMETSAT EUM P 42*, pages 307–314, Helsinki, Finland, June 2004.
13. A. Szantai and F. Desalmand. Basic information on msg images. fluid project 'www.fluid.irisa.fr', wp-1, report 1. Technical report, Laboratoire de Meteorologie Dynamique, 2005.
14. J. Weickert and C. Schnörr. Variational optic-flow computation with a spatio-temporal smoothness constraint. *J. Mathematical. Imaging and Vision*, 14(3):245–255, 2001.
15. J. Yuan, C. Schnoerr, and E. Memin. Discrete orthogonal decomposition and variational fluid flow estimation. *J. Mathematical Imaging and Vison*, (accepted for publication), 2006.
16. L. Zhou, C. Kambhamettu, and D. Goldgof. Fluid structure and motion analysis from multi-spectrum 2d cloud images sequences. In *Proc. Conf. Comp. Vision Pattern Rec.*, volume 2, pages 744–751, USA, 2000.

# Paretian Similarity for Partial Comparison of Non-rigid Objects

Alexander M. Bronstein, Michael M. Bronstein, Alfred M. Bruckstein, and Ron Kimmel

Department of Computer Science, Technion – Israel Institute of Technology, Haifa 32000, Israel
{bron,mbron,freddy,ron}@cs.technion.ac.il

**Abstract.** In this paper, we address the problem of partial comparison of non-rigid objects. We introduce a new class of set-valued distances, related to the concept of Pareto optimality in economics. Such distances allow to capture intrinsic geometric similarity between parts of non-rigid objects, obtaining semantically meaningful comparison results. The numerical implementation of our method is computationally efficient and is similar to GMDS, a multidimensional scaling-like continuous optimization problem.

## 1 Introduction

Analysis of non-rigid objects is an important field emerging in the pattern recognition community [16,25,14,28]. Such problems arise, for example, in face recognition [6,7], matching of articulated objects [27,21,30,24,5], image segmentation [20], texture mapping and morphing [9,3]. A central problem is defining a meaningful criterion of similarity between non-rigid objects. Such a criterion should be invariant to deformations, have metric properties and allow for consistent discretization and efficient computation.

Theoretically, many natural deformations of objects can be modeled as near-isometric (distance preserving) transformations. The problem in this setting is translated into finding intrinsic geometric similarity between the objects. Early attempts of approximate isometry-invariant comparison were presented by Elad and Kimmel [16]. The authors proposed representing the intrinsic geometry of objects in a common metric space with simple geometry, thereby allowing to undo the degrees of freedom resulting from isometries. Representations obtained in such a way were called the bending-invariant canonical forms and were computed using *multidimensional scaling* (MDS) [1]. The Elad-Kimmel method is not exactly isometry-invariant because of the inherent error introduced by such an embedding.

Mémoli and Sapiro [25] used the Gromov-Hausdorff distance, introduced in [18] for the comparison of metric spaces. This distance has appealing theoretical properties but its computation is NP-hard. The authors proposed an algorithm that approximates the Gromov-Hausdorff distance in polynomial time by computing a different distance related to it by a probabilistic bound. In follow-up

F. Sgallari, A. Murli, and N. Paragios (Eds.): SSVM 2007, LNCS 4485, pp. 264–275, 2007.

works, Bronstein *et al.* showed a different approach, according to which the computation of the Gromov-Hausdorff distance is formulated as a continuous MDS-like problem and solved efficiently using a local minimization algorithm [10,8]. This numerical framework was given the name of *generalized MDS* (GMDS).

Here, we address an even more challenging setting of the non-rigid object analysis problem – *partial comparison* of non-rigid objects. In this setting, we need to find similarity between non-rigid objects having similar subsets. Such a situation is very common in practice, for example, in three-dimensional face recognition, where due to imperfect data acquisition, use of eyeglasses, or changes in the facial hair, parts of the objects may be missing or differ substantially [2]. Attempts to cope with such artifacts were presented in [4]. In two-dimensional shape recognition, partial comparison is an underlying problem of many shape similarity methods, attempting to divide the objects into meaningful parts, compare the parts separately and then integrate the partial similarities together [26]. Psychophysical research suggests there is strong evidence that such a "recognition by parts" mechanism is employed by the human vision [19]. Unfortunately, we do not have a clear understanding of how our brain partitions the objects we see into meaningful parts, and therefore, cannot give a precise definition of a part [27]. The recent work of Latecki *et al.* [23] allows to avoid the ambiguous definition of a part by finding a simplification of shapes which minimizes some criterion of similarity.

The main contribution of this paper is a new class of set-valued distances, related to the concept of Pareto optimality in economics. Such distances allow to capture intrinsic geometric similarity between parts of non-rigid objects. We show that the Paretian similarity can be efficiently computed using numerics resembling the GMDS. This paper is organized as follows. In Section 2, we present the theoretical background and briefly overview the properties of the Gromov-Hausdorff distance. In Section 3 we introduce the Paretian similarity, and in Section 4 show how to efficiently compute it. Section 5 demonstrates some experimental results. Though we deal with three-dimensional objects, the method is generic and can be applied to two-dimensional non-rigid objects as well (see for example [24,5]). Section 6 concludes the paper. Due to space limitations, we do not prove our results here. The proofs will be published in the extended version of this paper.

## 2 Theoretical Background

We model the objects as two-dimensional smooth compact connected and complete Riemannian manifolds (surfaces), possibly with boundary. We will denote the space of such objects by $\mathbb{M}$. An object $\mathcal{S} \in \mathbb{M}$ is equipped with the metric $d_{\mathcal{S}} : \mathcal{S} \times \mathcal{S} \to \mathbb{R}$, induced by the Riemannian structure; $d_{\mathcal{S}}(s, s')$ is referred to as the *geodesic distance* between the points $s, s'$. The Riemannian structure of the surface also defines a measure $\mu_{\mathcal{S}}(\mathcal{S}')$, which measures the area of the set $\mathcal{S}' \subset \mathcal{S}$. We denote the corresponding $\sigma$-algebra (a collection of subsets of $\mathcal{S}$ closed under countable union and complement, on which the measure $\mu_{\mathcal{S}}$ is defined) by $\Sigma_{\mathcal{S}}$.

A property will be said to hold *almost everywhere* (abbreviated as *a.e.*) on $\mathcal{S}$ if it holds on a subset $\mathcal{S}' \subseteq \mathcal{S}$ with $\mu_{\mathcal{S}}(\mathcal{S}'^c) = \mu_{\mathcal{S}}(\mathcal{S} \setminus \mathcal{S}') = 0$.

The pair $(\mathcal{S}, d_{\mathcal{S}})$ can be thought of as a metric space. In a broad sense, we refer to the distance structure of $\mathcal{S}$ as to its *intrinsic geometry*, to distinguish it from the way in which the surface is embedded into the ambient space, which is called the *extrinsic geometry*. Given a subset $\mathcal{S}' \subset \mathcal{S}$, we have two meaningful ways to define a metric on it. One possibility is to restrict the metric $d_{\mathcal{S}}$ to $\mathcal{S}'$, i.e., $d_{\mathcal{S}}|_{\mathcal{S}'}(s, s') = d_{\mathcal{S}}(s, s')$ for all $s, s'$ in $\mathcal{S}^r$. Such metric is called the *restricted metric*. Another possibility is to derive the metric from the Riemannian structure of $\mathcal{S}'$; we call it the *induced metric* and denote it by $d_{\mathcal{S}'}$. $d_{\mathcal{S}'}$ coincides with $d_{\mathcal{S}}|_{\mathcal{S}'}$ if $\mathcal{S}'$ is *geodesically convex*. A subset $\mathcal{S}^r \subset \mathcal{S}$ with the restricted metric $d_{\mathcal{S}}|_{\mathcal{S}^r}$ is called an $r$-*covering* of $\mathcal{S}$ if $\mathcal{S} = \bigcup_{s \in \mathcal{S}^r} B_{\mathcal{S}}(s, r)$, where $B_{\mathcal{S}}(s_0, r) = \{s \in \mathcal{S} : d_{\mathcal{S}}(s, s_0) < r\}$ is a ball of radius $r$ around $s_0$ in $\mathcal{S}$. In practical applications, finite coverings of $\mathcal{S}$ are of particular interest. Such coverings always exist assuming that $\mathcal{S}$ is compact.

Two objects $\mathcal{S}$ and $\mathcal{Q}$ are said to be *isometric* if they are identical from the point of view of their intrinsic geometry. This implies the existence of a bijective bi-Lipschitz continuous distance preserving map called an *isometry*. In practice, genuine isometries rarely exist, and objects encountered in the real life may be only nearly-isometric. A map $f : \mathcal{S} \to \mathcal{Q}$ is said to have *distortion* $\epsilon$ if

$$\operatorname{dis} f = \sup_{s,s' \in \mathcal{S}} |d_{\mathcal{S}}(s, s') - d_{\mathcal{Q}}(f(s), f(s'))| = \epsilon. \tag{1}$$

We say that $\mathcal{S}$ is $\epsilon$-*isometrically embeddable* into $\mathcal{Q}$ if $\operatorname{dis} f \leq \epsilon$. Such $f$ is called an $\epsilon$-*isometric embedding*. If in addition $f$ is $\epsilon$-*surjectivie* (i.e. $d_{\mathcal{Q}}(q, f(\mathcal{S})) \leq \epsilon$ for all $q \in \mathcal{Q}$, where the set-to-point distance is defined as $d_{\mathcal{Q}}(q, f(\mathcal{S})) = \inf_{s \in \mathcal{S}} d_{\mathcal{Q}}(q, f(s))$), it is called an $\epsilon$-*isometry*, and $\mathcal{S}$ and $\mathcal{Q}$ are called $\epsilon$-*isometric*. In [18], Mikhail Gromov introduced a criterion of similarity between metric spaces, commonly known today as the *Gromov-Hausdorff distance*. For compact spaces, it can be written in the following way:

$$d_{\mathrm{GH}}(\mathcal{Q}, \mathcal{S}) = \frac{1}{2} \inf_{\substack{f:\mathcal{S} \to \mathcal{Q} \\ g:\mathcal{Q} \to \mathcal{S}}} \max\{\operatorname{dis} f, \operatorname{dis} g, \operatorname{dis}(f, g)\}, \tag{2}$$

where $\operatorname{dis}(f, g) = \sup_{s \in \mathcal{S}, q \in \mathcal{Q}} |d_{\mathcal{S}}(s, g(q)) - d_{\mathcal{Q}}(q, f(s))|$. The Gromov-Hausdorff distance is a metric on the quotient space $\mathbb{M} \setminus \mathrm{Iso}(\mathbb{M})$, the space in which a point represents the equivalence class of all the self-isometries of an object. Another property of $d_{\mathrm{GH}}$ will be of fundamental importance for us: (i) if $d_{\mathrm{GH}}(\mathcal{S}, \mathcal{Q}) \leq \epsilon$, then $\mathcal{S}$ and $\mathcal{Q}$ are $2\epsilon$-isometric; (ii) if $\mathcal{S}$ and $\mathcal{Q}$ are $\epsilon$-isometric, then $d_{\mathrm{GH}}(\mathcal{S}, \mathcal{Q}) \leq 2\epsilon$. Particularly, for $\epsilon = 0$ we have the *isometry invariance* property: $d_{\mathrm{GH}}(\mathcal{S}, \mathcal{Q}) = 0$ if and only if $\mathcal{S}$ and $\mathcal{Q}$ are isometric [13].

It may happen that $\mathcal{S}$ and $\mathcal{Q}$ are not $\epsilon$-isometric, but parts of them are. To describe such a situation, we introduce the notion of $(\lambda, \epsilon)$-*isometry*: $\mathcal{S}$ and $\mathcal{Q}$ are said to be $(\lambda, \epsilon)$-isometric if there exist $\mathcal{S}' \subseteq \mathcal{S}$ and $\mathcal{Q}' \subseteq \mathcal{Q}$ with $\max\{\mu_{\mathcal{S}}(\mathcal{S}'^c), \mu_{\mathcal{Q}}(\mathcal{Q}'^c)\} \leq \lambda$, such that $(\mathcal{S}', d_{\mathcal{S}'}|_{\mathcal{S}})$ and $(\mathcal{Q}', d_{\mathcal{Q}'}|_{\mathcal{Q}})$ are $\epsilon$-isometric.

## 3 Paretian Similarity

We can define partial similarity by saying that two objects are partially similar if they have large similar parts. What is implied by the words "similar" and "large" is a semantic question. Formally, we define a part $\mathcal{S}'$ as a subset of $\mathcal{S}$ belonging to the $\sigma$-algebra $\Sigma_{\mathcal{S}}$ (this condition is necessary in order for the part to be measurable). In our problem, it is natural to use intrinsic geometric similarity of parts, quantified by the Gromov-Hausdorff distance. The part size is quantified by the absolute or the normalized measure on the surface. We can give a more precise definition to partial similarity in the following way: two objects $\mathcal{S}$ and $\mathcal{Q}$ are partially similar if they have parts $\mathcal{S}' \in \Sigma_{\mathcal{S}}$ and $\mathcal{Q}' \in \Sigma_{\mathcal{Q}}$ of large measure $\mu_{\mathcal{S}}(\mathcal{S}')$ and $\mu_{\mathcal{Q}}(\mathcal{Q}')$, such that $(\mathcal{S}', d_{\mathcal{S}}|_{\mathcal{S}'})$ and $(\mathcal{Q}', d_{\mathcal{Q}}|_{\mathcal{Q}'})$ are nearly isometric. Note that we use the *restricted* metric on $\mathcal{S}'$ and $\mathcal{Q}'$; this fact will allow precompute the distances only ones and not recompute them every time for each $\mathcal{S}'$ and $\mathcal{Q}'$.

We will denote by $\epsilon(\mathcal{S}', \mathcal{Q}') = d_{\mathrm{GH}}(\mathcal{S}', \mathcal{Q}')$ the *similarity* of $\mathcal{S}'$ and $\mathcal{Q}'$, and by $\lambda(\mathcal{S}', \mathcal{Q}') = \max\{\mu_{\mathcal{S}}(\mathcal{S}'^c), \mu_{\mathcal{Q}}(\mathcal{Q}'^c)\}$ the *partiality*, representing the size of the region we crop off from the objects.[1] The computation of partial similarity can be formulated as a *multicriterion* optimization problem: among all the possible pairs $(\mathcal{S}', \mathcal{Q}') \in \Sigma_{\mathcal{S}} \times \Sigma_{\mathcal{Q}}$, find one that *simultaneously minimizes* $\epsilon$ and $\lambda$. In this formulation, our approach can be seen as a generalization of [23]. Obviously, in most cases it is impossible to bring both criteria to zero because they are competing. Each $(\mathcal{S}', \mathcal{Q}')$ can be represented as a point $(\lambda(\mathcal{S}', \mathcal{Q}'), \epsilon(\mathcal{S}', \mathcal{Q}'))$ in the plane. At certain points, improving one criterion inevitably compromises the other. Such solutions, representing the best tradeoff between the criteria, are called *Pareto optimal* in economics. This notion is closely related to rate-distortion analysis in information theory [15] and to receiver operation characteristics in pattern recognition [17]. We say that $(\mathcal{S}^*, \mathcal{Q}^*)$ is a Pareto optimum if at least one of the following holds,

$$\begin{aligned} \epsilon(\mathcal{S}^*, \mathcal{Q}^*) &\le \epsilon(\mathcal{S}', \mathcal{Q}'); \quad \text{or,} \\ \lambda(\mathcal{S}^*, \mathcal{Q}^*) &\le \lambda(\mathcal{S}', \mathcal{Q}'), \end{aligned} \tag{3}$$

for all $\mathcal{S}' \subseteq \mathcal{S}$ and $\mathcal{Q}' \subseteq \mathcal{Q}$. The set of all the Pareto optimal solutions is called the *Pareto frontier* and can be visualized as a planar curve (see Figures 1–2). Solutions below this curve do not exist.

The fundamental difference between the Paretian similarity and similarity in the traditional sense (which can be quantified by a scalar "distance" value), is the fact that we have a multitude of similarities, each corresponding to a Pareto optimum. We can think of the Pareto frontier as of a generalized, *set-valued distance*, which is denoted here by $d_{\mathrm{P}}$. Set-valued distance requires a redefinition of notions commonly associated with scalar-valued distances. For instance, it is usually impossible to establish a full order relation between the distances $d_{\mathrm{P}}(\mathcal{Q}, \mathcal{S})$ and $d_{\mathrm{P}}(\mathcal{Q}, \mathcal{R})$, since they may be mutually incompatible. We

[1] Partiality can be defined in other ways, for example, $\lambda(\mathcal{S}', \mathcal{Q}') = \mu_{\mathcal{S}}(\mathcal{S}'^c) + \mu_{\mathcal{Q}}(\mathcal{Q}'^c)$.

can only define point-wise order relations in the following way: if $(\lambda_0, \epsilon_0)$ is above $d_{\mathrm{P}}(\mathcal{Q}, \mathcal{S})$, we will write $d_{\mathrm{P}}(\mathcal{Q}, \mathcal{S}) < (\lambda_0, \epsilon_0)$; other strong and weak inequalities are defined in a similar way. The notation $(\lambda_0, \epsilon_0) \in d_{\mathrm{P}}(\mathcal{Q}, \mathcal{S})$ will be used to say that $(\lambda_0, \epsilon_0)$ is a Pareto optimum. Using this definition, we can summarize the properties of Paretian similarity as follows:

**Theorem 1 (Properties of $d_{\mathrm{P}}$).** *The distance $d_{\mathrm{P}}$ satisfies:*

*(P1)* Non-negativity: $d_{\mathrm{P}}(\mathcal{Q}, \mathcal{S}) \subseteq [0, \infty) \times [0, \infty)$.
*(P2)* Symmetry: $d_{\mathrm{P}}(\mathcal{Q}, \mathcal{S}) = d_{\mathrm{P}}(\mathcal{S}, \mathcal{Q})$.
*(P3)* Monotonicity: *If $d_{\mathrm{P}}(\mathcal{Q}, \mathcal{S}) \leq (\lambda, \epsilon)$, then $d_{\mathrm{P}}(\mathcal{Q}, \mathcal{S}) \leq (\lambda', \epsilon)$ for every $\lambda' \geq \lambda$, and $d_{\mathrm{P}}(\mathcal{Q}, \mathcal{S}) \leq (\lambda, \epsilon')$ for every $\epsilon' \geq \epsilon$.*
*(P4)* Partial similarity: *(i) If $d_{\mathrm{P}}(\mathcal{Q}, \mathcal{S}) \leq (\lambda, \epsilon)$, then $\mathcal{S}$ and $\mathcal{Q}$ are $(\lambda, 2\epsilon)$-isometric; (ii) if $\mathcal{S}$ and $\mathcal{Q}$ are $(\lambda, \epsilon)$-isometric, then $d_{\mathrm{P}}(\mathcal{Q}, \mathcal{S}) \leq (\lambda, 2\epsilon)$.*
*(P5)* Consistency to sampling: *If $\mathcal{S}^r$ and $\mathcal{Q}^r$ are finite $r$-coverings of two shapes of bounded curvature $\mathcal{S}$ and $\mathcal{Q}$, respectively, then $\lim_{r \to 0} d_{\mathrm{P}}(\mathcal{Q}, \mathcal{S}^r) = d_{\mathrm{P}}(\mathcal{Q}, \mathcal{S})$.*

Properties (P1)-(P5) follow from the properties of the Gromov-Hausdorff distance (see e.g. [8]). Due to space limitations, we do not give a formal proof of this theorem. $(0, 0) \in d_{\mathrm{P}}(\mathcal{Q}, \mathcal{S})$ if and only if there exists an a.e. isometry between $\mathcal{S}$ and $\mathcal{Q}$.

### 3.1 Converting Set-Valued Distances into Scalar-Valued Distances

In order to be able to compare similarities, we need to convert the set-valued distance into a traditional, scalar-valued one. The easiest way to do it is by considering a single point on the Pareto frontier. For example, we can fix the value of $\lambda$ and use the corresponding distortion $\epsilon$ as the distance. We obtain a scalar-valued distance, to which we refer as the *$\lambda$-Gromov-Hausdorff distance*:

$$d_{\mathrm{GH}}^{\lambda}(\mathcal{Q}, \mathcal{S}) = \frac{1}{2} \inf_{\substack{f:\mathcal{S}' \to \mathcal{Q}' \\ g:\mathcal{Q}' \to \mathcal{S}' \\ \mathcal{S}' : \mu_{\mathcal{S}}(\mathcal{S}'^c) \leq \lambda \\ \mathcal{Q}' : \mu_{\mathcal{Q}}(\mathcal{Q}'^c) \leq \lambda}} \max\{\mathrm{dis}\, f, \mathrm{dis}\, g, \mathrm{dis}\,(f, g)\}.$$

The particular case of $d_{\mathrm{GH}}^{0}(\mathcal{Q}, \mathcal{S})$ can be thought of as an a.e. Gromov-Hausdorff distance. Alternatively, we can fix the value of $\epsilon$; a scalar distance obtained this way may be useful in a practical situation when we know *a priory* the accuracy of surface acquisition and distance measurement. A third possibility is to take the area under the Pareto frontier as a scalar-valued distance.

We should note, however, that both of the above choices are rather arbitrary. A slightly more motivated selection of a single point out of the set of Pareto optimal solutions was proposed by Salukwadze [29] in the context of multicriterion optimization problems arising in control theory. Salukwadze suggested choosing a Pareto optimum, which is the closest (in sense of some norm) to some optimal, usually non-achievable point. In our case, such an optimal point is $(0,0)$. Given a Pareto frontier $d_{\mathrm{P}}(\mathcal{S},\mathcal{Q})$, we define the *Salukwadze similarity* as

$$d_{\mathrm{SAL}}(\mathcal{Q},\mathcal{S}) = \inf_{(\lambda,\epsilon)\in d_{\mathrm{P}}(\mathcal{S},\mathcal{Q})} \|(\lambda,\epsilon)\|. \tag{4}$$

Depending on the choice of the norm $\|\cdot\|$ in (4), we obtain different solutions, some of which have an explicit form. For instance, choosing the $L_p$-norm, we can define the $L_p$-*Salukwadze distance* as follows:

$$d^p_{\mathrm{SAL}}(\mathcal{Q},\mathcal{S}) = \inf_{\substack{f:\mathcal{S}'\to\mathcal{Q}'\\ g:\mathcal{Q}'\to\mathcal{S}'\\ \mathcal{S}',\mathcal{Q}'}} \left\{ \frac{1}{2^p}\max\{\mathrm{dis}\, f, \mathrm{dis}\, g, \mathrm{dis}\,(f,g)\}^p + \max\{\mu_{\mathcal{S}}(\mathcal{S}'^{c}), \mu_{\mathcal{Q}}(\mathcal{Q}'^{c})\}^p \right\}.$$

This formulation is a very intuitive interpretation of the multicriterion optimization problem: we are simultaneously minimizing $d_{\mathrm{GH}}(\mathcal{S}',\mathcal{Q}')$ and maximizing the measures of $\mathcal{S}'$ and $\mathcal{Q}'$. In order to avoid scaling ambiguity between the distortion and the measure, a normalization factor $\alpha \cdot d_{\mathrm{GH}}(\mathcal{S}',\mathcal{Q}')$ (where $\alpha$ has units of distance) can be used.

### 3.2 Relaxed Paretian Similarity

The computation of $d_{\mathrm{P}}(\mathcal{Q},\mathcal{S})$ requires optimization over $\Sigma_{\mathcal{S}} \times \Sigma_{\mathcal{Q}}$ and is impractical, since in the discrete case it gives rise to a combinatorial optimization problem with complexity growing exponentially in the sample size of $\mathcal{S}$ and $\mathcal{Q}$. However, the problem can be relaxed by resorting to *fuzzy* representation of parts as continuous *membership functions* $m_{\mathcal{S}} : \mathcal{S} \to [0,1]$ and $m_{\mathcal{Q}} : \mathcal{Q} \to [0,1]$. The value of $m_{\mathcal{S}}$ measures the degree to which a point belongs to the part of $\mathcal{S}$ (zero implies exclusion, one implies inclusion). Instead of $\lambda(\mathcal{S}',\mathcal{Q}')$, we define the *fuzzy partiality*

$$\tilde{\lambda}(m_{\mathcal{S}}, m_{\mathcal{Q}}) = \max\left\{ \int_{\mathcal{S}} (1 - m_{\mathcal{S}}(s))d\mu_{\mathcal{S}}, \int_{\mathcal{Q}} (1 - m_{\mathcal{Q}}(q))d\mu_{\mathcal{Q}} \right\} \tag{5}$$

and instead of $\epsilon(\mathcal{S}',\mathcal{Q}')$, use a fuzzy version of the Gromov-Hausdorff distance,

$$\tilde{\epsilon}(m_{\mathcal{S}}, m_{\mathcal{Q}}) = \frac{1}{2}\inf_{\substack{f:\mathcal{S}\to\mathcal{Q}\\ g:\mathcal{Q}\to\mathcal{S}}} \max \left\{ \begin{array}{l} \sup\limits_{s,s'\in\mathcal{S}} m_{\mathcal{S}}(s)m_{\mathcal{S}}(s')|d_{\mathcal{S}}(s,s') - d_{\mathcal{Q}}(f(s),f(s'))| \\ \sup\limits_{q,q'\in\mathcal{Q}} m_{\mathcal{Q}}(q)m_{\mathcal{Q}}(q')|d_{\mathcal{Q}}(q,q') - d_{\mathcal{S}}(g(q),g(q'))| \\ \sup\limits_{s\in\mathcal{S},q\in\mathcal{Q}} m_{\mathcal{S}}(s)m_{\mathcal{Q}}(q)|d_{\mathcal{S}}(s,g(q)) - d_{\mathcal{Q}}(f(s),q)| \\ \sup\limits_{s\in\mathcal{S}} D\,(1 - m_{\mathcal{Q}}(f(s)))m_{\mathcal{S}}(s) \\ \sup\limits_{q\in\mathcal{Q}} D\,(1 - m_{\mathcal{S}}(g(q)))m_{\mathcal{Q}}(q) \end{array} \right\},$$

where $D$ is some large constant. The computation of the relaxed partial similarity requires minimization of $(\tilde{\lambda}(m_{\mathcal{S}}, m_{\mathcal{Q}}), \tilde{\epsilon}(m_{\mathcal{S}}, m_{\mathcal{Q}}))$ on all the pairs of membership functions $(m_{\mathcal{S}}, m_{\mathcal{Q}})$, which is computationally tractable, as will be described in Section 4. The Pareto optimum of this problem is defined in the same way as in equation (3); we will henceforth denote the relaxed Pareto frontier by $\tilde{d}_{\mathrm{P}}$. The following relation between $d_{\mathrm{P}}$ and $\tilde{d}_{\mathrm{P}}$ holds:

**Theorem 2 (Relation of $d_{\mathrm{P}}$ and $\tilde{d}_{\mathrm{P}}$).** *Let $D = \max\{\mathrm{diam}\,\mathcal{S}, \mathrm{diam}\,\mathcal{Q}\}/\delta(1-\delta)$, for some $0 < \delta < 1$. Then, $\tilde{d}_{\mathrm{P}}(\mathcal{S}, \mathcal{Q}) \leq \left((1-\delta)^{-1}, \delta^{-2}\right) \cdot d_{\mathrm{P}}(\mathcal{S}, \mathcal{Q})$, where the inequality is interpreted in the vector sense.*

The proof is based on the Chebyshev inequality and is not given here due to space limitations.

# 4 Computational Framework

Practical computation of the Paretian similarity is performed on discretized objects. The surface $\mathcal{S}$ is represented as a triangular mesh, whose vertices constitute a finite $r$-sampling $\mathcal{S}_N = \{s_1, ..., s_N\}$. A point $s$ on $\mathcal{S}$ is represented as a pair $(t, u)$, where $t$ is the index of the triangular face enclosing it, and $u$ is the vector of barycentric coordinates with respect to the vertices of that triangle. The metric on $\mathcal{S}$ is discretized by numerically approximating the geodesic distances between the samples $s_i$ on the triangular mesh, using the *fast marching method* (FMM) [22]. Geodesic distances between two arbitrary points on the mesh are interpolated from $d_{\mathcal{S}}(s_i, s_j)$'s using the three-point interpolation approach presented in [8]. The measure on $\mathcal{S}$ is discretized as $\{\mu_{\mathcal{S}_N}(s_1), ..., \mu_{\mathcal{S}_N}(s_N)\}$, assigning to each $s_i \in \mathcal{S}_N$ the area of the corresponding Voronoi cell.

Given two discretized surfaces $\mathcal{S}_N$ and $\mathcal{Q}_M$, we compute the relaxed Paretian similarity as the solution to

$$\min_{\substack{q'_1,\dots,q'_N \in \mathcal{Q} \\ s'_1,\dots,s'_M \in \mathcal{S} \\ m_{\mathcal{S}_N}(s_1),\dots,m_{\mathcal{S}_N}(s_N)\in[0,1] \\ m_{\mathcal{Q}_M}(q_1),\dots,m_{\mathcal{Q}_M}(q_M)\in[0,1]}} \max \left\{ \begin{array}{l} \max\limits_{i,j} m_{\mathcal{S}_N}(s_i) m_{\mathcal{S}_N}(s_j) |d_{\mathcal{S}}(s_i, s_j) - d_{\mathcal{Q}}(q'_i, q'_k)| \\ \max\limits_{k,l} m_{\mathcal{Q}_M}(q_k) m_{\mathcal{Q}_M}(q_l) |d_{\mathcal{Q}}(q_k, q_l) - d_{\mathcal{S}}(s'_k, s'_l)| \\ \max\limits_{i,k} m_{\mathcal{S}_N}(s_i) m_{\mathcal{Q}_M}(q_k) |d_{\mathcal{S}}(s_i, s'_k) - d_{\mathcal{Q}}(q_k, q'_i)| \\ \max\limits_{i} D\,(1 - m_{\mathcal{S}_N}(s'_i)) m_{\mathcal{S}_N}(s_i) \\ \max\limits_{k} D\,(1 - m_{\mathcal{Q}_M}(q'_k)) m_{\mathcal{Q}_M}(q_k) \end{array} \right\}$$

$$\text{s.t.} \quad \sum m_{\mathcal{S}_N}(s_i) \mu_{\mathcal{S}_N}(s_i) \geq 1 - \lambda$$

$$\sum m_{\mathcal{Q}_M}(q_k) \mu_{\mathcal{Q}_M}(q_k) \geq 1 - \lambda, \qquad (6)$$

for a fixed set of values of $\lambda$, each $\lambda$ giving a different point on the Pareto frontier $\tilde{d}_{\mathrm{P}}$. Here, $m_{\mathcal{S}_N}(s_i)$ and $m_{\mathcal{Q}_M}(q_k)$ denote the discretized membership functions computed on $s_i$ and $q_k$, respectively. $m_{\mathcal{S}_N}(s)$ and $m_{\mathcal{Q}_M}(q)$ denote the interpolated weights for arbitrary points $s \in \mathcal{S}$ and $q \in \mathcal{Q}$. Note that the minimization over all mappings $f : \mathcal{S} \to \mathcal{Q}$ and $g : \mathcal{Q} \to \mathcal{S}$ is replaced by minimization over the images $q'_i = f(s_i)$ and $s'_k = g(q_k)$, in the spirit of multidimensional scaling.

The minimization problem (6) can be solved by alternatingly solving two smaller problems, namely the minimization of (6) with respect to $m_{\mathcal{S}}(s_i)$ and $m_{\mathcal{Q}}(q_k)$ for fixed $s'_k$ and $q'_i$, which can be cast as the following constrained minimization problem

$$\min_{\substack{\epsilon\geq 0\\ m_{\mathcal{S}_N}(s_1),\ldots,m_{\mathcal{S}_N}(s_N)\in[0,1]\\ m_{\mathcal{Q}_M}(q_1),\ldots,m_{\mathcal{Q}_M}(q_M)\in[0,1]}} \epsilon \text{ s.t. } \begin{cases} m_{\mathcal{S}_N}(s_i)m_{\mathcal{S}_N}(s_j)|d_{\mathcal{S}}(s_i,s_j)-d_{\mathcal{Q}}(q'_i,q'_k)|\leq\epsilon\\ m_{\mathcal{Q}_M}(q_k)m_{\mathcal{Q}_M}(q_l)|d_{\mathcal{Q}}(q_k,q_l)-d_{\mathcal{S}}(s'_k,s'_l)|\leq\epsilon\\ m_{\mathcal{S}_N}(s_i)m_{\mathcal{Q}_M}(q_k)|d_{\mathcal{S}}(s_i,s'_k)-d_{\mathcal{Q}}(q_k,q'_i)|\leq\epsilon\\ D(1-m_{\mathcal{S}_N}(s'_i))m_{\mathcal{S}_N}(s_i)\leq\epsilon\\ D(1-m_{\mathcal{Q}_M}(q'_k))m_{\mathcal{Q}_M}(q_k)\leq\epsilon\\ \sum m_{\mathcal{S}_N}(s_i)\mu_{\mathcal{S}_N}(s_i)\geq 1-\lambda\\ \sum m_{\mathcal{Q}_M}(q_k)\mu_{\mathcal{Q}_M}(q_k)\geq 1-\lambda, \end{cases} \tag{7}$$

and the minimization of (6) with respect to $s'_k$ and $q'_i$ for fixed $m_{\mathcal{S}}(s_i)$ and $m_{\mathcal{Q}}(q_k)$, which can be formulated as

$$\min_{\substack{\epsilon\geq 0\\ q'_1,\ldots,q'_N\in\mathcal{Q}\\ s'_1,\ldots,s'_M\in\mathcal{S}}} \text{ s.t. } \begin{cases} m_{\mathcal{S}_N}(s_i)m_{\mathcal{S}_N}(s_j)|d_{\mathcal{S}}(s_i,s_j)-d_{\mathcal{Q}}(q'_i,q'_k)|\leq\epsilon\\ m_{\mathcal{Q}_M}(q_k)m_{\mathcal{Q}_M}(q_l)|d_{\mathcal{Q}}(q_k,q_l)-d_{\mathcal{S}}(s'_k,s'_l)|\leq\epsilon\\ m_{\mathcal{S}_N}(s_i)m_{\mathcal{Q}_M}(q_k)|d_{\mathcal{S}}(s_i,s'_k)-d_{\mathcal{Q}}(q_k,q'_i)|\leq\epsilon\\ D(1-m_{\mathcal{S}_N}(s'_i))m_{\mathcal{S}_N}(s_i)\leq\epsilon\\ D(1-m_{\mathcal{Q}_M}(q'_k))m_{\mathcal{Q}_M}(q_k)\leq\epsilon \end{cases} \tag{8}$$

and solved using the multi-resolution approach proposed in [11,12,10,8] for the GMDS. Another, more efficient approach, is to solve a weighted $L_2$ approximation to (8) and use iterative re-weighting as a means of approximating the original $L_\infty$ problem.

### 4.1 Sensitivity to Noise

If the accuracy of geodesic distance measurement is $\delta$ (in FMM methods, $\delta$ is of order of the maximum edge length in the mesh), the accuracy of the Gromov-Hausdorff distance is bounded by $2\delta$. Using an $L_2$ criterion instead of the $L_\infty$ is advantageous in the case of noise, since it is less influenced by outliers. Like all the approaches based on the analysis of intrinsic geometry, our method may be sensitive to *topological noise*, or in other words, noise in extrinsic geometry that results in different topology of the surface.

## 5 Results

We tested our method on a set of partially overlapping objects, created from the Elad-Kimmel database [16]. Five objects (dog, spider, giraffe, man and crocodile) were used; a full version of each object appeared, in addition to four different deformations of which parts were cropped, resulting in five instances per object (total of 25 objects, see Figure 3). The resulting objects were partially overlapping (Figure 2, top).

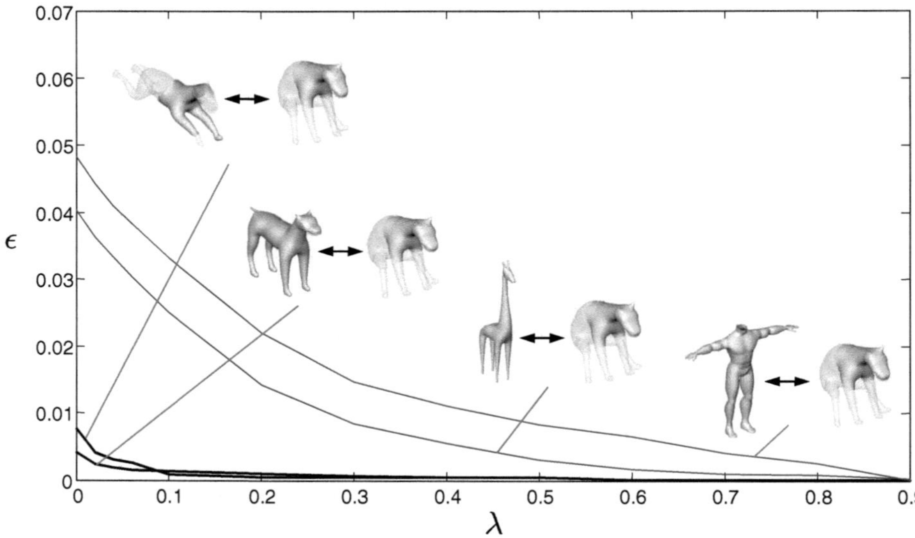


**Fig. 1.** Pareto similarity between different objects

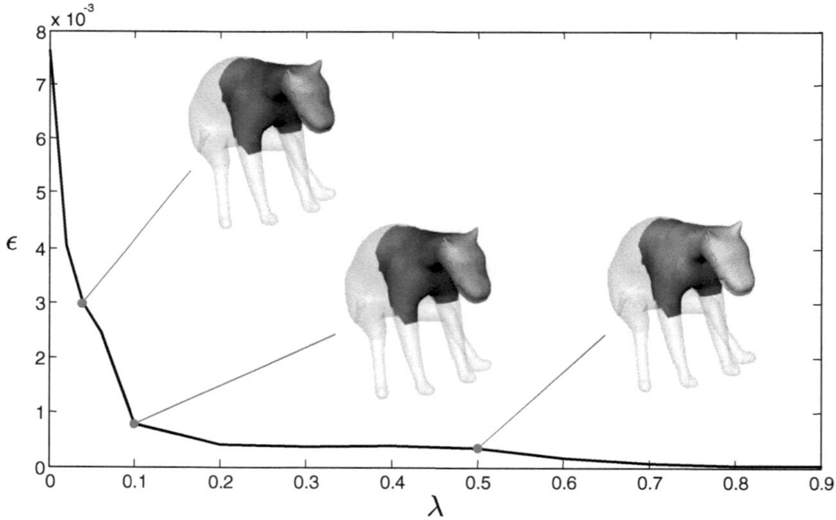


**Fig. 2.** Example of Paretian similarity. Shown is Pareto frontier corresponding to the set-valued distance between the dog objects. Colors encode the membership functions (red corresponding to 1).

**Fig. 3.** $L_1$-Salukwadze distance between partially missing objects, represented as Euclidean similarities

The objects were represented as triangular meshes and comprised between 1500 to 3000 points. The geodesic distances were computed using FMM. Set-valued distances were computed between all the objects using 13 values of $\lambda$. We used a multiresolution iteratively re-weighted scheme described in Section 4. Six resolution levels with 50 points at the finest level were used. The algorithms were implemented in MATLAB; the computation of the Pareto frontier took about a minute on a standard Intel Pentium IV computer.

Figure 2 shows the Pareto frontier corresponding to the set-valued distance between two instances of the dog object. Shades of red represent the values of the membership functions. The overlapping regions in the two objects are clearly visible. Figure 1 shows the Pareto frontiers arising from partial comparison of different objects. One can observe that the dog-man and the dog-giraffe comparisons (red) result in curves above those obtained for the comparison of different instances of the dog (black). Figure 3 depicts the $L_1$-Salukwadze distance (with scaling factor $\alpha = 200$) between the objects, represented as Euclidean similarities. Clusters corresponding to different objects are clearly distinguishable. For comparison, we refer the reader to [8], where the computation of the Gromov-Hausdorff distance between the full versions of the same objects presented here is shown.

## 6 Conclusions

We presented a method for partial comparison of non-rigid objects. Our approach suggest quantifying partial similarity as a tradeoff between the intrinsic geometric similarity and the area of a subset of the objects, using the formalism of Pareto optimality. Such a construction has a meaningful interpretation; the set-valued distances resulting from it have appealing theoretical and practical properties. For the efficient computation of our similarity criteria, we developed a numerical framework similar to the GMDS algorithm. Experimental results show that our method is able to recognize non-rigid objects even when large parts of them are missing or differ.

## Acknowledgement

This research was supported in part by United State-Israel Binational Science Foundation grant No. 2004274, the Israeli Science Foundation (ISF) grant No. 738/04 and the Horowitz fund.

## References

1. I. Borg and P. Groenen, *Modern multidimensional scaling - theory and applications*, Springer, 1997.
2. K. W. Bowyer, K. Chang, and P. Flynn, *A survey of 3D and multi-modal 3D+2D face recognition*, Dept. of Computer Science and Electrical Engineering Technical report, University of Notre Dame, January 2004.
3. A. M. Bronstein, A. M. Bronstein, and R. Kimmel, *Calculus of non-rigid surfaces for geometry and texture manipulation*, IEEE Trans. Visualization and Computer Graphics (2006), in press.
4. ———, *Robust expression-invariant face recognition from partially missing data*, Proc. ECCV, 2006, pp. 396–408.
5. A. M. Bronstein, M. M. Bronstein, A. M. Bruckstein, and R. Kimmel, *Matching two-dimensional articulated shapes using generalized multidimensional scaling*, Proc. AMDO, 2006, pp. 48–57.
6. A. M. Bronstein, M. M. Bronstein, and R. Kimmel, *Expression-invariant 3D face recognition*, Proc. AVBPA, Lecture Notes on Computer Science, no. 2688, Springer, 2003, pp. 62–69.

7. ———, *Three-dimensional face recognition*, IJCV **64** (2005), no. 1, 5–30.
8. ———, *Efficient computation of isometry-invariant distances between surfaces*, SIAM Journal Scientific Computing **28** (2006), 1812–1836.
9. ———, *Face2face: an isometric model for facial animation*, Proc. AMDO, 2006, pp. 38–47.
10. ———, *Generalized multidimensional scaling: a framework for isometry-invariant partial surface matching*, PNAS **103** (2006), no. 5, 1168–1172.
11. M. M. Bronstein, A. M. Bronstein, R. Kimmel, and I. Yavneh, *A multigrid approach for multidimensional scaling*, Copper Mountain Conf. Multigrid Methods, 2005.
12. ———, *Multigrid multidimensional scaling*, Numerical Linear Algebra with Applications **13** (2006), 149–171.
13. D. Burago, Y. Burago, and S. Ivanov, *A course in metric geometry*, Graduate studies in mathematics, vol. 33, American Mathematical Society, 2001.
14. G. Charpiat, O. Faugeras, and R. Keriven, *Approximations of shape metrics and application to shape warping and empirical shape statistics*, Found. Comput. Math. **5** (2005), no. 1, 1–58.
15. S. de Rooij and P. Vitanyi, *Approximating rate-distortion graphs of individual data: Experiments in lossy compression and denoising*, IEEE Trans. Information Theory (2006), submitted.
16. A. Elad and R. Kimmel, *Bending invariant representations for surfaces*, Proc. CVPR, 2001, pp. 168–174.
17. R. M. Everson and J. E. Fieldsend, *Multi-class ROC analysis from a multi-objective optimization perspective*, Pattern Recognition Letters **27** (2006), no. 8, 918–927.
18. M. Gromov, *Structures métriques pour les variétés riemanniennes*, Textes Mathématiques, no. 1, 1981.
19. D. Hoffman and W. Richards, *Visual cognition*, ch. Parts of recognition, MIT Press, Cambridge, 1984.
20. B.W. Hong, E. Prados, S. Soatto, and L. Vese, *Shape representation based on integral kernels: Application to image matching and segmentation*, Proc. CVPR, 2006, pp. 833–840.
21. D. Jacobs, D. Weinshall, and Y. Gdalyahu, *Class representation and image retrieval with non-metric distances*, IEEE Trans. PAMI **22** (2000), 583–600.
22. R. Kimmel and J. A. Sethian, *Computing geodesic on manifolds*, PNAS, vol. 95, 1998, pp. 8431–8435.
23. L.J. Latecki, R. Lakaemper, and D. Wolter, *Optimal partial shape similarity*, Image and Vision Computing **23** (2005), 227–236.
24. H. Ling and D. Jacobs, *Using the inner-distance for classification of articulated shapes*, Proc. CVPR, 2005.
25. F. Mémoli and G. Sapiro, *A theoretical and computational framework for isometry invariant recognition of point cloud data*, Foundations of Computational Mathematics **5** (2005), no. 3, 313–347.
26. A. Pentland, *Recognition by parts*, Proc. ICCV, 1987, pp. 612–620.
27. D. Geiger R. Basri, L. Costa and D. Jacobs, *Determining the similarity of deformable shapes*, Vision Research **38** (1998), 2365–2385.
28. M. Reuter, F.-E. Wolter, and N. Peinecke, *Laplace-Beltrami spectra as shape-DNA of surfaces and solids*, Computer-Aided Design **38** (2006), 342–366.
29. M. E. Salukwadze, *Vector-valued optimization problems in control theory*, Academic Press, 1979.
30. J. Zhang, R. Collins, and Y. Liu, *Representation and matching of articulated shapes*, Proc. CVPR, vol. 2, June 2004, pp. 342 – 349.

# Some Remarks on Perspective Shape-from-Shading Models

Emiliano Cristiani[1], Maurizio Falcone[2], and Alessandra Seghini[2]

[1] Dipartimento di Metodi e Modelli Matematici per le Scienze Applicate, Sapienza - Università di Roma
cristiani@dmmm.uniroma1.it
[2] Dipartimento di Matematica, Sapienza - Università di Roma
{falcone,seghini}@mat.uniroma1.it

**Abstract.** The Shape-from-Shading problem is a classical problem in image processing. Despite the huge amount of articles that deal with it, few real applications have been developed because the usual assumptions considered in the theory are too restrictive. Only recently two new PDE models have been proposed in order to include in the model the perspective deformation of the image, this allows to drop the unrealistic assumption requiring that the point of view is very far from the object. We compare these two models and present two semi-Lagrangian approximation schemes which can be applied to compute the solution. Moreover, we analyze the effect of various boundary conditions on the first order equations corresponding to the models. Some test problems on real and virtual images are presented.

## 1 Introduction

The Shape-from-Shading problem is a classical inverse problem in image processing. The goal is to reconstruct a three-dimensional object (the *shape*) from the brightness variation (the *shading*) in a greylevel photograph of that scene. In the classical model it is assumed that the surface $z = u(x)$, $x \in \Omega \subset \mathbb{R}^2$ is a graph. In order to characterize the solution of the problem several assumptions are needed, here are the most classical ones (see *e.g.* [3]):

$H_1$ - The image reflects the light uniformly, *i.e.* the albedo (ratio between energy reflected and energy captured) is constant.
$H_2$ - The material is Lambertian, *i.e.* the brightness function $I(x)$ is proportional to the scalar product $\eta(x) \cdot \omega(x)$ where $\eta(x)$ is the normal to the surface at the point $(x, z(x))$ and $\omega(x)$ is the direction of the light at the same point.
$H_3$ - The light source is unique and the rays of light which lighten the scene are parallel, *i.e.* $\omega(x)$ is constant.
$H_4$ - Multiple reflections are negligible.
$H_5$ - The aberrations of the objective are negligible.
$H_6$ - The distance between the scene and the objective is much larger than that between the objective and the CCD sensor.

F. Sgallari, A. Murli, and N. Paragios (Eds.): SSVM 2007, LNCS 4485, pp. 276–287, 2007.

$H_7$ - The perspective deformations are negligible.
$H_8$ - The scene is completely visible by the camera, *i.e.* there are not hidden regions.

Under those assumptions, if the light source is vertical the solution can be characterized in terms of an eikonal type equation

$$|\nabla u(x)| = q(x) \text{ for } x \in \Omega \quad \text{where} \quad q(x) = \sqrt{\frac{1 - I^2(x)}{I^2(x)}}. \tag{1}$$

Since the object can (obviously) have sharp sides there is no reason to assume that the solution of (1) is regular, the natural framework for this characterization is that of weak solutions in the viscosity sense (see *e.g.* [14,9]). However, the problem is not well-posed and suffers from the convex-concave ambiguity which makes impossible to prove uniqueness results without additional assumptions on the surface. A lot of work has been done to obtain uniqueness results in the framework of weak solution and new concepts have been introduced, such as that of maximal solutions (see [2] and references therein, [1,15] for some numerical applications and [8] for an up-to-date survay).

Recently, more realistic models have been proposed (see Prados and Faugeras [12], Tankus, Sochen and Yeshurun [16], Courteille, Crouzil, Durou and Gurdjos [4]). Since those papers, the Shape-from-Shading was finally applied to some real problems like reconstruction of faces ([11,13]), reconstruction of human organs ([17]) and the digitization of documents without scanners ([5,6,3]). We will examine in particular two models. The first, proposed in [4], drops the assumption $H_7$ so that it takes into account the perspective deformation due to the finite distance between the camera and the scene. In this model the distance of the light source is infinite so that all the rays are parallel (in the sequel this model will be denoted by $PSFS_\infty$). We perform a numerical approximation of the first order PDE related to the new problem via a semi-Lagrangian discretization and we discuss the effect of several possible boundary conditions for it.

In the second model, proposed by Prados and Faugeras [12], the assumptions $H_3$ and $H_7$ are dropped. The light source is placed at the optical center as in [10] so that this model is more realistic under flash lighting conditions (in the sequel this model will be denoted by $PSFS_r$). We present a semi-Lagrangian scheme also for this model.

The goal of this paper is to compare the two models and test them on real and synthetic images trying to sketch some conclusions which can be useful for future applications.

## 2 The $PSFS_\infty$ Problem

In this section we get rid of assumption $H_7$ so we will take into account the perspective deformation due to the fact that the camera is close to the scene.

Let us define the model adopting the same notations used in [5]. The point $O = (X_0, Y_0)$ is the principal point of the image, *i.e.* it is the intersection between

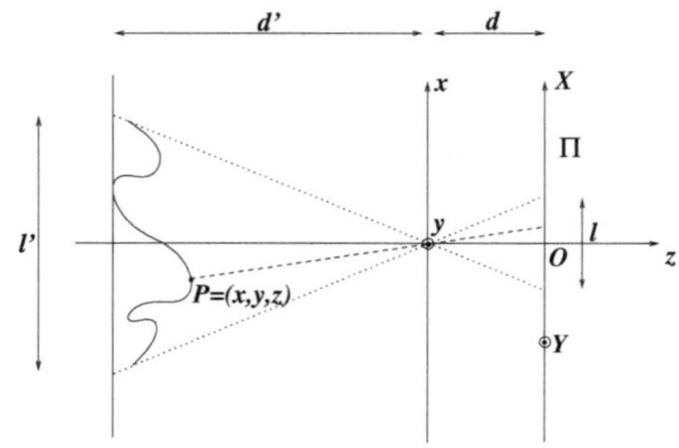


**Fig. 1.** The $\mathrm{PSFS}_\infty$ model

the perspective plane $\Pi$ and the $z$ axis, $d$ and $d'$ are respectively the distance of the objective from the perspective plane (the CCD sensor) and the distance of the objective from the (flat) background, $l$ and $l' = \frac{d'}{d} l$ are respectively the length of a generic segment in the perspective plane and the length of the real segment corresponding to it (see Figure 1 and [5] for more details). The representation of a point $P$ on the surface in terms of the $(X, Y)$ coordinates of a point in the perspective plane $\Pi$ is given by three parametric equations

$$x = r(X,Y), \; y = s(X,Y), \; z = t(X,Y) \tag{2}$$

where (see [5])

$$\begin{cases} r(X,Y) = \frac{X-X_0}{d} t(X,Y) \\ s(X,Y) = \frac{Y-Y_0}{d} t(X,Y). \end{cases} \tag{3}$$

Then the problem amounts to compute the third component $t$. This is the most difficult task since $t$ is the solution of the following eikonal type equation

$$\left(\frac{d}{\bar{t}(X,Y)}\right)^2 |\nabla t(X,Y)|^2 = \frac{I_{max}^2}{I'(X,Y)^2} - 1 \quad \text{in } \Omega \tag{4}$$

where $\Omega$ is the internal region bounded by the silhouette of the object ($\partial\Omega$ will denote its boundary) which is embedded in a rectangular domain $Q$,

$$\bar{t}(X,Y) = t(X,Y) + (X - X_0, Y - Y_0) \cdot \nabla t(X,Y), \tag{5}$$

$$I'(X,Y) = \frac{I(X,Y)\big((X-X_0)^2 + (Y-Y_0)^2 + d^2\big)^2}{d^4}, \tag{6}$$

and $I_{max}$ is a constant depending on parameters of the problem. The set $Q \setminus \Omega$ is the background.

Defining

$$f(X,Y) := \frac{1}{d^2}\left(\frac{I_{max}^2}{I'(X,Y)^2} - 1\right) \tag{7}$$

we can write (4) as

$$|\nabla t(X,Y)| = \sqrt{f(X,Y)}\;|\bar{t}(X,Y)|. \tag{8}$$

We want to write (8) in a fixed point form and construct an approximation scheme for this equation. To this end it is important to note that $\bar{t}$ has a sign. In fact, the exterior normal to the original surface at the point $P$ is given by $\hat{n}(P) = N(P)/|N(P)|$ where

$$N(P) := (d\,t_X(X,Y), d\,t_Y(X,Y), -\bar{t}(X,Y)) \tag{9}$$

and since $-\bar{t}$ must be positive (according to the orientation of the $z$ axis in Figure 1), $\bar{t}$ must be negative. This implies that (8) is in fact

$$|\nabla t(X,Y)| + \sqrt{f(X,Y)}\big((t(X,Y) + (X - X_0, Y - Y_0)\cdot \nabla t(X,Y)\big) = 0. \tag{10}$$

Equation (10) must be complemented with some boundary conditions, typically we will consider the Dirichlet boundary condition

$$t = g(X,Y) \quad \text{on } \partial\Omega, \quad \text{where } -d' \le g \le 0. \tag{11}$$

We will come back later on to this point.

The standard semi-Lagrangian scheme for (10)-(11) is

$$t(X,Y) = F[t](X,Y) := \frac{1}{1+h} \inf_{a\in B(0,1)} \{t\,(b_h(X,Y,a))\} \qquad \text{in } \Omega \tag{12}$$

where

$$b_h(X,Y,a) = (X,Y) + h\left(\frac{-a}{\sqrt{f}} - (X,Y)\right) \qquad (X,Y)\in\Omega,\ a\in B(0,1) \tag{13}$$

and $B(0,1)$ is the unit ball in $\mathbb{R}^2$. Let us examine the properties of the $F$ operator in order to guarantee convergence for the fixed point iteration. First, let us introduce the space $W = \{w : \Omega \to \mathbb{R},\ \text{such that } w|_{\partial\Omega} = g\}$.

**Lemma 1.** *[7] Under the above assumptions, the following properties hold true:*
*a) $F$ is a contraction mapping in $L^\infty(\Omega)$;*
*b) $F$ is monotone, i.e. $s \le t$ implies $F[s] \le F[t]$;*
*c) Let $V = \{w \in W : -d' \le w(X,Y) \le 0\}$, then $F : V \to V$;*

In practice, a variable step discretization $h$ can be applied to obtain more accurate results depending on $X, Y$ and $a$ in such a way that

$$h(X,Y,a)\left(\frac{-a}{\sqrt{f}} - (X,Y)\right) = \Delta x \quad \text{for all } X,Y,a$$

where $\Delta x$ is the space step discretization. Note that $h$ should be interpreted as a time step used to integrate along characteristics in the semi-Lagrangian scheme. This trick reduces the number of iterations needed to reach convergence.

Now let us examine the algorithm. Lemma 1 guarantees that, starting from any initial guess $t^{(0)}$ which satisfies the boundary conditions, the fixed point iteration

$$t^{(n+1)} = F[t^{(n)}] \tag{14}$$

converges to the unique solution $t^*$ (fixed point). The iterative algorithm stops when $\|t^{(n+1)} - t^{(n)}\|_\infty < \varepsilon$, where $\varepsilon$ is a given tolerance.

We note that a direct consequence of the above Lemma is that one can obtain a monotone increasing convergence just starting from any function below the final solution, *e.g.* choosing $t^{(0)} \equiv -d'$ in the internal nodes and imposing the Dirichlet boundary condition $t^{(0)} = g(X,Y)$ on $\partial\Omega$. Moreover, property *b*) guarantees that $\bar{t} < 0$ for all $(X,Y) \in \Omega$ *at every iteration*, so the equation associated to the problem is always (10).

## 3 The PSFS$_r$ Model

This model gets rid of assumptions $H_7$ and $H_3$. It takes into account the perspective deformation *and* the closeness of the light source which is now located at the optical center.

Let us define the model adopting the same notations used in [13]. Let $\Omega$ be an open set of $\mathbb{R}^2$. $\Omega$ represents the image domain. We denote by $f > 0$ the focal length and by $P$ a generic point on the surface. There exists a function $u : \Omega \to \mathbb{R}$ such that (see Fig. 2)

$$P = P(x,y) = u(x,y)m' \tag{15}$$

where

$$m' = \frac{f}{\sqrt{x^2+y^2+f^2}}\, m \qquad \text{and} \qquad m = (x,y,-f). \tag{16}$$

We also denote by $r(x,y)$ the distance between the light source and the point $P(x,y)$ on the surface. Note that $u(x,y) = \frac{r(x,y)}{f}$.

In [13] it was proved that $v = \ln(u)$ ($u$ is strictly positive since we assume that the scene is placed in front of the optical center) is the solution of the following equation

$$-\,e^{-2v(x,y)} + \sup_{a\in B(0,1)} \{-b(x,y,a)\cdot\nabla v(x,y) - l(x,y,a)\} = 0\,, \quad (x,y)\in\Omega \tag{17}$$

once we define

$$l(x,y,a) := -I(x,y)f^2\sqrt{1-|a|^2}\,,\; b(x,y,a) := -J(x,y)R^T(x,y)\,D(x,y)\,R(x,y)\,a\,,$$

$$D(x,y) := \begin{pmatrix} f & 0 \\ 0 & \sqrt{f^2+x^2+y^2} \end{pmatrix}, \qquad R(x,y) := \frac{1}{(x^2+y^2)^{1/2}}\begin{pmatrix} y & -x \\ x & y \end{pmatrix},$$

$$J(x,y) := I(x,y)f\sqrt{x^2+y^2+f^2}$$

where $R^T$ is the transpose of the matrix $R$. Note that $l(x,y,a) = 0$ on $\partial B(0,1)$, therefore in the numerical approximation we can not search only on $\partial B(0,1)$ but we have to discretize the unit ball entirely. We have the following

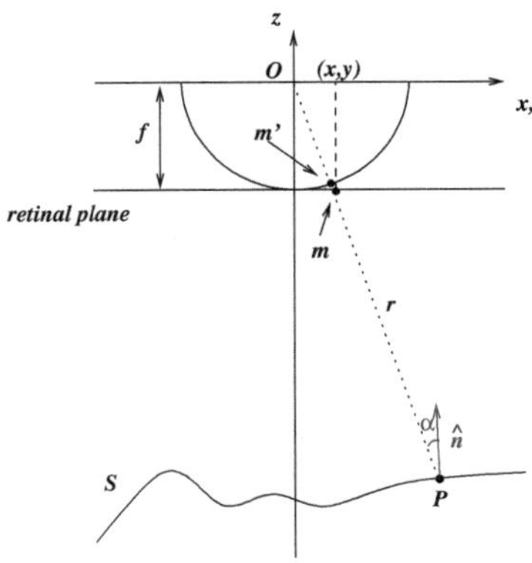


**Fig. 2.** The PSFS$_r$ model

**Theorem 1.** *Let $\Omega$ be bounded and smooth. If $I$ is differentiable and if there exist $\delta > 0$ and $M$ verifying $I \geq \delta$ and $|\nabla I| \leq M$, then equation (17) complemented with Dirichlet boundary condition $u = \phi$ on $\partial\Omega$ has a unique viscosity solution.*

**Theorem 2.** *Under the assumptions of Theorem 1, equation (17) complemented with state constraints boundary condition on $\partial\Omega$ has a unique viscosity solution.*

The proof of Theorems 1-2 can be found in [13]. These uniqueness results show that, under certain assumptions, the PSFS$_r$ model is well-posed. As we will see in the numerical experiments, the application of state constraint boundary condition (which is more convenient than Dirichlet's or Neumann's since it does not require a previous knowledge of any data) is typical of the PSFS$_r$ model due to its particular brightness attenuation and cannot be exported to the PSFS$_\infty$ model.

We present a semi-Lagrangian discretization for equation (17) which is simpler than that presented in [13] and which has a built-in up-wind correction. By standard arguments, we get, for $(x, y) \in \Omega$,

$$-v_h(x,y) + \inf_{a \in B(0,1)} \{v_h((x,y) + hb(x,y,a)) + hl(x,y,a)\} + he^{-2v_h(x,y)} = 0. \tag{18}$$

We want to solve equation (18) following a fixed point method. Note that once we compute the control $a^*$ where the infimum is attained we need some extra work to compute $v_h(x, y)$. In fact, let us define

$$c := v_h((x,y) + hb(x,y,a^*)) + hl(x,y,a^*)$$

and $t := v_h(x, y)$ for any $(x, y)$ fixed. At every iteration we have to solve the equation

$$g(t) := -t + c + he^{-2t} = 0.$$

This additional problem is solved by applying Newton's method as in [13] (note that $g'(t) < 0$ for all $t \in \mathbb{R}$).

We start from any supersolution $v_h^{(0)}$ of (18) and we compute its solution, iterating the procedure until $\|v_n^{(n+1)} - v_h^{(n)}\|_\infty < \varepsilon$, where $\varepsilon$ is a given tolerance. We choose the time step $h = h(x,y)$ in such a way $|h(x,y)b(x,y,a^*)| \leq \Delta x$ and we discretize the unit ball $B(0,1)$ by means of (*#directions* × *#circles*) points plus the central point.

**The Effect of Boundary Conditions**

As we have seen both models leads to a first order PDE which has to be complemented with a boundary condition in order to select a unique solution and run the approximation scheme. However, in practical applications boundary conditions on the surface are seldom known, so it useful to analyse in more detail the effect of different types of boundary conditions on the solution in order to define a minimal set of conditions which will allow to compute the exact solution.

In this section, we will briefly analyse the effect of Dirichlet, Neumann and state constraints boundary conditions on subsets of the boundary. Let us note first that boundary conditions should be imposed in a weak sense. The typical condition which defines a viscosity subsolution $u$ of an equation of the form $H(x, u, \nabla u) = 0$, $x \in \Omega$ requires that for any test function $\varphi \in C^1(\overline{\Omega})$ and $x_0 \in \partial\Omega$ local maximum point for $u - \varphi$

$$\min\{H(x_0, u(x_0), D\varphi(x_0)), B(x_0, u(x_0), D\varphi(x_0))\} \leq 0 \tag{19}$$

where the function $B$ is the operator describing the boundary conditions, *f.e.* $B(x, u, Du) = u - g$ for the Dirichlet condition. Similarly, the boundary condition for supersolutions requires that for any test function $\varphi \in C^1(\overline{\Omega})$ and $x_1 \in \partial\Omega$ local minimum point for $u - \varphi$

$$\max\{H(x_1, u(x_1), D\varphi(x_1)), B(x_1, u(x_1), D\varphi(x_1))\} \geq 0. \tag{20}$$

The effect of the Dirichlet condition is to impose a value on $u$ according to the above conditions, in particular the value $u(x) = g(x)$ is set at every point where $H(x, u(x), D\varphi(x)) \geq 0$ (for subsolutions) and $H(x, u(x), D\varphi(x)) \leq 0$ (for supersolutions). Neumann boundary condition corresponds to the operator $B(x, u, Du) = \partial u/\partial n(x) - m(x)$ where $n(\cdot)$ represents the outward normal to the domain $\Omega$. A typical use of it is when we know (or we presume) that the level curves of the surface are orthogonal to the boundary $\partial\Omega$ or to a subset of it where we simply choose $m(x) = 0$. The state constraints boundary condition is different from the above conditions since we do not impose neither a value for $u$ nor a value for its normal derivative $\partial u/\partial n(x)$ (cfr. [1]). In this respect it has been interpreted as a "no boundary condition" choice although this interpretation is rather sloppy. In fact, a real function $u$ bounded and uniformly continuous is said to be a *state constraints* viscosity solution if and only if it is a subsolution (in the viscosity sense) in $\Omega$ and a supersolution in $\overline{\Omega}$ (*i.e.* up to the boundary). It can be also stated as a Dirichlet boundary condition simply setting $g = C_g = constant$ provided $C_g > \max_{x \in \Omega} u(x)$.

## 4 Numerical Experiments

We will show some tests for the two schemes implemented in MATLAB v.7, compare the two methods and try to draw some conclusions.

**Numerical Experiments for PSFS$_\infty$**

We choose the following parameters: $X_0 = 0$, $Y_0 = 0$, $d = 1$, $d' = 4$, $l = 0.75$ and $l' = 3$. The computational procedure follows the steps described in the previous sections. We choose 16 controls for the discretization of the unit ball $B(0,1)$ (all of them are placed on the boundary $\partial B(0,1)$).

**Test 1: tent upside down**

In this test we consider a ridge tent upside down and its photograph (see Fig. 3). We choose $121 \times 121$ pixels grid and we solve the equation imposing Dirichlet boundary condition on the silhouette of the tent.

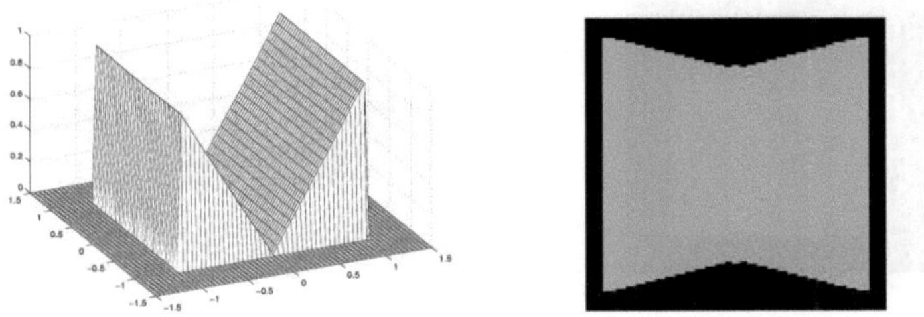

**Fig. 3.** Initial surface (left) and its photograph (right)

As we can see in Fig. 4, the reconstruction fails since the algorithm tries to compute the maximal solution instead of the correct solution. However, the shape of the domain (distorted in the photograph) is correctly straightened. Note that in Fig. 4-left MATLAB connects all points of the surface despite there is a hole in the domain of the reconstructed surface due to the fact that not all the background is visible by the camera. In Fig. 4-right the same surface is plotted by a slightly different point of view without interpolation.

**Test 2: Real image**

In this test we used a real photograph where the effect of perspective is visible. The surface is a sheet of paper with the shape of a roof tile. For this image the parameter values are: $l = 6.91\ mm$, $d = 5.8\ mm$, $l' = 200\ mm$, $d' = \frac{l'}{l}d = 167.87\ mm$, $\Delta x = 0.05\ mm$. We note that we performed the light correction (6) in the preprocessing step, so we can assume $I_{max} = 1$ during computation. Figure 5 shows the photograph ($128 \times 128$ pixels) and the surface reconstructed using Dirichlet boundary condition only on the left and right sides of the boundary and state constraints elsewhere (top and bottom sides).

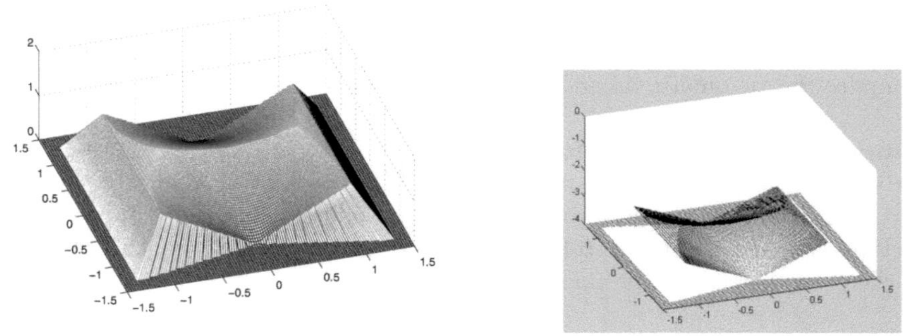


**Fig. 4.** Reconstructed surface with Dirichlet boundary condition. Interpolated (left) and non interpolated (right)

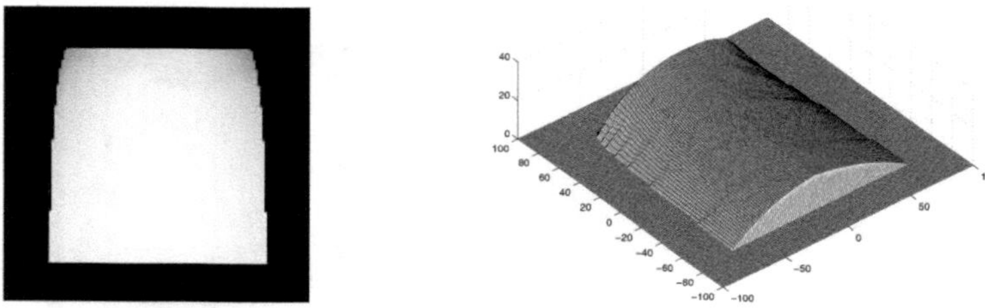

**Fig. 5.** The photograph, 128×128 pixels (left) and its reconstructed surface with mixed Dirichlet and state constraints boundary condition (right)

We can see that the solution is quite good considering the fact that light source (flash camera) is not far from the object and that direction of light source is not perfectly vertical as the mathematical model would have required.

We also tried to reconstruct the surface with two more practical boundary conditions. In the first case, we fixed a Dirichlet condition $t^0$ only on a vertical line in the center of the image (column 64) and then we turned over the computed surface with respect to the value $t^0$ (see Figure 6-left). Note that the solution is not very sensitive with respect to value $t^0$, so a rough knowledge of the behavior of the surface can be sufficient. We can see that the solution is quite good. We have a large maximum norm error on the boundary (17.7 $mm$, 41% of the maximum height of the tile), but not inside. In fact, assuming that the reconstructed surface in Figure 5-right is the exact solution, the average error on all nodes for Figure 6-left is about 1.2 $mm$.

In the second case (see Figure 6-right), we fixed a Dirichlet condition $t^0$ only on the point (64, 64) (at the center of the image) and then we turned over the computed surface as before. Note that in this case the solution has a shape very

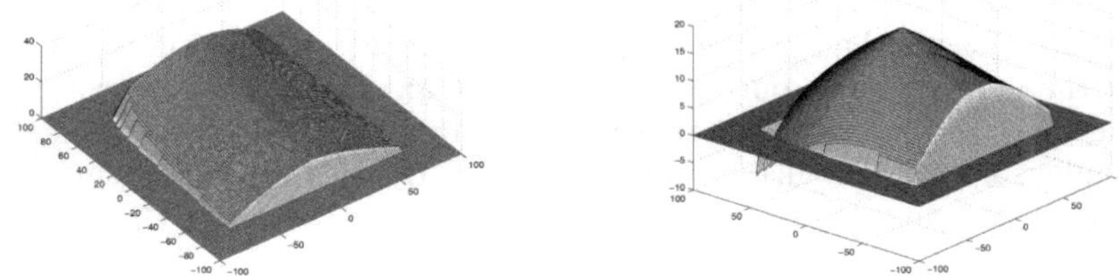

**Fig. 6.** Reconstructed surface with Dirichlet boundary condition on the center line (left) and on the center point (right, different scale)

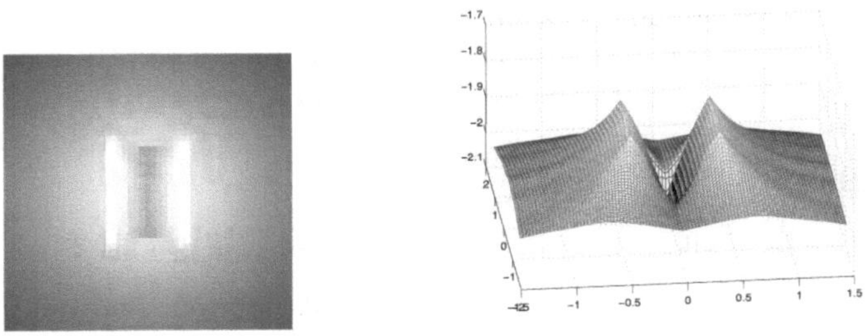

**Fig. 7.** The photograph (left) and its reconstructed surface with state constraints boundary condition (right)

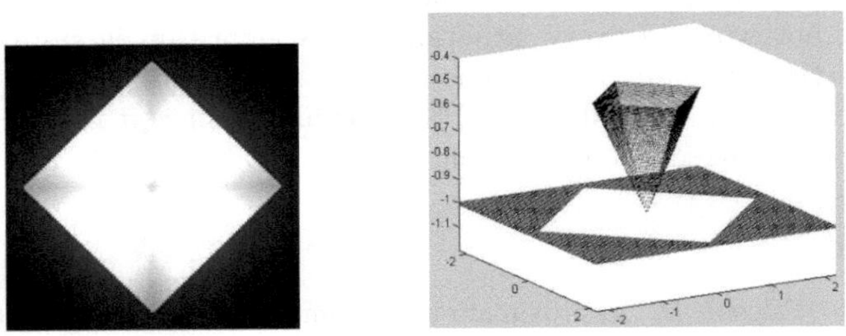

**Fig. 8.** The photograph (left) and its reconstructed surface with Dirichlet boundary condition (right)

different from the expected solution since it has a global maximum at the central point (64, 64). In these three tests the iterative procedure converges respectively in 167, 185 and 190 iterations, with $\varepsilon = 10^{-6}$.

**Numerical experiments for PSFS$_r$**

**Test 3: tent upside down**

In this test we consider a ridge tent upside down as in Test 1. We use a $100 \times 100$

pixels initial image and we choose $f = 1$ and $16 \times 2 + 1$ controls. Convergence is reached in 159 iterations. In Fig. 7 we show the result obtained by the $\mathrm{PSFS}_r$ algorithm imposing state constraints boundary condition on the boundary of the square (*i.e.* the background). In this case the reconstruction is definitively better than the previous one, considering that any boundary data was needed. On the other hand, we observe that the hole in the domain due to the regions in full shade is not reconstructed properly. In fact, the surfaces connecting the tent and the background are really computed and they are not due to the MATLAB's interpolation.

**Test 4: pyramid upside down**
In the second test we consider a pyramid upside down with the vertex standing on a flat background. We use a $128 \times 128$ pixels initial image and we choose $f = 1/4$. We impose Dirichlet boundary condition on the boundary of the pyramid (so we do not compute on the background). The initial image and the reconstructed surface are shown in Fig. 8. In this case we obtain a perfect result.

## 5 Conclusions

The above discussion allow us to draw some partial conclusions:

1. The $\mathrm{PSFS}_r$ is a more realistic model under flash lighting conditions. It allows to compute a solution without a previous knowledge of the surface just applying state constraints boundary conditions. However, this model has a more delicate initialization and the corresponding algorithm does not converge for any initial guess.
2. The $\mathrm{PSFS}_\infty$ requires additional information on the boundary (state constraints will not work in every situation). The algorithm is simpler and will converge starting from any initial guess.
3. The computational cost related to $\mathrm{PSFS}_r$ is higher since it requires a Newton iteration for every fixed point iteration.

## References

1. Camilli, F., Falcone, M.: Approximation of optimal control problems with state constraints: estimates and applications. B. S. Mordukhovic, H. J. Sussman (editors), "Nonsmooth analysis and geometric methods in deterministic optimal control", IMA Volumes in Applied Mathematics **78**, 23-57, Springer Verlag, 1996.
2. Camilli, F., Siconolfi, A.: Discontinuous solutions of a Hamilton-Jacobi equation with infinite speed of propagation. SIAM Journal on Mathematical Analysis, **28** (1997), 1420-1445.
3. Courteille, F.: Vision monoculaire: contributions théoriques et application à la numérisation des documents. Ph.D. thesis, Université Paul Sabatier, Toulouse, France, 2006.
4. Courteille, F., Crouzil, A., Durou, J.-D., Gurdjos, P.: Shape from shading en conditions réalistes d'acquisition photographique. Actes du $14^{ème}$ Congrès Francophone de Reconnaissance des Formes et Intelligence Artificielle (volume II), 925-934, Toulouse, France, 2004.

5. Courteille, F., Crouzil, A., Durou, J.-D., Gurdjos, P.: Towards shape from shading under realistic photographic conditions. Proceedings of the $17^{th}$ International Conference on Pattern Recognition (volume II), 277-280, Cambridge, England, 2004.
6. Courteille, F., Crouzil, A., Durou, J.-D., Gurdjos, P.: Shape from shading for the digitization of curved documents. Machine Vision and Applications, 2006 (to appear).
7. Cristiani, E., Falcone, M., Seghini, A.: Numerical solution of the perspective Shape from Shading problem. Proceedings of "Control Systems: Theory, Numerics and Applications" PoS (CSTNA2005) 008, `http://pos.sissa.it/`.
8. Durou, J.-D., Falcone, M., Sagona, M.: A survey of numerical methods for Shape from Shading. Rapport IRIT 2004-2-R, Université Paul Sabatier, Toulouse, France, 2004. Submitted to Computer Vision and Image Understanding, under revision.
9. Lions, P.-L., Rouy, E., Tourin, A.: Shape from shading, viscosity solution and edges. Numerische Mathematik, **64** (1993), 323-353.
10. Okatani, T., Deguchi, K.: Reconstructing Shape from Shading with a Point Light Source at the Projection Center: Shape Reconstruction from an Endoscope Image. Proceedings of the $13^{th}$ International Conference on Pattern Recognition (volume I), Vienna, Austria, 1996, 830-834.
11. Prados, E.: Application of the theory of the viscosity solutions to the Shape From Shading problem. Ph.D. thesis, Univ. of Nice - Sophia Antipolis, France, 2004.
12. Prados, E., Faugeras, O.: "Perspective Shape from Shading" and viscosity solutions. Proceedings of the $9^{th}$ IEEE International Conference on Computer Vision (volume II), 826-831, Nice, France, 2003.
13. Prados, E., Faugeras, O., Camilli, F.: Shape from Shading: a well-posed problem?. Rapport de Recherche n. 5297. INRIA Sophia Antipolis, 2004.
14. Rouy, E., Tourin, A.: A viscosity solutions approach to Shape-from-Shading. SIAM J. Numer. Anal., **29** (1992), 867-884.
15. Sagona, M.: Numerical methods for degenerate eikonal type equation and applications. Ph.D. thesis, Dipartimento di Matematica, Università di Napoli "Federico II", Italy, 2001.
16. Tankus, A., Sochen, N., Yeshurun, Y.: A new perspective [on] Shape-from-Shading. Proceedings of the $9^{th}$ IEEE International Conference on Computer Vision (volume II), 862-869, Nice, France, 2003.
17. Tankus, A., Sochen, N., Yeshurun, Y.: Reconstruction of medical images by perspective Shape-from-Shading. Proceedings of the $17^{th}$ International Conference on Pattern Recognition (volume III), 778-781, Cambridge, England, 2004.

# Scale-Space Clustering with Recursive Validation

Tomoya Sakai[1], Takuto Komazaki[2], and Atsushi Imiya[1]

[1] Institute of Media and Information Technology, Chiba University, Japan
{tsakai,imiya}@faculty.chiba-u.jp
[2] Graduate School of Science and Technology, Chiba University, Japan
komazaki@graduate.chiba-u.jp

**Abstract.** We present a hierarchical clustering method for a dataset based on the deep structure of the probability density function (PDF) of the data in the scale space. The data clusters correspond to the modes of the PDF, and their hierarchy is determined by regarding the nonparametric estimation of the PDF with the Gaussian kernel as a scale-space representation. It is shown that the number of clusters is statistically deterministic above a certain critical scale, even though the positions of the data points are stochastic. Such a critical scale is estimated by analysing the distribution of cluster lifetime in the scale space, and statistically valid clusters are detected above the critical scale. This cluster validation using the critical scale can be recursively employed according to the hierarchy of the clusters.

## 1 Introduction

Clustering is a pattern recognition technique for grouping a set of data samples with similar characteristics. Many clustering algorithms deal with data samples in a data space, which is called the feature space, to enable geometric approaches such as partitioning of the space with geometric criteria. Since the data points in the space can be considered as random points distributed according to an unknown probability density function (PDF), clustering is essentially a structural analysis of the PDF. In the case that a model of the PDF is not presumable, the PDF is estimated by a nonparametric approach. The details of the PDF structure, however, are controlled by the cardinality of the dataset. In other words, a finite number of data points represent the geometric structure of the PDF with some resolution. The scale-space theory is a powerful mathematical tool for the structural analysis of a multivariable function with respect to scale. We can clarify the PDF structure by applying the scale-space analysis to the nonparametric PDF estimation.

We interpret scale-space clustering [2,3,4] as an extraction of a hierarchical structure of the PDF on the basis of hierarchical relationships among the data points and subclusters. The modes are the local maxima of the PDF, and approximate barycentres of the clusters. A hierarchical clustering is achieved by determining the hierarchical relationships among the local extrema of the estimated PDF in scale space. We focus on an important property that the modes of

F. Sgallari, A. Murli, and N. Paragios (Eds.): SSVM 2007, LNCS 4485, pp. 288–299, 2007.

the estimated PDF are deterministic above a certain critical scale, even though the positions of the data points are stochastic. Such a critical scale can be estimated from the lifetime of the data points in the scale space. By selecting a suitable scale for clustering above the critical scale, we can obtain valid clusters without prior knowledge of the number of clusters or their locations.

In this paper, we first introduce the nonparametric PDF estimation in the scale space, and present a tree representation of its hierarchical structure. Second, we present a clustering method for the dataset and recursive validation of the clusters using the tree. Finally, we demonstrate the present method using artificial datasets.

## 2 Nonparametric Estimation of PDF from DataSet

Given a dataset $P = \{\mathbf{x}_i \in \mathbb{R}^N\}$, the sum of the Dirac delta functions at $\mathbf{x}_i$ is a primitive estimation of the PDF from the dataset.

$$f(P;\mathbf{x}) = \frac{1}{\mathrm{card}(P)} \sum_{\mathbf{x}_i \in P} \delta(\mathbf{x} - \mathbf{x}_i) \tag{1}$$

Here, $\delta$ is the Dirac delta function. $f(P;\mathbf{x})$ is normalised by the cardinality of $P$ so that $\int_{\mathbf{x}\in\mathbb{R}^N} f d\mathbf{x} = 1$. Since $\delta$ can be written as a limit of the Gaussian function $G(\mathbf{x}, \sigma)$, we have

$$\begin{aligned} f(P;\mathbf{x}) &= \frac{1}{\mathrm{card}(P)} \sum_{\mathbf{x}_i \in P} \lim_{\sigma\to 0} G(\mathbf{x} - \mathbf{x}_i, \sigma) \\ &= \lim_{\tau\to 0} f(P;\mathbf{x},\tau), \end{aligned} \tag{2}$$

where

$$\begin{aligned} f(P;\mathbf{x},\tau) &= \frac{1}{\mathrm{card}(P)} \sum_{\mathbf{x}_i \in P} G(\mathbf{x} - \mathbf{x}_i, \sqrt{2\tau}) \\ &= \frac{1}{\mathrm{card}(P)} \sum_{\mathbf{x}_i \in P} \frac{1}{\sqrt{4\pi\tau}^N} e^{-\frac{|\mathbf{x}-\mathbf{x}_i|^2}{4\tau}}. \end{aligned} \tag{3}$$

The one-parameter family of the PDF $f(P;\mathbf{x},\tau)$ is identical to the Parzen density estimation with a Gaussian kernel $G(\mathbf{x}, \sqrt{2\tau})$ [1]. Therefore, $f(P;\mathbf{x},\tau)$ converges to the true density function if the window width $\sigma = \sqrt{2\tau} \to 0$ when $\mathrm{card}(P) \to \infty$.

The use of the Gaussian kernel is justifiable from the viewpoints of the information theory and scale-space theory.

- The Gaussian distribution for each data point maximises the entropy

$$H(p) = -\int_{\mathbb{R}^N} p(\mathbf{x}) \log p(\mathbf{x}) d\mathbf{x} \tag{4}$$

for a given variance $\sigma^2 = 2\tau$.

- $f(P; \mathbf{x}, \tau)$ satisfies the diffusion equation

$$\frac{\partial f}{\partial \tau} = \Delta f. \tag{5}$$

  This property implies that the points disperse by Brownian motion. Initial positions at $\tau = 0$ are given by the dataset $P$, and a superposition of the Gaussian functions represents ambiguity of the locations of the points after the time $\tau$.
- $f(P; \mathbf{x}, \tau)$ satisfies the scale-space axioms [8,10,11,12]. The parameter $\tau$ can be regarded as the scale. Scale-space analyses are available for the PDF structure.

The modes of the PDF are its spatial critical points. The structure across the scale, or so-called deep structure [9,15,17], implies hierarchical relationships among the critical points. That is, the modes of the estimated PDF $f(P; \mathbf{x}, \tau)$ have hierarchical relationships among them. The modes are representative of clusters for the data points that are stochastically located in $\mathbb{R}^N$ according to $f(P; \mathbf{x}, \tau)$. We introduce a scale-space analysis for deriving an explicit representation of the hierarchical structure of $f(P; \mathbf{x}, \tau)$ from trajectories of the critical points in the scale space.

## 3 Gaussian Scale Space

The scale space is classically explained as blurred images by Gaussian filtering. A one-parameter family of a positive function $f(\mathbf{x}, \tau)$ is derived from resolution conversion, or a blurring operation of a grayscale image $f(\mathbf{x})$. The Gaussian scale-space image $f(\mathbf{x}, \tau)$ is defined as the convolution of the image $f(\mathbf{x})$ with the isotropic Gaussian function $G(\mathbf{x}, \sqrt{2\tau})$. The parameter $\tau$ is called the "scale", which can be regarded as an inversion of the image resolution. The isotropic Gaussian convolution satisfies the following axioms [8,10,11,12].

- Non-negative intensity.
- Linearity.
- Closedness under affine transformations.
- Associative (or semigroup) property.

The scale space $(\mathbf{x}, \tau)$ can also be defined as a space in which a spatial function is governed by a diffusion equation with respect to the scale $\tau$. Any function described as a convolution of the Gaussian function satisfies the linear diffusion equation (5). The PDF $f(P; \mathbf{x}, \tau)$ in (3) also satisfies the linear diffusion equation. Therefore, the Gaussian scale-space analysis is available for the estimation of the PDF structure if we equate $f(P; \mathbf{x}, \tau)$ as an image in the scale space. The PDF $f(P; \mathbf{x}) = \lim_{\tau \to 0} f(P; \mathbf{x}, \tau)$ can be considered as the image at the finest scale, and $f(P; \mathbf{x}, \tau)$ is the derived scale-space image at scale $\tau$.

### 3.1 Hierarchical Structure of Image

As the scale increases, an image is blurred, and its features are simplified. A remarkable feature of the image is a set of critical points where the spatial gradient of the image intensity vanishes.

$$\{\mathbf{x}(\tau)|\nabla f(\mathbf{x},\tau)=\mathbf{0}\} \tag{6}$$

The trajectories of the critical points observed in the scale space $(\mathbf{x},\tau)$ are called the critical curves. The critical curves (a.k.a stationary curves) are solutions to the equation

$$\mathbf{H}\frac{d\mathbf{x}(\tau)}{d\tau}=-\nabla\Delta f(\mathbf{x}(\tau),\tau), \tag{7}$$

where $\mathbf{x}(\tau)$ and $\mathbf{H}$ are the critical point and the Hessian matrix of image $f$, respectively [10]. Since the local extrema are representative of dark and bright regions of the image, the structures of the critical curves in scale space indicate the topological relationships among the regions.

The structure of an image indicated by the critical curves in scale space has been investigated by various authors [10,14,15,16,17]. The bifurcational properties of image features in the scale space imply that the image structure across the scale is hierarchical. A local extremum, for example, is annihilated when it comes across a saddle point with increasing scale. The annihilation point is a singular point where $\det\mathbf{H}=0$. Equation (7) indicates that the spatial velocity of the critical point with respect to scale becomes infinite at the annihilation point. Therefore, each top endpoint of the critical curves in scale space is a singular point, and does not have any connections via the critical curves to a higher scale, generically.

We have clarified the hierarchical structure on the basis of links of singular points, applying the catastrophe theory to the spatial gradient field of the image in the Gaussian scale space [18]. A singular point generically has a specific gradient field curve, which they call the antidirectional figure-flow curve. The antidirectional figure-flow curve defines the link of the singular point to another local extremum. Consequently, the hierarchical structure of the image can be derived from the critical curves across scales and the antidirectional figure-flow curves at fixed scales.

### 3.2 Hierarchical Structure of PDF

The PDF $f(P;\mathbf{x},\tau)$ consists of small card$(P)$ blobs in an isotropic Gaussian shape if the scale is sufficiently small. As the scale increases, the blobs merge into large ones, and local maxima at their peaks disappear one after another. This merging process is described as the trajectories of the local maxima in the scale space, that is, the critical curves of the local maxima. Each critical curve of the local maxima starts at one of the points in $P$ at $\tau=0$, and traces the peak of the blob with increasing scale. The critical curve terminates at the merging scale since the local maximum suddenly falls into another blob at that scale.

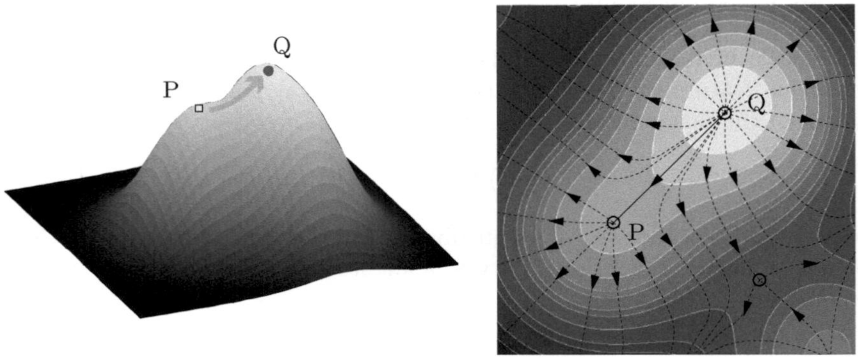


(a) (b)

**Fig. 1.** Merging blobs. (a) The annihilated local maximum P suddenly falls into a larger blob with the local maximum Q. (b) The contour map and gradient field curves at the merging scale. The antidirectional figure-flow curve (solid line) links P and Q.

This blob into which the local maximum falls is linked with an antidirectional figure-flow curve, as shown in Fig. 1.

As the result, a hierarchical clustering of the dataset $P$ is achieved by the hierarchical relationships among the local maxima and singular points of the PDF $f(P;\mathbf{x},\tau)$. $f(P;\mathbf{x},\tau)$ expresses dominant regions of the clusters at the scale $\tau$. This clustering method was proposed independently by Roberts [2], and Nakamura and Kehtarnavaz [3], and the scale-space theory on the hierarchy of critical points [18] mathematically underpins their method. Additionally, it is notable that the structure of the PDF discretely appears with a continuously decreasing scale. This property enables us to distinguish valid clusters from invalid ones. We will discuss this issue in later sections.

### 3.3 Mode Tree for Clustering

The hierarchical relationships among the points in $P$ are symbolically described as a tree [5]. For the construction of the tree, we need neither exact paths of the critical points nor forms of the antidirectional figure-flow curves. We require the relations of critical points between scales and at fixed scales. The algorithm of the tree construction is as follows.

**ALGORITHM I – Mode Tree**

1. Set card($P$) nodes with labels $k$ ($k = 1, \ldots, \mathrm{card}(P)$) to be leaves of a tree $T$.
2. Let $\hat{P} = P$ and $\tau = 0$.
3. Increase the scale $\tau$ by $\Delta\tau$, which is a small value so that $\sqrt{2\Delta\tau}$ is negligible compared with the space intervals of the points in $\hat{P}$.
4. For each point $\mathbf{p}_i \in \hat{P}$, update $\mathbf{p}_i$ by maximising $f(P;\mathbf{x},\tau)$ with $\mathbf{p}_i$ as the initial position. The mean shift [6] is available for this procedure.

5. If $\mathbf{p}_i$ falls into a local maximum of another blob corresponding to a point $\mathbf{p}_j \in \hat{P}$, remove $\mathbf{p}_i$ from $\hat{P}$, and add a new node with two branches attached to the nodes labelled $i$ and $j$ in $T$. The new node inherits the label $j$, and contains the values $\mathbf{p}_i$ and $\tau_i = \tau$.
6. If $\text{card}(\hat{P}) = 1$ then stop; otherwise go to Step 3.

# 4 Deterministic Structure of Estimated PDF

## 4.1 Cardinality and Critical Scale

A sample dataset $P$ can be regarded as an instance of a set of points stochastically located in a space according to the true PDF. If $f(P; \mathbf{x}, \tau)$ with a small scale $\tau$ well approximates the true PDF, the most probable distribution of points is the dataset $P$ itself, because each point approximately maximises the likelihood in its neighbourhood. However, if the structure of the true PDF is so complicated that the dataset $P$ cannot express the PDF in detail, the structure of the estimated PDF is stochastic at small scales. In that case, $f(P; \mathbf{x}, \tau)$ with a small scale is not a feasible estimation.

An important fact is that the cardinality of the dataset plays the role of the resolution of the PDF. The structure of the PDF is provided from coarse to fine by data samples, and a finite number of samples represent the PDF incompletely. Therefore, the PDF structure is established from top to bottom with increasing cardinality. There exists a critical lower bound of scale, above which the structure is deterministic and under which the structure is stochastic. Clustering should be employed above such a critical scale.

## 4.2 Detection of Valid Clusters

The number of clusters is determined by selecting a scale $\tau$ above the critical scale. All data points are classified into one cluster at the coarsest scale, which is represented by the root node of the scale-space tree. If the scale is sufficiently large, the position of one remaining local maximum converges to the barycentre of the cluster. As the scale decreases, new local maxima appear one after another. The appearance of the local maximum indicates that a cluster splits into subclusters, which are represented by the nodes of subtrees of the scale-space tree.

Detection of the critical scale is achieved by a statistical analysis of the lifetime of a data point. Recall that each critical curve of a local maximum starts at a data point at $\tau = 0$, and terminates at a singular point. We define the lifetime of a data point as follows.

**Definition 1.** *The lifetime of a data point* $\mathbf{p}_i \in P$ *is defined as the terminating scale* $\tau_i > 0$ *of a critical curve of local maximum whose starting point is* $(\mathbf{p}_i, 0)$ *in the scale space.*

Every node of the scale-space tree stores the lifetime. Since a local maximum corresponds to a cluster at any scale, the lifetime of a data point is regarded as that of the corresponding cluster, that is, the cluster lifetime.

We consider the distribution of the lifetime. If the data points are uniformly distributed, the number of local maxima of $f(P; \mathbf{x}, \tau)$ exponentially decays with increasing scale [2]. Accordingly, the distribution of the lifetime is exponential. Clustering of such data points with uniform distribution yields only invalid clusters due to spurious features of the estimated PDF. On the contrary, nonexponential decay of the lifetime with respect to scale implies nonuniformity of the distribution of data points. The exponential decay is allowed only under the small scale where the distribution seems to be locally uniform. Statistically deterministic features of the dataset emerge above the critical scale where the monotonic decay collapses.

In practice, the critical scale can be roughly estimated using the histogram of the lifetime. The critical scale is at the end of the exponential decay in the histogram. If there exist valid clusters, one can find outstanding lifetimes above the critical scale. If the data contain observation noise, it is preferable to select the clustering scale greatly above the critical scale so as to avoid noise infection for clustering.

### 4.3 Recursive Validation

Since the hierarchical relationships among the clusters are explicitly described as the scale-space tree, we can recursively validate whether a cluster can be split into deterministic subclusters. Construct a histogram of the lifetime stored in a subtree corresponding to the cluster. If a critical scale is found in that histogram, then there exist valid subclusters with outstanding lifetime values above the critical scale.

## 5 Experiments

### 5.1 Cardinality and Validity

We demonstrate the clustering for the artificial datasets with different cardinalities shown in Fig. 3. These datasets are generated from the PDF in Fig. 2(a). The PDF consists of five elliptic blobs, so the expected number of clusters is five. The critical curves of local maxima corresponding to the five blobs are found in the scale space, as shown in Fig. 2(b).

For each dataset, the critical curves of local maxima in scale space and the histogram of their lifetimes are shown in Figs. 4 and 5, respectively. For the dataset $P_{30}$, five critical curves of local maxima seem to represent the true five clusters. Their lifetimes, however, are not outstanding in the histogram in Fig. 5(a). Moreover, the hierarchy indicated by these five critical curves is different from that of the five blobs of the true PDF in Fig. 2. Therefore, the dataset $P_{30}$ is too poor to estimate the dominance of each cluster correctly.

Each dataset $P_{100}$, $P_{300}$ and $P_{1000}$ has a histogram with four outstanding lifetimes, as shown in Figs. 5(b), 5(c) and 5(d). They are well-detached and highly distinguishable from the others related to invalid clusters. These four

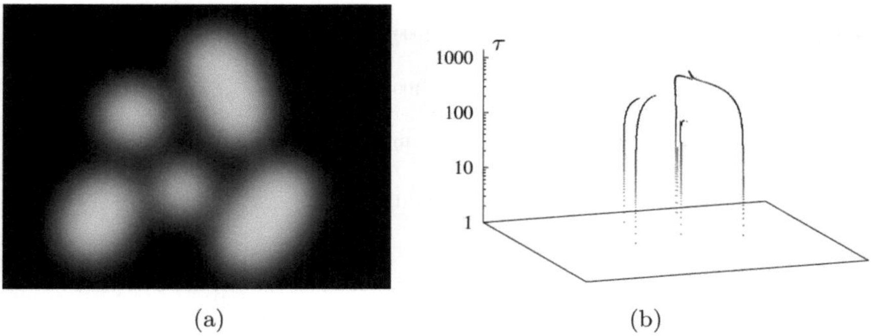


**Fig. 2.** PDF and its critical curves in scale space. (a) The PDF to be estimated. The brightness indicates the probability density of the data points. (b) Critical curves corresponding to local maxima of the five blobs.

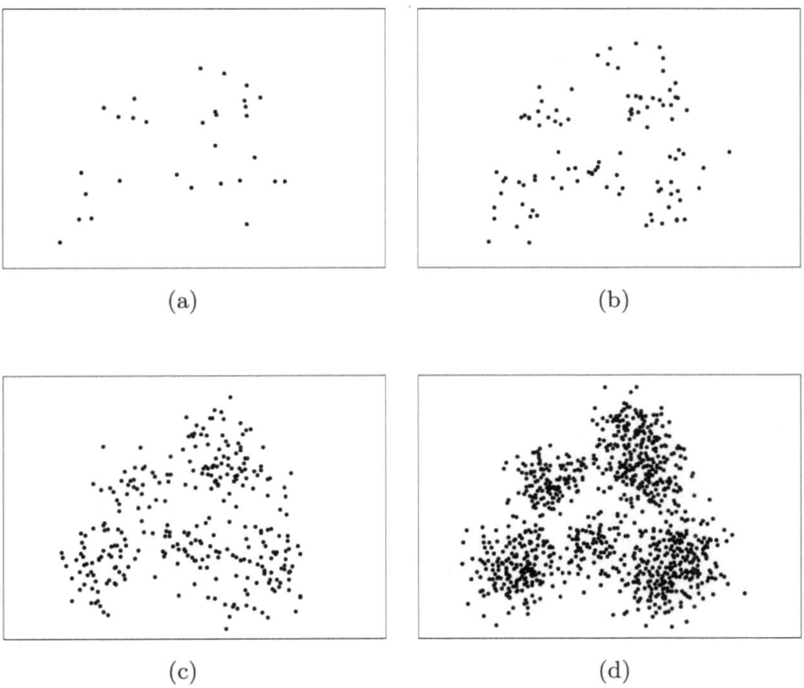


**Fig. 3.** Artifical dataset. (a) $P_{30}$, (b) $P_{100}$, (c) $P_{300}$, and (d) $P_{1000}$. $\mathrm{card}(P_{30}) = 30$, $\mathrm{card}(P_{100}) = 100$, $\mathrm{card}(P_{300}) = 300$, and $\mathrm{card}(P_{1000}) = 1000$.

maxima with the outstanding lifetimes and one remaining maximum correspond to the valid five clusters. The increase in the cardinality does not affect the number of valid clusters but clarifies the exponential decay under the critical scale.

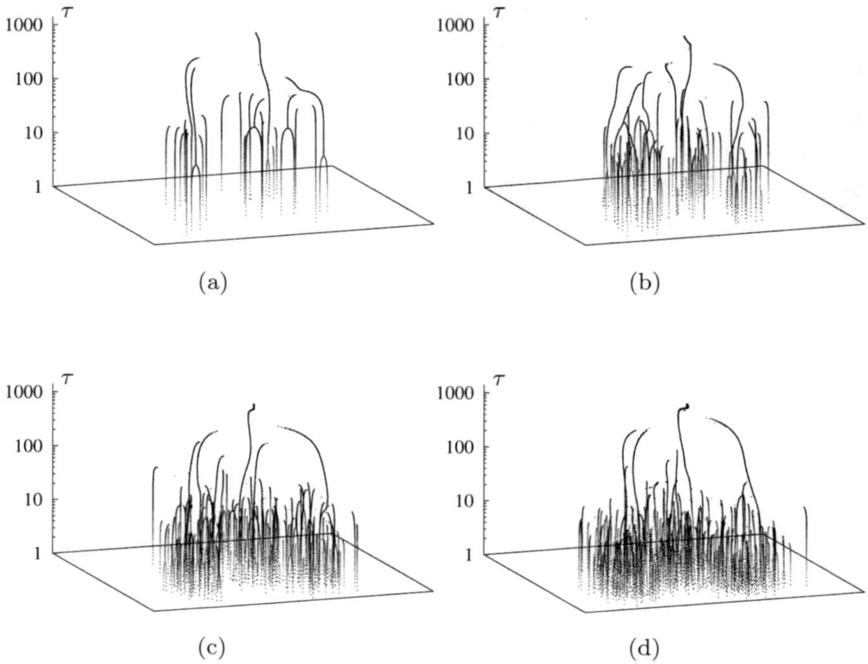


**Fig. 4.** Critical curves of local maxima for (a) $f(P_{30}; \mathbf{x}, \tau)$, (b) $f(P_{100}; \mathbf{x}, \tau)$, (c) $f(P_{300}; \mathbf{x}, \tau)$, and (d) $f(P_{1000}; \mathbf{x}, \tau)$

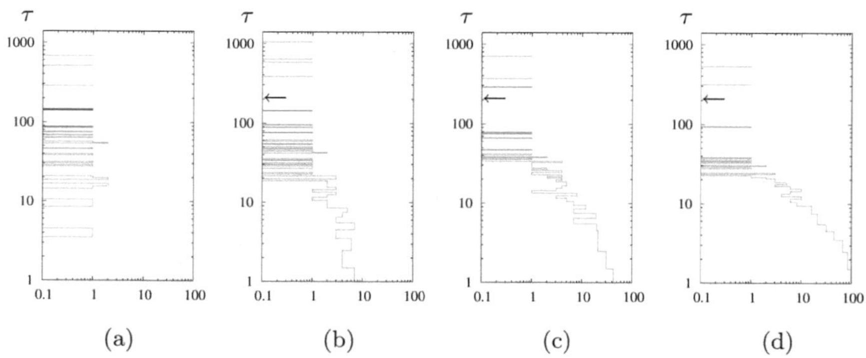


**Fig. 5.** Histograms of lifetimes for (a) $P_{30}$, (b) $P_{100}$, (c) $P_{300}$, and (d) $P_{1000}$. Roughly estimated critical scale is indicated by the arrows.

Nevertheless, any hierarchy of the valid clusters estimated by $P_{100}$ and $P_{300}$ disagrees with that of the five blobs in the true PDF. We confirm that $P_{1000}$ successfully estimates both the number of clusters and the hierarchy. The clustering result for $P_{1000}$ is shown in Fig. 6. A larger number of data points would be required for the estimation if the structure of the true PDF were more complicated.

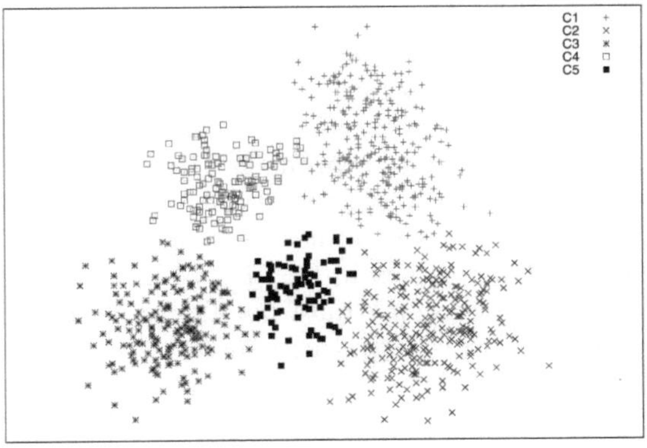


**Fig. 6.** Clustering result for $P_{1000}$

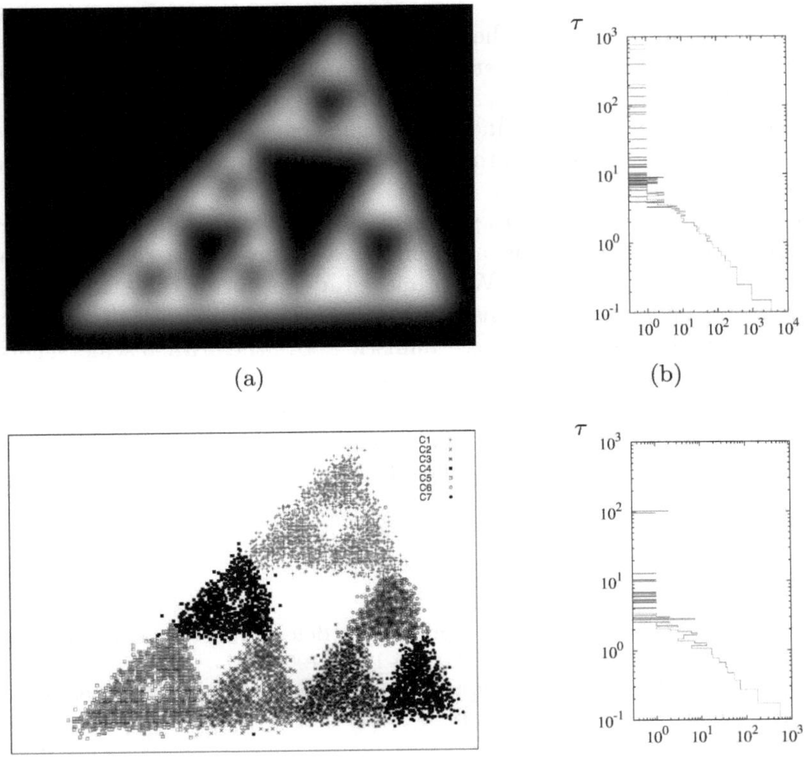


**Fig. 7.** Example of recursive validation of clusters. (a) The PDF to be estimated, (b) lifetime histogram for all $10^4$ data points, (c) clusters at $\tau = 200$, and (d) lifetime histogram for data points in the cluster C5.

### 5.2 Recursive Validation

We show, in Fig. 7, an example of the validation of detected clusters. Ten thousand data points are generated according to the PDF shown in Fig. 7(a), which is similar to a blurred version of Sierpinski's gasket. The decay in the lifetime histogram for all data points, as shown in Fig. 7(b), collapses at about $\tau = 10$. This implies the existence of valid clusters in the dataset, and every cluster with a lifetime of about 10 or more is identified as being valid. The clusters detected at $\tau = 200$, for instance, are shown in Fig. 7(c). Figure 7(d) shows the lifetime histogram for the cluster C5, whose data points are plotted as open squares in Fig. 7(c). We clearly see, in Fig.7(d), three outstanding lifetime values at about $10^2$, which indicate that the cluster C5 consists of at least three valid clusters.

## 6 Conclusions

We proposed and demonstrated a clustering method based on an estimation of the PDF of the data points in a scale space. The nonparametric PDF estimation using the Gaussian kernel can be regarded as an image in scale space obtained from the Gaussian filtering of the sum of the delta functions distributed at the data points. In the same manner as image analysis, the deep structure of the PDF can be estimated in the scale space. The geometric features of the PDF are established from coarse to fine in the scale space. The hierarchical structure of the PDF provides us with a top-down approach to identifying valid clusters of data points.

The scale-space analysis of the PDF clarifies how the statistically deterministic features of the dataset appear in higher scales even though the positions of the data points are stochastic. We proposed a concept of the critical scale for discriminating between deterministic and stochastic clusters. With a large cardinality, the number of clusters is deterministic above the critical scale. Therefore, clustering should be employed above the critical scale.

The present clustering method is effective for finding valid clusters at adequate scales. Although we only showed the experimental results for datasets in 2D space, our clustering scheme is applicable to arbitrary numbers of datasets in arbitrary dimensions.

## References

1. E. Parzen, "On the estimation of a probability density function and mode", Annals of Mathematical Statistics, vol. 33, 1065-1076, 1962.
2. S. J. Roberts, "Parametric and non-parametric unsupervised cluster analysis", *Pattern Recognition*, vol. 30, no. 2, pp. 261-272, 1997.
3. E. Nakamura and N. Kehtarnavaz, "Determining number of clusters and prototype locations via multi-scale clustering", Pattern Recognition Letters, vol. 19, no. 14, pp. 1265-1283, 1998.
4. Y. Leung, J.-S. Zhang, and Z.-B. Xu, "Clustering by scale-space filtering", *IEEE Trans. on Pattern Analysis and Machine Intelligence*, vol. 22, no. 12, pp. 1396-1410, 2000.

5. M. Minnotte and D. Scott, "The mode tree: A tool for visualization of nonparametric density features", *Journal of Computational and Graphical Statistics*, vol. 2, no. 1, pp. 51-68, 1993.
6. D. Comaniciu and P. Meer, "Mean shift: A robust approach toward feature space", *IEEE Trans. on Pattern Analysis and Machine Intelligence*, vol. 24, no. 5, pp. 603-619, 2002.
7. L. D. Griffin and M. Lillholm, "Mode estimation using pessimistic scale space tracking", *Lecture Notes in Computer Science*, vol. 2695, pp. 266-280, 2003.
8. A. P. Witkin, "Scale space filtering", *Proc. of 8th IJCAI*, pp. 1019-1022, 1983.
9. J. J. Koenderink, "The structure of images", *Biological Cybernetics*, vol. 50, pp. 363-370, 1984.
10. N.-Y. Zhao and T. Iijima, "Theory on the method of determination of view-point and field of vision during observation and measurement of figure", *IEICE Japan, Trans. D.*, vol. J68-D, pp. 508-514, 1985 (in Japanese).
11. J. Weickert, S. Ishikawa, and A. Imiya, "Linear scale-space has first been proposed in Japan", *Journal of Mathematical Imaging and Vision*, vol. 10, pp. 237-252, 1999.
12. J. Weickert, S. Ishikawa, and A. Imiya, "On the history of Gaussian scale-space axiomatics", *Gaussian Scale-Space Theory*, Computational Imaging and Vision Series, vol. 8, Kluwer Dordrecht, pp. 45-59, 1997.
13. T. Lindeberg, *Scale-Space Theory in Computer Vision*, Kluwer, Boston 1994.
14. P. Johansen, "On the classification of toppoints in scale space", *Journal of Mathematical Imaging and Vision*, vol. 4, no. 1, pp. 57-67, 1994.
15. L. D. Griffin and A. Colchester, "Superficial and deep structure in linear diffusion scale space: Isophotes, critical points and separatrices", *Image and Vision Computing*, vol. 13, no. 7, pp. 543-557, 1995.
16. L. M. J. Florack and A. Kuijper, "The topological structure of scale-space images", *Journal of Mathematical Imaging and Vision*, vol. 12, no. 1, pp. 65-79, 2000.
17. A. Kuijper, L. M. J. Florack, and M. A. Viergever, "Scale space hierarchy", *Journal of Mathematical Imaging and Vision*, vol. 18, no. 2, pp. 169-189, 2003.
18. T. Sakai and A. Imiya, "Scale-space hierarchy of singularities", *Lecture Notes in Computer Science*, vol. 3753, pp. 181-192, 2005.

# Scale Spaces on Lie Groups*

Remco Duits and Bernhard Burgeth

Eindhoven University of Technology, Dept. of Biomedical Engineering and Dept. Applied Mathematics and Computer Science, The Netherlands
R.Duits@tue.nl
Saarland University, Dept. of Mathematics and Computer Science, Germany
burgeth@mia.uni-saarland.de

**Abstract.** In the standard scale space approach one obtains a scale space representation $u : \mathbb{R}^d \rtimes \mathbb{R}^+ \to \mathbb{R}$ of an image $f \in \mathbb{L}_2(\mathbb{R}^d)$ by means of an evolution equation on the additive group $(\mathbb{R}^d, +)$. However, it is common to apply a wavelet transform (constructed via a representation $\mathcal{U}$ of a Lie-group $G$ and admissible wavelet $\psi$) to an image which provides a detailed overview of the group structure in an image. The result of such a wavelet transform provides a function $g \mapsto (\mathcal{U}_g\psi, f)_{\mathbb{L}_2(\mathbb{R}^2)}$ on a group $G$ (rather than $(\mathbb{R}^d, +)$), which we call a score. Since the wavelet transform is unitary we have stable reconstruction by its adjoint. This allows us to link operators on images to operators on scores in a robust way. To ensure $\mathcal{U}$-invariance of the corresponding operator on the image the operator on the wavelet transform must be left-invariant. Therefore we focus on left-invariant evolution equations (and their resolvents) on the Lie-group $G$ generated by a quadratic form $Q$ on left invariant vector fields. These evolution equations correspond to stochastic processes on $G$ and their solution is given by a group convolution with the corresponding Green's function, for which we present an explicit derivation in two particular image analysis applications. In this article we describe a general approach how the concept of scale space can be extended by replacing the additive group $\mathbb{R}^d$ by a Lie-group with more structure.[1]

## 1 Introduction

In the standard scale space approach one obtains a scale space representation $u : \mathbb{R}^d \rtimes \mathbb{R}^+ \to \mathbb{R}$ of a square integrable image $f : \mathbb{R}^d \to \mathbb{R}$ by means of an evolution equation on the additive group $(\mathbb{R}^d, +)$. It follows by the scale space axioms that the only allowable linear scale space representations are the so-called $\alpha$-scale space representations determined by the following linear system

$$\begin{cases} \partial_s u = -(-\Delta)^\alpha u, \qquad 0 < \alpha \leq 1 \\ u(\cdot, s) \in \mathbb{L}_2(\mathbb{R}^d) \text{ for all } s > 0 \text{ and } u(\cdot, s) \to 0 \text{ uniformly as } s \to \infty \\ u(\cdot, 0) = f \end{cases} \tag{1}$$

* The Dutch Organization for Scientific Research is gratefully acknowledged for financial support

[1] This article provides the theory and general framework we applied in [9],[5],[8].

F. Sgallari, A. Murli, and N. Paragios (Eds.): SSVM 2007, LNCS 4485, pp. 300–312, 2007.

including both Gaussian $\alpha = 1$ and Poisson scale space $\alpha = \frac{1}{2}$, [6]. By the translation invariance axiom these scale space representations are obtained via a convolution on the additive group $\mathbb{R}^d$. For example if $d = 2$, $\alpha = 1$ the evolution system (1) is a diffusion system and the scale space representation is obtained by $u(\mathbf{x}, s) = (G_s * f)(\mathbf{x})$ where $G_s(\mathbf{x}) = \frac{1}{4\pi s}e^{-\frac{\|\mathbf{x}\|^2}{4s}}$ denotes the Gaussian kernel. Its resolvent equation (obtained by Laplace transform with respect to scale) is

$$(-\Delta + \gamma I)u = f \Leftrightarrow u = (-\Delta + \gamma I)^{-1} f \tag{2}$$

the solution of which is given by $u(\mathbf{x}, \gamma) = (R_\gamma * f)(\mathbf{x})$, where the kernel

$$R_\gamma(\mathbf{x}) = -\frac{1}{2\pi\gamma^2} k_0(\gamma^{-1}\|\mathbf{x}\|), \qquad \mathbf{x} \in \mathbb{R}^2, \tag{3}$$

equals the Laplace transform of the Gaussian kernel $s \mapsto G_s(\mathbf{x})$ expressed in the well-known BesselK-function $k_0$. To this end we note that $\int_0^\infty e^{s\Delta} f \; e^{-\gamma s} \; \mathrm{d}s = -\gamma(\Delta - \gamma I)^{-1} f$. Although this explicit convolution kernel is not common in image analysis it plays an important mostly implicit role as it occurs in the minimization of a first order Sobolev norm

$$\mathcal{E}(u) = \|u\|^2_{\mathbb{H}^1(\mathbb{R}^d)} = \|u - f\|^2_{\mathbf{L}_2(\mathbb{R}^2)} + \|\nabla u\|^2_{\mathbf{L}_2(\mathbb{R}^2)}.$$

Indeed by some elementary variational calculus and partial integration one gets $\mathcal{E}'(u)v = ((\gamma I - \Delta)u - \gamma f, v)$ and $\mathcal{E}'(u)v = 0$ for all $v \in \mathbb{L}_2(\mathbb{R}^2)$ iff $u = \gamma(\gamma I - \Delta)^{-1} f$, where $R_\gamma = \gamma(\gamma I - \Delta)^{-1}\delta = \gamma \int_0^\infty e^{-\gamma s} e^{s\Delta}\delta \; \mathrm{d}s = \gamma \int_0^\infty e^{-\gamma s} G_s \; \mathrm{d}s$. The connection between a linear scale space and its resolvent equation is also relevant for stochastic interpretation. Consider $f$ as a probability density distribution of photons. Then its scale space representation evaluated at a point $(\mathbf{x}, s)$ in scale space, $u(\mathbf{x}, s)$, corresponds to the probability density of finding a random walker in a Wiener process at position $\mathbf{x}$ at time/scale $s > 0$. In such a process traveling time is negatively exponentially distributed.Now the probability density of finding a random walker at position $\mathbf{b}$ given the initial distribution $f$ equals

$$p(\mathbf{b}) = \int_0^\infty p(\mathbf{b}|T = t)p(T = t) \; \mathrm{d}t = \gamma \int_0^\infty e^{-\gamma t}(G_t * f) = \gamma(\gamma I - \Delta)^{-1} f.$$

In the remainder of this article we are going to repeat the above results for other Lie-groups than $(\mathbb{R}^d, +)$. Just like ordinary convolutions on $\mathbb{R}^d$ are the only translation invariant kernel operators, it is easy to show that the only left invariant operators on a Lie-group $G$ are $G$-convolutions, which are given by

$$(K *_G f)(g) = \int_G K(h^{-1}g) f(h) \; \mathrm{d}\mu_G(g), \tag{4}$$

where $\mu_G$ is the left invariant haar measure of the group $G$. However, if the Lie group $G$ is not commutative it is challenging to compute the analogues of the Gaussian and corresponding resolvent kernel.

**Definition 1.** *Let $H$ be a Hilbert space and let $G$ be a group with unit element $e$. Let $\mathcal{B}(H)$ denote the space of bounded operators on $H$. Then a mapping $\mathcal{R} : G \to \mathcal{B}(H)$ given by $g \mapsto \mathcal{R}_g$, where $\mathcal{R}_g$ is bounded for all $g \in G$, is a representation if $\mathcal{R}_g \mathcal{R}_h = \mathcal{R}_{gh}$ for all $g, h \in G$, with $\mathcal{R}_e = I$ . If for every $g \in G$ the operator $\mathcal{R}_g$ is unitary (so $\|\mathcal{R}_g f\| = \|f\|$ for all $f \in H$) the representation is called unitary.*

**Definition 2.** *An operator on $\mathbb{L}_2(G)$ is left invariant if it commutes with the left regular representation given by $(\mathcal{L}_g U)(h) = U(g^{-1}h)$, $\quad U \in \mathbb{L}_2(G), h, g \in G$. So an operator $\Phi$ on $\mathbb{L}_2(G)$ is left invariant if $\mathcal{L}_g \circ \Phi = \Phi \circ \mathcal{L}_g$ for all $g \in G$.*

The motivation for our generalization of scale space theory to arbitrary Lie-groups comes from wavelet-theory applications where one applies a wavelet transform $\mathcal{W}_\psi : \mathbb{L}_2(\mathbb{R}^d) \to \mathbb{L}_2(G)$ to the original image $f \in \mathbb{L}_2(\mathbb{R}^d)$ to provide more insight in the group structure of an image. Such a transform is usually given by

$$(\mathcal{W}_\psi f)(g) = (\mathcal{U}_g \psi, f)_{\mathbb{L}_2(\mathbb{R}^d)} \tag{5}$$

where the wavelet $\psi$ is admissible[2] and a unitary representation $\mathcal{U} : G \to \mathcal{B}(\mathbb{L}_2(\mathbb{R}^d))$ of a certain Lie-group $G$. Provided that $\psi$ is admissible such wavelet transform is an isometry from $\mathbb{L}_2(\mathbb{R}^d)$ into $\mathbb{L}_2(G)$ and thereby we have perfectly stable reconstruction by the adjoint wavelet transformation; $f = \mathcal{W}_\psi^*(\mathcal{W}_\psi f)$, allowing us to link scale operators on images and their scores $U_f := \mathcal{W}_\psi f$ in a stable manner. We must consider left invariant operators on $U_f$, see figure 1.

## 2 Scale Spaces on Lie-Groups: The General Recipe

In this section we provide a general recipe for scale spaces on Lie-groups. In contrast to other work on scale spaces via Lie-groups we consider the left-invariant vector fields on the Lie-group itself, rather than infinitesimal generators[3] of the representation $\mathcal{U}$ on $\mathbb{L}_2(\mathbb{R}^d)$, [15]. The main advantage over the infinitesimal generators is that these vector fields are defined on the group manifold $G$ rather than $\mathbb{R}^d$. Moreover, they give rise to left-invariant evolution equations on $\mathbb{L}_2(G)$.

First we compute the left-invariant vector-fields on a Lie-group. By definition those are vector fields on $G$, with unit element $e$, such that

$$X_g f = X_e(f \circ L_g), \tag{6}$$

for all $f \in C^\infty(\Omega_g)$ defined on some open neighborhood $\Omega_g$ of $g \in G$, where we note that $L_g h = gh$ denotes the left multiplication on the Lie-group. Note that by (6) the restriction $X_g$ of a left invariant vector field $X$ is always connected by its restriction to the unity element $X_e$. Consequently such left invariant vector

[2] this is a condition on $\psi$ to ensure that $\|\mathcal{W}_\psi f\|_{\mathbb{L}_2(G)} = \|f\|_{\mathbb{L}_2(\mathbb{R}^d)}$ for all $f \in \mathbb{L}_2(\mathbb{R}^d)$.

[3] The left invariant vector fields are obtained via the Lie-algebra $T_e(G) = \{A_1, \ldots A_n\}$ by means of the derivative of the right-regular representation on $\mathbb{L}_2(G)$ by $\{\mathrm{d}\mathcal{R}(A_i)\}_{i=1}^n$, whereas the infinitesimal generators are obtained by the derivative of some representation $\mathcal{U}$ on $\mathbb{L}_2(\mathbb{R}^d)$ by $\{\mathrm{d}\mathcal{U}(A_i)\}_{i=1}^n$.

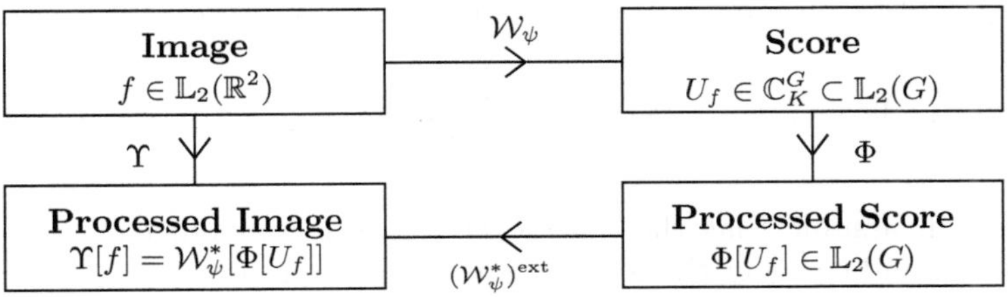


**Fig. 1.** The complete scheme; for admissible vectors $\psi$ the linear map $\mathcal{W}_\psi$ is unitary from $\mathbb{L}_2(\mathbb{R}^2)$ onto a closed subspace $V$ of $\mathbb{L}_2(G)$. So we can uniquely link a transformation $\Phi : V \to V$ on the wavelet domain to a transformation in the image domain $\Upsilon = (\mathcal{W}_\psi^*)^{ext} \circ \Phi \circ \mathcal{W}_\psi \in \mathcal{B}(\mathbb{L}_2(\mathbb{R}^d))$, where $(\mathcal{W}_\psi^*)^{ext}$ is the extension of the adjoint to $\mathbb{L}_2(G)$ given by $(\mathcal{W}_\psi^*)^{ext} U = \int_G \mathcal{U}_g \psi \, U(g) \, \mathrm{d}\mu_G(g)$, $U \in \mathbb{L}_2(G)$. It is easily verified that $\mathcal{W}_\psi \circ \mathcal{U}_g = \mathcal{L}_g \circ \mathcal{W}_\psi$ for all $g \in G$. As a result the net operator on the image domain $\Upsilon$ is invariant under $\mathcal{U}$ (which is highly desirable) if and only if the operator in the wavelet domain is left invariant, i.e. $\Upsilon \circ \mathcal{U}_g = \mathcal{U}_g \circ \Upsilon$ for all $g \in G$ if and only if $\Phi \circ \mathcal{L}_g = \mathcal{L}_g \circ \Phi$ for all $g \in G$. For more details see [4]Thm. 21 p.153. In our applications [8],[5], [4] we usually take $\Phi$ as a concatenation of *non-linear* invertible grey-value transformations and linear left invariant (anisotropic) scale space operators, for example $\Phi(U_f) = \gamma^{2/p}((Q(\underline{A}) - \gamma I)^{-1}(U_f)^p \; (Q(\underline{A}) - \gamma I)^{-1}(U_f)^p)^{1/p}$, for some sign-preserving power with exponent $p > 0$. A nice alternative, however, are non-linear adaptive scale spaces on Lie-groups as explored for the special case $G = \mathbb{R}^2 \rtimes \mathbb{T}$ in [9].

fields are isomorphic to the tangent space $T_e(G)$ at the unity element $e \in G$, also known as the Lie-Algebra of $G$. The isomorphism between $T_e(G)$ and the space of left invariant vector fields $\mathcal{L}(G)$ on $G$ (considered as differential operators) is

$$T_e(G) \ni A \leftrightarrow \mathcal{A} \in \mathcal{L}(G) \quad \Leftrightarrow \quad \mathcal{A}_g \phi = A(h \mapsto \phi(gh)), \text{ for all } \phi \in C^\infty(\Omega_g). \quad (7)$$

The Lie-product on $T_e(G)$ is given by $[A, B] = \lim_{t \downarrow 0} \frac{a(t)b(t)(a(t))^{-1}(b(t))^{-1} - e}{t^2}$, where $t \mapsto a(t)$ resp. $t \mapsto b(t)$ are *any* smooth curves in $G$ with $a(0) = b(0) = e$ and $a'(0) = A$ and $b'(0) = B$, whereas the Lie-product on $\mathcal{L}(G)$ is given by $[\mathcal{A}, \mathcal{B}] = \mathcal{A}\mathcal{B} - \mathcal{B}\mathcal{A}$. The mapping (7) is an isomorphism between $T_e(G)$ and $\mathcal{L}(G)$, so $A \leftrightarrow \mathcal{A}$ and $B \leftrightarrow \mathcal{B}$ imply $[A, B] \leftrightarrow [\mathcal{A}, \mathcal{B}]$.

Consider a Lie-group $G$ of finite dimension, with Lie-algebra $T_e(G)$. Let $\{A_1, \ldots, A_n\}$ be a basis within this Lie-algebra. Then we would like to construct the corresponding left invariant vector fields $\{\mathcal{A}_1, \ldots, \mathcal{A}_n\}$ in a direct way. This is done by computing the derivative $\mathrm{d}\mathcal{R}$ of the right-regular representation $\mathcal{R} : G \mapsto \mathcal{B}(\mathbb{L}_2(G))$. The right regular representation $\mathcal{R} : G \to \mathcal{B}(\mathbb{L}_2(G))$ is given by $(\mathcal{R}_g \Phi)(h) = \Phi(hg)$, for all $\Phi \in \mathbb{L}_2(G)$ and almost every $h \in G$. It is left-invariant and its derivative $\mathrm{d}\mathcal{R}$, which maps $T_e(G)$ onto $\mathcal{L}(G)$, is given by

$$(\mathrm{d}\mathcal{R}(A)\Phi)(g) = \lim_{t \to 0} \frac{(\mathcal{R}_{\exp(tA)}\Phi)(g) - \Phi(g)}{t}, \; A \in T_e(G), \Phi \in \mathbb{L}_2(G), g \in G. \quad (8)$$

So a basis for $\mathcal{L}(G)$ is given by

$$\{\mathcal{A}_1, \mathcal{A}_2, \ldots, \mathcal{A}_n\} := \{\mathrm{d}\mathcal{R}(A_1), \mathrm{d}\mathcal{R}(A_2), \ldots, \mathrm{d}\mathcal{R}(A_n)\}. \quad (9)$$

Now let $Q = Q^{\mathbf{D},\mathbf{a}}$ be some bilinear/quadratic form on $\mathcal{L}(G)$, i.e.

$$Q^{\mathbf{D},\mathbf{a}}(\mathcal{A}_1, \mathcal{A}_2, \ldots, \mathcal{A}_n) = \sum_{i=1}^{n} a_i \mathcal{A}_i + \sum_{j=1}^{n} D_{ij} \mathcal{A}_i \mathcal{A}_j, \qquad a_i, D_{ij} \in \mathbb{R}, \tag{10}$$

where we will assume that the matrix $D = [D_{ij}]$ is symmetric and positive semi-definite, and consider the following evolution equations

$$\boxed{\begin{cases} \partial_s W = Q(\mathcal{A}_1, \mathcal{A}_2, \ldots, \mathcal{A}_n)\, W\ , \\ \lim\limits_{s \downarrow 0} W(\cdot, s) = U_f(\cdot)\ , \end{cases}} \tag{11}$$

the solutions of which we call the $G$-scale space representation of initial condition $U_f$ (which is the score obtained from image $f$ by $\mathcal{W}_\psi[f]$). The corresponding $G$-Tikhonov regularization due to minimization of
$\mathcal{E}(u) = \|u\|^2_{\mathbb{H}^1(G)} = \|u - f\|^2_{\mathbf{L}_2(G)} + \sum_{i=1}^{d} D_{ii} \|\mathcal{A}_i u\|^2_{\mathbf{L}_2(G)}$ is again obtained by Laplace transform with respect to scale yielding the following resolvent equations

$$(-Q(\mathcal{A}_1, \mathcal{A}_2, \ldots, \mathcal{A}_n) + \gamma I)\, P_\gamma = \gamma\, U_f, \tag{12}$$

with $P_\gamma = \gamma \mathcal{L}(s \mapsto W(\cdot, s))(\gamma)$ and where traveling timeof a random walker in $G$ is assumed to be negatively exponentially distributed with $s \sim \mathrm{NE}(\gamma)$.

We distinguish between two types of scale space representations, the cases where $Q$ is non-degenerate and the cases where $Q$ is degenerate. If $Q$ is non-degenerate the principal directions in the diffusion span the whole tangent space in which case it follows that (11) gives rise to a strongly continuous semi-group, generated by a hypo-elliptic operator $A$, [12], such that the *left-invariant* operators $U_f \mapsto W(\cdot, s)$ and $U_f \mapsto P_\gamma$ are bounded operators on $\mathbb{L}_\infty(G)$ for all $\gamma, s > 0$ and by means of the Dunford-Pettis theorem [2] if follows that the solutions of (11) and (12) are given by $G$-convolutions with the corresponding smooth Green's functions

$$\begin{array}{ll} W(g, s) = (K_s *_G U_f)(g), & K_s \in C^\infty(G), s > 0 \\ P_\gamma(g) = (R_\gamma *_G U_f)(g), & R_\gamma \in C^\infty(G \backslash \{e\}), \end{array} \tag{13}$$

where $K_s$ and $R_\gamma$ are connected by Laplace transform $R_\gamma = \gamma \mathcal{L}(s \mapsto K_s)(\gamma)$.

The most interesting cases, however, arise if $Q$ is degenerate. If $Q$ is degenerate it follows by the general result by [11] the solutions of (11) and (13) are still given by group convolutions (13). The question though is whether the convolution kernels are to be considered in distributional sense only or if they are smooth functions. If $D = 0$ the convolution kernels are highly singular and concentrated at the exponential curves within the Lie-group. If $D \neq 0$ is degenerate diffusion takes place only in certain direction(s) and we can write $Q(\underline{\mathcal{A}}) = \sum_{j=1}^{d} \tilde{\mathcal{A}}_j^2 + \tilde{\mathcal{A}}_0,\ \ d = \mathrm{rank}(Q)$ where $\mathcal{A}_0$ is the convection part of $Q(\underline{\mathcal{A}})$ and where $\tilde{\mathcal{A}}_j$ are $d$-independent directions along which $Q$ is not degenerate. Now it is the question whether the non-commutativity of the vector fields results in a smoothing along the other directions. By employing the results of Hörmander[12] and Hebisch[11] we obtain the following necessary and sufficient conditions for smooth scale space representations of the type (13): Among the vector fields

$$\tilde{\mathcal{A}}_{j_1}, [\tilde{\mathcal{A}}_{j_1}, \tilde{\mathcal{A}}_{j_2}], \ldots [\tilde{\mathcal{A}}_{j_1}, [\tilde{\mathcal{A}}_{j_2}, [\tilde{\mathcal{A}}_{j_3}, \ldots, \tilde{\mathcal{A}}_{j_k}]]] \ldots, j_i = 0, 1, \ldots d \qquad (14)$$

there exist $n$ which are linearly independent.

## 3 Examples

### Spatial-Frequency Enhancement via left invariant scale spaces on Gabor Transforms: $G = H_3$, $Q(\underline{\mathcal{A}}) = D_{11}(\mathcal{A}_1)^2 + D_{22}(\mathcal{A}_2)^2$

Consider the Heisenberg group $H_{2d+1} = \mathbb{C}^d \rtimes \mathbb{R}$ with group product:

$$g\,g' = (\mathbf{z}, t)(\mathbf{z}', t') = (\mathbf{z} + \mathbf{z}', t + t' + 2\,\mathrm{Im}\{\sum_{j=1}^{d} z_j \overline{z_j'}\}), \quad \text{where } z_j = x_j + i\,\omega_j \in \mathbb{C}$$

and consider its representations $U^{\lambda}_{(\mathbf{x},\boldsymbol{\omega},t)}$ on $\mathbb{L}_2(\mathbb{R}^d)$:

$$(\mathcal{U}^{\lambda}_{(\mathbf{z},t)}\psi)(\boldsymbol{\xi}) = e^{i\lambda((\boldsymbol{\xi},\boldsymbol{\omega}) + \frac{t}{4} - \frac{1}{2}(\mathbf{x},\boldsymbol{\omega}))}\psi(\boldsymbol{\xi} - \mathbf{x}), \qquad \psi \in \mathbb{L}_2(\mathbb{R}^d), \mathbf{x}, \boldsymbol{\omega} \in \mathbb{R}^d, \lambda \in \mathbb{R}.$$

The corresponding wavelet transform is the windowed Fourier/Gabor transform:

$$(\mathcal{W}_{\psi}[f])(g) = (\mathcal{U}_g\psi, f)_{\mathbb{L}_2(\mathbb{R}^d)} = e^{-i\lambda(\frac{t}{4} - \frac{(\mathbf{x},\boldsymbol{\omega})}{2})} \int_{\mathbb{R}^d} \overline{\psi(\boldsymbol{\xi} - \mathbf{x})} f(\boldsymbol{\xi}) e^{-i\lambda(\boldsymbol{\xi},\boldsymbol{\omega})}\, \mathrm{d}\boldsymbol{\xi}$$

This is useful in practice as it provides a score of localized frequencies in signal $f$. Denote the phase subgroup of $H_{2d+1}$ by $\Theta = \{(\mathbf{0}, \mathbf{0}, t) \mid t \in \mathbb{R}\}$.

Now $\mathcal{U}^{\lambda}$ is a unitary, irreducible and square integrable representation with respect to $H_{2d+1}/\Theta$ with invariant measure $\mathrm{d}\mu_{H_{2d+1}/\Theta}(g) = \mathrm{d}\boldsymbol{\omega}\mathrm{d}\mathbf{x}$. Therefore by the theory of vector coherent states, [1], we employ that [4]

$$\int_{\mathbb{R}^d}\int_{\mathbb{R}^d} |\mathcal{W}_{\psi}[f](\mathbf{x}, \boldsymbol{\omega}, 0)|^2 \,\mathrm{d}\mathbf{x}\,\mathrm{d}\boldsymbol{\omega} = C_{\psi} \int_{\mathbb{R}^d} |f(\mathbf{x})|^2 \,\mathrm{d}\mathbf{x},$$

for all $f \in \mathbb{L}_2(\mathbb{R}^d)$ and for all $\psi \in \mathbb{L}_2(\mathbb{R}^d)$. As a result we obtain a perfectly stable reconstruction by means of the adjoint wavelet transform

$$\begin{aligned} f \quad &= \tfrac{1}{C_{\psi}} \mathcal{W}_{\psi}^{*}\mathcal{W}_{\psi} f = \tfrac{1}{C_{\psi}} \int_{H_{2d+1}/\Theta} \mathcal{W}_{\psi}[f](g)\,\mathcal{U}_g\psi \,\mathrm{d}\mu_{H_{2d+1}/\Theta}(g), \text{ i.e.} \\ f(\boldsymbol{\xi}) &= \tfrac{1}{C_{\psi}} \int_{\mathbb{R}^d}\int_{\mathbb{R}^d} (\mathcal{W}_{\psi} f)(\mathbf{x}, \boldsymbol{\omega}, t)\, e^{i\lambda[(\boldsymbol{\xi},\boldsymbol{\omega}) + (t/4) - (1/2)(\mathbf{x},\boldsymbol{\omega})]}\psi(\boldsymbol{\xi} - \mathbf{x})\,\mathrm{d}\mathbf{x}\,\mathrm{d}\boldsymbol{\omega}, \end{aligned}$$

for almost every $\boldsymbol{\xi} \in \mathbb{R}^d$ and all $f \in \mathbb{L}_2(\mathbb{R}^d)$.

Now that a stable connection between an image $f$ and its Gabor-transform $\mathcal{W}_{\psi}[f]$ is set we can think of left invariant scale spaces on the space of Gabor transforms which is embedded in $\mathbb{L}_2(H_{2d+1}/\Theta)$. Following the general recipe

[4] Note that $C_{\psi} = \int_{\mathbb{R}^d}\int_{\mathbb{R}^d} |(\mathcal{U}^{\lambda}_{(\mathbf{x}+i\boldsymbol{\omega},0)}\psi, \psi)|^2 \mathrm{d}\mathbf{x}\mathrm{d}\boldsymbol{\omega} = \frac{(2\pi)^d}{\lambda} \|\psi\|^4_{\mathbb{L}_2(\mathbb{R}^d)} < \infty$ for all $\psi \in \mathbb{L}_2(\mathbb{R}^2)$.

as described in Section 2 we compute the left invariant vector fields from the $2d+1$-dimensional Lie-algebra $T_e(G)$ spanned by

$$T_e(G)=\langle\{(\mathbf{e}_i,\mathbf{0},0),(\mathbf{0},\mathbf{e}_j,0),(\mathbf{0},\mathbf{0},1)\}_{i,j=1,\ldots d}\rangle =: \langle\{A_1,\ldots,A_d,A_{d+1},\ldots,A_{2d},A_{2d+1}\}\rangle$$

by means of $\mathcal{A}_i\psi = \mathrm{d}\mathcal{R}(A_i)\psi = \lim_{t\to 0}\frac{1}{t}\left(R(e^{tA_i})-I\right)\psi$. A straightforward calculation yields the following basis for $\mathcal{L}(H_{2d+1})$:

$$\mathcal{A}_i = \partial_{x_i}+2\omega_i\partial_t,\ \mathcal{A}_{d+i}=\partial_{\omega_i}-2x_i\partial_t,\ \text{for } i=1,\ldots,d,\ \text{and } \mathcal{A}_{2d+1}=\partial_t,$$

the commutators of which are given by

$$[\mathcal{A}_i,\mathcal{A}_j] = -4\,\delta_{j,i+d}\,\mathcal{A}_{2d+1},\ i,j=1,\ldots,2d,\ [\mathcal{A}_{2d+1},\mathcal{A}_j]=0,\ j=1,\ldots,2d+1. \tag{15}$$

Here we only consider (11) for the case where the quadratic form equals

$$Q(\underline{A}) = \sum_{j=1}^{d} D_{jj}\,(\mathcal{A}_j)^2 + D_{d+j,d+j}\,(\mathcal{A}_{d+j})^2. \tag{16}$$

Condition (14) is satisfied. In this case the scale space solutions $K_s^{\mathbf{D}}, P_\gamma^{\mathbf{D}}$ (13) initial condition $W(\cdot,\cdot,\cdot,0)=U_f\in\mathbb{L}_2(H_{2d+1})$ of (11) are group convolutions (4) with the corresponding Green's functions $K_s^{\mathbf{D}}$ and $R_\gamma$. For $d=1$ we get:

$$\begin{array}{l} W^{\mathbf{D}}(x,\omega,t,s) = (K_s^{\mathbf{D}} *_{H_3} U_f)(x,\omega,t) \\ = \int\limits_{\mathbb{R}}\int\limits_{\mathbb{R}}\int\limits_{\mathbb{R}^+} K_s^{\mathbf{D}}(x-x',\omega-\omega',t-t'-2(x\omega'-x'\omega))\,U_f(x',\omega',t')\,\mathrm{d}t\mathrm{d}\omega'\mathrm{d}x' \\ P_\gamma^{\mathbf{D}}(x,\omega,t) = (R_\gamma^{\mathbf{D}} *_{H_3} U_f)(x,\omega,t) \\ = \int\limits_{\mathbb{R}}\int\limits_{\mathbb{R}}\int\limits_{\mathbb{R}^+} R_\gamma^{\mathbf{D}}(x-x',\omega-\omega',t-t'-2(x\omega'-x'\omega))\,U_f(x',\omega',t')\,\mathrm{d}t'\mathrm{d}\omega'\mathrm{d}x'. \end{array} \tag{17}$$

Next we derive the Green's functions $K_s^{\mathbf{D}}$ and $R_\gamma$. First we note that in case $D_{jj}=D_{d+j,d+j}=\frac{1}{2}$ operator (16) coincides with Kohn's Laplacian, the fundamental solution of which is well-known [10]. As there exist several contradicting formulas for this Green's function, we summarize (for $d=1$) the correct derivation by Gaveau [10] which, together with the work of Lévy [13], provides important insight in the non-commutativity and the underlying stochastic process.

For $d=1$ the kernel $K_s^{\mathbf{D}}$ can be obtained by the Kohn Green's function $K_s := K_s^{D_{11}=\frac{1}{2},D_{22}=\frac{1}{2}}$ by means of a simple rescaling

$$\begin{array}{l} K_s^{\mathbf{D}}(x,y,t) = K_s(\sqrt{\frac{2}{D_{11}}}\,x,\sqrt{\frac{2}{D_{22}}}\,y,\frac{2}{\sqrt{D_{11}D_{22}}}\,t) \\ K_s(x,\omega,t) = s^{-2}K_1\left(\frac{x}{\sqrt{s}},\frac{\omega}{\sqrt{s}},\frac{t}{s}\right). \end{array} \tag{18}$$

Next we rewrite Kohn's $d$-dimensional Laplacian $\Delta_K$ in its fundamental form:

$$\begin{array}{rl} \Delta_K = & \sum\limits_{i=1}^{d}(\partial_{x_i})^2+(\partial_{\omega_i})^2+4\omega_i\,\partial_{x_i}\partial_t-4x_i\,\partial_{\omega_i}\partial_t+4|z_i|^2(\partial_t)^2 \\ = & \sum\limits_{i=1}^{2d+1}\sum\limits_{j=1}^{2d+1}\mathcal{A}_i g^{ij}\mathcal{A}_j = \sum\limits_{i=1}^{2d+1}\sum\limits_{j=1}^{2d+1}\mathcal{A}_i(\sigma^T\sigma)^{ij}\mathcal{A}_j = \sum\limits_{k=1}^{2d}\left(\sum\limits_{j=1}^{2d+1}\sigma^{kj}\mathcal{A}_j\right)^2, \end{array} \tag{19}$$

with $G = \sigma^T \sigma \in \mathbb{R}^{(2d+1)\times(2d+1)}$, $G = [g^{ij}]$, $\sigma = [\sigma^{kj}]$, $\sigma \in \mathbb{R}^{2d\times(2d+1)}$ given by

$$\sigma^{ij} = \begin{cases} \delta^{ij} & \text{if } i \leq 2d, j \leq 2d, \\ 2w_p & \text{if } i = 2p-1 \text{ and } j = 2d+1,\ p = 1,\dots,d \\ -2x_p & \text{if } i = 2p \text{ and } j = 2d+1,\ p = 1,\dots,d \end{cases}$$

where we recall that $z_j = x_j + i\,\omega_j$. By (19) the diffusion increments satisfy

$$(\mathrm{dx}^1,\dots,\mathrm{dx}^{\mathrm{d}},\mathrm{d}\omega^1,\dots,\mathrm{d}\omega^{\mathrm{d}},\mathrm{dt}) = (\mathrm{dx}^1,\dots,\mathrm{dx}^{\mathrm{d}},\mathrm{d}\omega^1,\dots,\mathrm{d}\omega^{\mathrm{d}})\sigma,$$

so that $\mathrm{d}t = 2\sum_{j=1}^{d} \omega_j \mathrm{d}x_j - x_j \mathrm{d}\omega_j$. So in case $G = H_{2d+1}$ the diffusion system (11) is the stochastic differential equation of the following stochastic process

$$\begin{cases} \mathbf{Z}(s) = \mathbf{X}(s) + i\,\mathbf{W}(s) = \mathbf{Z}_0 + \boldsymbol{\xi}\sqrt{s},\ \boldsymbol{\xi} = (\xi_1,\dots,\xi_d), \xi_j \sim \mathcal{N}(0,1) \\ T(s) = 2\sum\limits_{j=1}^{d}\int\limits_0^s W_j \mathrm{d}X_j - X_j \mathrm{d}W_j,\ s > 0 \end{cases} \tag{20}$$

so the random variable $\mathbf{Z} = (Z_1,\dots,Z_d)$ consists of $d$-independent Brownian motions in the complex plane The random variable $T(s)$ measures the deviation from a sample path with respect to a straight path $\mathbf{Z}(s) = \mathbf{Z}_0 + s(\mathbf{Z}(s) - \mathbf{Z}_0)$ by means of the *stochastic* integral $T(s) = 2\sum\limits_{j=1}^{d}\int\limits_0^s W_j \mathrm{d}X_j - X_j \mathrm{d}W_j$.

To this end we note that for[5] $s \mapsto (x(s),\omega(s)) \in C^\infty(\mathbb{R}^+,\mathbb{R}^2)$ such that the straight-line from $X_0$ to $X(s)$ followed by the inverse path encloses an oriented surface $\Omega \in \mathbb{R}^2$, we have by Stokes' theorem that

$$2\mu(\Omega) = -\int_0^s (-X'(t)W(t) + X(t)W'(t))\,\mathrm{d}t + 0 = \int_0^s W\mathrm{d}X - X\mathrm{d}W.$$

Now we compute the Fourier transform $\mathcal{F}_3 K_1$ of $K_1$ (with respect to $(x,\omega,t)$):

$$\begin{aligned}(\mathcal{F}_3 K_1)(\xi,\eta,\tau) &= \frac{1}{(2\pi)^{\frac{3}{2}}}\int_{\mathbb{R}^3} e^{-i(\xi x + \eta\omega + \tau t)} K_1(x,\omega,t)\ \mathrm{d}x\mathrm{d}\omega\mathrm{d}t \\ &= \frac{1}{\sqrt{2\pi}}\mathcal{F}_2\left((x,\omega) \mapsto e^{-\frac{\omega^2+x^2}{2}} E(e^{-i\tau T(1)} \mid X(1){=}x, W(1){=}\omega)\right),\end{aligned}$$

where $E(e^{-i\tau T(1)} \mid X(1) = x, W(1) = \omega)$ expresses the expectation of random variable $T(1)$, recall (20), given the fact that $X(1) = x$, $W(1) = \omega$. Now by the result of [13](formula 1.3.4) we have for $d = 1$

$$\begin{aligned}(\mathcal{F}_3 K_1)(\xi,\eta,\tau) &= \frac{1}{\sqrt{2\pi}}\mathcal{F}_2\left((x,\omega) \mapsto e^{-\frac{x^2+\omega^2}{2}} e^{+\frac{x^2+\omega^2}{2}} \frac{2\tau}{\sinh(2\tau)} e^{-|z|^2\tau\coth(2\tau)}\right)(\xi,\eta) \\ &= \frac{1}{(2\pi)^{\frac{3}{2}}}\frac{1}{\cosh(2\tau)} e^{-\frac{\xi^2+\eta^2}{2}\frac{\tanh(2\tau)}{2\tau}}.\end{aligned}$$

Now since $K_s > 0$ we have by (18) that $\|K_s\|_{\mathbb{L}_1(H_3)} = (2\pi)^{\frac{3}{2}} \lim\limits_{\tau\to 0}(\mathcal{F}_3 K_1)(\mathbf{0},\tau) = 1$. Application of inverse Fourier transform gives

$$K_1(x,\omega,t) = \frac{1}{(2\pi)^2}\int_{\mathbb{R}} \frac{2\tau}{\sinh(2\tau)} \cos(\tau t) e^{-\frac{|z|^2\tau}{\tanh(2\tau)}}\ \mathrm{d}\tau.$$

[5] A Brownian motion is a.e. not differentiable in the classical sense, nor does the integral in (20) make sense in classical integration theory.

Finally identities in (18) provide the general scale space kernel on $H_3$:

$$K_s^{\mathbf{D}}(x,\omega,t) = \frac{1}{(2\pi s)^2}\int_{\mathbb{R}} \frac{2\tau}{\sinh(2\tau)}\cos\left(\frac{2\,\tau\,t}{s\sqrt{D_{11}D_{22}}}\right)e^{-\frac{\left(\frac{x^2}{D_{11}s}+\frac{\omega^2}{D_{22}s}\right)\tau}{2\tanh(2\tau)}}\,d\tau, \quad (21)$$

which can be approximated with a one dimensional discrete cosine transform.

The corresponding resolvent kernel $R_\gamma(x,\omega,\tau)=\gamma\int_{\mathbb{R}^+} K_s^{\mathbf{D}}(x,\omega,\tau)\,e^{-\gamma s}\,\mathrm{d}s$ which is again a probability kernel, i.e. $R_\gamma > 0$ and $\|R_\gamma\|_{\mathbb{L}_1(H_3)} = 1$, is given by

$$R_\gamma(x,\omega,z) = \frac{2\gamma\sqrt{\gamma}}{\pi^2}\int_0^\infty \frac{\tau}{\sinh 2\tau}\,\mathrm{Re}\left(\frac{k_1\left(2\sqrt{\gamma}\sqrt{\frac{2\tau}{\tanh 2\tau}\left(\frac{x^2}{D_{11}}+\frac{\omega^2}{D_{22}}\right)-\frac{2i\,\tau t}{\sqrt{D_{11}D_{22}}}}\right)}{\sqrt{\frac{2\tau}{\tanh 2\tau}\left(\frac{x^2}{D_{11}}+\frac{\omega^2}{D_{22}}\right)-\frac{2i\,\tau t}{\sqrt{D_{11}D_{22}}}}}\right)\mathrm{d}\tau \quad (22)$$

with $k_1$ the 1st order BesselK-function.Formulae (22) and (21) are nasty for computation. The resolvent kernel with infinite lifetime is much simpler:

$$\lim_{\gamma\to 0}\gamma^{-1}R_\gamma(x,\omega,t) = \int_{\mathbb{R}^+} K_s^{\mathbf{D}}(x,\omega,t)\,\mathrm{d}s = \frac{1}{2\pi}\frac{1}{\sqrt{\left(\frac{x^2}{D_{11}}+\frac{\omega^2}{D_{22}}\right)^2+\frac{t^2}{D_{11}D_{22}}}}, \quad (23)$$

which follows by taking the limit $\gamma \to 0$ in (22) and substitution $v = \cosh(2\tau)$. It provides us the following left invariant metric $d^{\mathbf{D}} : H_3 \times H_3 \to \mathbb{R}^+$ given by

$$d^{\mathbf{D}}(g,h) = \sqrt{\left(D_{11}^{-1}(x-x')^2 + D_{22}^{-1}(\omega-\omega')^2\right)^2 + (D_{11}D_{22})^{-1}(t-t'-2(x\omega'-x'\omega)^2)^2},$$

with $g = (x,\omega,t), h = (x',\omega',t')$. Since (22) is not suitable for practical purposes if $\gamma < \infty$ and since $\tilde{R}^{\mathbf{D}}_{\gamma=\infty}$ decays slowly at infinity we propose to use

$$\tilde{R}^{\mathbf{D}}_\gamma(x,\omega,t) = \frac{4\gamma}{\pi^{\frac{3}{2}}D_{11}D_{22}}\frac{\exp\left\{-\left(\gamma\,d^{\mathbf{D}}(\,(x,\omega,t),e)\,\right)^2\right\}}{d^{\mathbf{D}}((x,\omega,t),e)}, \quad e = (0,0,0), \quad (24)$$

instead. Note that $\lim_{\gamma\to 0}\gamma\tilde{R}_\gamma \equiv \lim_{\gamma\to 0}\gamma R_\gamma$, $\tilde{R}_\gamma > 0$, $\|\tilde{R}_\gamma\|_{\mathbb{L}_2(H_3)} = 1$ for all $\gamma > 0$.

**Contour completion and enhancement via left invariant scale spaces on Orientation Scores: $G = \mathrm{SE}(2)$.**

Consider the Euclidean motion group $G = \mathrm{SE}(2) = \mathbb{R}^2 \rtimes \mathbb{T}$ with group product

$$(g,g') = (R_\theta\mathbf{b}' + \mathbf{b}, e^{i(\theta+\theta')}), \quad g = (\mathbf{b}, e^{i\theta}), g' = (\mathbf{b}, e^{i\theta'}) \in G = \mathbb{R}^2 \rtimes \mathbb{T},$$

which is (isomorphic to) the group of rotations and translations in $\mathbb{R}^2$. Then the tangent space at $e=(0,0,e^{i0})$ is spanned by $\{\mathbf{e}_x,\mathbf{e}_y,\mathbf{e}_\theta\}=\{(1,0,0),(0,1,0),(0,0,1)\}$ and again by the general recipe (9) we get the following basis for $\mathcal{L}(\mathrm{SE}(2))$:

$$\{\mathcal{A}_1,\mathcal{A}_2,\mathcal{A}_3\} = \{\partial_\theta,\partial_\xi,\partial_\eta\} = \{\partial_\theta, \cos\theta\,\partial_x + \sin\theta\,\partial_y, -\sin\theta\,\partial_x + \cos\theta\,\partial_y\}, \quad (25)$$

with $\xi = x\,\cos\theta + y\,\sin\theta$, $\eta = -x\,\sin\theta + y\,\cos\theta$.

The wavelet transform that maps an image $f \in \mathbb{L}_2(\mathbb{R}^2)$ to a so-called orientation score, [4], [5],[8] is given by

$$(\mathcal{W}_\psi f)(g) = (\mathcal{U}_g\psi, f)_{\mathbb{L}_2(\mathbb{R}^2)} = \int_{\mathbb{R}^2} \overline{\psi(R_\theta^{-1}(\mathbf{y}-\mathbf{x}))}\, f(\mathbf{y})\, \mathrm{d}\mathbf{y}, \quad \psi \in \mathbb{L}_2(\mathbb{R}^2)\cap\mathbb{L}_1(\mathbb{R}^2),$$

where $\psi$ is a suitable line-detecting wavelet in $\mathbb{L}_2(\mathbb{R}^2)$ and with representation $\mathcal{U} : \mathrm{SE}(2) \to \mathcal{B}(\mathbb{L}_2(\mathbb{R}^2))$ given by

$$\mathcal{U}_g\psi(\mathbf{y}) = \mathcal{T}_{\mathbf{y}}\mathcal{R}_\theta\psi(\mathbf{x}) = \psi(R_\theta^{-1}(\mathbf{x}-\mathbf{y})), \text{ with } R_\theta = \begin{pmatrix} \cos\theta & -\sin\theta \\ \sin\theta & \cos\theta \end{pmatrix} \in SO(2) \leftrightarrow$$
$e^{i\theta} \in \mathbb{T}$ and $\mathcal{R}_\theta\psi(\mathbf{x}) = \psi(R_\theta^{-1}\mathbf{x})$, $\mathcal{T}_{\mathbf{y}}\psi(\mathbf{x}) = \psi(\mathbf{x}-\mathbf{y})$.

Now $\mathcal{U}$ is reducible and the standard wavelet reconstruction theorems do not apply, nevertheless for proper choice of wavelets one can still obtain quadratic norm preservation. For details, see [8], [4] p.107-146. Now the wavelet transform maps $f$ to a so-called invertible orientation score $U_f$, which provides the initial condition $U_f \in \mathbb{L}_2(\mathrm{SE}(2))$ for our left invariant scale space representations on $SE(2)$ given by (11), generated by (10), which are computed by $G$-convolutions

$$\begin{aligned} W^{\mathbf{D},\mathbf{a}}(x,y,\theta,s) &= \int_{\mathbb{R}^2}\int_0^{2\pi} K_s^{\mathbf{D},\mathbf{a}}(e^{i(\theta-\theta')}, R_{\theta'}^{-1}(\mathbf{x}-\mathbf{x}'), e^{i(\theta-\theta')})\, U_f(\mathbf{x}', e^{i\theta'})\mathrm{d}\theta'\mathrm{d}\mathbf{x}' \\ P_\gamma^{\mathbf{D},\mathbf{a}}(x,y,\theta) &= \int_{\mathbb{R}^2}\int_0^{2\pi} R_\gamma^{\mathbf{D},\mathbf{a}}(e^{i(\theta-\theta')}, R_{\theta'}^{-1}(\mathbf{x}-\mathbf{x}'), e^{i(\theta-\theta')})\, U_f(\mathbf{x}', e^{i\theta'})\mathrm{d}\theta'\mathrm{d}\mathbf{x}'. \end{aligned} \tag{26}$$

For explicit formulae of our recently discovered exact convolution kernels $K_s^{\mathbf{D},\mathbf{a}}, R_\gamma^{\mathbf{D},\mathbf{a}} \in \mathbb{L}_1(\mathrm{SE}(2))$ we refer to our earlier work [7], with in particular the cases:

1. $Q(\underline{\mathcal{A}}) = D_{11}(\mathcal{A}_1)^2 + \mathcal{A}_2$, where the corresponding scale space representation is the Forward Kolmogorov equation of the direction process proposed by Mumford, [14] as a stochastic model for *contour completion*.
2. $Q(\underline{\mathcal{A}}) = D_{11}(\mathcal{A}_1)^2 + D_{22}(\mathcal{A}_2)^2$, where the corresponding scale space representation is the forward Kolmogorov equation of (up to scaling) the stochastic model for perceptual completion (*contour enhancement*) by Citti et al. [3].

Condition (14) is satisfied, since $\mathrm{span}\{\partial_\theta, [\partial_\theta, \partial_\xi], [\partial_\theta, [\partial_\theta, \partial_\xi]]\} = \mathrm{span}\{\partial_\theta, \partial_\eta, -\partial_\xi\} = \mathcal{L}(G)$. Although our exact solutions in [7], are simple, they consist of Mathieu-functions with disadvantages concerning computation time. Therefore it is worthwhile to replace the left invariant vector fields of $SE(2)$ (25) by the vector fields $\{\hat{A}_1, \hat{A}_2, \hat{A}_3\} = \{\partial_\theta, \partial_x + \theta\partial_y, -\theta\partial_x + \partial_y\}$. This leads to the following approximations[6] in case of contour completion, cf. [4]p.166-167 (for alternatives see [7]):

$$\begin{aligned} \hat{K}_s^{D_{11},a_1=1}(x,y,\theta) &= \delta(x-t)\frac{\sqrt{3}}{2\,D_{11}\pi x^2} e^{-\frac{3(x\theta-2y)^2 + x^2(\theta-\kappa_0 x)^2}{4x^3 D_{11}}} \\ \hat{R}_\gamma^{D_{11},a_1=1}(x,y,\theta) &= \alpha\,\frac{\sqrt{3}}{2\,D_{11}\pi x^2} e^{-\alpha x} e^{-\frac{3(x\theta-2y)^2 + x^2(\theta-\kappa_0 x)^2}{4x^3 D_{11}}} \tfrac{1}{2}(1+\mathrm{sign}(x)). \end{aligned}$$

---

[6] Note that the approximation $\hat{K}_s^{D_{11},a_1}$, in contrast to the exact solution(s) in [7], has singular behavior because of the violation of (14) as: $\mathrm{span}\{\partial_\theta[\partial_\theta, \partial_x + \theta\partial_y], [\partial_\theta, [\partial_\theta, \partial_x + \theta\partial_y]]\} = \mathrm{span}\{\partial_\theta, \partial_y\}$ which has dimension $2 < 3$.

The case of *contour enhancement* with $\tilde{Q}(\underline{A}) = D_{11}(\hat{A}_1)^2 + D_{22}(\hat{A}_2)^2$ requires a different approach. Here we apply a coordinates transformation

$\hat{K}_s^{D_{11},D_{22}}(x,y,\theta) = \tilde{K}_s(x',\omega',t') = \tilde{K}_s\left(\frac{x}{\sqrt{2D_{11}}}, \frac{\theta}{\sqrt{2D_{11}}}, \frac{2(y-\frac{x\theta}{2})}{\sqrt{D_{11}D_{22}}}\right)$ where we note

$$\partial_s \hat{K}_s^{D_{11},D_{22}} = \left(D_{11}\partial_\theta^2 + D_{22}(\partial_x + \theta\partial_y)^2\right)\hat{K}_s^{D_{11},D_{22}} \Leftrightarrow$$

$$\partial_s \tilde{K}_s^{D_{11},D_{22}} = \tfrac{1}{2}\left((\partial_{\omega'} - 2x'\partial_{t'})^2 + (\partial_{x'} + 2\omega'\partial_{t'})^2\right)\tilde{K}_s^{D_{11},D_{22}} = \tfrac{1}{2}\Delta_K \tilde{K}_s^{D_{11},D_{22}},$$

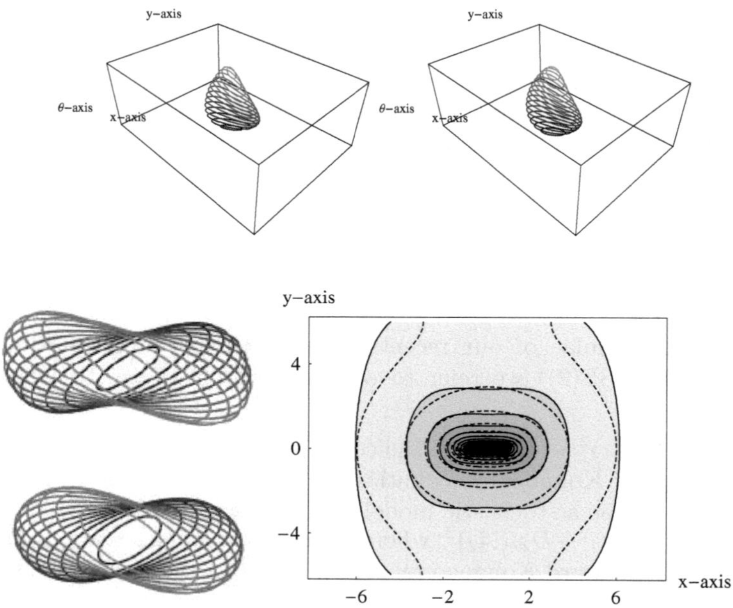


**Fig. 2.** A comparison between the exact Green's function of the resolvent diffusion process $(D_{11}\mathcal{A}_1^2 + D_{22}\mathcal{A}_2^2 - \gamma I)^{-1}\delta_e$, $\gamma = \frac{1}{30}$, $D_{11} = 0.1$, $D_{22} = 0.5$ which we explicitly derived in [7] and the approximate Green's function of the resolvent process with infinite lifetime $\lim_{\gamma\to 0}\gamma^{-1}(D_{11}\hat{A}_1^2 + D_{22}\hat{A}_2^2 - \gamma I)^{-1}\delta_e$, $D_{11} = 0.1$, $D_{22} = 0.5$ given by (27). Top row: a 3D view on a stack of spacial iso-contours with a 3D-iso-contour of the exact Green's function (right) and the approximate Green's function (left). Bottom-left; a close up on the same stacks of iso-contours but now viewed along the negative $\theta$-axis, with the approximation on top and the exact Green's function below. Bottom-right; an iso-contour-plot of the $xy$-marginal (i.e. Green's function integrated over $\theta$) of the exact Green's function with on top the corresponding iso-contours of the approximation in dashed lines. Note that the Green's functions nicely reflect the curvature of the Cartan-connection on $SE(2)$. The stochastic process corresponding to the approximation of the contour enhancement process is given by $X(s) + i\,\Theta(s) = X(0) + i\,\Theta(0) + \sqrt{s}(\epsilon_x + i\,\epsilon_\theta)$, where $\epsilon_x \sim \mathcal{N}(0, 2D_{11})$, $\epsilon_\theta \sim \mathcal{N}(0, 2D_{22})$ and, by (20), $Y(s) = \frac{X(s)\Theta(s)}{2} + \frac{1}{2}\int_0^s \Theta \mathrm{d}X - X\mathrm{d}\Theta = \int_0^s \Theta(t) - \Theta(0)\mathrm{d}t$.

which is exactly the evolution equation on $H_3$ generated by Kohn's Laplacian considered in the previous example ! As a result we have[7]

$$\begin{aligned} \hat{K}_s^{D_{11},D_{22}}(x,y,\theta) &= \frac{1}{2D_{11}D_{22}}\tilde{K}_s^{H_3}\left(\frac{x}{\sqrt{2D_{11}}}, \frac{\theta}{\sqrt{2D_{11}}}, \frac{2(y-\frac{x\theta}{2})}{\sqrt{D_{11}D_{22}}}\right) \\ &= \frac{1}{8D_{11}D_{22}\pi^2 s^2}\int_{\mathbb{R}} \frac{2\tau}{\sinh(2\tau)}\cos\left(\frac{2\tau(y-\frac{x\theta}{2})}{s\sqrt{D_{11}D_{22}}}\right) e^{-\frac{\left(\frac{x^2}{s\,D_{22}}+\frac{\theta^2}{s\,D_{11}}\right)\tau}{\tanh(2\tau)}}\, d\tau \\ \lim_{\gamma\to\infty}\gamma^{-1}\hat{R}_\gamma(x,y,\theta) &= \frac{1}{4\pi D_{11}D_{22}}\frac{1}{\sqrt{\frac{1}{16}\left(\frac{x^2}{D_{22}}+\frac{\theta^2}{D_{11}}\right)^2+\frac{(y-\frac{1}{2}x\theta)^2}{D_{11}D_{22}}}}. \end{aligned} \tag{27}$$

See Figure 2.

## 4 Conclusion

We derived a unifying framework for scale spaces (related to stochastic processes) on Lie-groups. These scale spaces are directly linked to operators on images by means of unitary wavelet transforms. To obtain proper invariance of these operators, the scale spaces must be left-invariant and thereby its solutions are $G$-convolutions with Green's functions. As this framework lead to fruitful applications on contour completion, contour enhancement and adaptive non-linear diffusion, see [9], [8], [5], in the special case $G = SE(2)$, this theory can be further employed for other groups, such as the Heisenberg group, $G = H_{2d+1}$.

## References


1. S.T. Ali, J.P. Antoine, and J.P. Gazeau. *Coherent States, Wavelets and Their Generalizations.* Springer Verlag, New York, Berlin, Heidelberg, 1999.
2. A.V. Bukhvalov and Arendt W. Integral representation of resolvent and semigroups. *Forum Math. 6*, 6(1):111–137, 1994.
3. G. Citti and A. Sarti. A cortical based model of perceptual completion in the roto-translation space. pages 1–27, 2004. Pre-print, available on the web `http://amsacta.cib.unibo.it/archive/00000822`.
4. R. Duits. *Perceptual Organization in Image Analysis.* PhD thesis, Eindhoven University of Technology, Dep. of Biomedical Engineering, The Netherlands, 2005.
5. R. Duits, M. Felsberg, G. Granlund, and B.M. ter Haar Romeny. Image analysis and reconstruction using a wavelet transform constructed from a reducible representation of the euclidean motion group. *IJCV.* Accepted for publication. To appear in 2007 Volume 72 issue 1, p.79–102.
6. R. Duits, L.M.J. Florack, J. de Graaf, and B. ter Haar Romeny. On the axioms of scale space theory. *Journal of Math. Imaging and Vision*, 20:267–298, 2004.
7. R. Duits and M.A. van Almsick. The explicit solutions of linear left-invariant second order stochastic evolution equations on the 2d-euclidean motion group. *Accepted for publication in Quarterly of Applied Mathematics, AMS*, 2007.
8. R. Duits and M.A. van Almsick. Invertible orientation scores as an application of generalized wavelet theory. *Image Processing, Analysis, Recognition and Understanding*, 17(1):42–75, 2007.


[7] Note that our approximation of the Green's function on the Euclidean motion group does not coincide with the formula by Citti in [3].

9. E. Franken, R. Duits, and B.M. ter Haar Romeny. Non-linear diffusion on the euclidean motion group. *Accepted for Publication in the Proceedings of SSVM-07, conf. on Scale Space and Var. Methods in Computer Vision.*, 2006.
10. B. Gaveau. Principe de moindre action, propagation de la chaleur et estimees sous elliptiques sur certains groupes nilpotents. *Acta mathematica*, 139:96–153, 1977.
11. W. Hebisch. Estimates on the semigroups generated by left invariant operators on lie groups. *Journal fuer die reine und angewandte Mathematik*, 423:1–45, 1992.
12. L. Hormander. Hypoellptic second order differential equations. *Acta Mathematica*, 119:147–171, 1968.
13. P. Lévy. Wiener random functions and other laplacian random functions. In *Proc. of the 2nd Berkely Symposium*, pages 171–187, USA, 1950. California Press.
14. D. Mumford. Elastica and computer vision. *Algebraic Geometry and Its Applications. Springer-Verlag*, pages 491–506, 1994.
15. C. Sagiv, N. A. Sochen, and Y. Y. Zeevi. Scale space generation via uncertainty principles. In *Scale Space and PDE Methods in Computer Vision 2005*, pages 351–362. Springer-Verlag.

# Convex Inverse Scale Spaces

Klaus Frick and Otmar Scherzer

Department of Computer Science, University of Innsbruck.
Technikerstrasse 21a, A 6020, Innsbruck, Austria
{klaus.frick,otmar.scherzer}@uibk.ac.at

**Abstract.** Inverse scale space methods are derived as asymptotic limits of iterative regularization methods. They have proven to be efficient methods for denoising of gray valued images and for the evaluation of unbounded operators.

In the beginning, inverse scale space methods have been derived from iterative regularization methods with squared Hilbert norm regularization terms, and later this concept was generalized to Bregman distance regularization (replacing the squared regularization norms); therefore allowing for instance to consider iterative total variation regularization. We have proven recently existence of a solution of the associated *inverse* total variation flow equation. In this paper we generalize these results and prove existence of solutions of inverse flow equations derived from iterative regularization with general convex regularization functionals.

We present some applications to filtering of color data and for the stable evaluation of the diZenzo edge detector.

## 1 Introduction

There are at least two evolutionary concepts based on partial differential equations for data filtering:

*Scale space methods* with time dependent partial differential equations approximate data $u^\delta$ (for instance images) by the solution of evolution equations (see e.g. [16]) at some time $t > 0$. The value of $t$ controls the amount of filtering.

*Inverse scale space methods* as introduced in [12] are defined as the semigroups corresponding to iterative regularization

$$\begin{aligned} u_{k+1} &= \operatorname*{argmin}_{u \in H_1} \left\{ \frac{1}{2\alpha} \left\| u - u^\delta \right\|_1^2 + \frac{1}{2} \left\| L\left(u - u_k\right) \right\|_2^2 \right\} \\ &= \operatorname*{argmin}_{u \in H_1} \left\{ \frac{1}{2\alpha} \left\| u - u^\delta \right\|_1^2 + \frac{1}{2} \left\| L(u) \right\|_2^2 - \left\langle L^* L u_k, u \right\rangle_1 \right\} \end{aligned} \quad (1)$$

where $L : H_1 \to H_2$ is a linear and densly defined operator between two Hilbert spaces $H_1$ and $H_2$ and $L^*$ denotes its adjoint. Here one typically initializes $u_0 = 0$ or $u_0 = \int_\Omega u^\delta \, dx$ and $u_{k+1}$ satisfies the Euler-Lagrange equation

$$u_{k+1} - u^\delta = \alpha \left( L^* L(u_{k+1}) - L^* L(u_k) \right). \quad (2)$$

F. Sgallari, A. Murli, and N. Paragios (Eds.): SSVM 2007, LNCS 4485, pp. 313–325, 2007.

Identifying the regularization parameter $\alpha$ and a time discretization $\Delta t$ via $\Delta t = \alpha^{-1}$ equation (2) can be considered as an implicit time step of the following flow equation

$$\begin{aligned} u - u^\delta &= (L^* L u)' \,, \\ u(0) &= u_0 \,. \end{aligned} \tag{3}$$

For instance, for $L = \nabla$ we have that $L^*L = -\triangle$ and (3) becomes *Showalter's method* which has been used successively for image denoising and for the stable evaluation of gradients (see e.g. [10]).

Two different approaches for generating *nonlinear* inverse scale spaces have been considered that allow a consistent generalization of the linear case:

In [12] nonlinear evolution equations have been derived from variational regularization techniques on reflexive Sobolev spaces.

In [14] the flow according to the iterative Bregman distance of the total variation semi-norm has been derived, which has been analyzed in [5]. Iterative Bregman distance regularization reads as follows:

- The first step consists in computing a minimizer $u_1$ of ROF functional

$$u_1 := \text{argmin} \frac{1}{2\alpha} \left\| u - u^\delta \right\|_{\mathrm{L}^2}^2 + |\mathrm{D}u|(\Omega).$$

- The $k+1$-th iterate is determined from

$$u_{k+1} = \operatorname*{argmin}_{u \in \mathrm{L}^2(\Omega)} \left\{ \frac{1}{2\alpha} \left\| u - u^\delta \right\|_{\mathrm{L}^2}^2 + |\mathrm{D}u|(\Omega) - \langle s, u \rangle \right\} , \tag{4}$$

  where $s$ is an element of the subgradient of the total variation semi norm at $u_k$.

Note that in the linear case (1) we always have that $L^*Lu$ is an element of the subgradient of $\frac{1}{2} \left\| L(\cdot) \right\|_2$ at $u$. This shows that replacing the squared regularization norm in the iterative method (1) by the *Bregman distance* (of the total variation semi-norm) gives a consistent definition of nonlinear inverse scale spaces. It has been shown in [5] that for $\alpha \to \infty$, the functions $u_\alpha : [0, \infty) \to \mathrm{L}^2(\Omega)$ with values $u_\alpha(t) = u_k$ for $k - 1 \leq \alpha t < k$ converge to the unique solution of the flow equation

$$v'(t) = u^\delta - u(t), \quad |\mathrm{D}u(t)|(\Omega) = \langle u(t), v(t) \rangle \,. \tag{5}$$

In this paper we generalize the results of [14] and prove results for variational regularization with Bregman distances of arbitrary convex functionals. Moreover, we prove existence of solutions of according flow equations. In particular, the results give existence of solutions of flow equations for denoising of vector valued data such as color images. Moreover, the techniques can be applied for the stable evaluation of unbounded operators such as the diZenzo edge detector.

## 2 Iterative Regularization with the Bregman Distance

In the sequel, without stating this explicitly, we always assume that $J : H \to \mathbb{R} \cup \{+\infty\}$ be a convex, lower semicontinuous and proper functional defined on an Hilbert space $H$. The norm on $H$ is denoted by $\|\cdot\|$ and is induced by the inner product $\langle \cdot, \cdot \rangle$. The *domain* $\mathrm{D}(J)$ of $J$ denotes the set of all $u \in \Omega$ such that $J(u) < +\infty$.

In order to analyze iterative Bregman regularization and the according gradient flows we review basic results from convex analysis.

**Definition 1.** *An element $s \in H$ is an element of the* subgradient $\partial J(u)$ *of $J$ at $u \in H$ if*

$$J(v) - J(u) - \langle s, v - u \rangle \geq 0 \textit{ for all } v \in H \, .$$

*The* Bregman distance *of $u, \tilde{u} \in H$ with respect to $J$ and $s \in \partial J(\tilde{u}) \subseteq H$ is defined by*

$$D_J^s(u, \tilde{u}) := J(u) - J(\tilde{u}) - \langle s, u - \tilde{u} \rangle \, . \tag{6}$$

Moreover, let

$$I(\alpha; u, \tilde{u}) := \left\{ \frac{1}{2\alpha} \left\| u - u^\delta \right\|^2 + D_J^{\tilde{v}}(u, \tilde{u}) \right\} \text{ for } \alpha > 0$$

With this notation iterative Bregman distance regularization reads as follows:

**Algorithm 1.** *Let $u^\delta \in H$.*

- *Choose $u_0 \in D(J)$ and $v_0 \in \partial J(u_0)$.*
- *For $k = 0, 1, \ldots$*

$$u_{k+1} := \underset{u \in H}{\mathit{argmin}} \, I(\alpha; u, u_k) \, .$$

$$v_{k+1} := v_k + \frac{1}{\alpha}(u^\delta - u_{k+1}) \, .$$

In the following we prove well-posedness of this algorithm and generalize some results in [6] for Bregman distance regularization with convex functionals $J$.

**Theorem 1.** *Assume that $u^\delta \in H$, $\alpha > 0$, $u_0 \in D(J)$ and $v_0 \in \partial J(u_0)$. Then for each $k \in \mathbb{N}$ there exists a unique minimizer $u_k \in H$ of $I(\alpha; \cdot, u_k)$ and a subgradient $v_k \in J(u_k)$ such that*

$$\alpha v_k + (u_k - u^\delta) = \alpha v_{k-1} \tag{7}$$

*and*

$$\left\| u_{k+1} - u^\delta \right\| \leq \left\| u_k - u^\delta \right\| . \tag{8}$$

*Proof.* Let $\tilde{u} \in H$ and $s \in \partial J(\tilde{u})$. We show weak lower semicontinuity and coercivity of $I(\alpha; \cdot, u_k)$. Then, existence of a minimizer follows from[7, Chap. 3,

Thm. 1.1]. Since both $u \to J(u)$ ($J$ is convex) and $u \to \langle s, u \rangle$ are weakly lower semicontinuous on $H$, the Bregman distance

$$u \mapsto D_J^s(u, \tilde{u})$$

is weakly lower semicontinuous. Therefore $I(\alpha; \cdot, u_k)$ is weakly lower semicontinuous on $H$ and proper. It remains to show that $I(\alpha; \cdot, u_k)$ is coercive on $H$, that is for every $k \in \mathbb{N}$ and every $\alpha > 0$ there exist constants $\lambda > 0$ and $\gamma \in \mathbb{R}$ such that

$$I(\alpha; u, u_k) > \lambda \|u\| + \gamma$$

for all $u$ in $H$. We verify the assertion for the functional $I(\alpha; \cdot, u_0)$. For $k > 1$ the assertion can be proven analogously taking into account that $u_k \in D(J)$ and $v_k \in \partial J(u_k)$. Since $J$ is convex we have that

$$J(u) \geq J(u_0) + \langle v_0, u - u_0 \rangle, \qquad \text{for all } u \in H.$$

Therefore,

$$\begin{aligned}
\frac{1}{2\alpha} \left\| u - u^\delta \right\|^2 + J(u) - \langle v_0, u \rangle &\geq \frac{1}{2\alpha} \left\| -u^\delta \right\|^2 + J(u_0) - \langle v_0, u_0 \rangle \\
&\geq \frac{1}{2\alpha} \left| \|u\| - \left\| u^\delta \right\| \right|^2 + J(u_0) - \langle v_0, u_0 \rangle .
\end{aligned}$$

for all $u \in H$ and hence we conclude that $I(\alpha; \cdot, u_0)$ is coercive. Thus we can apply [7, Chap. 3, Thm. 1.1] to obtain existence of a minimizer $u_1$ satisfying the Euler-Lagrange equation

$$v_1 := \frac{u^\delta - u_1}{\alpha} \in \partial J(u_1).$$

Uniqueness follows from the strict convexity of $\|\cdot\|^2$.

Moreover since the Bregman distance is nonnegative we have

$$\begin{aligned}
\frac{1}{2\alpha} \left\| u_{k+1} - u^\delta \right\|^2 &\leq \frac{1}{2\alpha} \left\| u_{k+1} - u^\delta \right\|^2 + D_J^{v_{k+1}}(u_{k+1}, u_k) \\
&= I(\alpha; u_{k+1}, u_k) \leq I(\alpha; u_k, u_k) = \frac{1}{2\alpha} \left\| u_k - u^\delta \right\|^2 .
\end{aligned}$$

This shows (8). □

As we will see in Sect. 3, the dual formulation of Algorithm 1 turns out to be the key ingredient in order to establish the corresponding continuous inverse scale space. The dual formulation is based on the Fenchel transform defined by:

**Definition 2.** *The* Legendre-Fenchel *conjugate of $J$ is the functional $J^* : H \to \mathbb{R} \cup \{+\infty\}$ defined by*

$$u^* \mapsto J^*(u^*) := \sup_{u \in H} \{\langle u^*, u \rangle - J(u)\} .$$

We consider the dual functional of $I$ with respect to $v \in H$ which is given as follows:

$$I^*(\alpha; v, \tilde{v}) := \frac{\alpha}{2} \|v - \tilde{v}\|^2 + J^*(v) - \langle u^\delta, v \rangle, \quad \tilde{v} \in H. \tag{9}$$

**Theorem 2.** *Assume that* $u^\delta \in H$, $\alpha > 0$, $u_0 \in D(J)$ *and* $v_0 \in \partial J(u_0)$. *Then* $v_k$ *as defined in Algorithm 1 satisfy*

$$v_k = \underset{v \in H}{\operatorname{argmin}}\, I^*(\alpha; v, v_{k-1}). \tag{10}$$

*Proof.* The functional $I^*$ is strictly convex and weakly lower semicontinuous with respect to $v$ and thus $I^*(\alpha; \cdot, v_{k-1})$ attains a unique minimizer $\tilde{v}_k$. It remains to show that $v_k = \tilde{v}_k$. From the definition of $v_k$ in Algorithm 1 and Theorem 1 it follows that

$$v_k = v_{k-1} - \frac{1}{\alpha}(u_k - u^\delta) \in \partial J(u_k). \tag{11}$$

Then, from the duality relation (see for instance [9]) we have that

$$u_k \in \partial J^* \left( v_{k-1} - \frac{1}{\alpha}(u_k - u^\delta) \right). \tag{12}$$

Moreover (11) is equivalent to $-\alpha(v_k - v_{k-1}) = u_k - u^\delta$ and this yields

$$\alpha(v_k - v_{k-1}) - u^\delta = -u_k. \tag{13}$$

Combination of (11), (12) and (13) shows that

$$0 \in \alpha(v_k - v_{k-1}) - u^\delta + \partial J^*(v_k) = \partial I^*(\alpha; v_k, v_{k-1}). \tag{14}$$

Therefore, $v_k$ minimizes the functional $I^*(\alpha; \cdot, v_{k-1})$, which together with the fact that the minimizer is unique implies that $v_k = \tilde{v}_k$. □

For the inverse total variation flow equation – i.e. $J(u) = |\mathrm{D}u|(\Omega)$ – $J^*$ is a barrier function. That is $J^*(v) = 0$ if and only if the $G$-norm (see [13], [3]) of $u \in \mathrm{BV}(\Omega)$ such that $|\mathrm{D}u|(\Omega) = \langle u, v \rangle$ is less than 1 (see e.g. [5]) and $+\infty$ else.

## 3 Continuous Inverse Scale Space Flow

In this section we show that the sequences $\{u_k\}$ and $\{v_k\}$ in Algorithm 1 can be considered as discrete approximations of the unique solution $(u, v)$ of

$$v'(t) = u^\delta - u(t), \qquad v(t) \in \partial J(u(t)), \tag{15a}$$

$$v(0) = v_0 \qquad u(0) = u_0. \tag{15b}$$

The following analysis uses results from [2] where a theory of gradient flows in metric spaces has been established. Moreover, we show that the solution $(u, v)$ of (15) satisfies the *inverse fidelity* axiom, that is

$$\lim_{t \to \infty} \left\| u(t) - u^\delta \right\| = 0$$

provided that $u^\delta \in \overline{D(J)}$. The last relation justifies to call (15) *inverse scale space method.*

For $\alpha > 0$, initial data $u_0 \in \mathrm{D}(J)$ and $v_0 \in \partial J(u_0)$ and $k \in \mathbb{N}$ the Bregman iterates $u_k$ and $v_k$ are extended to piecewise constants functions $\overline{U}_\alpha(t)$ and $\overline{V}_\alpha(t)$ satisfying

$$\overline{U}_\alpha(t) = u_k, \quad \overline{V}_\alpha(t) = v_k, \qquad \text{for } \frac{k}{\alpha} \le t < \frac{k+1}{\alpha} \tag{16}$$

Convergence of the functions $\{\overline{V}_\alpha(t)\}$ for $\alpha \to \infty$ follows from [2, Thm. 4.2.2] and reads as follows:

**Theorem 3.** *Assume that $u^\delta \in H$, $u_0 \in D(J)$ and $v_0 \in \partial J(u_0)$. Then there exists a absolutely continuous function $v : [0, \infty) \to H$ such that*

$$\lim_{\alpha\to\infty} \overline{V}_\alpha(t) = v(t)$$

*uniformly on every bounded* $[0, T]$.

*Proof.* The assertion is a consequence of [2, Thm. 4.2.2]. In order to apply this theorem the following two assumptions have to be verified.

**(A1)** The functional $\phi(v) = J^*(v) - \langle u^\delta, v\rangle$ is proper, lower semicontinuous and there exists $\tilde{v} \in \mathrm{D}(\phi)$, and $\tilde{r} > 0$ such that

$$\inf\{\phi(v) : \|v - \tilde{v}\| \le \tilde{r}\} > -\infty. \tag{17}$$

**(A2)** For every $v_0, v_1$ and $w$ in $\mathrm{D}(\phi)$ we have that

$$I^*(\alpha; (1-t)v_0 + tv_1, w) \\ \le (1-t)I^*(\alpha; v_0, w) + tI^*(\alpha; v_1, w) - \frac{\alpha}{2}t(1-t)\|v_0 - v_1\|^2$$

for all $t \in [0, 1]$ and $\alpha \ge 0$.

Since by assumption $J$ is lower semicontinuous and proper the same holds for $J^*$. To verify the coercivity of $J^*$ (assumption (17)) set $\tilde{v} = 0 \in \mathrm{D}(\phi)$ and $w = \operatorname{argmin}_{v\in H} I^*(\alpha; v, 0)$. Then

$$-\infty < \frac{\alpha}{2}\|w\|^2 + \phi(w) - \frac{\alpha\tilde{r}^2}{2} \le \phi(v),$$

for all $v \in H$ such that $\|v - 0\| \le \tilde{r}$. This shows **(A1)**. Moreover, for all $v_0, v_1$ and $w$ in $H$ we have

$$\|(1-t)v_0 + tv_1 - w\|^2 = (1-t)\|v_0 - w\|^2 + t\|v_1 - w\|^2 - t(1-t)\|v_0 - v_1\|^2.$$

This implies that for $v_0, v_1, w \in \mathrm{D}(\phi)$

$$\begin{aligned}\bar{I}^*(\alpha; (1-t)v_0 + tv_1, w) &= \frac{\alpha}{2}\|(1-t)v_0 + tv_1 - w\|^2 - \langle (1-t)v_0 + tv_1, w\rangle \\ &= \frac{\alpha}{2}\left((1-t)\|v_0 - w\|^2 + t\|v_1 - w\|^2\right) \\ &\quad - \frac{\alpha}{2}t(1-t)\|v_0 - v_1\|^2 - (1-t)\left\langle v_0, u^\delta\right\rangle - t\left\langle v_1, u^\delta\right\rangle \\ &= (1-t)\bar{I}^*(\alpha; v_0, w) + t\bar{I}^*(\alpha; v1, w) - \frac{\alpha}{2}t(1-t)\|v_0 - v_1\|^2.\end{aligned}$$

Therefore assumption **(A2)** holds. □

In [2] actually the stronger result has been shown, that for each partition of the interval $[0, \infty)$ the piecewise constant function converges to $v$ provided that the supremum of the stepsizes tends to zero.

In order to show that the function $v$ in Theorem 3 satisfies a gradient flow equation associated with $\phi$ we introduce the operator $\partial^0 \phi : D(\phi) \subseteq H \to H$ defined by

$$\begin{aligned} \partial^0 \phi(v) &= \operatorname*{argmin}_{u \in H} \{\|u\| \ : \ u \in \partial\phi\} \\ &= \operatorname*{argmin}_{u \in H} \left\{\left\|u - u^\delta\right\| \ : \ u \in \partial J^*(v)\right\} \end{aligned}$$

and the *slope function*

$$|\partial\phi| : v \mapsto \begin{cases} \left\|\partial^0 \phi(v)\right\| & \text{if } v \in D(\phi) \\ +\infty & \text{else.} \end{cases}$$

Since the norm in the Hilbert space $H$ is strictly convex, it follows that for every $v \in D(\phi)$ there exists a unique element $u$ in $\partial\phi$ with minimal norm. Thus the operator $\partial^0\phi(v)$ is single valued and well defined.

**Theorem 4.** *Under the assumptions of Theorem 3 the function $v$ is the unique solution of the gradient flow equation*

$$v'(t) = -\partial^0 \phi(v(t)), \qquad \textit{for a.e. } t \in [0, \infty) \tag{18a}$$

$$v(0) = v_0. \tag{18b}$$

*Proof.* The assertion the $v$ solves (18) follows directly from [2, Thm. 2.3.3] and [2, Prop. 1.4.1]. The results are applicable if **(A1)** and **(A2)** hold and the slope function $|\phi|(v)$ is lower semicontinuous on $H$. The later follows from the strong-weak closedness of $\partial\phi \subseteq H \times H$ ([2, Lem. 2.3.6]).

Uniqueness of $v$ can be proven with standard arguments as for instance from [4]. The mapping $v \to \phi(v)$ is convex and therefore $\partial\phi$ is monotone, that is

$$\langle u_1 - u_2, v_1 - v_2 \rangle \geq 0 \text{ for all } v_i \in H, u_i \in \partial\phi(v_i),\ i = 1, 2.$$

Assume that there exist two solutions $v, \hat{v}$ of (18), i.e.

$$-v'(t) \in \partial\phi(v(t)), \quad -\hat{v}'(t) \in \partial\phi(\hat{v}(t)).$$

From the monotonicity it follows that $\langle -v'(t) + \hat{v}'(t), v(t) - \hat{v}(t) \rangle \geq 0$ and therefore

$$\frac{\mathrm{d}}{\mathrm{d}t} \|v(t) - \hat{v}(t)\|^2 = \langle v'(t) - \hat{v}'(t), v(t) - \hat{v}(t) \rangle \leq 0.$$

This shows that $v = \hat{v}$. □

From definition it follows that the right hand side of (18) can be written as $u(t) - u^\delta$ for a (unique) function $u : [0, \infty) \to H$ satisfying $u(t) \in \partial J^*(v(t))$ for all $t \in [0, \infty)$. This is equivalent to $v(t) \in \partial J(u(t))$ such that $(u, v)$ is the unique solution of (15). In order to show the *inverse fidelity* property of the solution $(u, v)$ we cite one assertion of [2, Thm. 4.3.2]

**Theorem 5.** *The solution $(u, v)$ of (15) satisfies*

$$\left\|u(t) - u^\delta\right\|^2 \leq |\partial\phi|^2(\tilde{v}) + \frac{1}{t^2}\left\|\tilde{v} - v_0\right\| \tag{19}$$

*for all $\tilde{v} \in D(\phi)$.*

**Corollary 1.** *Let $u^\dagger$ denote the orthogonal projection of $u^\delta$ on to the closure of $D(J)$. Then we have that*

$$\limsup_{t\to\infty} \left\|u(t) - u^\delta\right\| \leq \left\|u^\dagger - u^\delta\right\| .$$

*In particular for $u^\delta \in \overline{D(J)}$ it follows that*

$$\lim_{t\to\infty} \left\|u(t) - u^\delta\right\| = 0 .$$

*Proof.* Let $\varepsilon > 0$. Then there exists $u_\varepsilon^\dagger \in \mathrm{D}(J)$ such that $\left\|u_\varepsilon^\dagger - u^\dagger\right\| < \varepsilon$. For $\tilde{v} \in \partial J(u_\varepsilon^\dagger)$ we have $u_\varepsilon^\dagger \in \partial J^*(\tilde{v})$ and consequently

$$|\partial\phi|(\tilde{v}) = \inf_{u\in H}\left\{\left\|u - u^\delta\right\| \;:\; u \in \partial J^*(\tilde{v})\right\} \leq \left\|u_\varepsilon^\dagger - u^\delta\right\| \leq \left\|u^\dagger - u^\delta\right\| + \varepsilon.$$

Using this inequality in (19) it follows that

$$\left\|u(t) - u^\delta\right\|^2 \leq \left(\left\|u^\dagger - u^\delta\right\| + \varepsilon\right)^2 + \frac{1}{t^2}\left\|\tilde{v} - v_0\right\| .$$

Taking into account that $\varepsilon > 0$ is arbitrary and taking $t \to \infty$ gives the assertion. □

The previous results show that $v_k$ as in Algorithm 1 approximates the function $v(t)$ in the flow equation (18).It remains to show that $\overline{U}_\alpha(t)$ approximates the primal function $u$ in (18) as well. We skip the proof since it uses exactly the same techniques as [5, Thm. 7].

**Theorem 6.** *Let $\overline{U}_\alpha(t)$ and $u$ be as in* (16) *and* (15), *respectively. Then*

$$\lim_{\alpha\to\infty} \overline{U}_\alpha(t) = u(t), \qquad \textit{almost everywhere in } [0, \infty) . \tag{20}$$

It is important to note that the generation of the (dual) flow equation (18) is independent of the minimizers of the primal variational problem in Algorithm 1. In other words, the function $u$ introduced above is established *artificially* and a connection to Algorithm 1 is a priori *not* obvious. Theorem 6 provides this relation.

**Corollary 2 (Monotonicity).** *Let $u^\delta \in H$. If $(u, v)$ is the solution of (15) we have*

$$\left\|u(s) - u^\delta\right\| \le \left\|u(t) - u^\delta\right\| \tag{21}$$

*for almost all $s, t$ in $[0, \infty)$ satisfying $s > t$.*

*Proof.* Let $\{\alpha_l\}_{l\in\mathbb{N}} \subseteq \mathbb{R}^+$ $\lim_{l\to\infty} \alpha_l = \infty$ and that (20) holds for $s$ and $t$. Then there exists an index $l_0$ such that for all $l > l_0$ we have that $\frac{1}{\alpha_l} < s - t$ and it consequently follows from (8) that

$$\left\|\overline{U}_{\alpha_l}(s) - u^\delta\right\| \le \left\|\overline{U}_{\alpha_l}(t) - u^\delta\right\|, \qquad \text{for all } l > l_0.$$

With this we obtain the estimate

$$\begin{aligned}
\left\|u(s) - u^\delta\right\| &\le \left\|\overline{U}_{\alpha_l}(s) - u^\delta\right\| + \left\|\overline{U}_{\alpha_l}(s) - u(s)\right\| \\
&\le \left\|\overline{U}_{\alpha_l}(t) - u^\delta\right\| + \left\|\overline{U}_{\alpha_l}(s) - u(s)\right\| \\
&\le \left\|\overline{U}_{\alpha_l}(t) - u(t)\right\| + \left\|\overline{U}_{\alpha_l}(s) - u(s)\right\| + \left\|u(t) - u^\delta\right\|.
\end{aligned}$$

Taking the limit $l \to \infty$ shows (21). □

## 4 Applications

In this section we highlight some applications of inverse scale spaces and iterative Bregman distance regularization.

### 4.1 Linear Inverse Scale Space

In [12] we introduced inverse scale spaces as methods for evaluation of unbounded operators. Let $L : D(L) \subseteq H_1 \to H_2$ be a linear, closed, densely defined and unbounded operator between two Hilbert spaces $H_1$ and $H_2$. Since $L$ is unbounded, computing $y = L(u)$ for a given $u \in H_1$ is ill posed. That is, for small perturbation $u^\delta$ of $u$ it might be that $u^\delta \notin D(L)$ or that $\left\|L(u^\delta) - L(u)\right\|$ might be significantly large.

In order to provide a stable method for evaluation of $L$ the *iterative Tikhonov-Morozov regularization* (1) can be used. In [12] it is shown that the discrepancy principle provides termination after a finite number of iterations. The discrepancy principle stops the iteration when for the first time the iteration error $\left\|u_k - u^\delta\right\|_1$ is below a given upper bound $\delta$ for the data error. As we have explained in Sect. 1 iterative Tikhonov Morozov regularization is equivalent to Bregman regularization with

$$J : u \mapsto \frac{1}{2}\left\|L(u)\right\|_2^2.$$

Note that linearity of $L$ implies convexity of $J$ and lower semicontinuity follows from the closedness of $L$. From the analysis presented in Sect. 3 it follows that iterative Bregman distance regularization can be considered the solution of an implicit time step of

$$(L^* L u)' = u^\delta - u, \quad u(0) = 0 \tag{22}$$

with time step size $\frac{1}{\alpha}$. Moreover since $L$ is densly defined we see from Corollary 1 that

$$\lim_{t\to\infty} u(t) = u^\delta \ .$$

Let $\Omega \subseteq \mathbb{R}^n$ and $m \geq 1$.

If $L$ denotes the gradient $D : \mathrm{H}^1(\Omega)^m \subseteq \mathrm{L}^2(\Omega)^m \to \mathrm{L}^2(\Omega)^{nm}$, then

$$L^* L(\boldsymbol{u}) = -\triangle \boldsymbol{u} := -\left(\triangle u_i\right)_{1\leq i \leq m} .$$

The according inverse scale space is the flow equation

$$(\triangle \boldsymbol{u})' = \boldsymbol{u} - \boldsymbol{u}^\delta, \quad \boldsymbol{u}(0) = 0 \ . \tag{23}$$

**Fig. 1. Left column.** top: data, middle: noisy data, bottom: solution of (22) at $t = 0.9$. **Middle column.** corresponding edge detectors $\lambda_1$ **right column.** corresponding eigenvectors of $z$ (zoom in some detail).

It can be used for stable evaluation of the derivative of a function $\boldsymbol{u} \in \mathrm{H}^1(\Omega)^m$ from noisy data $\boldsymbol{u}^\delta \in \mathrm{L}^2(\Omega)^m$. This is for instance useful for approximating the *diZenzo* edge detector [8] defined by

$$z(\boldsymbol{u})(x) = \left(D\boldsymbol{u}(x)\right)\left(D\boldsymbol{u}(x)\right)^T \in \mathbb{R}^{n\times n} \ . \tag{24}$$

The eigensystem of $z(\boldsymbol{u})$ at a point $x \in \Omega$ describes the geometry of the data. For $m = 1$ it is well known that the eigenvectors of $z(u)(x)$ are perpendicular and tangential to the level set of $u$ at $x$ and the corresponding eigenvalues are $|\nabla u|^2$ and 0 respectively.

Besides edge detection, the quantity $z(\boldsymbol{u})$ is also used as weight matrix in statistical models in image processing, e.g. in order to determine centers of gravities in (color) images (see e.g. [15]).

In the numerical experiments we have calculated the diZenzo edge detector with the inverse scale space method consisting in solving (23). Figure 1 shows the data (right column), the larger eigenvalue $\lambda_1$ (middle column) and the eigenvectors of $z(x)$ for a small subset of $\Omega$ (left column). The upper and middle row shows the original and noisy data respectively (additive Gaussian noise). The bottom row shows the solution of (22) at time $t = 0.9$.

## 4.2 Inverse Total Variation Denoising

Minimization of total variation based regularization functionals has turned out to be effective in image denoising applications.

**Fig. 2.** Inverse TV scale space: original image (upper left), solutions of (15) at times $t = 1, 5$ and 20 (from left to right)

**Fig. 3.** Difference images $|\boldsymbol{u}^\delta - \boldsymbol{u}(t)|_2$ for $t = 1, 5$ and 20

Let $\Omega \subseteq \mathbb{R}^n$ be an open and bounded set with piecewise Lipschitz boundary. We take $H = \mathrm{L}^2(\Omega)^m$, $m \geq 1$ and define for a vector valued function $\boldsymbol{u} \in \mathrm{L}^2(\Omega)^m$ the *total variation* semi-norm:

$$J(\boldsymbol{u}) := \left( \sum_{i=1}^{m} \left( \sup_{\substack{\phi \in C_c^1(\Omega) \\ |\phi(x)|_2 \leq 1}} \int_\Omega \mathrm{div}\phi \; u_i \; \mathrm{d}x \right)^2 \right)^{\frac{1}{2}} = \begin{cases} \left(\sum_{i=1}^m |\mathrm{D}u_i|(\Omega)^2\right)^{\frac{1}{2}} & \text{if } \boldsymbol{u} \in \mathrm{BV}(\Omega)^m \\ +\infty & \text{else} \end{cases} . \tag{25}$$

$J$ is convex and proper (since $\mathrm{BV}(\Omega)^m \cap \mathrm{L}^2(\Omega)^m \neq \emptyset$). Moreover $J$ is lower semicontinuous w.r.t. $\mathrm{L}^2(\Omega)^m$ norm (see e.g. [11, Chap. 5.2, Thm.1] or [1, Thm. 2.3]).

The numerical experiments in Fig. 2 show the multi-scale evolution of a color image by the inverse total variation flow with $J$ as in (25) and Fig. 3 shows the texture part of the images.

## 5 Conclusion

In this paper we have generalized the analysis of inverse Bregman distance total variation regularization to Bregman distance regularization of arbitrary convex functionals. Moreover, we have derived the according flow equations and analyzed them using general results from [2]. We applied the results for filtering of color data and for the stable evaluation of diZenzo's edge detector.

## Acknowledgment

This work has been supported by the Austrian Science Foundation (FWF) Projects Y-123INF, FSP 9203-N12 and FSP 9207-N12.

## References


1. R. Acar and C. Vogel. Analysis of bounded variation penalty methods for ill-posed problems. *Inverse Problems*, 10:1217–1229, 1994.
2. L. Ambrosio, N. Gigli, and G. Savaré. *Gradient flows.* Lectures in mathematics ETH Zürich. Birkhäuser Verlag, Basel, 2005.
3. G. Aubert and J.-F. Aujol. Modeling very oscillationg signals. application to image processing. *Appl. Math. Optim.*, 51(2):163–182, 2005.
4. H. Brézis. *Opérateurs maximaux monotones et semi-groupes de contractions dans les espaces de Hilbert.* Number 5 in North-Holland Mathematics Studies. North-Holland Publishing Co, Amsterdam-London, 1973.
5. M. Burger, K. Frick, Osher S., and O. Scherzer. Inverse total variation flow. *Multiscale Model. Simul.*, 2007. accepted.
6. M. Burger, G. Gilboa, S. Osher, and J. Xu. Nonlinear inverse scale space methods for image restoration. *Comm. Math. Sciences*, 4:179–212, 2006.
7. B. Dacorogna. *Direct methods in the calculus of variations.* Springer-Verlag, Berlin, 1989.
8. S. diZenzo. A note on the gradient of a multi-image. *Computer Vision, Graphics and Image Processing*, 33:116–126, 1986.
9. I. Ekeland and R. Temam. *Convex analysis and variational problems*, volume 1 of *Studies in Mathematics and its Applications.* North-Holland Publishing Co., Amsterdam-Oxford, 1976.
10. H. Engl, M. Hanke, and A. Neubauer. *Regularization of inverse problems.* Mathematics and its Applications. Kluwer Academic Publishers Group, Dordrecht, 1996.
11. L. C. Evans and R. F. Gariepy. *Measure theory and fine properties of functions.* Studies in Advanced Mathematics. CRC Press, Boca Raton, FL, 1992.
12. C. W. Groetsch and O. Scherzer. Non-stationary iterated Tikhonov - Morozov method and third-order differential equations for the evaluation of unbounded operators. *Math. Methods Appl. Sci.*, 23(15):1287–1300, 2000.
13. Y. Meyer. Oscillating patterns in image processing and nonlinear evolution equations. *University Lecture Series*, 22, 2001.
14. S. Osher, M. Burger, D. Goldfarb, J. Xu, and W. Yin. An iterative regularization method for total variation based image restoration. *Multiscale Model Simul.*, 4(2):460–489, 2005.
15. Förstner W. and Gülch E. A fast operator for detection and precise location of distinct points, corners and centers od circular features. pages 281–305, Interlaken, Swizerland, 1987.
16. J. Weickert. *Anisotropic Diffusion in Image Processing.* ECMI. B.G. Teubner, Stuttgart, 1998.

# Spatio-temporal Scale-Spaces

Daniel Fagerström

Computational Vision and Active Perception Laboratory (CVAP)
Department of Numerical Analysis and Computing Science
KTH (Royal Institute of Technology) SE-100 44 Stockholm, Sweden
danielf@nada.kth.se

**Abstract.** A family of spatio-temporal scale-spaces suitable for a moving observer is developed. The scale-spaces are required to be time causal for being usable for real time measurements, and to be "velocity adapted", i.e. to have Galilean covariance to avoid favoring any particular motion. Furthermore standard scale-space axioms: linearity, positivity, continuity, translation invariance, scaling covariance in space and time, rotational invariance in space and recursivity are used. An infinitesimal criterion for scale-spaces is developed, which simplifies calculations and makes it possible to define scale spaces on bounded regions. We show that there are no temporally causal Galilean scale-spaces that are semigroups acting on space and time, but that there are such scale-spaces that are semigroups acting on space and memory (where the memory is the scale-space). The temporally causal scale-space is a time-recursive process using current input and the scale-space as state, i.e. there is no need for storing earlier input. The diffusion equation acting on the memory with the input signal as boundary condition, is a member of this family of scale spaces and is special in the sense that its generator is local.

## 1 Introduction

While "spatial" scale-space theory has become the more or less canonical theory for low-level vision and its concepts, methods and algorithms now are part of the common toolbox in computer vision, there is this far no theory that can serve as a basis for low-level vision for moving images.

The spatio-temporal scale-space theory from Koenderink [11] is based on spatio-temporal convolution on the past image sequence, which makes it computationally heavy and less attractive for practical applications. From a conceptual point of view it is questionable to base a theory on spatio-temporal measurement on the past *signal*, a real-time system can only have access to the past *measurements*. The corresponding theory from Lindeberg [12] is more usable in an application setting due to its recursive formulation. On the other hand it is discrete in space-time, which makes it cumbersome to use analytical methods like differential geometry, methods which has been used for deriving a large part of the results in spatial scale-space theory. The spatio-temporal scale-space theory that we propose is both continuous and has a recursive formulation, and thus solves the problems described above.

F. Sgallari, A. Murli, and N. Paragios (Eds.): SSVM 2007, LNCS 4485, pp. 326–337, 2007.

In the rest of the article we will start by defining a spatio-temporal scale-space suitable for image motion measurements in the seemingly most natural way: It should be a semigroup on space-time, time causal, covariant with respect to spatial and temporal scale and Galilean boost [7]. We require Galilean covariance as for a moving observer no particular motion or lack of motion should be treated in a special way. It is relative motion that matters. This approach however doesn't work as we show that among the possible Galilean covariant, semigroups on space-time, there are no time causal ones. To simplify proofs and anticipating later needs we also develop infinitesimal criterion on scale-spaces.

After the failure of the "obvious" approach to time causal spatio-temporal scale-space, we perform a deeper analysis of the nature of temporal measurement. And observe that in addition to that a real time system cannot be dependent on future information it cannot be directly dependent on past input either, only its representation of past input, its *memory* [4]. Based on this insight, we instead define a temporally causal spatio-temporal scale space as a semigroups in space and *memory*, and derive a family of scale-spaces fulfilling this. If one requires the generator of the scale-space to be local, there is a unique scale-space that is generated by the heat equation.

## 2 Generalities

Our aim with defining a spatio-temporal scale-space is to develop a generic theory about image motion measurements. A measurement is done over an *extended point* (volume), it is *covariant* with the relevant symmetry group. It should have the *cascade property*, i.e. a measurement of a measurement should also be contained in the family of measurements [8]. As the structure of the Galilean similarity group is more complicated than the similarity group that is used in ordinary scale-space theory, some extra care is needed to handle the relation between covariance and the cascade property.

An image is defined as an integrable functions $u \in L_1(\Omega) = \Sigma$ over some connected subset of an Euclidean base manifold $\Omega \subset M$, $M = \mathbb{R}^d \ni x = (x_1, \ldots, x_d)$ for spatial images and $M = \mathbb{R}^{d+1} \ni x = (x_0, x_1, \ldots, x_d)$ for spatio-temporal images, where $x_0$ is a coordinate in the temporal direction.

**Definition 1.** *For any* $u, v \in \Sigma(= L_1(\Omega))$ *and* $\alpha, \beta \in \mathbb{R}$, $\Phi : \Sigma \to C^\infty(\Omega)$ *is a point measurement operator if it fulfills:*

**linearity.** $\Phi(\alpha u + \beta v) = \alpha\Phi(u) + \beta\Phi(v)$
**gray level invariance.** $\|\Phi u\|_{L_1} = \|u\|_{L_1}$
**positivity.** $u \geq 0 \Rightarrow \Phi u \geq 0$
**point.** *There is a sequence of operators* $\mathbb{R}_+ \ni s \mapsto \Phi_s$ *s.t.* $\lim_{s \to 0} \Phi_s u = u$

Image measurement should respect basic symmetries in the world that is measured. This is described in terms of Lie groups acting on images. Let $g \in G$, where $G$ is a Lie group, then the group act on the base space as $g \mapsto g \cdot x = T_g x$, where $T_g : \mathrm{Diff}(\Omega)$, (i.e a diffeomorphism on $\Omega$) and on images and measurements like $T_g u(x) = u(g \cdot x)$.

Ideally we would like point measurement operators that are *invariant*, $\Phi T_g = \Phi$, with respect to the group but for many groups (e.g. the scaling group) that is not possible. We have to be content with requiring *covariance*, i.e. we have a family $H \ni h \mapsto \Phi_h$ of measurement operators, fulfilling,

$$T_g \Phi_h = \Phi_{g \cdot h} T_g. \tag{1}$$

**Definition 2.** *We call such a family of point measurement operators a* G-covariant point measurement space.

From the definition it can easily be shown that the action $G \times H \ni (g,h) \mapsto g \cdot h = \sigma(g,h) = \sigma_g(h) \in H$ is a Lie group action on the set $H$. As a smaller set of measurement operators is easier to manage we prefer measurement spaces that are invariant to as large subgroup as possible.

For the spatio-temporal scale-space we will require covariance with respect to the $(n+1)$-dimmensional *Galilean similarity group* that in matrix form becomes:

$$\begin{pmatrix} t' \\ x' \end{pmatrix} = \begin{pmatrix} \tau & 0 \\ v & \sigma R \end{pmatrix} \begin{pmatrix} t \\ x \end{pmatrix} + a \tag{2}$$

where $x, v \in \mathbb{R}^n, t \in \mathbb{R}, R \in SO(n), \sigma, \tau \in \mathbb{R}_+$ and $a \in \mathbb{R}^{n+1}$.

Now we have a family of measurement devices $\Phi_h$ that is covariant under the chosen Lie group. But we still don't know how the actual measurements, $\Phi_h u$ are related. Considering that the result of a measurement can be measured in turn it is natural to require the family of measurements to be closed under composition,

$$\Phi_{h_1} \Phi_{h_2} u = \Phi_{h_1 \cdot h_2} u, \tag{3}$$

where $H \times H \ni (h_1, h_2) \mapsto h_1 \cdot h_2 \in H$, is an abstract Lie semigroup. Although there is a rich modern theory about Lie semigroups, it will for our needs be enough to consider Lie semigroups $H \subset G$ that are subsets of a Lie group $(G, \cdot)$ and closed under composition but not necessarily under inversion. A simple example is $(\mathbb{R}_+, +)$.

From (3) it can be seen that $\{\Phi_h | h \in H\}$ is an operator semigroup and furthermore as $H \ni h \mapsto \Phi_h$ is a Lie semigroup homomorphism, the operator semigroup must be a Lie semigroup as well. Combining $G$-covariance with the cascade property we finally get:

**Definition 3.** *A* G scale-space *is a minimal semigroup of G-covariant point measurements, i.e. a family of operators* $H \ni h \mapsto \Phi_h : \mathrm{L}(\Sigma, C^\infty(\Omega))$ *fulfilling*

$$T_g \Phi_h T_{g^{-1}} = \Phi_{g \cdot h}, \tag{4}$$

*a slight reorganization of (1) to emphasize how G act on the semi group of operators and (3), and where G is an abstract Lie group and H an abstract Lie semigroup.*

It can be seen that $G \times H \ni (g,h) \mapsto g \cdot h = \sigma(g,h) \in H$ is both a group homomorphism in the first argument and a semigroup homomorphism in the second. When the homomorphism is trivial in the first argument we have an *invariant operator*, $T_g \Phi_h T_{g^{-1}} = \Phi_h$.

# 3 Infinitesimal Generators

In preparation for later needs we derive infintesimal conditions for a $G$ scale-space. There are several advantages in doing that. It is more realistic in the sense that we mainly care for local symmetries. It allows for more general boundary conditions, e.g. for a bounded sensor and it simplifies calculations as it linearizes the problem.

The infinitesimal object that corresponds to a Lie group $G$ is a *Lie algebra*, $\mathcal{L}G = \mathfrak{g}$, which is the linear space of infinitesimal generators of the group at identity together with an anti-commutative bilinear operator $\mathfrak{g} \times \mathfrak{g} \ni (v, w) \mapsto [v, w] \in \mathfrak{g}$, the *Lie bracket* (see e.g. [14] for details). The transformation group $g \mapsto T_g$ corresponds to the Lie algebra of infinitesimal generators $\mathfrak{g} \ni v \mapsto A_v = A(v)$ where $A = dT(e) : \mathfrak{g} \to \mathrm{L}(\Sigma, \Sigma)$. Choosing a base $\{v_1, \ldots, v_n\} \subset \mathfrak{g}$ for $\mathfrak{g}$ the corresponding base for the infinitesimal generators of the transformation group is $\{A_1, \ldots, A_n\}$ where $A_j = A_{v_j}$

For a Lie semigroup $H$ the infinitesimal object is a *Lie wedge*, $\mathcal{L}H = \mathfrak{h}$, and is a closed cone in a Lie algebra. The properties for Lie algebras mentioned above also holds for Lie wedges with the restriction that it is not necessary for both $[v, w]$ and $[w, v]$ for $v, w \in \mathfrak{h}$ to be members of the wedge [9]. Only Lie semigroups that are generated by their Lie wedge will be considered, i.e. $\exp(\mathbb{R}_+\mathfrak{h}) = H$.

From a $G$ scale-space, using $\Phi(h)(x) = \Phi_h(x)$, for emphasizing the dependency on the semigroup $H$, the corresponding infinitesimal operator is $B = d\Phi(e) : \mathfrak{h} \to \mathrm{L}(\Sigma, \Sigma)$, $\mathfrak{h} \ni w \mapsto B_w = B(w)$. Choosing a base $\{w_1, \ldots w_m\} \subset \mathfrak{h}$ for the Lie wedge a corresponding base, $\{B_1, \ldots, B_m\}$ where $B_k = B_{w_k}$, of the Lie wedge of infinitesimal generators for the semi group is given.

## 3.1 Covariant Generators

The infinitesinal form of the covariance equation (4) is,

$$[A_v, B_w] = B_{C(v,w)}, \tag{5}$$

where $\mathfrak{g} \times \mathfrak{h} \ni (v, w) \mapsto C(v, w) \in \mathfrak{h}$ is the differential of $\sigma$ (from (4)) with respect to both arguments. This can be shown by calculating the differential to the equation both with respect to the group $G$ and the semigroup $H$. An operator $B_w$ fullfilling such an equation is called a *covariant tensor operator* [1] in mathematical physics. The operator $C$ is denoted the *covariance tensor* and is a Lie algebra representation in the first argument and a Lie wedge representation in the second.

In coordinate form the infinitesimal covariance equation becomes:

$$[A_j, B_k] = \sum_l C_{j,kl} B_l, \tag{6}$$

where $C(v_j, w_k) = C_j(w_k) = \sum_l C_{j,kl} w_l$. For an *infinitesimally invariant operator*, $[A_v, B_w] = 0$. Note that requiring an operator $B_w$ to be infinitesimally invariant with respect to an action $A_v$ means for the covariance tensor that $C(v, w) = 0$.

Now we take a look at the semigroup generated by a set of covariant tensor operators. A one-parameter Lie semigroup is generated by an element in the corresponding Lie wedge $w \in \mathfrak{h}$ by $\mathbb{R}_+ \ni s \mapsto \exp(sw) = h_w(s) \in H_w \subset H$. Let $u : H \times \Omega \to \mathbb{R}$ and set $\mathfrak{h} \ni w \mapsto u_w(s, x) = u(h_w(s), x)$,

$$\begin{cases} \partial_s u_w = B_w u_w \\ u_w(e, x) = f(x), \end{cases} \tag{7}$$

where $f \in \Sigma$. For an *abstract Cauchy problem*, like (7), the solution can be described in terms of a semigroup $u_w(s, x) = \exp(sB_w)f(x) = \Phi_{h_w(s)}f(x)$. And given that we have chosen semigroups $h \in H$, such that $h = \exp(sw)$ for some $s \in \mathbb{R}_+, w \in \mathfrak{h}$, we have [9],

$$u(h, x) = \Phi_h f(x). \tag{8}$$

Reconnecting to the $G$ scale-space axioms, Definition 3, the Cauchy problem (7) for a $\mathfrak{g}$ tensor operator generates a $G$-covariant semigroup. Furthermore, looking at Definition 1, the semigroup will be linear as long as the infinitesimal generator $B_w$ is independent of the evolution parameter ($s$ in (7)). Also the point property follows from the boundary condition in (7). What is left to give an infinitesimal characterization of is the positivity and gray level invariance properties.

### 3.2 Pseudo Differential Operators

For being able to continue we must be more concrete about the form of the operators in (5). The infinitesimal generators of the transformation group are on the form: $\sum_{j \leq n} a_j(x)\partial_j$, where $a_j : M \to \mathbb{R}$. In earlier work in scale-space theory the infinitesimal generator have been the inverse Fourier transform of some smooth function. By using *pseudodifferential operators*, $\Psi$DO, [3] we get a large enough class of operators to embed both of them. $\Psi$DO's are defined by

$$Au(x) = (2\pi)^{-n} \int e^{ix\cdot\xi} a(x, \xi)\tilde{u}(\xi)d\xi, \tag{9}$$

where $\tilde{u}(\xi) = \int e^{-ix\cdot\xi}u(x)dx$ and $a(x, \xi) = \sum_{|\alpha| \leq m} a_\alpha(x)\xi^\alpha$, and is called the *symbol* of $A$. The corresponding operator is denoted $a(x, D)$.

The composition of two symbols is:

$$c(x, \xi) = a(x, \xi) \circ b(x, \xi) = \sum_\alpha \frac{1}{\alpha!}[\partial_\xi^\alpha a(x, \xi)] \cdot [D_x^\alpha b(x, \xi)], \tag{10}$$

where $D_j = i^{-1}\partial_j$ and the commutator determines a Lie algebra structure on the symbols.

In this paper we will only consider translation invariant scale-spaces which further restricts the form of the symbol. A base for the Lie algebra of translations is, $\mathfrak{t}(n) = \{\partial_1, \ldots, \partial_n\}$, and the corresponding symbols are, $\partial_j \sim i^{-1}\xi_j$.

**Lemma 1.** *Translation invariant symbols are position independent,* $b(x,\xi) = b(\xi)$.

*Proof.* An infinitesimally translational invariant symbol $b$ fulfills, $[i^{-1}\xi_j, b(x,\xi)] = 0$, for $j \leq n$. Leading to, $\partial_{x_j} b(x,\xi) = 0$, for all $j \leq n$, i.e. the symbol is independent of $x$.

We can also see that any two translationally invariant symbols $b_i(\xi), b_j(\xi)$ commutes, $[b_i(\xi), b_j(\xi)] = 0$, which means that cones of translation invariant symbols automatically becomes Lie wedges.

### 3.3 Positivity and Gray Level Invariance

A positive translation invariant linear operator has a positive kernel. It can be shown that a kernel $\Omega \ni x \mapsto \phi(x)$, is positive iff its Fourier transform is *positive definite*, i.e. for each $k \in \mathbb{N}$ and each set of $\xi_1, \ldots, \xi_k \in \mathbb{R}^d$ the matrix $(\tilde{\phi}(\xi_i - \xi_j))_{i,j=1,\ldots,k}$ is positive Hermitian, (e.g. [10]). Furthermore for kernels that generate a semigroup $\mathbb{R}_+ \ni s \mapsto \phi_s$, their Fourier transform is $\tilde{\phi_s}(\xi) = e^{-sb(\xi)}$, where $\mathbb{R}^d \ni \xi \mapsto b(\xi) \in \mathbb{C}$.

**Definition 4.** *A symbol* $\mathbb{R}^d \ni \xi \mapsto b(\xi) \in \mathbb{C}$ *is* negative definite *if the matrix* $(b(\xi_i) + \overline{b(\xi_j)} - b(\xi_i - \xi_j))_{i,j=1,\ldots,k}$ *is positive Hermitian for any choice of* $k \in \mathbb{N}$ *and each set of* $\xi_1, \ldots, \xi_k \in \mathbb{R}^d$.

A kernel is positive definite iff its symbol is negative definite [10]. For negative definite symbols $b(\xi)$, $b(0) \geq 0$ and they have at most quadratic growth at $\infty$ i.e. $|b(\xi)| \leq k_b(1 + |\xi|^2)$ for some $k_b \in \mathbb{R}_+$.

For gray level invariance we use an adaption of a theorem from [10].

**Theorem 1.** *For a gray level invariant semigroup* $\Phi_s$ *generated by the* $B$ *with symbol* $b$ *the following holds:* $\Phi_s 1 = 1$ *for all* $s \geq 0$, $B1 = 0$ *and* $b(0) = 0$ *where* $\Omega \ni x \mapsto 1(x) = 1$.

A generator such that $B1 = 0$ is said to be *conservative.*

### 3.4 Infinitesimal $G$ Scale-Space

Combining our results about infinitesimal generators of a $G$ scale-space, we can now state:

**Definition 5.** *A* $\mathfrak{g}$ scale-space wedge, *is a minimal Lie wedge of negative definite conservative operators* $\mathfrak{h} \ni w \mapsto B_w : \mathrm{L}(\Sigma, \Sigma)$, *that is a covariant tensor operator (5), with respect on the Lie algebra action* $\mathfrak{g} \ni v \mapsto A_v : \mathrm{L}(\Sigma, \Sigma)$ *and the Lie algebra and Lie wedge representation* $(v, w) \mapsto C(v, w)$.

Summarizing the discussion above:

**Theorem 2.** *A* $G$ *scale-space is generated (using (7)) by its corresponding* $\mathfrak{g}$ *scale-space wedge.*

## 4 Scale-Space Generators

Now we have the tools we need for being able to study infinitesimal G-scale spaces. We will apply these tools for the affine line, Euclidean similarity spaces, and Galilean spaces with scaling in time and space.

### 4.1 The Affine Line

A one dimensional scale space measurement is invariant with respect to translation and covariant with respect to scaling. The affine line has the infinitesimal generators,

$$\mathfrak{gl}(1) = \mathfrak{t}(1) \cup \{x\partial_x\}, \tag{11}$$

commutator $[\partial_x, x\partial_x] = \partial_x$, and the symbol for scaling is, $i^{-1}x\xi$.

**Lemma 2.** *Translation invariant symbols $b : \mathbb{R} \to \mathbb{C}$ covariant with $\mathfrak{gl}(1)$ are on the form, $b(\xi) = k\xi^\alpha$ for any $k \in \mathbb{C}$ and $\alpha \in \mathbb{R}$.*

*Proof.* From translation invariance, the covariance tensor must obviously be trivial for the translation generator, for scaling the simplest non-trivial representation is $C(x\partial_x) = \alpha$, for any $\alpha \in \mathbb{R}$. Combining this with, (5) we have, $[i^{-1}x\xi, b(\xi)] = \alpha b(\xi)$, for the symbol $b(\xi)$. Using (10), we get $\xi b'(\xi) = \alpha b(\xi)$ which has solutions on the form given in the lemma.

Parameterizing $k$ as $k = ce^{i\theta\pi/2}$, where $c, \theta \in \mathbb{R}$ and disregarding the, in this context, uninteresting scaling parameter $c$, we have:

**Definition 6.** *The* Feller fractional derivative *[17] is defined as,*

$$D^\alpha_\theta(\xi) = -e^{i\theta\pi/2}\xi^\alpha = -|\xi|^\alpha e^{i\,\mathrm{sign}(\xi)\theta\pi/2}, \tag{12}$$

*where $0 < \alpha \leq 2, \alpha \neq 1$ and $|\theta| \leq \alpha$ for $0 < \alpha < 1$ and $|\theta| \leq 2-\alpha$ for $1 < \alpha \leq 2$.*

*The symmetric part $D^\alpha_0 = D^\alpha \sim |\xi|^\alpha$ is called the* Riesz fractional derivative. *For $\xi \in \mathbb{R}^d$ the notation $-(-\Delta)^{\alpha/2} = D^\alpha$, is also used for emphasizing that the derivative can be seen as a generalization of the Laplacian.*

*Decomposed as a linear combination of the two one sided operators, $D^\alpha_\theta(\xi) = -c_+(\alpha,\theta)D^\alpha_+(\xi) - c_-(\alpha,\theta)D^\alpha_-(\xi)$, the operator $D^\alpha_+$ ($D^\alpha_-$) is called the* left (right) sided Riemann Liouville fractional derivative *and is given by,*

$$D^\alpha_\pm(\xi) = (\mp i\xi)^\alpha = |\xi|^\alpha e^{\mp i\,\mathrm{sign}(\xi)\alpha\pi/2}, \quad c_\pm(\alpha,\theta) = \frac{\sin[(\alpha \mp \theta)\pi/2]}{\sin(\alpha\pi)}. \tag{13}$$

It can be shown that the Feller derivative is negative definite and conservative for the values of $\alpha, \theta$ given in the definition. And from this we have:

**Theorem 3.** *A $\mathfrak{gl}(1)$ scale-space wedge is generated by a Feller fractional derivative $D^\alpha_\theta$ with parameters according to the definition. The subspace of reflection symmetric wedges are generated by Riesz fractional derivatives $D^\alpha$ with $0 < \alpha \leq 2$ and temporally causal wedges by left sided Riemann Liouville fractional derivatives $D^\alpha_+$ with $0 < \alpha < 1$.*

The three parameter family of functions generated of the Feller derivative are called *stable densities* [5] and appear in generalizations of the central limit theorem. The stable density for $\alpha = 2, \theta = 0$ is the normal distribution, for $\alpha = 1, \theta = 0$ it is the Cauchy distribution and for $\alpha = \theta = 1/2$ is the solution of the signaling equation, i.e. diffusion on the half line with the signal as input at the end. Symmetric stable densities, $\theta = 0$ was the result of the scale space axiomatization in [15,2]. The maximally asymmetric, *extremal* stable density functions are one sided for $0 < \alpha < 1$ and $\theta = \pm\alpha$, these where the result of an axiomatization of scale spaces with temporal causality in [4].

In [15,2] these result are extended to the Euclidean similarity group, which on $\mathbb{R}^2$ consists of translation in the plane, rotation and scaling.

**Theorem 4.** *A Euclidean similarity* $\mathfrak{es}(2) = \mathfrak{t}(2) \cup \mathfrak{s}(2) \cup \mathfrak{so}(2)$ *scale-space wedge is generated for any* $0 < \alpha \leq 2$ *by the Riesz fractional derivative* $-(-\Delta)^{\alpha/2}$.

Where $\mathfrak{s}(2) = \{x_1\partial_1 + x_2\partial_2\}$ is the scaling generator is and $\mathfrak{so}(2) = \{x_2\partial_1 - x_1\partial_2\}$ is the rotation generator.

### 4.2 Galilean Similarity

The 1 + 1 dimensional Galilean similarity group, i.e. translation invariance in space and time, separate scaling in space and time and Galilean boost in space time, have the following set of infinitesimal generators,

$$\gamma\mathfrak{s}(2) = \mathfrak{t}(2) \cup \mathfrak{s}(1) \oplus \mathfrak{s}(1) \cup \gamma(1), \tag{14}$$

where $\gamma(1) = \{\gamma = x_0\partial_1\}$ is the Galilean boost that "skew" space-time and $\mathfrak{s}(1) \oplus \mathfrak{s}(1)$ is a direct sum of the scaling generator in space and time respectively. The non-zero commutators are $[\partial_j, x_j\partial_j] = \partial_j$, $[\partial_0, \gamma] = \partial_1$, $[x_0\partial_0, \gamma] = \gamma$ and $[x_1\partial_1, \gamma] = -\gamma$.

**Theorem 5.** *A* $\gamma\mathfrak{s}(2)$ *scale-space wedge is generated by* $\{\partial_0^2, \partial_0\partial_1, \partial_1^2\}$.

*Proof.* Requiring our Lie wedge separately covariant with respect to scaling booth in space and time and using the results from scale space on the affine line we can see that any wedge must contain two generators on the form $b_j(\xi_0, \xi_1) = k_j\xi_j^{\alpha_j}$, $j = 0, 1$. Applying the Galilean boost, which has the symbol $\gamma = i^{-1}x_0\xi_1$ on these the spatial generator disappears $[\gamma, k_1\xi_1^{\alpha_1}] = 0$, while repeated application of the Galilean boost on the temporal generator gives, $\mathrm{ad}(i^{-1}x_0\xi_1)^l(k_0\xi_0^{\alpha_0}) = k_0(\prod_{j=0}^{l-1}\alpha_0 - j)\xi_0^{\alpha_0 - l}\xi_1^l$, where $\mathrm{ad}(a)(b) = [a, b]$. To get a finite base of generators we must have $\alpha_0 \in \mathbb{Z}_+$, furthermore to generate a positive semigroup $\alpha_0, \alpha_1 \leq 2$. For $\alpha_0 = 2$ we get the set of symbols, $\{\xi_0^2, \xi_0\xi_1, \xi_1^2\}$, $k_0 \in \mathbb{R}$ for $\alpha = 2$ (then only $\theta = 0$ is allowed). This set of generators are booth closed under the Galilean similarity group and complete by choosing $\alpha_1 = 2, k_1 = 1$ for the spatial generator.

It should be noted that the generated scale-space is symmetric booth in time and space and thus no time causal scale-spaces are given with this axiomatization. And as $\gamma\mathfrak{s}(n), n \geq 2$ have $\gamma\mathfrak{s}(2)$ as a sub algebra, no time causal scale-spaces are possible for them either.

# 5 Time Causal Galilean Scale-Spaces

This far we have seen that Galilean scale-spaces as we have defined them have kernels that are symmetric in the temporal direction. This means that both the past and the future signal is used for the measurement, this is a disappointment if we want to use such a scale space for real time measurements.

**Definition 7.** *The history at time $t$ of the spatio-temporal signal $\mathbb{R} \times \mathbb{R}^d \ni (t, x) \mapsto f(t, x) \in \mathbb{R}$ is, $f(t, s, x) = f(t - s, x)$ when $s < t$ and 0 otherwise. A* time causal *measurement operator only depends on the history at time $t$ for a measurement at $t$.*

For a real time system temporal causality is obviously necessary. But as discussed in [4] defining temporal measurement in terms of a convolution or a evaluation equation on the history is to beg the question. It would require the measurement device to already have access to what it is supposed to measure.

This conceptual problem can be resolved by involving a *memory* of the history in the definition of temporal measurement. The measurement device should only have access to the current signal and its memory of previous measurements. As the memory is supposed to represent the history it is reasonable to make it as similar to the history as possible. This can be done by requiring the memory to be a half-space of the same dimensions as the history and subject to the same symmetry requirements.

We apply these considerations on the pure time causal scale-space (with no spatial dimensions).

**Definition 8.** *A* time causal *scale-space on the affine line $\mathbb{R} \times \mathbb{R}_+ \ni (t, \tau) \mapsto u(t, \tau)$, where $t$ is the temporal coordinate and $\tau$ the memory coordinate, is generated by the* signaling problem,

$$\begin{cases} \partial_t u = Bu \\ u(t, 0) = f(t), \end{cases} \tag{15}$$

*where the operator $B$ is independent of time and $f : \Sigma$ is the input signal. The measurement operator implicitly defined by $u(t, \tau) = \Phi_\tau f(t)$ is a $GL(1)$ covariant point measurement operator (but not necessarily a semigroup) and the infinitesimal generator $B$ is a $\mathfrak{gl}(1)$ scale-space wedge.*

**Theorem 6.** *A time causal scale-space on the affine line is generated by the right sided Riemann Liouville fractional derivative $D^\alpha_{\tau-}$, with $1 < \alpha \leq 2$ (where the suffix $\tau$ denotes that it is applied on the memory domain).*

*Proof.* The form of the generators of the $\mathfrak{gl}(1)$ scale-space wedge is given in Theorem 3. Of the generators given there, only $B = D^\alpha_{\tau-}$, $0 < \alpha \leq 2$ are translation invariant on the right (positive) half line, as all other generators have support on the left half line as well (for non integer differentiation order). In [4] it is shown that (15), with $B = D^\alpha_{\tau-}$ is equivalent to $\partial_\tau u = -D^{1/\alpha}_{t+} u$ with initial value $u(t, 0) = f(t)$, where $0 < 1/\alpha < 1$ that is $\alpha > 1$ for $\Phi_\tau$ to be a time causal scale space.

It is worth noticing that the only *local* generator for this family of scale spaces is $D^2_{\tau-} = \partial^2_\tau$, which means that in this case the scale space is generated by the heat equation, although with different boundary conditions compared to ordinary spatial scale space.

### 5.1 Galilean Similarity

Now we extend these results to Galilean space-time. The time causal Galilean scale-space is as discussed above a scale space on space and memory rather than on space and time. Something new compared to the the previously discussed scale-spaces is that for the generator of the scale space on memory $\partial_t = D^\alpha_{\tau-}$, the symmetry Lie algebra will not only act on the generator but on the time derivative $\partial_t$ as well.

**Definition 9.** *Let $\gamma\mathfrak{s}(d+1) = \mathfrak{es}(d) \oplus \mathfrak{gl}(1) \cup \gamma(d)$. The $d+1$-dimensional time causal Galilean scale-space $\mathbb{R}_+ \times \mathbb{R}^d \times \mathbb{R}_+ \times \mathbb{R} \times \mathbb{R}^d \ni (\sigma, v, \tau, t, x) \mapsto u(\sigma, v, \tau, t, x) \in \mathbb{R}$, where $\sigma$ is spatial scale, $v$ is velocity, $\tau$ is memory (and temporal scale) is a $\gamma\mathfrak{s}(d+1)$-covariant, point measurement space in space-time $(t, x)$ and a $\mathfrak{gl}(1)$-wedge in memory $\tau$.*

**Theorem 7.** *A $d+1$-dimensional time causal Galilean scale-space (for $d = 1, 2$) is generated by the evolution equation,*

$$\begin{cases} \partial_t u = -v \cdot \nabla_x u + D^{\alpha_0}_{\tau-} u \\ \partial_\sigma u = -(-\Delta_x)^{\alpha/2} u \\ u(0,0,0,t,x) = f(t,x), \end{cases} \tag{16}$$

*where $1 < \alpha_0 \leq 2$, $0 < \alpha \leq 2$, $\sigma$ is the spatial scale direction, $v \in \mathbb{R}^d$ the velocity vector, $\partial_s = \partial_t + v \cdot \nabla_x$ is the spatio-temporal direction, $\nabla_x = (\partial_1, \ldots, \partial_d)$ is the spatial gradient and $\Delta_x$ is the spatial Laplacian.*

*The equation,*

$$\begin{cases} \partial_t u = -v \cdot \nabla_x u + \partial^2_\tau u \\ \partial_\sigma u = \Delta_x u \\ u(0,0,0,t,x) = f(t,x), \end{cases} \tag{17}$$

*is unique in the family of evolution equations as it is the only one that has local generators.*

*Proof.* First it is shown in Theorem 4 that the spatial part of the scale space is generated by $\partial_\sigma u = -(-\Delta_x)^{\alpha/2} u$, $0 < \alpha \leq 2$ that besides being covariant with $\mathfrak{es}(2)$ is invariant with respect to temporal translation and scaling and spatio-temporal Galilean boost. For the temporal part of the scale-space we know from Theorem 6 that the generator $\partial_t u = D^{\alpha_0}_{\tau-} u$, $1 < \alpha \leq 2$ is a $\mathfrak{gl}(1)$-wedge. As $D^{\alpha_0}_{\tau-}$ is independent of space and time it is obviously invariant with the Galilean similarity group. But $\partial_t$ is not checking for the commutation relations with $\gamma\mathfrak{s}(d+1)$ the non zero commutators are for scaling $[\partial_t, t\partial_t] = \partial_t$ and for the Galilean boosts $[\partial_t, t\partial_j] = \partial_j$, $i = 1, 2$. Checking $\partial_j$ for all commutators no further generators are added, so $\{\partial_t, \partial_j\}$ is closed under $\gamma\mathfrak{s}(d+1)$. As a result, linear combinations of $\{\partial_j\}$, that is $v \cdot \nabla_x$, $v \in \mathbb{R}^d$ must be added to the temporal scale-space generator to make it closed under the required symmetries.

While there is no closed form for the general causal Galilean scale-space, it can be shown that (17) have the solution

$$u(\sigma, v, \tau, t, x) = \phi(\sigma, v, \tau, \cdot, \cdot) * f(t, x), \tag{18}$$

$$\phi(\sigma, v, \tau, t, x) = \frac{\tau \exp(-\frac{\tau^2}{4t} - \frac{(x-tv)\cdot(x-tv)}{4\sigma})}{\sqrt{4\pi} t^{3/2} (4\pi\sigma)^{d/2}}. \tag{19}$$

.

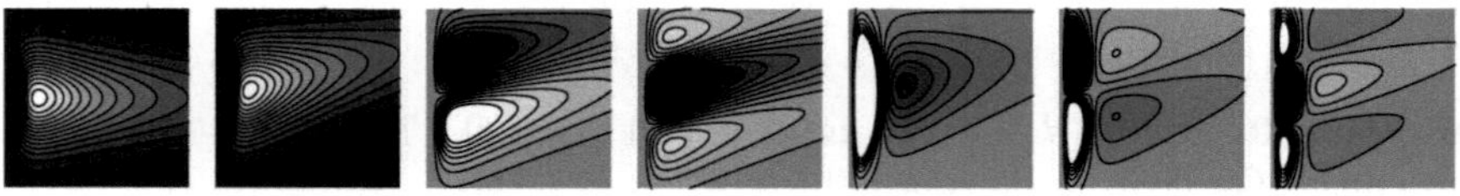

**Fig. 1.** Causal Galilean scale-space: space vertically and time horizontally, from left to right: two different velocity adaptations of $u$, then $u_x, u_{xx}, u_t, u_{xt}, u_{xxt}$

## 6 Discussion

The main result of this article is that we have shown (Theorem 7), that there is a reasonable spatio-temporal velocity adapted scale space for an active observer. It is booth time causal and recursively generated, in the sense that it only depends on current input and its memory, not the history of the input. The set of axioms is very close to modern scale-space axiomatizations [18], but with the main difference that we apply them on space and memory instead of space and time.

Comparing to earlier formulations of time causal Galilean scale spaces Lindeberg's [12] is close in the sense that it is a recursive formulation, but the formulation is discrete so covariances are only approximate and it is much harder analyze the properties. The approach from Florack [6] is based on a Gaussian Galilean scale-space that is made time causal by doing a logarithmic transformation of the time domain, it depends on the history of the signal and no recursive formulation has been suggested. Salden [16] have proposed a time causal spatio-temporal scale-space where the diffusion equation is applied (separately on the spatial and temporal domain) on the history of the signal so it is obviously dependent on history rather than recursive. Although the original formulation is not Galilean, it could easily be, by using a Galilean transformation on the generator.

For numerical implementation of the time causal Galilean scale-space the heat equation scheme (17) is most attractive as the fractional derivatives are integral operators and need to involve a much larger number of grid points for getting satisfying precision. For the heat equation it should be noted that it consists of two independent heat equations one in space and one in space time. The one in space can be computed with an explicit scheme for the heat equation with sub sampling as described in [13]. For the spatio-temporal part the scheme in [4] can be used. This makes the proposed scale-space highly efficient as only a two or three point derivative kernel needs to be applied in the temporal direction instead of a full temporal convolution kernel.

## References


1. A. O. Barut and R. Raczka. *Theory of Group Representaions and Applications.* World Scientific, 2 edition, 1986.
2. R. Duits, L. Florack, J. de Graaf, and B. ter Haar Romeny. On the axioms of scale space theory. *J. of Mathematical Imaging and Vision*, 20(3):267–298, May 2004.
3. Y. Egorov and S. M.A., editors. *Partial Differential Equations II*, volume 31 of *Encyclopaedia of Mathematical Sciences.* Springer Verlag, Berlin, 1994.
4. D. Fagerström. Temporal scale spaces. *International Journal of Computer Vision*, 64(2/3):97–106, 2005.
5. W. Feller. *An introduction to probability theory and its application*, volume 2. John Willey & Sons, Inc., 1966.
6. L. Florack. Visual representations embodying spacetime structure. Technical Report UU-CS-1999-07, Utrecht University, 1999.
7. L. Florack, B. ter Haar Romeny, J. Koenderink, and M. Viergever. Scale and the differential structure of images. *Image and Vision Computing*, 10:376–388, 1992.
8. L. M. J. Florack. *Image Structure.* Series in Mathematical Imaging and Vision. Kluwer Academic Publishers, Dordrecht, Netherlands, 1997.
9. E. Hille and R. S. Phillips. *Functional analysis and semi-groups.* American Mathematical Society, 1957.
10. N. Jacob. *Pseudo-Differential Operators and Markov Processes.* Number 94 in Mathematical Research. Akademie Verlag, Berlin, 1996.
11. J. Koenderink. Scale-time. *Biological Cybernetics*, 58:169–162, 1988.
12. T. Lindeberg. Time-recursive velocity-adapted spatio-temporal scale-space filters. In P. Johansen, editor, *Proc. 7th European Conference on Computer Vision*, volume 2350 of *Lecture Notes in Computer Science*, pages 52–67, Copenhagen, Denmark, May 2002. Springer Verlag, Berlin.
13. T. Lindeberg and L. Bretzner. Real-time scale selection in hybrid multi-scale representations. In L. Griffin, editor, *Proc. Scale-Space'03*, volume 2695 of *Lecture Notes in Computer Science*, pages 148–163, Isle of Skye, Scotland, June 2003. Springer-Verlag.
14. A. L. Onishchik, editor. *Lie Groups and Lie Algebra I*, volume 20 of *Encyclopaedia of Mathematical Sciences.* Springer-Verlag, 1993.
15. E. Pauwels, L. VanGool, P. Fiddelaers, and T. Moons. An extended class of scale-invariant and recursive scale space filters. *PAMI*, 17(7):691–701, July 1995.
16. A. Salden, B. ter Haar Romeny, and M. Viergever. Linear scale-space theory from physical principles. *JMIV*, 9(2):103–139, September 1998.
17. S. G. Samko, A. A. Kilbas, and O. I. Marichev. *Fractional integrals and derivatives : theory and applications.* Gordon and Breach Science Publishers, cop., Yverdon, 1992.
18. J. Weickert, S. Ishikawa, and A. Imiya. Linear scale-space has first been proposed in japan. *J. of Mathematical Imaging and Vision*, 10:237–252, 1999.

# A Scale-Space Reeb-Graph of Topological Invariants of Images and Its Applications to Content Identification

Jinhui Chao and Shintaro Suzuki

Department of Information and System Engineering, Chuo University
1-13-27, Kasuga, Bunkyo-ku, Tokyo, 112-8551, Japan

**Abstract.** In this paper, a new method of content identification using topological invariants is proposed. First, we show a Reeb-graph of topological invariants of images in a scale-space. Different from well-known scale-space trees of salient or critical points based on catastrophe or singularity theory, we use topologically stable blobs or primary sketches with nonzero lifetimes in scale and nonzero areas at each scale. The continuum of such blobs as a 3D manifold is featured by trees of topological invariants of the image called a Reeb graph. We show that this Reeb-graph representation is more robust against deformation attacks and perturbation such as numerical errors than traditional scale-space trees. A fast matching algorithm for the graphs is also presented.

## 1 Introduction

Robust identification and authentication of digital contents is one of the most important yet challenging tasks in digital rights management[1][2]. Currently used watermarking methods seems to be inevitably subjected to various deformation attacks in digital piracy and copyright infringement. These attacks include geometric transforms such as Euclidean, Affine transforms and other local or nonlinear deformations. Recent examples are replacement attacks and estimation-based attacks which aimed to erase the embedded watermarks. Besides, operations of image processing such as compression, filtering, gray scaling or histogram transforms could also significantly reduce detectability of watermarks. Many of these attacks have been implemented in e.g. the website of StirMark[21]. To defend against such attacks one could use stronger embedding but it always causes quality degradation of the original content. Until now, most proposals to defend against those attacks are effective only for one or two attacks which are known a priori. However, it seemed difficult for a watermark to survive under unknown and multiple attacks.

On another direction, "passive watermarking" or content identification approaches, instead of embedding foreign information into contents, try to identify illegal copies of the original content by extracting and matching their intrinsic features. These methods have no quality deterioration of the contents, and the intrinsic features of contents will not by easily erased.

F. Sgallari, A. Murli, and N. Paragios (Eds.): SSVM 2007, LNCS 4485, pp. 338–349, 2007.

In this paper, we first discuss general strategy to choose features and argue the importance to use stable and topological invariants in content identification and copyright protection. Then we review existing approaches on scale-space trees and discuss problems to apply them in content identification. A new scale-space tree called a scale-space Reeb-graph is proposed as features of contents. This scale-space Reeb-graph is specified by choice of scale filters, blobs and tree structure. Instead of using the unstable salient points in the images, we use topologically stable (explained in the next section) blobs which are primary sketches with nonzero lifetimes in scale and nonzero areas at each scale. We show that the new scale-space trees are robust against deformation and noise. An algorithm for fast matching of the scale-space Reeb-graphs using partial matching is presented. Finally, simulation is provided to examine the performance of the proposed method under various deformation attacks in the StirMark4.0. It is shown that this method could track the original image under these attacks and distinguishes it from the other images without a priori knowledge.

## 2 Stable and Topological Invariants for Content Identification

Content identification for copyright management is to find illegal copies of an original content. Comparing with image retrieving, these copies are usually more close to the original. At the same time, they are often maliciously deformed so that they are perceptually similar to but subtly different from the original. This means content identification needs more delicate and detailed search.

In fact, attacks in digital piracy and copyright infringement are consisted of minor and local deformations. An arbitrary deformation may not have the inverse map, but in the context of copyright infringement, these deformations should preserve the essential visual characteristics of the original content. Thus we can assume these deformations are bi-continuous maps or homeomorphisms which therefore form a subgroup of the topological transformation group acting on the contents.

Since it is difficult to formulate the subgroup of maps preserving visual characteristics and its action, we can use the invariants of its supergroup, i.e. the topological invariants or features which remain unchanged under bi-continuous transforms. In fact, it seemed to the authors the most effective way to defeat the deformation and other attacks is to use topological invariants as features of contents. This may seem at first sight unnecessarily restrict but is in fact reasonable considering the wide variety of potential unknown attacks in the future[18][17]. This formulation can also be extended to differentially topological group and invariants for fast computation.

Furthermore, we know that computation of features of an image, even topological ones could be sensitive to noise or perturbation. A feature which is unchanged under small perturbations is called topologically stable or simply stable (also structurally stable). Therefore, it is important to use stable topological invariants in content identification.

Another issue is that most active or passive watermarks are based on local features defined by e.g. DCT or wavelet transforms of certain blocks or meshes in the image. However, it is more desirable to use global features in order to resist deformation attacks e.g. containing a shift farther than the size of the blocks or meshes, since the shift distance of pixels under an attack is usually hard to predict.

## 3 Scale-Space Trees and Existing Approaches

It is known that "scale provides topology", i.e., topological features of a content can be obtained from the scale-space well known in image analysis and pattern recognition[7][10].

In particular, assume an image $I(\boldsymbol{x}), \boldsymbol{x} \in \mathbb{R}^n$ is smoothed by a filter $f_t$ of scale $t$, denote the smoothed image as $L(\boldsymbol{x}, t) := f_t(I(\boldsymbol{x}))$, then a scale-space representation is used to reveal the "deep structure" of an image, by considering all levels of scale $t$ simultaneously.

In fact, theory of embedding topology tells us that the continuum $\{L(\boldsymbol{x}, t), \forall \boldsymbol{x}, \forall t\}$ of a family of features for all scale $t$, as a 1-dimentionally higher object, has higher topological stability comparing with the stability at each scale $t$.

Besides, a tree structure especially a scale-space tree which represents the image in a descending order of resolution or scale can be convenient for fast searching in a large database. e.g., strategy of early abandonment can be used together with matching from low to high resolutions.

Furthermore, one of difficulties in content identification is due to the cropping attacks which embed the original image into another image. By matching between all sub-trees, it is possible to track illegal copies produced by these attacks.

Until now, one of the most important scale-space representations is the scale-space trees. Since the trajectory of a point on the image under scaling operation is a curve (may be branched) in the scale-space. The set of the trajectories of salient points of the image is called a scale-space tree. Scale-space trees or multi-resolution trees can be divided into the following categories.

1. Trees obtained by linear (e.g. Gaussian) filtering, which are consisted of trajectories of salient or critical points of Gauss curvatures of an image. Unfortunately, the properties of non-Morsian critical points are not structurally stable. Therefore these features are not robust again perturbations.
2. Trees obtained by nonlinear filtering
   Since linear scale-space filtering could blur edges or salient points in the image, two kinds of nonlinear filters are used to build a multi-resolution or scale-space tree:

   (a) Sieve methods [11][14] use morphological filters so that they can preserve edges and be robust against impulse noises.

(b) Critical point filters[12][13] preserve both positions and intensities of critical points in the original image. They are also based on morphological operations.

These nonlinear scale-space trees however could be noise sensitive due to discontinuous operations of the morphology filters. Since these discontinuous operations do not preserve topology, one can not guarantee uniqueness of the tree for an image under perturbation or attacks.

The reason to use nonlinear filters is that in traditional image processing, pattern recognition and image retrieving, performance under noisy environment is always a major concern. Especially, robustness against the impulse noise possesses a greater priority than smooth e.g. Gaussian noise. In content identification for copyright protection, noise is no longer a major issue but deformation attacks become the main threat. Therefore, the existing scale-space trees may not be able to provide stable topology.

## 4 A Scale-Space Reeb-Graph of Images

Below, we obtain trees of topological invariants in scale-space from the Morse theory.

### 4.1 Morse Theory and Reeb Graph

Here we recall briefly the Morse theory and Reeb graphs.

Consider a smooth function $f : M \to \mathbb{R}$ defined on a differentiable manifold $M$. A point $\boldsymbol{y} \in M$ at which the Jacobian of $f$ vanishes is called a critical point of $f$ and denoted as $\boldsymbol{p}$. A function is called a Morse function if its Hessian matrices at every critical points are nonsingular. The index of a Morse critical point is defined as the number of negative eigenvalues of the Hessian matrix at the critical point. It is easy to see that a Morse function has only isolated critical points.

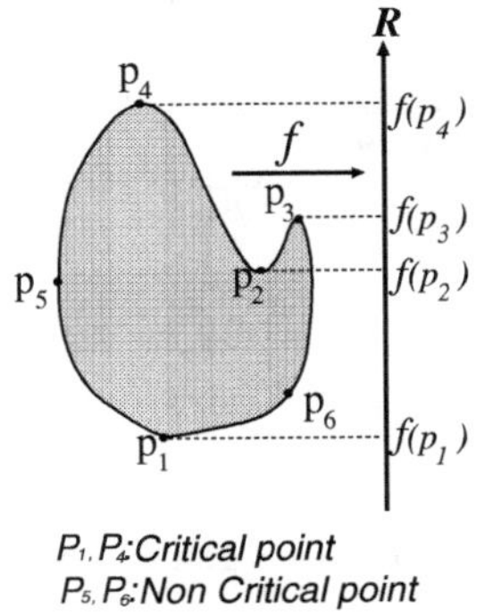


**Fig. 1.** Critical points of a Morse function $f$

Define the level set of a Morse function $f$ as

$$f^{-1}(c) := \{\boldsymbol{y} \in M | f(\boldsymbol{y}) = c\}.$$

then for $[a, b] \subset \mathbb{R}$, if $M_a^b := f^{-1}[a, b]$ contains no critical points of $f$, then $M_a := f^{-1}(-\infty, a]$ and $M_b := f^{-1}(-\infty, b]$ are diffeomorphic. This means that the level sets of a Morse function between two neighboring critical points are diffeomorphic to each other. In other words, they can be smoothly deformed from one to another so there is no change in properties of differential topology between them. On the other hand, topology only changes at the critical points.

It is known that topology of a smooth compact manifold can be efficiently described by the Morse theory in terms of CW homology In particular, the Morse theory provides a procedure for decomposition of the manifold into CW complexes. According to the theory, any smooth compact manifold $M$ can be constructed by repeatedly attaching cells or CW simlexes $e^{r_1}, e^{r_2}, \cdots, e^{r_k}$ of dimensions $r_1, r_2, \cdots, r_k$ at critical points $\boldsymbol{p}_1, \cdots, \boldsymbol{p}_k$ of a Morse function on $M$ . The dimension $r_i$ of the newly attached cell $e^{r_i}$ at $\boldsymbol{p}_i$ equals the index of $\boldsymbol{p}_i$(fig.1) [8].

$$M \simeq e^{r_1} \cup e^{r_2} \cup \cdots \cup e^{r_k}$$

Furthermore, any smooth compact manifold can be efficiently represented by a Reeb graph [6],[9]. A Reeb graph is derived by contracting a connected component of level sets of a Morse function into a point. On these graphs, each vertex corresponds to a critical point and information of its index is also labeled on the vertex( fig.2).

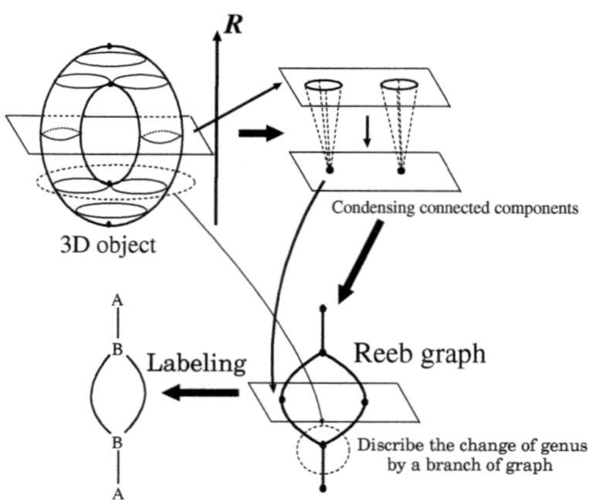


**Fig. 2.** Reeb Graph

### 4.2 Scale-Space Reeb Graph and Its Construction

To define the scale-space Reeb graph of an image, we regard the continuum $M := \{L(\boldsymbol{x}, t), \forall \boldsymbol{x}, \forall t\}$ of certain topological invariants in the smoothed images as a topological manifold.

The Morse function of the manifold $M$ is the scale function

$$\tau: \quad M \longrightarrow \mathbb{R}$$
$$L(\cdot,t) \longmapsto t$$

The blobs are the level sets of $\tau$:

$$\tau^{-1}(c) = L(\cdot,c), \quad c \in \mathbb{R}$$

The manifold $M$ in the scale space is then represented by a Reeb graph of the scale function, which we call as a scale space Reeb graph.

Notice that the scale function $t$ may not be a Morse function and the Reeb graph may not be structurally stable. Hence we need to build a scale-space Reeb-graph by careful choice of scaling filters and blobs as follows.

**Scaling filters:** It is clear that a continuous operator either a linear or nonlinear filter is desirable in order to preserve topological invariants. Here we use the simplest Gaussian scale linear filtering. Another reason to use Gaussian filtering is that it could be implemented by several fast algorithms [20],[19]. This could shorten the time of online extraction of the tree for application in a large database.

**Blobs:** We know that objects or features which survive longer during smoothing, i.e. with long lifetime are regarded as of higher visual significance. In our case this means these features have higher topological stability. THus, we use topologically robust features as blobs which keep appear in an interval of the scale with a certain length and also have enough area at each scale-level. i.e. This means these features possess large enough 3D volume in the scale-space.

In this way, we obtain a 3D continuum of the blobs in the scale space.

We then shrink each connected components of the blobs into a point at all scale $t$ to obtain a Reeb graph for this 3D continuum.

In this paper, we use the blobs as the positive areas of Gaussian curvatures of the tri-stimulus of the image.

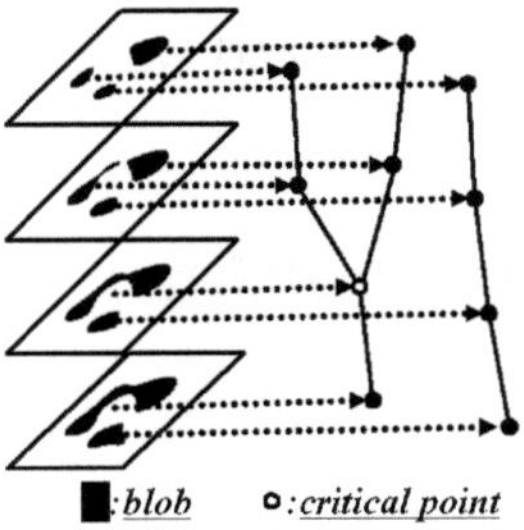


**Fig. 3.** Blobs with nonzero areas

In fact, we never extract the nodes as in the original Reeb-graph. Since we wish to avoid topological instability. In this extraction, only branches of the

Reeb graph are detected and recorded. According to Morse theory, the indices at critical points are equivalent to the topological change before and after the critical points. Therefore, this information alone will be enough to determine the topology of the manifold.

We actually use a weighted tree, i.e. each branch is labeled with information such areas of blobs and moments of the areas. This information can be used in fast matching of these graphs.

## 5 Fast Matching Algorithm

To compare the scale-space Reeb graphs of two contents, one can use the fast and parallel algorithms for graph-matching [15]. However, complete matching could still be time-consuming when the number of contents is large.

Below, we show a fast matching algorithm using partial matching of the trees.

### 5.1 Fast Matching of Nodes

In order to avoid exhaustive matching of all combinations of nodes at the same depth of the tree, the nodes are firstly matched using the information of blobs labeled to the nodes such as the areas and means $m$ of each blobs and their quadratic moments ( $\Psi_1$, $\Psi_2, \cdots$ , $\Psi_7$ ) etc. Here

$$
\begin{aligned}
\Psi_1 &= \eta_{20} + \eta_{02} \\
\Psi_2 &= (\eta_{20} - \eta_{02})^2 + 4\eta_{22}^2 \\
&\vdots \\
\Psi_7 &= (3\eta_{21} - \eta_{03})(\eta_{30} + \eta_{12})\{(\eta_{30} + \eta_{12})^2 - 3(\eta_{21} + \eta_{03})^2\} \\
&\quad +(3\eta_{12} - \eta_{30})(\eta_{21} + \eta_{03})\{(3\eta_{30} + \eta_{12})^2 - (\eta_{21} + \eta_{03})^2\} \\
\eta_{pq} &:= \frac{\mu_{pq}}{\mu_{00}^r} \\
\mu_{pq} &:= \sum_{x \in b_i} \sum_{y \in b_i} (x - \bar{x})^p (y - \bar{y})^q f(x, y) \\
r &:= (p + q + 2)/2.
\end{aligned}
$$

Nodes are represented as points in a vector space with coordinates $(m, \Psi_1, \Psi_2, \cdots, \Psi_7)$ (fig.4). We call this space a feature space and match nodes in it with the Euclidean distance between them for simplicity. We say two nodes $r_i, r'_j$ match if they are the nearest pair. .

**Remark**: To defend against cropping attack, or to find an embedded copy, we record also matching results between subsets of all nodes.

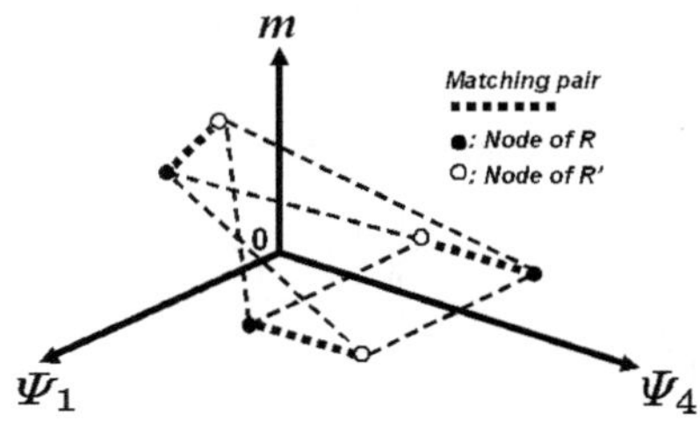


**Fig. 4.** Matching of nodes

### 5.2 Matching of Reeb Graphs

In order to fast match two scale-space Reeb graphs, we define a similarity $\sigma(R, R')$ between two Reeb-graphs $R, R'$ using only local graphic topology around nodes. It is described by the number of edges of the neighborhoods of the matching nodes, which is a variation of in [16].

$$\sigma(R, R') = S(R, R') \Big/ S(R, R).$$

$$S(R, R') = \sum_{r \in R, r' \in R'} s(r, r') - \frac{1}{2}\Big(\sum_{\bar{r} \in R} d_{\bar{r}} + \sum_{\bar{r'} \in R'} d_{\bar{r'}}\Big).$$

Here $L_r, L_{r'}$ are the number of preceding or parent nodes of nodes $r, r'$, $U_r, U_{r'}$ the numbers of subnodes of $r, r'$. $d_r$ the depth from the node $r$ upto the root node. $\bar{r}, \bar{r'}$ are the indices of non-matching nodes, $s(r, r')$ is the similarity between nodes $r, r'$ and is defined as

$$s(r, r') = 2^{-d_r}\Big\{w \cdot (1 - \frac{\|L_r - L_{r'}\|}{max(L_r, L_{r'})}) + (1 - w) \cdot (1 - \frac{\|U_r - U_{r'}\|}{max(U_r, U_{r'})})\Big\}.$$

$w$ is a weight used to adjust matching between $U_r, U_{r'}$ and $L_r, L_{r'}$. When $U_r = U_{r'} = 0, L_r = L_{r'} = 0$, we defines $max(U_r, U_{r'}) = 1, max(L_r, L_{r'}) = 1$.

## 6 Simulations

The image "Girl" is used as the original image and "Aerial, Baboon, Balloon, Car, Couple, Earth, F16, House, Jelly bean, Lady, Lena, Milk drop, Parrot, Pepper, Sail boat, Tiffany, Tree, Woman" are used as test images. in (fig.8)

The Gaussian curvatures $K_R, K_B, K_G$ for the tri-stimulus $R, G, B$ are firstly computed. Their scale-space representations as surfaces $\{L_{K_R}(\boldsymbol{x}, t)\}$, $\{L_{K_G}(\boldsymbol{x}, t)\}$, $\{L_{K_B}(\boldsymbol{x}, t)\}$ for $t \in \mathbb{R}_+$ are obtained by Gaussian filtering of $3 \times 3$ neighborhood with covariant $t$. The blobs are chosen as either the negative part or the positive

**Table 1.** Similarity: The original "Girl" vs deformed copies

| Attack | $\sigma$ | Attack | $\sigma$ |
|---|---|---|---|
| JPEG:100 | 0.9434 | JPG:90 | 0.9401 |
| JPEG:80 | 0.9248 | JPEG:70 | 0.9449 |
| JPEG:60 | 0.9330 | JPEG:50 | 0.9221 |
| JPEG:40 | 0.9231 | JPEG:35 | 0.9242 |
| JPEG:30 | 0.9186 | JPEG:25 | 0.9219 |
| JPEG:20 | 0.9167 | JPEG:15 | 0.9110 |
| AFFINE:1 | 0.9234 | AFFINE:2 | 0.8511 |
| AFFINE:3 | 0.9470 | AFFINE:4 | 0.8827 |
| AFFINE:5 | 0.9110 | AFFINE:6 | 0.9026 |
| AFFINE:7 | 0.9324 | AFFINE:8 | 0.9233 |
| ROTATION:0.25 | 0.9521 | ROTATION:0.50 | 0.9155 |
| ROTATION:0.75 | 0.9030 | ROTATION:0.90 | 0.8641 |
| ROTATION:1.00 | 0.8724 | ROTATION:5.00 | 0.8405 |
| ROTATION:10.0 | 0.8420 | ROTATION:15.0 | 0.8147 |
| ROTATION:30.0 | 0.8024 | ROTATION:45.0 | 0.7991 |
| ROTATION:90.0 | 0.9824 | RESCALE:0.50 | 0.8702 |
| RESCALE:0.75 | 0.8901 | RESCALE:0.90 | 0.9591 |
| RESCALE:1.10 | 0.9348 | RESCALE:1.50 | 0.9095 |
| RESCALE:2.00 | 0.9016 | PSNR:10 | 0.9849 |
| PSNR:20 | 0.9837 | PSNR:30 | 0.9753 |
| PSNR:40 | 0.9803 | PSNR:50 | 0.9618 |
| PSNR:60 | 0.9728 | PSNR:70 | 0.9801 |
| PSNR:80 | 0.9845 | PSNR:90 | 0.9797 |
| PSNR:100 | 0.9814 | - | - |

**Table 2.** Similarity: The original "Girl" vs other images

| Image | $\sigma$ | Image | $\sigma$ | Image | $\sigma$ |
|---|---|---|---|---|---|
| Aerial | 0.4036 | Baboon | 0.5721 | Balloon | 0.4797 |
| Car | 0.3457 | Couple | 0.4597 | Earth | 0.2079 |
| F16 | 0.3757 | House | 0.4905 | Jelley beans | 0.2612 |
| Lady | 0.4958 | Lenna | 0.7134 | Milk drop | 0.4210 |
| Parrot | 0.4454 | Pepper | 0.5185 | Sail boat | 0.3674 |
| Tiffany | 0.6266 | Tree | 0.3842 | Woman | 0.5509 |

part of the Gaussian curvatures, then transformed to binary images. In fig.5, fig.6 and fig.7 the Reeb graphs of $\{L_{K_R}(\boldsymbol{x},t)\}$, $\{L_{K_G}(\boldsymbol{x},t)\}$, $\{L_{K_B}(\boldsymbol{x},t)\}$ of the Girl are shown.

The proposed method is then applied to content identification under various attacks of StirMark4.0[21]. The matching results of similarity in percentage between the original image "Girl" and its deformed copies are shown in Table 1. "Attacks" denotes for the attacks to the original image by StirMark. In particular, these include JPEG compression with quality parameter from 15-100.

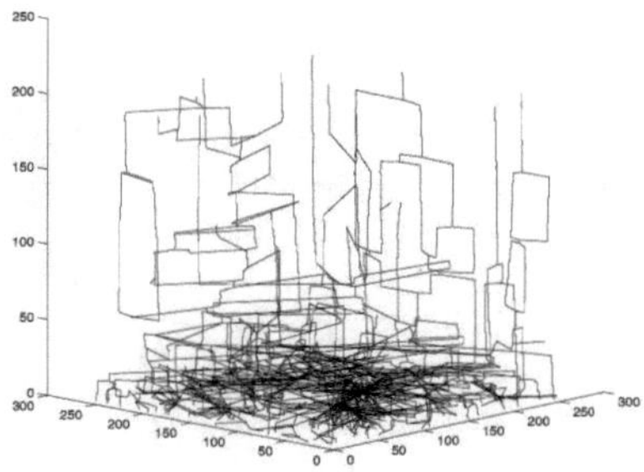

**Fig. 5.** Reeb Graph of curvature of Red in "Girl"

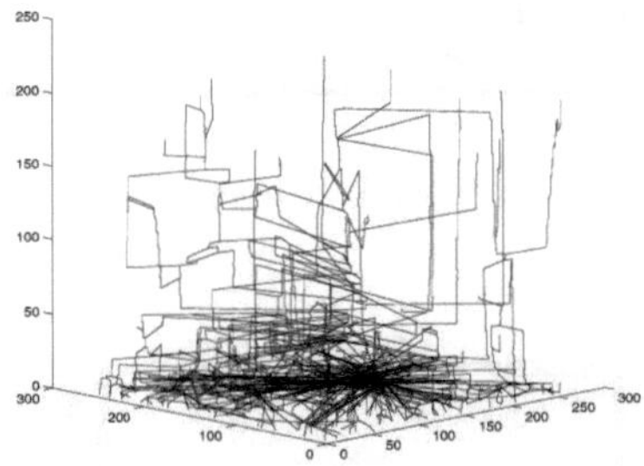

**Fig. 6.** Reeb Graph of curvature of Green in "Girl"

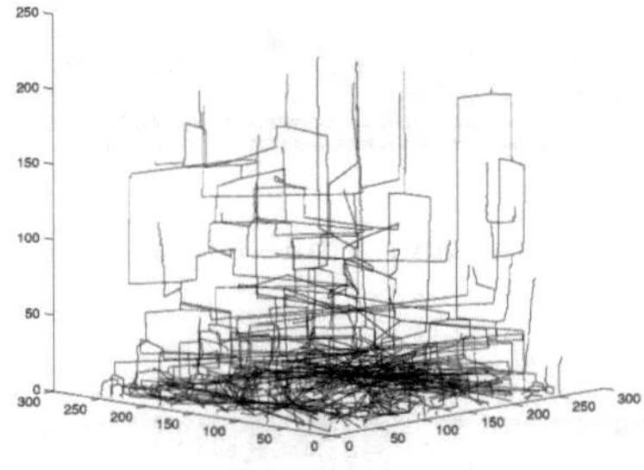

**Fig. 7.** Reeb Graph of curvature of Blue in "Girl"

eight random variations of AFFINE attacks, ROTATION from 0.25-90 degree, RESCALE with scaling factor from 0.5-2. PSNR transformations from 10-100dB. All the deformed or attacked images have similarities with the original image greater than 80%. On the other hand, matching between the original "Girl" and other images are shown in Table 2, the similarities are lower than 63 %.

**Fig. 8.** Test images

**Fig. 9.** Original and Affine-transformed Girls

**Fig. 10.** Original and Rotated Girls

## Acknowledgment

This research is partially supported by Institute of Science and Engineering, Chuo University.

## References

1. J.P.H.Huang, L.C.Jain "Intelligent watermarking techniques" World Scientific, 2004
2. M. Arnold, M. Schmucker, S.D. Wolthusen "Techniques and application of watermarking and content protection" Artech House INC. 2003
3. B. B. Zhu, M. D. Swanson, A. H. Tewfik, "When seeing isn't believing", IEEE Signal Processing Magazine, Vol.21, No2, pp40-49, Mar. 2004
4. C. S. Lu, C. Y. Hsu, S. W. Sun, P. C. Chang, "Robust Mesh-based Hashing for Copy Detection and Tracing of Images", Proceedings of ICME2004 ,Taiwan 2004.
5. J. Chao, S. Suzuki, "Copyright tracing using topological invariants of content" Proceedings of ICIP 2004, Singapore 2004.
6. A.T.Fomenko, T.L.Kunii, *"Topological Modeling for Visualization"*, Springer-Verlag, 1997.
7. T. Linderberg: *Scale-space theory in computer vision* Kluwer Academic Publishers.
8. J.Milnor, *Morse Theory, AMS:Vol 51*,Princeton Univ. Press,1963.
9. G, Reeb, *Surles points singuliers dúne form de pfaff completement integrable ou dúne fonction numerique"* Comptes Rendus Acad. Sciences Paris 222, 847-849.
10. L. Florack and A. Kuijper "The topological structure of scale space images" Journal of Mathematical Imaging and Vision, 12, 65-79, (2000)
11. J. A. Bangham, P. Chardaire, C. H. Pye, P. D. Ling "Multiscale nonlinear decomposition: the sieve decomposition theorem" IEEE Transactions on Pattern Analysis and Machine Intelligence, 18 (5): 529-539, 1996.
12. Y. Shinagawa, T. Kunii "Unconstrained automatic image matching using multiresolutional critical-point filters" IEEE Transactions on Pattern Analysis and Machine Intelligence, Vol.20, No.9, Sept. 1998.
13. K. Habuka and Y. Shinagawa, "Image interpolation using enhanced multiresolutional critical-point filters", Int. Journal of Computer Vision, 58 (1), 19-35, 2004.
14. R. Harvey, J. A. Bangham and A. Bosson "Scale-space filters and their robustness" In Proc. First Int. Conf. on Scale-space theory". P.341-344, Springer, 1977.
15. M.Karpinski, W.Rytter "Fast Parallel Algorithms for Graph Matching Problems: Combinatorial, Algebraic & Probabilistic Approach "Oxford Univ. Press 1998.
16. M. Hilaga, Y. Shinagawa, T. Kohmura, T.L. Kunii "Topology matching for fully automatic similarity estimation of 3D shapes", Proc. ACM SIGGRAPH 2001.
17. J. Kim, J. Chao "Topological invariants of color images and their application to copy right protection", Proc. of 2003 Symposium on Cryptography and Information Security, SCIS20 03. pp959-964, Jan. 2003.
18. M. Suzuki, J. Kim, J. Chao "Copyright protection using topological invariants of Image" Proc. of 2002 Symposium on Cryptography and Information Security, SCIS2002. pp.903-907.Jan.2002
19. L. J. van Villet, I. T. Young, P. W. Verbeek "Recursive Gaussian Derivative Filters", Proceedings of ICPR1998, Australia 1998.
20. I. T. Young, L. J. van Viliet "Recursive implementation of the Gaussian filter", Signal Processing, vol.44, p.139-151, 1995.
21. http://www.petitcolas.net/fabien/watermarking/stirmark/

# Salient Regions from Scale-Space Trees

Jose Roberto Perez Torres, Yuxuan Lan, and Richard Harvey

School of Computing Sciences, University of East Anglia, Norwich, NR4 7TJ, UK
{jrpt,yl,rwh}@cmp.uea.ac.uk

**Abstract.** Extracting regions that are noticeably different from their surroundings, so called *salient* regions, is a topic of considerable interest for image retrieval. There are many current techniques but it has been shown that SIFT and MSER regions are among the best. The SIFT methods have their basis in linear scale-space but less well known is that MSERs are based on a non-linear scale-space. We demonstrate the connection between MSERs and morphological scale-space. Using this connection, MSERs can be enhanced to form a *saliency tree* which we evaluate via its effectiveness at a standard image retrieval task. The tree out-performs scale-saliency methods. We also examine the robustness of the tree using another standard task in which patches are compared across images transformations such as illuminant change, perspective transformation and so on. The saliency tree is one of the best performing methods.

## 1 Introduction

In a variety of image processing tasks the concept of *saliency* [1] has become increasingly useful. Salient regions are ones that are somehow more informational than others. In [2], for example, salient regions are found in video frames, which are then used to retrieve new frames by selecting query regions. Since the human vision also seems to use saliency there is natural interest in automatic saliency algorithms. The most well known of recent algorithms is that due to Lowe [3]: the Scale Invariant Feature Transform (SIFT). SIFT uses a differences-of-Gaussian pyramid[1] to detect interesting regions (in much the same way as was firstly proposed in [5]) and then forms a 128-element feature vector to describe those regions.

An alternative approach is described in [1] which relies on local entropy:

$$H_{f,R} = -\sum_i P_{f,R}(f_i) \log_2 P_{f,R}(f_i) \tag{1}$$

where $P_{f,R}(f_i)$ is the probability of a descriptor $f$ taking the value $f_i$ in the local region $R$. Typically $R$ is a circle of varying scale centred around each point

[1] To properly preserve scale-space causality [4], these DoG filters $D(x, y, \sigma) = (G(x, y, k\sigma) - G(x, y, \sigma)) * I(x, y)$ should be formed from the discrete approximation to the Gaussian kernel [5] but, in practice, [3] uses a seven-point approximation to the kernel applied in a multi-resolution framework.

F. Sgallari, A. Murli, and N. Paragios (Eds.): SSVM 2007, LNCS 4485, pp. 350–361, 2007.

that one wishes to test. To assess which technique is most effective, a variety of local region detectors are evaluated in [6] using some benchmark images and several robustness scores. There are no universal winners for all tasks but the authors conclude that two methods have consistently better performance that the others: the SIFT method and another new technique known as Maximally Stable Extremal Regions (MSERs) [7].

MSERs are grayscale contours in the image that are likely to remain unchanged under typical transformations such as illuminant variation, perspective transformation, zoom and other distortions. In [8] it is shown that MSERs are a special case of a more general saliency detector known as Stable Salient Contours (SSCs) which we describe later.

Although the evaluation in [6] is quite comprehensive it is important to note that it is essentially a comparison of how reliably alike the selected salient regions are, rather than a test of how useful the regions are for particular tasks. In image retrieval this difference can be acute – a salient region that does not correspond to an object, or a semantically meaningful part of the image, is not likely to be useful. This problem has been tackled in [9] where the retrieval performance of difference-of-Gaussian detectors are evaluated on a standard test set, but they did not test SIFT, MSERs or SSCs so currently we do not know the most effective method for image retrieval.

This paper therefore attempts to measure the performance of a variety of scale-space and non-scale-space saliency algorithms for the task of image retrieval using the methodology developed in [9]. In Section 2 we briefly describe the saliency algorithms and derive an improved version of SSCs. Section 3 evaluates the reliability of the new SSC method using the methodology in [6] and in Section 4 we test several algorithms on an image retrieval task and give reasons for the difference in performance between entropy-based methods such as [1] and difference-based methods such as SIFT [3] and SSCs [8]. Section 5 discusses further properties of SSCs and their hierarchical representation which we call a *saliency tree.*

## 2 Algorithms

The SIFT algorithm is described in [3] and executables for a couple of operating systems are available from the authors' website. Here, however, we use public domain Matlab implementation[2] since it runs across multiple operating systems. The scale saliency algorithm [1] is available as an executable from the author's website[3]

Stable salient contours (SSCs) derive from a class of morphological scale-space processors called *sieves* that originated as one-dimensional filters that were later extended to two-dimensional graph morphology [10] and, more recently, colour [11]. They are often described as operations over a graph $G = (V, E)$

[2] `http://vision.ucla.edu/~vedaldi/code/sift/sift.html`

[3] Linux version 1.5: `http://www.robots.ox.ac.uk/~timork/salscale.html`.

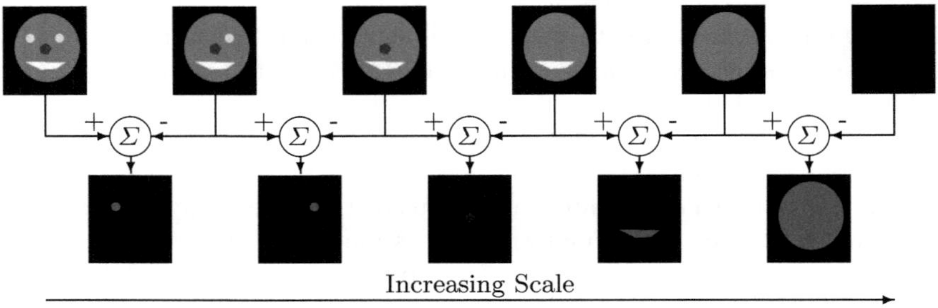


**Fig. 1.** Decomposition of a computer-generated image using the sieve algorithm. The top line shows images simplified with an $M$-sieve. The bottom line show the granules which are coded red for positive granules and blue for negative ones.

where $V$ is the set of vertices that label the pixels and $E$ are edges that indicate the adjacencies between pixels ([10] has definitions and references to graph morphology in which this notation is standard). There are four types of sieve corresponding to four grayscale graph morphology operators of open, $\mathcal{O}$, close, $\mathcal{C}$, $\mathcal{M}$- and $\mathcal{N}$-filters. $\mathcal{O}_s$, $\mathcal{C}_s$, $\mathcal{M}_s$ and $\mathcal{N}_s$ are defined for each integer scale, $s$, at pixel, $x$ as:

$$\mathcal{O}_s f(x) = \max_{\xi \in C_s(G,x)} \min_{u \in \xi} f(u), \quad \mathcal{C}_s f(x) = \min_{\xi \in C_s(G,x)} \max_{u \in \xi} f(u), \tag{2}$$

and $\mathcal{M}_s = \mathcal{O}_s\mathcal{C}_s$, $\mathcal{N}_s = \mathcal{C}_s\mathcal{O}_s$. Thus $\mathcal{M}_s$ is an opening followed by a closing, both of size $s$ and in any finite dimensional space. The sieves of a function, $f \in \mathbf{Z}^V$ are defined in as sequences $(f_s)_{s=1}^{\infty}$ with

$$f_1 = \mathcal{P}_1 f = f, \text{and } f_{s+1} = \mathcal{P}_{s+1} f_s \tag{3}$$

for, $s \geq 1$ where $\mathcal{P}_s$ is one of $\mathcal{O}_s$, $\mathcal{C}_s$, $\mathcal{M}_s$ and $\mathcal{N}_s$.

Sieves decompose an input image and remove its features by increasing the analysis scale from areas of size one (one pixel) to the size of the image area. Figure 1 shows a synthetic image that is progressively simplified by the removal of maxima and minima. The right "eye" is slightly larger than the left which is why it is removed at a larger scale. Here an $\mathcal{M}$-filter is used so the maxima and minima are removed together at each scale but, if the operator were an opening, then the image would be simplified by the progressive removal of maxima. Differences between successive scales are called *granule functions*, and these functions which exist in the *granularity domain* can be summed to give the original image. The method is thus a lossless transform. In [10] there is a proof that sieves preserve scale-space causality in the strict sense discussed in [12]. Since smaller granules are contained within larger ones, the method generates a tree [13] in which the vertices of the tree (red dots in Figure 2) represent granules and the edges (blue lines) represent containment. In Figure 2 the root of the tree is plotted on the image plane, nodes are plotted with their $(x, y)$ co-ordinate at the centroid of the granule, and their $z$ co-ordinate is tree depth.

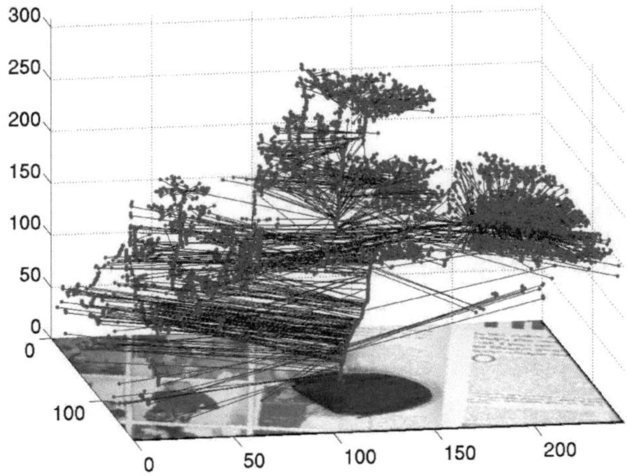


**Fig. 2.** Sieve tree computed using a morphological $\mathcal{M}$ filter on a 165 by 247 pixel image. Computing this tree, which has 6498 nodes, took 0.22 seconds on 996MHz Pentium III.

Via fast list algorithms [14], sieves and their corresponding trees can be computed quickly but, as can be seen from Figure 2, the trees can be complicated and hence subsequent processing can become unwieldy. This observation was the original motivation for simplifying the tree. However, independently [7], it was realized that contours could provide stable regions known as MSERs and in [8] it was shown that these regions are a subset of the nodes of opening and closing trees. Since it is known that $M$ and $N$-sieves are more robust than open/close trees [15] it has been postulated that contours extracted from $M$- and $N$-trees should be as stable, and possibly more stable, than MSERs which are known to be among the best performing regions in terms of repeatability. In [8] such regions are termed Stable Salient Contours or SSCs.

The algorithm in [8] (denoted SSC-A) can be improved. The new algorithm, SSC-B, uses a test on the intensity distributions measured in the parent and child regions. The test recommended in [8] is a Kolmogorov-Smirnov test which produces a score, denoted $\lambda$, that measures the similarity of two regions: large $\lambda$ tends to reject the null hypothesis that the two regions are drawn from the same intensity distribution. The original algorithm involved quite time-consuming parsing of the tree looking for long chains of nodes with low $\lambda$ scores whereas SSC-B is more efficient, as it avoids traversing the tree structure in the same way as SSC-A. More importantly, as will be shown later, it has a repeatability that is similar or better than the SSC-A but produces more salient regions. The algorithm has two inputs: $\mathcal{T}$, which is the tree, and $\theta$, which is a user-defined saliency threshold, and one output, $\mathcal{L}$, which is a list of SSC nodes. $R(i)$ is pixel set which supports the $i$th node in the image, $\Lambda(R_1, R_2)$ is the K-S $\lambda$ value of two regions $R_1$ and $R_2$. Each node $i$ in $\mathcal{T}$, is compared to its parent $j$ and the dissimilarity is measured by $\Lambda$. Also, the relative area difference between two

**Algorithm 1.** Stable salient contours of tree $\mathcal{T}$, Algorithm B

**function** $\mathcal{L} = SSC(\mathcal{T}, \theta)$
$\epsilon \leftarrow$ predefined value
$l \leftarrow \emptyset$
**for** $i := 1$ to $\|\mathcal{T}\|$ **do**
  $j \leftarrow \mathcal{T}.parent(i)$
  $\lambda \leftarrow \Lambda(R(i), R(j))$
  $a = (\|R(j)\| - \|R(i)\|)/\|R(j)\|$
  $p(i) = 1/(\lambda \times \sqrt{a})$ {$p$: saliency measurement }
  **if** $p(i) > \theta$ **then**
    $l \leftarrow [l, i]$
  **end if**
**end for**
$\mathcal{L} = groupnodes(l, \epsilon)$;

**return** $\mathcal{L}$

nodes is computed as $a$. A new saliency measurement, $p$, is introduced to be inverse-proportional to $\lambda$ and relative area $a$. Any node with $p$ exceeding the saliency threshold $\theta$ is a candidate of SSC. We set $\theta \propto \text{mean}(p)$.

A function *groupnodes* has the effect of merging several nodes into one if the regions they cover are very similar. In other words, if the overlapping error of these regions are below a threshold $\epsilon$ then they are merged. This step helps reduce the number of SSC regions produced but still retains the same amount of information. High $\epsilon$ values tend to suppress the large number of SSCs whereas, since large amount of nodes are merged into one, it may cause some information loss ($\epsilon = 0.2$ is the default). The effect is illustrated in Figure 3: having selected too many nodes we then have a "clean-up" phase where we remove overlaps. The effect of the algorithm is illustrated in Figure 4. On the left is an image of the same object and, on the right is the contour with the largest $p$ values.

## 3 Stability and Repeatability of SSCs

To evaluate the algorithm in a more general setting we use the standard framework of Mikolajczyk [6] in which regions are detected in sequences of images that have undergone various transformations. Since the transformations are known, it is possible to warp the detected regions onto the reference image and hence measure, firstly, the number of regions that overlap (indicating stability and repeatability) and, secondly, the number of regions detected indicating sensitivity. A full description of the methods is given in [6] although here we are interested only in the regions so we do not compute the SIFT-based matching scores which form another part of the evaluation.

Repeatability is defined as the ratio of the number of region-to-region correspondences to the smaller of the number of regions in the two images. A correspondence is declared if ellipses fitted to the regions overlap with a sufficiently

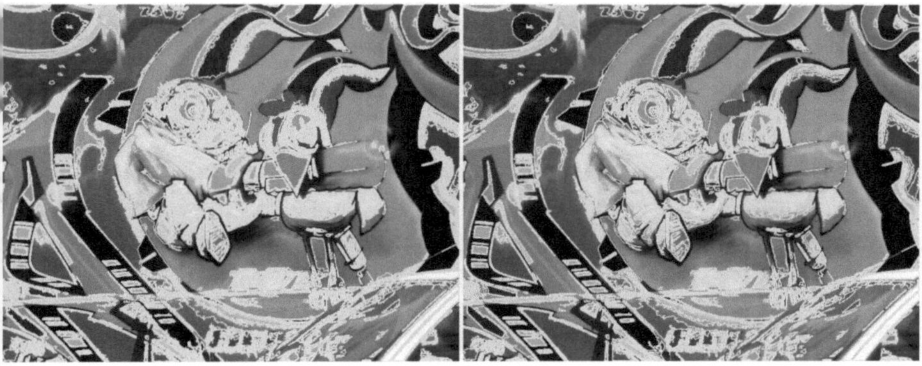

**Fig. 3.** SSCs before (left) and after (right) merging, $\epsilon = 0.2$. For the whole image, the number of SSCs is reduced from 11692 to 1665 afterwards, with no significant loss of information.

small error (we use 40% which is the value defined in the standard test harness in [6]). In Figure 5 the variation in repeatability between the methods does not vary much. As an example, in [6] the highest scoring method on the wall image is MSERs which here is almost identical to SSC-B. Given the strong similarities between SSCs and MSERs this perhaps not surprising. However SSC-B does appear to vary considerably in the number of correspondences generated. This is probably because both SSC-A and MSERs seek to identify long chains in the tree whereas SSC-B attempts to find stable salient contours directly.

## 4 Image Retrieval Performance

In [9] a retrieval benchmark based on the Washington ground-truth database is introduced. It has 1224 RGB images in 20 classes, each with different number of images. Some images have text descriptions of objects in the scene. Retaining

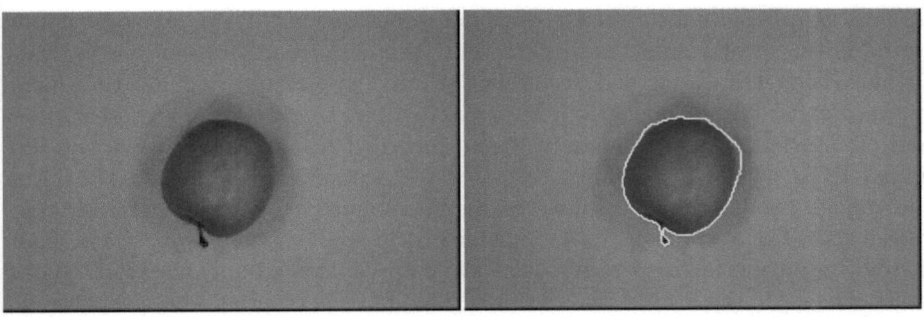

**Fig. 4.** Illustration of SSC-B algorithm showing how selecting contours with high $p$ value often corresponds to a segmentation or near-segmentation

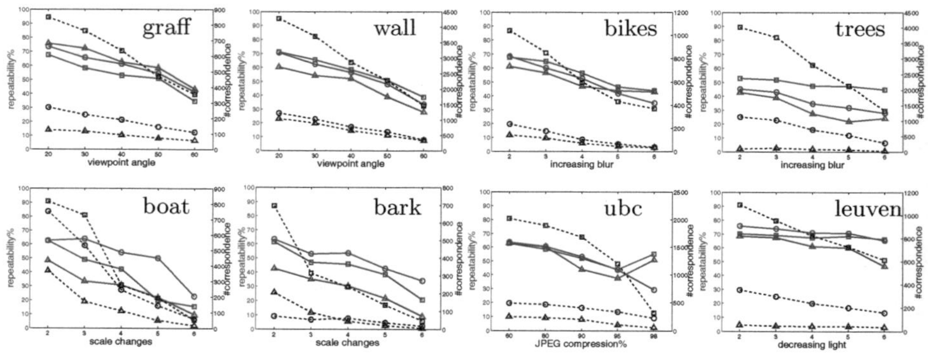


**Fig. 5.** Showing the repeatability (red solid lines) and number of correspondences (blue dashed lines) for SSC-B (square markers), SSC-A (triangles) and MSERS (circles). The code in the top right-hand corner of each plot refers to the test sequence in [6].

these, gives 697 images in 17 different classes. Each image is used as a query and the effectiveness of the retrieval is measured via *semantic relevance* [9], *SR*,

$$SR = \frac{\mid Q \bigcap R \mid}{\mid Q \mid} \tag{4}$$

where $Q$ is the set of words from the query image description, and $R$ is the set of words from a returned image. The final score is the mean of *SR* over all possible queries.

A simple retrieval method is to compute a histogram from each image and to rank retrievals by their Euclidean distance between histograms

$$d_E^2(\boldsymbol{F}_1, \boldsymbol{F}_2) = (\boldsymbol{F}_1 - \boldsymbol{F}_2)^2 = \sum_{k=1}^{K} |F_1(k) - F_2(k)|^2 \tag{5}$$

where $F_i(k)$ is the $k^{th}$ element of the $K$-element histogram $\boldsymbol{F}_i$. This method contains no segments or salient regions so is referred to in [9] as *global* and Table 1 gives results for a variety of colour quantizations. There are some minor errors in the results reported in [9]. Firstly the list of text descriptions appears to have been truncated and secondly the retrieved results include the query image. Rather than propagate these errors here, we use the full text descriptions and eliminate the query image from the returned results (For comparison, for global RGB 64-bin histograms, [9] reports a semantic relevance of 37.6% for the rank-1 result and 45.6% for the mean semantic relevance of the top five retrievals.) The results in Table 1 confirm the well known result that colour histograms outperform greyscale ones. We have considered only the RGB colour space since, for these data, it is known that more perceptual colour spaces do not substantially improve performance [9].

We now wish to compare these global methods to ones based on salient regions. We consider four methods based on saliency. The first is the standard

**Table 1.** Semantic relevance for global descriptors

| Feature | Space/Bins | Rank 1 Result | Mean Top 5 |
|---|---|---|---|
| Random | | 12.59% | 12.64% |
| Global | Grey/64 | 32.15% | 27.50% |
| Global | RGB/$64^3$ | 42.15% | 34.69% |
| Global | RGB/$5^3$ | 44.50% | 35.83% |

**Fig. 6.** Images from the Washington dataset with regions of saliency edged in yellow. Showing salient regions [1] (top row), salient sieve regions (middle row) and stable salient contours (bottom row). The middle row is plotted with slightly thinner lines so that the outline of the regions is visible.

scale-saliency method [1] which uses entropy computed in circular windows. The second is a modification in which entropy is still used, but the windows are taken from the sieve tree, and the third is SSCs which were described previously. The fourth is the SIFT method.

In [1] the entropy is considered as a function of the scale of a circular window. Salient regions are circles that have local entropy maxima as a function of scale. The top row of Figure 6 shows regions selected using this algorithm. It is possible to adapt this idea to the sieve tree. We consider every path from leaf to root. Each path generates a nested set of windows which can be selected by looking for windows that are entropy maxima. These nodes are then ranked by their entropy and, as recommended in [9] the fifty nodes with the highest entropy are retained. Example results are shown on the middle row of Figure 6. The regions are quite different to the scale saliency ones and there is now a tendency to select larger, more meaningful regions that lie within object boundaries. The final method, the SSCs, is shown on the bottom row of Figure 6 (we select only

**Table 2.** Semantic relevance for salient regions

| Feature | Space/Bins | Rank 1 Result | Mean Top 5 |
|---|---|---|---|
| SS | RGB/5 | 40.88% | 35.27% |
| ESC | Grey/256 | 30.73% | 25.26% |
| ESC | Grey/64 | 26.45% | 23.69% |
| ESC | RGB/$5^3$ | 36.33% | 30.55% |
| SSC | RGB/$5^3$ | 39.02% | 33.21% |
| SSC $\chi^2$ | RGB/$5^3$ | 49.31% | 39.91% |
| SSC B/G | RGB/$5^3$ | 43.21% | 35.95% |
| SIFT | Grey/256 | 16.45% | 15.55% |

SSCs that have areas greater than 500 pixels). Again we have selected the fifty strongest regions (highest $p$ score). Whereas entropy tends to favor textured regions (broad intensity histogram), SSCs favor smooth regions with structured intensity histograms. When using the SSC and sieve-based Entropy for retrieval we use the same histogram distance measure as in [9]

$$D_{12} = \sum_{j=1}^{N_1} \min k\{d_E(F_j, G_k)\} \tag{6}$$

where image 1 has $N_1$ histograms $F_j$, and image 2 has $N_2$ histograms $G_k$. $D$ is formed by finding the best matching feature from Image 1 to each feature in Image 2 and summing these best matches. Hence $D$ is non-symmetric.

We have not drawn the SIFT regions in Figure 6 because the SIFT technique works differently: the method identifies thousands of keypoints per image but not regions (the SIFT features are generated from square windows taken from blurred images around the keypoints – the scale selects the level of blurring). Every keypoint is a 128-dimensional feature vector and again the Euclidean distance is used to find the closest match to each keypoint. The high density of keypoints creates a further practical problem that has been addressed in [3]: the search time should be reduced by representing the points in a modified $k-d$ tree [16] and cutting-off the search after 200 nearest-neighbour candidates. Table 2 shows the results of this method as SIFT.

The retrieval results are shown in Table 2. The scale saliency method [1], denoted here SS, performs better than the modified Entropy Selected Contours (denoted ESC) which implies that entropy-selected regions will not be useful for windows that follow iso-intensity contours such as those produced by sieve. The middle row of Figure 6 implies that ESC is the worst of all methods: the regions appear to be quite noisy because searching for locally maximum entropy tends to force the regions to expand beyond object boundaries which has a consequence that the histogram derived from such regions is not particularly stable. The SSC method is slightly worse than the SS method and none of the methods can be said to be a substantial improvement on the global histogram.

Including an additional region, the background, which is the set difference between the image and the SSC regions, improves performance. Probably because

**Table 3.** Semantic relevance for salient regions with weighted histograms

| Feature | Space/Bins | Rank 1 Result | Mean Top 5 |
|---|---|---|---|
| SS | RGB/5 | 45.48% | 38.50% |
| ESC | Grey/256 | 29.63% | 25.76% |
| ESC | Grey/64 | 29.27% | 25.41% |
| ESC | RGB/$5^3$ | 44.46% | 36.51% |
| SSC | RGB/$5^3$ | 50.64% | 41.96% |
| SSC $\chi^2$ | RGB/$5^3$ | 52.35% | 43.82% |
| SSC B/G | RGB/$5^3$ | 50.50% | 42.22% |

it is a closer approximation to the global histogram which is the best performing method.

In [17] a large number of histogram methods were tested for image retrieval and it was shown that

$$\chi^2(F_1, F_2) = \sum_{k=1}^{K} \frac{(F_1(k) - e_{1k})^2}{e_{1k}} + \sum_{k=1}^{K} \frac{(F_2(k) - e_{2k})^2}{e_{2k}} \tag{7}$$

was almost always the best choice. In (7), $F_n(k)$ is the $k$th count in the $n$th histogram and $e_{ik} = \frac{N_i(F_1(k)+F_2(k))}{N_1+N_2}$. In Table 2 this is denoted SSC $\chi^2$ and, as predicted in [17], performance improves.

Nevertheless, the results in Table 2 are rather surprising and disappointing, given that the regions from the SSC methods look more meaningful. Our explanation stems from (5) which measures the Euclidean distance between histograms. For global histograms, formed from similarly-sized images, (5) is quite reasonable, but for SSCs there is quite a wide variation in area among regions and it is desirable for the distance measure to be invariant to the precise size of the regions in the image. This can be achieved by normalizing all histograms to sum to unity. Applying this new weighting gives the results in Table 3.

## 5 Conclusions

The new weightings improve most methods but particularly those with regions that correspond to homogeneous regions within the scene. The improvements on the ESC method are comparatively modest which we attribute to entropy preferring regions with intensity distributions that have high entropy. Such distributions arise when a region spans a boundary between two objects. While it may desirable for regions to span two objects if one wishes to correctly sample the intensity distribution of an image, it is highly undesirable if one wishes to correctly sample the intensity distributions of objects – we contend that it is mostly objects one needs to sample for effective image retrieval.

Generating salient regions has proved to be highly effective for many computer vision tasks. This paper has reviewed scale-space methods for generating salient

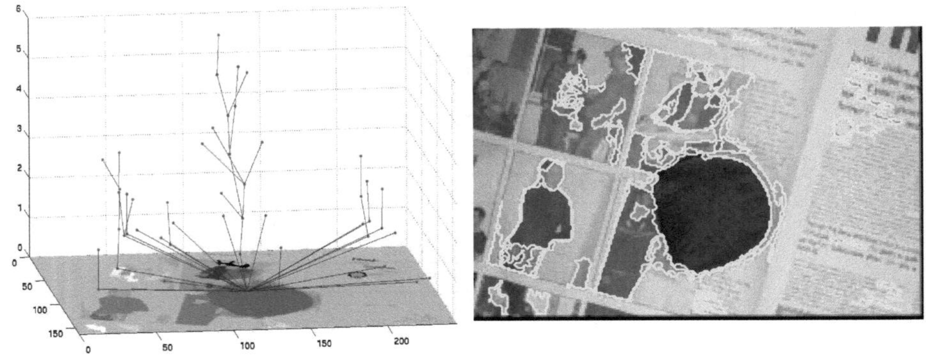

**Fig. 7.** Left: Saliency tree computed from the tree shown in Figure 2. Right: regions in the saliency tree which are regions selected from the sieve tree.

regions (SIFT and SSCs) and compared them to Scale saliency which is non-scale-space method that uses entropy. We have presented and improved version of the SSC and also a hybrid , ESC, which uses entropy to select the contours. These methods have been compared using a standard task. Our objective has been not only to evaluate these methods but also to understand the factors that govern the selection of effective regions for image retrieval.

Scale-saliency generates regions that often lie on the boundary of objects. For image retrieval this is a serious deficiency since objects are not locked to their background: as the viewpoint and context changes; so does the background. The SIFT method is know to perform very well at image matching and object recognition which implies that it is less prone to this problem. However, here the performance of SIFT was exceptionally poor even accounting for the fact that it ignores colour. There is some worsening due the use of the $k - d$ tree (although experiments using the full search imply that this is not serious) but the most likely factor is the way that SIFT forms its features: they are essentially weighted averages of the underlying intensities of patches around keypoints. For image matching this can be very effective since invariance to scale, rotation and illumination is inbuilt. But, for image retrieval the variation is not geometric, or photometric, it is variation within the class of objects: the descriptor "trees" can refer to many, many, instances of trees.

SSCs work the best in our experiments provided one accounts for the large variation in the size of SSC regions. Unlike the other methods, the SSC regions are much less likely to cross an object boundary. So the SSC features, which here are colour histograms, represent samples of object colours which often correlate well with text descriptors. It would be interesting to see if the SSC regions could be combined with the SIFT features to produce and even more effective system. The SSC regions form a data structure that we call a saliency tree which should be a good starting point for the analysis of images. An example saliency tree and its associated regions is shown in Figure 7. Compared to the original tree (Figure 2) the saliency tree is much simpler (50 nodes excluding the root) however the nodes that remain appear to be the outlines of useful regions in the image. Such regions inherit the good properties of MSERs and fast algorithms exist for their computation. The tree is controllable so that it is possible to adjust the number of regions selected from the sieve tree: in the limit one selects all the regions in the sieve tree in which case tree is a lossless transform of the image.

## References


1. Kadir, T., Brady, M.: Saliency, scale and image description. International Journal of Computer Vision **45** (2001) 83–105
2. Sivic, J., Zisserman, A.: Video Google: A text retrieval approach to object matching in videos. In: Proceedings of the International Conference on Computer Vision. Volume 2. (2003) 1470–1477
3. Lowe, D.G.: Distinctive image features from scale-invariant keypoints. International Journal of Computer Vision **60** (2004) 91–110
4. Koenderink, J.: The structure of images. Biological Cybernetics **50** (1984) 363–370
5. Lindeberg, T.: Scale-space theory in computer vision. Kluwer Academic Publishers (1994)
6. Mikolajczyk, K., Tuytelaars, T., Schmid, C., Zisserman, A., Matas, J., Schaffalitzky, F., Kadir, T., Van Gool, L.: A comparison of affine region detectors. International Journal of Computer Vision **65** (2005) 43–72
7. Matas, J., Chum, O., Urban, M., Pajdla, T. (In: Proceedings of the BMVC '02, Cardiff, England)
8. Lan, Y., Harvey, R., Perez Torres, J.R.: Finding stable salient contours. In: Proceedings of the BMVC '05, Oxford, England. (Volume 1.) 30–39
9. Hare, J.S., Lewis, P.H.: Salient regions for query by image content. In: The Challenge of Image and Video Retrieval: Third International conference. Volume 3115 of Lecture Notes in Computer Science., Springer-Verlag (2004) 317–325
10. Bangham, J.A., Harvey, R., Ling, P.D.: Morphological scale-space preserving transforms in many dimensions. Journal of Electronic Imaging **5** (1996) 283–299
11. Gimenez, D., Evans, A.: Colour morphological scale-spaces for image segmentation. In: Proceedings of the BMVC '05, Oxford, England. Volume 2. (2005) 909–918
12. Lifshitz, L.M., Pizer, S.M.: A multiresolution hierarchical approach to image segmentation based on intensity extrema. IEEE Transactions on Pattern Analysis and Machine Intelligence. **12** (1990) 529–540
13. Moravec, K., Harvey, R., Bangham, J.A.: Scale trees for stereo vision. IEE Proceedings-Vision Image and Signal Processing **147** (2000) 363–370
14. Tarjan, R.E.: Efficiency of a good but not linear set union algorithm. Journal of the ACM **22** (1975) 215–225
15. Harvey, R.W., Bangham, J.A., Bosson, A.: Some morphological scale-space filters and their properties. In: Proceedings Scale-Space '97. Volume 1252., Springer-Verlag (1997) 341–344
16. Arya, S., Mount, D.M.: Approximate nearest neighbor queries in fixed dimensions. In: SODA '93: Proceedings of the fourth annual ACM-SIAM Symposium on Discrete algorithms, Philadelphia, PA, USA, Society for Industrial and Applied Mathematics (1993) 271–280
17. Gibson, S., Harvey, R.: Morphological color quantization. In: Proceedings of CVPR (2). Volume 2. (2001) 525–530

# Generic Maximum Likely Scale Selection

Kim Steenstrup Pedersen[1], Marco Loog[1,2], and Bo Markussen[3]

[1] Department of Computer Science
University of Copenhagen
Copenhagen, Denmark
{kimstp,loog}@diku.dk
[2] Nordic Bioscience A/S
Herlev, Denmark
[3] Department of Natural Sciences
Royal Veterinary and Agricultural University
Frederiksberg, Denmark
bomar@kvl.dk

**Abstract.** The fundamental problem of local scale selection is addressed by means of a novel principle, which is based on maximum likelihood estimation. The principle is generally applicable to a broad variety of image models and descriptors, and provides a generic scale estimation methodology.

The focus in this work is on applying this selection principle under a Brownian image model. This image model provides a simple scale invariant prior for natural images and we provide illustrative examples of the behavior of our scale estimation on such images. In these illustrative examples, estimation is based on second order moments of multiple measurements outputs at a fixed location. These measurements, which reflect local image structure, consist in the cases considered here of Gaussian derivatives taken at several scales and/or having different derivative orders.

## 1 Introduction

The response of a feature detector is dependent on the scale at which it is applied. In order to fully exploit this scale dependency, it is necessary to perform some sort of scale selection or estimation procedure in order to fix, at every location in the image, the measurement scale to the appropriate one [1].

One of the well-known approaches to scale selection is the method proposed by Lindeberg [2,3]. His approach addresses the problem of automatic scale selection in a general way for a wide range of image descriptors. Generally, the idea is to study the evolution properties over scale of normalized Gaussian differential descriptors and to track local extrema over scale. The points thus detected are in a certain sense likely to correspond to interesting image structures. A more general setup of the ideas introduced by Lindeberg is presented in the work by Majer [4], in which the author, among other things, tries to formalize in a statistical way what is actually meant by 'interesting image structures'. In [5], a similar selection scheme based on a fractional Brownian image model draws the connection between the fraction parameter and the intrinsic scale of the feature to be detected.

F. Sgallari, A. Murli, and N. Paragios (Eds.): SSVM 2007, LNCS 4485, pp. 362–373, 2007.

Another recent variation on this theme is presented in [6]. It employs a statistical model selection framework in combination with cross-validation techniques to establish a powerful automatic scale estimation procedure. The framework is put up in such a way that it is quite easily applicable to nonlinear diffusion filtering as well. The work is reminiscent of the work by Majer [4] mentioned earlier.

A somewhat related approach to scale estimation may be found in [7]. In it, the authors derive a closed-form maximum likelihood estimate of the global scale for a broad class of Markov random field models. It is not directly clear how this relates to our estimation principle, however the global scale estimate in [7] may have a direct interpretation in terms of an averaging over local estimates as can be determined using our framework.

Although Lindeberg's proposal, and its variations, provide a unified framework for scale selection for different features — ranging from blobs, via corners, to edges and ridges, one may wonder how to perform generic scale estimation, i.e., how to carry out scale selection independent of the descriptor. In addition, it may be beneficial to formulate this task in terms of probabilities, making a probabilistic interpretation possible as well. In this work, we present a general probabilistic framework to do so.

The current exposition reviews and extends the preliminary work presented previously in [8]. In the latter, we introduced the general methodology and performed experiments on synthetic images. Here we refine the method, clarify the definition of intrinsic scale of images, and test the method both on natural and synthetic images.

To begin with, the general maximum likelihood approach to the scale selection problem is formalized [8]. Once the 'unknowns' in this formulation are filled in, actual scale estimation can be carried out. These unknowns are the probabilistic image model to use and the image measurements to base the local scale estimation on. Our main focus will be on the Brownian image model for its relation to natural images [9,10]. Subsequently, making the simplifying assumption that image filter responses can be described adequately by means of knowledge about the central moments of second order and, in addition, assuming that the local image measurements consist of Gaussian filters, or derivatives thereof, a general expression for the second order moments is given.

Following this formulation of the scale selection principle under the Brownian image model, some specific cases are illustrated. Examples on Brownian images and two natural images are provided. Additionally, we also discuss possible applications of our approach.

## 2 Maximum Likely Scale Selection

The general scale selection principle presented is based on maximum likelihood estimation, in which the maximum likely scale for every position in an image is determined based on local image measurements. To actually perform scale estimation, one has to specify the measurement types and the image model to use.

### 2.1 General Scale Selection Principle

Let $F_x^1, F_x^2, \ldots F_x^k$ be a set of $k$ filters at a location $x$. Typically, these filters are linear and assume the form of an inner product, i.e., for every $i \in \{1, \ldots, k\}$, there is a function $f_x^i$ defined on the domain of images such that

$$F_x^i[L] = \langle L, f_x^i \rangle$$

for an image $L$. $F_x := (F_x^1, F_x^2, \ldots F_x^k)^t$, that is, a $k$-dimensional vector of measurement apertures.

Given a particular stochastic image model, a $k$-dimensional probability distribution $\mathsf{p}_s$ of the filter responses coming from $F_x$ can be determined. Here the subscript $s$ indicates the dependency of $\mathsf{p}_s$ on the intrinsic image scale $s$. The maximum likelihood estimate for the scale at a specific location $x$, denoted by $\hat{s}(x)$, is then given by

$$\hat{s}(x) = \operatorname*{argmax}_{s \in (0,\infty)} \mathsf{p}_s(F_x[L]),$$

where the image $L$ is assumed to be a realization of the image model under consideration and $F_x[L]$ is the vector with filter responses at the location $x$ in the image $L$.

## 2.2 The Brownian Image Model

A Brownian image $B(x)$ is a Gaussian random function for which the increments $B(x + \Delta x) - B(x)$ are stationary independently, identically, and normally distributed with a variance proportional to the length of $\Delta x$ (see e.g. [11,9,12,10]). The Brownian image model is the least committed, scale invariant image model that adequately represents the first and second order structure in natural images [10]. In [8], based on the work in [10], assuming this Brownian image model, an expression for the second order moments of the output of localized Gaussian linear filters is established. Given an actual image and assuming that it is a realization of the Brownian image model, the local likelihood can be determined and our scale selection principle can be invoked.

More specifically, using the previous assumptions, the local likelihood over scale can be determined as follows. Firstly, for a zero-mean Brownian model at scale $s$ (with an initial zero-scale power spectrum $\beta/||\omega||^2$), one can derive an analytic expression for the second central moment $\mathsf{C}_{mn}^{\sigma\tau}(s)$, i.e., the covariance, of the responses of the linear Gaussian derivative filters $G_{m_1 n_1}^{\sigma}$ and $G_{m_2 n_2}^{\tau}$ located at a position $x$. In this, $m_1 n_1$ and $m_2 n_2$ indicate the order of differentiation for both filters. In the case considered in [10], $\sigma = \tau$ and $s = 0$, and the covariance is given by

$$\mathsf{C}_{mn}^{\sigma} = (-1)^{\frac{m+n}{2}+m_2+n_2} \frac{\beta m! n!}{2\pi \sigma^{m+n} 2^{m+n} (n+m) \frac{m}{2}! \frac{n}{2}!},$$

if both $m = m_1 + m_2$ and $n = n_1 + n_2$ are even, otherwise $\mathsf{C}_{mn}^{\sigma} = 0$. From the results in [10], it can be derived that generally for $s \in [0, \infty)$, if both $m = m_1 + m_2$ and $n = n_1 + n_2$ are even, the covariance is given by

$$\mathsf{C}_{mn}^{\sigma\tau}(s) = (-1)^{\frac{m+n}{2}+m_2+n_2} \frac{\beta m! n!}{2\pi \left( \sqrt{\frac{\sigma^2}{2} + \frac{\tau^2}{2} + s^2} \right)^{m+n} 2^{m+n} (n+m) \frac{m}{2}! \frac{n}{2}!}. \quad (1)$$

Again, $\mathsf{C}_{mn}^{\sigma\tau}(s) = 0$ otherwise.

Brownian images are scale invariant and as such do not have an intrinsic scale. By applying a scaling transformation $S$ to the domain $\mathbb{R}^2$ of a Brownian image we get the following distributional equivalence relation

$$B(x) \triangleq \frac{1}{\sqrt{\sqrt{|S|}}} B(S\,x) \triangleq (s_1 s_2)^{-1/4} B(S\,x) \tag{2}$$

where $\triangleq$ denotes equality in law and $|S|$ denotes the determinant of the scaling transformation defined by

$$S = \begin{pmatrix} s_1 & 0 \\ 0 & s_2 \end{pmatrix} . \tag{3}$$

The equivalence relation (2) can be generalized to differentiable mappings $h(x) : \mathbb{R}^2 \to \mathbb{R}^2$ of the image domain and reads

$$B(x) \triangleq \left| \frac{\partial h}{\partial x} \right|^{-1/4} B(h(x))$$

where $|\frac{\partial h}{\partial x}|$ denotes the determinant of the Jacobian of $h(x)$ and $\sqrt{|\frac{\partial h}{\partial x}|}$ is the effective isotropic scale of the possibly anisotropic scaling performed by $h(x)$.

Note that we can extend this definition to image sequences, i.e. including a time dimension, by considering Galilean type transformations of the domain $\mathbb{R}^2 \times \mathbb{R}$.

If we are to define an intrinsic scale of a Brownian image we need to do this relative to some reference or inner scale. By applying the Gaussian aperture $G_{00}^{s_0}$ of scale $s_0 > 0$ to a Brownian image $B(x)$ the resulting function $B_{s_0}(x) = G_{00}^{s_0}[B](x)$ is no longer scale invariant since we have introduced an inner scale $s_0$. To achieve a spatially varying scale of the Brownian image we can consider a spatially varying transformation $h(x)$ of the image domain. We are going to consider the following Brownian images with spatially varying scale

$$\tilde{B}_{s_0}(x) = \frac{1}{\sqrt{s_0 \sqrt{|\frac{\partial h}{\partial x}|}}} B_{s_0}(h(x)) = \frac{1}{\sqrt{s_0 \sqrt{|\frac{\partial h}{\partial x}|}}} G_{00}^{s_0}[B](h(x)) \tag{4}$$

and we define the isotropic intrinsic scale $s(x)$ of the Brownian image $\tilde{B}_{s_0}(x)$ as

$$s(x) = s_0 \sqrt{\left| \frac{\partial h}{\partial x} \right|} . \tag{5}$$

The rationale behind this choice is the combination of the scaling relation given in (2) and the well known scaling property of the linear scale space representation.

A problem with the forgoing analytic expression (1) for the covariance is that the parameter $\beta$ is not known a priori. The influence of this parameter on the eventual maximum likelihood estimation of the scale, is related to changing image intensities by a multiplicative factor, which can have a considerable impact. Our scale estimate should, of course, be invariant under intensity scaling and, therefore, intensities should be normalized in an explicit or implicit manner. Our choice is to take care of this explicitly by determining the global maximum likelihood estimate for $\beta$ together with the estimate for the local scale, assuming the $\beta$ to be constant over the whole image. We note that this choice appears to give more stable results than the, more ad hoc, implicit normalization employed in [8].

Because the scale estimation is based on the Brownian image model the distribution of the feature vector $F_x[L]$ is a multivariate Gaussian. Since all zeroth order moments equal zero, determining the maximum likely scale boils down to taking the scale which minimizes $\log|\mathsf{C}| + F_x[L]^t\mathsf{C}^{-1}F_x[L]$ for a feature vector $F_x[L]$. That is, under the Brownian image model the maximum likely estimate of the isotropic intrinsic scale $\hat{s}$ at position $x$ is given by

$$\hat{s}(x) = \underset{s\in(0,\infty)}{\operatorname{argmin}} \ \log|\mathsf{C}(s)| + F_x[L]^t\mathsf{C}^{-1}(s)F_x[L]\,. \tag{6}$$

The scale estimations performed in the next section are based on this approach.

## 3 Illustrative Examples and Results

Two Brownian images and two natural image patches are used to exemplify the performance and use of the scale estimation principle.

The Brownian images used in this section are constructed by sampling Brownian images using the Fourier technique, e.g. described in [10]. For each sample image we measure the image at scale $s_0$, apply an appropriate transformation $h(x)$ and compute the scaled image $\tilde{B}_{s_0}(x)$ using equation (4). We apply the scale estimation principle to $\tilde{B}_{s_0}(x)$ and expect to estimate scales close to the intrinsic scale as defined in (5).

The Brownian images $\tilde{B}_{s_0}(x)$ are all based on transformations of the form $h(x) = A(x)x$, where $A(x)$ is a spatially varying $2\times 2$ matrix. We consider the following transformations:

Slope:

$$A(x) = \begin{pmatrix} ax_1 + b & 0 \\ 0 & 1 \end{pmatrix}$$

Bar:

$$A(x) = \begin{pmatrix} \sigma g_w(x) + 1 & 0 \\ 0 & 1 \end{pmatrix}, \ g_w(x) = \exp(-\frac{x_1^2}{2w^2})$$

With the parameters $a = 0.15, b = 1, \sigma = 2, w = 2$ and $s_0 = 2$.

The corresponding Brownian images $\tilde{B}_{s_0}(x)$ obtained by applying these scaling transformations can be seen in the top left corner of Figure 2 - 3. The middle image in the top row of each figure illustrates the spatial effect of the transformation and the right image top row shows the correct intrinsic scale $s(x)$ as defined in (5). Finally, in Figure 1, the two natural image patches are displayed.

On these images, local scale estimation is performed using three different feature sets (behind the feature description, in parenthesis, is the number of filters and the position in the middle row of Figure 2 - 3 in which the corresponding image containing the estimated scale map $\hat{s}(x)$ is displayed):

1. the 5-jet at scale 2 (20 dimensions, left),
2. the 5-jet at scale 4 (20 dimensions, middle),
3. three 2-jets at scales 2, 4, and 8 (15 dimensions, right).

We note that in our setting an $n$-jet does not include the zeroth order measurement, because it does not contain any information about the image structure per se.

**Fig. 1.** The natural images used in the experiments

### 3.1 Behavior of Scale Estimates

Figures 2 to 3 show the outcome of the scale estimation procedure on the Brownian images based on the three different filter collections. The lighter the color that is displayed, the larger the estimated scale (with white the largest scale and black the smallest). In every figure, colors in the scale maps are scaled relative to each other.

The bottom row show 1-dimensional plots that gives the average of the estimated scale in the $y$-direction in the image directly besides it. The thin black lines give an indication of the variance around the mean and are plotted at one standard deviation from the mean. For comparison, the true intrinsic scale $s(x)$ is shown in the same plot as a dotted curve.

In general the estimation is inadequate when the underlying scaling transformation exhibits severe anisotropy. This is seen in both Figure 2 (underestimation for right most part of the slope image) and 3 (underestimation for the center part of the bar and overestimation for the two side valleys). Notice also the tendency of the standard deviation curves to move apart in the anisotropic regions, which further supports the observation.

Overestimation and to some extent underestimation seems to occur in regions with rapid scale changes, such as the sides of the bar image. This may be caused by the size of the filters $F_x[L]$ used in the estimation which is relatively large compared to the part of the image in which it actually takes on a particular scale, i.e., the scale varies relatively fast in the parts were the estimation failure occurs. For regions with nearly constant scale our estimation method produce good results such as in the left and right sides of the bar Figure 3.

The variance of the scale estimates seems to be affected by the scales used in the features $F_x[L]$. The results for the 5-jet at scale 2 in both Figure 2 and 3 produces the tightest standard deviation curves. This indicates that small scales should be used to obtain stable estimates for slowly varying scales, whereas the figures also show that this is not the case for regions with rapid scale changes.

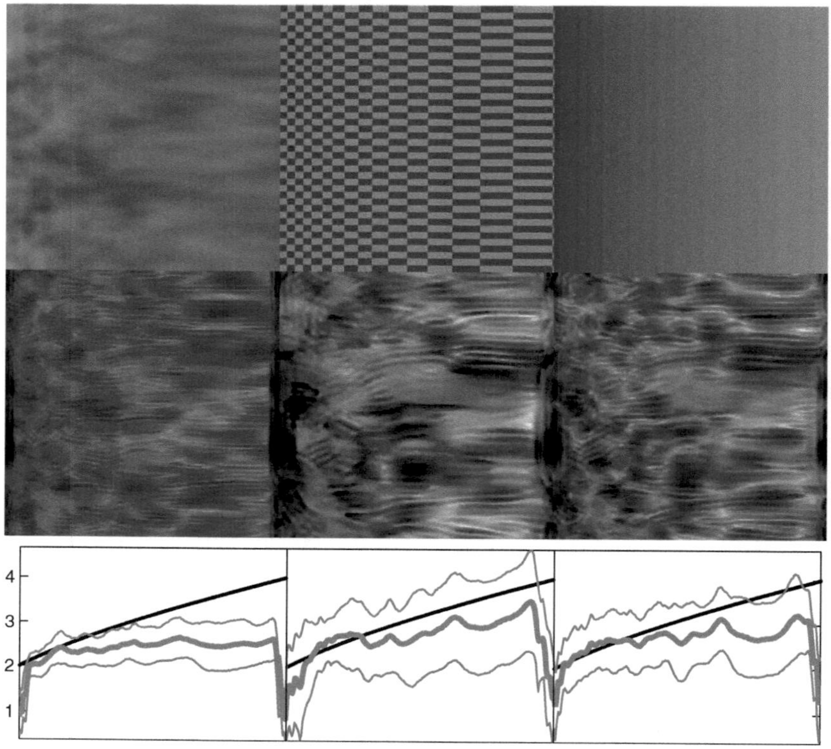


**Fig. 2.** Scale estimation for the Brownian slope image. (Top row) The slope image $\tilde{B}_{s_0}(x)$, illustration of the spatial warp, correct intrinsic scale $s(x)$. (Middle row) Estimated scale maps $\hat{s}(x)$ for the features 1.) 5-jet scale 2, 2.) 5-jet scale 4, 3.) three 2-jets at scales 2, 4, 8. (Bottom row) Scale estimates averaged across $y$ axis (fat curve), one standard deviation curve (thin curves), and true intrinsic scale $s(x)$ (dotted curve). Notice the underestimation when the anisotropy becomes to large.

Figure 4 provide the scale estimate maps for the 'leaves' (top row) and the 'house' image (bottom row). We readily observe that edges are typically estimated to be low scale phenomena while out-of-focus or flat areas are considered high scale. The latter behavior is easily explained by the fact that out-of-focus areas can be considered to be blurred regions and therefore should get a higher scale estimate than in-focus areas. Concerning the edges, it seems plausible to relate clearly discernible, sharp structures to low-scale, as high-scale would mean that they actually should have a blurred appearance.

In an additional test, it was investigated how blurring an image effects the scale estimation procedure. For a Brownian image, with scale image $s(x)$, it holds (by construction) that if the original image is blurred with a Gaussian of scale $\sigma$, the blurred image has a scale image $\sqrt{s^2(x) + \sigma^2}$. To a certain extent we expect such behavior also to hold true for the natural images. To give an indication to which extent this might actually hold, the leaves image was blurred with a Gaussian of scale 4 and the scale selection

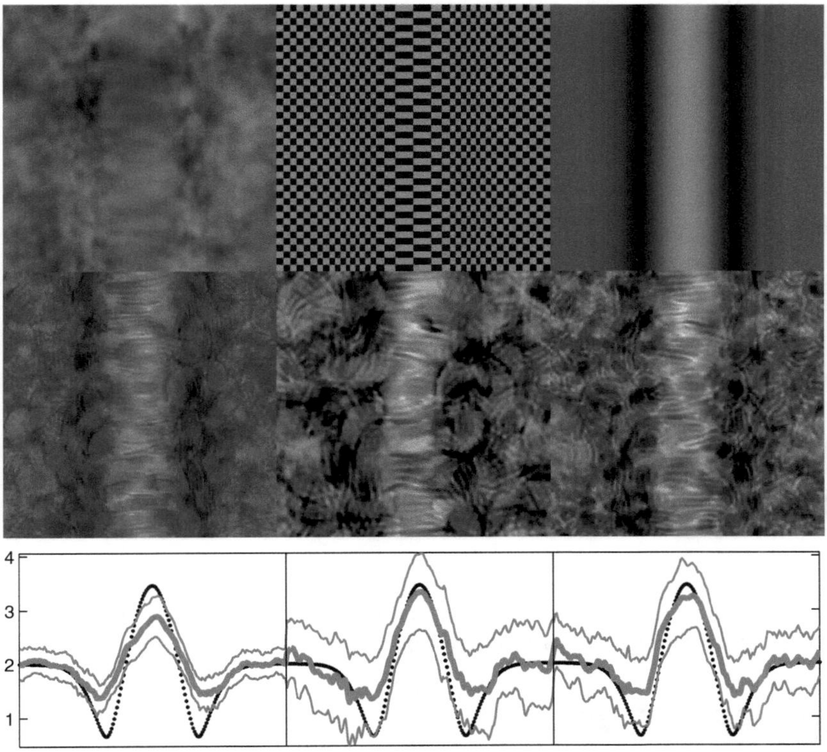


**Fig. 3.** Scale estimation for the Brownian bar image. (Top row) The bar image $\tilde{B}_{s_0}(x)$, illustration of the spatial warp, correct intrinsic scale $s(x)$. (Middle row) Estimated scale maps $\hat{s}(x)$ for the features 1.) 5-jet scale 2, 2.) 5-jet scale 4, 3.) three 2-jets at scales 2, 4, 8. (Bottom row) Scale estimates averaged across $y$ axis (fat curve), one standard deviation curve (thin curves), and true intrinsic scale $s(x)$ (dotted curve). Notice the over- and underestimation in regions of rapid scale change.

schemes, using the three different feature sets, were run on this image. Comparing the mean scale estimates for the blurred images (which are 4.8, 4.9, and 5.0) with the values derived from the original image (i.e., using $\sqrt{s^2(x) + 4^2}$; which gives 4.4., 4.7, and 4.7), we see that the estimates can be up to about 9% off. The second feature set, i.e., the 5-jet at scale 4, seems to perform best and the original estimated image is displayed together with the indirect and the direct scale estimate image for the blurred leaves image (see Figure 5).

## 4 Discussion

An interpretation of our scale estimation procedure is that it provides a scale map $\hat{s}(x)$ which correspond to one out of many possible choices of isotropic scaling transformations that will transform Brownian images into the image $L$ under investigation. The

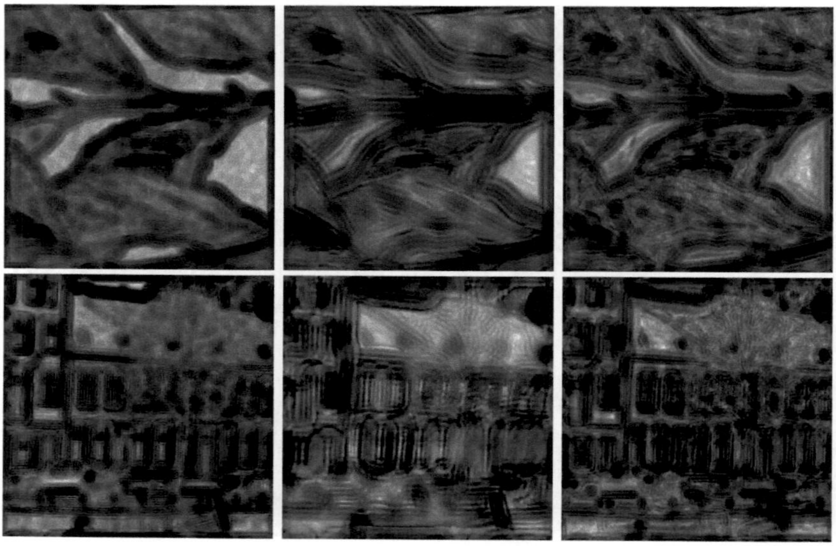

**Fig. 4.** Scale estimates for the 'leaves' (top row) and the 'house' image (bottom row) using the three different feature sets (in the same order) described in the beginning of Section 3. Edges have low intrinsic scale and out-of-focus areas have high intrinsic scales.

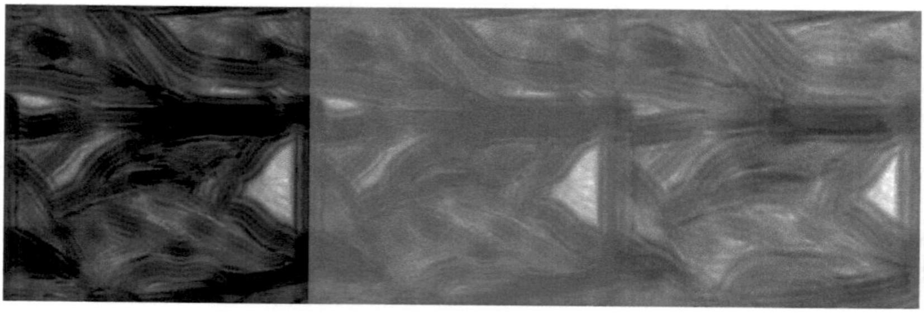

**Fig. 5.** From left to right: Original, unblurred scale image; the corresponding indirect scale estimate image, and the direct scale estimate image for the blurred-at-scale-4 leaves image. The estimates are based on the second feature set, i.e., the 5-jet at scale 4. Comparing the middle and right image a small underestimation may be noticed.

Brownian image model induces a probability distribution on the space of isotropic scaling transformations $s(x)$. Among these transformations our method picks the $\hat{s}(x)$ which is the local maximum likelihood estimate of the scale. Another natural choice would be to pick the estimate that is the global maximum likelihood (ML) transformation, i.e. pick the transformation function $\hat{s}(x)$ that maximizes the likelihood over the whole image. It is unclear to what extent our estimate approaches the global ML estimate.

Both artificial and natural images have been experimented with and from the experiments on the synthetic images, we conclude the method is promising. Estimates for Brownian images on regions of constant scale have a small amount of variance in the scale estimate and are reasonably accurate. On Brownian images having varying scale, although the general trend is clear, estimation failure can occur. It has to be investigated whether this can, for example, be remedied by extending the feature sets or improved statistical estimation techniques (e.g. by taking into account that neighboring pixels are not independent of each other). The best choice of features, however, may be hard to decide on. An obvious improvement to the estimation technique would be to do anisotropic scale estimation instead of isotropic as done here. This would clearly improve the estimation for regions of large anisotropic scaling.

Regarding the natural images, the scale estimation images seem interpretable and explainable up to a certain extent. However, here the differences between the different feature sets used becomes more apparent. Surely, as is the case for the artificial images, improvements are certainly possible, however it may be harder to realize them, not in the least because we do not have a corresponding ground truth scale image for them. A general improvement, may be to estimate the $\beta$ parameter, which in a certain way measures image contrast, in a local manner as well. The current assumption to keep $\beta$ constant over the whole image is presumably to restrictive.

One of the most notable conceptual differences between other selection schemes and the scheme proposed here is that, in principle, the former are employed in combination with a specific feature detection task, whereas the current approach allows for a scale estimate irrespective of actual features to be detected[1]. Therefore the scheme proposed here could be considered a way to obtain a local estimate of the *intrinsic* scale of the image. It would, on the other hand, be interesting to see under what assumptions for the maximum likelihood model, it is actually possible to exactly mimic the feature specific scale selection methods of Lindeberg [2,3]. As an additional positive aspect of the selection scheme, we should mention that it has a probabilistic basis and can therefore be easily incorporated in any statistical computer vision or image analysis framework. In this setting it might be more appropriate to estimate local scale distributions instead of selecting a particular scale. Finally, it is common in other techniques to select the scale by comparing features across scales. In our approach we can make do with measurements at a single scale which could mean a large reduction in the computational cost involved in performing the scale selection.

For applications such as object recognition and image retrieval a successful strategy is usually based on feature descriptors computed at a set of points called interest or salient points, see e.g. [13,14]. These points are selected such as to carry maximal information about the image content. Common techniques for selecting such points include detecting corners and blobs. Popular choices for the feature descriptors include SIFT [13] and collections of differential invariants [14] which are usually computed at some scale. The best performance have been reported for those cases that use some form of scale selection choosing, in some sense, a locally optimal scale for the descriptors. The usual approach is to select the scale identical to the scale of the interest point detector. However there are no particular arguments for doing this and we argue that one might as well apply our scale estimation principle on the feature descriptors to select the scale. This scale will be different from the scale selected by the interest point detector and

[1] Although it does of course depend on the actual feature set used for performing the estimation.

it would be interesting to investigate if our scale estimation principle can improve the performance of such systems.

Our method can also be used for solving the problem of estimating shape from texture. This can be done by estimating a general mapping $h(x)$ instead of only considering the isotropic scaling transformation. This requires that we assume that the 3 dimensional physical world consists of objects which are textured by Brownian images and that the scene is uniformly lit and without shadows. If we approximate the full perspective projection of the scene with an affine transformation $h(x)$ of the image domain we have to estimate 6 parameters. We can take advantage of the isotropy of the Brownian image model, as well as the fact that the filter responses obtained by using partial derivatives of the Gaussian aperture as features $F_x[L]$ are stationary, due to the stationarity of the increments of Brownian images. By locally estimating the transformation $h(x)$ we can produce an estimate of the 3 dimensional orientation of local planar object patches. An obvious extension would be to combine this estimation procedure with the additional information available in video sequences or stereo views.

## 5 Conclusions

A generic local scale estimation procedure employing maximum likelihood estimation has been proposed and demonstrated to work quite well in certain settings. The specific instance of the general framework we considered provides an estimate of the local intrinsic scale assuming the image to be from an underlying Brownian image model at a specific scale. Here scale is defined in terms of an inner scale $s_0$ and a spatially varying scaling transformation $h(x)$ of the image domain.

## References


1. ter Haar Romeny, B.: Front-end vision and multi-scale image analysis. Volume 27 of Computational Imaging and Vision Series. Kluwer Academic Publishers, Dordrecht, The Netherlands (2003)
2. Lindeberg, T.: Feature detection with automatic scale selection. International Journal of Computer Vision **30**(2) (November 1998) 79–116
3. Lindeberg, T.: Edge detection and ridge detection with automatic scale selection. International Journal of Computer Vision **30**(2) (November 1998) 117–154
4. Majer, P.: A Statistical Approach to Feature Detection and Scale Selection in Images. PhD thesis, University of Göttingen (2000)
5. Pedersen, K.S., Nielsen, M.: The Hausdorff dimension and scale-space normalisation of natural images. Journal of Visual Communication and Image Representation **11**(2) (2000) 266–277
6. Papandreou, G., Maragos, P.: Image denoising in nonlinear scale-spaces: Automatic scale selection through cross-validatory model selection. In: Proceeding of the International Conference on Image Processing (ICIP'05). IEEE, Genova, Italy (September 2005)
7. Bouman, C.A., Sauer, K.: Maximum likelihood scale estimation for a class of Markov random fields. In: Proceedings of ICASSP '94. IEEE International Conference on Acoustics, Speech and Signal Processing, IEEE (1994) 537–540

8. Loog, M., Pedersen, K.S., Markussen, B.: Maximum likely scale estimation. In Olsen, O.F., Florack, L.M.J., Kuijper, A., eds.: Deep structure, singularities and computer vision. Volume 3753 of LNCS., Springer Verlag (2005) 146–156
9. Pentland, A.P.: Fractal-based description of natural scenes. IEEE Transactions on Pattern Analysis and Machine Intelligence **6**(6) (1984) 661–674
10. Pedersen, K.S.: Properties of brownian image models in scale-space. In Griffin, L.D., Lillholm, M., eds.: Scale Space Methods in Computer Vision: Proceedings of the 4th Scale-Space conference. LNCS 2695, Isle of Skye, Scotland (June 2003) 281–296
11. Mandelbrot, B.B., van Ness, J.W.: Fractional brownian motions, fractional noises and applications. SIAM Review **10**(4) (October 1968) 422–437
12. Pesquet-Popescu, B., Vehel, J.L.: Stochastic fractal models for image processing. IEEE Signal Processing Magazine **19**(5) (September 2002) 48–62
13. Lowe, D.: Object recognition from local scale-invariant features. In: Proc. of 7th ICCV. (1999) 1150–1157
14. Schmid, C., Mohr, R., Bauckhage, C.: Evaluation of interest point detectors. International Journal of Computer Vision **37**(4) (2000) 151–172

# Combining Different Types of Scale Space Interest Points Using Canonical Sets

Frans Kanters[1], Trip Denton[2], Ali Shokoufandeh[2], Luc Florack[1,*], and Bart ter Haar Romeny[1]

[1] Eindhoven University of Technology,
Den Dolech 2, Postbus 513
5600 MB Eindhoven, the Netherlands
{F.M.W.Kanters,L.M.J.Florack,B.M.terHaarRomeny}@tue.nl
[2] Drexel university,
University Crossings 141, Philadelphia, USA
{tdenton,ashokouf}@cs.drexel.edu

**Abstract.** Scale space interest points capture important photometric and deep structure information of an image. The information content of such points can be made explicit using image reconstruction. In this paper we will consider the problem of combining multiple types of interest points used for image reconstruction. It is shown that ordering the complete set of points by differential (quadratic) TV-norm (which works for single feature types) does not yield optimal results for combined point sets. The paper presents a method to solve this problem using canonical sets of scale space features. Qualitative and quantitative analysis show improved performance over simple ordering of points using the TV-norm.

## 1 Introduction

It is known that from a sufficiently rich set of scale space interest points, a visually attractive reconstruction of the original image can be created [27,20,13,14]. Such a reconstruction can for example be used for image understanding or image editing. For robust reconstruction, many types of interest points can be used, e.g. top points, blobs, edges, corner points, ridge points, etc. These points are usually ordered by their strength [27,20] or by the differential (quadratic) TV norm [29,14] in the corresponding points. Lillholm and Nielsen showed that a combination of different types of interest points can improve the reconstruction quality substantially [27,20]. It is however not clear *how* exactly these points should be selected. Specifically, given a large set of different types of interest points, how should one pick points for an optimal subset? To motivate the problem, we show in section 2 a non optimal example of reconstruction where the differential TV norm is used to rank the combined set of scale space interest points. In this paper, we propose an optimization framework for selecting scale space interest points for image reconstruction using a canonical set of scale space features. We formulate the feature selection problem as a quadratic optimization and use a semi-definite program to approximate its solution. A quantitative analysis on an example

* The Netherlands Organisation for Scientific Research (NWO) is gratefully acknowledged for financial support.

F. Sgallari, A. Murli, and N. Paragios (Eds.): SSVM 2007, LNCS 4485, pp. 374–385, 2007.

image is presented in Section 3. In Section 4 we propose our method for selecting a good subset of combined scale space interest points for reconstruction using canonical sets [2,28] and finally conclusions are given in Section 5.

## 2 Problem Description

In a Gaussian scale space of an image, different types of special interest points can be extracted. Specifically, we will focus on 10 types of commonly used scale space interest points in Section 2.1 that we will use for image reconstruction in Section 2.2. In that section we will also present the results of reconstructions from combined point sets using the differential TV-norm for ordering the points. These results will show that a naive approach to combine multiple types of interest points is not optimal for image reconstruction.

### 2.1 Scale Space Interest Points

Consider a continuous signal $f : \mathbb{R}^d \to \mathbb{R}$. The linear scale space representation $u : \mathbb{R}^d \times \mathbb{R}_+ \to \mathbb{R}$ of $f$ is defined as the solution of the heat equation:

$$\begin{cases} \frac{\partial}{\partial s} u = \Delta u \\ \lim_{s \downarrow 0} u(\cdot, s) = f(\cdot) \end{cases} \tag{1}$$

In this equation $s$ denotes the scale. The unique solution to this equation leads to convolution with a Gaussian kernel. Spatial derivatives of the image can be calculated by convolution with a derivative of a Gaussian:

$$\partial_{\nu^1,\ldots,\nu^n} u(x, y, s) = (-1)^n (\partial_{\nu^1,\ldots,\nu^n} G_s * f)(x, y) \tag{2}$$

where $f$ is the original image and $G_s$ a Gaussian of scale $s$. The spatial indices $\nu^1, \ldots,$ $\nu^n$ indicate the $n$-th order partial derivative w.r.t. $x^{\nu^1}, \ldots, x^{\nu^n}$. Each index refers to either $x$ (value 1) or $y$ (value 2) for 2D images. For further reference we will use the short notation $u_{s,\nu^1 \ldots \nu^n}$ for $\partial_{\nu^1,\ldots,\nu^n} u(x, y, s)$. In this scale space representation of an image, several special types of interest points exist. In what follows, we will present 10 types of commonly used scale space interest points.

**Laplacian blobs.** Scale space blobs are defined as the positive local maxima (or negative local minima) in space and scale of the normalized Laplacian of the image [21]:

$$\underset{x,y,s}{\text{lmax}}\{|s^\gamma (u_{s,xx} + u_{s,yy})|\} \tag{3}$$

with lmax the local maximum and using $\gamma$-normalization with $\gamma = 1$. The $\gamma$-normalization with scale invariance is discussed in detail by Florack and Kuijper [4]. Blobs can be ordered in strength by the magnitude of the response of their respective filters [27,20].

**Hessian blobs.** Alternatively scale space blobs can be defined as the local maxima of the squared normalized determinant of the Hessian:

$$\underset{x,y,s}{\mathrm{lmax}}\{s^{4\gamma}(u_{s,xx}u_{s,yy} - u_{s,xy}^2)^2\}, \tag{4}$$

again using $\gamma$-normalization with $\gamma = 1$.

**Corner points.** Corner points in the image are defined as points with high curvature and high intensity gradient:

$$\underset{x,y,s}{\mathrm{lmax}}\{|s^{2\gamma}(2u_{s,x}u_{s,xy}u_{s,y} - u_{s,xx}u_{s,y}^2 - u_{s,x}^2u_{s,yy})|\} \tag{5}$$

using $\gamma$-normalization with $\gamma = 7/8$, following Lindeberg et al. [22]. Note that for ordering the corner points in strength, the magnitude of the corresponding filter response has to be normalized with $\gamma = 1$ to make magnitude values at different scales comparable.

**Edge points.** Edge points are defined by the following two constraints [20,27]:

$$\begin{cases} s^{\gamma}u_{ww} = 0 \\ \underset{s}{\mathrm{lmax}}\{s^{\gamma/2}u_w\} \end{cases} \tag{6}$$

using $\gamma$-normalization with $\gamma = 1/2$ [22]. Here $u_w$ is the first order derivative in the gradient direction and $u_{ww}$ the second order derivative in the gradient direction. For edge strength, the gradient magnitude is used re-normalized with $\gamma = 1$.

**Ridge points.** Ridge points are defined as the local extrema of the *square of the $\gamma$-normalized principal curvature difference* [9,22]:

$$\underset{x,y,s}{\mathrm{lmax}}\{s^{2\gamma}((u_{s,xx} - u_{s,yy})^2 + 4u_{s,xy}^2)\} \tag{7}$$

using $\gamma$-normalization with $\gamma = 3/4$. Note that for ordering the ridge points in strength, re-normalization with $\gamma = 1$ is necessary.

**Top points.** Top point are defined by:

$$\begin{cases} \nabla u_s = (u_{s,x}, u_{s,y})^T = \mathbf{0} \\ \det \mathcal{H}(u_s) = u_{s,xx}u_{s,yy} - u_{s,xy}^2 = 0 \end{cases} \tag{8}$$

where $\mathcal{H}(u_s)$ is the 2-nd order Hessian matrix defined by:

$$\mathcal{H}(u_s) = \begin{pmatrix} u_{s,xx} & u_{s,xy} \\ u_{s,xy} & u_{s,yy} \end{pmatrix} \tag{9}$$

Platel and Kanters showed that top points can be rank-ordered by a stability norm called differential (quadratic) TV-norm [29,14].

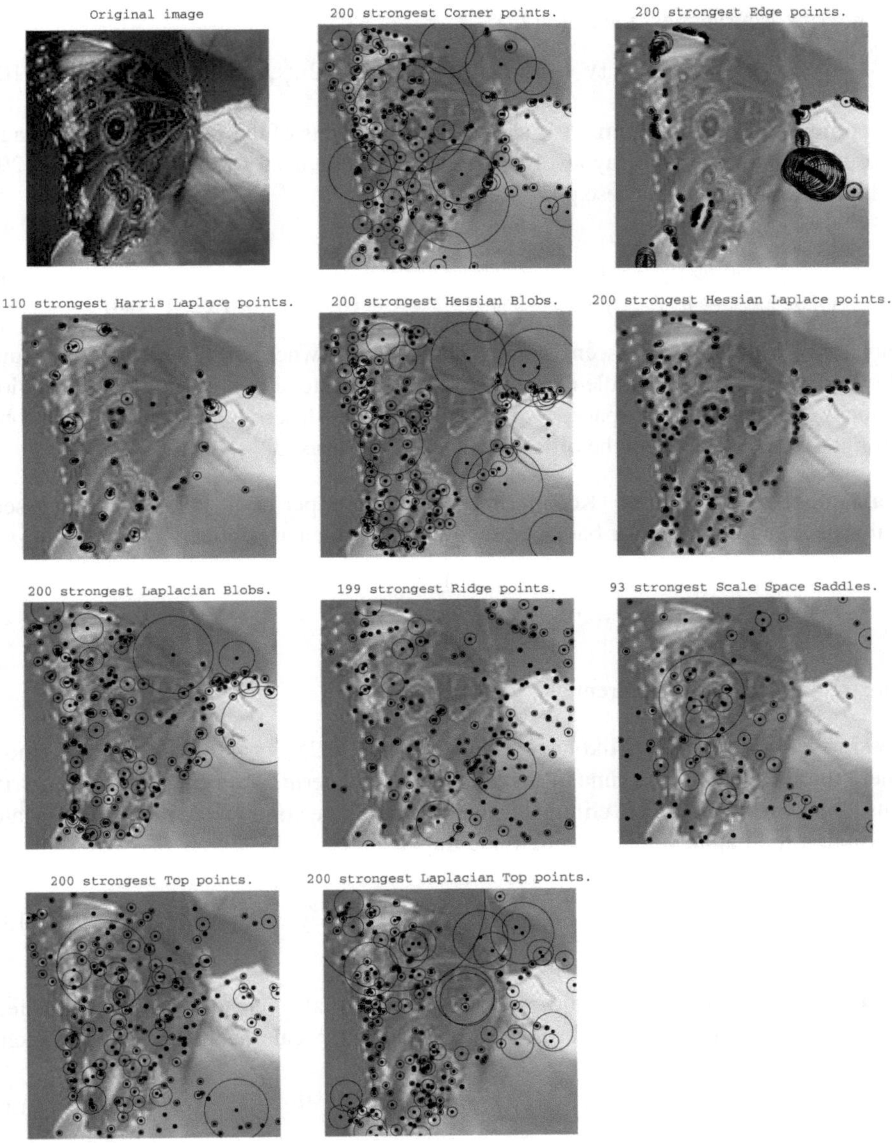


**Fig. 1.** Interest points of the *butterfly.jpg* image projected on the original image. The size of the circles represent the scale of the interest points. Maximum 200 points of each set are shown for the sake of clarity.

The amount of structure contained in a *spatial* area around a critical point can be quantified by the *total* (*quadratic*) *variation* (TV) norm over that area [1]. By using a spatial Taylor series around a considered critical point the TV-norm simplifies to Eqn. (10)

which is referred to as the *differential TV-norm* [29]. Note that this is also equal to Koenderink's curvedness measure [12].

$$\text{diff tv} = 4s^2(u_{s,xx}^2 + u_{s,yy}^2 + 2u_{s,xy}^2) \tag{10}$$

**Top points of the Laplacian.** It is shown that top points of the Laplacian of an image (versus top points of the gray level image itself) can be used for image matching [29] and reconstruction [14]. These points are defined by:

$$\begin{cases} u_{s,xxx} + u_{s,yyx} = 0 \\ u_{s,xxy} + u_{s,yyy} = 0 \\ (u_{s,xxxx} + u_{s,yyxx})(u_{s,xxyy} + u_{s,yyyy}) - (u_{s,xxxy} + u_{yyxy})^2 = 0 \end{cases} \tag{11}$$

Laplacian top points can be seen as points in scale space where one Laplacian extremum blob and one Laplacian saddle merge into one blob. Instead of describing the behavior of local extrema in scale space, Laplacian top points describe the behavior of blobs through scale. For strength the differential TV-Norm is used.

**Scale space saddle points.** Koenderink [16] and Kuijper et al. [17,18,19] discussed scale space saddle points (or balanced saddles [8]), which are defined by:

$$\begin{cases} u_{s,x} = 0 \\ u_{s,y} = 0 \\ u_{s,xx} + u_{s,yy} = 0 \end{cases} \tag{12}$$

For strength, again the differential TV-Norm is used.

**Hessian-Laplace points.** Mikolajczyk and Schmid [26,25] introduced a hybrid method where the local spatial maxima of the square of the determinant of the Hessian matrix (introduced by Lowe [23] to eliminate edge response) are combined with the local scale maxima of the Laplacian. For strength the Laplacian is used.

$$\begin{cases} \operatorname*{lmax}_{x,y}\{s^{4\gamma}(u_{s,xx}u_{s,yy} - u_{s,xy}^2)^2\} \\ \operatorname*{lmax}_{s}\{s^{2\gamma}(u_{s,xx} + u_{s,yy})\} \end{cases} \tag{13}$$

**Harris-Laplace points.** Mikolajczyk and Schmid [24] also introduced a scale-adapted version of the Harris corner detector [10]. Consider the scale-adapted second moment matrix:

$$\mu(s_D, s_I) = s_D G_{s_I} * \begin{pmatrix} u_{s_D,xx} & u_{s_D,xy} \\ u_{s_D,xy} & u_{s_D,yy} \end{pmatrix} \tag{14}$$

with $s_D$ the *differentiation* scale, $s_I$ the *integration* scale, and $G_{s_I}$ a Gaussian at scale $s_I$. The matrix describes the gradient variation in a local neighborhood of a point. The Harris measure [10] combines the trace and determinant of this matrix as a measure for cornerness. Scale selection is based on the local maxima over scale of the Laplacian. The scale-adapted Harris-Laplace points are defined as:

$$\begin{cases} \operatorname*{lmax}_{x,y}\{\det(\mu(s_D, s_I)) - \alpha \operatorname{trace}^2(\mu(s_D, s_I))\} \\ \operatorname*{lmax}_{s}\{s^{2\gamma}(u_{s,xx} + u_{s,yy})\} \end{cases} \tag{15}$$

with $\gamma = 1$. Note that in practice, $s_D = \beta s_I$ with $\beta$ a suitable constant. Again the Laplacian is used as strength measure. Figure 1 shows all different interest points projected on the original example image.

## 2.2 Reconstruction from Scale Space Interest Points

Using the Sobolev reconstruction algorithm by Janssen et al. [11] a reconstruction can be made using the calculated scale space interest points and the local $N$-jet in those points. Figure 2 shows the reconstruction results for the *butterfly.jpg* image. As can be seen from the reconstruction results in Fig. 2, the different types of interest points capture different aspects of image structure. Therefore, combining different types of interest points should improve reconstruction quality. The question is *how* to combine the different types of points. One possible option is to combine all types of points in a large set and re-order all points using their differential TV-norm or a similarly defined strength measure. The result of the 200 strongest combined interest points of the test image and the corresponding reconstruction is shown in Fig. 3.

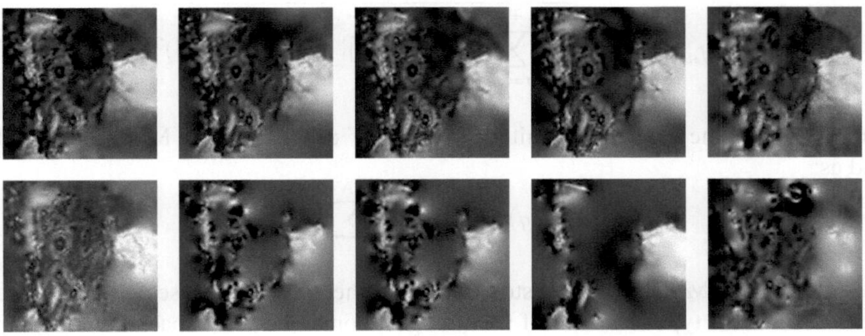

**Fig. 2.** Reconstructions of the *butterfly.jpg* image using the 200 strongest top points of the Laplacian, Hessian blobs, corner points, Laplacian blobs, top points, ridge points, Hessian Laplace points, Harris Laplace points, edge points and scale space saddles respectively

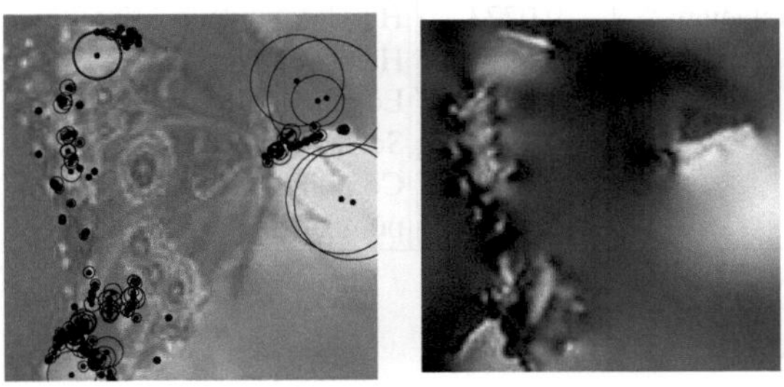

**Fig. 3.** Left: 200 strongest combined scale space interest points of the *butterfly.jpg* image ordered by differential TV-norm. Right: Reconstruction of the *butterfly.jpg* image using the 200 strongest combined scale space interest points ordered by differential TV-norm.

Note that the quality of the combined scale space interest point reconstruction is lower than some of the separate interest point reconstructions shown in Fig. 2. The reason for this is that in the combined point set many points that are close to each other will share a similar differential TV-norm (or other strength measure). Using strength measures as the sole criteria for ordering interest points may result in selecting points close to each other and consequently poor reconstruction results since these points will contain much redundant information.

## 3 Quantitative Analysis

In order to compare the results of the reconstruction experiments, an objective error measure is necessary. Ideally, this should reflect the human observer's notion of quality. The Multi Scale Differential Error (MSDE) is used for this purpose. A detailed motivation and evaluation of this error measure has been provided elsewhere [15]. For $R$ different scales, the gradient magnitude error map $\Psi_{f,g}$ between image $f$ and reconstructed image $g$ is defined as:

$$\Psi_{f,g}[i,j] = \sqrt{\frac{1}{R}\sum_{\sigma=\sigma_1}^{\sigma_R} (|\sigma\nabla_{f,\sigma}[i,j]| - |\sigma\nabla_{g,\sigma}[i,j]|)^2} \tag{16}$$

with $|\nabla_{f,\sigma}[i,j]|$ the gradient magnitude of image $f$ at scale $\sigma$. The MSDE is now defined as:

$$MSDE(f,g) = \frac{1}{MN}\sum_{i=1}^{M}\sum_{j=1}^{N}\Psi_{f,g}[i,j]. \tag{17}$$

Table 1 shows the MSDE for reconstructions from the 200 strongest scale space interest points of different types and 200 of the strongest combined scale space interest points using the TV-norm to rank the points.

**Table 1.** MSDE of reconstructions from separate scale space interest points

| Reconstruction from | MSDE | Reconstruction from | MSDE |
|---|---|---|---|
| Top points of the Laplacian | 0.0723 | Hessian Laplace points | 0.2527 |
| Hessian blobs | 0.0858 | Harris Laplace points | 0.2784 |
| Corner points | 0.0928 | Edge points | 0.2246 |
| Laplacian blobs | 0.0989 | Scale space saddles | 0.2460 |
| Top points | 0.0963 | Combined interest points using TV-norm | 0.1367 |
| Ridge points | 0.1493 | | |

## 4 Canonical Sets

Using only the strength measure of points can result in the selection from the combined point set of too many points that are close to each other. In order to take into account the spatial distance between points as well as the strength of the points, we select so-called canonical subsets of the combined point sets. Canonical sets are subsets of points with special properties, namely:

I. points in the canonical set are minimally similar;
II. points outside the canonical set are maximally similar to points in the canonical set;
III. points in the canonical set have high stability compared to elements outside the set.

The canonical set problem is formulated in terms of a quadratic integer programming optimization. Many problems of this type are known to be intractable [5], but they admit good approximation algorithms [7]. In what follows, we will present a brief overview of the problem formulation and its solution (for an in-depth treatment of the problem the reader is referred to [28,3,2]).

The input set to the canonical set problem consists of a set of points $\mathcal{P} = \{p_1, ..., p_n\}$, an associated set of strength (stability) measures $\{t_1, ..., t_n\}$, $t_i \in \mathbb{R}^+$, $1 \le i \le n$, and a similarity function $\mathcal{W} : \mathcal{P} \times \mathcal{P} \rightarrow \mathbb{R}_0^+$. In this work we let $t_i$ equal the differential TV-norm of point $p_i$ (See Eqn. 10), and define the similarity of two points, $p_i$ and $p_j$, as

$$\mathcal{W}_{ij} = \frac{1}{1 + d_{ij}},$$

where $d_{ij}$ denotes the Euclidean distance between the points $p_i$ and $p_j$.

The problem is formulated as a multi-objective quadratic integer program where the outcome of the optimization will determine whether each point is in the canonical set, $\mathcal{P}^*$, or not. Specifically, for each point, $p_i$, an indicator variable, $y_i$, is used that will be equal to 1 if $p_i \in \mathcal{P}^*$ and $-1$ otherwise. Using these indicator variables, it can be shown that the aforementioned properties I, II, and II can be stated as optimization objectives:

$$\text{Minimize } \frac{3}{4} \sum_{i,j} \mathcal{W}_{ij} y_i y_j + \frac{1}{2} \sum_{i=1}^{n} y_i \sum_{j=1}^{n} \mathcal{W}_{ij} + \frac{3}{4} \sum_{ij} \mathcal{W}_{ij} \tag{18}$$

$$\text{Minimize } \frac{1}{2} \sum_{i=1}^{n} t_i (1 - y_i) \tag{19}$$

$$\text{Subject to } y_i \in \{-1, +1\}, \forall\, 1 \le i \le n. \tag{20}$$

The optimal solution to this integer program is a vector $\mathbf{y} = [y_1, ..., y_n]^T$, indicating which points belong to the canonical set.

Vector labeling and lifting [6] are used to relax and reformulate the problem as a semidefinite program. To get rid of the linear terms we increase the dimension of the indicator vector by 1 and introduce a set indicator variable, $y_{n+1}$, which acts as a reference for membership in the canonical set. In an optimal solution, $p_i$ is a member of the canonical set only if $y_i = y_{n+1}$. The integrality constraints are removed by substituting a vector for each indicator variable, that is, we replace each indicator variable $y_i$, $1 \le i \le n+1$, with a vector $x_i \in S_{n+1}$, where $S_{n+1}$ is the unit sphere in $\mathbb{R}^{n+1}$. Let $w_\Sigma = \sum_{i,j} \mathcal{W}_{ij}$, define $\mathbf{d}$ as a column vector in $\mathbb{R}^n$ whose $i^{\text{th}}$ entry has value

$d_i = \sum_{j=1}^{n} \mathcal{W}_{ij}$, and let $t_\Sigma = \sum_{i=1}^{n} t_i$, and $\hat{\mathbf{0}}$ be an $n \times n$ matrix of zeros. The objectives are then encoded into matrices $\mathcal{C}$ and $\mathcal{T}$,

$$\mathcal{C} = \begin{bmatrix} \frac{3}{4}\mathcal{W} & \frac{1}{4}\mathbf{d} \\ \frac{1}{4}\mathbf{d}^T & \frac{3}{4}w_\Sigma \end{bmatrix}, \mathcal{T} = \begin{bmatrix} \hat{\mathbf{0}} & -\frac{1}{4}\mathbf{t} \\ -\frac{1}{4}\mathbf{t}^T & \frac{1}{2}t_\Sigma \end{bmatrix},$$

where $\mathbf{t}$ is a vector in $\mathbb{R}^n$ whose $i^{\text{th}}$ entry is $t_i$ (the strength of point $i$).

These objectives are combined using Pareto optimality by defining a parameter $\alpha \in [0, 1]$, and defining the matrix $\mathcal{Q}$ as a weighted convex combination of $\mathcal{C}$ and $\mathcal{T}$, where $\mathcal{Q} = \alpha\mathcal{C} + (1-\alpha)\mathcal{T}$. This implies that for any given $\alpha$ the combined objective is convex and a solution will be optimal for that $\alpha$. The semidefinite program formulation of the canonical set can be stated as

$$\begin{aligned} \textbf{(SDP):}\ \text{Minimize}\ & \mathcal{Q} \bullet \mathcal{X} \\ \text{Subject to}\ & \mathcal{D}_i \bullet \mathcal{X} \geq 0,\ \forall\, i = 1 \ldots, m, \\ & \mathcal{X} \succeq 0, \end{aligned}$$

where $m = n + 3$ and denotes the number of constraint matrices, and $\mathcal{X} \succeq 0$ denotes a positive semidefinite matrix. The notation $\mathcal{A} \bullet \mathcal{B}$ denotes the Frobenius inner product of matrices $\mathcal{A}$ and $\mathcal{B}$, i.e., $\mathcal{A} \bullet \mathcal{B} = \mathbf{Trace}(\mathcal{A}^T\mathcal{B})$. The first $n + 1$ constraint matrices, $\mathcal{D}_1, \mathcal{D}_1, \ldots, \mathcal{D}_{n+1}$, are all zeros with a single 1 that moves along the main diagonal, enforcing the $x_i^T x_i = 1$ constraints from Eqn. (20). The matrices $\mathcal{D}_{n+2}$ and $\mathcal{D}_{n+3}$, encode constraints that are used to bound the size of the canonical set.

Once the solution to this semidefinite program is computed, a rounding step is performed to obtain an approximate integer solution. This step identifies the set of values for indicator variables $y_1, ..., y_{n+1}$. We use a standard rounding scheme based on Cholesky decomposition and a multivariate normal hyperplane method [30].

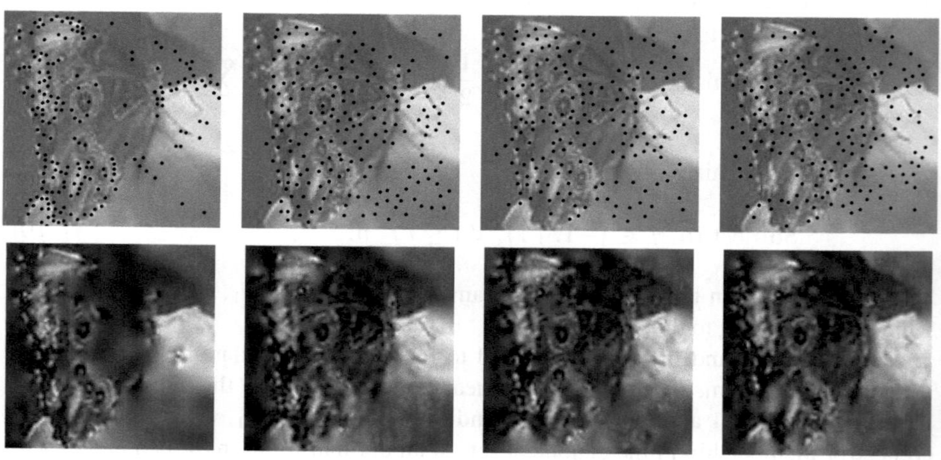

**Fig. 4.** Top: Canonical sets of the *butterfly.jpg* image using $\alpha = 0.0$, $\alpha = 0.1$, $\alpha = 0.3$ and $\alpha = 1.0$ respectively. Bottom: Reconstructions from canonical sets of the *butterfly.jpg* image using $\alpha = 0.0$, $\alpha = 0.1$, $\alpha = 0.3$ and $\alpha = 1.0$, respectively

**Table 2.** MSDE of reconstructions using canonical sets with various settings of $\alpha$

| $\alpha$ | 0.0 | 0.1 | 0.3 | 1.0 |
|---|---|---|---|---|
| MSDE | 0.0859 | 0.0854 | 0.0948 | 0.0971 |

The parameter $\alpha$ controls the relative significance of stability versus spatial distribution of the selected features. Specifically, in one extreme, setting $\alpha = 0$ will select the most stable features, and in the other extreme setting $\alpha = 1$ will select dispersed features from all across the image plane without regarding their stability. For our experiments an extra "minimum distance" constraint is used on the combined point set to reduce the size of the input set to the canonical set algorithm. Points that are within a radius of 3 pixels from a point that has a higher weight (TV-norm) are discarded. This ensures a minimum distance of 3 pixels between all points in the canonical set. Figure 4 shows some results of canonical sets for the example image for various settings of $\alpha$ and the corresponding reconstructions. Table 2 shows the corresponding MSDE. Note that visually, the reconstruction quality of canonical sets of combined scale space interest points is better than the reconstruction quality of combined scale space interest points ordered by TV-norm. The objective image quality measures support this conclusion, although the visual quality is more convincing than the MSDE.

## 5 Conclusions and Discussion

In this paper the problem of combining different types of scale space interest points for image reconstruction is addressed. It is shown that, in general, re-ordering the combined point set using the TV-norm (which works for single feature types) is not optimal. We have presented an optimization framework for selecting scale space interest points for image reconstruction that combines both stability and spatial distribution requirements in a single framework. The feature selection problem is presented as a quadratic optimization and an approximation algorithm for its solution is proposed. Our preliminary investigation indicates improved results using our canonical sets for combined scale space interest points over re-ordering through the TV-norm, visually as well as using an objective image quality measure. In future work we will conduct a more comprehensive set of experiments with comparative results, as well as, an evaluation of the $\alpha$ parameter on the reconstruction quality. The similarity measure currently used is based on Euclidean distance only. In future work we will consider similarity measures that also incorporate the scale and shape of the filters. The additional, somewhat ad-hoc minimum distance constraint, now performed before the canonical set algorithm, will also be incorporated within the canonical set formulation.

## References

1. T. Brox and J. Weickert. A tv flow based local scale measure for texture discrimination. In T. Pajdla and J. Matas, editors, *Proc. 8th European Conference on Computer Vision, Prague, Czech Republic.*, volume 2 of *Computer Vision - ECCV*, pages 578–590. Springer LNCS 3022, May 2004.

2. Trip Denton, Jeff Abrahamson, and Ali Shokoufandeh. Approximation of canonical sets and their application to 2d view simplification. In *CVPR*, volume 2, pages 550–557, June 2004.
3. Trip Denton, M. Fatih Demirci, Jeff Abrahamson, and Ali Shokoufandeh. Selecting canonical views for view-based 3-d object recognition. In *ICPR*, pages 273–276, August 2004.
4. L. Florack and A. Kuijper. The topological structure of scale-space images. *Journal of Mathematical Imaging and Vision*, 12(1):65–79, February 2000.
5. M. R. Gary and D. S. Johnson. *Computers and Intractability: A Guide to the Theory of NP-completeness*. Freeman, San Francisco, 1979. (ND2,SR1).
6. M. X. Goemans. Semidefinite programming in combinatorial optimization. *Mathematical Programming*, 79:143–161, 1997.
7. Michel X. Goemans and David P. Williamson. .878-approximation algorithms for max cut and max 2sat. In *Twenty-sixth Annual ACM Symposium on Theory of Computing*, pages 422–431, New York, 1994.
8. L. D. Griffin and A. C. F. Colchester. Superficial and deep structure in linear diffusion scale space: Isophotes, critical points and separatrices. *Image and Vision Computing*, 13(7):543–557, September 1995.
9. R.M. Haralick. Ridges and valleys on digital images. *Computer Vision, Graphics and Image Processing*, 22:28–38, 1983.
10. C. Harris and M. Stephens. A combined corner and edge detector. In *Proc. Alvey Vision Conf., Univ. Manchester*, pages 147–151, 1988.
11. B. Janssen, F. M. W. Kanters, R. Duits, L. M. J. Florack, and B. M. ter Haar Romeny. A linear image reconstruction framework based on Sobolev type inner products. *International Journal of Computer Vision*, 70(3):231–240, December 2006.
12. Koenderink J.J. and Doorn A.J. van. Local structure of gaussian texture. *Journal of the Institute of Electronics Information and Communication Engineers, Transactions on Information and Systems*, E86-D(7):1165–1171, 2003.
13. F. Kanters, L. Florack, B. Platel, and B. ter Haar Romeny. Image reconstruction from multiscale critical points. In L. D. Griffin and M. Lillholm, editors, *Scale-Space Methods in Computer Vision: Proceedings of the Fourth International Conference, Scale-Space 2003, Isle of Skye, UK*, volume 2695 of *Lecture Notes in Computer Science*, pages 464–478, Berlin, June 2003. Springer-Verlag.
14. F. M. W. Kanters, M. Lillholm, R. Duits, B. Janssen, B. Platel, L. M. J. Florack, and B. M. ter Haar Romeny. On image reconstruction from multiscale top points. In R. Kimmel, N. Sochen, and J. Weickert, editors, *Scale Space and PDE Methods in Computer Vision: Proceedings of the Fifth International Conference, Scale-Space 2005, Hofgeismar, Germany*, volume 3459 of *Lecture Notes in Computer Science*, pages 431–442, Berlin, April 2005. Springer-Verlag.
15. F.M.W Kanters, L.M.J. Florack, B. Platel, and B.M. ter Haar Romeny. Multi-scale differential error: A novel image quality assessment tool. In *Proc. of the 8th Int. conf. on Signal and Image Processing 2006,Honolulu, Hawaii.*, pages 188–194, August 2006.
16. J. J. Koenderink. A hitherto unnoticed singularity of scale-space. *IEEE Transactions on Pattern Analysis and Machine Intelligence*, 11(11):1222–1224, November 1989.
17. A. Kuijper and L. M. J. Florack. Hierarchical pre-segmentation without prior knowledge. In *Proceedings of the 8th International Conference on Computer Vision (Vancouver, Canada, July 9–12, 2001)*, pages 487–493. IEEE Computer Society Press, 2001.
18. A. Kuijper and L.M.J. Florack. Understanding and modeling the evolution of critical points under gaussian blurring. In *Proceedings of the 7th European Conference on Computer Vision (ECCV)*, volume LNCS 2350, pages 143–157, 2002.
19. A. Kuijper and L.M.J. Florack. The hierarchical structure of images. *IEEE-Transactions-on-Image-Processing*, 12(9):1067–1079, Sept. 2003.

20. M. Lillholm, M. Nielsen, and L. D. Griffin. Feature-based image analysis. *International Journal of Computer Vision*, 52(2/3):73–95, 2003.
21. T. Lindeberg. Scale-space behaviour of local extrema and blobs. *Journal of Mathematical Imaging and Vision*, 1(1):65–99, March 1992.
22. T. Lindeberg. Edge detection and ridge detection with automatic scale selection. *International Journal of Computer Vision*, 30(2):117–156, November 1998.
23. D. G Lowe. Distinctive image features from scale-invariant keypoints. *International Journal of Computer Vision*, 60(2):91–110, 2004.
24. K. Mikolajczyk and C. Schmid. Scale and affine invariant interest point detectors. *International Journal of Computer Vision*, 60(1):63–86, October 2004.
25. K. Mikolajczyk and C. Schmid. A performance evaluation of local descriptors. *IEEE Transactions on Pattern Analysis and Machine Intelligence*, 27(10):1615–1630, October 2005.
26. Krystian Mikolajczyk, Bastian Leibe, and Bernt Schiele. Local features for object class recognition. In *ICCV '05: Proceedings of the Tenth IEEE International Conference on Computer Vision (ICCV'05) Volume 2*, pages 1792–1799, Washington, DC, USA, 2005. IEEE Computer Society.
27. M. Nielsen and M. Lillholm. What do features tell about images? In M. Kerckhove, editor, *Scale-Space and Morphology in Computer Vision: Proceedings of the Third International Conference, Scale-Space 2001, Vancouver, Canada*, volume 2106 of *Lecture Notes in Computer Science*, pages 39–50. Springer-Verlag, Berlin, July 2001.
28. John Novatnack, Trip Denton, Ali Shokoufandeh, and Lars Bretzner. Stable bounded canonical sets and image matching. In *Energy Minimization Methods in Computer Vision and Pattern Recognition (EMMCVPR)*, pages 316–331, November 2005.
29. B. Platel, L. M. J. Florack, F. M. W. Kanters, and E. G. Balmachnova. Using multiscale top points in image matching. In *Proceedings of the 11th International Conference on Image Processing (Singapore, October 24–27, 2004)*, pages 389–392. IEEE, 2004.
30. Vijay V. Vazirani. *Approximation Algorithms*. Springer-Verlag, Berlin, Germany, second edition, 2003. ISBN 3-540-65367-8.

# Feature Vector Similarity Based on Local Structure*

Evgeniya Balmachnova, Luc Florack, and Bart ter Haar Romeny

Eindhoven University of Technology,
P.O. Box 513, 5600 MB Eindhoven, The Netherlands
{E.Balmachnova,L.M.J.Florack,B.M.terHaarRomeny}@tue.nl

**Abstract.** Local feature matching is an essential component of many image retrieval algorithms. Euclidean and Mahalanobis distances are mostly used in order to compare two feature vectors. The first distance does not give satisfactory results in many cases and is inappropriate in the typical case where the components of the feature vector are incommensurable, whereas the second one requires training data. In this paper a stability based similarity measure (SBSM) is introduced for feature vectors that are composed of arbitrary algebraic combinations of image derivatives. Feature matching based on SBSM is shown to outperform algorithms based on Euclidean and Mahalanobis distances, and does not require any training.

## 1 Introduction

Local descriptors evaluated at certain interest points in scale space are widely used in object-recognition and image retrieval due to their robustness under occlusion, certain image transformations (rotation, zooming, up to some extent view point changes) and distinctive power. First, the interest points should be localized in space and scale. There is wide range of interest points [1],[2],[3], such as Harris points, Harris-Laplace regions, Hessian-Laplace regions, DoG, Top-Points. The second step is to build a description of the interest point, which should be discriminative and invariant to certain image transformations.

There are many different techniques to describe local image properties. Mikolajczyk and Schmid [4] compared several local descriptors, such as steerable filters, differential invariants, moment invariants, complex filters, cross-correlation of different types of interest points, and SIFT [1]. In this paper we focus only on differential type of descriptors, particulary on improvements of discriminative power of descriptors by introducing a sensible distance measure for such descriptors.

Our approach is applicable to any feature vector constructed from Gaussian derivatives taken at the interest point and it shows improvement compared to Mahalanobis and Euclidean distances. The proposed *stability based similarity measure* (SBSM) is based on analysis of local structure at the interest point and therefore uses a more appropriate covariance matrix than in case of the global Mahalanobis distance. The symmetry property of a true distance function is lost, but this does not affect matching results.

* The Netherlands Organisation for Scientific Research (NWO) is gratefully acknowledged for financial support.

F. Sgallari, A. Murli, and N. Paragios (Eds.): SSVM 2007, LNCS 4485, pp. 386–393, 2007.

## 2 Feature Vectors

In this article we consider only differential features in scale-space, e.g. differential invariants proposed by Florack et al. [5] or steerable filters [6]. We collect local features with the same base point in scale-space into a "feature vector". Under perturbation (noise, rotation, jpeg-compression and so on) every feature vector behaves differently, depending on the local image structure at the interest point. Our intention is to take this fact into account when comparing features.

Local image structure can be captured by the so-called *local jet*[7], i.e. a set of image derivatives computed up to some order. As our features are differential descriptors feature vectors can be expressed as functions on the local jet $\{u_1, \ldots, u_n\}$ (for brevity, we have indicated the various image derivatives by $u_k$, $k = 1, \ldots, n$), as follows

$$d_i = d_i(u_1, \ldots, u_n),\ i = 1 \ldots m. \tag{1}$$

So each feature vector has $m$ components.

We construct the differential feature vectors in such a way that they are invariant to certain transformations, notably, rotation, zooming and linear intensity changes. In the experimental part we consider one particular set of differential invariants up to third order, evaluated at a top-point of the image Laplacian $\triangle u$, viz.

$$\mathbf{d} = \begin{pmatrix} \sigma\sqrt{(u_x^2 + u_y^2)}/u \\ \sigma(u_{xx} + u_{yy})/\sqrt{u_x^2 + u_y^2} \\ \sigma^2(u_{xx}^2 + 2u_{xy}^2 + u_{yy}^2)/(u_x^2 + u_y^2) \\ \sigma(u_x^2 u_{xx} + 2u_x u_y u_{xy} + u_y^2 u_{yy})/(u_x^2 + u_y^2)^{3/2} \\ \sigma^2((-3u_x^2 + u_y^2)u_y u_{xxx} + (-u_x^2 + 3u_y^2)u_x u_{yyy})/(u_x^2 + u_y^2)^2 \\ \sigma^2(-u_x^3 u_{xxx} + 3u_x u_{xxx} u_y^2 + 3u_x^2 u_y u_{yyy} - u_y^3 u_{yyy})/(u_x^2 + u_y^2)^2 \end{pmatrix} \tag{2}$$

One can show that this set is complete, in the sense that there exists no other third order invariants (at a Laplacian top-point) that is independent of the entries of $\mathbf{d}$, Eq. (2). Another set, used for testing, is the set of steerable filters up to third order as described by Freeman [6].

## 3 Distance Measure

The space of features is a vector space, but it is not obvious how to introduce a norm because of the incommensurability of the components. Similarity between descriptors is usually computed with either the Euclidean or the Mahalanobis distance measure. The Euclidean distance does not make much sense in case of Eq. (2), and indeed performs poorly in practice as we will illustrate in section 5. The Mahalanobis distance gives better matching results but has 3 disadvantages, viz

- it requires a covariance matrix estimated from training data;
- performance results will depend on the training set used;
- it is a global measure, not optimally adapted to the local structure at any feature point of interest.

We propose a more generic measure similar to the Mahalanobis distance that obviates training and takes into account the local structure at the feature points of interest. In this case the covariance matrix is obtained directly from the differential structure at each interest point. The matrix can be obtained in analytical form and reflects the actual behavior of the descriptor due to small perturbations. In the next section we present the details of this approach.

## 4 Feature Vector Perturbation

We use a perturbation approach for estimation of the covariance matrix for each feature vector as alluded to in the previous section.

We generically model changes in the image due to rendering artifacts induced by transformations, jpeg-compression effects and noise as additive random perturbations. This distribution of the random value is assumed pixel to be the same for all pixels. Due to linearity of scale-space the perturbed local jet in the point is

$$\{v_1, \ldots, v_n\} = \{u_1, \ldots, u_n\} + \{N_1, \ldots, N_n\}, \tag{3}$$

where the last term models the perturbation.

Let us rewrite Eq. (1) for the unperturbed image as

$$d_i = d_i(\mathbf{u}) \tag{4}$$

and for the perturbed image as

$$\tilde{d}_i = \tilde{d}_i(\mathbf{v}) \tag{5}$$

Then the difference between these two descriptors can be approximated by a Taylor expansion of Eq. (5) around $\mathbf{u}$ up to first order:

$$\Delta d_i = \tilde{d}_i - d_i = \sum_{k=1}^{n} \frac{\partial d_i}{\partial v_k}|_{v_k=u_k} N_k \tag{6}$$

Therefore, the approximate covariance matrix $\Sigma$ is given by ($\langle N_i \rangle$ is assumed to be zero)

$$\Sigma_{ij} =< \Delta d_i \Delta d_j >= \sum_{k=1}^{n} \sum_{l=1}^{n} \frac{\partial d_i}{\partial v_k}|_{v_k=u_k} \frac{\partial d_j}{\partial v_l}|_{v_l=u_l} < N_k N_l > \tag{7}$$

The only step left is to estimate this covariance matrix for the feature perturbation, which is the subject of the next section.

### 4.1 Gaussian Correlated Noise

The momentum $M^2_{m_x,m_y,n_x,n_y} =< N_{m_x,m_y} N_{n_x,n_y} >$ of Gaussian derivatives of orders $(m_x, m_y)$ and $(n_x, n_y)$ of correlated noise in case the spatial noise correlation distance $\tau$ is much smaller than scale $t$ is given by [8]

$$M^2_{m_x,m_y,n_x,n_y} \simeq < N^2 > \left(\frac{\tau}{2t}\right) \left(\frac{-1}{4t}\right)^{\frac{1}{2}(m_x+m_y+n_x+n_y)} Q_{m_x+n_x} Q_{m_y+n_y} \tag{8}$$

with $Q_k$ given by Table 1.

**Table 1.** Some values of $Q_n$ ($Q_n$=0 if $n$ is odd)

| $n$ | 0 | 2 | 4 | 6 |
|---|---|---|---|---|
| $Q_n$ | 1 | 1 | 3 | 15 |

Let us take the correlation kernel to be roughly of one pixel width corresponding to $\tau = \epsilon^2/4$, where $\epsilon$ denotes pixel size. For Gaussian derivatives of first and second order we obtain the following correlation matrix:

$$C = (\langle N_i N_j \rangle)_{ij} = \begin{pmatrix} 4t & 0 & 0 & 0 & 0 \\ 0 & 4t & 0 & 0 & 0 \\ 0 & 0 & 3 & 0 & 1 \\ 0 & 0 & 0 & 1 & 0 \\ 0 & 0 & 1 & 0 & 3 \end{pmatrix} \frac{\epsilon^2 \langle N^2 \rangle}{(4t)^3}, \tag{9}$$

where $(N_1, \ldots, N_5) = (N_x, N_y, N_{xx}, N_{xy}, N_{yy})$, and the matrix entries in Eq. (9) are labeled accordingly. This correlation matrix together with Eq. (7) gives an approximation of the covariance matrix of each local feature vector for given perturbation variance and pixel size.

### 4.2 Distance

We define the similarity between feature descriptors $\mathbf{u}$ and $\mathbf{u}_0$ in a similar way as the well-known Mahalanobis distance, except that for every point $u_0$ we insert its own covariance matrix

$$\rho(\mathbf{u}; \mathbf{u}_0) = (\mathbf{u} - \mathbf{u}_0)^T \Sigma_{\mathbf{u}_0}^{-1} (\mathbf{u} - \mathbf{u}_0) \tag{10}$$

Consequently, the function $\rho(\mathbf{u}; \mathbf{u}_0)$ is not symmetric, therefore it is not a distance in the strict sense. The reference image $u_0$ is considered to be the "ground truth". Covariance matrix and, as a consequence, distance are proportional to the constant $\epsilon^2 \langle N^2 \rangle$, i.e. the product of the noise variance and the pixel size. This constant is the same for all points of the reference image and hence does not change the ordering of distances from some point to the set of all points of the reference image, therefore the constant can be omitted.

## 5 Experiments

### 5.1 Database

For the experiments we use a data set containing transformed versions of 12 different magazine covers. The covers contain a variety of objects and text. The data set contains rotated, zoomed and noisy versions of these magazine covers as well as images with perspective transformations. For all transformations the ground truth is known, which enables us to verify the performance of different algorithms on the database. Mikolajczyk's data set used in [2,4] is not suitable for our purposes, as we require ground truth for genuine group transformations not confounded with other sources of image changes, such as changes in field of view or lighting conditions. To our knowledge Mikolajczyk's data set does not provide this.

**Fig. 1.** A selection of data set images. From left to right: unchanged, rotated, added noise, scaled, changed perspective.

### 5.2 Validation

We use a criterion proposed in [9]. It is based on the number of correct matches and the number of false matches obtained for an image pair. We couple interest points if the distance between their feature vectors is below a certain threshold $d$ (such a pair is called *a possible match*). Note that since we know the transformations we also know ground truth for the matches. Each feature vector from the reference image is compared to each vector from the transformed one, and the number of *correct matches* (according to the ground truth, out of all possible matches) as well as the number of *false matches*, is counted.

$$\text{\#possible matches} = \text{\#correct matches} + \text{\#false matches} \tag{11}$$

The threshold $d$ is varied to obtain curves as detailed in the next section. The results are presented with recall versus 1–precision. Recall is the number of correctly matched points relative to the number of ground truth correspondences between two images of the same scene:

$$\text{recall} = \frac{\text{\#correct matches}}{\text{\#correspondences}} \tag{12}$$

The number of false matches relative to the number of possible matches is presented by 1–precision

$$1 - \text{precision} = \frac{\text{\#false matches}}{\text{\#possible matches}} \tag{13}$$

The number of correct matches and correspondences is determined with the overlap error. The overlap error measures how well the points correspond under a transformation $H$. It is defined by the ratio of the intersection and union of two disks $S_1$ and $S_2$ with centers in the interest points, $x_1$ and $x_2$, and radii given by the scales of the points $\sigma_1$ and $\sigma_2$,

$$\varepsilon = 1 - \frac{S_2 \bigcap HS_1}{S_2 \bigcup HS_1}, \tag{14}$$

where $HS_1 = \{Hx | x \in S_1\}$.

In case of transformations closed to Scale-Euclidean ones, $HS_1$ can be also approximated by a disk, and areas of intersection and union can be computes analytically. A match is correct if the error $\varepsilon$ in the image area covered by two corresponding regions is less than 50% of the region union. The number of correspondences in order to compute recall in Eq. (12) is determined with the same criterion.

### 5.3 Results

In our experimental setting the distance between every point from the reference image and every point from the transformed one is calculated. Two points are considered to be matched if the distance $\rho$ between their feature vectors is below a threshold $d$. The result obtained by varying $d$ is presented by a curve. The curve presents 1-precision versus recall as a function of $d$. For experiments with the Mahalanobis distance training data are required. We use part of the data base for training and the rest for evaluation.

Experiments were conducted with different choices of image transformations, feature vectors and interest points. For every pair of images a recall vs 1–precision curve is built, and then the mean curve over 12 pairs is computed. In all the experiments usage of

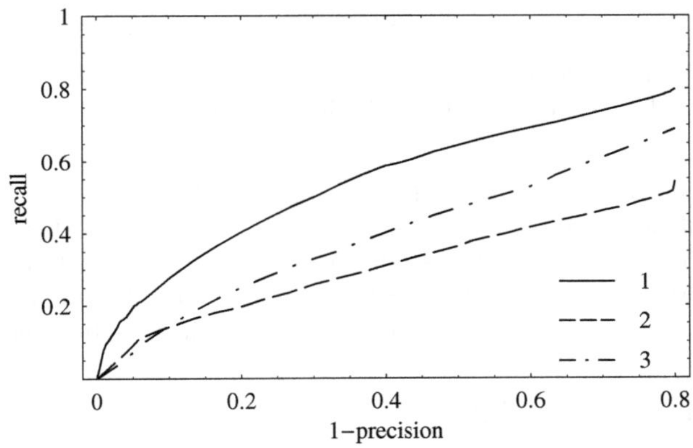


**Fig. 2.** Evaluation of different distances in case of DoG points, differential invariants for 5% noise. 1- SBSM, 2 - Euclidean distance, 3- Mahalanobis distance.

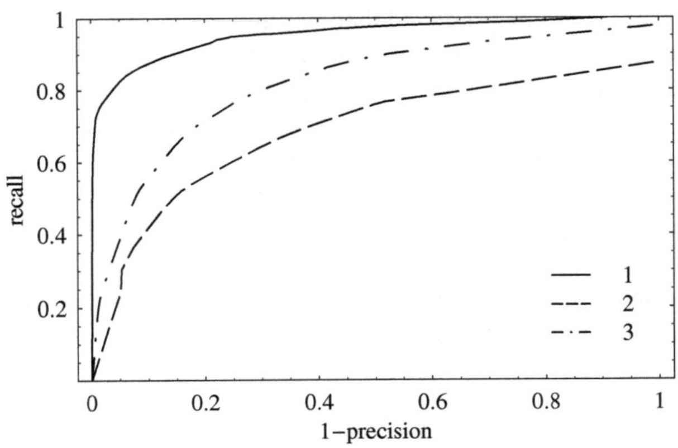


**Fig. 3.** Evaluation of different distances in case of Top-Points points, differential invariants and 45 degree rotation. 1- SBSM, 2 - Euclidean distance, 3- Mahalanobis distance.

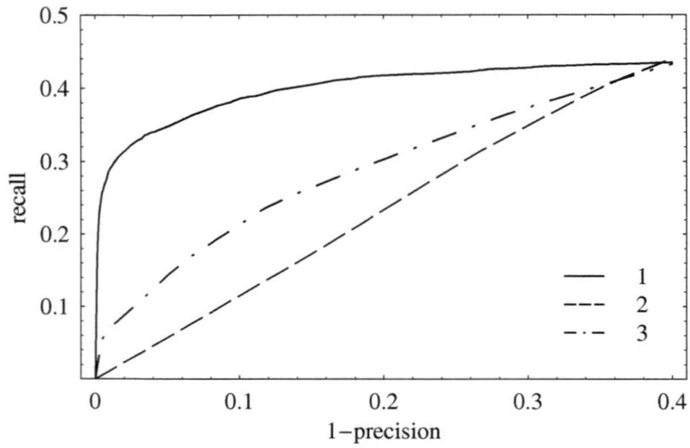


**Fig. 4.** Evaluation of different distances in case of Top-Points points, steerable filters and rotation+zooming. 1- SBSM, 2 - Euclidean distance, 3- Mahalanobis distance.

SBSM improved the performance. In the paper we present three examples. Figure 2 depicts SBSM, Euclidean and Mahalanobis curves in case of 5% noise, where differential invariants are used in Difference-of-Gaussian points. In Fig 3 Top-Points are used as interest points and differential invariants as features for 45 degree rotation case. Figure 4 depicts results of using steerable filters at Top-Points for image rotation and zooming.

## 6 Summary and Conclusions

In this paper we have introduced a new stability based similarity measure (SBSM) for feature vectors constructed by means of differential invariants. The algorithm is based on a perturbation approach and uses properties of noise propagation in Gaussian scale-space.

The advantage of this approach is that a local covariance matrix describing the stability of the feature vector can be predicted theoretically on the basis of the local differential structure, and that no training data are required. In fact the analytical noise model underlying the SBSM replaces the role of training. The drawback of using SBSM is a necessity to store a covariance matrix for every point of the reference image.

Another advantage of SBSM is possibility of using it in order to threshold interest points with very unstable and therefore unreliable feature vectors. One can think of eigenvalues of the covariance matrix as a criterium. This allows one to reduce the amount of data stored as well as computational time needed for matching.

The experiments show improvement in performance for different choices of interest points, different combinations of derivatives and several transformations.

## References

1. Lowe, D.G.: Distinctive image features from scale-invariant keypoints. Int. J. Comput. Vision **60** (2004) 91–110
2. Mikolajczyk, K., Schmid, C.: Scale and affine invariant interest point detectors. International Journal of Computer Vision **60** (2004) 63–86

3. Platel, B., Balmachnova, E.G., Florack, L.M.J., ter Haar Romeny, B.M.: Top-points as interest points for image matching. In: Proceedings of the Ninth European Conference on Computer Vision, Graz, Austria (2006) 418–429
4. Mikolajczyk, K., Schmid, C.: A performance evaluation of local descriptors. Submitted to PAMI (2004)
5. Florack, L.M.J., Haar Romeny, B.M.t., Koenderink, J.J., Viergever, M.A.: Cartesian differential invariants in scale-space. Journal of Mathematical Imaging and Vision **3** (1993) 327–348
6. Freeman, W., Adelson, E.: The design and use of steerable filters. IEEE Transactions on Pattern Analysis and Machine Intelligence **13** (1991) 891–906
7. Koenderink, J.J.: The structure of images. Biological Cybernetics **50** (1984) 363–370
8. Blom, J., Haar Romeny, B.M.t., Bel, A., Koenderink, J.J.: Spatial derivatives and the propagation of noise in Gaussian scale-space. Journal of Visual Communication and Image Representation **4** (1993) 1–13
9. Ke, Y., Sukthankar, R.: Pca-sift: A more distinctive representation for local image descriptors. In: Proceedings of the Conference on Computer Vision and Pattern Recognition. (2004) 511–517

# Maximum Likelihood Metameres for Local $2^{nd}$ Order Image Structure of Natural Images

Martin Lillholm and Lewis D Griffin

University College London, Department of Computer Science,
London - WC1E 6BT, United Kingdom
m.lillholm@cs.ucl.ac.uk

**Abstract.** We investigate the maximum likelihood metameres of local pure $2^{nd}$ order structure in natural images. Using the shape index, we re-parameterise the $2^{nd}$ order structure and gain a one-parameter index which offers a qualitative description of local pure $2^{nd}$ order image structure. Inspired by Koenderink and previous work within Geometric Texton Theory the maximum likelihood metameres are calculated for a quantised version of the shape index. Results are presented and discussed for natural images, Gaussian noise images, and Brownian or pink noise images. Furthermore, we present statistics for the shape index, principal direction, and curvedness of natural images. Finally, the results are discussed in the terms of their applicability to Geometric Texton Theory.

## 1 Introduction

The cornerstone of bottom-up feature-based image analysis is a representation of image structure in terms of qualitative feature descriptions such as edges, corners, lines, blobs, etc. Although Marr's work [24] set the bar, a complete realisation of his ideas has never be accomplished. There are of course many possible explanations for this — one simply is that research has progressed with much success using more direct methods based on quantitative descriptions of local image structure. Another possible explanation is that the mapping from local quantitative measurements of image structure to qualitative descriptions has never been fully explored. Although many feature types have been identified, they are typically the result of conscious design choices [10,1,22] rather than systematic discovery based on e.g. natural images statistics. It remains an interesting and unanswered question whether (natural) images can be locally described through a finite vocabulary of qualitatively distinct feature types. Inspired by Koenderink and previous work within Geometric Texton Theory, we explore iconic representatives of the local $2^{nd}$ order differential structure of natural images.

The remainder of this paper is organised as follows: The final two sections of the introduction describe quantitative measurements of local structure and a path to qualitative descriptions. The shape index descriptor for local $2^{nd}$ order structure is described in section 2.1. Based on this re-parameterisations the paper's main investigation is given and discussed in section 2. Finally, results and outlook is discussed in sections 3.1 and 4.

F. Sgallari, A. Murli, and N. Paragios (Eds.): SSVM 2007, LNCS 4485, pp. 394–405, 2007.

### 1.1 Quantitative Measurements of Local Image Structure

Traditionally, the output of V1 simple cells are modelled as the inner product of Receptive Field weighting functions (RFs) and the retinal irradiance [11]. Several models of V1 simple cells (ensembles) have been reported in the literature with Gabor filters as the dominant one [3,12].

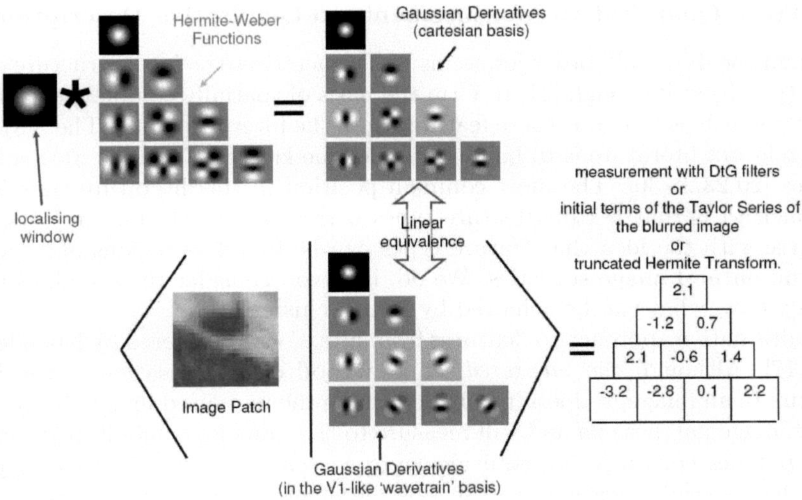


**Fig. 1.** Middle column: two alternative families of Gaussian Derivative (DtG) filters. Bottom row: how measurement with a DtG family gives rise to the local jet. Top row: equivalence between measurements with DtGs and computation of the Hermite Transform.

In this work we, however, prefer the Gaussian Derivatives (DtGs) as an uncommitted model to measure the quantitative local structure of images — see Figure 1. Although this choice may seem to counter reported trends, several arguments favour the DtGs. In terms of biological relevance DtGs up to $4^{th}$ or $5^{th}$ order provide an equally well fitting set of models (a simple linear combination brings the DtGs into the 'wave-train' basis very similar to Gabors [29,15]). Furthermore, the DtGs are well studied and possess many commendable properties [13,18,5]. The real strength of DtGs, however, lies in the fact that the measured quantities are much easier to interpret than corresponding Gabors. Local measurements of an image with DtGs up to some order $n$ is also referred to as the $n$-jet. The $n$-jet can be understood as the coefficients of truncated Taylor-series of the *blurred* image. Furthermore, the jet can also (Figure 1) be interpreted as isolating an image patch with a Gaussian window and then probing it with Hermite-Weber functions — not unlike a windowed Fourier transform.

The one-dimensional Gaussian kernel is given by $G(x) = \frac{1}{\sqrt{2\pi\sigma^2}} \exp \frac{-x^2}{2\sigma^2}$, where $\sigma$ is the standard deviation and controls the scale or *blur* at which the jet is calculated. Due to separability the two-dimensional Gaussian kernel is easily constructed as $G(x, y) = G(x)G(y)$. We'll use $G_x, G_{xy}$, etc. to denote partial derivatives of the two-dimension kernel and $L_{xy}$ to denote the inner product between an image patch $L$ and the suitable derivative of the Gaussian i.e. $L_{xy} = \langle G_{xy}, L\rangle$. The $n$-jet is the vector of inner products up to total order of derivation $n$ — the 2-jet is thus given by $(L, L_x, L_y, L_{xx}, L_{xy}, L_{Lyy})$.

### 1.2 From Quantitative Measurements to Qualitative Descriptions

Although the $4^{th}$ or $5^{th}$ order jet seems to characterises the local structure of an image (patch) well enough [21] to form the basis of spatially complex features, it is far obvious how to define these features from the filter responses. The majority of the relevant literature is to be found under the keyword 'textons' rather than features [20,23,28,30]. The most common position in this literature (implicitly or explicitly) is that textons/feature types correspond to clusters in jet space. We agree with the idea that feature types can be found as regions of jet space based on natural image statistics. We do, however, consider these regions more complex than what can be achieved by simple clustering.

An alternative approach to 'features from filters' was pioneered by Koenderink [14,16,17]: Although the *measured* jet is a good characterisation of the local structure in an image, it doesn't fully determine the measured image. In essence, several image patches can and will measure to the same jet or point in jet-space. These patches constitute an equivalence or *metamery* class under the applied filters. Koenderink suggested to use this metamerism and i) to associate with each point of jet space an iconic image from the metamery class of possible images, and (ii) the equivalence relation of 'qualitative identity of icons' gives rise to a partitioning of jet space into features. Koenderink further suggested (more or less) that these icons should be selected as the element in a metamery class with the smallest range of intensity values. We have earlier studied range minimisation and several other candidate rules for icon selection all based on minimisation of some measure of complexity — complexity was defined in terms of a norm either based on image luminance or image 'roughness'. A full description can be found in [8] where some of the main conclusions are: i) while low order luminance range minimisers are quite representative of natural structure, at higher order they are distinctly less so, (ii) roughness minimisers produce quite natural looking icons with the well-known Total Variation minimiser [26] as the most successful albeit results in 1D behave far better than corresponding 2D experiments. In summary, icons can be produced using complexity minimisers and partitions of jet space can be based on this. There are, however, two main problems with this approach: i) The choice of complexity minimiser can and will affect the icons and subsequently the partitioning of jet space, and ii) a complexity measure impose an underlying model distribution [21] of natural images that at worst is plain wrong and at best only models some aspects of the actual distribution of natural images.

To avoid these problems and as a refinement of Koenderink's original proposal, we proposed Geometric Texton Theory (GTT,[9]) where the icons are selected as the maximum likelihood (relative to natural image statistics) elements of metamery classes — that is the most likely or the modal patch in high-dimensional patch space. If successful, this essentially allows for a statistics driven approach to jet-space partitioning without underlying model assumptions and clustering schemes. Earlier published results for GTT can be found in section 3.1.

## 2 Maximum Likelihood Metameres for Local $2^{nd}$ Order Structure

This study focuses on the 'pure' $2^{nd}$ order structure of natural image patches measured as $L_{xx}, L_{xy}$, and $L_{yy}$. An idealised but also somewhat naïve application of the approach described above would be to attempt to calculate icons or maximum likelihood (ML) representatives for all points in response/jet space. For a number for fairly obvious reasons this wouldn't be tractable: i) There are an infinite number of points in the response space, ii) even circumventing this, we would need an unrealisable number of samples to make reliable mode estimates for each point, and iii) the study wouldn't be invariant to simple image transformations such a affine intensity changes and rotation.

We therefore, as in our previous studies [9,7,8], choose to factor out an affine component. To this end, the $0^{th}$ order $L$ is also included in our measurements and each patch is $p$ is transformed as $(p \rightarrow \frac{1}{\vartheta}p - L)$, where $\vartheta$ is the curvedness as described in 2.1. Furthermore, patches are extracted such that their 'x-axis' is parallel to the direction of main curvature. Measuring the $0^{th}$ and $2^{nd}$ order structure yields four degrees of freedom (DoF). The affine transformation and controlled extraction-angle soak up three of these leaving a one-dimensional factored jet-space [6]. As in our study of ML metameres for the 2-jet of natural image profiles, we seek a suitable re-parameterisation of this factored jet-space.

### 2.1 The Shape Index

We seek a re-parameterisation of, in this case, the pure second order structure $(L_{xx}, L_{xy}, L_{yy})$. Although the traditional invariants the Gaussian and mean curvature offer some insight into the local 'shape', we prefer the *shape index* as described in [2,19]. As for the Gaussian and mean curvature, the shape index for a point on a surface in $\mathbb{R}^3$ can be defined in terms of the eigenvalues $\kappa 1, \kappa 2$ of the local Hessian $H$:

$$H = \begin{pmatrix} L_{xx} & L_{xy} \\ L_{xy} & L_{yy} \end{pmatrix}$$

The shape index $\vartheta$ is now defined using the ratio of the sum and the difference of the eigenvalues $\kappa 1, \kappa 2$:

$$\vartheta = \arctan \frac{\kappa 1 + \kappa 2}{\kappa 1 - \kappa 2}$$

Per construction the shape index spans the interval $[-\pi/2, \pi/2]$ and is undefined for flat points ($\kappa 1 = \kappa 2 = 0$). At the extremes of the interval of the

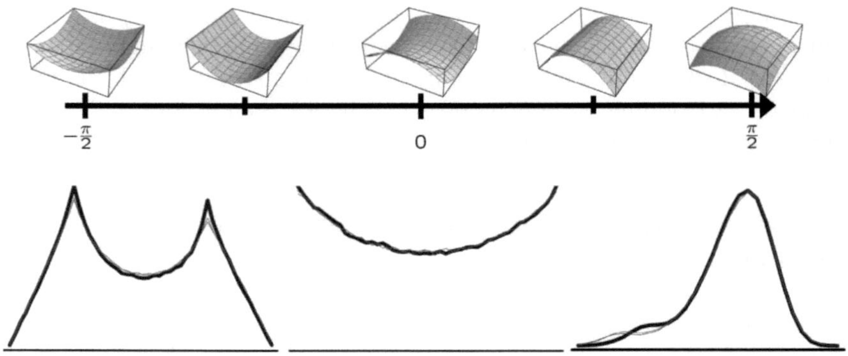


**Fig. 2.** The top row exemplifies the qualitative nature of the shape index. Bottom left is the histogram of shape indices in natural images — the axis spans $-\pi/2$ to $\pi/2$. Bottom middle is the histogram of natural image principal curvature directions from 0 (horizontal) to $\pi/2$ (vertical). Bottom right is the histogram of curvedness for natural images. Please note that the x-axis is logarithmic. For all three histograms, the main blue curved is the calculated at the scale used for experiments throughout the paper. The thinner red curves are the similar histograms calculated at two different scales indicating the approximate scale invariance expected in natural images [4].

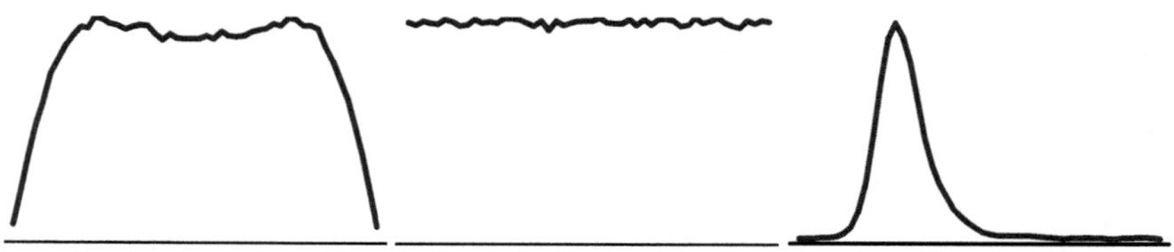

**Fig. 3.** The Gaussian noise version of Figure 2 except histograms are only calculated for one scale

shape index, we have umbilic like rotationally symmetric minima and maxima — typically found at isolated points. Parabolic or line(pass)-like points lie at the midpoints of the two half-intervals and at zero saddle-like points. This is illustrated in the top row of Figure 2.

Although the shape index in itself will be sufficient for our purposes, it is worth mentioning that the shape index form part of a complete re-parameterisation of the second order structure. The other two parts [2,19] are the *curvedness* $\lambda$ and the direction of principal curvature $\psi$ defined by:

$$\lambda = \frac{\sqrt{\kappa_1^2 + \kappa_2^2}}{2} \qquad \psi = \arctan \frac{L_{xy}}{\frac{L_{xx} - L_{yy}}{2}}$$

The curvedness can be interpreted as the amplitude of local $2^{nd}$ order structure and the principal direction as the direction in which the curvature of the normal section is maximal. As argued by Koenderink the triple $(\lambda, \vartheta, \psi)$ offers a much more intuitive decoupled description of the local $2^{nd}$ order structure than the corresponding Cartesian triple $(L_{xx}, L_{xy}, L_{yy})$.

In [19], Koenderink gave analytical and empirical densities for the shape index and the curvedness for Gaussian and Brownian noise. He furthermore gave empirical observations for a single 'natural' image. In Figure 2, we supplement these findings and give histograms for the shape index, curvedness, and principal direction calculated over a subset (see section 2.2) of the van Hateren database of natural images [27]. Some observations for natural images, beyond those given in [19], are as follows:

- The histogram of shape indices is slightly asymmetric compared to Koenderink's histogram for a single image. The bias is towards dark parabolic-like structures.
- The modes of the natural image shape index histogram are not only more kurtosed than the corresponding Gaussian noise (3) histograms but are centred on parabolic-like points whereas the corresponding modes in the Gaussian case are more or less halfway between the parabolic- and umbilic-like points. The modes of similar histograms (not shown) for Brownian noise lie in between those of natural images and Gaussian noise. An observation that is supported by the related densities over the $2^{nd}$ order local-image-structure solid in [6].
- The histogram of principal directions shows a bias towards horizontal and vertical structures. In the corresponding histograms for Gaussian noise (3) there is no directional bias. This bias for natural images is also observed for gradient orientation by among others [25] but is perhaps slightly more pronounced in the $2^{nd}$ order case.
- The histogram of curvedness over an ensemble of images is quite different from the single image histogram given in [19]. This suggests that the curvedness is less 'universal' than the other two components. Koenderink's single example image is quite structured whereas many of the van Hateren images have large almost flat areas of e.g. sky.
- The densities, over an ensemble of natural images, of the quantities $\lambda, \vartheta$, and $\psi$ exhibit the expected [4] approximate scale invariance.

### 2.2 Patch Extraction and Algorithm Details

The shape index offers the sought after re-parameterisation of the remaining DoF as discussed in the beginning of section 2 and constitutes our factored jet-space. The shape index spans a real interval and to make the study of $2^{nd}$ order metameres tractable we quantise the interval into 31 bins. This will pool different although neighbouring jets which is not optimal. We have, however, successfully used this approach in earlier studies.

For each bin $8 \times 10^6$, $17 \times 17$ patches are extracted. Each patch is extracted so that its 'x-axis' is parallel to the principal direction and then affinely normalised as described above. The jet is measured using scale $\sigma \approx 1.8$.

The patches are extracted from a subset of the van Hateren Natural Stimuli Collection [27] of 'natural images' — equally many from each image. The linear (.iml) version of the database originally consisted of approximately 4000 (1536×1024) images. A substantial part of the images does, however, suffer from pronounced visual artifacts in terms of either motion blur and/or partial oversaturation. This has led us to use only 1220 of the original 4000 images. The 8-bit depth of the images is in-fact pseudo 12-bit (see [27]) — this quantisation effect is reduced using the same technique as described in [9].

For comparison with the natural image study, we use two databases of artificial images — one set of Gaussian noise and one set of Brownian or pink noise. The methodology is otherwise the same for the control databases.

The maximum likelihood metamere for each bin is calculated as the mode of the $8 \times 10^6$ samples in $17 \times 17$ patch space. To this end we use a multi scale mean shift mode algorithm described in [8]. The algorithm tracks the mode to a single sample unless rare creation events close to sample scale are encountered. In this case, the mode estimate at the scale immediately preceding the catastrophe is returned. The mode estimates are always members of the 'fuzzy' metameric class defined by the current bin. The estimation of the mode for each bin (in each database) is repeated three times with new sets of patches. The presented results are the means of the three repetitions. Finally, initial results showed all ML metameres to be symmetric although noisy. In order to increase the reliability of the mode estimates through an increase of the number of samples the actual patches fed into the algorithm were averaged and of quarter size. The results shown in the next section are constructed from these quarter size ML estimates and are thus perfectly symmetric.

### 2.3 Results

The results for the three different images types can be found in Figure 4 (natural images), Figure 5 (Gaussian noise), and Figure 6 (Brownian noise). In all figures, the maximum likelihood metameres follow the binned shape index from top to bottom, left to right — the top left ML metamere corresponds to a shape index of approximately $-\frac{\pi}{2}$, the central to 0, and the bottom right to $\frac{\pi}{2}$. Individual patches are displayed as 10 level contour plots. Please note that due the principal direction dependent patch extraction, patches appear as if rotated ninety degrees as one passes through the central saddle-like bin.

In the following, we give our immediate comments to the results per se. In section 3.1, we further comment upon them in the context of Geometric Texton Theory.

Initially, we note that the Gaussian and the Brownian icons except for noise match the analytical predictions [9] based on corresponding bin centres. This confirms that the overall framework is working satisfactory and reinforce the validity of the natural image ML estimates. It should, however, be noted that the distribution of natural image patches is more complex than that of the symmetrical control experiments. In accordance with expectations and similar to our earlier studies, the Gaussian and Brownian ML estimates are more noisy than the corresponding natural image estimates.

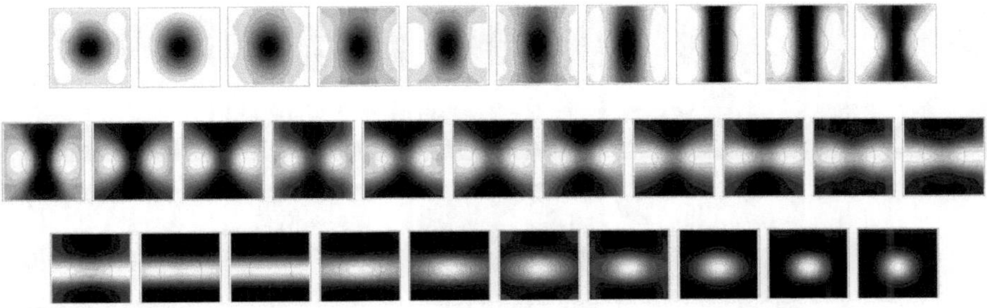

**Fig. 4.** The ML icons for natural images. Top left corresponds to shape index $-\pi/2$ and bottom right to $\pi/2$ .

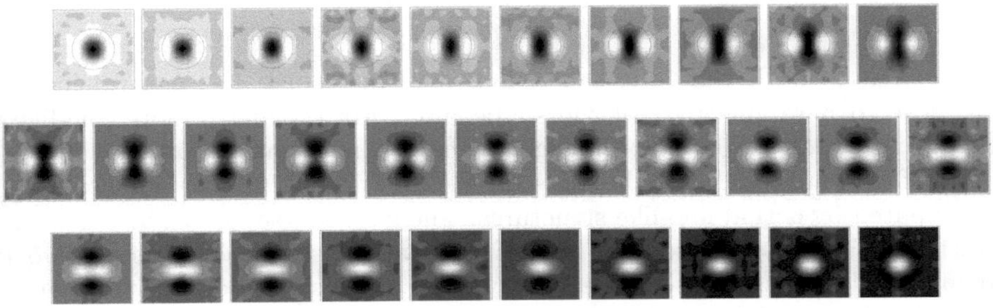

**Fig. 5.** The ML icons for Gaussian noise images

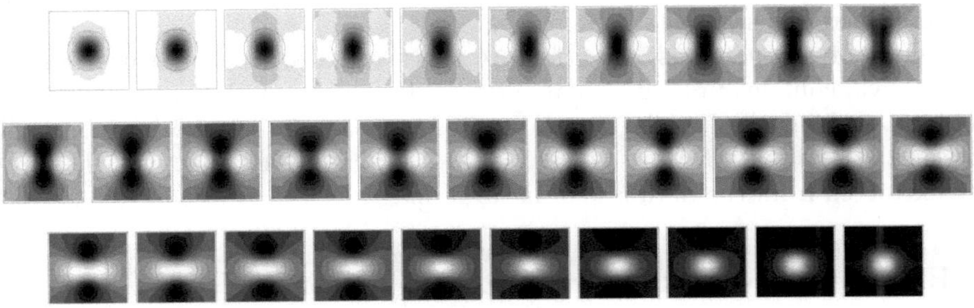

**Fig. 6.** The ML icons for Brownian noise images

We further remark that the natural images ML icons are different from both the Gaussian and Brownian control experiments making the result unique to natural images. The difference is not waste but nevertheless quite substantial given that all three experiments are otherwise identically constrained to the shape index.

Both the Gaussian and Brownian ML icons (although more noisy) come across as slightly more symmetrical across the central bin than the natural image ones. Although an essentially orthogonal and thus slightly speculative observation, this is also what we found for the shape index histograms (section 2.1).

**Fig. 7.** The context of 5 patches that were chosen as modes — top row is the large scale context and bottom row a zoom on the central part

Compared to both the Gaussian and Brownian results, the natural image icons close to parabolic part of the shape index are distinctly more line-like. Although all three shape index histograms have concentrations of mass in the vicinity of the parabolic parts the ML icons reveal that only natural images seem be dominated by actual line-like structures. The is of course fairly obvious from direct inspection of actual example images from each class but nevertheless reveal that marginal statistics based on e.g. the shape index can be deceptive. Again, in the slightly speculative department, we note that the strong continued lines for parabolic natural image icons could be explained by the fact that natural images normally do not have isolated locally line-like points by more typically long continued line segments.

In terms of distinct qualitative feature types revealed in the natural image icons, we will confine ourselves to naming five: dark blobs, dark bars, saddle-like, light bars, and light blobs. The remaining bins come across as smooth transitions between these primary feature types. Figure 7 shows where these five ML icons were extracted from the database of natural images.

## 3 Relevance to Geometric Texton Theory

In this section, we review previous studies using Geometric Texton Theory (GTT) and briefly comment on the natural image ML icons in this context. An in-depth review of GTT can be found in [8].

### 3.1 Previous Results

We have previously [9,7,8] applied GTT to 1-dimensional profiles and 2-dimensional image patches. As above, an affine (intensity) component is factored out of the profiles/patches. Figure 8 gives a summary of previous results.

The topmost left part shows the results for 1D natural image profiles, where the 1-jet is measured yielding two DoF. The affine transformation soaks up both DoF and results in one metamery class where the maximum likelihood metamere

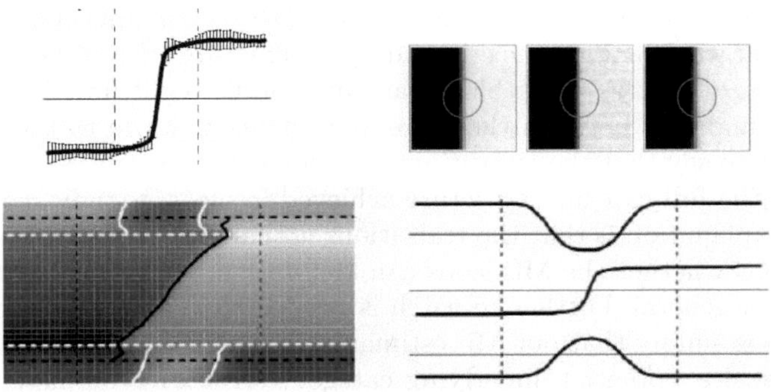

**Fig. 8.** A summary of previously published GTT results. Top left is the ML icon for $1^{st}$ order structure in natural image profiles. Top right 3 repetitions of the corresponding study for natural image patches. Bottom row depict the ML icons for full $2^{nd}$ order structure in natural image profiles — see main text for details.

or class representative is an approximate step edge. The topmost right part of Figure 8 shows three maximum likelihood metameres for the corresponding experiment for natural image patches. Measuring the 1-jet in 2D yields 3 DoF — two of which are soaked up by the affine transformation and one by the fact that all patches are extracted so that their 'x-axis' if parallel to the gradient direction. Again, the results is an approximate step-edge. In summary, if one looks at $1^{st}$ order structure in natural images (profiles or patches), the most likely representative for non-flat points is an approximate step-edge. Qualitatively speaking $1^{st}$ order structure in natural images is either flat or edge-like.

The bottom part of Figure 8 gives the results for profiles where the 2-jet is measured (3 DoF). Again, the affine transformation soaks up two of them — the one remaining DoF is re-parameterised as a phase-like [7] component and binned as described above. In the left part, each row of the tableau corresponds to a maximum likelihood metamere for a given phase bin. The right part of the figure shows the three dominant profiles. The interesting part of this result is that the 1D full $2^{nd}$ order natural image structure can be divided into one of three qualitative distinct feature types: Bar, edge, and pass. In the left tableau the transitions between these occur quite abruptly over a couple of bins which effectively enforces the category structure.

## 4 Discussion

We have added to the growing body of natural images statistics in two related ways: i) supplemented Koenderink's work on the shape index with statistics for a large ensemble of natural images, and ii) presented maximum likelihood metameres for the quantised shape index. The latter is also a step towards a more complete understanding of an icon based partitioning of jet-space for natural images.

Reviewing our current results in the light of 1D 2-jet example above, it is, however, clear that we do not get the crisp category structure offered there. Although, the five category centres (dark blob, dark line,saddle, light line, light blob) are clear, the boundaries between them are at best fuzzy. More to the point, it is probably fair to say that the transitions between category centres are extremely smooth and the full category structure achieved in our 1D study seems elusive. A possible explanation is that the transitions actually are smooth which in turn then questions whether the ML icons can facilitate a well-founded chopping up of jet-space in general. Further research is needed to answer this question. Another option is simply that our ML estimates are not accurate enough and thus does not reveal a potential underlying category structure completely — again further research is needed.

## References


1. J. Canny. A computational approach to edge detection. *IEEE Transactions on Pattern Analysis and Machine Intelligence*, 8(6):679–698, 1986.
2. F. Casorati. Nuova definitione della curvatura delle superficie e suo confronto con quella di gauss. *Rend. Maem. Accd. Lomb.*, 1867/1868.
3. J. G. Daugman. Uncertainty relations for resolution in space, spatial frequency, and orientation optimized by two-demensional visual cortical filters. *Journal of the Optical Society of America A*, pages 1160–1169, 1985.
4. D. J. Field. Relations between the statistics of natural images and the response proporties of cortical cells. *J. Optic. Soc. Am.*, 4(12):2379–2394, Dec 1987.
5. L. M. J. Florack, B. M. ter Haar Romeny, J. J. Koenderink, and M. A. Viergever. Cartesian differential invariants in scale-space. *Journal of Mathematical Imaging and Vision*, 3(4):327–348, November 1993.
6. L.D. Griffin and M. Lillholm. The 2nd order local-image-structure solid. *IEEE Transactions on Pattern Analysis and Machine Intelligence.* In Press.
7. L.D. Griffin and M. Lillholm. Image features and the 1-d, 2nd order gaussian derivative jet. *Proc. Scale Space 2005*, pages 26–37, 2005.
8. L.D. Griffin and M. Lillholm. Hypotheses for image features, icons and textons. *International Journal of Computer Vision*, 70(3):213–230, 2006.
9. L.D. Griffin, M. Lillholm, and M. Nielsen. Natural image profiles are most likely to be step edges. *Vision Research*, 44(4):407–421, February 2004.
10. C. Harris and M.J. Stephens. A combined corner and edge detector. In *Alvey88*, pages 147–152, 1988.
11. D. H. Hubel and T.N. Wiesel. Receptive fields and functional architecture of monkey striate cortex. *Journal of Physiology*, 195:215–243, 1968.
12. J. P. Jones and L. A. Palmer. The two-dimensional spatial structure of simple receptive fields in cat striate cortex. *Journal of Neurophysiology*, 58(6):1233–1258, 1987.
13. J. J. Koenderink. The structure of images. *Biological Cybernetics*, 50:363–370, 1984.
14. J. J. Koenderink. What is a feature? *Journal of Intelligent Systems*, 3(1):49–82, 1993.
15. J. J. Koenderink and A. J. van Doorn. Receptive field assembly specificity. *Journal of Visual Communication and Image Representation*, 3(1):1–12, 1992.

16. J. J. Koenderink and A. J. van Doorn. Metamerism in complete sets of image operators. In *Advances in Image Understading '96*, pages 113–129, 1996.
17. J. J. Koenderink and A. J. van Doorn. Local image operators and iconic structure. In G. Sommer and J. J. Koenderink, editors, *Algebraic Frames for the Perception-Action Cycle*, volume 1315 of *LNCS*, pages 66–93. Springer, 1997.
18. J.J. Koenderink and A. J. van Doorn. Receptive-field families. *Biological Cybernetics*, 63(4):291–297, 1990.
19. J.J. Koenderink and A.J. van Doorn. Local Structure of Gaussian Texture. *IEICE TRANSACTIONS on Information and Systems*, 86(7):1165–1171, 2003.
20. T. Leung and J. Malik. Representing and Recognizing the Visual Appearance of Materials using Three-dimensional Textons. *International Journal of Computer Vision*, 43(1):29–44, 2001.
21. M. Lillholm, M. Nielsen, and L. D. Griffin. Feature-based image analysis. *International Journal of Computer Vision*, 52(2/3):73–95, 2003.
22. T. Lindeberg. Feature detection with automatic scale selection. *International Journal of Computer Vision*, 30(2):79–116, 1998.
23. X. Liu and D.L. Wang. A spectral histogram model for texton modeling and texture discrimination. *Vision Research*, 42(23):2617–2634, 2002.
24. D. Marr. Vision: A computational investigation into the human representation and processing of visual information. In *W.H. Freeman*, 1982.
25. A.J. Nasrallah and L.D. Griffin. Gradient direction dependencies in natural images. *Spatial Vision*, Accepted Oct 2006.
26. L. I. Rudin, S. Osher, and E Fatemi. Nonlinear total variation based noise removal algorithms. *Physica D*, pages 259–268, 1992.
27. J. H. van Hateren and A. van der Schaaf. Independent components filters of natural images compared with simple cells in primary visual cortex. *Proc. R. Soc., London Series B*, 265:359–366, 1998.
28. M. Varma and A. Zisserman. A Statistical Approach to Texture Classification from Single Images. *International Journal of Computer Vision*, 62(1):61–81, 2005.
29. R. A. Young. The gaussian derivative theory of spatial vision: Analysis of cortical receptive field line-weighting profiles. *Gen.Motors Res.Tech.Rep, GMR-4920*, 1985.
30. S.C. Zhu, C. Guo, Y. Wang, and Z. Xu. What are Textons? *International Journal of Computer Vision*, 62(1):121–143, 2005.

# Fast and Accurate Gaussian Derivatives Based on B-Splines

Henri Bouma[1,⋆], Anna Vilanova[1], Javier Oliván Bescós[2], Bart M. ter Haar Romeny[1], and Frans A. Gerritsen[1,2]

[1] Biomedical Image Analysis, Technische Universiteit Eindhoven, The Netherlands
henri.bouma@tno.nl, {a.vilanova,b.m.terhaarromeny}@tue.nl
[2] Advanced Development, Healthcare Informatics, Philips Medical Systems, Best The Netherlands
{javier.olivan.bescos,frans.gerritsen}@philips.com

**Abstract.** Gaussian derivatives are often used as differential operators to analyze the structure in images. In this paper, we will analyze the accuracy and computational cost of the most common implementations for differentiation and interpolation of Gaussian-blurred multi-dimensional data. We show that – for the computation of multiple Gaussian derivatives – the method based on B-splines obtains a higher accuracy than the truncated Gaussian at equal computational cost.

## 1 Introduction

Computer vision aims at the automatic interpretation of structures in an image. The low-level image structure is often analyzed with differential operators, which are used to calculate (partial) derivatives. In mathematical analysis, the derivative expresses the slope of a continuous function at a point ($\frac{\partial f(x)}{\partial x} = \lim_{h\downarrow 0} \frac{f(x+h)-f(x)}{h}$). However, differentiation is an ill-posed operation, since the derivatives do not continuously depend on the input data [1]. The problem of an ill-posed differentiation on a discrete image $F$ is solved through a replacement of the derivative by a (well-posed) convolution with the derivative of a regularizing test function $\phi$ [2,3].

$$(\partial_{i_1\ldots i_n} F * \phi)(x) = (-1)^n \int_{-\infty}^{\infty} F(\xi)\, \partial_{i_1\ldots i_n}\phi(x+\xi)\,\mathrm{d}\xi$$
$$= \int_{-\infty}^{\infty} F(\xi)\, \partial_{i_1\ldots i_n}\phi(x-\xi)\mathrm{d}\xi = (F * \partial_{i_1\ldots i_n}\phi)(x) \qquad (1)$$

The Gaussian is the only regularizing function that is smooth, self-similar, causal, separable and rotation invariant [3,4]. The convolution of an image with a Gaussian is called blurring, which allows the analysis at a higher scale where small structures (e.g., noise) are removed.

⋆ Henri Bouma recently joined TNO, The Hague, The Netherlands.

F. Sgallari, A. Murli, and N. Paragios (Eds.): SSVM 2007, LNCS 4485, pp. 406–417, 2007.

Thanks to the mentioned properties, the Gaussian derivatives are often applied in the fields of image processing and computer vision as differential operators [5]. They are used to implement differential invariant operators – such as edge detectors, shape descriptors and motion estimators. In the medical field, the Gaussian derivatives are used to compute features in *huge* multi-dimensional images for a computer-aided interpretation of the data, sometimes even at multiple scales [6]. This processing requires an efficient and accurate implementation of the Gaussian derivatives.

The naive approach to obtain the blurred derivatives of an image, is to convolve a multi-dimensional image with a multi-dimensional truncated Gaussian (derivative) kernel. The same result can be obtain with lower computational cost by using separability, because the rotation-invariant multivariate Gaussian is equal to a product of univariate Gaussians. However, the cost of both approaches increases as the scale gets larger. Therefore, many techniques are proposed for an efficient implementation at large scales or at multiple scales.

The FFT [7] allows the replacement of an expensive convolution in the spatial domain by a cheaper multiplication in the Fourier domain. Usually, the cost of an FFT is only acceptable for large scales [8]. A recursive implementation [9,10,11] of the Gaussian (derivative) is even cheaper than the FFT [12], and the costs are – like the FFT – independent of the scale. However, this implementation lacks high accuracy, especially for small scales [13] and derivatives cannot be computed between voxels (e.g., for rendering) or locally at some voxels (e.g., to save time and memory for the computation of isophote curvature on a sparse surface). The low-pass pyramid technique [14,15] uses down-sampling at coarser scales to reduce the computational cost. Especially analysis at multiple or higher scales can benefit from this approach.

However, the use of large-scale Gaussian derivatives can be avoided because the Gaussian is a self-similar convolution operation. This means that a cascade application of two Gaussian kernels with standard deviation $\sigma_1$ and $\sigma_2$, results in a broader Gaussian function with $\sigma_{tot} = \sqrt{\sigma_1^2 + \sigma_2^2}$ (semi-group property). Therefore, Lindeberg [16] proposed to first blur an image once with a large Gaussian $G(\sigma_1)$, and then obtain all partial derivatives at lower cost with smaller Gaussian derivative kernels $g(\sigma_2)$. In this paper, we will compare the accuracy and computational cost of several approaches to obtain these derivatives.

Figure 1 shows four ways to obtain a Gaussian derivative. One way is to convolve an image in one pass with a truncated Gaussian derivative for each partial derivative. The second way is the approach of Lindeberg [16] that first blurs an image once and then obtains all the partial derivatives with small truncated Gaussian derivative kernels. Due to truncation, the Gaussian is not continuous and smooth anymore, although the error of this exponential function rapidly approaches zero. In the third way, which is similar to the second way, the small Gaussian derivative is replaced by a B-spline derivative [14,17,18]. The higher-order B-spline $\beta$ converges to a Gaussian as a consequence of the central-limit theorem. An advantage of the B-spline of order $n$ is that it is a compact kernel that guarantees $C^{n-1}$ continuity. The fourth way to compute the Gaussian

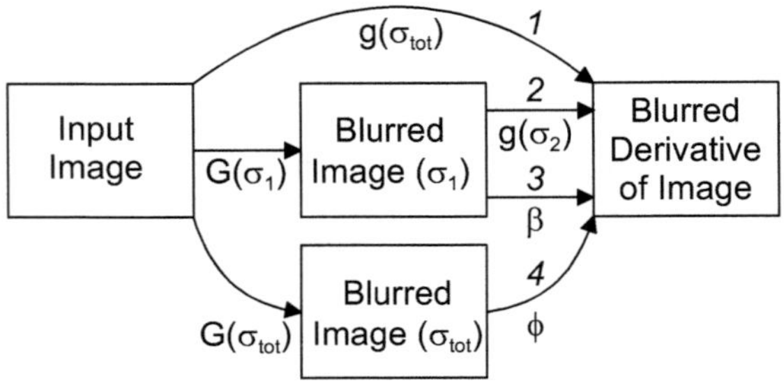


**Fig. 1.** Four ways to obtain the blurred derivative of an image. The first way performs one convolution with the derivative of a Gaussian $g(\sigma_{tot})$. The second and third way convolve the image with a Gaussian $G(\sigma_1)$ for most of the blurring and then with a smaller derivative of a Gaussian $g(\sigma_2)$ or a B-spline derivative $\beta$; where $\sigma_{tot} = \sqrt{\sigma_1^2 + \sigma_2^2}$. The fourth way convolves the image with a Gaussian $G(\sigma_{tot})$ for all the blurring and then with the derivative of an interpolator $\phi$ for differentiation.

derivatives makes a separation between blurring and differentiation. After blurring the image – an operation that can benefit from the mentioned optimizations – the derivative is computed *without* blurring.

Many operators have been proposed to compute the derivative in an image (e.g., the Roberts, Prewitt and Sobel operators [19,20,21]). However, they do not compute the derivative without adding extra blur and they are very inaccurate.

The unblurred derivative of an image can be computed as a convolution with the derivative of an interpolation function $\phi$ (4th way in Fig. 1). A quantitative comparison of several interpolation methods can be found in papers by Meijering *et al.* [23,24], Jacobs *et al.* [25] and Lehmann *et al.* [26]. The comparisons show that for each of the methods for differentiation and interpolation there is a trade-off between accuracy, continuity and kernel size, and that *B-spline interpolation* [27,28] appears to be superior in many cases. Therefore, we used the derivative of a B-spline interpolator to implement the unblurred derivative (4th way in Fig. 1).

In the last decades, we have seen a growing competition between Gaussian- and spline-based image-analysis techniques, which are both frequently used. To our knowledge, a comparison between the truncated Gaussian and the approaches based on B-spline approximation and B-spline interpolation (Fig. 1) for a fast and accurate implementation of Gaussian derivatives has not been published before. In this paper, we will compare the accuracy (Sec. 2) and computational cost (Sec. 3) of the four strategies.

## 2 Accuracy of Methods

In this section, the true Gaussian derivatives are compared to their approximations to analyze the accuracy of these approximations on one-dimensional data.

Thanks to the separability of the Gaussian $\mathcal{G}$, this analysis is also valid for higher dimensions.

$$\mathcal{G}(x,\sigma) = \frac{1}{\sigma\sqrt{2\pi}}\, e^{-\frac{x^2}{2\sigma^2}} \tag{2}$$

The error $\epsilon$ of an approximation $\tilde{y}$ of the true continuous signal $y$ is computed as the normalized RMS-value, which is directly related to the energy.

$$\epsilon = \frac{\sqrt{\int_{-\infty}^{\infty} |\tilde{y}(x) - y(x,\sigma)|^2 \,\mathrm{d}x}}{\sqrt{\int_{-\infty}^{\infty} |y(x,\sigma)|^2 \mathrm{d}x}} \tag{3}$$

For the normalized RMS-value, the error of the impulse response of a Gaussian-derivative kernel that is truncated at $x = a\sigma$ is independent of the standard deviation. For example, the normalized RMS-error of a Gaussian is $\epsilon = \sqrt{1 - \mathrm{erf}(a)}$, and for a first-order Gaussian derivative: $\epsilon = \sqrt{1 + 2\,a\,e^{-a^2}/\sqrt{\pi} - \mathrm{erf}(a)}$.

## 2.1 Aliasing and Truncation of the Gaussian

Small-scale Gaussian operators may suffer from sampling artifacts. According to the Nyquist theorem, the sampling frequency must be at least twice the bandwidth in order to avoid overlap of the copied bands (aliasing) [5]. If the copies do not overlap, then perfect reconstruction is possible. For a small amount of blurring (e.g., $\sigma < 1$ pixel) limitation of the bandwidth of the reconstructed signal is not guaranteed and serious aliasing artifacts may be the result. Band-limitation is enforced by a convolution with a sinc function [13]. If high-frequencies are (almost) absent in the stop band, then aliasing is negligible and the signals with and without band-limitation ($g * \phi_{sinc}(x)$ and g(x) respectively) are (approximately) equal. Figure 2a shows that sampling causes serious aliasing artifacts for a small-scale zeroth-order derivative of a Gaussian, and it shows that a first- or second- order derivative requires even more blurring for the same reduction of the aliasing effects. To avoid aliasing artifacts, second-order derivatives are often computed at $\sigma = 2.0$ pixels. Therefore, we will mainly focus further analysis on this amount of blurring ($\sigma_{tot} = 2.0$ px).

For a fixed kernel size $N$, small scales will lead to aliasing artifacts but large scales will lead to truncation artifacts. The optimal trade-off between aliasing and truncation is selected by minimizing the difference between a band-limited Gaussian ($g * \phi_{sinc}$) and a truncated Gaussian. Figure 2b shows that the error is minimal at $\sigma \approx (N/6.25)^{0.50} \approx \sqrt{N/6}$. If the truncated kernel is applied to a blurred input signal – e.g. blurred by the PSF or pre-filtering with $\sigma_1$ – so that the total blurring is $\sigma_{tot} = \sqrt{\sigma_1^2 + \sigma_2^2} = 2.0$ px, the optimal scale can even be reduced to approximately $\sigma_2 \approx (N/9.9)^{0.56} \approx \sqrt{N/10}$, as shown in Figure 2c. The scale with a minimal error is used to implement the second approach in Figure 1.

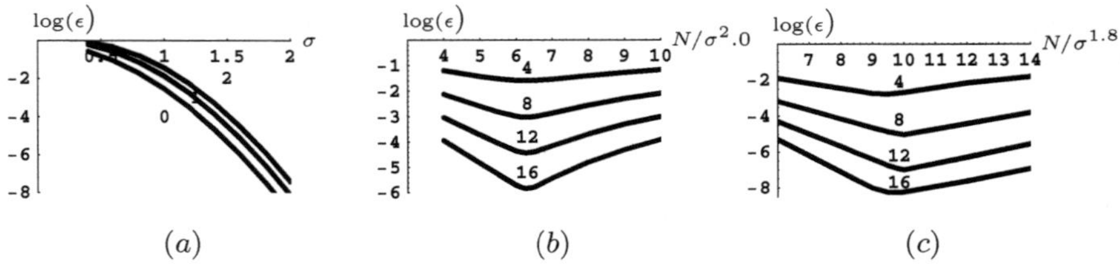


**Fig. 2.** (a) The normalized RMS-error $\epsilon$ due to aliasing for a zeroth-, first- and second-order derivative of a Gaussian at $\sigma = [0.5 - 2.0]$. (b) The difference between a band-limited Gaussian and a truncated Gaussian is minimal at $\sigma = (N/6.25)^{0.50}$, where the kernel size $N = [4, 8, 12, 16]$. (c) On a blurred signal, the normalized RMS-error is minimal at $\sigma = (N/9.9)^{0.56}$.

## 2.2 B-Spline Approximation

The B-spline approximator is used to implement the third approach in Figure 1. A high-order B-spline [18], or a cascade application of kernels [17,14], will converge to a Gaussian (central-limit theorem). The B-spline approximator $\beta^n(x)$ of order $n$ is:

$$\beta^n(x) := \frac{1}{n!} \sum_{i=0}^{n+1} \binom{n+1}{i} (-1)^i \, \mu^n\left(x - i + \frac{n+1}{2}\right) \tag{4}$$

where $\mu^n(x)$ is $x^n$ for $x \geq 0$ and zero for other values, and where $\binom{n+1}{i}$ is the binomial coefficient. The derivatives of the B-spline can be obtained analytically in a recursive fashion based on the following property:

$$\frac{\partial \beta^n(x)}{\partial x} = \beta^{n-1}\left(x + \frac{1}{2}\right) - \beta^{n-1}\left(x - \frac{1}{2}\right) \tag{5}$$

The z-transform [29] is commonly used in digital signal processing to represent filters in the complex frequency domain. For example, for cubic ($n = 3$) spline filtering, the z-transform is:

$$B^3(z) = \frac{1\,z^{-1} + 4\,z^0 + 1\,z^1}{6} \quad \Leftrightarrow \quad y_i = \frac{1}{6}x_{i-1} + \frac{4}{6}x_i + \frac{1}{6}x_{i+1} \tag{6}$$

The output $y_i$ of this digital filter only depends on the inputs $x$, which makes it a *finite impulse response* (FIR) filter.

The derivative of a B-spline approximator $\beta^n(x)$ can be used as a small-scale Gaussian derivative. Figure 3a shows the normalized RMS-error between a Gaussian and a B-spline is minimal for the standard deviation $\sigma = \sqrt{N/12}$ [30]. Although the B-spline converges to a Gaussian for higher orders, the error is not reduced for higher orders (Fig. 3b) when it is applied to a blurred signal (to obtain $\sigma_{tot} = 2.0$ px). The scale with a minimal error is used to analyze the accuracy of this approach.

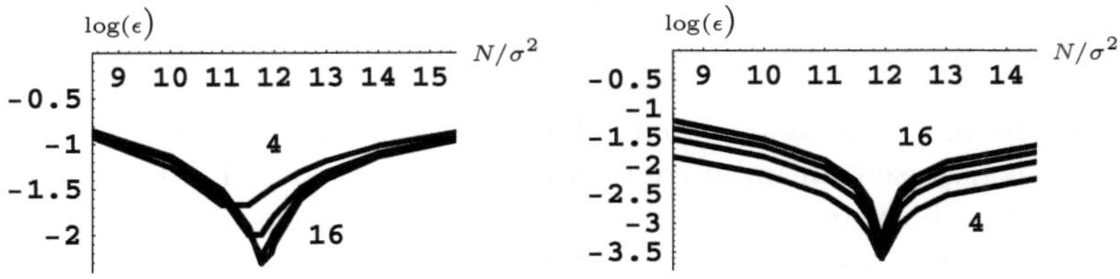


**Fig. 3.** The normalized RMS-error between a Gaussian and a B-spline approximator is minimal at $\sigma = \sqrt{N/12}$, for kernel size $N = 4, 8, 12, 16$. (b) The same relation can be found on a blurred signal.

### 2.3 B-Spline Interpolation

The B-spline interpolator is used to implement the fourth approach in Figure 1. In order to perform B-spline interpolation of the blurred image $H$ with the approximating B-spline kernels ($\beta^n$ in Eq. 4), an inverse operation $B^n_{inv}$ is required.

$$\tilde{h} = H \ast B^n_{inv} \ast \beta^n \tag{7}$$

The inverse operator can easily be calculated in the z-domain as $B^n(z)^{-1}$. To obtain a stable filter, this inverse operator can be decomposed by its negative roots with magnitude smaller than one [27]. For example, the root of the inverse of a cubic B-spline (Equations 6 and 8) is $\lambda = -2 + \sqrt{3}$.

$$B^3(z)^{-1} = \frac{1}{B^3(z)} = \frac{6}{z^1 + 4 + z^{-1}} = -6\lambda \frac{1}{(1 - \lambda z^{-1})} \frac{1}{(1 - \lambda z^1)} \tag{8}$$

Multiplication of two parts in the z-domain is equivalent to a cascade convolution with both parts in the spatial domain. The last part in Equation (8), with $z^1$, can be applied backward, so that it also becomes a $z^{-1}$ operation. This results in a stable and fast filter, which should be applied forward and backward:

$$\frac{1}{1 - \lambda z^{-1}} \quad \Leftrightarrow \quad y_i = x_i + \lambda\, y_{i-1} \tag{9}$$

The output $y_i$ of this digital filter does not only depend on the input $x_i$, but also on the output $y_{i-1}$, which makes it a recursive – or *infinite impulse response* (IIR) – filter. The recursive inverse operation makes the B-spline interpolator computationally more expensive than the B-spline approximator at equal order $n$. For more information about B-Spline interpolation, we refer to the work of Unser *et al.* [27,28].

### 2.4 Comparison of Accuracy

An experiment was performed to estimate the normalized RMS-error between the impulse response of a continuous Gaussian derivative ($\sigma_{tot} = 2.0$ px to avoid aliasing) and each of the four approaches (Fig. 1). Measuring for each approach the error of the impulse response gives an indication of the accuracy in general, because a discrete image can be modelled as a sum of impulses with varying amplitude. The first approach, which is based on a one-pass truncated Gaussian of $\sigma = 2.0$ pixels, used an unblurred impulse as input signal. The second and third approach, which are based on a small-scale truncated Gaussian and on a B-spline approximator, used a sampled Gaussian as an input signal to obtain a total blurring of $\sigma = 2.0$ pixels. The fourth approach, which is based on B-spline interpolation, used a sampled Gaussian of $\sigma = 2.0$ pixels as input signal.

Truncation of the one-pass Gaussian is often performed at $3\sigma$ or $4\sigma$, which corresponds to a kernel size of 12 or 16 pixels for $\sigma = 2.0$ pixels. Figure 4 shows that for these kernel sizes the normalized RMS-error in the second-order derivative is $5.0 \cdot 10^{-2}$ or $2.4 \cdot 10^{-3}$ respectively. The results show that B-spline approximation requires much smaller kernels to obtain the same accuracy as the truncated Gaussian (4 or 6 px respectively). The figure also shows that B-spline interpolation and cascade application of small-scale Gaussians may be interesting if higher accuracies are required, but for most applications the approach based on B-spline approximation will be sufficiently accurate.

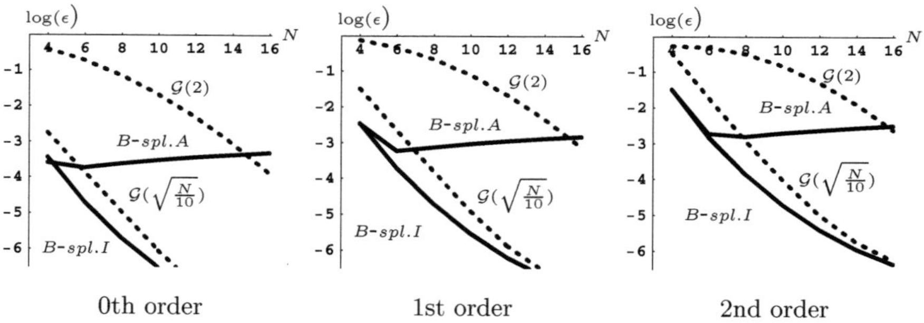


**Fig. 4.** The normalized RMS-error $\epsilon$ in estimating the zeroth-, first- and second-order Gaussian derivative ($\sigma = 2.0$px) for the four approaches based on the one-pass truncated Gaussian ($\mathcal{G}(2)$, dashed), the cascade application Gaussians ($\mathcal{G}(\sqrt{N/10})$, dashed), the B-spline approximator (*B-spl.A*, solid) and the B-spline interpolator (*B-spl.I*, solid). The first approach requires much larger kernels than the others to obtain the same accuracy.

## 3 Computational Cost

Our comparison of computational cost will focus on the calculation of first- and second-order derivatives at a low scale ($\sigma = 2.0$ px) in three-dimensional (3D) data, because these derivatives are frequently used in the medical field. For these parameters, we will show that – in most cases – it is beneficial to use the B-spline approximator. For larger scales, more derivatives or higher-dimensionality it will be even more beneficial to make a separation between the blurring and differentiation. Therefore, our analysis can easily be extended to the computation of an arbitrary number of derivatives at higher scales on multi-dimensional data.

Figure 4 showed that the truncated Gaussian requires 12 or 16 pixels to obtain the same accuracy as the B-spline approximator of 4 or 6 pixels respectively. For these sizes the B-spline approximator (*B-spl.A*) is more accurate than the cascaded Gaussians ($\mathcal{G}(\sqrt{N/10})$) and computationally cheaper than the B-spline interpolator (*B-spl.I*) because no inverse is required (Eq. 7). Therefore, we will focus on the comparison of the B-spline approximator with the truncated Gaussian. Despite its small kernel, the B-spline is not always cheaper than the truncated Gaussian because it requires preprocessing to obtain the same amount of blur. The computational cost of this global blurring step can be reduced – especially for large scales – by using a recursive implementation [11].

The estimation of the computational cost $C$ is based on the number of multiplications, which is equal to the kernel size. Three approaches are distinguished to analyze the performance for different purposes (Table 1). In the first approach, all volume-elements (voxels) are processed in a 3D volume. In the second, the derivatives are computed at some voxel-locations, and in the third, interpolation and differentiation is allowed at arbitrary (sub-voxel) locations in the volume. Finally, our estimation of the computational cost is verified with an experiment.

**Table 1.** The computational cost $C$ in a 3D volume of $d$ derivatives based on the truncated Gaussian (kernel size $k$) and B-spline approximation (order $n$)

| | Blur | All Voxels | Some Voxels | Some Points |
|---|---|---|---|---|
| Trunc. Gauss | – | $3\,d\,(k+1)$ | $d\,(k+1)^3$ | $d\,(k)^3$ |
| B-spline approx. | $3\,(k+1)$ | $3\,d\,(n)$ | $d\,(n)^3$ | $d\,(n+1)^3$ |

### 3.1 Cost of Differentiation on All Voxels

The computation of Gaussian derivatives on all voxels allows the use of a separable implementation with discrete one-dimensional filters. The continuous B-spline of order $n$ with kernel size $n+1$ is zero at the positions $-(n+1)/2$ and $(n+1)/2$. Therefore, the number of non-zero elements in a discrete B-spline kernel is $n$. The truncated Gaussian with kernel size $k$ is not zero at its end points and therefore it requires $k+1$ elements in the discrete kernel to avoid the loss of accuracy.

For a 'fast' computation ($n = 3$, $k = 12$) of three first-order derivatives on all voxels, the B-spline approximator is 1.8 times faster than the truncated Gaussian despite the required preprocessing. For the nine first- and second-order derivatives, the B-spline is 2.9 times faster. For a 'more-accurate' computation ($n = 5$, $k = 16$) of three or nine derivatives, the B-spline approximator is 1.6 resp. 2.5 times faster than the truncated Gaussian (horizontal lines in Fig. 5).

### 3.2 Cost of Differentiation on Some Voxels

If only a small percentage $p$ of the volume needs to be processed (e.g, to compute shape descriptors on the surface of an object) – or if storage of multiple derivatives of the whole image consumes too much memory – the non-separable implementation may be more efficient to compute the derivatives than the separable implementation. However, in 3D data, the cost of a non-separable *local* operation increases with a *power* of three instead of a *factor* of three (Tab. 1).

The non-separable implementation is more efficient than the separable for the 'fast' B-spline approximator ($n = 3$) if less than $p = 33\%$ of the volume is processed, and for the 'more-accurate' ($n = 5$) if less than $p = 12\%$ is processed (Fig. 5).

Figure 5 also shows that the B-spline implementation ($n = 3$, $d = 3$) is more efficient than the truncated Gaussian if more than $p = 0.6\%$ of the voxels is processed ($d = 9$ reduces the trade-off point to $p = 0.2\%$). For example, the B-spline ($n = 3$, $d = 9$) is 8 times faster than the truncated Gaussian at $p = 2.0\%$.

If we would have assumed that the blurring for the B-splines was incorporated in the preprocessing, then the B-spline approximator would even have been 81 times faster than the truncated Gaussian for each voxel.

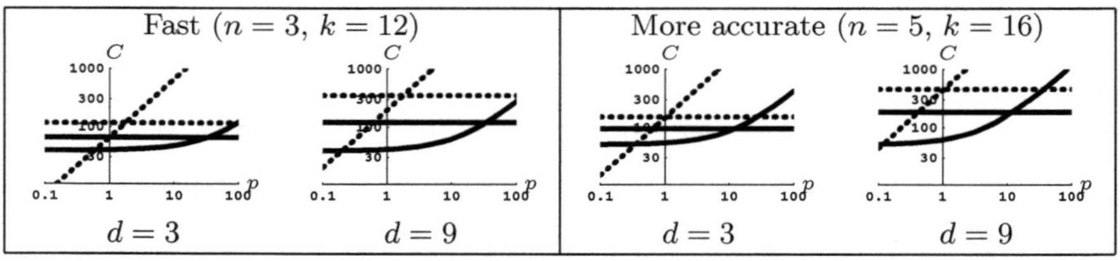


**Fig. 5.** The curves show the computational cost $C$ for processing a percentage $p$ of the voxels with a non-separable implementation in a 3D volume with a truncated Gaussian (dashed) and the B-spline (solid) for $d$ derivatives. The horizontal lines show the cost of processing all voxels with a separable implementation. The plots show that the B-spline is expected to be more efficient if more than $p = 0.6\%$ of the data is processed.

### 3.3 Cost of Interpolation and Differentiation on Arbitrary Points

To interpolate and differentiate at arbitrary (sub-voxel) points in the volume continuous kernels are needed and a separable implementation cannot be used. The $n$-th order B-spline has a continuous kernel size of $n + 1$ (Table 1).

Figure 6 shows that if more than $p = 0.8\%$ of the data is processed the B-spline is more efficient than the truncated Gaussian. For example, if nine derivatives are computed at a number of points that equals $p = 10\%$ of the voxels, the B-spline (n=3) is more than 16 times faster than the truncated Gaussian ($k = 12$).

### 3.4 Validation of Cost of Differentiation on Voxels

To validate our estimation of the computational cost, we measured the time that was required to compute the nine first- and second-order derivatives on a 3D volume of 512x512x498 voxels with a Pentium Xeon 3.2 GHz processor. In this experiment, we compared the implementations based on the truncated

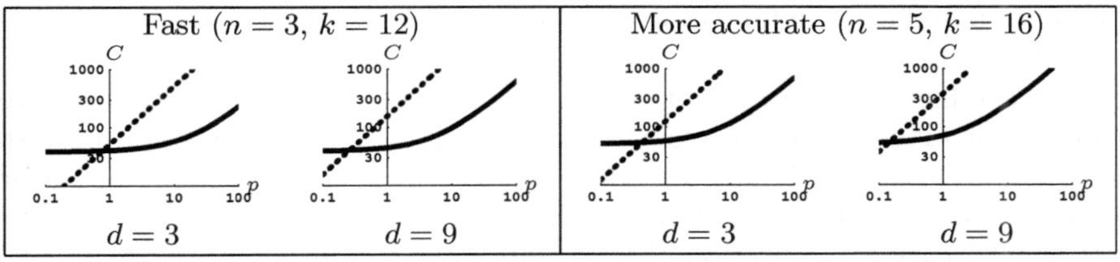


**Fig. 6.** The computational cost $C$ for processing arbitrary points as a percentage $p$ of the voxels in a 3D volume with a truncated Gaussian (dashed) and the B-spline (solid) for $d$ derivatives. The plots show that the B-spline is expected to be more efficient if more than $p = 0.8\%$ of the data is processed.

Gaussian ($k = 12$) and the B-spline approximator ($n = 3$) as an example to show that our assumptions are valid. The measured results in Figure 7 are in good agreement with our analysis. The measurements show that the B-spline is more efficient if more than 0.3% of the data is processed (estimated 0.2%). The B-spline appears to be 6 times faster than the truncated Gaussian if 2% of the volume is processed with a non-separable implementation (estimated 8 times faster). And if all voxels are processed with a separable implementation the B-spline appears to be 2.1 times faster (estimated 2.9 times faster).

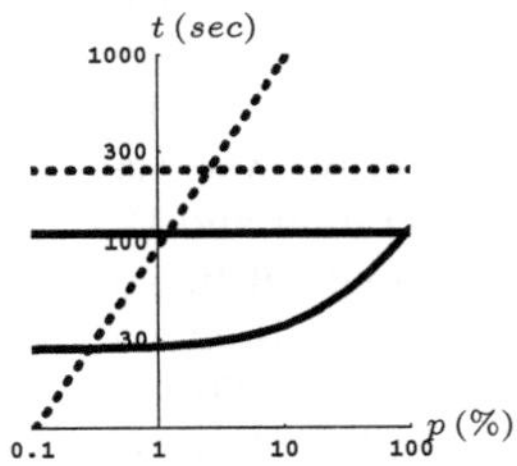


**Fig. 7.** The measured computation time $t$ in seconds for processing a percentage $p$ of the voxels in a 3D volume (512x512x498 voxels) with a truncated Gaussian ($k = 12$, dashed) and a B-spline approximator ($n = 3$, solid) for 9 derivatives. The horizontal lines show the cost of processing all voxels with a separable implementation. The plot shows that, for equivalent accuracy, the B-spline is more efficient if more than $p = 0.3\%$ of the data is processed.

## 4 Conclusions

We analyzed the accuracy and computational cost of several common implementations for differentiation and interpolation of Gaussian blurred multi-dimensional data. An efficient implementation is extremely important for all fields that use Gaussian derivatives to analyze the structure in data. A comparison between an implementation based on the truncated Gaussian and alternative approaches based on B-spline approximation and B-spline interpolation has not been published before, to the best of our knowledge.

If the vesselness or isophote curvature of a data set needs to be computed (requiring six or nine derivatives respectively), the B-spline approach will perform much faster than the approach based on truncated Gaussians. These operators are very important in the field of medical imaging for shape analysis. Our analysis shows that, for the computation of first- and second-order Gaussian derivatives on three-dimensional data, the B-spline approximator is faster than the truncated Gaussian at equal accuracy, provided that more than 1% of the data is processed. For example, if 2% of a 3D volume is processed, B-spline approximation is more than 5 times faster than the truncated Gaussian at equal accuracy. Our analysis can be extended easily to an arbitrary number of derivatives on multi-dimensional data.

Higher accuracy will not always lead to better results. However, in many cases, the same accuracy can be obtained at lower computational cost, as was shown in this paper. Another advantage of the B-spline of order $n$ is that $C^{n-1}$ continuity is guaranteed, whereas the truncated Gaussian is not even $C^0$ continuous.

## References


1. J. Hadamard: Sur les problèmes aux Dérivées Partielles et leur Signification Physique. Bulletin, Princeton University **13** (1902) 49–62
2. L. Schwartz: Théorie des distributions. In: Actualités Scientifiques et Industrielles, Institut de Mathématique, Université de Strasbourg. Vol. 1,2. (1951) 1091–1122
3. L.M.J. Florack: Image Structure. Kluwer Academic Publ., The Netherlands (1997)
4. R. Duits, L.M.J. Florack, J. de Graaf and B.M. ter Haar Romeny: On the axioms of scale-space theory. J. Mathematical Imaging and Vision **20**(3) (2004) 267–298
5. B.M. ter Haar Romeny: Front-End Vision and Multi-Scale Image Analysis. Kluwer Academic Publ., The Netherlands (2003)
6. Y. Masutani, H. MacMahon and K. Doi: Computerized detection of pulmonary embolism in spiral CT angiography based on volumetric image analysis. IEEE Trans. Medical Imaging **21**(12) (2002) 1517–1523
7. M. Frigo and S.G. Johnson: An FFT compiler. Proc. IEEE **93**(2) (2005) 216–231
8. L.M.J. Florack: A spatio-frequency trade-off scale for scale-space filtering. IEEE Trans. Pattern Analysis and Machine Intelligence **22**(9) (2000) 1050–1055
9. M. Abramowitz and I.A. Stegun: Handbook of Mathematical Functions. Dover, New York, USA (1965)
10. R. Deriche: Fast algorithms for low-level vision. IEEE Trans. Pattern Analysis and Machine Intelligence **12**(1) (1990) 78–87
11. L.J. van Vliet, I.T. Young and P.W. Verbeek: Recursive Gaussian derivative filters. In: Proc. Int. Conf. Pattern Recognition (ICPR). Vol. 1. (1998) 509–514
12. I.T. Young and L.J. van Vliet: Recursive implementation of the Gaussian filter. Signal Processing, Elsevier **44** (1995) 139–151
13. R. van den Boomgaard and R. van der Weij: Gaussian convolutions, numerical approximations based on interpolation. In: Proc. Scale Space. LNCS 2106 (2001) 205–214
14. P.J. Burt and E.H. Adelson: The Laplacian pyramid as a compact image code. IEEE Trans. Communications **31**(4) (1983) 532–540

15. J.L. Crowley and R.M. Stern: Computation of the difference of low-pass transform. IEEE Trans. Pattern Analysis and Machine Intelligence **6**(2) (1984) 212–222
16. T. Lindeberg: Discrete derivative approximations with scale-space properties: A basis for low-level feature extraction. J. Mathematical Imaging and Vision **3**(4) (1993) 349–376
17. M. Wells: Efficient synthesis of Gaussian filters by cascaded uniform filters. IEEE Trans. Pattern Analysis and Machine Intelligence **8**(2) (1986) 234–239
18. Y.P. Wang and S.L. Lee: Scale space derived from B-splines. IEEE Trans. Pattern Analysis and Machine Intelligence **20**(10) (1998) 1040–1055
19. I.E. Abdou and W.K. Pratt: Quantitative design and evaluation of enhancement/thresholding edge detectors. Proc. IEEE **67**(5) (1979) 753–763
20. V.S. Nalwa and T.O. Binford: On detecting edges. IEEE Trans. Pattern Analysis and Machine Intelligence **8**(6) (1986) 699–714
21. V. Torre and T.A. Poggio: On edge detection. IEEE Trans. Pattern Analysis and Machine Intelligence **8**(2) (1986) 147–163
22. S.R. Marschner and R.J. Lobb: An evaluation of reconstruction filters for volume rendering. In: Proc. IEEE Visualization. (1994) 100–107
23. E.H.W. Meijering, W.J. Niessen and M.A. Viergever: The sinc-approximating kernels of classical polynomial interpolation. In: Proc. IEEE Int. Conf. Image Processing. Vol. 3. (1999) 652–656
24. E.H.W. Meijering, W.J. Niessen and M.A. Viergever: Quantitative evaluation of convolution-based methods for medical image interpolation. Medical Image Processing **5**(2) (2001) 111–126
25. M. Jacob, T. Blu and M. Unser: Sampling of periodic signals: A quantitative error analysis. IEEE Trans. Signal Processing **50**(5) (2002) 1153–1159
26. T.M. Lehmann, C. Gönner and K. Spitzer: Survey: Interpolation methods in medical image processing. IEEE Trans. Medical Imaging **18**(11) (1999) 1049–1075
27. M. Unser, A. Aldroubi and M. Eden: B-spline signal processing: Part I: Theory, and Part II: Efficient design and applications. IEEE Trans. Signal Processing **41**(2) (1993) 821–848
28. M. Unser: Splines, a perfect fit for signal and image processing. IEEE Signal Processing Magazine (1999) 22–38
29. E.I. Jury: Theory and Application of the Z-Transform Method. John Wiley and Sons, New York, USA (1964)
30. M. Unser, A. Aldroubi and M. Eden: On the asymptotic convergence of B-spline wavelets to Gabor functions. IEEE Trans. Inf. Theory **38**(2) (1992) 864–872

# Uniform and Textured Regions Separation in Natural Images Towards MPM Adaptive Denoising

Noura Azzabou[1,2], Nikos Paragios[1], and Frédéric Guichard[2]

[1] Laboratoire MAS, Ecole Centrale de Paris, Grande Voie des Vignes, France
noura.azzabou@ecp.fr, nikos.paragios@ecp.fr
[2] DxOLabs, 3 Rue Nationale, Boulogne Billancourt, France
fguichard@dxo.com

**Abstract.** Natural images consist of texture, structure and smooth regions and this makes the task of filtering challenging mainly when it aims at edge and texture preservation. In this paper, we present a novel adaptive filtering technique based on a partition of the image to "noisy smooth zones" and "texture or edge + noise" zones. To this end, an analysis of local features is used to recover a statistical model that associates to each pixel a probability measure corresponding to a membership degree for each class. This probability function is then encoded in a new denoising process based on a MPM (Marginal Posterior Mode) estimation technique. The posterior density is computed through a non parametric density estimation method with variable kernel bandwidth that aims to adapt the denoising process to image structure. In our algorithm the selection of the bandwidth relies on a non linear function of the membership probabilities. Encouraging, experimental results demonstrate the potential of our approach.

## 1 Introduction

In spite of the progress made in the field of image denoising, we can claim that it is still an open research problem. Traditional techniques of image enhancement and noise reduction rely on the assumption that image is homogeneous at local scale. Natural images consist of smooth and patterned regions like texture and therefore the use of such a simplistic denoising techniques deteriorate the quality of the reconstruction in regions with texture. Texture, refers to regions with repetitive patterns and structure at various scale and orientations. In this paper our goal is to propose a technique that takes into account the particularities of such patterns and design an adaptive enhancement technique able to preserve texture while removing noise.

State of the art techniques in image enhancement refers to local methods, image decomposition in orthogonal spaces, partial differential equations as well as complex mathematical models. Filters and morphological operators are the most prominent local approaches [13,19,22,5] and exploit homogeneity of the image through convolution. Global methods represent images through a set of invertible transformations of an orthogonal basis [14,8,12] where noise is removed

F. Sgallari, A. Murli, and N. Paragios (Eds.): SSVM 2007, LNCS 4485, pp. 418–429, 2007.

through the modification of the coefficients with limited importance in the reconstruction process. Partial differential equations [23,17,1,2,20] like the heat equation, anisotropic diffusion , etc. incorporate more structure in the denoising process where the noise-free images corresponds to the steady state solution of the PDE. Last, but not least global approaches [16,18,11] recover the noise-free image through the lowest potential of a cost function that aims to separate image structure from noise.

Presence of texture often violates the fundamental assumption considered in the image enhancement field that is local homogeneity and despite numerous provisions of the above methods, denoising of texture images is still an open problem. Separating structure from texture is the most prominent technique to deal with such limitation and has gained significant attention in the past years [4,21,15]. These techniques model images as a mixture of a uniform and an oscillatory component which are computed through optimization of a specifically designed cost. In spite of their performance, these methods fail sometimes to separate noise from texture. A more recent work introduced in [3] addresses such limitation but the proposed approach is highly dependent on the noise model.

One can conclude that, traditional/state-of-the art techniques make the assumption that an image is a mixture of a piecewise constant and an oscillatory component is relative to the noise. It is clear that this hypothesis is not satisfied because texture is also an oscillatory pattern and many examples of natural images show that it's sometimes difficult to distinguish between noisy regions and textured ones. An effort to address this limitation was carried out in [10] where an adaptive total variation algorithm was proposed. This algorithm selects the coefficient of the fidelity-to-data term according to the presence or not of texture. The texture characterization is only based on local variance information which often is not sufficient to separate texture from noise.

In the present paper, we propose a novel denoising technique which relies in a soft pre-classification step that aims to (i) identify regions where the assumption of local smoothness is valid and (ii) patterned regions that contain texture, edges and other image details. To this end, we propose an automatic technique of image partition into two classes: "homogeneous regions + noise" and " (texture, edges, details) + noise". This partition is performed through a local feature analysis and assigns to each pixel a probability measure that reflects its degree of membership to noise or texture. This classification is then integrated in a non parametric image model with variable bandwidth kernel to perform an MPM (Marginal Posterior Mode) estimation of the original image.

The reminder of the paper is organized as follows; section 2 is devoted to the soft image classification. In section 3, we present our new non-parametric image model to perform image denoising. Validation and comparisons with the state of the art methods are presented in section 4. Finally, we conclude in section 5.

## 2 Texture and Homogeneous Region Classification

Analysis of texture has been a long term research topic in computer vision. While most of the existing techniques aim to separate different texture patterns, we focus on a simpler problem that is separation of texture and smooth regions (with noise patterns being present). In other words, if we consider a local image patch, we want to know whether it corresponds to a texture pattern or to a uniform one altered because of noise. To this end, relative image moments (the mean is subtracted) at local scale are computed. If we assume that noise is independent from signal and independent from image position, we expect that all image patches that correspond to the noise will produce similar local descriptors, that is not the case of edges as well as textured regions. We consider a fairly simple set of feature vectors that consists of the variance, the skewness, the kurtosis and the entropy. With this set of features we can capture local behavior for each pixel and thus can discriminate between patterns that correspond to noise and the ones with important deviations that refer to texture or edges and small details. Intuitively, we expect that image patches that correspond to noise will have similar variance (the noise variance), with a low skewness value because noise have a symmetric behavior. Once local descriptors are computed, our main concern is to define a proper statistical model to interpret them towards classification.

The dimension of the feature vector and the small number of present samples (number of pixels in the image) makes density estimation in this space rather impractical. In order to reduce the dimensionality of our problem, we perform a principle component analysis (PCA), towards a linear transformation of the feature space that retains the largest amount of variation within the data. Such a selection is a reasonable compromise between computing complexity and discrimination power. Thus, we decrease the dimensionality of the classification problem by considering the projection of the feature vector on the first mode of variation.

Once the feature space has been determined, classification consists of assigning to each pixel of the image a probability value according to its membership to a textured pattern or a noisy patch. We recall that our observation space is the projection of the local descriptor on the first eigen vector. Now, our concern is to find a statistical model that is able to describe the distribution of the observation and provides an automatic classification tool of them. To this end, we use the gaussian mixture model which is a very popular tool to approximate a probability density function of the projected samples. Considering that three population are present in an image that are: edges, texture and smooth region, we can consider that each gaussian in the model describes a population.

Let's call $O = (o_1, o_2, ..., o_n)$ the $n$ unlabeled observations corresponding to the feature vectors of the image pixels. We can consider in the case of three component:

$$p(o_x|\Theta) = P_{edge}\ p_{edge}(o_x) + P_{tex}\ p_{tex}(o_x) + P_{smooth}\ p_{smooth}(o_x)$$

with $\Theta$ being the parameter vector and $P_{edge}, P_{tex}, P_{smooth}$ are respective conditional probabilities of edges, texture and noise. Recovering the prior marginals $(P_{edge}, P_{tex}, P_{smooth})$ and the parameters of each gaussian is done according to the maximum likelihood principle and the Expectation Maximization (EM) algorithm [9]. Some results of unsupervised image partitions based on gaussian mixture model is shown in [Fig(1)].

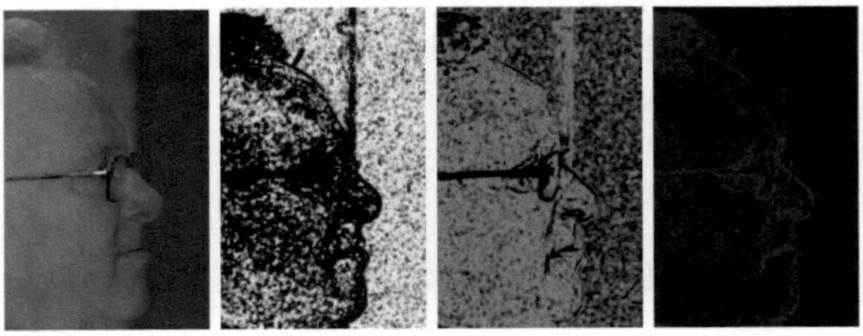

(a) (b) (c) (d)

**Fig. 1.** Example of partition of an image: (a) original image, and conditional probability function relative to (b) "smooth component" $p(o_x|smooth)$ , (c) "texture" $p(o_x|tex)$ and (d) "edges" $p(o_x|edge)$

It is important to point out the fact that we perform a fully unsupervised classification where each gaussian is representative of one image component. Thus, we need to assign a label to each gaussian component and more explicitly we want to know which one of the computed gaussian is representative of the smooth component in the image. Under the assumption that an image consists mostly of smooth regions we can use the prior density of each gaussian to assign labels to them according to the following relation $P_{smooth} > P_{tex} > P_{edge}$. Parallel to that, knowing that pixels that belong to a uniform noisy patch have similar descriptor we expect that the gaussian which represents the noise will have a weak variance compared to the others, an assumption that has been validated by experimental results. For the example shown in [Fig.(1)], the gaussian component with the smallest variance and its dominance over the other hypotheses corresponds to the column (b). It is clear that this component refers to the class describing the uniform assumption with noise.

## 3 Non Parametric Model and Adaptive Denoising

These memberships now can be encoded in the denoising process towards adjusting the behavior of the algorithm according the pixel classification. In this section we will focus on the definition of our denoising model as well as the use of the partition step to perform adaptive denoising.

### 3.1 Non Parametric Model

To introduce our denoising method, let us consider $I$, $U$ and $N$ three random variables defined on a discrete partition $\Omega \in \mathbb{Z}^2$ relative to the image domain related according to:

$$I = U + N$$

with $N$ being an additive noise independent from the signal, $I$ is the observed noisy image and $U$ is the noise-free image. Given such a model, we consider a method based on Marginal Posterior Mode estimation. It consists in estimating the intensity of a given pixel by maximizing its conditional probability relative to the whole observed noisy image. This estimation is done in an independent manner for each image pixel. Thus, the estimate $\hat{U}_x$ of the original observation at a given position $x$ satisfies,

$$\hat{U}_x = \text{argmax}_{U_x} \left[ p\left(U_x | I\right) \right]$$

Considering that $U$ and $I$ are random markov fields in $\Omega$, we can define the conditional probability accordong to the observations in the neighborhood of the pixel instead of the whole image. Thus if we note $\mathbf{N}_x$, a local neighborhood of $x$ the marginal posterior is defined as:

$$p\left(U_x | I\right) = p\left(U_x | I\left(\mathbf{N}_x\right)\right) = \frac{p\left(U_x, I\left(\mathbf{N}_x\right)\right)}{p\left(I\left(\mathbf{N}_x\right)\right)}$$

To perform the MPM estimation, one has to determine a model relative to the posterior. To this end, we consider a non parametric density function based on multidimensional gaussian kernels with variable bandwidth. The set of samples used to perform this estimation is extracted from the local neighborhood of the observed noisy image.

$$p\left(U_x, I\left(\mathbf{N}_x\right)\right) = \frac{1}{M} \sum_{y \in R_x} G_y\left(\|(U_x, I(\mathbf{N}_x)) - (I_y, I(\mathbf{N}_y))\|_2\right)$$

with $M$ being the total number of observations and $R_x$ the local neighborhood system relative to $x$. $G_y$ is a multidimensional isotropic gaussian kernel with zero mean and a covariance matrix $\Sigma = \sigma_y^2 \mathbf{I_n}$ where $\mathbf{I_n}$ is the identity matrix. The bandwidth of these kernels depends on the image content associated with the given pixel position and plays a critical role in the denoising process. The selection of this parameter using the image partition introduced in the beginning of this work will be discussed later. The posterior can now be expressed as follows:

$$p\left(U_x | I\left(\mathbf{N}_x\right)\right) = \frac{\sum_{y \in R_x} G_y\left(\|(U_x, I(\mathbf{N}_x)) - (I_y, I(\mathbf{N}_y))\|_2\right)}{M p\left(I\left(\mathbf{N}_x\right)\right)} \tag{1}$$

the maximum of this probability density function corresponds to the value that provides the optimal numerator since the denominator is constant with respect to $U$. This function penalizes high photometric distances between similar neighbouring pixels which reduces the amount of noise in the image.

A calculus of variation and a gradient descent algorithm is used to compute an estimate of the mode of the marginal posterior probability. If we note, $E$ the numerator of the previous expression, its gradient with respect to $U_x$ in case of gaussian kernel is equal to:

$$\frac{\partial E}{\partial U_x} = -\frac{\partial}{\partial U_x} \sum_{y \in R_x} G_y\left(\|(U_x, I(\mathbf{N}_x)) - (I_y, I(\mathbf{N}_y))\|_2\right)$$

$$\frac{\partial E}{\partial U_x} = \sum_{y \in R_x} \left( \frac{I_y - U_x}{\sigma_y^2} \right) G_y \left( \| (U_x, I(\mathbf{N}_x)) - (I_y, I(\mathbf{N}_y)) \|_2 \right)$$

If we introduce $w_{xy}$ that are weights reflecting the image content agreement between local neighborhood around $x$ and $y$ and defined as:

$$w_{xy} = G_y \left( \| (U_x, I(\mathbf{N}_x)) - (I_y, I(\mathbf{N}_y)) \|_2 \right)$$

The gradient energy expression becomes:

$$\frac{\partial E}{\partial U_x} = \sum_{y \in R_x} \left( \frac{I_y - U_x}{\sigma_y^2} \right) w_{xy}$$

According to the gradient descent algorithm the estimated intensity is updated according to:

$$\begin{aligned} U_x^{t+1} &= U_x^t - dt \sum_{y \in R_x} \left( \frac{I_y - U_x^t}{\sigma_y^2} \right) w_{xy} \\ &= U_x^t - dt \left[ \sum_{y \in R_x} \frac{w_{xy}}{\sigma_y^2} I_y - \sum_{y \in R_x} \frac{w_{xy}}{\sigma_y^2} U_x^t \right] \end{aligned} \tag{2}$$

Where $dt$ is the time step and $U_x^t$ is the intensity value at time $t$.

We point out that such an expression bears certain similarities to the well known mean shift filtering algorithm [7] where the update is proportional to the mean shift value which is defined by the distance between the weighted mean of samples using the kernel $G$ and $U_x^t$ the center of the kernel window.

$$m_G(U_x^t) = U_x^t - \frac{\sum_{y \in R_x} \left( w_{xy} / \sigma_y^2 \right) I_y}{\sum_{y \in R_x} \left( w_{xy} / \sigma_y^2 \right)}$$

The main difference between our approach and the mean shift filtering algorithm lies in the fact that we consider in the mean computation a local neighborhood and we don't restrict the observation to only a pixel level. Therefore, we are able to improve the performance in the case of textured regions because the process goes beyond simple pixel-wise comparisons between pixels to be extended to the entire local neighborhood.

An important parameter of the proposed denoising approach is the kernel bandwidth. A main contribution of the present paper is to consider variable bandwidth gaussian, to model the posterior probability density function. The selection of this parameter will be the focus of the following section.

### 3.2 Bandwidth Selection

The role of the bandwidth in the density approximation is to guarantee the proper use of samples when constructing the pdf. In the case of smooth regions we can assume that all samples have equal contribution on the pdf and therefore, their bandwidth should reflect such condition. Towards decreasing the importance of variation in the kernel, we can consider the increase of their bandwidth. All the samples will inherit equal importance from such a selection within the pdf approximation process. In case of textured regions or edges such a choice will over smooth the small details. That's why in case of sparse distribution we will use smaller bandwidth values which lead to a multi modal density function which is coherent with the fact that in natural image the local histograms in textured regions and edges are often multi modal. Such a selection relies on the implicit assumption that only a small portion of samples express the pdf. The selection of these samples is purely based on the photometric matching between the associated patches. In the other hand, using small bandwidth in case of texture and edges, enable a better selection of the neighborhood samples that will be used in the intensity estimation.

To satisfy such a demand, it is more appropriate to use kernels with variable bandwidth than fixed ones in order to guarantee at the same time detail preservation and good denoising. To this end, we introduce a new function to determine the gaussian kernel bandwidth using the conditional probability values obtained after the partition step. Such a function should be monotonically decreasing with respect to the conditional probability relative to textured region or edges. One possible choice for such a function is defined as follows:

$$\sigma_y = \sigma_0 \left( \frac{p_{smooth}(o_y)}{p_{smooth}(o_y) + p_{tex}(o_y) + p_{edge}(o_y)} + c \right) \tag{3}$$

we recall that $o_y$ is the observation relative to the feature vector in the position $y$ and $p_{smooth}$,$p_{tex}$,$p_{edge}$ are respectively the conditional probability for $o_y$ to be in a noisy smooth region, textured region and edges. $\sigma_0$ and $c$ are parameters to be fixed according to noise level. With such a choice, we adopt for pixels that belong to smooth regions with high value of $p_{smooth}$, high kernel bandwidth. For image component identified as texture or edges, $p_{smooth}$ tends to be close to zero and implies smaller bandwidth values [1].

# 4 Experimental Results and Validation

One can now use the theoretical framework introduced in the previous section for image enhancement. This denoising method is based on a non parametric model estimation of marginal posterior mode. This model is based on an automatic and unsupervised partition of the image on local smooth regions and textured one. Parameters that are involved in our denoising approach are the size of the two neighborhood $R_x$ and $\mathbf{N_x}$ as well as the bandwidth of the kernels that are selected according to expression (3). $R_x$ is the size of the neighborhood used for the posterior estimation. One can think that using an important size of $R_x$ allows a better estimation, but such a choice will involve many irrelevant samples in the

[1] In order to account for error in classification due to noise a morphological filtering approach is used to smooth the obtained probability map.

estimation process. Experimental validation has shown that choosing $R_x = 9 \times 9$ gives good results while remaining computationally efficient. $\mathbf{N_x}$ is the size of the noisy patch around the pixel that we want to recover, in our experiments $\mathbf{N_x}$ is set to $7 \times 7$. Finally the choice of the photometric bandwidth is dependent on noise level. In case of additive gaussian noise with standard deviation $\sigma_n = 20$, we considered this couple of parameter ($\sigma_0$=4, $c = 4$). The contribution of the use a variable kernel bandwidth ($\text{MPM}_{var}$) towards fixed one ($\text{MPM}_{fix}$) was also evaluated through our tests.

In order to evaluate the performance of our method, we have used natural images corrupted by a synthetic gaussian noise ($\sigma_n$=10,20) as well as digital images corrupted by real camera noise. We compared our approach to well known filtering techniques such as the bilateral filter [19], the Non Local Mean approach [6], the total variation [18] and the anisotropic filtering [17] using an edge stopping function of the type $(1 + |\nabla I|^2 / K^2)^{-1}$. The parameters of the considered methods were tuned to get a good balance between texture preserving and noise suppression as well as the highest possible PSNR value. As far as subjective criteria are concerned, we adopt the whole aspect of the image in term of noise suppression and small detail preservation. Visual comparison results of denoising [Fig.(2), (3)] show that the total variation, the anisotropic diffusion and the bilateral filter, fail to preserve small detail and image texture. Furthermore, one can observe structured noise component in the presence of texture. We explain such behavior by the local nature of these methods where comparisons and structure information are considered only on a pixel level and not at local patches. Better quality of denoising was reached using the proposed approach, since the residual images contain less image structure compared to other techniques [Fig.(3)]. When considering real digital camera noise [Fig.(4,5,6)], we noticed a better restoration using our method with variable bandwidth kernels. In [Fig.(4,5)], we noticed that in case of fixed bandwidth the texture skin is over smoothed to lead to an artificial appearance when the variable bandwidth model is able to suppress the same amount of noise in smooth regions while preserving the texture. In fact this region was identified by our classification algorithm as textured region leading thus to more adapted smoothing constraints [Fig.(4-b)].

As far as quantitative validation is concerned we used the Peak Signal to Noise Ratio criterion defined by

$$PSNR = 10 log_{10} \frac{255^2}{MSE} \quad MSE = \frac{1}{\|\Omega\|} \sum_{\mathbf{x} \in \Omega} (U_x - \hat{U}_x)^2$$

where $U$ is the noise free ideal image and $\hat{U}$ its estimation by the denoising process.

In table 1 and 2, we report experimental validation results for the different methods on a set of image with various content corrupted by additive gaussian noise. PSNR values, confirm the subjective results and show that our non parametric estimation technique outperforms prior state-of-the art methods. Nevertheless for some examples, the classification step fails to capture fine scale details and texture is considered as noise. It tends to be a binary classification leading to

**Table 1.** PSNR values for denoised image (The PSNR of the image corrupted by gaussian noise of std=10 is equal to 28.11)

| | barbara | Boat | FingerPrint | House | Lena | baboon |
|---|---|---|---|---|---|---|
| TV | 29.60 | 32.17 | **30.65** | 33.86 | 33.83 | 27.81 |
| AD | 30.85 | 31.92 | 29.02 | 33.72 | 33.36 | 28.11 |
| Bilateral | 31.05 | 31.52 | 28.81 | 33.40 | 33.01 | 29.31 |
| NLmean | 32.96 | 32.49 | 30.60 | 34.66 | 34.65 | 29.54 |
| $MPM_{fix}$ | **33.07** | **32.57** | 30.58 | **34.8** | **34.77** | **29.7** |
| $MPM_{var}$ | 32.61 | 32.43 | 30.43 | 34.67 | 34.48 | 29.62 |

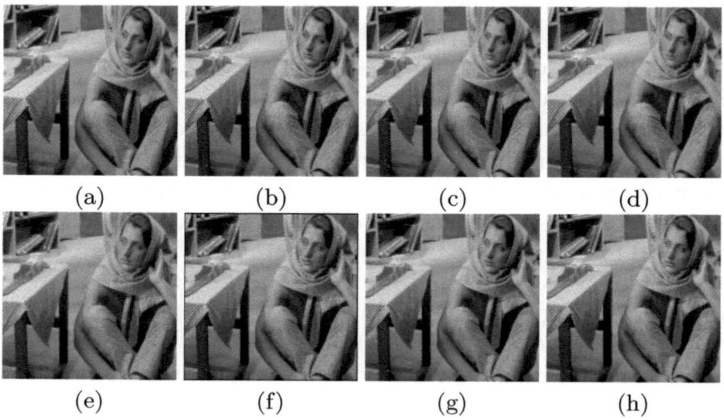

**Fig. 2.** Results of denoising on barbara image corrupted by gaussian noise with std=10: (a) original (b) noisy image (c) total variation (d) Anisotropic filtering (e) Bilateral filter (f) Non local mean (g) $MPM_{fix}$ (h)$MPM_{var}$

an important variation in the choice of the kernel bandwidth and thus affecting the quality of reconstruction.

## 5 Conclusion

In this paper we have proposed a new model to image denoising that exploits information of the image context. Such a method first decomposes the image domain into smooth and patterned regions. To this end, we associate to each image location a feature vector that refers to a statistical descriptor of local patches. Then, an analysis on these features provides measures of "smoothness". The obtained measures are then encoded in a new denoising process through an MPM estimation based on a non parametric model of the posterior density. As shown in tables (1,2) our method outperforms in all cases the existing ones. In spite of the marginal improvement of the statistical denoising process while considering

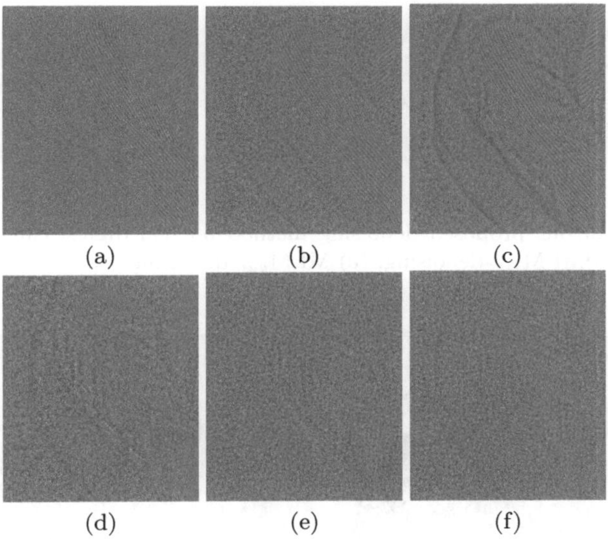

Fig. 3. Zoom on the residual of the different tested methods: (a) Total variation (b) Anisotropic filtering (c) Bilateral filter (d) Non local mean (e) $MPM_{fix}$ (f)$MPM_{var}$

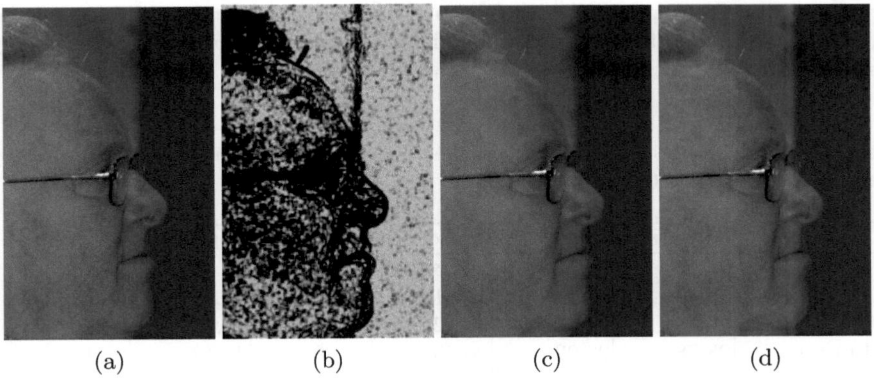

Fig. 4. Results of our proposed denoising method on real digital camera Noise, (a) original image (b) variable bandwidth function (low intensity ( $\sigma_x$=2 ), high intensity ($\sigma_x$=4)) (c)$MPM_{fix}$ denoising, (d) $MPM_{var}$ denoising

the image partition, we believe that this idea is promising and a better integration of the image structure in the denoising model should be studied. In fact, more appropriate selection of the kernels as well as their bandwidth could also improve the performance of the method, and is a direction that we are willing to address in the coming future. Furthermore, better extraction of local features using higher order projections is the most prominent direction to improve the initial step of our approach.

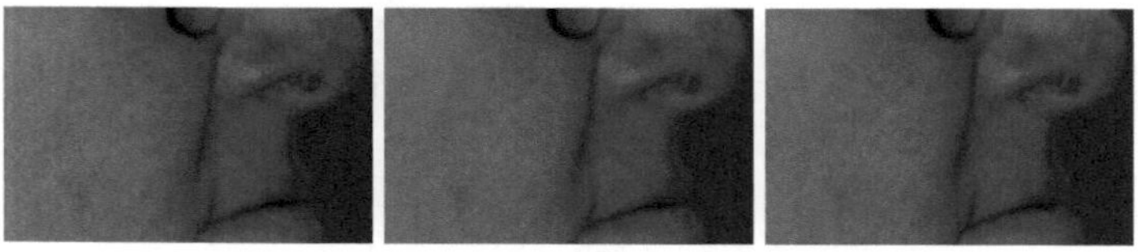

(a) (b) (c)

**Fig. 5.** Results of our proposed denoising method on real digital camera Noise, (a) original image (b)$\mathrm{MPM}_{fix}$ denoising, (c) $\mathrm{MPM}_{var}$ denoising

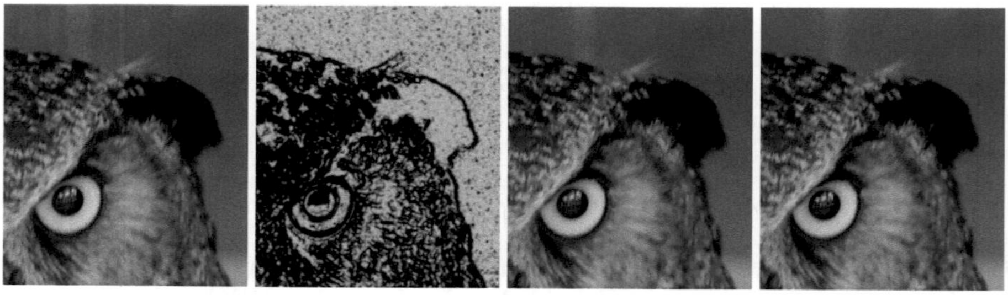

(a) (b) (c) (d)

**Fig. 6.** Results of our proposed denoising method on real digital camera Noise, (a) original image (b) variable bandwidth function (low intensity ( $\sigma_x$=2 ), high intensity ($\sigma_x$=4)), (c)$\mathrm{MPM}_{fix}$ denoising, (d) $\mathrm{MPM}_{var}$ denoising

**Table 2.** PSNR values for denoised image (The PSNR of the image corrupted by gaussian noise of std=20 is equal to 22.15)

| | barbara | Boat | FingerPrint | House | Lena | baboon |
|---|---|---|---|---|---|---|
| TV | 26.18 | 27.72 | 26.08 | 28.43 | 28.45 | 25.18 |
| AD | 26.45 | 28.06 | 24.81 | 29.41 | 29.27 | 23.68 |
| Bilateral | 26.75 | 27.82 | 24.12 | 29.18 | 29.28 | 24.95 |
| NLmean | 28.78 | 28.92 | 26.45 | 30.86 | 31.13 | 25.18 |
| $\mathrm{MPM}_{fix}$ | **29.18** | 28.84 | 26.38 | **31.16** | 31.19 | 25.26 |
| $\mathrm{MPM}_{var}$ | 28.9 | **29.11** | **26.68** | 31.02 | **31.25** | **25.39** |

## References


1. L. Alvarez, F. Guichard, P-L. Lions, and J-M. Morel. Axioms and fundamental equations of image processing. *Archive for Rational Mechanics*, 123:199–257, 1993.
2. G. Aubert and P. Kornprobst. *Mathematical Problems in Image Processing: Partial Differential Equations and the Calculus of Variations*. Springer-Verlag, 2001.
3. JF. Aujol and Chambolle A. Dual norms and image decomposition models. *International Journal of Computer Vision*, 63(1):85–104, 2005.

4. JF. Aujol, G. Gilboa, T. Chan, and S. Osher. Structure-texture image decomposition - modeling, algorithms, and parameter selection. *International Journal of Computer Vision*, 67(1):111–136, 2006.
5. N. Azzabou, N. Paragios, and F. Guichard. Random walks, constrained multiple hypotheses testing and image enhancement. In *ECCV*, pages 379–390, 2006.
6. A. Buades, B. Coll, and J-M. Morel. A non-local algorithm for image denoising. In *CVPR*, pages 60–65, 2005.
7. D. Comaniciu and P. Meer. Mean shift: A robust approach toward feature space analysis. *IEEE Transactions on pattern analysis and machine intelligence*, 15(5):603–619, 2002.
8. M. Do and M. Vetterli. Pyramidal directional filter banks and curvelets. In *ICIP01*, pages III: 158–161, 2001.
9. R.O. Duda and P. Hart. *Pattern Classification And Scene Analysis.* WileyInterscience Publication, 1973.
10. G. Gilboa, N. Sochen, and Zeevi Y.Y. Variational denoising of partly-textured images by spatially varying constraints. *IEEE Transactions on Image Processing*, 24(8):2281–2289, 2006.
11. R. Kimmel, R. Malladi, and N. Sochen. Image processing via the beltrami operator. In *ACCV*, pages 574–581, 1998.
12. E. Le Pennec and S. Mallat. Sparse geometric image representations with bandelets. *IEEE Transactions on Image Processing*, pages 423– 438, 2005.
13. S. Lee. Digital image smoothing and the sigma filter. *CVGIP*, 24(2):255–269, November 1983.
14. S. Mallat. A theory for multiscale signal decomposition: The wavelet representation. *IEEE Transactions on Pattern and Machine Intelligence*, pages 674–693, 1989.
15. Y. Meyer. *Oscillating Patterns in Image Processing and Nonlinear Evolution Equations: The Fifteenth Dean Jacqueline B. Lewis Memorial Lectures.* American Mathematical Society, Boston, MA, USA, 2001.
16. D. Mumford and J. Shah. Optimal Approximation by Piecewise Smooth Functions and Associated Variational Problems. *Communications on Pure and Applied Mathematics*, 42:577–685, 1989.
17. P. Perona and J. Malik. Scale space and edge detection using anisotropic diffusion. *IEEE Transactions on Pattern and Machine Intelligence*, 12:629–639, 1990.
18. L. Rudin, S. Osher, and E. Fatemi. Nonlinear Total Variation Based Noise Removal. *Physica D*, 60:259–268, 1992.
19. C. Tomasi and R. Manduchi. Bilateral filtering for gray and color images. In *ICCV*, pages 839–846, 1998.
20. D. Tschumperle and R. Deriche. Vector-valued image regularization with pde's : A common framework for different applications. *IEEE Transactions on Pattern Analysis and Machine Intelligence*, pages 506–517, 2005.
21. L. Vese and S. Osher. Modeling Textures with Total Variation Minimization and Oscillating Patterns in Image Processing. *Journal of Scientific Computing*, 19: 553–572, 2003.
22. L. Vincent. Morphological grayscale reconstruction in image analysis: Applications and efficient algorithms. *IEEE Transactions on Image Processing*, 2:176–201, 1993.
23. J. Weickert, B.M. ter Haar Romeny, and M. Viergever. Efficient and reliable schemes for nonlinear diffusion filtering. *IEEE Trans. lmage Processing*, 7(2):398, 1998.

# The Variational Origin of Motion by Gaussian Curvature

Niels Chr. Overgaard and Jan Erik Solem

Applied Mathematics Group
School of Technology and Society
Malmö University, Sweden
{nco,jes}@ts.mah.se

**Abstract.** A variational formulation of an image analysis problem has the nice feature that it is often easier to predict the effect of minimizing a certain energy functional than to interpret the corresponding Euler-Lagrange equations. For example, the equations of motion for an active contour usually contains a mean curvature term, which we know will regularizes the contour because mean curvature is the first variation of curve length, and shorter curves are typically smoother than longer ones. In some applications it may be worth considering Gaussian curvature as a regularizing term instead of mean curvature. The present paper provides a variational principle for this: We show that Gaussian curvature of a regular surface in three-dimensional Euclidean space is the first variation of an energy functional defined on the surface. Some properties of the corresponding *motion by Gaussian curvature* are pointed out, and a simple example is given, where minimization of this functional yields a nontrivial solution.

**Keywords:** Total mean curvature, Gaussian curvature, gradient descent flow, level set methods, Euler characteristic.

## 1 Introduction

For almost two decades, following the publication of the seminal papers by Kass, Witkin, and Terzopoulos [10] and Mumford and Shah [13], variational principles have been both popular and powerful tools in the inventory of the image analysts' toolbox. The level set method of Osher and Sethian [15] has made it considerably easier to implement and visualize curve and surface evolutions such as *geometric active contours* [2], *geodesic active contours* [3], *active contours without edges* [4], and notably *motion by mean curvature* (MMC), see e.g. Brakke [1]. Existence and uniqueness of viscosity solutions of the level set equations for MMC was established simultaneously by Chen, Giga, and Goto [5] and Evans and Spruck [7].

The present paper focuses on *motion by Gaussian curvature* (MGC). By MGC we mean a differentiable one-parameter family of regular surfaces $I \ni t \mapsto \Gamma(t) \subset \mathbf{R}^3$ in three-dimensional Euclidean space, $I$ being an open interval containing $t = 0$, which solves the initial value problem,

F. Sgallari, A. Murli, and N. Paragios (Eds.): SSVM 2007, LNCS 4485, pp. 430–441, 2007.

$$\frac{d}{dt}\Gamma(t) = -K_{\Gamma(t)} \ , \quad \Gamma(0) = \Gamma_0 \ , \tag{1}$$

for some given initial surface $\Gamma_0$. Here $(d/dt)\Gamma(t)$ is the (scalar) normal velocity of the evolving surface, and $K = K_{\Gamma(t)} = K_{\Gamma(t)}(\mathbf{x})$ denotes the Gaussian curvature at $\mathbf{x} \in \Gamma(t)$.

Motion by Gaussian curvature has not received nearly as much attention as MMC. One of the first papers on the subject is that of Firey [8] who constructed an idealized model of the wearing process of a convex stone on the beach, assuming that the local rate of wear is proportional to the Gaussian curvature. Oliker [14] studied the MGC for a surfaces which are graphs, $\Gamma(t) : z = u(x,y,t)$, where the function $u : \overline{U} \times [0,\infty) \to \mathbf{R}$ is defined on bounded, strictly convex subset $\overline{U} \subset \mathbf{R}^2$ with smooth boundary $\partial U$. Since the Gaussian curvature of such a surface is given by, see do Carmo [6, p. 163],

$$K = \frac{u_{xx}u_{yy} - u_{xy}^2}{(1 + u_x^2 + u_y^2)^2} \ , \tag{2}$$

and the normal velocity of $t \mapsto \Gamma(t)$ is $(d/dt)\Gamma(t) = u_t/(1 + u_x^2 + u_y^2)^{1/2}$, substitution into (1) gives the PDE

$$u_t = \frac{u_{xx}u_{yy} - u_{xy}^2}{(1 + u_x^2 + u_y^2)^{3/2}} \quad \text{in } U \times [0,\infty),$$

which is solved with homogeneous Dirichlet boundary conditions on $\partial U$.

In image analysis MGC has been used only in a few cases, most recently by Lee and Seo [11]. One way to understand this lack of use is to observe that MMC enters into the existing variational segmentation models as a *regularizing term* alongside the *fidelity term*. This application is based on a variational principle, namely that MMC is the gradient descent motion for the minimization of the surface area functional

$$E_{\mathrm{A}}(\Gamma) = \int_\Gamma dA \ ,$$

where $dA$ is the element of surface area on $\Gamma$. A typical segmentation functional has the form $E = E_F + \lambda E_{\mathrm{A}}$, where $\lambda > 0$ is a parameter. The first term $E_F$ is the fidelity term, which contains all the information about the input image, and the second is the area term $\lambda E_{\mathrm{A}}$, which is included as a smoothness prior. Smoothness is achieved as a trade-off between a good fit of the model to the input image, on one hand, and a small surface area of the interface between object and background, on the other. When gradient descent minimization is applied to $E$, the resulting evolution equation contains a mean curvature term. The variational interpretation of MMC as a minimizing flow of the surface area enables us to predict the regularizing nature of this mean curvature term. For MGC the corresponding variational interpretation is not well-known, making it harder to see the effects of including a Gaussian curvature term into an evolution equation. This may be one of the reasons why MGC has not been used so much.

In this paper we show that MGC is the gradient descent evolution for the minimization of an energy functional defined on the evolving surface. In fact, if we consider the *total mean curvature* of a a regular surface $\Gamma$ in $\mathbf{R}^3$,

$$E_{\mathrm{H}}(\Gamma) := \int_{\Gamma} H \, dA \;, \tag{3}$$

where $H = H(\mathbf{x})$ denotes the mean curvature of the surface at $\mathbf{x} \in \Gamma$, then we prove that the *first variation* (or Gâteaux derivative or directional derivative) of $E_{\mathrm{H}}$ is

$$dE_{\mathrm{H}}(\Gamma)(v) = \int_{\Gamma} K v \, dA \;, \tag{4}$$

for all normal variations $v : \Gamma \to \mathbf{R}$ of the surface. Loosely speaking, Gaussian curvature $K$ is the first variation of the total mean curvature.

One way of proving (4) is to use that any regular surface may locally be considered as the graph $\Gamma = \Gamma(u) : z = u(x, y)$ of a smooth function $u : U \to \mathbf{R}$ defined on some bounded, open subset $U$ of $\mathbf{R}^2$. In this representation the mean curvature of the surface is given by [6, p. 163],[1]

$$H = -\frac{1}{2} \frac{(1 + u_y^2)u_{xx} - 2u_x u_y u_{xy} + (1 + u_x^2)u_{yy}}{(1 + u_x^2 + u_y^2)^{3/2}} \;.$$

Since the area element on $\Gamma(u)$ is $dA = (1 + u_x^2 + u_y^2)^{1/2} \, dxdy$, the total mean curvature becomes a functional in $u$ of the form,

$$E(u) := E_{\mathrm{H}}(\Gamma(u)) = -\int_U \frac{(1 + u_y^2)u_{xx} - 2u_x u_y u_{xy} + (1 + u_x^2)u_{yy}}{2(1 + u_x^2 + u_y^2)} \, dxdy \;.$$

If $\varphi \in C_0^\infty(U)$ is a test function, then $t \mapsto u + t\varphi$ is a variation of $u$ which corresponds to a local smooth deformation of the surface $\Gamma$. The first variation of the total mean curvature thus becomes

$$\begin{aligned} dE(u)(\varphi) &= \frac{d}{dt} E(u + t\varphi)\Big|_{t=0} \\ &= \int_U \frac{u_{xx}u_{yy} - u_{xy}^2}{(1 + u_x^2 + u_y^2)^2} \, \varphi \; dxdy = \int_{\Gamma} K v \, dA \;, \end{aligned}$$

where $v = \varphi / (1 + u_x^2 + u_y^2)^{1/2}$ is the normal variation of $t \mapsto \Gamma(u + t\varphi)$ at $t = 0$. We have also used the formula (2) for the Gauss curvature $K$, and the expression for the area element $dA$ on $\Gamma(u)$. This identity proves (4).

This proof is both straight-forward and reliable, but at the same time tedious and dull, because of the long routine calculations involved. The method of proof that we are going propose below is more geometrical in nature, and applies to more general situations without extra work. In fact, we are going to formulate a

[1] Notice our sign convention: The sphere of radius $R$ has positive mean curvature $H = 1/R$. In [6] and [12] the opposite sign is used.

somewhat more general result, Theorem 1 in Sect. 3, which has (4) as a simple corollary.[2] After some geometrical preliminaries in Sect. 4, the proof is sketched in Sect. 5. Motion by Gaussian curvature is considered briefly in Sect. 6, and finally, surfaces of revolution with minimal total mean curvature in Sect. 7.

## 2 Volume, Surface Area, and Total Mean Curvature

Let $\Gamma$ denote a compact regular surface in $\mathbf{R}^3$. Notice that $\Gamma$'s complement consists of exactly two components, one of which is bounded. This bounded component is called the *inside of* $\Gamma$, and will be denoted $\Omega$. Let $w \in C^\infty(\mathbf{R}^3)$ be an arbitrary smooth function, and consider the following weighted surface functionals: First of all the **weighted volume**,

$$E_{\mathrm{V}}(w, \Gamma) := \int_\Omega w \, d\mathbf{x} \ , \tag{5}$$

then the **weighted surface area**,

$$E_{\mathrm{A}}(w, \Gamma) := \int_\Gamma w \, dA \ , \tag{6}$$

and the **weighted total mean curvature**,

$$E_{\mathrm{H}}(w, \Gamma) := \int_\Gamma H w \, dA \ . \tag{7}$$

Finally, we also consider the **weighted total Gaussian curvature**,

$$E_{\mathrm{K}}(w, \Gamma) := \int_\Gamma K w \, dA \ . \tag{8}$$

Notice that the map $C^\infty(\mathbf{R}^3) \ni w \mapsto E_{\mathrm{A}}(w, \Gamma) \in \mathbf{R}$ defines a (Schwartz-) distribution with compact support, that is $E_{\mathrm{A}}(\cdot, \Gamma) \in \mathcal{E}'(\mathbf{R}^3)$. On the other hand, $\Gamma \mapsto E_{\mathrm{A}}(w, \Gamma)$ defines a surface functional, in the usual sense. This holds for the other functionals $E_{\mathrm{V}}, E_{\mathrm{H}}$, and $E_{\mathrm{K}}$, as well. If $w \equiv 1$ we write

$$E_{\mathrm{V}}(\Gamma) := E_{\mathrm{V}}(1, \Gamma), \quad E_{\mathrm{A}}(\Gamma) := E_{\mathrm{A}}(1, \Gamma) \ ,$$

$$E_{\mathrm{H}}(\Gamma) := E_{\mathrm{H}}(1, \Gamma), \quad \text{and} \quad E_{\mathrm{K}}(\Gamma) := E_{\mathrm{K}}(1, \Gamma) \ ,$$

corresponding to the volume of $\Omega$, the surface area of $\Gamma$, and total mean and Gaussian curvatures of $\Gamma$, respectively. One of the most famous results in classical global differential geometry, namely *Gauss-Bonnet's Theorem*, tells us that the value of the total Gaussian curvature $E_{\mathrm{K}}(\Gamma)$ is entirely determined by the topological type of $\Gamma$,

[2] It has been brought to our attention that the results in (4) and Theorem 1 may be found in a more general version in [9, 1.6 Scholia, p. 82]. However, since the results do not seem to be commonly known, and our proof is new and simple, we believe this paper is still of interest to members of the image analysis community.

$$E_{\mathrm{K}}(\Gamma) = \int_\Gamma K\, d\sigma = 2\pi\chi(\Gamma) \ , \tag{9}$$

where $\chi(\Gamma)$ denotes the *Euler characteristic* of the surface $\Gamma$, i.e. $\chi(\Gamma) = 2 - 2g$, where $g$ is the *genus* of the surface, cf. [6, p.273].

## 3 The First Variation of Total Mean Curvature

The first variation, or Gâteaux derivative, of a surface functional $E = E(\Gamma)$ is a mapping $v \mapsto dE(\Gamma)(v)$ defined by the derivative

$$dE(\Gamma)(v) = \frac{d}{dt} E(\Gamma(t))\Big|_{t=0} \ ,$$

where $t \mapsto \Gamma(t)$ is an evolving surface satisfying $\Gamma(0) = \Gamma$ and $(d/dt)\Gamma(t) = v$. The latter means that the *normal velocity*, or *normal variation*, of the surface evolution is given by the function $v : \Gamma \to \mathbf{R}$. The first variation of a surface functional is homogeneous of degree one, by definition, but not necessarily additive, hence generally not a linear mapping. If the first variation $dE$ happens to be linear, then we call it the *differential of* $E$. Two such differentials, which are extensively used in image analysis, are that of the weighted volume,

$$dE_{\mathrm{V}}(w, \Gamma)(v) = \int_\Gamma wv\, dA = E_{\mathrm{A}}(wv, \Gamma) \ , \tag{10}$$

and weighted surface area,

$$dE_{\mathrm{A}}(w, \Gamma)(v) = \int_\Gamma w_n v\, dA + 2\int_\Gamma Hwv\, dA = E_{\mathrm{A}}(w_n v, \Gamma) + 2E_{\mathrm{H}}(wv, \Gamma), \tag{11}$$

where $w_n$ denotes the normal derivative on $\Gamma$ of the function $w \in C^\infty(\mathbf{R}^3)$, and $v$ is any normal variation of the surface. Readers will recognize (11) as the differential of the *geodesic active contours* [3]. Missing from the above list is the first variations of the total mean curvature functional $E_{\mathrm{H}}$ and the Gaussian curvature functional $E_{\mathrm{K}}$. They are provided by the main result of this paper:

**Theorem 1.** *Let $\Gamma$ be a compact regular surface in $\mathbf{R}^3$ and $w \in C^\infty(\mathbf{R}^3)$. The first variations of the weighted total mean curvature (7) and the weighted Gaussian curvature (8) are given by*

$$dE_{\mathrm{H}}(w, \Gamma)(v) = E_{\mathrm{H}}(w_n v, \Gamma) + E_{\mathrm{K}}(wv, \Gamma) \ , \tag{12}$$

*and*

$$dE_{\mathrm{K}}(w, \Gamma)(v) = E_{\mathrm{K}}(w_n v, \Gamma) \ , \tag{13}$$

*for any normal variation $v : \Gamma \to \mathbf{R}$, and $w_n$ denoting the normal derivative of $w$. Both variations are linear functionals of $v$, hence they are the differentials of $E_{\mathrm{H}}$ and $E_{\mathrm{K}}$, respectively.*

If $w$ is identically equal to one then, as an easy corollary of the theorem, we find that $dE_{\mathrm{H}}(\Gamma)(v) = E_{\mathrm{K}}(v, \Gamma)$ and $dE_{\mathrm{K}}(\Gamma)(v) = 0$. The first identity is exactly the assertion in (4), and the second is a consequence of Gauss-Bonnet's theorem (9).

## 4 Some Geometric Preliminaries

We prepare for the proof of Theorem 1 in the next section by recalling some facts from differential geometry. Suppose $\Gamma$ is a compact regular surface in $\mathbf{R}^3$, $\mathbf{x}_0$ a point on $\Gamma$, and let $\mathbf{x} = \mathbf{x}(u, v)$ be a local parametrization of a neighbourhood of $\mathbf{x}_0$, with parameters $(u, v) \in U$ where $U \subset \mathbf{R}^2$ is an open set. In the parameterized patch $\mathbf{x}(U)$ the Euclidean surface area element $dA$ is given by the formula,

$$dA = |\mathbf{x}_u \wedge \mathbf{x}_v| \, dudv \ , \tag{14}$$

where $\mathbf{x}_u$ and $\mathbf{x}_v$ are the partial derivatives of $\mathbf{x}$ with respect to $u$ and $v$, respectively, and "$\wedge$" denotes the vector product in $\mathbf{R}^3$. The principal curvatures of the surface at a point $\mathbf{x}$ on $\Gamma$ are denoted $\kappa_1 = \kappa_1(\mathbf{x})$ and $\kappa_2 = \kappa_2(\mathbf{x})$, respectively. The *mean curvature* $H = H(\mathbf{x})$ and the *Gaussian curvature* $K = K(\mathbf{x})$ are then given by

$$H = \frac{\kappa_1 + \kappa_2}{2} \qquad \text{and} \qquad K = \kappa_1 \kappa_2 \ .$$

Denote by $\mathbf{n} = \mathbf{n}(\mathbf{x})$, $\mathbf{x} \in \Gamma$, the *outward unit normal* on $\Gamma$ (Recall that we have a well-defined "inside".) The principal curvatures at $\mathbf{x}$ are the eigenvalues of the differential $D\mathbf{n}(\mathbf{x})$ which maps the tangent space at $\mathbf{x}$ into itself. The surface $\Gamma$ is said to be locally parameterized by *lines of principal curvature* if the parametrization $\mathbf{x} = \mathbf{x}(u, v) : U \to \mathbf{R}^3$ satisfies

$$\mathbf{n}_u(u, v) = \kappa_1(u, v)\mathbf{x}_u(u, v) \quad \text{and} \quad \mathbf{n}_v(u, v) = \kappa_2(u, v)\mathbf{x}_v(u, v) \ , \tag{15}$$

that is, the coordinate directions $\mathbf{x}_u$ and $\mathbf{x}_v$ are eigenvectors of the differential $D\mathbf{n}(\mathbf{x})$ at $\mathbf{x} = \mathbf{x}(u, v)$.

For $t \in \mathbf{R}$ define $\Gamma(t) = \{\mathbf{x} \in \mathbf{R}^3 : d(\mathbf{x}, \Gamma) = t\}$, where $d(\cdot, \Gamma)$ is the signed distance to the surface $\Gamma$. Since $\Gamma$ is assumed to be compact there exists a real number $\epsilon > 0$ such that if $t \in (-\epsilon, \epsilon)$, then the set $\Gamma(t)$ is again a compact regular surface, called a *parallel surface* to $\Gamma$. If $\mathbf{x} = \mathbf{x}(u, v)$ is a local parametrization of $\Gamma$, then each member the one-parameter family of parallel surfaces $t \mapsto \Gamma(t)$, $t \in (-\epsilon, \epsilon)$, can be parameterized locally by

$$\mathbf{x}^t = \mathbf{x}^t(u, v) := \mathbf{x}(u, v) + t\mathbf{n}(u, v) \ . \tag{16}$$

Notice that $t \mapsto \Gamma(t)$ is the surface evolution satisfying the initial value problem,

$$\frac{d}{dt}\Gamma(t) = 1 \quad \text{on } \Gamma(t), \text{ and} \quad \Gamma(0) = \Gamma \ . \tag{17}$$

The area element on the parallel surface $\Gamma(t)$ can be expressed in terms of the area element on $\Gamma$ and its curvatures:

**Lemma 1.** *The Euclidean area element $dA^t$ on the parallel surface $\Gamma(t)$ with the local parametrization (16) is given by*

$$dA^t = (1 + 2tH + t^2K) \, dA, \quad (-\epsilon < t < \epsilon) \ ,$$

*where $dA = dA^0$ is the area element on $\Gamma$.*

*Proof.* We prove the lemma under the simplifying assumption that the local parametrization can be chosen such that the coordinate lines $u \mapsto \mathbf{x}(u,v)$ and $v \mapsto \mathbf{x}(u,v)$ are *lines of principal curvature*, in which case it follows that

$$\begin{aligned}\mathbf{x}_u^t(u,v) &= \mathbf{x}_u(u,v) + t\mathbf{n}_u(u,v)\\ &= \mathbf{x}_u(u,v) + t\kappa_1(u,v)\mathbf{x}_u(u,v) = (1+t\kappa_1(u,v))\mathbf{x}_u(u,v) \ ,\end{aligned}$$

and similarly $\mathbf{x}_v^t(u,v) = (1+t\kappa_2(u,v))\mathbf{x}_v(u,v)$. Using (14) we find that

$$\begin{aligned}dA^t &= |\mathbf{x}_u^t \wedge \mathbf{x}_v^t|\, dudv\\ &= |(1+t\kappa_1)\mathbf{x}_u \wedge (1+t\kappa_2)\mathbf{x}_v|\, dudv\\ &= (1+t\kappa_1)(1+t\kappa_2)|\mathbf{x}_u \wedge \mathbf{x}_v|\, dudv = (1+t(\kappa_1+\kappa_2)+t^2\kappa_1\kappa_2)\, dA \ ,\end{aligned}$$

which is the desired result. For a complete proof, see [12, p. 145]. □

A point $\mathbf{x}$ whose distance to $\Gamma$ is less than $\epsilon$ is said to belong to a tubular neighbourhood $T_\epsilon$ of $\Gamma$. Any $\mathbf{x} \in T_\epsilon$ has a unique representation of the form $\mathbf{x} = \mathbf{x}_0 + t\mathbf{n}(\mathbf{x}_0)$ for some $\mathbf{x}_0 \in \Gamma$ and some $t \in (-\epsilon, \epsilon)$. The point $\mathbf{x}_0$ is called $\mathbf{x}$'s *projection onto* $\Gamma$, and $t$ is the *signed distance* of $\mathbf{x}$ to $\Gamma$. This representation is used in the proof of the following result:

**Corollary 1.** *For any weight $w \in C^\infty(\mathbf{R}^3)$, and the one-parameter family of parallel surfaces $t \mapsto \Gamma(t)$ defined by (16), we have the Taylor expansion,*

$$\begin{aligned}E_{\mathrm{A}}(w,\Gamma(t)) = E_{\mathrm{A}}(w,\Gamma) &+ t\big(E_{\mathrm{A}}(w_n,\Gamma) + 2E_{\mathrm{H}}(w,\Gamma)\big) +\\ &+ \frac{1}{2}t^2\big[E_{\mathrm{A}}(w_{nn},\Gamma) + 4E_{\mathrm{H}}(w_n,\Gamma) + 2E_{\mathrm{K}}(w,\Gamma)\big] + O(t^3) \ ,\end{aligned}$$

*as $t \to 0$, where $w_n$ and $w_{nn}$ denote the first and second derivatives of $w$ in the direction normal to the surface $\Gamma$.*

*Proof.* For each point $\mathbf{x} \in \Gamma$ fixed, the function $t \mapsto w(\mathbf{x} + t\mathbf{n}(\mathbf{x}))$ has the following Taylor expansion,

$$w(\mathbf{x} + t\mathbf{n}(\mathbf{x})) = w(\mathbf{x}) + tw_n(\mathbf{x}) + \frac{1}{2}t^2 w_{nn}(\mathbf{x}) + O(t^3) \ ,$$

which, in combination with the formula for $dA^t$ in the Lemma 1, gives the desired result. □

Any smooth function $v : \Gamma \to \mathbf{R}$ has a smooth extension to a tubular neighbourhood $T_\epsilon$ of $\Gamma$ which is constant along rays normal to the surface. This extension, which is also denoted $v$, is given by the formula

$$v(\mathbf{x}) = v(\mathbf{x}_0), \quad (\mathbf{x} \in T_\epsilon) \tag{18}$$

where $\mathbf{x}_0$ is the unique projection of $\mathbf{x}$ onto $\Gamma$. This extension is convenient in the formulation of the lemma below, and will play an important role in the proof of Theorem 1.

Let $s \mapsto \Gamma(s)$ be a surface evolution defined for $s \in I$, where $I$ is an open interval containing $s = 0$. For $s \in I$ fixed, let $t \mapsto \Gamma(s)(t) := \Gamma(s,t)$ denote the family of parallel surfaces of $\Gamma(s)$. Then we have,

**Lemma 2.** *If the normal velocity of $s \mapsto \Gamma(s)$ at $s=0$ is given by the scalar function $(d/ds)\Gamma(s)|_{s=0} = v$, then, for $t$ fixed, the normal velocity at $s=0$ of the evolution $s \mapsto \Gamma(s,t)$ of a parallel surface is*

$$\frac{d}{ds}\Gamma(s,t;\mathbf{x})\bigg|_{s=0} = v(\mathbf{x}) \quad (\text{for } \mathbf{x} \in \Gamma(0,t)) \ ,$$

*where $v$ is the extension (18) of the normal velocity $v : \Gamma(0) \to \mathbf{R}$ to a tubular neighbourhood of $\Gamma(0)$.*

*Proof.* The proof is carried out in a local parametrization $\mathbf{x} = \mathbf{x}(u,v,s)$ of the evolution $s \mapsto \Gamma(s)$. The corresponding parametrization of the parallel surfaces $s \mapsto \Gamma(s,t)$ is then given by

$$\mathbf{x}^t = \mathbf{x}^t(u,v,s) = \mathbf{x}(u,v,s) + t\mathbf{n}(u,v,s) \ ,$$

where $\mathbf{n} = \mathbf{n}(u,v,s)$ is the parametrization of the outward unit normal of $\Gamma(s)$. Using the notation $\dot{} = d/ds$ we find that

$$\begin{aligned}\frac{d}{ds}\Gamma(s,t;\mathbf{x}^t)\bigg|_{s=0} &= \mathbf{n}(u,v,0)\cdot\dot{\mathbf{x}}^t(u,v,0)\\ &= \mathbf{n}(u,v,0)\cdot\dot{\mathbf{x}}(u,v,0) + t\mathbf{n}(u,v,0)\cdot\dot{\mathbf{n}}(u,v,0)\\ &= \mathbf{n}(u,v,0)\cdot\dot{\mathbf{x}}(u,v,0) = \frac{d}{ds}\Gamma(s;\mathbf{x})\bigg|_{s=0} = v(\mathbf{x}) \ ,\end{aligned}$$

because $0 = (d/ds)|\mathbf{n}(u,v,s)|^2 = 2\mathbf{n}(u,v,s)\cdot\dot{\mathbf{n}}(u,v,s)$ for all $(u,v,s)$. This proves the lemma because $\mathbf{x}$ is the projection of $\mathbf{x}^t$ onto $\Gamma(0)$. □

## 5 Proof of the Main Theorem

We now come to the proof of Theorem 1 itself. Again, let $t \to \Gamma(t)$ denote the family of surfaces parallel to $\Gamma$, and observe that equations (10) and (17) imply that

$$\frac{d}{dt}E_{\mathrm{V}}(w,\Gamma(t)) = dE_{\mathrm{V}}(w,\Gamma(t))(\frac{d}{dt}\Gamma(t)) = dE_{\mathrm{V}}(w,\Gamma(t))(1) = E_{\mathrm{A}}(w,\Gamma(t)) \ .$$

The right hand side of this identity is known from Corollary 1, so by integrating we find the following Taylor expansion of the weighted volume functional on the parallel surface $\Gamma(t)$ as $t \to 0$:

$$E_{\mathrm{V}}(w,\Gamma(t)) = E_{\mathrm{V}}(w,\Gamma) + tE_{\mathrm{A}}(w,\Gamma) + \frac{1}{2}t^2\big[E_{\mathrm{A}}(w_n,\Gamma) + 2E_{\mathrm{H}}(w,\Gamma)\big] + O(t^3) \,. \tag{19}$$

Now, the idea is to use the fact that the (19) holds for any surface $\Gamma$ and its parallel surface $\Gamma(t)$, for any fixed sufficiently small $t$. We begin by computing the differential with respect to normal variations $v$ of $\Gamma$ on both sides of the equality sign. Using Lemma 2 to find the normal variation of $\Gamma(t)$ in terms $v$ in the left hand side of (19) it follows that,

$$dE_{\mathrm{V}}(w,\Gamma(t))(v) = dE_{\mathrm{V}}(w,\Gamma)(v) + t dE_{\mathrm{A}}(w,\Gamma)(v) + \\ + \frac{1}{2}t^2\big(dE_{\mathrm{A}}(w_n,\Gamma)(v) + 2dE_{\mathrm{H}}(w,\Gamma)(v)\big) + O(t^3) \ .$$

Substituting the formulas for $dE_{\mathrm{V}}$ and $dE_{\mathrm{A}}$ in (10) and (11) into this identity gives

$$E_{\mathrm{A}}(wv,\Gamma(t)) = E_{\mathrm{A}}(wv,\Gamma) + t\big[E_{\mathrm{A}}(w_n v,\Gamma) + 2E_{\mathrm{H}}(wv,\Gamma)\big] + \\ + \frac{1}{2}t^2\big[E_{\mathrm{A}}(w_{nn}v,\Gamma) + 2E_{\mathrm{H}}(w_n v,\Gamma) + 2dE_{\mathrm{H}}(w,\Gamma)(v)\big] + O(t^3). \tag{20}$$

Now, replace the test function $w$ by in Corollary 1 by the product $wv$, where $v$ is the extension of the normal velocity $v : \Gamma \to \mathbf{R}$ defined in (18). Since $v$ is constant along rays normal to the surface, $(wv)_n = w_n v$ and $(wv)_{nn} = w_{nn} v$, so we get

$$E_{\mathrm{A}}(wv,\Gamma(t)) = E_{\mathrm{A}}(wv,\Gamma) + t\big[E_{\mathrm{A}}(w_n v,\Gamma) + 2E_{\mathrm{H}}(wv,\Gamma)\big] + \\ + \frac{1}{2}t^2\big[E_{\mathrm{A}}(w_{nn}v,\Gamma) + 4E_{\mathrm{H}}(w_n v,\Gamma) + 2E_{\mathrm{K}}(wv,\Gamma)\big] + O(t^3) \ ,$$

If we compare the coefficients in this expansion with those found in the Taylor expansion (20) we find that

$$dE_{\mathrm{H}}(w,\Gamma)(v) = E_{\mathrm{H}}(w_n v,\Gamma) + E_{\mathrm{K}}(wv,\Gamma) \ ,$$

which is the desired formula for the differential of the weighted total mean curvature. The differential for the weighted total Gaussian curvature can be obtained in a similar manner by including third order terms in the expansions. The details are left to the reader. □

## 6 Some Properties of Motion by Gaussian Curvature

In this section we want to point to some interesting properties of the motion by Gaussian curvature, $t \mapsto \Gamma(t)$, defined by the initial value problem (1). Consider the volume of the interior $\Omega(t)$ of a surface $\Gamma(t)$,

$$V(t) := E_{\mathrm{V}}(\Gamma(t)) = \int_{\Omega(t)} dx \ .$$

It follows from (10) with $w \equiv 1$, and the definition (1) of MGC, that

$$V'(t) = dE_{\mathrm{V}}(\Gamma(t), \frac{d}{dt}\Gamma(t)) = dE_{\mathrm{V}}(\Gamma(t), -K_{\Gamma(t)}) = -E_{\mathrm{K}}(\Gamma(t)) \ , \tag{21}$$

**Fig. 1.** From left to right, a comparison between motion by mean curvature (top) and motion by Gaussian curvature (bottom) for the standard torus $T^2$

so, in view of Gauss-Bonnet's theorem (9), we find the following differential equation

$$V'(t) = -2\pi\chi(\Gamma(t)) \ , \tag{22}$$

where $\chi(\Gamma)$ is the Euler characteristic of $\Gamma$. This equation has some interesting consequences. First of all, (22) seems to suggest that the surface does not change topological type as it evolves. This is true as long as $V(t)$ is continuously differentiable because the Euler characteristic is an integer, so a change of topological type would lead to a jump in the right hand side of the equation. Secondly, (22) shows that the volume of $\Omega(t)$ changes at a constant rate. For instance, if $\Gamma_0$ is homeomorphic to the two-sphere $S^2$, then so is $\Gamma(t)$ for all sufficiently small $t > 0$, and since $\chi(S^2) = 2$, cf. [6, p. 273], it follows that

$$V'(t) = -4\pi \quad (\text{for } \Gamma_0 \text{ homeomorphic to } S^2).$$

In particular $\Gamma(t)$ ceases to exist after a certain *extinction time* $t_*$ given by

$$t_* = \frac{V(0)}{4\pi} \ .$$

If $\Gamma_0$ is homeomorphic to the standard torus $T^2$, then $\chi(\Gamma_0) = 0$ ([6, p. 273]), implying that

$$V(t) = V(0) \quad (\text{for } \Gamma_0 \text{ homeomorphic to } T^2),$$

that is, MGC preserves the volume of $\Omega(t)$. Finally, is $\Gamma_0$ is a surface of higher genus than the sphere or the torus (i.e. $g \geq 2$), then $\chi(\Gamma_0) < 0$, and we conclude that the volume $V(t)$ increases at a constant rate.

In Fig. 1 a comparison between MMC and MGC is shown for $T^2$. MMC decreases surface area and leads to shrinking in contrast to MGC which does not change the volume but moves the surface closer to the symmetry axis.

# 7 Surfaces of Revolution with Minimal Total Mean Curvature

Let $u : [a, b] \to \mathbf{R}$ be a twice continuously differentiable function, such that $y(x) > 0$ for all $x \in [a, b]$, and $\Gamma(u) = \{(x, y, z) \in \mathbf{R}^3 \mid y^2 + z^2 = u(x)^2\}$ be the surface of revolution obtained by rotating $u$'s graph through an angle of 360° about the $x$-axis. We are now going to determine the surfaces or revolution which minimizes total mean curvature. The mean curvature of $\Gamma(u)$ is given by the formula

$$H = \frac{1 + (u')^2 - uu''}{2u(1 + (u')^2)^{3/2}} ,$$

and the surface area element by $dA = 2\pi u(1 + (u')^2)^{1/2}\, dx$, so the total mean curvature of $\Gamma(u)$ is the functional of $u$ given by the formula,

$$E_{\mathrm{H}}(u) := E_{\mathrm{H}}(\Gamma(u)) = \pi \int_a^b \frac{1 + (u')^2 - uu''}{1 + (u')^2}\, dx \ . \tag{23}$$

Let $A, B > 0$ and set

$$\mathcal{A} = \{u \in C^2([a, b]) \mid u(a) = A,\ u(b) = B \text{ and } u(x) > 0 \text{ for all } a < x < b\}.$$

The task is to find an admissible function $u = u_0$ such that

$$u_0 \in \mathcal{A} : \quad E_{\mathrm{H}}(u_0) \le E_{\mathrm{H}}(u) \quad \text{ for all } u \in \mathcal{A}.$$

Assume that such a function $u_0$ exists, and pick a test function $\varphi \in C_0^2(a, b)$. If $\epsilon > 0$ is sufficiently small, then the function $u_0(x) + t\varphi(x) \in \mathcal{A}$ for all $t \in (-\epsilon, \epsilon)$. The necessary condition for a minimum is,

$$0 = \frac{d}{dt} E_{\mathrm{H}}(u_0 + t\varphi)\Big|_{t=0} = 2\pi \int_a^b \frac{-u_0''}{(1 + (u_0')^2)^2}\, \varphi\, dx \ .$$

The right hand side was obtained by differentiation with respect to $t$ under the integral sign, followed by integration by parts, and some simplifications. Since the test function $\varphi$ is arbitrary, the minimizer $u_0$ must satisfy $u_0''(x) = 0$ for all $x \in (a, b)$, hence

$$u_0(x) = (A(b - x) + B(x - a))/(b - a)$$

which is the straight line segment connecting the fixed endpoints $(a, A)$ and $(b, B)$. The corresponding surface of revolution $\Gamma(u_0)$ is therefore a part of a circular cone. Whether the solution is a local minimum or just a stationary point is at present not known to us.

Although this example is simple, it shows that minimization problems for the total mean curvature, in the presence of boundary conditions or constraints, may yield nontrivial and meaningful results.

## 8 Conclusion

We have seen that motion by Gaussian curvature is the gradient descent flow for a geometric surface functional, namely the total mean curvature of the surface. This functional can be used in applications as an alternative to the area functional which leads to the frequently used mean curvature motion. Some properties of the Gaussian curvature motion were pointed out and minimization of the total mean curvature functional subject to boundary conditions was considered briefly in a simple case. More work remains to be done in the area and it will be interesting to see if the theory for motion by Gaussian curvature will become as rich as the one for motion by mean curvature.

## References


1. Brakke, K.A.: The Motion of a Surface by its Mean Curvature. Volume 20 of Mathematical Notes. Princeton University Press (1978)
2. Caselles, V., Catte, F., Coll, T., Dibos, F.: A Geometric Model for Active Contours. Numerische Mathematik **66** (1993) 1–31
3. Caselles, V., Kimmel, R., Sapiro, G.: Geodesic active contours. Int. Journal of Computer Vision (1997)
4. Chan, T., Vese, L.: Active contours without edges. IEEE Transactions on Image Processing **10** (2001) 266–277
5. Chen, Y.G., Giga, Y., Goto, S.: Uniqueness and Existence of Viscosity Solutions of Generalized Mean Curvature Flow Equations. J. Diff. Geometry **33** (1991) 749–786
6. do Carmo, M.: Differential Geometry of Curves and Surfaces. Prentice-Hall (1976)
7. Evans, L.C., Spruck, J.: Motions of level sets by mean curvature, I. J. Diff. Geometry **33** (1991) 635–681
8. Firey, W.J.: Shapes of worn stones. Mathematika **21** (1974) 1–11
9. Giaquinta, M., Hildebrandt, S.: Calculus of Variation I. Grundlehren der mathematischen Wissenschaften **310**. Springer-Verlag (1996)
10. Kass, M., Witkin, A., Terzopoulos, D.: Snakes: Active contour models. Int. J. Computer Vision **1** (1987) 321–331
11. Lee, S.H., Seo, J.K.: Noise Removal with Gauss Curvature-Driven Diffusion. IEEE Transactions on Image Processing **14** (2005)
12. Montiel, S., Ros, A.: Curves and Surfaces. Volume 69 of Graduate Studies in Mathematics. American Mathematical Society & Real Soceidad Mathmática Española (2005)
13. Mumford, D., Shah, J.: Optimal approximation by piecewise smooth functions and associated variational problems. Comm. Pure Appl. Math. **42** (1989) 577–685
14. Oliker, V.: Evolution of Nonparametric Surfaces with Speed Determined by Curvature, I. the Gauss Curvature Case. Indiana University Mathematical Journal **40** (1991) 237–258
15. Osher, S., Sethian, J.A.: Fronts propagating with curvature-dependent speed: Algorithms based on Hamilton-Jacobi formulations. Journal of Computational Physics **79** (1988) 12–49

# A Variational Method with a Noise Detector for Impulse Noise Removal

Shoushui Chen and Xin Yang

Institute of Image Processing and Pattern Recognition
Shanghai Jiao Tong University, Shanghai 200030, P.R. China
{sschen,yangxin}@sjtu.edu.cn

**Abstract.** In this paper we propose a combined method for removing impulse noise. In the first phase, we use an efficient detector, called the *Statistics of Ordered Difference Detector* (SODD) to identify pixels which are likely to be corrupted by impulse noise. The proposed SODD can yield very high noise detection accuracy at high noise density. This noise detection phase is crucial for the following noise removal. In the second phase, only these noise candidates are restored using the variational method. Edges and noise free pixels of images filtered by our combined method are preserved. Simulation results indicate that the proposed method is significantly better than those using other impulse noise reduction filters.

## 1 Introduction

As we know, images are frequently corrupted by impulse noise during acquisition, transformation and so on. Impulse noise is characterized by replacing a portion of pixel values of the image with intensity values drawn from some distribution [1]. For images corrupted by impulse noises, noisy images is related to the original image by

$$u_{ij} = \begin{cases} u_{ij}^s, \texttt{with probability } 1-pr \\ \left.\begin{matrix} p_1, \texttt{with probability } pr_1 \\ p_2, \texttt{with probability } pr_2 \\ \vdots \\ p_n, \texttt{with probability } pr_n \end{matrix}\right\} u_{ij}^n \end{cases} \qquad (1)$$

where $u_{ij}^s$ represents the original image (signal), $u_{ij}$ denotes the observed noisy image, and $u_{ij}^n$ signifies the signal-independent impulse noise. Of course, the following restriction must be valid: $pr_1 + pr_2 + ... + pr_n = pr$. In the case of salt and pepper impulse noise, there are only two values $p_1$ and $p_2$, which are the maximum or the minimum pixel value of the considered integer interval. (in our case respectively 255 and 0). Many different filtering methods have been proposed for the removal of impulse noise from digital images. A great majority of these methods are based on standard median filter (SMF)[2] and its modifications [3][4][5][6], which utilize the rank order information of the pixels contained in the filtering window. Adaptive center-weighted median filter (ACWMF)[7] gives

F. Sgallari, A. Murli, and N. Paragios (Eds.): SSVM 2007, LNCS 4485, pp. 442–450, 2007.

the current pixel a large weight and the final output is chosen between the median and the current pixel value. Switching methods, such as, detail-preserving median based filter [8] and EDPA [9], use a noise detector to determine whether a pixel is noise or not. Then, noise reduction process is applied only to noise candidates. A novel filtering scheme using a signal-dependent shape for the moving window is proposed in [10]. These methods mentioned above can achieve good result at low noise density. However, their denoising performances are unsatisfactory at high noise density. Recently, a detail-preserving variational method (DPVM) has been proposed in [11] to reduce impulse noise. It uses smooth data-fitting term along with edge-preserving regularization.

In this paper, inspired by the work in [11], we propose a combined impulse noise removal method which combines the variational method with an efficient impulse noise detector. Firstly, impulse noise candidates are precisely detected by the proposed SODD, and then these candidates are removed by the variational method. Since image details are preserved for the noise candidates and image pixels which are noise free are unchanged, the performance of our combined method is much better than other state-of-art methods. The rest of the paper is organized as follows. Section 2 introduces the proposed SODD for impulse noise detection and the variational method for impulse noise reduction. Section 3 introduces some numerical simulations and experiments. The conclusions of the paper are listed in Section 4.

## 2 SODD and the Variational Method

### 2.1 SODD Algorithm for Impulse Detection

Assuming any noisy image of size $m \times n$ , and let $u_{ij}$ be its pixel value at position $(i,j)$, for$(i,j) \in I \equiv \{1,...,m\} \times \{1,...,n\}$. Let $W_{ij}^{u}(h)$ denote the window of size $(2h+1) \times (2h+1)$ centered about $u_{ij}$ , i.e., $W_{ij}^{u}(h) = \{u_{i+k1,j+k2} | -h \leq k1, k2 \leq h\}$ is the set of points in $(2h+1) \times (2h+1)$ neighborhood centered at $u_{ij}$ for some positive integer $h$. In the following discussion, let us only consider $h \geq 1$. We define

$$\Phi^0(p) = \Phi_p(h) \backslash \{p\} \tag{2}$$

as the set of points in $W_{ij}^{u}(h)$ deleted center pixel $p$. We define relativity measure $d(p,q)$ in intensity of pixels between the center pixel p and its neighbors $q_i$ as

$$d(p,q) = \sum_{j=\lfloor \alpha t \rfloor}^{t} \exp(s_j) \tag{3}$$

where $t = (2h+1) \times (2h+1) - 1$, $\alpha$ is the trimming parameter and is assumed between 0 and 1, $\lfloor . \rfloor$ is the floor function and $s_j$ is the jth data item in the increasingly ordered samples of$|q_{(1)} - p| \leq |q_{(2)} - p| \leq ... \leq |q_{((2h+1)\times(2h+1)-1)} - p|$ ($q_i$ $\in \phi^0(p)$). The relativity measure $d(p,q)$ provides a measure of how close a pixel is to its neighbors. It is verified that impulse intensity value varies greatly from most or all of its neighboring pixels, whereas other pixels' neighbors composing pixels of similar intensity, even pixels of image details. Fig.1 and table 1 show an example from the Lena image [12] comparing an impulse noise pixel to an edge pixel. We set $h = 3$,$\alpha = 0.6$ in this test and select three typical points $(P_a, P_b, P_c)$ from edge points, impulse noise and smoothing region points respectively. Table 1 gives their positions and $d(p,q)$ values and demonstrates that $d(p,q)$ can make clear difference between the edge pixel and the impulse noise pixel.

**Table 1.** Three pixels $P_a(40, 80)$ $P_b(120, 80)$ $P_c(150, 10)$ selected from edge points, impulse noises and smoothing region points and their d(p,q) values

| Point (x, y) | $P_a(40, 80)$ | $P_b(120, 80)$ | $P_c(150, 10)$ |
|---|---|---|---|
| d(p,q) | 28.314 | 49.224 | 25.182 |

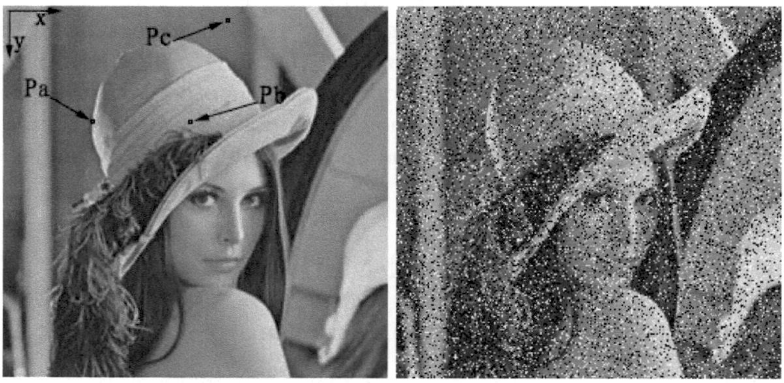


(a) (b)

**Fig. 1.** Three typical pixels (Pa, Pb, Pc) from Lena image, (a) Three pixels (Pa, Pb, Pc), (b) 20% salt and pepper noise

According to above discussion, we can get relativity measure matrix $d_{ij}(p,q)$ of the whole image. Then this matrix is divided into $g \times g$ blocks, which are neighboring, but do not overlap one another. Let $rms$ be root mean square value of each block respectively. Firstly we define an $g \times g$ zero matrix $M$. In each block, we make mark as follows,

$$M_{ij} = 1, \texttt{if} \;\; d_{ij}(p,q) > rms \tag{4}$$

If any pixel value $u_{ij}$ in each block is equal to the pixel value of point that we mark in the equation (4), we set $M_{ij} = 1$ also. Then, we combine each neighboring block $M_{g \times g}$ and get the whole mark matrix $M$. If an element of this mark matrix is 1, $u_{ij}$ is a noise candidate pixel and if it is 0, $u_{ij}$ is noise free pixel.

Since the noise detection plays the key role on noise reduction, it would be insightful to evaluate the performance of noise detection. In every noise image, let

$\Omega_a$ denote a set of all actual corrupted pixels, $\Omega_d$ denote a set of pixels, which are regarded as contaminated pixels by our proposed detector. Test indices include $T = \tau_n/\tau_d$ and $V = \tau_n/\tau_a$ , where $\tau_n$ denotes the number of pixels in $\Omega_a \cap \Omega_d$ , $\tau_d$ is that of pixels in $\Omega_d$, and $\tau_a$ is that of pixels in $\Omega_a$. We experiment on Lena and gold hill images [12] and test results are plotted in the table 2. It is showed that the accuracy of our detector can satisfy $V$=1 even in high noise density and only a little number of uncorrupted pixels are misclassified as noise candidates. The larger detection window size is, the more accurate noise detector is, and however, larger window size leads to higher computation complexity.

**Table 2.** T and V values of different noise densities on salt and pepper noise detection. (a) Lena; (b) gold hill.

| Noise density | 30% | 35% | 40% | 45% | 50% | 55% | 60% | 65% | 70% | 75% | 80% |
|---|---|---|---|---|---|---|---|---|---|---|---|
| T | 1 | 1 | 1 | 0.995 | 1 | 1 | 1 | 1 | 1 | 0.999 | 0.994 |
| V | 1 | 1 | 1 | 1 | 1 | 1 | 1 | 1 | 1 | 1 | 1 |

(a)

| Noise density | 30% | 35% | 40% | 45% | 50% | 55% | 60% | 65% | 70% | 75% | 80% |
|---|---|---|---|---|---|---|---|---|---|---|---|
| T | 0.916 | 0.944 | 0.989 | 0.995 | 0.991 | 0.992 | 1 | 1 | 1 | 0.999 | 0.997 |
| V | 1 | 1 | 1 | 1 | 1 | 1 | 1 | 1 | 1 | 1 | 1 |

(b)

The proposed SODD for impulse noise ensures that almost all of the noise pixels are detected even at a high noise density. It is noted that the noise candidates are restored by the variational method in the second phase, which we will discuss in the next section, while the remaining pixels are left unaltered.

### 2.2 The Variational Method for Impulse Noise Removal

In [11], images corrupted by impulse noise are restored by minimizing a convex objective function of the form $F_y : R^{m\times n} \to R$ .

$$F_y(u) = \sum_{(i,j)\in I} |u_{ij} - y_{ij}| + \frac{\beta}{2} \sum_{(i,j)\in I} \sum_{(v,w)\in N(i,j)} \varphi(u_{ij} - u_{vw}) \tag{5}$$

where $\varphi$ is an edge-preserving potential function and $N(i,j)$ is composed of those neighbors of (i,j). DPVM [11] method furnishes a new framework for the processing of image with different kinds of impulse noise and preserves edges during impulse noise reduction. However, this method alters all pixels in the image, including those that are not corrupted by the impulse noise and also has problem in detecting noisy patches. To avoid the drawback of DPVM method, we do noise reduction in two phases. In the first phase, impulse noise candidates are detected by the proposed SODD. We get noise candidate set $\Omega_d$ of image set $I$. Then these candidates are removed by DPVM in the following phase. That is to say, for a noise candidate $u_{ij}$. restore the noisy image by minimizing the following function over $\Omega_d$,

$$F_y(u) = \sum_{(i,j)\in\Omega_d} |u_{ij} - y_{ij}| + \frac{\beta}{2} \sum_{(i,j)\in\Omega_d} \sum_{(v,w)\in N(i,j)} \varphi(u_{ij} - u_{vw}) \tag{6}$$

In the following, we calculate the minimization of the equation (6) [11],

(1) Initialize $z_{ij}^0 = 0$ for each $(i,j) \in \Omega_d$.

(2) At each iteration $k$, for each $(i,j) \in \Omega_d$ , we calculate

$$\xi_{ij}^k = \beta \sum_{(v,w)\in N_{ij}} \varphi'(y_{ij} - z_{vw} - y_{vw})$$

where $z_{vw} \in N_{ij}$ and $\varphi'$ is the derivative of $\varphi$.

(3) If $\left|\xi_{ij}^k\right| \leq 1$ , then set $z_{ij}^k = 0$,

If $\left|\xi_{ij}^k\right| > 1$, then find $z_{ij}^k$ by solving

$$\beta \sum_{(v,w)\in N_{ij}} \varphi'(z_{ij}^k + y_{ij} - z_{vw} - y_{vw}) = sign(\xi_{ij}^k) \tag{7}$$

As suggested in [11], we choose $\varphi(t) = |t|^\alpha$ ,$\alpha = 1.41$ and $\beta = 10$. The procedure is repeated until convergence, $z^k$ can converges to $\hat{z} = \hat{u} - y$ and the restored image $\hat{u}$ is the minimization of $F_y(u)$. In each iteration, equation (7) is solved using Newton's method.

## 3 Numerical Simulations and Experiments

In this section, we compare our algorithm with SMF[2], ACWMF[7], EDPA[9] and DPVM[11]. Extensive experiments are conducted on a variety of standard gray-scale test images with distinctly different features, including peppers, mandrill and man [12]. In all experiments, we set $h = 3$,$\alpha = 0.6$,$g = 4$ in the noise candidate detection phase. The 256-by-256 images of peppers and mandrill are used as the original image in subjective experiments. Then two images are corrupted by 40% and 80% impulse noise respectively. Fig.2,3,4 and 5 show that our method achieves better performance than other filters do. It is obvious that SMF($7 \times 7$), ACWMF, DPVM and the proposed method gets better results in noise reduction than SMF($3\times3$) and EDPA do. But images filtered by SMF($7\times7$) and ACWMF have noise patches and DPVM fails in noise reduction at noise density of 80%.

To assess the effectiveness of our method in processing various images, we try another 256-by-256 gray scale image man. The parameters chosen are the same as in the previous simulations. The results in terms of mean absolute error (MAE),peak signal-to-noise ratio (PSNR) and structural similarity (SSIM)[13] are used as performance indices. Simulations are carried out for a wide range of

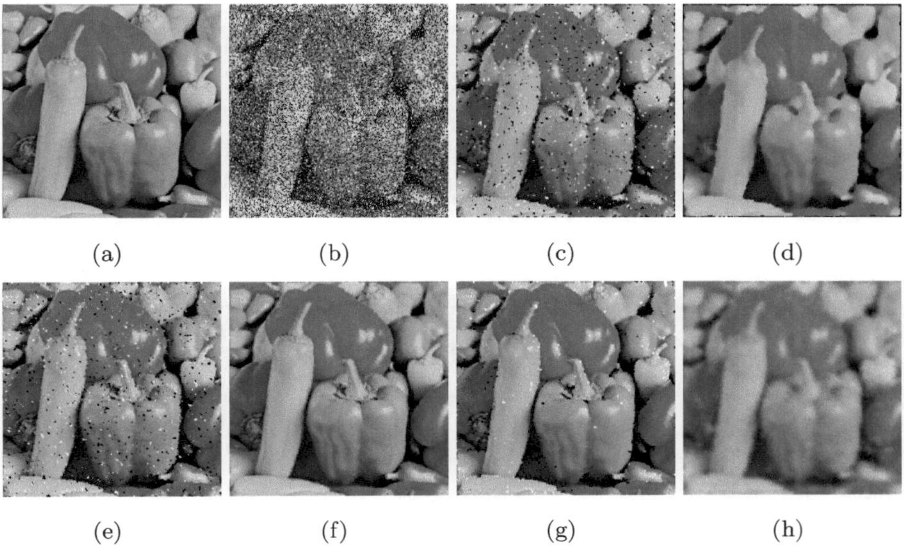


**Fig. 2.** Results of different filters for the Peppers image (with 40% salt and pepper noise),(a) Original image, (b) 40% noise, (c) SMF ($3 \times 3$), (d) SMF ($7 \times 7$), (e) EDPA, (f) Proposed, (g) ACWMF, (h) DPVM

**Fig. 3.** Results of different filters for the Peppers image (with 80% salt and pepper noise) (a) Original image, (b) 80% noise, (c) SMF ($3 \times 3$), (d) SMF ($7 \times 7$), (e) EDPA, (f) Proposed, (g) ACWMF, (h) DPVM

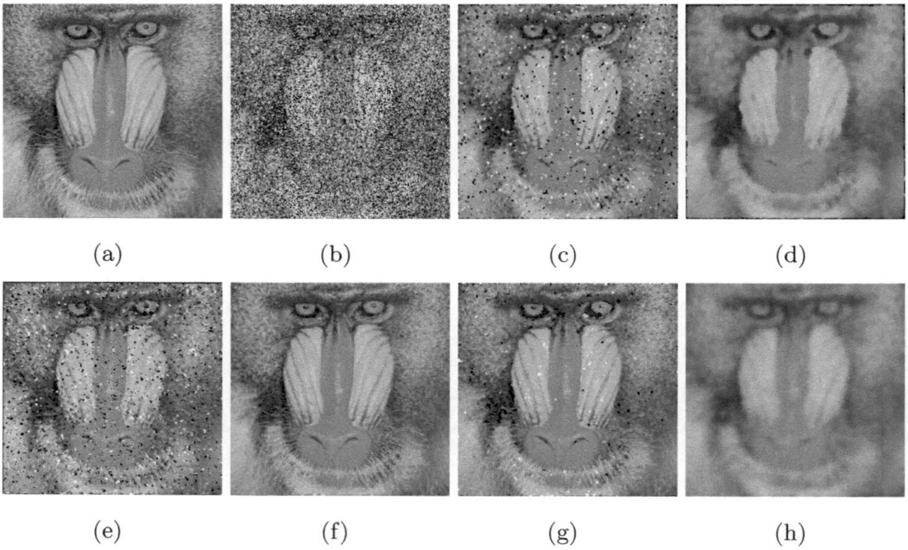

**Fig. 4.** Results of different filters for the mandrill image (with 40% salt and pepper noise) (a) Original image, (b) 40% noise, (c) SMF (3 × 3), (d) SMF (7 × 7), (e) EDPA, (f) Proposed, (g) ACWMF, (h) DPVM

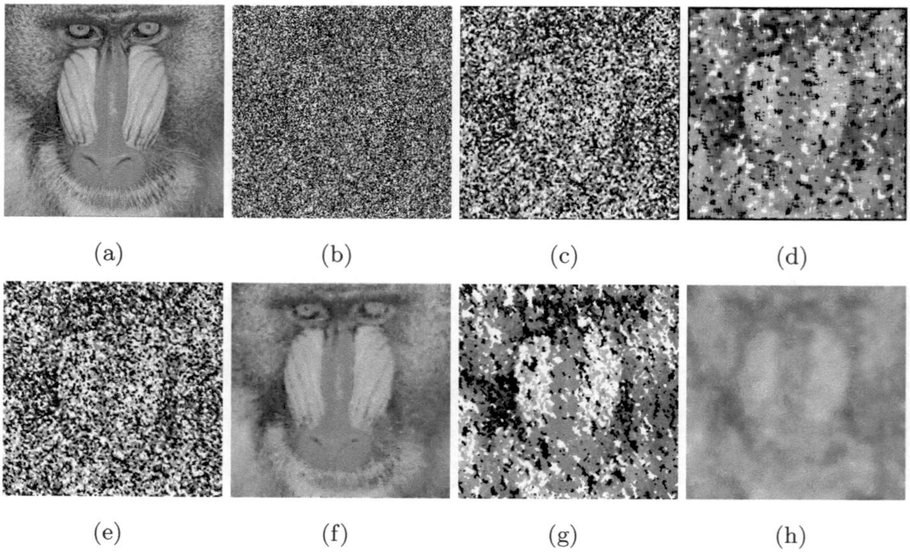

**Fig. 5.** Results of different filters for the mandrill image (with 80% salt and pepper noise) (a) Original image, (b) 80% noise, (c) SMF (3 × 3), (d) SMF (7 × 7), (e) EDPA, (f) Proposed, (g) ACWMF, (h) DPVM

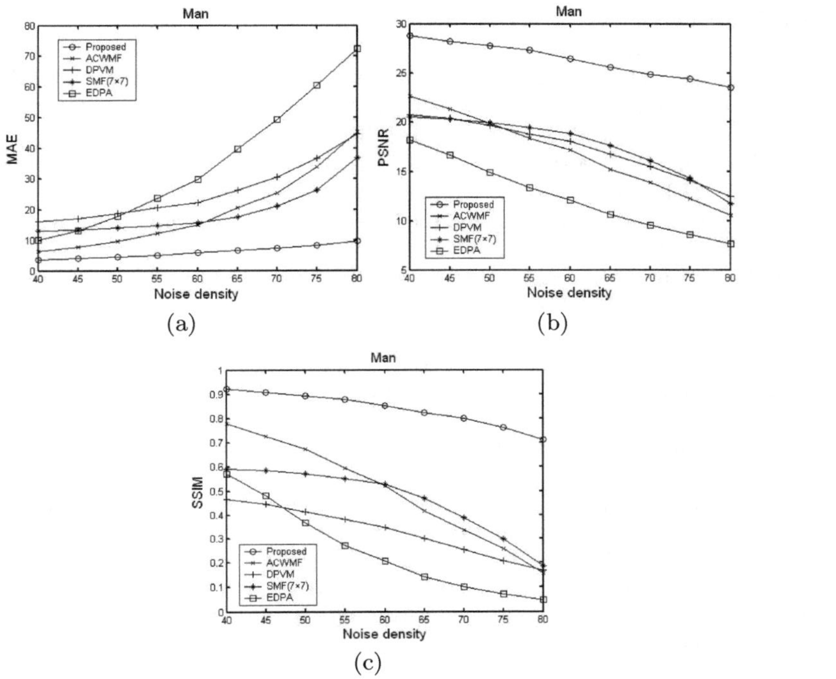


**Fig. 6.** MAE, PSNR and SSIM performance of different filters for the man image, (a) MAE (b) PSNR, (c) SSIM

noise density levels-40% $\leq p \leq$ 80% with an increment step of 5%. The comparisons between our method, ACWMF, EDPA, DPVM and SMF are shown in Fig.6 (a),(b) and (c). From the plots, we see that our proposed method consistently achieves significantly higher PSNR, SSIM and lower MAE than other filters do. The reason is that our method is mainly based on the accurate noise detection and detail-preserving variational method.

## 4 Conclusions

In this paper, a combined method based on the Statistics of Ordered Difference Detector and the variational method is proposed for removing impulse noise from images. To demonstrate the superior performance of the proposed method, extensive experiments have been conducted on a variety of standard test images to compare our method with many other well-known filters. Experimental results indicate that the proposed method performs significantly better than many other existing filters, especially at high noise density corruption.

**Acknowledgments.** This paper has been partially supported by China Natural Science Foundation (60572154). The authors would like to thank the referees for their valuable suggestions.

## References


[1] S. Schulte, M. Nachtegael, V. De Witte, D. Van der Weken and E.E. Kerre, "A fuzzy impulse noise detection and reduction method," IEEE Trans. Image Process., vol. 15, Issue 5, pp. 1153 - 1162, May. 2006.

[2] I. Pitas and A. N. Venetsanopoulos, "Order statistics in digital image processing," Proc. IEEE, vol. 80, no. 12, pp. 1893-1921, Dec. 1992.

[3] D. R. K. Brownrigg, "The weighted median filter," Commun. ACM, vol. 27, no. 8, pp. 807-818, Aug. 1984.

[4] G.R. Arce and R. E. Foster, "Detail-preserving ranked-order based filters for image processing," IEEE Trans. Acoustics, Speech, Signal Processing. , vol. 37, no. 1, pp. 83 - 98, Jan. 1989.

[5] H.G. Senel, R.A. II Peters and B. Dawant, "Topological median filters," IEEE Trans. Image Process. , vol. 10, no. 12, pp. 89 - 104, Dec. 2002.

[6] S.-J. Ko and Y. H. Lee, "Center weighted median filters and their applications to image enhancement," IEEE Trans. Circuits Syst., vol. 38, no. 9, pp. 984-993, Sep. 1991.

[7] Tao Chen and Hong Ren Wu"Adaptive impulse detection using center-weighted median filters," IEEE Signal Processing Letters., vol. 8, pp. 1-3, Jan. 2001.

[8] T. Sun and Y. Neuvo, "Detail-preserving median based filters in image processing," Pattern Recognit. Lett., vol. 15, no. 4, pp. 341-347, Apr. 1994.

[9] Wenbin, Luo, "An efficient detail-preserving approach for removing impulse noise in images," IEEE signal processing letters. , vol. 13, no. 7, pp. 413 - 416, July. 2006.

[10] Mikhail Mozerov and Vitaly Kober, "Impulse noise removal with gradient adaptive neighborhoods," Opt. Eng. , 45, pp. 067003. 1- 3, Jun. 2006.

[11] M. Nikolova, "A variational approach to remove outliers and impulse noise," J. Math. Imag, Vis. , vol. 20, pp.99-120, 2004.

[12] http://sipi.usc.edu/database/index.html.

[13] Z. Wang, A. C. Bovik, H. R. Sheikh, and E. P. Simoncelli, "Image quality assessment: From error visibility to structural similarity, " IEEE Trans. on Image Process. , vol. 13, no. 4, pp.600-612, Apr. 2004.

# Detection and Completion of Filaments: A Vector Field and PDE Approach

Alexis Baudour[1,2], Gilles Aubert[1], and Laure Blanc-Féraud[2]

[1] Laboratoire Jean-Alexandre Dieudonné UNSA - Faculté des Sciences
Parc Valrose, 06108 - NICE Cedex 02,France
[2] Project ARIANA, CNRS/INRIA/UNSA
2004 route des Lucioles, BP 93 06902 SOPHIA ANTIPOLIS Cedex, France
gaubert@math.unice.fr, alexisbaudour@gmail.com, blancf@sophia.inria.fr

**Abstract.** Our goal in this paper is to propose a new algorithm for the detection and the completion of thin filaments in noisy and blurred 2D or 3D images. The detection method is based on the construction of a 3D vector field whose singularities (vorticity points) correspond to the filaments. The completion is then obtained by solving a Ginzburg-Landau system which is well-adapted to the study of such singularities. Numerical results for 2D images are given.

## 1 Introduction

The goal of this paper is to propose a new algorithm for the detection and the completion of thin filaments in noisy and blurred 2D or 3D images. For example, think about actin filaments in confocal microscopic images of cells. We mean by filament, structures of codimension $n-1$ in an ambient space of dimension $n$. For example, in 3D, a filament is a discontinuity line within an homogeneous constant intensity volume. Generally, in pre-detection methods, filaments have a non nul thickness due to the blur introduced during the acquisition process. Moreover, some parts are missing (due to intensity cancellation, noise, ...). Our aim in this paper is twofold: firstly, we want to refine the pre-detection in order to recover the original thin filaments and secondly, we want to complete the missing parts. The main tool to detect a filament in a 2D or a 3D image, relies on the construction of a 3D vector field $\overrightarrow{\mathcal{B}}$ lying locally in the orthogonal plane to the filament. These vectors spin locally around the filament which then correspond to their singularity set (called vorticity set). In fact, filaments can be viewed as current lines and we use the associated magnetic field to construct $\overrightarrow{\mathcal{B}}$.

The method we propose for the completion relies on the minimization of a Ginzburg-Landau (GL) energy. GL models are dedicated to the study of singularities of codimension $k$ for function $u$ from $\mathbb{R}^{n+k}$ into $\mathbb{R}^k$ [6,3] .

More precisely in the 3D case, we search for a vector $u : \Omega \subset \mathbb{R}^3 \to \mathbb{R}^2$ minimizing

F. Sgallari, A. Murli, and N. Paragios (Eds.): SSVM 2007, LNCS 4485, pp. 451–460, 2007.

$$E_\epsilon(u) = \int_\Omega \left( \|\nabla u\|^2 + \frac{(1-|u|^2)^2}{\epsilon^2} \right) dx$$

with some prescribed values $g$ on the boundary $\partial\Omega$. The completion zone $\Omega$ and $g$ are defined from the vector field $\overrightarrow{\mathcal{B}}$.

It can be shown in [6] that the sequence $(u_\epsilon)_{\epsilon>0}$ of minimizers of $E_\epsilon$ converges as $\epsilon \longrightarrow 0$ to a vector field $u_*$ whose discontinuity set gives the completed filament.

Computing numerically $u_\epsilon$ is generally, for 3D images, a difficult task for small $\epsilon$. Therefore we use a two step algorithm proposed in [9] which consists in alternately solving a diffusion equation for a short time and then applying a renormalization process.

Note that there exist few methods in the literature dedicated to the detection and completion of fine structures in an image. Most of them rely on tubular structures and not on curves [7,8].

The paper is organized as follows. In section 1, we set the notations and give our definition of a filament. Then we first describe a method based on the derivatives of the image to localize the filaments. In a second step, we construct a vector field $\overrightarrow{\mathcal{B}}$ whose singularities give the "skeleton" of the filament. In section 2 we introduce the Ginzburg-Landau system and we show how this PDE based system allows us to complete missing part of the filaments. One important point is the orientation of the filaments to be completed and this point is examined in detail. Finally in section 3 numerical results are given on synthetic images. Though our theory applies for 3D or 2D images, we only present results in two dimensions.

## 2 Filament Detection

We define a filament as a piecewise $C^1$−curve in the support $D \subset \mathbb{R}^3$ of an original image $I_r : D \longrightarrow \mathbb{R}$. For example in confocal microscopy, the observed image $I$ can be approximated as the convolution of the original image $I_r$ and a Gaussian kernel [10]. Consequently the observed filaments in $I$ can appear thick and blurry. Our goal is to recover the original filaments which can be seen as the "skeleton" of the thick filaments.

### 2.1 Filament Localisation

The observed 3D image $I$ is a twice differentiable function since we suppose that it is the result of a convolution with a Gaussian kernel $G_\sigma$, i.e. $I = I_r * G_\sigma$. Filaments can be defined as local extrema of $I$, without jump. For example, if the backround is uniform and the intensity is constant along the filaments, $I_r$ can be modeled as a binary image: $I_r = 0$ in the backround and $I_r = 1$ on the filaments. $\nabla I(x)$ has a zero crossing for $x$ belonging to the filaments. It is then necessary to use second order differential operators to detect filaments, e.g. the Hessian matrix. We denote by $\lambda_1$, $\lambda_2$, $\lambda_3$ the eigenvalues (with $|\lambda_1| \leq |\lambda_2| \leq |\lambda_3|$) and $\overrightarrow{v_1}$, $\overrightarrow{v_2}$, $\overrightarrow{v_3}$ the eigenvectors of the Hessian matrix of $I$. It can be shown, when filaments are present in an image, that $|\lambda_2|$ and $|\lambda_3|$ tend to $+\infty$ as $\sigma$ tends to zero. On the other hand we have $|\lambda_1| << |\lambda_2|$. This leads us to define the following set $M_\alpha$

$$M_\alpha = \{(x, y, z); |\lambda_2 \lambda_3| > \alpha\}$$

where $\alpha$ is a given threshold. The set $M_\alpha$ is useful to perform a pre-detection of the filament, by selecting points where the image $I$ has high variations in at least two directions ($\overrightarrow{v_2}$ and $\overrightarrow{v_3}$). Therefore $M_\alpha$ contains curves, whose directions are given by $\overrightarrow{v_1}(x)$, or isolated points. In general, there are some false alarms and the filament detection is not thin (see Figure 1). In this figure the directions of $\overrightarrow{v_1}(x)$ for $x \in M_\alpha$ are quite good and we see that $M_\alpha$ gives a first localisation of the filament. We present in the next section a method to refine in $M_\alpha$ this coarse pre-detection.

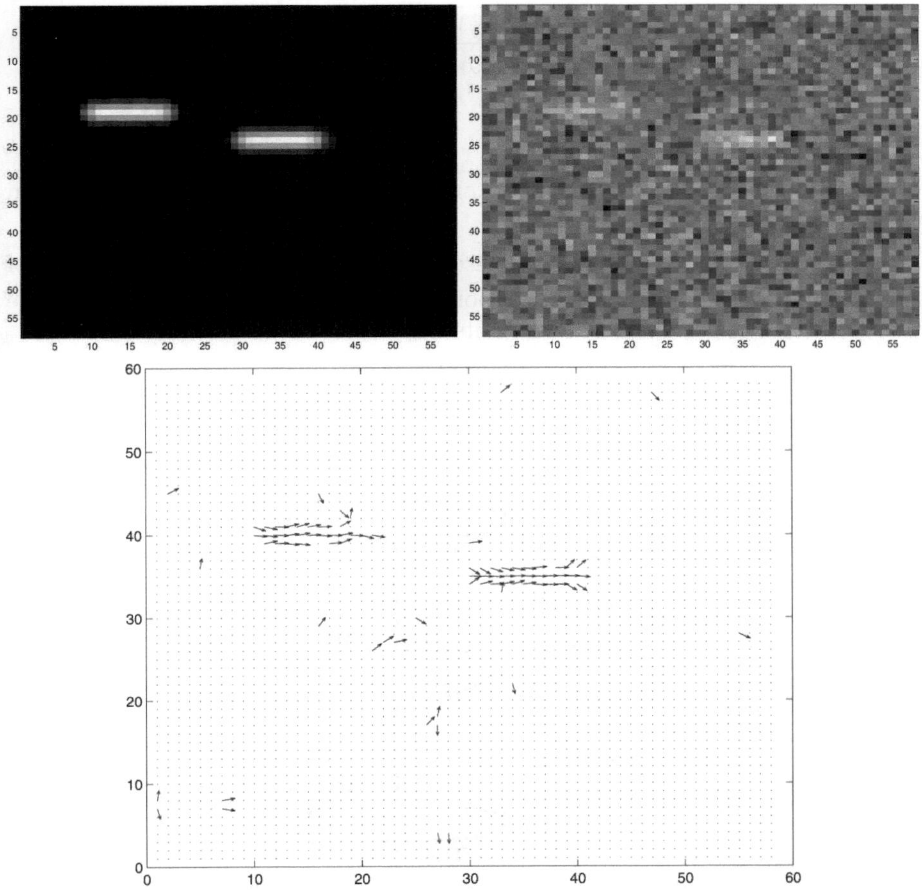


**Fig. 1.** Top left: Blurred image, top right: Blurred and noisy image, bottom: $v_1(x)$ for $x \in M_\alpha$, PSNR=7.06 Db

### 2.2 Definition and Computation of the Skeleton of the Filament

Our aim is to refine our pre-detection. With the same notations we denote by $\overrightarrow{v} : D \rightarrow \mathbb{R}^3$ the vector field which represents a pre-detection of the filaments:

$$\overrightarrow{v} = \begin{cases} \overrightarrow{v_1} \text{ if } |\lambda_2\lambda_3| > \alpha \\ \overrightarrow{0} \text{ otherwise} \end{cases}$$

$M_\alpha$ is the support of the vector field $\overrightarrow{v}$, and gives a coarse representation of the thick filaments.

In order to model a filament, we draw our inspiration from magnetostatic. The filament is considered as an electric current and we construct the associated potential $\overrightarrow{A}$ and magnetic field $\overrightarrow{B}$ on D.

$$\overrightarrow{A(x)} = \int_{M_\alpha} \frac{\overrightarrow{v(y)}}{||\overrightarrow{xy}||} dy \qquad \forall x \in D$$

$$\overrightarrow{B(x)} = \overrightarrow{rot}\,\overrightarrow{A} \qquad \forall x \in D$$

We also introduce the vector field $\overrightarrow{B'(x)}$, the normalized projection of $\overrightarrow{B(x)}$ in the orthogonal plane to $\overrightarrow{A(x)}$.

In $M_\alpha$, vector field $\overrightarrow{A}$ is a regularized version of $\overrightarrow{v}$ since it is the sum of the contributions of all pre-detected points. Vector $\overrightarrow{B'}$ spin locally around the filament which in turn corresponds to its singularity set (see Figure 2).

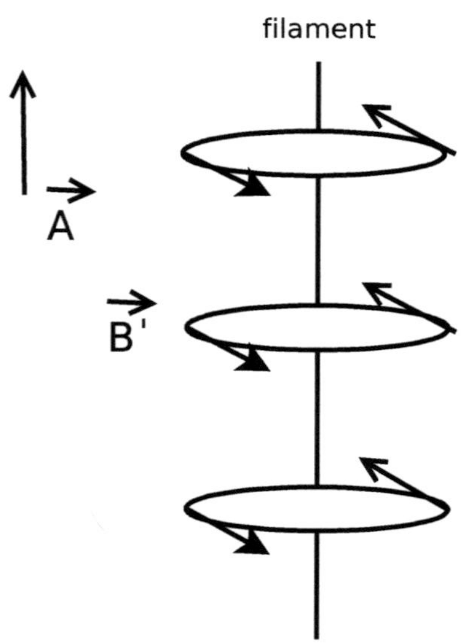


**Fig. 2.** Vector field B' and A

For example, in the case of a rectilinear filament, vector field $\overrightarrow{B'}$ is in the orthogonal plane to the filament and so equals vector field $\overrightarrow{B}$. In this plane the cartesian coordinates of $\overrightarrow{B'}$ are $(-sin\theta, cos\theta)$ where $\theta$ is the polar angle. We have $||\nabla\overrightarrow{B'}||^2 = ||\frac{\partial\overrightarrow{B'}}{\partial r}||^2 + \frac{1}{r^2}||\frac{\partial\overrightarrow{B'}}{\partial\theta}||^2$. As $\overrightarrow{B'}$ does not depend on the radial component $r$, we have $\frac{\partial\overrightarrow{B'}}{\partial r} = \overrightarrow{0}$ and $||\nabla\overrightarrow{B'}||^2 = \frac{1}{r^2}||\frac{\partial\overrightarrow{B'}}{\partial\theta}||^2$. Thus we get $||\nabla\overrightarrow{B'}||^2 = \frac{1}{r^2}$. Therefore it is clear that $||\nabla\overrightarrow{B'}||^2$ tends to $+\infty$ as $r$ tends to 0. This particular case leads us to the definition of the skeleton of a filament:

**Definition 1.** *The skeleton of the set of filaments is defined as*

$$S = \{x \in D, \lim_{y\longrightarrow x} ||\nabla\overrightarrow{B'}(y)|| = +\infty\}.$$

In practice to remove false alarms we only keep in the skeleton points where $||\overrightarrow{A}|| \geq \beta$ for a given $\beta$. Note that our definition of skeleton is not the one usually given in mathematical morphology. However it allows us to get an intuitive idea of what can be a skeleton of a thick filament.

We can see an illustration of our method in Figure 3, where we have displayed the skeleton of a filament in an image corrupted by blur and Gaussian noise (see Figure 1). We have computed the vector field $\overrightarrow{v_1}$ and the associated vector fields $\overrightarrow{A}$, $\overrightarrow{B}$ and $\overrightarrow{B'}$. The skeleton $S$ gives the original filament. This method is available in all cases, even if the Point Spread Function of the blur ant the noise are unperfectly known or unknown, which is usualy the case in practical applications.

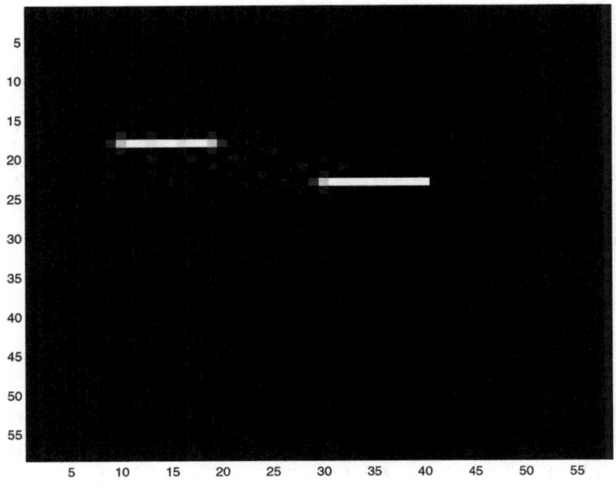


**Fig. 3.** Skeleton of filaments computed from Figure 1

# 3 Filament Completion

## 3.1 The Ginzburg-Landau Model

Due to degradation and occlusion missing parts can occur in our previous detection. We cannot use classical inpainting methods [1,2] to perform a completion. In fact the main idea of these methods consists in connecting level lines which are, in general, jump sets in an image. But unfortunately in an image of filaments there are no level lines. Thus we propose a new method of completion based on the Ginzburg-Landau theory.

Ginzburg-Landau models are dedicated to the study of singularities of codimension $k$ for function $u$ from $\mathbb{R}^{n+k}$ into $\mathbb{R}^k$ [6] . In our case we look for filaments in $\mathbb{R}^3$, so $k = 2$ and we need to use a vector field with values into $\mathbb{R}^2$.

A Ginzburg-Landau energy is of the type:

$$E_\epsilon(u) = \int_\Omega \left( \|\nabla u\|^2 + \frac{(1-|u|^2)^2}{\epsilon^2} \right) dx$$

with some prescribed values on the boundary $\partial\Omega$.

We remark that the term $\frac{(1-|u|^2)^2}{\epsilon^2}$ in the Ginzburg-Landau energy obliges, in a minimization process, the norm of $\overrightarrow{u}$ to be close to 1 for small $\epsilon$.

In our application we choose $\Omega = D - E_\delta$ where $E_\delta = \{x \in D; ||\overrightarrow{A(x)}|| \geq \delta\}$ and the boundary condition is $u = \overrightarrow{B'}$ on $\partial E_\delta$ (see [6]). In fact the set $E_\delta$ can be viewed as a tubular neighborhood of a regularized version of the set $M_\alpha$ (the support of the thick filaments, see Figure 1) where isolated points and bad oriented vectors $\overrightarrow{v}$ have been removed (see Figure 4). Let us precise that the computation of $\overrightarrow{B'}$ depends on the orientation of the vector field $\overrightarrow{v}$. This point will be developed in section 3.3.

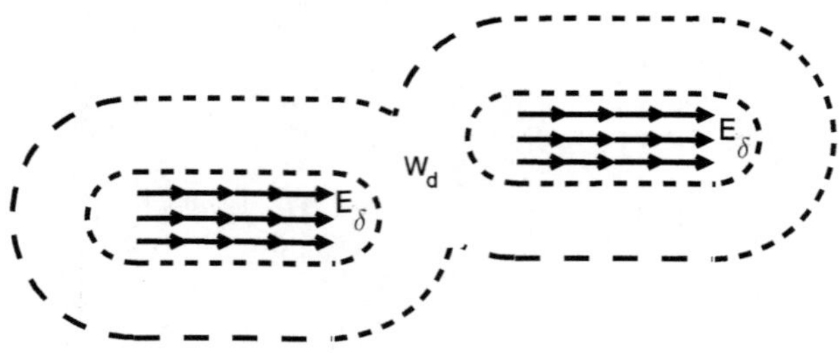


**Fig. 4.** $E_\delta$ and $W_d$

For $\epsilon > 0$ fixed, it is easy to show that $E_\epsilon(u)$ admits a minimizer $u_\epsilon$ on the Sobolev space $H^1(\Omega;\mathbb{R}^2) = \{u \in L^2(\Omega)^2; \nabla u \in L^2(\Omega)^6\}$. As $\epsilon \to 0$ we can prove that $u_\epsilon \to u_*$. Thanks to the general theory for Ginzburg-Landau systems, we can conjecture that the singularity set of $u_*$ is a set of codimension two which creates connection of minimal length between the extremities of the skeleton, i.e. the desired completion. In fact, $u_*$ spins around the completed filament.

### 3.2 The Completion Algorithm

In practice, in order to avoid the connection of farther filaments, we perform our algorithm in a smaller domain than $D$. We thus define the set $W_d = \{x \in D; ||\overrightarrow{A(x)}|| \geq d\}$, where $d << \delta$ (see Figure 4). The set $\Omega$ in the Ginzburg-Landau energy will be then $\Omega = W_d - E_\delta$.

Computing numerically $u_\epsilon$ is generally, for 3D image, a difficult task for small $\epsilon$. Multigrid methods may be used [4]. In this work we use a two step algorithm proposed in [9] which consists in alternately solving a diffusion equation for a short time and then applying a renormalization process. We initialize the algorithm with $\overrightarrow{u_0} = \overrightarrow{B'}$.

$$\frac{v_{n+1} - u_n}{\delta t} = \Delta v_{n+1} \qquad \text{in } \Omega$$

$$u_{n+1} = \frac{v_{n+1}}{|v_{n+1}|} \qquad \text{in } \Omega$$

$$u_{n+1}(x) = \overrightarrow{B'} \qquad \text{on } \partial\Omega$$

### 3.3 Orientation of the Filaments

The completion of two filaments will be possible if the orientation of the pre-detected filaments (i.e. the vector field $\overrightarrow{v}$) are compatible. We mean by compatible the fact that the winding numbers [5] $\eta_1$ and $\eta_2$ of the vector field $\overrightarrow{B'}$ associated to filament $F_1$ (resp. $F_2$) are of opposite sign (see Figure 5). In fact, for two filaments, $\eta_1$ and $\eta_2$ are equal to $\pm 1$. We cannot find a global orientation which allows the completion. For example on Figure 5, the two filaments are well-oriented for a completion in the left handside image, but they are opposite in the right handside image.

To choose a local orientation for a filament $F_1$, it is necessary to consider the orientation of the other filaments in a neighborhood of filament $F_1$. We give below an algorithm to find a local orientation.

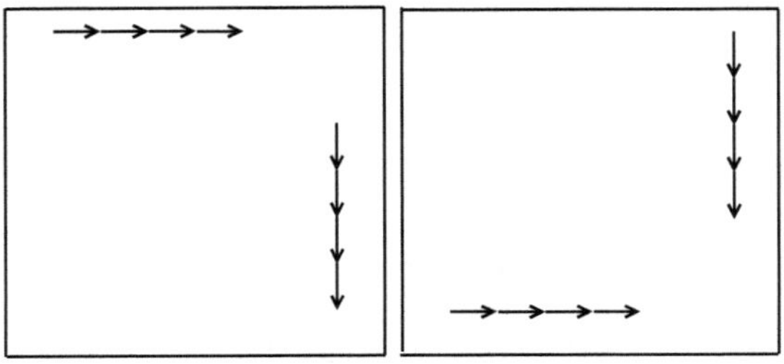

**Fig. 5.** Left: good orientation, right: bad orientation

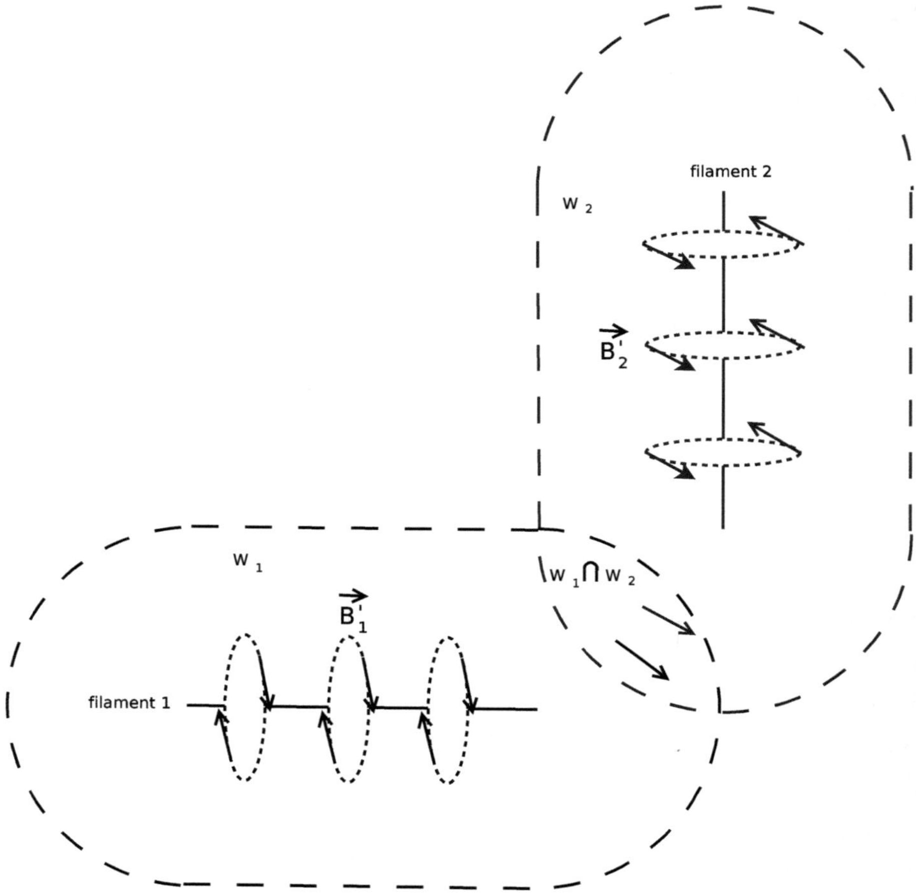


**Fig. 6.** Orientation algorithm

The idea is that the two magnetic fields associated to the two filaments have nearby directions if the orientation is compatible (cf. Figure 6). We explain how to find a compatible orientation of two filaments $F_1$ and $F_2$ which are connected components of $M_\alpha$.

We denote by $W_d^i = \{x \in D,\ ||\overrightarrow{A_i(x)}|| \geq d\}$ where $\overrightarrow{A_i}$ is the potential vector associated to filament $F_i$, $i = 1, 2$. We compute the magnetic fields $\overrightarrow{B_1'}$ and $\overrightarrow{B_2'}$ on $W_d^1$ and $W_d^2$ associated to $F_1$ and $F_2$. Let $\overrightarrow{B'_{2*}}$ be the magnetic field associated to $F_2$ with the reverse orientation. Then we compare $I_2$ and $J_2$:

$$I_2 = \int_{W_d^1 \cap W_d^2} ||\overrightarrow{B_1'} - \overrightarrow{B_2'}|| \qquad J_2 = \int_{W_d^1 \cap W_d^2} ||\overrightarrow{B_1'} - \overrightarrow{B'_{2*}}||$$

If $J_2 < I_2$ we change the orientation of $F_2$. We can see in Figure 6 that with the good orientation $\overrightarrow{B_1'}$ and $\overrightarrow{B_2'}$ are quite similar in $W_d^1 \cap W_d^2$.

For many filaments, we use an iterative algorithm that selects the orientation of neighbouring filaments two by two.

## 4 Results

Figure 7 is the result of the completion of filaments of Figure 1. We remind that $\overrightarrow{B'}$ defines the boundary conditions in the Ginzburg-Landau system and that $\Omega = W_d - E_\delta$ (see section 2.2). The filament is extracted from vector field $\overrightarrow{u_*}$ (the solution of GL) by selecting points $x$ such as $||\overrightarrow{\nabla u(x)}|| > \gamma$.

We also give a result for a triple junction (see Figure 8). In this case the winding number of one of the three filament should be equal to two and the two others to $\pm 1$. We also have to adapt the orientation algorithm.

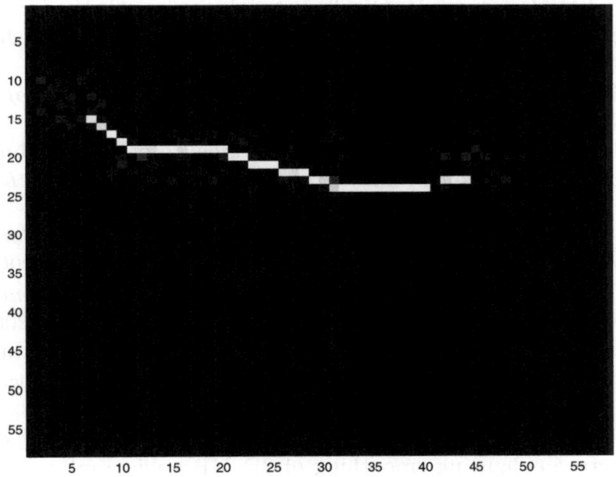

**Fig. 7.** Completion of filaments computed from Figure 1

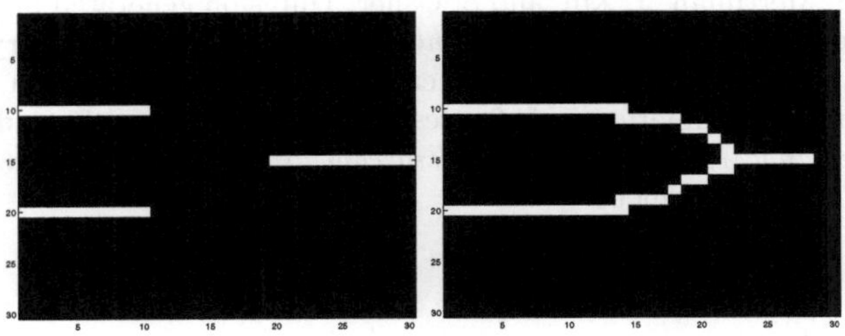

**Fig. 8.** Result for a triple junction

Note that the Ginzburg-Landau algorithm is also applicable from the filament skeleton as defined in section 1.2. However in this case we must compute again a new set $M_\alpha$ and a new vector field $\overrightarrow{B'}$.

## 5 Conclusion

In this paper we have proposed a simple and robust method to detect filaments in 2D or 3D images. The algorithm is based on the computation of the eigenvectors of the Hessian matrix of the image and on the representation of the filaments by a vector field $\overrightarrow{B'}$. Since there are missing parts in the detection we use this vector field and a Ginzburg-Landau energy to complete the filaments. We have presented 2D numerical results but in a future work we plan to test our algorithm on 3D real images.

## References


1. C. Ballester, J.M Bertalmio, V. Caselles, and J. Sapiro, G. Verdera. Filling-in by joint interpolation of vector fields and gray levels. *Transactions on Image Processing, IEEE*, 10:1200–1211, August 2001.
2. M. Bertalmio, V Caselles, G. Haro, and G. Sapiro. *The Handbook of Mathematical Models of Computer Vision.* Springer Verlag, 2005.
3. F. Bethuel, G. Orlandi, and D. Smets. Convergence of the parabolic Ginzburg-Landau equation to motion by mean curvature. *Ann. of Math. 163*, (1):37–163, 2006.
4. A. Borzi and K. Kunisch. *A Multigrid Method for the Optimal Control of Time-Dependent Reaction Diffusion Processes*, volume 69. Birkhäuser, 2001.
5. H. Brezis. Degree theory: Old and new. *in M. Matzeu and A. Vignoli eds., Topological Nonlinear Analysis II. Degree, Singularity and Variations*, 2005.
6. F. H. Lin and T. Rivière. Complex Ginzburg-Landau equation in high dimension and codimension two area minimizing currents. *J.Eur.Math.Soc.*, (1):237–311, 1999.
7. O. Nemitz, M. Rumpf, T. Tasdizen, and R. Whitaker. Anisotropic curvature motion for structure enhancing smoothing of 3D MR angiography data. *Journal of Mathematical Imaging and Vision.* to appear.
8. M. Rochery, I.H. Jermyn, and J. Zerubia. Higher order active contours. *International Journal of Computer Vision*, 69(1):27–42, August 2006.
9. S. Ruuth, B. Merriman, J. Xin, and S. Osher. Diffusion generated motion by mean curvature for filaments. *Journal of Nonlinear Science*, 11(6), January 2001.
10. Bo Zhang, Josiane Zerubia, and Jean-Christophe Olivo-Marin. A study of Gaussian approximations of fluorescence microscopy psf models. *Three-Dimensional and Multidimensional Microscopy: Image Acquisition and Processing XIII*, 6090(1):60900K, 2006.

# Nonlinear Diffusion on the 2D Euclidean Motion Group

Erik Franken*, Remco Duits**, and Bart ter Haar Romeny

Eindhoven University of Technology, Dept. of Biomedical Engineering, The Netherlands
{E.M.Franken,R.Duits,B.M.terHaarRomeny}@tue.nl

**Abstract.** Linear and nonlinear diffusion equations are usually considered on an image, which is in fact a function on the translation group. In this paper we study diffusion on *orientation scores*, i.e. on functions on the Euclidean motion group SE(2). An orientation score is obtained from an image by a linear invertible transformation. The goal is to enhance elongated structures by applying nonlinear left-invariant diffusion on the orientation score of the image. For this purpose we describe how we can use Gaussian derivatives to obtain regularized left-invariant derivatives that obey the non-commutative structure of the Lie algebra of SE(2). The Hessian constructed with these derivatives is used to estimate local curvature and orientation strength and the diffusion is made nonlinearly dependent on these measures. We propose an explicit finite difference scheme to apply the nonlinear diffusion on orientation scores. The experiments show that preservation of crossing structures is the main advantage compared to approaches such as coherence enhancing diffusion.

## 1 Introduction

A scale space of a scalar-valued image is obtained by solving an evolution equation on the additive group $(\mathbb{R}^n, +)$, i.e. the translation group. The most widely used evolution equation is the diffusion equation, which in the linear case leads to the Gaussian scale space [1] [2]. In the nonlinear case with an isotropic diffusion tensor it leads to a nonlinear scale space of Perona and Malik type [3]. An anisotropic diffusion tensor leads to edge- or coherence-enhancing diffusion [4].

Recently, processing of *tensor images* gains attention, for instance in Diffusion Tensor Imaging (DTI). A related type of data are *orientation scores* [5] [6], where orientation is made an explicit dimension. Orientation scores arise naturally in high angular resolution diffusion imaging, but can also be created out of an image by applying a wavelet transform [6]. Both tensor images and orientation scores have in common that they contain richer information on local orientation. They

* The project was financially supported by the Dutch BSIK program entitled Molecular Imaging of Ischemic heart disease (project number BSIK 03033).

** The Netherlands Organisation for Scientific Research (NWO) is gratefully acknowledged for financial support.

F. Sgallari, A. Murli, and N. Paragios (Eds.): SSVM 2007, LNCS 4485, pp. 461–472, 2007.

can both be considered as functions on the Euclidean motion group $SE(2) = \mathbb{R}^2 \rtimes \mathbb{T}$, i.e. the group of all 2D rotations and translations. This richer structure is often overlooked, e.g. if one applies component-wise nonlinear diffusion on tensor images [7]. When processing tensor images or orientation scores, it is actually *more natural to define the evolution equation on the Euclidean motion group*, leading to scale spaces on the Euclidean motion group.

In this paper we will introduce the analogue of nonlinear diffusion on the Euclidean motion group, with the goal to enhance oriented structures or patterns in two-dimensional images. We will start with introducing orientation scores and (nonlinear) diffusion in orientation scores in more detail. We will propose nonlinear conductivity functions to enable a coherence enhancing diffusion operation in orientation scores, which can handle crossings and adapts to the curvature of line structures. An explicit numerical finite difference scheme will be presented that has good rotational invariance. Finally we will show examples of coherence enhancing diffusion in orientation scores on applications with crossing and curved line structures.

This paper focusses on nonlinear diffusion on SE(2) and how to operationalize this. Scale spaces on Lie groups in general are treated in [8].

## 2 Orientation Scores

An orientation score is a function $U \in \mathbb{L}_2(SE(2))$. Such a function has one additional dimension compared to the original image, which explicitly encodes information on local orientations in the image. An example is shown in Figure 1a-b. The domain of the orientation score can be parameterized by the group elements $g = (\mathbf{x}, \theta)$ where $\mathbf{x} = (x, y) \in \mathbb{R}^2$ are the two spatial variables that label the domain of the image $f$, and $\theta \bmod 2\pi$ is the orientation angle that captures the orientation of structures in image $f$. The group product and group inverse of elements in $SE(2)$ are given by

$$g\,g' = (\mathbf{x}, \theta)\,(\mathbf{x}', \theta') = (\mathbf{x} + \mathbf{R}_\theta \mathbf{x}', \theta + \theta' \bmod 2\,\pi), \quad g^{-1} = (-\mathbf{R}_\theta^{-1}\mathbf{x}, -\theta) \tag{1}$$

We will use both short notation $g$ and explicit notation $(\mathbf{x}, \theta)$ for group elements.

An orientation score $U_f : \mathbb{R}^2 \rtimes \mathbb{T} \to \mathbb{C}$ of an image [1] $f \in \mathbb{L}_2(\mathbb{R}^2)$ is obtained by convolving the image with an anisotropic convolution kernel $K \in \mathbb{L}_2(\mathbb{R}^2)$,

$$U_f(\mathbf{x}, \theta) = (K^\theta * f)(\mathbf{x}) = \int_{\mathbb{R}^2} K(\mathbf{R}_\theta^{-1}(\mathbf{x} - \mathbf{x}'))f(\mathbf{x}')d\mathbf{x}', \tag{2}$$

where $K(\mathbf{x})$ is the kernel with orientation $\theta = 0$, and $\mathbf{R}_\theta$ is the rotation matrix $\mathbf{R}_\theta = \begin{pmatrix} \cos\theta & -\sin\theta \\ \sin\theta & \cos\theta \end{pmatrix}$. For some choices of $K$ there exists a stable *inverse transformation* [6], which is obtained by either convolving $U(\cdot, \theta)$ with the mirrored conjugate kernel of $K$ followed by integration over $\theta$, or simply by $f = \int_0^{2\pi} U(\mathbf{x}, \theta) d\theta$.

[1] The space of orientation scores of images $V = \{U_f | f \in \mathbb{L}_2(\mathbb{R}^2)\}$ is a vector subspace of $\mathbb{L}_2(SE(2))$. Note that the operations in $\mathbb{L}_2(SE(2))$ described in the rest of this paper do not leave $V$ invariant. From a practical point of view, however, this is not a prolbem since the inverse transformation implicitely projects on $V$. For mathematical details, see [8] where $V = \mathbb{C}_K^G$.

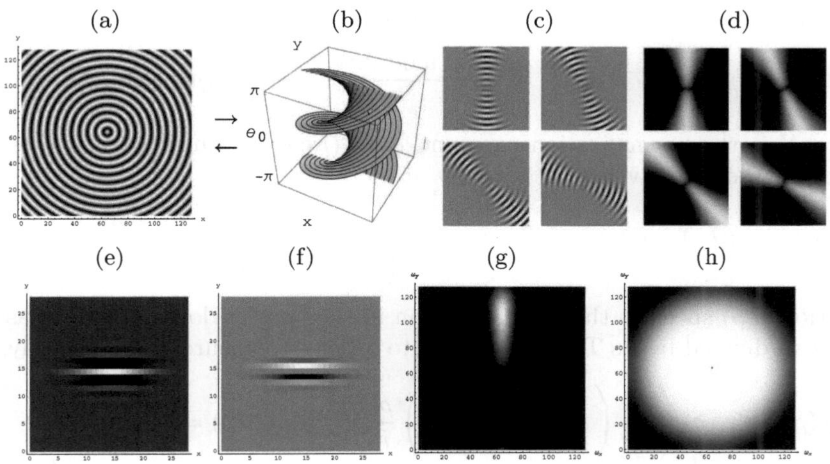


**Fig. 1.** (a) Example of an image with concentric circles. (b) The structure of the corresponding orientation score. The circles become spirals and all spirals are situated in the same helicoid-shaped plane. Note that the $\theta$-dimension is periodic. (c) Real part of the orientation score $U$ displayed for 4 different orientations. (d) The absolute vale $|U|$ yields a phase-invariant response displayed for 4 orientations. (e) Real part of the kernel with $\theta = 0$ and parameter values $k = 2$, $q = 8$, $t = 400$, $s = 10$, $n_\theta = 64$. (f) Imaginary part. (g) Fourier tranform. (h) Fourier transform of the net operation, i.e. orientation score transformation followed by the inverse orientation score transformation.

For our purpose, invertibility is required to be able to obtain an enhanced image after applying non-linear diffusion in the orientation score of an image.

In practice the $\theta$-dimension is sampled with steps $\frac{2\pi}{n_\theta}$ where $n_\theta$ the number of samples. To emphasize discretization we will use the notation $U[\mathbf{x}, l] = U(\mathbf{x}, l \cdot s_\theta)$ with $\mathbf{x} \in [0, 1, \ldots, N_x - 1] \times [0, 1, \ldots, N_y - 1]$, $l \in [0, 1, \ldots n_\theta - 1]$, and $s_\theta = \frac{2\pi}{n_\theta}$.

Note that if the operation in performed in the orientation score is linear, the net operation is just a linear filter operation on the original image. Therefore it is very natural to consider *non*linear evolution equations on orientation scores.

## 2.1 An Invertible Orientation Score Transformation

To transform images to orientation scores using (2) for the purpose of nonlinear diffusion we need a kernel $K$ with the following properties

1. A finite number of orientations.
2. Reconstruction by summing all orientations.
3. Directional kernel, i.e. the kernel should be a convex cone in the Fourier domain [9].
4. Localization in the spatial domain.
5. Quadrature property [10]. This is especially useful since the absolute value $|U|$ of the resulting complex-valued orientation score will render a *phase invariant* signal responding to both edges and ridges.

Based on these properties we propose the following kernel

$$K(\mathbf{x}) = \frac{1}{N}\mathcal{F}^{-1}\left[\boldsymbol{\omega} \mapsto B^k\left(\frac{(\varphi \bmod 2\pi) - \pi/2}{s_\theta}\right) f(\rho)\right](\mathbf{x})\, G_s(\mathbf{x}) \tag{3}$$

where $N$ is the normalization constant, $\boldsymbol{\omega} = (\rho\cos\varphi, \rho\sin\varphi)$, $B^k$ denotes the $k$th order B-spline given by

$$B^k(x) = (B^{k-1} * B^0)(x), \qquad B^0(x) = \begin{cases} 1 & \text{if } -1/2 < x < +1/2 \\ 0 & \text{otherwise} \end{cases}. \tag{4}$$

Function $f(\rho)$ specifies the radial function in the Fourier domain, chosen as the Gaussian divided by its Taylor series up to order $q$ to ensure a slower decay, i.e.

$$f(\rho) = G_t(\rho)\left(\sum_{i=0}^{q}\left(\frac{d}{d\rho'}G_t(\rho')\bigg|_{\rho'=0}\right)\frac{\rho^i}{i!}\right)^{-1}, \quad G_t(\rho) = \frac{1}{2\sqrt{\pi t}}e^{-\frac{\rho^2}{4t}}. \tag{5}$$

Function $G_s$ in (3) is a Gaussian kernel with scale $s$, which ensures spatial locality. Figure 1 shows an example of this orientation score transformation.

## 3 Diffusion on the Euclidean Motion Group

### 3.1 Left-Invariant Derivatives

We want to construct the diffusion equation from *left-invariant* differential operators on orientation scores. An operator $\Upsilon : \mathbb{L}_2(SE(2)) \to \mathbb{L}_2(SE(2))$ is left-invariant if $\mathcal{L}_g \Upsilon U = \Upsilon \mathcal{L}_g U$, for all $g \in G$ and for all $U \in \mathbb{L}_2(SE(2))$, where $\mathcal{L}_g : \mathbb{L}_2(SE(2)) \to \mathbb{L}_2(SE(2))$ is given by $(\mathcal{L}_g U)(h) = U(g^{-1}h)$. This property is important because a left-invariant operator in an orientation score implies that the net operation on the corresponding image is *rotation invariant.* The differential operators $\partial_x$ and $\partial_y$ (we will consistently use the shorthand notation $\partial_x$ for derivative operators corresponding to partial derivative $\partial/\partial x$) on the orientation score are not left-invariant and are therefore unsuitable. However, the differential operators $\{\partial_\xi, \partial_\eta, \partial_\theta\}$, where

$$\partial_\xi(g) = \cos\theta\,\partial_x + \sin\theta\,\partial_y, \quad \partial_\eta(g) = -\sin\theta\,\partial_x + \cos\theta\,\partial_y, \quad \partial_\theta(g) = \partial_\theta, \tag{6}$$

with $g = (\mathbf{x}, \theta)$, are all left-invariant, see [11] for a derivation. Consequently, all combinations of the operators $\{\partial_\xi, \partial_\eta, \partial_\theta\}$ are also left-invariant. The tangent space at $g$ is spanned by $\{\partial_\xi, \partial_\eta, \partial_\theta\}$. To distinguish between the derivative operator at $g$ and the basis of the tangent space at $g$ we will use the following notation for the latter

$$\{\mathbf{e}_\xi(g), \mathbf{e}_\eta(g), \mathbf{e}_\theta(g)\} = \{\cos\theta\,\mathbf{e}_x + \sin\theta\,\mathbf{e}_y, -\sin\theta\,\mathbf{e}_x + \cos\theta\,\mathbf{e}_y, \mathbf{e}_\theta\}. \tag{7}$$

For notational simplicity the dependency on $g$ is omitted further on.

It is very important to note that not all the derivatives $\{\partial_\xi, \partial_\eta, \partial_\theta\}$ commute. The nonzero commutators (definition $[A, B] = AB - BA$) are given by

$$[\partial_\theta, \partial_\xi] = \partial_\eta, \qquad [\partial_\theta, \partial_\eta] = -\partial_\xi. \tag{8}$$

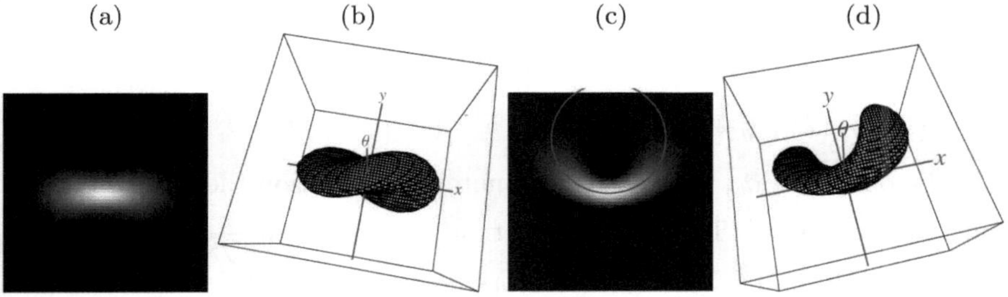


**Fig. 2.** Illustrations of Green's functions for different parameter values, obtained using an explicit iterative numerical scheme (Section 7) with end time $t = 70$. (a) Shows the effect of a nonzero $D'_{11}$ in the spatial plane, i.e. all orientations are summed. Parameters $D'_{11} = 0.003$, $D_{22} = 1$, $D_{33} = 0$ and $\kappa = 0$. (b) Isosurface of (a) in the orientation score. (c) Shows the effect of nonzero $\kappa$. The superimposed circle shows the curvature. Parameters $D'_{11} = 0$, $D_{22} = 1$, $D_{33} = 0$ and $\kappa = 0.06$. (d) Isosurface of (c) in the orientation score, showing the typical spiral shape of the Green's function.

### 3.2 Diffusion Equation

The general diffusion equation for orientation scores using left-invariant derivative operators is

$$\partial_t u = \begin{pmatrix} \partial_\theta & \partial_\xi & \partial_\eta \end{pmatrix} \begin{pmatrix} D_{11} & D_{12} & D_{13} \\ D_{21} & D_{22} & D_{23} \\ D_{31} & D_{32} & D_{33} \end{pmatrix} \begin{pmatrix} \partial_\theta \\ \partial_\xi \\ \partial_\eta \end{pmatrix} u = \mathcal{A}\, u \tag{9}$$

with $u(\mathbf{x}, \theta; 0) = U(\mathbf{x}, \theta)$, and $u(\mathbf{x}, 0; t) = u(\mathbf{x}, 2\pi; t)$. This equation constitutes a scale space on SE(2) [8]. The solution can be written as $u(\cdot, \cdot; t) = e^{t\,\mathcal{A}} U$.

In practice, it makes no sense to consider the full diffusion tensor. If we want the diffusion to be optimal for straight lines with any orientation, we only have to consider the diagonal elements. In that case $D_{22}$ determines the diffusion along the line structure, $D_{33}$ determines the diffusion orthogonal to the line structure, and $D_{11}$ accounts for diffusion between different orientations. For curved lines, diffusion with a diagonal diffusion tensor is not optimal. We can obtain a diffusion process with a curvature $\kappa$ by replacing $\partial_\xi$ in the diagonal diffusion equation by $\partial_\xi + \kappa\, \partial_\theta$ (i.e. the generator of a curved line), yielding

$$\partial_t u = \begin{pmatrix} \partial_\theta & \partial_\xi & \partial_\eta \end{pmatrix} \begin{pmatrix} D'_{11} + D_{22}\kappa^2 & D_{22}\kappa & 0 \\ D_{22}\kappa & D_{22} & 0 \\ 0 & 0 & D_{33} \end{pmatrix} \begin{pmatrix} \partial_\theta \\ \partial_\xi \\ \partial_\eta \end{pmatrix} u. \tag{10}$$

When $\kappa$ is nonzero, the resulting kernels will be curved in the image plane. Figure 2 shows examples of Green's functions of linear evolutions of this type.

## 4 Using Gaussian Derivatives in Orientation Scores

Regularized derivatives on the orientation score are operationalized by $\mathcal{D}e^{t\mathcal{A}}\, u$ where $\mathcal{D}$ is a derivative of any order constructed from $\{\partial_\xi, \partial_\eta, \partial_\theta\}$. The order of the regularization operator and differential operators matters in this case, i.e. the diffusion should come first.

In this paper we restrict ourselves to $D_{22} = D_{33}$ and $\kappa = 0$, leading to

$$\partial_t u = \left(D_{11}\partial_\theta^2 + D_{22}\left(\partial_\xi^2 + \partial_\eta^2\right)\right) u = \left(D_{11}\partial_\theta^2 + D_{22}\left(\partial_x^2 + \partial_y^2\right)\right) u \tag{11}$$

Since the operators $\partial_\theta$, $\partial_x$, and $\partial_y$ commute, this equation the same as the diffusion equation in $\mathbb{R}^3$. The Green's function is a Gaussian $\frac{1}{8\sqrt{\pi^3 t_s^2 t_o}} e^{-\frac{x^2+y^2}{4t_s} - \frac{\theta^2}{4t_o}}$ where $t_o = t\,D_{11}$ and $t_s = t\,D_{22}$. In this special case we can use standard (separable) implementations of Gaussian derivatives, but we have to be careful because of the non-commuting operators. A normal $(i, j, k)$th order Gaussian derivative implementation for a 3D image $f$ is

$$\partial_x^i \partial_y^j \partial_z^k e^{t(\partial_x^2+\partial_y^2+\partial_z^2)} f = \partial_x^i e^{t\,\partial_x^2} \partial_y^j e^{t\,\partial_y^2} \partial_z^k e^{t\,\partial_z^2} f, \tag{12}$$

where the equality between the left and right side is essential, since it implies separability along the three dimensions. We want to use the same implementations to construct Gaussian derivatives in the orientation scores, meaning that we have to ensure that the same permutation of differential operators and regularization operators is allowed. By noting that

$$\begin{aligned} \partial_\xi^i\, \partial_\eta^j\, \partial_\theta^k\, e^{t_o \partial_\theta^2 + t_s(\partial_\xi^2+\partial_\eta^2)} &= \partial_\xi^i \partial_\eta^j\, e^{t_s(\partial_x^2+\partial_y^2)}\, \partial_\theta^k\, e^{t_o \partial_\theta^2}, \\ \partial_\theta^k\, \partial_\xi^i\, \partial_\eta^j\, e^{t_o \partial_\theta^2 + t_s(\partial_\xi^2+\partial_\eta^2)} &\neq \partial_\theta^k\, e^{t_o \partial_\theta^2}\, \partial_\xi^i\, \partial_\eta^j\, e^{t_s(\partial_x^2+\partial_y^2)}, \end{aligned} \tag{13}$$

we conclude that we always should ensure a certain *ordering* of the derivative operators, i.e. one should first calculate the orientational derivative $\partial_\theta$ and then the commuting spatial derivatives $\{\partial_\xi, \partial_\eta\}$, which are calculated from the Cartesian derivatives $\{\partial_x, \partial_y\}$ using (6). The commutator relations of (8) allow to rewrite the derivatives in this canonical order. For instance, the derivative $\partial_\xi \partial_\theta$ can be calculated directly with Gaussian derivatives, while $\partial_\theta \partial_\xi$ should be operationalized as $\partial_\xi \partial_\theta + \partial_\eta$.

Note that one has to be careful with the sampled $\theta$ dimensions of the orientation score. One should ensure to make both the scale $t_s$ and the derivatives $\partial_\theta^k$ dimensionless and consequently independent on the sampling step $s_\theta$.

## 5 Curvature Estimation in Orientation Scores

Before turning to nonlinear diffusion, we first discuss how to estimate curvature from orientation scores. Our procedure to measure this is inspired by van Ginkel [12]. Suppose we have at position $g_0$ a tangent vector $\mathbf{v}(g_0) = v_\theta \mathbf{e}_\theta + v_\xi \mathbf{e}_\xi + v_\eta \mathbf{e}_\eta$. Similar to the concept of a tangent line in $\mathbb{R}^3$ we can define a "tangent spiral" in an orientation score by means of the exponential map. The parametrization $\mathbf{h} : \mathbb{R} \to \mathbb{R}^2 \rtimes \mathbb{T}$ of this spiral at $g_0 = (\mathbf{x}_0, \theta_0)$ with tangent vector $\mathbf{v}(g_0)$ is given by (if $v_\theta \neq 0$)

$$\mathbf{h}(t) = e^{t\mathbf{v}(g_0)} = \left(\mathbf{x}_0 + \frac{1}{v_\theta}\left(\mathbf{R}_{v_\theta t+\theta_0-\pi/2} - \mathbf{R}_{\theta_0-\pi/2}\right)\begin{pmatrix} v_\xi \\ v_\eta \end{pmatrix}, v_\theta t + \theta_0\right) \tag{14}$$

We are interested in the curvature on the spatial plane, so we project $\mathbf{h}(t)$ to the $\mathbb{R}^2$ plane $\mathbf{x}(t) = \mathbb{P}_{\mathbb{R}^2}\mathbf{h}(t)$. The curvature in this plane is given by

$$\boldsymbol{\kappa}(t) = \frac{d^2}{ds^2}\mathbf{x}(t(s)) = \frac{v_\theta}{v_\eta^2 + v_\xi^2}\begin{pmatrix} v_\eta \cos(tv_\theta + \theta_0) + v_\xi \sin(tv_\theta + \theta_0) \\ v_\xi \cos(tv_\theta + \theta_0) - v_\eta \sin(tv_\theta + \theta_0) \end{pmatrix} \tag{15}$$

where $s$ is the parameterization such that $||\frac{d}{ds}x(t(s))|| = 1$. The signed norm of the curvature vector is

$$\kappa = ||\boldsymbol{\kappa}|| \operatorname{sign}(\boldsymbol{\kappa} \cdot \mathbf{e}_\eta) = \frac{-v_\theta}{\sqrt{v_\eta^2 + v_\xi^2}} \tag{16}$$

This result has an intuitive interpretation: the curvature is equal to the slope at which the curve in the orientation score meets the spatial plane spanned by $\{\mathbf{e}_\xi, \mathbf{e}_\eta\}$.

Ideally, $v_\eta = 0$ because by construction oriented structures are orthogonal to $\mathbf{e}_\eta$. In practice, however, assuming $v_\eta = 0$ leads to a biased curvature estimate if the orientation $\theta$ deviates from the true orientation of an oriented structure, which occurs frequently since an oriented structure will always cause a response within a certain range of orientations.

How to find the vector field $\mathbf{v}$ from an orientation score $u$? A curve or oriented pattern appears in the phase-invariant representation of the orientation score cf. Section 2.1 as a ridge. Therefore we calculate the Hessian, which is defined by

$$\mathbf{H}(u) = \begin{pmatrix} \partial_\theta^2|u| & \partial_\xi\partial_\theta|u| & \partial_\eta\partial_\theta|u| \\ \partial_\theta\partial_\xi|u| & \partial_\xi^2|u| & \partial_\eta\partial_\xi|u| \\ \partial_\theta\partial_\eta|u| & \partial_\xi\partial_\eta|u| & \partial_\eta^2|u| \end{pmatrix} = \begin{pmatrix} \partial_\theta^2|u| & \partial_\xi\partial_\theta|u| & \partial_\eta\partial_\theta|u| \\ \partial_\xi\partial_\theta|u| + \partial_\eta|u| & \partial_\xi^2|u| & \partial_\xi\partial_\eta|u| \\ \partial_\eta\partial_\theta|u| - \partial_\xi|u| & \partial_\xi\partial_\eta|u| & \partial_\eta^2|u| \end{pmatrix}, \tag{17}$$

where Gaussian derivatives are used with scales $t_s$ and $t_o$ using the canonical ordering in the expression on the right. Note that the Hessian matrix is not symmetric because of the torsion of the space, implying that we can get complex-valued eigenvalues and eigenvectors. However, we can still find local and global extrema of $\{||\mathbf{H}\,\mathbf{a}||^2 \mid ||\mathbf{a}||^2 = 1\}$ with $\mathbf{a} = (x, y, \theta)$. Now by Lagrange these extrema satisfy $\nabla_{\mathbf{a}}||\mathbf{H}\,\mathbf{a}||^2 = \nabla_{\mathbf{a}}(\mathbf{a}^{\mathrm{T}}\mathbf{H}^{\mathrm{T}}\mathbf{H}\,\mathbf{a}) = 2\,\mathbf{H}^{\mathrm{T}}\mathbf{H}\,\mathbf{a} = 2\,\lambda||\mathbf{a}||$. Therefore we apply eigen analysis on $\mathbf{H}^{\mathrm{T}}\mathbf{H}$ rather than $\mathbf{H}$. An oriented structure will lie approximately within the 2D plane spanned by $\{\mathbf{e}_\xi, \mathbf{e}_\theta\}$, so the two eigenvectors of $\mathbf{H}^{\mathrm{T}}\mathbf{H}$ that are closest to the plane are selected by leaving out the eigenvector with the largest $\mathbf{e}_\eta$ component. From these two eigenvectors the one with the smallest eigenvalue is tangent to the oriented structure and is used to estimate the curvature with (16).

## 6 Conductivity Functions for Nonlinear Diffusion

At positions in the orientation score with a strongly oriented structure we only want to diffuse tangent to this structure, i.e. $D_{22}$ should be large, $D'_{11}$ and $D_{33}$ should be small, and the curvature measurement of the previous section should be taken into account. If there is no strong orientation, the diffusion should be isotropic in the spatial plane, i.e. $D_{22} = D_{33}$ should be large as well as $D'_{11}$. Curvature is not defined at such positions so $\kappa = 0$.

If an oriented structure is present at a position in the orientation score, one eigenvalue of the Hessian of $|u|$ will have a large negative real part. Therefore we propose as measure for the presence of oriented structures

$$s(\mathbf{x}, \theta) = \mathrm{Max}(-\mathrm{Re}(\lambda_1(\mathbf{x}, \theta)), 0) \tag{18}$$

where $\lambda_1$ denotes the largest eigenvalue of the Hessian at every position. In the equation for the Hessian (17) we substitute $\partial_\theta \leftarrow \gamma\, \partial_\theta$ where $\gamma$ is a parameter with unit 1/pixel that is necessary to make the units of all Hessian components 1/pixel$^2$. For the conductivity functions we propose

$$\begin{aligned} D_{33}(\mathbf{x}, \theta) &= \exp\left(-\frac{s(\mathbf{x}, \theta)}{c}\right); \quad D'_{11}(\mathbf{x}, \theta) = \epsilon_{11} D_{33}(\mathbf{x}, \theta); \\ \kappa(\mathbf{x}, \theta) &= \left(1 - \exp\left(-\left(\frac{d_\kappa}{D_{33}(\mathbf{x}, \theta)}\right)^4\right)\right) \kappa_{\mathrm{est}}(\mathbf{x}, \theta); \quad D_{22}(\mathbf{x}, \theta) = 1. \end{aligned} \tag{19}$$

where the nonlinear function for $D_{33}$ makes the separation between isotropic and oriented regions stronger. The function is chosen such that the result is always between 0 and 1 for $s \geq 0$. The nonlinear function for $\kappa$ is chosen such that it puts a soft threshold determining whether to include the curvature estimate $\kappa_{\mathrm{est}}$ depending on the value of $D_{33}$. There are six parameters involved: $c$ controls the behavior of the nonlinear $e$-power, $\gamma$ controls the weight factor of the $\theta$ derivatives, $\epsilon_{11}$ controls the strength of the diffusion in $\theta$ direction in isotropic regions, $d_\kappa$ determines the soft threshold on including curvature, and $t_s$ and $t_o$ are the two scale parameters.

## 7 Numerical Scheme

We propose an explicit finite difference scheme to solve diffusion equation (10). Since the PDE on the orientation score is highly anisotropic we require good rotational invariance. Many efficient numerical schemes proposed in literature, e.g. the AOS (additive operator splitting) scheme [4], are therefore discarded since they show poor rotation invariance. The LSAS scheme [13] has good rotational invariance, but it is not straightforward to make a 3D version. The scheme in [14] suffers from checkerboard artefacts.

An important property of the differential operators $\partial_\xi$, $\partial_\eta$, and $\partial_\theta$ is their left-invariance. The performance of a numerical scheme will therefore be more optimal if this left-invariance is carried over to the finite differences that are used. To achieve this we should define the spatial finite differences in the directions defined by the left-invariant $\mathbf{e}_\xi$, $\mathbf{e}_\eta$ tangent basis vectors, instead of the sampled $\mathbf{e}_x, \mathbf{e}_y$ grid. In effect, the principal axes of diffusion in the spatial plane are always aligned with the finite differences.

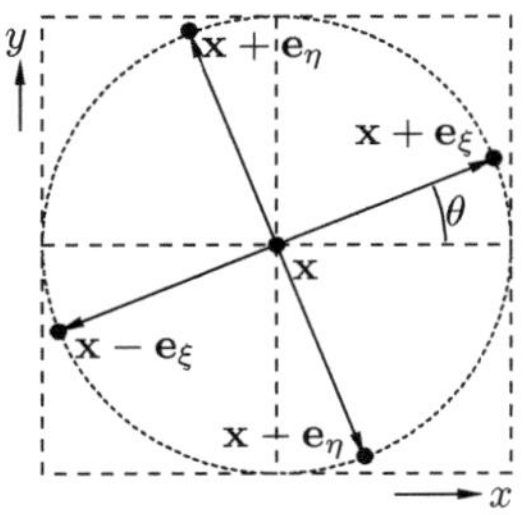


$$\partial_\theta u \approx \frac{1}{2s_\theta}\left(u(\mathbf{x}, l+1) - u(\mathbf{x}, l-1)\right)$$

$$\partial_\theta^2 u \approx \frac{1}{s_\theta^2}\left(u(\mathbf{x}, l+1) - 2u(\mathbf{x}, l) + u(\mathbf{x}, l-1)\right)$$

$$\partial_\xi u \approx \frac{1}{2}\left(u(\mathbf{x} + \mathbf{e}_\xi^l, l) - u(\mathbf{x} - \mathbf{e}_\xi^l, l)\right)$$

$$\partial_\xi^2 u \approx u(\mathbf{x} + \mathbf{e}_\xi^l, l) - 2u(\mathbf{x}, l) + u(\mathbf{x} - \mathbf{e}_\xi^l, l)$$

$$\partial_\eta u \approx \frac{1}{2}\left(u(\mathbf{x} + \mathbf{e}_\eta^l, l) - u(\mathbf{x} - \mathbf{e}_\eta^l, l)\right)$$

$$\partial_\eta^2 u \approx u(\mathbf{x} + \mathbf{e}_\eta^l, l) - 2u(\mathbf{x}, l) + u(\mathbf{x} - \mathbf{e}_\eta^l, l)$$

$$\partial_\xi \partial_\theta u \approx \frac{1}{4s_\theta}\left(u(\mathbf{x} + \mathbf{e}_\xi^l, l+1) - u(\mathbf{x} + \mathbf{e}_\xi^l, l-1) - u(\mathbf{x} - \mathbf{e}_\xi^l, l+1) + u(\mathbf{x} - \mathbf{e}_\xi^l, l-1)\right)$$

$$\partial_\theta \partial_\xi u \approx \frac{1}{4s_\theta}\left(u(\mathbf{x} + \mathbf{e}_\xi^{l+1}, l+1) - u(\mathbf{x} + \mathbf{e}_\xi^{l+1}, l-1) - u(\mathbf{x} - \mathbf{e}_\xi^{l-1}, l+1) + u(\mathbf{x} - \mathbf{e}_\xi^{l-1}, l-1)\right)$$

**Fig. 3.** Illustration of the spatial part of the stencil of the numerical scheme. The horizontal and vertical dashed lines indicate the sampling grid, which is aligned with $\{\mathbf{e}_x, \mathbf{e}_y\}$. The stencil points, indicated by the black dots, are aligned with the rotated coordinate system cf. (7) with $\theta = l\, s_\theta$.

For our numerical scheme we apply the chain rule on the right-hand side of the PDE (10) (i.e., analog to 1D: $\partial_x D\, \partial_x u = D\, \partial_x^2 u + (\partial_x D)(\partial_x u)$) and the derivatives are replaced by the finite differences defined in Figure 3. In time direction we use the first order forward finite difference, i.e. $(u^{k+1} - u^k)/\tau$ where $k$ is the discrete time and $\tau$ the time step. Interpolation is required at spatial positions $\mathbf{x} \pm \mathbf{e}_\xi$ and $\mathbf{x} \pm \mathbf{e}_\eta$. For this purpose we use the algorithms for B-spline interpolation proposed by Unser et al. [15] with B-spline order 3. This interpolation algorithm consists of a prefiltering step with a separable IIR filter to determine the B-spline coefficients. The interpolation images such as $u^k(\mathbf{x} \pm \mathbf{e}_\xi)$ can then be calculated by a separable convolution with a shifted B-spline. The examples in Figure 2 and all experiments in the next section are obtained with this numerical scheme.

The drawback of this explicit scheme is the numerical stability. An analysis on stability is difficult due to the interpolation step. From experiments, we conclude that one should choose $\tau \leq 0.25$ to ensure numerical stability.

## 8 Experiments

In this section we compare the results of coherence enhancing diffusion in the orientation score (CED-OS) with results obtained by the normal coherence enhancing diffusion (CED) approach [4] where we use the LSAS numerical scheme with [13] since this has particularly good rotation invariance.

In all experiments we construct orientation scores with period $2\pi$ with $n_\theta = 64$. The following parameters are used for the orientation score transformation (Section 2.1): $k = 2$, $q = 8$, $t = 1000$, and $s = 200$. These parameters are chosen such that the reconstruction is visually indistinguishable from the original. Since computational speed was not our main concern, we use a small time step of $\tau = 0.1$ to ensure numerical stability. The parameters for the nonlinear diffusion

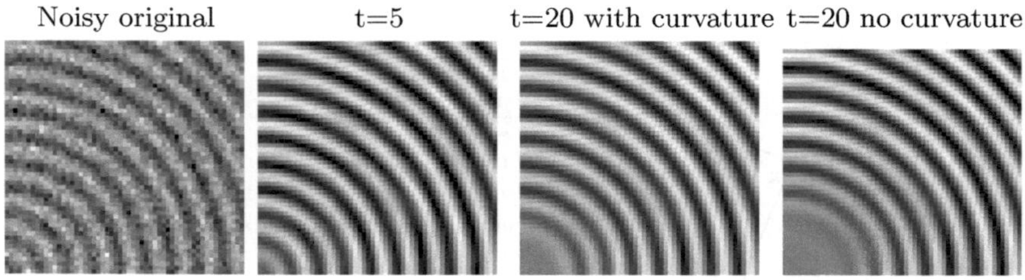


**Fig. 4.** Shows the effect of including curvature on a noisy test image in CED-OS. At $t = 5$ the results with and without curvature are visually indistinguishable. At $t = 20$ the effect is visible: higher curvatures are better preserved when curvature is included.

in $SE(2)$ for all experiments are: $\epsilon_{11} = 0.001$, $t_s = 12$, $t_o = 0.04$, $\gamma = 0.05$, $c = 0.08$, and $d_\kappa = 0.13$. Note that the resulting images we will show of CED-OS do not represent the evolving orientation score, but only the reconstructed image (i.e. after summation over all orientations).

The parameters that we used for CED are (see [4]): $\sigma = 1$, $\rho = 1$ (artificial images) or $\rho = 6$ (medical image), $C = 1$, and $\alpha = 0.001$. The artificial images all have a size of $56 \times 56$ and a range of 0 to 255.

Figure 4 shows CED-OS with and without including curvature. As expected, the noise is removed while the line structures are well-preserved. At time $t = 5$ no visible differences are observed in the resulting image reconstructions so only the result with curvature is shown. At $t = 20$, however, the difference is visible: when curvature is included the preservation of the high-curvature inner circles is better. Still, in all cases the smallest circles are blurred isotropically. This is

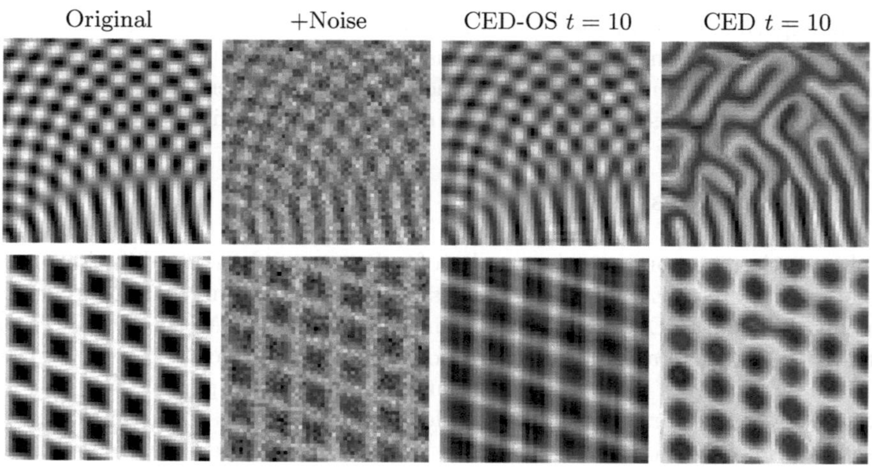


**Fig. 5.** Shows the typical different behavior of CED-OS compared to CED. In CED-OS crossing structures are better preserved.

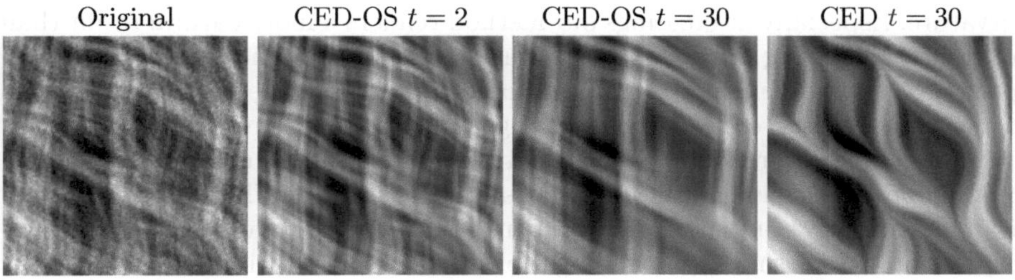


**Fig. 6.** Shows results on an image constructed from two rotated 2-photon images of collagen tissue in the heart. At $t = 2$ we obtain a nice enhancement of the image. Comparing with $t = 30$ a nonlinear scale-space behavior can be seen. For comparison, the right column shows the behavior of CED.

due to smaller response of the Hessian on curved lines, causing the value for $D_{33}$ to be larger on high-curvature circles.

Note that CED will also perform good on the image in Figure 4. The difference in behavior becomes apparent if we consider images with crossing line structures. This is shown in Figure 5. The upper image shows an additive superimposition of two images with concentric circles. Our method is able to preserve this structure, while CED can not. The same holds for the lower image with crossing straight lines, where it should be noted that our method leads to amplification of the crossings, which is because the lines in the original image are not superimposed linearly.

Figure 6 shows the results on an image of collagen fibres obtained using 2-photon microscopy. These kind of images are acquired in tissue engineering research, where the goal is to create artificial heart valves. The image shows an artificial superposition of the same image with two different rotations, for the purpose of this experiment. This is not entirely artificial, since there exist collagen structures with this kind of properties. The parameters during these experiments were set the same as the artificial images. The image size is $160 \times 160$.

## 9 Conclusions

In this paper we introduced nonlinear diffusion on the Euclidean motion group. Starting from a 2D image, we constructed a three-dimensional orientation score using rotated versions of a directional quadrature filter. We considered the orientation score as a function on the Euclidean motion group and defined the left-invariant diffusion equation. We showed how one can use normal Gaussian derivatives to calculate regularized derivatives in the orientation score. The nonlinear diffusion is steered by estimates for oriented feature strength and curvature that are obtained from Gaussian derivatives. Furthermore, we proposed to use finite differences that approximate the left-invariance of the derivative operators.

The experiments show that we are indeed able to enhance elongated patterns in images and that including curvature helps to enhance lines with large curvature. Especially at crossings our method renders a more natural result than coherence enhancing diffusion. The diffusion shows the typical nonlinear scale-space behavior when increasing time: blurring occurs, but the important features of images are preserved over a longer range of time. Furthermore we showed that including curvature renders better results on curved line structures.

Some problems should still be addressed in future work. The numerical algorithm is currently computationally expensive due to the small time step and interpolation. Furthermore, embedding the nonlinear diffusion in orientation scores in the variational framework may lead to better control on the behavior of the evolution equations. Finally, it would be interesting to extend this approach to the similitude group, i.e. to use multi-scale *and* multi-orientation simultaneously to resolve the problem of selecting the appropriate scale.

## References


1. Iijima, T.: Basic theory of pattern observation. Papers of Technical Group on Automata and Automatic Control, IECE, Japan (1959) (in Japanese).
2. Witkin, A.: Scale-space filtering. In: 8th Int. Joint Conf. Artificial Intelligence. (1983) 1019–1022
3. Perona, P., Malik, J.: Scale-space and edge detection using anisotropic diffusion. IEEE Trans. Pattern Analysis and Machine Intelligence **12** (1990) 629–639
4. Weickert, J.: Coherence-enhancing diffusion filtering. Int. J. Comp. Vision **31**(2–3) (1999) 111–127
5. Kalitzin, S.N., ter Haar Romeny, B.M., Viergever, M.A.: Invertible orientation bundles on 2D scalar images. In ter Haar Romeny, B.M., Florack, L., Koenderink, J., Viergever, M., eds.: Scale Space Theory in Computer Vision. (1997) 77–88
6. Duits, R., Felsberg, M., Granlund, G., ter Haar Romeny, B.: Image analysis and reconstruction using a wavelet transform constructed from a reducible representation of the euclidean motion group. Int. J. Comp. Vision **72**(1) (2007) 79–102
7. Brox, T., Weickert, J., Burgeth, B., Mrázek, P.: Nonlinear structure tensors. Image and Vision Computing **24**(1) (2006) 41–55
8. Duits, R., Burgeth, B.: Scale spaces on Lie groups. In: Conference on Scale Space and Variational Methods in Computer Vision (SSVM-07). (2007)
9. Antoine, J.P., Murenzi, R.: Two-dimensional directional wavelets and the scale-angle representation. Signal Processing **52**(3) (1996) 241–272
10. Granlund, G.H., Knutsson, H.: Signal Processing for Computer Vision. Kluwer Academic Publisher (1995)
11. Duits, R., van Almsick, M.: The explicit solutions of linear left-invariant second order stochastic evolution equations on the 2D-euclidean motion group. To appear in: Quarterly of Applied Mathematics, American Mathetical Society (2007)
12. van Ginkel, M.: Image Analysis using Orientation Space based on Steerable Filters. PhD thesis, Technische Universiteit Delft, The Netherlands (2002)
13. Welk, M., Weickert, J., Steidl, G.: From tensor-driven diffusion to anisotropic wavelet shrinkage. In: ECCV 2006. (2006) 391–403
14. Weickert, J., Scharr, H.: A scheme for coherence-enhancing diffusion filtering with optimized rotation invariance. Journal of Visual Communication and Image Representation (2002) 103–118
15. Unser, M.: Splines: A perfect fit for signal and image processing. IEEE Signal Processing Magazine **16**(6) (1999) 22–38

# A TV-Stokes Denoising Algorithm

Talal Rahman[1], Xue-Cheng Tai[1], and Stanley Osher[2]

[1] Department of Mathematics, University of Bergen
CIPR, Allégt. 41, 5007 Bergen, Norway
talal.rahman@mi.uib.no,tai@mi.uib.no
[2] Department of Mathematics, UCLA, California, USA
sjo@math.ucla.edu

**Abstract.** In this paper, we propose a two-step algorithm for denoising digital images with additive noise. Observing that the isophote directions of an image correspond to an incompressible velocity field, we impose the constraint of zero divergence on the tangential field. Combined with an energy minimization problem corresponding to the smoothing of tangential vectors, this constraint gives rise to a nonlinear Stokes equation where the nonlinearity is in the viscosity function. Once the isophote directions are found, an image is reconstructed that fits those directions by solving another nonlinear partial differential equation. In both steps, we use finite difference schemes to solve. We present several numerical examples to show the effectiveness of our approach.

## 1 Introduction

A digital image $d$ is a function defined on a two dimensional rectangular domain $\Omega \subset \mathbf{R}^2$ where $d(\mathbf{x})$ represents the gray-level value of the image, associated with the pixel at $\mathbf{x} = (x, y) \in \Omega$. Let $d_0$ be the observed image (the given data) which contains some additive noise $\eta$, in other words,

$$d_0(\mathbf{x}) = d(\mathbf{x}) + \eta(\mathbf{x}), \tag{1}$$

where $d$ is the true image. The problem is then to recover the true image from the given data $d_0$. This is a typical example of an inverse problem, a solution of which is normally sought through the minimization of an energy functional consisting of a fidelity term and a regularization term (a smoothing term). The classical Tikhonov regularization involving the $H^1$ seminorm, is quite effective for smooth functions, but behaves poorly when the function $d(\mathbf{x})$ includes discontinuities or steep gradients, like edges and textures. The famous model based on the TV-norm regularization, proposed by Rudin-Osher-Fatemi (ROF) in [14], has proven to be quite effective for such functions, removing noise without causing excessive smoothing of the edges. However, it is well known that the TV-norm regularization suffers from the so-called stair-case effect, which may produce undesirable blocky images. Several methods have been proposed since the ROF model, see for instance in [6,8,9,11,12,13,16].

F. Sgallari, A. Murli, and N. Paragios (Eds.): SSVM 2007, LNCS 4485, pp. 473–483, 2007.

Recently, a two step method has been proposed by Lysaker-Osher-Tai (LOT) in [11] involving a smoothing of the normal vectors $\nabla d_0/|\nabla d_0|$ in the first step, and then finding a surface to fit the smoothed normals in the second step based on ideas from [4,2,16]. In this paper we use the same two-step approach, but we modify the first step being motivated by the observation that tangent directions to the isophote lines (lines along which the intensity is constant) correspond to an incompressible velocity field, see [3,15] where this observation was used to develop effective algorithms for the image inpainting. The aim of this paper is to extend this idea further into developing an effective algorithm for the image denoising. Instead of smoothing the normal field in the first step, we smooth the tangential field imposing the constraint that the field is divergence free (incompressible). This results into an algorithm that generates smooth isophote lines turning noisy images into smooth and visually pleasant denoised images, and preserves the edges quite well.

As the algorithm is still in its early stage of research, so far, we have only been interested in its qualitative nature, and not much in the convergence speed. As a result, we have only been using straight forward explicit schemes for the discrete solution. Search for a faster algorithm constitutes part of our future plans.

The paper is organized as follows. In Section 2, we present our two-step algorithm, and include a brief description of the numerical explicit scheme involved in each step. Numerical experiments showing its performance are presented in Section 3.

## 2 The Denoising Algorithm

Given an image $d$, the normal and the tangential vectors of the level curves (or the isophote lines) are given by $\mathbf{n} = \nabla d(\mathbf{x}) = (d_x, d_y)^T$ and $\tau = \nabla^{\perp} d = (-d_y, d_x)^T$. The vector fields then satisfy the following conditions: $\nabla \cdot \tau = 0$ and $\nabla \times \mathbf{n} = 0$, the first one being called the incompressibility condition in the fluid mechanics, a natural condition to use in our algorithm.

Let the noisy image $d_0$ be given. We compute $\tau_0 = \nabla^{\perp} d_0$. The algorithm is then defined in two steps. In the first step, we solve the following minimization problem.

$$\min_{\tau} \int_{\Omega} |\nabla \tau| \, d\mathbf{x} \; + \; \frac{\delta}{2} \int_{\Omega} |\tau - \tau_0|^2 \, d\mathbf{x} \qquad \text{subject to} \qquad \nabla \cdot \tau = 0, \tag{2}$$

where $\delta$ is a constant which is used to balance between the smoothing of the tangent field and the fidelity to the original tangent field. The gradient matrix and its norm, of the tangent vector $\tau = (v, u)$, are defined as

$$\nabla \tau = \begin{pmatrix} \nabla v \\ \nabla u \end{pmatrix}, \qquad |\nabla \tau| = \sqrt{v_x^2 + v_y^2 + u_x^2 + u_y^2}, \tag{3}$$

respectively.Once we have the smoothed tangent field, we can get the corresponding normal field $\mathbf{n} = (u, -v)$. In the second step, we reconstruct our image by fitting it to the normal field through solving the following minimization problem.

$$\min_{d} \int_{\Omega} \left( |\nabla d| - \nabla d \cdot \frac{\mathbf{n}}{|\mathbf{n}|} \right) d\mathbf{x} \qquad \text{subject to} \qquad \int_{\Omega} (d - d_0)^2 \, d\mathbf{x} = \sigma^2, \tag{4}$$

where $\sigma^2$ is the estimated noise variance. This can be estimated using statistical methods. If the exact noise variance cannot be obtained, then an approximate value may be used. In which case, a larger value would result in over-smoothing and a smaller value would result in under-smoothing.

For the discretization, we use a staggered grid, see [15] for some more details. Each vertex of the rectangular grid corresponds to the position of a pixel or pixel center where the image intensity variable $d$ is defined. Let the horizontal axis and the vertical axis represent the x-axis and the y-axis, respectively. The variables $v$ and $u$, i.e. the components of the tangential vector $\tau$, corresponding to the $-d_y$ and $d_x$, are then defined respectively along the vertical and the horizontal edges of the grid. Further, we approximate the derivatives by finite differences using the standard forward/backward difference operators $D_x^{\pm}$ and $D_y^{\pm}$, and the centered difference operators $C_x^h$ and $C_y^h$ respectively in the $x$ and $y$ direction where $h$ correspond to the $h-$spacing.

## 2.1 Step 1: Tangent Field Smoothing

A method of augmented Lagrangian [7] is used for the solution of (2), where we use a Lagrange multiplier to deal with the constraint $\nabla \cdot \tau = 0$, and include a penalty term associated with the same constraint. The corresponding Lagrange functional takes the following form.

$$\mathcal{L}(\tau, \lambda) = \int_{\Omega} |\nabla \tau| \, d\mathbf{x} + \frac{\delta}{2} \int_{\Omega} |\tau - \tau_0|^2 \, d\mathbf{x} + \int_{\Omega} \lambda \nabla \cdot \tau \, d\mathbf{x} + \frac{r}{2} \int_{\Omega} (\nabla \cdot \tau)^2 \, d\mathbf{x}, \quad (5)$$

where $\lambda$ is the Lagrange multiplier and $r$ is a penalty parameter. The optimality condition for the saddle point is the following set of Euler-Lagrange equations,

$$-\nabla \cdot \left( \frac{\nabla \tau}{|\nabla \tau|} \right) + \delta(\tau - \tau_0) - \nabla \lambda - r \nabla (\nabla \cdot \tau) = 0 \qquad \text{in } \Omega, \quad (6)$$

$$\nabla \cdot \tau = 0 \qquad \text{in } \Omega, \quad (7)$$

with the following boundary condition,

$$\left( \frac{\nabla \tau}{|\nabla \tau|} + \lambda I \right) \cdot \nu = \mathbf{0} \qquad \text{on } \partial\Omega, \quad (8)$$

where $\nu$ is the unit outward normal and $I$ is the identity matrix. For the solution we use the method of gradient-descent requiring to solve the following equation to steady-state.

$$\frac{\partial \tau}{\partial t} - \nabla \cdot \left( \frac{\nabla \tau}{|\nabla \tau|} \right) + \delta(\tau - \tau_0) - \nabla \lambda - r \nabla (\nabla \cdot \tau) = 0 \qquad \text{in } \Omega, \quad (9)$$

$$\frac{\partial \lambda}{\partial t} - \nabla \cdot \tau = 0 \qquad \text{in } \Omega, \quad (10)$$

with (8) being the boundary condition, and $t$ being the artificial time variable.

The discrete approximation of (8)-(10) now follows. Let the Lagrange multiplier $\lambda$ be defined at the centers of the rectangles of the grid. We first determine the tangential vector $\tau^0$ as $(v^0, u^0)^T = (-D_y^- d_0, D_x^- d_0)^T$, and take $h$ to be equal to one. The values of the variables $u, v$ and $\lambda$ at step $n+1$ are then calculated from

$$\frac{v^{n+1} - v^n}{\Delta t} = D_x^- \left( \frac{D_x^+ v^n}{T_1^n} \right) + D_y^- \left( \frac{D_y^+ v^n}{T_2^n} \right) - \delta \left( v^n - v_0 \right) \\ + D_x^- \left( \lambda^n + Div(\tau^n) \right), \quad (11)$$

$$\frac{u^{n+1} - u^n}{\Delta t} = D_x^- \left( \frac{D_x^+ u^n}{T_2^n} \right) + D_y^- \left( \frac{D_y^+ u^n}{T_1^n} \right) - \delta \left( u^n - u_0 \right) \\ + D_y^- \left( \lambda^n + Div(\tau^n) \right), \quad (12)$$

$$\frac{\lambda^{n+1} - \lambda^n}{\Delta t} = D_x^+ v^n + D_y^+ u^n, \quad (13)$$

where $Div(\tau^n) = D_x^+ v^n + D_y^+ u^n$ is a discrete divergence operator. For the terms $T_1$ and $T_2$, we introduce two average operators $A_x$ and $A_y$ by defining $A_x w = (w(x,y) + w(x+h,y))/2$ and $A_y w = (w(x,y) + w(x,y+h))/2$. Then

$$T_1 = \sqrt{\left(A_x(C_y^h v^n)\right)^2 + \left(D_x^+ v^n\right)^2 + \left(D_y^+ u^n\right)^2 + \left(A_y(C_x^h u^n)\right)^2 + \epsilon}, \quad (14)$$

$$T_2 = \sqrt{\left(A_y(C_x^h v^n)\right)^2 + \left(D_y^+ v^n\right)^2 + \left(D_x^+ u^n\right)^2 + \left(A_x(C_y^h u^n)\right)^2 + \epsilon}. \quad (15)$$

## 2.2 Step 2: Image Reconstruction

Once we have the value of $\tau = (v, u)^T$ from Step 1 of the algorithm, we use them here to reconstruct our image $d$. Using a Lagrange multiplier $\mu$ for the constraint in (4) we get the following Lagrange functional.

$$\mathcal{L}(d, \mu) = \int_\Omega \left( |\nabla d| - \nabla d \cdot \frac{\mathbf{n}}{|\mathbf{n}|} \right) \, d\mathbf{x} + \mu \left( \int_\Omega (d - d_0)^2 \, d\mathbf{x} - \sigma^2 \right). \quad (16)$$

The corresponding set of Euler-Lagrange equations for the saddle point is

$$-\nabla \cdot \left( \frac{\nabla d}{|\nabla d|} - \frac{\mathbf{n}}{|\mathbf{n}|} \right) + \mu(d - d_0) = 0 \qquad \text{in } \Omega, \quad (17)$$

$$\int_\Omega \left( \frac{d - d_0}{\sigma} \right)^2 d\mathbf{x} = 1, \quad (18)$$

with the Neumann boundary condition

$$\left( \frac{\nabla d}{|\nabla d|} - \frac{\mathbf{n}}{|\mathbf{n}|} \right) \cdot \nu = 0 \qquad \text{on } \partial\Omega. \quad (19)$$

One way to calculate the Lagrange multiplier $\mu$ is to make use of the condition (18), see [11] for detail,

$$\mu = -\frac{1}{\sigma^2} \int_\Omega \left( \frac{\nabla d}{|\nabla d|} - \frac{\mathbf{n}}{|\mathbf{n}|} \right) \cdot \nabla (d - d_0) \, d\mathbf{x}. \quad (20)$$

Introducing an artificial time variable $t$ we get the following time dependent problem, which needs to be solved to steady-state,

$$\frac{\partial d}{\partial t} - \nabla \cdot \left( \frac{\nabla d}{|\nabla d|} - \frac{\mathbf{n}}{|\mathbf{n}|} \right) + \mu(d - d_0) = 0 \qquad \text{in } \Omega, \tag{21}$$

with the Neumann boundary condition (19), and $\mu$ is given by the equation (20). If we replace the unit vector $\frac{\mathbf{n}}{|\mathbf{n}|}$ with the zero vector $\mathbf{0}$, then the method reduces to the classical TV denoising algorithm of Rudin, Osher and Fatemi [14].

Noting that $\mathbf{n} = (u, -v)$, the discrete formulation of the image reconstruction step takes the following form.

$$\frac{d^{n+1} - d^n}{\Delta t} = D_x^- \left( \frac{D_x^+ d^n}{T_3^n} - n_1 \right) + D_y^- \left( \frac{D_y^+ d^n}{T_4^n} - n_2 \right) - \mu^n \left( d^n - d_0 \right), \tag{22}$$

where $\mu^n$ is approximated as

$$\begin{aligned} \mu^n = & -\frac{1}{\sigma^2} \sum \left( \frac{D_x^+ d^n}{T_3^n} - n_1 \right) D_x^+ \left( d^n - d_0 \right) \\ & -\frac{1}{\sigma^2} \sum \left( \frac{D_y^+ d^n}{T_4^n} - n_2 \right) D_y^+ \left( d^n - d_0, \right) \end{aligned} \tag{23}$$

with $T_3$ and $T_4$ being defined as

$$T_3 = \sqrt{\left(D_x^+ d^n\right)^2 + \left(A_x(C_y^h d^n)\right)^2 + \epsilon}, \tag{24}$$

$$T_4 = \sqrt{\left(D_y^+ d^n\right)^2 + \left(A_y(C_x^h d^n)\right)^2 + \epsilon}, \tag{25}$$

and $n_1$ and $n_2$ as

$$n_1 = \frac{u}{\sqrt{u^2 + \left(A_x(A_y v)\right)^2 + \epsilon}}, \qquad n_2 = \frac{-v}{\sqrt{v^2 + \left(A_y(A_x u)\right)^2 + \epsilon}}. \tag{26}$$

## 3 Numerical Experiments

Several experiments with the proposed two-step algorithm have been performed, we present a few of them in this section.

As in each step of the algorithm the minimal of an energy functional is being sought, it is reasonable to use the energy functional as an objective measure for the stopping criterion in the corresponding step. However, since the minimization problems are subject to constraints, it is not enough just to use the energy functionals as stoping criterion. It has been observed through the image reconstruction step that even after the energy has nearly stabilized at some numerical minimum the image continues to improve, and the image becomes visually super when both the energy is minimum and the constraint is satisfied accurate enough. We have therefore included the constraints as additional measures in

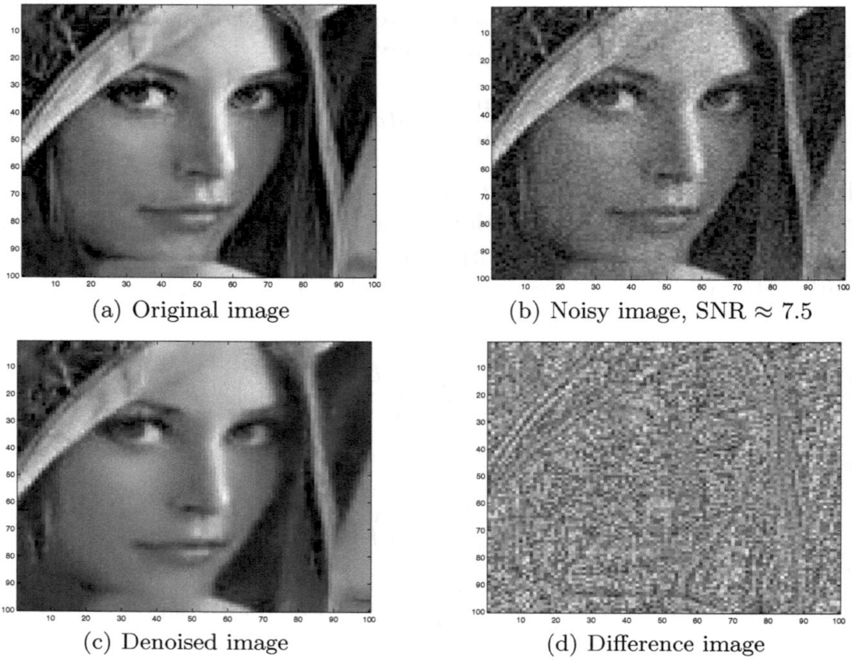

(a) Original image

(b) Noisy image, SNR $\approx 7.5$

(c) Denoised image

(d) Difference image

**Fig. 1.** The Lena image, denoised using the TV-Stokes algorithm

determining when to terminate the iterations. For the vector smoothing step we use the value of $\left(\int_\Omega |\nabla \cdot \tau|^2 \, d\mathbf{x}\right)^{\frac{1}{2}}$, and for the reconstruction step we compare the value of $\left(\int_\Omega (d - d_0)^2 \, d\mathbf{x}\right)^{\frac{1}{2}}$ with the given noise level $\sigma$.

Experiments have shown that the TV-Stokes can very often give smooth images that are visually very pleasant, specially in the smooth areas of the image. At places where the texture changes very rapidly, TV-Stokes seems to smear out the image and thereby loose the fine scale details.

For the presentation of our results, we have chosen three images with gray-level values in the range between 0 (black) and 255 (white). In all our experiments, we expose our image to random noise with zero mean, and apply different denoising algorithms on the noisy image. The value of $\epsilon$ is set equal to $10^{-11}$ in all cases.

The first image is the well known Lena image, cf. Figure 1. The TV-Stokes algorithm is applied first to the noisy image. In Figure 1 we can see the result of one such test where the parameter $\delta$ is equal to 0.07. As we can see from the denoised image, it is evident that the TV-Stokes algorithm does a very good job recovering the smooth area and yet keep the edges quite well.

To understand the behavior of the TV-Stokes algorithm as the iteration continues we have plotted the energy and the corresponding constraint measure, as shown in Figure 2, for both the normal smoothing and the image reconstruction step. It is clear from the plots in both cases, that the energies may stabilize long before the constraints are met with some accuracy.

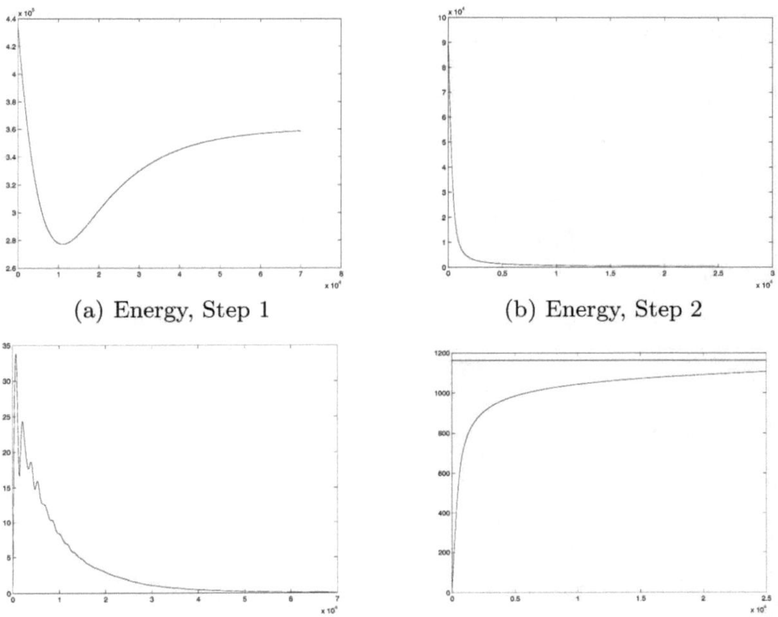

(a) Energy, Step 1 (b) Energy, Step 2

(c) Discrete $L^2$-norm of $\nabla \cdot \tau$ (d) Noise level $\sigma$

**Fig. 2.** Plot of energy during tangent field smoothing (Step 1) and during image reconstruction (Step 2), followed by plots of their corresponding constraint measures. The solid line in (d) indicates the true noise level $\sigma$.

The experiment of Figure 1 and 2 has been performed using fixed time steps $\Delta t$ equal to $10^{-3}$ and $5 \times 10^{-3}$ respectively for the first and second steps of the algorithm. It is well known that with large time step $\Delta t$ the algorithm may become unstable and not converge to steady state. It is then necessary to choose a reasonably smaller time step. It is usual to choose $\Delta t$ by trial and error or from experience. The choice of $\Delta t$ depends on the parameter $\epsilon$, for a large $\epsilon$ $\Delta t$ can be large, but for smaller $\epsilon$ it is necessary to use a smaller $\Delta t$, which ofcourse slows down the algorithm in reaching the steady state. However, with large $\epsilon$ the algorithm will result in an image which may not be sharp, but the image gets sharper as the parameter is reduced. In several occasions, we have exploited this situation in reducing the number of iterations by first running the algorithm with a large $\epsilon$ and a corresponding time step, and then gradually decreasing their values as the iteration continues.

For the comparison, we include the results of applying the classical TV denoising scheme [14] and the LOT algorithm of [11] on our noisy image of Figure 1. The denoised images and the corresponding difference images are shown in Figure 3. As seen from the difference images, the LOT algorithm seems to have preserved the edges best, while the proposed algorithm shows a performance which is very close to the LOT algorithm and much better than the TV algorithm in preserving edges. From the recovered images, however, we see that

(a) Denoised using LOT

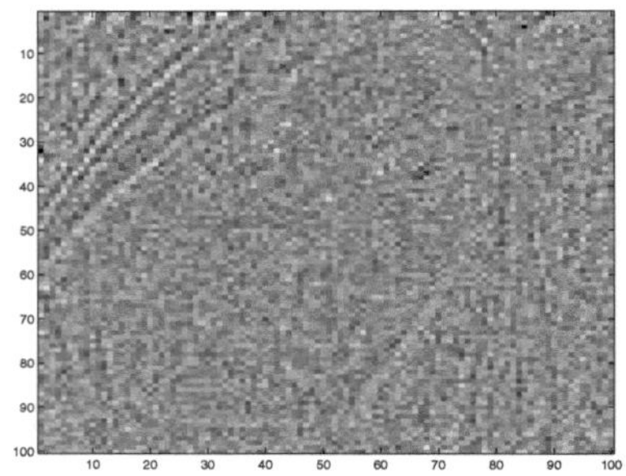

(b) Difference image, LOT

(c) Denoised using TV

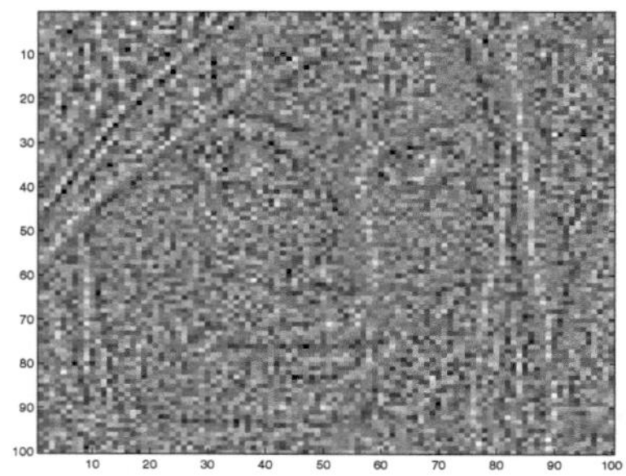

(d) Difference image, TV

**Fig. 3.** Denoised Lena image using the LOT algorithm of [11] the classical TV algorithm of [14]

(a) Denoised using TV-Stokes

(b) Denoised using TV

**Fig. 4.** Comparison of the results of the two methods: the TV-Stokes on the left, and the TV on the right the image created by the TV-Stokes algorithm is the smoothest and visually the most pleasant. Moreover, the stair-case effect of the TV algorithm does not exist in the new algorithm, see Figure 4 for a comparison, where the TV method shows clear evidence of a blocky image.

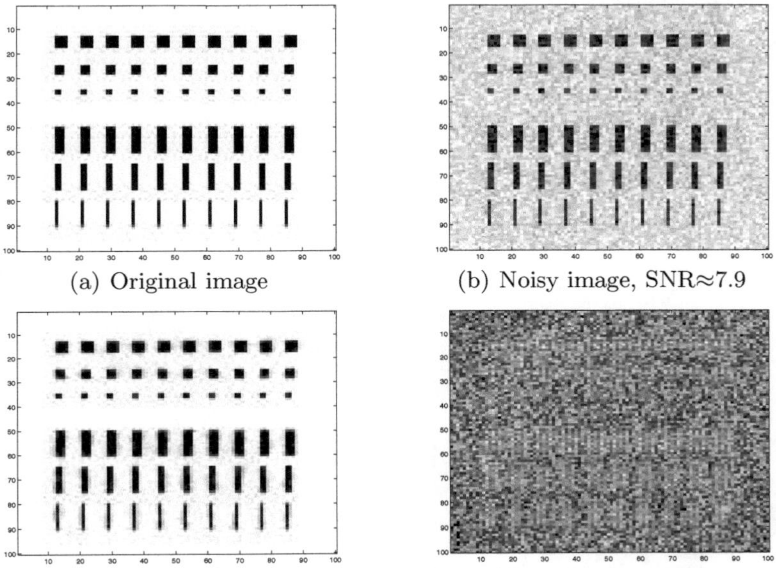

(a) Original image

(b) Noisy image, SNR≈7.9

(c) Denoised using TV-Stokes

(d) Difference image

**Fig. 5.** TV-Stokes algorithm on a blocky image

The next image we consider is a commonly used blocky image on which the TV algorithm is known to perform the best. It is not easy to smooth as well as preserve the edges. For this particular experiment the parameter $\delta$ is chosen equal to 0.2, and the time steps for the first and second steps of the algorithm are chosen equal to $5 \times 10^{-3}$ and $10^{-2}$, respectively. The denoised image and the difference image obtained by using the TV-Stokes algorithm are shown in Figure 5, illustrating that the TV-Stokes algorithm has managed to suppress the noise sufficiently well, and at the same time it has maintained the edges.

As the final image we consider the Cameraman image, cf. Figure 6, consisting of a smooth background (the sky), a relatively weak skyline, and very random grass texture. This image is considered to be difficult for most algorithms including the TV algorithm, the LOT algorithm of [11] and the TV-Stokes algorithm. However, as seen from the recovered images in this case, the TV-Stokes algorithm performs better than the LOT algorithm preserving the shapes of objects far away. Moreover, the TV-Stokes algorithm results in a much smoother image compared to the other one, cf. Figure 6. Here, the parameter $\delta$ is equal to 0.575, and the time steps are equal to $5 \times 10^{-3}$ and $10^{-2}$ respectively for the first and the second steps of the TV-Stokes algorithm.

(a) Original image
(b) Noisy image, SNR=8.21
(c) Denoised using LOT
(d) Denoised using TV-Stokes

**Fig. 6.** Comparing the TV-Stokes algorithm with the LOT algorithm of [11]

We close this section with a comment on the choice of the parameter $\delta$. The choice of $\delta$ is crucial for the TV-Stokes algorithm to succeed. For $\delta$ sufficiently small the algorithm performs normally quite well. The recovered image may however become too smooth causing it to loose edges. The image can in most cases be improved by tuning up the parameter. However, as we gradually increase the parameter $\delta$, the quality of the restored image may decrease. For instance, in case of the Lena image it has been observed that the algorithm restores the image perfectly well for $\delta$ around 0.06, but for $\delta$ around 0.1 the restored image still contains some noise.

## References

1. Gilles Aubert and Pierre Kornprobst, *Mathematical Problems in Image Processing: Partial Differential Equations and the Calculus of Variations*, Applied Mathematical Sciences 147, 2002, Springer Verlag, New York.

2. C. Ballaster, M. Bertalmio, V. Caselles, G. Sapiro, and J. Verdera, *Filling in by Joint Interpolation of Vector Fields and Gray Levels*, IEEE Trans. Image Processing, No. 10, 2000, pp. 1200–1211.
3. M. Bertalmio, A. L. Bertozzi, and G. Sapiro, *Navier-Stokes, Fluid Dynamics and Image and Video Inpainting*, In Proc. Conf. Comp. Vision Pattern Rec., 2001, pp. 355–362.
4. P. Burchard, T. Tasdizen, R. Whitaker, and S. Osher, *Geometric Surface Processing via Normal Maps*, Tech. Rep. 02-3, Applied Mathematics, 2002, UCLA.
5. T.F. Chan and J. Shen, *Image Processing and Analysis: Variational, PDE, Wavelet, and Stochastic Methods*, 2005, SIAM, Philadelphia.
6. C. Frohn-Schauf, S. Henn, and K. Witsch, *Nonlinear Multigrid Methods for Total Variation Image Denoising*, Comput. Visual. Sci., Vol. 7, 2004, pp. 199–206.
7. R. Glowinski and P. LeTallec, *Augmented Lagrangian and Operator-Splitting Methods in Nonlinear Mechanics*, SIAM Studies in Applied Mathematics, Vol. 9, 1989, SIAM, Philadelphia.
8. D. Goldfarb and W. Yin, *Second Order Cone Programming Methods for Total Variation-Based image Restoration*, SIAM J. Sci. Comput., Vol. 27, No. 2, 2005, pp. 622–645.
9. S. Kindermann, S. Osher, and J. Xu, *Denoising by BV-duality*, J. Sci. Comput., Vol. 28, Sept. 2006, pp. 414–444.
10. D. Krishnan, P. Lin, and X.C. Tai, *An Efficient Operator-Splitting Method for Noise Removal in Images*, Commun. Comput. Phys., Vol. 1, 2006, pp. 847–858.
11. M. Lysaker, S. Osher, and X.C. Tai, *Noise Removal Using Smoothed Normals and Surface Fitting*, IEEE Trans. Image Processing, Vol. 13, No. 10, October 2004, pp. 1345–1357.
12. S. Osher, A. Sole, and L. Vese, *Image Decomposition and Restoration Using Total Variation Minimization and the $H^{-1}$ norm*, Multiscale Modelling and Simulation, A SIAM Interdisciplinary J., Vol. 1, No. 3, 2003, pp. 1579–1590.
13. S. Osher, M. Burger, D. Goldfarb, J. Xu, and W. Yin, *An Iterative Regularization Method for Total Variation Based Image Restoration*, Multiscale Modelling and Simulation, Vol. 4, No. 2, 2005, pp. 460–489.
14. L.I. Rudin, S. Osher, and E. Fatemi, *Nonlinear Total Variation Based Noise Removal Algorithms*, Physica D., Vol. 60, 1992, pp. 259–268.
15. X.C. Tai, S. Osher, and R. Holm, *Image Inpainting using TV-Stokes equation*, in: Image Processing based on partial differential equations, 2006, Springer, Heidelberg.
16. L. Vese and S. Osher, *Numerical Methods for P-Harmonic Flows and Applications to Image Processing*, SIAM J. Numer. Anal., Vol. 40, No. 6, December 2002, pp. 2085–2104.

# Anisotropic $\alpha$-Kernels and Associated Flows

Micha Feigin[1], Nir Sochen[1,*], and Baba C. Vemuri[2,**]

[1] School of Mathematics Tel Aviv University
sochen@post.tau.ac.il, michf@post.tau.ac.il
[2] Department of Computer & Information Science & Engineering
University of Florida Gainesville, Fl. 32611, USA
vemuri@cise.ufl.edu

**Abstract.** The Laplacian raised to fractional powers can be used to generate scale spaces as was shown in recent literature by Duits et al. In this paper, we study the anisotropic diffusion processes by defining new generators that are fractional powers of an anisotropic scale space generator. This is done in a general framework that allows us to explain the relation between a differential operator that generates the flow and the generators that are constructed from its fractional powers. We then generalize this to any other function of the operator. We discuss important issues involved in the numerical implementation of this framework and present several examples of fractional versions of the Perona-Malik flow along with their properties.

## 1 Introduction

Scale space analysis of images is fundamental to image understanding and has been a very active area of research in image processing and computer vision. There are several ways to generate scale-spaces or diffusion flows. One of the most widely used scale spaces is the linear scale-space generated using the heat equation, which was first suggested by Ijima [1] and then by Witkin [2] and Koenderink [3]. This scale-space has a special status as it was the first to be discovered, linear and easy to generate. It was re-derived by many researchers using different approaches. One approach puts the emphasis on the semi-group properties and shows that this equation is the unique semi-group that satisfies several invariance properties (see [4] and references therein). Another approach derives it as a gradient descent of a functional, and yet another approach derives it from random walk considerations [5,6]. This equation was first generalized by Perona and Malik[1] [8] that suggested an inhomogeneous (also known as anisotropic) diffusion. It was then generalized by several approaches in various forms with

* This research was supported by MUSCLE Multimedia Understanding through Semantic, Computation and Learning, an European Network of Excellence funded by the EC 6th Framework IST Programme, and by the Adams super center for Brain Research.
** BCV was partially supported by the NIH grant EB007082.
[1] And was suggested independently in other domains of application by Rosenau [7].

F. Sgallari, A. Murli, and N. Paragios (Eds.): SSVM 2007, LNCS 4485, pp. 484–495, 2007.

numerous applications. A different approach was taken in [9] and recently developed in [10,11] for bounded domains and in [12,13] for unbounded domains. In their work, the generator i.e. the Laplacian, of the linear flow (or the semigroup) is considered. New flows were produced by taking *fractional* powers of the Laplacian as generators of the diffusion-like flows.

In this paper we use this framework and generalize it from fractional powers of the Laplacian to fractional powers, and even other functions, of other diffusion operators. Of particular interest are the operators that generate the anisotropic flows. In [12,13], the flows that are generated by fractional powers of the Laplacian were given explicitly via a convolution with a kernel. We follow this approach and derive general kernels for the flows generated by the fractional powers of the "anisotropic Laplacian".

The kernels are computed via a spectral decomposition and are given as summations of terms that involve the eigenvalues and eigenfunctions (eigenvectors in the discrete setting). These summations can be re-cast in to the form given in [12,13] for the case of the linear Laplacian. No such analytic expression can be derived in the anisotropic case. Instead, a numerical scheme is suggested for the computation of the eigenvectors and eigenvalues such that the summation can be correctly approximated. For $N$ by $N$ images the operators become after discretization $N^2$ by $N^2$ matrices. Care must be exercised in the discretization process in order to achieve manageable computations.

The rest of the paper is organized as follows: In Section 2 we recall the basic definitions of $\alpha$ scale-spaces, based on the Laplacian, and generalize it to other linear operators and to more general functions of the operator. In Section 3 we discuss the anisotropic case and present our anisotropic $\alpha$-kernels. Implementation and numerical issues are discussed in Section 4. Section 5 is devoted to presentation of the results and we conclude in Section 6 with a discussion.

## 2 General Linear $\alpha$-Scale Space

### 2.1 Laplacian $\alpha$-Scale Space

Let us first recall few basic ideas from the $\alpha$-scalespace framework. We start with linear scale-space

$$I_t = \Delta I \tag{1}$$

where $\Delta$ is the Laplacian operator. The new flows are written formally as

$$I_t = \Delta^\alpha I \tag{2}$$

or more exactly,

$$I_t = -\left(-\Delta\right)^\alpha I \ . \tag{3}$$

It has recently been defined and studied for $0 < \alpha < 1$. The way this equation should be understood is via the Fourier transform. Since $\mathcal{F}\{\Delta I\} = -k^2\mathcal{F}\{I\}$ equation (1) becomes

$$\mathcal{F}\{I\}_t = -k^2\mathcal{F}\{I\} \ . \tag{4}$$

Applying on both sides of the equation the inverse Fourier transform leads to

$$I_t = \mathcal{F}^{-1}\{-k^2\mathcal{F}\{I\}\} = -\{\mathcal{F}^{-1}k^2\mathcal{F}\}I \ . \tag{5}$$

The generalization to fractional powers arises from the observation that in the Fourier space the differential operator in question, the Laplacian, is diagonalized and is expressed as a multiplication operator. In this form the linear scale-space is easily generalized to fractional powers by defining the new equation

$$\mathcal{F}\{I\}_t = -(k^2)^\alpha\mathcal{F}\{I\} \tag{6}$$

Transforming back to the spatial domain we find that

$$I_t = \mathcal{F}^{-1}\{-k^{2\alpha}\mathcal{F}\{I\}\} = -\{\mathcal{F}^{-1}k^{2\alpha}\mathcal{F}\}I \ . \tag{7}$$

This is an interesting result. First notice that the differential operator that defines the linear scale-space is replaced by an integral operator. Second, we may interpret the Fourier transform and the powers of the frequency in a different way. Indeed, the exponentials $e^{-ikx}$ are the eigenfunctions of the Laplacian and the frequencies are the eigenvalues. The Fourier transform is nothing but summation over all eigenfunctions of the Laplacian operator. This is our starting point for the generalization.

### 2.2 Kernels of Linear Operators

In order to understand the relation between an operator and its powers, or more general functions of it, we have to understand its spectral decomposition. Consider the following equation

$$I_t = L(x)I$$

where $-L$ is a linear, positive definite and time independent differential operator. Let us further assume that the differential operator $L$ is self-adjoint and that it acts in a Hilbert space. Upon discretization of the spatial domain the differential operator is approximated by a symmetric matrix. Let $\phi_i(x)$ be a basis for this Hilbert space. Upon discretization the basis function $\phi_i(x)$ is approximated by vector with elements $(V_i)_j = \phi_i(x_j)$. We will use also the "bra - ket" notation, that is $|\phi_i\rangle$ is the function $\phi_i(x)$ and $\langle\phi_i|$ is the linear functional over that Hilbert space $\int \phi_i^*(x)(\ \cdot\ )dx$, where $\phi_i^*$ is the complex conjugate of $\phi$. The linear functional acts on the space of images and takes values in $\mathbb{R}$. Thus, if $I(x)$ is an image, the value of the functional evaluated on this image is $\langle\phi_i|I\rangle = \int \phi_i^*(x)I(x)dx$.

If we discretize the spatial domain the ket $|\phi_i\rangle$ is approximated, as we saw, by the vector $V_i$ and the bra $\langle\phi_i|$ is approximated by the transposed vector $V_i^t$. Clearly $V^t$ applied to vectors returns scalars and is a linear functional in the corresponding finite dimensional space.

The index $i \in Ind$ belongs to an index set $Ind$ that may be continuous. Since the functions $|\phi\rangle$ span the Hilbert space we get $I = \sum_{i\in Ind} c_i|\phi_i\rangle$, where summation should be understood as integration when the indices assume continuous value. The operator's matrix elements are given by $D_{ij} = \langle \phi_i | L \phi_j \rangle$ where $|L\phi_i\rangle$ are functions obtained when the operator $L$ is applied to $|\phi_i\rangle$. In these terms the basic diffusion equation (1) becomes

$$\sum_{i \in Ind} \frac{\partial c_i}{\partial t} |\phi_i\rangle = \sum_{i \in Ind} c_i |L\phi_i\rangle \tag{8}$$

and by taking the inner product with $\langle \phi_j |$ we get

$$\langle \phi_j | \phi_i \rangle \frac{\partial c_i}{\partial t} = \sum_{i \in Ind} D_{ji} c_i$$

A clever choice of the basis can simplify these equations. In particular, since the operator $-L$ is self adjoint and positive definite, we may choose the eigenfunctions of $-L$ as the desired basis. The eigenfunctions satisfy the relation

$$L|\phi_i\rangle = -\lambda_i |\phi_i\rangle \ .$$

We normalize the eigenfunctions to unit magnitude. Under these assumptions we have the following relations:

$$\begin{aligned} \langle \phi_j | \phi_i \rangle &= \delta_{ij} \\ D_{ji} &= -\lambda_i \delta_{ij} \ . \end{aligned} \tag{9}$$

Using the eigenfunctions we may think on the decomposition formula $I = \sum_{i \in Ind} c_i |\phi_i\rangle$ as the "inverse Fourier" transform, and the coefficients $c_i = \langle \phi_i | I \rangle$ are the "Fourier" transform of the image $I$. When the differential operator is the Laplacian these expressions coincide with the usual definitions of the Fourier transform and its inverse.

Using these concepts the diffusion-like equation takes the following form

$$\frac{\partial c_i}{\partial t} = -\lambda_i c_i \quad \forall i \in Ind \ . \tag{10}$$

This equation is the analog of eq. (4). The "Fourier" transform diagonalizes the differential equation and the differential operator acts via multiplication. This insight will be used below in the generalization of the flow. This simple equation is readily solved by

$$c_i(t) = c_i(0) e^{-\lambda_i t} = \langle \phi_i | I_0 \rangle e^{-\lambda_i t} \ . \tag{11}$$

We finally find the solution

$$I(x,t) = \sum_{i \in Ind} c_i(t) \phi_i(x) = \sum_{i \in Ind} |\phi_i\rangle e^{-\lambda_i t} \langle \phi_i | I_0 \rangle \tag{12}$$

This is a solution based on a kernel $I(x, t_1 + t_2) = \int K(x, x'; t_2) I(x', t_1) dx'$ where the kernel is given by

$$K(x, x'; t) = \sum_{i \in Ind} \phi_i(x) e^{-\lambda_i t} \phi_i^*(x') \ . \tag{13}$$

Now that we understand how the solution of the diffusion-like equation is expressed via the eigenfunctions and the eigenvalues we can take the next step and suggest new flows.

### 2.3 $\alpha$-Kernels for Linear Operators

Our suggestion is to consider functions of the operator $L$ as generators for new flows. Note that for the flow to be stable we should consider functions $f : \mathbb{R}_+ \to \mathbb{R}_+$. Such functions will map positive definite operators to positive definite operators. Since the operator $L$ is negative definite we define the new flows as

$$I_t = -f(-L)I \ , \tag{14}$$

were the first minus sign makes sure that the equation is stable and the second ensures that $f(-L)$ is positive definite. In general a function of an operator $f(-L)$ can be expressed in the basis of eigenfunctions of the operator $-L$. The resulting (Pseudo) differential operator has the the same eigenfunctions but with the the eigenvalues changed $\lambda_i \to f(\lambda_i)$. Equivalently one can define the flow in the "Fourier" space by taking equation (10) and applying the function $f(-L)$ on the corresponding multiplication operator

$$\frac{\partial c_i}{\partial t} = -f(\lambda_i)c_i \quad \forall i \in Ind \ . \tag{15}$$

The solution of this newly defined equation (15) is obtained via the kernel method as before. The kernel is constructed from the same eigenfunctions of the original operator but with the new eigenvalues. It reads therefore

$$K(x, x'; t) = \sum_{i \in Ind} \phi_i(x) e^{-f(\lambda_i)t} \phi_i^*(x') \tag{16}$$

Before proceeding to the anisotropic diffusion equations let us work out the $\alpha$-kernels that were studied in [10,11]. Consider the linear Laplacian $\Delta$ operator. The eigenfunctions of the Laplacian are the complex exponentials $\exp -ik \cdot x$ and the eigenvalues of $-\Delta$ are $k^2$.

The kernel for the non-fractional case is constructed by this method as

$$\begin{aligned} K(x, x'; t) &= \sum_i \phi_i(x) e^{-\lambda_i t} \phi_i^*(x') = \int e^{-ik\cdot x} e^{-k^2 t} e^{ik \cdot x'} dk \\ &= \int e^{-k^2 t} e^{-ik(x-x')} dk = \frac{\pi}{t} e^{-\frac{(x-x')^2}{4t}} \end{aligned} \tag{17}$$

Such that the usual kernel of the linear diffusion equation is recovered.

The kernel of the linear $\alpha$-kernels is given by the following expression

$$K(x, x'; t) = \sum_i \phi_i(x) e^{-\lambda_i^\alpha t} \phi_i^*(x') = \int e^{-k^{2\alpha} t} e^{-ik(x-x')} dk \tag{18}$$

where the analytic description of the kernel in the spatial domain is more complicated and is omitted here. The kernels for few values of $\alpha$ are presented graphically in Fig. (1).

We can see that as $\alpha$ decreases the corresponding kernel is localized more sharply around zero but is developing a heavier tail.

Next we consider anisotropic diffusion equations and their generalizations.

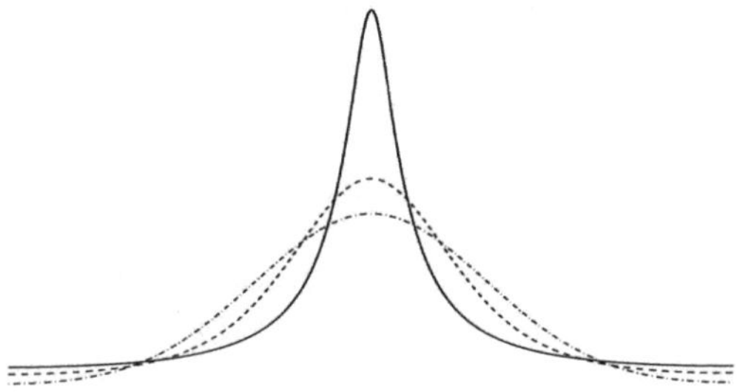

**Fig. 1.** Sample kernels for the $\Delta^\alpha$ case at $\alpha = 1$ (dash-dot), 0.7 (dashes), 0.4 (solid)

## 3 Anisotropic α-Kernels

Our starting point in this section is the anisotropic diffusion equation

$$I_t = Div(g(|\nabla I|)\nabla I) \equiv \Delta_g I \ . \tag{19}$$

Note that the differential operator that acts on the image $I$ is self adjoint and negative definite yet it is certainly not linear and the analysis in the previous section is not applicable as is. We linearize this equation by using the lagged-diffusivity technique i.e., lagging the nonlinear terms by one iteration. We thus solve the following equation:

$$I_t^{n+1} = Div(g(|\nabla I^n|)\nabla I^{n+1}) \equiv \Delta_{g^n} I^{n+1} \tag{20}$$

for a short time interval $(t, t+\delta t)$ with an initial condition $I^{n+1}(t) = I^n(t)$. Acting on $I^{n+1}$ the operator $Div\,(g\,(|\nabla I^n|)\,\nabla\,(\cdot))$ is linear and self adjoint. Choosing the diffusivity function $g(s)$ properly makes this operator negative definite as well.

The solution for the kernels of the anisotropic operator $\Delta_{g^n}$ is simply given in the kernel notation with the eigenvalues of $\Delta_{g^n}$ as

$$K(x, x'; t) = \sum_{i \in Ind} \phi_i(x) e^{-(\lambda_g)_i^\alpha t} \phi_i^*(x') \tag{21}$$

where now we don't have analytic expressions neither for the eigenfunctions nor for the eigenvalues. Nevertheless, these expressions can be computed numerically. Thus, our main result in this paper can be written in the following form:

$$I(x, t_0 + t) = \int K(x, x'; t) I(x', t_0) dx' = \sum_i |\phi_i(x)\rangle e^{-(\lambda_g)_i^\alpha t} \langle \phi_i | I(t_0)\rangle \ . \tag{22}$$

It is important to understand that the equation is linearized for each small time interval. In practice it means that the operator, which depends on the image in the previous time step is changing from one iteration to another. This implies that we have to recompute in each iteration the new eigenvectors and eigenvalues. This is not very computationally efficient. The advantage is that to the same order of accuracy, we can get longer time steps without loosing stability.

If we take the limiting case of this approach , we may ask, how does the flow look if the kernel depends only on the initial image. In fact, recently Nadler et al [14] used exactly this type of reasoning and took the Beltrami flow with a fixed metric to define "diffusion maps" which are used as data clustering techniques.

In the next sections we elaborate on implementation issues and discuss the results.

## 4 Numerical Implementation for the Perona and Malik Operator

Whenever one makes the move from a continuous to a discrete operator, some of the qualities of the original operator are lost. The discretization scheme is chosen according to the properties important for the problem at hand. For the eigenvector decomposition, one of the most important qualities of a self adjoined operator is symmetry. This has both implementational and computational efficiency ramifications.

The eigenvectors of a symmetric (self adjoint) operator form a complete orthogonal decomposition of the appropriate Hilbert space. Orthogonality is a prerequisite for both efficiency and stability in the computation of the (generalized) Fourier coefficients which can be calculated independently using an inner product. Another good quality of self adjoint real operators is that all eigenvalues and eigenvectors are real. Using a truncated series provides a real-valued approximation. In fact, due to the size of the resulting matrices and the proximity and multiplicity of the eigenvalues, complex eigenvector/eigenvalue pairs usually occur in practice for non-symmetric schemes.

The efficiency implications arise from the details of the eigenvalue/eigenvector algorithms. The most common method currently used for large scale problems (and the one used by Matlab for the eigs function), is the ARPACK package which is based on the implicitly restarted Arnoldi/Lanczos methods [15,16]. First the Hesenberg form is calculated [17], and then successive iterations are used to approximate the diagonal form. In the case of symmetric matrices, the Hesenberg form reduces to a tri-diagonal matrix.

### 4.1 Numerical Scheme

Here we describe an approach to discretize the Perona and Malik flow in a way that preserves symmetry. For the continuous case, the flow is given by

$$I_t = Div\left(c\left(I\left(x,y,t\right)\right)\nabla I\right) \tag{23}$$

where $c\left(I\left(x,y,t\right)\right)$ is dependent on the image (it is a measure of whether a given pixel is an object border), or more frequently, the image gradient. In the classic case, $c\left(I\left(x,y,t\right)\right)$ is taken to be

$$c(I(x,y,t)) = g(|\nabla I(x,y,t)|) = \frac{1}{1+|\nabla I/k|^2} \tag{24}$$

with $k$ being the scale parameter. Circumventing the $Div, \nabla$ notation this translates to

$$I_t = \frac{\partial}{\partial x} c \frac{\partial}{\partial x} I + \frac{\partial}{\partial y} c \frac{\partial}{\partial y} I \tag{25}$$

($\frac{\partial}{\partial x} c \frac{\partial}{\partial x} I$ and $\frac{\partial}{\partial y} c \frac{\partial}{\partial y} I$ should very rarely be developed further for both stability and symmetry reasons).

As described earlier, the operator is linearized by setting $c$ to be dependent on $I^n$ and not on the input. We denote it as $c(|\nabla J^n|)$ to make the distinction where $J^n = I^n$

Denote by $i$, $j$ the matrix indices and h the step size. Calculating $c$ at a half step offset, the spacial discretization is given by

$$\begin{aligned} \frac{\partial}{\partial t} I_{i,j} &= \left(\Delta_{x-} c_{i+\frac{1}{2},j} \Delta_{x+} + \Delta_{y-} c_{i,j+\frac{1}{2}} \Delta_{y+}\right) I_{i,j} \\ &= \frac{1}{h^2}\left[c_{i-\frac{1}{2},j} I_{i-1,j} - \left(c_{i-\frac{1}{2},j} + c_{i+\frac{1}{2},j}\right) I_{i,j} + c_{i+\frac{1}{2},j} I_{i+1,j}\right] + \\ &\quad \frac{1}{h^2}\left[c_{i,j-\frac{1}{2}} I_{i,j-1} - \left(c_{i,j-\frac{1}{2}} + c_{i,j+\frac{1}{2}}\right) I_{i,j} + c_{i,j+\frac{1}{2}} I_{i,j+1}\right] \end{aligned} \tag{26}$$

where $\Delta_{x+}, \Delta_{x-}, \Delta_{y+}, \Delta_{y-}$ denote the forward and backward difference operators for $x$ and $y$.

To calculate $c$ at the intermediate (half step) point one needs to interpolate. In each case the derivative in the same direction of the half step is easy, but the complement requires some additional work. We thus get

$$c_{i+\frac{1}{2},j} = \frac{1}{1+\left(J_x^2 + J_y^2\right)/k^2} \tag{27}$$

$$= \frac{1}{1+\frac{\frac{1}{h^2}(J_{i+1,j}-J_{i,j})^2 + \frac{1}{(2h)^2}\left(\frac{1}{2}((J_{i,j+1}-J_{i,j-1})+(J_{i+1,j+1}-J_{i+1,j-1}))\right)^2}{k^2}} \tag{28}$$

$$c_{i-\frac{1}{2},j} = \frac{1}{1+\left(J_x^2 + J_y^2\right)/k^2} \tag{29}$$

$$= \frac{1}{1+\frac{\frac{1}{h^2}(J_{i,j}-J_{i-1,j})^2 + \frac{1}{(2h)^2}\left(\frac{1}{2}((J_{i-1,j+1}-J_{i-1,j-1})+(J_{i,j+1}-J_{i,j-1}))\right)^2}{k^2}} \tag{30}$$

Where $J_x$ at $i+\frac{1}{2}$ is calculated using symmetric difference with steps $h/2$ which results with the normal forward difference. $J_y$ is calculated using symmetric difference at both $i$ and $i+1$ and these are averaged to get the results at $i+1/2$. The values at $c_{i,j+1/2}$ and $c_{i,j-1/2}$ are calculated in a similar fashion.

### 4.2 Evaluating the Operator (Eigenvectors/Eigenvalues)

Calculating the flow is a two step operation. First, the eigenmodes of

$$-L(\cdot) = -Div\left(c\left(\nabla I^n\right)\nabla\cdot\right) \tag{31}$$

are calculated. Next, $I^{n+1}$ is calculated using

$$\bar{I}^{n+1} = V\cdot\exp\left(-\left(\Lambda\right)^{\alpha}\cdot\Delta t)\right]\cdot V^T\cdot\bar{I}^n \tag{32}$$

where $\bar{I}^n$ and $\bar{I}^{n+1}$ are the vectors formed column wise from the image. $V$'s columns are the eigenvectors of $-L$ and $\Lambda$ is a diagonal matrix who's elements are the matching eigenvalues.

A truncated series is used for the eigenvector decomposition as it is unstable and impractical to compute all of the vectors. The longer the time step, the less vectors that are needed.

The size of the time step depends on the stability in time of the eigenvectors. For the Perona-Malik operator, the eigenvectors for the small eigenvalues represent large scale structure which changes little over scale and thus evolve very slowly. This allows for relatively large time steps.

Experimentation has shown that we can keep $c$ constant, i.e: $c\left(I\left(x,y,t\right)\right) = c\left(I\left(x,y,0\right)\right)$ for rather large time steps. This changes the evolution speed but leaves the results mostly unchanged (see figure 3). In most cases, the eigenmodes need to recalculated at most once, if at all.

## 5 Experimental Results

Figure 2 shows a comparison for changing the value of $\alpha$. As can be seen in the figure (eyes, nose, scarf), decreasing $\alpha$ results with more loss of smaller detail while preserving larger objects. This is expected due to the heavier tail and narrower main section of the kernel. We also see less aliasing effects.

The example was produce with $c\left(I\left(x,y,t\right)\right) = c\left(I\left(x,y,0\right)\right)$. This may seem to be inappropriate as $c$ to changes quickly small $t$. Figure 3 shows that against expectations, this a rather good approximation. It shows a comparison of for recalculation the eigenmodes at each step compared to keeping $c$ constant.

This is explained by the face that the change is $c$ over time mainly effects finer detail. This finer detail (high frequency detail) is represented in eigenvectors belonging to larger magnitude eigenvalues and these are highly attenuated (or dropped completely). The lower magnitude eigenvectors represent large structures which are already rather smooth. The results is that the lower the magnitude of the eigenvalue, the stabler the eigenvector is over time. Thus, the less eigenvectors used, the larger the time step over which $c$ can be kept constant.

Lastly, figure 4 shows some of the dominant eigenvectors in the decomposition. As mentioned previously, large structures and dominant edges appear mostly in eigenvectors corresponding to small magnitude eigenvalues. The larger the eigenvalue is, the finer the detail represented in the eigenvector.

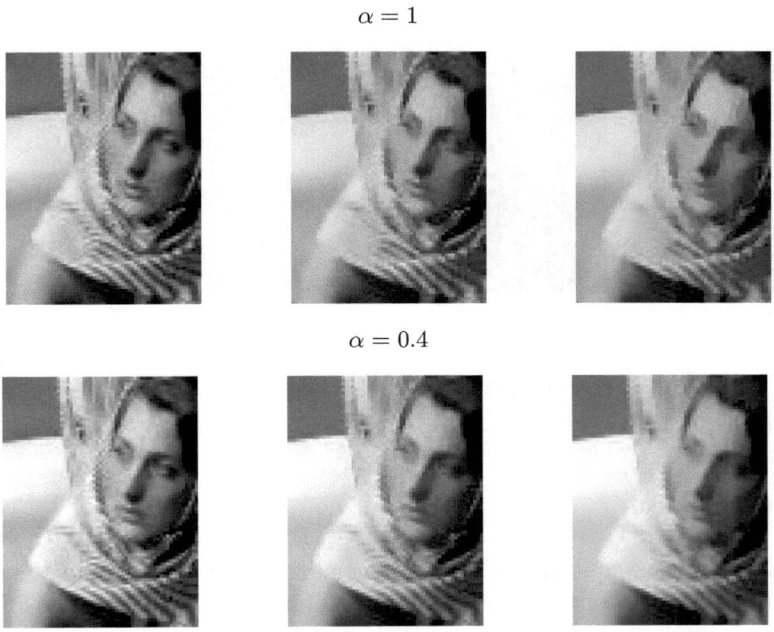


**Fig. 2.** Perona-Malik ($k = 6$) for $\alpha = 1, 0.4$

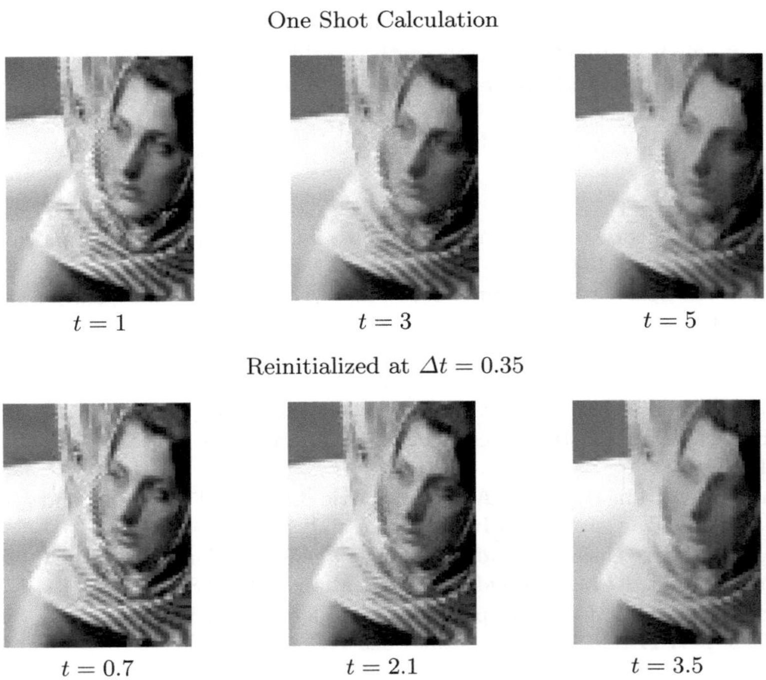


**Fig. 3.** Reinitialization vs. One shot Calculation of the Eigenvectors for $\alpha = 0.4$

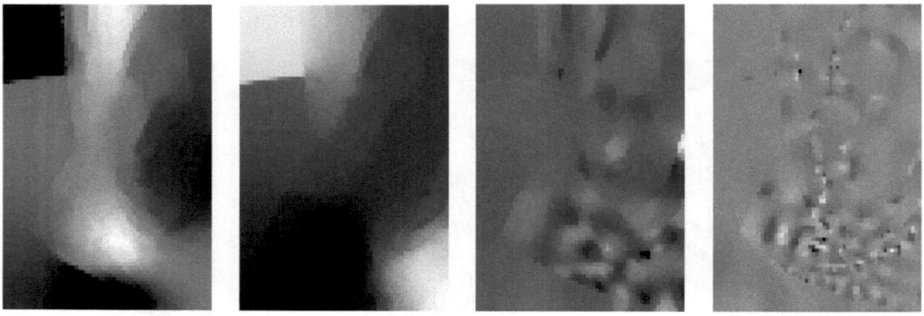

$\lambda = -0.0009$ $\lambda = -0.00353$ $\lambda = -0.01$ $\lambda = -0.03252$

**Fig. 4.** Feature size depending on eigenvalue magnitude

## 6 Conclusion

We studied in this paper new scale-space generating flows. These flows are generated from fractional powers, or more general functions, of anisotropic diffusion operator. This is a direct generalization of the $\alpha$-kernels that were based on the linear scale-space flow. The generalization is being done via a spectral decomposition of the anisotropic operators. Since analytic expressions can not be obtained in these cases a numerical approach was employed. We discussed discretization schemes and truncation methods. The results of different such flows are demonstrated on real images.

## References


1. Ijima, T.: Theory of pattern recognition. Electronics and Communications in Japan (1963) 123–134
2. Witkin, A.: Scale-space filtering. In: Proc. of Eighth Int. Joint Conf. on Artificial Intelligence. (Volume 2.)
3. Koenderink, J.J.: The structure of images. Biol. Cybern. (1984) 363–370
4. Weickert, J., Ishikawa, S., Imiya, A.: Linear scale-space has first been proposed in japan. Journal of Mathematical Imaging and Vision **10** (1999) 237–252
5. Sochen, N.A.: Stochastic processes in vision: From langevin to beltrami. (In: ICCV 2001) 288–293
6. Unal, G.B., Krim, H., Yezzi, A.J.: Stochastic differential equations and geometric flows. IEEE Transactions on Image Processing **11** (2002) 1405–1416
7. Rosenau, P.: Extension of landau-ginzburg free-energy functions to high-gradients domains. Phys. Rev. A (Rapid Communications) **39** (1989) 6614–6617
8. Perona, P., Malik, J.: Scale-space and edge detection using anisotropic diffusion. IEEE T PATTERN ANAL **12** (1990) 629–639
9. Pauwels, E.J., Gool, L.J.V., Fiddelaers, P., Moons, T.: An extended class of scale-invariant and recursive scale space filters. PAMI **17** (1995) 691–701

10. Duits, R., Felsberg, M., Florack, L., Platel, B.: $\alpha$ scale spaces on a bounded domain. In Griffin, L.D., Lillholm, M., eds.: Proc. Scale Space 2003. Volume 2695 of LNCS., Springer, Heidelberg (2003) 494–510
11. Felsberg, M., Duits, R., Florack, L.: The monogenic scale space on a bounded domain and its applications. In: Proc. Scale-Space 2003. (2003) 209–224
12. Duits, R., Florack, L.M.J., de Graaf, J., ter Haar Romeny, B.M.: On the axioms of scale space theory. JMIV **20** (2004) 267–298
13. Felsberg, M., Sommer, G.: The monogenic scale-space: A unifying approach to phase-based image processing in scale-space. JMIV **21** (2004) 5–26
14. Nadler, B., Lafon, S., Coifman, R.R., Kevrekidis, I.G.: Diffusion maps, spectral clustering and eigenfunctions of fokker-planck operators. Neural Information Processing Systems (NIPS) (2005)
15. Lehoucq, R., Sorensen, D., Yang, C.: ARPACK User's Guide: Solution of Large Scale Eigenvalue Problems with Implicitly Restarted Arnoldi Methods. (1997)
16. Sorense, D.: Implicitly restarted arnoldi/lanczos methods for large scale eigenvalue calculations. Technical Report TR-96-40, Department of Computational and Applied Mathematics, Rice University (1995)
17. Trefethen, L.N., Bau, D.: Numerical linear algebra. SIAM (1997)

# Bounds on the Minimizers of (nonconvex) Regularized Least-Squares

Mila Nikolova

CMLA, ENS Cachan, CNRS, PRES UniverSud
61 Av. President Wilson, F-94230 Cachan, France
nikolova@cmla.ens-cachan.fr

**Abstract.** This is a theoretical study on the minimizers of cost-functions composed of an $\ell_2$ data-fidelity term and a possibly nonsmooth or non-convex regularization term acting on the differences or the discrete gradients of the image or the signal to restore. More precisely, we derive general *nonasymptotic* analytical bounds characterizing the local and the global minimizers of these cost-functions. We provide several bounds relevant to the observation model. For edge-preserving regularization, we exhibit a tight bound on the $\ell_\infty$ norm of the residual (the error) that is independent of the data, even if its $\ell_2$ norm is being minimized. Then we focus on the smoothing incurred by the (local) minimizers in terms of the differences or the discrete gradient of the restored image (or signal).

**Keywords:** restoration, regularization, variational methods, bounds, $\ell_\infty$ norm, edge preservation, inverse problems, non-convex analysis, non-smooth analysis

## 1 Introduction

We consider the classical inverse problem of the finding of an estimate $\hat{x} \in \mathbb{R}^p$ of an unknown image or signal $x \in \mathbb{R}^p$ based on data $y \in \mathbb{R}^q$ corresponding to $y = Ax+\text{noise}$, where $A \in \mathbb{R}^{q \times p}$ models the data-acquisition system. For instance, $A$ can be a point spread function (PSF) accounting for optical blurring, a distortion wavelet in seismic imaging and non-destructive evaluation, a Radon transform in X-ray tomography, a Fourier transform in diffraction tomography, or it can be the identity in denoising and segmentation problems. Such problems are customarily solved using regularized least-squares methods: the solution $\hat{x} \in \mathbb{R}^p$ minimizes a cost-function $\mathcal{F}_y : \mathbb{R}^p \to \mathbb{R}$ of the form

$$\mathcal{F}_y(x) = \|Ax - y\|^2 + \beta\Phi(x) \tag{1}$$

where $\Phi$ is the regularization term and $\beta > 0$ is a parameter which controls the trade-off between the fidelity to data and the regularization [2,6,1]. The role of $\Phi$ is to push $\hat{x}$ to exhibit some a priori expected features, such as the presence of edges and smooth regions. Since [2,8], regularization functions are usually of the form

$$\Phi(x) = \sum_{i \in J} \varphi(\|G_i x\|), \quad J = \{1, \ldots, r\}, \tag{2}$$

F. Sgallari, A. Murli, and N. Paragios (Eds.): SSVM 2007, LNCS 4485, pp. 496–507, 2007.

where $G_i \in \mathrm{R}^{s \times p}$, $i \in J$, are linear operators with $s \geq 1$, $\|.\|$ is the $\ell_2$ norm and $\varphi : \mathrm{R}_+ \to \mathrm{R}$ is called a potential function. When $x$ is an image, the usual choices for $G_i$ are either that $s = 2$ and $\{G_i\}$ is the discrete approximation of the gradient operator, or that $s = 1$ and $\{G_i x\}$ are the first-order difference between each pixel and its 4 or 8 nearest neighbors. In the following, the letter $G$ will denote the $rs \times p$ matrix obtained by vertical concatenation of the matrices $G_i$ for $i \in J$, i.e. $G = [G_1^T, G_2^T, \ldots, G_r^T]^T$ where $^T$ means transposed. A basic requirement to have regularization is

$$\ker(A) \cap \ker(G) = \{0\}. \tag{3}$$

Many different potential functions (PFs) have been used in the literature. The most popular PFs are given in Table 1. Although PFs differ in convexity,

**Table 1.** Commonly used PFs $\varphi$ where $\alpha > 0$ is a parameter

| Convex PFs | | | |
|---|---|---|---|
| $\varphi(\|t\|)$ is smooth at zero | | $\varphi(\|t\|)$ is nonsmooth at zero | |
| (f1) $\varphi(t) = t^\alpha,\ 1 < \alpha \leq 2$ | [4] | (f3) $\varphi(t) = t$ | [2,20] |
| (f2) $\varphi(t) = \sqrt{\alpha + t^2}$ | [22] | | |
| **Nonconvex PFs** | | | |
| $\varphi(\|t\|)$ is smooth at zero | | $\varphi(\|t\|)$ is nonsmooth at zero | |
| (f4) $\varphi(t) = \min\{\alpha t^2, 1\}$ | [15,3] | (f8) $\varphi(t) = t^\alpha,\ 0 < \alpha < 1$ | [21] |
| (f5) $\varphi(t) = \dfrac{\alpha t^2}{1 + \alpha t^2}$ | [10] | (f9) $\varphi(t) = \dfrac{\alpha t}{1 + \alpha t}$ | [9] |
| (f6) $\varphi(t) = \log(\alpha t^2 + 1)$ | [11] | (f10) $\varphi(t) = \log(\alpha t + 1)$ | |
| (f7) $\varphi(t) = 1 - \exp(-\alpha t^2)$ | [13,19] | (f11) $\varphi(0) = 0,\ \varphi(t) = 1$ if $t \neq 0$ | [13] |

boundedness, differentiability, etc., they share some common features. Based on these common features, we systematically assume the following:

**H1** *$\varphi$ increases on $\mathrm{R}_+$ so that $\varphi(0) = 0$ and $\varphi(t) > 0$ for any $t > 0$.*

According to the smoothness of $\varphi(|t|)$ at zero, we will consider either H2 or H3:

**H2** *$\varphi$ is $\mathcal{C}^m$ with $m \geq 1$ on $\mathrm{R}_+ \setminus T$ where the set $T = \{t > 0 : \varphi'(t^-) > \varphi'(t^+)\}$ is at most finite and $\varphi'(0) = 0$.*

The conditions on $T$ in this assumption allows us to address the PF in (f4) which is the discrete version of the Mumford-Shah functional.

**H3** *$\varphi$ is $\mathcal{C}^m$ with $m \geq 1$ on $\mathrm{R}_+$ and $\varphi'(0) > 0$.*

Under this assumption $t \to \varphi(|t|)$ is nonsmooth at zero since $\varphi'(0) > 0$. Notice that our assumptions address convex and nonconvex functions $\varphi$. A particular attention is devoted to *edge-preserving* functions $\varphi$ because of their ability to give rise to solutions $\hat{x}$ involving sharp edges and homogeneous regions. Based on various conditions for edge-preservation in the literature [9,20,4,14,1,18], a common requirement is that for $t$ large, $\varphi$ can be upper bounded by a nearly affine function.

The aim of this paper is to give *nonasymptotic* analytical bounds on the local and the global minimizers $\hat{x}$ of $\mathcal{F}_y$ in (1)-(2) that hold for all functions $\varphi$ described above. To our knowledge, similar questions have mainly been considered in particular situations, such as $A$ the identity, or a particular $\varphi$, or when $y$ is a special noise-free function, or in asymptotic conditions when one of the terms in (1) vanishes. In a statistical setting, the mean and the variance of the minimizers $\hat{x}$ for strictly convex and differentiable functions $\varphi$ has been explored in [7].

## 2 Bounds Relevant to the Data Term

In this section we give bounds that characterize the data term at a local minimizer $\hat{x}$ of $\mathcal{F}_y$. Before to get into the heart of the matter, we will get sure that even if $\varphi$ is non-smooth in the sense specified in H2, the function $\mathcal{F}_y$ is smooth on a neighborhood of each one of its local minimizers.

**Proposition 1.** *Let $\varphi$ satisfy H1 and H2 where $T \neq \emptyset$. If $\hat{x}$ is a (local) minimizer of $\mathcal{F}_y$, we have $\|G_i\hat{x}\| \neq \tau$, for all $i \in J$, for every $\tau \in T$. Moreover, $\hat{x}$ satisfies $D\mathcal{F}_y(\hat{x}) = 0$.*

The proof of this Proposition can be found in [17].

### 2.1 The Observation Equation

The theorem below corroborates a very intuitive statement.

**Theorem 1.** *Let $\mathcal{F}_y : \mathrm{R}^p \to \mathrm{R}$ read as in (1)-(2) where (3) holds and $\varphi$ satisfies H1. Let $\varphi$ also satisfy either H2 or H3 for $m = 1$. In addition, if* $\mathrm{rank}(A) < p$ *we assume that $\varphi'(t) > 0$ for all $t > 0$.*
*If $\mathcal{F}_y$ reaches a (local) minimum at $\hat{x}$ then $\|A\hat{x}\| \leq \|y\|$.*

When $A$ is the identity and $x$ is defined on a convex subset of $\mathrm{R}^d$, $d \geq 1$, it is shown in [1] that $\|\hat{x}\| \leq \sqrt{2}\|y\|$. So we provide here a bound which is sharper and more general for images on discrete grids and which holds for any regularization term satisfying the assumptions.

The proof of the theorem relies on the lemma stated below whose proof is outlined in a forthcoming paper.

**Lemma 1.** *Given $A \in \mathrm{R}^{q\times p}$ and $G \in \mathrm{R}^{r\times p}$, assume that (3) holds. Let $\theta \in \mathrm{R}^r_+$. If* $\mathrm{rank}(A) < p$ *assume also that $\theta[i] > 0$, for all $1 \leq i \leq r$. Then the $q\times q$ matrix $C$ below*

$$C = A\Big(A^T A + G^T \mathrm{diag}(\theta) G\Big)^{-1} A^T \tag{4}$$

*is well defined and its spectral norm satisfies $\|C\|_2 \leq 1$.*

In (4), $\mathrm{diag}(\theta)$ denotes the diagonal matrix whose main diagonal is $\theta$. For any $\nu \geq 0$, $\|.\|_\nu$ will denote the $\ell_\nu$ norm. For simplicity, we systematically write $\|.\|$ in place of $\|.\|_2$. For any matrix $C$, $\|C\|_\nu$ is the matrix norm associated with the vector norm $\|.\|_\nu$, namely $\|C\|_\nu = \sup_{\|u\|_\nu=1} \|Cu\|_\nu$. Let us remind that in particular $\|C\|_2 = \max\{\sqrt{\lambda} : \lambda \text{ is an eigenvalue of } C^T C\} = \sup_{\|u\|=1} \|Cu\|$. Since $C$ in (4) is symmetric and positive semi-definite, the Lemma means that all the eigenvalues of $C$ are contained in $[0, 1]$.

For every $i \in J$, the rows of $G_i \in \mathbb{R}^{s\times p}$ will be denoted $G_i^j \in \mathbb{R}^{1\times p}$, $1 \leq j \leq s$, its columns $G_i[n]$, $1 \leq n \leq p$, while its entries $G_i^j[n]$.

*Proof of Theorem 1.* The cases corresponding to H2 and H3 are considered separately.

*$\varphi$ satisfying H1 and H2.* The function $z \to \varphi(\|z\|)$ is smooth at $z = 0$, so using Proposition 1 if necessary, we can write that $D\mathcal{F}_y(\hat{x}) = 0$ where

$$D\mathcal{F}_y(x) = 2A^T Ax + \beta\nabla\Phi(x) - 2A^T y. \tag{5}$$

Noticing that $\|G_i x\| = \sqrt{\sum_{j=1}^s (G_i^j x)^2}$, for any $x$ such that $\|G_i x\| \notin T$ for all $i$, the entries $\partial_n\Phi(x) = \partial\Phi(x)/\partial x[n]$ of $\nabla\Phi(x)$ read

$$\partial_n\Phi(x) = \sum_{i\in J} \frac{\varphi'(\|G_i x\|)}{\|G_i x\|} \sum_{j=1}^{s} G_i^j x \, G_i^j[n] = \sum_{i\in J} \big(G_i[n]\big)^T \frac{\varphi'(\|G_i x\|)}{\|G_i x\|} G_i x. \tag{6}$$

If $s = 1$, we have $G_i^1 = G_i$ and (6) is simplified to

$$\partial_n\Phi(x) = \sum_{i\in J} \varphi'(|G_i x|)\mathrm{sign}(G_i x) G_i[n]$$

where we set $\mathrm{sign}(0) = 0$. Let $\theta \in \mathbb{R}^{rs}$ be defined for $\hat{x}$ by

$$\forall i = 1, \ldots, r,$$
$$\theta\big[(i-1)s+1\big] = \cdots = \theta\big[(i-1)s+s\big] = \begin{cases} \dfrac{\varphi'(\|G_i\hat{x}\|)}{\|G_i\hat{x}\|} & \text{if } \|G_i\hat{x}\| \neq 0, \\ 1 & \text{if } \|G_i\hat{x}\| = 0. \end{cases} \tag{7}$$

Since $\varphi$ is increasing on $[0, \infty)$, (7) shows that $\theta[i] \geq 0$ for every $1 \leq i \leq rs$. Moreover, if the assumption (a) holds, then $\theta[i] > 0$ for every $1 \leq i \leq rs$. Using that the value of a $\theta_i$ such that $G_i\hat{x} = 0$ can be any, we can write that $\nabla\Phi(\hat{x}) = G^T \mathrm{diag}(\theta) G\hat{x}$. Introducing this into (5) yields

$$\Big(A^T A + \frac{\beta}{2} G^T \mathrm{diag}(\theta) G\Big)\hat{x} = A^T y. \tag{8}$$

Using Lemma 1 we can write that $A\hat{x} = Cy$ where $C$ is the matrix given in (4). By this lemma, $\|A\hat{x}\|_2 \leq \|C\|_2 \, \|y\|_2 \leq \|y\|_2$.

*$\varphi$ satisfying H1 and H3.* Now $z \to \varphi(\|z\|)$ is nonsmooth at $z = 0$. Given $\hat{x}$, let us introduce the subsets

$$J_0 = \{i \in J : \|G_i\hat{x}\| = 0\} \quad \text{and} \quad J_1 = J \setminus J_0. \tag{9}$$

If $J_0$ is empty then $\mathcal{F}_y$ is differentiable at $\hat{x}$ and the result follows from the previous paragraph. Consider next that $J_0$ is nonempty. Since $\mathcal{F}_y$ has a (local) minimum at $\hat{x}$, for every $v \in \mathbf{R}^p$, the one-sided derivative of $\mathcal{F}_y$ at $\hat{x}$ in the direction of $v$,

$$\delta\mathcal{F}_y(\hat{x})(v) = 2v^TA^T(A\hat{x} - y) + \beta\sum_{i\in J_1}\frac{\varphi'(\|G_i\hat{x}\|)}{\|G_i\hat{x}\|}(G_i\hat{x})^T\, G_iv + \beta\varphi'(0)\sum_{i\in J_0}\|G_iv\|, \tag{10}$$

satisfies $\delta\mathcal{F}_y(\hat{x})(v) \geq 0$. Let $\mathcal{K}_0 \subset \mathbf{R}^p$ be subspace

$$\mathcal{K}_0 = \{v \in \mathbf{R}^p : G_iv = 0,\ \forall i \in J_0\}. \tag{11}$$

Since the last term in the right side of (10) vanishes if $v \in \mathcal{K}_0$, we can write that

$$\forall v \in \mathcal{K}_0, \quad v^T\left(A^TA\hat{x} + \frac{\beta}{2}\sum_{i\in J_1}G_i^T\,\frac{\varphi'(\|G_i\hat{x}\|)}{\|G_i\hat{x}\|}G_i\hat{x} - A^Ty\right) = 0. \tag{12}$$

Let $k \leq p$ denote the dimension of $\mathcal{K}_0$ and $B_0$ be a $p \times k$ matrix whose columns form an orthonormal basis of $\mathcal{K}_0$. Then (13) is equivalent to

$$B_0^T\left(A^TA\hat{x} + \frac{\beta}{2}\sum_{i\in J_1}G_i^T\,\frac{\varphi'(\|G_i\hat{x}\|)}{\|G_i\hat{x}\|}G_i\hat{x}\right) = B_0^TA^Ty. \tag{13}$$

Let $\theta \in \mathbf{R}^{rs}$ be defined as in (7). Using that $\|G_i\hat{x}\| = 0$ for all $i \in J_0$ we can write that

$$\sum_{i\in J_1}G_i^T\,\frac{\varphi'(\|G_i\hat{x}\|)}{\|G_i\hat{x}\|}G_i\hat{x} = \sum_{i\in J_1}G_i^T\,\theta[is]G_i\hat{x} + \sum_{i\in J_0}G_i^T\,\theta[is]G_i\hat{x} = G^T\mathrm{diag}(\theta)G\hat{x}.$$

Then (13) is equivalent to

$$B_0^T\left(A^TA + \frac{\beta}{2}G^T\mathrm{diag}(\theta)G\right)\hat{x} = B_0^TA^Ty. \tag{14}$$

Since $\hat{x} \in \mathcal{K}_0$, there is a unique $\tilde{x} \in \mathbf{R}^k$ such that

$$\hat{x} = B_0\tilde{x}. \tag{15}$$

Define $A_0 = AB_0 \in \mathbf{R}^{q\times k}$ and $G_0 = GB_0 \in \mathbf{R}^{r\times k}$. Then (14) reads

$$\left(A_0^TA_0 + \frac{\beta}{2}G_0^T\mathrm{diag}(\theta)G_0\right)\tilde{x} = A_0^Ty. \tag{16}$$

If $A^TA$ is invertible, $A_0^TA_0$ is invertible as well. Otherwise, the assumption (a) ensures that $\theta[i] > 0$ for all $i \in J$. In both cases, we can write down

$$A_0\tilde{x} = C_0 y$$
$$C_0 = A_0 \left( A_0^T A_0 + \frac{\beta}{2} G_0^T \mathrm{diag}(\theta) G_0 \right)^{-1} A_0^T.$$

By Lemma 1, $|||C_0||| \le 1$. Using (15), we deduce that $\|A\hat{x}\| = \|A_0\tilde{x}\| \le \|y\|$.
If $A$ is unitary, then $\|A\hat{x}\| = \|\hat{x}\|$ hence the result. The proof is complete.

## 2.2 The Residuals

The next theorem focuses on edge-preserving functions $\varphi$ for which a current assumption is that $\|\varphi'\|_\infty = \sup_{0\le t<\infty} |\varphi'(t)|$ is finite. A look at Table 1 shows that this condition is satisfied by all the functions given there except for (f1).

**Theorem 2.** *Let $\mathcal{F}_y : \mathrm{R}^p \to \mathrm{R}$ read as in (1)-(2) where (3) holds. Let $\varphi$ satisfy H1 and only one of the assumptions H2 or H3. Assume also that*

*(a)* $\mathrm{rank}(A) = q \le p$.
*(b)* $\|\varphi'\|_\infty$ *is finite.*

*For every $y \in \mathrm{R}^q$, if $\mathcal{F}_y$ reaches a (local) minimum at $\hat{x}$ then*

$$\|y - A\hat{x}\|_\infty \le \frac{\beta}{2} \|\varphi'\|_\infty \; |||(AA^T)^{-1}A^T|||_\infty |||G|||_1. \tag{17}$$

*If $A$ is the identity and $\|\varphi'\|_\infty = 1$, and if either $G_i \approx \nabla_{x_i}$, $\forall i \in J$ or if $\{G_i\}$ yields the first-order differences between each pixel and its 4 nearest neighbors, then*

$$\|y - \hat{x}\|_\infty \le 2\beta. \tag{18}$$

*Remark 1.* Let us emphasize that (17) and (18) provide tight bounds that are *independent of $y$ and hold for any local or global minimizer $\hat{x}$ of $\mathcal{F}_y$.*

It may be useful to remind that for any matrix $C$, we have $|||C|||_1 = \max_j \sum_i |C[i,j]|$ and $|||C|||_\infty = \max_i \sum_j |C[i,j]|$, see e.g. [12,5].

**Proof.** The cases relevant to H2 and H3 are analyzed separately.

*$\varphi$ satisfying H1 and H2.* From the first-order necessary condition for a minimum $\nabla\mathcal{F}_y(\hat{x}) = 0$,

$$2A^T(y - A\hat{x}) = \beta \nabla \Phi(\hat{x})$$

By (a), $AA^T$ is invertible. Multiplying both sides of the above equation by $\frac{1}{2}(AA^T)^{-1}A$ yields

$$y - A\hat{x} = \frac{\beta}{2}(AA^T)^{-1}A \, \nabla\Phi(\hat{x})$$

and then

$$\|y - A\hat{x}\|_\infty \le \frac{\beta}{2} \left\|\left|(AA^T)^{-1}A\right\|\right|_\infty \|\nabla\Phi(\hat{x})\|_\infty.$$

The entries of $\nabla\Phi(\hat{x})$ are given in (6). Introducing into(6) the assumption (b) and the observation that for all $i$ we have $\frac{|G_i^j x|}{\|G_i x\|} \le 1$, $1 \le j \le s$, leads to

$$\left|\partial_n \Phi(x)\right| \le \sum_{i \in J} \left|\varphi'(\|G_i x\|)\right| \sum_{j=1}^{s} \frac{\left|G_i^j x\right|}{\|G_i x\|} \left|G_i^j[n]\right| \le \|\varphi'\|_\infty \sum_{i \in J} \sum_{j=1}^{s} \left|G_i^j[n]\right|$$

Observe that the double sum on the right is the $\ell_1$ norm of the $n$th column of $G$. It follows that

$$\|\nabla\Phi(\hat{x})\|_\infty = \max_{n \in J} \left|\partial_n \Phi(x)\right| \le \|\varphi'\|_\infty \max_{n \in J} \sum_{i \in J} \sum_{j=1}^{s} \left|G_i^j[n]\right| = \|\varphi'\|_\infty \|\!|G|\!\|_1.$$

Suppose $x$ corresponds to an $m \times n$ image, $mn = p$, whose pixels are concatenated column by column. If $G_i \approx \nabla_{x_i}$, we have $G_i \in \mathrm{R}^{2\times p}$. For definiteness, let us consider a forward discretization for $\{\nabla_{x_i}\}$. Then the only non-zero entries of $G_i$ are $G_i^1[i] = 1$, $G_i^1[i+1] = -1$ and $G_i^2[i] = 1$, $G_i^2[i+m] = -1$. Whatever the boundary conditions, each column of $G$ has at most 4 non-zero entries with values in $\{-1, 1\}$. The same is true if $G$ corresponds to the first-order differences between each pixel and its 4 adjacent neighbors. In both cases,

$$\|\!|G|\!\|_1 = 4,$$

hence (18).

*$\varphi$ satisfying H1 and H3.* Using that $\delta\mathcal{F}_y(\hat{x})(v) \ge 0$ and $\delta\mathcal{F}_y(\hat{x})(-v)$ for every $v \in \mathrm{R}^p$, where $\delta\mathcal{F}_y(\hat{x})(v)$ is given in (10), we obtain that

$$\left|2(A\hat{x} - y)^T Av + \beta \sum_{i \in J_1} \frac{\varphi'(\|G_i\hat{x}\|)}{\|G_i\hat{x}\|} (G_i\hat{x})^T G_i v\right| \le \beta\varphi'(0) \sum_{i \in J_0} \|G_i v\|,$$

where $J_0$ and $J_1$ are defined in (9). Using that $\|\varphi'\|_\infty = \varphi'(0)$,

$$\begin{aligned}
\left|2(A\hat{x} - y)^T Av\right| &\le \beta \left| \sum_{i \in J_1} \varphi'(\|G_i\hat{x}\|) \sum_{j=1}^{s} \frac{G_i^j \hat{x}}{\|G_i\hat{x}\|} G_i^j v \right| + \beta\varphi'(0) \sum_{i \in J_0} \|G_i v\| \\
&\le \beta \sum_{i \in J_1} \varphi'(\|G_i\hat{x}\|) \sum_{j=1}^{s} \frac{\left|G_i^j \hat{x}\right|}{\|G_i\hat{x}\|} \left|G_i^j v\right| + \beta\varphi'(0) \sum_{i \in J_0} \|G_i v\| \\
&\le \beta\varphi'(0) \sum_{i \in J_1} \sum_{j=1}^{s} \left|G_i^j v\right| + \beta\varphi'(0) \sum_{i \in J_0} \|G_i v\| \\
&= \beta\varphi'(0) \sum_{i \in J_1} \|G_i v\|_1 + \beta\varphi'(0) \sum_{i \in J_0} \|G_i v\|.
\end{aligned}$$

Since $\|u\|_2 \leq \|u\|_1$ for every $u$, we can write that for every $v \in \mathbb{R}^p$,

$$\left|(A\hat{x} - y)^T Av\right| \leq \frac{\beta}{2}\varphi'(0)\sum_{i\in J}\|G_i v\|_1 = \frac{\beta}{2}\varphi'(0)\|Gv\|_1. \tag{19}$$

Let $\{e_n, 1 \leq n \leq q\}$ denote the canonical basis of $\mathbb{R}^q$. For any $n = 1, \ldots, q$, we apply (19) for

$$v = A^T(AA^T)^{-1}e_n.$$

Then $\left|(A\hat{x} - y)^T Av\right| = \left|(A\hat{x} - y)[n]\right|$ and (19) yields

$$\left|(A\hat{x} - y)[n]\right| \leq \frac{\beta}{2}\varphi'(0)\|GA^T(AA^T)^{-1}e_n\|_1 \leq \frac{\beta}{2}\varphi'(0)\|\!|G\|\!|_1 \left\|A^T(AA^T)^{-1}e_n\right\|_1$$

It follows that

$$\|A\hat{x} - y\|_\infty = \max_{1\leq n\leq q}\left|(A\hat{x} - y)[n]\right| \leq \frac{\beta}{2}\|\varphi'\|_\infty \|\!|G\|\!|_1 \max_{1\leq n\leq q}\left\|A^T(AA^T)^{-1}e_n\right\|_1.$$

Since $\left\|A^T(AA^T)^{-1}e_n\right\|_1$ is the $\ell_1$ norm of the $n$th column of $A^T(AA^T)^{-1}$, we find that $\left\|A^T(AA^T)^{-1}\right\|_\infty = \max_{1\leq n\leq q}\left\|A^T(AA^T)^{-1}e_n\right\|_1$. Hence (17) and the particular case of (18).

*Remark 2.* In a statistical setting, the quadratic data-fidelity term $\|Ax - y\|^2$ in (1) corresponds to white Gaussian noise on the data [2,6,1]. Such a noise is unbounded, even if its $\ell_2$ norm is finite. It may seem surprising to realize that whenever $\varphi$ is edge-preserving with $\|\varphi'\|_\infty$ bounded, the minimizers $\hat{x}$ of $\mathcal{F}_y$ give rise to residuals $(y - A\hat{x})[i]$, $1 \leq i \leq q$—i.e. estimates of the noise—that are tightly bounded as stated in (17) and (18). The proof of the theorem reveals that this behavior is due to the boundedness of the gradient of the regularization term.

Theorem 2 addresses observation matrices $A$ that have full row rank, i.e. which are surjective. Let us remark that the observation equation can always be rewritten with such a matrix.

When $x$ is an image defined on a convex subset of $\mathbb{R}^2$ and $\varphi$ is applied to $\nabla x$, and when $A$ is the identity, it is well known that the solution $\hat{x}$ and the data $y$ have the same mean [1]. This is an important property because the noise corrupting the data is usually zero-mean. Below we consider this question in our context where $x$ is defined on a discrete grid and $A$ is no longer the identity.

**Proposition 2.** *Assume that* $\mathbb{1} \in \ker G$ *and that* $A\mathbb{1} = c\mathbb{1}$ *for some* $c \neq 0$. *Then*

$$\mathbb{1}^T(A\hat{x} - y) = 0. \tag{20}$$

**Proof.** When $\varphi$ is smooth at zero (H1 and H2), multiplying (8) on the left by $\mathbb{1}^T$ leads to the result. Under H1 and H3, we notice that $\mathbb{1} \in \mathcal{K}_0$ where $\mathcal{K}_0$ is defined in (11). Applying (13) with $v = \mathbb{1}$ leads to the result. ◇

When $A$ is the identity, we find the classical result that $\frac{1}{p}\sum_{i=1}^{p}\hat{x}_i = \frac{1}{p}\sum_{i=1}^{p} y_i$. The assumption $A\mathbb{1} = c\mathbb{1}$ holds in the case of shift-invariant blurring under the periodic boundary condition since then $A$ is block-circulant. However, this assumption is very restrictive, so for general operators $A$ (20) does not hold.

## 3 Bounds on the Differences

The regularization term $\Phi$ being defined on the discrete gradients or differences $Gx$ we wish to find bounds characterizing how they behave in the solution $\hat{x}$. This problem is quite intricate. In this section we systematically suppose that

**H4** *$A^TA$ is invertible.*

The considerations are slightly different according to the smoothness of $\varphi(|t|)$.

### 3.1 $\varphi(|t|)$ Is Smooth at Zero

When the data $y$ involve a linear transform $A$ from the original $x$, it does not make sense to consider the (discrete) gradient or the differences of $y$. Instead, we can compare $G\hat{x}$ with a solution that is built based only on the data, without any prior. We will compare $G\hat{x}$ with $G\hat{z}$ where $\hat{z}$ is the least-squares solution, i.e. the minimizer of $\|Ax - y\|^2$:

$$\hat{z} = (A^TA)^{-1}A^Ty. \tag{21}$$

**Proposition 3.** *Assume that H1, H2 and H4 hold. For every $y \in \mathrm{R}^q$, if $\mathcal{F}_y$ has a (local) minimum at $\hat{x}$, there is a linear operator $H_y \in \mathcal{L}(\mathrm{R}^r;\mathrm{R}^r)$ such that*

$$\begin{cases} G\hat{x} = H_y\, G\, \hat{z}, \\ \text{Spectral Radius}(H_y) \leq 1. \end{cases} \tag{22}$$

*If $A$ is orthonormal, then $\|G\hat{x}\| \leq |\!|\!|G|\!|\!|\ \|y\|, \quad \forall y \in \mathrm{R}^q$.*

The result in (22) can be seen as a bound on the smoothness of the solution $\hat{x}$. Indeed, for any $\varepsilon > 0$ there is at least one norm $\|.\|$ such that $\|G\hat{x}\| \leq (1+\varepsilon)\|G\hat{z}\|$ (see e.g. [5]). Let us notice that this results holds even if $\varphi$ is nonconvex.

**Proof.** Multiplying both sides of (8) by $G(A^TA)^{-1}$ yields

$$G\hat{x} + \frac{\beta}{2}G(A^TA)^{-1}G^T\mathrm{diag}(\theta)G\hat{x} = G\hat{z}$$

Then the operator $H_y$ introduced in (22) reads

$$H_y = \left(I + \frac{\beta}{2}G(A^TA)^{-1}G^T\Theta\right)^{-1}. \tag{23}$$

Let $\lambda \in \mathrm{R}$ be an eigenvalue of $H_y$ and let the relevant eigenvector be $u \in \mathrm{R}^r$, $\|u\| = 1$. Starting with $H_y u = \lambda u$ we derive

$$u = \lambda u + \lambda \frac{\beta}{2} G(A^T A)^{-1} G^T \Theta u.$$

If $\Theta u = 0$, we have $\lambda = 1$, hence (22). Consider now that $\Theta u \neq 0$ then $u^T \Theta u > 0$. Multiplying both sides of the above equation from the left by $u^T \Theta$ yields

$$\lambda = \frac{u^T \Theta u}{u^T \Theta u + \frac{\beta}{2} u^T \Theta G(A^T A)^{-1} G^T \Theta u}.$$

Using that $u^T \Theta G(A^T A)^{-1} G^T \Theta u \geq 0$ shows that $\lambda \leq 1$.

In the case when $A$ is orthonormal, $G\hat{x} = G\big(I + \frac{\beta}{2} G^T \mathrm{diag}(\theta) G\big)^{-1} y$ and hence

$$\|G\hat{x}\| \leq \|G\|_2 \, \left\|\big(I + \frac{\beta}{2} G^T \mathrm{diag}(\theta) G\big)^{-1} y\right\| \leq \|G\|_2 \, \|y\|$$

where the last inequality is obtained by applying Lemma 1.

### 3.2 $\varphi(|t|)$ Is Nonsmooth at Zero

It is well known that if $\varphi'(0) > 0$ in (2), the minimizers $\hat{x}$ of $\mathcal{F}_y$ typically satisfy $G_i \hat{x} = 0$ for a certain (possibly large) subset of indexes $i \in J$ [16,17]. With any local minimizer $\hat{x}$ of $\mathcal{F}_y$ we associate the subsets $J_0$ and $J_1$ defined in (9) as well as the subspace $\mathcal{K}_0$ given in (11), $k = \dim K_0$, along with its orthonormal basis given by the columns of $B_0 \in \mathrm{R}^{k \times p}$.

**Proposition 4.** *Assume that H1, H3 and H4 hold. For every $y \in \mathrm{R}^q$, if $\mathcal{F}_y$ has a (local) minimum at $\hat{x}$, there is a linear operator $H_y \in \mathcal{L}(\mathrm{R}^r; \mathrm{R}^r)$ such that*

$$\begin{cases} G\hat{x} = H_y \, G\hat{z}_0, \\ \text{Spectral Radius}(H_y) \leq 1, \end{cases} \tag{24}$$

*where $\hat{z}_0$ is the least-squares solution constrained to $\mathcal{K}_0$, i.e. the point yielding*

$$\min_{x \in \mathcal{K}_0} \|Ax - y\|^2. \tag{25}$$

*If $A$ is the identity, then $\|G\hat{x}\| \leq \|G\|_2 \, \|y\|, \quad \forall y \in \mathrm{R}^q$.*

**Proof.** The least-squares solution constrained to $\mathcal{K}_0$ is of the form $\hat{z}_0 = B_0 \tilde{z}$ where $\tilde{z} \in \mathrm{R}^k$ yields the minimum of $\|AB_0 z - y\|^2$. Let us denote

$$A_0 = AB_0.$$

Then we can write that $\tilde{z} = (A_0^T A_0)^{-1} A_0^T y$ and hence

$$\hat{z}_0 = B_0 (A_0^T A_0)^{-1} A_0^T y.$$

Let $\tilde{x} \in \mathrm{R}^k$ be the unique element such that $\hat{x} = B_0\tilde{x}$. Multiplying both sides of (16) on the left by $(A_0^T A_0)^{-1}$ and using the expression for $\tilde{z}$ yields

$$\tilde{x} + \frac{\beta}{2}(A_0^T A_0)^{-1} G_0^T \mathrm{diag}(\theta) G_0 \tilde{x} = \tilde{z}. \tag{26}$$

Multiplying both sides of the last equation on the left by $GB_0$, then using the expression for $\hat{z}_0$ and reminding that $GB_0\tilde{x} = G\hat{x}$ and that $G_0 = GB_0$ shows that

$$\left(I + \frac{\beta}{2} G_0 (A_0^T A_0)^{-1} G_0^T \mathrm{diag}(\theta)\right) G\hat{x} = G\hat{z}_0.$$

The operator $H_y$ introduced in (24) reads $H_y = \left(I + \frac{\beta}{2} G_0 (A_0^T A_0)^{-1} G_0^T \mathrm{diag}(\theta)\right)^{-1}$. Its structure is similar to (23). By the same arguments, it is found that (24) holds.

When $A$ is the identity on $\mathrm{R}^p$, by using (26) we obtain

$$\left(I + \frac{\beta}{2} G_0^T \mathrm{diag}(\theta) G_0\right)\tilde{x} = B_0^T y$$

and then $\tilde{x} = \left(I + \frac{\beta}{2} G_0^T \mathrm{diag}(\theta) G_0\right)^{-1} B_0^T y$. Multiplying on the left by $GB_0$ yields

$$G\hat{x} = GB_0\left(I + \frac{\beta}{2} G_0^T \mathrm{diag}(\theta) G_0\right)^{-1} B_0^T y.$$

Using that $B_0$ is an orthonormal basis,

$$\begin{aligned}
\|G\hat{x}\| &\le \|G\|_2 \left\| B_0\left(I + \frac{\beta}{2} G_0^T \mathrm{diag}(\theta) G_0\right)^{-1} B_0^T y \right\| \\
&\le \|G\|_2 \left\| \left(I + \frac{\beta}{2} G_0^T \mathrm{diag}(\theta) G_0\right)^{-1} y \right\| \le \|G\|_2 \, \|y\|.
\end{aligned}$$

As in Proposition 3, the result in (24) is a bound on the smoothness of the solution $\hat{x}$ since for every $\varepsilon > 0$ there is at least one norm $\|.\|$ such that $\|G\hat{x}\| \le (1+\varepsilon)\|G\hat{z}_0\|$.

## 4 Conclusion

We provide simple bounds characterizing the minimizers of regularized least-squares. These bounds are for arbitrary signals and images of a finite size and they hold for regular values of $\beta$, and for possibly nonsmooth or nonconvex regularization terms.

## References


1. G. Aubert and P. Kornprobst, *Mathematical problems in images processing*, Springer-Verlag, Berlin, 2002.
2. J. E. Besag, *Digital image processing : Towards Bayesian image analysis*, J. of Applied Statistics, 16 (1989), pp. 395–407.

3. A. Blake and A. Zisserman, *Visual reconstruction*, The MIT Press, Cambridge, 1987.
4. C. Bouman and K. Sauer, *A generalized Gaussian image model for edge-preserving* MAP *estimation*, IEEE Trans. on Image Processing, 2 (1993), pp. 296–310.
5. P. G. Ciarlet, *Introduction à l'analyse numérique matricielle et à l'optimisation*, Collection mathématiques appliquées pour la maîtrise, Dunod, Paris, 5e ed., 2000.
6. G. Demoment, *Image reconstruction and restoration : Overview of common estimation structure and problems*, IEEE Trans. on Acoustics Speech and Signal Processing, ASSP-37 (1989), pp. 2024–2036.
7. F. Fessler, *Mean and variance of implicitly defined biased estimators (such as penalized maximum likelihood): Applications to tomography*, IEEE Trans. on Image Processing, 5 (1996), pp. 493–506.
8. D. Geman, *Random fields and inverse problems in imaging*, vol. 1427, École d'Été de Probabilités de Saint-Flour XVIII - 1988, Springer-Verlag, lecture notes in mathematics ed., 1990, pp. 117–193.
9. D. Geman and G. Reynolds, *Constrained restoration and recovery of discontinuities*, IEEE Trans. on Pattern Analysis and Machine Intelligence, PAMI-14 (1992), pp. 367–383.
10. S. Geman and D. E. McClure, *Statistical methods for tomographic image reconstruction*, in Proc. of the 46-th Session of the ISI, Bulletin of the ISI, vol. 52, 1987, pp. 22–26.
11. T. Hebert and R. Leahy, *A generalized EM algorithm for 3-D Bayesian reconstruction from Poisson data using Gibbs priors*, IEEE Trans. on Medical Imaging, 8 (1989), pp. 194–202.
12. R. Horn and C. Johnson, *Matrix analysis*, Cambridge University Press, 1985.
13. Y. Leclerc, *Constructing simple stable description for image partitioning*, International J. of Computer Vision, 3 (1989), pp. 73–102.
14. S. Li, *Markov Random Field Modeling in Computer Vision*, Springer-Verlag, New York, 1 ed., 1995.
15. D. Mumford and J. Shah, *Boundary detection by minimizing functionals*, in Proc. of the IEEE Int. Conf. on Acoustics, Speech and Signal Processing, 1985, pp. 22–26.
16. M. Nikolova, *Local strong homogeneity of a regularized estimator*, SIAM J. on Applied Mathematics, 61 (2000), pp. 633–658.
17. ——, *Weakly constrained minimization. Application to the estimation of images and signals involving constant regions*, J. of Mathematical Imaging and Vision, 21 (2004), pp. 155–175.
18. ——, *Analysis of the recovery of edges in images and signals by minimizing nonconvex regularized least-squares*, SIAM J. on Multiscale Modeling and Simulation, 4 (2005), pp. 960–991.
19. P. Perona and J. Malik, *Scale-space and edge detection using anisotropic diffusion*, IEEE Trans. on Pattern Analysis and Machine Intelligence, PAMI-12 (1990), pp. 629–639.
20. L. Rudin, S. Osher, and C. Fatemi, *Nonlinear total variation based noise removal algorithm*, Physica, 60 D (1992), pp. 259–268.
21. S. Saquib, C. Bouman, and K. Sauer, *ML parameter estimation for Markov random fields, with appications to Bayesian tomography*, IEEE Trans. on Image Processing, 7 (1998), pp. 1029–1044.
22. C. R. Vogel and M. E. Oman, *Iterative method for total variation denoising*, SIAM J. on Scientific Computing, 17 (1996), pp. 227–238.

# Numerical Invariantization for Morphological PDE Schemes

Martin Welk[1], Pilwon Kim[2], and Peter J. Olver[3]

[1] Mathematical Image Analysis Group
Faculty of Mathematics and Computer Science
Saarland University, 66041 Saarbrücken, Germany
welk@mia.uni-saarland.de
http://www.mia.uni-saarland.de
[2] Department of Mathematics
Ohio State University
Columbus, Ohio 43210, U.S.A.
pwkim@math.ohio-state.edu
[3] School of Mathematics
University of Minnesota
Minneapolis, MN 55455, U.S.A.
olver@math.umn.edu
http://www.math.umn.edu/~olver

**Abstract.** Based on a new, general formulation of the geometric method of moving frames, invariantization of numerical schemes has been established during the last years as a powerful tool to guarantee symmetries for numerical solutions while simultaneously reducing the numerical errors. In this paper, we make the first step to apply this framework to the differential equations of image processing. We focus on the Hamilton–Jacobi equation governing dilation and erosion processes which displays morphological symmetry, i.e. is invariant under strictly monotonically increasing transformations of gray-values. Results demonstrate that invariantization is able to handle the specific needs of differential equations applied in image processing, and thus encourage further research in this direction.

## 1 Introduction

Image filters based on partial differential equations play an important role in contemporary digital image processing. The field therefore has a need for efficient and accurate numerical algorithms for solving the PDEs that arise in applications.

The method of *invariantization* provides a general framework for designing numerical schemes for (ordinary and partial) differential equations [17,12,10] that preserve symmetries of the continuous-scale differential equation. The method is based on a new approach to the Cartan method of moving frames [4] that applies to completely general group actions, and has been extensively developed in the last few years [7,18]. The invariantization process is based on a choice of cross-section to the symmetry group orbits, and careful selection of the cross-section can produce a more robust numerical scheme that is better able to handle rapid variations and singularities. So far, the invariantization technique has been studied for standard numerical schemes for ordinary

F. Sgallari, A. Murli, and N. Paragios (Eds.): SSVM 2007, LNCS 4485, pp. 508–519, 2007.

differential equations [10], as well as for a number of partial differential equations including the heat equation, the Korteweg–deVries equation, and Burgers' equation [11], with encouraging results.

In this paper, we will investigate the applicability of the invariantization framework in the context of image processing. This field of application poses special needs in that it requires in particular an accurate representation of sharp discontinuity-type structures such as edges. A variety of partial differential equations with discontinuity-preserving properties has been developed over the years but often numerical dissipation adversely affects their favorable theoretical properties. We are therefore especially interested in whether invariantization can contribute to reducing numerical dissipation effects and thereby improve the treatment of edges in images.

For our investigation, we select the Hamilton–Jacobi equations governing the morphological processes of dilation and erosion [1,6]. They offer the advantage of combining formal simplicity with high relevance for image analysis – mathematical morphology being one of the oldest and most successful techniques in the field [15,21] – and a particularly attractive symmetry property, namely the so-called morphological invariance. The latter is also a characteristic of many other image processing PDEs such as mean curvature motion [8,9,2], and the affine invariant morphological scale space [1,20]. Thus, our present results can be viewed as a proof of concept for a wider application of the invariantization idea in this field.

## 2 Morphological PDEs

Dilation and erosion are the basic operations of mathematical morphology. Let $S$ be a closed connected convex set containing zero. Dilation of a gray-value image $u$ with $S$ as structuring element then comes down to taking at each location the maximum of gray-values within the translated structuring element while erosion uses the minimum instead:

$$\begin{aligned} \text{dilation:} \qquad & (u \oplus S)(x) := \max_{y \in S} u(x+y) \ , \\ \text{erosion:} \qquad & (u \ominus S)(x) := \min_{y \in S} u(x+y) \ . \end{aligned} \tag{1}$$

Dilation and erosion with disk-shaped structuring elements are closely related to the Hamilton–Jacobi partial differential equation

$$u_t = \pm |\nabla u| \tag{2}$$

where $\nabla u$ denotes the spatial gradient of $u$, i.e. $\nabla u = u_x$ in the 1D case, or $\nabla u = (u_x, u_y)^{\mathrm{T}}$ in the 2D case: Given the initial image $u_0$ at time $t = 0$, we evolve via (2) up to time $t$. In the case of the positive sign in (2) the resulting image $u$ will be the dilation of $u_0$ with the disk $S = D_t = \{x \mid |x| \le t\}$ as structuring element while in the case of the negative sign an erosion with the same structuring element results.

### 2.1 The Upwind Scheme

In spite of the simplicity of the PDE (2), its numerical evaluation remains a challenge. In image processing, one is particularly interested in the correct treatment of steep gradients which represent image edges. Under the Hamilton–Jacobi flow, these should propagate in space at constant speed without being blurred. Moreover, the partial differential equation (2) obeys a maximum–minimum principle which is also essential in image processing applications.

The simplest approach, a forward Euler discretization, with central spatial differences, generates oscillations in the vicinity of edges that violate the maximum–minimum principle; this is another manifestation of the general Gibbs phenomena observed in numerical approximations to discontinuous solutions, [14]. They can only be reduced but not eliminated by choosing very small time step sizes. Moreover, edges are smeared out as the number of iterations increases, and so the problem becomes even worse with smaller time steps.

An alternative scheme that avoids the oscillatory behavior and obeys the maximum–minimum principle is the upwind scheme [22]. Its idea is to discretize the first-order derivatives on the right-hand side of (2) by one-sided difference and switch between their possible directions depending on the local gradient direction, and hence on the information flow direction. In the case of 1D dilation, $u_t = |u_x|$, one step of the resulting explicit scheme with spatial grid size $h$ and time step size $\tau$ then reads

$$u_i^{k+1} = u_i^k + \frac{\tau}{h} \max\{u_{i+1}^k - u_i^k, u_{i-1}^k - u_i^k, 0\} . \tag{3}$$

For time step sizes $\tau \leq h$ this scheme respects the maximum–minimum principle.

There are several ways to adapt this idea to the two-dimensional case. We defer these considerations until Subsection 3.3.

## 3 Morphological Invariantization

In general, given a freely acting $r$-parameter transformation group acting on an $m$-dimensional space, one defines a moving frame by the choice of a cross-section to the group orbits, [7,18]. In practice, one begins by writing out the group transformations as explicit functions of the coordinates $z = (z_1, \ldots, z_m)$ and the group parameters $\lambda = (\lambda_1, \ldots, \lambda_r)$. One then normalizes $r$ of these expressions by equating them to well-chosen constants – typically either 0 or 1 – and solving for the group parameters in terms of the coordinates: $\lambda = \rho(z)$, which defines the moving frame map. The invariantization of any function, numerical scheme, etc. is then found by first writing out its transformed version and then replacing the group parameters by their moving frame formula. In particular, the invariantization of the coordinates $z_i$ yields the fundamental invariants $I_i(z)$, with those corresponding to the $r$ normalization coordinates being constant. The invariantization of any other function $F(z_1, \ldots, z_m)$ is then found by replacing each $z_i$ by its corresponding invariant (constant or not), leading to the invariantized function $I(z) = F(I_1(z), \ldots, I_m(z))$. In particular, invariantization does not change a function that is already invariant under the group. This so-called Replacement Rule makes it particularly easy to convert (both mathematically and in pre-existing software packages) numerical schemes into invariant numerical schemes. The resulting schemes are guaranteed to be consistent with the underlying differential equations, since invariantization preserves consistency of numerical schemes. In fact, one of the key benefits of the invariantization method is that it enables one to modify and tune existing schemes without affecting their consistency. In numerical applications, one selects the normalization coordinates and constants so as to try to eliminate as many of the error terms as possible; see [12,10,18] for further details.

### 3.1 Symmetry Group

The Hamilton–Jacobi PDE (2) that governs the processes of dilation and erosion displays one outstanding symmetry: It is invariant under *any* (differentiable) strictly monotonically increasing gray-value transformation [3]. This specific symmetry which is shared by a class of other PDEs relevant for image processing like mean curvature motion and affine curvature flow is called *morphological invariance*. PDEs with this symmetry can be re-formulated into intrinsic level set evolutions, i.e. curve or hypersurface evolutions of the level sets which depend on nothing else but the geometry of the evolving level sets themselves [19,3]. Infinitesimal generators for this symmetry are given by $f(u)\partial_u$ for arbitrary differentiable functions $f(u)$.

From the viewpoint of the invariantization of numerical schemes, the morphological symmetry is special in that it involves the function values only, in contrast to the symmetries of many other differential equations that involve both the independent and the dependent variables. Moreover, it is a very rich symmetry since the group of strictly increasing differentiable maps of $\mathbb{R}$ is an infinite-dimensional Lie pseudogroup. Though an extension of the invariantization framework for the Lie pseudogroup case has been recently developed, [18], to simplify the constructions, we will restrict our attention to a particular one-dimensional subgroup.

To this end, we use the strictly monotonically increasing transformations

$$\tau_\lambda : [0,1] \longrightarrow [0,1], \qquad u \longmapsto \frac{\lambda u}{1+(\lambda-1)u} \tag{4}$$

where $\lambda \in \mathbb{R}^+$ is the group parameter. This family of functions on $[0,1]$ forms a one-parameter Lie group with infinitesimal generator $u(1-u)\partial_u$, satisfying the group laws $\tau_\mu \circ \tau_\lambda = \tau_{\lambda\mu}$, $(\tau_\lambda)^{-1} = \tau_{1/\lambda}$.

### 3.2 The One-Dimensional Case

We want now to use the invariantization idea in order to improve the accuracy of numerical schemes for the 1D Hamilton–Jacobi equation $u_t = |u_x|$. With respect to image processing applications we are particularly interested in reducing numerical dissipation at edges.

The one-parameter Lie group selected in the previous subsection allows us to impose one equality constraint on the local numerical data. A closer look reveals that both the forward Euler scheme with central spatial differences and the upwind scheme are exact if the function $u$ is linear in $x$. We want therefore to annihilate locally the second derivative $u_{xx}$.

While this idea is easy to carry out for the central difference scheme, it turns out that the numerical dissipation is in no way reduced. Thus, we turn our attention to the upwind scheme. Since this scheme uses one-sided difference approximations for the first derivatives, the question arises which approximation of the second derivative should be used in the constraint that is to be enforced by invariantization. Since the first order derivative approximations can be considered as central differences located at inter-pixel positions $i \pm 1/2$, thus providing higher accuracy at these locations, we decide to use a four-pixel stencil centred at the same location for the second derivative.

Let us consider without loss of generality the case $u_x > 0$ in which the upwind scheme uses the right-sided derivative approximation. As approximation of the second derivative we then use $(u_{xx})_i \approx u_{i+2} - u_{i+1} - u_i + u_{i-1}$. For the invariantization at pixel $i$ in time step $k$, we linearly transform the pixel values $u_j^k$, $j = i-1, i, i+1, i+2$, to $[0,1]$ which gives $\tilde{u}_j^k$, and apply (4) to obtain $v_j^k = \tau_\lambda(\tilde{u}_j^k)$. Herein, the parameter $\lambda = \lambda_i^k > 0$ is to be determined, using the invariantization condition

$$v_{i+2}^k - v_{i+1}^k - v_i^k + v_{i-1}^k = 0\,. \tag{5}$$

Inserting (4) into (5) gives

$$\begin{aligned}0 = \lambda\Big((\lambda+1)^2(&-\tilde{u}_{i+2}^k\tilde{u}_{i+1}^k\tilde{u}_i^k + \tilde{u}_{i+2}^k\tilde{u}_{i+1}^k\tilde{u}_{i-1}^k + \tilde{u}_{i+2}^k\tilde{u}_i^k\tilde{u}_{i-1}^k - \tilde{u}_{i+1}^k\tilde{u}_i^k\tilde{u}_{i-1}^k)\\ &+ 2(\lambda+1)(\tilde{u}_{i+2}^k\tilde{u}_{i-1}^k - \tilde{u}_{i+2}^k\tilde{u}_i^k) + (\tilde{u}_{i+2}^k - \tilde{u}_{i+1}^k - \tilde{u}_i^k + \tilde{u}_{i-1}^k)\Big)\,.\end{aligned} \tag{6}$$

This equation has exactly one positive solution if the sequence $u_{i-1}^k, u_i^k, u_{i+1}^k, u_{i+2}^k$ is strictly monotonic. If this is not the case, our one-parameter transformation group in fact does not contain a transformation that satisfies (5). Instead, $\lambda = 0$ is then calculated as largest solution of (6). We select therefore a small $\varepsilon > 0$ and use $\tilde{\lambda} = \max\{\lambda, \varepsilon\}$ in the algorithm. Whenever $\lambda < \varepsilon$, our invariantization is therefore imperfect, and the second derivative error term not completely annihilated. Still, the numerical error is reduced in these cases.

One time step for pixel $i$ of a 1D signal reads therefore as follows.

1. Compute the one-sided derivative approximations $\Delta_{i,+}^k := u_{i+1}^k - u_i^k$, $\Delta_{i,-}^k := u_i^k - u_{i-1}^k$. If $\max\{\Delta_{i,+}^k, -\Delta_{i,-}^k, 0\} = 0$, let $u_i^{k+1} = u_i^k$ and finish. Otherwise, if $\Delta_{i,+}^k \geq -\Delta_{i,-}^k$, let $\sigma = +1$, else $\sigma = -1$. Let $\widehat{u}_j := u_{i+j\sigma}^k$ for $j = -1, 0, 1, 2$.
2. Let
$$\begin{aligned} m &:= \min\{\widehat{u}_j \mid j \in \{-1,0,1,2\}\}\,,\\ M &:= \max\{\widehat{u}_j \mid j \in \{-1,0,1,2\}\}\,,\\ \tilde{u}_j &:= \frac{\widehat{u}_j - m}{M-m}\,, \quad j \in \{-1,0,1,2\}\,. \end{aligned} \tag{7}$$
3. Compute the coefficients
$$\begin{aligned} a &:= \tilde{u}_2\tilde{u}_1\tilde{u}_0 - \tilde{u}_2\tilde{u}_1\tilde{u}_{-1} - \tilde{u}_2\tilde{u}_0\tilde{u}_{-1} + \tilde{u}_1\tilde{u}_0\tilde{u}_{-1}\,,\\ b &:= \tilde{u}_2\tilde{u}_{-1} - \tilde{u}_1\tilde{u}_0\,,\\ c &:= \tilde{u}_2 - \tilde{u}_1 - \tilde{u}_0 + \tilde{u}_{-1} \end{aligned} \tag{8}$$
and the transformation parameter
$$\lambda := 1 + \frac{b + \sqrt{b^2 + 4ac}}{a}\,. \tag{9}$$

Bound the transformation parameter via $\tilde{\lambda} := \max\{\lambda, \varepsilon\}$.

4. Transform the pixel values by

$$v_j := \tau_{\tilde{\lambda}}(\tilde{u}_j)\,, \quad j \in \{-1, 0, 1, 2\}\,. \tag{10}$$

5. Perform one step of the upwind scheme on the transformed data:

$$\tilde{v}_0 := v_0 + \frac{\tau}{h}(v_1 - v_0)\,. \tag{11}$$

6. Transform back:

$$u_i^{k+1} := m + (M - m)\tau_{1/\tilde{\lambda}}(\tilde{v}_0)\,. \tag{12}$$

It is easy to see that as for the unmodified upwind scheme, the maximum–minimum principle is guaranteed for the modified algorithm if the time step size fulfills $\tau < 1$.

### 3.3 The Two-Dimensional Case

In the two-dimensional situation there is a continuum of possible "upwind" directions. This adds complication to the discretization of first and second derivatives. While in the original upwind scheme an approximation of the gradient magnitude based on one-sided difference approximations of $u_x$ and $u_y$ works reasonably, experiments show that the invariantization via second derivatives is highly sensitive to misestimations of the second derivatives in gradient direction.

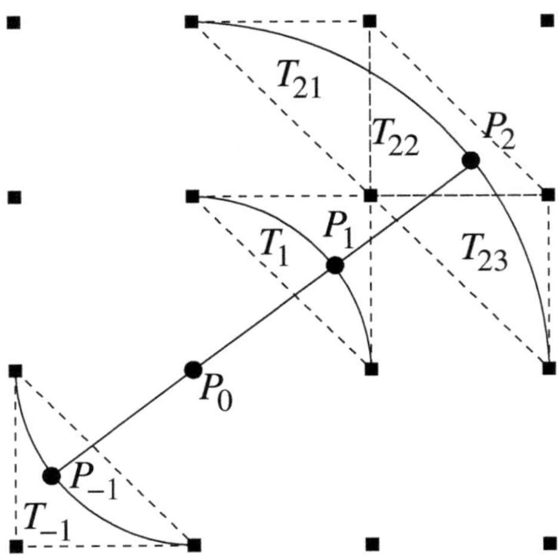


**Fig. 1.** Interpolation of a local 1-D subsample in gradient direction consisting of function values at $P_{-1}$, $P_0$, $P_1$, and $P_2$. The points $P_{-1}$, $P_1$, $P_2$ are located on circular arcs around $P_0$. $P_{-1}$ is linearly interpolated within the triangle $T_{-1}$, $P_1$ within $T_1$, and $P_2$ within one of the triangles $T_{21}, T_{22}, T_{23}$

However, since the 2D Hamilton–Jacobi flow at every single location is essentially a 1D process, we can directly build on our 1D algorithm in the following way. First, we compute via $u_x$ and $u_y$ approximations in the spirit of classical 2D upwind scheme implementations the gradient direction. Then, we resample the needed pixels along this direction to obtain a 1D section that represents the problem at the given location. While in principle this could be done via bilinear interpolation within grid squares, we choose an interpolation within isosceles right triangles of side length 1 that experimentally represents the local features of the 1D section slightly better (see Fig. 1). To interpolate $u$ for a point $P$ on a 1D section through $(i,j)$ in gradient direction, we use the triangle of grid points that encloses $P$ and whose vertex has either maximal or minimal distance to $(i,j)$ among the three corner points.

One time step for pixel $(i,j)$ then reads as follows.

1. Compute
$$\begin{aligned} \Delta^k_{i,j;x+} &:= u^k_{i+1,j} - u^k_{i,j}\,, & \Delta^k_{i,j;x-} &:= u^k_{i,j} - u^k_{i-1,j}\,, \\ \Delta^k_{i,j;y+} &:= u^k_{i,j+1} - u^k_{i,j}\,, & \Delta^k_{i,j;y-} &:= u^k_{i,j} - u^k_{i,j-1}\,. \end{aligned} \tag{13}$$
If $\max\{\Delta^k_{i,j;x+}, -\Delta^k_{i,j;x-}, 0\} = 0$, let $s_x := 0$, $\Delta_x := 0$, else if $\Delta^k_{i,j;x+} \ge -\Delta^k_{i,j;x-}$, let $s_x := +1$, $\Delta_x := \Delta^k_{i,j;x+}$, else let $s_x := -1$, $\Delta_x := -\Delta^k_{i,j;x-}$. Proceed analogously to determine $s_y$ and $\Delta_y$.
2. If $\Delta_x = \Delta_y = 0$, let $u^{k+1}_{i,j} = u^k_{i,j}$ and finish. Otherwise, let
$$\sigma_x := \frac{s_x}{\sqrt{s_x^2 + s_y^2}}\,, \qquad \sigma_y := \frac{s_y}{\sqrt{s_x^2 + s_y^2}}\,. \tag{14}$$
3. Compute
$$\widehat{u}_l := u^k_{i+l\sigma_x, j+l\sigma_y}\,, \qquad l = -1, 0, 1, 2\,, \tag{15}$$
where inter-pixel values of $u$ are linearly interpolated between three neighboring grid locations.
4. Apply steps 2–6 of the 1D algorithm to the 1D signal $\widehat{u}$, and assign the resulting value to $u^{k+1}_{i,j}$.

Though the calculation on the resampled 1D subsample involves inter-pixel sample values which are not present in the previous time step of the image, the maximum–minimum principle is still obeyed because the linear interpolation itself satisfies the maximum–minimum principle.

# 4 Experiments

## 4.1 One-Dimensional Case

To illustrate the effect of invariantization on a 1D example, Figure 2 shows the dilation of a single peak by the upwind scheme and our invariantized modification together with the theoretical solution. The higher sharpness of the invariantized scheme is clearly visible. We note that comparing with the theoretical result the propagation of the edge is slightly accelerated, an undesired effect that even increases for smaller time step sizes. The reason is that our scheme in its present form does not compensate for the bias in the treatment of regions of opposite curvature which is introduced by the use of one-sided derivative approximations. Since experimentally the effect is much smaller in the 2D case, we do not discuss remedies here.

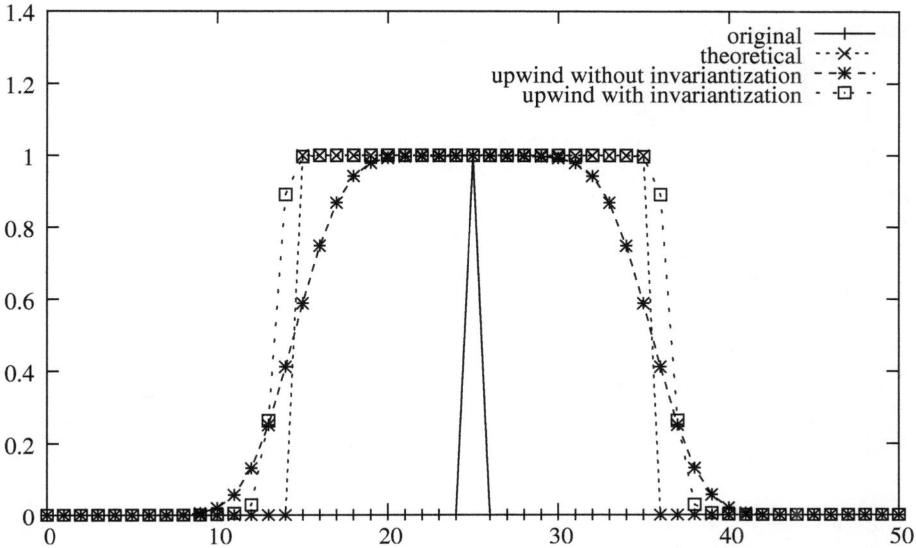


**Fig. 2.** 1D dilation of a single peak, 20 iterations with $\tau = 0.5$ of upwind scheme without and with invariantization. For comparison, the theoretical dilation result at evolution time $t = 10$ is also included.

## 4.2 Two-Dimensional Case

We demonstrate the 2D version of our algorithm with two experiments. First, Fig. 3 shows a test image featuring three discs, together with two stages of dilation evolution, for both the upwind scheme and our method. It is evident that the sharp boundaries of the expanding discs are preserved better by the invariantized scheme. The second stage of evolution demonstrates the correct handling of the merging between the objects.

At the same time, one can observe the reasonable degree of rotational invariance achieved by our method. This has been supported by choosing a smaller time step size than in the 1D case. Still, a close look suggests that a small amount of additional blur is added in diagonal directions due to the interpolation procedure used to obtain the 1D subsample.

A 1D section from the 2D evolution (slightly above the horizontal diameter of one circle, as indicated in Fig. 3) is shown in Fig. 4. The increased sharpness of the invariantized scheme is again visible; the interface between the bright and dark region attains a width of approx. four to five pixels, which is in accordance with the effective region of influence of each time step. This degree of edge blur remains essentially unchanged even after many more time steps.

The position of the expanded contour under an exact dilation with equal evolution time is also shown. Here, the speed of expansion of the bright regions is in good agreement with the theoretically derived speed, even with the smaller time step size. Besides this, the maximum–minimum stability is confirmed by Fig. 4.

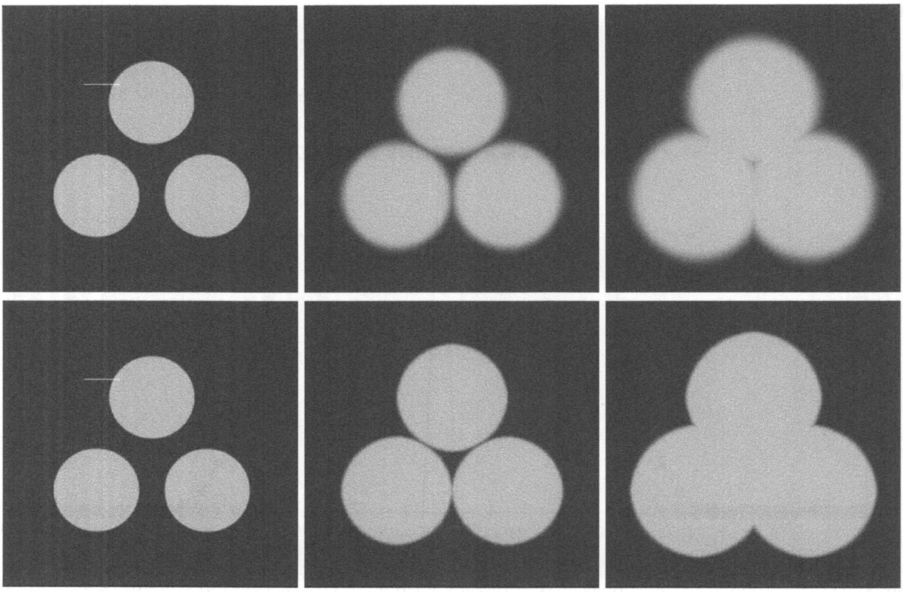

**Fig. 3. Top, left:** Original image (256 × 256 pixels) showing three discs. White line marks a 1D section shown in Fig. 4. **Top, middle:** Dilation by upwind scheme without invariantization, 100 iterations, time step $\tau = 0.1$. **Top, right:** Same but with 200 iterations. **Bottom row:** Same as above but with invariantized upwind scheme.

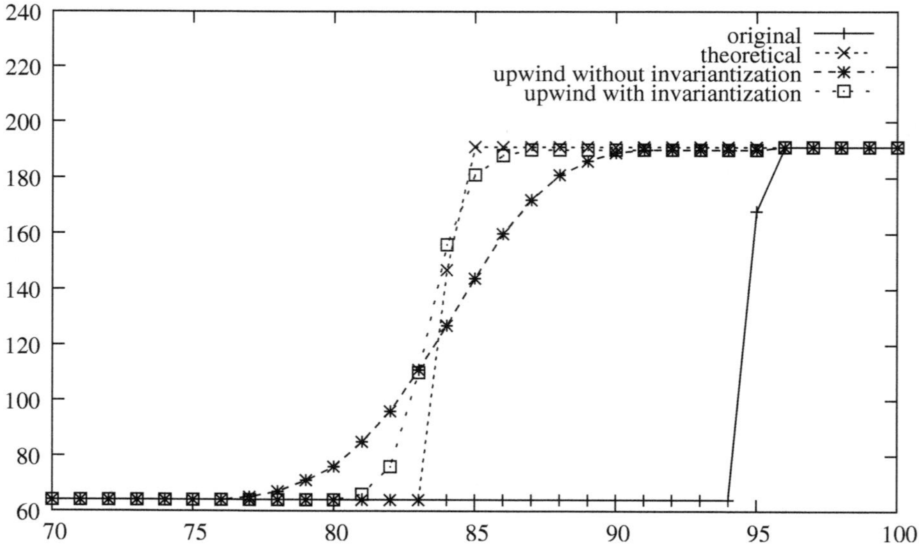


**Fig. 4.** Profiles of 2D dilation results along the line marked in Fig. 3. Original image, theoretical result of dilation at time $t = 10$, upwind scheme without and with invariantization, 100 iterations with $\tau = 0.1$.

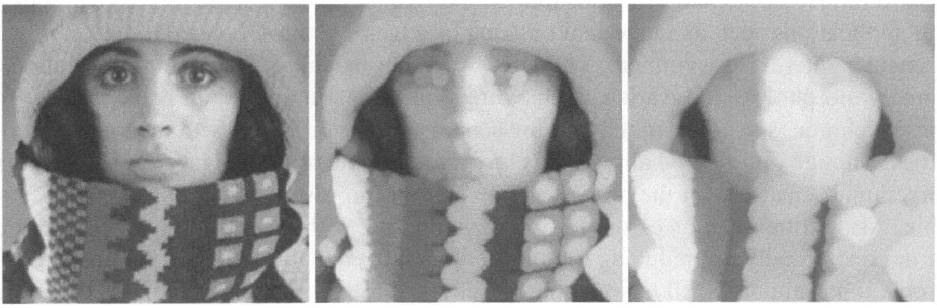

**Fig. 5. Left:** Original image (256 $\times$ 256 pixels). **Middle:** Dilation with invariantized upwind scheme, 50 iterations with $\tau = 0.1$ **Right:** Same with 150 iterations.

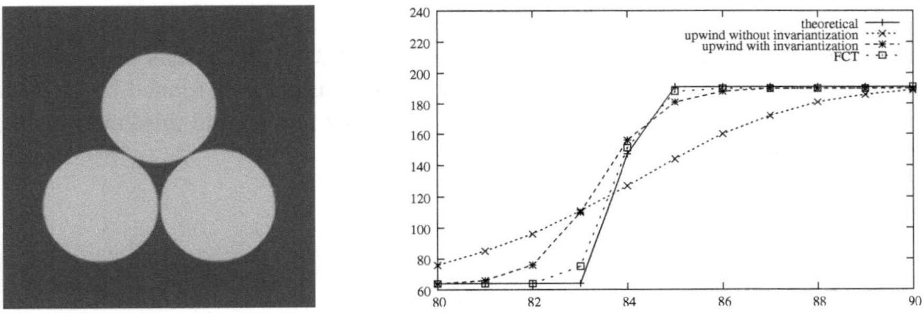


**Fig. 6. Left:** Dilation of Fig. 3 by flux-corrected transport (FCT) scheme, 20 iterations with $\tau = 0.5$ (provided by M. Breuß). **Right:** Central part of the dilated profiles from Fig. 4 and corresponding profile of the FCT result.

Fig. 5 finally demonstrates the dilation process of a natural halftone image by our algorithm.

A comparison with another state-of-the-art numerical method for evaluating the Hamilton–Jacobi equation of dilation is shown in Fig. 6. The flux-corrected transport (FCT) scheme by Breuß and Weickert [5] relies on a direct modelling of, and compensation for, the numerical viscosity of the upwind scheme. Thereby, it achieves a higher degree of sharpness, with an interface width of only one to two pixels. Note that for the FCT scheme a larger time step size has been used. Since the two approaches exploit different aspects of the process, it will be worth conducting future research to look for ways how their respective advantages can be combined.

Erosion is equivalent to dilation of an inverted image and can therefore be performed in a completely analogous fashion by our method. Due to space limitations, we have not included an erosion example here.

## 5 Conclusion

We have demonstrated that the invariantization technique can be applied to the numerics of PDE-based image filters. It allows to raise the accuracy of numerical schemes and also to reduce numerical problems that are particularly troublesome in image processing applications such as numerical blurring of edges. We have concentrated here on a particular interesting symmetry of PDEs occurring in image processing applications, namely morphological invariance. One direction of ongoing research is the transfer of these techniques to other image filtering schemes based on PDEs with invariance properties. Though our method already displays a reasonable rotational invariance, the high directional sensitivity of the process makes further improvements in this respect desirable. Also, combinations of the invariantization idea with conservation properties are of interest. Finally, by reducing the morphological symmetry to a one-parameter subgroup, it has not been fully used so far; a better exploitation of its potential is therefore also a topic of continued research.

## Acknowledgement

The authors thank the Institute for Mathematics and its Applications in Minnesota, where this project was initiated. The work of the first author was funded by Deutsche Forschungsgemeinschaft under grant We 3563/2-1. The second and third authors were supported in part by NSF Grant DMS 05–05293.

## References

1. L. Alvarez, F. Guichard, P.-L. Lions, and J.-M. Morel. Axioms and fundamental equations in image processing. *Archive for Rational Mechanics and Analysis*, 123:199–257, 1993.
2. L. Alvarez, P.-L. Lions, and J.-M. Morel. Image selective smoothing and edge detection by nonlinear diffusion. II. *SIAM Journal on Numerical Analysis*, 29:845–866, 1992.
3. F. Cao. *Geometric Curve Evolution and Image Processing*, volume 1805 of *Lecture Notes in Mathematics*. Springer, Berlin, 2003.
4. É. Cartan. *La méthode du repère mobile, la théorie des groupes continus, et les espaces généralisés.* Exposés de Géométrie no. 5, Hermann, Paris, 1935.
5. M. Breuß and J. Weickert. A shock-capturing algorithm for the differential equations of dilation and erosion. *Journal of Mathematical Imaging and Vision*, in press.
6. R.W. Brockett and P. Maragos. Evolution equations for continuous-scale morphological filtering. *IEEE Transactions on Signal Processing*, 42:3377–3386, 1994.
7. M. Fels and P.J. Olver. Moving coframes. II. Regularization and theoretical foundations. *Acta Appl. Math.*, 55:127–208, 1999.
8. M. Gage and R. S. Hamilton. The heat equation shrinking convex plane curves. *Journal of Differential Geometry*, 23:69–96, 1986.
9. M. Grayson. The heat equation shrinks embedded plane curves to round points. *Journal of Differential Geometry*, 26:285–314, 1987.
10. P. Kim. *Invariantization of Numerical Schemes for Differential Equations Using Moving Frames.* Ph.D. Thesis, University of Minnesota, Minneapolis, 2006.
11. P. Kim. *Invariantization of the Crank-Nicholson Method for Burgers' Equation.* Preprint, University of Minnesota, Minneapolis.
12. P. Kim and P.J. Olver. Geometric integration via multi-space. *Regular and Chaotic Dynamics*, 9(3):213–226, 2004.
13. R. Kimmel. *Numerical Geometry of Images.* Springer, Berlin 2004.
14. P.D. Lax. Gibbs Phenomena. *Journal of Scientific Computing*, 28:445–449, 2006.

15. G. Mathéron. *Random Sets and Integral Geometry.* Wiley, New York 1975.
16. P.J. Olver. *Applications of Lie Groups to Differential Equations.* Springer, New York 1986.
17. P.J. Olver. Geometric foundations of numerical algorithms and symmetry. *Applicable Algebra in Engineering, Communication and Computing*, 11:417–436, 2001.
18. P.J. Olver. A survey of moving frames. In H. Li, P.J. Olver, and G. Sommer, eds., *Computer Algebra and Geometric Algebra with Applications.* Volume 3519 of *Lecture Notes in Computer Science*, 105–138, Springer, Berlin, 2005.
19. G. Sapiro. *Geometric Partial Differential Equations and Image Analysis.* Cambridge University Press, Cambridge, UK, 2001.
20. G. Sapiro and A. Tannenbaum. Affine invariant scale-space. *International Journal of Computer Vision*, 11:25–44, 1993.
21. J. Serra. *Image Analysis and Mathematical Morphology.* Volume 1. Academic Press, London 1982.
22. J.A. Sethian. *Level Set Methods.* Cambridge University Press, Cambridge, UK, 1996.

# Bayesian Non-local Means Filter, Image Redundancy and Adaptive Dictionaries for Noise Removal

Charles Kervrann[1,3], Jérôme Boulanger[1,3], and Pierrick Coupé[2]

[1] INRIA, IRISA, Campus de Beaulieu, 35 042 Rennes, France
[2] Université de Rennes 1, IRISA, Campus de Beaulieu, 35 042 Rennes, France
[3] INRA - MIA, Domaine de Vilvert, 78 352 Jouy-en-Josas, France

**Abstract.** Partial Differential equations (PDE), wavelets-based methods and neighborhood filters were proposed as locally adaptive machines for noise removal. Recently, Buades, Coll and Morel proposed the *Non-Local (NL-) means filter* for image denoising. This method replaces a noisy pixel by the weighted average of other image pixels with weights reflecting the similarity between local neighborhoods of the pixel being processed and the other pixels. The *NL-means filter* was proposed as an intuitive *neighborhood filter* but theoretical connections to diffusion and non-parametric estimation approaches are also given by the authors. In this paper we propose another bridge, and show that the *NL-means filter* also emerges from the Bayesian approach with new arguments. Based on this observation, we show how the performance of this filter can be significantly improved by introducing adaptive local dictionaries and a new statistical distance measure to compare patches. The new *Bayesian NL-means filter* is better parametrized and the amount of smoothing is directly determined by the noise variance (estimated from image data) given the patch size. Experimental results are given for real images with artificial Gaussian noise added, and for images with real image-dependent noise.

## 1 Introduction

Denoising (or restoration) is still a widely studied and an unsolved problem in image processing. Many methods have been suggested in the literature, and a recent outstanding review of them can be found in [4]. Some of the more advanced methods are based on PDEs [28,29,33,37] and aim at preserving the image details and local geometries while removing the undesirable noise; in general, an initial image is progressively approximated by filtered versions which are smoother or simpler in some sense. Other methods incorporate a neighborhood of the pixel under consideration and perform some kind of averaging on the gray values. One of the earliest examples for such filters has been presented by Lee [23] and a recent evolution is the so-called *bilateral filter* [34] with theoretical connections to local mode filtering [38], non-linear diffusion [3,5] and nonlocal regularization approaches [27,12].

However, natural images often contain many structured patterns which can be misclassified either as details to be preserved or noise, when usual neighborhood filters are applied. Very recently, the so-called *NL-means filter* has been proposed by Buades *et al.*

F. Sgallari, A. Murli, and N. Paragios (Eds.): SSVM 2007, LNCS 4485, pp. 520–532, 2007.

[4] that can deal with such a "structured" noise: for a given pixel, the restored gray value is obtained by the weighted average of the gray values of all pixels in the image; each weight is proportional to the similarity between the local neighborhood of the pixel being processed and the neighborhood corresponding to the other image pixels. A similar patch-based regularization approach based on the key idea of iteratively increasing a window at each pixel and adaptively weighting input data has been also proposed in [21] with excellent results on a commonly-used image database [31]. The success of the *NL-means filter* (see [21,22,26,25,7,16,2]), inspired by the Efros and Leung's exemplar-based approach for texture synthesis [11], is mainly related to image redundancy. A similar idea was early and independently proposed for Gaussian [9] and impulse noise [36,39] removal in images, and more recently for image inpainting [8]. Similarities between image patches have been also used in the early 90's for texture segmentation [15,20]. More recently, other recent denoising methods demonstrated that representations based on local image patches outperform the best denoising methods of the state-of-the-art [1,21,10,18,13]; in [32], patch-based Markov random field (MRF) models and learning techniques have been introduced to capture non-local pairwise interactions, and were successfully applied in image denoising.

In this paper, we present a new Bayesian motivation for the *NL-means filter* briefly described in Section 2. In Section 3, we adopt a blockwise (vectorial) representation and introduce spatially adaptive dictionaries in the modeling for better contrast restoration (Section 4). Using the proposed Bayesian framework, we revise the usual Euclidean distance used for patch comparison, yielding to a filter which is better parametrized, and with a higher performance. In Section 4, we also show how smooth parts in the image can be are better recovered if the restored image is "recycled" once. Experimental results on artificial and real images are presented in Section 5, and the performance is very close to the most competitive denoising methods. It is worth noting that the proposed modeling framework is general and could be used to restore images corrupted by non-Gaussian noises in applications such as biomedical imaging (microscopy, ultrasound imagery, ...) or remote sensing.

## 2 Image Denoising with the NL-Means Filter

In this section, a brief overview of the *NL-means* method introduced in [4] is presented. Consider a gray-scale image $z = (z(x))_{x\in\Omega}$ defined over a bounded domain $\Omega \subset \mathbb{R}^2$, (which is usually a rectangle) and $z(x) \in \mathbb{R}_+$ is the noisy observed intensity at pixel $x \in \Omega$. The *NL-means filter* is defined as

$$NL\, z(x) = \frac{1}{C(x)} \sum_{y\in\Omega} w(x,y)\, z(y) \tag{1}$$

where $NL\, z(x)$ at pixel $x$ is the weighted average of all gray values in the image and $C(x)$ is a normalizing factor, i.e. $C(x) = \sum_{y\in\Omega} w(x,y)$. The weights $w(x,y)$ defined as

$$w(x,y) = \exp\left(-\frac{1}{h^2}\int_{\mathbb{R}^2} G_a(t)|z(x+t)-z(y+t)|^2 dt\right) := \exp -\frac{\|\mathbf{z}(x)-\mathbf{z}(y)\|_{2,a}^2}{h^2} \tag{2}$$

express the amount of similarity between the vectorized image patches $\mathbf{z}(x)$ and $\mathbf{z}(y)$ (or neighborhoods) of each pair of pixels $x$ and $y$ involved in the computation. The decay parameter $h \approx 12\sigma$ acts as a filtering parameter. A Gaussian kernel $G_a(\cdot)$ of standard deviation $a$ is used to take into account the distance between the central pixel and other pixels in the patch. In (2), the pixel intensities of a $\sqrt{n} \times \sqrt{n}$ square neighborhood $B(x)$ centered at pixel $x$, are taken and reordered lexicographically to form a $n$-dimensional vector $\mathbf{z}(x) := (z(x_k), x_k \in B(x)) \in \mathbb{R}^n$. In [4], it was shown that $7 \times 7$ patches are able to take care of the local geometries and textures seen in the image while removing undesirable distortions. The range of the search space in the *NL-means* algorithm can be as large as the whole image. In practice, it is necessary to reduce the total number of computed weights – $|\Omega|$ weights for each pixel – to improve the performance of the algorithm. This can be achieved by selecting patches in a *semi-local* neighborhood corresponding to a search window $\Delta(x)$ of $21 \times 21$ pixels. The *NL-means filter* we will now consider, is then defined as

$$NL_h z(x) = \frac{1}{C(x)} \sum_{y \in \Delta(x)} e^{-\|\mathbf{z}(x)-\mathbf{z}(y)\|^2/h^2} z(y), \quad C(x) = \sum_{y \in \Delta(x)} e^{-\|\mathbf{z}(x)-\mathbf{z}(y)\|^2/h^2} \quad (3)$$

where, for the sake of simplicity, $\|\cdot\|$ denotes the usual $\ell^2$-norm. In practice, it is just required to set the $\sqrt{n} \times \sqrt{n}$ patch size, the search space $\Delta(x)$ and the filtering parameter $h$. Buades *et al.* showed that this filter substantially outperforms the *bilateral filter* [34] and other iterative approaches [33].

Since, several accelerated versions of this filter have been proposed [26,7]. In [4], Buades *et al.* recommended the *vectorial (or block-based) NL-means filter* defined as

$$NL_h \mathbf{z}(x) = \frac{1}{C(x)} \sum_{y \in \Delta(x)} e^{-\|\mathbf{z}(x)-\mathbf{z}(y)\|^2/h^2} \mathbf{z}(y), \quad C(x) = \sum_{y \in \Delta(x)} e^{-\|\mathbf{z}(x)-\mathbf{z}(y)\|^2/h^2}, \quad (4)$$

which amounts to simultaneously restore pixels of a whole patch $\mathbf{z}(x)$ from nearby patches. The restored value at a pixel $x$ is finally obtained by averaging the different estimators available at that position [4]. This filter can been considered as a fast implementation of the *NL-means filter*, especially if the blocks are picked up on a subsampled grid of pixels. In this paper, the proposed filter is inspired by this intuitive *vectorial NL-means filter* [4], but also by other recent accelerated versions [26,7], as explained in the next sections. The related Bayesian framework enables to establish the relationships between these algorithms, to justify some underlying statistical assumptions and to give keys to set the control parameters of the *NL-means filter*. It is worth noting that this framework could be also used to remove noise in applications for which the noise distribution is assumed to be known and non-Gaussian.

## 3 A Bayesian Risk Formulation

In a probabilistic setting, the image denoising problem is usually solved in a discrete setting. The estimation problem is then to guess a $n$-dimensional patch $\mathbf{u}(x)$ from its noisy version $\mathbf{z}(x)$ observed at point $x$. Typically, the unknown vectorized image patch

$\mathbf{u}(x)$ is defined as $\mathbf{u}(x) := (u(x_k), x_k \in B(x)) \in \mathbb{R}^n$ where $B(x)$ defines the $\sqrt{n} \times \sqrt{n}$ square neighborhood of point $x$ and the pixels in $\mathbf{u}(x)$ are ordered lexicographically. Let us suppose now that $\mathbf{u}(x)$ is unknown but we can observe $\mathbf{z}(x) = f(\mathbf{u}(x), \mathbf{v}(x))$ where $\mathbf{z}(x) := (z(x_k), x_k \in B(x))$, $\mathbf{v}(x)$ represents noise and $f(\cdot)$ is a linear or a non-linear function related to the image formation process. The noise $\mathbf{v}(x)$ is a random vector which components are iid, and $\mathbf{u}(x)$ is considered as a stochastic vector with some unknown probability distribution function (pdf).

**Conditional mean estimator.** To compute the optimal Bayesian estimator for the vector $\mathbf{u}(x)$, it is necessary to define an appropriate loss function $L(\mathbf{u}(x), \widehat{\mathbf{u}}(x))$ which measures the loss associated with choosing an estimator $\widehat{\mathbf{u}}(x)$ when the true vector is $\mathbf{u}(x)$. The optimal estimator $\widehat{\mathbf{u}}_{opt}(x)$ is found by minimizing the posterior expected loss

$$\mathbb{E}[L(\mathbf{u}(x), \widehat{\mathbf{u}}(x))] = \sum_{\mathbf{u}(x) \in \Lambda} L(\mathbf{u}(x), \widehat{\mathbf{u}}(x))\, p(\mathbf{u}(x)|\mathbf{z}(x)),$$

taken with respect to the posterior distribution $p(\mathbf{u}(x)|\mathbf{z}(x))$ and $\Lambda$ denotes the large space of all configurations of $\mathbf{u}(x)$ (e.g $|\Lambda| = 256^n$ if $u(x) \in \{0, \cdots, 255\}$). The loss function used in most cases is $L(\mathbf{u}(x), \widehat{\mathbf{u}}(x)) = 1 - \delta(\mathbf{u}(x), \widehat{\mathbf{u}}(x))$ where the $\delta$ function equals 1 if $\mathbf{u}(x) = \widehat{\mathbf{u}}(x)$ and 0 otherwise. Minimizing $\mathbb{E}[L(\mathbf{u}(x), \widehat{\mathbf{u}}(x))]$ is then equivalent to choose $\widehat{\mathbf{u}}_{opt}(x) = \arg\max p(\mathbf{u}(x)|\mathbf{z}(x))$, with the motivation that this should correspond to the most likely vector given the observation $\mathbf{z}(x)$. However, this loss function $L(\mathbf{u}(x), \widehat{\mathbf{u}}(x))$ may not be the most appropriate since it assigns 0 cost only to the perfect solution and unit cost to all other estimators. Another thought would be to use a cost function that depends on the number of pixels that are in error such as $L(\mathbf{u}(x), \widehat{\mathbf{u}}(x)) = \|\mathbf{u}(x) - \widehat{\mathbf{u}}(x)\|^2$. Assuming this quadratic loss function, the optimal Bayesian estimator is then

$$\widehat{\mathbf{u}}_{opt}(x) = \arg\min_{\widehat{\mathbf{u}}(x)} \sum_{\mathbf{u}(x)} \|\mathbf{u}(x) - \widehat{\mathbf{u}}(x)\|^2\, p(\mathbf{u}(x)|\mathbf{z}(x)) = \sum_{\mathbf{u}(x)} \mathbf{u}(x)\, p(\mathbf{u}(x)|\mathbf{z}(x)).$$

Referred as the conditional mean estimator, $\widehat{\mathbf{u}}_{opt}(x)$ can be also written as

$$\widehat{\mathbf{u}}_{opt}(x) = \sum_{\mathbf{u}(x)} \mathbf{u}(x) \frac{p(\mathbf{u}(x), \mathbf{z}(x))}{p(\mathbf{z}(x))} = \frac{\sum_{\mathbf{u}(x)} \mathbf{u}(x) p(\mathbf{z}(x)|\mathbf{u}(x)) p(\mathbf{u}(x))}{\sum_{\mathbf{u}(x)} p(\mathbf{z}(x)|\mathbf{u}(x)) p(\mathbf{u}(x))} \tag{5}$$

by using the Bayes' and marginalization rules, and $p(\mathbf{z}(x)|\mathbf{u}(x))$ and $p(\mathbf{u}(x))$ respectively denote the distribution of $\mathbf{z}(x)|\mathbf{u}(x)$ and the prior distribution of $\mathbf{u}(x)$.

**Bayesian filter and image redundancy.** Ideally, we would like to know the pdfs $p(\mathbf{z}(x)|\mathbf{u}(x))$ and $p(\mathbf{u}(x))$ to compute $\widehat{\mathbf{u}}_{opt}(x)$ for each point $x$ in the image from a large number of "repeated" observations (i.e. images). Unfortunately, we have only one image at our disposal, meaning that we have to adopt another way of estimating these pdfs. Due to the fact that the pdfs $p(\mathbf{z}(x)|\mathbf{u}(x))$ and $p(\mathbf{u}(x))$ cannot be obtained from a number of observations at the same point $x$, we choose to use the observations at a number of neighboring points taken in a semi-local neighborhood (or window) $\Delta(x)$. The window $\Delta(x)$ needs to be not too large since the remote samples are likely less significant and can originate from other spatial contexts. We then assume that this set of nearby samples may be considered as a set of samples from $p(\mathbf{u}(x)|\mathbf{z}(x))$.

More formally, we suppose that $p(\mathbf{u}(x)|\mathbf{z}(x))$ is unknown, but we have a set $\{\mathbf{u}(x_1), \mathbf{u}(x_2), \cdots, \mathbf{u}(x_{N(x)})\}$ of $N(x)$ posterior samples taken in $\Delta(x)$. In what follows, $|\Delta(x)|$ is fixed for all the pixels but the size $N(x) \leq |\Delta(x)|$ is spatially varying since irrelevant and unlikely samples in $\Delta(x)$ are preliminarily discarded. From this set, we start by examining the prior distribution $p(\mathbf{u}(x))$. A first natural ideal would be to introduce MRFs and Gibbs distributions to capture interactions between pixels in the image patch, but the MRF framework involves the computationally intensive estimation of additional *hyperparameters* which must be likely adapted to each spatial position. Due to the huge domain space $\Lambda$, a computational alternative is then to set $p(\mathbf{u}(x))$ to uniform, i.e. $p(\mathbf{u}(x)) = 1/N(x)$. This means there is no preference to choose a vector $\mathbf{u}(x_i)$ in the set $\{\mathbf{u}(x_1), \cdots, \mathbf{u}(x_{N(x)})\}$ assumed to be composed of $N(x)$ preliminarily selected "similar" patches. Then, we have the following approximations (see [17]):

$$\frac{1}{N(x_i)} \sum_{j=1}^{N(x_i)} \mathbf{u}(x_j) p(\mathbf{z}(x_i)|\mathbf{u}(x_j)) \xrightarrow{P} \sum_{\mathbf{u}(x)} \mathbf{u}(x) p(\mathbf{z}(x)|\mathbf{u}(x)) p(\mathbf{u}(x)),$$

$$\frac{1}{N(x_i)} \sum_{j=1}^{N(x_i)} p(\mathbf{z}(x_i)|\mathbf{u}(x_j)) \xrightarrow{P} \sum_{\mathbf{u}(x)} p(\mathbf{z}(x)|\mathbf{u}(x)) p(\mathbf{u}(x)),$$

and we can propose a reasonable estimator $\widehat{\mathbf{u}}_N(x_i)$ for $\widehat{\mathbf{u}}_{opt}(x)$:

$$\widehat{\mathbf{u}}_N(x_i) = \frac{\frac{1}{N(x_i)} \sum_{j=1}^{N(x_i)} \mathbf{u}(x_j) p(\mathbf{z}(x_i)|\mathbf{u}(x_j))}{\frac{1}{N(x_i)} \sum_{j=1}^{N(x_i)} p(\mathbf{z}(x_i)|\mathbf{u}(x_j))}. \tag{6}$$

Nevertheless, we do not have the set $\{\mathbf{u}(x_1), \cdots, \mathbf{u}(x_{N(x)})\}$, but only a spatially varying dictionary $\mathcal{D}(x) = \{\mathbf{z}(x_1), \cdots, \mathbf{z}(x_{N(x)})\}$ composed of noisy versions is available. A way to solve this problem will be then to substitute $\mathbf{z}(x_j)$ to $\mathbf{u}(x_j)$ in (6) as shown in the next section. In a second step, this estimator will be refined by substituting the "aggregated" estimator computed from $\widehat{\mathbf{u}}_N(x_j)$ (see below) to $\mathbf{u}(x_j)$. Indeed, the restored patch at pixel $x_j$ is a better approximation of $\mathbf{u}(x_j)$ than the noisy patch $\mathbf{z}(x_j)$ used as a *"pilot"* estimator, and the performance will be improved at the second iteration.

**Aggregation of estimators.** The estimator (6) requires spatially sliding windows over the whole image for image reconstruction. Hence, a set of $L$ (constant for uniform image sub-sampling) concurrent scalar values $\widehat{u}_{N,1}(x_i), \widehat{u}_{N,2}(x_i), \cdots, \widehat{u}_{N,L}(x_i)$ is calculated for the same pixel $x_i$ due to the overlapping between patches. This set of competing estimators must be fused or *aggregated* into the one final estimator $\widetilde{u}(x_i)$ at pixel $x_i$. Actually, when no estimator is a "clear winner", one may prefer to combine the different estimators $\widehat{u}_{N,1}(x_i), \widehat{u}_{N,2}(x_i), \cdots, \widehat{u}_{N,L}(x_i)$ and a natural approach, well-grounded in statistics [6], is to use a convex or linear combination of estimators [19,10]. Here, our *aggregate* estimator $\widetilde{u}_N(x_i)$ is simply defined as the average of competing estimators:

$$\widetilde{u}_N(x_i) = \frac{1}{L} \sum_{l=1}^{L} \widehat{u}_{N,l}(x_i). \tag{7}$$

In practice, patches are picked up on a sub-sampled grid (e.g. factor 3) of pixels to speed up the algorithmic procedure (e.g. factor 8), while preserving a good visual quality.

# 4 Bayesian NL-Means Filter

As explained before, to compute $\widehat{\mathbf{u}}_N(x_i)$, we first substitute $\mathbf{z}(x)$ to $\mathbf{u}(x)$ in (6). This yields the following estimator

$$\frac{\frac{1}{N(x_i)}\sum_{j=1}^{N(x_i)} p(\mathbf{z}(x_i)|\mathbf{z}(x_j))\mathbf{z}(x_j)}{\frac{1}{N(x_i)}\sum_{j=1}^{N(x_i)} p(\mathbf{z}(x_i)|\mathbf{z}(x_j))} \approx \widehat{\mathbf{u}}_N(x_i) \tag{8}$$

which can be computationally calculated provided the pdfs are known. It is worth noting that $p(\mathbf{z}(x_i)|\mathbf{z}(x_j))$ is not well defined if $x_i = x_j$ and it could be recommended to set $p(\mathbf{z}(x_i)|\mathbf{u}(x_i)) \approx \max_{x_j \neq x_i} p(\mathbf{z}(x_i)|\mathbf{z}(x_j))$ in (8) (see [22]). The central data point involved in the computation of its own average is then re-weighted to get the higher weight. Actually, the probability to detect an exact copy of $\mathbf{z}(x_i)$ corrupted by noise in the neighborhood tends to 0 because the space of $\sqrt{n} \times \sqrt{n}$ patches is huge and, to be consistent, it is necessary to limit the influence of the central patch.

In the remainder of this section, we shall now consider the usual image model

$$\mathbf{z}(x) = \mathbf{u}(x) + \mathbf{v}(x) \tag{9}$$

where $\mathbf{v}(x)$ is an additive white Gaussian noise with variance $\sigma^2$. We will further assume that the likelihood can be factorized as $p(\mathbf{z}(x_i)|\mathbf{z}(x_j)) = \prod_{k=1}^{n} p(z(x_{i,k})|z(x_{j,k}))$ with $x_{i,k} \in B(x_i)$ and $x_{j,k} \in B(x_j)$. It follows that $\mathbf{z}(x_i)|\mathbf{z}(x_j)$ follows a multivariate normal distribution, $\mathbf{z}(x_i)|\mathbf{z}(x_j) \sim \mathcal{N}(\mathbf{z}(x_j), \sigma^2\mathbf{I}_n)$ where $\mathbf{I}_n$ is the $n$-dimensional identity matrix. From (8), the filter adapted to white Gaussian noise is then given by

$$\frac{1}{C(x_i)}\sum_{j=1}^{N(x_i)} e^{-\|\mathbf{z}(x_i)-\mathbf{z}(x_j)\|^2/(2\sigma^2)}\mathbf{z}(x_j) \text{ with } C(x_i) = \sum_{j=1}^{N(x_i)} e^{-\|\mathbf{z}(x_i)-\mathbf{z}(x_j)\|^2/(2\sigma^2)}. \tag{10}$$

If we arbitrarily set $N(x_i) = N$ to a constant value and $h^2 = 2\sigma^2$, this filter is nothing else than the *vectorial NL-means filter* given in (4). However, it is recommended to set $h \approx 12\sigma$ in [4] to produce satisfying denoising results. In our experiments, it is also confirmed that $h \approx 5\sigma$ is good choice if we use (3) and (4) for denoising. Actually, the filtering parameter $h$ is actually set to a higher value than the expected value $\sqrt{2}\sigma$ in practical imaging. In the next sections, we shall see how this parameter can be better interpreted and theoretically estimated.

**Spatially adaptive dictionaries .** The filter (10) can be refined if the adaptive dictionary $\mathcal{D}(x_i) = \{\mathbf{z}(x_1), \cdots, \mathbf{z}(x_{N(x_i)})\}$ around $x_i$ is reliably obtained using an *off-line* procedure. Since $\mathcal{D}(x_i)$ is assumed to be composed of noisy versions of the more likely samples from the posterior distribution $p(\mathbf{u}(x_i)|\mathbf{z}(x_i))$, the irrelevant image patches in $\Delta(x_i)$ must be discarded in a preliminary step. Consequently, the size $N(x_i)$ is adaptive according to local spatial contexts and a simple way to detect these unwanted samples can be based on local statistical features between images patches. In our experiments, we have consider two basic features, that is the mean $m(\mathbf{z}(x)) = n^{-1}\sum_{k=1}^{n} z(x_k)$ and the variance $\text{var}(\mathbf{z}(x)) = n^{-1}\sum_{k=1}^{n}(z(x_k) - m(\mathbf{z}(x)))^2$ of a vectorized patch $\mathbf{z}(x) := (z(x_k), x_k \in B(x))$.

Intuitively, $\mathbf{z}(x_j)$ will be discarded from the local dictionary $\mathcal{D}(x_i)$ if the mean $m(\mathbf{z}(x_j))$ is too 'far" from the mean $m(\mathbf{z}(x_i))$ when $\mathbf{z}(x_j)$ and $\mathbf{z}(x_i)$ are compared. More formally, if $|m(\mathbf{z}(x_j)) - m(\mathbf{z}(x_i))| > \lambda_\alpha \sigma/\sqrt{n}$, where $\lambda_\alpha \in \mathbb{R}_+$ is chosen as a quantile of the standard normal distribution, the hypothesis that the two patches belong to the same "population" is rejected. Hence, setting $\lambda_\alpha = 3$ given $\mathbb{P}(|m(\mathbf{z}(x_j)) - m(\mathbf{z}(x_i))| \leq \lambda_\alpha \sigma/\sqrt{n}) = 1 - \alpha$, yields to $\alpha = 2(1 - \Phi(\lambda_\alpha/\sqrt{2})) = 0.034$ where $\Phi$ means the Laplace distribution.

Similarly, the variance $\text{var}(\mathbf{z}(x_j))$ is expected to be close to the variance $\text{var}(\mathbf{z}(x_i))$ for the central patch $\mathbf{z}(x_i)$. A $F$-test[1] is used and the ratio $F = \frac{\max(\text{var}(\mathbf{z}(x_j)), \text{var}(\mathbf{z}(x_i)))}{\min(\text{var}(\mathbf{z}(x_j)), \text{var}(\mathbf{z}(x_i)))}$ is compared to a threshold $T_{\beta,n-1}$ to determine if the value falls within the zone of acceptance of the hypothesis that the variances are equal. The threshold $T_{\beta,n-1}$ is the critical value for the $F$-distribution with $n-1$ degrees of freedom for each patch and a significance level $\beta$. Typically, when $7 \times 7$ patches are compared, we have $\mathbb{P}(F > T_{0.05,48} = 1.6) = 0.05$. If the ratio $F$ exceeds the value $T_{\beta,n-1}$, the sample $\mathbf{z}(x_j)$ is discarded from the local dictionary $\mathcal{D}(x_i)$.

This formal description is related to the intuitive approach proposed in [26,7,16] to improve the performance of the *NL-means filter*.

**New statistical distance measure for patch comparison .** In this section, we propose to revise the distance used for patch comparison, yielding to a *NL-means filter* which is better parametrized. In (3) and (4), it is implicitly assumed that $\mathbf{z}(x_i)|\mathbf{z}(x_j) \sim \mathcal{N}(\mathbf{z}(x_j), \frac{1}{2}h^2\mathbf{I}_n)$. Actually, this hypothesis is valid only for non-overlapping and statistically independent patches, but most of patches overlapped in $\Delta(x)$ since $\Delta(x)$ is not so large (e.g $21 \times 21$ pixels). At the opposite, if $\mathbf{z}(x_j)$ is horizontally (or vertically) shifted by one pixel from the location of $\mathbf{z}(x_j)$, $\mathbf{z}(x_i)|\mathbf{z}(x_j)$ is expected to follow a multivariate Laplacian distribution. However, this statistical hypothesis does not hold true for arbitrary locations of overlapping patches in $\Delta(x)$. The adjustment of the decay parameter $h \approx 5\sigma$ in (3) to a value higher that the expected value $\sqrt{2}\sigma$ is probably related to the fact that the two compared patches are not independent. Note that some pixel values are in common in the two vectors but at different locations.

Hence, $p(\mathbf{z}(x_i)|\mathbf{z}(x_j))$ must be carefully examined and we propose the following definition for the likelihood: $p(\mathbf{z}(x_i)|\mathbf{z}(x_j)) \propto e^{-\phi(\|\mathbf{z}(x_i)-\mathbf{z}(x_j)\|)}$. Typically, we can choose $\phi(t) = t^2$ or $\phi(t) = |t|$ (or a scaled version) to compare patches. Here, we examine the distribution of $\|\mathbf{z}(x_i) - \mathbf{z}(x_j)\|$ from the local dictionary $\mathcal{D}(x_i)$ to determine $\phi$. First, it can be observed that $\mathbb{E}[\|\mathbf{z}(x_i) - \mathbf{z}(x_j)\|]$ is non-zero in most cases and the probability to find an exact copy of $\mathbf{z}(x_i)$ in $\Delta(x)$ tends to 0, especially if $\Delta(x)$ is large. The maximum of the assumed zero-mean multivariate Gaussian distribution in (3) should be then "shifted" to $\mathbb{E}[\|\mathbf{z}(x_i){-}\mathbf{z}(x_j)\|]$. However, this training step could be hard in practice since it must adapted to each spatial position, and we propose instead to use asymptotic results. Actually, we have already assumed $(z(x_{i,k}){-}z(x_{j,k})) \sim \mathcal{N}(0, 2\sigma^2)$ when two pixels taken in $\mathbf{z}(x_i)$ and $\mathbf{z}(x_j) \in \mathcal{D}(x_i)$ are compared. Hence, the normalized distance $\text{dist}(\mathbf{z}(x_i), \mathbf{z}(x_j)) = \|\mathbf{z}(x_i) - \mathbf{z}(x_j)\|^2/(2\sigma^2)$ follows a chi-square $\chi^2_n$ distribution with $n$ degrees of freedom. For $n$ large ($n \geq 25$), it can be proved that

[1] The $F$-distribution is used to compare the variance of two independent samples from a normally distributed population.

$\sqrt{2\,\mathrm{dist}(\mathbf{z}(x_i), \mathbf{z}(x_j))}$ is approximately normally distributed with mean $\sqrt{2n-1}$ and unit variance:

$$p\left(\sqrt{2\,\mathrm{dist}(\mathbf{z}(x_i), \mathbf{z}(x_j))}\right) \propto \exp -\frac{1}{2}\left(\sqrt{2\,\mathrm{dist}(\mathbf{z}(x_i), \mathbf{z}(x_j))} - \sqrt{2n-1}\right)^2 \quad (11)$$
$$\propto \exp -\left(\frac{\|\mathbf{z}(x_i) - \mathbf{z}(x_j)\|^2}{2\sigma^2} - \frac{\|\mathbf{z}(x_i) - \mathbf{z}(x_j)\|}{(\sigma/\sqrt{2n-1})} + \frac{2n-1}{2}\right).$$

Accordingly, we define the likelihood as $p(\mathbf{z}(x_i)|\mathbf{z}(x_j)) \propto \exp -\phi(\|\mathbf{z}(x_i) - \mathbf{z}(x_j)\|)$ and choose $\phi(t) = at^2 - b|t| + c$ with $a = 1/(2\sigma^2)$, $b = \sqrt{2n-1}/\sigma$ and $c = (2n-1)/2$ depending only on the patch size $n$ and the noise variance $\sigma^2$. From our experiments, it was confirmed that no additional adjustment parameter is necessary provided the noise variance is robustly estimated and the performance is maximum for the true noise variance as expected. Figure 1 (bottom-right) shows the functions $e^{-t^2/h^2}$ and $e^{-(|t|/\sigma - \sqrt{2n-1})^2/2}$ by setting $n = 49$, $\sigma^2 = 1$ and $h = 5\sigma$, and then illustrates how data points are currently weighted when the original *NL-means filter* and the so-called *Adaptive NL-means filter* (*ANL*) defined as

$$ANL_{\sigma,n}\mathbf{z}(x_i) = \frac{\sum_{j=1}^{N(x_i)} \exp -\frac{1}{2}\left(\frac{\|\mathbf{z}(x_i) - \mathbf{z}(x_j)\|}{\sigma} - \sqrt{2n-1}\right)^2 \mathbf{z}(x_j)}{\sum_{j=1}^{N(x_i)} \exp -\frac{1}{2}\left(\frac{\|\mathbf{z}(x_i) - \mathbf{z}(x_j)\|}{\sigma} - \sqrt{2n-1}\right)^2}. \quad (12)$$

where $N(x_i) = \#\{\mathcal{D}(x_i)\}$, are applied. Note that the data at point $x_i$ should participate significantly to the weighted average. Accordingly, since $p(\mathbf{z}(x_i)|\mathbf{z}(x_j))$ tends to 0 when $x_i = x_j$ in (12), we arbitrarily decide to set $p(\mathbf{z}(x_i)|\mathbf{z}(x_i)) \approx \max_{x_j \neq x_i} p(\mathbf{z}(x_i)|\mathbf{z}(x_j))$ as explained before (see also [22]).

**Bayesian NL-means filter and plugin estimator.** In the previous sections, $\mathbf{z}(x_j)$ was substituted to $\mathbf{u}(x_j)$ in (6) to give (8) and further (12). Now, we are free to substitute the vector $\widetilde{\mathbf{u}}_{ANL}(x_j)$ of aggregated estimators (computed from the set of restored blocks $\{ANL_{\sigma,n}\mathbf{z}(x_j)\}$ , see (7)) to $\mathbf{u}(x_j)$. This *plugin ANL* estimator defined as

$$ANL_{\sigma,n}\widetilde{\mathbf{u}}_{ANL}(x_i) = \frac{\sum_{j=1}^{N(x_i)} \exp -\frac{1}{2}\left(\frac{2\,\|\mathbf{z}(x_i) - \widetilde{\mathbf{u}}_{ANL}(x_j)\|}{\sigma} - \sqrt{2n-1}\right)^2 \widetilde{\mathbf{u}}_{ANL}(x_j)}{\sum_{j=1}^{N(x_i)} \exp -\frac{1}{2}\left(\frac{2\,\|\mathbf{z}(x_i) - \widetilde{\mathbf{u}}_{ANL}(x_j)\|}{\sigma} - \sqrt{2n-1}\right)^2} \quad (13)$$

is expected to improve the restored image since $\widetilde{\mathbf{u}}_{ANL}(x_j)$ is a better approximation of $\mathbf{u}(x_j)$ than $\mathbf{z}(x_j)$. In (13) the restored image is recycled but the weights is a rescaled function (theoretically by a factor 2) of the distance between the *"pilot"* estimator $\widetilde{\mathbf{u}}_{ANL}(x_j)$ given by (12) and the input vector $\mathbf{z}(x_i)$. The estimators are finally aggregated to produce the final restored image (see (7)).

## 5 Experiments

In this section, we evaluate the performance of different versions of the Bayesian filter and the original *NL-means filter* using the peak signal-to-noise ratio (PSNR in db) defined as PSNR $= 10\log_{10}(255^2/\text{MSE})$ with MSE $= |\Omega|^{-1}\sum_{x\in\Omega}(z_0(x) - \widehat{u}(x))^2$ where $z_0$ is the noise-free original image. We also use the *"method noise"* described in [4] which consists in computing the difference between the input noisy image and its denoised version. The *NL-means filter* (see (3)) was applied with $h = 5\sigma$ and our experiments have been performed with $7 \times 7$ patches and $15 \times 15$ search windows, corresponding to the best visual results and the best PSNR values. For all the presented results, we set $T_{0.05,n-1} = 1.6$ and $\lambda_{0.034} = 3$ to build spatially adaptive dictionaries.

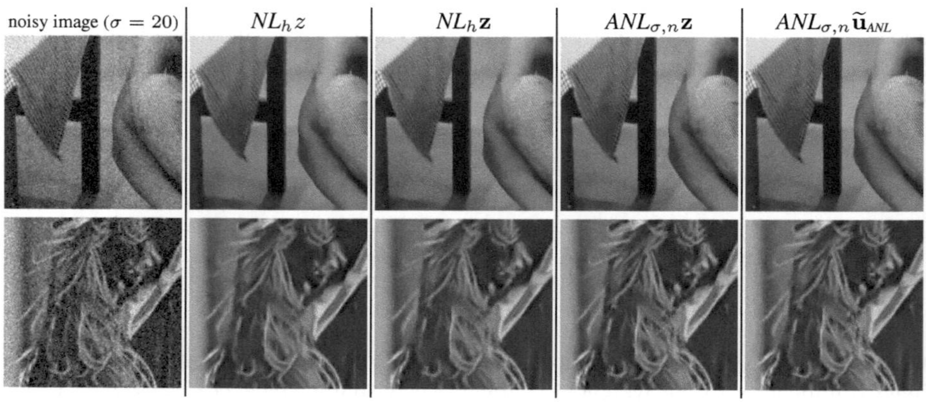


| | Timings | *Lena* $512 \times 512$ | *Barbara* $512 \times 512$ |
|---|---|---|---|
| $NL_h z$ | 58.1 sec | 31.85 db | 30.27 db |
| $NL_h \mathbf{z}$ | 96.2 sec | 32.04 db | 30.49 db |
| $ANL_{\sigma,n}\mathbf{z}$ | 75.2 sec | 32.51 db | 30.79 db |
| $ANL_{\sigma,n}\widetilde{\mathbf{u}}_{ANL}$ | 173.3 sec | **32.63** db | **30.88** db |
| Fast $ANL_{\sigma,n}\mathbf{z}$ | 10.6 sec | 32.36 db | 30.61 db |
| Fast $ANL_{\sigma,n}\widetilde{\mathbf{u}}_{ANL}$ | 21.2 sec | 32.49 db | 30.71 db |

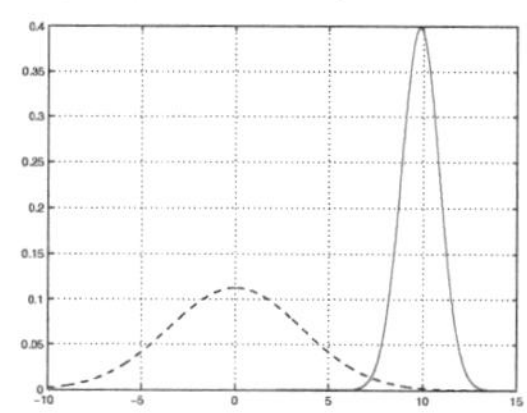

**Fig. 1.** Comparisons of different *NL-means filters*. From top to bottom and from left to right: part of noisy images (*Barbara, Lena*, $\sigma = 20$), *NL-means filter*, *vectorial NL-means filter*, *Adaptive NL-means filter*, plugin *Adaptive NL-means filter*; numerical results for each filter; comparison of exponential weights for the original *NL-means filter* (blue dashed line) and for the *Adaptive NL-means filter* (solid green line) (see text).

The potential of the estimation method is mainly illustrated with the $512 \times 512$ *Lena* and *Barbara* images corrupted by an additive white-Gaussian noise (WGN) (PSNR = 22.13 db, $\sigma = 20$). We compared the original *NL-means* algorithm with the proposed modifications and Fig. 1 shows the denoising results using several filters ($n = 7 \times 7$ and $N = 15 \times 15$): *i)* the *NL-means filter* $NL_h z$ when the similarity is only measured by the Euclidean distance (see (3)); *ii)* the *vectorial NL-means filter* $NL_h \mathbf{z}$ with

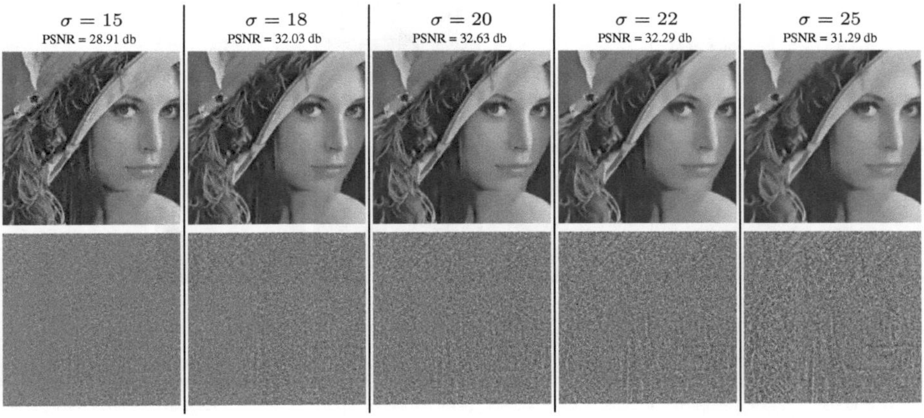


**Fig. 2.** Denoising with the plugin *Adaptive NL-means filter* $ANL_{\sigma,n}\widetilde{\mathbf{u}}_{ANL}$ of the artificially corrupted *Lena* image (WGN, $\sigma = 20$) with different estimations of the noise variance and experiments with the *"method noise"* (bottom)

sliding blocks but no spatial sub-sampling (see (4)); *iii)* our *Adaptive NL-means filter* $ANL_{\sigma,n}\mathbf{z}$ which includes adaptive dictionaries and a similarity measure based on the new distance (see (12)); *iv)* the plugin *Adaptive NL-means filter* $ANL_{\sigma,n}\widetilde{\mathbf{u}}_{ANL}$ (see (13)). In most cases, the PSNR values are slightly affected if a spatial sub-sampling (by a factor 3) is used in the implementation, but the time computing is drastically reduced (speed is 8 times less than before): the implementation of the *fast Adaptive NL-means filter* took 10 sec on a single-CPU PC 2.0 Ghz running linux, and the full *Adaptive NL-means filter* (with no spatial sub-sampling) took 75 sec for denoising a $512 \times 512$ image (see table in Fig. 1). In these practical examples, the use of spatially adaptive dictionaries enables to enhance contrasts. Note that the residual noise in flat zones is more reduced with no additional blur, when $ANL_{\sigma,n}\widetilde{\mathbf{u}}_{ANL}$ is applied. In Fig. 2, we modified the estimation of noise variance to assess the sensitivity of this parameter on filtering results. In general, the PSNR value is maximum with the true value (e.g. $\sigma = 20$ in Fig. 2) but decreases if this value is under-estimated or over-estimated. In Fig. 2, the estimated noise component is similar to a simulated white Gaussian noise but contains few

**Table 1.** Left: performance of different methods when applied to test noisy (WGN) images (*NL-means filter* $NL_h z$ is implemented as in (3) and the maximum weighting $w(x,y)$ at $x = y$ is given by $w(x,x) = max_{x \neq y} w(x,y)$) ; right: performance of the plugin *Adaptive NL-means filter* $ANL_{\sigma,n}\widetilde{\mathbf{u}}_{ANL}$ for different signal-to-noise ratios (WGN)

| Image<br>$\sigma$/PSNR | *Lena*<br>20/22.13 | *Barbara*<br>20/22.18 | *Boats*<br>20/22.17 | *House*<br>20/22.11 | *Peppers*<br>20/22.19 |
|---|---|---|---|---|---|
| $ANL_{\sigma,n}\widetilde{\mathbf{u}}_{ANL}$ | **32.63** | **30.88** | **30.16** | **33.24** | **30.75** |
| *NL-means filter* | 31.85 | 30.27 | 29.42 | 32.24 | 29.86 |
| Dabov *et al.* [10] | 33.03 | 31.77 | 30.65 | 33.54 | 30.87 |
| Elad *et al.* [13] | 32.38 | 30.83 | 30.36 | 33.20 | 30.82 |
| Kervrann *et al.* [21] | 32.64 | 30.37 | 30.12 | 32.90 | 30.59 |
| Portilla *et al.* [31] | 32.66 | 30.32 | 30.38 | 32.39 | 30.31 |
| Roth *et al.* [32] | 31.92 | 28.32 | 29.85 | 32.17 | 30.58 |
| Rudin *et al.* [33] | 30.48 | 27.07 | 29.02 | 31.03 | 28.51 |

| $\sigma$/PSNR | *Lena*<br>$512^2$ | *Barbara*<br>$512^2$ | *Boats*<br>$512^2$ | *House*<br>$256^2$ | *Peppers*<br>$256^2$ |
|---|---|---|---|---|---|
| 5 / 34.15 | 37.98 | 36.93 | 36.39 | 38.89 | 37.13 |
| 10 / 28.13 | 35.25 | 33.82 | 33.18 | 35.67 | 33.87 |
| 15 / 24.61 | 33.68 | 32.21 | 31.45 | 34.23 | 32.06 |
| 20 / 22.11 | 32.63 | 30.88 | 30.16 | 33.24 | 30.75 |
| 25 / 20.17 | 31.55 | 29.77 | 29.11 | 32.30 | 29.77 |
| 50 / 14.15 | 27.51 | 24.91 | 25.13 | 27.64 | 23.84 |

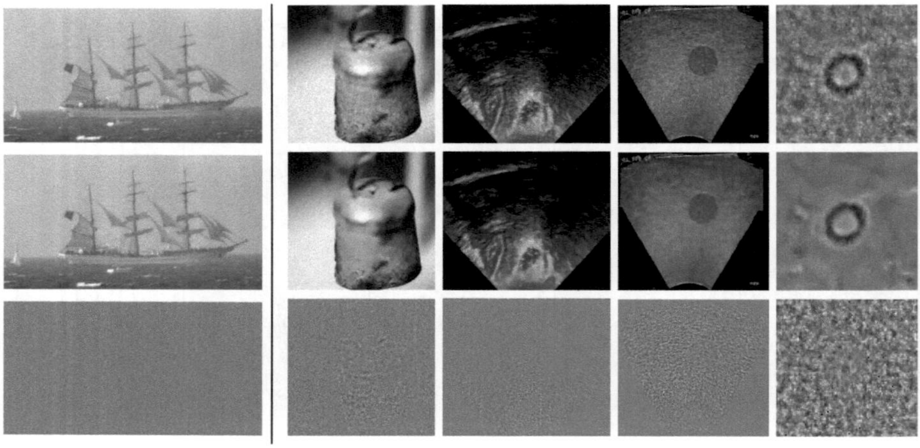

**Fig. 3.** Denoising of real noisy images. From top to bottom: original images, denoised images, *"method noise"* experiments. The old photograph (left) has been denoised using an usual additive noise and the four other images (right) are denoised using the image-dependent noise model ($\gamma = 0.5$, see text).

geometric structures if we under-estimate or over-estimate the noise variance. To evaluate the performance of those filters, we reported the PSNR values for different versions of the *NL-means filter*. In table 1, the numerical results are improved using our filter, with performance very close to competitive patch-based denoising methods. Note that the best results (PSNR values) were recently obtained by filtering in 3D transform domain and combining sliding-window transform processing with block-matching [10].

In the second part of experiments, we have applied $ANL_{\sigma,n}\widetilde{\mathbf{u}}_{ANL}$ to restore an old photograph (Fig. 3 - left column); in that case, the noise variance is estimated from image data (see [21]). Nevertheless, in real digital imaging, images are better described by the following model $z(x) = u(x) + u^{\gamma}(x)\,\varepsilon(x)$ where the sensor noise $u^{\gamma}(x)\,\varepsilon(x)$ is defined as a function of $u(x)$ and $\varepsilon(x)$ is a zero-mean Gaussian noise of variance $\sigma^2$. Accordingly, the noise in bright regions is higher than in dark regions (see Fig. 2 - second column). From experiments on real digital images [14], it was confirmed that $\gamma \approx 0.5$ ($\gamma = 0$ corresponds to WGN in the previous experiments). Accordingly, we modify the normalized distance as follows: $\text{dist}(\mathbf{z}(x_i), \mathbf{z}(x_j)) = \|\mathbf{z}(x_i) - \mathbf{z}(x_j)\|^2/(2\sigma^2\mathbf{z}(x_j))$ with $\sigma \in [1.5, 3]$. Moreover, this model $z(x) = u(x) + \sqrt{u(x)}\,\varepsilon(x)$ has already been considered to denoise log-compressed ultrasonic images [24]. Preliminary results of $ANL_{\sigma,n}\widetilde{\mathbf{u}}_{ANL}$ with this model is shown in Fig. 3, when applied to two log-compressed ultrasonic images and a cryo-Electronic Microscopy image (cross-section of a microtubule (10-30 nm)) where brights areas are smoother than dark areas.

## 6 Conclusion

We have described a Bayesian motivation for the *NL-means filter* and justify some intuitive adaptations described in previous papers [4,26,7]. The proposed framework yields to a filter which is better parametrized: the size of the adaptive dictionary and the noise variance are computed from the input image, and the patch size must be large enough. Our Bayesian approach has been used to remove image-dependent noise and could be adapted in applications with appropriate noise distributions. A more thoroughly evaluation with other methods [1,2,35,16], and recent improvements of the *NL-means filter* described in [5], would be also desirable.

**Acknowledgments.** We thank B. Delyon and P. Pérez for fruitful conversations and comments.

## References


1. Awate, S.P., Whitaker, R.T.: Higher-order image statistics for unsupervised, information-theoretic, adaptive, image filtering. CVPR'05, San Diego (2005)
2. Azzabou, N., Paragios, N., Guichard, F.: Random walks, constrained multiple hypothesis and image enhancement. ECCV'06, Graz (2006)
3. Barash, D., Comaniciu, D.: A Common framework for nonlinear diffusion, adaptive smoothing, bilateral filtering and mean shift. Image Vis. Comp. **22** (2004) 73-81
4. Buades, A., Coll, B., Morel, J.M.: A review of image denoising methods, with a new one. Multiscale Modeling and Simulation. **4** (2005) 490-530
5. Buades, A., Coll, B., Morel, J.M.: The staircasing effect in neighborhood filters and its solution. IEEE T. Image Process. **15** (2006)
6. Bunea, F., Nobel, A.B.: Sequential procedures for aggregating arbitrary estimators of a conditional mean *(under revision)* (2005)
7. Coupé P., Yger, P. Barillot, C.: Fast non-local means denoising for 3D MR images. MICCAI'06, Copenhagen (2006)
8. Criminisi, A., Pérez, P., Toyama, K.: Region filling and object removal by exemplar-based inpainting. IEEE T. Image Process. **13** (2004) 1200-1212
9. De Bonet, J.S.: Noise reduction through detection of signal redundancy. Rethinking Artificial Intelligence, MIT AI Lab (1997)
10. Dabov, K., Foi, A., Katkovnik, V., Egiazarian, K.: Image denoising with block-matching and 3D filtering. In Electronic Imaging'06, Proc. SPIE 6064, no. 6064A-30, San Jose, California USA, 2006.
11. Efros, A., Leung, T.: Texture synthesis by non-parametric sampling. ICCV'99, Kerkyra (1999)
12. Elad, M.: On the bilateral filter and ways to improve it. IEEE T. Image Process. **11** (2002) 1141-1151
13. Elad, M. Aharon, M.: Image denoising via learned dictionaries and sparse representation. CVPR'06, New-York (2006)
14. Faraji, H. and MacLean, J.W.: CCD noise removal in digital images. IEEE T. Image Process. **15**(9) (2006) 2676-2685
15. Geman, D., Geman, S., Graffigne, C., Dong, P: Boundary detection by constrained optimization. IEEE T. Patt. Anal. Mach. Intell. **12** (1990), 609-628
16. Gilboa, G., Osher, S.: Nonlocal linear image regularization and supervised segmentation. UCLA CAM Report 06-47 (2006)
17. Godtliebsen, F., Spjotvoll, E., Marron, J.S.: A nonlinear Gaussian filter applied to images with discontinuities. J. Nonparametric Stat. **8** (1997) 21-43
18. Hirakawa, K., Parks, T.W.: Image denoising using total least squares. IEEE T. Image Process. **15**(9) (2006) 2730-2742

19. Katkovnik, V., Egiazarian, K., Astola, J.: Adaptive window size image denoising based on intersection of confidence intervals (ICI) rule. J. Math. Imag. Vis. **16** (2002) 223-235
20. Kervrann, C., Heitz, F.: A Markov random field model-based approach to unsupervised texture segmentation using local and global spatial statistics. IEEE T. Image Process. **4** (1995) 856-862
21. Kervrann, C., Boulanger, J. Unsupervised patch-based image regularization and representation. ECCV'06, Graz (2006)
22. Kinderman, S., Osher, S. Jones, P.W.: Deblurring and denoising of images by nonlocal functionals. Multiscale Modeling and Simulation. **4** (2005) 1091-1115
23. Lee, J.S.: Digital image smoothing and the sigma filter. Comp. Vis. Graph. Image Process. **24** (1983) 255-269
24. Loupas, T., McDicken, W.N., Allan, P.L.: An adaptive weighted median filter for speckle suppression inmedical ultrasonic images. IEEE T. Circ. Syst. **36** (1989) 129-135
25. Lukin, A.: A multiresolution approach for improving quality of image denoising algorithms. ICASSP'06, Toulouse (2006)
26. Mahmoudi, M.; Sapiro, G.: Fast image and video denoising via nonlocal means of similar neighborhoods. IEEE Signal Processing Letters. **12** (2205) 839-842
27. Mrazek, P., Weickert, J., Bruhn, A.: On robust estimation and smoothing with spatial and tonal kernels. Preprint 51, U. Bremen (2004)
28. Mumford, D., Shah, J.: Optimal approximations by piecewise smooth functions and variational problems. Comm. Pure and Appl. Math. **42** (1989) 577-685
29. Perona. P., Malik, J.: Scale space and edge detection using anisotropic diffusion. IEEE T. Patt. Anal. Mach. Intell. **12** (1990) 629-239
30. Polzehl, J., Spokoiny, V.: Adaptive weights smoothing with application to image restoration. J. Roy. Stat. Soc. B **62** (2000) 335-354
31. Portilla, J., Strela, V., Wainwright, M., Simoncelli, E.: Image denoising using scale mixtures of Gaussians in the wavelet domain. IEEE T. Image Process. **12** (2003) 1338-1351
32. Roth, S., Black, M.J.: Fields of experts: a framework for learning image priors with applications. CVPR'05, San Diego (2005)
33. Rudin, L., Osher, S., Fatemi, E.: Nonlinear Total Variation based noise removal algorithms. Physica D (2992) **60** (1992) 259-268
34. Tomasi, C., Manduchi, R.: Bilateral filtering for gray and color images. ICCV'98, Bombay (1998)
35. Tschumperlé, D.: Curvature-preserving regularization of multi-valued images using PDE's. ECCV'06, Graz (2006)
36. Wang, Z., Zhang, D.: Restoration of impulse noise corrupted images using long-range correlation. IEEE Signal Processing Letters. **5** (1998) 4-6
37. Weickert, J.: Anisotropic Diffusion in Image Processing. Teubner-Verlag, Stuttgart (1998)
38. van de Weijer, J., van den Boomgaard, R.: Local mode filtering. CVPR'01, Kauai (2001)
39. Zhang, D., Wang, Z.: Image information restoration based on long-range correlation. IEEE T. Circ. Syst. Video Technol. **12** (2002) 331-341

# Restoration of Images with Piecewise Space-Variant Blur

Leah Bar[1], Nir Sochen[2], and Nahum Kiryati[3]

[1] Department of Electrical and Computer Engineering,
University of Minnesota, Minneapolis, MN 55455, USA
[2] Dept. of Applied Mathematics
[3] School of Electrical Engineering
Tel Aviv University, Tel Aviv 69978, Israel

**Abstract.** We address the problem of space-variant image deblurring, where different parts of the image are blurred by different blur kernels. Assuming a region-wise space variant point spread function, we first solve the problem for the case of known blur kernels and known boundaries between the different blur regions in the image. We then generalize the method to the challenging case of unknown boundaries between the blur domains. Using variational and level set techniques, the image is processed globally. The space-variant deconvolution process is stabilized by a unified common regularizer, thus preserving discontinuities between the differently restored image regions. In the case where the blurred subregions are unknown, a segmentation procedure is performed using an evolving level set function, guided by edges and image derivatives.

## 1 Introduction

Most image deblurring methods rely on the standard model of a space invariant kernel and additive noise

$$g = h * f + n. \tag{1}$$

Here $h$ denotes a known space-invariant blur kernel, $f$ is an ideal version of the observed image $g$, and $n$ is noise. Yet, the assumption of space-invariance is not accurate in real photographic images. For example, when multiple objects move at different velocities and in different directions in a scene, one gets space-variant motion blur. Likewise, when a camera lens is focused on one specific object, other objects nearer or farther away from the lens are not as sharp. In such situations, different blur kernels degrade different areas of the image. Other applications include astronomy [12] and medical (SPECT) imaging [21]. In the space-variant case, assuming a linear blur, Eq. (1) takes the form

$$g(x) = \int h(x, x-u) f(u) du + n,$$

where $x$ is a location in the observed image and $u$ is a location in the ideal image $f$.

F. Sgallari, A. Murli, and N. Paragios (Eds.): SSVM 2007, LNCS 4485, pp. 533–544, 2007.

Most of the space-variant deconvolution literature considers the space-variant problem under the assumption that the blur function is known. For example, Nagy and O'leary developed a fast deconvolution algorithm via matrix-vector multiplication involving banded Toeplitz matrices [14]. The blurring kernels are assumed to have small domain of support. Lauer [12] dealt with astronomical images and modeled the point spread function (PSF) as a sum of orthogonal functions. Welk *et al.* [20] presented a variational deblurring method such that the recovered image $f$ was the minimizer of an objective functional that consists of a robust fidelity term and a Perona-Malik regularizer [16].

Few works address the blind space-variant deconvolution issue. You and Kaveh [22] presented a variational approach to this problem. In their method, a parametric space-variant PSF with piecewise smooth space variance constraint is utilized. In the case of motion blur, for example, the velocity $v(x)$ was calculated. This approach, however, suffers from severe multiple local minima problems.

Favaro and Soatto [7] presented a variational approach to segmentation and restoration of scenes containing multiple moving objects. From a collection of motion blurred images, they calculate the motion field, depth map and radiance. In their method, the blur kernel is *implicitly* formulated in the objective functional. The idea of simultaneous variational restoration and segmentation was first introduced by Kim *et al.* [10]; they used a variant of the Mumford-Shah functional [13] together with a curve evolution technique, but assumed a *known space-invariant* blur kernel.

A significant contribution to the related *shape from defocus* problem was given by Favaro *et al.* The 3D shape of a scene was estimated from a collection of defocused images [8,9]. In the case of a moving scene, the depth map, radiance and motion was inferred via an anisotropic diffusion model from a collection of defocused and motion-blurred images [4].

Schechner *et al.* [18] presented an algorithm for the separation of two transparent layers. Nonetheless, their work assumes that two images are available such that in each image one layer is in focus.

Here we propose a novel variational approach to space-variant restoration. The underlying model of this work assumes a space-variant PSF with piecewise location dependence. This means that every sub-domain in the image is blurred by a different kernel. The regions are separated by contours that are expressed as zero level set functions. First we will solve the case where the contours and blur kernels are known in advance. The contribution of the proposed non-blind space-variant restoration approach relies on the usage of a global regularizer, which eliminates the requirement of dealing with region boundaries. As a result, the gray levels continuity of the recovered image is inherent. This method does not limit the number of sub-regions, their geometrical shape nor the kernel support size. In addition, since the restoration is global, the convolution operation can be efficiently implemented by using Fast Fourier Transform (FFT) multiplications.

In the second part of the paper, we will address the blur segmentation given a *single* image. We assume two sub-regions and known or easily estimated blur kernels but an unknown contour between the sub-regions. The blur level difference between the regions is assumed to be visually significant. This problem is approached via a sequential process. We first segment the observed image into its blurred sub-regions. The boundaries are determined using an evolving level set function, guided by edges and image derivatives. Then we use the suggested non-blind space-variant restoration method given the sub-regions and estimated blur kernels. The advantage of this approach over a one-step process is the additional regularization of this ill-posed problem. In the contours detection stage, some assumptions are made regarding the contour properties. Pre-detected edges and derivatives provide additional important information on this process. In such a manner, the solutions space is reduced. The suggested method accommodates arbitrary blur functions. It is because the contour detection process depends essentially on the blur level rather on the blur type.

## 2 Non-blind Space-Variant Restoration

Consider an image that consists of known sub-domains blurred by known kernels. It is well known [14] that region-wise deblurring may yield boundary discontinuities. This problem is illustrated in Fig. 1. The $256 \times 256$ *Lena* image was blurred by an 8 pixels horizontal motion blur within the marked rectangular region. This region was recovered by the Total Variation deconvolution method [19] and put back into its place in the observed image. The outcome of the region-wise procedure is shown on the right. It can be easily seen that although the regional recovery is satisfying, the gray levels on both sides of the boundary are not compatible. To solve this problem, some blending constraints have to be added to the boundary conditions. Moreover, if the region shape is more complex, dealing with boundaries requires additional algorithmic effort.

In the current study, we suggest a variational setting in which a global regularizer automatically takes care of gray level continuity in region boundaries. Furthermore, there are no limitations on the number and shapes of the blur regions.

**Fig. 1.** Failure of the region-wise image deconvolution algorithm. *Left:* Spatially variant blurred image. *Right*: Restoration using a region-wise algorithm.

**Fig. 2.** Non-blind space-variant restoration. *Left column*: Spatially variant motion blurred images. *Right column:* The corresponding recovered images using the suggested method.

Let us define $\Omega$ as an open bounded set in $\mathbb{R}^n$, and an image $f(x) : \Omega \rightarrow \mathbb{R}^n$. The open non-overlapping subsets $w_i \subset \Omega$ denote regions that are blurred by kernels $h_i$ respectively. In addition, $\Omega \setminus \cup \overline{w_i}$, denotes the background region blurred by the background kernel $h_b$, and $\overline{w_i}$ stands for the closure of $w_i$. The region boundaries are designated by $\partial w_i$. Deconvolution is known to be an ill-posed inverse problem that has to be regularized. In the variational framework, the recovered image is the minimizer of an objective functional that consists of a fidelity term $\mathcal{T}$ and a regularizer $\mathcal{R}$ that reflects some *a priori* knowledge about the image, such as smoothness. Boundary discontinuities can be prevented by using a global regularizer. This means that the smoothness constraint is applied to the whole image. Formally, the fidelity term is given by

$$\mathcal{T} = \frac{1}{2} \sum_i \eta_i \int_{w_i} (h_i * f - g)^2 dx + \frac{\eta_b}{2} \int_{\Omega \setminus (\cup \overline{w_i})} (h_b * f - g)^2 dx, \qquad (2)$$

where $\eta_i$ and $\eta_b$ are positive scalars.

Using the level set framework [15], the region boundaries $\partial w_i$ can be implicitly represented by Lipschitz functions $\phi_i : \Omega \rightarrow \mathbb{R}$, such that

$$C_i \doteq \partial w_i = \{x \in \Omega : \phi_i(x) = 0\}.$$

Following the formulation of Chan and Vese [6], the domains $w_i$ can be replaced by the Heaviside function $H(\phi_i)$, where

$$H(\phi_i) = \begin{cases} 1, & \phi_i > 0, \\ 0, & \phi_i \leq 0. \end{cases} \tag{3}$$

The fidelity term $\mathcal{T}$ can, therefore, be rewritten as

$$\begin{aligned} \mathcal{T} &= \frac{1}{2} \sum_i \eta_i \int_\Omega (h_i * f - g)^2 H(\phi_i) dx \\ &+ \frac{\eta_b}{2} \int_\Omega (h_b * f - g)^2 \left(1 - \sum_i H(\phi_i)\right) dx. \end{aligned} \tag{4}$$

Several edge preserving regularizers are known to be effective in image restoration and denoising. The Total Variation [17] and Perona-Malik [16] stabilizers, for example, are widely used in the scientific literature. Here we favor the recently proposed Mumford-Shah [13] based regularizer that is used for the deblurring process [2]. In this approach, the image is modeled as a piecewise smooth function, that is the observed image can be decomposed into smooth regions by well-behaved contours. The regularizer is expressed as

$$\mathcal{R}^{MS} = \beta \int_{\Omega \setminus K} |\nabla f|^2 dx + \alpha \int_K d\sigma$$

where $K$ denotes the unknown edge set and $\int_K d\sigma$ is the total edge length. The Mumford-Shah regularizer, which leads to a free discontinuity problem, can be approximated by a regular functional in the $\Gamma$-convergence framework [3]. Ambrosio and Tortorelli [1] approximated the irregular regularizer by a sequence $\mathcal{R}_\epsilon$ where $\epsilon \to 0$. The edge set $K$ is represented by the characteristic function $(1 - \chi_K)$, which in turn is approximated by a smooth auxiliary function $v$ such that $v \approx 0$ across the edges and $v \approx 1$ within the segments. The approximated regularizer thus takes the form

$$\mathcal{R}_\epsilon^{MS} = \beta \int_\Omega v^2 |\nabla f|^2 dx + \alpha \int_\Omega \left( \epsilon |\nabla v|^2 + \frac{(v-1)^2}{4\epsilon} \right) dx. \tag{5}$$

Hence, the objective functional $\mathcal{F}_\epsilon^{MS} \doteq \mathcal{T} + \mathcal{R}_\epsilon^{MS}$ is given by

$$\begin{aligned} \mathcal{F}_\epsilon^{MS}(f, v) &= \frac{1}{2} \sum_i \eta_i \int_\Omega (h_i * f - g)^2 H(\phi_i) dx \\ &+ \frac{\eta_b}{2} \int_\Omega (h_b * f - g)^2 \left(1 - \sum_i H(\phi_i)\right) dx \\ &+ \beta \int_\Omega v^2 |\nabla f|^2 dx + \alpha \int_\Omega \left( \epsilon |\nabla v|^2 + \frac{(v-1)^2}{4\epsilon} \right) dx. \end{aligned} \tag{6}$$

Whenever the contours $C_i$ are known *a priori*, the $\phi_i$ functions can easily be determined. The recovered image $f$ and edge map $v$ are alternately obtained via the solution of the following Euler-Lagrange equations [2]:

$$\frac{\delta \mathcal{F}_\epsilon^{MS}}{\delta f} = \sum_i \eta_i h_i(-x) * [(h_i * f - g) H(\phi_i)] \quad (7)$$

$$+ \eta_b h_b(-x) * \left[ (h_b * f - g) \left( 1 - \sum_i H(\phi_i) \right) \right] - 2\beta \nabla \cdot (v^2 \nabla f) = 0,$$

$$\frac{\delta \mathcal{F}_\epsilon^{MS}}{\delta v} = 2\beta v |\nabla f|^2 + \alpha \cdot \frac{v-1}{2\epsilon} - 2\epsilon\, \alpha\, \nabla^2 v = 0. \quad (8)$$

**Fig. 3.** Non-blind space-variant restoration with different blur kernels in each region. *Left:* Blurred image. *Right:* Restoration using the suggested method.

Fig. 2 demonstrates the performance of the suggested algorithm. The two images in the left column were synthetically blurred by an 8 pixels length horizontal motion blur within the marked rectangle (top) and circle (bottom). The corresponding recovered images are shown in the right column. Since there was no background blur, $h_b$ was as a delta function. The parameter set used was: $\eta_{1,b} = 1$, $\beta = 10^{-3}$, $\alpha = 10^{-8}$ and $\epsilon = 10^{-3}$. The superiority of this method over the region-wise approach (Fig. 1) is evident.

Fig. 3 demonstrates a more complex case. The ellipse and curve sub-regions were blurred by pill boxes of radius 4.9 and 2.1 pixels, respectively. The background blur was again a delta function. The image was recovered again with $\eta_{1,2,b} = 1$ and the same parameters set as before.

## 3 Blur Segmentation and Space-Variant Restoration

In this section, we address the problem of deblurring an image consisting of two *unknown* sub-regions blurred by different kernels. We assume that the blur level

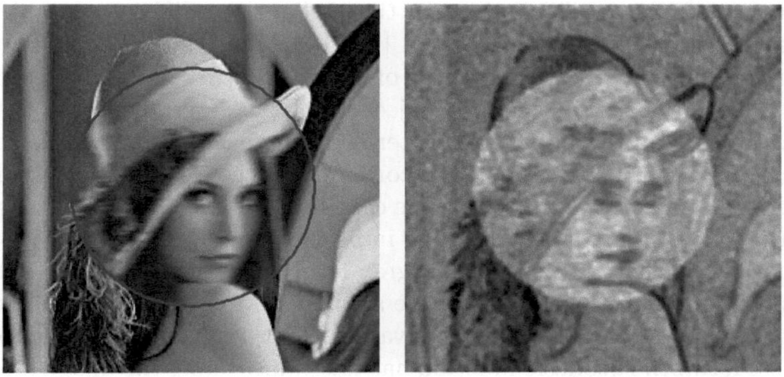

**Fig. 4.** *Left:* Blurred image. *Right:* The corresponding $E$ function.

difference between the regions is significant, and that both PSFs are known or can be easily estimated. We present a novel approach to detect the boundary between the sub-regions by the evolution of a level set function.

The key idea is to distinguish between different blur levels in the observed image. It is readily apparent that highly blurred regions accommodate low spatial derivatives. Edge information provides additional useful knowledge for the discrimination process. Let us define a function $E$ such that

$$E \doteq \left|\log\left(\left|\nabla^2 g\right| * B_r\right)\right|, \tag{9}$$

and $B_r$ is a ball of radius $r$. This function represents a smoothed version of the Laplacian of the observed image $g$. Fig. 4 illustrates $E$ with $r = 2$ pixels. The two sub-regions are visibly distinguishable by their different gray levels. A closer look at $E$ shows that the gray levels of the edges in both regions are darker than the segments, but the average value of the edges in the blurred sub-region is lighter than in the sharp sub-region. This observation motivated us to perform the contour detection by the difference of gray levels of $E$.

Let $C$ denote the separating contour, $K$ the edge set of the observed image, and $\overrightarrow{c} \doteq \{c_1, c_2, c_3, c_4\}$ the average gray levels of the edges inside the contour, the edges outside the contour, the segments inside the contour and the segments outside the contour respectively. Ideally, the separating contour $C$ is the minimizer of the functional:

$$\begin{aligned}\mathcal{F}(C, \overrightarrow{c}) = {} & \frac{\lambda_1}{2}\int_{\text{inside}(C)\cap K}(E - c_1)^2 dx + \frac{\lambda_2}{2}\int_{\text{outside}(C)\cap K}(E - c_2)^2 dx \\ & + \frac{\lambda_3}{2}\int_{\text{inside}(C)\setminus K}(E - c_3)^2 dx + \frac{\lambda_4}{2}\int_{\text{outside}(C)\setminus K}(E - c_4)^2 dx \\ & \underbrace{- \oint_C |\langle \nabla E, \overrightarrow{n}\rangle| ds}_{\text{Robust Alignment}} \quad + \quad \underbrace{\oint \mathcal{G}(C(s)ds}_{\text{Geodesic Active Contour}} \quad .\end{aligned} \tag{10}$$

The first four terms are fidelity terms that reflect the different gray levels in $E$. The fifth term is a robust alignment [11]. It integrates the absolute value of the inner product between the gradient of $E$ and the curve normal $\vec{n}$ along the contour. This measure reflects the projection of the $E$ gradients on the curve normals. Minimizing the minus of this term aligns these two vectors together.

The last term is the geodesic active contour [5], which is an integration of an *inverse* edge indicator function along the contour. A curve $C$ that minimizes this term is the curve along which the edge indicator function is a local maximum. Fig. 4 shows that there is a significant gray level difference between the blurred sub-regions, which in is this case can be seen to be a circle. Thus, the geodesic active contour function must have low values along the separating contours. In its classical form, the geodesic function has low values along the image gradients, but in our case, the contribution from the edges needs to be eliminated. As a result, the active contour will eventually follow the separating contour. This is accomplished by detecting the edges in advance. We use the Mumford-Shah based non-blind space-invariant restoration method [2]. As noted before, in this framework edges are represented by a smooth function $v(x)$ where $v \approx 0$ across the edges and $v \approx 1$ within the segments. In an iterative process, both $f$ and $v$ are alternately calculated as the minimizers of the following objective functional

$$\begin{aligned}\mathcal{F}(f,v) &= \frac{1}{2}\int_{\Omega}(g * B_r - h_b * f)^2 + \beta\int_{\Omega} v^2|\nabla f|^2 dx \\ &+ \alpha\int_{\Omega}\left(\epsilon|\nabla v|^2 + \frac{(v-1)^2}{4\epsilon}\right)dx.\end{aligned} \tag{11}$$

The $g$ function was smoothed a little in order to get thicker edges. The blur kernel was selected to be the weaker of the two kernels (the background kernel in our example) to avoid over-deconvolution artifacts. Detailed minimization techniques can be found in [2]. Now, let

$$v_T = \begin{cases} 1, & v > T, \\ 0, & v \leq T, \end{cases} \tag{12}$$

be a binary edge map, where $v_T = 1$ in the segments and $v_T = 0$ across the edges.

We proceed to introducing the $\mathcal{G}$ function (for Eq. 10):

$$\mathcal{G}(\nabla E, v_T) = \frac{\mu}{1 + (v_T^2|\nabla E|^2)/\gamma} + \nu. \tag{13}$$

The constants $\gamma$, $\mu$, $\nu$ are positive scalars. This inverse edge indicator function has low values along the separating contour because edge contributions are now eliminated by $v_T$.

As already noted, a contour $C$ can be implicitly represented as the zero level set function $C = \{x \in \Omega : \phi(x) = 0\}$. In terms of a level set function [6]

$$\text{Length}\{C\} = \oint_C ds = \int_{\Omega}|\nabla H(\phi(x))|dx = \int_{\Omega}\delta(\phi(x))|\nabla\phi(x)|dx,$$

and the curve normal $\overrightarrow{n} = \nabla\phi/|\nabla\phi|$ [11]. Therefore, Eq. (10) can be rewritten as

$$\begin{aligned}\mathcal{F}(\phi,\overrightarrow{c}) = &\frac{\lambda_1}{2}\int_\Omega (1-v)^2(E-c_1)^2 H(\phi)dx + \frac{\lambda_2}{2}\int_\Omega (1-v)^2(E-c_2)^2(1-H(\phi))dx \\ &+ \frac{\lambda_3}{2}\int_\Omega v^2(E-c_3)^2 H(\phi)dx + \frac{\lambda_4}{2}\int_\Omega v^2(E-c_4)^2(1-H(\phi))dx \\ &- \underbrace{\int_\Omega \left|\left\langle \nabla E, \frac{\nabla\phi}{|\nabla\phi|}\right\rangle\right| |\nabla H(\phi)|dx}_{\text{Robust Alignment}} + \underbrace{\int_\Omega \mathcal{G}(\nabla E, v_T)|\nabla H(\phi)|dx}_{\text{Geodesic Active Contour}}.\end{aligned} \tag{14}$$

When $v \approx 0$ (across edges), the third and forth terms vanish, while the first and second terms vanish within the segments whenever $v \approx 1$.

Using the alternate minimization method, we first calculate the level set function $\phi$ while keeping the constants $\overrightarrow{c}$ fixed. Then we keep $\phi$ fixed and minimize the functional with respect to each of the four constants. The implicit level set evolution takes the form

$$\begin{aligned}\phi_t = -\frac{\delta\mathcal{F}}{\delta\phi} = -\Big\{ &\lambda_1(1-v)^2(E-c_1)^2 - \lambda_2(1-v)^2(E-c_2)^2 \\ &+ \lambda_3 v^2(E-c_3)^2 - \lambda_4 v^2(E-c_4) \\ &- \operatorname{sign}(\langle\nabla E, \nabla\phi\rangle)\,\nabla^2 E - \nabla\cdot\left(\mathcal{G}(\nabla E, v_T)\frac{\nabla\phi}{|\nabla\phi|}\right)\Big\}\,\delta(\phi).\end{aligned} \tag{15}$$

See Ref. [11] for more details on the derivation of the fifth and sixth terms. Keeping $\phi$ fixed and minimizing $\mathcal{F}(\phi,\overrightarrow{c})$ with respect to $c_1$ yields

$$c_1 = \frac{\int(1-v)^2 E H(\phi)}{\int(1-v)^2 H(\phi)}. \tag{16}$$

The constants $c_2, c_3$ and $c_4$ are calculated in the same manner. Following Chan and Vese [6], the Heaviside function $H$ is approximated by the $C^\infty(\overline{\Omega})$ function

$$H_\varepsilon(z) = \frac{1}{2}\left(1 + \frac{2}{\pi}\arctan\left(\frac{z}{\varepsilon}\right)\right),$$

hence

$$\delta_\varepsilon(z) = \frac{d}{dz}H_\varepsilon(z) = \frac{1}{\pi}\frac{\varepsilon}{\varepsilon^2+z^2}.$$

In our implementation we use this approximation with $\varepsilon = 1$.

Fig. 5 shows experimental results with a real image degraded by space variant defocus blur. The top-left $250 \times 333$ image was obtained with deliberate defocus blur using a Canon Power Shot G5 digital camera. The foreground object was located $1.2\,m$ in front of the background poster. The picture was taken such that only the background was in focus. The top-right image shows the recovered image using the Mumford-Shah based non-blind *space-invariant* method [2] such

**Fig. 5.** Real space-variant restoration. *Top-left:* Observed image, background is in focus while foreground is out of focus. *Top-right:* Space-invariant image restoration using a single out of focus blur kernel. *Bottom-left:* Recovered image using the suggested method. *Bottom-right:* Separating contour (red) on the background of the $E$ function and image edges (pink).

that the whole image was restored by the foreground kernel. The failure of this recovery is evident.

Better results were obtained using our space-variant method. At the first stage, the separating contour was calculated by the minimization of Eq. (14). The $E$ function together with the binary edge map $v_T$ and final contour, are shown bottom-right. In this case the parameter set was $\lambda_1 = \lambda_2 = 3$, $\lambda_3 = \lambda_4 = 0.1$, $\nu = 0.95$, $\mu = 14$, $\gamma = 0.6$, $T = 0.93$ and $\Delta t = 0.1$ as the gradient descent time step. The ball radius was set to 8 pixels. The background PSF was assumed to be a delta function and the foreground PSF was manually tuned to be a pill-box of radius 4. In the second stage of the restoration process, the non-blind space-variant method was employed using the calculated $\phi$ function with $\beta = 0.001$, $\alpha = 0.1$, $\epsilon = 0.1$, $\eta_1 = 1$, and $\eta_b = 30$. In the bottom-left is the reconstruction using the method suggested in this paper. The quality of the image obtained using the proposed method shows its applicability to real-world life situations.

## 4 Discussion

We presented a novel approach to space-variant deblurring of a *single* image, where region-wise space-variant point spread function is the underlying model of this research. In the non-blind space-variant restoration case, we proposed an efficient algorithm that globally recovers the image with no limitations on the number and shapes of the sub-regions. Special handling of the region boundaries is not necessary. Later, we introduced a segmentation method to determine the blur regions. The separating contour between two blurred sub-regions was represented as an evolving level set function and the blur kernels are assumed to be known or easily estimated. Promising experimental results of real and synthetic images demonstrate the potential of the suggested method to recover space-variant blurred images.

This study suggests additional challenging topics for future research. In the segmentation process, the number of blurred regions can be greater than two. The detection of the blur kernels, analysis of the algorithm in the presence of noise, convergence issues and parameters estimation can be also considered. The fully blind restoration problem is highly ill-posed, therefore sophisticated regularizers will have to be employed.

## Acknowledgment

This research was supported by the A.M.N. Foundation and by MUSCLE: Multimedia Understanding through Semantics, Computation and Learning, a European Network of Excellence funded by the EC 6th Framework IST Programme.

## References

1. L. Ambrosio and V.M. Tortorelli. Approximation of functionals depending on jumps by elliptic functionals via $\Gamma$-convergence. *Comm. Pure Appl. Math.*, 43(8):999–1036, 1990.

2. L. Bar, N. Sochen, and N. Kiryati. Semi-blind image restoration via Mumford-Shah regularization. *IEEE Trans. Image Processing*, 15(2):483–493, 2006.
3. A. Braides. *Approximation of Free-Discontinuity Problems*, volume 1694 of *Lecture Notes in Mathematics*, pages 47–51. Springer, 1998.
4. P. Favaro M. Burger and S. Soatto. Scene and motion reconstruction from defocused and motion-blurred images via anisotropic diffusion. In *Proc. of 8th European Conference on Computer Vision*, pages 257–269, 2004.
5. V. Caselles, R. Kimmel, and G. Sapiro. Geodesic active contours. *International Journal of Computer Vision*, 22:61–79, 1997.
6. T. Chan and L. Vese. Active contours without edges. *IEEE Trans. Image Processing*, 10:266–277, 2001.
7. P. Favaro and S. Soatto. A variational approach to scene reconstruction and image segmentation from motion-blur cues. In *Proc. of IEEE Conference on Computer Vision and Pattern Recognition*, pages 631–637, 2004.
8. P. Favaro and S. Soatto. A geometric approach to shape from defocus. *IEEE Trans. Pattern Analysis and Machine Intelligence*, 27:406–417, 2005.
9. H. Jin and P. Favaro. A variational approach to shape from defocus. In *Proc. of 7th European Conference on Computer Vision*, pages 18–30, 2002.
10. J. Kim, A. Tsai, M. Cetin, and A.S. Willsky. A curve evolution-based variational approach to simultaneous image restoration and segmentation. In *Proc. of IEEE International Conference on Image Processing*, volume 1, pages 109–112, 2002.
11. R. Kimmel. Fase edge integration. In S. Osher and N. Paragios, editors, *Geometric Level Set Methods in Imaging Vision and Graphics*. Springer-Verlag, 2003.
12. T. Lauer. Deconvolution with a spatially-variant PSF. In *Proc. of the SPIE, Astronomical Data Analysis II.*, pages 167–173, 2002.
13. D. Mumford and J. Shah. Optimal approximations by piecewise smooth functions and associated variational problems. *Comm. Pure Appl. Math.*, 42:577–685, 1989.
14. J.G. Nagy and D.P. Oĺeary. Restoring images degraded by spatially-variant blur. *Siam Journal on Scientific Computing*, 19:1063–1082, 1998.
15. S. Osher and J.A. Sethian. Fronts propagating with curvature-dependent speed: Algorithms based on hamilton-jacobi formulation. *Journal of Computational Physics*, 79:12–49, 1988.
16. P. Perona and J. Malik. Scale-space and edge detection using anisotropic diffusion. *IEEE Trans. Pattern Analysis and Machine Intelligence*, 12:629–639, 1990.
17. L.I. Rudin, S. Osher, and E. Fatemi. Non linear total variatrion based noise removal algorithms. *Physica D*, 60:259–268, 1992.
18. Y. Schechner, N. Kiryati, and R. Basri. Separation of transparent layers using focus. *International Journal of Computer Vision*, 39:25–39, 2000.
19. C.R. Vogel and M.E. Oman. Fast, robust total variation-based reconstruction of noisy, blurred images. *IEEE Trans. Image Processing*, 7:813–824, 1998.
20. M. Welk, D. Theis, and J. Weickert. Variational deblurring of images with uncertain and spatially variant blurs. In *Pattern Recognition, 27th DAGM Symposium*, volume 3663 of *LNCS*, pages 485–492, 2005.
21. W. Xia, R. M. Lewitt, and P. R. Edholm. Fourier correction for spatially variant collimator blurring in SPECT. *IEEE Trans. Medical Imaging*, 14:100–115, 1995.
22. Y. You and M. Kaveh. Blind image restoration by anistropic regularization. *IEEE Trans. Image Processing*, 8:396–407, 1999.

# Mumford-Shah Regularizer with Spatial Coherence

Erkut Erdem[1], Aysun Sancar-Yilmaz[2], and Sibel Tari[1]

[1] Middle East Technical University, Department of Computer Engineering, Ankara, TR-06531
erkut@ceng.metu.edu.tr, stari@metu.edu.tr
[2] Aselsan Inc., Microwave and System Technologies Division, Ankara, TR-06172
asancar@aselsan.com.tr

**Abstract.** As recently discussed by Bar, Kiryati, and Sochen in [3], the Ambrosio-Tortorelli approximation of the Mumford-Shah functional defines an extended line process regularization where the regularizer has an additional constraint introduced by the term $\rho|\nabla v|^2$. This term mildly forces some spatial organization by demanding that the edges are smooth. However, it does not force spatial coherence such as edge direction compatibility or edge connectivity, as in the traditional edge detectors such as Canny. Using the connection between regularization and diffusion filters, we incorporate further spatial structure into the regularization process of the Mumford-Shah model. The new model combines smoothing, edge detection and edge linking steps of the traditional approach to boundary detection. Importance of spatial coherence is best observed if the image noise is salt and pepper like. Proposed approach is able to deal with difficult noise cases without using non-smooth cost functions such as $L^1$ in the data fidelity or regularizer.

## 1 Introduction

Mumford and Shah [13] formulated image segmentation process as a functional minimization via which a piecewise smooth approximation of a given image and an edge set are to be recovered simultaneously. The Mumford-Shah energy is:

$$E_{MS}(u,\Gamma) = \beta \int_R (u-g)^2 dx + \alpha \int_{R\setminus\Gamma} |\nabla u|^2 dx + length(\Gamma) \tag{1}$$

where

- $R \subset \Re^2$ is connected, bounded, open subset representing the image domain,
- $g$ is an image defined on $R$,
- $\Gamma \subset R$ is the edge set segmenting $R$,
- $u$ is the piecewise smooth approximation of $g$,
- $\alpha$, $\beta$ are the scale space parameters of the model.

F. Sgallari, A. Murli, and N. Paragios (Eds.): SSVM 2007, LNCS 4485, pp. 545–555, 2007.

The first term in $E_{MS}(u, \Gamma)$ is the data fidelity term, which forces the solution $u$ to be as close as to the original image $g$. The other two terms are the regularization terms, which give preference to piecewise smooth images with simple edge sets. The unknown edge set $\Gamma$ makes the minimization mathematically difficult. A convenient approximation is suggested by Ambrosio and Tortorelli [2] following the $\Gamma$ convergence framework [7]. The basic idea is to introduce a smooth edge indicator function $v$ which is more convenient than the original edge indicator represented by the characteristic function $1-\chi_\Gamma$. The function $v$ depends on a parameter $\rho$, and as $\rho \to 0$, $v \to 1-\chi_\Gamma$. That is, $v(x) \approx 0$ if $x \in \Gamma$ and $v(x) \approx 1$ otherwise. Moreover, the cardinality of the edge set $\Gamma$ can be approximated by $\frac{1}{2}\left(\rho|\nabla v|^2 + \frac{(1-v)^2}{\rho}\right)$. The new functional is as follows:

$$E_{AT}(u, v) = \int_R \beta(u-g)^2 + \alpha(v^2|\nabla u|^2) + \frac{1}{2}\left(\rho|\nabla v|^2 + \frac{(1-v)^2}{\rho}\right) dx \qquad (2)$$

Notice that as $v \to 0$, the smoothness constraint in the piecewise smooth model is switched off. It is possible to interpret $v^2$ as an analog form of the line process introduced by Geman and Geman [12]. As shown by Bar et al. [3] and Teboul et al. [18], the Ambrosio-Tortorelli approximation of the Mumford Shah functional defines an extended line process regularization where the regularizer has an additional constraint introduced by the term $\rho|\nabla v|^2$. This term mildly forces some spatial organization by demanding that the edges are smooth. However, it does not force spatial coherence such as edge direction compatibility or edge connectivity. On the other hand, in the traditional approach, segmentation is defined as a sequential bottom-up process composed of the following three steps:

- smoothing,
- edge detection,
- edge linking.

The purpose of the last step is to force global consistency to locally detected edges in order to come up with a coherent edge set. Interestingly, this last step is what Mumford-Shah model or its Ambrosio-Tortorelli approximation lacks. The importance of spatial coherence can be best observed when the image contains impulse noise (Fig 1). Some works use Mumford-Shah regularizer or its modification [17,1] for restoration in the presence of impulse noise. In [3,4], Bar et al. present a very promising approach. However, the success of their method stems mostly from the use of robust data fidelity by replacing $L^2$ norm with $L^1$. Similarly, in [17], Shah uses $L^1$ norm for both the data fidelity and the regularizer. In fact, the use of non-smooth cost functions such as $L^1$ for the data fidelity term in order to deal with outliers and impulse noise is well motivated both theoretically and experimentally [8,14,17]. Teboul et al. [18] present a modification to (2), by replacing the quadratic cost $|\nabla v|^2$ with $L^1$ cost, which leads to singular diffusivity. Numerical difficulties are the cons of singular diffusivities [9]. The cost function choice in [18] also leads to directional smoothing. As explored by Weickert [19], directional smoothing can offer significant feature preserving

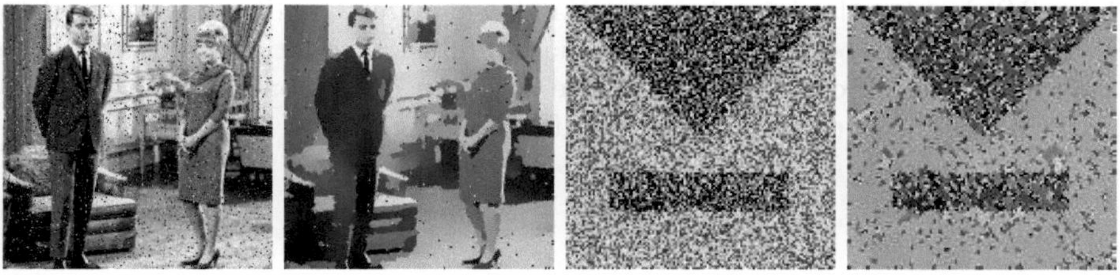

(a) (b) (c) (d)

**Fig. 1.** Denoising cases which can not be handled by the Ambrosio-Tortorelli. (a)-(b) couple image corrupted with 5% salt and pepper noise and its reconstruction using the Ambrosio-Tortorelli. (c)-(d) A noisy test image –70% of the pixels are degraded with uniform noise– and its reconstruction using the Ambrosio-Tortorelli.

capabilities. However, the models get complicated and numerics is not as simple as in the case of isotropic diffusion.

In this work, we propose a modification to the Ambrosio-Tortorelli approximation of the Mumford-Shah functional, turning it into an edge-preserving regularization with spatial coherence. Key to our approach is the link between edge preserving regularization and diffusion filters [15,16,6]. Proposed model is a set of coupled linear diffusion equations. Hence, it is easy to implement.

We experimentally demonstrate denoising and edge preserving abilities of the proposed method. It can handle impulse noise and fill boundary gaps. Moreover, it can produce sharper results. Smoothed images obtained by the proposed method are qualitatively comparable to that are obtained by singular diffusion equations [9].

In the next section, we review the Ambrosio Tortorelli approximation and analyze it's behavior relevant to our developments given in § 3. We present and discuss experimental results in § 4. Finally, § 5 is the summary and the conclusion.

## 2 Gradient Descent Equations for Ambrosio-Tortorelli Energy

Gradient descent equations for the Ambrosio-Tortorelli functional yield the following coupled PDEs:

$$\frac{\partial u}{\partial t} = \nabla \cdot (v^2 \nabla u) - \frac{\beta}{\alpha}(u - g); \quad \left.\frac{\partial u}{\partial n}\right|_{\partial R} = 0 \tag{3}$$

$$\frac{\partial v}{\partial t} = \nabla^2 v - \frac{2\alpha |\nabla u|^2 v}{\rho} - \frac{(v-1)}{\rho^2}; \quad \left.\frac{\partial v}{\partial n}\right|_{\partial R} = 0 \tag{4}$$

where $\partial R$ denotes the boundary of $R$ and $n$ denotes the direction normal to $\partial R$. By alternating between these two biased diffusion equations, smooth image $u$ and the edge indicator function $v$ are simultaneously computed. Keeping $v$ fixed, (3) minimizes a convex quadratic functional given by

$$\int_R \alpha v^2 |\nabla u|^2 + \beta(u-g)^2 \tag{5}$$

While the bias term in (3) or equivalently in (5) forces $u$ to be as close as to the original image $g$, the first term acts as an edge preserving regularizer. It smoothes the image with a smoothing radius proportional to the value of the function $v$ and $\frac{\alpha}{\beta}$. If there is an edge ($v \approx 0$), no smoothing (diffusion) is applied.

On the other hand, keeping $u$ fixed, (4) minimizes a convex quadratic functional given by

$$\int_R \rho |\nabla v|^2 + \frac{1+2\alpha\rho|\nabla u|^2}{\rho}\left(\frac{1}{1+2\alpha\rho|\nabla u|^2} - v\right)^2 \tag{6}$$

The reciprocal relationship between $v$ and $|\nabla u|^2$ can be best observed in (6). Clearly, it asserts that $v$ function is nothing but a smoothing of

$$\frac{1}{1+2\alpha\rho|\nabla u|^2} \tag{7}$$

with a blurring radius proportional to $\rho$ and reciprocal to $|\nabla u|$. Ignoring the smoothing, by letting $\rho \to 0$ [17,11],

$$v \approx \frac{1}{1+2\alpha\rho|\nabla u|^2} \tag{8}$$

## 3 Regularizer with Spatial Coherence

The regularization of $v$ via $\nabla^2 v$ term in (4) or equivalently $|\nabla v|^2$ term in (6) imposes some mild spatial organization [3,18], by forcing $v$ to be smooth. However, this regularization does not force edge direction compatibility or edge continuity. Key contribution of our work is to modify the PDE given in (3) in a way that these constraints are incorporated. Our essential idea is to introduce a spatially varying function $c$, which increases for unpreferred spatial organization of the edges and decreases for the preferred ones. The value of the function appears as a multiplier in the diffusivity function giving (9).

$$\frac{\partial u}{\partial t} = \nabla \cdot ((c.v)^2 \nabla u) - \frac{\beta}{\alpha}(u-g); \quad \left.\frac{\partial u}{\partial n}\right|_{\partial R} = 0 \tag{9}$$

Function $(cv)^2$ can be seen as a modified edge indicator, which we compute directly without explicitly computing the scalar $c$. Regularization of $u$ is influenced by the modified indicator whereas the edge indicator $v$ itself remains the same.

In the following subsections, we present two possible choices for the scalar function $c$, considering the edge coherency by means of directional consistency and edge continuity, respectively. Since the edge coherency corresponds a multiplier for the diffusivity function, it is possible to combine these two proposed functions into a single framework by taking the function $c$ as the product. The resulting framework considers the coherency of the edges by means of both the directional consistency and the edge continuity.

### 3.1 Directional Consistency

In the edge linking step of traditional boundary detection, edge pixels detected based on the magnitude of gradient are linked to give a connected edge set if their gradient directions are in agreement. Unlinked edge pixels are discarded. We induce such an effect in our diffusion model by increasing the relative persistence of the edge pixels, which are consistent with their neighbors, by increasing the diffusivity at inconsistent ones. We consider a coherency function $\phi(u)$ such that $\phi(u) \to 1$ on the preferred configurations and $\phi(u) \to 0$ on the incoherent configurations, and let $c$ has the following form:

$$c = 1 + (1 - \phi(u))\frac{1 - v}{v} \tag{10}$$

First, notice that $c$ increases in proportion to the image gradient $|\nabla u|$, which is proportional to $\frac{1-v}{v}$ (See (8)). Second, notice that the overall diffusivity coefficient $(c.v)^2$ can be estimated as follows, without explicitly computing the variable $c$:

$$(cv)^2 = (\phi(u)v + (1 - \phi(u))\,1)^2 \tag{11}$$

The new diffusion coefficient is the square of the convex combination of $v$ and 1. Value of the diffusivity is bounded by 1, attaining maximum value as $\phi(u) \to 0$, and decaying as $\phi(u) \to 1$ to a value determined by the edge indicator function $v$. We consider the following $\phi$ function, which simply measures the coherency as a function of edge directions.

$$\phi(u_i) = exp\left[\varepsilon\left(\frac{1}{|\eta_s|}\sum_{j\in\eta_s} \nabla u_i \cdot \nabla u_j - 1\right)\right] \tag{12}$$

where $\eta_s$ represents the neighborhood of pixel $i$ having $s$ neighbors. We define $\eta_s$ as $\pm s$ pixels along the orthogonal edge direction $\nabla {u_i}^{\perp}$. The parameter $\varepsilon$ is a scalar, which determines the decay rate of the $\phi$ function. If the neighboring pixels are coherent (having similar edge directions), then the average angle between $\nabla u_i$ and $\nabla u_j$'s is close to 0 making $\phi \to 1$.

### 3.2 Edge Continuity

The principle of edge continuity is used to eliminate streaking or breaking up of an edge contour due to noise or changing contrast. It is commonly referred as *hysteresis* due to successful application of threshold retardation in Canny edge detector [10]. In our diffusion model, we lower the diffusivity at pixels that correspond to broken parts of boundary segments to favor edge formation. There may be various choices for the selection of $c$. The important point is to decrease the modified diffusivity $(cv)^2$ if the neighboring site supports formation of an edge i.e having a low $v$ value. Recall (8) that gives the reciprocal relationship between $v$ and $|\nabla u|$. Decreasing diffusivity can be achieved by increasing the estimate of the image gradient, which is used in estimating the diffusivity. Therefore, a natural choice is to add an offset $h \in [0,1]$ indicating a support in favor of edge formation to the gradient term in the diffusivity estimate:

$$(cv)^2 = \left(\frac{1}{1 + h + 2\alpha\rho|\nabla u|^2}\right)^2 \tag{13}$$

Such a choice yields

$$c = \frac{1}{1 + hv} \tag{14}$$

In the discrete implementation of (9), diffusivities are estimated at mid-grid points. Hence, $h$ should be computed as a support from a suitably chosen neighbor. For example, modified diffusivity $(cv)^2_{i+0.5,j}$ at a mid point between $(i,j)$ and $(i+1,j)$ may receive support in the form of either $(1 - v_{i+0.5,j-1})$ or $(1 - v_{i+0.5,j+1})$. Notice that lower the value of edge indicator at a neighboring site, higher the support it provides.

Adding spatial organization to energies defining regularization with line process has been previously proposed by Black and Rangarajan [5]. They define a local interaction energy that favors formation of unbroken contours. In [6], Black et al. derives the necessary update equations. If we let $\frac{v_k^2+v_l^2}{2}$ define a line process between site $k$ and site $l$, then our development becomes equivalent to that of Black et al. Thus, solving new coupled equations are qualitatively equivalent to modifying the Mumford-Shah with an additive term favoring unbroken contours as in Black and Rangarajan [5].

## 4 Experimental Results

The importance of directional consistency is best observed if the image contains impulse noise. Processing of the noisy couple image, shown in Fig 1(a), using the Ambrosio-Tortorelli and the new method are illustrated in Fig 2. Fig 2(a) and (b) illustrate smoothing obtained using the Ambrosio-Tortorelli with 400 iterations with different smoothing radius, $\frac{\alpha}{\beta}$. The result in Fig 2(a) is obtained with $\alpha = 1$, $\beta = 0.01$, $\rho = 0.01$. When we increase the smoothing radius by choosing $\beta = 0.001$, diffusion is highly severe that we even lose the head of the lady (Fig 2(b)). Yet, the noise is still present. If we use a regularization term which forces spatial coherence of the edges by means of the directional consistency, as discussed in § 3.1, the image is denoised without blurring (Fig 2(c) and (d)). The perceptual difference between Fig 2(c) and (d) is in the sharpness level. The result in Fig 2(c) is obtained with the segmentation parameters

(a) (b) (c) (d)

**Fig. 2.** Considering directional consistency eliminates impulse noise. (a)-(b) Reconstructions using the Ambrosio-Tortorelli with two different smoothing radius. Notice that the noise is still present even when we lose the head of the lady. (c)-(d) Reconstructions with directional consistency with two different sharpness levels. Notice that at comparable scales noise is completely eliminated.

specified for Fig 2(a) and the coherency parameters $s = 2$ and $\varepsilon = 0.25$ (this set of parameters are used for all of the experiments reported in the paper unless otherwise stated) with 50 iterations. For the result given in Fig 2(d), we use the same parameters except $\varepsilon = 0.02$ and 300 iterations. The variable $\varepsilon$ determines the decay rate of the coherency function used in the segmentation process and therefore specifies the level of sharpness. For large $\varepsilon$ value, the decay rate is high and the edges are more smoothed out depending on the coherency. Hence, as observed, the resulting image is smoother. On the other hand, for small $\varepsilon$ values, we get sharper results.

Increasing the value of $\alpha$ while keeping $\frac{\alpha}{\beta}$ fixed means decreasing the penalty of the length term, yielding more detailed reconstruction. In Fig 3, the proposed modification is again tested with *couple* image with 5% salt and pepper noise (Fig 1(a)), however, forcing the reconstruction to be more detailed by the proper choice of parameters. Fig 3(a) is the outcome of the proposed modification after 20 iterations. On the other hand, Fig 3(c) is obtained by performing 50 iterations

(a) (b) (c) (d)

**Fig. 3.** $u$ and $1 - v$ functions computed with $\alpha = 1$, $\beta = 0.01$ and $\alpha = 4$, $\beta = 0.04$ respectively. Even in more detailed reconstructions, modified scheme is able to remove noise completely.

with the same parameters except $\alpha = 4$, $\beta = 0.04$. The corresponding edge indicator functions are also shown in Fig 3(b) and (d) respectively. As they demonstrate, even the detailed reconstruction with $\alpha = 4$ is noise free.

The example presented in Fig 4 illustrates the effect of edge continuity as described in § 3.2. The results are obtained with 100 iterations. The reconstructions of the *venice* image shown in Fig 4(a) are presented in Fig 4 (b) and (c) together with the corresponding edge indicator functions. Fig 4(b) illustrates the outcome of the Ambrosio-Tortorelli whereas Fig 4(c) illustrates the result obtained by considering edge continuity. As it can be clearly seen from the zoomed indicator functions given in Fig 4(d), the modified scheme eliminates broken contours.

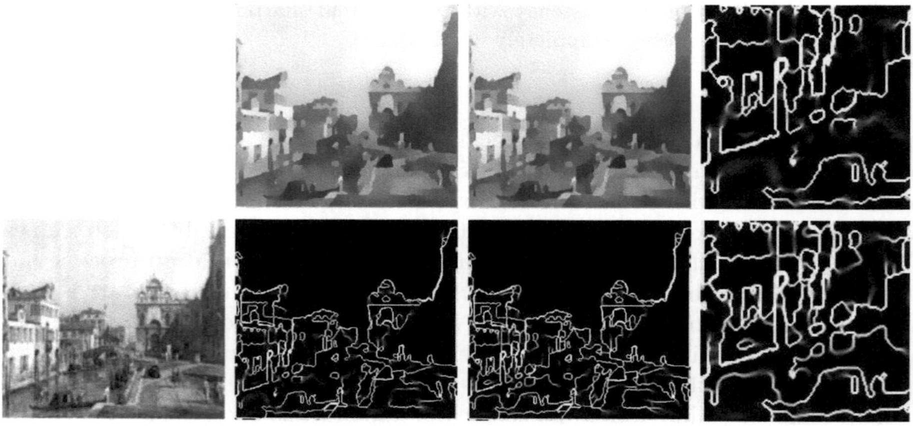

(a) (b) (c) (d)

**Fig. 4.** Considering edge continuity eliminates broken contours. (a) input image. (b) Reconstruction using the Ambrosio-Tortorelli ($u$ and $1 - v$). (c) Reconstruction with modified scheme forcing edge continuity ($u$ and $1-v$) (d) Details from the edge indicator functions given in (b) and (c) respectively.

In Fig 5, we demonstrate the results obtained with a regularization considering both the directional consistency and the edge continuity via the product of individual $c$ functions. The reconstruction results of *venice* image with 10% salt and pepper noise (Fig 5(a)) after 100 iterations are given in Fig 5(b)-(d). Fig 5(b) is the result obtained with edge continuity. As it can be clearly seen, the noise is not eliminated. Fig 5(c) is obtained with the modification which considers the directional consistency. Finally, Fig 5(d) is the outcome of the combined framework which is not only noise free and but also having stronger edges.

In Fig 6, the combined framework is tested with a noisier image (Fig 1(c)). Fig 6(a) is the outcome of the Ambrosio-Tortorelli approximation after 500 iterations. Fig 6(b) is obtained with 150 iterations by using the modification,

(a) (b) (c) (d)

**Fig. 5.** Considering a combined framework eliminates both noise and the broken contours. (a) *venice* image corrupted with 10% salt and pepper noise. (b) Reconstruction with edge continuity. (c) Reconstruction with directional consistency. (d) Reconstruction using both edge continuity and directional consistency.

which considers only the directional consistency. Fig 6(c) is the outcome of the combined framework again after 150 iterations. Both reconstructions are qualitatively comparable to the ones obtained by means of singular diffusivities [9]. Notice that the two results are visually similar. This is due to the fact that the contrast is almost constant in the image. Hence, broken lines do not occur.

Mumford-Shah regularizer gives preference to piecewise smooth images with simple edge sets, without directly forcing edge direction compatibility or edge connectivity. Our final experiment demonstrates the potential of the modified model for textured images where the piecewise smooth assumption fails. As shown in Fig 7, directional consistency can also be used for further smoothing of the inhomogeneous textured regions that results in more coherent texture boundaries. Fig 7(a) is the input image *sunflower*. Fig 7(b) is the edge indicator function obtained by the Ambrosio-Tortorelli after 100 iterations with $\alpha = 4$, $\beta = 0.04$ and $\rho = 0.001$. Even though the outer boundary separating leaves and seeds start to vanish, the inner boundary (small circles due to seeds) is clearly visible. Increasing $\alpha$ worsens the situations. On the other hand, when we consider the directional consistency of the edges with the parameters $\alpha = 8$, $\beta = 0.08$,

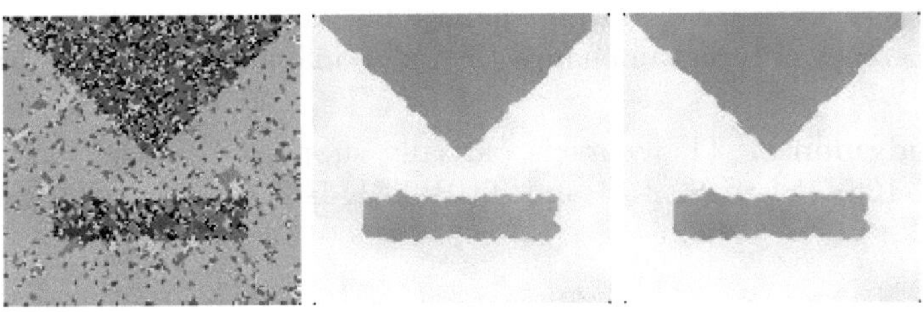

(a) (b) (c)

**Fig. 6.** A difficult denoising case. (a) Reconstruction using Ambrosio-Tortorelli. (b) Reconstruction with directional consistency. (c) Reconstruction using the combined framework.

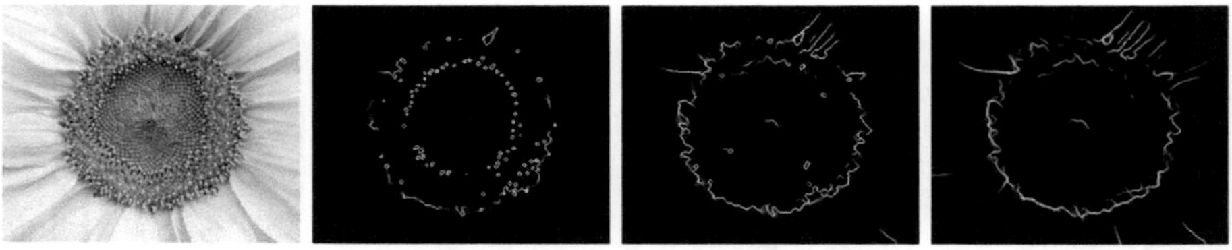

(a) (b) (c) (d)

**Fig. 7.** An application to an image which violates the piecewise smooth assumption. (a) input image. (b) Edge indicator function computed using the Ambrosio-Tortorelli. Notice that when the outer boundary separating leaves and seeds are smoothed, the small circles due to texture gradient are still present. (c)-(d) Edge indicator functions computed with the directional consistency and the combined framework respectively. Observe that the edges due to texture gradient disappeared, yet the outer boundary is clearly visible.

$\rho = 0.001$, $s = 2$ and $\varepsilon = 0.25$, the inner boundary is smoothed out and the outer boundary is present (Fig 7(c)). A reasonable result is also obtained by using the combined framework with the same parameters except $\varepsilon = 2$ (Fig 7(d)).

## 5 Summary and Conclusion

Mumford-Shah model and its Ambrosio-Tortorelli approximation unify image smoothing and edge detection via coupling of two functions $u$ and $v$ representing smooth image and the edge indicator respectively. The edge indicator function $v$ defines an analog line process and its regularization imposes smoothness of the edge set. However, the model does not directly enforce spatial coherence as in the edge linking step of the traditional processing. We modify Ambrosio-Tortorelli model in its coupled diffusion equations form such that the regularization of $u$ is steered by the coherent edges. Our experiments demonstrate that the new regularization is able to remove difficult noise types and produce almost segmentation like results without using directional or singular diffusivities that arose from $L^1$ norm in the cost functions. In our experiments, we consider spatial coherency in terms of edge direction compatibility and edge continuity. However, further coherency criteria can be investigated, remaining in the same framework.

**Acknowledgments.** This work is partially supported by the research grant TUBITAK-105E154 to S. Tari and TUBITAK-BAYG PhD scholarship to E. Erdem.

## References

1. R. Alicandro, A. Braides, and J. Shah. Free-discontinuity problems via functionals involving the $L^1$-norm of the gradient and their approximation. *Interfaces and Free Boundaries*, 1(1):17–37, 1999.

2. L. Ambrosio and V. Tortorelli. On the approximation of functionals depending on jumps by elliptic functionals via $\Gamma$-convergence. *Commun. Pure Appl. Math.*, 43(8):999–1036, 1990.
3. L. Bar, N. Kiryati, and N. Sochen. Image deblurring in the presence of impulsive noise. *Int. J. Comput. Vision*, 70(3):279–298, 2006.
4. L. Bar, N. Sochen, and N. Kiryati. Image deblurring in the presence of salt-and-pepper noise. In *Scale-Space*, pages 107–118, 2005.
5. M. J. Black and A. Rangarajan. On the unification of line processes, outlier rejection, and robust statistics with applications in early vision. *Int. J. Comput. Vision*, 19(1):57–91, 1996.
6. M. J. Black, G. Sapiro, D. H. Marimont, and D. Heeger. Robust anisotropic diffusion. *IEEE Trans. Image Processing*, 7(3):421–432, March 1998.
7. A. Braides. *Approximation of Free-discontinuity Problems.* Lecture Notes in Mathematics, Vol. 1694. Springer-Verlag, 1998.
8. T. Brox, A. Bruhn, N. Papenberg, and J. Weickert. High accuracy optical flow estimation based on a theory for warping. In *ECCV (4)*, pages 25–36, 2004.
9. B. Burgeth, J. Weickert, and S. Tari. Minimally stochastic schemes for singular diffusion equations. In X.-C. Tai, K.-A. Lie, T. F. Chan, and S. Osher and, editors, *Image Processing Based on Partial Differential Equations*, Mathematics and Visualization, pages 325–339. Springer Berlin Heidelberg, 2006.
10. J. Canny. A computational approach to edge detection. *IEEE Trans. Pattern Anal. Mach. Intell.*, 8(6):679–698, 1986.
11. T. Chan and L. Vese. Variational image restoration and segmentation models and approximations. *UCLA, CAM-report 97-47*, September, 1997.
12. S. Geman and D. Geman. Stochastic relaxation, gibbs distributions, and the bayesian restoration of images. *IEEE Trans. Pattern Anal. Mach. Intell.*, 6: 721–639, 1984.
13. D. Mumford and J. Shah. Optimal approximations by piecewise smooth functions and associated variational problems. *Commun. Pure Appl. Math.*, 42(5):577–685, 1989.
14. M. Nikolova. A variational approach to remove outliers and impulse noise. *J. Math. Imaging Vis.*, 20(1-2):99–120, 2004.
15. O. Scherzer and J. Weickert. Relations between regularization and diffusion filtering. *J. Math. Imaging Vis.*, 12(1):43–63, 2000.
16. J. Shah. Segmentation by nonlinear diffusion. In *CVPR*, pages 202–207, 1991.
17. J. Shah. A common framework for curve evolution, segmentation and anisotropic diffusion. In *CVPR*, pages 136–142, 1996.
18. S. Teboul, L. Blanc-Fraud, G. Aubert, and M. Barlaud. Variational approach for edge preserving regularization using coupled pde's. *IEEE Trans. Image Processing*, 7(3):387–397, March 1998.
19. J. Weickert. Coherence-enhancing diffusion filtering. *Int. J. Comput. Vision*, 31 (2-3):111–127, 1999.

# A Generic Approach to the Filtering of Matrix Fields with Singular PDEs

Bernhard Burgeth*, Stephan Didas**, Luc Florack*, and Joachim Weickert

Saarland University, Dept. of Mathematics and Computer Science,
Germany
burgeth@mia.uni-saarland.de
Eindhoven University of Technology, Dept. of Biomedical Engineering
The Netherlands
L.M.Florack@tue.nl

**Abstract.** There is an increasing demand to develop image processing tools for the filtering and analysis of matrix-valued data, so-called matrix fields. In the case of scalar-valued images parabolic partial differential equations (PDEs) are widely used to perform filtering and denoising processes. Especially interesting from a theoretical as well as from a practical point of view are PDEs with singular diffusivities describing processes like total variation (TV-)diffusion, mean curvature motion and its generalisation, the so-called self-snakes. In this contribution we propose a generic framework that allows us to find the matrix-valued counterparts of the equations mentioned above. In order to solve these novel matrix-valued PDEs successfully we develop truly matrix-valued analogs to numerical solution schemes of the scalar setting. Numerical experiments performed on both synthetic and real world data substantiate the effectiveness of our matrix-valued, singular diffusion filters.

## 1 Introduction

Matrix-fields are used, for instance, in civil engineering to describe anistropic behaviour of physical quantities. Stress and diffusion tensors are prominent examples. The output of diffusion tensor magnetic resonance imaging (DT-MRI) [14] are symmetric $3 \times 3$-matrix fields as well. In medical sciences this image acquisition technique has become an indispensable diagnostic tool in recent years. Evidently there is an increasing demand to develop image processing tools for the filtering and analysis of such matrix-valued data.

d-dimensional scalar images $f : \Omega \subset \mathbb{R}^d \to \mathbb{R}$ have been denoised, segmented and/or enhanced successfully with various filters described by nonlinear parabolic PDEs. In this article we focus on some prominent examples of PDEs used in image processing and which can serve as a proof-of-concept:

* The financial support of the Dutch Organization for Scientific Research NWO is gratefully acknowledged.

** The financial support of the German Organization for Scientific Research DFG is gratefully acknowledged.

F. Sgallari, A. Murli, and N. Paragios (Eds.): SSVM 2007, LNCS 4485, pp. 556–567, 2007.

- Total-Variation (TV)-Diffusion (p=1), [3,10] and balanced-forward-backward (BFB)-diffusion (p=2), [13],

$$\partial_t u = \operatorname{div}\left(\frac{\nabla u}{\|\nabla u\|^p}\right), \tag{1}$$

- Mean curvature motion (MCM), [2],

$$\partial_t u = \|\nabla u\| \operatorname{div}\left(\frac{\nabla u}{\|\nabla u\|}\right), \tag{2}$$

- Self-Snakes involving a Perona-Malik type diffusivity $g$, [15],

$$\partial_t u = \|\nabla u\| \operatorname{div}\left(g(\|\nabla u\|^2)\,\frac{\nabla u}{\|\nabla u\|}\right), \tag{3}$$

where we impose the initial condition $u(x,0) = f(x)$ for $x \in \Omega$ in all cases. TV-type diffusion filters require no tuning of parameters but have shape-preserving qualities [6] and a finite extinction time [4]. Even arbitrary exponents have been considered, [1,17]. Extensions of curvature-based PDEs to matrix fields have been proposed in [11] and more recently in [16], based on generalisations of the so-called structure tensor for scalar images to matrix fields. The research on these structure-tensor concepts has been initiated by [7,19]. The approaches to matrix field regularisation suggested in [9] are based on differential geometric considerations. Comprehensive survey articles on the analysis of matrix fields using various techniques can be found in [20].

In this article we will proceed along a different path. We will develop a generic framework for deriving matrix-valued counterparts for scalar PDEs. This does not just mean that we derive systems of PDEs which can be written in matrix form. Instead we will exploit the operator-algebraic properties of (symmetric) matrices to establish truly matrix-valued PDEs. For this work we concentrate on the matrix-valued analogs of the singular PDEs (1)–(3) as particularly interesting equations. It is also worth mentioning that in contrast to [11] and [16] our framework does not rely on a notion of structure tensor. Nevertheless, the proposed concept ensures an appropriate and desirable coupling of channels. The methodology to be developed will also enable us to transfer numerical schemes from the scalar to the matrix valued setting.

The article is structured as follows: The subsequent Section 2 contains the basic definitions necessary for our framework, such as functions of a matrix, partial derivatives, and generalised gradient of a matrix field. In Section 3 we turn first to the simple linear diffusion for matrix fields for the sake of later comparison. After introducing a symmetrised multiplication for symmetric matrices we then formulate the matrix-valued counterparts of the singular equations mentioned above. By considering the already rather complicated one-dimensional case, first properties of the matrix-valued diffusion processes are inferred. The transition from scalar numerical solution schemes to matrix-valued algorithms for the solutions of the new diffusion equations is discussed in Section 4. Example applications on synthetic and real DT-MRI data are presented in Section 5, followed by concluding remarks in the last Section 6.

## 2 Matrix-Valued PDEs: A Generic Framework

This section contains the key definitions for the formulation of matrix-valued PDEs. The underlying idea is that to a certain extent symmetric matrices can be regarded as a generalisation of real numbers. In that spirit we would like to generalise notions like functions of matrices, derivatives and gradients of such functions to the matrix-valued setting as instigated in [8]. We juxtapose the corresponding basic definitions in Table 1, and comment on them in the subsequent remarks. We start with clarifying notation. A matrix field is considered as a mapping $F : \Omega \subset \mathbb{R}^d \longrightarrow M_n(\mathbb{R})$, from a $d$-dimensional image domain into the set of $n \times n$-matrices with real entries, $F(x) = (f_{p,q}(x))_{p,q=1,\ldots,n}$. Important for us is the subset of symmetric matrices $\mathrm{Sym}_n(\mathbb{R})$. The set of positive (semi-) definite matrices, denoted by $\mathrm{Sym}_n^{++}(\mathbb{R})$ (resp., $\mathrm{Sym}_n^{+}(\mathbb{R})$), consists of all symmetric matrices $A$ with $\langle v, Av\rangle := v^\top Av > 0$ (resp., $\geq 0$) for $v \in \mathbb{R}^n \setminus \{0\}$. This set is of special interest since DT-MRI produces data with this property. Note that at each point the matrix $F(x)$ of a field of symmetric matrices can be diagonalised yielding $F(x) = V(x)^\top D(x)V(x)$, where $x \mapsto V(x) \in O(n)$ is a

**Table 1.** Extensions of elements of scalar valued calculus **(middle)** to the matrix-valued setting **(right)**

| Setting | scalar valued | matrix-valued |
|---|---|---|
| function | $h : \begin{cases} \mathbb{R} \longrightarrow \mathbb{R} \\ x \mapsto h(x) \end{cases}$ | $h : \begin{cases} \mathrm{Sym}_n(\mathbb{R}) \longrightarrow \mathrm{Sym}_n(\mathbb{R}) \\ U \mapsto V^\top \mathrm{diag}(h(\lambda_1), \ldots, h(\lambda_n))V \end{cases}$ |
| partial derivatives | $\partial_\omega u$, $\omega \in \{t, x_1, \ldots, x_d\}$ | $\overline{\partial}_\omega U := (\partial_\omega u_{ij})_{ij}$, $\omega \in \{t, x_1, \ldots, x_d\}$ |
| higher derivatives | $\partial_\omega^k u$, $\omega \in \{t, x_1, \ldots, x_d\}$ | $\overline{\partial}_\omega^k U := \left(\partial_\omega^k u_{ij}\right)_{ij}$, $\omega \in \{t, x_1, \ldots, x_d\}$ |
| Laplacian | $\Delta u := \sum_{i=1}^{d} \partial_{x_i}^2 u$ | $\overline{\Delta} U := \sum_{i=1}^{d} \overline{\partial}_{x_i}^2 U$ |
| gradient | $\nabla u(x) := (\partial_{x_1} u(x), \ldots, \partial_{x_d} u(x))^\top$, $\nabla u(x) \in \mathbb{R}^d$ | $\overline{\nabla} U(x) := (\overline{\partial}_{x_1} U(x), \ldots, \overline{\partial}_{x_d} U(x))^\top$, $\overline{\nabla} U(x) \in (\mathrm{Sym}_n(\mathbb{R}))^d$ |
| divergence | $\mathrm{div}\ (a(x))^\top := \sum_{i=1}^{d} \partial_{x_i} a_i(x)$, $a(x) := (a_1(x), \ldots, a_d(x))$ | $\overline{\mathrm{div}}\ (A(x))^\top := \sum_{i=1}^{d} \overline{\partial}_{x_i} A_i(x)$, $A(x) := (A_1(x), \ldots, A_d(x))$ |
| length | $\lvert w\rvert_p := \sqrt[p]{\lvert w_1\rvert^p + \cdots + \lvert w_d\rvert^p}$, $\lvert w\rvert_p \in [0, +\infty[$ | $\lvert W\rvert_p := \sqrt[p]{\lvert W_1\rvert^p + \cdots + \lvert W_d\rvert^p}$, $\lvert W\rvert_p \in \mathrm{Sym}_n^+(\mathbb{R})$ |
| multiplication | $a \cdot b$ | $A^{\frac{1}{2}} B A^{\frac{1}{2}}$ |

matrix field of orthogonal matrices, while $x \mapsto D(x)$ is a matrix field of diagonal matrices. In the sequel we will denote $n \times n$ - diagonal matrices with entries $\lambda_1, \ldots, \lambda_n \in \mathbb{R}$ from left to right simply by $\text{diag}(\lambda_i)$. $O(n)$ stands for the matrix group of orthogonal $n \times n$-matrices. We will also assume that the matrix field $U(x)$ to be diagonalisable with $U = (u_{i,j})_{i,j} = V^\top \text{diag}(\lambda_1, \ldots, \lambda_n)V$, where $V \in O(n)$ and $\lambda_1, \ldots, \lambda_n \in \mathbb{R}$.

**Remarks 1:**

1. **Functions of matrices.** The definition of a function $h$ on $\text{Sym}_n(\mathbb{R})$ is standard [12]. As an important example, $|U|$ denotes the matrix-valued equivalent of the *absolute value* of a real number, $|U| = V^\top \text{diag}(|\lambda_1|, \ldots, |\lambda_n|)V \in \text{Sym}_n^+(\mathbb{R})$, not to be confused with the determinant $\det(U)$ of $U$.

2. **Partial derivatives.** The *componentwise* definition of the partial derivative for matrix fields is a natural extension of the scalar case:

$$\begin{aligned}\overline{\partial}_\omega U(\omega_0) &= \lim_{h \to 0} \frac{1}{h}\left(U(\omega_0 + h) - U(\omega_0)\right) = \left(\lim_{h \to 0} \frac{u_{ij}(\omega_0 + h) - u_{ij}(\omega_0)}{h}\right)_{i,j} \\ &= \left(\partial_\omega u_{ij}(\omega_0)\right)_{i,j} ,\end{aligned}$$

   where $\overline{\partial}_\omega$ stands for a spatial or temporal derivative. By iteration, higher order partial differential operators such as the Laplacian or other more sophisticated operators find their natural counterparts in the matrix-valued framework.
   It is worth mentioning that for the operators $\overline{\partial}_\omega$ a *product rule* holds:

$$\overline{\partial}_\omega (A(x) \cdot B(x)) = (\overline{\partial}_\omega A(x)) \cdot B(x)) + A(x) \cdot (\overline{\partial}_\omega B(x)) .$$

   Observe that positive definiteness in general is *not* preserved through derivation $\overline{\partial}_\omega$.

3. **Generalized gradient of a matrix field.** The definition of a generalised gradient is somewhat different from one that might be expected when viewing a matrix as a tensor (of second order). The rules of differential geometry would tell us that derivatives are tensors of third order. Instead, we adopt a more operator-algebraic point of view: The matrices are self-adjoint operators that can be added, multiplied with a scalar, and concatenated. Thus, they form an algebra, and we aim at consequently replacing the field $\mathbb{R}$ by the algebra $\text{Sym}_n(\mathbb{R})$ in the scalar, that is, $\mathbb{R}$-based formulation of PDEs used in image processing. Hence, the generalised gradient $\overline{\nabla} U(x)$ at a voxel $x$ is regarded as an element of the module $(\text{Sym}_n(\mathbb{R}))^d$ over $\text{Sym}_n(\mathbb{R})$ in close analogy to the scalar setting where $\nabla u(x) \in \mathbb{R}^d$.
   In the sequel we will call a mapping from $\mathbb{R}^d$ into $(\text{Sym}_n(\mathbb{R}))^d$ a module field rather than a vector field.

4. **Generalised divergence of the module field.** The generalization of the divergence operator div acting on a vector field to an operator $\overline{\text{div}}$ acting

on a module field $A$ is straightforward, and is in accordance with the formal relation $\overline{\Delta}U = \overline{\mathrm{div}}\,\overline{\nabla}U = \overline{\nabla}.\overline{\nabla}U$ known in its scalar form from standard vector analysis.

5. **Generalised Length in $(\mathrm{Sym}_n(\mathrm{I\!R}))^d$.** Considering the formal definition in Table 1 the length of an element of a module field $A$ is close at hand. It results in a positive semidefinite matrix from $\mathrm{Sym}_n^+(\mathrm{I\!R})$ the direct counterpart of a nonnegative real number as the length of a vector in $\mathrm{I\!R}^d$.
6. **Symmetrised Multiplication in $\mathrm{Sym}_n(\mathrm{I\!R})$.** The scalar TV-diffusion equation (1) requires the multiplication of the components of a vector (namely $\nabla u$) with a scalar (namely $\frac{1}{\|\nabla u\|}$). In the matrix-valued setting the components of $\overline{\nabla}U$, that is, $\overline{\partial}_{x_i}U$, $i = 1, \ldots, d$, and (the inverse of) its generalised length $|\overline{\nabla}U|_2 =: |\overline{\nabla}U|$ are symmetric matrices. However, the product of two symmetric matrices $A, B \in \mathrm{Sym}_n(\mathrm{I\!R})$ is not symmetric unless the matrices commute. Among the numerous options to define a symmetrised matrix product we focus on one that is inspired from pre-conditioning of symmetric linear equation systems [12]. We define
$$A \bullet B := A^{\frac{1}{2}} B A^{\frac{1}{2}}$$
as the *symmetrised multiplication* of symmetric matrices.

For the sake of future comparison we first consider the matrix-valued version of the linear diffusion equation on $\mathrm{I\!R}^d \times [0, \infty[$ in the next section.

# 3 Diffusion Equations for Matrix-Fields

## 3.1 Matrix-Valued Linear Diffusion

The linear diffusion equation $\partial_t u = \sum_{i=1}^{d} \partial_{x_i}\partial_{x_i} u = \sum_{i=1}^{d} \partial_{x_i x_i} u = \overline{\Delta} u$ on $\mathrm{I\!R}^d \times [0, \infty[$ is directly extended to the matrix valued setting:

$$\overline{\partial}_t U = \sum_{i=1}^{d} \overline{\partial}_{x_i}\overline{\partial}_{x_i} U = \sum_{i=1}^{d} \overline{\partial}_{x_i x_i} U = \overline{\Delta} U \tag{4}$$

with initial condition $U(x, 0) = F(x)$. The diffusion process described by this equation acts on each of the components of the matrix independently. It is proven in [11] that positive (semi-)definiteness of the initial matrix field $F$ is indeed bequeathed to $U$ for all times.

## 3.2 Matrix-Valued Singular Diffusion Equations

In Section 2, Remark 1, (6) we set $A \bullet B := A^{\frac{1}{2}} B A^{\frac{1}{2}}$ for a symmetric multiplication of symmetric matrices. It is easily verified that this product is neither associative, nor commutative, and distributive only in the second argument. However, if $A$ is non-singular, the so-called *signature* $s = (s_+, s_-, s_0)$ of $B$ is preserved, where $s_+$, $s_-$, and $s_0$, stand for the number of positive, negative, and vanishing eigenvalues of $B$, respectively. This implies in particular that the positive definiteness of $B$ is preserved. Furthermore, for commuting matrices $A, B$ we have $A \bullet B = A \cdot B$. Another even more prominent candidate for a symmetrised multiplication would be the so-called Jordan product $A \bullet_J B := \frac{1}{2}(AB + BA)$, which is neither associative nor distributive, but commutative. The reason we disregard it in this article lies in the fact that it does not preserve positive (semi-)definiteness as the following simple example shows:

$$\begin{pmatrix} 2 & 0 \\ 0 & 0 \end{pmatrix} \bullet_J \begin{pmatrix} 1 & 1 \\ 1 & 1 \end{pmatrix} = \frac{1}{2}\left(\begin{pmatrix} 2 & 2 \\ 0 & 0 \end{pmatrix} + \begin{pmatrix} 2 & 0 \\ 2 & 0 \end{pmatrix}\right) = \begin{pmatrix} 2 & 1 \\ 1 & 0 \end{pmatrix} \quad \text{but} \quad \det \begin{pmatrix} 2 & 1 \\ 1 & 0 \end{pmatrix} = -1\,.$$

**Remark 2:** It should be mentioned that the logarithmic multiplication introduced in [5] and given by $A \bullet_L B := \exp(\log(A) + \log(B))$ is defined only for positive definite matrices. However, the matrix-valued PDE-based filtering proposed here requires the symmetric multiplication to be able to cope with at least one factor matrix being indefinite. Furthermore, matrix fields that are not necessarily positive semidefinite should also be within the reach of our PDE-based methods. Hence the logarithmic multiplication is not suitable for our purpose.

With these definitions we are now in the position to state the matrix-valued counterparts for the PDEs (1)-(3) mentioned above. For the sake of brevity we concentrate on the most general one, the self-snakes:

$$\begin{aligned} \partial_t u &= |\overline{\nabla} U| \bullet \operatorname{div}\left(\frac{g(|\overline{\nabla} U|^2)}{|\overline{\nabla} U|} \bullet \overline{\nabla} U\right) \qquad (5) \\ &= \sqrt{|\overline{\nabla} U|} \cdot \left[\sum_{i=1}^{d} \overline{\partial}_{x_i}\left(\sqrt{\frac{g(|\overline{\nabla} U|^2)}{|\overline{\nabla} U|}} \cdot (\overline{\partial}_{x_i} U) \cdot \sqrt{\frac{g(|\overline{\nabla} U|^2)}{|\overline{\nabla} U|}}\right)\right] \cdot \sqrt{|\overline{\nabla} U|}\,, \end{aligned}$$

where we used the notation

$$\frac{g(|\overline{\nabla} U|^2)}{|\overline{\nabla} U|} := g(|\overline{\nabla} U|^2) \cdot |\overline{\nabla} U|^{-1} = |\overline{\nabla} U|^{-1} \cdot g(|\overline{\nabla} U|^2) = |\overline{\nabla} U|^{-1} \bullet g(|\overline{\nabla} U|^2)\,.$$

Specifying $g = 1$ we regain the matrix-valued PDE for mean curvature motion of matrix fields, while neglecting the factor $|\overline{\nabla} U|$ and setting $g(s^2) = \frac{1}{|s|}$ in equation (5) produces the equation for BFB-diffusion, for instance.

### 3.3 Matrix-Valued Signals

In this section we investigate matrix-valued TV-related diffusion processes, mean curvature motion and self-snakes in the case of one space dimension. We restrict ourselves to the one-dimensional case ($d = 1$), $U : \mathbb{R} \longrightarrow \mathrm{Sym}_n(\mathbb{R})$, since then simplifications occur. Only one spatial derivative appears, and the expressions containing the matrix $\overline{\partial}_x$ commute. Hence, in those expressions the symmetric multiplication "$\bullet$" collapses to "$\cdot$", facilitating the analysis. The equation for the matrix-valued self-snakes in one space dimension simplifies to

$$\overline{\partial}_t U = |\overline{\partial}_x U| \bullet \overline{\partial}_x \left( \frac{g((\overline{\partial}_x U)^2)}{|\overline{\partial}_x U|} \cdot \overline{\partial}_x U \right) .$$

However, even in this simplified setting this type of data exhibits directional (through eigenvectors) as well as shape information (through eigenvalues) which allows for the appearance of new phenomena. The partial derivative $\overline{\partial}_x$ of a signal $U$ of symmetric matrices results again in symmetric matrices, $\overline{\partial}_x U(x) \in \mathrm{Sym}_n(\mathbb{R})$. Hence we have $\overline{\partial}_x U(x) = \tilde{V}^\top(x)\mathrm{diag}(\tilde{\lambda}_i(x))\tilde{V}(x)$ with $\tilde{V}(x) \in O(n)$ for all $x \in \Omega$. We observe that $\frac{g((\overline{\partial}_x U)^2)}{|\overline{\partial}_x U|}$ is also diagonalised by $\tilde{V}$,

$$\frac{g((\overline{\partial}_x U)^2)}{|\overline{\partial}_x U|} = \tilde{V}^\top \mathrm{diag}\left( \frac{g(\tilde{\lambda}_i^{\,2})}{|\tilde{\lambda}_i|} \right) \tilde{V} ,$$

and introducing the abbreviation $h(s^2) := \frac{g(s^2)}{\sqrt{s^2}}$ it follows that $h((\overline{\partial}_x U)^2)\cdot\overline{\partial}_x U = \tilde{V}^\top \mathrm{diag}\left(h(\tilde{\lambda}_i^{\,2}) \cdot \tilde{\lambda}_i\right) \tilde{V}$. We introduce a flux function $\Phi$ by $\Phi(s) := s \cdot h(s^2)$ which gives $\frac{d\Phi}{d\,s}(s) = \Phi'(s) = 2s^2 h'(s^2) + h(s^2)$ at least for $s \neq 0$. In order to treat the singularity at $s = 0$ it is customary to regularise $h$ in one way or the other to make $h$ differentiable in $[0, +\infty[$. Keeping numerical issues in mind we also adopt this point of view, rather than interpreting the derivatives in the following calculations in the distributional sense. The product rule for matrix-valued functions and incorporating $\Phi$ then yields, if we suppress the explicit dependence of $V$ and $\lambda_i$ on $x$ notationally, the following matrix-valued version of the self-snakes equation

$$\overline{\partial}_t U = |\overline{\partial}_x U| \bullet \Big( \overline{\partial}_x \tilde{V} \mathrm{diag}(h(\tilde{\lambda}_i^{\,2})\cdot\tilde{\lambda}_i)\tilde{V}^\top + \tilde{V}\mathrm{diag}(h(\tilde{\lambda}_i^{\,2})\cdot\tilde{\lambda}_i)\overline{\partial}_x\tilde{V}^\top \tag{6}$$

$$+ \tilde{V}\mathrm{diag}(\Phi'(\tilde{\lambda}_i) \cdot \partial_x\tilde{\lambda}_i]) \;\tilde{V}^\top \Big) \tag{7}$$

We infer that the matrix-valued data allow for a new phenomenon: unlike in the scalar setting, a matrix carries *directional* information conveyed through the eigenvectors as well as *shape* information mediated via eigenvalues. The evolution process described in (6) and (7) displays a coupling between shape and directional information by virtue of the simultaneous occurrence of terms containing $\overline{\partial}_x \tilde{V}(x)$ in (6) and $\partial_x \tilde{\lambda}(x)$ in (7). Clearly there is no equivalent for this in the scalar setting.

## 4 Matrix-Valued Numerical Schemes

In the previous sections the guideline to infer matrix-valued PDEs from scalar ones was, roughly speaking, analogy by making a transition from the real field $\mathbb{R}$ to the vector space $\mathrm{Sym}_n(\mathbb{R})$ endowed with some 'symmetric' product "$\bullet$".

We follow this guideline also in the issue of numerical schemes for matrix-valued PDEs.

For the sake of brevity we restrict ourselves to the TV-type diffusion, which means $h(s^2) = \frac{1}{\sqrt{s^2}}$ (or in its regularised form $h(s^2) = \frac{1}{\sqrt{\varepsilon^2+s^2}}$ with $0 \leq \varepsilon \ll 1$) and two space dimensions ($d = 2$). The necessary extensions to self-snakes in dimensions $d \geq 3$ are immediate. A possible space-discrete scheme for the scalar TV-diffusion can be cast into the form

$$\begin{aligned}\frac{du(i,j)}{dt} = &\frac{1}{\tau_1}\left(h(i+\frac{1}{2},j)\cdot\frac{u(i+1,j)-u(i,j)}{\tau_1} - h(i-\frac{1}{2},j)\cdot\frac{u(i,j)-u(i-1,j)}{\tau_1}\right)\\ &+\frac{1}{\tau_2}\left(h(i,j+\frac{1}{2})\cdot\frac{u(i,j+1)-u(i,j)}{\tau_2} - h(i,j-\frac{1}{2})\cdot\frac{u(i,j)-u(i,j-1)}{\tau_2}\right),\end{aligned}$$

where $h(i,j)$ and $u(i,j)$ are samples of the (regularised) diffusivity $h$ and of $u$ at pixel $(i,j)$ and, for example, $h(i \pm \frac{1}{2}, j) := \frac{h(i\pm1,j)+h(i,j)}{2}$. According to our preparations in Section 2 its matrix-valued extension to solve the TV-diffusion equation in the matrix setting reads

$$\begin{aligned}\frac{dU(i,j)}{dt} = &\frac{1}{h_1}\left(H(i+\frac{1}{2},j)\bullet\frac{U(i+1,j)-U(i,j)}{h_1} - H(i-\frac{1}{2},j)\bullet\frac{U(i,j)-U(i-1,j)}{h_1}\right)\\ &+\frac{1}{h_2}\left(H(i,j+\frac{1}{2})\bullet\frac{U(i,j+1)-U(i,j)}{h_2} - H(i,j-\frac{1}{2})\bullet\frac{U(i,j)-U(i,j-1)}{h_2}\right).\end{aligned}$$

The arithmetic mean $H(i \pm \frac{1}{2}, j) := \frac{H(i\pm1,j)+H(i,j)}{2} \in \mathrm{Sym}_n(\mathrm{I\!R})$ approximates the diffusivity $H(|\overline{\nabla} U|^2)$ between the pixels $(i \pm 1, j)$ and $(i, j)$. However, for the numerical treatment of MCM and self-snakes the usage of the properly defined *harmonic mean* instead of the arithmetic mean is advised. In the scalar setting this was already observed and put to work in [18].

## 5 Experiments

In our experiments we used a 3-D DT-MRI data set of a human head consisting of a $128 \times 128 \times 38$-field of positive definite matrices. The data are represented as ellipsoids via the level sets of the quadratic form $\{x^\top A^{-2} x = const. : x \in \mathrm{I\!R}^3\}$ associated with a matrix $A \in \mathrm{Sym}^+(3)$. By using $A^{-2}$ the length of the semi-axes of the ellipsoid correspond directly with the three eigenvalues of the matrix. However, for a better judgement of the denoising qualities of the smoothing processes we utilise also artificial data sets.

In Figure 1 below we compare the results of matrix-valued TV- and BFB-diffusion. The noise is removed while the edge is preserved, in very good agreement with the well-known denoising properties of their scalar predecessors.

Another set of artificial data, depicted in Figure 2, is used to demonstrate exemplarily the denoising capabilities of matrix-valued self-snakes, see Figure 3.Figure 4 juxtaposes matrix-valued linear diffusion, and smoothing with MCM and self-snakes. The smoothing as well as the convexifying and shrinking of image objects to circular structures known as features of scalar mean curvature motion

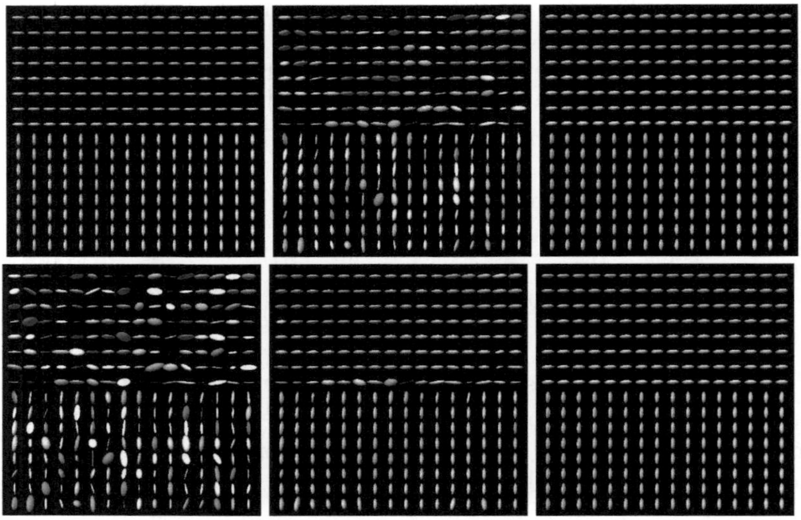

**Fig. 1. (a) Top row, from left to right:** Original matrix field. TV-diffusion on the noisy image after $t = 5$, and $t = 100$. **(b) Bottom row, from left to right:** Original polluted additively with a random matrix field $R$. The eigenvalues of $R$ stem from a Gaussian distribution with vanishing mean and standard deviation 100, its normalised eigenvectors have uniform spatial distribution. Then BFB-diffusion on the noisy image after $t = 0.5$, and $t = 10$.

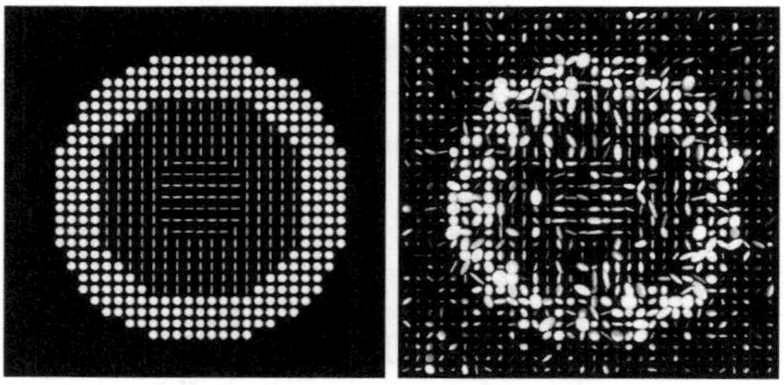

**Fig. 2. Left:** Original matrix field. **Right:** Original polluted additively with a random matrix field $R$ as in Figure 1.

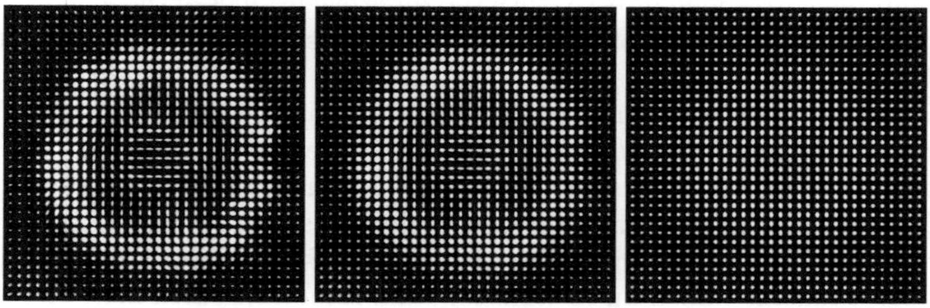

**Fig. 3. From left to right:** Filtering results for the polluted image of Figure 2 with self-snakes ($\lambda = 2000$) after $t = 5$, $t = 10$, and $t = 100$

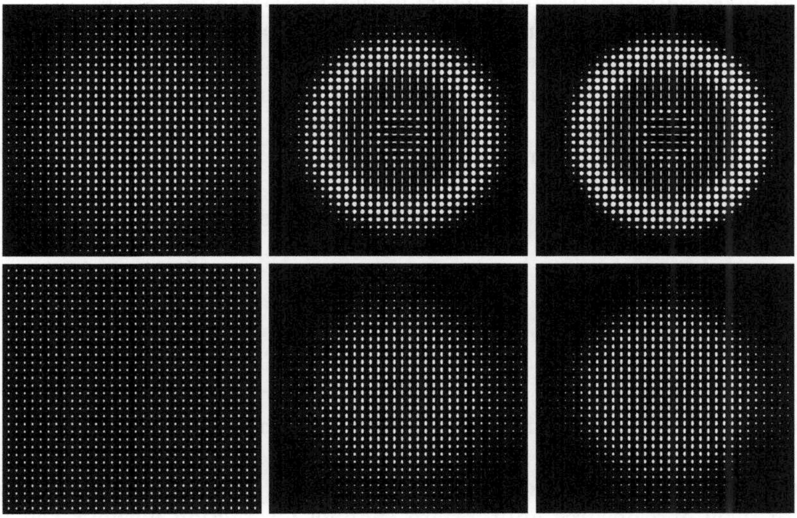

**Fig. 4.** Smoothing of image (a) in Figure 2. **(a) First column, top to bottom:** Linear Diffusion. Stopping times $t = 10$, and $t = 100$. **(b) Second column, top to bottom:** Mean curvature motion. Stopping times $t = 10$, and $t = 100$. **(c) Third column, top to bottom:** Self-snakes with $\lambda = 2000$. Stopping times $t = 10$, and $t = 100$.

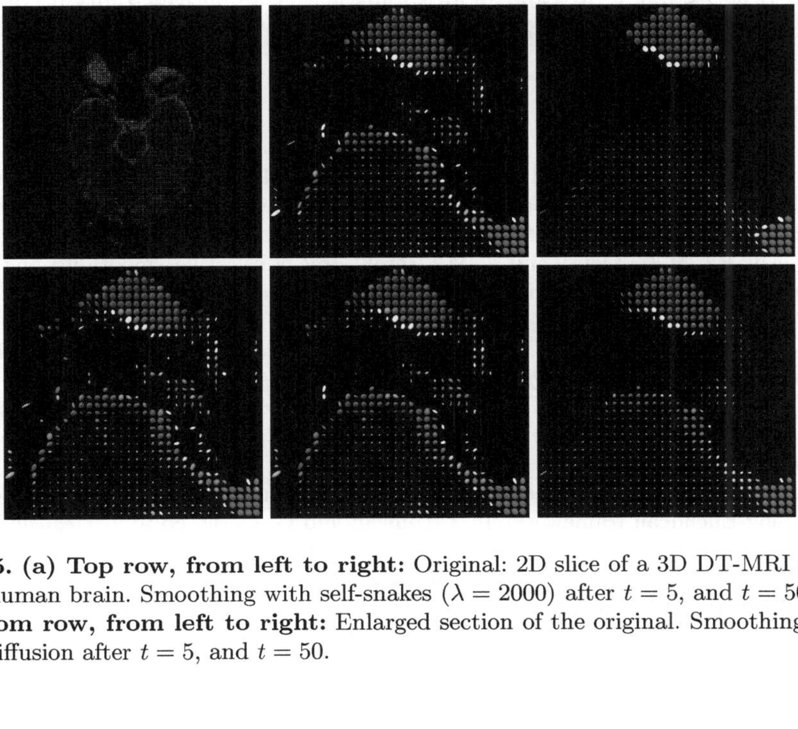

**Fig. 5. (a) Top row, from left to right:** Original: 2D slice of a 3D DT-MRI image of a human brain. Smoothing with self-snakes ($\lambda = 2000$) after $t = 5$, and $t = 50$. **(b) Bottom row, from left to right:** Enlarged section of the original. Smoothing with TV-diffusion after $t = 5$, and $t = 50$.

and self-snakes are clearly discernable in our matrix-valued setting. Finally, in Figure 5 the smoothing and enhancing properties of matrix-valued self-snakes and TV-diffusion are juxtaposed while acting on a 2-D slice of a real 3-D DT-MRI data set. The matrix-valued extensions inherit the filtering capabilities of their scalar counterparts.

It is worth mentioning that the results are in good agreement with the results in [11] and [16]. However, the framework presented here is generic, hence more general, and does not rely on any notion of a potentially parameter-steered structure tensor.

## 6 Conclusion

In this article we have presented a novel and generic framework for the extension of singular PDEs to symmetric matrix fields in any spatial dimension. We focused on the extension of scalar TV/BFB-diffusion, mean curvature motion, and self-snakes as leading examples. The approach takes an operator-algebraic point of view and ensures appropriate channel interaction without the use of a structure tensor. Experiments on positive semidefinite DT-MRI and artificial data illustrate that the matrix-valued methods inherit desirable characteristic properties of their scalar valued predecessors, e.g. very good denoising capabilities combined with feature preserving qualities, and the absence of tuning parameters. In future work we will investigate how this framework can help to extend other scalar PDEs and more sophisticated numerical solution concepts in image processing to the matrix-valued setting.

## References


1. L. Alvarez, F. Guichard, P.-L. Lions, and J.-M. Morel. Axioms and fundamental equations in image processing. *Archive for Rational Mechanics and Analysis*, 123:199–257, 1993.
2. L. Alvarez, P.-L. Lions, and J.-M. Morel. Image selective smoothing and edge detection by nonlinear diffusion. II. *SIAM Journal on Numerical Analysis*, 29: 845–866, 1992.
3. F. Andreu, C. Ballester, V. Caselles, and J. M. Mazón. Minimizing total variation flow. *Differential and Integral Equations*, 14(3):321–360, March 2001.
4. F. Andreu, V. Caselles, J. I. Diaz, and J. M. Mazón. Qualitative properties of the total variation flow. *Journal of Functional Analysis*, 188(2):516–547, February 2002.
5. V. Arsigny, P. F., X. Pennec, and N. Ayache. Fast and simple calculus on tensors in the log-Euclidean framework. In J. Duncan and G. Gerig, editors, *Medical Image Computing and Computer-Assisted Intervention - MICCAI 2005, Part I*, volume 3749 of *LNCS*, pages 115–122. Springer, 2005.
6. G. Bellettini, V. Caselles, and M. Novaga. The total variation flow in $R^N$. *Journal of Differential Equations*, 184(2):475–525, 2002.
7. T. Brox, J. Weickert, B. Burgeth, and P. Mrázek. Nonlinear structure tensors. *Image and Vision Computing*, 24(1):41–55, 2006.

8. B. Burgeth, A. Bruhn, S. Didas, J. Weickert, and M. Welk. Morphology for matrix-data: Ordering versus PDE-based approach. *Image and Vision Computing*, 2006.
9. C. Chefd'Hotel, D. Tschumperlé, R. Deriche, and O. Faugeras. Constrained flows of matrix-valued functions: Application to diffusion tensor regularization. In A. Heyden, G. Sparr, M. Nielsen, and P. Johansen, editors, *Computer Vision – ECCV 2002*, volume 2350 of *Lecture Notes in Computer Science*, pages 251–265. Springer, Berlin, 2002.
10. F. Dibos and G. Koepfler. Total variation minimization by the Fast Level Sets Transform. In *Proc. First IEEE Workshop on Variational and Level Set Methods in Computer Vision*, pages 145–152, Vancouver, Canada, July 2001. IEEE Computer Society Press.
11. C. Feddern, J. Weickert, B. Burgeth, and M. Welk. Curvature-driven PDE methods for matrix-valued images. *International Journal of Computer Vision*, 69(1):91–103, August 2006.
12. R. A. Horn and C. R. Johnson. *Matrix Analysis.* Cambridge University Press, Cambridge, UK, 1990.
13. S. L. Keeling and R. Stollberger. Nonlinear anisotropic diffusion filters for wide range edge sharpening. *Inverse Problems*, 18:175–190, January 2002.
14. C. Pierpaoli, P. Jezzard, P. J. Basser, A. Barnett, and G. Di Chiro. Diffusion tensor MR imaging of the human brain. *Radiology*, 201(3):637–648, December 1996.
15. G. Sapiro. *Geometric Partial Differential Equations and Image Analysis.* Cambridge University Press, Cambridge, UK, 2001.
16. T. Schultz, B. Burgeth, and J. Weickert. Flexible segmentation and smoothing of DT-MRI fields through a customizable structure tensor. In *Proceedings of the International Symposium on Visual Computing*, Lecture Notes in Computer Science. Springer, Berlin, 2007.
17. V. I. Tsurkov. An analytical model of edge protection under noise suppression by anisotropic diffusion. *Journal of Computer and Systems Sciences International*, 39(3):437–440, 2000.
18. J. Weickert. Applications of nonlinear diffusion in image processing and computer vision. *Acta Mathematica Universitatis Comenianae*, 70(1):33–50, 2001.
19. J. Weickert and T. Brox. Diffusion and regularization of vector- and matrix-valued images. In M. Z. Nashed and O. Scherzer, editors, *Inverse Problems, Image Analysis, and Medical Imaging*, volume 313 of *Contemporary Mathematics*, pages 251–268. AMS, Providence, 2002.
20. J. Weickert and H. Hagen, editors. *Visualization and Processing of Tensor Fields.* Springer, Berlin, 2006.

# Combining Curvature Motion and Edge-Preserving Denoising

Stephan Didas and Joachim Weickert

Mathematical Image Analysis Group
Department of Mathematics and Computer Science
Saarland University, Building E1 1
66041 Saarbrücken
{didas,weickert}@mia.uni-saarland.de

**Abstract.** In this paper we investigate a family of partial differential equations (PDEs) for image processing which can be regarded as isotropic nonlinear diffusion with an additional factor on the right-hand side. The one-dimensional analogues to this filter class have been motivated as scaling limits of one-dimensional adaptive averaging schemes. In 2-D, mean curvature motion is one of the most prominent examples of this family of PDEs. Other representatives of the filter class combine properties of curvature motion with the enhanced edge preservation of Perona-Malik diffusion. It becomes appearent that these PDEs require a careful discretisation. Numerical experiments display the differences between Perona-Malik diffusion, classical mean curvature motion and the proposed extensions. We consider, for example, enhanced edge sharpness, the question of morphological invariance, and the behaviour with respect to noise.

## 1 Introduction

Mean curvature motion and nonlinear diffusion filtering are classical methods in image processing for which feature directions in the image are important. Usually the two prominent directions for the local geometry in the image are the direction of the level line or isophote (along an edge) and its orthogonal direction, the flowline (across an edge). Choosing the amount of diffusion along these two directions appropriately gives a various range of different methods [1,2,3]. The earliest examples go back to the 1960s when Gabor proposed a method for deblurring of electron microscopy images [4,5].

A prominent example where we only have diffusion along edges is mean curvature motion (MCM). Its theoretical properties have first been investigated by Gage, Hamilton and Huisken [6,7,8] in the 1980s. In the context of image processing, it first appeared in [9,10]. Two nonlinear extensions of MCM are proposed by Alvarez et al. [11], or by Sapiro with the so-called self-snakes [12].

Nonlinear diffusion has first been proposed for image processing by Perona and Malik in 1990 [13]. Especially in its regularised variant by Catté et al. [14] it has become one of the standard tools for image denoising and simplification

F. Sgallari, A. Murli, and N. Paragios (Eds.): SSVM 2007, LNCS 4485, pp. 568–579, 2007.

in the meantime. The decomposition of the Perona-Malik approach in diffusion along gradient and normal direction is used to clearify its properties [15].

A general formulation for such filters which respect more general feature directions is given by Carmona and Zhong [16]. They discuss different ways to determine the local feature directions in practice, for example with second order derivatives or Gabor filters. The corresponding techniques are not only applied to grey value images, but there are also extensions to the vector-valued case by Tschumperlé and Deriche [17]. They use so-called oriented Laplacians, that means weighted sums of second derivatives of an image in two orthogonal directions.

The goal of the present paper is to investigate a class of PDEs for image processing which combine denoising capabilities of Perona-Malik filtering with curve shrinkage properties of mean curvature motion. The proposed model is the 2-D generalisation of a one-dimensional equation considered in [18] as scaling limit of adaptive averaging schemes. We are going to relate this filter class to previously known techniques and analyse its properties with numerical experiments.

The paper is organised as follows: The following Section 2 reviews some of the classical directional depending filters mentioned above to introduce the topic. Section 3 then introduces our model, which will be called *generalised mean curvature motion* (GMCM). In Section 4, we discuss some aspects of discretisation as this turns out to be a crucial question for this filter class. We display practical properties of all presented methods in Section 5 with several numerical examples and compare them to classical models. Section 6 concludes the paper with a summary and some questions of ongoing research.

## 2 Classical Directional Depending Filters

In this section, we review some well known filters which depend on the local feature directions. Fig. 1 gives an impression of the properties of several filtering methods. Nonlinear diffusion filtering as introduced by Perona and Malik [13]

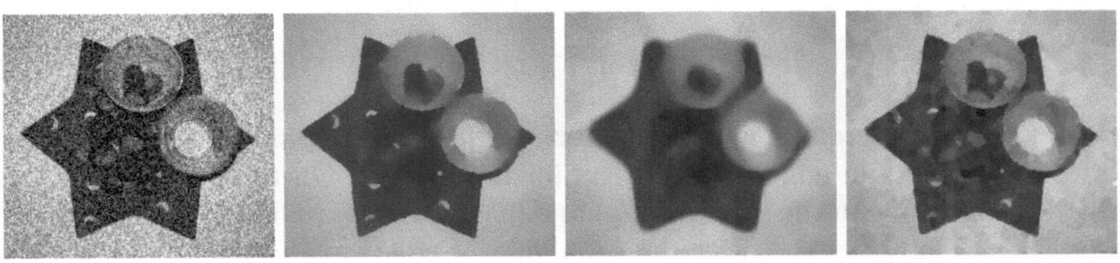

**Fig. 1.** Classical PDE-based filters. *Left:* Original image, 300×275 pixels, with Gaussian noise (standard deviation $\sigma = 50$). *Second left:* Perona-Malik filtering, $\lambda = 6$, $t = 75$. *Second right:* Mean curvature motion, $t = 50$. *Right:* Self-snakes, $\lambda = 10$, $t = 50$.

uses the evolution of an image $u$ under the PDE

$$\begin{aligned} u(\cdot,0) &= f \ , \\ \partial_t u &= \operatorname{div}\left(g(|\nabla u|)\nabla u\right) \end{aligned} \tag{1}$$

where the given initial image is denoted by $f$ and the evolving one by $u$. Usually homogeneous Neumann boundary conditions $\partial_n u = 0$ are considered, i. e. the derivative in the normal direction of the boundary is zero. For all PDE methods in this paper, we have the same initial and boundary conditions, and we therefore do not state them explicitly for each equation in the following. Following the ideas in [15,16] we decompose the diffusion into two parts acting in orthogonal directions. We consider the two local orientations of an image:

$$\eta := \frac{\nabla u}{|\nabla u|} = \frac{1}{\sqrt{u_x^2+u_y^2}} \begin{pmatrix} u_x \\ u_y \end{pmatrix} \tag{2}$$

is the direction of the gradient or steepest ascent, that means across an edge in the image. Orthogonal to the gradient we have the direction of the level set

$$\xi := \frac{\nabla u^\perp}{|\nabla u^\perp|} = \frac{1}{\sqrt{u_x^2+u_y^2}} \begin{pmatrix} -u_y \\ u_x \end{pmatrix} \ , \tag{3}$$

which points locally along an edge. In the following considerations we want to decompose a diffusion process into diffusion along and across image edges. For this reason we need the second derivatives of the image in the directions $\xi$ and $\eta$, namely

$$u_{\xi\xi} = \frac{u_{xx}u_y^2 - 2u_xu_yu_{xy} + u_{yy}u_x^2}{u_x^2+u_y^2} \quad \text{and} \tag{4}$$

$$u_{\eta\eta} = \frac{u_{xx}u_x^2 + 2u_xu_yu_{xy} + u_{yy}u_y^2}{u_x^2+u_y^2} \ . \tag{5}$$

With these equations we follow Alvarez et al. [15] and decompose the Perona-Malik equation into two diffusion components acting in direction $\xi$ and $\eta$:

$$\partial_t u = g(|\nabla u|)\, u_{\xi\xi} + \left(g(|\nabla u|) + g'(|\nabla u|)|\nabla u|\right) u_{\eta\eta} \tag{6}$$

We see that on the one hand, the factor $g(|\nabla u|)$ can reduce the velocity of the diffusion close to an edge (when the gradient is large). On the other hand, the first derivative of $g$ in the second summand makes backward diffusion in direction $\eta$ possible. This gives the filter the capability of edge enhancement.

Starting from (6), Carmona and Zhong [16] proposed a more general evolution equation

$$\partial_t u = c(a u_{\eta\eta} + b u_{\xi\xi}) \tag{7}$$

where the function $c$ controls the whole amount of smoothing, and $a$ and $b$ weight this smoothing between the two feature directions. Carmona and Zhong let the functions $a, b$, and $c$ as given by the Perona and Malik equation (6) and focus on different ways to choose the local feature directions $\xi$ and $\eta$. For example, they use eigenvectors of the Hessian of $u$ or Gabor filters.

In contrast to their approach, we are going to modify the function $c$ and leave $\eta$ and $\xi$ as given in (2) and (3). We will not focus on different choices of $\xi$ and $\eta$ in this paper, although it is clear that this could also be combined with our modified functions $c$. We are going to see with numerical examples that even small changes in $c$ can change the filtering behaviour significantly.

Although this was not mentioned by Carmona and Zhong, mean curvature motion (MCM) can be obtained by choosing special parameters in their general filter class. It only performs smoothing in the direction $\xi$ of the isophotes in the image:

$$\partial_t u \;=\; u_{\xi\xi} \;=\; |\nabla u| \operatorname{div}\left(\frac{\nabla u}{|\nabla u|}\right) \quad . \tag{8}$$

There are some very useful properties of this technique [6,7,8,9,10]: First it is contrast invariant. Furthermore, it makes non-convex shapes convex and obeys a shape inclusion principle. Convex shapes are shrunken to circular points and finally vanish. In filtering time $t = \frac{1}{2}r^2$, a circle of radius $r$ and everything inside has vanished. Nevertheless, mean curvature motion has the disadvantage to blur the edges during the evolution.

In the context of segmentation, Sapiro [12] proposed the so-called self-snakes which can be understood as nonlinear extension of mean curvature motion. The corresponding equation is

$$\partial_t u = |\nabla u| \operatorname{div}\left(g(|\nabla u|)\frac{\nabla u}{|\nabla u|}\right) \tag{9}$$

and allows for sharper edges. This can be explained by a decomposition of the equation in a curvature motion term and a shock filtering term making edges sharper.

## 3 Generalised Mean Curvature Motion

After reviewing some classical filtering methods in the last section, we now introduce the approach we will focus on in this paper. First we spend some words on the derivation of the general model, coming from adaptive averaging schemes. Then we interpret the appearing PDE with respect to the classical models. The study of important special cases obtained by several choices for the diffusivity-type function $g$ in the model concludes this section.

### 3.1 Derivation of the Model

The starting point for our derivations in this section is the consideration of adaptive averaging schemes

$$u_i^0 = f_i \ ,$$

$$u_i^{k+1} = \frac{g\left(\left|\frac{u_{i+1}^k - u_i^k}{h}\right|\right) u_{i+1}^k + g\left(\left|\frac{u_{i-1}^k - u_i^k}{h}\right|\right) u_{i-1}^k}{g\left(\left|\frac{u_{i+1}^k - u_i^k}{h}\right|\right) + g\left(\left|\frac{u_{i-1}^k - u_i^k}{h}\right|\right)} \tag{10}$$

in [18]. It is explained in detail there that a scaling limit of this averaging scheme leads to the so-called accelerated nonlinear diffusion equation

$$\partial_t u = \frac{1}{g(|\partial_x u|)} \partial_x \left(g(|\partial_x u|)\partial_x u\right) \quad . \tag{11}$$

The only difference of this equation to a classical Perona-Malik model is the factor $\frac{1}{g(|\partial_x u|)}$ on the right-hand side. This factor is understood as acceleration of the process in the following sense: If we are in an almost constant region, the derivative of $u$ is small, and the factor is close to 1. This does not change the evolution very much. On the other hand, at the position of an edge, we have a large derivative of $u$, and the factor is becoming much larger than 1. This leads to a higher evolution velocity close to an edge.

There is no unique way to generalise (11) to two or more dimensions. As described in [18], considering a 2-D weighted averaging scheme as starting point and taking the scaling limit leads to an anisotropic diffusion equation including a diffusion tensor.

In this paper, we generalise (11) in a different way to 2-D: We replace the first derivative of $u$ in (11) by a gradient and the outer derivative by the divergence. This directly leads to the PDE

$$\partial_t u = \frac{1}{g(|\nabla u|)} \operatorname{div}\left(g(|\nabla u|)\nabla u\right) \quad , \tag{12}$$

which will be called *generalised mean curvature motion (GMCM)* here. To justify this name, consider the special case $g(s) = \frac{1}{s}$ to obtain the standard mean curvature motion equation (8). This already indicates that the additional factor on the right-hand side changes the behaviour compared to the Perona-Malik model more than in the 1-D case.

## 3.2 Interpretation

From the decomposition of the Perona-Malik filter in (6), we immediately derive that generalised mean curvature motion (12) can be decomposed as

$$\partial_t u = u_{\xi\xi} + \left(1 + \frac{g'(|\nabla u|)|\nabla u|}{g(|\nabla u|)}\right) u_{\eta\eta} \tag{13}$$

$$= u_{\xi\xi} + a(|\nabla u|)\, u_{\eta\eta} \quad . \tag{14}$$

That means we have a mean curvature motion equation with an additional diffusive component in orthogonal direction $\eta$ which is steered by the factor $a(s) := 1 + \frac{g'(s)s}{g(s)}$. As argument $s$, the factor depends on the norm of the gradient $|\nabla u|$. We will discuss later how the choice of $g$ influences the behaviour of this factor $a(s)$. The basic idea is that the filter performs shrinkage of level lines in the sense of mean curvature motion while the second term keeps edges sharp during the evolution.

There is also another way of understanding the process: Having the equation $\Delta u = u_{\xi\xi} + u_{\eta\eta}$ in mind, we can rewrite this as

$$\partial_t u \;=\; \Delta u + \frac{g'(|\nabla u|)|\nabla u|}{g(|\nabla u|)}\, u_{\eta\eta} \;. \tag{15}$$

In this form, we see that the generalised mean curvature motion can be understood as linear diffusion with an additional shock term for edge enhancement. While classical Perona-Malik filtering slows down the linear diffusion part near edges by the factor $g(|\nabla u|)$, the velocity of this part is constant for generalised mean curvature motion.

### 3.3 Choices for the Function $g$

After specifying the general framework, we now focus on several choices of the function $g$ and give some first insight in the expected behaviour of the corresponding methods.

**Perona-Malik diffusivity.** Let us choose the classical diffusivity function $g(s) = \left(1 + \frac{s^2}{\lambda^2}\right)^{-1}$ proposed by Perona and Malik [13]. This diffusivity is especially interesting because it is capable of switching between forward and backward diffusion adaptively. In this case we have

$$a(s) \;=\; 1 + \frac{g'(s)s}{g(s)} \;=\; 1 - 2\frac{s^2}{s^2+\lambda^2} \tag{16}$$

which immediately shows that $-1 \le a(s) \le 1$ for all $s \in \mathbb{R}$. In a region where $|\nabla u|$ is small, we have forward diffusion. That means the whole process (13) acts like linear diffusion there. Close to an edge, we have forward diffusion along the edge and backward diffusion across the edge. This explains the edge-preserving behaviour which can be observed at the practical results in Section 5.

**An example with unbounded backward diffusion.** Another frequently used diffusivity function is $g(s) = \exp\left(-\frac{s^2}{2\lambda^2}\right)$ which has also been proposed by Perona and Malik [13]. In the classical nonlinear diffusion approach, it has the same properties as the function discussed above. In our case we obtain $a(s) = 1 - \frac{s^2}{\lambda^2}$. We have $a(s) \le 1$ for all $s \in \mathbb{R}$, but $a$ is not bounded from below. That means in theory there would be no limit for the amount of backward diffusion in a pixel where $|\nabla u|$ is very large. We see that similar diffusion properties do not imply a similar behaviour in the corresponding GMCM model. Nevertheless, this special example is of rather theoretical interest, since for realistic values of $|\nabla u|$ and $\lambda$, the values $\exp\left(|\nabla u|^2/\lambda^2\right)$ and $\exp\left(-|\nabla u|^2/\lambda^2\right)$ differ by so many orders of magnitude that a numerical treatment gets very complicated.

**Special case: Constant diffusion velocity in both directions.** So far, we have chosen the amount of diffusion in direction $\eta$ adaptively depending on the gradient magnitude of the evolving image $|\nabla u|$. Now we consider the case that the diffusion in direction $\eta$ has a constant velocity. This is equivalent to

$$a(s) \;=\; 1 + \frac{g'(s)s}{g(s)} \;=\; c \in \mathbb{R}\ . \tag{17}$$

We see that this condition is satisfied for the family of functions $g(s) = \frac{1}{s^p}$ for $p > 0$ where we have $a(s) = 1 - p$. The corresponding equation is given by

$$\partial_t u \;=\; |\nabla u|^p \operatorname{div}\left(\frac{\nabla u}{|\nabla u|^p}\right) \;=\; u_{\xi\xi} + (1-p)\, u_{\eta\eta}\ . \tag{18}$$

For $p = 1$, we have the special case of mean curvature motion. In the experimental section, we are going to take a closer look at the behaviour of this filtering family for several integer values of $p$.

**Historical remark.** A certain special case of this family of methods has been proposed already in 1965 by Gabor in the context of electron microscopy [4]. Later on, his approach has been reviewed and brought into the context of contemporary image analysis [5]. Rewriting his approach in our notation gives the equation

$$u \;=\; f - \frac{\mu^2}{2}\left(f_{\eta\eta} - \frac{1}{3} f_{\xi\xi}\right) \tag{19}$$

for an initial image $f$ and a filtered version $u$. The quantity $\mu$ is derived from the application. We rewrite this equation as

$$\frac{6}{\mu^2}(u - f) \;=\; f_{\xi\xi} - 3 f_{\eta\eta}\ . \tag{20}$$

The left-hand side of (20) can be interpreted as finite difference approximation of a time-derivative. That means, Gabor's approach (19) can be seen as one step in an Euler forward time-discretisation of (18) with $p = 4$ and time step size $\tau = \frac{\mu^2}{6}$. Due to limited computational tools, the approach was rather theoretically motivated than used in practice at that time.

## 4 Discretisation

This section describes one possibility of discretising generalised mean curvature motion (13). In our first practical examples, it turns out that the question of finding a suitable discretisation is very important for this kind of equations.

Let $h_1, h_2 > 0$ be the pixel distance in $x$- and $y$-direction and $\mathcal{N}_d(i)$ the indices of the direct neighbouring pixels in direction $d$ to the pixel with index $i$. Let $u_i^k$ denote the grey value of pixel $i$ at the time step $k$. We start with the grey values of the given initial image $u_i^0 := f_i$. Let further $g_i^k \approx g(|\nabla u(x_i)|)$ denote an approximation to the weighting function evaluated at pixel $i$. We have used the approximation

$$|\nabla u(x_i)| \;\approx\; \sqrt{\sum_{d=1}^{2} \sum_{j \in \mathcal{N}_d(i)} \frac{(u_j - u_i)^2}{2h_d^2}}\ . \tag{21}$$

This approximation yields better results for (13) than standard central differences. Similar to [19] we use a finite difference scheme with harmonic averaging of the diffusivity approximations:

$$u_i^{k+1} = \begin{cases} u_i^k & \text{if } g_i^k = 0 \\ u_i^k + \frac{1}{g_i^k} \sum_{d=1}^{2} \sum_{j \in \mathcal{N}_d(i)} \frac{2}{\frac{1}{g_j^k} + \frac{1}{g_i^k}} \frac{u_j^k - u_i^k}{h_d^2} & \text{else} \end{cases} \tag{22}$$

Why this scheme is very stable in practice can be seen by a simple equivalent reformulation of the scheme for $g_i^k \neq 0$:

$$u_i^{k+1} = u_i^k + \sum_{d=1}^{2} \sum_{j \in \mathcal{N}_d(i)} \frac{2g_j^k}{g_j^k + g_i^k} \frac{u_j^k - u_i^k}{h_d^2} . \tag{23}$$

Under the assumption that $g$ is a non-negative function, we have $0 \leq \frac{g_j^k}{g_j^k + g_i^k} \leq 1$. This allows us to see that for sufficiently small time step size $\tau \leq \frac{1}{8}$ the iteration step (23) yields a convex combination of grey values from the old time step. We conclude that $\min_j f_j \leq u_i^k \leq \max_j f_j$ for all $k \in \mathbb{N}$ and all $i$, i. e. the process satisfies a maximum-minimum principle:

## 5 Numerical Experiments

In this section we study the properties of generalised mean curvature motion with some practical examples. In Fig. 2 we compare the results of Perona-Malik filtering with mean curvature motion and generalised mean curvature motion. It is clearly visible that GMCM offers a combination of the properties of MCM with Perona-Malik filtering: On the one hand, the contrast parameter $\lambda$ gives us the opportunity to distinguish between smoothed and preserved sharp edges as known from Perona-Malik filtering. On the other hand, the objects are shrunken to points and vanish at finite time as known from mean curvature motion.

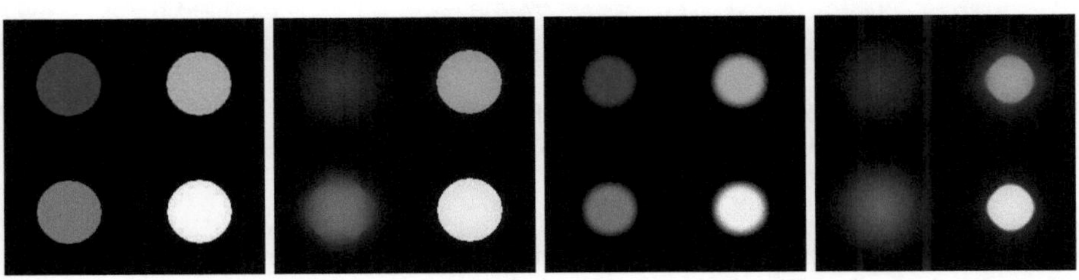

**Fig. 2.** Comparison of different filtering techniques. *Left:* Original image, 256×256 pixels. *Second left:* Perona-Malik filtering, $\lambda = 10$. *Second right:* Mean curvature motion. *Right:* Generalised mean curvature motion (12) with $g(s) = \left(1 + s^2/\lambda^2\right)^{-1}$, $\lambda = 10$. Stopping time in all examples: $t = 200$.

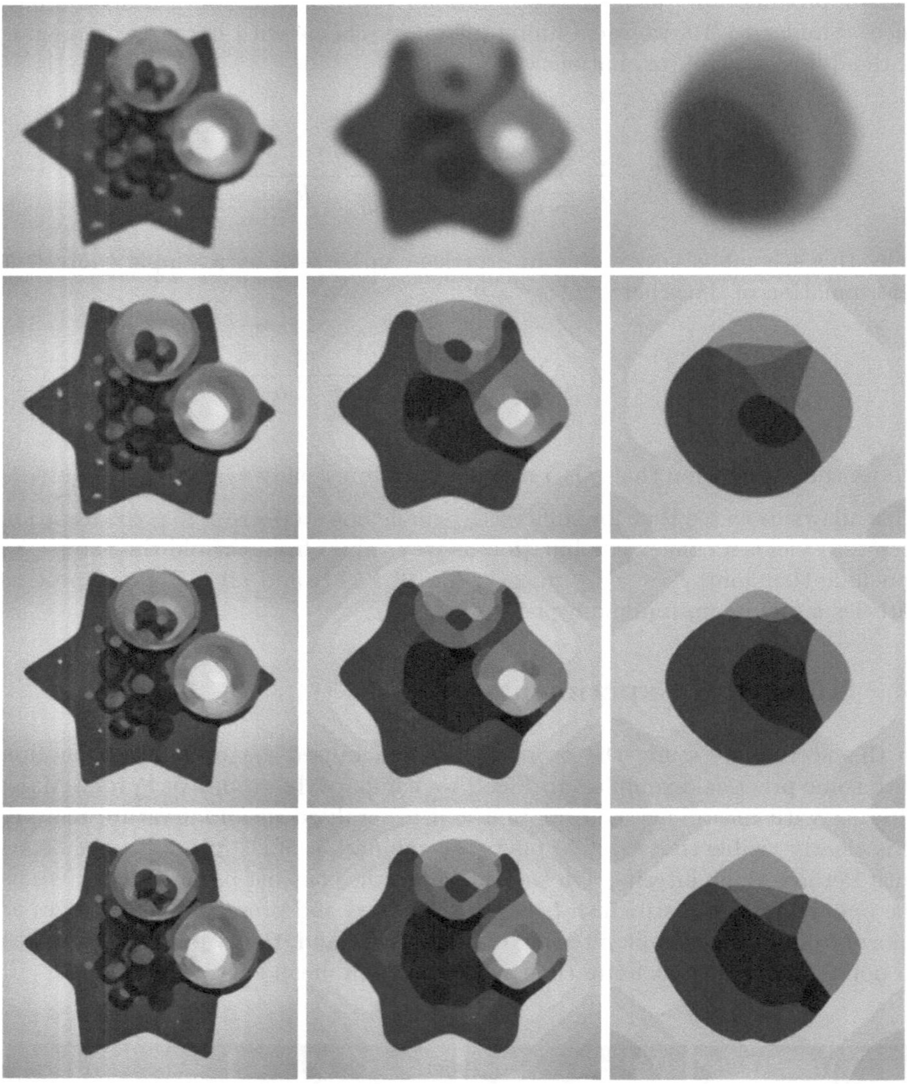

**Fig. 3.** Comparison of the evolution (18) for different values of $p$. *Rows from top to bottom:* $p = 1, 2, 4, 10$. *Left column:* $t = 5$. *Middle column:* $t = 100$. *Right column:* $t = 1000$.

In our second experiment, we compare the behaviour of equation (18) for different values of $p$. Fig. 3 shows the results of the application to the same test image. We see that $p = 1$ yields blurred results while $p \geq 2$ leads to sharp edges. Some basic properties of mean curvature motion are also satisfied here: At this example we see that non-convex objects are getting convex and shrink in finite time. Further, for larger $p$ it is possible that corners are also kept longer in the

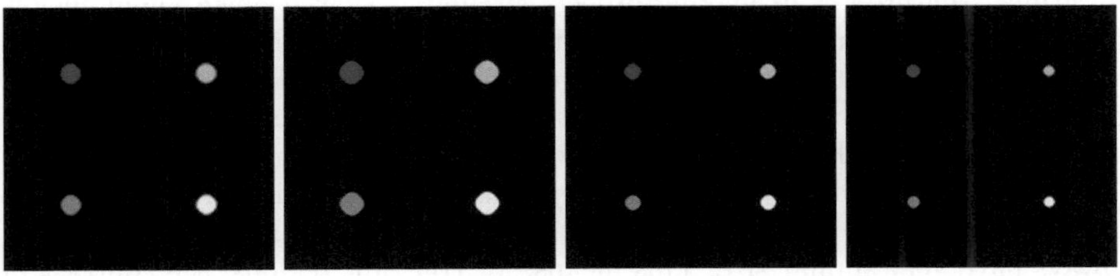

**Fig. 4.** Contrast dependency of a discrete version for the constant generalised mean curvature motion (18). *Left:* $p = 2, t = 375$. *Second left:* $p = 4, t = 312.5$. *Second right:* $p = 6, t = 375$. *Right:* $p = 10, t = 625$.

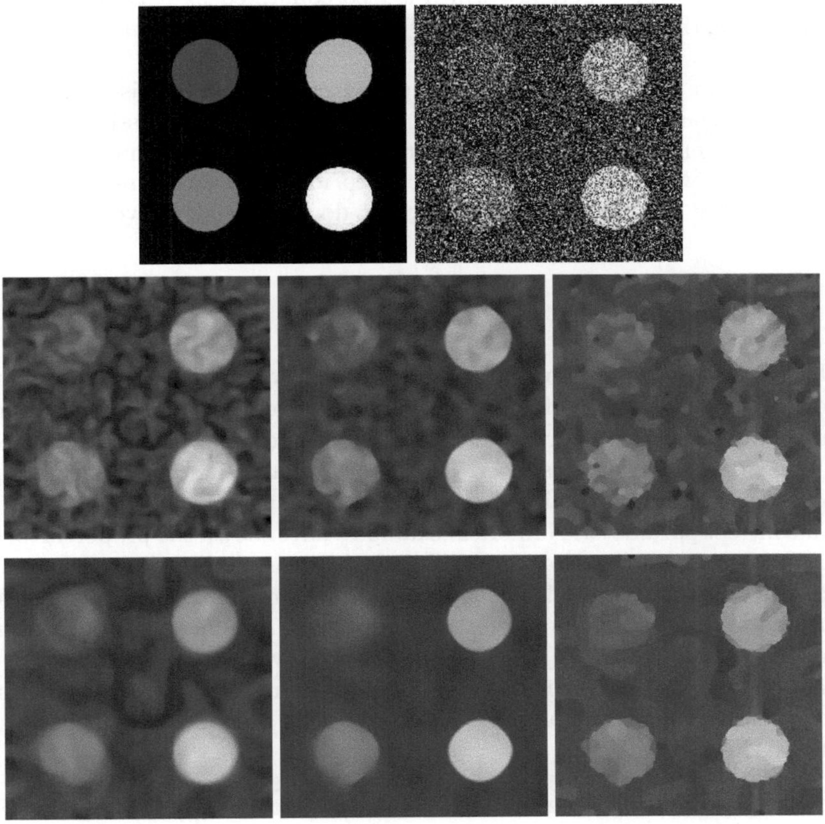

Mean curvature motion Generalised MCM Self-snakes

**Fig. 5.** Joint denoising and curvature motion. *Top left:* Original image, $256 \times 256$ pixels. *Top right:* Original image with Gaussian noise, standard deviation $\sigma = 200$. *First column:* MCM, $t = 12.5, 50$. *Second column:* Generalised mean curvature motion (12), $g(s) = \left(1 + s^2/\lambda^2\right)^{-1}$, $t = 12.5, 50$. *Third column:* Self-snakes, $\lambda = 10, t = 50, 100$.

iterations, the process of making objects circular is slowed down. That means objects are getting smaller during evolution while the shape is preserved longer than for mean curvature motion. We have already mentioned that one important property of mean curvature motion is the morphological invariance. We use a test image composed out of four circles with different contrast to the background (see Fig. 2) to determine the contrast dependence of generalised mean curvature motion (18). We see that for $p = 2, 4, 6, 10$ the four circles in one filtering result always have very similar size. This means, for constant regions, the contrast in this example does hardly influence the shrinkage time. We know from Fig. 3 that these processes tend to segment images into constant regions after a few steps. Further we notice that the stopping times for shrinkage of the circles changes strongly with $p$. Our experience which is also confirmed by a larger number of experiments is that the stopping time is smallest for $p = 4$ and increases rapidly for larger $p$.

In Fig. 5, we see how joint denoising and curve shrinking is possible with generalised mean curvature motion. In this example, it is possible to obtain sharp edges even for a highly noisy test image. At the same time, the process shrinks circles with a comparable velocity to mean curvature motion. We see that self-snakes also denoise the image, but do not shrink it even for a larger stopping time.

## 6 Conclusions

We have investigated a family of partial differential equations which is motivated by the consideration of adaptive averaging schemes. This family comprises mean curvature motion as prominent special case and earns properties related to level line shrinkage from this filter. On the other hand, its close relationship to Perona-Malik filtering explains that it is capable of smoothing edges selectively with a contrast parameter. Numerical examples have shown the properties of several representants of the filter family. It is clearly visible that linear generalised mean curvature motion yields much sharper results than classical mean curvature motion and keeps its interesting properties. Nonlinear generalised mean curvature motion combines Perona-Malik filtering with curve shrinkage. Questions of ongoing research include other ways to discretise the equations without the harmonic mean as well as theoretical properties such as shape inclusion and shrinkage times.

**Acknowledgements.** We gratefully acknowledge partly funding by the *Deutsche Forschungsgemeinschaft (DFG)*, project WE 2602/2-2.

## References

1. Sapiro, G.: Geometric Partial Differential Equations and Image Analysis. Cambridge University Press, Cambridge, UK (2001)
2. Kimmel, R.: Numerical Geometry of Images: Theory, Algorithms, and Applications. Springer, New York (2003)

3. Cao, F.: Geometric Curve Evolution and Image Processing. Volume 1805 of Lecture Notes in Mathematics. Springer, Berlin (2003)
4. Gabor, D.: Information theory in electron microscopy. Laboratory Investigation **14** (1965) 801–807
5. Lindenbaum, M., Fischer, M., Bruckstein, A.: On Gabor's contribution to image enhancement. Pattern Recognition **27** (1994) 1–8
6. Gage, M.: Curve shortening makes convex curves circular. Inventiones Mathematicae **76** (1984) 357–364
7. Gage, M., Hamilton, R.S.: The heat equation shrinking convex plane curves. Journal of Differential Geometry **23** (1986) 69–96
8. Huisken, G.: Flow by mean curvature of convex surfaces into spheres. Journal of Differential Geometry **20** (1984) 237–266
9. Osher, S., Sethian, J.A.: Fronts propagating with curvature-dependent speed: Algorithms based on Hamilton–Jacobi formulations. Journal of Computational Physics **79** (1988) 12–49
10. Kimia, B.B., Tannenbaum, A., Zucker, S.W.: Toward a computational theory of shape: an overview. In Faugeras, O., ed.: Computer Vision – ECCV '90. Volume 427 of Lecture Notes in Computer Science. Springer, Berlin (1990) 402–407
11. Alvarez, L., Lions, P.L., Morel, J.M.: Image selective smoothing and edge detection by nonlinear diffusion ii. SIAM Journal on Numerical Analysis **29** (1992) 845–866
12. Sapiro, G.: Vector (self) snakes: a geometric framework for color, texture and multiscale image segmentation. In: Proc. 1996 IEEE International Conference on Image Processing. Volume 1., Lausanne, Switzerland (1996) 817–820
13. Perona, P., Malik, J.: Scale space and edge detection using anisotropic diffusion. IEEE Transactions on Pattern Analysis and Machine Intelligence **12** (1990) 629–639
14. Catté, F., Lions, P.L., Morel, J.M., Coll, T.: Image selective smoothing and edge detection by nonlinear diffusion. SIAM Journal on Numerical Analysis **29** (1992) 182–193
15. Alvarez, L., Guichard, F., Lions, P.L., Morel, J.M.: Axioms and fundamental equations of image processing. Archive for Rational Mechanics and Analysis **123** (1993) 199–257
16. Carmona, R.A., Zhong, S.: Adaptive smoothing respecting feature directions. IEEE Transactions on Image Processing **7** (1998) 353–358
17. Tschumperlé, D., Deriche, R.: Vector-valued image regularization with PDE's: A common framework for different applications. IEEE Transactions on Image Processing **27** (2005) 1–12
18. Didas, S., Weickert, J.: From adaptive averaging to accelerated nonlinear diffusion filtering. In Franke, K., ed.: Pattern Recognition. Volume 4174 of Lecture Notes in Computer Science., Springer, Berlin (2006) 101–110
19. Weickert, J.: Applications of nonlinear diffusion in image processing and computer vision. Acta Mathematica Universitatis Comenianae **LXX** (2001) 33–50

# Coordinate-Free Diffusion over Compact Lie-Groups

Yaniv Gur and Nir Sochen

Department of Applied Mathematics
Tel-Aviv university
Ramat-Aviv, Tel-Aviv 69978, Israel
{yanivg,sochen}@post.tau.ac.il

**Abstract.** We have seen in recent years a need for regularization of complicated feature spaces: Vector fields, orientation fields, color perceptual spaces, the structure tensor and Diffusion Weighted Images (DWI) are few examples. In most cases we represent the feature space as a manifold. In the proposed formalism, the image is described as a section of a fiber bundle where the image domain is the base space and the feature space is the fiber. In some distinguished cases the feature space has algebraic structure as well. In the proposed framework we treat fibers which are compact Lie-group manifolds (e.g., $O(N)$, $SU(N)$). We study here this case and show that the algebraic structure can help in defining a sensible regularization scheme. We solve the parameterization problem of compact manifold that is responsible for singularities anytime that one wishes to describe in one coordinate system a compact manifold. The proposed solution defines a coordinate-free diffusion process accompanied by an appropriate numerical scheme. We demonstrate this framework in an example of $S^1$ feature space regularization which is known also as orientation diffusion.

## 1 Introduction

Denoising of tensor-valued images via diffusion-like equation generally leads to a constrained partial differential equation (PDE) that generate a flow where the tensor properties have to be preserved [3,12,16,18]. When the tensors are elements of a matrix Lie-group, the solution of the PDE has to evolve on the Lie-group manifold [4]. Generally, PDEs on manifolds and in particular Lie-group manifolds may be defined with respect to the local coordinates on the manifold (i.e., the parameterization of the group). However, this approach is somewhat problematic. By using explicitly the coordinates on the manifold one has to use different charts to cover the manifold in order to avoid coordinates singularities [7]. Another approach embed the manifold in a higher dimensional Euclidean space and operate on the external coordinates. This is problematic since the flow may be drifted away from the manifold due to numerical errors and it needs to be constantly projected on the manifold [15].

Alternatively, a PDE on a Lie-group manifold may be defined in a coordinate-free manner. In this way the PDE is defined with respect to the Lie-group element

F. Sgallari, A. Murli, and N. Paragios (Eds.): SSVM 2007, LNCS 4485, pp. 580–591, 2007.

directly *without any parameterization.* This approach leads to elegant coordinate-free numerical implementations [5,6]. Working directly on the Lie-group matrices was done before in [3,16] for the $SO(3)$ case but with a parameterization and using a unique feature of that group.

In this paper we define PDEs on compact Lie-group in a coordinate free manner via the mathematical concept of fiber bundles. The space of "images" may be described as a fiber bundle. The image spatial domain is the base space $B$ and the fiber is the feature space $F$ (e.g., gray-scale value, color, vector field, etc.). The total space is described locally as $M = B \times F$ where in the context of image processing, $M$ is the spatial-feature manifold. A specific image is a section in the fiber bundle space, that is, a special map from the base manifold to the feature space. The section is referred here as the image manifold. In these terms for an example, the spatial-feature manifold of a gray-scale image may be described locally by $\mathbb{R}^2 \times \mathbb{R}^1$. More complex feature space objects (fibers) include vector spaces [1,17,19] and even Riemannian manifolds [8,9].

The feature space objects we treat here are compact Lie-groups. In the case where the feature space is a Lie-group manifold, the fiber bundle is a principle bundle. The image manifold is embedded in the principle bundle $\mathbb{R}^2 \times G$ where $G$ denotes the Lie-group. Lie-group manifolds have the advantage that there is a strong interplay between the geometry and the algebra such that many geometric objects of interest may be expressed in algebraic form. Thus, for example, one may define a Riemannian metric on a Lie-group manifold via the Killing form over the corresponding Lie-algebra. This structure is combined with the Euclidean metric for the spatial part and extended to the whole fiber bundle. This enables to avoid both explicit and implicit parameterizations. The algebraic structure permits as well the definition of a functional over sections, e.g. images, that takes into account the Riemannian structure. By varying this functional with respect to the Lie-group element, we obtain the equations of motions. The gradient descent equation, directly on the group elements, defines the desired diffusion equation on the Lie-group manifold. The anisotropic behavior of this equation is determined by the structure of the spatial-feature manifold via the induced metric. Since the parameters of the group are not manifested in the PDE, this equation defines a coordinate-free flow over the group manifold. The derivation of such a coordinate-free feature and structure preserving equation and its numerical discretization are the main results of this study.

This framework may be applied to any compact Lie-group. Here we examplify it by treating the well known problem of orientation diffusion over $S^1$ vector fields via principle bundles, in a coordinate-free manner. In this formulation each vector orientation is described by $2 \times 2$ rotation matrix which form the $SO(2)$ feature space, thus, the feature space objects are the rotation matrices and not the orientations. Then, regularization of this field is done by applying the proposed PDE directly to the matrices field.

The plan of this paper is as follows: in section 2 we give a short and non-formal introduction to Lie-groups and Lie-algebra. A discussion on metrics on Lie-group manifold is given in section 3. The calculus of variations in principle bundles is discussed in section 4. The coordinate-free numerical scheme will be discussed in section 5. A formulation of direction diffusion via the principle bundle framework is given in section 6. In this section the framework is presented by few examples. Finally, concluding remarks are given in section 7. Readers who are not familiar with fiber bundles and Lie-groups may go to the explicit examples which we hope that clarify the following discussion.

## 2 Matrix Lie-Groups and Lie-Algebras

The discussion is this paper demands basic knowledge in Lie algebras and Lie groups. In the following subsection we give a non-formal introduction to matrix Lie-groups and Lie-algebras.

Basically, a matrix Lie-group is a set of $N \times N$ matrices which have a structure of a group under the matrix multiplication and matrix inversion operations. The identity element is the identity matrix. Few concrete examples of matrix Lie-groups will help to clarify the definition:

- The set of all real $N \times N$ and nonsingular matrices is the general linear group which is denoted $GL(N)$.
- The set of all $N \times N$ orthogonal matrices with determinant one is the special orthogonal group $SO(N)$. This group is actually the group of rotations in N-dimensional space. A well known group which is very common in computer graphics and image processing is the $SO(3)$ group which is the group of rotations in three-dimensional space. As an example we give the matrix

$$R_x = \begin{pmatrix} 1 & 0 & 0 \\ 0 & cos(\theta) & -sin(\theta) \\ 0 & sin(\theta) & cos(\theta) \end{pmatrix} \tag{1}$$

  which represent a rotation by $\theta$ about the x-axis.
- The set of all $2N \times 2N$ real matrices $P$ which obey the relation $PJP^T = J$ where

$$J = \begin{pmatrix} 0_N & I_N \\ -I_N & 0_N \end{pmatrix}, \tag{2}$$

  $0_N$ and $I_N$ are the $N \times N$ zeros and identity matrices, respectively is the real symplectic group denoted by $Sp(N)$.

The Lie algebra is a linear vector space equipped with the Lie bracket (the commutator) operation $[A, B] = AB - BA$. This operation is bilinear, skew-symmetric ($[A, B] = -[B, A]$), and satisfies the Jacobi identity

$$[A, [B, C]] + [C, [A, B]] + [B, [C, A]] = 0. \tag{3}$$

The commutator operation is also denoted by $ad_A B$.

Every matrix Lie-group has its corresponding Lie-algebra. The basis vectors of the Lie-algebra vector space are called the generators of the Lie-group. Every element in the Lie-algebra may be spanned by these generators. Also, one can map elements of the Lie algebra to the corresponding elements in the Lie-group and vice versa using several maps. The common map is the exponential mapping $expm : \mathfrak{g} \mapsto G$ where $\mathfrak{g}$ denote the Lie-algebra and $G$ denote the Lie-group.

As an example we give the set of skew-symmetric matrices which form the Lie algebra of $SO(N)$ and is denoted by $\mathfrak{so}(N)$. An element of $\mathfrak{so}(3)$ for an example is the matrix

$$r = \begin{pmatrix} 0 & 0 & 0 \\ 0 & 0 & -\theta \\ 0 & \theta & 0 \end{pmatrix} \tag{4}$$

whereas one can check that

$$expm(r) = \begin{pmatrix} 1 & 0 & 0 \\ 0 & cos(\theta) & -sin(\theta) \\ 0 & sin(\theta) & cos(\theta) \end{pmatrix} \tag{5}$$

which is a rotation by $\theta$ about the x-axis and belong to $SO(3)$. We use also the inverse of the matrix exponential which is the matrix logarithm. It maps elements in the Lie-group to elements in the Lie-algebra $logm : G \mapsto \mathfrak{g}$.

To end this discussion it is important to mention that a matrix Lie-group is also a manifold equipped with the matrices multiplication operation. A flow which lies on this manifold has to preserve the matrices properties (e.g., determinant, orthogonality, etc) at all times.

# 3 Metrics on Lie-Group Manifolds

The metric over the image domain is Euclidean. In order to write the metric over the principle bundle, i.e. the total space, we have to find also the metric over the fiber which is the Lie-group manifold.

Recalling Riemannian Geometry, the metric over a Riemannian manifold $M$ is given by a symmetric, positive-definite bilinear form $g : \langle V, W \rangle \mapsto \mathbb{R}$ where $V, W \in T_pM$ (vectors in the tangent vector space of the manifold $M$ at the point $p \in M$). In terms of the local coordinates on the manifold, this is written as $g(V, W) = g_{ij}V^iW^j$ where we sum over repeated indices. Distances on the manifold are computed using the line element $ds^2 = g_{ij}dx^idx^j$.

The Lie-algebra is the tangent vector space to the Lie-group manifold $G$ at the identity. This is denoted by $T_eG$. Thus, in analogy, we may define a symmetric, positive-definite bi-invariant form on a Lie-group manifold with respect to the Lie-algebra elements, $w : \langle A, B \rangle \mapsto \mathbb{R}$ where $A, B \in T_eG$. The existence and uniqueness of a bi-invariant Riemannian metric is assured when the Lie-group is compact and connected [2]. We use a bi-invariant metric in the form $\langle A, B \rangle = \frac{1}{2}Tr(AB^T)$ where $A, B \in \mathfrak{g}$. The exact form of $A$ and $B$ may be given in terms of the Lie-group elements via the Maurer-Cartan form $g^{-1}dg \in \mathfrak{g}$. Thus, the explicit form for the bi-invariant metric over a compact Lie-group manifold will be given by the expression

$$ds^2 = \frac{1}{2}Tr\left[(g^{-1}dg)(g^{-1}dg)^T\right]. \tag{6}$$

Every Lie-algebra element may be spanned in terms of the basis vectors of the Lie-algebra (the generators of the Lie-group). Thus, $g^{-1}dg = f^i(x, dx)T_i$ where the coefficients $f(x, dx)$ are one-forms i.e. linear combination of the differentials $dx^i$ with coefficients which are functions of the group parameters (the local coordinates on the manifold). The Lie group generators $T_i$ are the basis vectors. We choose an orthonormal basis for the Lie algebra. Thus the generators of the Lie-group obey $\langle T_i, T_j\rangle = Tr(T_i T_j^T) = 2\delta_{ij}$ where $\delta_{ij}$ is the Kronecker delta function. We may write the metric over the Lie-group manifold in terms of the coefficients

$$ds^2 = \frac{1}{2}Tr\left[f^i(x, dx)T_i f^j(x, dx)T_j\right] = \sum_i f^i(x, dx)^2. \qquad (7)$$

As a simple example we give the group of one-dimensional rotation in 2D, $SO(2)$. A general rotation in 2D space is given by the matrix

$$g = \begin{pmatrix} cos\theta & sin\theta \\ -sin\theta & cos\theta \end{pmatrix}. \qquad (8)$$

Calculation of the Maurer-Cartan form $g^{-1}dg$ yields a skew symmetric matrix as expected (the elements of the Lie algebra of the orthogonal matrix Lie-groups are skew symmetric matrices):

$$g^{-1}\frac{dg}{d\theta}d\theta = \begin{pmatrix} 0 & -d\theta \\ d\theta & 0 \end{pmatrix} = d\theta\begin{pmatrix} 0 & -1 \\ 1 & 0 \end{pmatrix}, \qquad (9)$$

where $T = \begin{pmatrix} 0 & -1 \\ 1 & 0 \end{pmatrix}$ is the only basis vector which span the $\mathfrak{so}(2)$ Lie algebra. We finally get the metric over the $SO(2)$ Lie-group manifold

$$ds^2 = \frac{1}{2}Tr\left[(g^{-1}dg)(g^{-1}dg)^T\right] = \frac{1}{2}Tr\left[\begin{pmatrix} 0 & -1 \\ 1 & 0 \end{pmatrix}\begin{pmatrix} 0 & 1 \\ -1 & 0 \end{pmatrix}\right]d\theta^2 = d\theta^2. \qquad (10)$$

Note that $d\theta^2$ is the metric over the unit-circle $S^1$ which is the manifold of $SO(2)$.

## 4 Calculus of Variations in Principle Bundles

In the proposed formalism the image manifold (the section) is embedded in the spatial-feature manifold (the principle-bundle). In order to define a meaningful functional over sections in the principle bundle we use ideas from the Beltrami framework [11]. The Beltrami framework has been intensively discussed in the literature (e.g., [7,11,13,14]). Let us briefly review its main ideas.

### 4.1 The Polyakov Action

Denote by $(\Sigma, \gamma)$ the image manifold and its metric and by $(M, h)$ the spatial-feature manifold and its metric, then, the section of interest is expressed locally

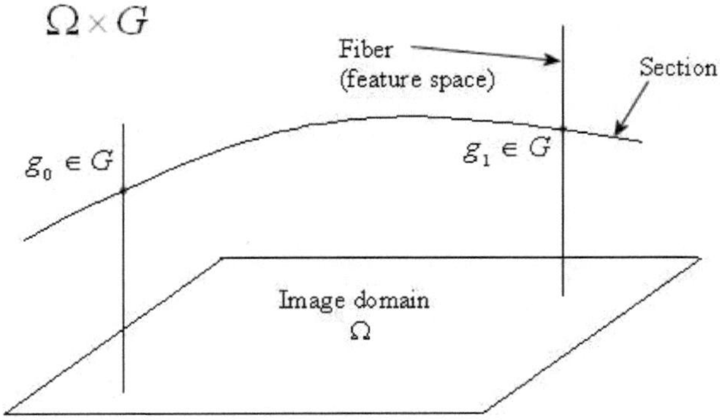


**Fig. 1.** Description of a matrix-valued image as a principal fiber bundle. The base manifold $B$ is the image domain which is denoted by $\Omega$, the fiber $F$ is the feature space (e.g., graylevel value, Riemannian manifold). Here we take the image domain to be Euclidean $\Omega = R^2$. Locally the bundle is a trivial bundle, described as a direct product $M = B \times F$ where $M$ is the total space. When the fiber is a Lie-group, the fiber bundle is a principle bundle. A trivial principle bundle is described as $\Omega \times G$ where $G$ denotes the Lie-group. A particular selection of a Lie-group element to each point of the base manifold is called a section. The section is referred here as the image manifold.

by the map $X : \Sigma \mapsto M$. A functional over the space of sections is given in local coordinates by the following expression

$$S = \int_{\Sigma} d^2x \sqrt{|\gamma|} \gamma^{\mu\nu} \partial_\mu X^i \partial_\nu X^j h_{ij}(X) \tag{11}$$

which is known in the literature as the Polyakov action [10]. The integration is taken over the two-dimensional image manifold, $|\gamma|$ is the determinant of the image metric, $\gamma^{\mu\nu}$ is the inverse of the metric tensor and $h_{ij}$ is the metric of the embedding space (the spatial-feature fiber bundle manifold). The coordinates in the spatial-feature space are denoted by $X^i$. The values of $\mu$ and $\nu$ ranges from 1 to dimension $\Sigma$ (the dimension of the image manifold) and the values of the $i$ and $j$ indices ranges from 1 to the dimension of the embedding space.

## 4.2 The Induced Metric

In the Beltrami framework the object of interest is the geometry of the image manifold (the section) and not the geometry of the image domain (the base manifold). Therefore one has to take the induced metric from the principal bundle to the section.

Let $M = \Omega \times G$ be the spatial-feature manifold and denote by $y^i$, $i = 1, ...dim\, M$ its coordinates, then the metric on $M$ is given by $ds^2_M = h_{ij} dy^i dy^j = ds^2_\Omega + ds^2_G$. The metric over the image manifold is given by $ds^2 = g_{\mu\nu} dx^\mu dx^\nu$

where the coordinates on the image manifold are denoted by $x^\mu$. The assumption is that infinitesimal distances on the spatial-feature manifold equal to infinitesimal distances on the image manifold. By using the chain rule for the metric of the spatial-feature manifold and comparing the solution to the image manifold metric, the expression for the induced metric is easily obtained

$$g_{\mu\nu} = h_{ij}\partial_\mu dy^i \partial_\nu dy^j. \tag{12}$$

As an example let us calculate the induced metric for $SO(2)$ group. The metric over the $SO(2)$ group manifold was already calculated in section 3: $ds_G^2 = d\theta^2$. Thus $ds_M^2 = ds_\Omega^2 + ds_G^2 = dx^2 + dy^2 + d\theta^2$. Using the chain rule for $\theta$ we have $d\theta = \theta_x dx + \theta_y dy$. Substituting this result into $ds_M^2$ and arranging the terms we get $ds_M^2 = (1+\theta_x^2)dx^2 + (1+\theta_y^2)dy^2 + 2\theta_x\theta_y dxdy$. Thus, the induced metric tensor is given by

$$g_{\mu\nu} = \begin{pmatrix} 1+\theta_x^2 & \theta_x\theta_y \\ \theta_x\theta_y & 1+\theta_y^2 \end{pmatrix}. \tag{13}$$

This tensor may be easily obtained by using directly Eq. (12) where the metric tensor of the embedding space is $h_{ij} = \delta_{ij}$. In this case the induced metric tensor reduces to $g_{\mu\nu} = \partial_\mu dy^i \partial_\nu dy^i$.

### 4.3 Coordinate-Free Formulation

There are two critical observations that enable to work in a coordinate-free manner. First we notice that in terms of the induced metric eq. (12) we may re-write the Polyakov action (11) as such

$$S = \int_\Sigma d^2x \sqrt{|\gamma|}\gamma^{\mu\nu} g_{\mu\nu} \quad . \tag{14}$$

Second the induced metric may be written in general form in terms of the group element $g$:

$$g_{\mu\nu} = \delta_{\mu\nu} + \frac{1}{2}Tr\left[(g^{-1}\partial_\mu g)(g^{-1}\partial_\nu g)^T\right]. \tag{15}$$

For an example, let us calculate $g_{xx}$ where $g \in SO(2)$:

$$g_{xx} = 1 + \frac{1}{2}Tr\left[(g^{-1}\partial_x g)(g^{-1}\partial_x g)^T\right] = 1 + \theta_x^2. \tag{16}$$

Combining these observations, the energy action in Eq. (14) may be written in a modified form, by using the induced metric

$$S = \int_\Sigma d^2x \sqrt{|\gamma|}\gamma^{\mu\nu}\{\delta_{\mu\nu} + \frac{1}{2}Tr\left[(g^{-1}\partial_\mu g)(g^{-1}\partial_\nu g)^T\right]\} . \tag{17}$$

This is the (fiber) coordinate-free proposed functional over sections of a principle bundle.

Using calculus of variation, we may vary the energy with respect to the group element $g$. This yields the equations of motion

$$0 = -\partial_\mu \left( \sqrt{|\gamma|}\gamma^{\mu\nu} g^{-1}\partial_\nu g \right) . \tag{18}$$

Since $g^{-1}\partial_\nu g \in \mathfrak{g}$, the right hand side of this equation lies in the Lie-algebra. Therefore, the left hand side of this equation should lie in the Lie-algebra as well. Thus, the gradient descent equation may be written in the from

$$g^{-1}\partial_t g = \partial_\mu \left( \sqrt{|\gamma|}\gamma^{\mu\nu} g^{-1}\partial_\nu g \right) . \tag{19}$$

To obtain the PDE flow on the fiber (the Lie-group manifold) we multiply both sides of Eq. (19) by $g$ from the left such that

$$\partial_t g = g\partial_\mu \left( \sqrt{|\gamma|}\gamma^{\mu\nu} g^{-1}\partial_\nu g \right) . \tag{20}$$

Up to now the metric $\gamma_{\mu\nu}$ was not specified. Varying the functional with respect to this metric and solving the Euler-Lagrange equations determines it. The solution happens to be $\gamma_{\mu\nu} = \delta_{\mu\nu} + \frac{1}{2}Tr\left[(g^{-1}\partial_\mu g)(g^{-1}\partial_\nu g)^T\right]$ which is the $2 \times 2$ induced metric tensor and $|\gamma|$ is its determinant. The amount of local regularization thus is determined by the structure of the spatial-feature manifold via the induced metric.

## 5 Numerical Integration on Lie-Group Manifolds

In this work we treat matrix Lie-groups. Since the group operation for matrix Lie-groups is matrix multiplication, the sum or the difference of two group elements $g_1, g_2 \in G$ does not belong, in general, to the group. This implies that we cannot use classical iterative schemes. Also, derivatives on the Lie-group manifold cannot be estimated by using finite-difference for example. However, this problem is solved by estimating the derivatives in the Lie-algebra which is a linear space and mapping the solution back to the Lie-group via the exponential mapping.

Therefore, in order to estimate numerically the term $g^{-1}\partial_\mu g$ we substitute $g = exp(a)$. Then, the spatial derivative of the group element reads $\partial_\mu exp(a)$ where now we have to evaluate the derivative of the exponent. For the scalar case $a \in \mathbb{R}$ and for Abelian groups (where $a, a' \in \mathfrak{g}$ commute) the formula for the derivative of the exponent is $\frac{d}{dx}exp(a(x)) = a'(x)exp(a(x))$ where $a'$ is the derivative with respect to the spatial coordinate. However, this formula does not hold for non-Abelian groups where $[a, a'] \neq 0$. The correct formula, for the non-Abelian case, may be written in terms of the *dexp* function such that

$$\begin{aligned} &\frac{\partial}{\partial x} exp(a(x,t)) \\ &= dexp_{a(x,t)} a'(x,t) exp(a(x,t)) \\ &= exp(a(x,t)) dexp_{-a(x,t)} a'(x,t), \end{aligned} \tag{21}$$

where the *dexp* function is defined as a power series as follows [6]

$$dexp_A B = B + \frac{1}{2!}[A,B] + \frac{1}{3!}[A,[A,B]] \qquad (22)$$
$$+\frac{1}{4!}[A,[A,[A,B]]] + \ldots = \sum_{k=0}^{\infty} \frac{1}{(k+1)!} ad_A^k B.$$

Finally, we multiply this equation from the left by $g^{-1} = exp(-a(x,t))$ to obtain the numerical estimation

$$g^{-1}\partial_\mu g \approx dexp_{-a(x,t)}\partial_\mu a(x,t). \qquad (23)$$

Since $a$, the *dexp* series and $\partial_\mu dexp$ are all in the Lie-algebra and since the Lie-algebra is a linear space, the partial derivatives may be evaluated using e.g. the forward finite difference scheme. Mapping from the Lie-group to the Lie-algebra is done via the logarithmic mapping, $logm : G \mapsto \mathfrak{g}$.

For the time forward step operator we take the Lie-group version of the forward Euler operator [6]. It reads

$$g_{n+1} = g_n expm(dt\, a(g_n, t_n)), \qquad (24)$$

where $dt$ is the time step and $a$ is the element of the algebra. In terms of group elements our time step operator reads

$$g_{n+1} = g_n\, expm\left(dt\, \partial_\mu\, dexp_{-a(x,t)}(\partial^\mu a(x,t))\right) \qquad (25)$$
$$= g_n\, expm\left(dt\, \partial_\mu\, dexp_{-logm\, g(x,t)}(\partial^\mu logm\, g(x,t))\right).$$

Although on each iteration we have to calculate the *dexp* power series, this calculation is efficient since this power series converges very fast. Thus for example $\frac{||M_{100}-M_7||_2}{||M_{100}||_2} \approx 10^{-7}$ where $M_{100}$ and $M_7$ represents the value of the 100 terms and 7 terms *dexp* series, respectively. The norm estimation is with respect to the matrix 2-norm.

## 6 Direction Diffusion on $S^1$ Via Lie-Groups

In the following section we demonstrate regularization of oriented unit-norm vectors which are attached to the image domain. The orientations field may describe image quantities such as gradient directions, ridge directions and hue values as well as physical quantities such as turbulence, wind flows and more. The feature space here is described by the unit-circle, $S^1$ manifold.

Let us consider the $SO(2)$ Lie-group. It is the group of one-dimensional rotations in two-dimensional space where its element is an orthogonal two by two rotation matrix with determinant one (see Eq. 8). The rotation angle $\theta \in [0, 2\pi)$ parameterize the unit-circle $S^1$. Therefore, the unit-circle is the group manifold of $SO(2)$ and $U(1)$ which is the one-dimensional unitary group. These two groups

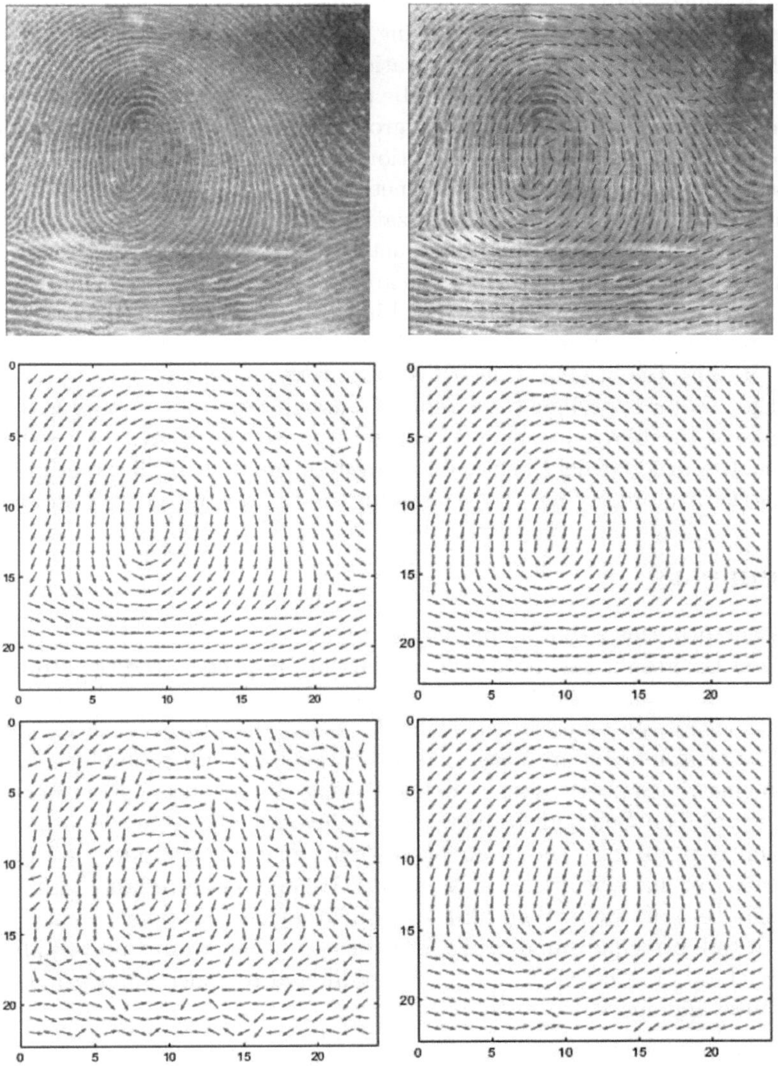

**Fig. 2.** Top left: Original fingerprint image. Right: The image has been blurred by using a Gaussian filter. The ridge directions are plotted on the image. Middle left: Ridge directions of the blurred image. Right: Ridge directions after regularization with the PDE flow on $S^1$. Bottom left: Gaussian noise of variance 0.15 has been added to the ridge directions of the blurred image. Right: The directions after denoising with the PDE flow.

are isomorphic where $U(1)$ corresponds to the embedding of $S^1$ in $\mathbb{C}$ by exponentiating the angle such that $g = e^{i\theta}$, and $SO(2)$ corresponds to the embedding of $S^1$ in $\mathbb{R}^2$. The principal bundle here may be described as $\mathbb{R}^2 \times S^1$. The metric over the image domain is Euclidean and the metric over $S^1$ is simply $ds^2_{S^1} = d\theta^2$. This was calculated in section 3 by using the group element of $SO(2)$.

In the proposed formalism, each rotation will be described by the proper rotation matrix, where the diffusion will be performed with respect to the rotation matrices. Thus, we diffuse over the Lie-group manifold via the Lie-group element and not the coordinate $\theta$. As was mentioned in the introduction, this approach solves the problems of manifolds parameterization.

In figure 2 we demonstrate regularization of ridge directions of a fingerprint image. On the first experiment we extracted ridge directions of a blurred fingerprint image (top row). These directions are unit-norm vectors which are oriented in the 2D space. Then, we applied the Lie-group PDE flow in a coordinate-free manner to the orientations field of the blurred image. At the end of the regularization process, the original ridge directions were recovered.

On the second experiment we added Gaussian noise of variance 0.15 to the directions of the blurred image. Again, we applied the PDE flow to the noisy field and the original field has been recovered. Since in both experiments the flow is structure preserving, the vectors unit-norm has not been changed.

## 7 Summary

We presented in this paper a new way to regularize tensor fields that take their values in a Lie group. The proposed regularization respects the Lie group structure and is done in a *coordinate-free manner*. This enables to avoid numerical problems such as parameterization singularities and/or projections on the manifold. The tensor fields are described as sections of a fiber bundle and a Riemannian metric over the fiber is defined using the algebraic structure i.e. the Killing form. The Beltrami framework was invoked and a coordinate-free new functional is proposed. The gradient descent equations are derived and a new coordinate-free numerical scheme is devised. We demonstrated the framework on the simple case of orientation diffusion which is from our point of view section of the $SO(2)$ principle bundle. Applications to other groups and applications (notably the diffusion tensor imaging) are differed to other publications due to lack of time and space.

**Acknowledgments.** This research was supported by MUSCLE Multimedia Understanding through Semantic, Computation and Learning, an European Network of Excellence funded by the EC 6th Framework IST Programme, and by the Adams super center for Brain Research.

## References

1. Blomgren, P. and Chan, T. F.: Color TV: Total Variation Methods for Restoration of Vector Valued Images. IEEE Trans. on Image processing **7** (1998) 304–309.
2. Boothby, W. M.: An Introduction to Diffrentiable Manifolds and Riemannian Geometry, Academic Press (1975).

3. Chefd'hotel, C., Tschumperlé, D., Deriche, R. and Faugeras, O.: Regularizing Flows for Constrained Matrix-Valued Images. Journal of Mathematical Imaging and Vision **20** (2004) 147–162.
4. Gur, Y. and Sochen, N.: Denoising Tensors via Lie-Group Flows. Lecture Notes in Computer Science **3752** (2005) 13–24.
5. Hairer, E., Lubich, C. and Wanner, G.: Geometric Numerical Integration. Springer-Verlag, Berlin Heidelberg (2002).
6. Iserles, A., Munthe-Kaas, H. Z., Nørsett, S. P. and Zanna, A.: Lie-Group Methods. Acta Numerica **9** (2000) 215–365.
7. Kimmel, R. and Sochen, N.: Orientation Diffusion or How to Comb a Porcupine. Journal of Visual Communication and Image Representation **13** (2001) 238–248.
8. Pennec, X.: Probabilities and Statistics on Riemannian Manifolds: A Geometric approach. Research Report 5093, INRIA, January (2004).
9. Pennec, X., Fillard, P. and Ayache, N.: A Riemmannian Framework for Tensor Computing. International Journal of Computer Vision **66** (2006) 41–66.
10. A. M. Polyakov, Quantum Geometry of Bosonic Strings, Phys. Lett. B, **103** (1981) 207–210.
11. Sochen, N., Kimmel, R. and Malladi, R.: From High Energy Physics to Low Level Vision. Scale-Space Theories in Computer Vision (1997) 236–247.
12. Sapiro, G. and Ringach, D.: Anisotropic Diffusion of Multivalued Images with Application to Color Filtering. IEEE Trans. On Image Processing **5** (1996) 1582–1586.
13. Shafrir, D. Sochen, N., Deriche, R.: Regularization of Mappings between Implicit Manifolds of Arbitrary Dimension and Codimension, VLSM '05: Proceedings of the 3rd IEEE Workshop on Variational, Geometric and Level-set methods in Computer Vision (2005).
14. Sochen, N., Deriche, R. and Lopez-Perez, L.: The Beltrami Flow over Implicit Manifolds, ICCV '03: Proceedings of the Ninth IEEE International Conference on Computer Vision (2003) 832-839.
15. Tang, B., Sapiro, G. and Caselles, V.: Direction Diffusion. International Conference on Computer Vision (1999).
16. Tschumperlé, D.: PhD thesis: PDE's Based Regularization of Multivalued Images and Applications. Universite de Nice-Sophia, Antipolis (2002).
17. Tschumperlé, D. and Deriche, R.: Orthonormal Vector Sets Regularization with PDE's and Application. International Journal of Computer Vision **50** (2002) 237–252.
18. Weickert, j., Feddern, C., Welk, M., Burgeth, B. and Brox, T.: PDEs for Tensor Image Processing. In: Weickert, J. and Hagen, H. (Eds.): Visualization and Processing of Tensor Fields. Springer, Berlin (2005).
19. Whitaker, R. and Gerig, G.: Vector-Valued Diffusion. In: B. M. ter Haar Romeny (ed.): Geometry Driven Diffusion in Computer Vision, Kluwer Academic Publishers, The Netherlands, (1994) 93–134.

# Riemannian Curvature-Driven Flows for Tensor-Valued Data

Mourad Zéraï and Maher Moakher

Laboratory for Mathematical and Numerical Modeling in Engineering Science
National Engineering School at Tunis
ENIT-LAMSIN, B.P. 37, 1002 Tunis Belvédère, Tunisia
mourad.zerai@lamsin.rnu.tn, maher.moakher@enit.rnu.tn

**Abstract.** We present a novel approach for the derivation of PDE modeling curvature-driven flows for matrix-valued data. This approach is based on the Riemannian geometry of the manifold of symmetric positive-definite matrices $\mathcal{P}(n)$. The differential geometric attributes of $\mathcal{P}(n)$ −such as the bi-invariant metric, the covariant derivative and the Christoffel symbols− allow us to extend scalar-valued mean curvature and snakes methods to the tensor data setting. Since the data live on $\mathcal{P}(n)$, these methods have the natural property of preserving positive definiteness of the initial data. Experiments on three-dimensional real DT-MRI data show that the proposed methods are highly robust.

## 1 Introduction

With the introduction of diffusion tensor magnetic resonance imaging (DT-MRI) [4], there has been an ever increasing demand on rigorous, reliable and robust methods for the processing of tensor-valued data such as the estimation, filtering, regularization and segmentation. Many well established PDE-based methods used for the processing of scalar-valued data have been extended in various ways to the processing of multi-valued data such as vector-valued data and smoothly constrained data [5, 12, 22, 23, 24, 25]. Recently, some efforts have been directed toward the extension of these methods to tensor fields [3, 6, 9, 8, 14, 17, 20, 26, 27]. The generalization of the methods used for scalar- and vector-valued data to tensor-valued data is being pursued with mainly three formalisms: the use of geometric invariants of tensors like eigenvalues, determinant, trace; the generalization of Di Zenzo's concept of a structure tensor for vector-valued images to tensor-valued data; and recently, differential-geometric methods.

The aim of the present paper is to generalize the total variation (TV) flow, mean curvature motion (MCM), modified mean curvature flow and self snakes to tensor-valued data such as DT-MRI. The key ingredient for these generalizations is the use of the Riemannian geometry of the space of symmetric positive-definite (SPD) matrices. The remainder of this paper is organized as follows. In Section 2 we give a compilation of results that gives the differential geometry of the Riemannian manifold of symmetric positive-definite matrices. We explain in Section 3 how to describe a DT-MR image by differential-geometric concepts. Section 4 is the key of our paper in which we extend several mean curvature-based flows for the denoising and segmentation from the scalar and vector setting to the tensor one. In Section 5 we present some numerical results.

F. Sgallari, A. Murli, and N. Paragios (Eds.): SSVM 2007, LNCS 4485, pp. 592–602, 2007.

## 2 Differential Geometry of $\mathcal{P}(n)$

Positive-definite matrices are omnipresent in many engineering and physical contexts. They play important roles in various disciplines such as control theory, continuum mechanics, numerical analysis, covariance analysis, signal processing, etc. Recently, they gained an increasing attention within the diffusion tensor magnetic resonance imaging (DT-MRI) community as they are used as an encoding for the principal diffusion directions and strengths in biological tissues.

We here recall some differential-geometric facts about the space of symmetric positive-definite matrices, see [28] for a comprehensive treatment. We denote by $\mathcal{S}(n)$ the vector space of $n \times n$ symmetric matrices. A matrix $A \in \mathcal{S}(n)$ is said to be positive semidefinite if $\boldsymbol{x}^T A \boldsymbol{x} \geq 0$ for all $\boldsymbol{x} \in \mathbb{R}^n$, and positive definite if in addition $A$ is invertible. The space of all $n \times n$ symmetric, positive-definite matrices will be denoted by $\mathcal{P}(n)$. We note that the set of positive-semidefinite matrices is a pointed convex cone in the linear space of $n \times n$ matrices, and that $\mathcal{P}(n)$ is the interior of this cone. The space $\mathcal{P}(n)$ is a differentiable manifold endowed with a Riemannian structure. The tangent space to $\mathcal{P}(n)$ at any of its points $P$ is the space $T_P\mathcal{P}(n) = \{P\} \times \mathcal{S}(n)$, which for simplicity is identified with $\mathcal{S}(n)$. On each tangent space $T_P\mathcal{P}(n)$ we introduce the base point-dependent inner product defined by $\langle A, B\rangle_P := \operatorname{tr}(P^{-1}AP^{-1}B)$.

This inner product leads to a natural Riemannian metric on the manifold $\mathcal{P}(n)$ that is given at each $P$ by the differential

$$ds^2 = \operatorname{tr}\left(P^{-1}dPP^{-1}dP\right), \tag{1}$$

where $dP$ is the symmetric matrix with elements $(dP_{ij})$. We note that the metric (1) is invariant under congruent transformations: $P \to LPL^T$ and under inversion $P \to P^{-1}$.

For an $n \times n$ matrix $A$, we denote by $\operatorname{vec} A$ the $n^2$-column vector that is obtained by stacking the columns of $A$. If $A$ is symmetric, then $\frac{1}{2}n(n-1)$ elements of $\operatorname{vec}(A)$ are redundant. We will denote by $\upsilon(A)$ the $d = \frac{1}{2}n(n+1)$-vector that is obtained from $\operatorname{vec}(A)$ by eliminating the redundant elements, e.g., all supradiagonal elements of $A$. We note that there are several ways to arrange the independent elements of $\operatorname{vec}(A)$ into $\upsilon(A)$. In any case, there exists a unique $n^2 \times \frac{1}{2}n(n+1)$ matrix, called the duplication matrix and denoted by $D_n$, that by duplicating certain elements, reconstructs $\operatorname{vec} A$ from $\upsilon(A)$, i.e., is the matrix such that

$$\operatorname{vec} A = D_n \upsilon(A). \tag{2}$$

The duplication matrix $D_n$, which has been studied extensively by Henderson and Searle [10], and by Magnus and Neudecker [15], has full column rank $\frac{1}{2}n(n+1)$. Hence, $D_n^T D_n$ is non-singular and it follows that the duplication matrix $D_n$ has a Moore-Penrose inverse denoted by $D_n^+$ and is given by

$$D_n^+ = \left(D_n^T D_n\right)^{-1} D_n^T.$$

It follows from (2) that

$$\upsilon(A) = D_n^+ \operatorname{vec} A. \tag{3}$$

By using the vector $\upsilon(P)$ as a parametrization of $P \in \mathcal{P}(n)$ we obtain the matrix of components of the metric tensor associated with the Riemannian metric (1) is given explicitly by [28]

$$G(P) = D_n^T (P^{-1} \otimes P^{-1}) D_n, \tag{4}$$

where $\otimes$ denotes the Kronecker product. For differential-geometric operators on $\mathcal{P}(n)$ it is important to obtain the expression of the inverse of the metric and that of its determinant. The matrix of components of the inverse metric tensor is given by [28]

$$G^{-1}(P) = D_n^+ \, (P \otimes P) \, D_n^{+T}, \tag{5}$$

and the determinant of $G$ is

$$\det(G(P)) = 2^{n(n-1)/2} \left((\det(P))^{(n+1)}\right). \tag{6}$$

In the coordinate system $(p^\alpha)$, the Christoffel symbols are given by [28]

$$\Gamma_{\alpha\beta}^{\gamma} = -[D_n^T \left(P^{-1} \otimes E^\gamma\right) D_n]_{\alpha\beta}, \quad 1 \leq \alpha, \beta, \gamma \leq d,$$

where $E^\gamma$ is the dual basis associated with the local coordinates $(p^\alpha)$. As the elements of $E^\gamma$ and $D_n$ are either 0, 1, or $\frac{1}{2}$, it follows from the above theorem that each non-vanishing Christoffel symbol is given by an element of $P^{-1}$ or half of it.

Let $P$ be an element of $\mathcal{P}(3)$ and let $dP$ be a (symmetric) infinitesimal variation of it

$$P = \begin{bmatrix} p^1 & p^4 & p^6 \\ p^4 & p^2 & p^5 \\ p^6 & p^5 & p^3 \end{bmatrix}, \qquad dP = \begin{bmatrix} dp^1 & dp^4 & dp^6 \\ dp^4 & dp^2 & dp^5 \\ dp^6 & dp^5 & dp^3 \end{bmatrix}.$$

Hence, the complete and reduced vector forms of $P$ are respectively,

$$\mathrm{vec}(P) = [p^1 \, p^4 \, p^6 \, p^4 \, p^2 \, p^5 \, p^6 \, p^5 \, p^3]^T, \quad \upsilon(P) = [p^1 \, p^2 \, p^3 \, p^4 \, p^5 \, p^6]^T.$$

The components of the inverse metric tensor and the Christoffel symbols are given explicity in Appendix A.

## 3 Diffusion-Tensor MRI Data as Isometric Immersions

A volumetric tensor-valued image can be described mathematically as an isometric immersion $(x^1, x^2, x^3) \mapsto \phi = (x^1, x^2, x^3; P(x^1, x^2, x^3))$ of a three-dimensional domain $\Omega$ in the trivial fiber bundle $\mathbb{R}^3 \times \mathcal{P}(3)$, which is a nine-dimensional manifold. (The reader is referred to Appendix B for a succinct review of immersions and mean curvature.)

We denote by $(M, \gamma)$ the image manifold and its metric and by $(N, g)$ the target manifold and its metric. Here $M = \Omega$ and $N = \mathbb{R}^3 \times \mathcal{P}(3)$. Consequently, a tensor-valued image is a trivial section of this fiber bundle. The metric $\hat{g}$ of $N$ is then given by

$$d\hat{s}^2 = ds^2_{\mathrm{spatial}} + ds^2_{\mathrm{tensor}}. \tag{7}$$

The target manifold $N$, in this context is also called the *space-feature manifold* [25]. We can rewrite the metric defined by (7) as the quadratic form

$$d\hat{s}^2 = (dx^1)^2 + (dx^2)^2 + (dx^3)^2 + (d\boldsymbol{p})^T D_n^T (P^{-1} \otimes P^{-1}) D_n (d\boldsymbol{p}),$$

where $\boldsymbol{p} = (p^i) = v(P)$. The corresponding metric tensor is

$$\hat{g} = \begin{pmatrix} I_3 & 0_{3,6} \\ 0_{6,3} & g \end{pmatrix},$$

where $g$ is the metric tensor of $\mathcal{P}(3)$ as defined in Section 2.

Since the image is an isometric immersion, we have $\gamma = \phi^* \hat{g}$. Therefore

$$\gamma_{\alpha\beta} = \delta_{\alpha\beta} + g_{ij} \partial_\alpha p^i \partial_\beta p^j, \quad \alpha, \beta = 1, \ldots, m, \quad i, j = 1, \ldots, d. \tag{8}$$

We note that $d = n - m$ is the codimension of $M$. In compact form, we have

$$\gamma = I_m + (\nabla \boldsymbol{p})^T G(\phi) \nabla \boldsymbol{p}, \tag{9}$$

where $G$ is given by (4). (We take $m = 2$ for a slice and $m = 3$ for a volumetric DT-MRI image.)

## 4 Geometric Curvature-Driven Flows for Tensor-Valued Data

The basic concept in which geometric curvature-driven flows are based is the mean curvature of a submanifold embedded in a higher dimensional manifold. Here we generalize the scalar mean curvature flow to mean curvature flow in the space-feature manifold $\Omega \times \mathcal{P}(3)$. For this, we embed the Euclidean image space $\Omega$ into the Riemannian manifold $\Omega \times \mathcal{P}(3)$, and use some classical results from differential geometry to derive the *Riemannan Mean Curvature* (RMC). We then use the RMC to generalize mean curvature flow to tensor-valued data. Given the expression of the mean curvature vector $\boldsymbol{H}$, we can establish some PDEs based tensor-image filtering. Especially, we are interested of the so called level-set methods, which relay on PDEs that modify the shape of level sets in an image.

### 4.1 Riemannian Total Variation Flow

The total variation norm (TV) method introduced in [21] and its reconstructions have been successfully used in reducing noise and blurs without smearing sharp edges in grey-level, color and other vector-valued images [7, 13, 18, 22, 23, 24]. It is then natural to look for the extension of the TV norm to tensor-valued images.

The TV norm method is obtained as a gradient-decent flow associated with the $L^1$-norm of the tensor field. This yields the following PDE that express the motion by the mean curvature vector $\boldsymbol{H}$

$$\partial_t \phi^i = H^i. \tag{10}$$

This flow can be considered as a deformation of the tensor field toward *minimal immersion*. Indeed, it derives from variational setting that minimize the *volume* of the embedded image manifold in the space-feature manifold.

### 4.2 Riemannian Mean Curvature Flow

The following flow was proposed for the processing of scalar-valued images

$$\partial_t u = |\nabla u| \operatorname{div} \frac{\nabla u}{|\nabla u|}, \quad u(0,x,y) = u_0(x,y), \tag{11}$$

where $u_0(x,y)$ is the grey level of the image to be processed, $u(t,x,y)$ is its smoothed version that depends on the scale parameter $t$.

The "philosophy" of this flow is that the term $|\nabla u| \operatorname{div} \frac{\nabla u}{|\nabla u|}$ represents a degenerate diffusion term which diffuses $u$ in the direction orthogonal to its gradient $\nabla u$ and does not diffuse at all in the direction of $\nabla u$. This formulation has been proposed as a "morphological scale space" [2] and as more numerically tractable method of solving total variation [16].

The natural generalization of this flow to tensor-valued data is

$$\partial_t \phi^i = |\nabla^\gamma \phi|_g H^i, \quad i = 1, \ldots, d, \tag{12}$$

where

$$|\nabla^\gamma \phi|_g = \gamma^{\alpha\beta} g_{ij} \partial_\alpha \phi^i \partial_\beta \phi^j.$$

We note that several authors have tried to generalize curvature-driven flows for tensor-valued data in different ways. We think that the the use of differential-geometric tools and concepts yield the correct generalization.

### 4.3 Modified Riemannian Mean Curvature Flow

To denoise highly degraded images, Alvarez et al. [1] have proposed a modification of the mean curvature flow equation (11) that reads

$$\partial_t \phi = c\left(|K \star \nabla\phi|\right) |\nabla \phi| \operatorname{div} \frac{\nabla \phi}{|\nabla \phi|}, \quad \phi(0,x,y) = \phi_0(x,y), \tag{13}$$

where $K$ is a smoothing kernel (a Gaussian for example), $K \star \nabla\phi$ is therefore a local estimate of $\nabla\phi$ for noise elimination, and $c(s)$ is a nonincreasing real function which tends to zero as $s$ goes to infinity. We note that for the numerical experiments we have used $c(|\nabla\phi|) = k^2/(k^2 + |\nabla\phi|^2)$.

The generalization of the modified mean curvature flow to tensor-field processing is

$$\partial_t \phi^i = c\left(|K \star \nabla^\gamma \phi|_g\right) |\nabla^\gamma \phi|_g H^i, \quad \phi^i(0, \Omega) = \phi^i_0(\Omega). \tag{14}$$

The role of $c(\cdot)$ is to reduce the magnitude of smoothing near edges. In the scalar case, this equation does not have the same action as the Perona-Malik equation of enhancing edges. Indeed, the Perona-Malik equation has a variable diffusivity function and has been shown to selectively produce a "negative diffusion" which can increase the contrast of edges. Equation of the form (13) have always positive or forward diffusion, and the term $c$ merely reduces the magnitude of that smoothing. To correct this situation, Sapiro have proposed the self-snakes formalism [22], which we present in the next subsection and generalize to the matrix-valued data setting.

### 4.4 Riemannan Self-Snakes

The method of Sapiro, which he names *self-snakes* introduces an edge-stopping function into the mean curvature flow

$$\begin{aligned}\partial_t \phi &= |\nabla \phi| \operatorname{div}\left(c\,(K \star |\nabla \phi|)\,\tfrac{\nabla \phi}{|\nabla \phi|}\right) \\ &= c\,(K \star |\nabla \phi|)\,|\nabla \phi| \operatorname{div}\left(\tfrac{\nabla \phi}{|\nabla \phi|}\right) + \nabla c\,(K \star |\nabla \phi|) \cdot \nabla \phi.\end{aligned} \tag{15}$$

Comparing equation (15) with (13), we observe that the term $\nabla c\,(K \star |\nabla \phi|) \cdot \nabla \phi$ is missing in the previous model. This is due to the fact that the Sapiro model takes into account the image structure. Indeed, equation (15) can be re-written as

$$\partial_t \phi = \mathcal{F}_{\text{diffusion}} + \mathcal{F}_{\text{shock}}, \tag{16}$$

where

$$\begin{aligned}\mathcal{F}_{\text{diffusion}} &= c\,(K \star |\nabla \phi|)\,|\nabla \phi| \operatorname{div}\left(\frac{\nabla \phi}{|\nabla \phi|}\right), \\ \mathcal{F}_{\text{shock}} &= \nabla c\,(K \star |\nabla \phi|) \cdot \nabla \phi.\end{aligned}$$

The term $\mathcal{F}_{\text{diffusion}}$ is as in the anisotropic flow proposed in [1]. The second term in (16), i.e., $\nabla c \cdot \nabla \phi$, increases the attraction of the deforming contour toward the boundary of "objects" acting as the shock-filter introduced in [19] for deblurring. Therefore, the flow $\nabla c \cdot \nabla \phi$ is a shock filter acting like the backward diffusion in the Perona-Malik equation, which is responsible for the edge-enhancing properties of self snakes. See [22] for detailed discussion on this topic.

We are now interested in generalizing the self-snakes method for the case of tensor-valued data. We will start the generalization from equation (16) in the following manner

$$\partial_t \phi = \mathcal{F}_{\mathit{diffusion}} + \mathcal{F}_{\mathit{shock}}, \tag{17}$$

where

$$\begin{aligned}\mathcal{F}_{\mathit{diffusion}} &= c\,(K \star |\nabla^\gamma \phi|_g)\,|\nabla^\gamma \phi|_g H^i, \\ \mathcal{F}_{\mathit{shock}} &= \nabla c\,(K \star |\nabla^\gamma \phi|_g) \cdot \nabla^\gamma \phi^i.\end{aligned}$$

This decomposition is not artificial, since the covariant derivative on manifolds follow the same chain rule as the Euclidean directional derivative: let $V$ a vector field on $M$ which components are $v^i$, and let $\rho$ a scalar function. From classical differential geometry we have

$$\nabla_i^\gamma (\rho v^i) = \rho \nabla_i^\gamma v^i + v^i \nabla_i^\gamma \rho, \tag{18}$$

and in compact form

$$\operatorname{div}_\gamma(\rho V) = \rho \operatorname{div}_\gamma V + V \cdot \operatorname{grad}_\gamma \rho. \tag{19}$$

## 5 Numerical Experiments

In Fig. 1-a, we give a slice of a 3D tensor field defined over a square in $\mathbb{R}^2$. We note that a symmetric positive-definite $3 \times 3$ matrix $P$ is represented graphically by an ellipsoid whose principal directions are parallel to the eigenvectors of $P$ and whose axes are proportional to the eigenvalues of $P^{-1}$. Figure 1-b shows this tensor field after the

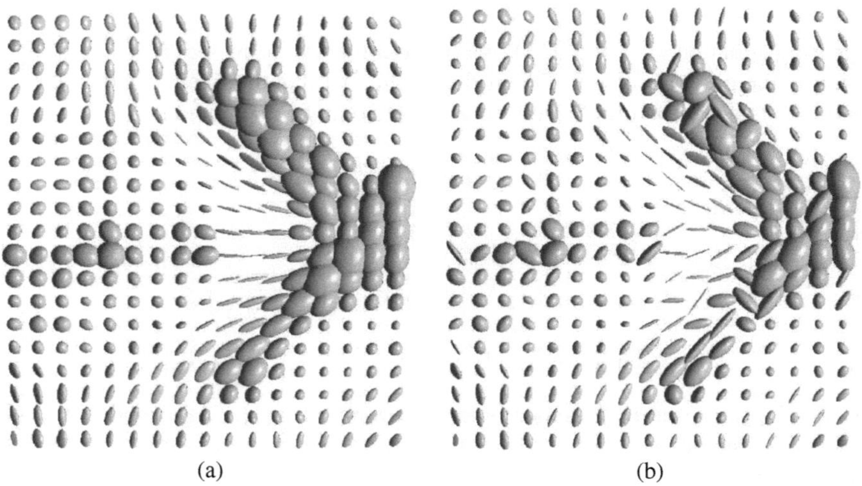

**Fig. 1.** Original tensor field (a) and that with added noise (b)

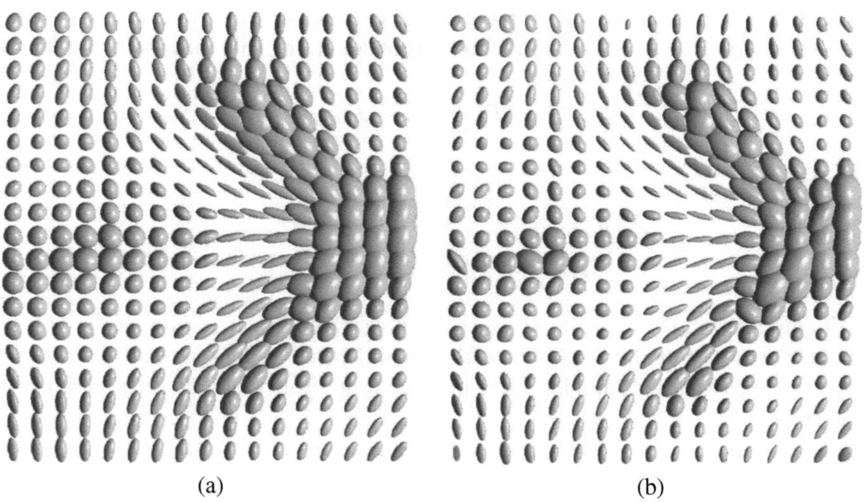

**Fig. 2.** The tensor field after 50 iterations of smoothing by: the TV flow (a) and by the Riemannian mean curvature flow (b)

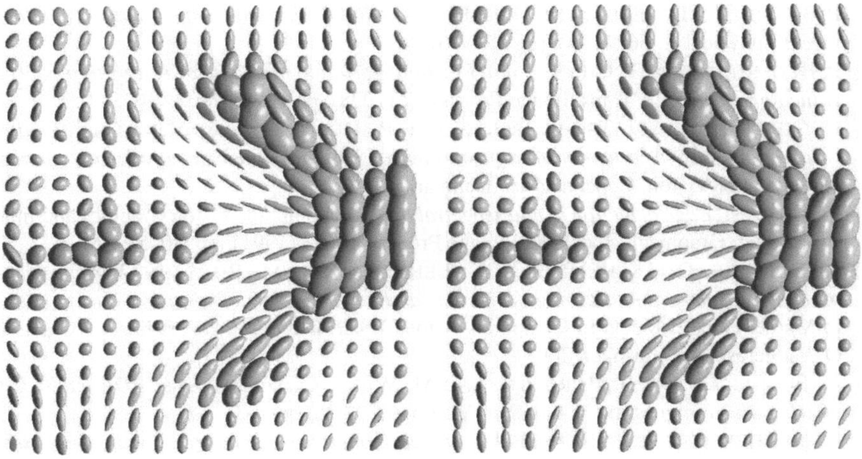

(a) (b)

**Fig. 3.** The tensor field after 50 iterations of smoothing by: the modified Riemannian mean curvature flow (a) and by the Riemannian self snake flow (b)

addition of noise. The resultant tensor field $P_0(x^1, x^2, x^3)$ is used as an initial condition for the partial differential equations (10), (12), (14) and (17) which we solve by a finite difference scheme with Neumann boundary conditions. We used 50 time steps of 0.01 each for these different flows. Figure 2-a represents the tensor smoothed by (10) and Figure 2-b shows the tensor smoothed by (12). Figure 3-a depicts the tensor smoothed by (14) and Figure 3-b shows the tensor smoothed by (17).

In this paper we generalized several curvature-driven flows of scalar- and vector-valued data to tensor-valued data. The use of the differential-geometric tools and concepts yields a natural extension of these well-known scalar-valued data processing methods to tensor-valued data processing. The preliminary numerical results presented show that in the TV flow there is a swelling effect usually observed in Euclidean methods. For the other flows, we note that edges between different homogeneous regions are preserved and there is no swelling effect. A detailed analysis of the numerical results and comparison with other methods are beyond the scope of this paper and will be published elsewhere.

## References

[1] L. Alvarez, P.-L. Lions, and J.-M. Morel, *Image selective smoothing and edge detection by nonlinear diffusion (II)*, SIAM J. Num. Anal., 29 (1992), pp. 845–866.

[2] L. Alvarez and J.-M. Morel, *Morphological Approach to Multiscale Analysis: From Principles to Equations*, Kluwler Academic Publishers, 1994.

[3] V. Arsigny, P. Fillard, X. Pennec, and N. Ayache, *Fast and simple calculus on tensors in the Log-Euclidean framework*, in Proc. 8th Int. Conf. on Medical Image Computing and Computer-Assisted Intervention - MICCAI 2005, Part I, J. Duncan and G. Gerig, eds., LNCS, Palm Springs, CA, 2005, Springer Verlag, pp. 115–122.

[4] P. Basser, J. Matiello, and D. LeBihan, *MR diffusion tensor spectroscopy and imaging*, Biophysical J., 66 (1994), pp. 259–267.
[5] P. Blomgren and T. Chan, *Color TV: Total variation methods for restoration of vector valued images*, IEEE Trans. Image Process., 7 (1998), pp. 304–378.
[6] C. Chefd'hotel, D. Tschumperlé, and R. Deriche, *Constrained flows of matrix-valued functions: Application to diffusion tensor regularization*, in European Conference on Computer Vision, Copenhaguen, Denmark, 2002.
[7] A. Cumani, *Edge detection in multispectral images*, Computer Vision Graphics and Image Processing: Graphical Models and Image Processing, 53 (1991), pp. 40–51.
[8] R. Deriche, D. Tschumperlé, C. Lenglet, and M. Rousson, *Variational approaches to the estimation, regularization and segmentation of diffusion tensor images*, in Mathematical Models in Computer Vision: The Handbook, N. Paragios, Y. Chen, and O. Faugeras, eds., Springer, 2005.
[9] C. Feddern, J. W. B. Burgeth, and M. Welk, *Curvature-driven PDE methods for matrix-valued images*, Int. J. Comput. Vision, 69 (2006), pp. 93–107.
[10] H. Henderson and S. Searle, *Vec and vech operators for matrices, with some uses in Jacobians and multivariate statistics*, Canad. J. Statist., 7 (1979), pp. 65–81.
[11] J. Jost, *Riemannian Geometry and Geometric Analysis*, Springer, 2nd ed., 1998.
[12] R. Kimmel, R. Malladi, and N. Sochen, *Images as embedded maps and minimal surfaces: movies, color, texture, and volumetric medical images*, Int. J. Compt. Vision, 39 (2000), pp. 111–129.
[13] H.-C. Lee and D. Cok, *Detecting boundaries in a vector field*, IEEE Trans. Signal Proc., 39 (1991), pp. 1181–1194.
[14] C. Lenglet, M. Rousson, R. Deriche, and O. Faugeras, *Statistics on the manifold of multivariate normal distributions: Theory and application to diffusion tensor MRI processing*, J. Mathematical Imaging and Vision, 25 (2006), pp. 423–444.
[15] J. Magnus and H. Neudecker, *The elimination matrix: some lemmas and applications*, SIAM J. Alg. Disc. Meth., 1 (1980), pp. 422–449.
[16] A. Marquina and S. Osher, *Explicit algoritms for a new time dependent model based on level set motion for non-linear deblurring and noise removal*, SIAM J. Sci. Compt., 22 (2000), pp. 378–405.
[17] M. Moakher and P. Batchelor, *The symmetric space of positive definite tensors: From geometry to applications and visualization*, in Visualization and Processing of Tensor Fields, J. Weickert and H. Hagen, eds., Springer, Berlin, 2006, ch. 17, pp. 285–298.
[18] R. Neviata, *A color edge detector and its use in scene segmentation*, IEEE Trans. Syst. Man. Cybern., 7 (1977), pp. 820–826.
[19] S. Osher and L. Rudin, *Feature-oriented image enhancement using shock filters*, SIAM J. Numer. Anal., 27 (1990), pp. 919–940.
[20] X. Pennec, P. Fillard, and N. Ayache, *A Riemannian framework for tensor computing*, Int. J. Compt. Vision, 66 (2006), pp. 41–66.
[21] L. Rudin, S. Osher, and E. Fatemi, *Nonlinear total variation based noise removal algorithms*, Physica D, 60 (1992), pp. 259–268.
[22] G. Sapiro, *Color snakes*, Tech. Rep. HPL-95-113, Hewlett Packard Computer Peripherals Laboratory, 1995.
[23] ——, *Vector-valued active contours*, in Proceedings of Computer Vision and Pattern Recognition, 1996, pp. 520–525.
[24] G. Sapiro and D. Ringach, *Anisotropic diffusion of multivalued images with applications to color filtering*, IEEE Trans. Image Process., 5 (1996), pp. 1582–1586.
[25] N. A. Sochen, R. Kimmel, and R. Malladi, *A general framework for low level vision*, IEEE Trans. Image Process., 7 (1998), pp. 310–318.

[26] Z. WANG, D. VEMURI, Y. CHEN, AND T. MARECI, *A constrained variational principle for direct estimation and smoothing of the diffusion tensor field from complex DWI*, IEEE Trans. Medical Imaging, 23 (2004), pp. 930–939.

[27] J. WEICKERT AND H. HAGEN, eds., *Visualization and Processing of Tensor Fields*, Springer, Berlin, 2006.

[28] M. ZÉRAÏ AND M. MOAKHER, *The Riemannian geometry of the space of positive-definite matrices and its application to the regularization of diffusion tensor MRI data*, Submitted to: J. Mathematical Imaging and Vision, (2006).


## Appendix A: Explicit Formulae for the Geometry of $\mathcal{P}(3)$

We give here the explicit form of the inverse metric tensor and Christoffel symbols for the Riemannian metric on $\mathcal{P}(3)$. The components of the inverse metric tensor are given by

$$G^{-1} = \begin{bmatrix} (p^1)^2 & (p^4)^2 & (p^6)^2 & p^1p^4 & p^4p^6 & p^1p^6 \\ (p^4)^2 & (p^2)^2 & (p^5)^2 & p^2p^4 & p^2p^5 & p^4p^5 \\ (p^6)^2 & (p^5)^2 & (p^3)^2 & p^6p^5 & p^5p^3 & p^6p^3 \\ p^1p^4 & p^2p^4 & p^6p^5 & \frac{1}{2}(p^1p^2+(p^4)^2) & \frac{1}{2}(p^4p^5+p^6p^2) & \frac{1}{2}(p^1p^5+p^4p^6) \\ p^4p^6 & p^2p^5 & p^5p^3 & \frac{1}{2}(p^4p^5+p^6p^2) & \frac{1}{2}((p^5)^2+p^2p^3) & \frac{1}{2}(p^6p^5+p^4p^3) \\ p^1p^6 & p^4p^5 & p^6p^3 & \frac{1}{2}(p^1p^5+p^4p^6) & \frac{1}{2}(p^6p^5+p^4p^3) & \frac{1}{2}(p^1p^3+(p^6)^2) \end{bmatrix}.$$

The determinant of $P$ is $\rho = p^1p^2p^3 + 2p^4p^5p^6 - p^1(p^5)^2 - p^2(p^6)^2 - p^3(p^4)^2$. Let $\boldsymbol{s} := [s^1, s^2, s^3, s^4, s^5, s^6]^T = \upsilon(\mathrm{adj}(P))$, where $\mathrm{adj}(P) = \rho P^{-1}$ is the adjoint matrix of $P$. The Christoffel symbols are arranged in the following six symmetric matrices:

$$\Gamma^1 = \frac{-1}{\rho}\begin{bmatrix} s^1 & 0 & 0 & s^4 & 0 & s^6 \\ 0 & 0 & 0 & 0 & 0 & 0 \\ 0 & 0 & 0 & 0 & 0 & 0 \\ s^4 & 0 & 0 & s^2 & 0 & s^5 \\ 0 & 0 & 0 & 0 & 0 & 0 \\ s^6 & 0 & 0 & s^5 & 0 & s^3 \end{bmatrix}, \qquad \Gamma^2 = \frac{-1}{\rho}\begin{bmatrix} 0 & 0 & 0 & 0 & 0 & 0 \\ 0 & s^2 & 0 & s^4 & s^5 & 0 \\ 0 & 0 & 0 & 0 & 0 & 0 \\ 0 & s^4 & 0 & s^1 & s^6 & 0 \\ 0 & s^5 & 0 & s^6 & s^3 & 0 \\ 0 & 0 & 0 & 0 & 0 & 0 \end{bmatrix},$$

$$\Gamma^3 = \frac{-1}{\rho}\begin{bmatrix} 0 & 0 & 0 & 0 & 0 & 0 \\ 0 & 0 & 0 & 0 & 0 & 0 \\ 0 & 0 & s^3 & 0 & s^5 & s^6 \\ 0 & 0 & 0 & 0 & 0 & 0 \\ 0 & 0 & s^5 & 0 & s^2 & s^4 \\ 0 & 0 & s^6 & 0 & s^4 & s^1 \end{bmatrix}, \qquad \Gamma^4 = \frac{-1}{2\rho}\begin{bmatrix} 0 & s^4 & 0 & s^1 & s^6 & 0 \\ s^4 & 0 & 0 & s^2 & 0 & s^5 \\ 0 & 0 & 0 & 0 & 0 & 0 \\ s^1 & s^2 & 0 & 2s^4 & s^5 & s^6 \\ s^6 & 0 & 0 & s^5 & 0 & s^3 \\ 0 & s^5 & 0 & s^6 & s^3 & 0 \end{bmatrix},$$

$$\Gamma^5 = \frac{-1}{2\rho}\begin{bmatrix} 0 & 0 & 0 & 0 & 0 & 0 \\ 0 & 0 & s^5 & 0 & s^2 & s^4 \\ 0 & s^5 & 0 & s^6 & s^3 & 0 \\ 0 & 0 & s^6 & 0 & s^4 & s^1 \\ 0 & s^2 & s^3 & s^4 & 2s^5 & s^6 \\ 0 & s^4 & 0 & s^1 & s^6 & 0 \end{bmatrix}, \qquad \Gamma^6 = \frac{-1}{2\rho}\begin{bmatrix} 0 & 0 & s^6 & 0 & s^4 & s^1 \\ 0 & 0 & 0 & 0 & 0 & 0 \\ s^6 & 0 & 0 & s^5 & 0 & s^3 \\ 0 & 0 & s^5 & 0 & s^2 & s^4 \\ s^4 & 0 & 0 & s^2 & 0 & s^5 \\ s^1 & 0 & s^3 & s^4 & s^5 & 2s^6 \end{bmatrix}.$$

# Appendix B: Immersions and Mean Curvature

In this appendix, we recall some facts about immersions between Riemannian manifolds and their mean curvature that we use in Section 3. Let $(M, \gamma)$ and $(N, g)$ be two connected Riemanian manifolds of dimensions $m$ and $n$, respectively. We consider a map $\phi : M \to N$ that is of class $C^2$, i.e., $\phi \in C^2(M, N)$. Let $\{x^\alpha\}_{1 \leq \alpha \leq m}$ be a local coordinate system of $x$ in a neighborhood of a point $p \in M$ and let $\{y^i\}_{1 \leq i \leq n}$ be a local coordinate system of $y$ in a neighborhood of $\phi(p) \in N$.

The mapping $\phi$ induces a metric $\phi^* g$ on $M$ defined by $\phi^* g\ (X_p, Y_p) = g\ (\phi_*(X_p), \phi_*(Y_p))$ . This metric is called the *pull-back* metric induced by $\phi$, as it maps the metric in the opposite direction of the mapping $\phi$.

An isometry is a diffeomorphism $\phi : M \to N$ that preserves the Riemannian metric, i.e., if $g$ and $\gamma$ are the metrics for $M$ and $N$, respectively, then $\gamma = \phi^* g$. It follows that an isometry preserves the length of curves, i.e., if $c$ is a smooth curve on $M$, then the curve $\phi \circ c$ is a curve on $N$ of the same length. Also, the image of a geodesic under an isometry is again a geodesic.

A mapping $\phi : M \to N$ is called an *immersion* if $(\phi_*)_p$ is injective for every point $p$ in $M$. We say that $M$ is immersed in $N$ by $\phi$ or that $M$ is an immersed submanifold of $N$. When an immersion $\phi$ is injective, it is called an *embedding* of $M$ into $N$. We then say that $M$ is an *embedded submanifold*, or simply, a *submanifold* of $N$.

Now let $\phi : M \to N$ be an immersion of a manifold $M$ into a Riemannian manifold $N$ with metric $g$. The first fundamental form associated with the immersion $\phi$ is $h = \phi^* g$. Its components are $h_{\alpha\beta} = \partial_\alpha \phi^i \partial_\beta \phi^j g_{ij}$ where $\partial_\alpha \phi^i = \frac{\partial \phi^i}{\partial x^\alpha}$. The total covariant derivative $\nabla d\phi$ is called the second fundamental form of $\phi$ and is denoted by $II^M(\phi)$. The second fundamental form $II^M$ takes values in the normal bundle of $M$. The mean curvature vector $\boldsymbol{H}$ of an isometric immersion $\phi : M \to N$ is defined as the trace of the second fundamental form $II^M(\phi)$ divided by $m = \dim M$ [11]

$$\boldsymbol{H} := \frac{1}{m} \operatorname{tr}_\gamma II^M(\phi).$$

In local coordinates, we have [11]

$$mH^i = \Delta_M \phi^i + \gamma^{\alpha\beta}(x)^N \Gamma^i_{jk} \left(\phi(x)\right) \frac{\partial \phi^j}{\partial x^\alpha} \frac{\partial \phi^k}{\partial x^\beta},$$

where ${}^N\Gamma^i_{jk}$ are the Christoffel symbols of $(N, g)$ and $\Delta_M$ is the Laplace-Beltrami operator on $(M, \gamma)$ given by

$$\Delta_M = \frac{1}{\sqrt{\det \gamma}} \frac{\partial}{\partial x^\alpha} \left( \sqrt{\det \gamma} \gamma^{\alpha\beta} \frac{\partial \phi^i}{\partial x^\beta} \right).$$

# A Variational Framework for Spatio-temporal Smoothing of Fluid Motions

Nicolas Papadakis and Étienne Mémin

IRISA/INRIA, Campus de Beaulieu, 35042 Rennes Cedex, France
{Nicolas.Papadakis,Etienne.Memin}@irisa.fr

**Abstract.** In this paper, we introduce a variational framework derived from data assimilation principles in order to realize a temporal Bayesian smoothing of fluid flow velocity fields. The velocity measurements are supplied by an optical flow estimator. These noisy measurement are smoothed according to the vorticity-velocity formulation of Navier-Stokes equation. Following optimal control recipes, the associated minimization is conducted through an iterative process involving a forward integration of our dynamical model followed by a backward integration of an adjoint evolution law. Both evolution laws are implemented with second order non-oscillatory scheme. The approach is here validated on a synthetic sequence of turbulent 2D flow provided by Direct Numerical Simulation (DNS) and on a real world meteorological satellite image sequence depicting the evolution of a cyclone.

## 1 Introduction

The analysis and control of complex fluid flows is a major scientific issue. In that prospect, flow visualization and extraction of accurate kinetic or dynamical measurements are of the utmost importance. For several years, the study of dynamic structures and the estimation of dense velocity fields from fluid image sequences have received great attention from the computer vision community [3,7,8,15,18,26]. Application domains range from experimental visualization in fluid mechanical to geophysical flow analysis in environmental sciences. In particular, accurate measurement of atmospheric flow dynamics is of the greatest importance for weather forecasting, climate prediction or analysis, etc...

The analysis of motion in such sequences is particularly challenging due to abrupt and sudden changes of the luminance function in image sequences. For these reasons, motion analysis techniques designed for computer vision application and quasi-rigid motions, are not well adapted in this context. Recently, methods for fluid-dedicated dense estimation have been proposed to characterize fluid motion [3,4,5,12,24,25]. However, these motion estimators are still using only a small set of images and thus may suffer from a temporal inconsistency from frame to frame. The set of motion fields provided may not respect fluid mechanics conservation laws. The design of appropriate methods enabling to take into account the underlying physics of the observed flow constitutes a widely

F. Sgallari, A. Murli, and N. Paragios (Eds.): SSVM 2007, LNCS 4485, pp. 603–615, 2007.

open domain of research. We are here interested in using the *vorticity-velocity* formulation of *Navier-Stokes* equations which describes accurately the evolution of vorticity transported by the flow for the filtering of noisy motion fields.

The approach we propose in this work is related to variational data assimilation principles used for instance in meteorology [1,6,23]. Such techniques enable, in the same spirit as a Kalman filter, a temporal smoothing along the whole image sequence. As does a Bayesian smoother, it combines a dynamical evolution law of state variables representing the target of interest with the whole set of available noisy measurements related to this target. Nevertheless, unlike Kalman filtering and stochastic Bayesian filtering approaches such as particle filtering, variational assimilation techniques allows to cope with state spaces of very large dimension.

The technique we devise allows us to incorporate in the whole set of motion fields a dynamical consistency along the image sequence. The approach is expressed as the minimization of a global spatio-temporal functional stemming from a Bayesian formulation. The optimization process is led through the introduction of an adjoint evolution model. This method has the advantage to provide an efficient numerical approximation of the gradient functional without resorting to the complete analytical expressions of Euler-Lagrange equations. This is particularly interesting when dealing with high order differential operators.

## 2 Data Assimilation

### 2.1 Introduction

Data Assimilation is a technique related to optimal control theory which allows estimating over time the state of a system of variables of interest. This method enables a smoothing of the unknown variables according to an initial state of the system, a dynamic law and noisy measurements of the system's state. Let the vector of variables $\boldsymbol{X} \in \Xi$ represents the state of the system. The evolution of the system is assumed to be described through a (possibly non linear) differential dynamical model $\mathbb{M}$:

$$\begin{cases} \partial_t \boldsymbol{X} + \mathbb{M}(\boldsymbol{X}, U) = 0 \\ \boldsymbol{X}(t_0) = \boldsymbol{X}_0 \end{cases} \tag{1}$$

This system is monitored by a control variable $v = (U, \boldsymbol{X}_0)$ defined in control space $P$. This control variable may be set to the initial condition $\boldsymbol{X}_0$ and/or to any free parameters $U$ of the evolution law.

We then assume that observations $Y \in O_{obs}$ are available. These observations may live in a different space (a reduced space for instance) from the state variable. We will nevertheless assume that there exists a differential operator $\mathbb{H}$, that goes from the variable space to the observation space. A least squares estimation of the control variable regarding the whole sequence of measurements available within a considered time range comes to minimize with respect to the control variable $v \in P$, a cost function of the following form:

$$J(v) = \frac{1}{2} \int_{t_0}^{t_f} ||Y - \mathbb{H}(\boldsymbol{X}(v))||^2 \, dt. \tag{2}$$

A first approach consists in computing the functional gradient through finite differences:

$$\nabla_v J \simeq \lim_{\epsilon \longrightarrow 0} \frac{J(v + \epsilon e_k) - J(v)}{\epsilon},$$

where $\epsilon \in \mathbb{R}$ is an infinitesimal perturbation and $\{e_k, k = 1, \dots, N\}$ denotes the unitary basis vectors of the control space of dimension $N$. Such a computation is impractical for control space of large dimension since it requires $N$ integrations of the evolution model for each required value of the gradient functional.

*Adjoint models* as introduced first in meteorology by Le Dimet and Talagrand in [6] authorize the computation of the gradient functional in a single backward integration of an adjoint variable. The value of this adjoint variable at the initial time provides the value of the gradient at the desired point. This first approach is widely used in environmental sciences for the analysis of geophysical flows [6,23]. However, these methods rely on a perfect dynamical modeling of the system evolution. Such modeling seems to us irrelevant in image analysis since the different models on which we can rely on are usually inaccurate due for instance to 3D-2D projections, varying lighting conditions, completely unknown boundary conditions, etc ... Considering imperfect models, defined up to a Gaussian noise comes to an optimization problem where the control variable is constituted by the whole trajectory of the state variable. This is the kind of problem we are facing in this work.

### 2.2 Data Assimilation with Imperfect Model

The ingredients of the new data assimilation problem are now composed by an imperfect dynamic model of the target system (without parameter $U$), an initialization of the system's state and an observation equation which relates the system variables to some measurements:

$$\begin{cases} \partial_t \boldsymbol{X} + \mathbb{M}(\boldsymbol{X}) = \boldsymbol{\eta}(\boldsymbol{x}, t) \\ \boldsymbol{X}(t_0) = \boldsymbol{X}_0 + \boldsymbol{\eta}_i(\boldsymbol{x}) \\ Y(t) = \mathbb{H}(\boldsymbol{X}) + \eta_o(\boldsymbol{x}, t), \end{cases} \tag{3}$$

where $\boldsymbol{\eta}$, $\boldsymbol{\eta}_i$ and $\eta_o$ are time varying zero mean Gaussian noise vector functions. They are respectively associated to covariance matrices $Q(\boldsymbol{x}, t)$, $B(\boldsymbol{x}, \boldsymbol{x}')$ and $R(\boldsymbol{x}, t)$. The state variable $\boldsymbol{X}$ is defined on the image plan $\Omega$. The noise functions represent the errors involved in the different components of the system (i.e model, initialization and measurement errors) and are assumed to be uncorrelated in time. The system of equations (3) could be specified describing the three Gaussian conditional probability densities $p(\boldsymbol{X}|\boldsymbol{X}(0))$ , $p(\boldsymbol{X}(0)|\boldsymbol{X}_0)$ and $p(Y|\boldsymbol{X})$ which relates respectively the state trajectory $\boldsymbol{X}$ along time to the initial state value $\boldsymbol{X}(0)$, the initial state value to the initial condition $\boldsymbol{X}_0$ and $Y$, and the complete set of measurements to the state $\boldsymbol{X}$. As in any stochastic filtering problem, we aim at estimating the conditional expectation of the state trajectory given the whole set of available observations. As all the pdf involved here are Gaussian, it becomes estimating the mode of the *a posteriori* distribution $p(\boldsymbol{X}|Y, \boldsymbol{X}_0)$.

**Penalty function.** The goal is thus to minimize the new functional:

$$J(\boldsymbol{X}) = \frac{1}{2}\int_{t_0}^{t_f} ||\partial_t \boldsymbol{X} + \mathbb{M}(\boldsymbol{X})||_Q^2 dt + \frac{1}{2}\,||\boldsymbol{X}(t_0) - \boldsymbol{X}_0||_B^2 + \frac{1}{2}\int_{t_0}^{t_f} ||Y - \mathbb{H}(\boldsymbol{X})||_R^2\, dt, \quad (4)$$

where the norms are the Mahalanobis distance defined by the inverse matrices associated to $Q$, $B$ and $R$ (the information matrices) and the dot product of $L^2(\Omega)$. The minimization has to be done according to variable $\boldsymbol{X}$. It is the complete trajectory of the state variable that constitutes the control variable of the associated problem.

**Minimization of the functional.** A minimizer $\boldsymbol{X}$ of functional $J$ is also a minimizer of a perturbed function $J(\boldsymbol{X} + \beta\boldsymbol{\theta}(\boldsymbol{x}, t))$, where $\boldsymbol{\theta}(\boldsymbol{x}, t)$ belongs to a space of admissible function and $\beta$ is a positive parameter. In other words, $\boldsymbol{X}$ must cancel out the directional derivative:

$$\delta J_{\boldsymbol{X}}(\boldsymbol{\theta}) = \lim_{\beta \to 0} \frac{dJ(\boldsymbol{X} + \beta\boldsymbol{\theta}(\boldsymbol{x}, t))}{d\beta} = 0.$$

The functional $J$ at point $\boldsymbol{X} + \beta\boldsymbol{\theta}(\boldsymbol{x}, t))$ reads:

$$\begin{aligned} J = &\frac{1}{2}\int_{\Omega} (\boldsymbol{X} + \beta\boldsymbol{\theta} - \boldsymbol{X}_0)^\top B^{-1}(\boldsymbol{X} + \beta\boldsymbol{\theta} - \boldsymbol{X}_0) d\boldsymbol{x} \\ &+ \frac{1}{2}\int_{\Omega,T} (\partial_t \boldsymbol{X} + \beta\partial_t\boldsymbol{\theta} + \mathbb{M}(\boldsymbol{X} + \beta\boldsymbol{\theta}))^\top Q^{-1}(\partial_t \boldsymbol{X} + \beta\partial_t\boldsymbol{\theta} + \mathbb{M}(\boldsymbol{X} + \beta\boldsymbol{\theta}))\, dtd\boldsymbol{x} \\ &+ \frac{1}{2}\int_{\Omega,T} (Y - \mathbb{H}(\boldsymbol{X} + \beta\boldsymbol{\theta}))^\top R^{-1}(Y - \mathbb{H}(\boldsymbol{X} + \beta\boldsymbol{\theta}))dtd\boldsymbol{x}, \end{aligned} \quad (5)$$

where integration with respect to $\Omega$ denotes spatial integration on the image domain and subscript $T$ stands for temporal integration between an initial time $t_0$ and a final instant $t_f$

**Adjoint variable.** In order to perform an integration by part – to factorize this expression by $\boldsymbol{\theta}$ – we introduce an "adjoint variable" $\boldsymbol{\lambda}$ defined by:

$$\boldsymbol{\lambda} = Q^{-1}\left(\partial_t \boldsymbol{X} + \mathbb{M}(\boldsymbol{X})\right), \quad (6)$$

as well as *linear tangent operators* $\partial_{\boldsymbol{X}}\mathbb{M}$ and $\partial_{\boldsymbol{X}}\mathbb{H}$ defined by:

$$\lim_{\beta \to 0} \frac{d\mathbb{M}(\boldsymbol{X} + \beta\boldsymbol{\theta})}{d\beta} = \partial_{\boldsymbol{X}}\mathbb{M}\boldsymbol{\theta}. \quad (7)$$

Such linear operators correspond to the Gâteaux derivative at point $\boldsymbol{X}$ of the operators $\mathbb{M}$ and $\mathbb{H}$. Let us note that the derivative of a linear operator is the operator itself. By taking the limit $\beta \to 0$, and applying integrations by parts, we can get rid of the partial derivatives of the admissible function $\theta$. We also have to introduce the adjoint operators $\partial_X\mathbb{M}^*$ and $\partial_X\mathbb{H}^*$ as compact notation of the integration by parts of the associated linear tangent operator. Considering the dot product $\langle \phi, \psi \rangle = \int_{\Omega,T} \phi(\boldsymbol{x},t)\psi(\boldsymbol{x},t)d\boldsymbol{x}dt$ associated to $L^2(\Omega, T \in [t_0, t_f])$ such operators are defined as:

$$\langle \partial_{\boldsymbol{X}} \mathbb{M}\boldsymbol{X}_1, \boldsymbol{X}_2 \rangle_{\Xi} = \langle \boldsymbol{X}_1, \partial_{\boldsymbol{X}} \mathbb{M}^* \boldsymbol{X}_2 \rangle_{\Xi} \qquad \langle \partial_{\boldsymbol{X}} \mathbb{H}\boldsymbol{X}, Y \rangle_{O_{obs}} = \langle \boldsymbol{X}, \partial_{\boldsymbol{X}} \mathbb{H}^* Y \rangle_{\Xi}. \tag{8}$$

Gathering all the elements we have so far, equation (5) can be rewritten as:

$$\begin{aligned} \lim_{\beta \to 0} \frac{dJ}{d\beta} = & \int_{\Omega,T} \boldsymbol{\theta}^\top \left[ (-\partial_t \boldsymbol{\lambda} + \partial_{\boldsymbol{X}} \mathbb{M}^* \boldsymbol{\lambda}) - \partial_{\boldsymbol{X}} \mathbb{H}^* R^{-1} (Y - \mathbb{H}(\boldsymbol{X})) \right] dt d\boldsymbol{x} \\ & + \int_{\Omega} \boldsymbol{\theta}^\top (\boldsymbol{x}, t_0) \left[ \left( B^{-1}(\boldsymbol{X}(\boldsymbol{x}, t_0) - \boldsymbol{X}_0(\boldsymbol{x})) - \boldsymbol{\lambda}(\boldsymbol{x}, t_0) \right) d\boldsymbol{x}' \right] d\boldsymbol{x} \\ & + \int_{\Omega} \boldsymbol{\theta}^\top (\boldsymbol{x}, t_f) \boldsymbol{\lambda}(\boldsymbol{x}, t_f) d\boldsymbol{x} = 0. \end{aligned} \tag{9}$$

**Forward/backward equations.** Since the functional derivative must be null for any arbitrary independent admissible functions in the three integrals of expression (9), all the other members appearing in the three integral terms must be identically null. It follows a coupled system of forward and backward PDE's with initial and final conditions:

$$\boldsymbol{\lambda}(\boldsymbol{x}, t_f) = 0, \tag{10}$$

$$-\partial_t \boldsymbol{\lambda} + \partial_{\boldsymbol{X}} \mathbb{M}^* \boldsymbol{\lambda} = \partial_{\boldsymbol{X}} \mathbb{H}^* R^{-1} (Y - \mathbb{H}(\boldsymbol{X})), \tag{11}$$

$$\boldsymbol{\lambda}(\boldsymbol{x}, t_0) = \left( B^{-1}(\boldsymbol{X}(\boldsymbol{x}, t_0) - \boldsymbol{X}_0(\boldsymbol{x}) \right), \tag{12}$$

$$\partial_t \boldsymbol{X} + \mathbb{M}(\boldsymbol{X}) = Q\boldsymbol{\lambda}. \tag{13}$$

The forward equation (13) corresponds to the definition of the adjoint variable (6) and has been obtained introducing $Q$, the pseudo-inverse of $Q^{-1}$ [1]. Let us remark that if the model is assumed to be perfect, we would have $Q = 0$ and retrieve the case of a perfect dynamical state model associated to an initial state control problem.

Otherwise, equation (10) constitutes an explicit end condition for the adjoint evolution model equation (11). This adjoint evolution model can be integrated backward from the end condition assuming the knowledge of an initial guess for $\boldsymbol{X}$ to compute the discrepancy $Y - \mathbb{H}(\boldsymbol{X})$. This backward integration provides the gradient of the associated functional. To perform this integration, an expression of the *adjoint evolution operator* is required. Let us recall that this operator is defined from an integration by part of the *linear tangent operator* associated to the evolution law operator. The analytic expression of such an operator is obviously not accessible in general. Nevertheless, it can be noticed that a discrete expression of this operator can be obtained from the discretization of the linear tangent operator. Thus, the adjoint of the linear tangent operator discretized as a matrix simply consists of the transpose of that matrix. Knowing a first solution of the adjoint variable, an initial update condition for the state variable can be obtained from (12) and a pseudo inverse expression of the covariance matrix $B$. From this initial condition, equation (13) can be finally integrated forward and supply a new right hand part for the adjoint equation (11) and so forth.

**Incremental state function.** The previous system can be slightly modified to rely on an adequate initial guess for the state function. Considering a function of state increments linking the state function and an initial condition function, $\delta \boldsymbol{X} = \boldsymbol{X} - \boldsymbol{X}_0$, and linearizing the operator $\mathbb{M}$ around the initial condition function $\boldsymbol{X}_0$, as $\mathbb{M}(\boldsymbol{X}) = \mathbb{M}(\boldsymbol{X}_0) + \partial_{\boldsymbol{X}_0}\mathbb{M}(\delta \boldsymbol{X})$ enables to split equation (13) into two pde's with an explicit initial condition:

$$\boldsymbol{X}(\boldsymbol{x}, t_0) = \boldsymbol{X}_0(\boldsymbol{x}), \tag{14}$$

$$\partial_t \boldsymbol{X}_0 + \mathbb{M}(\boldsymbol{X}_0) = 0, \tag{15}$$

$$\partial_t \delta \boldsymbol{X} + \partial_{\boldsymbol{X}_0}\mathbb{M}\delta \boldsymbol{X} = Q\boldsymbol{\lambda}. \tag{16}$$

Combining equations (10-12) and (14-16) leads to the final tracking algorithm.

The method first consists of a forward integration of the initial condition $\boldsymbol{X}_0$ with the system's dynamical model equation (15). The current solution is then corrected by performing a backward integration (10, 11) of the adjoint variable. The evolution of $\boldsymbol{\lambda}$ is guided by a discrepancy measure between the observation and the estimate: $Y - \mathbb{H}(\boldsymbol{X})$. The initial condition is then updated through equation (12) and a forward integration of the increment $\delta \boldsymbol{X}$ is realized through the equation (16). The overall process is iteratively repeated until convergence.

## 3 Application to Fluid Motion Tracking

We aim here at applying the previous framework for a consistent tracking along time of fluid motion velocity fields. For fluid flows, the Navier-Stokes equation provides a universal general law for predicting the evolution of the flow. The purpose will be thus to incorporate into a data assimilation process such a dynamical model together with noisy velocity measurements.

### 3.1 Basic Definitions

In this work, the formulation of the Navier-Stokes on which we will rely on uses the vorticity $\xi = \boldsymbol{\nabla}^{\perp} \cdot \boldsymbol{w} = v_x - u_y$ and on the divergence $\zeta = \boldsymbol{\nabla} \cdot \boldsymbol{w} = u_x + v_y$ of a bidimensional motion field $\boldsymbol{w} = [u, v]^{\top}$. The vorticity is related to the presence of a rotating motion, whereas the divergence is related to the presence of sinks and sources in a flow. Assuming $\boldsymbol{w}$ vanishes at infinity[1], the vector field is decomposed using the orthogonal Helmholtz decomposition, as a sum of two potential functions gradient $\boldsymbol{w} = \boldsymbol{\nabla}^{\perp}\Psi + \boldsymbol{\nabla}\Phi$. The stream function $\Psi$ and the velocity potential $\Phi$ respectively correspond to the solenoidal and the irrotational part of the vector field $\boldsymbol{w}$. They are linked to the divergence and vorticity maps through two Poisson Equations: $\xi = \Delta\Psi$, $\zeta = \Delta\Phi$. Expressing the solution of both equations as a convolution product with the Green kernel

[1] A divergence and curl free global transportation component is removed from the vector field. This field is estimated on the basis of a Horn and Schunck estimator associated to a high smoothing penalty [3].

$G = \frac{1}{2\pi}\ln(|x|)$ associated to the 2D Laplacian operator: $\Psi = G * \xi$, $\Phi = G * \zeta$, the whole velocity field can be recovered knowing its divergence and vorticity:

$$\boldsymbol{w} = \boldsymbol{\nabla}^{\perp} G * \xi + \boldsymbol{\nabla} G * \zeta. \tag{17}$$

This computation can be very efficiently implemented in the Fourier domain.

### 3.2 Fluid Motion Evolution Equation

In order to consider a tractable expression of the Navier-Stokes equation for the tracking problem, we rely in this work on the 2D vorticity-velocity formulation of the 3D incompressible Navier-Stokes equation, as obtained in the shallow water model:

$$\partial_t \xi + \boldsymbol{w} \cdot \boldsymbol{\nabla} \xi + \xi\zeta - \nu \Delta \xi = 0. \tag{18}$$

This formulation states roughly that the vorticity is transported by the velocity field and is diffused along time. Modeling the vorticity divergence product as a zero mean Gaussian random variable, we end up with an imperfect 2D incompressible vorticity-velocity formulation.

Concerning the divergence map, it is more involved to exhibit any conservation law. We will assume here that it behaves like a noise. More precisely we will assume that the divergence map is a function of a Gaussian random variable, $X_t$, with stationary increments (a Brownian motion) starting at points, $\boldsymbol{x}$. It can be shown through Ito formula and Kolmogorov's backward equation, that the expectation at time $t$ of such a function, $u(t, \boldsymbol{x}) = \mathbb{E}[\zeta(X_t)]$ obeys to a heat equation [20]:

$$\begin{aligned} &\partial_t u - \nu_\zeta \Delta u = 0, \\ &u(0, \boldsymbol{x}) = \zeta(\boldsymbol{x}). \end{aligned} \tag{19}$$

According to this equation, we indeed make the assumption that the divergence at any time of the sequence is a solution of a heat equation (i.e. it can be recovered from a smoothing of the initial motion field divergence map with a Gaussian function of standard deviation $2\sqrt{\nu_\zeta}$ ).

As the curl and divergence maps completely determine the underlying velocity field, equations (18) and (19) allow us to write the following imperfect dynamical model for the fluid motion field:

$$\partial_t \begin{bmatrix} \xi \\ \zeta \end{bmatrix} + \underbrace{\begin{bmatrix} \boldsymbol{w} \cdot \boldsymbol{\nabla} - \nu_\xi \Delta & 0 \\ 0 & -\nu_\zeta \Delta \end{bmatrix} \begin{bmatrix} \xi \\ \zeta \end{bmatrix}}_{\mathbb{M}(\xi,\zeta)} = \boldsymbol{\eta}(t). \tag{20}$$

The noise function $\boldsymbol{\eta}(t)$ is a Gaussian random vector modeling the errors of our evolution law.

### 3.3 Fluid Motion Observations

With regards to the velocity measurements, we will assume that an observation motion field $\boldsymbol{w}_{obs}$ is available. This motion field can be provided by any dense motion estimator. In this work, a dense motion field estimator dedicated to fluid flows is used [3] to supply the velocity measurements. Taking the vorticity and divergence of these optical-flow motion fields as observation, provides measurements in the state variable space and consequently $\mathbb{H} = Id$.

## 4 Discretization of the Vorticity-Velocity Equation

The discretization of the vorticity-velocity equation 18 must be done cautiously. In particular, the advective term $\boldsymbol{\nabla}\xi \cdot \boldsymbol{w}$ must be treated specifically. A lot of non-oscillatory schemes for conservation laws have been developed to solve this problem [11,14,19]. Such schemes consider a polynomial reconstruction of the sought function on cells and discretize the intermediate value of this function at the cell's boundaries. The involved derivatives of the transported quantity are computed with high orders accurate difference scheme. The value of these derivatives are attenuated through limiting function (so called slope limiters). This prevents from inappropriate numerical error amplifications. The ENO (Essentially non-oscillatory) or WENO (Weighted ENO) constitute the most used schemes of such family[16,21].

To achieve an accurate and stable discretization of the advective term, one must use conservative numerical scheme. Such schemes exactly respect the conservation law within the cell by integrating the flux value at cell boundaries. Total Variation Diminishing (TVD) scheme (which are monotonicity preserving flux) prevents from an increase of oscillations over time and enables to transport shocks. All these methods are well detailed in [21].

**Reconstruction of the vorticity.** In our work, the reconstruction of the vorticity on the cell boundaries is realized through a second order accurate method [17] based on a *Min-Mod limiter* on the regular spatial grid $(i\Delta_x, j\Delta_y)$ :

$$\xi^+_{i+\frac{1}{2},j} = \xi_{i+1,j} - \frac{\Delta_x}{2}(\xi_x)_{i+1,j} \qquad \text{and} \qquad \xi^-_{i+\frac{1}{2},j} = \xi_{i,j} + \frac{\Delta_x}{2}(\xi_x)_{i,j},$$

$$\text{with } (\xi_x)_{i,j} = Minmod\left(2\frac{(\xi_{i,j} - \xi_{i-1,j})}{\Delta_x}, \frac{\xi_{i+1,j} - \xi_{i-1,j}}{2\Delta_x}, 2\frac{(\xi_{i+1,j} - \xi_{i,j})}{\Delta_x}\right),$$

$$\text{and } Minmod(x_1, \cdots, x_n) = \begin{cases} \inf_i(x_i) \text{ if } x_i \geq 0 \ \forall i \\ \sup_i(x_i) \text{ if } x_i \leq 0 \ \forall i \\ 0 \qquad \text{otherwise.} \end{cases}$$

The intermediate values $\xi^+_{i,j+\frac{1}{2}}$ and $\xi^-_{i,j+\frac{1}{2}}$ are computed in the same way. As the Mid-Mod limiter provides the smallest slope, the reconstructed values of the vorticity on the cell boundaries attenuate amplification effect due to spatial discontinuities.

**Vorticity-velocity scheme.** To deal with the advective term, we use the following semidiscrete central scheme [13,14]:

$$\partial_t \xi_{i,j} = -\frac{H^x_{i+\frac{1}{2},j}(t) - H^x_{i-\frac{1}{2},j}(t)}{\Delta_x} - \frac{H^y_{i,j+\frac{1}{2}}(t) - H^y_{i,j-\frac{1}{2}}(t)}{\Delta_y} + \nu_\xi D_{i,j}, \tag{21}$$

with a numerical convection flux derived from the monotone Lax-Friedricks flux:

$$
\begin{aligned}
H^x_{i+\frac{1}{2},j}(t) &= \frac{u_{i+\frac{1}{2},j}(t)}{2}\left[\xi^+_{i+\frac{1}{2},j}+\xi^-_{i+\frac{1}{2},j}\right]-\frac{|u_{i+\frac{1}{2},j}(t)|}{2}\left[\xi^+_{i+\frac{1}{2},j}-\xi^-_{i+\frac{1}{2},j}\right]\\
H^y_{i,j+\frac{1}{2}}(t) &= \frac{v_{i,j+\frac{1}{2}}(t)}{2}\left[\xi^+_{i,j+\frac{1}{2}}+\xi^-_{i,j+\frac{1}{2}}\right]-\frac{|v_{i,j+\frac{1}{2}}(t)|}{2}\left[\xi^+_{i,j+\frac{1}{2}}-\xi^-_{i,j+\frac{1}{2}}\right].
\end{aligned}
\tag{22}
$$

This resulting second order, semidiscrete central scheme is TVD [13,17] and not very dissipative. The intermediate values of the velocities are computed with a fourth-order averaging:

$$
\begin{aligned}
u_{i+\frac{1}{2},j}(t) &= \frac{-u_{i+2,j}(t)+9u_{i+1,j}(t)+9u_{i,j}(t)-u_{i-1,j}(t)}{16}\\
v_{i,j+\frac{1}{2}}(t) &= \frac{-v_{i,j+2}(t)+9v_{i,j+1}(t)+9v_{i,j}(t)-v_{i,j-1}(t)}{16}.
\end{aligned}
\tag{23}
$$

The linear viscosity $\Delta\xi$ is approximated by the fourth-order central differencing:

$$
\begin{aligned}
D_{i,j}(t) = &\frac{-\xi_{i+2,j}(t)+16\xi_{i+1,j}(t)-30\xi_{i,j}(t)+16\xi_{i-1,j}(t)-\xi_{i-2,j}(t)}{12\Delta_x^2}\\
+&\frac{-\xi_{i,j+2}(t)+16\xi_{i,j+1}(t)-30\xi_{i,j}(t)+16\xi_{i,j-1}(t)-\xi_{i,j-2}(t)}{12\Delta_y^2}.
\end{aligned}
\tag{24}
$$

The time integration is realized with a third-order Runge Kutta scheme, which also respect the TVD property [21]. The divergence is integrated with a stable implicit discretization. The motion field is updated at each time step in the Fourier domain with equation (17). With this whole non-oscillatory scheme, the vorticity-velocity and divergence equations can be integrated in the image area.

**Adjoint discretization.** Variational data assimilation assumes that the adjoint operator is the exact numerical adjoint of the direct operator [22]. Thus, the adjoint computation must be done according to the previously described vorticity simulation method. For large scale applications involving several coupled state variables of huge dimension and where for each of them a specific dynamical model is discretized accordingly, automatic differentiation programs [9] are used to compute the adjoint model. In our case, as only two variables are involved, it is possible to derive an explicit version of the discretized adjoint operator and a backward Runge-Kutta integration can be realized [10].

## 5 Results

In order to assess the benefits of our technique for the tracking of fluid motion, we first applied it on a synthetic sequence of particles images of a 2D divergence free turbulence obtained through a direct numerical simulation of the Navier-Stokes equation [2]. In this sequence composed of 52 images, we compare in figure 1 the vorticity map of the actual, the observed and the assimilated motion fields. It

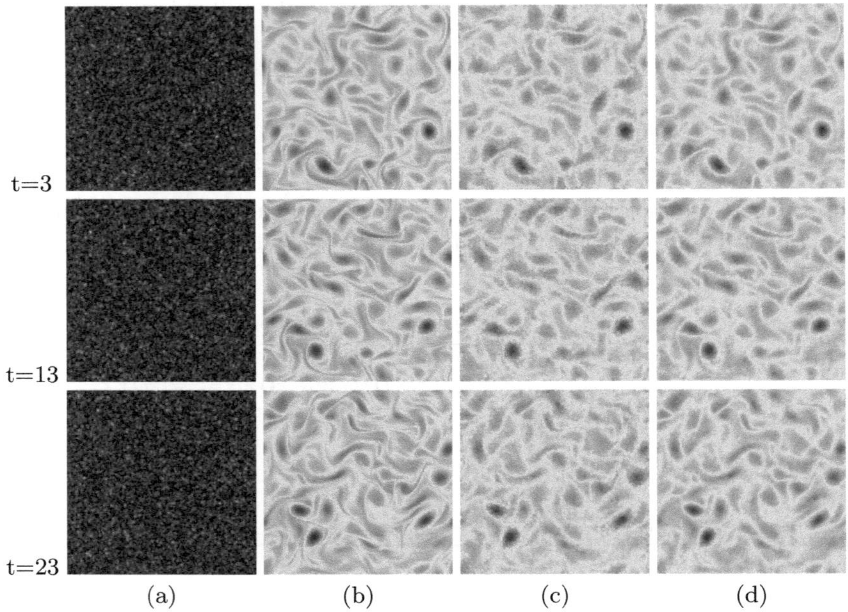


**Fig. 1. 2D Direct Numerical Simulation.** *a)* Particle images sequence. *b)* True vorticity. *c)* Vorticity observed by optic flow estimator. *d)* Recovered vorticity.

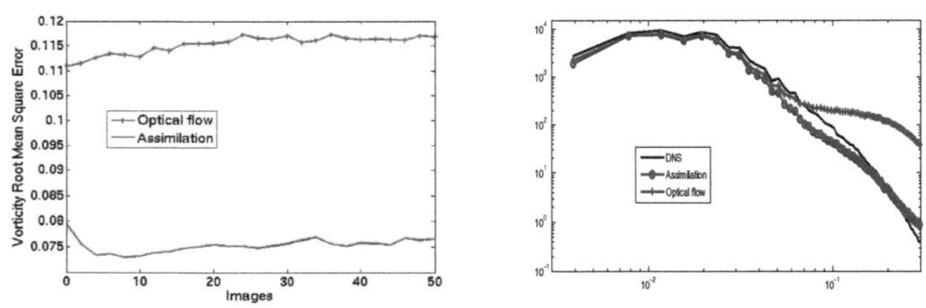


**Fig. 2. Comparison of errors.** On the left figure, the red curve outlines the mean square error of the vorticity computed by the optic flow technique on the 25 images of the particle sequence. The green curve exhibits the error obtained at the end of the assimilation process. The assimilated vorticity is then closer to the reality than the observed one. The actual mean absolute value of the vorticity is about 0.43. On the right, a spectral analysis of the energy of the row average vorticity is represented in the log-log scale. Contrarily to the observed vorticity, the assimilated vorticity recovers high frequencies.

can be observed that the proposed technique not only denoises the observations, but also enables to recover small scales structures that were smoothed out in the original velocity fields. These observations were obtained from a dedicated optic flow estimator [3]. To give some quantitative evaluation results, we present the comparative errors in figure 2. As can be seen, we have been able to significantly improve the quality of the recovered motion field (about 30%). A spectral analysis of the energy of the row average vorticity is also realized in order to show that our assimilation process recovers the high frequencies of the flow, which correspond to the small scales spatial structures.

Finally, we applied our technique on a Infra-red meteorological sequence showing Vince cyclone over north atlantic. The sequence is composed by 20 satellite images acquired the 9 october 2005 from 00:00 up to 5:00 am[2]. Complete results in term of curves and vorticity maps are presented on figure 3. Line (b) exhibits the different vorticity maps of the initial motion fields used as noisy measurements. These motion observations present temporal inconsistencies and discontinuities, whereas the recovered vorticity maps shown on line (c) are more compliant with the vorticity conservation law.

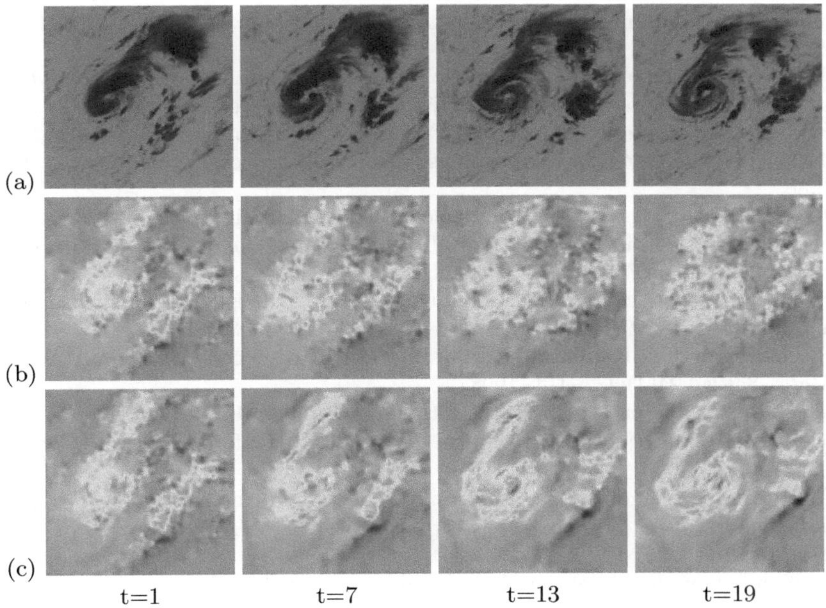


**Fig. 3. Cyclone sequence.** *(a)* Cyclone sequence. *(b)* Sample of observed vorticity maps. *(c)* Vorticity maps corresponding to the recovered motion fields.

## 6 Conclusion

In this work a variational framework for the tracking of fluid flows has been introduced. This approach relies on variational data assimilation principles. The proposed method allows to recover the state of an unknown function on the basis

[2] We thank the Laboratoire de Météorologie Dynamique for providing us this sequence.

of an imperfect dynamical models and noisy measurement of this function. This technique has been applied to the tracking of fluid motion from image sequences.

**Acknowledgments.** This work was supported by the European Community through the IST FET Open FLUID project (`http://fluid.irisa.fr`).

## References

1. A.F. Bennet. *Inverse Methods in Physical Oceanography.* Cambridge University Press, 1992.
2. J. Carlier and B. Wieneke. Report on production and diffusion of fluid mechanics images and data. Technical report, Fluid Project deliverable 1.2, 2005.
3. T. Corpetti, E. Mémin, and P. Pérez. Dense estimation of fluid flows. *IEEE Trans. Pattern Anal. Machine Intell.*, 24(3):365–380, 2002.
4. T. Corpetti, E. Mémin, and P. Pérez. Extraction of singular points from dense motion fields: an analytic approach. *J. Math. Imaging and Vision*, 19(3):175–198, 2003.
5. A. Cuzol and E. Mémin. A stochastic filter for fluid motion tracking. In *Int. Conf. on Computer Vision (ICCV'05)*, Beijing, China, October 2005.
6. F.-X. Le Dimet and O. Talagrand. Variational algorithms for analysis and assimilation of meteorological observations: theoretical aspects. *Tellus*, pages 97–110, 1986.
7. J.M. Fitzpatrick. A method for calculating velocity in time dependent images based on the continuity equation. In *Proc. Conf. Comp. Vision Pattern Rec.*, pages 78–81, San Francisco, USA, 1985.
8. R.M. Ford, R. Strickland, and B. Thomas. Image models for 2-d flow visualization and compression. *Graph. Mod. Image Proc.*, 56(1):75–93, 1994.
9. R. Giering and T. Kaminski. Recipes for adjoint code construction. *ACM Trans. Math. Softw.*, 24(4):437–474, 1998.
10. M. Giles. On the use of runge-kutta time-marching and multigrid for the solution of steady adjoint equations. Technical Report 00/10, Oxford University Computing Laboratory, 2000.
11. A. Harten, B. Engquist, S. Osher, and S. R. Chakravarthy. Uniformly high order accurate essentially non-oscillatory schemes, 111. *J. of Comput. Phys.*, 71(2): 231–303, 1987.
12. T. Kohlberger, E. Mémin, and C. Schnörr. Variational dense motion estimation using the helmholtz decomposition. In *Int. conf on Scale-Space theories in Computer Vision (Scale-Space'03)*, volume 2695, pages 432–448, Isle of Skye, UK, june 2003.
13. A. Kurganov and D. Levy. A third-order semidiscrete central scheme for conservation laws and convection-diffusion equations. *SIAM J. Sci. Comput.*, 22(4):1461–1488, 2000.
14. A. Kurganov and E. Tadmor. New high-resolution central schemes for nonlinear conservation laws and convetion-diffusion equations. *J. Comput. Phys.*, 160(1): 241–282, 2000.
15. R. Larsen, K. Conradsen, and B.K. Ersboll. Estimation of dense image flow fields in fluids. *IEEE trans. on Geosc. and Remote sensing*, 36(1):256–264, Jan. 1998.
16. D. Levy, G. Puppo, and G. Russo. A third order central weno scheme for 2d conservation laws. *Appl. Num. Math.: Trans. of IMACS*, 33(1–4):415–421, 2000.

17. D. Levy and E. Tadmor. Non-oscillatory central schemes for the incompressible 2-d euler equations. *Math. Res. Let*, 4:321–340, 1997.
18. E. Mémin and P. Pérez. Fluid motion recovery by coupling dense and parametric motion fields. In *Int. Conf. on Computer, ICCV'99*, pages 620–625, 1999.
19. H. Nessyahu and E. Tadmor. Non-oscillatory central differencing for hyperbolic conservation laws. *J. of Comput. Phys.*, 87(2):408–463, 1990.
20. B. Oksendal. *Stochastic differential equations.* Spinger-Verlag, 1998.
21. C.-W. Shu. Essentially non-oscillatory and weighted essentially non-oscillatory schemes for hyperbolic conservation laws. *Advanced Numerical Approximation of Nonlinear Hyperbolic Equations, Lecture Notes in Mathematics*, 1697:325–432, 1998.
22. O. Talagrand. *Variational assimilation. Adjoint equations.* Kluwer Academic Publishers, 2002.
23. O. Talagrand and P. Courtier. Variational assimilation of meteorological observations with the adjoint vorticity equation. I: Theory. *J. of Roy. Meteo. soc.*, 113:1311–1328, 1987.
24. J. Yuan, P. Ruhnau, E. Mémin, and C. Schnörr. Discrete orthogonal decomposition and variational fluid flow estimation. In *5th Int. Conf. on Scale-Space and PDE methods in Computer Vision (Scale-Space'05), Hofgeismar, Germany*, April 2005.
25. J. Yuan, C. Schnörr, and E. Mémin. Discrete orthogonal decomposition and variational fluid flow estimation. *Accepted for publication in J. of Math. Imaging and Vision*, 2006.
26. L. Zhou, C. Kambhamettu, and D. Goldgof. Fluid structure and motion analysis from multi-spectrum 2D cloud images sequences. In *Proc. Conf. Comp. Vision Pattern Rec.*, volume 2, pages 744–751, Hilton Head Island, USA, 2000.

# Super-Resolution Using Sub-band Constrained Total Variation

Priyam Chatterjee[1], Vinay P. Namboodiri[2], and Subhasis Chaudhuri[2]

[1] University of California, Santa Cruz,
California 95064, USA
priyam@soe.ucsc.edu
[2] Indian Institute of Technology, Bombay,
Powai, Mumbai 400 076, India
{vinaypn,sc}@ee.iitb.ac.in

**Abstract.** Super-resolution of a single image is a severely ill-posed problem in computer vision. It is possible to consider solving this problem by considering a total variation based regularization framework. The choice of total variation based regularization helps in formulating an edge preserving scheme for super-resolution. However, this scheme tends to result in a piece-wise constant resultant image. To address this issue, we extend the formulation by incorporating an appropriate sub-band constraint which ensures the preservation of textural details in trade off with noise present in the observation. The proposed framework is extensively evaluated and the experimental results for the same are presented.

## 1 Introduction

Super-resolution is the process of increasing the spatial details in an image by computational means. In certain applications it is often not possible to obtain an image with a high level of detail. In such cases super-resolution methods become extremely necessary to provide a better observation from one or more degraded available images of the scene. Image super-resolution finds a variety of applications in video and image quality improvement (HDTV conversion), health diagnosis (from X-ray or sonographic images) and as a preprocessing step for any application where a better quality input picture is a requirement.

The problem of super-resolution can formally be stated as follows. There are $p$ observed images $y_m$ $(m = 1...p)$, each of size $M_1 \times M_2$ which are the shifted, decimated, blurred and noisy versions of a single high resolution image $z$ of size $N_1 \times N_2$ where $N_1 = qM_1$ and $N_2 = qM_2$. If $y_m$ is the $M_1M_2 \times 1$ lexicographically ordered vector containing pixels from the low resolution image then a vector $z$ of size $q^2M_1M_2 \times 1$ containing pixels of the high resolution image can be formed by placing each of the $q \times q$ pixel neighborhoods sequentially so as to maintain the relationship between a low resolution pixel and its corresponding high resolution pixel. After incorporating the blur matrix and the noise vector, the image formation model is written as

$$Y_m = D(H_m * z) + n_m, m = 1, ..., p \tag{1}$$

F. Sgallari, A. Murli, and N. Paragios (Eds.): SSVM 2007, LNCS 4485, pp. 616–627, 2007.

where $D$ is the decimation matrix of size $M_1M_2 \times q^2M_1M2$, $H$ is the blurring point spread function (PSF) and $n_m$ is the $M_1M_2 \times 1$ noise vector and $p$ is the number of low resolution observations. Here we assume the blur kernel to be shift invariant and $*$ denotes the convolution operation.

The various approaches towards solving this problem can be broadly classified as being based on super-resolution from multiple images, from a single image based on learning low-level features and super-resolution of a single image based on interpolation. In this paper, we propose a technique for interpolation based super-resolution of a single image using total variation regularization with a stronger data term which ensures that the resulting image has better texture and edge preserving properties. This is achieved using spectral decomposition followed by sub-band correlation between the reconstructed and the given LR image.

In the next section we discuss the related work. In Sec. 3 we present the proposed technique. In Sec. 4 we discuss the implementation details and the technique is validated in Sec. 5 by comparison with various other techniques. We finally conclude in Sec. 6.

## 2 Related Work

A vast collection of literature is available where researchers aim to perform resolution enhancement using a wide variety of techniques. Here we discuss the different approaches.

### 2.1 Super-Resolution from Multiple Images

In cases where multiple degraded observations are available the low resolution (LR) images must first be registered to determine the inter-pixel shifts before placing them onto a higher dimensional grid. A survey of different registration methods is provided in [1]. The reconstruction of the image in the higher dimensional grid using maximum likelihood estimation (MLE), Maximum a-priori (MAP) estimation with priors like Gaussian Markov random field (MRF) and Huber MRFs has been demonstrated by Capel *et al.* in [2]. Using a MAP estimator and blur as a cue Rajan *et al.* perform super-resolution in [3],[4]. In [5], [6] Joshi *et al.* use zoom as a cue. They consider the linear dependency of pixels in a neighborhood and model it as a simultaneous auto-regressive process which is then used as a prior term in a regularization framework. Regularization based techniques have also been used by Chan *et al.* in [7]. A total variation (TV) based regularization method is used to perform simultaneous inpainting and deblurring to obtain a super-resolved image from multiple LR video frames. A combination of TV and bilateral filter named Bilateral TV (BTV) has been used as a regularizing term for super-resolution by Farsiu *et al.* in [8], [9]. As opposed to the $L_2$ norm used as a data fidelity term in most cases, the authors use the $L_1$ norm. They present a two step algorithm where they first use the median filter to build a high resolution grid from multiple LR images. Regularization is then done to perform an iterative interpolation to deblur as well as inpaint missing pixels in the HR grid. In [8] the authors make use of the temporal information between frames of a LR video. SR is then performed under a control theoretic approach using an approximation of the Kalman Filter along with the previous framework.

### 2.2 Learning Based Super-Resolution

Considerable amount of work has also been done in the domain of super-resolution from a single image by making use of a image database where LR-HR image pairs are provided. One of the foremost papers in this approach is the work by Freeman *et al.* [10]. Here the authors' model have a generative model for scenes and their rendered image, with a Markovian relationship between them. Bayesian belief propagation is used to estimate the posterior probability of the scene given an image. The priors are learnt through a database of LR-HR images. A similar approach has been proposed by Baker and Kanade [11], where low level features are recognized and the corresponding high resolution features are "hallucinated". In [12], the authors suggest a method to obtain an approximate fast one pass solution to the Markov network where again the low-level features are learnt using patches from LR-HR image database.

### 2.3 Interpolation Based Super-Resolution

Some researchers have also applied different methods to address the issue of zooming into an image when only a single low resolution version of the scene is available. The main challenge here is to preserve edges that are present in any natural image. A variety of linear and non-linear tools are available which try to address this issue. A detailed mathematical analysis of regularization based schemes has been provided by Malgouyres and Guichard in [13]. The authors in [14] provide an interpolation method under the total variation regularization scheme. In their paper they start off with a higher resolution image formed by zero-padded interpolation of the LR image. A constrained gradient descent algorithm is presented where the authors minimize the gradient energy of the image which conforms to a linear smoothing and sampling process. The use of TV for super-resolution has also been demonstrated by Aly and Dubois in [15]. In their method they modify the data fidelity term to closely model the assumed image acquisition model. Their iterative algorithm then makes use of the back projection technique introduced by Irani and Peleg [16] for data fidelity in a regularization framework. The resultant image depends upon the choice of the image formation model and the dependence on the mathematical model that captures the downsampling process for such regularization based methods has been discussed in [17]. Jiji *et al.* in [18] propose an interpolation technique where the aliasing present in the LR image is used. They assume knowledge of the bandwidth and the amount of aliasing in a given observation and use a signal processing approach to perform super-resolution. In this paper we restrict ourselves to the case where super-resolution is performed from a single observation without the use of any such database. Our work falls in the general category of hybrid image restroration models proposed by Coifman *et al.* [19] and Malgouyres [20].

## 3 Super-Resolution Using TV Approach

The image formation model for the low resolution image from a high resolution image is given as

$$y(x) = d(x)(h * z) + n(x) \tag{2}$$

Here $d(x)$ is the decimation matrix, $h$ is the blur point spread function, $z$ is the high resolution image and $n(x)$ is the noise function. Given an approximation $u$ to the high resolution image $z$, and the image $u_0$ which is the upsampled version of the observed low resolution image, the residual error is given as

$$r(x) = u_0(x) - (h * u). \tag{3}$$

Based on the error function an objective function can be formulated, the minimization of which gives the high resolution image. The objective function is given as

$$E(u) = \int ((r(x))^2 + \alpha|\nabla u|) \tag{4}$$

The solution for super-resolution from a single image can be given in terms of the following objective function. Here the first term is the data term and the second term is the $L_1$ (TV) regularization term. The choice of TV norm has found favor in the image restoration community because it allows discontinuities in its solution. As opposed to the $L_2$ norm it does not smoothen the image across edges. Our motivation for the use of TV based regularization stems from its edge preserving property which is vital for super-resolution. However, the current formulation of data term and regularization term results in a solution that preserves strong edges but the finer details of texture are lost in the solution of the above objective function. This can be easily understood by considering the following argument. If there exists a weak edge (the magnitude of gradient is small), then the regularization constraint gives it a low weight. The data fidelity constraint, due to the averaging nature of the blurring kernel, would also not enforce the preservation of the edge. In the iterative energy minimization approach these finer details are therefore lost. This can be expected of the cost function because it tries to minimize the variations without paying much regard to the fact whether the variations are due to thin edges, texture or noise in smooth regions. This motivates us to search for an additional term which will penalize variations at the finer edges and texture regions more than those over smoother regions. In other words we are more interested in retaining the texture and finer details in an image. For this we need to separate the finer edges and textural details in the image and use data fidelity terms that are able to retain the desired details. This we formulate as additional correlation constraints over various sub-bands of the spectrum of an image.

The objective function we use is then

$$J(u) = \int_{\Omega} |\nabla u| \, dxdy + \frac{1}{2}\alpha \int_{\Omega} (u * h - u_0)^2 dxdy + \frac{1}{2} \sum_k \lambda_k \int_{\Omega_k} (\tilde{U}_k - U_{0k})^2 \, d\nu d\omega \tag{5}$$

where $u$ denotes the HR restored image, $h$ is the blurring kernel, $u_0$ is the interpolated version of the input LR image, $\tilde{U}_k$ denotes the $k^{th}$ spectral sub-band of the estimated LR image formed under the known decimation model, $U_{0k}$ is the $k^{th}$ sub-band of the input LR image and $\lambda_k$ is the corresponding weighing term for the regularizer. This simple band splitting framework allows us to differentiate the smoother regions of the image from textures and sharp edges and allows us to weigh the data fidelity between the different spectral bands separately.

## 4 Implementation Details

The objective function given in Eqn. (5) is minimized by an iterative gradient descent technique as done commonly in the literature [21]. The corresponding Euler Lagrange equation for the objective function is given by

$$\begin{aligned} \frac{\partial}{\partial x}\left(\frac{u_x}{\sqrt{{u_x}^2 + {u_y}^2}}\right) + \frac{\partial}{\partial y}\left(\frac{u_y}{\sqrt{{u_x}^2 + {u_y}^2}}\right) - \alpha(u * h - u_0) \\ -D^{-1}\mathscr{F}^{-1} \sum_k \lambda_k (\tilde{U}_k - U_{0k}) = 0, \ x, y \in \Omega \ . \end{aligned} \tag{6}$$

$$\frac{\partial u}{\partial n} = 0 \text{ on the boundary of } \ \Omega = \partial \Omega \tag{7}$$

The resulting iterative updation process is then given by

$$u^{(n+1)} = u^{(n)} + \Delta t \left( \nabla . \frac{\nabla u}{|\nabla u|} + \alpha(u_0 - u * h) + D^{-1}\mathscr{F}^{-1} \sum_k \lambda_k (U_{0k} - \tilde{U}_k) \right)^{(n)} \tag{8}$$

where $D^{-1}$ is the upsampling process and $\mathscr{F}^{-1}$ implies the inverse Fourier transform. Under this framework, we assign different weights ($\lambda_k$) to existing errors terms in different spectral bands to weigh different bands separately. As opposed to other schemes that we have discussed before, the additional constraint term is calculated and weighed in the spectral domain. The inverse Fourier transform is then applied and finally it is scaled to match the high resolution image dimensions. It may be argued here that it follows from Parseval's theorem that calculating error power in the spatial domain and the frequency domain should be equivalent. However, the operation in the spectral domain makes it easy to split an image into separable components based on spectral contribution. The number of spectral bands $k$ does affect the quality of super-resolution. In general, the higher the number of spectral bands, more the flexibility for preserving details. But this also increases the number of tunable parameters ($\lambda_k$) linearly. Also, if the image does not have enough energy in each decomposed band, the weighing parameter of that band does not have much effect on the restored image. We are thus faced with a tradeoff between the number of bands and computational efficiency and free parameters. We have experimentally tried the method with 2 and 4 spectral bands. Under the absence of noise, we use a higher weight factor ($\lambda_k$) for the higher spectral bands which capture the finer details and the edges of the image. Using such a model it is then possible for us to enforce that more importance is given to data fidelity at the edges. This should ensure that the image that is formed is a sharper super-resolved image of the input LR observation under the known image decimation model. On the other hand, noise in an image can be expected to be captured in the higher frequency sub-bands, which necessitates the use of smaller weights for higher sub-bands when the input image is noisy. An appropriate choice of $\lambda_k$s would ensure a proper trade-off between the sharpness of the super-resolved image and the accentuation of the noise present. Due to different considerations under different conditions, the choice of $\lambda$ values unfortunately have to be manually determined. We have experimented with both noisy and noiseless cases and the results are discussed in Sec. 5.

## 5 Results

For our experiment we take the initial starting image as the bicubic interpolation of the input LR image. The images at every iteration of the restoration process are decomposed into two bands and based on the theory presented above we apply a higher weight to the higher frequency components of the image. For the results shown here using a decomposition into only 2 bands we use the values $\alpha = 0.7$, $\lambda_1 = 0.6$ and $\lambda_2 = 0.8$ where a higher index of $\lambda$ value implies a higher frequency band. In this experiment the higher 40% of the spectrum was assumed to capture most of the edge information of the image. We also demonstrate results when a 4 band decomposition is done. In this case the frequency spectrum is equally divided into 4 bands. The parameter values in this case are $\alpha = 0.8$, $\lambda_1 = 0.4$, $\lambda_2 = 0.6$, $\lambda_3 = 1$ and $\lambda_4 = 1.2$. The results obtained using these parameter values are shown in Fig. 1 and Fig. 2.

In Fig. 1 we can see that the total variational deblurring performed on the bicubic reconstruction sharpens the image at edges but at the cost of loss of the texture. This is not the case for Fig. 1(d) & (e). This can be noted from the presence of texture in the hat and the hair, even though the overall reconstruction remains sharp. This is specifically what we wanted to achieve by our method. A similar effect can be seen in Fig. 2 where the result from our method yields a better texture than that of TV based deblurring. This is visible at the terrain and finer details on the tank. This proves that band splitting and differential weighing of the bands indeed perform better as far as restoration of texture is concerned.

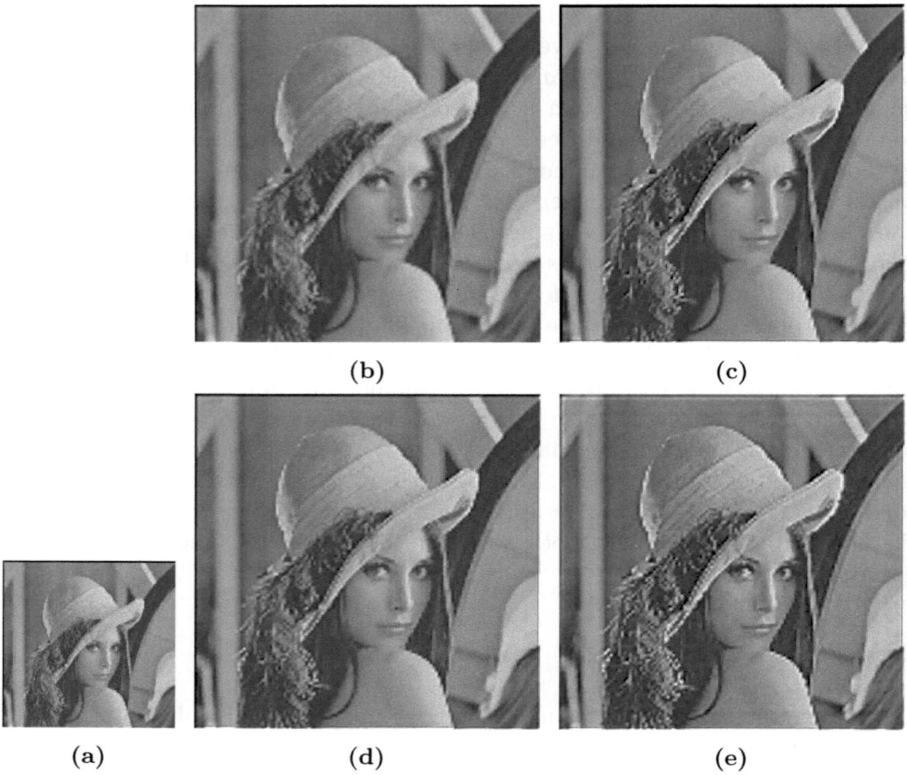

**Fig. 1.** SR using proposed TV based approach for 2× zoom : (a) Input LR image, (b) bicubic interpolated image used as the initial estimate, (c) TV based deblurring of (b), (d) SR using modified TV based approach using only 2 bands, (e) reconstruction using 4 bands

We also compare our results with the alias-free interpolation method [18] proposed by Jiji *et al.* and the results are presented in Fig. 3. The alias-free interpolation method performs super-resolution by introducing high frequency components. Since, this method does not assume any priors or additional data set, the assumptions are comparable to our method. Here, it is assumed that the aliasing is present in the top ten percent of the spectrum and it performs reconstruction based on samples from other parts of the spectrum. The input image is a $64 \times 64$ LR image and we perform 2× zoom using both the methods. The parameters used for TV based method are same as those used in the previous experiment on a $128 \times 128$ LR image. We show the results for the proposed method using two bands and four bands respectively. The results shown in Fig. 3 (c) and (d), show that the proposed method adds more coherent high frequency components and the resultant images are sharper as compared to the alias free

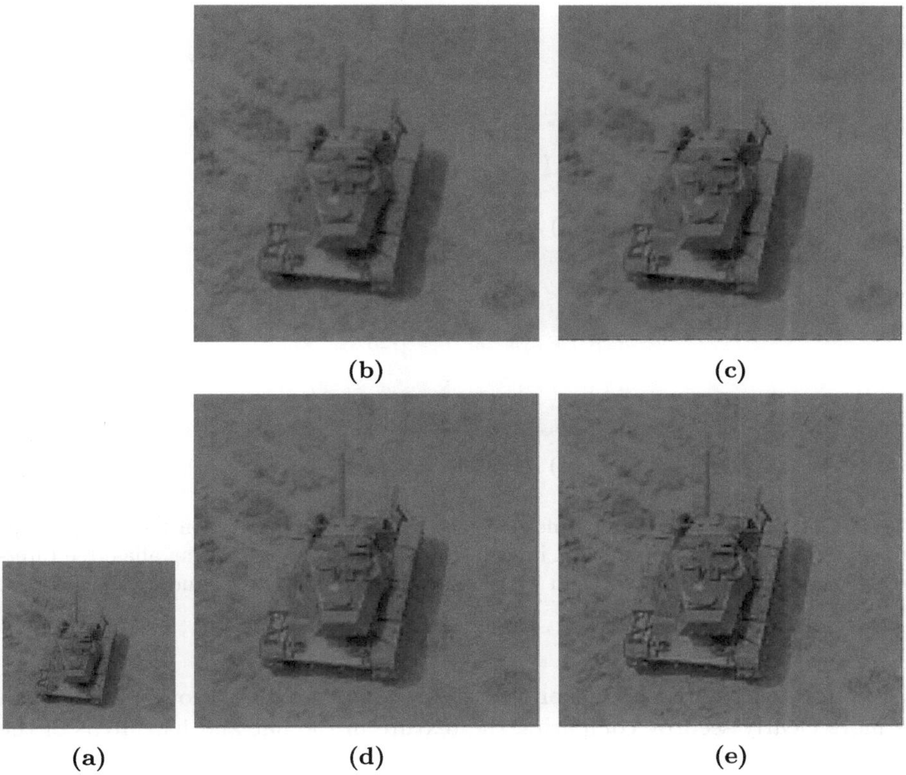


**Fig. 2.** SR using proposed TV based approach for 2× zoom : (a) Input LR image, (b) bicubic interpolated image used as the initial estimate, (c) TV based deblurring of (b), (d) SR using modified TV based approach using only 2 bands, (e) reconstruction using 4 bands

interpolation method. The result with four bands are better as compared to the result with two frequency bands, indicating the effectiveness of multiple bands.

We next compare our results with image super-resolution methods described in [22] and [23]. In [22], the authors suggest a multi-image super-resolution method. However, the method based on delaunay triangulation based approximation can also be used for interpolation with a single image. In [23], the authors consider the use of kernel regression for image upscaling. We compare our proposed method with these methods and the results are shown in Fig. 4. Fig. 4(a) shows the bicubic interpolated Lena image for 3× zoom factor which is used as the initial condition in our algorithm. Fig. 4(b) shows the result from the delaunay triangulation based method [22] and Fig. 4(c) shows the result from the kernel regression based method [23]. Fig. 4(d) shows the result from the proposed approach. It can be seen that the results from the proposed approach

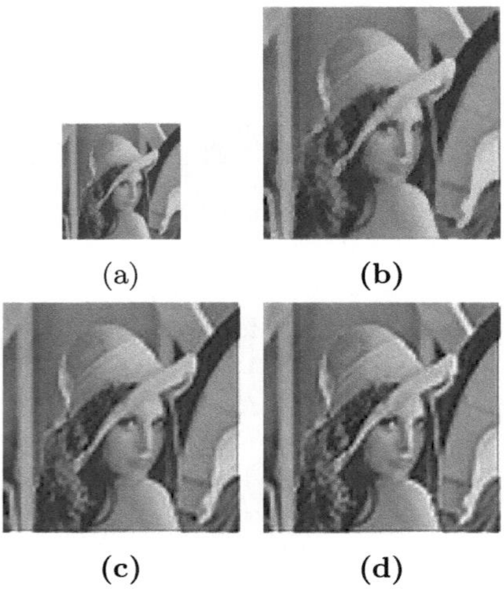

**Fig. 3.** Comparison of proposed method with alias-free interpolation method [18] approach for 2× zoom : (a) input LR image, (b) image interpolated using alias-free Interpolation method [18], (c) SR from proposed approach using a 2 band decomposition, (e) reconstruction using 4 bands

has better texture preserving properties as compared to the others. This can be more clearly seen by comparing the texture in the hat and hair areas of the image.

We also attempt to try our method where the image is corrupted by zero mean additive noise. We use a Gaussian noise of variance 25 ($[0, 255]$ being the range of pixel values in the image). In our reconstruction process we do not make use of any information about the nature of the noise. Our theory builds on the assumption that this noise remains limited to higher frequency bands of the image. Hence we use a lower scaling factor for the higher frequency band. The results obtained using a 2 band decomposition with parameter values of $\alpha = 0.7$, $\lambda_1 = 0.8$ and $\lambda_2 = 0.6$ are shown in Fig. 5(d). For the case where we make use of a 4 band decomposition the parameters used for denoising are $\alpha = 0.6$, $\lambda_1 = 1$, $\lambda_2 = 0.8$, $\lambda_3 = 0.6$ and $\lambda_4 = 0.4$. The result obtained using this configuration is shown in Fig. 5(e).

It can be seen from Fig. 5(b) that the bicubic interpolation does not perform any denoising, as is expected. Total variation based regularization performed on the bicubic interpolated image, in Fig. 5(c), reduces a lot of noise but the resultant image is smooth and lacks texture. The modified approach yields a sharper reconstruction but the presence of noise is clearly visible. Apart from an enhancement of details, an improvement in the PSNR is also observed for the proposed method. More denoising will lead to a smoother reconstruction, as can be expected from the method. The reason for this is that spectral bands which

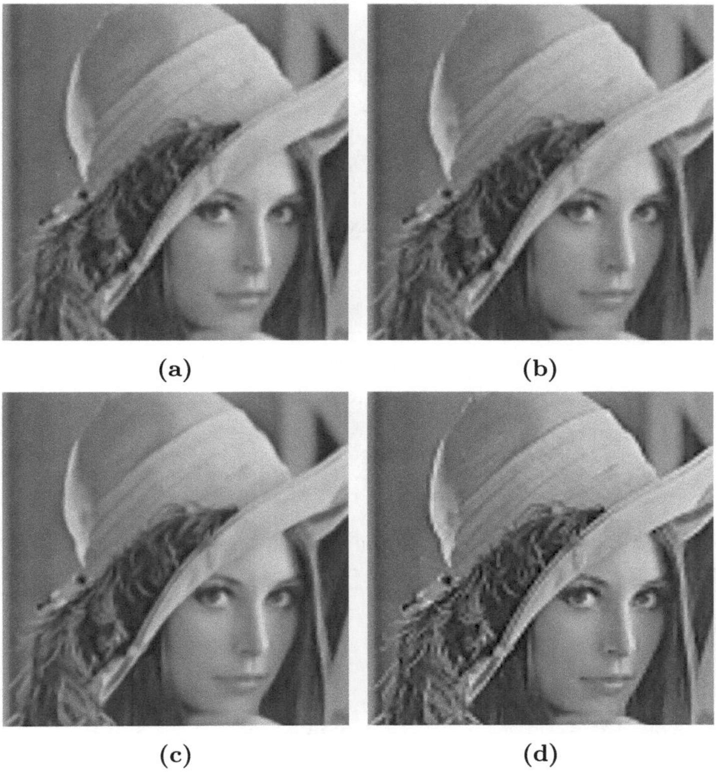

**Fig. 4.** SR using proposed TV based approach for 3× zoom : (a) bicubic interpolated image used as the initial estimate, (b) image interpolated using delaunay triangulation [22], (c) image interpolated using kernel regression method [23] (d) reconstruction using proposed method with 4 bands

capture sharp edges and texture will also contain most of the noise and hence it becomes difficult to distinguish between noise and texture using the present method.

## 6 Conclusion

The total variation based regularization scheme has been widely used by researchers for image restoration. It has been known to be a very good tool for denoising and deblurring. In our work we make use of the advantages of this method and apply it to perform super-resolution. We show that though in itself the method results in a good interpolated image, it fails to reconstruct texture and other finer details in an image. This is obtained through decomposition of the image into multiple spectral bands and enforcing differential data fidelity in each band. As shown and discussed before, in the presence of noise this method

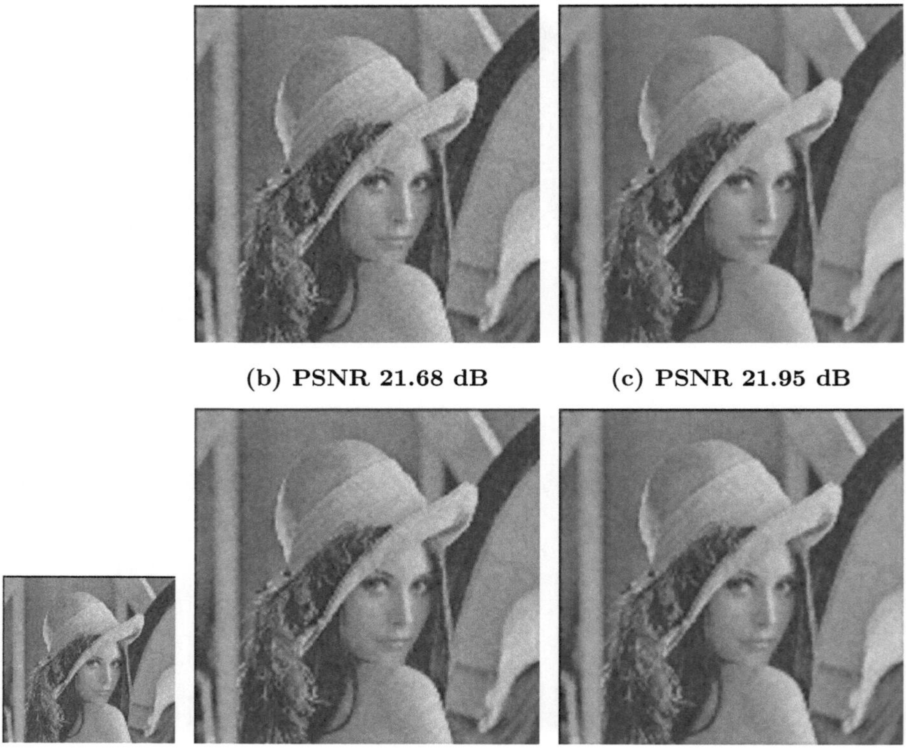

(b) **PSNR 21.68 dB** (c) **PSNR 21.95 dB**

(a) (d) **PSNR 22.36 dB** (e) **PSNR 22.13 dB**

**Fig. 5.** SR using proposed TV based approach for 2× zoom for a noisy case: (a) input LR image, (b) bicubic interpolated image, (c) TV based denoising of (b), (d) SR using modified TV based approach using only 2 bands, (e) reconstruction using 4 bands

provides a trade-off between the desired sharpness of the edges and the amount of denoising achieved. This is due to the inherent limitation of the TV denoising process which cannot tell finer texture apart from noise. Further investigation is thus necessary to make the method more robust in the presence of noise.

## References

1. Zitova, B., Flusser, J.: Image registration methods : A survey. Image and Vision Computing **21** (2003) 977–1000
2. Capel, D., Zisserman, A.: Computer vision applied to super resolution. IEEE Signal Processing Magazine **20** (2003) 75–86
3. Rajan, D., Chaudhuri, S., Joshi, M.V.: Multi-objective super resolution: Concepts and examples. IEEE Signal Processing Magazine **20** (2003) 49–61

4. Rajan, D., Chaudhuri, S.: Simultaneous estimation of super-resolved scene and depth map from low resolution defocused observations. IEEE Transactions on Pattern Analysis and Machine Intelligence **25** (2003) 1102–1117
5. Joshi, M., Chaudhuri, S., Rajkiran, P.: Super-resolution imaging: Use of zoom as a cue. Image and Vision Computing **22** (2004) 1185–1196
6. Joshi, M., Chaudhuri, S., Rajkiran, P.: A learning based method for image super-resolution from zoomed observations. IEEE Trans. Systems, Man & Cybernetics, Part-B **35** (2005) 527–537
7. Chan, T.F., Ng, M.K., Yau, A.C., Yip, A.M.: Superresolution image reconstruction using fast inpainting algorithms. Technical report, University of California, Los Angeles (2006)
8. Farsiu, S., Elad, M., Milanfar, P.: A practical approach to super-resolution. In: Proc. of the SPIE: Visual Communications and Image Processing. Volume 6077., San Jose, USA (2006) 24–38
9. Farsiu, S., Robinson, D., Elad, M., Milanfar, P.: Advances and challenges in super-resolution. International Journal of Imaging Systems and Technology **14** (2004) 47–57
10. Freeman, W.T., Pasztor, E.C., Carmichael, O.T.: Learning low-level vision. International Journal of Computer Vision **40** (2000) 25–47
11. Baker, S., Kanade, T.: Limits on super-resolution and how to break them. IEEE Transactions on Pattern Analysis and Machine Intelligence **24** (2002) 1167–1183
12. Freeman, W., Jones, T., Pasztor, E.: Example-based super-resolution. IEEE Transactions on Computer Graphics and Applications **22** (2002) 56–65
13. Malgouyres, F., Guichard, F.: Edge direction preserving image zooming: A mathematical and numerical analysis. SIAM Journal on Numerical Analysis **39** (2002) 1–37
14. Guichard, F., Malgouyres, F.: Total variation based interpolation. In: Proceedings of the European Signal Processing Conference. Volume 3., Amsterdam, The Netherlands, The Netherlands, Elsevier North-Holland, Inc. (1998) 1741–1744
15. Aly, H., Dubois, E.: Image up-sampling using total-variation regularization. IEEE Transaction on Image Processing **14** (2005) 1646–1659
16. Irani, M., Peleg, S.: Improving resolution by image registration. CVGIP: Graphical Model and Image Processing **53** (1991) 231–239
17. Aly, H., Dubois, E.: Specification of the observation model for regularized image up-sampling. IEEE Transaction on Image Processing **14** (2005) 567–576
18. Jiji, C., Neethu, P., Chaudhuri, S.: Alias-free interpolation. In: Proceedings of 9th ECCV - Part IV. Lecture Notes in Computer Science, Graz, Austria, Springer (2006) 255–266
19. Coifman, R.R., Sowa, A.: Combining the calculus of variations and wavelets for image enhancement. Applied and Computational Harmonic Analysis **9** (2000) 1–18
20. Malgouyres, F.: Minimizing the total variation under a general convex constraint for image restoration. IEEE Trans. on Image Processing **11** (2002) 1450–1456
21. Rudin, L.I., Osher, S., Fatemi, E.: Nonlinear Total Variation Based Noise Removal Algorithms. Physica D **60** (1992) 259–268
22. Lertrattanapanich, S., Bose, N.K.: High resolution image formation from low resolution frames using delaunay triangulation. IEEE Transactions on Image Processing **11** (2002) 1427–1441
23. Takeda, H., Farsiu, S., Milanfar, P.: Kernel regression for image processing and reconstruction. IEEE Transactions on Image Processing **16** (2007) 349–366

# Non-negative Sparse Modeling of Textures

Gabriel Peyré

Ceremade, Université Paris Dauphine,
Place du Marchal De Lattre De Tassigny,
75775 Paris Cedex 16 France
gabriel.peyre@ceremade.dauphine.fr
http://www.ceremade.dauphine.fr/~peyre/

**Abstract.** This paper presents a statistical model for textures that uses a non-negative decomposition on a set of local atoms learned from an exemplar. This model is described by the variances and kurtosis of the marginals of the decomposition of patches in the learned dictionary. A fast sampling algorithm allows to draw a typical image from this model. The resulting texture synthesis captures the geometric features of the original exemplar. To speed up synthesis and generate structures of various sizes, a multi-scale process is used. Applications to texture synthesis, image inpainting and texture segmentation are presented.

## 1 Statistical Models for Texture Synthesis

The characterization of textures is a central topic in computer vision and graphics, mainly approached from a probabilistic point of view.

*Spatial domain modeling.* The works of both Efros and Leung [1] and Wei and Levoy [2] pioneered a whole area of greedy approaches to texture synthesis. These methods copy pixels one by one, enforcing locally the consistence of the synthesized image with the exemplar. Recent approaches such as the method of Lefebvre and Hoppe [3] are fast, multiscale and give impressive results.

*Transformed domain modeling.* Julesz [4] stated simple axioms about the probabilistic characterization of textures. A texture is described as a realization of a random process characterized by the marginals of responses to a set of linear filters. Zhu, Wu and Mumford [5] setup a Gibbs energy to learn both the filters and the marginals. They use a Gibbs sampler to draw textures from this model.

A fast synthesis can be obtained by fixing the analyzing filters to be steerable wavelets as done by Heeger and Bergen [6]. The resulting textures are similar to those obtained by Perlin [7]. They exhibit isotropic cloud-like structures and fail to reproduce long range anisotropic features. This is because wavelets decompositions represent sparsely point wise singularities but do not compress enough long edge features. Higher order statistics such as local correlations are used by Portilla and Simoncelli [8] to synthesize high quality textures.

F. Sgallari, A. Murli, and N. Paragios (Eds.): SSVM 2007, LNCS 4485, pp. 628–639, 2007.

*Sparse image decompositions.* Representing a complex image with few meaningful elements is at the core of the visual processing made by the human cortex. Atteneave [9] and Barlow [10] first stated that efficient high level computations should be performed over a representation of reduced complexity.

This biological processing suggests a sparse description of a patches $y \in \mathbb{R}^N$ of $N$ pixels extracted from a natural image as

$$y[n] = \sum_{k=0}^{p-1} x^k \, d_k[n] \qquad \text{where} \qquad \|x\|_{\ell^0} \overset{\text{def.}}{=} \#\left\{k \setminus x^k \neq 0\right\} \leqslant \tau \ll N, \tag{1}$$

where $x = [x^0, \ldots, x^{p-1}]$ are the coefficients of the decomposition, and where typically $p \geqslant N$. This simple linear model uses as prior a dictionary $D = [d_0, \ldots, d_{p-1}] \in \mathbb{R}^{N \times p}$ of atoms.

This generative process is appealing as it involves $p$ degrees of freedom to perform statistical modeling while generating a large amount of different images. Classical works in computational harmonic analysis describe the set of candidate images $y$ as belonging to some functional space and study the sparsity of the decomposition of $y$ in some fixed basis or dictionary. Fourier decomposition is only suitable for smooth images and a wavelet decomposition [11] allows to model sparsely data with pointwise singularities. Images with geometrical singularities such as edges should be analyzed with more involved constructions such as the frame of curvelets [12] or a bandelet best basis [13].

These tools however are not efficient to capture the complex geometry of textures, which might include turbulent structures or complex overlapping junctions. To sparsely represent these structures, we use a exemplar-based approach where the dictionary $D$ is learned from a single input texture.

*Learning the dictionary.* Given a set of $m$ typical patches $Y = [y_0, \ldots, y_{m-1}] \in \mathbb{R}^{N \times m}$, one needs to compute both the dictionary $D$ and the coefficients $X = [x_0, \ldots, x_{m-1}]$ of the decomposition $Y \approx DX$ that leads to the sparsity $\|x_j\|_{\ell^0} \leqslant \tau$ of each column $x_j$ of X required by equation (1).

Olshausen and Field [14] first proposed a learning scheme to build a dictionary of atoms to represent each signal of a data set using few elements. For natural images, edge filters emerge to efficiently represent the geometry of images. Algorithms have been proposed in signal processing such as the K-SVD of Aharon et al [15] and the tight frames construction of Tropp et al. [16].

ICA and sparse dictionaries have been applied in texture modeling mainly for features extraction in classification [17,18]. An ICA decomposition is used as a post-processing step by Manduchi and Portilla [19] to enhance the synthesis results of Heeger and Bergen multiscale approach [6].

An alternative way to perform sparse coding is to enforce positivity of both the coefficients $X$ and the dictionary $D$. Non-negative factorization, as proposed by Lee and Seung [20] tends to to decompose an image into its meaningful parts, see the theoretical study of Donoho and Stodden [21].

*Contributions.* In this paper, we propose a signal-processing approach to the sparse modeling of textures. It is based on the following ingredients:

- Only marginal responses of a decomposition are used. It provides a simple modeling of textures ensembles in the transformed domain.
- An additive generative model based on positive atoms is used. Although our method works with traditional linear decompositions, positive atoms are more localized and decompose the exemplar texture into it constitutive elements.
- A fast signal-processing based method is used to learn the set of atoms adapted to a given texture. This is motivated by the sparse description of textures and does not require complex solvers such as the one of [5].
- A fast iterative scheme is used to sample a typical texture that matches the same marginal statistics. It does not ensure a sampling with maximum entropy distribution [5] but initializing the iterations with a random noise is good enough in typical applications, as noticed in [6,8].

## 2 Non-negative Atoms for Texture Decomposition

*Non-negative matrix factorization.* The process of learning a dictionary $D$ of atoms $D = [d_0, \ldots, d_{p-1}]$ to represent accurately a set $Y = [y_0, \ldots, y_{m-1}]$ of exemplars is equivalent to performing a factorization $Y = DX$. This problem is underconstrained and the non-negative factorization [20] enforces positivity of both $X$ and $D$ to constrained the learning problem.

The problem of learning $D$ and $X$ requires to minimize the reconstruction error $\|Y - DX\|$ subject to the constraints $D, X \succeq 0$. This minimization is not convex on both $D$ and $X$, but following [22], an alternate minimization on each matrix can be carried using iterations of the following two steps

$$X_{ab} \leftarrow X_{ab} \frac{\sum_i D_{ia} Y_{ib}/(DX)_{ib}}{\sum_i D_{ia}} \quad \text{and} \quad D_{ia} \leftarrow D_{ia} \frac{\sum_b X_{ab} Y_{ib}/(DX)_{ib}}{\sum_b X_{ab}},$$

which converge to a local minimum of $\|Y - DX\|$. To enforce the dictionary elements $y_i$ to be of unit norm, the columns of $D$ are normalized at each iteration. An explicit sparsity prior can be added into this process [23] to enforce the constraints $\|x_j\|_{\ell^0} \leqslant \tau$ required by equation (1). In practice, this did not result in a noticeable enhancement for texture modeling.

*Patch-based decomposition.* Our sparse modeling of textures is based on the assumption of local ergodicity common to many previous works, see for instance [8]. It assumes that the texture is a realization of a random vector whose statistics are invariant under translation.

Starting from an exemplar texture $f_e \in \mathbb{R}^N$ of $N$ pixels given by the user, one works at a fixed scale $w > 0$ that sets the size of the typical structures present in the texture. Section 4 describes how a multiscale procedure can alleviate the issue of using a fixed scale. If $n$ is a pixel in the image, we denote by $p_n = p_n(f_e)$ the patch of $w \times w$ pixels centered at $n$ in $f_e$.

The ergodicity assumption leads us to model each patch $p_n(f_e)$ as being sparsely represented in some dictionary $D$. In practice, a set of $m$ square patches

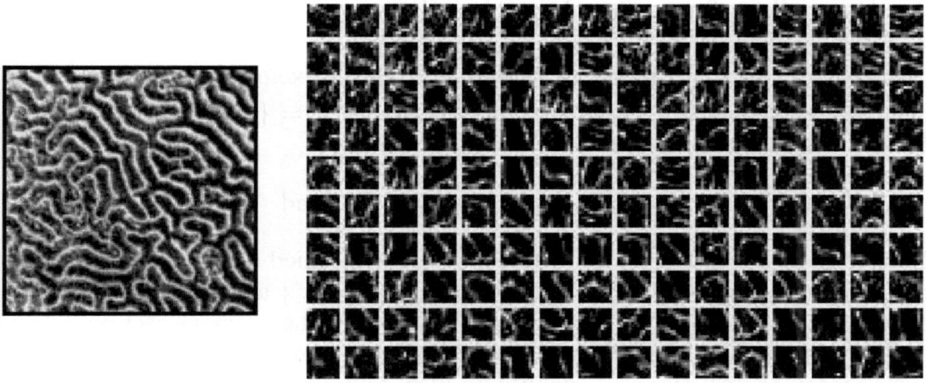

**Fig. 1.** Example of dictionary for patches of $12 \times 12$ pixels. We used $m = 20w^2$ random patches for the learning stage.

$Y = [y_0, \ldots, y_{m-1}]$ of $N = w \times w$ pixels is extracted at random from the exemplar $f_e$. The non negative factorization $Y = DX$ is used to learn the dictionary from the exemplar. Figure 1 shows some examples of atoms $d_j$ of $D$ learned from a texture.

## 3 Parametric Modeling over a Learned Dictionary

Once a dictionary $D$ is learned to efficiently represent the patches $Y$ of size $w \times w$ extracted from $f_e$, a statistical model is built using the marginal of the decomposition coefficients $X$ such that $Y = DX$. Ergodicity allows to use the sets of coefficients $x^k = \{X_{k,j}\}_{j=0}^{m-1}$ to estimate the marginals of the underlying probabilistic model that generates the patches of $f_e$.

The marginal distributions are both on-sided and highly concentrated near 0. We thus keep track of the empirical variance and kurtosis of the decomposition of $Y$ onto the dictionary $D$ defined by

$$\sigma^k(Y) \stackrel{\text{def.}}{=} \sigma(x^k) \stackrel{\text{def.}}{=} M_2(x^k), \quad \text{and} \quad \kappa^k(Y) \stackrel{\text{def.}}{=} \kappa(x^k) \stackrel{\text{def.}}{=} \frac{M_4(x^k)}{(M_2(x^k))^2}, \tag{2}$$

$$\text{where} \qquad M_s(x^k) \stackrel{\text{def.}}{=} \frac{1}{N} \sum_{j=1}^{m} (x^k[j])^s. \tag{3}$$

A texture $f_e$ is characterized, at a scale $w$, by
- its adapted dictionary $D$,
- the empirical variance $\sigma^k(Y)$ and kurtosis $\kappa^k(Y)$.

Both are computed from the decomposition of a large enough set of patches $Y$ extracted from $f_e$.

# 4 Sampling from the Positive Texture Model

*Texture ensembles.* A dictionary $D$ learned from the exemplar defines an equivalence relationship $\overset{D}{\sim}$ between two sets of patches $P = \{p_n\}_n$ and $Q = \{q_n\}_n$ that shares the same statistics

$$P \overset{D}{\sim} Q \quad \Longleftrightarrow \quad \forall k, \quad \sigma^k(P) = \sigma^k(Q) \quad \text{and} \quad \kappa^k(P) = \kappa^k(Q).$$

Note that these statistics are defined over a transformed domain, which means that one first has to factor the matrix $P = [p_0, p_1, \ldots]$ as $P = DX$ and then extract the statistics of the rows $x^k$, as explained in the previous section.

The set of images $f$ of $N_1$ pixels whose local decompositions in $D$ share the same marginal statistics as $f_e$ defines an ideal texture ensemble

$$\widetilde{\mathcal{T}}(f_e) \overset{\text{def.}}{=} \left\{ f \setminus \{p_n(f)\}_{n\in\mathcal{P}} \overset{D}{\sim} \{p_n(f_e)\}_{n\in\mathcal{P}} \right\} \subset \mathbb{R}^{N_1}.$$

where $\mathcal{P}$ denotes the set of pixels. Note that although $f_e$ contains $N$ pixels, the texture ensemble can be defined for any size $N_1$. We use a model based on overlapping patches to have a translation invariant description that avoids blocking artifact.

Imposing simultaneously all the statistical constraints that define $\mathcal{T}(f_e)$ is a complex non-linear process due to the overlapping between patches $p_n(f)$. We approximate this texture ensemble using the following decomposition

$$\widetilde{\mathcal{T}}(f_e) \approx \mathcal{T}(f_e) \overset{\text{def.}}{=} \bigcap_{\delta\in\Delta} \mathcal{T}_\delta(f_e) \qquad \text{where} \qquad \Delta \overset{\text{def.}}{=} \{0, \ldots, w-1\}^2. \tag{4}$$

Each $\mathcal{T}_\delta(f_e)$ imposes constraints on a sub-set of non-overlapping patches

$$\mathcal{T}_\delta(f_e) \overset{\text{def.}}{=} \left\{ f \setminus \{p_n(f)\}_{n\in\mathcal{P}_\delta} \overset{D}{\sim} \{p_n(f_e)\}_{n\in\mathcal{P}_\delta} \right\}$$

where $\mathcal{P}_\delta$ with $\delta = (\delta_1, \delta_2)$ is the sub-lattice of pixels $n$ defined by

$$\mathcal{P}_\delta = \{n = (n_1, n_2) \setminus n_1\%w = \delta_1 \quad \text{and} \quad n_2\%w = \delta_2\},$$

where % is the modulo operator.

*Projection on the texture ensemble.* The approximation of the texture ensemble of equation (4) describes $\mathcal{T}(f_e)$ as the intersection of convex sets $\mathcal{T}_\delta(f_e)$. Following [8], an image $f$ can be approximately projected onto $\mathcal{T}(f_e)$ by iterating projections onto each $\mathcal{T}_\delta(f_e)$. Since each set $\mathcal{T}_\delta(f_e)$ involves constraints on independent patches of $f$, this projection can be carried by local ajustements on the decomposition of each patch $p_n(f)$ for $n \in \mathcal{P}_\delta$.

We denote by $\pi_\delta(f)$ the approximate projection of $f$ onto $\mathcal{T}_\delta(f_e)$, which is computed using the following steps

– The set of patches $P = \{p_n(f)\}_{n\in\mathcal{P}_\delta}$ are gathered from $f$.

- The positive factorisation $P = DX$ is performed using the algorithm described in section 2. We now adjust the statistics of each vector of coefficients $x^k = \{X_{k,j}\}_j$ representing the decomposition on a single atom $d_k$ of $D$.
- The projection on variance constraints is performed by $\widetilde{x}^k \leftarrow x^k \sigma^k(f_e)/\sigma(x^k)$.
- As done in [8], enforcing the kurtosis is performed using a gradient descent of the potential $\widetilde{x}^k \mapsto |\kappa(\widetilde{x}^k) - \kappa^k(f_e)|^2$ while keeping $\sigma(\widetilde{x}^k)$ constant.
- The updated patches $\widetilde{P}$ are reconstructed using the dictionary $\widetilde{P} = D\widetilde{X}$ where the rows of $\widetilde{X}$ are the new coefficients $\widetilde{x}^k$ for $k = 0, \ldots, p-1$. The projection $\pi_\delta(f)$ is computed by rearranging the non-overlapping patches of $\widetilde{P}$.

*Sampling from the texture ensemble.* The set $\mathcal{T}(f_e)$ is compact and the uniform distribution on $\mathcal{T}(f_e)$ thus defines the probability measure with maximum entropy. Zhu et al. [5] sample this distribution in order to synthesize textures without bias.

Rather than performing an exact sampling of the uniform distribution on $\mathcal{T}(f_e)$, we follow [8] and use a sampling strategy that finds a point in $\mathcal{T}(f_e)$ by iterating projections on each of set $\mathcal{T}_\delta(f_e)$. Starting from an initial point with high entropy such as a gaussian noise ensures that the set of generated images span $\mathcal{T}(f_e)$ with a minimum bias. This leads to the following synthesis algorithm:

- Preprocessing: extract $m$ random patches $Y = [y_0, \ldots, y_{m-1}]$ of width $w \times w$ from $f_e$. Compute the positive factorization $Y = DX$ and record the parameters $\sigma^k(f_e)$ and $\kappa^k(f_e)$ of the marginals defined in equation (2).
- Initialization: set $f$ as a realization of a gaussian white noise on $N_1$ pixels.
- Repeat for random shift $\delta \in \{0, \ldots, w-1\}^2$ until convergence: $f \leftarrow \pi_\delta(f)$.

## 5 Texture Synthesis

*Mono-scale synthesis.* Starting from an exemplar $f_e$ of $N$ pixels, the mono-scale texture synthesis process consists in computing an image $f \in \mathcal{T}(f_e)$ of $N_1$ pixel. This sampling is carried out using the iterative projection method exposed in the previous section.

Note that this method generates textures of arbitrary size. Furthermore, we use cyclic boundary conditions when extracting patches $\{p_n(f)\}_{n \in \mathcal{P}_\delta}$ from the output texture, which results in periodic textures that tile the plane. Figure 2 shows two exemples of synthesis. The short range structures are well synthesized, but the algorithm fails to capture long range fiber-like structures.

Color texture are synthesized by applying the algorithm on each channel independently. Moving from the RGB color representation to the HSV representation improves synthesis quality since the intensity channel tends to have more distinct structures than the remaining channels. Figure 2 compares our method with the method of [6] and [8]. Both the input texture $f_e$ and output $f$ are of size $256 \times 256$ pixels. The multiscale histogram matching of Heeger and Bergen [6] is not able to reproduce textures with geometric features. In contrast, both the higher order

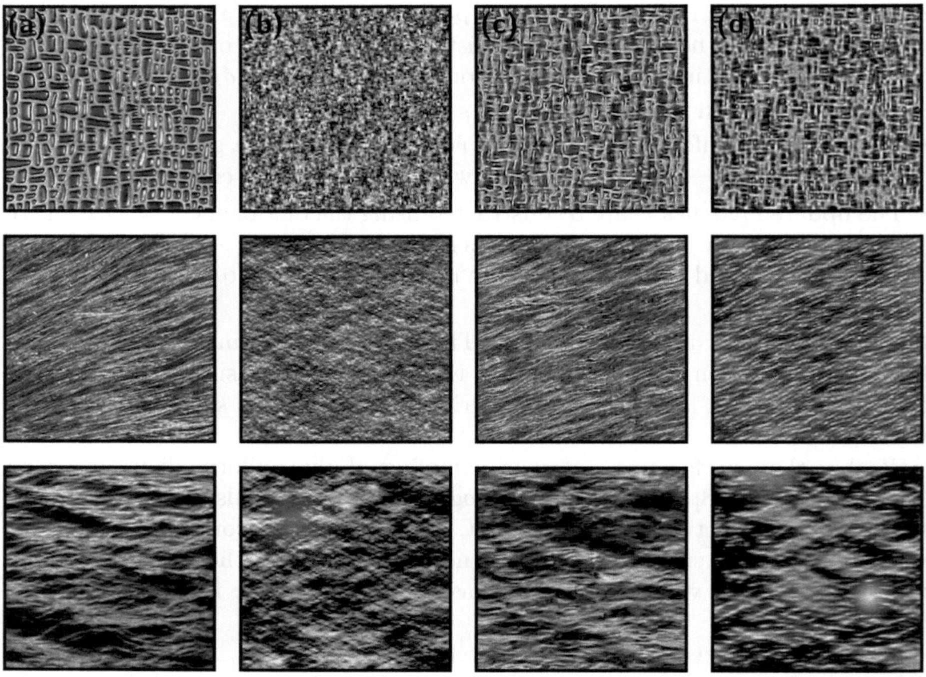


**Fig. 2.** (a) Original texture. (b) Textures synthesized with the method of Heeger and Bergen [6] (c) Textures synthesized with the method of Portilla and Simoncelli [8]. (d) Textures synthesized with our method.

model of Portaill and Simoncelli [8] and our method can synthesize textures with complex structures.

*Multiscale synthesis.* In order to cope with the fixed scale $w$ used in previous section, one can use a multiscale synthesis strategy. We use a fixed number of pixels $w_0$ but consider textures with increasing resolutions. This allows to capture first elongated low frequencies structures and then fine scale geometric details. A simple interpolation is used to switch between the various resolutions. At each scale, the synthesis algorithm manipulates only small patches of size $w_0 \times w_0$. This leads to the following algorithm that handles $J$ scales.

- Initialization: Set $j = J$ to be the coarser scale. Initialize the synthesis with a random noise $f$ of $N_1/2^J \times N_1/2^J$ pixels.
- Step 1: Set $w = 2^j w_0$. Smooth the exemplar $f_e^j = f_e * h_j$ where $h_j$ is a gaussian kernel of width $2^j$ pixels. Extract a set of $m$ patches $Y_j$ from $f_e^j$. Sub-sample these squares by a factor $2^j$ so that vectors in $Y_j$ are of size $w_0 \times w_0$.
- Step 2: Perform the mono-scale synthesis algorithm using patches $Y_j$ to train the dictionary and with the current $f$ as initialization.

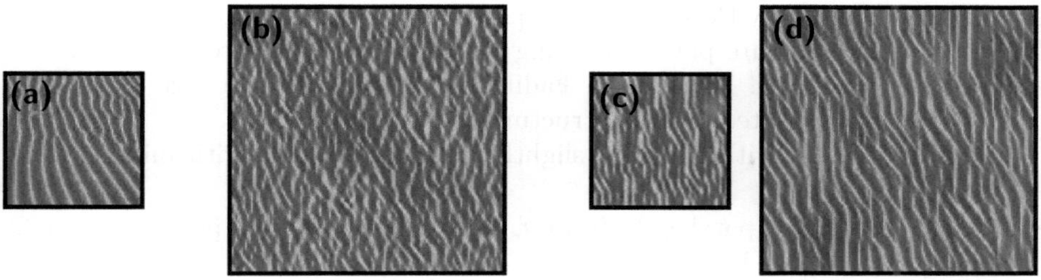


**Fig. 3.** (a) Original texture $f_e$. (b) Texture synthesized with the mono-scale procedure with patches of width $w = 8$. (c) Texture synthesized at scale $2^j = 2$ with $w = 16$. (d) Texture synthesized at scale $2^j = 1$ with $w = 8$.

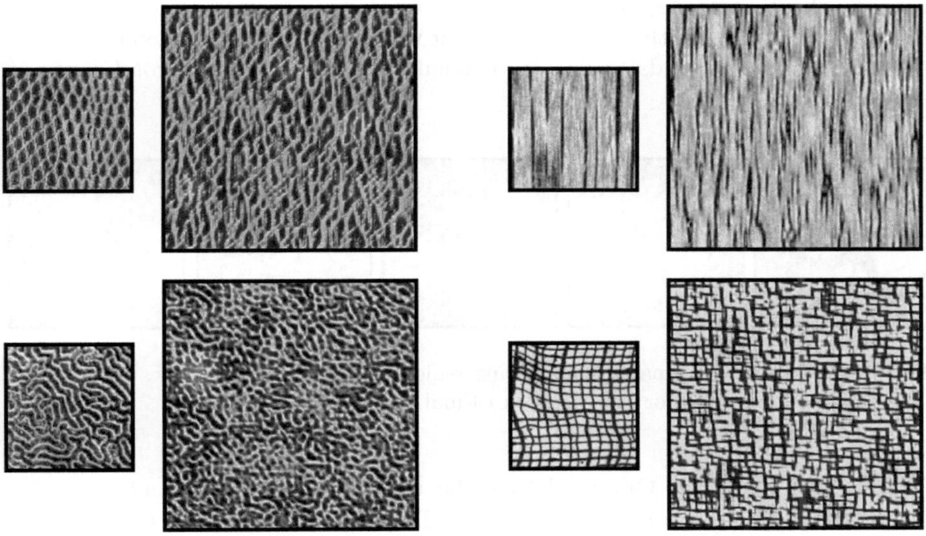

**Fig. 4.** Examples of multiscale texture synthesis

– If $j = 0$ then stop the algorithm. Otherwise, upsample the current synthesized texture using linear interpolation from $N/2^j \times N/2^j$ pixels to $2N/2^j \times 2N/2^j$ pixels. Set $j \to j - 1$ and go back to step 1.

In our implementation, we have used 2 scales $j = 0, 1$ and a base width $w_0 = 8$. Figure 3 compares the fixed scale synthesis and the multiscale synthesis which is able to creates elongated singularities. Figure 4 shows additional synthesis results.

## 6 Texture Inpainting

The inpainting problem consists in filling a set of missing pixels $\Omega$ in a given image $f_e$. This problem has been approached using evolution equation derived from fluid dynamics by Bertalmio et al. [24] however diffusion-based approaches fail to reproduce texture patterns. Using a sparsity prior in a set of fixed bases such as Curvelets and local DCT, Fadili and Starck [25] are able to inpaint oscillatory and elongated texture structures.

Our synthesis algorithm can be slightly modified to cope with missing data as follow.

- Extract a set of $m$ patches $Y$ from $f_e$ that are as close as possible from $\Omega$ without intersecting it.
- Set as initial inpainted image $f$ the original $f_e$ with values at random inside $\Omega$.
- Step 1: for a random shift $\delta$, perform one step of synthesis $f \leftarrow \pi_\delta(f)$. The projection needs only to be performed for patches $p_n(f) \cap \Omega \neq \emptyset$.
- Step 2: impose the known values, $\forall n \notin \Omega, \quad f[n] \leftarrow f_e[n]$. Go back to step 1.

Figure 5 shows some step of this inpainting process and figure 6 shows additional results. A limitation of this method is that it works well for homogeneous textures and the inpainting tends to give poor results if $\Omega$ intersects a broad range of structures.

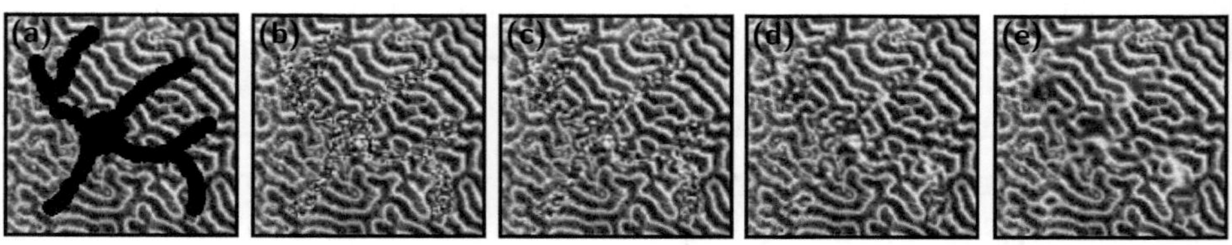


**Fig. 5.** (a) Texture to inpaint, the missing region $\Omega$ is depicted in black. (b,c,d) Evolution of the inpainting for step 1,2,4. (e) Final result.

*Texture segmentation.* Our model can be used to perform segmentation of a given texture $f$ into components corresponding to patterns similar to exemplars $\{f_e^1, \ldots, f_e^s\}$. Learned dictionaries have already been used for segmentation [18], and we recast this approach into our patch-based non-negative model. The idea is to project the texture $f$ onto each texture ensemble $\mathcal{T}(f_e^\ell)$ and select locally the class $\ell$ that generates the least deviation from $f$. This leads to the following algorithm.

- Learn the statistical model $\mathcal{T}(f_e^\ell)$ for each class $\ell$.
- Compute the projection $f_\ell$ of $f$ on each ensemble $\mathcal{T}(f_e^\ell)$ using the algorithm of section 4.
- Compute the class-wise error for each pixel and smooth it with a gaussian kernel $G_{s_0}$ of with $s_0$

$$\widetilde{E}_\ell[n] \overset{\text{def.}}{=} |f[n] - f_\ell[n]|^2 \qquad \text{and} \qquad E_\ell = \widetilde{E}_\ell * G_{s_0}.$$

  The smoothing removes estimation noise and reflects the prior knowledge that class boundary should be smooth curves.

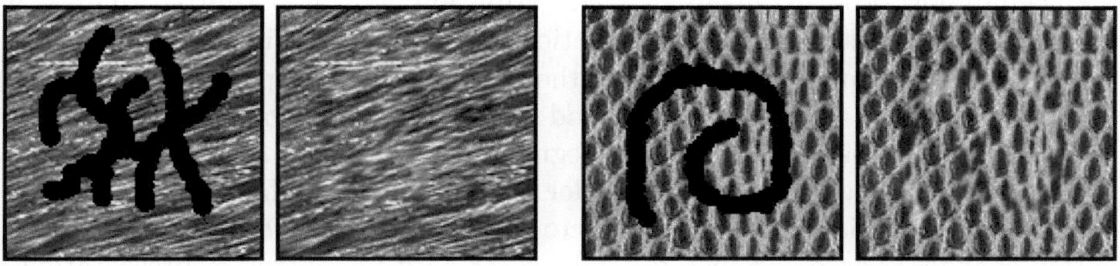

**Fig. 6.** Examples of inpainting

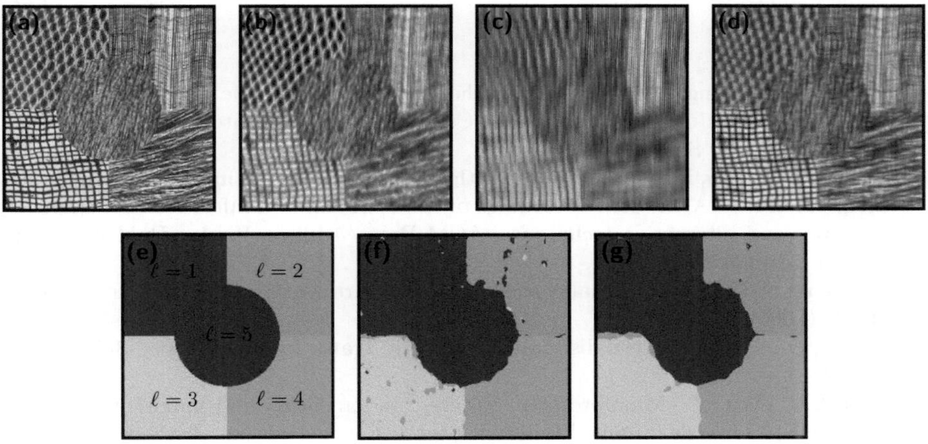


**Fig. 7.** (a) Original texture $f$. (b) Projected texture $f_1$ of $f$ onto $\mathcal{T}(f_e^1)$. Note how the upper left corner is well preserved. (c) Projected texture $f_2$. (d) Projected texture $f_3$. (e) Ground trust segmentation. (f) Segmentation $\ell[n]$ computed with $s_0 = 3$ pixels. (g) Segmentation computed with $s_0 = 6$ pixels.

– Compute the segmentation into classes using

$$\ell[n] = \underset{\ell}{\operatorname{argmin}}\ E_\ell[n].$$

We have tested this segmentation using a set of $s = 5$ exemplar textures. The input image $f$ of $256 \times 256$ pixels is a patchwork of five textures extracted from the upper left corner of the original images $\widetilde{f}_e^\ell$ of $512 \times 512$ pixels. The exemplars $f_e^\ell$ are extracted from the lower right corner of $\widetilde{f}_e^\ell$. Figure 7 shows the segmentation process.

## 7 Conclusion

We have proposed a statistical model for textures built out of marginal distributions of a positive decomposition. The statistical model is parametric since we use only low order moments of the distribution. This is permitted thanks to the sparsity provided by a learned dictionary. Such a positive dictionary captures with few atoms the structures of the textures. This simple model allows to perform multiscale texture synthesis and can be fitted into various applications such as texture inpainting or texture segmentation.

An important parameter of our model is the redundancy factor $p/N$. Redundancy brings invariances to various factors such as translations or local illumination changes. These parameters are however hard to control. Representations that can explicitly capture this invariances include bilinear decompositions [26] that could improve our model.

## References


1. Efros, A.A., Leung, T.K.: Texture synthesis by non-parametric sampling. In: ICCV '99: Proceedings of the International Conference on Computer Vision-Volume 2, IEEE Computer Society (1999) 1033
2. Wei, L.Y., Levoy, M.: Fast texture synthesis using tree-structured vector quantization. In: SIGGRAPH '00: Proceedings of the 27th annual conference on Computer graphics and interactive techniques, ACM Press/Addison-Wesley Publishing Co. (2000) 479–488
3. Lefebvre, S., Hoppe, H.: Parallel controllable texture synthesis. ACM Trans. Graph. **24**(3) (2005) 777–786
4. Julesz, B.: Visual pattern discrimination. IRE Trans. Inform. Theory **8**(2) (1962) 84–92
5. Zhu, S.C., Wu, Y., Mumford, D.: Filters, random fields and maximum entropy (FRAME): Towards a unified theory for texture modeling. Int. J. Comput. Vision **27**(2) (1998) 107–126
6. Heeger, D.J., Bergen, J.R.: Pyramid-Based texture analysis/synthesis. In Cook, R., ed.: SIGGRAPH 95 Conference Proceedings. Annual Conference Series, ACM SIGGRAPH, Addison Wesley (1995) 229–238
7. Perlin, K.: An image synthesizer. In: SIGGRAPH '85: Proceedings of the 12th annual conference on Computer graphics and interactive techniques, New York, NY, USA, ACM Press (1985) 287–296
8. Portilla, J., Simoncelli, E.P.: A parametric texture model based on joint statistics of complex wavelet coefficients. Int. J. Comput. Vision **40**(1) (2000) 49–70
9. Attneave, F.: Some informational aspects of visual perception. Psychological Review **61** (1954) 183–193
10. Barlow, H.B.: Possible principles underlying the transformation of sensory messages. In Rosenblith, W.A., ed.: Sensory Communication, MIT Press (1961) 217–234
11. Mallat, S.: A Wavelet Tour of Signal Processing. Academic Press, San Diego (1998)
12. Candès, E., Donoho, D.: New tight frames of curvelets and optimal representations of objects with piecewise $C^2$ singularities. Comm. Pure Appl. Math. **57**(2) (2004) 219–266
13. Le Pennec, E., Mallat, S.: Bandelet Image Approximation and Compression. SIAM Multiscale Modeling and Simulation **4**(3) (2005) 992–1039
14. Olshausen, B.A., Field, D.J.: Emergence of simple-cell receptive-field properties by learning a sparse code for natural images. Nature **381**(6583) (1996) 607–609

15. Aharon, M., Elad, M., Bruckstein, A.: The k-svd: An algorithm for designing of overcomplete dictionaries for sparse representation. IEEE Trans. On Signal Processing (to appear) (2006)
16. Tropp, J., Dhillon, I., Heath, R., Strohmer, T.: Designing structured tight frames via an alternating projection method. IEEE Trans. on Information Theory **51**(1) (2005) 188–209
17. Zeng, X.Y., Chen, Y.W., van Alphen, D., Nakao, Z.: Selection of ica features for texture classification. In: ISNN (2). (2005) 262–267
18. Skretting, K., Husoy, J.: Texture classification using sparse frame based representations. EURASIP Journal on Applied Signal Processing, to appear (2006)
19. Manduchi, R., Portilla, J.: Independent component analysis of textures. In: ICCV. (1999) 1054–1060
20. Lee, D.D., Seung, H.S.: Learning the parts of objects by non-negative matrix factorization. Nature **401** (1999) 788–791
21. Donoho, D., Stodden, V.: When does non-negative matrix factorization give a correct decomposition into parts? In Thrun, S., Saul, L., Schölkopf, B., eds.: Advances in Neural Information Processing Systems 16, Cambridge, MA, MIT Press (2004)
22. Lee, D.D., Seung, H.S.: Algorithms for non-negative matrix factorization. In: Advances in Neural Information Processing Systems 13, Cambridge, MA, MIT Press (2001)
23. Hoyer, P.O.: Non-negative matrix factorization with sparseness constraints. Journal of Machine Learning Research **5** (2004) 1457–1469
24. Bertalmio, M., Sapiro, G., Caselles, V., Ballester, C.: Image inpainting. In: Siggraph 2000. (2000) 417–424
25. Fadili, M., Starck, J.L.: Em algorithm for sparse representation-based image inpainting. In: IEEE International Conference on Image Processing, Vol. II. (2005) 61–63
26. Grimes, D.B., Rao, R.P.N.: Bilinear sparse coding for invariant vision. Neural Computation **17**(1) (2005) 47–73

# Texture Synthesis and Modification with a Patch-Valued Wavelet Transform

Gabriel Peyré

Ceremade, Université Paris Dauphine,
Place du Marchal De Lattre De Tassigny,
75775 Paris Cedex 16 France
gabriel.peyre@ceremade.dauphine.fr
http://www.ceremade.dauphine.fr/~peyre/

**Abstract.** This paper models a texture as a 2D mapping onto a nonlinear manifold representing the local structures of the image. This manifold is learned from the set of local patches from an exemplar texture. A multiscale decomposition of this manifold valued representation is computed that mimics the orthogonal wavelet transform. The key ingredient of this decomposition is a geometric association field that drives the computations along the manifold. Iterated predictions leads to the computation of details coefficients over the features manifold. The resulting transform is invertible, non-linear and represents efficiently the local geometric structures of the exemplar. The multiscale coefficients of this transform are used to perform analysis and synthesis of textures.

**Keywords:** Manifold of patches, non linear wavelet transform, image geometry, texture synthesis, texture modification.

## 1 Geometric Modeling of Images

*Multiscale decompositions.* The classical isotropic wavelet transform [10] cannot capture efficiently geometrical structures of textures. The local anisotropy of edges requires specific constructions such as the curvelets frame of Candès and Donoho [1] or the bandelets framework of Le Pennec and Mallat [6].

*Geometric structure propagation.* The local geometry of images can be described as points on a curved manifold. Lee et al. [7] show how the edge manifold emerges from the set of patches extracted from natural images. To handle more complex texture features we propose to learn this manifold from an exemplar image. Computation are performed along this manifold using a local connexion describing how structures propagate in the image plane. Mumford first introduced this notion for edges propagation with the elastica model [12]. Computer vision scientists such as Williams and Jacobs [17] proposed in some cases efficient approximations of this edge propagation field.

F. Sgallari, A. Murli, and N. Paragios (Eds.): SSVM 2007, LNCS 4485, pp. 640–651, 2007.

*Texture synthesis.* Pioneer work of Julesz [5] states filtering rules for the probabilistic characterization of textures. The wavelet domain modeling of Heeger and Bergen [4] synthesizes cloudy textures.

Most successful approaches for texture synthesis in graphics are based on non-parametric copy of small patches from an original texture, see for example Efros and Leung [3]. Recent approaches such as the method of Lefebvre and Hoppe in [8] have re-casted this non-parametric patches sampling into a multiscale framework. A random shuffling of patches coordinates allows to deviate from the original texture. These modifications can be seen roughly as wavelet details that are added at various scales during the synthesis.

Our manifold description of textures paves the way between multiscale texture descriptions [4] and more complex, non parametric sampling [3,8].

*Manifold valued analysis.* Modeling data using a manifold structure can performed using non-linear estimation algorithms such as Isomap [15] or LLE [13]. These dimensionality reduction procedures enable the analysis of libraries of images such as the ones studied by Donoho and Grimes [2].

Processing of manifold valued functions is studied by Ur-Raman et al. [16] who propose a wavelet-like decomposition for such data. Our multiscale feature valued analysis presented in section 4 is inspired by this work but is adapted to geometric images and textures. In particular, our scheme takes into account the relative position of two features in the image, which enables the generation of anisotropic geometries such as elongated features.

Recent approaches to image synthesis in computed graphics use manifold modeling of textures. Matusik et al. define a manifold from a set of textures [11]. Lefebvre and Hoppe introduce a mapping of an image into a higher dimensional appearance space [9]. This embedding allows a synthesis with high fidelity and spacial variations.

## 2 Manifolds as Image Models

We model the local structure of textures using a set of manifolds $\{\mathcal{M}^w\}_w$ computed from an exemplar texture $f_e$. The manifold $\mathcal{M}^w$ is composed of patches of fixed size $w$ that are extracted from the original texture $f_e$

$$\mathcal{M}^w \stackrel{\text{def.}}{=} \{p_t^w \setminus t \in [0,1]^2\} \subset \mathrm{L}^2([-w/2, w/2]^2) \tag{1}$$

$$\text{where} \quad \forall\, x \in [-w/2, w/2]^2, \quad p_t^w(x) = f_e(x-t). \tag{2}$$

As explained by Grimes and Donoho [2], the set $\mathcal{M}^w$ might have a complex structure and be non-differentiable even for a simple images $f_e$. Figure 1 shows some examples of patches extracted from a discretized texture. In numerical applications, the image $f_e$ is composed of $n \times n$ pixels, so $\mathcal{M}^w$ is a set of patches considered as vectors of $\mathbb{R}^{m \times m}$ with $m = wn$.

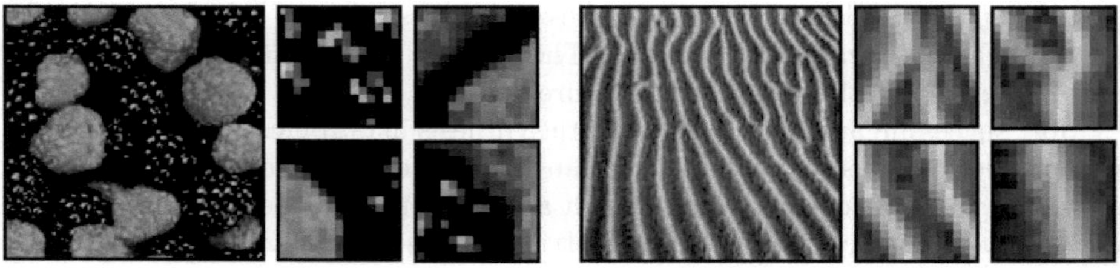

**Fig. 1.** Two examples of textures and some local geometric structures

*Models and estimation.* A gray scale image is a mapping $f : \{0,\ldots,n-1\}^2 \rightarrow \mathbb{R}$ from pixels to real numbers. Our manifold description of images corresponds to a factorization

$$f : \{0,\ldots,n-1\}^2 \xrightarrow{\tilde{f}} \mathcal{M}^{w_0} \xrightarrow{\Phi_{w_0}} \mathbb{R} \tag{3}$$

where the manifold $\mathcal{M}^{w_0}$ is estimated from an exemplar $f_e$ (that might be different from $f$) and the scale parameter $w_0 > 0$ represents the smallest size of a typical feature. Note that the mapping $\Phi_{w_0}$ depends only on $\mathcal{M}^{w_0}$ and thus on $f_e$ and can be computed independently of $f$. Factorization (3) extends to color images by considering each channel independently.

In this paper, we are interested by the mapping $\tilde{f}$ which locates an image feature $\tilde{f}(x) \in \mathcal{M}^{w_0}$ of size $w_0$ at location $x$ in the image plane. The function $\Phi_{w_0}$ controls the display of each feature, and we assume a simple sampling process

$$\Phi_{w_0} : p \in \mathcal{M}^{w_0} \mapsto p(0) \in \mathbb{R}.$$

We note that recent works in harmonic analysis has focussed on the processing of such function $\Phi_{w_0} \in \mathrm{L}^2(\mathcal{M})$ defined over a manifold domain $\mathcal{M}$. In particular Fourier analysis of this function can be carried using eigenvector of the Laplace Beltrami operator. Szlam et al. [14] use this spectral decomposition to perform a non-linear filtering.

In this paper, instead of modifying $\Phi_{w_0}$, we propose to modify $\tilde{f}$ which involves two key ingredients.

- Estimating $\tilde{f}$ such that factorization (3) hold at least approximately. To that end we perform a simple local best fit

$$\tilde{f}[k] \stackrel{\text{def.}}{=} \underset{g \in \mathcal{M}^{w_0}}{\operatorname{argmin}} \sum_{t_1=-nw_0/2}^{nw_0/2} \sum_{t_2=-nw_0/2}^{nw_0/2} |f[k+t] - g[t]|^2. \tag{4}$$

  The minimization of equation (4) extends to color images. In practice a color equalization is performed on $f_e$ prior to the matching as shown in image 2.
- Performing signal processing for manifold-valued functions. The general frameworks of Ur Uaman et al. [16] is not suited for processing patches that have some spacial connexions. We develop our own computation model and derive a multiscale decomposition in this paper.

*Examples of texture representations.* In the case of a texture manifold $\mathcal{M}^{w_0}$ learned from an exemplar $f_e$ using equation (1), the estimation of $\tilde{f}$ from a given texture $f$ defines a 2D-valued mapping

$$\varphi : [0,1]^2 \to [0,1]^2, \qquad \text{where} \qquad \forall x \in [0,1]^2, \quad \tilde{f}(x) = p^{w_0}_{\varphi(x)} \in \mathcal{M}^{w_0},$$

where $p^w_t$ is the patch extracted from $f_e$ at some position $t$, as defined by equation (1). A similar coordinate mapping $\varphi$ is generated by the texture synthesis algorithm of Lefebvre and Hoppe [8,9].

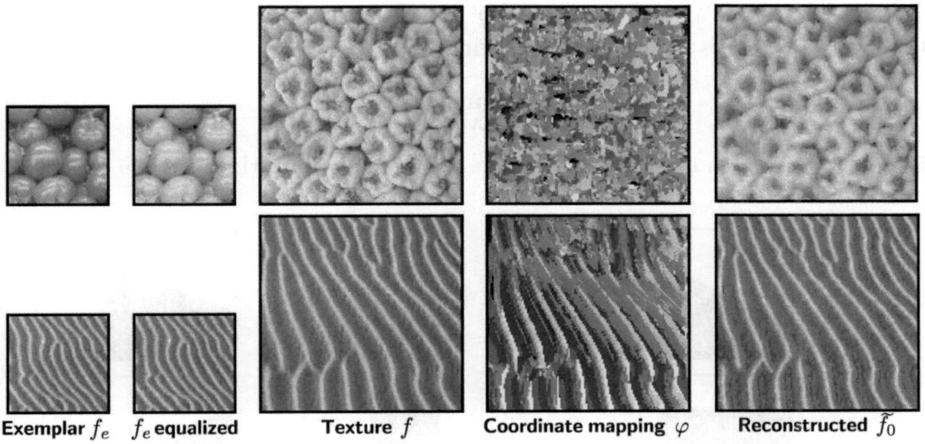


**Fig. 2.** Example of coordinate mapping $\varphi$ estimated for two different pairs of textures $(f_e, f)$. The mapping $\varphi(x) = (\varphi_1(x), \varphi_2(x)) \in \mathbb{R}^2$ is depicted using a red color for $\varphi_1$ and green for $\varphi_2$. Prior to computing $\varphi$, the colors of $f_e$ are equalized to match those of $f$. The reconstructed texture is $\Phi_{w_0} \circ \tilde{f}$.

Figure 2 shows two examples of mappings $\varphi$ estimated for target textures $f$ of a size twice bigger (in pixels) than the size of the exemplar $f_e$. For natural images, the mapping $\varphi$ is usually piecewise linear and exhibit jump discontinuities. The right column of figure 2 shows the function $\Phi_{w_0} \circ \tilde{f} \approx f$ reconstructed from the manifold representation $\tilde{f}$. This reconstruction satisfies $(\Phi_{w_0} \circ \tilde{f})(x) = f_e(\varphi(x))$. If the learned manifold $\mathcal{M}^{w_0}$ does not contains the local geometric structures of $f$, the reconstruction might differ from the original texture. On the example of figure 2 (upper row), one can see that the green line features of the pepper image are lacking.

## 3 Association Field over the Patch Manifold

The mean computation of two features $g, g' \in \mathcal{M}^w$ must take into account the relative positions of the features in the image to process. The two features are thus assumed to be samples of some manifold valued function $\tilde{f}$ such that $\tilde{f}(x) = g \in \mathcal{M}^w$ and $\tilde{f}(x') = g' \in \mathcal{M}^w$. For the purpose of our multiscale processing framework, one only needs to compute such average for horizontal $\omega = (1,0)$ and vertical $\omega = (0,1)$ alignements of features. We thus assume that $x - x' = w\omega$. We describe the computation of the mean mapping in the horizontal case, the vertical one being similar. Starting from two feature elements $g, g' \in \mathcal{M}^w$, a new patch is computed by mixing the right side of $g$ with the left side of $g'$

$$\mu^\omega(g, g')(x) \stackrel{\text{def.}}{=} \begin{cases} g(x_1 + w/2, x_2) & \text{if} \quad x_1 < 0, \\ g'(x_1 - w/2, x_2) & \text{if} \quad x_1 \geqslant 0. \end{cases}$$

The mean $m^\omega(g, g') \in \mathcal{M}^w$ of $g$ and $g'$ in the direction $\omega$ is then computed as the feature that best matches this mixing image

$$m^\omega(g, g') \stackrel{\text{def.}}{=} \underset{h \in \mathcal{M}^w}{\operatorname{argmin}} \|\mu^\omega(g, g') - h\|.$$

Figure 3 illustrates the process of mean features computations $\mathcal{M}^w$.

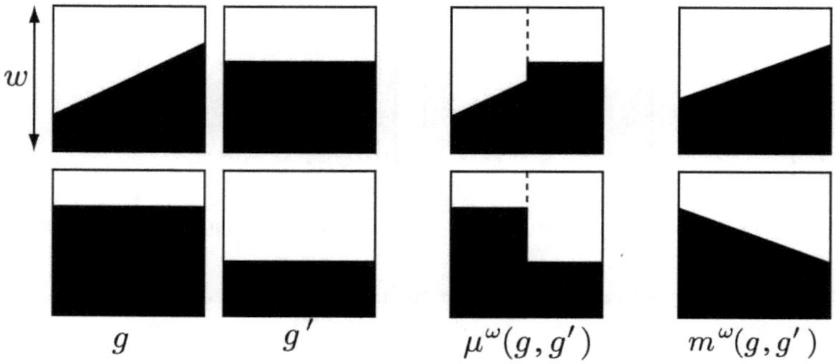


**Fig. 3.** Computation of the mean $m^\omega(g, g')$ of two edge features $g, g' \in \mathcal{M}^w$ along direction $\omega = (1, 0)$

This numerical algorithm is a simple way to estimate an association field between feature elements. Similar ideas are used in graphics for texture synthesis using copy of patches, see for instance [3,8]. More complex approaches relying on a biological or variational modeling of the association field could be used [12,17].

## 4 Multiscale Feature-Valued Transform

The multiscale transform decomposes a given patch valued image $\tilde{f} : [0,1]^2 \mapsto \mathcal{M}^{w_0}$ of $n \times n$ values into a set of $n^2$ coefficients $\{d_j\}_{j=1}^J \cup \{\tilde{f}_J\}$ that encodes at each scale $j \leqslant J \stackrel{\text{def.}}{==} \log_2(n^2)$ the details needed to reconstruct the features of the original texture. This transform is inspired by the interpolating manifold wavelet decomposition of Ur-Raman et al. [16] but with important differences that make it suitable for the analysis of local image features.

*Manifold parameterization.* In order to carry computations along the manifold $\mathcal{M}^w$ we assume that it is globally parameterized by a bijective mapping

$$\gamma_w : \Omega \subset \mathbb{R}^p \to \mathcal{M}^w. \tag{5}$$

Such a parameterization is hardly available in practice. In numerical applications, we compute an approximate mapping using the Locally Linear Embedding [13]. The goal of this dimensionality reduction is to compute wavelet coefficients as differences in the parameter domain $d = \gamma_w(g) - \gamma_w(g') \in \mathbb{R}^p$ of two patches $g, g' \in \mathcal{M}^w$. Ur-Raman et al. [16] define a wavelet difference $d$ as a vector in the tangent plane of the manifold that is tangent to the geodesic joining $g$ and $g'$ and whose magnitude is the length of the geodesic. We use a parametric approach to cope with our patch-valued data. The value of the dimensionality $p$ should reflect the complexity of the local geometry of $f_e$ and we use $p = 5$ in our numerical applications.

**Table 1.** Pseudo code for the forward feature valued transform

```
Function [{d_j^ω}_{j,ω}, f̃_J] = multiscale_transform_fwd(f̃).

Input : feature valued image f̃,
Output : detail coefficients d_j^ω for j = 1,...,J and ω ∈ {(1,0),(0,1)},
         coarse approximation at scale J, f_J.

Initialization : set f̃_0 = f̃.
For j = 1,...,J = log_2(n),
  | Set w_j = 2^j/n.
  | (Expand) Consider f_j[k] = f_{j-1}[k] ∈ M^{w_j}.
  | For ω = (1,0) and then ω = (0,1)
  |   | (Split) Compute even and odd set of coefficients:
  |   |     f_j^e[k_1,k_2] = f_j[2k_1,k_2]   and   f_j^o[k_1,k_2] = f_j[2k_1+1,k_2].
  |   | (Extract) Compute μ = m^ω(f_j^e[k_1,k_2], f_j^e[k_1+1,k_2]) and set
  |   |     d_j^ω[k_1,k_2] = γ_{w_j}(μ) − γ_{w_j}(f_j^o[k_1,k_2])   and   f_j[k_1,k_2] = f_j^e[k_1,k_2].
  |   | (Swap) Switch lines and columns : f_j[k_1,k_2] = f_j[k_2,k_1].
  |   End
  End
```

Reconstruction from the wavelet coefficients requires the computation of the inverse mapping $\gamma_w^{-1}$. In numerical applications where $\mathcal{M}^w$ is estimated using a discrete set of points, we use a nearest neighbor search

$$\forall x \in \mathbb{R}^d, \qquad \gamma_w^{-1}(x) \overset{\text{def.}}{=} \underset{g \in \mathcal{M}}{\operatorname{argmin}} \|\gamma_w(g) - x\|_{\ell^2}^2.$$

*Feature-valued Wavelet Decomposition* In the following, we use cyclic boundary conditions, which means that we formally write $v[k] = v[k+n]$ for a vector $v \in \mathbb{R}^n$. For the analysis of non-periodic textures, we use an extension by symmetrization along the boundaries of the image.

The algorithm starts with an initial feature valued mapping $\tilde{f}_0 = \tilde{f}$ of $n \times n$ values. Each step generates successively coarser features valued images $\tilde{f}_j[k] \in \mathcal{M}^{w_j}$ of $n/2^j \times n/2^j$ coefficients, with $w_j = 2^j/n$.

Each column of the array $\tilde{f}_j$ is processed and we drop the column index to simplify the notations. The same computation is then performed on the rows of the array, as described in the pseudo code 1. The current vector $\tilde{f}_j$ is split into even and odd coefficients

$$\forall\, k = 0, \ldots, n/2^j - 1, \quad \tilde{f}_j^e[k] \stackrel{\text{def.}}{=} \tilde{f}_j[2k] \qquad \text{and} \qquad \tilde{f}_j^o[k] \stackrel{\text{def.}}{=} \tilde{f}_j[2k+1].$$

The even coefficients are used to predict the values of the odd coefficients using the averaging operator described in section 3

$$\tilde{f}_j^o[k] \approx \mu \stackrel{\text{def.}}{=} m^\omega(\tilde{f}_j^e[k], \tilde{f}_j^e[k+1]) \qquad \text{where} \qquad \omega = (1,0).$$

The wavelet decomposition at scale $j$ along direction $\omega$ encodes the prediction error, measured using differences computed over the parameter domain

$$d_j^\omega[k] \stackrel{\text{def.}}{=} \gamma_{w_j}(\mu) - \gamma_{w_j}(\tilde{f}_j^o[k]) \in \mathbb{R}^p.$$

The even coefficients are then copied $\tilde{f}_j = \tilde{f}_j^e$ in order to be further processed along the rows. Once $\tilde{f}_j$ as been processed, the resulting feature valued image of $n/2^{j+1} \times n/2^{j+1}$ values is copied into $\tilde{f}_{j+1}$ considered as taking values into the coarser manifold $\mathcal{M}^{w_{j+1}}$. Note that at each iteration of the analysis, the assignment $\tilde{f}_j \leftarrow \tilde{f}_{j-1}$ assumes a switch from elements $\tilde{f}_{j-1}[k] \in \mathcal{M}^{w_{j-1}}$ to elements $\tilde{f}_j[k] \in \mathcal{M}^{w_j}$.

**Table 2.** Pseudo code for the backward feature valued transform

Function $\tilde{f} =$ `multiscale_transform_bwd`$(\{d_j^\omega\}_{j,\omega}, f_J)$.

*Input* : detail coefficients $d_j^\omega$ for $j = 1, \ldots, J$ and $\omega \in \{(1,0),(0,1)\}$,
coarse approximation at scale $J$, $f_J$.
*Output* : feature valued image $\tilde{f}$.

*Initialization* : set $\tilde{f}_J = \tilde{f}$.
For $j = J = \log_2(n), \ldots, 1$,
  Set $w_j = 2^j/n$.
  For $\omega = (0,1)$ and then $\omega = (1,0)$
    (Reconstruct) Compute $\mu = m^\omega(f_j[k_1,k_2], f_j[k_1+1,k_2])$ and set
      $f_j^o[k_1,k_2] = \gamma_{w_j}^{-1}(\mu - d_j^\omega[k_1,k_2])$ and $f_j^e[k_1,k_2] = f_j[k_1,k_2]$.
    (Join) Compute coefficients:
      $f_j[2k_1,k_2] = f_j^e[k_1,k_2]$ and $f_j[2k_1+1,k_2] = f_j^o[k_1,k_2]$.
    (Swap) Switch lines and columns : $f_j[k_1,k_2] = f_j[k_2,k_1]$.
  End
  (Contract) Consider $f_{j-1}[k] = f_j[k] \in \mathcal{M}^{w_{j-1}}$.
End
Output $\tilde{f} = \tilde{f}_0$.

The resulting wavelet transform can be inverted by retrieving the odd coefficients $\tilde{f}_j^o$ from the available details $d_j^\omega$ and even coefficients $\tilde{f}_j^e = \tilde{f}_j$

$$\tilde{f}_j^o[k] = \gamma_{w_j}^{-1}(\gamma_{w_j}(\tilde{f}_j^e[k]) - d_j^\omega[k]).$$

The pseudo code 1 implements the forward transform, whereas the pseudo code 2 implements the reverse transform. Figure 7 (b) shows the magnitude of the wavelet coefficients packed similarly to the orthogonal wavelet coefficients [10].

# 5 Texture Analysis and Synthesis

In this section, an input texture $f_e$ is used to provide an estimation of the feature manifolds $\mathcal{M}^w$ for dyadic sizes $w$. The multiscale transform is used with these manifolds to perform texture synthesis and modification.

## 5.1 Texture Synthesis

An ideal interpolation of a coarse scale sets of coefficients can be constructed using the inverse multiscale transform. The user defines the four coefficients composing $\tilde{f}_{j_0}$ for the scale $j_0 = 1$ (they are chosen at random in our examples). The wavelet coefficients $\{d_j^\omega\}_{j,\omega}$ are set to zero in order to deviate as little as possible from the exemplar texture $f_e$. Coefficients at scale $j = 1$ define four corners of the texture and successive steps in the backward transform (pseudo-code 2) perform dyadic refinements which compute feature valued functions $\tilde{f}_j$ of increasing dyadic sizes. The corresponding images $f_j \stackrel{\text{def.}}{=} (\Phi_{w_j} \circ \tilde{f}_j)$ have $2^j \times 2^j$ pixels and can be seen as a non linear interpolation. The final synthesized texture is $f \stackrel{\text{def.}}{=} (\Phi_{w_J} \circ \tilde{f})$. The pixels of these intermediate images are sampled from the exemplar texture via a coordinate mapping $f_j[k] = f_e[\varphi_j[k]]$ and the mapping $\varphi_j$ is refined through the iterations. Figure 4 shows the iterations of the synthesis algorithm which refines $f_j$ and $\varphi_j$.

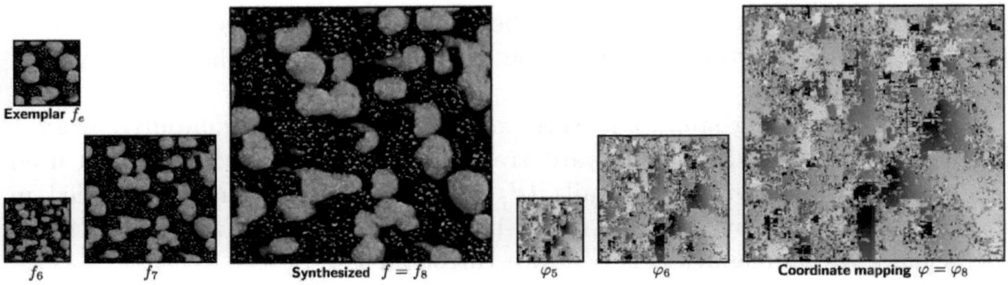


**Fig. 4.** Progression of the synthesis algorithm

Figure 5 shows two examples of texture synthesis. Note that the size of the synthesized textures can be arbitrary large. We use periodic boundary conditions for the backward transform so that the resulting texture is periodic (although the exemplar needs not be) and tiles the plane.

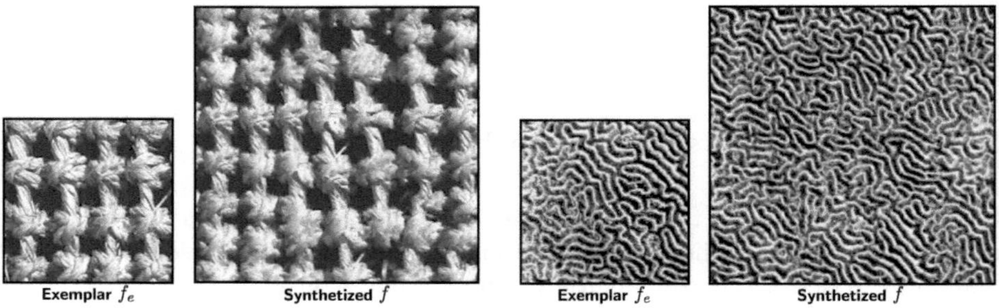


**Fig. 5.** Two examples of texture synthesis using zero padding in the multiscale domain

In order to deviate from the original texture, one can use non-zero wavelet coefficients. More precisely, we define a parametric texture model over the wavelet coefficients using

$$\forall j > 0, \quad \forall \omega, \quad \forall k \in \{0, \ldots, n/2^j - 1\}^2 \qquad d_j^\omega[k] \sim X(\sigma_j),$$

where $\sim$ means that each coefficient is drawn from a gaussian random variable $X(\sigma_j)$ of mean 0 and variance $\sigma_j$ which allows a user control of the synthesis. Figure 6 shows different examples of texture synthesis using various spectral content $\sigma_j$.

### 5.2 Texture Modification

In contrast to texture synthesis, texture modification requires both the forward and backward multiscale transform. The modification process takes as input an original texture $f$, which is to be modified according to some exemplar texture $f_e$. This exemplar texture is used as the exemplar model for the feature valued multiscale transform.

In a pre-processing step, a feature valued mapping $\tilde{f}$ is computed from $f$ following equation (4). The forward transform described in listing 1 is used to decompose $\tilde{f}$ as wavelets details $\{d_j^\omega\}_{j,\omega}$. The texture can be modified in the wavelet domain by applying a non-linear thresholding that removes small coefficients whose magnitude is bellow a threshold $T$

$$d_{j,T}^\omega[n] \stackrel{\text{def.}}{=} \begin{cases} d_j^\omega[n] & \text{if} \quad \|d_j^\omega[n]\| \leqslant T, \\ 0 & \text{otherwise.} \end{cases}$$

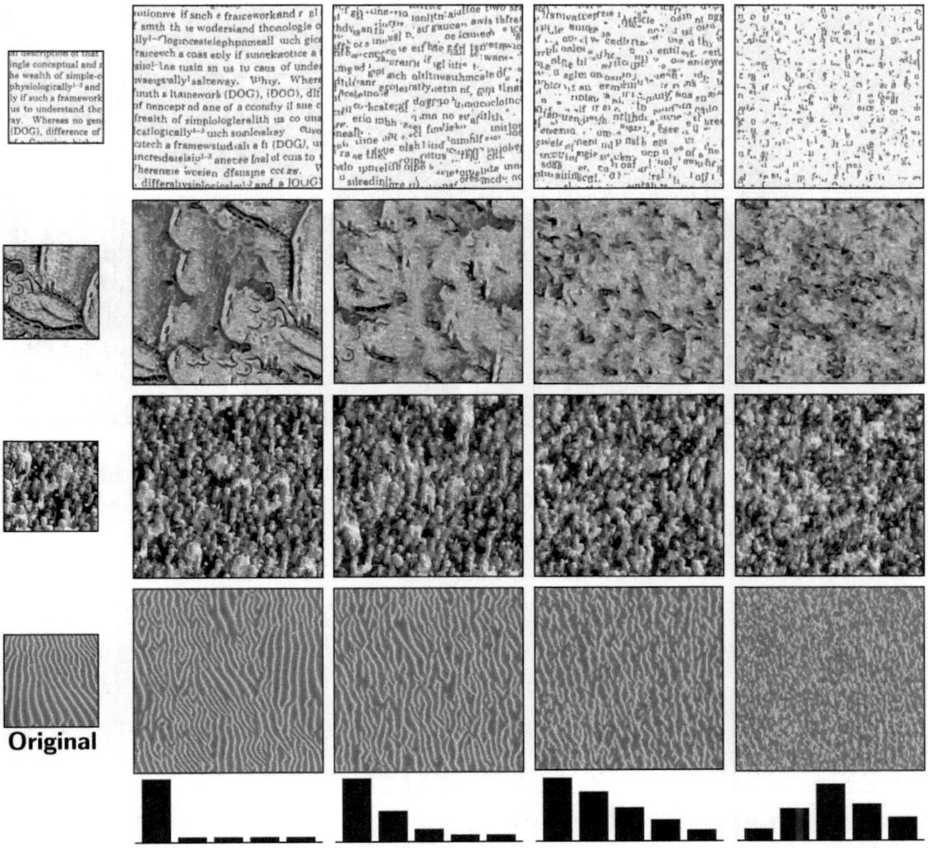


**Fig. 6.** Texture generation with varying multiscale spectral distributions. The bottom line shows the spectral variation $\sigma_j$ (coarse scales on the left). A small band indicate little deviation with respect to an ideal synthesis for the considered scale.

A more complex thresholding strategy could be used to include a normalization of the wavelet coefficients but we do not use it in our numerical experiments. Note that each wavelet detail $d_j^\omega[n]$ is a $p$-dimensional vector so its magnitude $\|d_j^\omega[n]\|$ is computed using the usual norm in $\mathbb{R}^p$. The thresholded wavelet coefficients $d_{j,T}^\omega$ are used to synthesize a modified feature valued function $\tilde{f}_T$ using the backward transform described in listing 2. The modified texture is computed as $f_T \stackrel{\text{def.}}{=} (\Phi_{2^J} \circ \tilde{f}_T)$.

On figure 7, bottom row, one can see how the wavelets coefficients are progressively shrinked toward zero. When $T = \max(d_j^\omega)$, the modification algorithm performs a pure texture synthesis since all wavelets coefficients are set to zero. On figure 8 one can see how the modification process smoothly interpolates between the two textures by progressively adding and removing texture structures when wavelet coefficients are thresholded to zero.

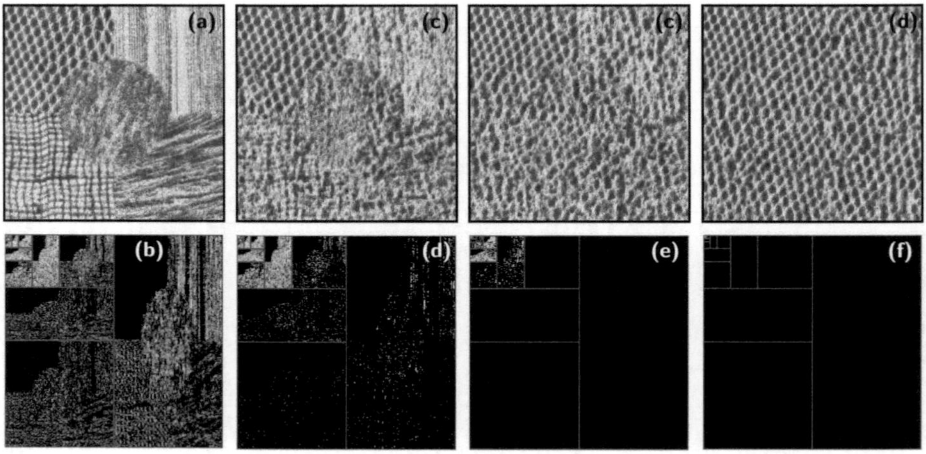


**Fig. 7.** Progressive modification of a texture. (b,d,f,h) Magnitude $S$ of the thresholded wavelets coefficients for $T/\max d_j^\omega = 0, 0.1, 0.5$ and 1 respectively. (a,c,e,g) Corresponding reconstructed textures.

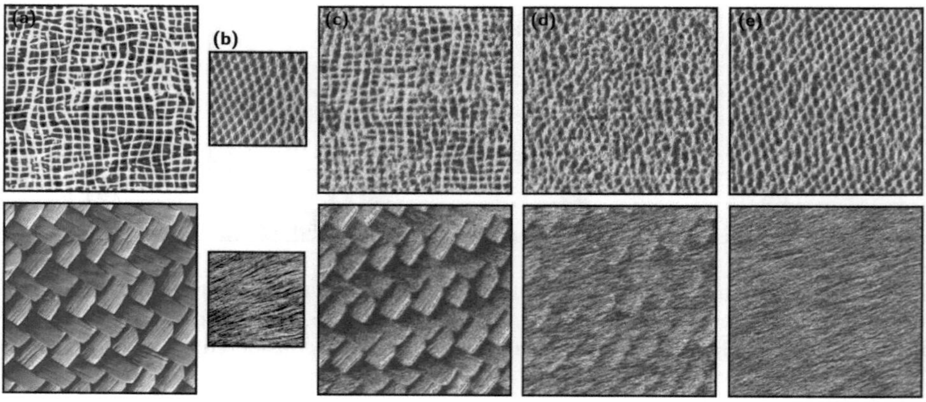


**Fig. 8.** Two examples of texture modification. (a) Input texture $f$. (b) Exemplar texture $f_e$. (c,d,e) Reconstruction from the thresholded wavelet coefficients for thresholds $T/\max(d_j^\omega) = 0.1, 0.5$ and 1 respectively.

## 6 Conclusion

In this paper, we have proposed a new texture model as a mapping into a non-linear features manifold. A multiscale decomposition of such a mapping is performed by using a pairwise association field. This non-linear analysis and synthesis framework allows to perform feature preserving texture generation and modification. This new model is promising for capturing multiple structures that exists in non-homogeneous textures using several target manifolds. Taking into account non-manifold constraints such as symmetries is another avenue for future work.

## References


1. E. Candès and D. Donoho. New tight frames of curvelets and optimal representations of objects with piecewise $c_2$ singularities. *Comm. Pure Appl. Math.*, 57(2):219–266, 2004.
2. D.L. Donoho and C. Grimes. Image manifolds isometric to euclidean space. *J. Math. Imaging Computer Vision*, 23, July 2005.
3. A. A. Efros and T. K. Leung. Texture synthesis by non-parametric sampling. In *ICCV '99: Proceedings of the International Conference on Computer Vision-Volume 2*, page 1033. IEEE Computer Society, 1999.
4. D. J. Heeger and J. R. Bergen. Pyramid-Based texture analysis/synthesis. In Robert Cook, editor, *SIGGRAPH 95 Conference Proceedings*, Annual Conference Series, pages 229–238. ACM SIGGRAPH, Addison Wesley, August 1995.
5. B. Julesz. Visual pattern discrimination. *IRE Trans. Inform. Theory*, 8(2):84–92, 1962.
6. E. Le Pennec and S. Mallat. Bandelet Image Approximation and Compression. *SIAM Multiscale Modeling and Simulation*, page to appear, 2005.
7. A.B. Lee, K.S. Pedersen, and D. Mumford. The nonlinear statistics of high-contrast patches in natural images. *International Journal of Computer Vision*, 54(1-3):83–103, August 2003.
8. S. Lefebvre and H. Hoppe. Parallel controllable texture synthesis. *ACM Trans. Graph.*, 24(3):777–786, 2005.
9. S. Lefebvre and H. Hoppe. Appearance-space texture synthesis. *ACM Trans. Graph.*, 2006.
10. S. Mallat. *A Wavelet Tour of Signal Processing*. Academic Press, San Diego, 1998.
11. W. Matusik, M. Zwicker, and F. Durand. Texture design using a simplicial complex of morphable textures. *ACM Trans. Graph.*, 24(3):787–794, 2005.
12. D. Mumford. Elastica and computer vision. *In C. L. Bajaj (Ed.), Algebraic geometry and its applications*, pages 491–506, 1994.
13. S. Roweis and L. Saul. Nonlinear Dimensionality Reduction by Locally Linear Embedding. *Science*, 290(5500):2323–2326, Dec. 2000.
14. A. Szlam, M. Maggioni, and R. R. Coifman. A general framework for adaptive regularization based on diffusion processes on graphs. *Yale technichal report*, July 2006.
15. J. B. Tenenbaum, V. de Silva, and J. C. Langford. A Global Geometric Framework for Nonlinear Dimensionality Reduction. *Science*, 290(5500):2319–2323, Dec. 2000.
16. I. Ur-Raman, I. Drori, V. Stodden, and P. Schroeder. Multiscale representations of manifold-valued data. *to appear in SIAM Multiscale Modeling and Simulation*, 2005.
17. L.R. Williams and D. W. Jacobs. Stochastic completion fields: a neural model of illusory contour shape and salience. *Neural Comput.*, 9(4):837–858, 1997.

# A Variational Framework for the Simultaneous Segmentation and Object Behavior Classification of Image Sequences

Laura Gui[1], Jean-Philippe Thiran[1], and Nikos Paragios[2]

[1] Signal Processing Institute (ITS), Ecole Polytechnique Fédérale de Lausanne (EPFL), Lausanne, Switzerland
[2] Laboratoire MAS - Ecole Centrale de Paris, Chatenay-Malabry, France

**Abstract.** In this paper, we advance the state of the art in variational image segmentation through the fusion of bottom-up segmentation and top-down classification of object behavior over an image sequence. Such an approach is beneficial for both tasks and is carried out through a joint optimization, which enables the two tasks to cooperate, such that knowledge relevant to each can aid in the resolution of the other, thereby enhancing the final result. In particular, classification offers dynamic probabilistic priors to guide segmentation, while segmentation supplies its results to classification, ensuring that they are consistent with prior knowledge. The prior models are learned from training data and updated dynamically, based on segmentations of earlier images in the sequence. We demonstrate the potential of our approach in a hand gesture recognition application, where the combined use of segmentation and classification improves robustness in the presence of occlusion and background complexity.

## 1 Introduction

Image segmentation is one of the most basic yet most challenging problems of computer vision. Segmentation requires finding in an image semantically salient regions (or their bounding contours) associated with "objects". Behavior classification in image sequences is an important higher level task towards comprehensive visual perception. By "behavior" of an object in an image sequence, we mean the temporal evolution of its attributes (such as position, orientation, shape, color, texture, etc.) apparent in the image sequence. The classification of object behavior refers to assigning one of several behavior class labels to each of its temporal evolution instances. For example, we would like to classify object motion (e.g., car turn directions at an intersection), classify motion and deformation (e.g., hand gestures, body motions), or classify intensity changes in a brain activation map for clinical purposes.

Conventionally, segmentation and behavior classification are solved separately and sequentially: one segments the image sequence, extracts the relevant features, and finally classifies their time evolution. However, the task of behavior

F. Sgallari, A. Murli, and N. Paragios (Eds.): SSVM 2007, LNCS 4485, pp. 652–664, 2007.

classification can be facilitated if segmentation information is available. Reciprocally, segmentation can greatly benefit from considering the expected behavior of the targeted object(s) (related to, e.g., its shape, color, texture), which is usually built into classification tasks as a priori models of the behavior classes to be distinguished. Therefore, benefits should accrue from a collaboration between image segmentation and behavior classification.

Our contribution in this paper is a coupled solution to image sequence segmentation and classification of object behavior, which enables the information related to each of them to enhance the results of both. To this end, we develop a new variational framework that smoothly integrates the two main sources of information: the target image sequence and the prior behavior models, which adapt dynamically to the latest segmented images through the classification strategy.

Variational methods underlie the mathematical formulation of numerous computer vision problems. The image segmentation problem has been formulated in terms of energy minimization, where one can seamlessly introduce various criteria describing the desired solution, such as smoothness, region homogeneity, edge correspondence, etc. Starting with the original active contour (snakes) model [1], continuing with the Mumford-Shah model [2], the introduction of the level set approach [3] and geodesic active contours [4,5,6], recent work has yielded versatile segmentation approaches such as [7,8]. Statistical shape priors were introduced into active contours [9] and also into level set active contours [10,11,12,13] and the Mumford-Shah segmentation [14,15,16]. These techniques have made it possible to successfully segment a familiarly-shaped object in difficult cases. Variational methods for contour evolution have also been adopted for object tracking (e.g., [1,17,14]). Coherence between frames has been exploited by approaches based on Kalman filtering [18], particle filtering [19], and autoregressive models [20].

Our proposed framework deals simultaneously with the issues of image sequence segmentation and object behavior classification, leading to increased chances of success for both tasks through cooperation and information sharing. On the one hand, segmentation is improved by guidance towards the target object via probabilistic priors, offered by classification. On the other hand, classification is improved from the consideration of segmentation results captured from new images, while maintaining consistency with prior knowledge and with previous segmentations in the sequence. To our knowledge, the fusion of segmentation and behavior classification over image sequences is novel in the domain of variational image analysis, while it of course capitalizes on existing experience in the use of shape priors. The idea of combining segmentation and object recognition has previously yielded good results in the case of single images, both in variational [15] and non-variational [21,22,23,24] settings. Our work makes a significant contribution in that we address *image sequences* and the temporal problem of *object behavior classification.* To tackle this problem, we introduce a variational framework that incorporates dynamic probabilistic priors automatically obtained via a machine learning approach. We illustrate the potential of our proposed approach in a gesture recognition application, where the combination of segmentation and classification dramatically increases the tolerance to occlusion and background complexity present in the input image sequence.

We propose a *general framework* for the joint resolution of the two tasks—segmentation and behavior classification—whose components can be adapted to suit the needs of a wide range of applications. The next section details the collaborating halves of our general framework, first behavior classification and then segmentation. A particular implementation of the framework is proposed in Section 3, which employs a specific image term and dynamic prior component, for the purposes of gesture recognition. Experimental results are presented at the end of Section 3. Section 4 concludes the paper.

## 2 Formulation of the Variational Framework

Our goal is to segment an image sequence and classify it in terms of object behavior. As illustrated in Fig. 1, the key idea in our framework is to interweave the classification and segmentation processes while iterating through the given image sequence. This enables them to collaborate in exploiting the available prior knowledge and to improve each other by sharing partial results obtained throughout the image sequence. More concretely, for each image in the sequence, classification offers dynamic probabilistic attribute priors to guide segmentation. These priors, which are based on training, adapt in time according to knowledge gained from new segmentations. In turn, segmentation detects, and supplies to classification, object attributes that best explain the image evidence consistently with the prior knowledge. These object attributes are used in the subsequent step of the classification, and so on, until the entire sequence is segmented and classified.

Note that by "attribute" we generically designate a visual property of the object of interest, representable as a functional $A(C, I)$ of the image $I$ and of the object's segmenting contour $C$ ($A$ is assumed to be differentiable with respect to $C$). The palette of such attributes is quite large, including all properties computable with boundary-based and/or region-based functionals, such as position, orientation, average intensity/color, or higher-order statistics describing texture.

### 2.1 Classification and Its Cooperation with Segmentation

The behavior classification task aims at estimating, for a given time instance of an image sequence, the behavior class of the object, based on its observed attributes. Supposing for the moment that the attribute values are known, we need only find the generating behavior classes. We solve this problem using the machine learning concept of generative models [25], in particular Hidden Markov Models (HMMs) [26], where the observations are attribute values and the hidden states are the unknown behavior classes. Once trained on typical attribute evolution sequences, an HMM classifies new attribute sequences by estimating the most likely state sequence generating them.

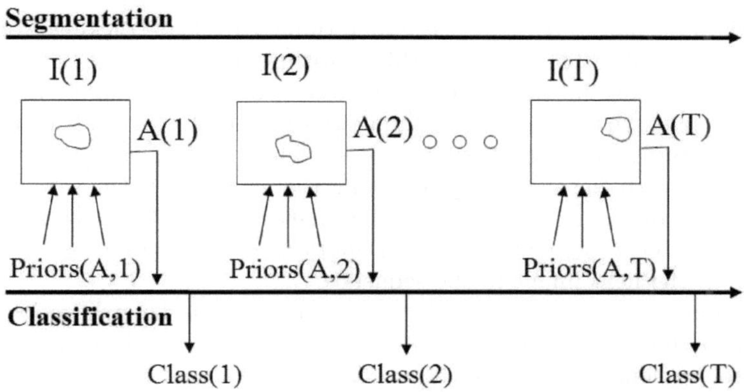


**Fig. 1.** Our approach: cooperation of segmentation and classification along the image sequence

We denote the states of the HMM (each corresponding to a behavior class) by $S = \{S_1, S_2, \ldots, S_M\}$, the state at time $t$ by $q_t$, and the attribute value at time $t$ by $A(t)$. The HMM parameters are:

1. the initial state distribution $\pi = \{\pi_i\}$, with $\pi_i = P(q_1 = S_i), \quad i = 1..M$,
2. the state transition probability distribution $T = \{t_{ij}\}$, with $t_{ij} = P(q_{t+1} = S_j | q_t = S_i), \quad i, j = 1..M$, and
3. the state observation probability distributions (class likelihoods)

$$P(A(t) \,|\, q_t = S_i) = P_i(A(t)),\ i = 1..M. \tag{1}$$

To support cooperation with the segmentation process, we require that these class likelihood functions $P_i(A(t))$ be differentiable with respect to $A(t)$.

Once having estimated the set $\lambda$ of HMM parameters from training data, the HMM can be used to classify new attribute sequences. In order to assign a behavior class to each observation in a new sequence $A_{1..T} = \{A(1), A(2), \ldots, A(T)\}$, we estimate the state sequence $q_{1..T}^{\text{opt}} = \{q_1, q_2, \ldots, q_T\}^{\text{opt}}$ that best explains the observation sequence

$$q_{1..T}^{\text{opt}} = \arg\max_{q_{1..T}} P(q_{1..T} | A_{1..T}, \lambda) = \arg\max_{q_{1..T}} P(q_{1..T}, A_{1..T} | \lambda), \tag{2}$$

using the Viterbi algorithm [26]. At each time step $t$ and for each state $S_i$, the Viterbi algorithm calculates the quantity

$$\delta_t(i) = \max_{q_1, q_2, \ldots, q_{t-1}} P(q_{1..t-1}, q_t = S_i, A_{1..t} | \lambda), \tag{3}$$

representing the highest probability along a state sequence, at time $t$, which explains the first $t$ observations and ends in state $S_i$. This quantity is computed by initializing the $\delta$s and then using the following recursion:

$$\delta_t(i) = (\max_j \delta_{t-1}(j)\, t_{ji})\; P_i(A(t) \,|\, \lambda). \tag{4}$$

Finally, the optimal state sequence is determined by backtracking from these maximization results. Thus, the Viterbi algorithm iterates through the attribute sequence, computing its best estimate for the probability of different generating classes, given the knowledge accumulated in the HMM. We can use these estimates to guide the segmentation process. The idea is to run this algorithm synchronously with the segmentation, using the attribute of the segmented object as the next observation as soon as it becomes available. Then, we incorporate the algorithm's best momentary class estimations as attribute priors for the segmentation of the next image in the sequence.

Now, suppose we have completed step $t-1$ of both the segmentation and the Viterbi algorithm, so that attributes $A_{1..t-1}$ and $\delta_{t-1}(j)$, $j = 1..M$, are available. In order to segment $I(t)$, we use the maximum available *a priori* knowledge:

1. the predictions of each class $i$ for the next attribute $A(t)$; i.e., the likelihood functions $P_i(A(t) \mid \lambda)$, $i = 1..M$
2. our relative confidence in the prediction of each class $i$, given by the Viterbi algorithm; i.e., the maximum probability of reaching state $S_i$ at time step $t$, after having observed attributes $A_{1..t-1}$:

$$w_t(i) = \max_{j=1..M} \delta_{t-1}(j) t_{ji} = \max_{q_1, q_2, \ldots, q_{t-1}} P(q_{1..t-1}, q_t = S_i, A_{1..t-1} | \lambda). \quad (5)$$

As prior information offered by each behavior class $i$, we shall use the product of these two quantities, which according to (4) is actually

$$\delta_t(A(t), i) = w_t(i)\, P_i(A(t) \mid \lambda), \;\; i = 1..M; \quad (6)$$

i.e., $\delta_t$ as a function of the unknown attribute $A(t)$. Next, we explain how to introduce these class contributions into the segmentation framework.

### 2.2 Segmentation and Its Cooperation with Classification

We take a variational approach to segmentation that incorporates the dynamic probabilistic priors offered by classification. For an image $I(t)$, these priors consist of the delta functions of the object attribute corresponding to each class $i$; i.e., $\delta_t(A(t), i)$. We introduce these class contributions into the segmentation model by means of a competition mechanism, since we are searching for a single "winning" class that best accounts for the generation of the next observation.

To create a "competition" during segmentation among the priors associated with different classes, we employ a labeling mechanism motivated by [15]. For each prior $i$, we use one label $L_i$, a scalar variable that varies continuously between 0 and 1 during energy minimization and converges either to 0 or 1. The value of the set of labels $\mathcal{L} = (L_1, \ldots, L_M)$ after convergence designates a "winner" among the attribute priors, corresponding to the probability which has been maximized through segmentation. Each of the prior terms carries a label factor equal to $L_i^2$. Competition is enforced by constraining the label factors to sum up to 1 through the addition of the term $(1 - \sum_{i=1}^{M} L_i^2)^2$ to the segmentation energy.

Once having run our joint segmentation/classification framework on the first $t-1$ frames of an image sequence, we segment $I(t)$ by minimizing with respect to the contour $C$ and the labels $\mathcal{L}$ the following energy functional:

$$E(C,\mathcal{L},I(t)) = E_{\text{data}}(C,I(t)) + \alpha E_{\text{prior}}(C,\mathcal{L},I(t)), \tag{7}$$

where $\alpha$ is a positive weight parameter. Here $E_{\text{data}}(C,I(t))$ can be any boundary-based or region-based segmentation energy, suitable to the application at hand (e.g., the energy proposed in [27]). The energy due to the priors is

$$E_{\text{prior}}(C,\mathcal{L},I(t)) = -\sum_{i=1}^{M} \log\left(\delta_t(A(C,I(t)),i)\right) L_i^2 + \beta\left(1-\sum_{i=1}^{M} L_i^2\right)^2, \tag{8}$$

where $\beta$ is a positive constant and the $\delta$ function is defined in (6). The only assumptions regarding energy (8) are that the likelihood functions $P_i(A(C,I(t))$ are differentiable with respect to the attribute $A(C,I(t)$ and the attribute is a differentiable functional of the contour $C$.

The minimization of (7) simultaneously with respect to the segmenting contour $C$ and the label vector $\mathcal{L}$ is performed via the calculus of variations and gradient descent. The contour $C$ is driven by image forces due to $E_{\text{data}}(C)$, and by the $M$ attribute priors due to $E_{\text{prior}}(C,\mathcal{L})$:

$$\frac{\partial C}{\partial \tau} = -\frac{\partial E_{\text{data}}(C,I(t))}{\partial C} - \alpha\frac{\partial E_{\text{prior}}(C,\mathcal{L},I(t))}{\partial C}. \tag{9}$$

Here $\partial E_{\text{data}}(C,I(t))/\partial C$ can be derived through the calculus of variations for the particular chosen form of $E_{\text{data}}(C,I(t))$. The second term can be written as:

$$\begin{aligned} \frac{\partial E_{\text{prior}}(C,\mathcal{L},I(t))}{\partial C} &= -\sum_{i=1}^{M} \frac{L_i^2}{\delta_t(A(C,I(t)),i)} \cdot \frac{\partial \delta_t(A(C,I(t)),i)}{\partial A} \cdot \frac{\partial A(C,I(t))}{\partial C}, \\ \text{with } \frac{\partial \delta_t(A(C,I(t)),i)}{\partial A} &= w_t(i)\,\frac{\partial P_i(A(C,I(t)) \mid \lambda)}{\partial A}. \end{aligned} \tag{10}$$

Derivatives $\partial P_i/\partial A$ and $\partial A(C,I(t))/\partial C$ are computed according to the particular likelihood function and attribute employed. In parallel with contour evolution, the labels compete to maximize the probability of the most likely prior given the image evidence:

$$\frac{\partial L_i}{\partial \tau} = \sum_{i=1}^{M} \delta_t(A(C,I(t)),i)\,L_i - \beta\,L_i\left(1-\sum_{i=1}^{M} L_i^2\right). \tag{11}$$

The effect of these equations is that the label $L_i$ corresponding to the maximum $\delta_t(A(C,I(t)),i)$ is driven towards 1 – i.e., the maximum $\delta_t$ is extremized – while the other labels are driven to 0.

In probabilistic terms, the minimization of our proposed energy using competing priors amounts to the maximization of the probability $\delta_t(A(t),i)$ with respect

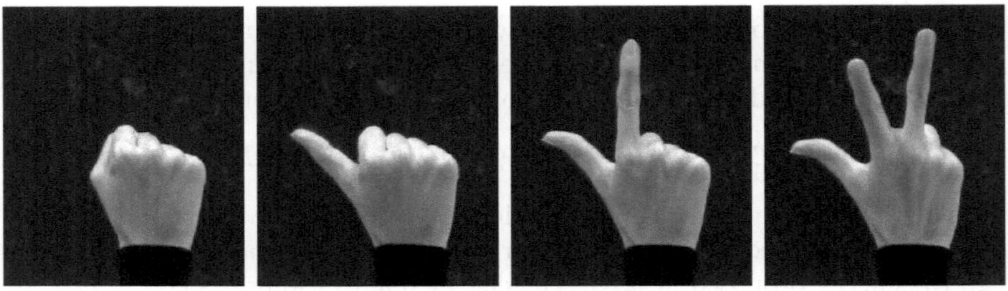

(a) Class 0 (b) Class 1 (c) Class 2 (d) Class 3

**Fig. 2.** Samples from the four gesture classes that we use in our application

to both the attribute $A(t)$ and class $i$, subject to image-based constraints. Then, the segmentation of image $I(t)$ can be regarded as the joint estimation of the attribute value $A^*(t)$ and the class $i^*$ as

$$\begin{aligned}(A^*(t), i^*) &= \arg\max_{A(t),i} \delta_t(A(t), i), \\ &\text{subject to image constraints } (A(t), I(t)).\end{aligned} \tag{12}$$

Thus, segmentation works concurrently towards the same goal as classification: maximizing the joint probability of the class and the observation at time $t$, while remaining consistent with previous observations, according to prior knowledge (through the HMM), and incorporating new information from image $I(t)$.

The segmentation of $I(t)$ yields $A(t)$, enabling the Viterbi algorithm to estimate $\delta_t(i)$ and $w_{t+1}(i)$, so that we can continue by segmenting $I(t+1)$ and repeat the cycle to the end of the image sequence. Finally, we obtain the classification of the image sequence as the most probable state sequence given the observations, by backtracking from the results of the Viterbi algorithm.

## 3 A Specific Implementation of Our Framework for Hand Gesture Recognition

We now demonstrate the strength of our framework of Section 2 in a hand gesture recognition application. After describing the problem that we wish to address, we detail the particular implementation of our general framework, including the specific model that we use. Finally, we present the results obtained.

### 3.1 Application

In our application, we identify four gesture classes consisting of a right hand going through four finger configurations, exemplified in Fig. 2: fist (Class 0), thumb extended (Class 1), thumb and index finger extended (Class 2), and thumb, index, and middle finger extended (Class 3). Given an image sequence of such gestures, our goal is to perform joint segmentation and classification; i.e., for each image, extract the segmenting contour of the hand and determine the gesture class to which it belongs. Note that our gesture image sequence depicts finger-counting from 1 to 3 and back to 1, ending with the initial fist position; i.e., the following succession of gesture classes: 0,1,2,3,2,1,0. Our strategy is to train a 4-class HMM with such sequences and then to incorporate the HMM into our framework in order to segment and classify new sequences of this sort. In an application with more complicated gesture scenarios, the optimal number of gesture sub-states could be determined as in [28].

### 3.2 Solution Using the Proposed Framework

For this application, the object attribute employed within our framework is the contour segmenting the hand $A(C, I) = C$. Using the level set approach [3], we represent the contour by the level set function (LSF) $\phi : \Omega \to \mathbb{R}$, chosen to be the signed distance function to the contour, so that $C \equiv \{(x, y) : \phi(x, y) = 0\}$.

Regarding the segmentation energy (7), as a data term we use the piecewise constant Mumford-Shah model as in [27]:

$$E_{\text{data}}(\phi) = \iint_{\Omega} (I - \mu_+)^2 H(\phi) dx dy + \iint_{\Omega} (I - \mu_-)^2 (1 - H(\phi)) dx dy \\ + \nu \iint_{\Omega} |\nabla H(\phi)| dx dy, \tag{13}$$

where $H$ is the Heaviside function and $\mu_+$, $\mu_-$ are mean image intensities over the positive, respectively negative regions of $\phi$. The prior energy is given by

$$E_{\text{prior}}(\phi, \mathcal{L}) = -\sum_{i=1}^{M} \log\left(\delta_t(\phi, i)\right) L_i^2 + \beta \left(1 - \sum_{i=1}^{M} L_i^2\right)^2, \tag{14}$$

where $\delta_t(\phi, i) = w_t(i)\, P_i(\phi)$. We use a local Gaussian probability model for each class $i$:

$$p_i^{(x,y)}(\phi) = \frac{1}{\sqrt{2\pi}\sigma_i((x, y))} e^{-\dfrac{(\phi(x, y) - \phi_i(x, y))^2}{2\sigma_i^2(x, y)}}, \tag{15}$$

where $(x, y) \in \Omega$ is an image location, $\phi_i$ is the average LSF of class $i$, and the variance $\sigma_i(x, y)$ models the local variability of the level set at $(x, y)$. Assuming densities independent across pixels, the likelihood of an LSF $\phi$, offered by class $i$, is the product of these densities over the image domain:

$$P_i(\phi) = \prod_{(x,y)\in\Omega} p_i^{(x,y)}(\phi). \tag{16}$$

Substituting likelihoods $P_i(\phi)$ in (14) and augmenting by similarity transformations (including translation, rotation, and scale) that align each prior $i$ with contour $\phi$, the prior energy becomes:

$$E_{\text{prior}}(\phi, \mathcal{L}, \boldsymbol{\tau}^{i=1..M}) = \sum_{i=1}^{M} \left( -\log w_t(i) + \iint_{\Omega} \left( \frac{(\phi(x,y) - \phi_i(h_{\boldsymbol{\tau}^i}(x,y)))^2}{2\sigma_i^2(h_{\boldsymbol{\tau}^i}(x,y))} \right. \right.$$
$$\left. \left. + \log \sigma_i(h_{\boldsymbol{\tau}^i}(x,y)) \right) dxdy \right) L_i^2 + \beta \left( 1 - \sum_{i=1}^{M} L_i^2 \right)^2 . \quad (17)$$

Here, $\boldsymbol{\tau} = \{s, \theta, T_x, T_y\}$ are the parameters of a similarity transformation

$$h_{\boldsymbol{\tau}}\left([x \;\; y]^T\right) = s \begin{pmatrix} \cos\theta & \sin\theta \\ -\sin\theta & \cos\theta \end{pmatrix} \begin{bmatrix} x \\ y \end{bmatrix} + \begin{bmatrix} T_x \\ T_y \end{bmatrix} . \quad (18)$$

Parameters $\boldsymbol{\tau}^i$ for each class $i$ evolve during segmentation according to their corresponding gradient descent equations, leading to the minimization of (17).

### 3.3 Training the Model

In the training phase, we estimated the parameters of the HMM (see, e.g., [26]) using a labeled, rigidly aligned sequence of LSFs corresponding to a manual segmentation of the mentioned gesture sequence (0,1,2,3,2,1,0). We used the method in [13] to obtain smooth estimates of the mean $\phi_i$ and variance $\sigma_i$ for each gesture class $i$.

### 3.4 Results

In the testing phase, we run classification and segmentation jointly on new image sequences of a hand performing the same succession of gestures in front of a complex background, in the presence of occlusions and noise. By virtue of the prior information supplied by classification, segmentation is able to cope with severe occlusions, as can be seen in Fig. 3(a)–(d), (i)–(l). Figure 3(e)–(h), (m)–(p) shows that conventional segmentation of the same sequences is clearly inferior, failing to recover the desired shape of the object because of the occlusions.

Fig. 4 shows the classification results for the first test sequence, which correctly follow our understanding of the executed gestures. Moreover, the final classification yielded by the Viterbi algorithm corresponds to the partial classification results used to guide segmentation throughout the sequence. This can be seen in Fig. 4, which exhibits, as functions of time (frame), (a) the final classification, (b) the delta functions of each class, and (c) the prior confidence of each class (the $w$ function) used as input to the segmentation. The $w$ values are scaled with respect to their maximum value for every frame.

Even though our chosen test application might not seem especially challenging, the proposed framework can potentially be applied to much more complicated scenarios. Its power lies in its flexibility: it allows a large variety of implementations, capitalizing on existing expertise in both probabilistic learning and variational segmentation.

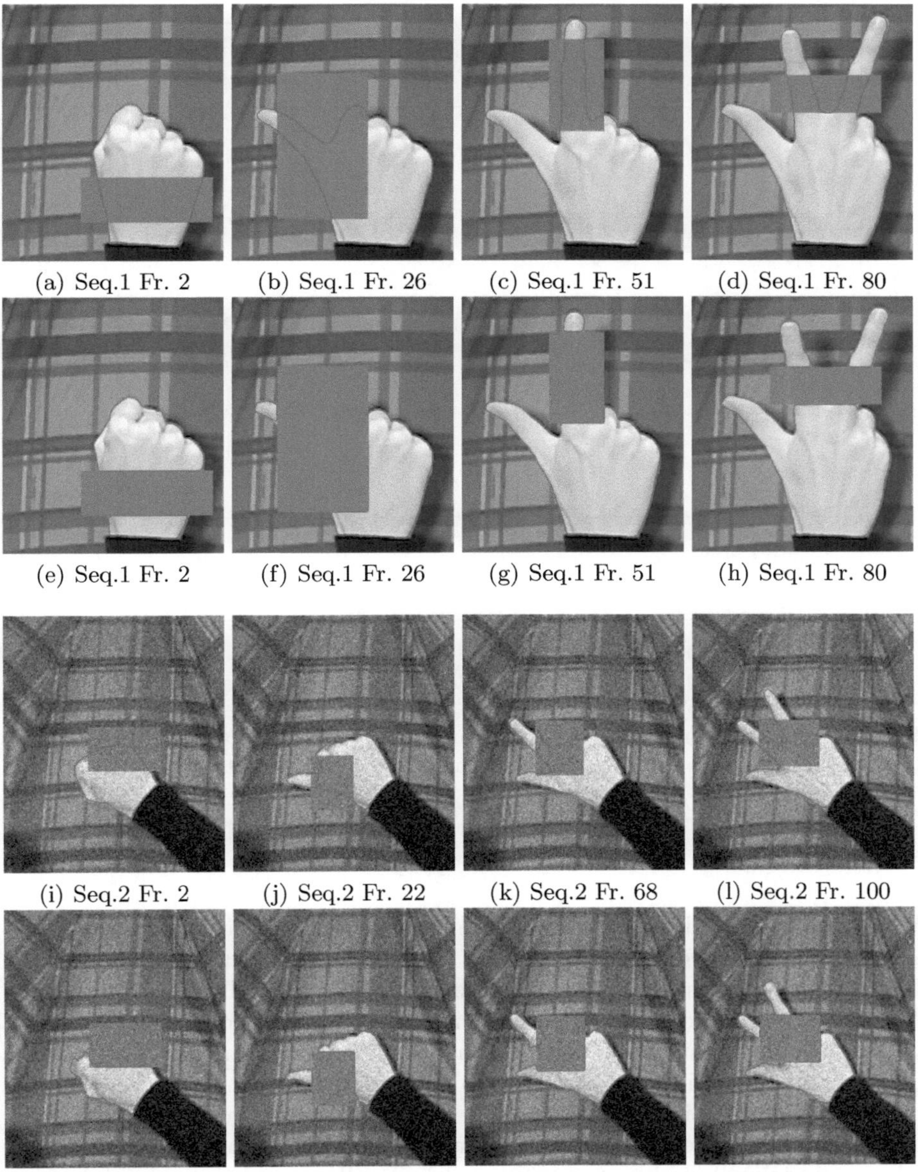

(a) Seq.1 Fr. 2 (b) Seq.1 Fr. 26 (c) Seq.1 Fr. 51 (d) Seq.1 Fr. 80

(e) Seq.1 Fr. 2 (f) Seq.1 Fr. 26 (g) Seq.1 Fr. 51 (h) Seq.1 Fr. 80

(i) Seq.2 Fr. 2 (j) Seq.2 Fr. 22 (k) Seq.2 Fr. 68 (l) Seq.2 Fr. 100

(m) Seq.2 Fr. 2 (n) Seq.2 Fr. 22 (o) Seq.2 Fr. 68 (p) Seq.2 Fr. 100

**Fig. 3.** (a)–(d), (i)–(l) Segmentation with the proposed framework of two image sequences in the presence of occlusion, background complexity and noise (second sequence). (e)–(h), (m)–(p) Conventional segmentation of the same image sequences

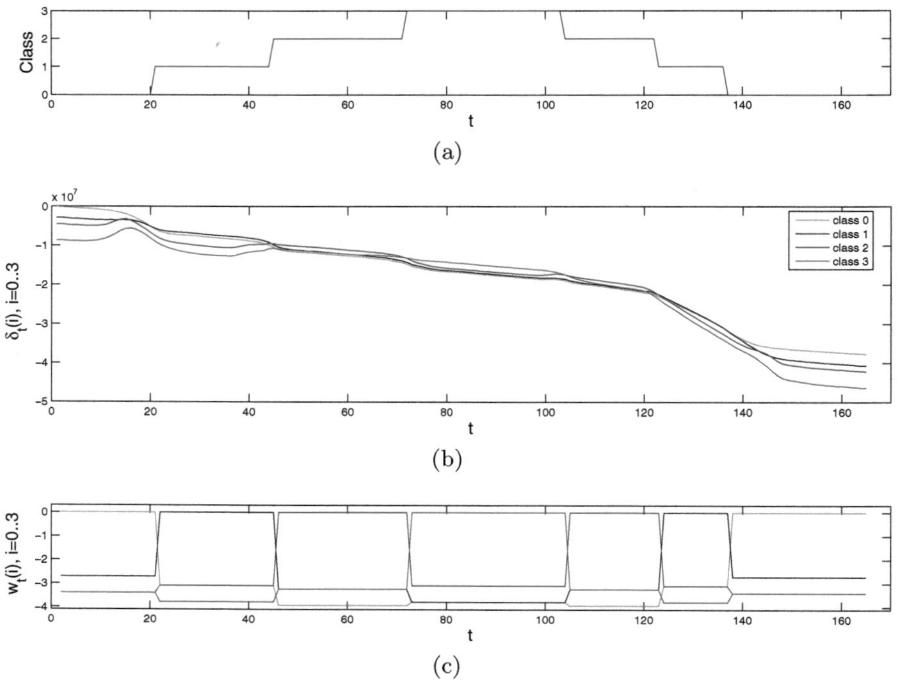


**Fig. 4.** Classification results plotted per frame. (a) Final classification. (b) Delta functions of each class. (c) Prior confidence of each class used as input to the segmentation.

## 4 Conclusion

We have introduced a novel variational framework for the simultaneous segmentation and object behavior classification of image sequences. Cooperation between the segmentation and classification processes facilitates a mutual exchange of information, which is beneficial to their joint success. In particular, we employed a classification strategy based on generative models that provides dynamic probabilistic attribute priors to guide image segmentation. These priors allow the segmentation process to work towards the same goal as classification, by outlining the object that best accounts for both image data and prior knowledge encapsulated in the generative model. We illustrated the potential of our general framework in a hand gesture analysis application, where we successfully segmented and classified image sequences of a gesturing hand before a complex background, in the presence of occlusions and noise. Future directions of our work will include the use of more complex attribute priors that would be better suited to challenging, under-constrained problems in high-dimensional spaces, such as the inference of 3D hand pose and behavior from monocular images.

## Acknowledgements

The collaborative work between the authors of this paper has been made possible via the PAI De Stael 08392QA French-Swiss program. The authors would like to thank Xavier Bresson for fruitful discussions supportive of this paper.

## References

1. Kass, M., Witkin, A., Terzopoulos, D.: Snakes: Active contour models. International Journal of Computer Vision **1** (1987) 321–331
2. Mumford, D., J.Shah: Optimal approximations by piecewise smooth functions and associated variational problems. Communications in Pure and Applied Mathematics **42** (1989) 577–685
3. Osher, S., Sethian, J.: Fronts propagating with curvature-dependent speed: Algorithms based on the Hamilton-Jacobi formulation. Journal of Computational Physics **79** (1988) 12–49
4. Caselles, V., Kimmel, R., Sapiro, G.: Geodesic active contours. In: Proc. IEEE Intl. Conf. on Comp. Vis., Boston, USA (1995) 694–699
5. Kichenassamy, S., Kumar, A., Olver, P., Tannenbaum, A., Yezzi, A.: Gradient flows and geometric active contour models. In: Proc. IEEE Intl. Conf. on Comp. Vis. (1995) 810–815
6. Malladi, R., Sethian, J., Vemuri, B.: Shape modeling with front propagation: A level set approach. IEEE PAMI **17** (1995) 158–175
7. Vese, L., Chan, T.: A multiphase level set framework for image segmentation using the Mumford and Shah model. International Journal of Computer Vision **50(3)** (2002) 271–293
8. Paragios, N., Deriche, R.: Geodesic active regions and level set methods for supervised texture segmentation. International Journal of Computer Vision **46(3)** (2002) 223–247
9. Cootes, T., Beeston, C., Edwards, G., Taylor, C.: Unified framework for atlas matching using active appearance models. Intl Conf. Inf. Proc. in Med. Imaging (1999) 322–333
10. Leventon, M., Grimson, W., Faugeras, O.: Statistical shape influence in geodesic active contours. In: IEEE International Conference on Computer Vision and Pattern Recognition. (2000) 316–323
11. Tsai, A., Yezzi, A., Wells, W., Tempany, C., Tucker, D., Fan, A., Grimson, W., Willsky, A.: Model-based curve evolution technique for image segmentation. In: IEEE International Conference on Computer Vision and Pattern Recognition. (2001) 463–468
12. Chen, Y., Tagare, H., Thiruvenkadam, S., Huang, F., Wilson, D., Gopinath, K., Briggs, R., Geiser, E.: Using prior shapes in geometric active contours in a variational framework. International Journal of Computer Vision **50(3)** (2002) 315–328
13. Paragios, N., Rousson, M.: Shape priors for level set representations. In: European Conference in Computer Vision. Volume 2. (2002) 78–92
14. Cremers, D., Osher, S., Soatto, S.: Kernel density estimation and intrinsic alignment for knowledge-driven segmentation: Teaching level sets to walk. Pattern Recognition **3175** (2004) 36–44
15. Cremers, D., Sochen, N., Schnör, C.: Multiphase dynamic labeling for variational recognition-driven image segmentation. In: European Conference on Computer Vision. Volume 3024. (2004) 74–86

16. Bresson, X., Vandergheynst, P., Thiran, J.P.: A variational model for object segmentation using boundary information and shape prior driven by the Mumford-Shah functional. International Journal of Computer Vision **28**(2) (2006) 145 – 162
17. Paragios, N., Deriche, R.: Geodesic active regions and level set methods for motion estimation and tracking. Computer Vision and Image Understanding **97** (2005) 259–282
18. Terzopoulos, D., Szeliski, R.: Tracking with Kalman snakes. Active vision (1993) 3–20
19. Rathi, Y., Vaswani, N., Tannenbaum, A., Yezzi, A.: Particle filtering for geometric active contours with application to tracking moving and deforming objects. In: Proc. CVPR. Volume 2. (2005) 2–9
20. Cremers, D., Funka-Lea, G.: Dynamical statistical shape priors for level set based tracking. In LNCS, S., ed.: 3rd. Workshop on Variational, Geometric and Level Set Methods in Computer Vision. Volume 3752. (2005) 210–221
21. Tu, Z., Chen, X., Yuille, A., Zhu, S.: Image parsing: Segmentation, detection, and recognition. In: ICCV. (2003) 18–25
22. Leibe, B., Leonardis, A., Schiele, B.: Combined object categorization and segmentation with an implicit shape model. In: ECCV Workshop on SLCV. (2004)
23. Ferrari, V., Tuytelaars, T., Gool, L.V.: Simultaneous object recognition and segmentation by image exploration. In: ECCV. (2004)
24. Kokkinos, I., Maragos, P.: An Expectation Maximization approach to the synergy between image segmentation and object categorization. In: ICCV. (2005) 617–624
25. Bishop, C.: Pattern Recognition and Machine Learning. Springer (2006)
26. Rabiner, L.: A tutorial on Hidden Markov Models and selected applications in speech recognition. Proceedings of the IEEE **77(2)** (1989)
27. Chan, T., Vese, L.: Active contours without edges. IEEE Transactions on Image Processing **10(2)** (2001) 266–277
28. Stenger, B., Ramesh, V., Paragios, N., Coetzee, F., Bouhman., J.: Topology-free Hidden Markov Models: Application to background modeling. In: ICCV. (2001) 294–301

# Blur Invariant Image Priors

Marco Loog[1,2] and Francois Lauze[2]

[1] Department of Computer Science
University of Copenhagen
Copenhagen, Denmark
[2] Nordic Bioscience A/S
Herlev, Denmark

**Abstract.** We introduce functionals on image spaces that can be readily interpreted as image priors, i.e., probability distributions expressing one's uncertainty before having observed any (image) data. However, as opposed to previous work in this area, not the actual images are considered but their *observed* version, *i.e.*, we assume the images are obtained be means of a linear aperture, which is typically taken to be Gaussian. More specifically, we consider those functionals that are invariant under blurring of the observed images and the main aim is to fully describe the class of admissible functionals under these assumptions. As it turns out, this class of 'priors' is rather large and adding additional constraints may be considered to restrict the possible solutions.

## 1 Introduction

Image statistics have been widely studied, especially from an empirical point of view. After the seminal paper by Field [1], many other publications in this field followed of which the reader can find an extensive overview in [2] (see also [3]).

Less attention has been given to obtaining insight in image statistics through theoretical considerations, reasoning from first, invariance, or physical principles. Examples of the latter approach can be found in [4,5,6,7,8,9]. In these works, often the notion of scale invariance—or at least some self-similarity assumptions or underlying power-law behavior—plays a major role. As pointed out in [7], defining priors on image spaces which exhibit a scale invariant structure poses some serious problems, the main one being that there exists no scale invariant measures on functions spaces and one needs to consider images as generalized functions (see also [10]). All in all, there seem to remain some serious flaws to what are reasonable prior model for images. For instance, histograms of empirical measurements in images do not fit the marginals of the suggested theoretical models particularly well [7].

This paper relaxes the stringent scale invariance requirement and studies, what we call, blur invariance (*cf.* [6]) of the prior probability measure. The notion, however, still comes rather close to that of scale invariance and states that a distribution on images should not change when the whole ensemble of images is blurred by a kernel of the same scale, as is explained in more detail shortly. In a sense, this work also deviates from other's in that *observed* images are considered and not the underlying real physical process, *i.e.*, images are obtained by convolving a, possibly generalized, function with (say) a Gaussian kernel.

F. Sgallari, A. Murli, and N. Paragios (Eds.): SSVM 2007, LNCS 4485, pp. 665–674, 2007.

### 1.1 Outline

Next section starts by providing description of the way functionals can be interpreted as improper priors on function spaces. Subsequently, for comparison purposes, a definition of scale invariant priors is provided, which is followed by a formulation of the notion of blur invariance. Section 3 provides the basic theoretical manipulations that aim at fully describing the class of admissible functionals that are blur invariant. This class turns out to be huge and Section 4 dwells on imposing additional constraints to further reduce the class of functionals of interest. Section 5 provides some further remarks and concludes this paper.

## 2 Prior Functionals and Invariance

We interpret a functional $\mathscr{L}$ on a space of images $I : \mathbb{R}^d \to \mathbb{R}$ as a—possibly improper —prior on this image space if $\mathscr{L}[I] \geq 0$ for all $I$. The idea is that, the absolute value that $\mathscr{L}[I]$ takes on does not matter. We want to have the opportunity to compare the 'probabilities' of two images, say $I$ and $J$, and are therefore merely interested in the quantity $q$ providing their relative probability

$$q(I, J) = \frac{\mathscr{L}[I]}{\mathscr{L}[J]}, \tag{1}$$

or, equivalently, $\log q(I, J) = \log \mathscr{L}[I] - \log \mathscr{L}[J]$. With this construct, we avoid issues concerning the normalization of distributions on infinite dimensional function spaces, while keeping a possibility of devising computational models based on this methodology.

### 2.1 Scale and Blur Invariance

We are now in the position to define scale invariance and blur invariance of functionals on images.

Let $\mathscr{S}_c$ be the scaling operator that scales images $I : \mathbb{R}^d \to \mathbb{R}$ with a constant $c > 0$ in the following way

$$\mathscr{S}_c[I](x) = c^a I(cx), \tag{2}$$

for some exponent $a$. A functional $\mathscr{L}$ is said to be (relatively) invariant under scaling of images if the following holds for every $c \in \mathbb{R}^+$ and every image $I$ (*cf.* [7,8]):

$$q(\mathscr{S}_c[I], \mathscr{S}_c[J]) = q(I, J). \tag{3}$$

Blur invariance, on the other hand, is not defined through a rescaling of the images as such, but through a scaling of the aperture by which images are observed. In this work, we restrict our attention to Gaussian apertures [11] and consider blurred version $I_\sigma$ of images $I$, and the blurring operator $\mathscr{B}_\sigma$ may be defined as:

$$\mathscr{B}_\sigma[I](x) = I_\sigma(x) = \int_{\mathbb{R}^d} g_\sigma(x - y) I(y) dy, \tag{4}$$

in which $g_\sigma$ is a Gaussian kernel at scale $\sigma$. Now blur invariance of a functional $\mathscr{L}$ is defined as

$$q(\mathscr{B}_\sigma[I], \mathscr{B}_\sigma[J]) = q(I, J)\,, \tag{5}$$

which should hold for *every* choice of $\sigma$ in which negative values formally indicate deblurring.

## 2.2 The Idea Behind Blur Invariance

The previous requirement basically describes that we do not want the overall properties of the *observed* image space we are working in to change with changing aperture. That this can indeed be stated as above is explained in more detail here.

Consider a standard probability density function $P_X : \mathbb{R}^d \to \mathbb{R}_0^+$ on a finite dimensional space. Applying an invertible *linear* operator $L$ to the elements $x$ in this space, such that $y = L(x)$, we can express the probability density of the transformed random variables $y$ in terms of the initial probability density function and the absolute determinant $\lambda = |\det L|$ of the transform. The change of variables formula [13] immediately gives

$$P_Y(y) = \lambda^{-1} P_X(L^{-1}(y))\,. \tag{6}$$

Now, requiring that the overall statistics in $\mathbb{R}^d$ do not get changed by the transform $L$, can be readily stated as the condition that

$$P_Y = P_X\,, \tag{7}$$

*i.e.*, $P_Y(x) = P_X(x)$ for every instantiation $x$. As $P_Y = \lambda^{-1} P_X(L^{-1}(\cdot))$, this obviously leads to the requirement that

$$\lambda^{-1} P_X(L^{-1}(x)) = P_X(x) \tag{8}$$

for all $x$, or equivalently, assuming $L$ to be sufficiently regular,

$$\lambda^t P_X(L^t(x)) = P_X(x)\,, \tag{9}$$

for every power $t \in \mathbb{R}$.

With the blur invariance condition in Equation (5), we want to express the fact that the way the *perceived* visual world looks like statistically, is independent of the scale at which it is observed; the overall statistics should remain the same. Noting that the blurring operation $\mathscr{B}_\sigma$ is linear, we see that we are essentially in the same situation as in the finite dimensional space described above, which is characterized by Equation (7) or, equivalently, Equation (9) for all $t$. The actual requirement is simply slightly reformulated and the normalizing inverse of the absolute determinant of the linear transform $\lambda$ is removed by considering the quotient of the functionals $\mathscr{L}$ defining the prior on the image space, *i.e.*, we consider the quotient

$$\frac{\mathscr{L}[\mathscr{B}_\sigma[I]]}{\mathscr{L}[\mathscr{B}_\sigma[J]]} = \frac{\mathscr{L}[I]}{\mathscr{L}[J]}\,, \tag{10}$$

which is Equation (5).

### 2.3 Blurring vs. Scaling

Although the condition of blur invariance somehow resembles that of scale invariance to a certain extent, they are different in an important way. It turns out that they have quite distinct implications concerning necessary requirements on the underlying function spaces (*cf.* [10,7,3]). In this work, we restrict ourselves to merely providing an illustration of this issue and only concern ourselves with the actual functional on these spaces. We do, however, want to emphasize that the topic is an important one in need of further investigation.

The illustration: Let the sinusoidal image $I : \mathbb{R}^d \to \mathbb{R}$ be defined by

$$I(x) = \sin\langle\omega, x\rangle \tag{11}$$

with $\omega \neq 0$ a vector of frequencies for the various direction. Assume that $I$ is an element of the image space under consideration. Imposing scale invariance basically means that every function of the form $c^\alpha \sin(c\langle\omega, x\rangle)$, $c > 0$ should also be an element in the image space, *i.e.*, functions having arbitrarily rapid oscillations should be part of this space with the limiting function being rather ill-defined.

On the other hand, imposing blur invariance, $I$ being an element of the image space merely implies that all function of the form $c \sin\langle\omega, x\rangle$, with $c > 0$, should also be an element of the space. This follows from the fact that $I$ is an eigenfunction of the blurring operator [14] (see also [15]) and therefore only the amplitude is being altered. In a sense, over the whole (de)blurring range, the class of functions stays rather well-behaved as compared to the one encountered when imposing scale invariance. The latter exhibits arbitrarily rapid oscillations as part of its solution space, which—especially in a more realistic setting, in which many more functions are part of image space—may be harder to deal with than the behavior induced by blur.

## 3 Construction of Blur Invariant Functionals

Equation (5) formulates the restriction we want to impose on functionals and in this section we aim to describe the complete class of functionals that fulfill the requirement. It is again illustrative to switch back to the finite dimensional setting and solve the problem there. Afterward, we can reformulate the results obtained for the infinite dimensional case and switch back to functionals on image spaces.

### 3.1 Finite Dimensional Considerations

Consider an invertible positive definite matrix $A$, which has an eigenvalue decomposition $VDV^{-1}$ with diagonal matrix

$$D = \begin{pmatrix} d_1 & & \\ & \ddots & \\ & & d_n \end{pmatrix} . \tag{12}$$

Let $H : \mathbb{R}^n \to \mathbb{R}$ be a homogeneous function of order $\alpha$, that is,

$$H(cx) = c^\alpha H(x) \tag{13}$$

for all $x \in \mathbb{R}^n$ and $c \in \mathbb{R}^+$ [16]. Furthermore, define the transformation $T : \mathbb{R}^n \to \mathbb{R}^n$, which transforms every element from $\mathbb{R}^n$ by the inverted eigenvector matrix and takes every $i$th component to the power of $\log d_i$, the logarithm of the $i$th eigenvalue $d_i$. That is,

$$T_i(x) \mapsto (V^{-1}x)_i^{\frac{1}{\log d_i}} \tag{14}$$

for all $i \in \{1, \ldots, n\}$.

As is demonstrated now, the function $\Phi : \mathbb{R}^n \to \mathbb{R}$, defined as $\Phi := H \circ T$, is invariant—in the sense of Equation (9)—under the transformation group $A^t$, where $t \in \mathbb{R}$ and

$$A^t = VD^tV^{-1} = V\begin{pmatrix} d_1^t & & \\ & \ddots & \\ & & d_n^t \end{pmatrix}V^{-1}. \tag{15}$$

First note that for all $i$

$$\begin{aligned} T_i(A^tx) &= T_i(VD^tV^{-1}x) = (V^{-1}VD^tV^{-1}x)_i^{\frac{1}{\log d_i}} \\ &= (D^tV^{-1}x)_i^{\frac{1}{\log d_i}} = (d_i^t(V^{-1}x)_i)^{\frac{1}{\log d_i}} \\ &= d_i^{\frac{t}{\log d_i}}(V^{-1}x)_i^{\frac{1}{\log d_i}} = e^t(V^{-1}x)_i^{\frac{1}{\log d_i}} = e^tT_i(x)\,. \end{aligned} \tag{16}$$

From the previous, we derive that for all $t \in \mathbb{R}$

$$\begin{aligned} \Phi(A^t(x)) &= H(T(A^t(x))) \\ &= H((e^t(V^{-1}x)_1^{\frac{1}{\log d_1}}, \ldots, e^t(V^{-1}x)_n^{\frac{1}{\log d_n}})^\mathsf{T}) \\ &= H(e^t((V^{-1}x)_1^{\frac{1}{\log d_1}}, \ldots, (V^{-1}x)_n^{\frac{1}{\log d_n}})^\mathsf{T}) \\ &= e^{\alpha t}H(((V^{-1}x)_1^{\frac{1}{\log d_1}}, \ldots, (V^{-1}x)_n^{\frac{1}{\log d_n}})^\mathsf{T}) = e^{\alpha t}H(T(x)) = \lambda_t\Phi(x)\,, \end{aligned} \tag{17}$$

with $\lambda_t = e^{\alpha t}$. This proofs our claim.

### 3.2 Blur and Invariant Functionals on Infinite Dimensional Spaces

In this subsection, we simply take the results from the previous and reformulate them in terms of infinite dimensional spaces. Additionally, the operator $A$ is now restricted to be the blurring operator, which we denoted by $\mathscr{B}_\sigma$. Exponentiating $A$ with $t$ readily corresponds to blurring at a certain scale, say, $\sigma$ and we may identify $A^t$ with $\mathscr{B}_\sigma$. As already indicated, $\sigma$ may also be taken to be negative indeed, which corresponds to a deblurring of the images.

Similar to the function $H$ from Equation (13), take $\mathscr{H}$ to be an $\alpha$-homogeneous functional on images, i.e,

$$\mathscr{H}(cI) = c^\alpha\mathscr{H}(I) \tag{18}$$

for all $I$ and $c \in \mathbb{R}^+$.

We also define the analogous form of the transformation $T$ as in Equation (14). The function $\mathscr{T}$ is given by

$$\mathscr{T}[I](\xi) \mapsto (\mathscr{F}[I](\xi))^{-\frac{2}{\|\xi\|^2}} \tag{19}$$

with $\xi$ in the image domain $\mathbb{R}^d$. In this equation, $\mathscr{F}$ denotes the Fourier transform, which diagonalizes the blurring operator and the $-\frac{2}{\|\xi\|^2}$ are the operator's spectral values, $e^{-\frac{1}{2}\|\xi\|^2}$, of which the logarithm and the reciprocal are taken (*cf.* $\frac{1}{\log d_i}$ in Equation (14)).

The combination of both operators $\mathscr{H} \circ \mathscr{T}$, which we denote by $\mathscr{P}$,

$$\mathscr{P}[I] = \mathscr{H}\left[(\mathscr{F}[I])^{-\frac{2}{\|\cdot\|^2}}\right], \tag{20}$$

is blur invariant in the sense of Equation (5), as can be checked by substituting the solution from Equation (20) into Equation (5):

$$\begin{aligned}\frac{\mathscr{P}[\mathscr{B}_\sigma[I]]}{\mathscr{P}[\mathscr{B}_\sigma[J]]} &= \frac{\mathscr{H}\left[(\mathscr{F}[\mathscr{B}_\sigma[I]](x))^{-\frac{2}{\|x\|^2}}\right]}{\mathscr{H}\left[(\mathscr{F}[\mathscr{B}_\sigma[J]](x))^{-\frac{2}{\|x\|^2}}\right]} = \frac{\mathscr{H}\left[\left(\hat{I}(x)e^{-2\sigma^2\|x\|^2}\right)^{-\frac{2}{\|x\|^2}}\right]}{\mathscr{H}\left[\left(\hat{J}(x)e^{-2\sigma^2\|x\|^2}\right)^{-\frac{2}{\|x\|^2}}\right]} \\ &= \frac{\mathscr{H}\left[e^{-\sigma^2}\left(\hat{I}(x)\right)^{-\frac{2}{\|x\|^2}}\right]}{\mathscr{H}\left[e^{-\sigma^2}\left(\hat{J}(x)\right)^{-\frac{2}{\|x\|^2}}\right]} = \frac{\mathscr{H}\left[\left(\hat{I}(x)\right)^{-\frac{2}{\|x\|^2}}\right]}{\mathscr{H}\left[\left(\hat{J}(x)\right)^{-\frac{2}{\|x\|^2}}\right]} = \frac{\mathscr{P}[I]}{\mathscr{P}[J]},\end{aligned} \tag{21}$$

where the second to last equality follows from the homogeneity of $\mathscr{H}$.

As a rather modest example: $\mathscr{P}[g_{\sqrt{\sigma}}]$ equals $ce^{\alpha\sigma}$ for some constant $c$, which in turn leads to $q(g_{\sqrt{\sigma}}, g_{\sqrt{\tau}})$ being equal to $e^{\alpha(\sigma-\tau)}$, i.e., the probability is merely dependent on the difference in (squared) scale.

### 3.3 On Homogeneous Functions

Every single homogeneous function $\mathscr{H}$ used in Subsection 3.2 leads to a different prior functional on images. To get a feel for the richness of this class of functions, and thus for the family of admissible functionals, consider first the $\alpha$-homogeneous functions $F$ on finite dimensional spaces again [16]. For this, define the intermediate functions $A : \mathbb{R}^d \to S^{d-1}$, where $S^{d-1}$ denotes the $d-1$-dimensional sphere in $\mathbb{R}^d$, and we have $A(x) = \frac{x}{\|x\|}$ and $R : \mathbb{R}^d \to \mathbb{R}_0^+$, with $R(x) = \|x\|$.

Given an arbitrary function $F : S^{d-1} \to \mathbb{R}_0^+$, the following function $H$ now provides a general homogeneous function with index $\alpha$: $H(x) = R^\alpha(x)F(A(x))$, with $x = 0$ mapping to 0.

Firstly, we note that $R^\alpha$ leads to a power-law behavior depending on the length of the vector $x$ for which $F \circ A$ provides the amplitude of the power-low corresponding to every unit 'direction' $\frac{x}{\|x\|}$. Secondly, we point out that blur invariance does not seem to impose very heavy restrictions on the actual form of functionals. In a sense, it takes out merely a single degree of freedom, *i.e.*, because of the necessary homogeneity of the function $H$, we are not free to choose an arbitrary function on $\mathbb{R}^d$, however we are only restricted to $S^{d-1}$.

Now, considering $\mathscr{H}$, similar observations can be made. However, in order to find mappings equivalent to $A$ and $R$, we need the infinite dimensional image space to be normed such that a norm $||I||$ is readily available. In the latter case one can define mappings $\mathscr{A}$ and $\mathscr{R}$ on function spaces in analogy with the functions $A$ and $R$ introduced above. If the space is not normed, one has to work directly with the definition in Equation (18). One way or the other, the possible choices for $\mathscr{H}$ are huge and we may want to restrict the class further by imposing additional (invariance) constraints. Two such restrictions are briefly discussed in the next section.

# 4 Two Additional Restrictions

In addition to blur invariance, we can also impose translation invariance, *i.e.*, stationarity, or rotation invariance—both sensible requirements in image processing and analysis tasks.

## 4.1 Translation Invariance

Observing that translating an image $I$ by means of vector $\xi$, simply means that the Fourier transform employed in Equation (20) is multiplied by $e^{i\langle\xi,x\rangle}$. One can easily get rid of this by taking the absolute value of the Fourier transform, *i.e.*, the power spectrum is taken. In short, to get translation invariance, one can simply consider functionals of the form

$$\mathscr{P}'[I] = \mathscr{H}\left[|\mathscr{F}[I]|^{-\frac{2}{\|\cdot\|^2}}\right]. \tag{22}$$

Note however, that it is not directly clear, that this approach gives all possible translation invariant forms. In addition, in is not clear how we can describe this requirement as some easy to handle equivalent requirement on the functions $A : \mathbb{R}^d \to S^{d-1}$ or $R : \mathbb{R}^d \to \mathbb{R}^+$, or their corresponding infinite dimensional forms $\mathscr{A}$ and $\mathscr{R}$ from the previous section.

## 4.2 Rotation Invariance

Stating what is meant by rotation invariance is rather straightforward, however finding an explicit form for all functionals that adhere to it is rather difficult. For this reason, we here only demonstrate the possibility of incorporating both blur and rotation invariance starting from a fairly restricted class of rotation invariant functionals, which is given by the set of all functionals of the form

$$\mathscr{K}[I] = \int_0^\infty \left( \int_{\|x\|=r} f_r(I(x))\, dx \right) dr\,. \tag{23}$$

The function $f_r : \mathbb{R} \to \mathbb{R}$ described how gray values at a distance $r$ from the origin are being transformed. Through this function, image values in positions $x$ that have an equal distance from the origin are all treated in a similar way and the actual position does not play a role in this. It indeed can be checked easily that Equation (23) defines rotation invariance.

Imposing blur invariance, in addition, we have to find a prior functional $\mathscr{P}$, that, besides Equation (23), also respects the general form in Equation (20). Invariance under image rotation can also be handled by imposing invariance under rotation of the Fourier transformed image $\mathscr{F}[I]$, and therefore of its exponentiated form $(\mathscr{F}[I])^{-\frac{2}{\|\cdot\|^2}}$, and so one may consider rotational invariant functionals of the form

$$\mathscr{K}'[I] = \mathscr{K}\left[(\mathscr{F}[I])^{-\frac{2}{\|\cdot\|^2}}\right] . \tag{24}$$

The previous expression is similar to the one for general blur invariant functionals. However, in order to make it blur invariant, $\mathscr{K}$ should be homogeneous. Considering the form of $\mathscr{K}$ given in Equation (23), this can only be enforced if the function $f_r$ is a monomial of order $\alpha$ for all $r$:

$$f_r(x) = c(r)x^{\alpha} . \tag{25}$$

Note that the coefficient may still depend on $r$, however $\alpha$ is not allowed to vary. All in all, this leads to the following blur and rotation invariant prior functionals

$$\mathscr{P}[I] = \int_0^{\infty} c(r) \int_{\|\xi\|=r} \hat{I}(\xi)^{-\frac{2\alpha}{r^2}} \, d\xi \, dr . \tag{26}$$

This example nicely illustrates that taking into account blur invariance, in addition to rotation invariance, does seem to impose some considerable restrictions on the admissible prior functional. However, this might be due to the fact that the set of functionals considered had already been restricted from the outset.

Also remark that based on this case, functionals may be easily constructed that adhere to blur, rotation, and translation invariance simultaneously.

## 5 End Remarks

Through the identification of functionals on image spaces with, improper, priors, we were able to study a form of invariance somewhat related to, but different from, scale invariance. We demonstrated that the class of prior functionals that fulfill this so-called blur invariance requirement is huge, implying that additional constraints should be imposed before a workable class of functionals would be obtained. This situation is, however, not necessarily different from any other in which a single type of invariance constraint has been imposed. Subsequently, two moderate examples of classes of functionals on which an additional type of invariance was imposed, besides blur invariance, were briefly considered. Although we suspect that the translation invariant case may be described in its full generality, the general case in which rotation invariance is assumed is hard to describe in an explicit form and we merely considered a, probably, small example class of such functionals.

Although not discussed in depth here, one of the main topics for future investigations should be to study the aforementioned possibility of combining various invariance properties in order to restricting the set of admissible functionals further. In this, one may actually also consider scale invariance as an additional requirement. Considering this constraint in the light of *observed* images, the ones we considered in this work, may provide the opportunity to deal with it in another way then has been done up to know. In direct relation there is the, in a sense, fundamental question about the actual function spaces on which all these operations are carried out. This, we briefly mentioned in an earlier section. Clearly, regarding blur invariance, the blurring (and more even, the deblurring) imposes some rather specific restrictions on the allowable function spaces. Further invariance requirements, of course, only complicate the situation further.

To illustrate that deblurring *ad infinitum* does not necessarily have to be an odd and ill-posed operation, we point out that such can be readily done on, for instance, functions that have a finitely supported Fourier transform (e.g. see [15]).

In addition to all necessary theoretical developments, one may also consider a more empirical approach to further investigations. The analysis of actual natural images or the simulation of images from the proposed priors may give additional insight into the validity, consequences, and quirks of a blur invariance assumption. We, however, consider such studies more important at a later stage and would argue to focus on more theoretical understanding first.

## Acknowledgements

The reviewers of our manuscript are kindly acknowledged for the comments, remarks, and suggestions provided.

## References

1. Field, D.J.: Relations between the statistics of natural images and the response properties of cortical cells. Journal of the Optical Society of America. A **4** (1987) 2379–2394
2. Srivastava, A., Lee, A.B., Simoncelli, E.P., Zhu, S.C.: On advances in statistical modeling of natural images. Journal of Mathematical Imaging and Vision **18** (2003) 17–33
3. Mumford, D.: Empirical Statistics and Stochastic Models for Visual Signals. In: New Directions in Statistical Signal Processing. From Systems to Brains. The MIT Press, Cambridge, MA (2006)
4. Geusebroek, J.M., Smeulders, A.W.M.: A physical explanation for natural image statistics. In: Proceedings of the 2nd International Workshop on Texture Analysis and Synthesis (Texture 2002). (2002) 47–52
5. Grenander, U., Srivastava, A.: Probability models for clutter in natural images. IEEE Transactions on Pattern Analysis and Machine Intelligence **23** (2001) 424–429
6. Markussen, B., Pedersen, K.S., Loog, M.: A scale invariant covariance structure on jet space. In: Proceeding of the DSSCV Workshop. Lecture Notes in Computer Science, Maastricht, The Netherlands (2005)
7. Mumford, D., Gidas, B.: Stochastic models for generic images. Quarterly of Applied Mathematics **59** (2001) 85–111
8. Mumford, D.: Gaussian models for images. as yet unpublished (2001)
9. Pedersen, K.S.: Properties of brownian image models in scale-space. In: Proceedings of the 4th Scale-Space conference. Volume 2695 of Lecture Notes in Computer Science., Springer-Verlag (2003) 281–296
10. Florack, L.M.J.: Image Structure. Volume 10 of Computational Imaging and Vision. Kluwer, Dordrecht . Boston . London (1997)

11. Koenderink, J.J., van Doorn, A.J.: Receptive field families. Biological Cybernetics **63** (1990) 291–297
12. Lindeberg, T.: Feature detection with automatic scale selection. International Journal of Computer Vision **30** (1998) 79–116
13. Billingsley, P.: Probability and Measure. Wiley, New York (1986)
14. Widder, D.V.: The Heat Equation. Academic Press (1975)
15. Johansen, P., Skelboe, S., Grue, K., Andersen, J.D.: Representing signals by their toppoints in scale-space. In: Proceedings of the 8th International Conference on Pattern Recognition, Paris (1986) 215–217
16. Aczél, J.: Lectures on Functional Equations and their Applications. Academic Press Inc., New York (1966)

# A Variational Framework for Adaptive Satellite Images Segmentation*

Olfa Besbes[1,2], Ziad Belhadj[2], and Nozha Boujemaa[1]

[1] IMEDIA project, INRIA Rocquencourt, 78153 Le Chesnay, France
[2] URISA - SUP'COM, 2088 Ariana, Tunisia
Olfa.Besbes, Nozha.Boujemma@inria.fr, ziad.belhadj@supcom.rnu.tn

**Abstract.** In this paper, we present an adaptive variational segmentation algorithm of spectral / texture regions in satellite images using level set. Satellite images contain both textured and non-textured regions, so for each region spectral and texture cues are integrated according to their discrimination power. Motivated by Fisher-Rao linear discriminant analysis, two region weights are defined to code respectively the relevance of spectral and texture cues. Therefore, regions with or without texture are processed in an unified framework. The obtained segmentation criterion is minimized via curves evolution within an explicit correspondence between the interiors of evolving curves and regions in the segmentation. The shape derivation principle is used to derive the system of coupled evolution equations in such a way that we consider the region weights and the statistical parameters variability. Experimental results on both natural and satellite images are shown.

## 1 Introduction

As the volume of satellite imagery continues to grow exponentially, effective querying and browsing in these image databases are becoming a serious challenge. In particular, Region-Based Image Retrieval [16] is a powerful tool since it allows to search images containing similar objects of a reference image. It requires the satellite image to be segmented into a number of regions. Segmentation consists of partitioning the image into non-overlapping regions that are homogeneous with regards to some characteristics such as spectral and texture. Remote sensed images contain both textured and non-textured regions. This is even more true today with high resolution images such as IKONOS, SPOT-5 and QuickBird data. Different segmentation approaches have been developed to deal with natural images having both textured and non-textured regions. As region growing algorithm, [7] performs a color quantization and measures the confidence of pixels to be boundaries or interiors of color/texture regions. In a graph partitioning framework, [9] analyzes gray-level images using contour and texture cues and gates their contribution by a local measure of texturedness. [19] proposes a stochastic method to segment natural images by estimating the region type and selecting its appropriate model. A speed improvement is obtained

* This work is partially supported by QuerySat project and INRIA STIC project.

F. Sgallari, A. Murli, and N. Paragios (Eds.): SSVM 2007, LNCS 4485, pp. 675–686, 2007.

by the generalized cluster algorithm [3]. For this method, the segmentation is only carried on intensity images. Recently, variational approaches have shown their ability to integrate various cues. Only few of them cope cope with images containing both textured and non-textured regions. For instance, in [18] the segmentation is performed by integrating the Edgeflow vector field to the curve evolution. It utilizes a predictive coding model to identify the direction of changes in color/texture features and construct an edge flow vector. The main limitation of this method is the scale parameter selection. A fixed global scale is inappropriate for images which contain multiple scale information. To detect meaningful boundaries in these images, a local scale parameter depending on the local color/texture properties is required. A coarse to fine multi-scale edge detection framework is applied implicitly to overcome this limitation [17]. But, as an edge-based method it has over-segmentation results. However, in [2] a region-based variational image segmentation algorithm is presented. It's based on the image decomposition into a geometrical component and a textured component. The segmentation is carried out by combining the information from both channels with a logical framework. This approach is restricted to gray-level images. Besides, its results depend on the image decomposition reliability and the supervised definition of the region's logical model. The others existing variational methods are specifically devoted to only one-kind-region images segmentation: non-textured [13] or textured [11,14]. In particular [4], since color and texture are combined similarly they must be discriminative simultaneously for the set of regions in a natural image to be well segmented. On the other hand, natural image segmentation requires either different models to describe various region types or an adaptive cue combination according to region types. It's for the latter that we contribute to image segmentation by level set.

In this paper, we propose an adaptive variational segmentation method of spectral/texture images using active curves evolution via level set [15]. Inspired from [11,13,6], we apply a multi-dimensional statistical model to describe regions. In order to cope with content heterogeneity of remote sensed data, we evaluate spectral and texture features relevance according to each region in the image. The idea is as following: For a non-textured region, color is the most discriminant information since non-textured regions have similar response in the texture feature space. However, for a textured region we evaluate the coherence between spectral and texture cues in order to discriminate it from other regions in the image. Furthermore, we use the multiregion competition algorithm [10] to guarantee an unambiguous segmentation that is a partition of the image domain into N fixed but arbitrary regions.

## 2 Adaptive Multiple Region Segmentation

Let $I : \Omega \rightarrow \mathbb{R}^C$ be the multispectral image to be segmented, defined on $\Omega \subset \mathbb{R}^2$. We aim to find a partition of $I$ into $N$ fixed but arbitrary homogeneous regions with respect to spectral and texture characteristics. Let $\Re = \{R_i\}_{i=1}^N$ be a partition of $\Omega$ such that $\cup_{i=1}^N R_i = \Omega$ and $R_i \cap R_k = \emptyset$, if $i \neq k$.

## 2.1 Energy Functional

Let $\mathbf{U} : \Omega \rightarrow \mathbb{R}^M$ be the computed feature vector of the image $I$. It consists of $C$ spectral channels and 4 texture features where a coupled edge preserving smoothing process is applied. This so-called nonlinear diffusion [20] deals with outliers in the data, closes structures and synchronizes all channels in order to facilitate the segmentation process. The texture features [4] are given by the second order matrix and the local scale feature. The latter is deduced from the speed of a diffusion process based on the TV flow [5]. The resultant feature vector $\mathbf{U}$ describes each region by its spectral (or color) property, the magnitude, the orientation and the local scale of its texture. Unlike the Gabor filters, this method leads to a significantly reduced number of features. In figure 1, the feature vector components of a panchromatic satellite image are shown. Let $p_{ij}(x)$

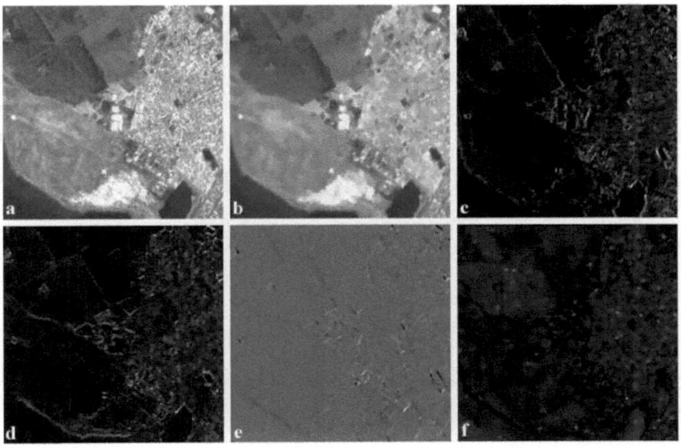


**Fig. 1.** Original image {a} and its feature channels ({b} spectral feature, {c, d, e} nonlinear structure tensor components and {f} local scale feature)

be the conditional probability density function of a value $\mathbf{U}_j(x)$ to be in a region $R_i$. The feature channels are assumed to be independent. Assuming all partitions to be equally probable and the pixels within each region to be independent, the adaptive segmentation problem consists of finding the partition $\hat{\Re}$ that minimizes the following Bayes-derived energy functional :

$$E(\Re) = -\sum_{i=1}^{N} \int_{R_i} w_{is} \sum_{j=1}^{C} \log p_{ij}(\mathbf{U}_j(x))\, dx + \int_{R_i} w_{it} \sum_{j=C+1}^{M} \log p_{ij}(\mathbf{U}_j(x))\, dx, \qquad (1)$$

where $w_{is}$ and $w_{it}$ are two weights which code respectively the discrimination power of spectral features and texture features in each region $R_i$. Thanks to this new cost function, we ensure an adaptive segmentation method that deals with images without texture, images with only textured regions and general images that contain regions of both kinds.

### 2.2 Defining the Region Weights

We use the well-known Fisher-Rao statistical criterion [12], to determine the different weights $\{w_{is}, w_{it}\}_{i=1}^{N}$. It consists on maximizing the ratio of the between-class variance $\mathbf{B}$ to the within-class variance $\mathbf{W}$ in any particular data set thereby guaranteeing the maximal separability. This linear discrimination technique has widely used for the feature space dimensionality reduction [8]. A generalization into the kernel version was developed for nonlinear discriminative analysis [21]. For a $N$-classes case, $\mathbf{B}$ and $\mathbf{W}$ are defined below:

$$\mathbf{B} = \sum_{i=1}^{N-1} \sum_{k=i+1}^{N} p_i p_k \left(\mu_i - \mu_k\right)\left(\mu_i - \mu_k\right)^T \qquad \mathbf{W} = \sum_{i=1}^{N} p_i \mathbf{W_i}, \tag{2}$$

where $\mu_i$ is the mean vector of a class $i$, $p_i$ is its *a priori* probability and $\mathbf{W_i}$ is its within-class covariance. Assuming equally probable regions, $p_i$ equals $\frac{1}{N}$. For a defined feature space, the $\mathbf{W}^{-1}\mathbf{B}$ eigenvalues code the relevance of their corresponding feature channels. Furthermore, $\mathrm{trace}(\mathbf{W}^{-1}\mathbf{B})$ measures the feature space relevance. The higher value of this inter-intra criterion, the more informative is the feature space. In order to measure the relevance of a feature space to discriminate a given region $R_i$ from others regions $R_{k|k\neq i}$, we only consider the $R_i$'s between-class variance. Therefore, we define $\mathbf{B_i} = \frac{1}{N^2} \sum_{k=1|k\neq i}^{N} (\mu_i - \mu_k)(\mu_i - \mu_k)^T$. Then the spectral and texture weights of a region $R_i$ can be written as :

$$w_{is} = \frac{\sum_{j=1}^{C} \left[\mathbf{W}^{-1}\mathbf{B_i}\right]_{jj}}{\sum_{j=1}^{M} \left[\mathbf{W}^{-1}\mathbf{B_i}\right]_{jj}},\ w_{it} = \frac{\sum_{j=C+1}^{M} \left[\mathbf{W}^{-1}\mathbf{B_i}\right]_{jj}}{\sum_{j=1}^{M} \left[\mathbf{W}^{-1}\mathbf{B_i}\right]_{jj}},\ \text{where} \left[\mathbf{W}^{-1}\mathbf{B_i}\right]_{jj} = \frac{\sum_{k=1}^{N} (\mu_{ij} - \mu_{kj})^2}{N \sum_{k=1}^{N} \sigma_{kj}^2}. \tag{3}$$

Noting that $\{w_{is}, w_{it}\} \in [0,1]$ and $w_{is} + w_{it} = 1, \forall i \in [1, N]$.

### 2.3 Representation of a Partition

In order to find the partition $\hat{\Re} = \left\{\hat{R}_i\right\}_{i=1}^{N}$ that minimizes the energy functional (1), we use active curves evolution. We consider a family $\{\vec{\gamma}_i : [0,1] \rightarrow \Omega\}_{i=1}^{N-1}$ of plane curves parameterized by the arc parameter $s \in [0,1]$. As proposed in [10], we use an explicit correspondence between the regions $R_{\vec{\gamma}_i}$ enclosed by the curves $\vec{\gamma}_i$ and the regions of the partition. With this choice of a partition representation and with the addition of a regularization term related to the length of curves, the energy functional (1) becomes:

$$\begin{aligned} E\left(\{R_i\}_{i=1}^{N}\right) &= \int_{R_{\vec{\gamma}_1}} \xi_1(x)\,dx + \int_{R^c_{\vec{\gamma}_1} \cap R_{\vec{\gamma}_2}} \xi_2(x)\,dx + \ldots \\ &+ \int_{R^c_{\vec{\gamma}_1} \cap R^c_{\vec{\gamma}_2} \cap \ldots \cap R^c_{\vec{\gamma}_{k-1}} \cap R_{\vec{\gamma}_k}} \xi_k(x)\,dx + \ldots \\ &+ \int_{R^c_{\vec{\gamma}_1} \cap R^c_{\vec{\gamma}_2} \cap \ldots \cap R^c_{\vec{\gamma}_{N-2}} \cap R^c_{\vec{\gamma}_{N-1}}} \xi_N(x)\,dx + \lambda \sum_{i=1}^{N-1} \oint_{\vec{\gamma}_i} ds, \end{aligned} \tag{4}$$

where $\lambda > 0$ weights the contribution of the regularization term. This partition representation guarantees an unambiguous image segmentation to $N$ arbitrary number of regions without adding a constraint term to the energy functional [11]. For a region $R_i$, we have $\xi_i(x) = w_{is}\xi_{is}(x) + w_{it}\xi_{it}(x)$. Using a Gaussian model for all feature channels to describe $R_i$, $\xi_{is}(x) = \sum_{j=1}^{C} \log(2\Pi\sigma_{ij}^2) + \frac{(\mathbf{U}_j(x)-\mu_{ij})^2}{\sigma_{ij}^2}$ and $\xi_{it}(x) = \sum_{j=C+1}^{M} \log(2\Pi\sigma_{ij}^2) + \frac{(\mathbf{U}_j(x)-\mu_{ij})^2}{\sigma_{ij}^2}$.

## 3 Level Set Segmentation with Multiple Regions

The energy functional (4) minimum is obtained via the gradient descent with respect to the positions of the $N-1$ curves. We directly differentiate this functional using the shape derivative tool introduced in [1]. Here, we extend [13] to the multiple region segmentation task. Starting with $\vec{\gamma}_1$, we rewrite the energy functional as:

$$E(\Re) = \int_{R_{\vec{\gamma}_1}} \xi_1(x)\, dx + \int_{R^c_{\vec{\gamma}_1}} \phi_1(x)\, dx + \lambda \oint_{\vec{\gamma}_1} ds + \lambda \sum_{i=2}^{N-1} \oint_{\vec{\gamma}_i} ds. \tag{5}$$

We define, for $1 \leq i, n \leq N$, $\phi_n(x) = \sum_{i=n+1}^{N} \xi_i(x)\, \chi_i^n(x)$ and $\chi_i^n$ as:

$$\begin{cases} \chi_i^n(x) = 0 & \text{if } \; i < n \\ \chi_i^n(x) = -1 & \text{if } \; i = n \\ \chi_i^n(x) = \chi_{R^c_{\vec{\gamma}_{n+1}}}(x) \dots \chi_{R^c_{\vec{\gamma}_{i-1}}}(x)\, \chi_{R_{\vec{\gamma}_i}}(x) & \text{otherwise} \end{cases}. \tag{6}$$

The statistical parameters for $j \in [1, M]$ are also rewritten as follows :

$$\begin{array}{ll} \mu_{1j} = \frac{1}{|R_1|} \int_{R_{\vec{\gamma}_1}} \mathbf{U}_j(x)\, dx & \sigma_{1j}^2 = \frac{1}{|R_1|} \int_{R_{\vec{\gamma}_1}} (\mathbf{U}_j(x) - \mu_{1j})^2\, dx \\ \mu_{kj} = \frac{1}{|R_k|} \int_{R^c_{\vec{\gamma}_1}} \mathbf{U}_j(x)\, \chi_k^1(x)\, dx & \sigma_{kj}^2 = \frac{1}{|R_k|} \int_{R^c_{\vec{\gamma}_1}} (\mathbf{U}_j(x) - \mu_{kj})^2\, \chi_k^1(x)\, dx \end{array}, \tag{7}$$

where $|R_1| = \int_{R_{\vec{\gamma}_1}} dx$ and $|R_k| = \int_{R^c_{\vec{\gamma}_1}} \chi_k^1(x)\, dx, \forall k = 2, \dots, N$ are the region areas. The integrals $D_1 = \int_{R_{\vec{\gamma}_1}} \xi_1(x) dx$ and $D_2 = \int_{R_{\vec{\gamma}_1^c}} \phi_1(x) dx$ may be differentiated with respect to the position of $\vec{\gamma}_1$. In particular, the spectral (textural) weight Gâteaux derivative is determined to compute the region integral shape derivative.

### 3.1 The Weight's Gâteaux Derivative

The spectral (textural) weight is adaptive and changes during the segmentation process. Thus, we need to compute its Gâteaux derivative to derive the $\vec{\gamma}_1$'s evolution equation. Obviously, $\langle (w_{1t})', V \rangle = -\langle (w_{1s})', V \rangle$.

As defined in (3), $Z_{1j} = \frac{\sum_{i=1}^{N}(\mu_{1j}-\mu_{ij})^2}{\sum_{i=1}^{N}\sigma_{ij}^2}$ and its Gâteaux derivative is $\langle (Z_{1j})', V\rangle = \sum_{i=1}^{N}\left(\frac{\partial Z_{1j}}{\partial \mu_{ij}}\left\langle (\mu_{ij})', V\right\rangle + \frac{\partial Z_{1j}}{\partial \sigma_{ij}^2}\left\langle (\sigma_{ij}^2)', V\right\rangle\right)$. Using the following equalities :

$$\frac{\partial Z_{1j}}{\partial \mu_{1j}} = \frac{2\sum_{i=2}^{N}(\mu_{1j}-\mu_{ij})}{\sum_{i=1}^{N}\sigma_{ij}^2},\quad \frac{\partial Z_{1j}}{\partial \mu_{kj}}\Big|_{k\in[2,N]} = \frac{-2\left(\mu_{1j}-\mu_{kj}\right)}{\sum_{i=1}^{N}\sigma_{ij}^2},\quad \frac{\partial Z_{1j}}{\partial \sigma_{kj}^2}\Big|_{k\in[1,N]} = -\frac{\sum_{i=1}^{N}(\mu_{1j}-\mu_{ij})^2}{\left(\sum_{i=1}^{N}\sigma_{ij}^2\right)^2} \tag{8}$$

$$\begin{aligned}
&\left\langle (\mu_{1j})', V\right\rangle = -\tfrac{1}{|R_1|}\textstyle\int_{\vec{\gamma}_1}(\mathbf{U}_j(x)-\mu_{1j})(V(x).N_1(x))\,da(x)\\
&\left\langle (\mu_{kj})', V\right\rangle\big|_{k\in[2,N]} = \tfrac{1}{|R_k|}\textstyle\int_{\vec{\gamma}_1}(\mathbf{U}_j(x)-\mu_{kj})\chi_k^1(x)(V(x).N_1(x))\,da(x)\\
&\left\langle (\sigma_{1j}^2)', V\right\rangle = -\tfrac{1}{|R_1|}\textstyle\int_{\vec{\gamma}_1}\left((\mathbf{U}_j(x)-\mu_{1j})^2-\sigma_{1j}^2\right)(V(x).N_1(x))\,da(x)\\
&\left\langle (\sigma_{kj}^2)', V\right\rangle\big|_{k\in[2,N]} = \tfrac{1}{|R_k|}\textstyle\int_{\vec{\gamma}_1}\left((\mathbf{U}_j(x)-\mu_{kj})^2-\sigma_{kj}^2\right)\chi_k^1(x)(V(x).N_1(x))\,da(x)
\end{aligned} \tag{9}$$

we obtain $\left\langle Z_{1j}', V\right\rangle = \int_{\vec{\gamma}_1} k_{1j}^1(x)\,(V(x).N_1(x))\,da(x)$, where :

$$\begin{aligned}
k_{1j}^1(x) = &\frac{-2}{\sum_{i=1}^{N}\sigma_{ij}^2}\sum_{k=2}^{N}\left[(\mu_{1j}-\mu_{kj})\left(\frac{\mathbf{U}_j(x)-\mu_{1j}}{|R_1|}+\frac{\mathbf{U}_j(x)-\mu_{kj}}{|R_k|}\chi_k^1(x)\right)\right]\\
&+\frac{\sum_{i=1}^{N}(\mu_{1j}-\mu_{ij})^2}{\left(\sum_{i=1}^{N}\sigma_{ij}^2\right)^2}\left[\frac{(\mathbf{U}_j(x)-\mu_{1j})^2-\sigma_{1j}^2}{|R_1|}-\sum_{k=2}^{N}\left(\frac{(\mathbf{U}_j(x)-\mu_{kj})^2-\sigma_{kj}^2}{|R_k|}\chi_k^1(x)\right)\right].
\end{aligned} \tag{10}$$

Then, the Gâteaux derivative of $w_{1s}$ in the direction of $V$ is :

$$\left\langle (w_{1s})', V\right\rangle = \frac{1}{\sum_{j=1}^{M} Z_{1j}}\int_{\vec{\gamma}_1}\left[w_{1t}\sum_{j=1}^{C}k_{1j}^1(x) - w_{1s}\sum_{j=C+1}^{M}k_{1j}^1(x)\right](V(x).N_1(x))\,da(x). \tag{11}$$

### 3.2 The Curve Evolution Equations

Now, we can compute the Gâteaux derivative of the functional $D_1 = \int_{R_{\vec{\gamma}_1}}\xi_1(x)dx$ in the direction of $V$. It equals $\int_{R_{\vec{\gamma}_1}}\xi_1^{sh}(x, R_{\vec{\gamma}_1}, V)\,dx - \int_{\vec{\gamma}_1}\xi_1(x)\,(V(x).N_1(x))\,da(x)$, where $\xi_1^{sh}$ is the shape derivative of $\xi_1$ and $N_1(x)$ is the unit normal vector to the curve $\vec{\gamma}_1$ at a point $x$. Using the result of the weight Gâteaux derivative (11), we obtain :

$$\left\langle D_1', V\right\rangle = \int_{\vec{\gamma}_1}\left[A_1\left(w_{1t}\sum_{j=1}^{C}k_{1j}^1(x) - w_{1s}\sum_{j=C+1}^{M}k_{1j}^1(x)\right) - \xi_1(x)\right](V(x).N_1(x))\,da(x). \tag{12}$$

Similarly, we calculate the Gâteaux derivative of $D_2 = \int_{R^c_{\vec{\gamma}_1}} \phi_1(x)dx$ in the direction of $V$. Since $\left\langle D_2^{'}, V \right\rangle = \sum_{i=2}^{N} \int_{R^c_{\vec{\gamma}_1}} [\xi_i(x)\, \chi_i(x)]^{sh} dx + \int_{\vec{\gamma}_1} \phi_1(x)(V(x)\,.N_1(x))\, da(x)$, we have :

$$\left\langle D_2^{'}, V \right\rangle = \int_{\vec{\gamma}_1} \left[ \sum_{i=2}^{N} A_i \left( w_{it} \sum_{j=1}^{C} k_{ij}^1(x) - w_{is} \sum_{j=C+1}^{M} k_{ij}^1(x) \right) + \phi_1(x) \right] (V(x)\,.N_1(x))\, da(x). \tag{13}$$

In the general case, we have the energy term $\Psi_n(x) = \zeta_n(x) + \phi_n(x)$. The first new term is added due to our adaptive energy functional formulation (4). However, the second term is classical and obtained for non-adaptive segmentation case where the region spectral/textural weights are constant and equal for all regions. Furthermore, the adaptive derived energy term is defined as followed : $\zeta_n(x) = \sum_{i=1}^{N} A_i \left( w_{it} \sum_{j=1}^{C} k_{ij}^n(x) - w_{is} \sum_{j=C+1}^{M} k_{ij}^n(x) \right)$, where $\forall \{i,n\} = \{1,\ldots,N\}$ $A_i = \dfrac{|R_i| \left( (2C-M)(1+\log(2\Pi)) + \sum_{j=1}^{C} \log(\sigma_{ij}^2) - \sum_{j=C+1}^{M} \log(\sigma_{ij}^2) \right)}{\sum_{j=1}^{M} Z_{ij}}$ and

$$\begin{aligned} k_{ij}^n(x) = & \frac{2(\mu_{ij}-\mu_{nj})}{\sum_{l=1}^{N} \sigma_{lj}^2} \frac{\mathbf{U}_j(x)-\mu_{nj}}{|R_n|} + \frac{2}{\sum_{l=1}^{N} \sigma_{lj}^2} \sum_{k=1}^{N} (\mu_{ij} - \mu_{kj}) \frac{\mathbf{U}_j(x)-\mu_{ij}}{|R_i|} \chi_i^n(x) \\ & - \frac{2}{\sum_{l=1}^{N} \sigma_{lj}^2} \sum_{k=n+1}^{N} \left[ (\mu_{ij} - \mu_{kj}) \frac{\mathbf{U}_j(x)-\mu_{kj}}{|R_k|} \chi_k^n(x) \right] \\ & + \frac{\sum_{k=1}^{N} (\mu_{ij}-\mu_{kj})^2}{\left( \sum_{l=1}^{N} \sigma_{lj}^2 \right)^2} \left[ \frac{(\mathbf{U}_j(x)-\mu_{nj})^2 - \sigma_{nj}^2}{|R_n|} - \sum_{k=n+1}^{N} \left( \frac{(\mathbf{U}_j(x)-\mu_{kj})^2 - \sigma_{kj}^2}{|R_k|} \chi_k^n(x) \right) \right] \end{aligned} \tag{14}$$

Therefore, we can write the $\vec{\gamma}_1$'s evolution equation as :

$$\frac{\partial \vec{\gamma}_1}{\partial t}(x) = -\left[ \xi_1(x) - \Psi_1(x) + \lambda \kappa_1 \right] N_1, \tag{15}$$

where $\kappa_1$ is the mean curvature function of $\vec{\gamma}_1$. As the same manner, we compute the $\vec{\gamma}_2$'s evolution equation. We consider that the $R_{\vec{\gamma}_1}$ statistical parameters $\left( \{\mu_{1j}, \sigma_{1j}^2\}_{j \in [1,M]}, |R_1| \right)$ are constant during the evolution of $\vec{\gamma}_2$. Therefore, the $\vec{\gamma}_2$'s evolution equation is :

$$\frac{\partial \vec{\gamma}_2}{\partial t}(x) = [-\chi_{R^c_{\vec{\gamma}_1}}(x)\, (\xi_2(x) - \Psi_2(x)) + \lambda \kappa_2] N_2. \tag{16}$$

Proceeding similarly, a curve $\vec{\gamma}_n$ has an evolution equation given by:

$$\frac{\partial \vec{\gamma}_n}{\partial t}(x) = -[\chi_1^c(x) \ldots \chi_{n-1}^c(x)\, (\xi_n(x) - \Psi_n(x)) + \lambda \kappa_n] N_n. \tag{17}$$

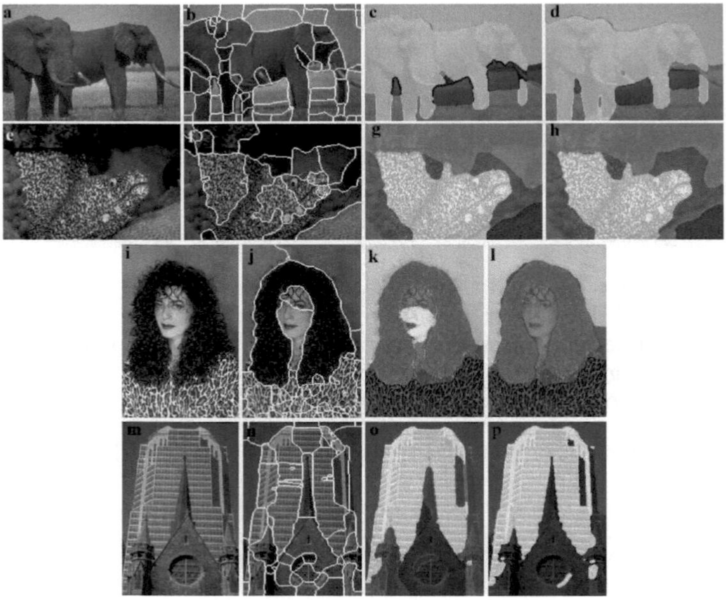


**Fig. 2.** {a, e, i, m} Natural images. Their corresponding segmentation results superimposed on the original images : {b, f, j, n} Edgeflow-based method results, {c, g, k, o} Non-adaptive method results and {d, h, l, p} our adaptive method results.

Finally, the minimization of the adaptive multiregion competition functional (4) is achieved through the following system of coupled curves evolution equations:

$$\begin{cases} \frac{\partial \vec{\gamma}_1}{\partial t}(x) = -[\xi_1(x) - \Psi_1(x) + \lambda\kappa_1]N_1 \\ \frac{\partial \vec{\gamma}_n}{\partial t}(x) = -[\chi_{R^c_{\vec{\gamma}_1}}(x)\dots\chi_{R^c_{\vec{\gamma}_{n-1}}}(x)\,(\xi_n(x) - \Psi_n(x)) + \lambda\kappa_n]N_n, \; . \\ n = 2, \dots, N-1 \end{cases} \quad (18)$$

### 3.3 Level Set Implementation

The system of curve evolution equations (18) is implemented via level set formalism [15]. Thus, the curves $\{\vec{\gamma}_i\}_{i=1}^{N-1}$ are implicitly represented by the zero level set of a function $u_i : \mathbb{R}^2 \rightarrow \mathbb{R}$. The region $R_{\vec{\gamma}_i}$ inside $\vec{\gamma}_i$ corresponds to $u_i > 0$. This implicit representation has several advantages. First, the region membership is explicitly maintained. Second, it allows topological changes and can be implemented by stable numerical schemes. The system of curve evolution equations leads to the following system :

$$\begin{cases} \frac{\partial u_1}{\partial t}(x) = -[\xi_1(x) - \Psi_1(x) + \lambda\kappa_{u_1}]\|\vec{\nabla} u_1\| \\ \frac{\partial u_n}{\partial t}(x) = -[\chi_{\{u_1(x,t)\le 0\}}\cdots\chi_{\{u_{n-1}(x,t)\le 0\}}\,(\xi_n(x) - \Psi_n(x)) + \lambda\kappa_{u_n}]\|\vec{\nabla} u_n\|, \\ n = 2, \dots, N-1 \end{cases} \quad (19)$$

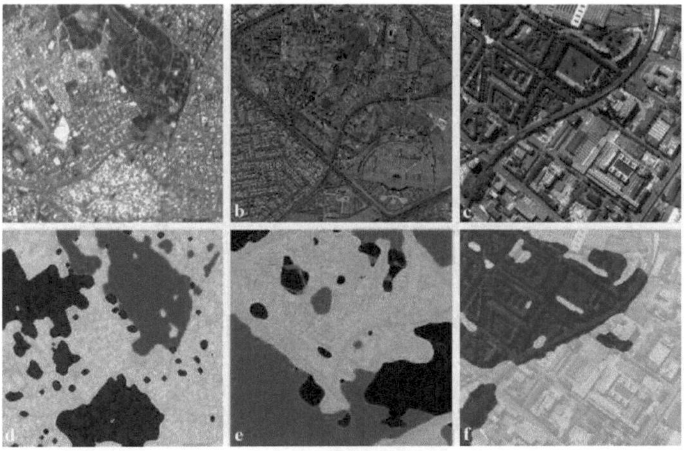

**Fig. 3.** {a, b, c} Original images with only textured regions. {d, e, f} Their corresponding segmentation results superimposed on the original images.

where $\chi_{\{u_i(x,t)\leq 0\}} = 1$ if $u_i(x,t) \leq 0$ and 0 otherwise. $\kappa_{u_i} = -div\left(\frac{\vec{\nabla} u_i}{\|\vec{\nabla} u_i\|}\right)$ is the $u_i$'s curvature. At convergence, the final segmentation is given by the family $\{R_{u_1}, R^c_{u_1} \cap R_{u_2}, \ldots, \cap_{i=1}^{N-1} R^c_{u_i}\}$ where $R_{u_i} = \{x \in \Omega \,|\, u_i(x,\infty) > 0\}$, $i = 1, \ldots, N-1$.

## 4 Experimental Results

In our implementation, the same initialization with fixed-size circles was used. These circles were labeled automatically with k-means algorithm using the multivariate Gaussian and the Bhattacharya distance. Such an initialization has been already used in the past because of its ability to detect easily holes and improve the convergence speed. Moreover, the time step and the regularization parameters were fixed to the same value 0.1 for all the tests. Only the numbers of regions and nonlinear diffusion iterations were set according to the image. The diffusion time has great influence on the results: if there is too much diffusion there is too much information lost for the statistics. On the other hand, if there is not enough diffusion the risk to hit a local optimum increases. Therefore, an adequate value of this parameter was selected experimentally by taking into account the present amount of texture in the image. Concerning the convergence speed, it depends greatly on the size of the image and the number of segmented regions.

In figure (2), we illustrate the segmentation results obtained on natural images with our adaptive method (2d, 2h, 2l and 2p), the non-adaptive method by setting all region weights to 0.5 (2c, 2g, 2k and 2o) and the Edgeflow-based method[1] (2b, 2f, 2j and 2n) [17]. For the latter method, its three parameters were fixed as followed: the texture contribution was 50%, the minimum scale was 10% and

[1] An on-line demo of this method is at http://aakash.ece.ucsb.edu/imdiffuse/segment.aspx

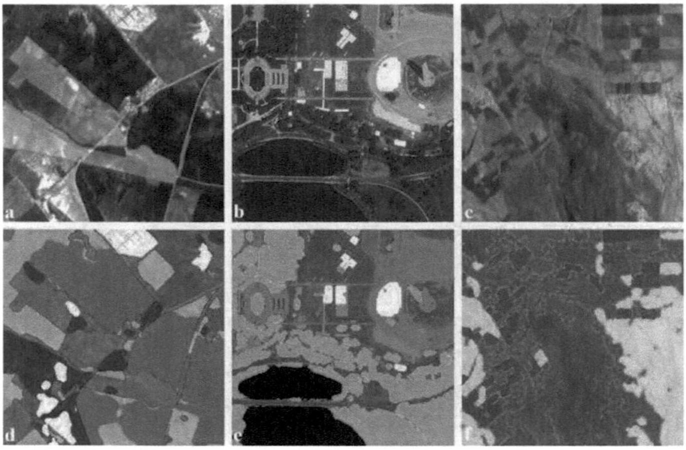

**Fig. 4.** {a, b, c} Original images without textured regions. {d, e, f} Their corresponding segmentation results superimposed on the original images.

a balanced merge. In contrast to these two methods of which color and texture cues are similarly combined, the boundaries are better located and the different smooth/textured regions are better segmented with our adaptive segmentation method. Furthermore, we demonstrate in figure 2 the ability of our method to deal with images without textured regions (2a), images with only textured regions (2e) and images containing both kinds of regions (2i and 2m). In all the cases, the regions are separated from each other thanks to adaptive cues combination of color and texture. For instance, the lady lips and eyes (2l) are segmented as the face parts due to a high applied diffusion time. A finer segmentation can be obtained with a less diffusion time and so a higher number of regions.

An interesting application of our method is satellite image segmentation. Combining cues of spectral and texture according to their discrimination power provides a powerful framework to cope with satellite images. We applied our algorithm on various panchromatic and multi-spectral images acquired by SPOT-3 (4c), SPOT-5 (4a,3a,5a), IKONOS (5c) and QuickBird (4b,3b, 3c, 5b) satellites. Results for images without texture are illustrated in figure 4. Smooth regions like sea, agricultural area, urban in low resolution, green area and ground are cleanly segmented. Figure 3 shows segmentation results for textured region images. Urban with different densities, vegetation area and ground are well segmented. Finally, figure 5 illustrates the capabilities of our approach on images which contain both non-textured regions (eg. agricultural areas, ground, river) and textured regions (eg. mountains, urban).

## 5 Conclusion

In this paper, we have presented an adaptive variational segmentation method for satellite images. It is based on combining the spectral and texture cues according

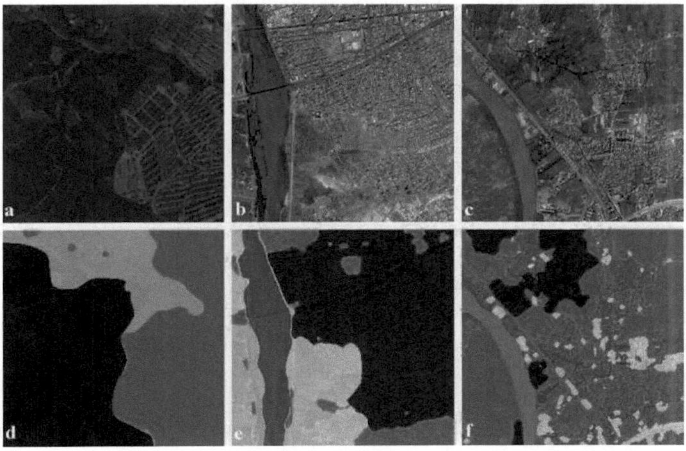


**Fig. 5.** {a, b, c} Original images containing both non-textured and textured regions. {d, e, f} Their corresponding segmentation results superimposed on the original images.

to their discrimination power. Two weights, motivated by Fisher-Rao's linear discriminant analysis, are defined to code respectively the relevance of each cue. Therefore, smooth regions, textured regions and both kinds regions images are processed similarly with neither a multi-model regions description nor a prior selection of relevant features for the segmentation nor a prior fixed weighting of color and texture features. Promising results have been obtained on both natural and satellite images.

In a future work, we intend to develop a coarse-to-fine multi-resolution approach coupled with a hierarchical splitting. This proceeding is useful not only to estimate the optimal number of regions but also to obtain suitable initializations and increase the convergence speed. Moreover, we aim to use the constructed nonlinear scale-space to provide a multiscale satellite image segmentation. Our aim is to decompose satellite image content into a hierarchy of attributed regions describing semantically topological relations and properties.

## References

1. G. Aubert, M. Barlaud, O. Faugeras, and S. Jehan-Besson. Image segmentation using active contours: Calculus of variations of shape gradients? *SIAM JAM*, 63(6):2128–2154, 2003.
2. J. F. Aujol and T. F. Chan. Combining geometrical and textured information to perform image classification. *J. Visual Communication and Image Representation*, 17(5):1004–1023, 2006.
3. Adrian Barbu and Song-Chun Zhu. Generalizing swendsen-wang to sampling arbitrary posterior probabilities. *IEEE Trans. PAMI*, 27(8):1239–1253, 2005.
4. T. Brox, M. Rousson, R. Deriche, and J. Weickert. Unsupervised segmentation incorporating colour, texture and motion. Technical Report 4760, INRIA, 2003.

5. T. Brox and J. Weickert. A TV flow based local scale measure for texture discrimination. In *8th ECCV*, volume 2, pages 578–590, Parague, 2004.
6. D. Cremers, M. Rousson, and R. Deriche. A review of statistical approaches to level set segmentation: Integrating color, texture, motion and shape. *IJCV*, 72(2):195–215, 2007.
7. Y. Deng and B. S. Manjunath. Unsupervised segmentation of color-texture regions in images and video. *IEEE Trans. PAMI*, 23(8):800–810, 2001.
8. M. Loog, R. P. W. Duin, and R. Haeb-Umbach. Multiclass linear dimension reduction by weighted pairwise Fisher criteria. *IEEE Trans. PAMI*, 23(7):762–766, 2001.
9. J. Malik, S. Belongie, T. Leung, and J. Shi. Contour and texture analysis for image segmentation. *IJCV*, 43(1):7–27, 2001.
10. A.-R. Mansouri, A. Mitiche, and C. Vazquez. Multiregion competition: A level set extension of region competition to multiple region image partitioning. *CVIU*, 101(3):137–150, March 2006.
11. N. Paragios and R. Deriche. Geodesic active regions and level set methods for supervised texture segmentation. *IJCV*, 46(3):223–247, 2002.
12. C. R. Rao. The utilization of multiple measurements in problems of biological classification. *Journal of the Royal Statistical Society*, 10(B):159–203, 1948.
13. M. Rousson and R. Deriche. A variational framework for active and adaptive segmentation of vector valued images. In *IEEE WMVC*, pages 56–61, Florida, 2002.
14. B. Sandberg, T. Chan, and L. Vese. A level-set and gabor-based active contour algorithm for segmenting textured images. Technical Report 39, Math. Dept. UCLA, USA, 2002.
15. J. A. Sethian. *Level set methods and fast marching methods.* Cambridge University Press, 1999.
16. A. W. M. Smeulders, M. Worring, S. Santini, A. Gupta, and R. Jain. Content-based image retrieval at the end of the early years. *IEEE Trans. PAMI*, 22(12):1349–1380, 2000.
17. B. Sumengen and B. S. Manjunath. Edgeflow-driven variational image segmentation: Theory and performance evaluation. Technical report, VRL, ECE, UCSB, May 2005.
18. B. Sumengen, B. S. Manjunath, and C. Kenney. Image segmentation using curve evolution and flow fields. In *IEEE ICIP*, pages 105–108, 2002.
19. Zhuowen Tu and Song-Chun Zhu. Image segmentation by data-driven markov chain monte carlo. *IEEE Trans. PAMI*, 24(5):657–673, 2002.
20. J. Weickert, B. M. Romeny, and M. A. Viergever. Efficient and reliable schemes for nonlinear diffusion filtering. *IEEE Trans. IP*, 7(3):398–410, 1998.
21. J. Yang, A. F. Frangi, J. Y. Yang, D. Zhang, and Z. Jin. KPCA plus LDA: A complete kernel Fisher discriminant framework for feature extraction and recognition. *IEEE Trans. PAMI*, 27(2):230–244, 2005.

# Piecewise Constant Level Set Method for 3D Image Segmentation

Are Losnegård, Oddvar Christiansen, and Xue-Cheng Tai

Department of Mathematics, University of Bergen, Johannes Brunsgate 12, 5008 Bergen, Norway

**Abstract.** Level set methods have been proven to be efficient tools for tracing interface problems. Recently, some variants of the Osher- Sethian level set methods, which are called the Piecewise Constant Level Set Methods (PCLSM), have been proposed for some interface problems. The methods need to minimize a smooth cost functional under some special constraints. In this paper a PCLSM for 3D image segmentation is tested. The algorithm uses the gradient descent method combined with a Quasi-Newton method to minimize an augmented Lagrangian functional. Experiments for medical image segmentation are shown on synthetic three dimensional MR brain images. The efficiency of the algorithm and the quality of the obtained images are demonstrated.

## 1 Piecewise Constant Level Set Methods

In many applications one wants to divide an image into subsections based on intensity values, e.g. extract the gray matter, white matter and cerebrospinal fluid from a brain MRI, recognize the letters on a car plate or isolate any interesting objects in an image. There are many ways to achieve this segmentation. In the last decade, methods based on level sets have been quite popular [1,2,3,4,5].

In the traditional level set methods [6,7] the idea is to define a function $\phi(x)$, whose zero level set represents an interface $\Gamma$. This will divide a domain $\Omega$ into 2 subdomains, i.e.

$$\begin{aligned} \phi(x) &> 0, \quad \text{x inside } \Gamma, \\ \phi(x) &= 0, \quad \text{x on } \Gamma, \\ \phi(x) &< 0, \quad \text{x outside } \Gamma. \end{aligned}$$

Extensions of these methods results in the possibilities of separating the domain $\Omega$ into $2^N$ disjunct regions, using $N$ level set functions.

Recently, methods based on Piecewise Constant Level Set Methods (PCLSM) have been introduced for multiphase segmentation [8,9,10]. Instead of using $N$ level set functions one uses a single piecewise constant level set function $\phi$, defined as

$$\phi = i \text{ in } \Omega_i, \quad i = 1, 2, \ldots, N. \tag{1}$$

The discontinuities of $\phi$ give the curves that separate the regions. This extend the traditional multiphase level set method of [11,12]. In [13], the layers for the

F. Sgallari, A. Murli, and N. Paragios (Eds.): SSVM 2007, LNCS 4485, pp. 687–696, 2007.

$\phi$ functions are used to identify the phases. Another approach for multiphase segmentation is related to the phase-field models and the so-called binary level set methods, see [14,15,16,17].

In this work we will extend the approach introduced in [9], where a novel method for multiphase image segmentation is proposed and tested on 2d images. In this work we extend the code to 3d images and test the method on synthetic MRI data. We shall note that the level set idea has been applied [18] to the geodesic active surface model for gray matter segmentation. In [19] they combine the Chan-Vese [7] model with a variant of the Haralick/Canny edge detector and the geodesic active surface model for thin structures segmentation in volumetric data.

This work is organized in the following way. We start by introducing the minimization problem in section 2. In section 3 we show how the minimization problem can be solved using the gradient descent method and a quasi-Newton method, and in section 4 we show numerical results for the method.

## 2 The Minimization Problem

To segment an image $u_0$ based on intensity values, Mumford and Shah [20] proposed to approximate the function $u_0$ by a piecewise constant function $u$ by solving:

$$\inf_{u,\Gamma}\left\{ F^{MS}(u,\Gamma) = \sum_i \int_{\Omega_i} |c_i - u_0|^2 \mathrm{d}x + \nu|\Gamma| \right\}, \tag{2}$$

where $\Omega = \cup_i \Omega_i \cup \Gamma$, i.e. they tried to find a decomposition $\Omega_i$ of $\Omega$, where $u = c_i$ (constant) inside each connected component $\Omega_i$. The length of the curve $\Gamma$ is controlled by a positive parameter $\nu$. For a fixed $\Gamma$, we see that (2) is minimized when $c_i = mean(u_0)$ in $\Omega_i$. One challenge when solving (2) is to find a unique representation of the parametrized curve $\Gamma$.

In [8] they propose to solve the segmentation problem by using a PCLSM, thus solving the following constrained minimization problem:

$$\min_{\substack{c,\phi \\ K(\phi)=0}} \left\{ F(c,\phi) = \frac{1}{2}\int_\Omega |u-u_0|^2 \;\mathrm{d}x + \beta \sum_{i=1}^{N} \int_\Omega |\nabla \psi_i| \;\mathrm{d}x + \nu \int_\Omega |\nabla\phi| \;\mathrm{d}x \right\}, \tag{3}$$

in order to segment an image $u_0$ into $N$ phases. Above, The functions $\psi_i$ are defined as

$$\psi_i = \frac{1}{\alpha_i} \prod_{j=1, j\neq i}^{N} (\phi - j), \quad \alpha_i = \prod_{k=1,k\neq i}^{N} (i-k) \tag{4}$$

and the constraint $K$ is defined as

$$K(\phi) = (\phi-1)(\phi-2)\cdots(\phi-N) = \prod_{\phi=i}^{N} (\phi - i). \tag{5}$$

The piecewise constant image $u$ is a linear combination of the characteristic functions

$$u = \sum_{i=0}^{N} c_i \psi_i. \tag{6}$$

We see that large approximation errors will be penalized by the fidelity term $\frac{1}{2}\int_{\Omega} |u - u_0|^2$ and the last two terms suppress oscillation, whereas the regularization parameters $\beta > 0$, $\nu > 0$ control the effect of the two terms.

## 3 Steepest Descent and Quasi-Newton

In [8] the augmented Lagrangian method was used to solve the constrained minimization problem (3), and they defined this as

$$L(c, \phi, \lambda) = F(c, \phi) + \int_{\Omega} \lambda K(\phi)\, \mathrm{d}x + \frac{r}{2}\int_{\Omega} |K(\phi)|^2\, \mathrm{d}x, \tag{7}$$

where $\lambda \in L^2(\Omega)$ is the Langrange-multiplier and $r > 0$ is a penalty parameter. We normally choose the penalization parameter $r$ large, compared to the other parameters. To minimize (3), we have to find the saddle points for $L$. The saddle points are found by minimizing $L$ with respect to $\phi$ and $c$, and maximizing with respect to $\lambda$. By minimizing with respect to $\phi$ and $c$, we ensure that $F(c, \phi)$ is minimized, and by maximizing with respect to $\lambda$, the constraint must be fulfilled at convergence, otherwise the Lagrangian term of (7) will not vanish. The result is the following algorithm:

**Algorithm 1.** *Choose initial values for $\phi^0$, $c^0$ and $\lambda^0$. For $k = 1, 2, \ldots$, do:*

1. Find $c^k$ from
$$L(c^k, \phi^{k-1}, \lambda^{k-1}) = \min_{c} L(c, \phi^{k-1}, \lambda^{k-1}). \tag{8}$$
2. Use (6) to update $u = \sum_{i=1}^{N} c_i^k \psi_i(\phi^{k-1})$.
3. Find $\phi^k$ from
$$L(c^k, \phi^k, \lambda^{k-1}) = \min_{\phi} L(c^k, \phi, \lambda^{k-1}). \tag{9}$$
4. Use (6) to update $u = \sum_{i=1}^{N} c_i^k \psi_i(\phi^k)$.
5. Update the Langrange-multiplier by
$$\lambda^k = \lambda^{k-1} + rK(\phi^k). \tag{10}$$

The minimizer for (9) is solved here by the gradient descent method [8]. We introduce an artificial time variable and look for a steady-state solution to the PDE

$$\phi_t = -\frac{\partial L}{\partial \phi}. \tag{11}$$

We use the following numerical approximation to solve this problem:

$$\phi^{new} = \phi^{old} - \Delta t \frac{\partial L}{\partial \phi}(c, \phi^{old}, \lambda^{k-1}). \tag{12}$$

In order to compute this, we need the Gateux derivative:

$$\frac{\partial L}{\partial \phi} = (u-u_0)\frac{\partial u}{\partial \phi} - \beta \sum_{i=1}^{N} \nabla \cdot \left( \frac{\nabla \psi_i}{|\nabla \psi_i|} \right) \frac{\partial \psi_i}{\partial \phi} - \nu \nabla \cdot \left( \frac{\nabla \phi}{|\nabla \phi|} \right) + \lambda K'(\phi) + rK(\phi)K'(\phi) \tag{13}$$

and the step size $\Delta t$ is fixed during the whole iterative procedure.

As always, the derivative of $L$ with respect to $\lambda$ recovers the constraint

$$\frac{\partial L}{\partial \lambda} = K(\phi). \tag{14}$$

Because $u$ is linear with respect to the $c_i$ values, we see that $L$ is quadratic with respect to $c_i$. Thus the minimization problem (8) can be solved exactly. We see that

$$\frac{\partial L}{\partial c_i} = \int_\Omega \frac{\partial L}{\partial u} \frac{\partial u}{\partial c_i} = \int_\Omega (u - u_0)\psi_i \mathrm{d}x \quad \text{for } i = 1, 2, \ldots, N. \tag{15}$$

Therefore, the minimizer of (8) satisfies a linear system of equations $Ac^k = b$:

$$\sum_{j=1}^{n} \int_\Omega (\psi_j \psi_i) c_j^k \mathrm{d}x = \int_\Omega u_0 \psi_i \mathrm{d}x, \quad \text{for } i = 1, 2, \ldots, N. \tag{16}$$

Tests have shown [9] that this algorithm alone converges very slowly, i.e. it would require thousands of iterations for a brain MRI to achieve convergence. The strategy is to terminate it after a certain number of iterations and then use a quasi-Newton method to make it converge fast. The reason for choosing a quasi-Newton method, is to avoid inverting a huge linear algebraic system due to the regularization terms in (3). So we define

$$Q(c, \phi, \lambda) = \frac{1}{2} \int_\Omega |u - u_0|^2 \, \mathrm{d}x + \int_\Omega \lambda K(\phi) \, \mathrm{d}x. \tag{17}$$

Numerical experiments also reveals that it is not necessary to use the penalization term with the quasi-Newton updatings. Thus, we also define

$$L_0(c, \phi, \lambda) = F(c, \phi) + \int_\Omega \lambda K(\phi) \, \mathrm{d}x. \tag{18}$$

We see that $L_0$ is equal to $L$ if we take $r = 0$ in (7). Also, the functional $L_0$ reduces to $Q$ if we take $\beta = \nu = 0$. Thus the Hessian matrix for $Q$ is a good approximation for the Hessian matrix of $L_0$, using the fact that $\beta$ and $\nu$ are normally very small. So we arrive at the next algorithm:

**Algorithm 2.** *Choose initial values $\phi^0$, $c^0$ and $\lambda^0$. For $k = 1, 2, \ldots$, do:*

1. Find $c^k$ from

$$L(c^k, \phi^{k-1}, \lambda^{k-1}) = \min_c L(c, \phi^{k-1}, \lambda^{k-1}). \tag{19}$$

2. Update $u = \sum_{i=1}^N c_i^k \psi_i(\phi^{k-1})$.
3. Find $\phi^k$, $\lambda^k$ from

$$\begin{pmatrix} \frac{\partial^2 Q}{\partial \phi^2} & \frac{\partial^2 Q}{\partial \phi \partial \lambda} \\ \frac{\partial^2 Q}{\partial \phi \partial \lambda} & 0 \end{pmatrix} \begin{pmatrix} \phi^k - \phi^{k-1} \\ \lambda^k - \lambda^{k-1} \end{pmatrix} = - \begin{pmatrix} \frac{\partial L_0}{\partial \phi} \\ \frac{\partial L_0}{\partial \lambda} \end{pmatrix} \tag{20}$$

4. Update $u = \sum_{j=1}^N c_j^k \psi_j\left(\phi^k\right)$
5. If converged, end the loop. Otherwise go to step 1.

The updating of $c$ in (19) can be done in the same way as for algorithm 1. To solve (20), we need

$$\frac{\partial^2 Q}{\partial \phi^2} = \left(\frac{\partial u}{\partial \phi}\right)^2 + (u - u_0)\frac{\partial^2 u}{\partial \phi^2} + \lambda K''(\phi), \quad \frac{\partial^2 Q}{\partial \phi \partial \lambda} = \frac{\partial^2 Q}{\partial \lambda \partial \phi} = K'(\phi), \tag{21}$$

where $\lambda = \lambda^{k-1}$, $\phi = \phi^{k-1}$ and $u = u(c^k, \phi^{k-1})$. In addition we use that $\frac{\partial L_0}{\partial \lambda} = K(\phi^{k-1})$ and $\frac{\partial L_0}{\partial \phi}$ can be found from (13) by setting $r = 0$.

## 4 Numerical Results

We have applied our methods to some synthetic 3D brain MRI, and in this section we present some of the results. First we discuss how to choose the initial values, before looking into the tested images.

To find $\phi^0$ we have scaled the image, which normally takes the values between 0 and 255, to take the values between 1 and $n$, where $n$ is the number of phases for the image, i.e.

$$\phi^0(x) = 1 + \frac{u_0(x) - min_{x \in \Omega} u_0}{max_{x \in \Omega} u_0 - min_{x \in \Omega} u_0} \cdot (n-1). \tag{22}$$

To find initial $c^0$, we have used the function *kmeans* in MatLab and numerical tests have shown that the results are improved by not updating $c$ in (8) and (19). Initial value $\lambda^0$ is set to 0.

The tests are run on a computer with a 2 Opteron 270 dualcore processor and 8 GB memory.

In the following, we will look at some results of a synthetic brain image obtained from [21]. This is an image of size $217 \times 181 \times 181$ with 20% inhomogenity and 7% noise.

In the numerical experiments for this image we have achieved the best results with 1000 iterations of algorithm 1 followed by 30 iterations of algorithm 2. The parameters are $r = 500$, $\nu = 500$, $\beta = 0$, $dt = 1e^{-6}$ and we have used 3 phases.

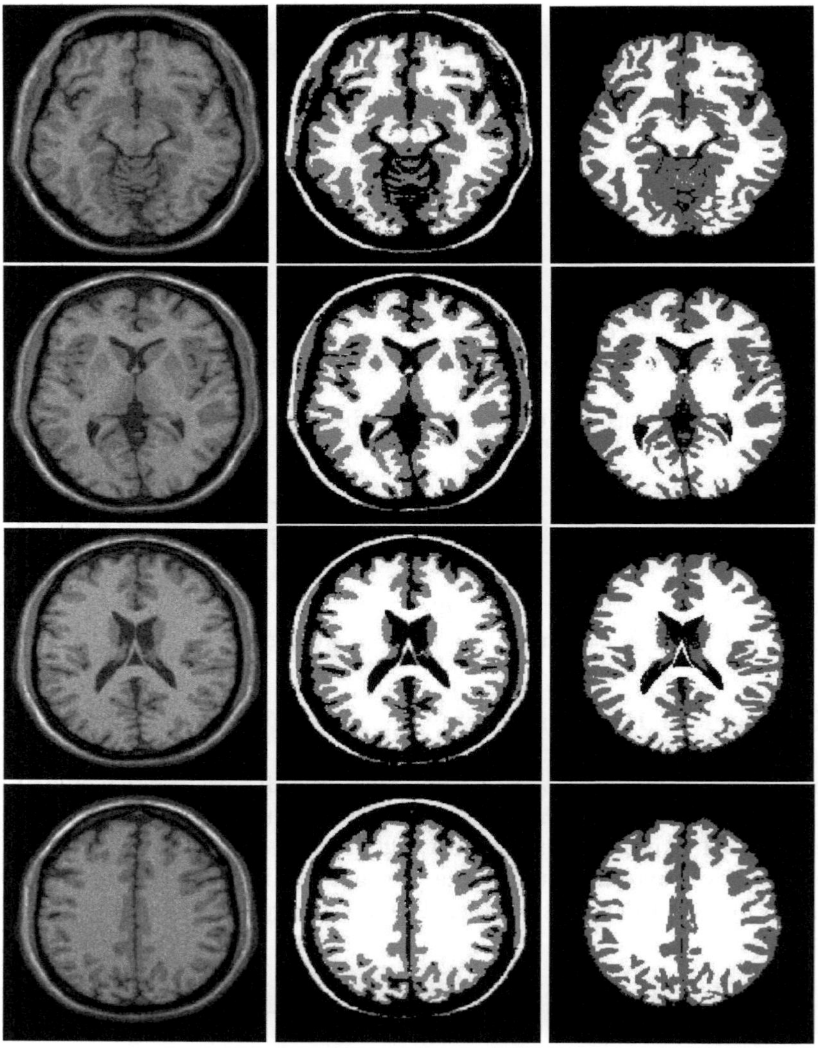

**Fig. 1.** Comparing the synthetic image, axial view. From left to right we display: The original image, the result from our method and the result from brainweb [21]. From top to bottom we show slice nr.: 61, 76, 91 and 106.

In figure 1 and 2 we try to compare the original image, the image segmented by our method and the image from brainweb [21] indicating the 3 phases. This is done by looking at some of the slices from the brain. In figure 3 we take a closer look at slice 91, comparing the original image and the result from our method. Finally, in figure 4 we present the white matter extracted from our segmented image, and in figure 5 we also show the white matter, but now sliced and with slice 91 displayed in gray.

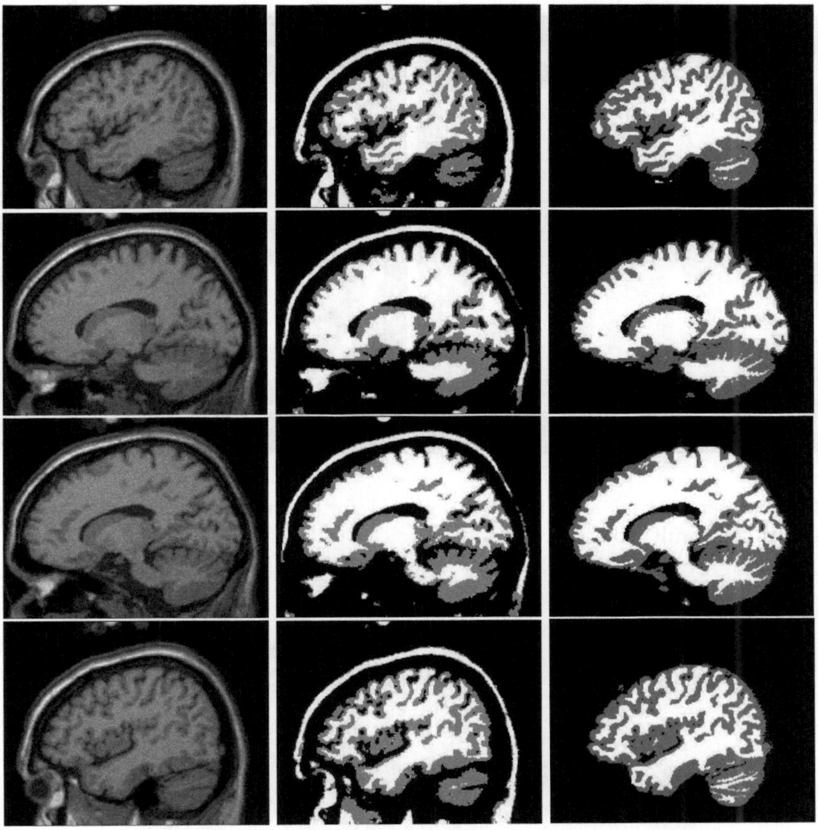

**Fig. 2.** Comparing the synthetic image, sagittal view. From left to right we display: The original image, the result from our method and the result from brainweb [21]. From top to bottom we show slice nr.: 45, 75, 105 and 135.

By looking at figure 1 and 2, we can see that our results are very good compared to the images in the 3rd column. By looking at the slices in the 1st row in figure 1, we can observe that there is a difference in the gray matter in the lower middle part of the images. In the 2nd row, we have the same tendency and there are also minor differences with respect to the white matter in the same part of the image. In figure 2 we can observe that there is a difference in the lower right parts of the images with respect to the white matter. Also small differences with respect to the gray matter are observable in this rather complex part of the image, probably due to noise and inhomogeneity.

## 5 Summary

In this paper we have extended the Piecewise Constant Level Set Methods (PCLSM) and tested it on complex 3D medical MR images. The segmentation

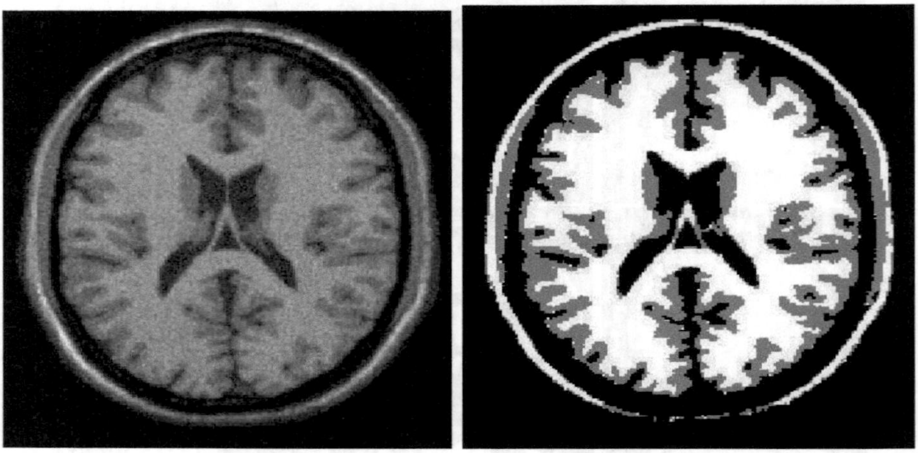

**Fig. 3.** A closer look at slice 91 from figure 1 for, from top to bottom, the original image and the result from our method

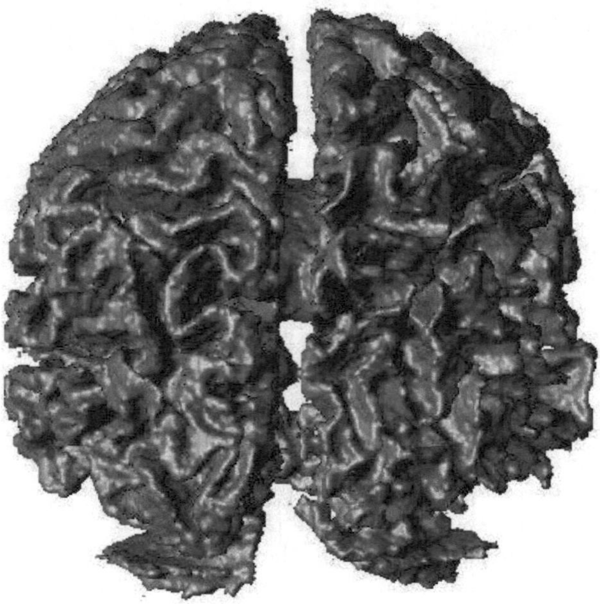

**Fig. 4.** We have extracted the white matter from our test image and used the built-in MatLab function *bwareaopen* to strip the skull together with a little gaussian smoothing

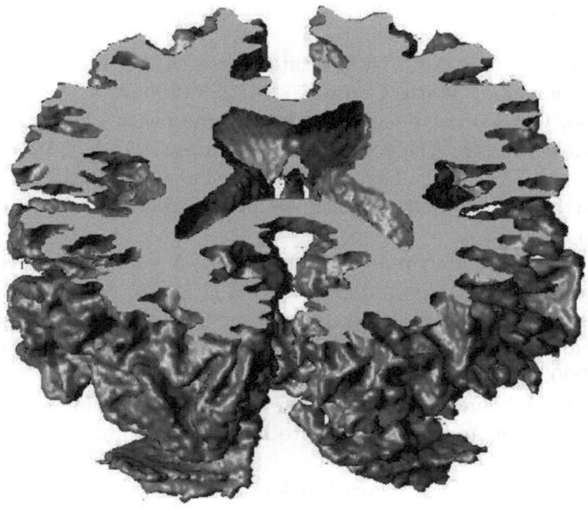

**Fig. 5.** Lower part of the white matter, where slice 91 is displayed in gray

results are very promising and we believe that it has a great potential for these kinds of applications.

Manual segmentation is very time consuming and we believe that our method has a potential of becoming quicker than it is today, but the convergence naturally depends on the size and the complexity of the image.

In future work we will test our method on 3D clinical MRI data which is an even harder task.

## References


1. Osher, S., Fedkiw, R.: An overview and some recent results. J.Comput.Phys, 169 No. 2:463-502 (2001)
2. Tai, X.C., Chan, T.F.: A survey on multiple level set methods with applications for identifying piecewise constant functions. Int. J. Numer. Anal. Model, 1(1):25-47 (2004)
3. Osher, S., Burger, M.: A survey on level set methods for inverse problems and optimal design. Cam-report-04-02, UCLA, Applied Mathematics (2004)
4. Vese, L.A., Chan, T.F.: A new multiphase level set framework for image segmentation via the Mumford and Shah model. International Journal of Computer Vision **50** (2002.) 271–293
5. Chan, T.F., Vese, L.A.: Image segmentation using level sets and the piecewise constant mumford-shah model. Technical report, CAM Report 00-14, UCLA, Math. Depart. (April 2000) revised December 2000.
6. Osher, S., Sethian, J.: Fronts propagating with curvature dependent speed: algorithms based on Hamilton-Jacobi formulations. J. Comput. Phys., vol. 79, no. 1 (1988)

7. Chan, T., Vese, L.: Active contours without edges. IEEE Image Proc., 10, pp. 266-277 (2001)
8. Lie, J., Lysaker, M., Tai, X.C.: A variant of the level set method and applications to image segmentation. Math. Comp. 75 no. 255 (2006)
9. Tai, X.c., Yao, C.h.: Image segmentation by piecewise constant Mumford-Shah model without estimating the constants. J. Comput. Math. **24**(3) (2006) 435–443
10. Christiansen, O., Tai, X.C.: Fast implementation of piecewise constant level set methods. In: Image processing based on partial differential equations, Springer Verlag (2006) 253–272
11. Zhao, H.K., Chan, T., Merriman, B., Osher, S.: A variational level set approach to multiphase motion. J. Comput. Phys. **127**(1) (1996) 179–195
12. Vese, L., Chan, T.: A new multiphase level set framework for image segmentation via the Mumford and Shah model. International Journal of Computer Vision, 50, pp.271-293 (2002)
13. Chung, J., Vese, A.: Energy minimization based segmentation and denoising using a multilayer level set approach. Lecture notes in computer science **3757** (2005) 439–455
14. Song, B., Chan, T.F.: Fast algorithm for level set segmentation. UCLA CAM report 02-68 (2002)
15. Gibou, F., Fedkiw, R.: A fast hybrid k-means level set algorithm for segmentation. 4th Annual Hawaii International Conference on Statistics and Mathematics, pp. 281-291 (2005)
16. Esedoḡlu, S., Tsai, Y.H.R.: Threshold dynamics for the piecewise constant Mumford-Shah functional. J. Comput. Phys. **211**(1) (2006) 367–384
17. Shi, Y., Karl, W.C.: A fast level set method without solving pdes. In: ICASSP'05. (2005)
18. Goldenberg, R., Kimmel, R., Rivlin, E., Rudzsky, M.: Cortex Segmentation: A Fast Variational Geometric Approach. IEEE Transactions on medical imaging, Vol. 21, No. 2 (2002)
19. Holtzman-Gazit, M., Kimmel, R., Peled, N., Goldsher, D.: Segmentation of Thin Structures in Volumetric Medical Images. IEEE Transactions on image processing, Vol. 15, No. 2 (2006)
20. Mumford, D., Shah, J.: Optimal approximation by piecewise smooth functions and associated variational problems. Comm. Pure Appl. Math, 42 (1989)
21. McConnell Brain Imaging Centre, Montréal Neurological Institute, M.U.: Brainweb. http://www.bic.mni.mcgill.ca/brainweb/ (12.6.2006)

# Histogram Based Segmentation Using Wasserstein Distances

Tony Chan[1], Selim Esedoglu[2], and Kangyu Ni[3]

[1] Department of Mathematics, UCLA
chan@math.ucla.edu
[2] Department of Mathematics, University of Michigan
esedoglu@umich.edu
[3] Department of Mathematics, UCLA
kni66@math.ucla.edu

**Abstract.** In this paper, we propose a new nonparametric region-based active contour model for clutter image segmentation. To quantify the similarity between two clutter regions, we propose to compare their respective histograms using the *Wasserstein* distance. Our first segmentation model is based on minimizing the Wasserstein distance between the object (resp. background) histogram and the object (resp. background) reference histogram, together with a geometric regularization term that penalizes complicated region boundaries. The minimization is achieved by computing the gradient of the level set formulation for the energy. Our second model does not require reference histograms and assumes that the image can be partitioned into two regions in each of which the local histograms are similar everywhere.

**Keywords:** image segmentation, region-based active contour, Wasserstein distance, clutter.

## 1 Introduction

Parametric region-based active contour models have been widely used in image segmentation. One of their advantages is that they incorporate region information with boundary information. For example, the Chan-Vese model is able to carry out foreground and background segmentation without any explicit reference to edges [5]. However, the standard Chan-Vese model is based on the assumption that the foreground (resp. background) intensity is fairly homogeneous, i.e. the probability density functions of object intensities and background intensities are both Gaussian with the same variance. This can be a significant restriction in applications. Other parametric region-based active contours models, including certain generalizations of the Chan-Vese model, assume that the histogram of image intensities in different regions of the segmentation are Gaussian. For example, in [18], the segmentation models distinguish the object from the background by intensity means and/or variances of image regions.

The purpose of this work is to segment images consisting of clutter features. There are also many other image models that are not within our scope, such as

F. Sgallari, A. Murli, and N. Paragios (Eds.): SSVM 2007, LNCS 4485, pp. 697–708, 2007.

shape, texture and smooth regions. An effective method of incorporating different image models can be found in [17]. Clutter features are often found in natural scenes, such as trees and grass. They are highly nonhomogeneous in intensity and their corresponding histograms do not necessarily have particular statistical structure – for example, they may not obey a Gaussian distribution. They also usually do not have a particular geometric content. Therefore, parametric methods are not suitable for segmentation of cluttered regions. In this work, we use image intensity histograms to drive the segmentation process, which makes no simplifying assumptions about the statistics of the image intensity values. It also does not rely on any geometric content found in the regions. We thus segment images purely based on histogram information found within its various regions. Image histograms have been used extensively in PDE-based image processing. For instance, [4] and [16] are variational methods in histogram modification for contrast enhancement.

There are a number of nonparametric segmentation models in the literature that are closely related to our work. In [10,7], the authors propose to maximize the mutual information between the region labels and the image intensities. In [3,1], the proposed model is to minimize the chi-2 comparison function between the object (resp. background) histogram and the object (resp. background) reference histogram. Their experimental results show effectiveness in segmenting slightly-textured images, e.g. human faces. However, the chi-2 comparison function is not a metric and is not suitable for comparing histograms in many situations. As a simple demonstration, the chi-2 distance between two delta functions with disjoint supports is the same no matter how far apart the supports are; this is a situation that arises often in segmentation applications, since for example images consisting of two objects with approximately constant but different intensities would fall into this category. To overcome this issue, we propose to use the Wasserstein distance (Monge-Kantorovich distance) to compare histograms. The Wasserstein distance between two normalized histograms is the least work required to move the region lying under the graph of one histogram to that of the other. It extends as a metric to measures such as the delta function. We believe this to be the more natural and appropriate way to compare histograms, since it does not suffer from the shortcoming mentioned above concerning pointwise metrics such as the standard $L^p$ norms or the chi-2 comparison function. Experimental results show that there is indeed a significant benefit using the Wasserstein distance to compare histograms, and that it is quite effective in segmenting images consisting of cluttered regions. Optimal transport ideas have been used in other context in image processing, such as [8] on image registration and morphing and many others [2], [6] and [15].

The layout of the paper is as follows. Section II presents facts from optimal transportation theory used in this paper; in particular, we describe briefly the Monge-Kantorovich problem and how to solve it. Section III consists of two subsections, each one devoted to one of proposed new models. Also, level set formulations of these new models and their associated optimality conditions and gradient descent equations are given here. Section IV shows the algorithms and discretization for solving the proposed models. Section V shows experimental results and comparison with other methods in both synthetic and real images.

## 2 Wasserstein Distance

The original Monge-Kantorovich problem was first posed in 1781 by G. Monge in [11]: what is the minimum work required to move a pile of dirt into a hole with the same volume? The original mathematical formulation turned out to be a difficult problem because it requires that no mass be split. Kantorovich proposed a relaxed version in [9], which we summarize in the following. Let $(X, \mu)$ and $(Y, \nu)$ be two probability measure spaces. Let $\pi$ be a probability measure on the product space $X \times Y$ and $\Pi(\mu, \nu) = \{\pi \in P(X \times Y) : \pi[A \times Y] = \mu[A]$, and $\pi[X \times B] = \nu[B]$ hold for all measureable sets $A \in X$ and $B \in Y\}$ be the set of admissible transference plans. For a given cost function $c : X \to Y$, the total transportation cost, associated to $\pi \in \Pi(\mu, \nu)$, is $I[\pi] = \int_{X \times Y} c(x, y) d\pi(x, y)$. The optimal transportation cost between $\mu$ and $\nu$ is $T_c(\mu, \nu) = inf_{\pi \in \Pi(\mu,\nu)} I[\pi]$. More detail can be found in [19] and [14], which is a good exposition on this subject.

In this paper, we are interested in the case when the probability is on the real line. Let $\mu$ and $\nu$ be two probability measures on $\mathbb{R}$, with respective cumulative distribution functions $F$ and $G$. Then, it is known that for a convex cost function $c(x, y)$, the optimal transportation cost is $T_c(\mu, \nu) = \int_0^1 c(F^{-1}(t), G^{-1}(t))dt$. In particular, the optimal transportation cost for the linear cost function $c(x, y) = |x - y|$ is $T_1(\mu, \nu) = \int_0^1 |F^{-1}(t) - G^{-1}(t)|dt$ and by Fubini's Theorem, $T_(\mu, \nu) = \int_0^1 |F(t) - G(t)|dt$.

In the proposed models, we use the Wasserstein distance to determine the similarities between two normalized image histograms. Let $P_a(y)$ and $P_b(y)$ be two normalized histograms and let $F_a(y)$ and $F_b(y)$ be their corresponding cumulative distributions. The linear Wasserstein distance ($W_1$ distance) between $P_a(y)$ and $P_b(y)$ is defined by

$$W_1(P_a, P_b) = T_1(P_a, P_b) = \int_0^1 |F_a(y) - F_b(y)|dy \, . \tag{1}$$

An important consequence of this definition is that, unlike chi-2 function, Wasserstein distance is a metric. If two $\delta$-functions are close by, the Wasserstein distance between them is small, because the area between their corresponding cumulative distribution functions is small.

## 3 Proposed Models

In this paper, we propose two segmentation energy models using the $W_1$ distance. By minimizing these energies, we hope to find an optimal region such that the region boundaries match the clutter boundaries. The first proposed model requires reference object (resp. background) histograms as inputs; this is the same setting as in [1]. The first model is to minimize the Wasserstein distance between object (resp. background) histogram and the object (resp. background) reference histogram, together with a geometric regularization term on the interface. The second model do not require any reference histograms and assumes that the local histograms within the object region (resp. background region) are similar everywhere. We assign each pixel a neighborhood histogram, the histogram of a small neighborhood around that pixel. This model is to find an optimal region such that the object (resp. background) histogram is similar to all the neighborhood histograms inside (resp. outside) the region.

Given a grey scale image $I : \Omega \to [0, 255]$, the normalized image histogram restricted on the region $\Sigma$ and the associated cumulative distribution function can be written in the following level set representation

$$P_{\Sigma}(y) = \frac{\int_{\Omega} H(\phi(z))\delta(y - I(z))dz}{\int_{\Omega} H(\phi(z))dz} \tag{2}$$

and

$$F_{\Sigma}(y) = \frac{\int_{\Omega} H(\phi(z))H(y - I(z))dz}{\int_{\Omega} H(\phi(z))dz} , \tag{3}$$

where $y \in [0, 255]$ is an intensity value, $\phi$ is a level set function [13] such that $\Sigma = \{x \in \Omega : \phi(x) > 0\}$, and $\delta$ and $H$ are the Dirac and Heaviside function, respectively. Similarly, using the same $\phi$ for outside the region $\Sigma^c$, we have

$$P_{\Sigma^c}(y) = \frac{\int_{\Omega}[1 - H(\phi(z))]\delta(y - I(z))dz}{\int_{\Omega}[1 - H(\phi(z))]dz} \tag{4}$$

and

$$F_{\Sigma^c}(y) = \frac{\int_{\Omega}[1 - H(\phi(z))]H(y - I(z))dz}{\int_{\Omega}[1 - H(\phi(z))]dz} . \tag{5}$$

We use the level set method [13], because it allows changes of topology, such as merging and splitting. We normalize histograms because two identical clutters of different sizes should have zero distance.

### 3.1 Histogram Segmentation with Reference Histograms

For the first segmentation model, we are given a foreground reference histogram $P_f(y)$ and a background reference histogram $P_b(y)$. The model is

$$\inf_{\Sigma} E_1(\Sigma) = \text{Per}(\Sigma) + \lambda\{W_1(P_{\Sigma}, P_f) + W_1(P_{\Sigma^c}, P_b))\} ,$$

where $W_1$ is the $W_1$ distance described in (1). The first term is the length of the boundary of $\Sigma$, as a regularization term. The second (resp. third) is a fitting term between the object (resp. background) histogram and object (resp. background) reference histogram. The level set formulation of (6) is

$$\inf_{\phi} E_1(\phi) = \int_{\Omega} |\nabla H(\phi(x))| dx + \quad \lambda \quad \{ \int_0^{255} |F_{\Sigma}(y) - F_f(y)| dy$$
$$+ \int_0^{255} |(F_{\Sigma^c}(y) - F_b(y)| dy\} ,$$

where we plug in $F_{\Sigma}$ and $F_{\Sigma^c}$ by (3) and (5), respectively.

To minimize the energy, we derive the associated Euler-Lagrange equation. The gradient descent for $\phi$ is given by the following evolution equations

$$\phi_t = \delta(\phi) \left[ \nabla \cdot \left( \frac{\nabla \phi}{|\nabla \phi|} \right) - \lambda(A - B) \right] , \tag{6}$$

where

$$A = \frac{1}{Area(\Sigma)} \int_0^{255} \frac{F_{\Sigma}(y) - F_f(y)}{|F_{\Sigma}(y) - F_f(y)|} [H(y - I(x)) - (F_{\Sigma}(y))] dy$$

and

$$B = \frac{1}{Area(\Sigma^c)} \int_0^{255} \frac{F_{\Sigma^c}(y) - F_b(y)}{|F_{\Sigma^c}(y) - F_b(y)|} [H(y - I(x)) - (F_{\Sigma^c}(y))] dy .$$

### 3.2 Histogram Segmentation with Neighborhood Histograms

We modify the first segmentation model (6) so that input reference histograms are not required. For simplicity, we assume that the image of interest has two regions, object and background region, each of which has the same histograms locally (e.g. clutter features). The histogram restricted on a small region (neighborhood histogram) is similar to either the object histogram or the background histogram. Therefore, we compare the object (resp. background) histogram with all the neighborhood histograms in the object (resp. background) region.

For each point $x \in \Omega$, we compute the neighborhood cumulative distribution function

$$F_{x,r}(y) = \frac{Area(\{x \in B_r(x) : I(x) \leq y\})}{Area(\{B_r(x)\})} .$$

The size $r$ of the neighborhood is chosen according to the clutter features in an image. It needs to be greater than or equal to the size of the clutter feature. For an accurate result, it should not be too large. In this paper, the selection of the size is specified by the user. The proposed model is

$$\inf_{\Sigma} E_2(\Sigma) = Per(\Sigma) + \lambda \{ \int_{\Sigma} W_1(P_1, P_{x,r}) dx + \int_{\Sigma^c} W_1(P_2, P_{x,r}) dx \} . \tag{7}$$

In a level set formulation, (7) becomes

$$\inf_{\Sigma} E_2(\Sigma) = \int_{\Omega} |\nabla H(\phi(x))| dx$$

$$+\lambda\{\int_\Omega H(\phi(x)) \int_0^{255} |F_1(y) - F_{x,r}(y)| dy dx$$
$$+ \int_\Omega [1 - H(\phi(x))] \int_0^{255} |F_2(y) - F_{x,r}(y)| dy dx\} . \quad (8)$$

Note that $F_{x,r}(y)$'s need to be computed only once before optimization. $F_1(y)$ and $F_2(y)$ are two constant cumulative distribution to be determined, independent of $\phi$.

To minimize this energy, we first fix $\phi$ and minimize with respect to $F_1(y)$ and $F_2(y)$, respectively. Then, we fix $F_1(y)$ and $F_2(y)$ and minimize with respect to $\phi$. The evolution equations are

$$F_1(y) = \frac{\int_\Omega H(\phi(x)) F_{x,r}(y) dx}{\int_\Omega H(\phi(x)) dx}$$

$$F_2(y) = \frac{\int_\Omega [1 - H(\phi(x))] F_{x,r}(y) dx}{\int_\Omega [1 - H(\phi(x))] dx}$$

$$\phi_t = \delta(\phi) \left[ \nabla \cdot \left( \frac{\nabla \phi}{|\nabla \phi|} \right) - \lambda \int_0^1 (|F_1(y) - F_{x,r}(y)| - |F_2(y) - F_{x,r}(y)|) \, dy \right] \quad (9)$$

As the evolution equations suggest, the object (resp. background) cumulative distribution function $F_1$ (resp. $F_2$) is the average of all the neighborhood cumulative distribution functions $F_{x,r}$ inside (resp. outside) the curve. The minimization forces the 0-level curve of $\phi$ to move toward the boundaries of the object, so that the object (resp. background) cumulative distribution function is similar to all the neighborhood cumulative distribution histograms inside (resp. outside) the curve.

## 4 Numerical Method

For numerical implementation, we use a $C^\infty$ regularized Heaviside function and the corresponding regularized Dirac function as follows

$$H_\varepsilon(z) = \frac{1}{2}\left(1 + \frac{2}{\pi} \arctan\left(\frac{z}{\varepsilon}\right)\right) , \text{ and } \delta_\varepsilon(z) = \frac{1}{\pi} \frac{\varepsilon}{\varepsilon^2 + z^2} .$$

The evolution equations (6) and (9) for both proposed models have the following form

$$\phi_t = \delta(\phi) \left[ \nabla \cdot \left( \frac{\nabla \phi}{|\nabla \phi|} \right) + \lambda A(\phi) \right] .$$

We compute $\phi$ by the following discretization

$$\frac{\phi^{n+1} - \phi^n}{\Delta t} = \delta_\epsilon(\phi^n)\left[\Delta_x^-\left(\frac{\Delta_x^+\phi^n}{|\nabla\phi^n|}\right) + \Delta_y^-\left(\frac{\Delta_y^+\phi^n}{|\nabla\phi^n|}\right) + \lambda A(\phi^n)\right] ,$$

where

$$|\nabla\phi^n| = \sqrt{\left(\Delta_x^+\phi^n\right)^2 + \left(\Delta_y^+\phi^n\right)^2 + \epsilon} ,$$
$$\Delta_x^-\phi_{i,j} = \phi_{i,j} - \phi_{i-1,j} , \Delta_x^+\phi_{i,j} = \phi_{i+1,j} - \phi_{i,j} ,$$
$$\Delta_y^-\phi_{i,j} = \phi_{i,j} - \phi_{i,j-1} , \Delta_y^+\phi_{i,j} = \phi_{i,j+1} - \phi_{i,j} .$$

In the evolution equation (6), the corresponding $A(\phi)$ term can be written as

$$\int B(y)\left[H(y - f(x)) - C(y)\right]dy = \int B(y)C(y)dy + \int B(y)H(y - f(x))dy , \quad (10)$$

for some functions $B(y)$ and $C(y)$.
Note that the first term is independent of $x$, while the second term can be simplified as

$$\int B(y)H(y - f(x))dy = \int_{f(x)}^{255} B(y)dy .$$

Now, we only need to compute once

$$G(i) = \int_i^{255} A(y)dy$$

for $i \in \{0, 1, ..., 255\}$. Then, the second term of the right hand side of (10) can be obtained fast by looking up $G(f(x))$ and by linear interpolation.

## 5 Experimental Results

We show and compare the proposed segmentation methods with some of existing methods. Figure 1 shows a $144 \times 144$ synthetic image, which has three regions with different distributions, as shown in Fig. 2. The inner region and the middle region look distinct, as well as their corresponding histograms, even though the histograms overlap 50 percents. On the other hand, the middle region and the outer region look similar, as well as their corresponding histograms, even though the histograms do not overlap at all. In both cases, the degree of similarity in image regions agree with the degree of similarity in their corresponding histograms. Figure 3 shows results of proposed and existing segmentation methods. The first row is the final contour, corresponding histograms, and cumulative distributions (from left to right) of proposed segmentation with reference histograms. The foreground and background reference histograms are obtained by

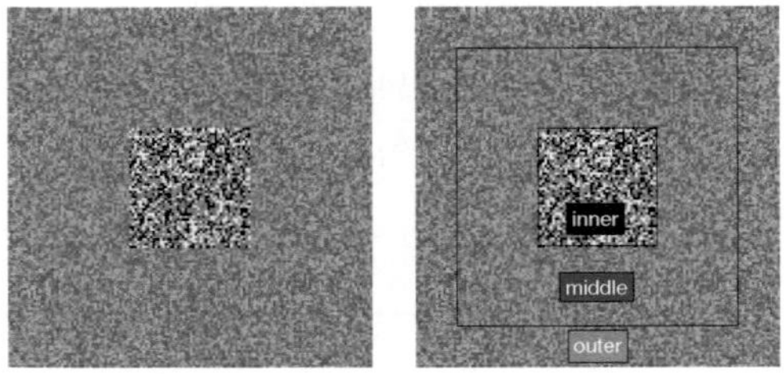


**Fig. 1.** Left: synthetic image. Right: boundaries between inner, middle, and outer regions.

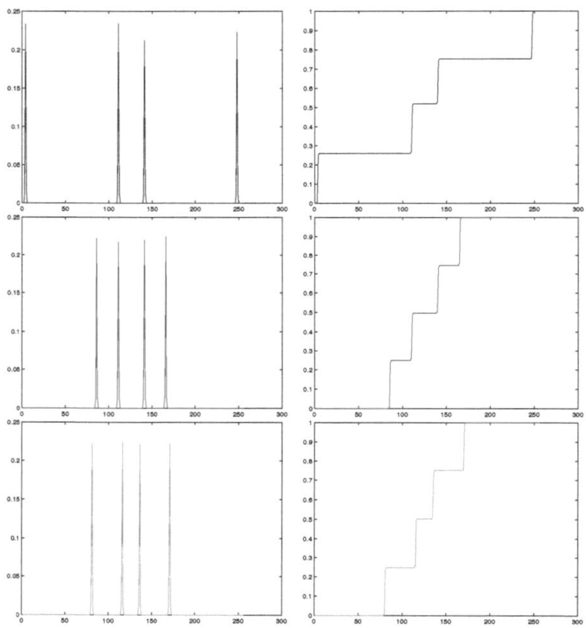

**Fig. 2.** From top to bottom. Left column: inner, middle, and outer region histograms. Right column: inner, middle, and outer region cumulative distributions.

calculating on the inner and outer region, respectively. The final contour shows that the proposed model is able to segment the middle and the outer region as background. The second row is the final contours, corresponding histograms, and cumulative distributions of proposed segmentation model with neighborhood histograms. This proposed method is also able to distinguish the foreground (inner region) from the background (middle and outer region). This shows that the $W_1$ distance is effective in histogram segmentation. The third row is the final contour, corresponding histograms, and cumulative distributions of segmentation with reference histograms using chi-2 function. Since this model strongly favors

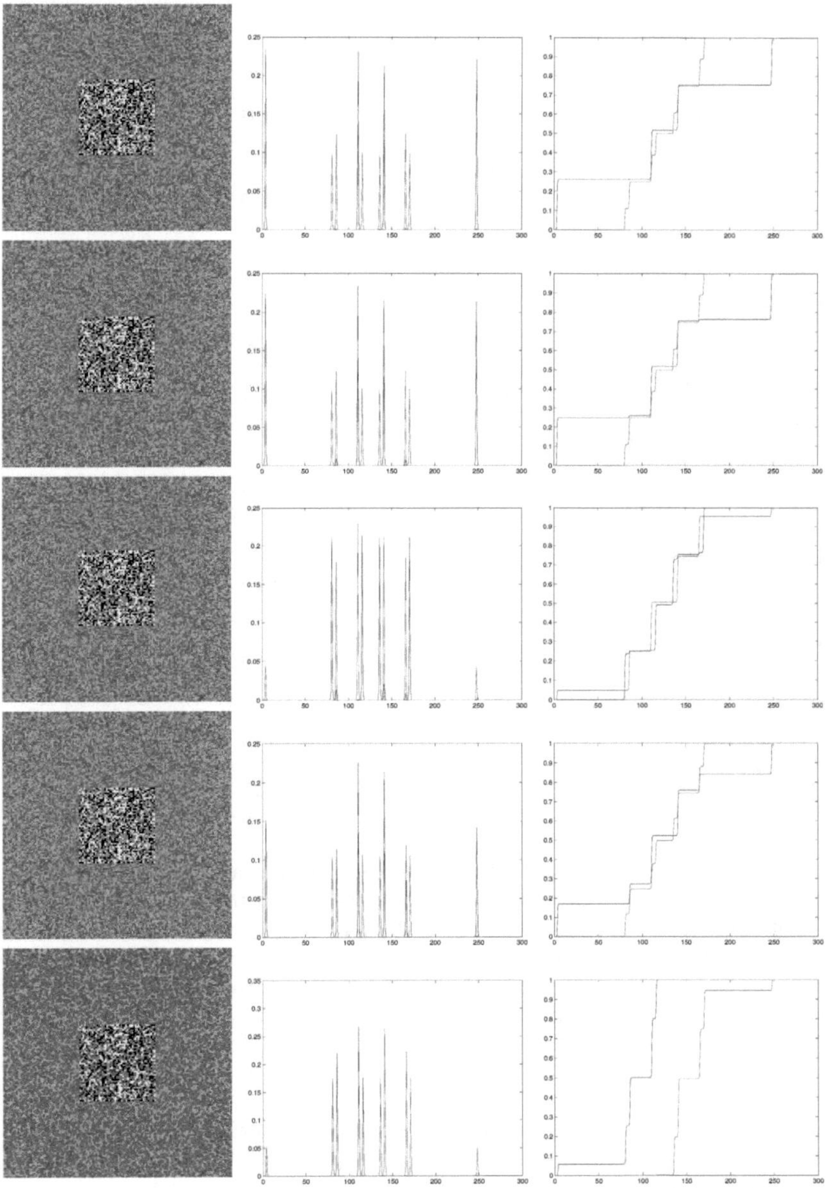

**Fig. 3.** Comparison of proposed histogram segmentation and existing methods. The first column is the final contour of different segmentation methods. The second (resp. third) columns are corresponding foreground a nd background histograms (resp. cumulative distributions). First row: proposed histogram segmentation with reference histograms. Second row: proposed histogram segmentation with neighborhood histograms. Third row: histogram segmentation using chi-2 function with reference histograms. Fourth row: Chan-Vese segmentation. Fifth row: Chan-Vese segmentation.

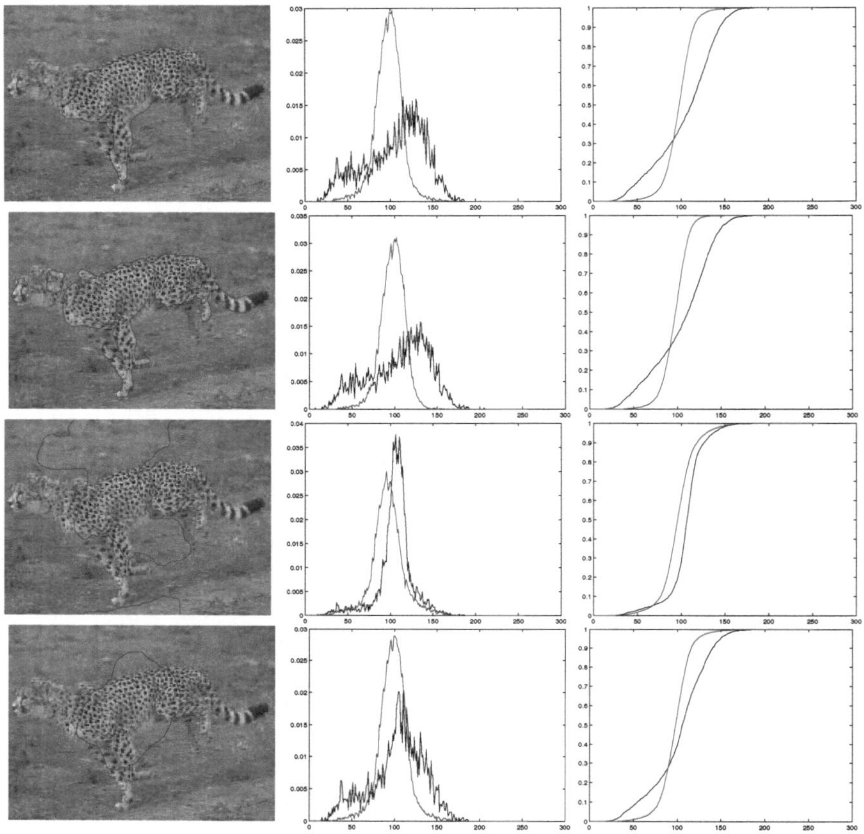

**Fig. 4.** Comparison of proposed histogram segmentation and existing methods. The first column is the final contour of different segmentation methods. The second (resp. third) columns are corresponding foreground a nd background histograms (resp. cumulative distributions). First row: proposed histogram segmentation with reference histograms. Second row: proposed histogram segmentation with neighborhood histograms. Third row: histogram segmentation using chi-2 function with reference histograms. Fourth row: Chan-Vese segmentation.

overlapping histograms, the middle region is segmented falsely as foreground. The fourth and fifth columns are the final contours, corresponding histograms, and cumulative distributions of Chan-Vese segmentation, with different fidelity parameters. The fourth row shows that the Chan-Vese segmentation is not able to come close to a correct segmentation. The fifth row shows that the Chan-Vese segmentation, with larger fidelity parameters, segments at a pixel level, in order to distinguish foreground and background intensity values. In any case, the standard Chan-Vese segmentation fails the task because the average intensity of any region in this image is the same.

Figure 4 shows segmentation results of various methods for a $135 \times 175$ real image. This image has complicated features, i.e. cheetah patterns, that are impossible to segment by existing methods. The first row is the final contour, corresponding histograms, and cumulative distributions of proposed histogram segmentation with reference histograms. The given foreground reference histogram is obtained by calculating the histogram on a small patch of the cheetah. The second row is the final contours, corresponding histograms, and cumulative distributions by segmentation with neighborhood histograms. As the results show, our proposed models proficiently segment the cheetah patterns. The third row is the final contour, corresponding histograms, and cumulative distributions of segmentation with reference histograms using chi-2 function. The fourth row is the final contour, corresponding histograms, and cumulative distributions of Chan-Vese segmentation. Both the existing segmentation methods fail to segment the cheetah pattern from the background. In this example, our proposed models outperform in segmenting clutters.

## 6 Conclusion

In this work, we propose a novel nonparametric region-based active contour model for segmenting clutter images. It is based on the use of Wasserstein mass transfer metrics in comparing histograms of different regions in the image. Our numerical results corroborate that these metrics are more suitable for histogram comparisons than what has been utilized previously in the existing literature, and lead to substantially better segmentations. Wasserstein metrics can be incorporated into a variety of histogram and curve evolution based segmentation models; we give two examples of such in this paper in order to substantiate our claims.

## References


1. G.Aubert, M.Barlaud, O.Faugeras, S.Jehan-Besson, Image Segmentation Using Active Contours: Calculus of Variations or Shape Gradients?, SIAM APPL. Math. Vol. 1, No. 2, pp.2128-2145, 2005.
2. R.E.Broadhurst, Statistical estimation of histogram variation for texture classification, In Texture 2005: Proceedings of the 4th Internation Workshop on Texture Analysis and Synthesis, pp. 25-30, 2005.
3. S.Jehan-Besson, M.Barlaud, G.Aubert, O.Faugeras, Shape Gradients for Histogram Segmentation Using Active Contours, in Proc. Int. Conf. Computer Vision, Nice, France, 2003, pp.408-415.
4. V.Caselles,J-L.Lisani,J-M.Morel, and G.Sapiro, Shape preserving local histogram modification, IEEE Trans. Image Proc. 8, pp. 220-230, 1999.
5. T.F.Chan, L.A.Vese, Active contours without edges, IEEE Transactions on Image Processing. 10:2(2001), pp. 266-277.
6. R.Chartrand, K.Vixie, B.Wohlber, E.Bollt, A Gradient Descent Solution to the Monge-Kantorovich Problem.

7. A.Herbulot, S.Jehan-Besson, M.Barlaud, G.Aubert, Shape Gradient for Image Segmentation using Information Theory, in ICASSP, may 2004, Vol. 3, pp.21-24.
8. S.Haker, L.Zhu, and A.Tannnenbaum, Optimal Mass Transport for Registration and Warping, International Journal of Computer Vision 60(3), 225-240,2004.
9. L.V.Kantorovich, On the translocation of masses, C.R.(Doklady) Acad. Sci. URSS(N.S.), 37(1942),pp.199-201.
10. J.Kim, J.W.Fisher, A.Yezzi, M.Cetin, and A.S.Willsky, Nonparametric methods for image segmentation using information theory and curve evolution, in ICIP, 2002, pp. III:797-800.
11. G.Monge, Mémoire sur la théorie des déblais at des remblais, Histoire de l'Académie Royale des Sciences de Paris, pp.666-704, 1781.
12. D.Mumford and J.Shah, Optimal approximation by piecewise smooth functions and associated variational problems, Commun. Pure Appl. Math, vol. 42, pp. 577-685,1989.
13. S.Osher and J.A.Sethian, Fronts propagation with curvature-dependent speed: Algorithms based on Hamilton-Jacobi Formulation, J. Comput. Phys., vol. 79, pp. 12-49, 1988.
14. S.Rachev and L.Rüschendorf, Mass transportation problems. Vol.I: Theory, Vol.II: Applications.Probability and its applications. Springer-Verlag, New York, 1998.
15. Y.Rubner, C.Tomasi, and L.J.Guibas, A metric for distributions with applications to image databases, In IEEE International Conference on Computer Vision, pp. 59-66, Bombay, India, January 1998.
16. G.Sapiro and V.Caselles, Histogram modification via differential equations, Journal of Differential Equations 135:2, pp. 238-268, 1997.
17. Z.Tu and S.Zhu, Image segmentation by data-driven Markov Chain Monte Carlo, in IEEE transactions on pattern analysis and machine intelligenc, Vol.24, No.5, May 2002.
18. A.Yezzi,Jr.,A.Tsai, and A.Willsky, A statistical approach to snakes for bimodal and trimodal imagery, In Int. Conf. on Computer Vision, pp.898-903, 1999.
19. C.Villani, Topics in Optimal Transportation. Graduate Studies in Mathematics, Vol. 58, American Mathematical Society, Providence Rhode Island, 2003.

# Efficient Segmentation of Piecewise Smooth Images

Jérome Piovano[1], Mikaël Rousson[2], and Théodore Papadopoulo[1]

[1] Odyssée team, INRIA/ENPC/ENS, 2004 route des Lucioles, BP 93, 06902 Sophia-Antipolis, France
{Jerome.Piovano,Theodore.Papadopoulo}@sophia.inria.fr

[2] Siemens Corporate Research, 755 College Road East, Princeton, NJ 08540, USA
Mikael.Rousson@siemens.com

**Abstract.** We propose a fast and robust segmentation model for piecewise smooth images. Rather than modeling each region with global statistics, we introduce local statistics in an energy formulation. The shape gradient of this new functional gives a contour evolution controlled by local averaging of image intensities inside and outside the contour. To avoid the computational burden of a direct estimation, we express these terms as the result of convolutions. This makes an efficient implementation via recursive filters possible, and gives a complexity of the same order as methods based on global statistics. This approach leads to results similar to the general Mumford-Shah model but in a faster way, without solving a Poisson partial differential equation at each iteration. We apply it to synthetic and real data, and compare the results with the piecewise smooth and piecewise constant Mumford-Shah models.

## 1 Introduction

The extraction of piecewise smooth regions from an image is of great interest in different domains, and still remains a challenging task. For example, this is very useful in medical imaging where organs or structures of interest are often characterized by smooth intensity regions. This problem has been formulated as the minimization of an energy by Mumford and Shah in [9]:

$$E^{MS}(u,\Gamma) = \mu^2 \int_{\Omega} (u_0 - u)^2 dx + \int_{\Omega\setminus\Gamma} |\nabla u|^2 dx + \nu|\Gamma| \tag{1}$$

where $u$ is the piecewise smooth function, $\Gamma$ the interface between smooth regions, and $u_0$ the original image. The interpretation of the three term is straightforward: the first one is the usual mean-square data term; the second one means that we want to extract smooth regions; the third one means that we want to extract regions with smooth boundaries.

The minimizer of this so-called *Mumford-Shah functional* gives a boundary that separates the image domain in smooth regions. A very interesting property of this approach is that it solves two common image-processing tasks simultaneously: image denoising and image segmentation. However, finding the minimizer

F. Sgallari, A. Murli, and N. Paragios (Eds.): SSVM 2007, LNCS 4485, pp. 709–720, 2007.

is not straightforward and remains an issue. Being non-convex, the functional is most of time minimized using gradient descent techniques which are subject to local minima. For example, in [16,14], the optimization process alternates between the evolution of one or two level set functions [10,13,5] and the resolution of Poison partial differential equations. This process is computational expensive and requires a very good initialization to avoid being stuck into a local minima.

To relax this problem, one can consider a restriction of $E^{MS}$ to piecewise constant functions. Let $\Omega_i$ be the open subsets delimited by $\Gamma$, the piecewise constant Mumford-Shah functional writes:

$$E_0^{MS}(\Gamma) = \sum_i \int_{\Omega_i} (u_0 - \text{mean}_{\Omega_i}(u_0))^2 dx + \nu|\Gamma|. \tag{2}$$

This functional was shown in [9] to be a limit functional of (1) as $\mu \to 0$. A level set implementation of this functional known as the Chan-Vese model was proposed in [16]. While this simplified functional is easier to minimize, it also makes a very strong assumption on the image by assuming implicitly a Gaussian intensity distribution for each region $\Omega_i$ [12]. Other papers model this distribution with Gaussian mixtures [11,6] or with nonparametric distributions [8] but they all make the assumption of a global distribution over each region. In many real images, these global intensity models are not valid. This is often the case in medical imaging, especially in MR images where an intensity bias can be observed.

Several approaches are available to overcome the limitation of global techniques. One is to consider image gradients by fitting the contour to image discontinuities. This is generally referred to as edge-based methods, and it is the basis of the Geodesic Active Contours [2,7][1]. Edge-based methods are also well-known for their high sensitivity to noise and for the presence of local minima in the optimization [3]. Another alternative was briefly discussed in [15] where the function $u$ of the Mumford-Shah functional is restricted to a linear function of the spatial location $x$: $u(x) = a.x + b$. Even though this last one is promising, it is still restricted to very particular spatial distributions of the intensity.

In this paper, a general approach for extracting piecewise smooth regions is proposed. Instead of minimizing the distance between the intensity and the average intensity of the region like in (2), the distance between the intensity and local averaging inside the region is minimized. This gives a model able to approximate piecewise smooth functions like the original Mumford-Shah functional (1), but with a complexity closer to that of the piecewise constant model.

Section 2 explains in detail our model and how it can be linked to the Mumford-Shah model. The minimized energy as well as its derivative using the shape gradient are expressed. In Section 2.2, the level set method is used to compute the evolution of the interface, and each term of the derivative is expressed as the result of a convolution. The importance of the fast recursive filter is

[1] A third functional was also introduced in the seminal work of Mumford and Shah [9]. This functional is the integral along $\Gamma$ of a *generalized Finsler metric* and leads indeed to the first geodesic active contour (before [2,7]).

briefly explained. In Section 3, some results on synthetic and real data are shown and compared with the piecewise smooth and piecewise constant Mumford-Shah Model.

## 2 Piecewise Smooth Approximation Through Gaussian Convolutions

The Mumford-Shah model approximates the image by a piecewise smooth function by penalizing the high gradient of an "ideal" cartoon image ($u$ in (1)). The estimation of this image at the same time as the segmentation makes the functional difficult to minimize, and computationally expensive, as it's the solution of a poisson equation.

Here we approach the problem differently by fixing the cartoon image to a smoothing of the image intensity inside each subset $\Omega_i$. for the 2 Dimensional case, the smooth function in $\Omega_i$ is then defined as,

$$u_\sigma(x, \Omega_i) = \frac{\int_{\Omega_i} g_\sigma(x-y) u_0(y) dy}{\int_{\Omega_i} g_\sigma(x-y) dy}, \quad g_\sigma(v) = \frac{1}{\sqrt{2\pi}\sigma} \exp\left(-\frac{v^2}{2\sigma^2}\right). \tag{3}$$

The denominator of this expression is a normalization factor which is important for voxels that are close to the border of $\Omega_i$, i.e. when the Gaussian kernel does not overlap completely with $\Omega_i$. This can be also interpreted as a local weighted averaging of the image intensity around the voxel $x$ inside the region $\Omega_i$ (Figure 1).

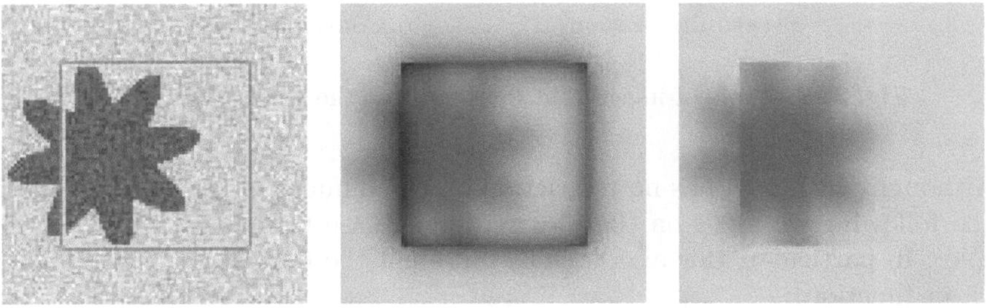

(a) Original Image (b) Without the normalization factor (c) With the normalization factor

**Fig. 1.** Importance of the denominator in the formulation of smooth regions

Let $\chi_i$ be a characteristic function of $\Omega_i$ such that $\chi_i(x) = 1$ if $x \in \Omega_i$ and 0 otherwise. We can express the overall piecewise smooth approximation of the image as

$$u_\sigma(x, \Gamma) = \sum_i \chi_i \, u_\sigma(x, \Omega_i). \tag{4}$$

With this approximation, $u_\sigma$ is a piecewise smooth function that is given analytically with respect to the boundary $\Gamma$, and we do not need anymore the regularization term on $u$ present in (1). This leads us to a new functional:

$$E(\Gamma) = \mu^2 \int_S (u_0 - u_\sigma(\Gamma))^2 dx + \nu|\Gamma| = \mu^2 \sum_i \int_{\Omega_i} (u_0 - u_\sigma(\Omega_i))^2 dx + \nu|\Gamma| \quad (5)$$

Interestingly, when the variance $\sigma$ of the Gaussian goes to infinity, this functional becomes equivalent to the piecewise constant model. This limit model has become very popular in its level set formulation (Chan-Vese model[16]) because it performs very well for regions that are characterized by quite different global means. However it is not able to discriminate regions with nearly the same global intensity distributions (Figure 2). With a different choice of $\sigma$, our model becomes more local and can segment a wider set of images where image regions differs only in their local intensity distributions. Hence, tuning the parameter $\sigma$ permits to control the locality of intensity statistics inside each region .

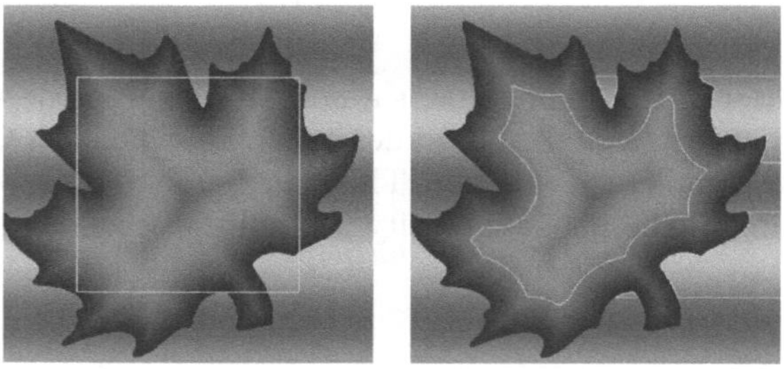

(a) Initialization (b) Convergence

**Fig. 2.** Example of image that does not suit the Chan-Vese model

Although our model has no restriction on the number of regions to segment, in the following we focus on the bi-partitioning case to make the explanations simpler. In particular, this allows us to represent the boundary $\Gamma$ with a single level set function[2].

## 2.1 Energy Minimization

In the case of bi-partitioning, the contour $\Gamma$ separates a region $\Omega$ from its complementary $\overline{\Omega}$, and the energy (5) becomes:

$$E(\Gamma) = \mu^2 \sum_{\mathcal{D}=\{\Omega,\overline{\Omega}\}} \int_{\mathcal{D}} \left( u_0(x) - \frac{\int_{\mathcal{D}} g_\sigma(x-y) u_0(y) dy}{\int_{\mathcal{D}} g_\sigma(x-y) dy} \right) dx + \nu|\Gamma| \quad (6)$$

[2] Several extensions using multiple level set functions have been proposed to segment an arbitrary number of regions [16].

To minimize this energy, we use the shape gradient tools developed in [1]. The detailed derivation is presented in appendix A. It leads to the following evolution of the boundary $\Gamma$:

$$\frac{\partial \Gamma}{\partial t}(x) = \left( \left(u_0(x) - u_\sigma(x, \overline{\Omega})\right)^2 + q_\sigma(x, \Omega) - \left(u_0(x) - u_\sigma(x, \Omega)\right)^2 - q_\sigma(x, \overline{\Omega}) \right) \mathbf{N}(x),$$

$$\text{with } q_\sigma(x, \Omega) = \int_\Omega \frac{2\left(u_0(y) - u_\sigma(y, \Omega)\right)\left(u_0(x) - u_\sigma(y, \Omega)\right) g_\sigma\left(y - x\right)}{\int_\Omega g_\sigma(y - z)\, dz}\, dy. \tag{7}$$

Where $\mathbf{N}(x)$, the outward normal vector to $\Gamma$ at the point $x$. The first two terms of this evolution equation are similar to the ones that can be found in the usual piecewise smooth and constant Mumford-Shah cases. Their interpretation is quite straightforward: the contour will locally move to include the current image voxel in the region it is the most similar to. The other terms are unique to our formulation and come from the analytical expression of the piecewise smooth image as a function of the boundary.

## 2.2 Level Set Implementation

Any curve representation can be used to implement the evolution described in (7). Here we present how to do it with a level set representation. In particular, this allows us to give an implementation that is valid in any dimension.

Let $\phi$ be the signed distance function to the boundary $\Gamma$, positive in $\Omega$ and negative in $\overline{\Omega}$. We introduce $H_\alpha$, regularized versions of the Heaviside functions. Equation (7) becomes:

$$\frac{\partial \phi}{\partial t} = \left( \left(u_0 - \overline{u}_\sigma(\phi)\right)^2 - \left(u_0 - u_\sigma(\phi)\right)^2 - \overline{q}_\sigma(\phi) + q_\sigma(\phi) \right) |\nabla \phi|, \tag{8}$$

All four different terms $u_\sigma(\phi)$, $\overline{u}_\sigma(\phi)$, $q_\sigma(\phi)$ and $\overline{q}_\sigma(\phi)$ can be computed with convolutions by the Gaussian kernel $g_\sigma$:

$$\begin{cases} u_\sigma(\phi) = \dfrac{g_\sigma * H_\alpha(\phi)\, u_0}{g_\sigma * H_\alpha(\phi)} \\ \overline{u}_\sigma(\phi) = \dfrac{g_\sigma * (1 - H_\alpha(\phi))\, u_0}{g_\sigma * (1 - H_\alpha(\phi))} \\ q_\sigma(\phi) = u_0 \left( g_\sigma * \dfrac{2(u_0 - u_\sigma) H_\alpha(\phi)}{g_\sigma * H_\alpha(\phi)} \right) - g_\sigma * \dfrac{2(u_0 - u_\sigma) u_\sigma H_\alpha(\phi)}{g_\sigma * H_\alpha(\phi)} \\ \overline{q}_\sigma(\phi) = u_0 \left( g_\sigma * \dfrac{2(u_0 - \overline{u}_\sigma)(1 - H_\alpha(\phi))}{g_\sigma * (1 - H_\alpha(\phi))} \right) - g_\sigma * \dfrac{2(u_0 - \overline{u}_\sigma)\overline{u}_\sigma (1 - H_\alpha(\phi))}{g_\sigma * (1 - H_\alpha(\phi))} \end{cases} \tag{9}$$

Each one of these terms need to be updated at each evolution of the level set. Even though these expressions seem complicated, their estimation is quite straightforward since they are the results of convolutions by a Gaussian kernel (more details are given in appendix B). This is a good advantage because it can be implemented very effectively with a recursive filter [4]. Hence, for a $d$ dimensional image with $N$ voxels, the complexity of each convolution $O(d.N)$, and only six convolutions are needed to compute the four terms. If we compare this complexity to the piecewise constant case, where the means inside each region also need to be recomputed at each iteration, it is of the same order.

## 3 Results and Comparisons

We applied our model on several images :

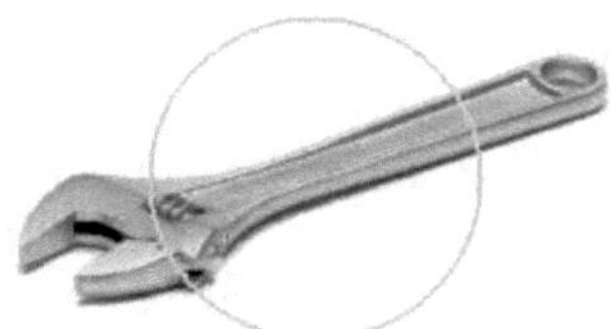

(a) Initialization

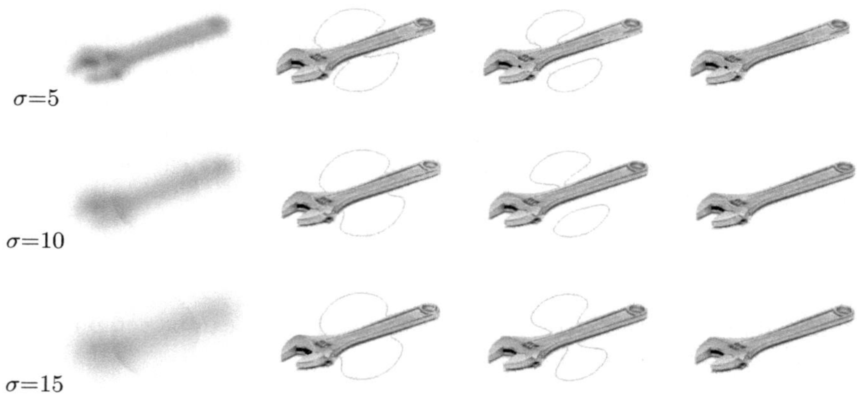


**Fig. 3.** Role of the variance in evolution : small variance extract thinner details whereas large variance suit well for large homogeneous shape

In Figure 3 we can see the role of the variance parameter $\sigma$. As pointed out in section 2, when $\sigma$ goes to infinity, the functional become equivalent to the Chan-Vese model. However, for small variances, the initial contour has to be close to edges to evolve, but is able to extract much thinner details. Actually, the model behave like the geodesic active contour model without any balloon force : it drives the front toward edges in the image, and makes it evolve via a mean curvature flow in homogeneous regions. More generally, the evolution of the front “follow” the edges that cross the initial contour. Our model is thus very dependent to the initialization.

(a) Initialization

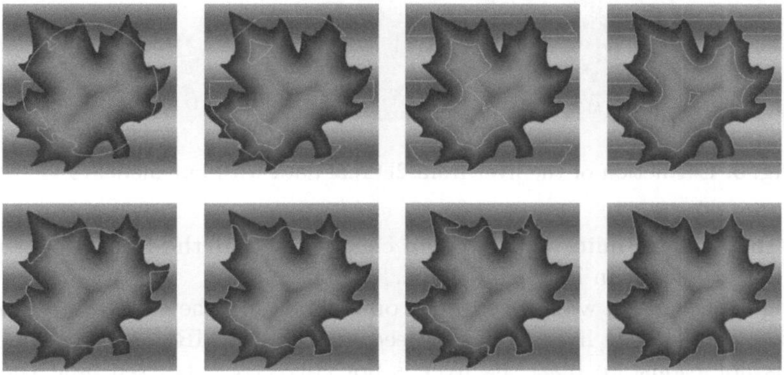

**Fig. 4.** Limitation of the Chan-Vese model. first row : Chan-Vese model; second row : our model.

In Figure 4 we applied our model in the case of two distinct regions characterized by same global statistic. In the first row, we pointed out the limitations of the Chan-Vese model, that separate the white regions from the black, and do not extract accurately the leaf in the image. We can see in the second row our

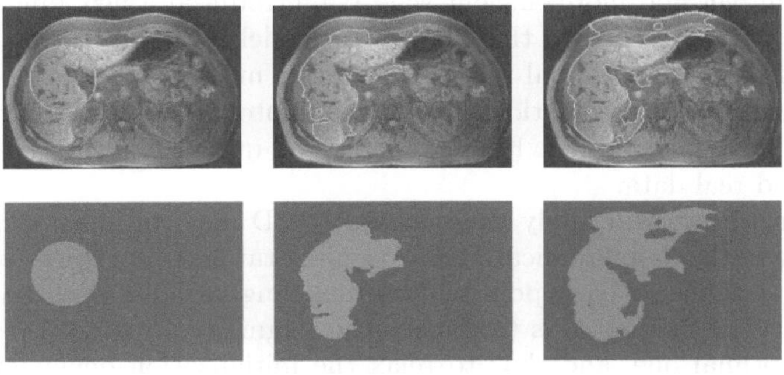

**Fig. 5.** Extraction of the liver from 2D real data with the Chan-Vese model

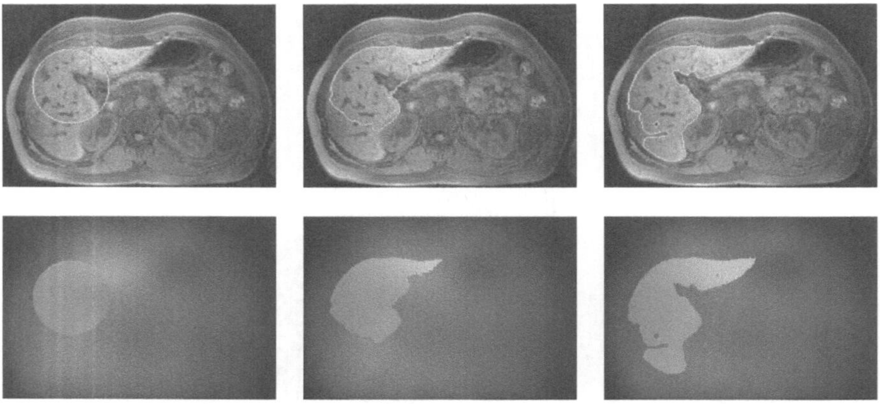

**Fig. 6.** Extraction of the liver from 2D real data with our model ($\sigma = 16$)

model, that behave quite well. We can clearly see that the front "follows" the edges that crossed it in its initial shape.

In Figure 5 and 6, we applied both our model and the Chan-Vese model in order to extract a 2D liver from a biased anatomical MRI. As the Chan-Vese model is not robust to bias, the liver is not correctly extracted, and the front "leak" in the part of the image where intensity is close to the global mean inside the front. We can see that our model behave quite well, as the liver is correctly delimited at the local level. We just have to make the initial front cross the edges of the liver, and make the front evolve, thus following the edges of the liver , and finally extract it almost completely.

## 4 Conclusion and Future Works

In this paper, we presented a new model for extracting smooth regions from image data. This model is based on the Mumford-Shah functional, but is formulated in a simpler and more efficient way. We introduced a new functional, and showed that it was linked to the Chan-Vese model, by representing regions as local average instead of global mean. One of the most interesting point is that the minimization of this functional can be computed in a very fast way, thanks to the Deriche recursive filter. Finally, we showed quite promising results on 2D synthetic and real data.

In the future, we will apply this method to 3D medical images, in order to segment organs from anatomical MRI. We also started some works to estimate a different variance in each points, thus modeling regions as a space-varying convolution. The main goal is to remove the "sigma parameter" by estimating a varying optimal one, and thus to relax the initialization dependency of the model.

## References


1. G. Aubert, M. Barlaud, O. Faugeras, and S. Jehan-Besson. Image segmentation using active contours: Calculus of variations or shape gradients? *SIAM Journal of Applied Mathematics*, 63(6):2128–2154, 2003.
2. V. Caselles, R. Kimmel, and G. Sapiro. Geodesic active contours. *International Journal of Computer Vision*, Vol 22:61–79, 1997.
3. D. Cremers, M. Rousson, and R. Deriche. A review of statistical approaches to level set segmentation: integrating color, texture, motion and shape. *International Journal of Computer Vision*, 2006.
4. R. Deriche. Using Canny's criteria to derive a recursively implemented optimal edge detector. *The International Journal of Computer Vision*, 1(2):167–187, May 1987.
5. A. Dervieux and F. Thomasset. A finite element method for the simulation of Raleigh-Taylor instability. *Springer Lect. Notes in Math.*, 771:145–158, 1979.
6. O. Juan, R. Keriven, and G. Postelnicu. Stochastic motion and the level set method in computer vision: Stochastic active contours. *The International Journal of Computer Vision*, 69(1):7–25, 2006.
7. S. Kichenassamy, A. Kumar, P. Olver, A. Tannenbaum, and A. Yezzi. Gradient flows and geometric active contour models. In *Proceedings of the 5International Conference on Computer Vision*, pages 810–815, Boston, MA, June 1995. IEEE Computer Society Press.
8. J. Kim, J. Fisher, A. Yezzi, M. Cetin, and A. Willsky. Nonparametric methods for image segmentation using information theory and curve evolution. In *IEEE International Conference on Image Processing*, pages 797–800, September 2002.
9. D. Mumford and J. Shah. Optimal approximations by piecewise smooth functions and associated variational problems. *Comm. Pure Appl. Math.*, 42:577–685, 1989.
10. S. Osher and J. Sethian. Fronts propagation with curvature dependent speed: Algorithms based on Hamilton–Jacobi formulations. *J. of Comp. Phys.*, 79:12–49, 1988.
11. N. Paragios and R. Deriche. Geodesic active regions and level set methods for supervised texture segmentation. *The International Journal of Computer Vision*, 46(3):223–247, 2002.
12. M. Rousson and R. Deriche. A variational framework for active and adaptative segmentation of vector valued images. In *Proc. IEEE Workshop on Motion and Video Computing*, pages 56–62, Orlando, Florida, December 2002.
13. J. Sethian. *Level Set Methods and Fast Marching Methods: Evolving Interfaces in Computational Geometry, Fluid Mechanics, Computer Vision, and Materials Sciences.* Cambridge Monograph on Applied and Computational Mathematics. Cambridge University Press, 1999.
14. A. Tsai, A. Jr. Yezzi, and A.S. Willsky. Curve evolution implementation of the mumford-shah functional for image segmentation, denoising, interpolation, and magnification. *IEEE Transactions on Image Processing*, 10(8):1169–1186, August 2001.
15. L. Vese. Multiphase object detection and image segmentation. In S. Osher and N. Paragios, editors, *Geometric Level Set Methods in Imaging, Vision and Graphics*, pages 175–194. Springer Verlag, 2003.
16. L. Vese and T. Chan. A multiphase level set framework for image segmentation using the mumford and shah model. *International Journal of Computer Vision*, 50:271–293, 2002.

# A Derivation of the Energy

In section 2, we defined an energy that is composed of two domain integrals. Here, we present a detailed derivation using shape gradients. We consider the following energy :

$$E(\Omega,\overline{\Omega}) = \int_{\Omega} f(x,\Omega)dx + \int_{\overline{\Omega}} f(x,\overline{\Omega})dx,$$

where

$$f(x,\Omega) = (u_0(x) - u_\sigma(x,\Omega))^2 = \left(u_0(x) - \frac{G_1(x,\Omega)}{G_2(x,\Omega)}\right)^2, \tag{10}$$

and

$$\begin{cases} G_1(x,\Omega) = \int_{\Omega} H_1(x,y)dy \text{ and } H_1(x,y) = u_0(y)g_\sigma(x-y) \\ G_2(x,\Omega) = \int_{\Omega} H_2(x,y)dy \text{ and } H_2(x,y) = g_\sigma(x-y) \end{cases} \tag{11}$$

We recall the main theorem of shape gradient method presented in [1]:

**Theorem 1.** *The Gâteaux derivative of a functional* $J(\Omega) = \int_\Omega g(x,\Omega)dx$ *in the direction of a vector field* $\mathbf{V}$ *is:*

$$\langle J'(\Omega),\mathbf{V}\rangle = \int_{\Omega} g_s\left(x,\Omega,\mathbf{V}\right)d\mathbf{x} - \int_{\partial\Omega} g\left(x,\Omega\right)\left(\mathbf{V}(x).\mathbf{N}(x)\right)d\mathbf{a}(x)$$

*Where* $g_s\left(x,\Omega,\mathbf{V}\right)$ *is the shape derivative of* $g\left(x,\Omega\right)$ *in the direction of* $\mathbf{V}$, $\partial\Omega$ *is the boundary of* $\Omega$, $\mathbf{N}$ *is the unit inward normal to* $\partial\Omega$ *and* $\mathbf{da}$ *its area element.*

In our case, for the bi-partitioning problem, we have,

$$\begin{aligned}\left\langle E'(\Omega,\overline{\Omega}),\mathbf{V}\right\rangle = &\int_{\Omega} f_s\left(x,\Omega,\mathbf{V}\right)dx + \int_{\overline{\Omega}} f_s\left(x,\overline{\Omega},\mathbf{V}\right)dx \\ &-\int_{\Gamma}\left(f\left(x,\Omega\right) - f\left(x,\overline{\Omega}\right)\right)\left(\mathbf{V}(x).\mathbf{N}(x)\right)d\mathbf{a}(x)\end{aligned}$$

with

$$f_s\left(x,\Omega,\mathbf{V}\right) = f_{G_1}\left(x,\Omega,G_1,G_2\right)\left\langle G_1'(x,\Omega).\mathbf{V}\right\rangle + f_{G_2}\left(x,\Omega,G_1,G_2\right)\left\langle G_2'(x,\Omega).\mathbf{V}\right\rangle,$$

where $f_{G_1}$ and $f_{G_2}$ denote the partial derivative of (10) with respect to $G_1$ and $G_2$. They can be expressed as:

$$\begin{cases} f_{G_1}\left(x,\Omega,G_1,G_2\right) = -\dfrac{2}{G_2(x,\Omega)}\left(I(x) - \dfrac{G_1(x,\Omega)}{G_2(x,\Omega)}\right) \\ f_{G_2}\left(x,\Omega,G_1,G_2\right) = \dfrac{2G_1(x,\Omega)}{G_2(x,\Omega)^2}\left(I(x) - \dfrac{G_1(x,\Omega)}{G_2(x,\Omega)}\right) \end{cases} \tag{12}$$

and by using Theorem 1,

$$\begin{cases} \langle G_1'(x,\Omega).\mathbf{V}\rangle = \int_\Omega H_{1s}(x,y,\mathbf{V})dy - \int_\Gamma H_1(x,y)(\mathbf{V}(y).\mathbf{N}(y))d\mathbf{a}(y) \\ \langle G_2'(x,\Omega).\mathbf{V}\rangle = \int_\Omega H_{2s}(x,y,\mathbf{V})dy - \int_\Gamma H_2(x,y)(\mathbf{V}(y).\mathbf{N}(y))d\mathbf{a}(y) \end{cases} \tag{13}$$

Since $H_1$ and $H_2$ do not depend on $\Omega$, we obtain $H_{1s} = 0$ and $H_{2s} = 0$.
Putting all the term together we find

$$f_s(x,\Omega,\mathbf{V}) = f_{G_1}\langle G_1'.V\rangle + f_{G_2}\langle G_2'.V\rangle \tag{14}$$

$$= \frac{2\,(u_0(x) - u_\sigma(x,\Omega))}{\int_\Omega g_\sigma(x-y)dy}\int_\Gamma g_\sigma(x-y)\,(u_0(y) - u_\sigma(x,\Omega))\,(\mathbf{V}(y).\mathbf{N}(y))\,d\mathbf{a}(y) \tag{15}$$

and at last we obtain, by changing the order of integration,

$$\begin{aligned}\left\langle E'(\Omega,\overline{\Omega}),\mathbf{V}\right\rangle &= \int_\Omega f_s(x,\Omega,\mathbf{V})\,dx + \int_{\overline{\Omega}} f_s\left(x,\overline{\Omega},\mathbf{V}\right)dx \\ &\quad - \int_\Gamma \left(f(\mathbf{y},\Omega) - f\left(\mathbf{y},\overline{\Omega}\right)\right)(\mathbf{V}(y).\mathbf{N}(y))\,d\mathbf{a}(y) \\ &= \int_\Gamma \Big(q(y,\Omega) - q(y,\overline{\Omega}) - (u_0(y) - u_\sigma(y,\Omega))^2 \\ &\quad + \left(u_0(y) - u_\sigma(y,\overline{\Omega})\right)^2\Big)(\mathbf{V}(y).\mathbf{N}(y))\,d\mathbf{a}(y)\end{aligned}$$

with

$$q(y,\Omega) = \int_\Omega \frac{2\,(u_0(x) - u_\sigma(x,\Omega))\,g_\sigma(x-y)\,(u_0(y) - u_\sigma(x,\Omega))}{\int_\Omega g_\sigma(x-z)dz}dx \tag{16}$$

We finally get the following gradient descent:

$$\boxed{\frac{\partial\Gamma}{\partial\tau}(x) = \left[\left(q(x,\Omega) - q(x,\overline{\Omega}) - (u_0(x) - u_\sigma(x,\Omega))^2 + \left(u_0(x) - u_\sigma(x,\overline{\Omega})\right)^2\right)\right]\mathbf{N}(x)} \tag{17}$$

## B Implementation

Each integral term in the gradient descent can be seen as a convolution by a kernel of variance $\sigma$. Therefore, the computation of the evolution can be done in a very fast way in two separate steps :

– first we make several convolutions via a fast recursive filter (ref deriche)

– then we compute the speed of each points in the narrow-band by using the previously computed blurred images.

We have,

$$u_\sigma(x,\Omega) = \frac{\int_\Omega g_\sigma(x-y)u_0(y)dy}{\int_\Omega g_\sigma(x-y)dy} = \frac{(g_\sigma * u_0)\,|_\Omega(x)}{(g_\sigma * 1)\,|_\Omega(x)} \tag{18}$$

and

$$\begin{aligned} q(y,\Omega) &= \int_\Omega \frac{2\left(u_0(x)-u_\sigma(x,\Omega)\right) g_\sigma\left(x-y\right)\left(u_0(y)-u_\sigma(x,\Omega)\right)}{\int_\Omega g_\sigma(x-z)dz} dx \\ &= u_0(y)\int_\Omega \frac{2\left(u_0(x)-u_\sigma(x,\Omega)\right)}{(g_\sigma * 1)\,|_\Omega(x)} g_\sigma\left(x-y\right)dx \\ &\quad - \int_\Omega \frac{2\left(u_0(x)-u_\sigma(x,\Omega)\right) u_\sigma(x,\Omega)}{(g_\sigma * 1)\,|_\Omega(x)} g_\sigma\left(x-y\right)dx \\ &= u_0(y)\left(g_\sigma * q_1\right)|_\Omega(y) - \left(g_\sigma * q_2\right)|_\Omega(y) \end{aligned}$$

with

$$q_1(x,\Omega) = \frac{2\left(u_0(x)-u_\sigma(x,\Omega)\right)}{(g_\sigma * 1)\,|_\Omega(x)} \quad \text{and} \quad q_2(x,\Omega) = \frac{2\left(u_0(x)-u_\sigma(x,\Omega)\right) u_\sigma(x,\Omega)}{(g_\sigma * 1)\,|_\Omega(x)} \tag{19}$$

We can compute the domain-convolution by using the Heaviside $H_\sigma$ of the level-set function :

$$\begin{cases} (g_\sigma * f)\,|_\Omega(x) = (g_\sigma * H_\sigma f)\,(x) \\ (g_\sigma * f)\,|_{\overline{\Omega}}(x) = (g_\sigma * (1-H_\sigma)f)\,(x) \end{cases} \tag{20}$$

# Space-Time Segmentation Based on a Joint Entropy with Estimation of Nonparametric Distributions

Ariane Herbulot[1], Sylvain Boltz[1], Eric Debreuve[1], Michel Barlaud[1], and Gilles Aubert[2]

[1] Laboratoire I3S, CNRS, Université de Nice-Sophia Antipolis, France
[2] Laboratoire Dieudonné, CNRS, Université de Nice-Sophia Antipolis, France

**Abstract.** This paper deals with video segmentation based on motion and spatial information. Classically, the nucleus of the motion term is the motion compensation error (MCE) between two consecutive frames. Defining a motion-based energy as the integral of a function of the MCE over the object domain implicitly results in making an assumption on the MCE distribution: Gaussian for the square function, Laplacian for the absolute value, or other parametric distributions for functions used in robust estimation. However, these assumptions are generally false. Instead, it is proposed to integrate a function of (an estimation of) the MCE distribution. The function is taken such that the integral is the Ahmad-Lin entropy of the MCE, the purpose being to be more robust to outliers. Since a motion-only approach can fail in homogeneous areas, the proposed energy is the joint entropy of the MCE and the object color. It is minimized using active contours.

**Keywords:** space-time segmentation, joint entropy, active contour.

## 1 Introduction

Segmentation of moving objects in video sequences is a challenging problem. Let us first consider the problem of motion estimation alone. The motion of a given object domain $\Omega$ can be computed by choosing a motion model and finding the motion parameters that minimize a function of the motion compensation error (MCE) over the object domain. At a pixel level, making the assumption of brightness constancy, the MCE is classically the following residual

$$e_n(x, v(x)) = I_n(x) - I_{n+1}(x + v(x)) \quad (1)$$

where $x$ is a pixel of the object domain, $I_n$ is the $n^{\text{th}}$ grayscale or color frame of the sequence, $v(x)$ is the optical flow [1,2] (*i.e.*, the apparent motion related to the assumption of brightness constancy) between $I_n$ and $I_{n+1}$ at $x$, and, ideally , $e_n(x, v(x))$ is equal to zero up to some noise. In grayscale, there is only one equation ((1) = 0) for two unknowns (the components of $v(x)$) and, both in grayscale and color, it is likely that several pixels have the same intensity

F. Sgallari, A. Murli, and N. Paragios (Eds.): SSVM 2007, LNCS 4485, pp. 721–732, 2007.

value(s). As a consequence, the motion estimation problem cannot be solved as is. It must be constrained. A possible way to make it solvable is to suppose that the motion is coherent with a chosen model inside $\Omega$ [3]. Let $v$ denotes a vector of motion parameters related to the model. For simplicity, let us consider a translation model. Then, (1) is replaced with

$$e_n(v, x) = I_n(x) - I_{n+1}(x + v), \ x \in \Omega \ . \tag{2}$$

The motion estimate $\hat{v}$ can be computed as

$$\hat{v} = \arg\min_v \sum_{x \in \Omega} \varphi(e_n(v, x)) \tag{3}$$

or, in the continuous framework,

$$\hat{v} = \arg\min_v \int_\Omega \varphi(e_n(v, x)) \, \mathrm{d}x \tag{4}$$

where $\varphi$ can be, for example, the square function, the absolute value, or a function typical of the robust estimation framework [4].

The motion-based segmentation of frame $I_n$ can be formulated as the largest domain $\Omega$ inside which the motion is coherent with model (4), formally,

$$\begin{cases} \hat{\Omega} = \arg\min_\Omega \int_\Omega \varphi(e_n(v(\Gamma), x)) \, \mathrm{d}x \\ v(\Gamma) = \arg\min_v \int_\Omega \varphi(e_n(v, x)) \, \mathrm{d}x \end{cases} \tag{5}$$

where $\Gamma$ is the boundary $\partial\Omega$ of $\Omega$. Note that writing $v(\Gamma)$ or $v(\Omega)$ is only a matter of notation since $\Omega$ is completely determined by $\Gamma$ and conversely. Let us denote by $E_t$, where $t$ stands for *temporal*, the following domain energy

$$E_t(\Gamma) = \int_\Omega \varphi(e_n(v(\Gamma), x)) \, \mathrm{d}x \ . \tag{6}$$

The choice of $\varphi$ results in making an assumption on the distribution of residual $e_n$ in $\Omega$. For example, if $\varphi$ is equal to the square function, the motion estimation is performed based on the assumption that the distribution is Gaussian and if $\varphi$ is equal to the absolute value, the distribution is assumed to be Laplacian. However, these assumptions are false in general. In particular, the presence of outliers in the residual (*e.g.*, due to occlusions, mismatch between the chosen motion model and the actual motion, complex noise characteristics, variation of luminance...) may result in a complex, multimode distribution. As a consequence, the motion estimator in (5) may be biased, leading to a loss of accuracy of the motion-based segmentation. Moreover, motion only-based segmentation may fail in homogeneous areas or, say, in insufficiently textured areas.

The approach using energy (6) will be referred to as a parametric approach since $\varphi$ is characterized by a small set of parameters, *e.g.*, the scaling of the abscissas. In Sect. 2.1, this parametric assumption on the residual distribution will be discarded by involving through entropy an estimation of the true residual distribution, which will be referred to as a nonparametric approach in the sense that the estimated distribution will not follow any particular parametric scheme [5,6,7]. In Sect. 2.2, the proposed motion-based energy will be combined with a spatial energy in order to deal with less textured areas. Spatio-temporal video segmentation and tracking have already been approached as in [8,9,10] but we present here a method in a full nonparametric context.

# 2 Proposed Segmentation Energy

## 2.1 Nonparametric, Entropy-Based Motion Energy

To account for the true distribution of the residual $e_n$, it is proposed to make the domain energy $E_t$ depend on an estimation $f$ of the residual distribution rather than on the residual itself as in (6). The proposed domain energy is then

$$E_t(\Gamma) = \int_{\mathbb{R}} \psi(f(e_n(v(\Gamma)))(r)) \, \mathrm{d}r \tag{7}$$

where $f$ is an estimation of the distribution of $e_n$ inside $\Omega$ and $v(\Gamma)$ is equal to

$$v(\Gamma) = \arg\min_{v} \int_{\mathbb{R}} \psi(f(e_n(v))(r)) \, \mathrm{d}r \ . \tag{8}$$

Note that the domain of definition of the residual distribution is $\mathbb{R}$ in grayscale (illustrated in (7) for simplicity) or $\mathbb{R}^3$ in color.

It is proposed to choose $\psi$ such that (7) is the Shannon entropy of the residual $\psi(r) = -\, r \ \log(r)$. Indeed, entropy is a measure of dispersion. If the motion-based segmentation is optimal, the residual should be around zero with a minimal dispersion. Moreover, entropy coincides locally asymptotically with likelihood at the optimum. Noting that maximum likelihood is optimal when the distribution of data is parametric, a minimum entropy criterion should have similar performances as a maximum likelihood criterion in such cases while being able to adapt to a nonparametric distribution. In particular, entropy appears less sensitive to outliers in practice. Explicit estimation of the residual distribution $f$ is not necessary to compute its entropy [11]. However, as will be shown in Sect. 3, $f$ will be needed since it appears explicitly in the proposed segmentation method. Parzen windowing is a usual distribution estimation procedure. It provides a smooth estimate of $f$

$$f(e_n(v(\Gamma)))(r) = \frac{1}{|\Omega|} \int_{\Omega} K_\sigma(e_n(v(\Gamma), x) - r) \, \mathrm{d}x \tag{9}$$

where $|\Omega|$ is the measure of $\Omega$ and $K_\sigma$ is a Gaussian function with a mean equal to zero and a standard deviation equal to $\sigma$ called the Parzen kernel. It is usual to adapt $\sigma$ to the data [12].

The proposed energy (7) is an integral over $\mathbb{R}$ whereas the classical energy (6) is an integral over $\Omega$. For unifying both approaches in a common framework, the expression of entropy (7) the following approximation

$$E_t(\Gamma) = \int_{\Omega} \psi(f(e_n(v(\Gamma), x)))\ \mathrm{d}x \tag{10}$$

where $\psi(r) = -\frac{1}{|\Omega|}\ \log(r)$. As a consequence, one can say that the parametric approach (6) is extended to nonparametric distributions by changing $\varphi(e_n)$ into $\psi(f(e_n))$, resulting in the following motion-based energy

$$\begin{cases} E_t(\Gamma) = -\dfrac{1}{|\Omega|} \displaystyle\int_{\Omega} \log f(e_n(v(\Gamma), x))\ \mathrm{d}x \\ v(\Gamma) = \arg\min\limits_{v} -\dfrac{1}{|\Omega|} \displaystyle\int_{\Omega} \log f(e_n(v, x))\ \mathrm{d}x \end{cases} . \tag{11}$$

Some similarity can be seen with the Ahmad-Lin entropy [13]. The domain $\hat{\Omega}$ minimizing (11) represents a motion-based segmentation of frame $I_n$.

### 2.2 Spatial Energy

Energy (11) is well suited for segmenting objects over a textured background. However, it might cause segmentation to include homogeneous or quasi-homogeneous areas of the background. Indeed, this type of areas has a low residual even if compensated with the estimated motion of the object, at least as long as the motion-compensated object domain remains in the homogeneous area. Therefore, energy (11) might increase only negligibly when expanding in such areas. Since the notions of object and background are arbitrary and can be swapped for one another, one can note that an equivalent undersegmentation phenomenon can occur if the object contains homogeneous areas near its boundary.

The proposed solution is to combine spatial and temporal information. In order to conform to the proposed framework, the entropy will be changed into a joint entropy of the residual and the object intensity or color (see Sect. 2.3). Intuitively, the entropy of the object color increases if the object domain includes some background since it adds new colors to the object[1] and, therefore, increases the dispersion of its color distribution. Consequently, the joint entropy also increases.

Note that this approach defines a general framework for multimodal segmentation: the joint entropy allows to combine an arbitrary number of modalities. In practice, though, the number of modalities that can be combined is limited by the number of samples available, *i.e.*, the number of pixels of the image or sequence frame. If the samples fill the distribution space too sparsely, then the entropy cannot be approximated accurately.

[1] If the background has the same color as the object near the boundary, there is no objective information to find the object boundary.

### 2.3 Combining Spatial and Temporal Energies

As mentioned in Sect. 2.2, spatial and temporal information are combined by means of the joint entropy of the residual and the object intensity or color. Then, the spatio-temporal energy is

$$E(\Gamma) = -\frac{1}{|\Omega|} \int_{\Omega} \log f(e_n(v(\Gamma), x), I_n(x)) \, dx \tag{12}$$

where $f(e_n, I_n)$ is the joint distribution of the residual and the image color inside object domain $\Omega$.

In order to simplify energy (12), the residual and the color are supposed to be independent. Let us consider the following sequence model

$$I_{n+1}(x) = I_n(T(x)) + n(x) \tag{13}$$

where $T$ is a transformation and $n$ is a Gaussian white noise. The residual is equal to

$$e_n(x, v(x)) = I_n(x) - I_{n+1}(v(x)) \; . \tag{14}$$

If the motion is perfectly estimated, then $v$ is equal to $T^{-1}$ and $e_n(x, v(x)) = n(T^{-1}(x))$, which is independent of $I_n$. If the transformation is not correctly estimated, then the residual is correlated with $I_n$. However, the correlation is only partial. In fact, model (13) is not realistic. There is not such transformation $T$, frame $I_{n+1}$ being a projection on a two-dimensional plane of a three-dimensional scene. In general, some parts of objects in $I_n$ become visible in $I_{n+1}$ while others become invisible. Therefore, frame $I_{n+1}$ cannot be deduced entirely from $I_n$. In other words, the residual contains information unpredictable given $I_n$. In addition, if the transformation, represented by motion $v(\Gamma)$, is fairly well estimated, then the assumption of independence is reasonable. With this assumption, energy (12) can be rewritten as the sum of the marginal distributions

$$\begin{aligned} E(\Gamma) &= -\frac{1}{|\Omega|} \int_{\Omega} \log f(e_n(v(\Gamma), x)) \, dx - \frac{1}{|\Omega|} \int_{\Omega} \log f(I_n(x)) \, dx & (15) \\ &= E_t(\Gamma) + E_s(\Gamma) & (16) \end{aligned}$$

where $s$ in $E_s$ stands for *spatial* and, as a reminder, $v(\Gamma)$ is defined in (11).

Note that, by making the assumption of independence, one get a sum of two energies, meeting the philosophy usually adopted when one want to simultaneously minimize several energies (although, in general, weighting parameters are introduced to tune the influence of the respective energies).

## 3 Segmentation Using Active Contours

### 3.1 Shape Gradient of the Energy

Minimization of energy (15) requires the computation of its derivative with respect to $\Gamma$. There exists an infinite number of ways of deforming $\Gamma$. The shape derivative [14] of (15) can be interpreted as the derivative in a direction $F$, a vector field defined on $\Gamma$. It can be shown that it is equal to

$$\begin{aligned} \mathrm{d}E_t(\Gamma, F) = \frac{1}{|\Omega|} \int_\Gamma \Bigg[ & \log f(e_n(v(\Gamma), s)) - 1 + E_t(\Gamma) \\ & + \frac{1}{|\Omega|} \int_\Omega \frac{K_\sigma(e_n(v(\Gamma), s) - e_n(v(\Gamma), x))}{f(e_n(v(\Gamma), x))} \mathrm{d}x \Bigg] N(s) \cdot F(s) \, \mathrm{d}s \, . \end{aligned} \tag{17}$$

where $N$ is the inward unit normal of $\Gamma$. Note that the distribution $f$ appears as itself in (17), hence the necessity to estimate it, as mentioned in Sect. 2.1.

The expression of derivative $\mathrm{d}E_s(\Gamma, F)$ is similar: it suffices to replace $e_n(v(\Gamma), .)$ with $I_n(.)$. Then, the shape derivative of (15) is equal to

$$\mathrm{d}E(\Gamma, F) = \mathrm{d}E_t(\Gamma, F) + \mathrm{d}E_s(\Gamma, F) \tag{18}$$

The shape derivative (17) has the following form

$$\mathrm{d}E_t(\Gamma, F) = \int_\Gamma (\alpha_t(s) \; N(s)) \cdot F(s) \, \mathrm{d}s = \langle \alpha_t \; N, F \rangle_{L_2} \tag{19}$$

where $\langle , \rangle_{L_2}$ is the $L_2$-inner product on $\Gamma$. Therefore, $\alpha_t \; N$ is, by definition, the gradient of (11) at $\Gamma$. The gradient of $E_s$ is obtained in a similar way and put in the form $\alpha_s \; N$.

### 3.2 Evolution Equation

Based on the notion of gradient defined in Sect. 3.1, energy (15) can be minimized using a steepest descent procedure in the space of contours. The contour evolution process is known as the active contour technique [15,16,10,17]: an initial contour[2] is iteratively deformed until a convergence condition is met. Here, at each iteration, the contour should be deformed in the opposite direction of the gradient. The evolution equation of the active contour is written as follows

$$\begin{cases} \Gamma(\tau = 0) = \Gamma_0 \\ \dfrac{\partial \Gamma}{\partial \tau} = -(\alpha_t + \alpha_s) \; N \end{cases} . \tag{20}$$

where $\tau$ is the evolution parameter. Classically, the convergence condition corresponds to a gradient equal to zero, *i.e.*, $\alpha_t + \alpha_s = 0$.

### 3.3 Region Competition

Energy (15) is positive or equal to zero. In practice, due to approximations and round-off errors, it is never equal to zero. Unless $\Gamma$ gets stuck in a local minimum, evolution (20) will make the contour smaller and smaller until it disappears.

[2] For example, a user-defined contour.

A common solution is known as region competition: the energy of the background is added to the energy of the object (15). As a result, the segmentation will represent a trade-off between the minimization of the object energy and the minimization of the background energy. It can also be interpreted as the maximal separation between object and background descriptors, here, the respective joint distributions.

To account for the relative areas of the object and the background, or, in other words, to account for the probability of a pixel to belong to either of them, the following weighted sum will be used

$$E_{\rm rc}(\Gamma) = |\Omega|\ E(\Gamma) + |\Omega^c|\ E(\Gamma^c) \tag{21}$$

where $\Omega^c$ is the complement of $\Omega$ in $D$, the image domain, and $\Gamma^c$ is its boundary $\partial\Omega^c$. Let us remind that $E$ is given by 18. Note that $\Gamma$ and $\Gamma^c$ are identical up to a change of orientation. In particular, the inward unit normal $N^c$ of $\Gamma^c$ is equal to $-N$.

Energy (21) can be divided by a positive number without influencing the segmentation

$$\frac{E_{\rm rc}(\Gamma)}{|D|} = \frac{|\Omega|}{|D|}\ E(\Gamma) + \frac{|\Omega^c|}{|D|}\ E(\Gamma^c) \tag{22}$$

$$= p(C=1)\ E(\Gamma) + p(C=0)\ E(\Gamma^c) \tag{23}$$

where $C$ is the characteristic function of the object and $p(C = 1)$ denotes the probability of the event $C = 1$. As defined in (12), energy $E(\Gamma)$ is (an approximation of) the joint entropy of the residual and the color conditional on $C = 1$. Let us denote it by $H(e_n, I_n|C = 1)$. Equivalently, $E(\Gamma^c)$ is equal to $H(e_n, I_n|C = 0)$. Then, (23) is equal to

$$\frac{E_{\rm rc}(\Gamma)}{|D|} = \sum_{i\in\{0,1\}} p(C=i)\ H(e_n, I_n|C=i) \tag{24}$$

$$= H(e_n, I_n|C)\ . \tag{25}$$

Therefore, energy (21) is equal, up to a multiplicative constant, to the conditional, joint entropy of the residual and the color $H(e_n, I_n|C)$.

The gradient of (21) can be obtained by applying the traditional differentiation rule $(u\ v)' = u'\ v + u\ v'$, and using (17) and the following shape derivative

$$\mathrm{d}(|\Omega|)(\Gamma, F) = -\int_\Gamma N(s)\cdot F(s)\ \mathrm{d}s\ . \tag{26}$$

# 4 Experimental Results

## 4.1 Test Conditions

As a reminder, the proposed segmentation energy has the following form

$$E_{\rm rc}(\Gamma) = |\Omega|\ E(\Gamma) + |\Omega^c|\ E(\Gamma^c) \tag{27}$$

where (see (15))

$$E(\Gamma) = E_t(\Gamma) + E_s(\Gamma) \ . \tag{28}$$

For comparison purposes, energy (27) will also be used in two incomplete forms: when $E_s$ is removed from the definition of $E$ in (28), the energy will be called temporal energy; when $E_t$ is removed from the definition of $E$, the energy will be called spatial energy. In its complete form, it is called spatio-temporal energy.

The tests were performed on synthetic and natural sequences. The input data of our method are the two color channels of the image and the residual (Y and U in the color space YUV). The statistics of these data are complex: on the color channels, as the object is moving, hidden parts of the objects and the background become visible and visible parts are occluded. On the residual, the motion is estimated using a global model, it is disturbed by a lot of outliers, on both homogeneous or textured objects. We present segmentation and tracking results on synthetic sequences and on three real sequences.

To estimate the density of the data over a region $\Omega$, the distribution is estimated using a normal kernel method. An optimal kernel bandwidth is computed using the standard deviation of the residual over the region [12]

$$\sigma = 0.9 \min(\hat{\sigma}, \hat{p}/1.34) \ |\Omega|^{-1/5} \tag{29}$$

This estimation of the bandwidth is called plug-in [12] as the standard deviation $\hat{\sigma}$, or the interquartile range $\hat{p}$, of the data estimated over the region $\Omega$ is *plugged* into the bandwidth estimator.

The motion estimation has been implemented by using a model allowing translation, to use a fast suboptimal search method (the diamond search) that allows in our case an amplitude of motion of +12/-12 pixels in both directions.

### 4.2 Synthetic Sequences

We created sequences with different textures and quite homogeneous zones to show the benefit of combining spatial and temporal terms. An object is moving in translation over a background moving also in translation but in the opposite direction. The following images show an homogeneous object on a homogeneous background (Fig. 1) and on a textured background (Fig. 2), a textured object on a textured background (Fig. 3). The results tends to show that when we are in presence of textures, the temporal energy seems to be efficient. On the contrary, the spatial energy seems more reliable if homogeneous zones occur. In any case, the combination of temporal and spatial information appears as a good candidate for a segmentation energy when the images contain both homogeneous and textured zones.

### 4.3 Standard Test Sequences

In this section, we evaluate the two energies separately and observe what happens when they are combined on the standard sequence `'Flowers and garden'` (Fig. 4). Segmentation on this sequence is a difficult problem due to the complex

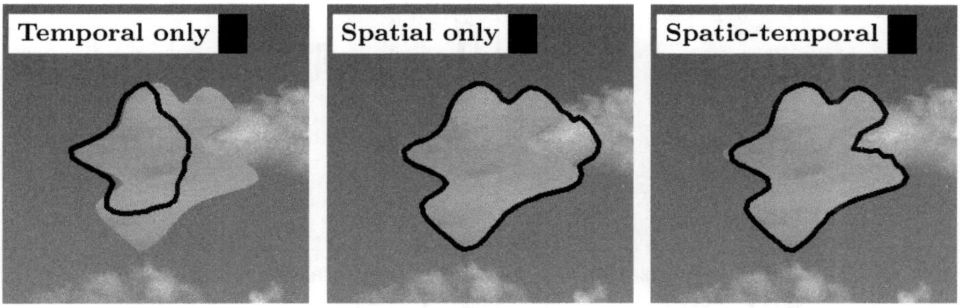


**Fig. 1.** Segmentation of an homogeneous object on an homogeneous background

**Fig. 2.** Segmentation of an homogeneous object on a texture background

statistics and the homogeneous sky whose motion cannot be estimated. The segmentation initialization is a small circle at the bottom of the image. The test is first run only with the temporal entropy. As explained in Sect. 5 when the contour reaches the sky, the temporal entropy decreases and an oversegmentation taking all the sky is observed. The test is then run only on the spatial entropy, the segmentation process fails again as some parts of the houses in the background have similar colors as the tree. Finally, the spatio-temporal entropy has

**Fig. 3.** Segmentation of a textured object on a textured background

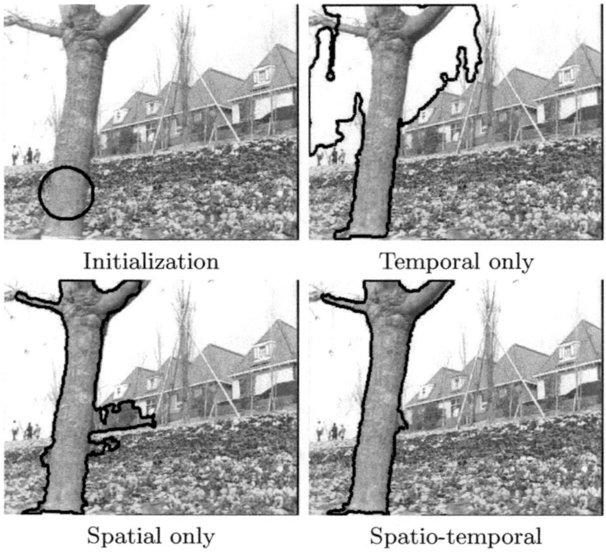


**Fig. 4.** Segmentation of the tree on the sequence `'Flowers and garden'`

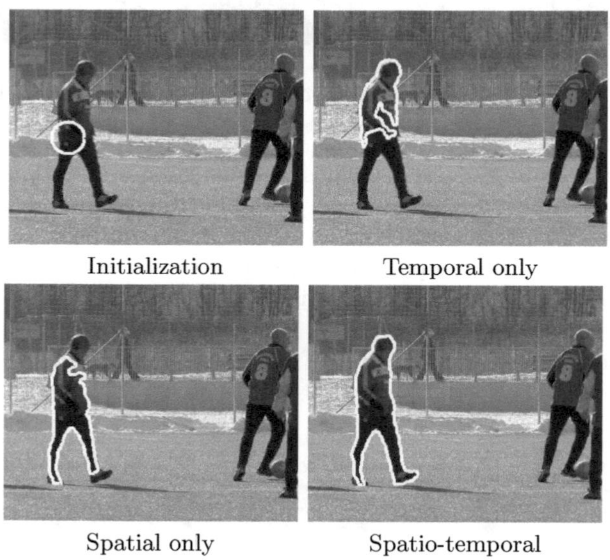


**Fig. 5.** Segmentation of a player on the sequence `'Soccer'`

good properties as it does not oversegment the sky thanks to the spatial entropy. It also does not oversegment same color zones thanks to the temporal entropy.

Next test is run on sequence `'Soccer'` (Fig. 5). This sequence is difficult because the object, here a soccer player, is not performing a rigid translation as

**Fig. 6.** Tracking of a player on the sequence `'Soccer'`

allowed in our model but an articulate motion. The colors of his head and foot are also quite similar to some colors of the background. Temporal energy only captures the rigid part of the body while spatial energy does not capture the head. Spatio-temporal energy performs well in both segmentation and tracking, although it sometimes miss a foot of the player, as both temporal (the motion is different than the rigid part of the body) and spatial (the color of the shoe is more similar to background colors) fails. However, we can observe that the spatial energy helps the temporal term when the motion is articulated, and the temporal energy helps the spatial term when the color is not discriminating.

Finally we present some results in tracking (Fig. 6), the initialization is the result of the segmentation presented above. The initialization of each frame is then the result of the segmentation on the previous frame, compensated by the motion of the object. In spite of the articulate motion and the large deformation of the player due to the motion of the legs, the method succeeds in tracking the player over the sequence.

## 5 Conclusion

We have presented a method for video segmentation combining motion and intensity information with estimation of nonparametric distributions. Joint entropy seems a good candidate for being robust to outliers when data like motion prediction error do not have parametric distributions. The results on synthetic and real sequences show that a combination of motion and intensity outperforms the use of one modality only. Comparison with existing methods is out of scope of this paper. Nevertheless, on sequence `'Flowers and garden'`, our results are comparable to those of a recent spatio-temporal segmentation method [18] (see Fig. 4). This method was originally proposed in [10]. Future works will focus on the study of other energies in the nonparametric framework.

## References


1. Weickert, J., Schnörr, C.: Variational optic flow computation with a spatio-temporal smoothness constraint. Journal of Mathematical Imaging and Vision **14** (2001) 245–255
2. Alvarez, L., Weickert, J., Sánchez, J.: A scale-space approach to nonlocal optical flow calculations. In: Int. Conf. on Scale-Space Theories in Computer Vision. (1999) 235–246
3. Wu, S.F., Kittler, J.: A gradient-based method for general motion estimation and segmentation. Journal of Visual Com. Image Repr. **4** (1993) 25–38
4. Black, M.J., Anandan, P.: The robust estimation of multiple motions: parametric and piecewise-smooth flow fields. Comput. Vis. Image Underst. **63** (1996) 75–104
5. Comaniciu, D., Meer, P.: Mean shift analysis and applications. In: International Conference on Computer Vision. Volume 2. (1999) 1197–1203
6. Kim, J., Fisher, J.W.F., Yezzi, A., Çetin, M., Willsky, A.S.: A nonparametric statistical method for image segmentation using information theory and curve evolution. IEEE Trans. on Image Processing **14**(10) (2005) 1486–1502
7. Mittal, A., Paragios, N.: Motion-based background subtraction using adaptive kernel density estimation. In: Computer Vision and Pattern Recognition. Volume 2. (2004) 302–309
8. Mitiche, A., El-Feghali, R., Mansouri, A.R.: Motion tracking as spatio-temporal motion boundary detection. Robotics and Autonomous Sys. **43** (2003) 39–50
9. Brox, T., Rousson, M., Deriche, R., Weickert, J.: Unsupervised segmentation incorporating colour, texture, and motion. In: Computer Analysis of Images and Patterns. Volume 2756 of LNCS., Springer (2003) 353–360
10. Cremers, D., Soatto, S.: Motion competition: A variational framework for piecewise parametric motion segmentation. International Journal of Computer Vision **62**(3) (2005) 249–265
11. Goria, M., Leonenko, N., Mergel, V., Inverardi, P.N.: A new class of random vector entropy estimators and its applications in testing statistical hypotheses. Journal of Nonparametric Statistics **17**(3) (2005) 277–297
12. Silverman, B.: Density Estimation for Statistics and Data Analysis. Chapman and Hall, London (1986)
13. Ahmad, I., Lin, P.: A nonparametric estimation of the entropy for absolutely continuous distributions. In: IEEE Trans. on Information Theory. Volume 36. (1989) 688–692
14. Delfour, M.C., Zolésio, J.P.: Shapes and geometries: Analysis, differential calculus and optimization. Advances in Design and Control. Society for Industrial and Applied Mathematics, Philadelphia (2001)
15. Caselles, V., Kimmel, R., Sapiro, G.: Geodesic active contours. International Journal of Computer Vision **22**(1) (1997) 61–79
16. Hintermuller, M., Ring, W.: A second order shape optimization approach for image segmentation. SIAM Journal on Applied Mathematics **64**(2) (2004) 442–467
17. Cremers, D., Rousson, M., Deriche, R.: A review of statistical approaches to level set segmentation: integrating color, texture, motion and shape. International Journal of Computer Vision **72**(2) (2007)
18. Schoenemann, T., Cremers, D.: Near real-time motion segmentation using graph cuts. In: Pattern Recognition (Proc. DAGM). Volume 4174 of LNCS., Springer (2006) 455–464

# Region Based Image Segmentation Using a Modified Mumford-Shah Algorithm

Jung-ha An[1] and Yunmei Chen[2]

[1] Institute for Mathematics and its Applications (IMA), University of Minnesota, USA
[2] Department of Mathematics, University of Florida, USA

**Abstract.** The goal of this paper is to develop region based image segmentation algorithms. Two new variational PDE image segmentation models are proposed. The first model is obtained by minimizing an energy function which depends on a modified Mumford-Shah algorithm. The second model is acquired by utilizing prior shape information and region intensity values. The numerical experiments of the proposed models are tested against synthetic data and simulated normal human-brain MR images. The preliminary experimental results show the effectiveness and robustness of presented models against to noise, artifact, and loss of information.

## 1 Introduction

Segmentation techniques have been developed to capture the object boundary by several different approaches; edge-based methods mainly using active contour models, region-based methods, or the combination of the two by using Geodesic Active Region models.

The most celebrating region based image segmentation model is introduced by Mumford and Shah in 1989 [15]. In this model, an image is decomposed into a set of regions within the bounded open set $\Omega$ and these regions are separated by smooth edges $\Gamma$. Ambrosio and Tortorelli approximated the measurement of an edge $\Gamma$ length term in the Mumford-Shah model by a quadratic integral of an edge signature function in 1990 [1]. Chan and Vese proposed a piecewise constant Mumford-Shah model in [4,5] by using a level set formulation [16]. Developments of variational level set implementation techniques based Mumford-Shah model are followed by [8,9,13]. The segmentation is represented by characteristics functions using phase fields in [8,13]. The details of phase field theory can be found in [2,14,18,19,20].

Recently, a soft (fuzzy) Mumford-Shah segmentation of mixture image patterns is introduced by [19]. This technique is defined by using the Bayesian rationale and the MAP estimator. It has been shown that hard segmentation technique is the special case of soft segmentation in [19]. The first proposed model in this paper is motivated by [8,18,19]. The region based variational partial differential equation (PDE) image segmentation model is suggested. The model is obtained by minimizing an energy function which depends on a modified Mumford-Shah algorithm. A fuzzy segmentation in our model is similar to Shen [18,19], but our model is variational which differs from Esedoglu [8].

Yet these algorithms have a limit to obtain an efficient segmentation result, if images have strong noise, occlusion, or loss of information. The prior shape

F. Sgallari, A. Murli, and N. Paragios (Eds.): SSVM 2007, LNCS 4485, pp. 733–742, 2007.

information has been incorporated into the segmentation process to overcome these problems [3,6,7,10,11,12,17]. The second suggested model in this paper is acquired by using prior shape information and region intensity values. The prior shape information is attained by using a modified Mumford-Shah model.

Two new region based variational partial differential equation (PDE) models for an image segmentation are presented with an application to synthetic data and simulated human brain MR images. This paper is organized as follows: In section 2, a modified Mumford-Shah model is described. Euler-Lagrange equations of the suggested model are presented in this section as well. Experimental results of the model which were applied to synthetic data and simulated human brain MR images are shown. In section 3, the prior shape information based image segmentation technique is presented. Numerical results to the noisy synthetic data is also shown in this section. Finally, in section 4, the conclusion follows and future work is stated.

## 2 A Modified Mumford-Shah Model Based Image Segmentation

In this section, a region based image segmentation is introduced. The segmentation is attained by using a modified Mumford-Shah segmentation technique. Two phases are assumed for the simplicity of the problem in our model. The presented model is followed in a similar way from [18] with $z(\bar{x}) = \frac{2}{\pi}arctan(\frac{\bar{x}}{\epsilon})$. The model is aimed to find $\theta$, $c_1$, and $c_2$ by minimizing the energy functional:

$$E(\theta, c_1, c_2) =$$

$$\lambda_1 \int_\Omega \{\frac{1}{2}(1+\frac{2}{\pi}\arctan(\frac{\theta}{\epsilon}))\}^2(I(\bar{x})-c_1)^2 + \{\frac{1}{2}(1-\frac{2}{\pi}\arctan(\frac{\theta}{\epsilon})\}^2$$

$$(I(\bar{x})-c_2)^2 d\bar{x} + \lambda_2 \int_\Omega \frac{9\epsilon_1|\nabla(\frac{\theta}{\epsilon})|^2}{\pi^2(1+(\frac{\theta}{\epsilon})^2)^2} + \frac{(\pi^2 - 4\arctan^2(\frac{\theta}{\epsilon}))^2}{64\epsilon_1\pi^4} d\bar{x}, \qquad (2.1)$$

where $I$ is a given image, $\Omega$ is domain, $\epsilon$ and $\epsilon_1$ are positive parameters, and $\lambda_i > 0,\ \ i = 1, 2$ are parameters balancing the influences from the two terms in the model. Here, $c_1(\theta)$=average($I$) in $\{\theta \geq 0\}$ and $c_2(\theta)$=average($I$) in $\{\theta < 0\}$. As $\epsilon \to 0$, the approximation $H_\epsilon(\theta) = \frac{1}{2}(1+\frac{2}{\pi}\arctan(\frac{\theta}{\epsilon}))$ converges to the heaviside function $H(\theta) = 1,\ \ if\ \theta \geq 0$ and $H(\theta) = 0,\ \ if\ \theta < 0$ as in [4]. The square of $H_\epsilon$ is used in our model for the computational stability. In the second term of Equation (2.1), $\epsilon_1 \ll 1$ controls the transition bandwidth. As $\epsilon_1 \to 0$, the first term is to penalize unnecessary interfaces and the second term forces the stable solution of $z(\theta)$ to take one of the two phase field values 1 or -1 in a similar way from [18]. The proposed model is similar to [4], but is different by using the $\Gamma$ approximation to the piecewise-constant Mumford-Shah model.

The first term forces $\{\frac{1}{2}(1+\frac{2}{\pi}\arctan(\frac{\theta}{\epsilon}))\}^2$, towards 0 if $I(\bar{x})$ is different from $c_1$ and towards 1 if $I(\bar{x})$ is close to $c_1$, for every $\bar{x} \in \Omega$. In a similar way, $\{\frac{1}{2}(1-\frac{2}{\pi}\arctan(\frac{\theta}{\epsilon}))\}^2$, towards 0 if $I(\bar{x})$ is different from $c_2$ and towards 1 if $I(\bar{x})$ is close to $c_2$, for every $\bar{x} \in \Omega$.

In the theory of $\Gamma$-convergence, the measuring an edge $\Gamma$ length term in the Mumford-Shah model can be approximated by a quadratic integral of an edge signature function $p(x)$ such that

$$\int_\Gamma dS = \int_\Omega (\epsilon_1 |\nabla p|^2 + \frac{(p-1)^2}{4\epsilon_1}) d\bar{x}, \epsilon_1 \ll 1$$

by Ambrosio and Tortorelli in 1990 [1]. This model is combined with double-well potential function $W(p) = p^2(1-p)^2$ which is quadratic around its minima and is growing faster than linearly at infinity, where $p \in H^1(\Omega)$. In [18], the following is suggested for two phase model:

$$\int_\Omega (9\epsilon_1 |\nabla p|^2 + \frac{(1-p^2)^2}{64\epsilon_1}) d\bar{x},$$

where the range of $p$ is restricted within $[-1, 1]$. Here $\epsilon_1 \ll 1$ controls the transition bandwidth. As $\epsilon_1 \to 0$, the first term is to penalize unnecessary interfaces and the second term forces the stable solution to take one of the two phase field values 1 or -1. The second term in our model is followed from [18]. For the details of phase field models and double-well potential functions, please refer [18,19,20].

## 2.1 Euler-Lagrange Equations of the Model

The evolution equations associated with the Euler-Lagrange equations in Equation (2.1) are

$$\frac{\partial \theta}{\partial t} = -\lambda_1 \{(1 + \frac{2}{\pi}\arctan(\frac{\theta}{\epsilon}))(\frac{\epsilon^2}{\pi(\epsilon^2+\theta^2)})(I(\bar{x}) - c_1)^2$$

$$-(1 - \frac{2}{\pi}\arctan(\frac{\theta}{\epsilon}))(\frac{\epsilon^2}{\pi(\epsilon^2+\theta^2)})(I(\bar{x}) - c_2)^2\}$$

$$+\lambda_2\{\frac{72\epsilon_1}{\pi^2}(div(\frac{\nabla\theta}{(1+\theta^2)^2}) + \frac{2\theta|\nabla\theta|^2}{(1+\theta^2)^3})$$

$$+\frac{(\pi^2 - 4\arctan^2(\frac{\theta}{\epsilon}))\arctan(\frac{\theta}{\epsilon})}{4\epsilon_1\pi^4(1+(\frac{\theta}{\epsilon})^2)}\}, \quad in\ \Omega$$

$$\frac{\partial \theta}{\partial n} = 0, \quad on\ \partial\Omega, \tag{2.2}$$

where the optimal means $c_1$ and $c_2$ are attained by

$$c_1 = \frac{\int_\Omega \{\frac{1}{2}(1 + \frac{2}{\pi}\arctan(\frac{\theta}{\epsilon}))\}^2 I(\bar{x}) d\bar{x}}{\int_\Omega \{\frac{1}{2}(1 + \frac{2}{\pi}\arctan(\frac{\theta}{\epsilon}))\}^2 d\bar{x}}, \tag{2.3}$$

$$c_2 = \frac{\int_\Omega \{\frac{1}{2}(1 - \frac{2}{\pi}\arctan(\frac{\theta}{\epsilon}))\}^2 I(\bar{x}) d\bar{x}}{\int_\Omega \{\frac{1}{2}(1 - \frac{2}{\pi}\arctan(\frac{\theta}{\epsilon}))\}^2 d\bar{x}}. \tag{2.4}$$

## 2.2 Numerical Results

In this part, the numerical results with applications to synthetic data and simulated normal human brain MR images. The Equation (2.1) was solved by finding a steady state solution of the evolution equations. The evolution equations are associated with the Euler-Lagrange equations of the Equation (2.1). A finite difference scheme and the gradient descent method is applied to discretize the evolving equations.

The Figure 1 showed the proposed model segmentation results to synthetic data. Figure 2 showed the segmentation results by [4]. The first figure in Figure 1 and Figure 2 is the given synthetic image $I$ with an initial contour as a solid line. In Figure 1 and Figure 2, the second image is the segmented image and the third one is the segmented contour as a solid line with $I$. Our model performs better than [4] to capture the boundary of the region which has similar intensity. The numerical results with an application to the simulated human brain MR image are shown from Figure 3 to Figure 6. The data is obtained from *http://www.bic.mni.mcgill.ca/brainweb*. The simulated normal human brain image with the ground truth white matter image is shown in Figure 3. In Figure 3, the first figure is the simulated brain image $I$, the second one is an image $I$ with an initial contour as a solid line, and the third one is the ground truth brain white matter. From Figure 4 to Figure 5, the first image is the given image $I$ with an initial contour as a solid line, the second image is the segmented image results by our proposed model and [4] each, and the third figure is the segmented contour in $I$. The numerical results of segmented image by the proposed model and [4] compared to the ground truth white matter brain image which is in the first figure are in Figure 6. Our proposed model captures the boundary of the brain image better than [4]. The second image in Figure 1, Figure2, Figure 4, and Figure 5 is obtained by $sign(\theta)$, where $sign(\theta) = 1, if\ \ \theta \geq 0$ and $sign(\theta) = -1, if\ \ \theta < 0$.

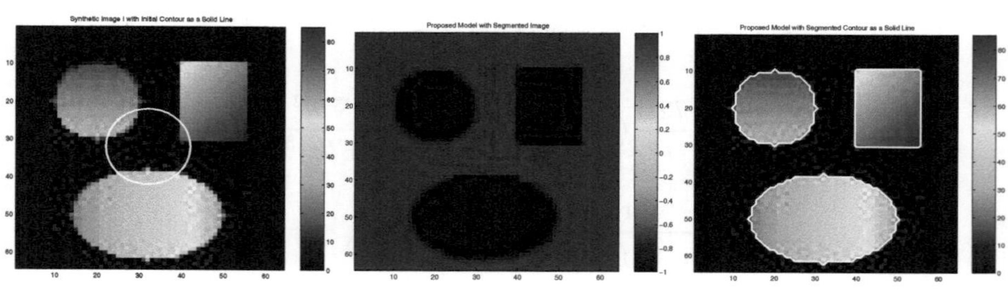

**Fig. 1.** Left : A Given Image $I$ with an Initial Contour, Middle : Segmented Synthetic Image using the Proposed Model, Right : Segmented Synthetic Image with a Segmented Contour as a Solid Line

# 3 Image Segmentation Using a Prior Shape Information

A new variational image segmentation model based on the region intensity information utilizing prior shape knowledge is suggested. The goal of the model is to develop

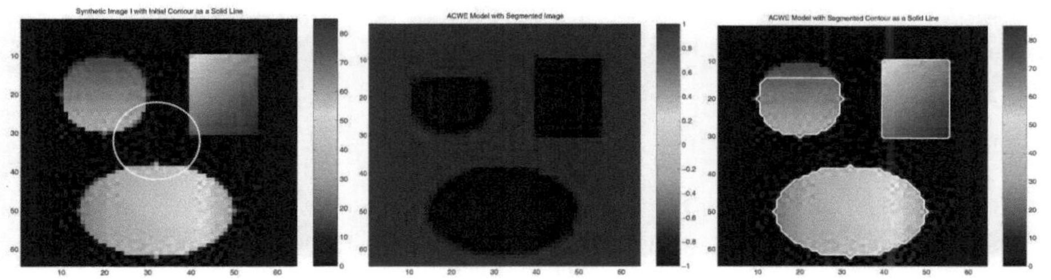

**Fig. 2.** Left : A Given Image $I$ with an Initial Contour, Middle : Segmented Synthetic Image using Active Contour Without Edges Model, Right : Segmented Synthetic Image with a Contour as a Solid Line

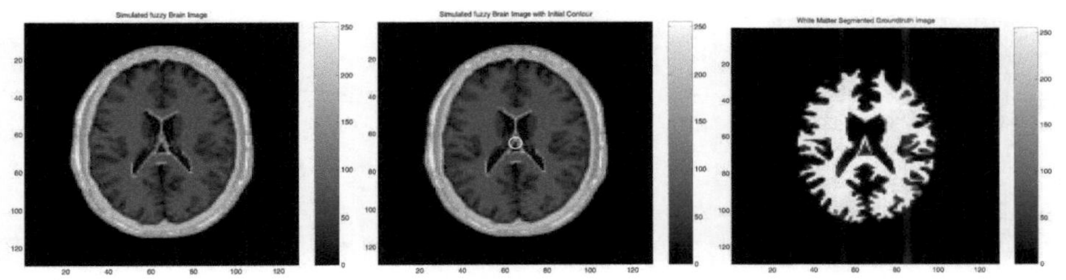

**Fig. 3.** Left : A Given Image $I$, Middle : Brain Image with an Initial Contour, Right : Ground Truth Brain White Matter

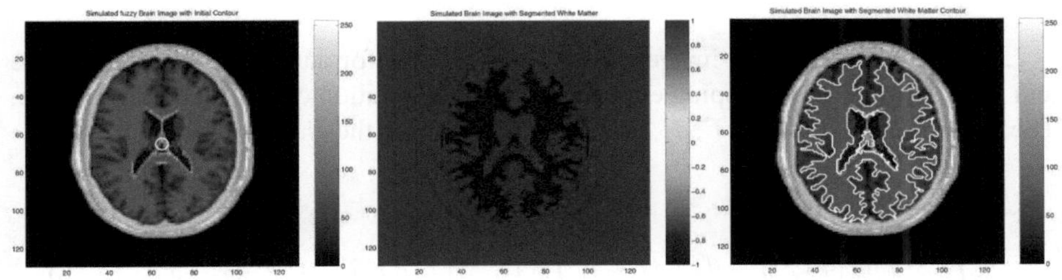

**Fig. 4.** Left : Brain Image with an Initial Contour, Middle : Segmented Simulated White Matter Brain Image using the Proposed Model, Right : Segmented Simulated Brain Image with a Solid Line

the prior shape information based image segmentation technique. The shape prior is obtained by using a modified Mumford-Shah segmentation technique from section 2. Then the heaviside function was applied to $\theta$ to get binary image, where $H(\theta) = 1, \ \ if \ \theta \geq 0$ and $H(\theta) = 0, \ \ if \ \theta < 0$. In a similar way from Section 2, two phases are assumed for the simplicity of the problem in our model.

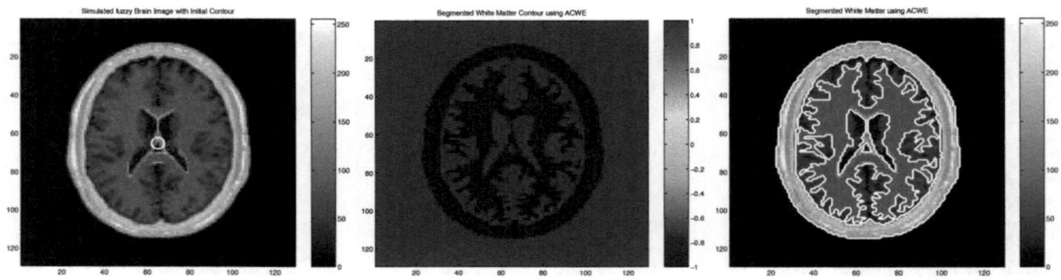

**Fig. 5.** Left : Brain Image with an Initial Contour, Middle : Segmented Simulated White Matter Brain Image using a Active Contour Without Edges Model, Right : Segmented Simulated Brain Image with a Solid Line

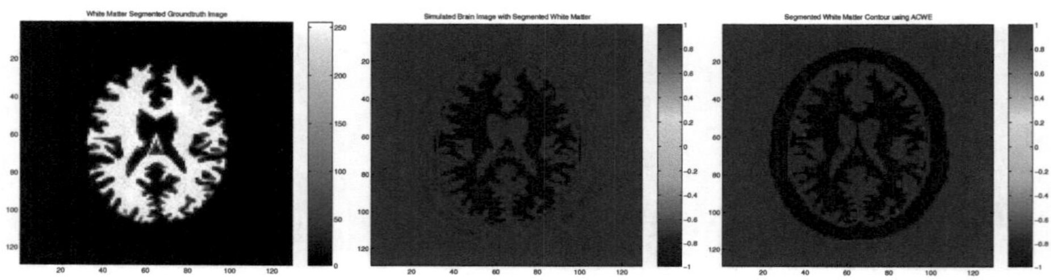

**Fig. 6.** Left : Ground Truth Brain White Matter Segmented Image, Middle : Segmented Simulated White Matter Brain Image by our Proposed Model, Right : Segmented Simulated White Matter Brain Image using an Active Contour without Edges Model

Let $S$ be a given binary image, called shape prior information. $I$ ia a given image and $\Upsilon$ is a registration mapping either a rigid transformation or nonrigid deformation. The model is aimed to find $\Upsilon$, $d_1$, and $d_2$ by minimizing the energy functional:

$$E(\Upsilon, d_1, d_2) =$$
$$\int_{\Omega} S(\bar{x})(I(\Upsilon(\bar{x})) - d_1)^2 d\bar{x} + \lambda \int_{\Omega} (1 - S(\bar{x}))(I(\Upsilon(\bar{x})) - d_2)^2 d\bar{x}. \quad (3.1)$$

A shape prior image $S$, a given image $I$, and $\Omega$ as a domain are given and $\lambda$ is a positive parameter balancing the influences from two terms in the model. $d_1$ and $d_2$ are average intensity values of $I(\Upsilon)$ in $\{S\}$ and $\{1 - S\}$ respectively. In our numerical experiments, only rigid transformation is considered, but the model can be generalized to non-rigid deformation as well. Therefore the rigid transformation vector $\Upsilon(\bar{x}) = \mu R\bar{x} + T$, where $\mu$ is a scaling, $R$ is a rotation matrix with respect to an angle $\theta$, and $T$ is a translation.

For each $\bar{x} \in \Omega$, the first term is forcing $I(\Upsilon(\bar{x}))$ to be close to $d_1$ using prior shape information $S(\bar{x})$. In a similar way, the second term is compelling $I(\Upsilon(\bar{x}))$ to be close to $d_2$ utilizing prior shape information $\{1 - S(\bar{x})\}$. After minimizing these three terms, the best $\Upsilon$, $d_1$, and $d_2$ are obtained. The Equation (3.1) is similar to [4], but is different by using the prior shape information $S$ and the rigid transformation $\Upsilon$. During the image segmentation process, the shape knowledge information supports the robustness to loss of information and image noise and the rigid transformation $\Upsilon$ helps to find the correct correspondence to the transformed image.

### 3.1 Euler-Lagrange Equations of the Proposed Model

The evolution equations associated with the Euler-Lagrange equations in Equation (3.1) are

$$\frac{\partial \mu}{\partial t} = -\int_\Omega S(\bar{x})(I(\Upsilon(\bar{x})) - d_1)(\nabla I(\Upsilon(\bar{x})))R\bar{x}d\bar{x} - \lambda \int_\Omega (1 - S(\bar{x}))(I(\Upsilon(\bar{x})) - d_2)(\nabla I(\Upsilon(\bar{x})))R\bar{x}d\bar{x}, \tag{3.2}$$

$$\frac{\partial \theta}{\partial t} = -\int_\Omega S(\bar{x})(I(\Upsilon(\bar{x})) - d_1)(\nabla I(\Upsilon(\bar{x})))\mu \frac{dR}{d\theta}\bar{x}d\bar{x} - \lambda \int_\Omega (1 - S(\bar{x}))(I(\Upsilon(\bar{x})) - d_2)(\nabla I(\Upsilon(\bar{x})))\mu \frac{dR}{d\theta}\bar{x}d\bar{x}, \tag{3.3}$$

$$\frac{\partial T}{\partial t} = -\int_\Omega S(\bar{x})(I(\Upsilon(\bar{x})) - d_1)(\nabla I(\Upsilon(\bar{x})))d\bar{x} - \lambda \int_\Omega (1 - S(\bar{x}))(I(\Upsilon(\bar{x})) - d_2)(\nabla I(\Upsilon(\bar{x})))d\bar{x}, \tag{3.4}$$

$$\frac{\partial d_1}{\partial t} = \int_\Omega S(\bar{x})(I(\Upsilon(\bar{x})) - d_1)d\bar{x}, \tag{3.5}$$

$$\frac{\partial d_2}{\partial t} = \lambda \int_\Omega (1 - S(\bar{x}))(I(\Upsilon(\bar{x})) - d_2)d\bar{x}, \tag{3.6}$$

where $R$ is the rotation matrix in terms of the angle $\theta$.

### 3.2 Numerical Results

The Equation (3.1) was solved by finding a steady state solution of the evolution equations. The evolution equations are associated with the Euler-Lagrange equations of the Equation (3.1). A finite difference scheme and the gradient descent method is applied to discretize the evolving equations. Initial value of $d_1$ and $d_2$ are 0.95 and 0.01 respectively. Here $\lambda = 1$ is used in the numerical experiments. Figure 7 showed original image, the binary image $S$ as a prior shape information, and synthetic image $I$ with noise, rotation, and loss of information. The numerical results of the proposed model using the prior shape information are shown in Figure 8. In Figure 8, the first one is the binary image as prior shape information and the second one is the given image $I$. The segmented image and segmented contour result by using the Equation (3.1) are the third and fourth figure in Figure 8. The first figure in Figure 9 and Figure 10 is the given image $I$. The second and the third figure in Figure 9 showed the numerical results of

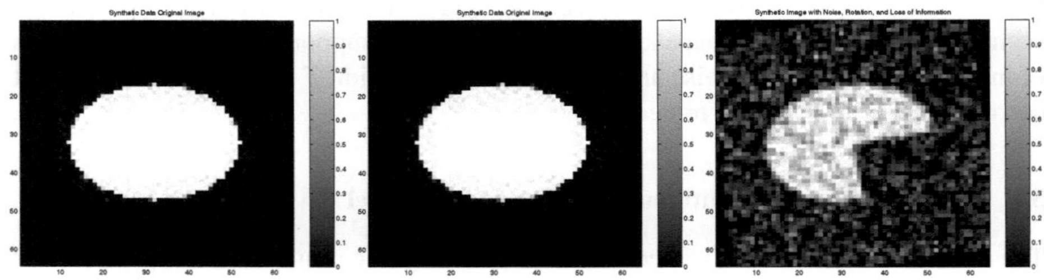


**Fig. 7.** Left: Original Image, Middle : A Binary Image $S$ as a Prior Shape Information, Right : Given Synthetic Image $I$ with Noise, Loss of Information, and Rotation

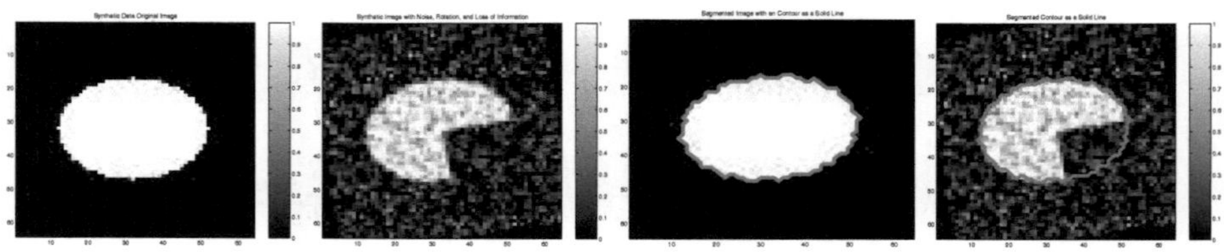

**Fig. 8.** First : A Binary Image $S$ as a Prior Shape Information, Second: Given Synthetic Image $I$, Third : Segmented Image with a contour as a Solid Line, Fourth : Given Image $I$ with Segmented Contour as a Solid Line

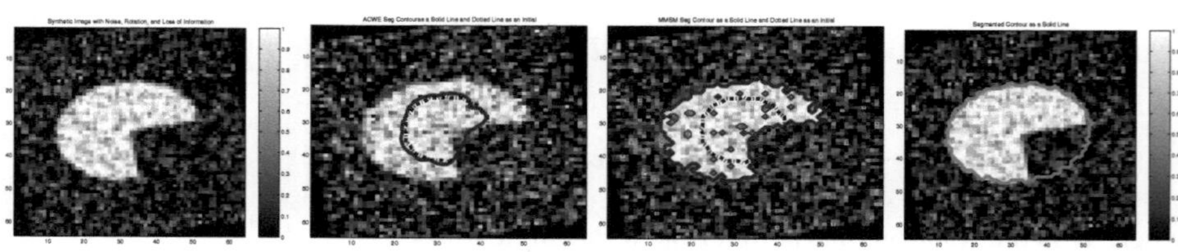

**Fig. 9.** First : A Given Image $I$, Second : Segmented Image using an Active Contour without Edges with Dotted Line as Initial Contour and Solid Line as Segmented Contour, Third : Segmented Image using our Proposed Model in Section 2 with Dotted Line as Initial Contour and Solid Line as Segmented Contour, Fourth : Segmented Image using the Prior Shape Information $S$ with Solid Line Contour

segmented contour by [4] and the Equation (2.1) to a given Image $I$. Due to the noise, rotation, and loss of information, only using the model by [4] or the Equation (2.1) was not sufficient to get desired segmentation results. Hence the prior shape information is necessary in the segmentation process. The fourth image in Figure 9 is the segmented contour result using the Equation (3.1). In Figure 10, the second and the third figure showed the numerical results of segmented image by [4] and the Equation (2.1) to a given Image $I$. The fourth image in Figure 10 is the segmented image result using the Equation (3.1) with a contour as a solid line. The optimal solution $\mu$, $R$ associated with

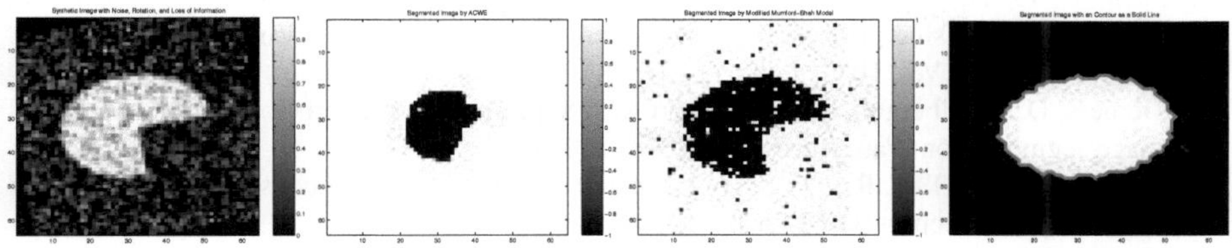

**Fig. 10.** First : A Given Image $I$, Second : Segmented Synthetic Image by an Active Contour without Edges, Third : Segmented Synthetic Image by our Proposed Model in Section 2, Fourth : Segmented Synthetic Image with a Solid Line Contour using the Prior Shape Information

$\theta$, and $T$ from the Equation (3.1) is applied to the Prior Shape Information $S$ to get the segmented image and segmented contour as a solid line from Figure 8 to Figure 10.

## 4 Conclusions and Future Work

Two new region based variational partial differential equation (PDE) models for an image segmentation are proposed with an application to synthetic and simulated human brain MR images. The first model utilizes a modified piecewise constant Mumford-Shah model. Even though this model performs better than existing model with fuzzy images, this algorithm has some limits with strong noise, rotation, and loss of information. Therefore, the second model is obtained using a prior shape information and region intensity value. Numerical Results show the effectiveness and robustness of the presented model against to noise, rotation, loss of information, and artifact. In the future work, the research will be focused on the improvements of the first model with the robustness to the choice of the initial contour. In addition, the generalization of the second model to non-rigid deformation will be developed and the numerical experiments on simulated human brain MR images of the second proposed model will be performed.

**Acknowledgement.** Chen is partially supported by NSF CCF-0527967 and NIH R01 NS052831-01 A1.

## References

1. Ambrosio, L. and Tortorelli, V. : Approximation of functionals depending on jumps by elliptic functionals via $\Gamma$-convergence . Comm. on Pure and Applied Math. Vol. **43** (1990) 999–1036
2. Baldo, S. : Minimal interface criterion for phase transitions in mixtures of Cahn-Hilliard fluids. Annals. Inst. Henri Poincare. Vol. **7** (1990)67–90
3. Bresson, X., Vandergheynst, P., and Thiran, J. : A variational model for object segmentation using boundary information and shape prior driven by the Mumford-Shah functional Int. J. Comp. Vis. Vol. **68**(2) (2006) 145–162
4. Chan, T. and Vese, L. : Active contours without edges. IEEE Trans. Image Proc. **10** (2) (2001) 266–277
5. Chan, T. and Vese, L. : A level set algorithm for minimizing the Mumford-Shah functional in image processing. Proc. 1st IEEE Workshop Varia. Level Set Meth. Comp. Vis. **13** Vancouver B.C. Canada (2001) 161–168

6. Chen, Y., Tagare, H., Thiruvenkadam, S., Huang, F., Wilson, D., Gopinath, K., Briggs, R., and Geiser, E. : Using prior shapes in geometric active contours in a variational framework. Int. J. Comp. Vis. Vol. **50** (3) (2002) 315–328
7. Cremers, D., Kohlberger, T., and Schnörr, C. : Shpe statistics in kernel space for variational image segmentation. Patt. Recog. 36 (2003) 1929–1943
8. Esedoglu, S. and Tsai, R. : Threshold dynamics for the piecewise constant Mumford-Shah fuctional. CAM Report 04-63 UCLA (2004)
9. Gibou, F., and Fedkiw, R. : A fast hybrid k-means level set algorithm for segmenttion. Stanford Technical Report (2002)
10. Huang, X., Li, Z., and Metaxas, D. : Learning Coupled Prior Shape and Appearance Models for Segmentation. Proc. 7th. Ann. Int. Conf. on Med. Image Comp. Computer-Assi. Interv. MICCAI04 Vol. **I** LNCS-3216 (2004) 60-69.
11. Huang, X., Metaxas, D., and Chen, T. : MetaMorphs: Deformable Shape and Texture Models. Proc. IEEE Comp. Soc. Conf. Comp. Vis. Pat. Recog. CVPR04, Vol. **I** (2004) 496–503
12. Leventon, M., Grimson, E., and Faugeras, O. : Statistical Shape Influence in Geodesic Active Contours. Proc. IEEE Conf. CVPR (2000) 316–323
13. Lie, J., Lysaker, M., and Tai, X. : A binary level set model and some aplications to Mumford-Shah segmentation. CAM Report. Vol. **31** (2004)
14. Modica, L. : The gradient theory of phase transitions and the minimal interface criterion. Arch. Rational Mech. Anal. Vol. **98** (1987) 123–142
15. Mumford, D. and Shah, J. : Optimal approximations by piecewise smooth functions and associated variational problems. Comm. on Pure and Applied Math. Vol. **42** (1989) 577–685
16. Osher, S. and Fedkiw, R. : Level set methods and dynamic implicit surfaces. Springer Verlag, New York, (2003)
17. Rousson, M. Paragios, N.: Shape prior for level set representations. Comp. Vis. ECCV2002 7th. Eur. Conf. Comp. Vis. Copenhgen Demark Proc. (2002) 78–92
18. Shen. J.: $\Gamma$-Convergence Approximation to Piecewise Constant Mumford-Shah Segmentation. Lec. Notes Comp. Sci., 3708 (2005) 499–506
19. Shen. J.: A stochastic-variational model for soft Mumford-Shah segmentation . Int. J. Biomed. Imaging, special issue on "Recent Advances in Mathematical Methods for the Processing of Biomedical Images," Vol. **2006** (2006) 1–14
20. Wang, M. and Zhou, S. : Phase Field: A Variational Method for Structural Topology Optimization . Computer Modeling in Engineering & Sciences. Vol. **6** no.6 (2004) 547–566

# Total Variation Minimization and Graph Cuts for Moving Objects Segmentation

Florent Ranchin[1], Antonin Chambolle[2], and Françoise Dibos[3]

[1] CEA Saclay, LIST-LCEI, Gif-sur-Yvette
[2] CMAP, Ecole Polytechnique, Palaiseau, France
[3] LAGA & L2TI , Universit Paris 13, 93430 Villetaneuse

**Abstract.** In this paper, we are interested in the application to video segmentation of the discrete shape optimization problem

$$\lambda J(\theta) + \sum_i (\alpha - f_i)\theta_i \tag{1}$$

incorporating a data $f = (f_i)$ and a total variation function $J$, and where the unknown $\theta = (\theta_i)$ with $\theta_i \in \{0, 1\}$ is a binary function representing the region to be segmented and $\alpha$ a parameter. Based on the recent works [1], and Darbon and Sigelle [2,3], we justify the equivalence of the shape optimization problem and a weighted TV regularization in the case where $J$ is a "weighted" total variation. For solving this problem, we adapt the projection algorithm proposed in [4] to this case. Another way of solving (1) investigated here is to use graph cuts. Both methods have the advantage to lead to a global minimum.

Since we can distinguish moving objects from static elements of a scene by analyzing norm of the optical flow vectors, we choose $f$ as the optical flow norm. In order to have the contour as close as possible to an edge in the image, we use a classical edge detector function as the weight of the weighted total variation. This model has been used in the former work [5]. We also apply the same methods to a video segmentation model used by Jehan-Besson, Barlaud and Aubert. In this case, it is a direct but interesting application of [1], as only standard perimeter is incorporated in the shape functional.

**Keywords:** total variation, motion detection, active contour models.

## 1 Introduction

Segmentation of moving objects from a video sequence is an important task whose applications cover domains such like video compression, video surveillance or object recognition. In video compression, the MPEG-4 video coding standard is based on the representation of the scene as different shapes-objects. This representation simplifies the scene and is used for the encoding of the sequence.

There are different ways to perform moving objects segmentation, using different mathematical techniques. For Markov Random Fields based methods, we refer to the works of Bouthemy [6] and for maximum likelihood based methods,

F. Sgallari, A. Murli, and N. Paragios (Eds.): SSVM 2007, LNCS 4485, pp. 743–753, 2007.

to the works of Deriche and Paragios ([7]). For variational techniques, we refer to the works of Deriche *et al.* ([8]) and Barlaud *et al.* ([9]). At last, mathematical morphology has been more and more used these last ten years, see the works of Salembier, Serra and their teams ([10]).

In this paper, based on the former work [5] concerning moving object segmentation, we focus on two different techniques, the first one relying on the recent result of [1] (the same results were derived independently, and previously, by Darbon and Sigelle [2,3] in a probabilistic setting) and the second one is the use of graph cuts (Boykov, Veksler, and Zabih [11], Kolmogorov and Zabih [12]). The result of [1] states that solving the Rudin-Osher-Fatemi Total Variation regularization problem [13] and thresholding the result at the level $\alpha$ gives the region that is solution of the shape optimization problem 7. It is related to former works like [14] and [15].

In this paper, we use the framework of [1] in the case of a non-homogeneous total variation functional, corresponding to a weighted anisotropic perimeter like the one studied in [5]. The outline is the following : in the first part we present the energy used to segment moving objects in the image in the second part and we expose formal mathematic arguments for the use of TV regularization. It is followed by a mathematical part about TV regularization and results about the equivalence with solving a class of shape optimization problems, and by a part where we present graph cuts and their use for our functional. It is ended by an experimental part where we show the results obtained.

## 2 A Shape Optimization Problem for Moving Object Detection

### 2.1 The Functional

Once we have determined the optical flow $\mathbf{v}$, we keep it for the segmentation purpose. We will denote $\Omega$ the moving region and $D$ the image domain. As a moving object should be characterised by a sufficiently large flow magnitude, it seems natural to incorporate $\int_\Omega \alpha - |\mathbf{v}|(x)\,dx$ to the energy we want to minimize, where $\alpha - |\mathbf{v}|(x)$ have to take different signs on the image domain, otherwise the solution of the shape optimization problem will be trivial. As we want the boundary of $\Omega$ to remain stable in the presence of noise or spurious variations, we also penalize the total length of this boundary (that is, the perimeter of $\Omega$) in our functional. Finally, as thresholding the optical flow will not give exact object contours (due to the temporal integration), we add a weighted perimeter which integrates a function of the gradient (here $g_I = \frac{1}{1+|\nabla I|^2}$) along the boundary. It gives the functional

$$\int_\Omega \alpha\,dx + \int_{D\setminus\Omega} |\mathbf{v}|\,dx + \int_{\partial\Omega} g(x)\,dS, \tag{2}$$

$dS$ denoting the arclength variation along the boundary and $g = \lambda g_I + \mu$. Within the framework of shape sensitivity analysis (see Delfour and Zolesio [16]), one can compute the *shape derivative* of this functional and obtain the steepest gradient descent. Combining it to the famous level set method (Osher, Sethian, [17]), we would obtain

$$\frac{\partial u}{\partial t} = |\nabla u| \left( |\mathbf{v}| - \alpha + \operatorname{div}\left( g \frac{\nabla u}{|\nabla u|} \right) \right). \tag{3}$$

That was done in [5], unfortunately, if we want to adjust the value of $\alpha$ in a suitable way, we have to recompute the result by this partial differential equation as many times as necessary. We overcome this problem by using the equivalence between solving the ROF model with a weighted total variation and solving of (2) for all the possible values of $\alpha$. Another advantage of this method is to converge towards a global minimum of the energy, which is not guaranteed by solving (3).

In [1], functionals do not involve standard perimeter but a different anisotropic one. This is for theoretical reasons explained in [1] : the discrete total variation does not satisfy the coarea formula which is needed in the main result of [1]. In fact, the theory can be developped with the isotropic total variation in the continuous setting, and results could still be (approximately) computed.

Thus we slightly modify the functional to fit in the framework given in [1] ($\nu$ denotes the outside normal to the boundary and $|\cdot|_1$ the 1-norm : $|(a,b)|_1 = |a| + |b|$, $R_{\frac{\pi}{4}}$ denotes the rotation of angle $\frac{\pi}{4}$)

$$E(\Omega) = \int_\Omega \alpha \, dx + \int_{D\setminus\Omega} |\mathbf{v}| \, dx + \frac{1}{2} \int_{\partial\Omega} g(x)(|\nu|_1 + |R_{\frac{\pi}{4}}(\nu)|_1) \, dS. \tag{4}$$

This is a change of metric : the standard length and its weighted counterpart are replaced by an isotropic version of what it is usually called “Manhattan” or “taxicab” length $\int_{\partial\Omega} g(x)|\nu|_1 \, dS$. At last, we rewrite our functional in discrete setting, as this will be in the rest of the paper. We introduce the notations

$$(\nabla^x \theta)_{i,j} = \begin{cases} \theta_{i+1,j} - \theta_{i,j} & if\ 0 \le i \le N-1, 0 \le j \le N \\ 0 & if \quad i = N, 0 \ge j \le N \end{cases}$$

$$(\nabla^y \theta)_{i,j} = \begin{cases} \theta_{i,j+1} - \theta_{i,j} & if\ 0 \le i \le N, 0 \le j \le N-1 \\ 0 & if \quad 0 \ge i \le N, j = N \end{cases}$$

$$(\nabla^{xy} \theta)_{i,j} = \begin{cases} \theta_{i+1,j+1} - \theta_{i,j} & if\ 0 \le i \le N-1, 0 \le j \le N-1 \\ 0 & if \quad i = N\ or j = N \end{cases}$$

$$(\nabla^{yx} \theta)_{i,j} = \begin{cases} \theta_{i+1,j-1} - \theta_{i,j} & if\ 0 \le i \le N-1, 1 \le j \le N \\ 0 & if \quad i = N\ or j = 0 \end{cases}$$

$$\begin{aligned} E(\theta) = {} & \textstyle\sum_{i,j} (\alpha - |\mathbf{v}|_{i,j})\theta_{i,j} + \frac{1}{2} \sum_{i,j} g_{i,j} \left( |(\nabla^x \theta)_{i,j}| + |(\nabla^y \theta)_{i,j}| \right) \\ & + \textstyle\frac{1}{2\sqrt{2}} \sum_{i,j} g_{i,j} \left( |(\nabla^{xy} \theta)_{i,j}| + |(\nabla^{yx} \theta)_{i,j}| \right). \end{aligned} \tag{5}$$

Let us observe that the weight $g_{i,j}$ could be different on each edge (connecting two neighboring pixels) of the grid and that the choice we have made is quite arbitrary. However, we did not observe a significant change in the output when weighing the edges in a different way.

## 3 On the Equivalence of Total Variation Regularization and a Class of Shape Optimization Problems

In this section, we will use the following notations : $|\cdot|$ denotes the euclidean norm $|(a,b)| = \sqrt{a^2+b^2}$, $|\cdot|_p$ denotes the $p$-norm $|(a,b)|_p = (|a|^p + |b|^p)^{1/p}$ and $|\cdot|_\infty$ denotes the $\infty$-norm $|(a,b)|_\infty = \sup(|a|,|b|)$. The scalar product between two vectors $u$ and $v$ is denoted by $u \cdot v$.

### 3.1 Settings

In this section, we recall the main results obtained in [1]. The problem considered is

$$\min_{\theta \in X, \theta_i \in \{0,1\}} \lambda J(\theta) + \sum_i (\alpha - f_i)\theta_i \qquad (P_\alpha)$$

where $X$ is the space of functions defined on the $N$ pixels of the image grid ($i$ denotes the pixel index and $f$ is still a data function). We consider the Rudin-Osher-Fatemi TV regularization problem

$$\min_{u \in X} J(u) + \frac{1}{2\lambda}\|u - f\|^2 \tag{6}$$

and the discrete shape optimization problem

$$\min_{\theta \in X, \theta_i \in \{0,1\}} \lambda J(\theta) + \sum_i (\alpha - f_i)\theta_i \tag{7}$$

The main theorem of [1] states an equivalence between solving (6) and thresholding the result at threshold $\alpha$ and solving (7). As we are concerned only with solving (7), we give only the part of the theorem which states that thresholding the solution of the discretized ROF model gives a solution of the shape optimisation problem.

**Theorem 1.** ([1]) *Let $w$ solve (6) and $J$ satisfying suitable properties (see [1]). Then, for any $s \in \mathbb{R}$, both $w_i^s = \mathbf{1}_{w_i > s}$ and $\bar{\bar{w}}_i^s = \mathbf{1}_{w_i > s}$ solve (7). If $w^s = \bar{\bar{w}}^s$, then the solution of (7) is unique.*

With the notations introduced in the previous section, the energy (2) involves the function

$$TV_{1,\frac{\pi}{4},g}(u) = \frac{1}{2}\sum_{i,j} g_{i,j}\left(|(\nabla^x\theta)_{i,j}| + |(\nabla^y\theta)_{i,j}|\right) + \frac{1}{2\sqrt{2}}\sum_{i,j} g_{i,j}\left(|(\nabla^{xy}\theta)_{i,j}| + |(\nabla^{yx}\theta)_{i,j}|\right). \tag{8}$$

It derives from the discrete Manhattan total variation $TV_1(u) = \sum_{i,j}|u_{i+1,j} - u_{i,j}| + |u_{i,j+1} - u_{i,j}|$ and its modifications: $TV_{1,\frac{\pi}{4}}(u) = TV_1(u) + \frac{1}{2\sqrt{2}}\sum_{i,j}|u_{i+1,j+1} - u_{i,j}| + |u_{i-1,j+1} - u_{i,j}|$ (more isotropic version), $TV_{1,g}(u) = \sum_{i,j} g_{i,j}\left(|u_{i+1,j} - u_{i,j}| + |u_{i,j+1} - u_{i,j}|\right)$ ("weighted" Manhattan total variation). As shown in [1], the Manhattan total variation satisfies the conditions exposed in the previous section. It is straightforward to show that all these modifications, including $TV_{1,\frac{\pi}{4},g}$ satisfie them also.

### 3.2 The Projection Algorithm of [4]

In [4], a new algorithm for computing the solution of (6) was proposed. It is based on duality results and consists in finding the projection of $f$ onto a convex set. Let us describe how it works on the energy we are interested in. Here we follow the calculus of [1] which generalize well to the g-weighted Manhattan TV.

We denote $\nabla w = (\nabla^x w, \nabla^y w)$ and $\nabla' w = (\nabla^{xy} w, \nabla^{yx} w)$, with these notations

$$TV_{1,\frac{\pi}{4},g}(u) = \frac{1}{2}\sum_{i,j} g_{i,j}|(\nabla u)_{i,j}|_1 + \frac{1}{2}\sum_{i,j} g_{i,j}|(\nabla' u)_{i,j}|$$

From discrete gradients, we get the definition of discrete divergence div $= -\nabla^*$

$$(\text{div}\xi, w)_X = -(\xi, \nabla w)_{X\times X}, \ \forall w \in X, \xi \in X \times X,$$

and similarly with is rotated counterpart div$' = -(\nabla')^*$

$$(\text{div}'\xi, w)_X = -(\xi, \nabla' w)_{X\times X}, \ \forall w \in X, \xi \in X \times X.$$

Using these duality principles, we can demonstrate that

$$TV_{1,\frac{\pi}{4},g}(w) = \sup_{|\xi|_\infty \le 1} (w, \text{div}(g\,\xi))_X + \sup_{|\eta|_\infty \le 1} \left(w, \text{div}'(g\,\eta)\right)_X = \sup_{v\in K} (w, v)_X\,,$$

where (the overlining denotes the closure)

$$K = \overline{\{\frac{1}{2}(\text{div}(g\,\xi) + \text{div}'(g\,\eta))|\,(\xi,\eta) \in A^2\}}$$

and

$$A = \{p = (p^x, p^y) \in X \times X,\ |p^x_{i,j}| \le 1,\ |p^y_{i,j}| \le 1\}.$$

Exactly in the same manner than in [1], it can be established that the solution of the ROF problem is given by the orthogonal projection of $w$ onto $\lambda K$ This is for constraints simplicity. Finally, the solution of

$$\min_{w\in X} TV_{1,\frac{\pi}{4},g}(w) + \frac{1}{2\lambda}\|w - w_0\|^2 \tag{9}$$

is given by $\bar{w} = w_0 - \frac{1}{2}\left(\lambda \text{div}(g\,\bar{\xi}) + \lambda \text{div}'(g\,\bar{\eta})\right)$ where $(\bar{\xi},\bar{\eta})$ is a solution to

$$\min_{(\xi,\eta)\in A^2} \|\frac{1}{2}\left(\lambda \text{div}(g\,\xi) + \lambda \text{div}'\,(g\eta)\right) - w_0\|^2 \tag{10}$$

In the same way than in [1], we obtain the Euler-Lagrange equations and a fixed-point algorithm from the gradient descent

$$\begin{aligned}
w^n &= \tfrac{1}{2}\left(\lambda \text{div}(g\,\xi_n) + \lambda \text{div}'(g\,\eta_n)\right) - w_0\\
(\xi^{n+1}_{i,j})^x &= \frac{(\xi^n_{i,j})^x + g_{i,j}\frac{\tau}{\lambda}(\nabla^x w^n)_{i,j}}{1 + g_{i,j}\frac{\tau}{\lambda}|(\nabla^x w^n)_{i,j}|}\\
(\xi^{n+1}_{i,j})^y &= \frac{(\xi^n_{i,j})^y + g_{i,j}\frac{\tau}{\lambda}(\nabla^y w^n)_{i,j}}{1 + g_{i,j}\frac{\tau}{\lambda}|(\nabla^y w^n)_{i,j}|}\\
(\eta^{n+1}_{i,j})^x &= \frac{(\eta^n_{i,j})^x + g_{i,j}\frac{\tau}{\lambda}(\nabla^{xy} w^n)_{i,j}}{1 + g_{i,j}\frac{\tau}{\lambda}|(\nabla^{xy} w^n)_{i,j}|}\\
(\eta^{n+1}_{i,j})^y &= \frac{(\eta^n_{i,j})^y + g_{i,j}\frac{\tau}{\lambda}(\nabla^{yx} w^n)_{i,j}}{1 + g_{i,j}\frac{\tau}{\lambda}|(\nabla^{yx} w^n)_{i,j}|}
\end{aligned}$$

Following the convergence proof of [4], we obtain the convergence theorem

**Theorem 2.** *Let* $\tau \leq \frac{1}{8\max_{i,j} g_{i,j}}$. *Then,* $\frac{\lambda}{2}(\mathrm{div}(g\xi^n) + \mathrm{div}'(g\eta^n))$ *converges to the orthogonal projection of* $w_0$ *onto the convex set* $\lambda K$ *as* $n \to \infty$, *and* $w^n$ *converges to the solution of (9).*

# 4 How to Minimize the Energies with Graphcuts

## 4.1 Principle

Greig, Portehous and Seheult proved in [18] that discrete energy minimization can be exactly performed. Graphcuts have been introduced in computer vision by Y. Boykov and his collaborators in [11] as an algorithm for this type of minimization. They have been extended to many areas : stereovision [19], medical imaging [20]... The idea is to add a "source" and a "sink" in such a way that to each point in the image grid a link is created to either the source or the sink. A cost is assigned to the links so that the global cost be related to the energy. Finally, solving the energy minimization problem is equivalent to find a cut of minimal cost along the graph (source-points-sink). This is achieved by finding a "maximal flow" along the edges of the graph, due to a duality between min-cut and max-flow problems, first observed by Ford and Fulkerson.

## 4.2 Construction

We recall the energy is (we replace $\lambda + \mu g$ by $g$ for simplicity)

$$J(\theta) = \sum_{(i,j)} (\alpha - |\mathbf{v}|_{i,j})\theta_{i,j}$$
$$+\frac{1}{2}\sum_{i,j} g_{i,j}\left(|(\nabla^x\theta)_{i,j}| + |(\nabla^y\theta)_{i,j}| + \frac{\sqrt{2}}{2}|(\nabla^{xy}\theta)_{i,j}| + \frac{\sqrt{2}}{2}|(\nabla^{yx}\theta)_{i,j}|\right)$$

which gives, with simpler notations (we denote a pixel $x = (i,j)$)

$$J(\theta) = \sum_x (\alpha - |\mathbf{v}|_x)\theta_x + \sum_{x,y} w_{x,y}|\theta_y - \theta_x|$$

The coefficients $w_{x,y}$ are given by $w((i,j),(i\pm1,j)) = w((i,j),(i,j\pm1)) = g_{(i,j)}$ and $w((i,j),(i\pm1,j\pm1)) = w((i,j),(i\mp1,j\pm1)) = \frac{\sqrt{2}}{2}g_{(i,j)}$.

One can see that the weights $w_{x,y}$ are nonsymmetric : $w_{x,y} \neq w_{y,x}$ due to the presence of $g$ which has a dependency with respect to the pixel. This is not the problem (indeed, the conditions found by Kolmogorov and Zabih for an energy invoving an interaction term which satisfies a suitable property techniques does not rule out such cases), and we found unnecessary to symmetrize the weight in the implementation.

Then, we build the graph $\mathcal{G} = (\mathcal{V}, \mathcal{E})$ made of vertices $\mathcal{V} = \{i, i = 1, ..., N\} \cup \{t\} \cup \{s\}$ and whose edges are

$$\mathcal{E} = \{(x,y)|\, w_{x,y} > 0\} \cup \{(s,x)|\, 1 \leq x \leq N\} \cup \{(x,t)|\, 1 \leq x \leq N\}.$$

As a cut of this graph define a partition $(\mathcal{V}_s, \mathcal{V}_t)$ of the graph into two sets, the first one containing the source and the second one containing the sink, the global cost of a cut is given by

$$E(\mathcal{V}_s, \mathcal{V}_t) = \sum_{\substack{e=(a,b)\in\mathcal{E} \\ a\in\mathcal{V}_s, b\in\mathcal{V}_t}} C(e).$$

So what we would like to realize is $E(\mathcal{V}_s, \mathcal{V}_t) = J(\theta)$. The construction is given by Kolmogorov in [12], it consists in assigning the weight $w_{x,y}$ to an edge $e = (x, y) \in \mathcal{E}$ in the image grid, the weight $\alpha + \max_i G_i$ to the edges $(s, x)$ and $\max_i G_i - G_i$ to the edges $(x, t)$, then the equality between the global cost and the energy holds.

## 5 Experimental Results

All the experiments whose results are presented here were performed on a laptop equiped with a 1.8GHz Pentium 4 and 1 Gb of RAM.

### 5.1 Experiments with Optical Flow

For the implementation, we have used the `maxflow-v2.1` and `energy-v2.1` graphcuts implementation of V. Kolmogorov, available at `http://www.cs.cornell.edu/People/vnk/software.html`. Type of capacities has been set to `double`, though `short` or `int` leads to faster computation when quantized quantities are chosen in input.

The optical flow is computed by the Weickert and Schnrr method [21] with a multiresolution procedure (see [22]). As optical flow computation has been improved since the Weickert and Schnrr spatiotemporal model (using mixed model combining local and global information, using intensity or gradient intensity...), we emphasize that our purpose is not to obtain a very precise estimation of the optical flow but to show how we can improve this with the $g$-weighted term and thus to obtain a segmentation as close as possible to the image edges. Figure 1 shows results obtained successively with $TV_{1,g}$ and $TV_{1,g,\frac{\pi}{4}}$. Parameters are chosen from previous computations with classical snakes (see [5]). The values are set in relation with the range of value of the optical flow amplitude. For $TV_{1,g,\frac{\pi}{4}}$, the computational time is of 0.11, 0.12 or 0.13 second. Same times are obtained with $TV_{1,g}$, though it can reach 0.09 or 0.10 second on some images.

The method could also be applied in the same way to a functional that was used by Jehan-Besson, Barlaud and Aubert in [9] for video segmentation purpose (actually it even inspired the work [5])

$$J(\Omega) = \int_\Omega \alpha\, dx + \int_\Omega |B - I|(x)\, dx + \lambda \int_{\partial\Omega} dS$$

where $B$ represent a background image and $I$ the current image in the movie. In the discrete formalism which is used in this paper, it gives

$$\sum_i (\alpha - |B - I|(i))\theta_i + \lambda TV_1(\theta).$$

**Fig. 1.** Results obtained with graphcuts with the energy involving $TV_{1,g}$ (first image on top left) and $TV_{1,g,\frac{\pi}{4}}$ (top right). The initial data is the optical flow norm $\mathbf{v}$. Parameters are $\alpha = 0.6$, $\lambda = 0.2$ and $\mu = 10$.

**Fig. 2.** Results obtained (10th image of the sequence) with total variation minimisation (Manhattan with horizontal, vertical and diagonal neighbors $TV_{1,g,\frac{\pi}{4}}$) for the Jehan-Besson–Aubert–Barlaud model (initial data: $|B - I|$). Parameters are $\lambda = 50$. From left to right: $\alpha = 10, 15$.

In this case it is a direct application of the previous work [1] (as before we have to modify the perimeter to a Manhattan perimeter). The background can be computed using more or less sophisticated methods. We tried time median filter and the method proposed by Kornprobst, Deriche and Aubert [8]. Some results are shown on figure 2 for $\alpha = 10, 15$ and $\lambda = 50$.

Let us explain some technical aspects. Iterations were stopped when the maximum of the two residues between $\xi^n$ and $\xi^{n+1}$ and between $\eta^n$ and $\eta^{n+1}$ become lower than 0.002, a maximum of 2000 iterations being set to prevent the algorithm to become too slow. The time step is $\tau = 0.1$. Such a value could be quite high, as we have indicated the time step should be lower than $\frac{1}{8 \max_{i,j} g_{i,j}}$, but a simple trick is to write $g = \tilde{g} \max g$, which changes the regularization parameter

**Fig. 3.** Results obtained (10th image of the sequence) with $TV_{1,g}$ and $TV_{1,g,\frac{\pi}{4}}$ and the optical flow norm as initial data. Parameters are $\alpha = 0.6$, $\lambda = 0.2$ and $\mu = 10$. Notice the smoothness of the result on the right image in comparison to the one on the left image.

from 1 to $\max g$, and thus has no incidence over the time step condition, as this one does not depend on the regularization parameter. Typical computational time is of $300s$. The results are shown on figure 3.

## 6 Conclusion

In this paper, we have extended the main result of [1] in order to handle shape optimization functionals involving weighted anisotropic perimeter. It states that all the solutions of some shape optimization problems depending on a parameter $\alpha$ are $\alpha$-level sets of the solution of the Rudin-Osher-Fatemi problem. Thus the algorithm used for total variation regularization — as in [1] — allows to compute all the solutions for different values of $\alpha$ in one pass. This is in our opinion the main advantage over classical snakes methods like Chan and Vese one in this particular type of shape optimization.

On the other hand, we have also minimised the discrete version of the functional with graph cuts techniques. The main advantage of this method is that it is very fast, and whenever the advantage of the TV-minimization algorithm does not occur when we employ graph cuts, even a great number of computations of the algorithm leads to a very competitive computational time (close to a single computation of a classical continuous snake algorithm).

We have used these both methods on two video segmentation models : one introduced in [5] in which weighted perimeter is involved and a previous one introduced by Jehan-Besson, Barlaud and Aubert [9]. We would like to emphasize that the general model studied in the theoretical part of the paper covers many applications. One could think for example about segmentation with shape priors, using a perimeter weighted by a distance to the prior. Such models have been used by Freedman and Zhang [23], or by Gastaud, Jehan-Besson, Barlaud and Aubert [24]...

**Acknowledgements.** Antonin Chambolle is partially funded by the A.C.I. "MULTIM" (Fonds National de la Science).

## References


1. Chambolle, A.: Total variation minimization and a class of binary mrf models. In Rangarajan, A., Vemuri, B., Yuille, A.L., eds.: Proceedings of the 5th International Workshop on Energy Minimization Methods in Computer Vision and Pattern Recognition. Volume 3757 of LNCS. (2005) 136–152
2. Darbon, J., Sigelle, M.: Exact optimization of discrete constrained total variation minimization problems. In Klette, R., Zunic, J., eds.: Tenth International Workshop on Combinatorial Image Analysis. Volume 3322 of LNCS. (2004) 548–557
3. Darbon, J., Sigelle, M.: A fast and exact algorithm for total variation minimization. In Marques, J.S., de la Blanca, N.P., Pina, P., eds.: 2nd Iberian Conference on Pattern Recognition and Image Analysis. Volume 3522 of LNCS. (2005) 351–359
4. Chambolle, A.: An algorithm for total variation minimization and applications. Journal of Mathematical Imaging and Vision **20** (2004) 89–97
5. Ranchin, F., Dibos, F.: Segmentation des objets en mouvement par utilisation du flot optique. ORASIS 2005 Proceedings (2005)
6. Bouthemy, P., Heitz, F., Prez, P.: Multiscale minimization of global energy functions in some visual recovery problems. CVGIP : Image understanding **59** (1989) 191–212
7. Deriche, R., Paragios, N.: Geodesic active regions for motion estimation and tracking. Proceedings of the Int. Conf. in Computer Vision (1999) 224–240
8. Aubert, G., Deriche, R., Kornprobst, P.: Image sequence analysis via partial differential equations. Journal of Mathematical Imaging and Vision **11** (1999) 5–26
9. Aubert, G., Barlaud, M., Jehan-Besson, S.: Video object segmentation using eulerian region-based active contours. In: International Conference in Computer Vision Proceedings, Vancouver, Canada (2001)
10. Bouchard, L., Corset, I., Jeannin, S., Marqus, F., Meyer, F., Morros, R., Pards, M., Marcotegui, B., Salembier, P.: Segmentation-based video coding system allowing the manipulation of objects. IEEE Transactions on Circuits and Systems for Video Technology, (RACE/MAVT and MORPHECO Projects) **7** (1997) 60–74
11. Boykov, Y., Veksler, O., Zabih, R.: Fast approximate energy minimization via graph cuts. IEEE Transactions on Pattern Analysis and Machine Intelligence **23** (2001) 1222–1239
12. Kolmogorov, V., Zabih, R.: What energy functions can be minimized via graph cuts? IEEE Transactions on Pattern Analysis and Machine Intelligence (PAMI) (2004)
13. Rudin, L., Osher, S., Fatemi, E.: Non linear total variation based noise removal algorithms. Physica D **60** (2002) 259–268
14. Dibos, F., Koepfler, G.: Global total variation minimization. SIAM Journal of Numerical Analysis **37** (2000) 646–664
15. Chan, T., Esedoglu, S., Nikolova, M.: Algorithms for finding global minimizers of image segmentation and denoising models. Technical Report CAM 04-07, UCLA (2004)
16. Delfour, M., Zolsio, J.P.: Shapes and Geometries. SIAM, Philadelphia, PA, Advances in Design and Control (2001)

17. Osher, S., Sethian, J.: Fronts propagating with curvature dependent speed: Algorithms based on the hamilton-jacobi formulation. Journal of Computational Physics **79** (1990) 12–49
18. Greig, D., Portehous, B., Seheult, A.: Exact maximum a posteriori estimation for binary images. Journal of the Royal Statistical Society **51** (1989) 271–279
19. Kolmogorov, V., Zabih, R., Gortler, S.: Generalized multi-camera scene reconstruction using graph cuts. EMMCVPR 03 Proceedings (2003)
20. Boykov, Y., Jolly, M.P.: Interactive organ segmentation using graph cuts. Medical Image Computing and Computer-Assisted Intervention (2000) 276–286
21. Weickert, J., Schnrr, C.: Variational optic flow computation with a spatio-temporal smoothness constraint. Journal of Mathematical Imaging and Vision **14** (2001) 245–255
22. Mémin, E., Prez, P.: A multigrid approach for hierarchical motion estimation. In: Proceedings of the 6th International Journal of Computer Vision, IEEE Computer Society Press (1998) 933–938
23. Freedman, D., Zhang, T.: Interactive graph cut based segmentation with shape priors. IEEE Computer Society Conference on Computer Vision and Pattern Recognition (CVPR) **1** (2005) 755–762
24. Gastaud, M., Jehan-Besson, S., Barlaud, M., Aubert, G.: Region-based active contours using geometrical and statistical features for image segmentation. In: Proceedings of the IEEE International Conference in Image Processing. Volume II. (2003) 643–646

# Curve Evolution in Subspaces

Aditya Tatu[1], François Lauze[2], Mads Nielsen[1,2], and Ole Fogh Olsen[3]

[1] DIKU, University of Copenhagen, Universitetsparken 1, 2100 Copenhagen East, Denmark
{aditya,francois,madsn}@diku.dk
[2] Nordic Bioscience Imaging A/S, Herlev Hovegade 207, 2370 Herlev, Denmark
{francois,madsn}@nordicbioscience.com
[3] IT University of Copenhagen, Rued Langgaardsvej 7, 2300 Copenhagen South, Denmark
fogh@itu.dk

**Abstract.** Curve evolution forms the basis of active contour algorithms used for image segmentation. In many applications the curve under evolution needs to be restricted to the shape space given by some example shapes, or some linear space given by a set of basis vectors. Also, when a curve evolution is carried out on a computer, the evolution is approximated by some suitable discretization. Here too, the evolution is implicitly carried out in some subspace and not in the space of all curves. Hence it is important to study curve evolution in subspace of all curves. We look at a formulation that describes curve evolution restricted to subspaces. We give numerical methods and examples of a formulation for curvature flow for curves restricted to the B-spline subspace.

## 1 Introduction

Curve evolution methods have been used for several applications like propagating interfaces, object segmentation, image inpainting etc. [5,9]. They arise in minimization problems for which an objective function $\mathbf{C} \mapsto \mathcal{E}(\mathbf{C})$ cannot be directly optimized, and we need an evolution scheme to reach an optimal point of ths cost function via say, a gradient descent approach:

$$\frac{\partial \mathbf{C}}{\partial t} = \pm \frac{\partial \mathcal{E}}{\partial \mathbf{C}}. \tag{1}$$

Whether arising from a descent formulation or from some sort of "direct design", the most common type of evolution equation studied has the form

$$\frac{\partial \mathbf{C}}{\partial t}(p,t) = V(p,t)\mathbf{N}(p,t) \tag{2}$$

where the scalar function $V(p,t)$ is the velocity giving the amount of deformation of the curve along the inner unit normal $\mathbf{N}(p,t)$ to the closed curve $\mathbf{C}$, $p$ is the parameter of the curve and $t$ is the time of evolution [5]. $V$ is chosen depending on the task to be performed. Depending on the flow chosen, the initial curve

F. Sgallari, A. Murli, and N. Paragios (Eds.): SSVM 2007, LNCS 4485, pp. 754–764, 2007.

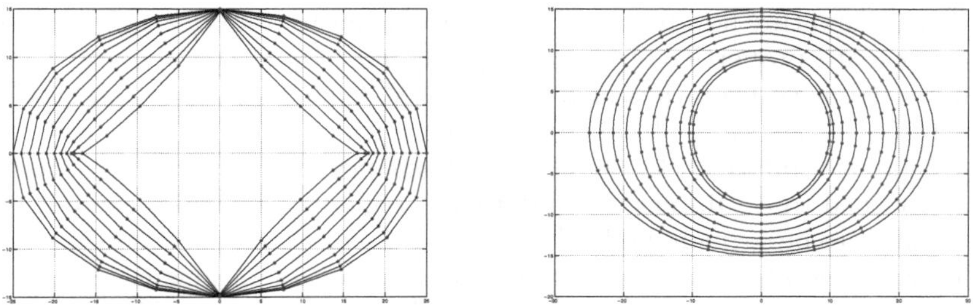

**Fig. 1.** (left) Curvature flow of points on an ellipse. The ellipse is approximated as a polygon. The properties of the curve like the curvature are computed using the central difference derivatives. (right) Curvature flow of the same points on the ellipse. The curve is approximated by $3^{rd}$ order B-splines and evolution is carried out using our approach. We can see that the flow shown in the left, is erroneous, and the one to the right is as expected, evolving towards a round point.

changes to other intermediate curves in some "space of all curves with a given regularity" $\Gamma$, assumed large enough to be able to define generic evolutions.

This generic space $\Gamma$ is often not clearly specified, and, a somehow paradoxical situation, the evolution equations are restricted to to a given much more specific subspace of the generic space $\Gamma$: some application may require all intermediate curves to lie in some given subspace specified by some set of shapes, for instance the space of all curves of a human right hand boundary and its simplifications [3]. In fact, in any case, while implementing the partial differential equation (PDE) (2) on a computer, one has to discretize it, and hence the evolution is implicitly carried out in some *finite dimensional* (linear or not) subspace of the generic space $\Gamma$. One often used approach is to approximate the curve evolution and neglect the fact that the intermediate curves obtained in the discrete implementation may be different than the actual intended curve evolution. The subspace, obviously depends on the type of discretization used. The evolution results too. This fact is illustrated in Figure 1.(left), which depicts the well known (see [1]) mean curvature flow, starting here with an ellipse, and where curvature is approximated by central difference derivatives, and the curve is approximated by a polygon with given vertices. On its right, the mean curvature flow of the same set of points, considering a $3^{rd}$ order B-spline interpolation, with our approach is shown. Note that, in both cases, we evolve just the vertex or control points marked as '*'. Clearly, it advocates study of curve evolution in subspaces.

In this paper, we formulate the curve evolution equations, given the fact that we need to restrict the curves to some subspace of all curves. Here, we work with finite dimensional linear subspaces with a Hilbertian structure. The subspace may be specified by giving some example shapes from the subspace, or by specifying the basis vectors of the subspace. We do not impose the restriction of orthogonality on the basis vectors.

The main object of this paper is not new. Several proposal in order to restrict curve evolution to some particular subspace are given in [10], where the authors restrict the curvature flow to curves with a given constant area, [7] gives methods to approximate properties of curves represented as planar polygons, in order to carry out curve evolution. There is related work in the area of statistical shape priors in active contours. In [2], the authors restrict the curve in the linear shape space formed by some example shapes. In [14], the active contours are constrained to a subspace having desired regional properties, similar to the object that is to be segmented. In this work, we focus howver on a certain numver of points: a general formulation, some time-step considerations in the case of a descent equation as well as a rigorous analysis in the case of uniform B-splines.

This paper is organized as follows. In section 2, we describe the problem in general. We then discuss in Section 3 time steps in actual implementations when the evolution is the gradient descent from a cost function, restricted to a finite dimensional subspace. In section 4 we take up an example of a particular subspace, namely the B-spline subspace, and give details of our method. We give experimental results in section 5 and conclude in section 6.

## 2 Evolution in Subspaces

We consider the "space of all reasonable closed planar curves" $\Gamma$, we could consider the space of continuous maps $\mathbf{C} : \mathbb{S}^1 \to \mathbb{R}^2$ where $\mathbb{S}^1$ is the unit circle such that we can define a inner unit vector along $\mathbf{C}$ almost everywhere, we endow it with the standard scalar product

$$\langle \mathbf{C}_1, \mathbf{C}_2 \rangle_\Gamma = \int_{\mathbb{S}^1} \langle \mathbf{C}_1(p), \mathbf{C}_2(p) \rangle_{\mathbb{R}^2} \, dp$$

where $\langle -, - \rangle_{\mathbb{R}^2}$ is the usual scalar product on $\mathbb{R}^2$ and silently ignore completness questions[1]. Instead of parameterizing closed curves $\mathbf{C}$ on $\mathbb{S}^1$, they are often parameterized on closed ranges $[\alpha, \beta]$ with the added requirement that they take equal values at endpoints $\mathbf{C}(\alpha) = \mathbf{C}(\beta)$.

Let $\Delta \subset \Gamma$ a given subspace. Then in general, for a curve $\mathbf{C}_0 = \mathbf{C}(-, t_0) \in \Delta$, the solution $\mathbf{C}(-, t)$ of equation (2) is not guarantied to remain in $\Delta$ unless at each time $t$, the deformation $V\mathbf{N}(-, t)$ is tangent to $\Delta$ at $\mathbf{C}(-, t)$. Hence it is clear that it is not possible to exactly follow the given curve evolution in a subspace, unless the velocity vector forces the curve to stay in the given subspace. So the best alternative is to follow the flow as accurately as possible, i.e., to project a deformed curve or back into the desired subspace, or the velocity vector in its tangent subspace. The Classical Projection theorem [6] states that the best approximation to a given vector is its orthogonal projection into the desired subspace, when this subspace is linear. We follow this principle in our formulation. Given a curve evolution scheme as in equation (2), our approach is to carry out the corresponding projected evolution given as

$$\frac{\partial \mathbf{C}}{\partial t} = proj(V\mathbf{N}) \tag{3}$$

[1] Completion in $L^2$ will lead to non continuous curves, although restriction to final dimensional subspaces makes the problem essentially void.

where $proj(.)$ is the projection operator on the tangent space $T\Delta$ of the given subspace $\Delta$.

The projection operator could depend on the curve $\mathbf{C}(-,t)$ if the subspace is non-linear. In this paper, we will nevertheless only consider the case where $\Delta$ is a linear subspace of $\Gamma$ and thus equal to its tangent space. Using Forward difference scheme for the evolution time $t$, letting $C^i$ denote $C(-,i\delta t)$ where $\delta t$ is a small time step, we can write a discrete counterpart to equation (3) as

$$\mathbf{C}^{i+1} = \mathbf{C}^i + proj(V^i\mathbf{N}^i)\delta t. \tag{4}$$

We either project the velocity $V^i\mathbf{N}^i$ onto our subspace or allow our curve to deform and move out of the subspace at every iteration, then we project it back into the desired subspace, the two operations are equivalent in the linear case. Let $\{B_k\}_{k=0,1..N-1}$ be the set of basis vectors of the subspace. A curve $C$ in the subspace is represented as $\mathbf{C} = \sum_{k=0}^{N-1} a_k\ B_k$. In order to evolve this curve according to equation (2) in the subspace, we use,

$$\mathbf{a}^{i+1} = \mathbf{a}^i + G^{-1}.P_B(V^i\mathbf{N}^i)^t\delta t \tag{5}$$

where $P_B(V^i\mathbf{N}^i) = [< V^i\mathbf{N}^i, B_0 > < V^i\mathbf{N}^i, B_1 > \ \ldots \ < V^i\mathbf{N}^i, B_{N-1} >]$, $\mathbf{a}^i = [a_0^i a_1^i ... a_{N-1}^i]$ are the parameters of the curve at iteration $i$, $G^{-1}$ is the inverse Gram matrix [6] (also explained in section 4), required in case our basis vectors $B_k$ are not orthogonal, $< .,. >$ is the inner product and $\delta t$ is a suitable time step ($P_B^t$ indicates matrix transpose of $P_B$). The deformation is projected onto the tangent space of the desired subspace of all curves. If the subspace is linear then the tangent space can be identified with the subspace. The new curve is given as

$$C^{i+1} = \sum_{k=0}^{N-1} a_k^{i+1}\ B_k.$$

This problem is similar to constrained optimization problem over linear subspaces. It is shown in [4] that solving an optimization problem restricted to linear subspaces using Lagrange multipliers, is equivalent to solving the unconstrained problem and then projecting the solution orthogonally into the required subspace. The projection evolution will have a stationary point in case the evolution occurs in a direction orthogonal to the subspace. Two different curves trace out different evolution paths in the generic space $\Gamma$ when evolved according to the original equation, but may cross each other's trajectory when the evolution is carried out in some subspace, because of the projection. In case, we are given example shapes lying in a subspace, we compute the basis vectors by finding the eigenvectors of the covariance matrix, after aligning the shapes.

## 3 Time Step and Cost Function Descent

Tuning the time-step $\delta t$ is extremely important for numerical stability. It may give rise to very complex problems and adhoc solutions. In the rather common situation where the given curve evolution is a gradient descent on a cost/energy function, say $\mathcal{E}(\mathbf{C}(\mathbf{a}))$ where $\mathbf{C}$ is the curve, $\mathbf{a}$ are the coefficients of the basis vectors of the subspace $\Delta$, we propose an adaptive choice for this time step based on the second order structure of the cost hypersurface. A standard and very important example the mean curvature flow which is a descent on the length of the curve. Then the curve evolution equation can be written as

$$\frac{\partial \mathbf{C}}{\partial t} = -\frac{\partial \mathcal{E}(\mathbf{C}, t)}{\partial \mathbf{C}} = F(\mathbf{C}, t)$$

Using the Forward difference scheme,

$$\mathbf{C}^{i+1} = \mathbf{C}^{i} + F(\mathbf{C}, t)\delta t$$

Now if the cost function is purely quadratic, then the gradient is linear and one can fix an optimal value of the time step for the evolution. But for example in the curvature flow, the cost function is highly non-linear. The time-step has to be adapted to the shape of the cost function at each iteration. The time-step suggested in such cases is,

$$\delta t = \frac{1}{max(V\mathbf{N})}$$

When employed in our approach to curve evolution, it gives rise to oscillations in the curve at certain stage in the evolution. To overcome this, we use the information about the shape(curvature) of the cost function. Hessian of the cost function gives the information about the curvature of the cost function(surface). We would like to take small steps where the curvature of the cost function is high, and larger steps where it is low. The reason being that there are cost surfaces with different curvatures but similar gradient values near the optima. In such cases, if the curvature is not taken into account for the time-step the evolution may jump over the point of maxima/minima. So, one way to achieve this is to make the time step dependant on the inverse of hessian, $Hess(\mathcal{E}(\mathbf{C}, t))$. But the cost surface may not have the same curvature in all directions. Taking a conservative approach, we use the maximum curvature at the current point in evolution, to decide the time-step. The eigenvalues of the hessian matrix are the principle curvatures of the cost surface [8]. Hence we use the time-step given by

$$\delta t = \frac{1}{|\lambda_{max}|}. \tag{6}$$

In a somehow geometric formulation, we limit the evolution displacements to a displacement smaller than (equal to) the radius of the smallest osculatory circle along one of the principal curvature directions.

## 4 B-Spline Subspace

In this section, we derive the equations of curve evolution restricted to a given subspace, in details and apply it to the $3^{rd}$ order B-spline subspace. We take an example of a particular subspace of interest and a particular curve evolution equation and derive our formulation for it. The following formulation applies equally well to other examples of linear subspaces and other curve evolution equations. We restrict the curvature flow to the space of curves generated by $3^{rd}$ order B-spline basis functions. The $n^{th}$ order spline basis functions are given by

$$\beta^n(x) = \underbrace{\beta^0(x) * \cdots * \beta^0(x)}_{n \text{ times}}$$

where $\beta^0(x)$ is the $0^{th}$ order spline function, equal to 1 in $[-\frac{1}{2}, \frac{1}{2}]$ and 0 otherwise, and $*$ denotes convolution. Note that the spline basis are not orthogonal. If $n^{th}$ order spine is used, then the curve obtained belongs to $\mathcal{C}^{n-1}$ class of functions $n-1$ times continuously differentiable. This is important in cases where we need to carry out evolution of the curve with a velocity vector $V$ that depends on some differential property of the curve. For example, to carry out the curvature flow of a curve, it becomes important that we should be able to compute the first and second derivatives of the curve with respect to its parameters, analytically. For details regarding B-spline functions, we refer the reader to [11,12,13]. We now derive the evolution equation for curvature flow of a curve, restricted to $3^{rd}$ order B-spline subspace. The curvature flow being the gradient descent of the arclength

$$L(\mathbf{C}) = \int_{\alpha}^{\beta} |\mathbf{C}'(p)| \, dp$$

we will be able to use the results of the previous section. Let us denote the curvature flow by

$$\frac{\partial \mathbf{C}}{\partial t}(p,t) = \kappa \mathbf{N} \tag{7}$$

Let the initial curve be given as

$$\mathbf{C}^i(p) = \sum_{k=0}^{N-1} a_k^i B_k^n(p) = A^i.[B_0^n B_1^n .. B_{N-1}^n] \tag{8}$$

where $a_k$ are the spline basis coefficients and $B_k^n(p)|_{k=0,1,..N-1} = \beta^n(p-k)$ are the spline basis, $n$ representing the order of the spline. The curve is specified by the user with $N$ node points. Let $y$ represent the curve obtained by evolving our curve according to the curvature flow for one iteration, i.e.

$$y = \mathbf{C}^i(p) + \kappa \, \mathbf{N}\delta t$$

Now, our task is to project the curve $y$, into the subspace orthogonally. Let the best approximate of $y$ be represented as $\hat{y} = p_0 B_0^n + p_1 B_1^n + ... + p_{N-1} B_{N-1}^n$. We need to compute the coefficients $p_0, p_1, ..., p_{N-1}$. From the projection theorem we know that the difference vector $(y - \hat{y})$ is orthogonal to the B-spline subspace. Therefore $< (y - \hat{y}), B_i^n|_{i=0,..,N-1} >= 0$. These conditions give us $N$ linear equations, which can be written as,

$$\begin{pmatrix} <y|B_0^n> \\ . \\ . \\ <y|B_{N-1}^n> \end{pmatrix} = \begin{pmatrix} <B_0^n|B_0^n> & .. & <B_0^n|B_{N-1}^n> \\ . & .. & . \\ . & .. & . \\ <B_{N-1}^n|B_0^n> & .. & <B_{N-1}^n|B_{N-1}^n> \end{pmatrix} \begin{pmatrix} p_0 \\ . \\ . \\ p_{N-1} \end{pmatrix}$$

i.e.,

$$P_B(y) = G.P$$

Therefore,

$$P = G^{-1}.P_B(y)$$

The matrix $G$ is known as the *Gram matrix*. The elements of this matrix, for the B-spline subspace are given as

$$G(i,j) = \int_{-\infty}^{\infty} \beta^n(p-i)\beta^n(p-j)\ dp = \int_{-\infty}^{\infty} \beta^n(p-(i-j))\beta^n(p)\ dp.$$

$G$ can be computed beforehand, given the number of node points and the order of spline and is a symmetric matrix. $\beta^n(p) = 0$ for $|p| > \frac{n}{2}$. So for computing $G$, we can take into account the small support of the spline basis functions (if the basis vectors were orthonormal then then Gram matrix $G$ would be the identity matrix, this is for instance the case with Fourier descriptors). The vector $P_B(y)$ can be computed as

$$P_{B_k}(y)|_{k=0,1...N-1} = \int_{-\infty}^{\infty} y(p)B_k^n(p)|_{k=0,1...N-1}dp$$

(integration of course in reality limited to the support of $B_k^n$). In our case, $B_k^n(p) = \beta^n(p-k)$ and $y(p) = \kappa\mathbf{N}(p)$. The projection gives the change in the spline coefficients of the curve

$$A^{i+1} = A^i + P\delta t \tag{9}$$

where $\delta t$ is the time-step. From the new spline coefficients, we can compute the evolved curve at each iteration using equation (8). This method incorporates curvature information in a neighborhood of each node point, thereby giving a better approximation. Although we have explicitly used the spline subspace, this formulation will clearly apply to any finite dimensional linear subspace.

The next step is to decide the time-step in equation (9). We have tried with the well known time-step

$$\delta t < \frac{1}{max(\kappa\mathbf{N})}. \tag{10}$$

But, as we will show in the next section, this leads to oscillations in the curve evolution. As we reduce the time-step, the oscillations enter into our evolution at a later time. In order to overcome this problem, we use the Hessian based approach that we proposed in the previous section. The curvature flow is the steepest descent of the arclength, which for a spline curve is given as

$$\mathcal{E}(C(\mathbf{a},p)) = \int_{-(n+1)/2}^{(n+1)/2} dp \sqrt{(\sum_k a_{x_k}(B_k^n(p))')^2 + (\sum_k a_{y_k}(B_k^n(p))')^2}$$

where **a** are the spline coefficients, $(a_x, a_y)$ are the coefficients for the x and y co-ordinates respectively, $(B_k^n(p))'$ are the first order derivatives of the B-spline basis with respect to the parameter $p$. For details we refer the reader to [12]. If we minimize this cost function with respect to the spline coefficients **a**, we get, as expected,

$$\left[\frac{\partial \mathcal{E}}{\partial a_{x_m}}, \frac{\partial \mathcal{E}}{\partial a_{y_m}}\right] = \int_{-(n+1)/2}^{(n+1)/2} \kappa \mathbf{n}[B_m^n(p), B_m^n(p)]dp =< \kappa \mathbf{n}, B_m^n >$$

where **n** is the normal vector to the curve. This is consistent with our projection scheme. The hessian of the cost function with respect to the spline coefficients, is given as (dropping intgration bounds to keep notations simple)

$$Hess(\mathcal{E}) = \begin{pmatrix} \int \frac{C_x'^2(p)B_i'(p)B_j'(p)}{|C'(p)|^3} & \int \frac{-C_x'(p)C_y'(p)B_i'(p)B_j'(p)}{|C'(p)|^3} \\ \int \frac{-C_x'(p)C_y'(p)B_i'(p)B_j'(p)}{|C'(p)|^3} & \int \frac{C_y'^2(p)B_i'(p)B_j'(p)}{|C'(p)|^3} \end{pmatrix}$$

where, $C_y'(p), C_x'(p)$ and $B_j'(p)$ are first derivatives of x and y components of the curve and that of the B-spline basis with respect to the parameter $p$ of the curve and $i, j = 0, 1, ...N - 1$. They can be computed as shown in [12]. If the curve has $N$ node points, the Hessian is a $2N \times 2N$ matrix. Let $\lambda_{max}$ denote the maximum eigenvalue of the Hessian matrix. The time-step is given as,

$$\delta t = \frac{1}{|\lambda_{max}|} \tag{11}$$

In the Newton scheme, the gradient is multiplied by the inverse of the hessian matrix of the cost function. We take a more conservative approach by using the maximum eigenvalue of the hessian matrix. We take the lowest bound on the step in all directions. This scheme works much better then the one given by equation (10), as can be seen in the results in section 5. For numerical computations of all integrations, we use Riemann sums.

## 5 Experiments

First we consider the scheme where we evolve only the node points of the curve according to the curvature and normal at those points, in the B-spline subspace. As one can observe in Figure 5, the node points tend to lie on a straight line segment, thereby stopping further evolution.

Next we apply the projection algorithm, in which the time-step is computed according to equation (10). We see that oscillations creep into the curve at some stage. We use two different time steps, but still the oscillations creep into the curves. Figure 3 is obtained with time step $\delta t = \frac{1}{2\ max(\kappa \mathbf{N})}$ and Figure 4 is obtained with time step $\delta t = \frac{1}{5\ max(\kappa \mathbf{N})}$.

In Figure 5, we use the time step given by equation (11). Here the oscillations occur after evolving the curve for a long time as compared to the first scheme.

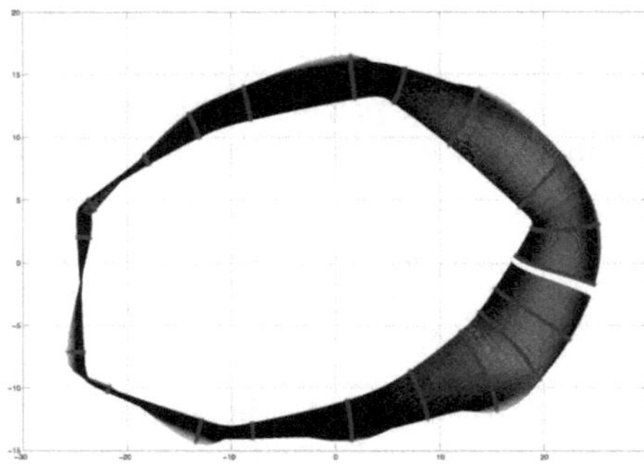

**Fig. 2.** Curvature flow, by evolving only the node points of the spline curve. Here the curve is represented by $3^{rd}$ B-splines, hence the properties of the curve like curvature etc., are accurately computed. But still the evolution takes the node points to lie on a straight line, after which the evolution practically stops. This happens because the curvature information on the entire curve is not taken into account. In our formulation, the projection operator takes into account the curvature information in the neighborhood of the node points, as shown in the previous section.

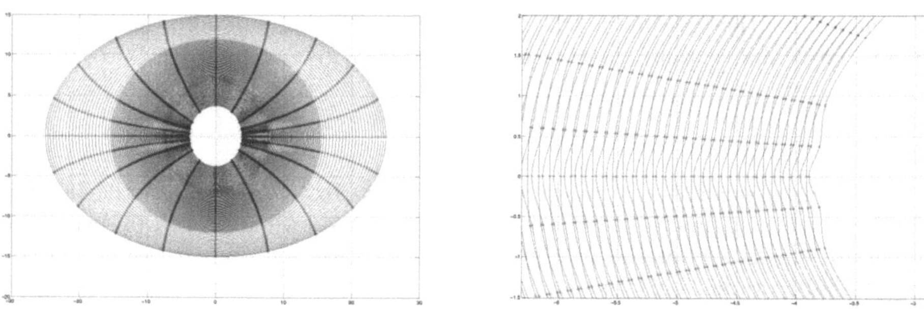

**Fig. 3.** (left) Curvature flow of points on an ellipse, using time step given in Equation 10. (right) Magnified view of the figure on left, note the oscillations in the curve.

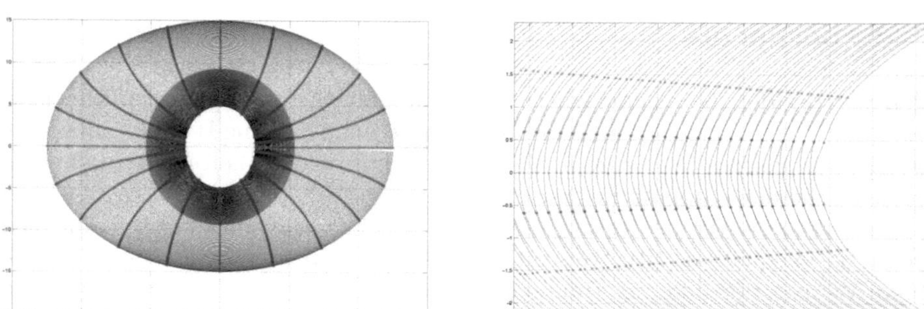

**Fig. 4.** (left) Curvature flow of points on an ellipse, using time step given in Equation 10. (right) Magnified view of the figure on left, note the oscillations. Here the time step is still smaller then the one used in Figure. 3

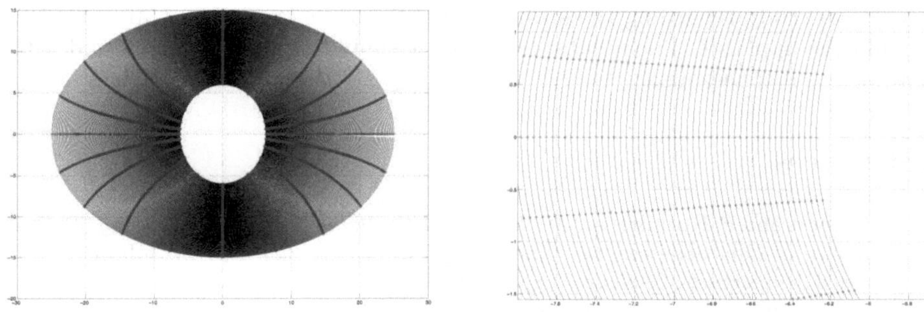

**Fig. 5.** (left) Curvature flow of points on an ellipse, using time step given in Equation 11. (right) Magnified view of the figure on left, note that there are no oscillations.

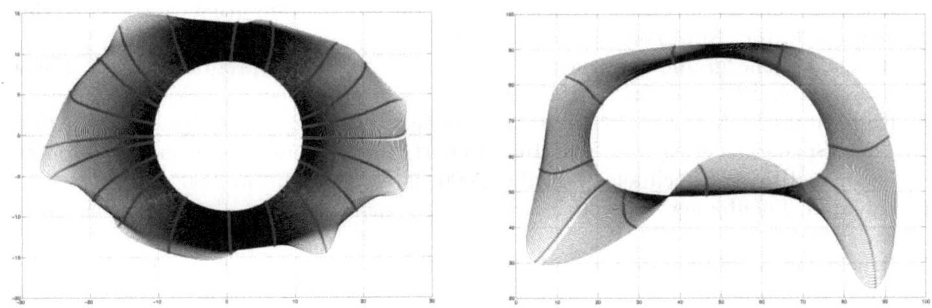

**Fig. 6.** (left) Curvature flow of a curve, using our approach, using time step given in Equation 11. (right) Curvature flow of a curve, using our approach, using time step given in Equation 11. These are couple of examples of curves having convex and concave parts.

Also, in this case, the oscillations may be because of the fact that the node points on the curve come too close to each other, as discussed in [2].We show couple of other examples in Figure 6.

## 6 Conclusion

We have taken a different view on curve evolution. Generally, people try to approximate the continuous PDE of curve evolution while implementing it, but we consider the fact that the PDE cannot be accurately followed and thus we try to best approximate the flow. This approach is also useful when there is a constraint desiring the curve to remain in a specified subspace. We have given experimental results for the B-spline subspace. The results show that the numerical problems with discretization have been taken care of to a large extent. We also have given a numerical scheme for the time step for the curve evolution in the B-spline subspace, which is a variant of the Newton scheme. Further theoretical study regarding the stability of the scheme is needed. Currently the formulation applies to linear subspaces, but it should be possible to extend it to non-linear subspaces also.

## References


1. G. Aubert and P. Kornprobst. *Mathematical Problems in Image Processing: Partial Differential Equations and the Calculus of Variations, 2nd Edition*, volume 147 of *Applied Mathematical Sciences*. Springer-Verlag, 2006.
2. Daniel Cremers, Florian Tischhauser, Joachim Weickert, and Christoph Schnorr. Diffusion snakes: Introducing statistical shape knowledge into the mumford-shah functional. *International Journal of Computer Vision*, 50(3):295–313, December 2002.
3. U. Grenander, Y. Chow, and D. M. Keenan. *Hands: A Pattern Theoretic Study of Biological Shapes*. Springer Research Notes In Neural Computing. Springer-Verlag New York, Inc. New York, NY, USA, 1991.
4. Rosen J. The gradient projection method for nonlinear programming, i, linear constraints. *SIAM J.*, 8:181–217, 1960.
5. Ron Kimmel. *Numerical Geometry of Images, Theory, Algorithms and Applications*. Springer-Verlag New York,Inc., Department of Computer Science, Technion-Israel Institute of Technology, Haifa 32000, Israel, 1st edition, 2004.
6. David G. Luenberger. *Optimization by vector space methods*. John Wiley & Sons, Inc., 1969.
7. Bruckstein A. M., Sapiro G., and Shaked D. Evolutions of planar polygons. *International Journal of Pattern Recognition and Artificial Intelligence*, 9(6):991–1014, 1995.
8. Andrew Pressley. *Elementary Differential Geometry*. Springer Undergraduate Mathematics Series. Springer-Verlag London Ltd., Department of Mathematics, King's College, The Strand, London WC2R 2LS, UK, first edition, 2002.
9. James A. Sethian. *Level Set Methods*. Cambridge Monograph on Applied and Computational Mathematics. Cambridge University Press, 1st edition, 1996.
10. B. M. ter Haar Romeny, editor. *Geometry-Driven Diffusion in Computer Vision*. Kluwer Academic Publishers, 1994.
11. Michael Unser. Splines: A perfect fit for signal and image processing. *IEEE Signal Processing Magazine*, pages 22–38, November 1999.
12. Michael Unser, Akram Aldroubi, and Murray Eden. B-spline signal processing: Part 1 - theory. *IEEE Transactions on Signal Processing*, 41(2):821–832, February 1993.
13. Michael Unser, Akram Aldroubi, and Murray Eden. B-spline signal processing: Part 2 - efficient design and applications. *IEEE Transactions on Signal Processing*, 41(2):834–848, February 1993.
14. Wang X., He L., and Wee W. G. Deformable contour method: A constrained optimization approach. *International Journal of Computer Vision*, 59(1):87–108, August 2004.

# Identification of Grain Boundary Contours at Atomic Scale

Benjamin Berkels[1], Andreas Rätz[2], Martin Rumpf[1], and Axel Voigt[2,3]

[1] Institut für Numerische Simulation,
Rheinische Friedrich-Wilhelms-Universität Bonn,
Nussallee 15, 53115 Bonn, Germany
{benjamin.berkels,matrin.rumpf}@ins.uni-bonn.de
http://numod.ins.uni-bonn.de/
[2] Crystal Growth Group, Research Center caesar
Ludwig-Erhard-Allee 2, 53175 Bonn, Germany
[3] Institut für Wissenschaftliches Rechnen,
Technische Universität Dresden,
01062 Dresden, Germany

**Abstract.** Nowadays image acquisition in materials science allows the resolution of grains at atomic scale. Grains are material regions with different lattice orientation which are typically not in equilibrium. At the same time, new microscopic simulation tools allow to study the dynamics of such grain structures. Single atoms are resolved in the experimental and in the simulation results. A qualitative study of experimental images and simulation results and the comparison of simulation and experiment requires the robust and reliable extraction of mesoscopic properties from these microscopic data sets. Based on a Mumford–Shah type functional, grain boundaries are described as free discontinuity sets at which the orientation parameter for the lattice jumps. The lattice structure itself is encoded in a suitable integrand depending on the local lattice orientation. In addition the approach incorporates solid–liquid interfaces. The resulting Mumford–Shah functional is approximated with a level set active contour model following the approach by Chan and Vese. The implementation is based on a finite element discretization in space and a step size controlled gradient descent algorithm.

## 1 Introduction

The goal of this paper is to present a method for joint segmentation and orientation classification in materials science. For many problems in materials science on an atomic microscale, it is essential to link the underlying atomic structure to the material properties (electrical, optical, mechanical, etc.). The actual material properties are usually determined on a mesoscopic length scale on which non–equilibrium structures exist, which form and evolve during material processing. For example, the yield strength of a polycrystal varies with the inverse square of the average grain size. Grains are material regions with different lattice orientation which are typically not in equilibrium. Experimental tools such as TEM

F. Sgallari, A. Murli, and N. Paragios (Eds.): SSVM 2007, LNCS 4485, pp. 765–776, 2007.

**Fig. 1.** On a TEM image (left), light dots render atoms from a single atom layer of aluminum, in particular this image shows a $\Sigma 11(113)/[\bar{1}00]$ grain boundary [12] (courtesy of Geoffrey H. Campbell, Lawrence Livermore National Laboratory). Nowadays, there are physical models like the phase field crystal model which enable the numerical simulations of grains. Indeed, a time step from a numerical simulation (right) on the microscale shows an similar atomic layer. In both images, grain boundaries are characterized by jumps in the lattice orientation.

(transmission electron microscopy) [12] today allow measurements down to an atomic resolution (cf. Fig.1). A reliable extraction of grains and grain boundaries from these TEM-images is essential for an efficient material characterization. On the other hand, recent numerical simulation tools have been developed for physical models of grain formation and grain dynamics on the atomistic scale. Concerning such simulations, we refer to numerical results obtained from a phase field crystal (PFC) model [10] derived from the density function theory (DFT) of freezing [23]. Its methodology describes the evolution of the atomic density of a system according to dissipative dynamics driven by free energy minimization. The resulting highly nonlinear partial differential equation of sixth order is solved applying a finite element discretization [20]. These simulations in particular will allow a validation of the physical models based on the comparison of mesoscopic properties such as the propagation speed of grain boundaries. The formation of grains from an undercooled melt happens on a much faster time scale than their subsequent growing and coalescence. The evolution of grain boundaries at later stages of the process is of particular interest. Fig. 1 and 2 show a comparison of experimental (TEM) and numerically simulated (PFC) single grain boundaries on the atomic scale and the nucleation of grains, respectively. As material properties result from microstructures the segmentation of these structures in experimental or simulation results is of utmost importance to material scientists. Currently, the post processing of experimental images and the pattern analysis is mostly based on local, discrete Fourier filtering. In this paper, we aim at a reliable extraction of grain boundaries and in addition of interfaces between the liquid and the solid phase. Thus, we apply a variational approach based on the description of the interfaces by level sets. Our focus is on the post processing of phase field simulation results, but we will as well demonstrate the applicability of our approach to experimental images.

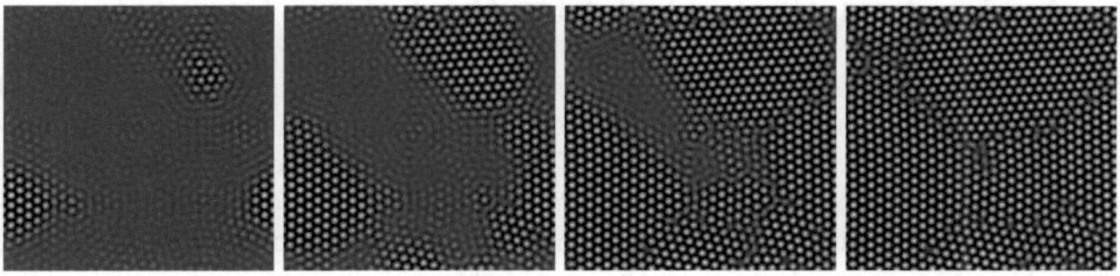

**Fig. 2.** Nucleation of grains in a phase field crystal simulation

Image classification has extensively been studied in the last decades. It consists of assigning a label to each point in the image domain and is one of the basic problems in image processing. Classification can be based on geometric and on texture information. Many models have been developed either based on region growing [24,19,6], on statistical approaches [4,5,14,15], and in particular recently on variational approaches [3,1,8,16,26].

The boundaries of the classified regions can be considered as free discontinuity sets of classification parameters, which connects the problem with the Mumford–Shah approach [17] to image segmentation and denoising. A robust and efficient approximation of the Mumford–Shah functional has been presented by Chan and Vese [7] for piecewise constant image segmentation and extended to multiple objects segmentation based on a multiphase approach [25]. Thereby, the decomposition of the image domain is implicitly described by a single or multiple level set functions (for a review on level sets we refer to [18,22]). In [21], their approach has been further generalized for the texture segmentation using a directional sensitive frequency analysis based on Gabor filtering. Texture classification based on the energy represented by selected wavelet coefficients is investigated in [2]. Inspired by the work of Meyer [16] on cartoon and texture decomposition, the classification of geometrical and texture information has been investigated further in [3]. There a logic classification framework from [21] has been considered to combine texture classification and geometry segmentation. A combination of level set segmentation and filter response statistics has been considered for texture segmentation in [11]. For a variational texture segmentation approach in image sequences based on level sets we refer to [9].

Our method to be presented here differs to the best of our knowledge significantly from other variational approaches in the literature. Our focus is not on a general purpose texture classification and segmentation tool but on the specific application in materials science. Texture segmentation can be regarded as a two–scale problem, where the microscale is represented by the structure of the texture and the macroscale by the geometric structure of interfaces between differently textured regions. In this sense, we have strong a priori knowledge on the geometric structure of the texture on the microscale and incorporate this directly into the variational approach on the macroscale. Thus, the scale separation is more direct than in other approaches based on a local, direction sensitive frequency analysis.

## 2 A Mumford–Shah Model for the Lattice Orientation

We consider a single atom layer resolved on the microscale. In the phase field simulation results as well as in the experimental images, single atoms are represented by blurry, dot-like structures. These dots are either described via the image intensity of the TEM image or the phase field function from the simulation. Let us denote this intensity or phase field by a function $u : \Omega \subset \mathbb{R}^2 \to \mathbb{R}$, where $\Omega$ is the image domain or the computational domain, respectively. Furthermore, we introduce the lattice orientation as a function $\alpha : \Omega \subset \mathbb{R}^2 \to \mathbb{R}$. In addition we consider the decomposition of the domain $\Omega$ into a solid phase $\Omega_S$ and a liquid phase $\Omega_L$. The solid domain $\Omega_S$ is further partitioned into grains, each of them characterized by a constant lattice orientation $\alpha$. The grain boundaries form the jump set of the orientation function $\alpha$. In the following section, we will first introduce a Mumford–Shah type model for the segmentation of grains and then expand this model to incorporate liquid–solid interfaces as well.

### 2.1 Segmenting Grain Boundaries

As already discussed in the introduction, grains are characterized by a homogeneous lattice orientation. At first, let us suppose that there is no liquid phase. Thus, the whole domain $\Omega$ is partitioned into grain subdomains. The lattice is uniquely identified by a description of the local neighborhood of a single atom in the lattice. In a reference frame with an atom at the origin, the neighboring atoms are supposed to be placed at positions $q_i$ for $i = 1, \cdots, m$, where $m$ is the number of direct neighbors in the lattice. In case of a hexagonal packing each atom has six direct neighbors at equal distances and we obtain

$$q_i := d\left(\cos\left(i\tfrac{\pi}{3}\right), \sin\left(i\tfrac{\pi}{3}\right)\right) \quad i = 1, \cdots, 6\,.$$

Here $d > 0$ denotes the distance between two atoms. If the lattice from the reference configuration is now rotated by an angle $\alpha$ and translated to a position $x$, the neighboring atoms are located at the positions $x + M(\alpha)q_i$ where $M(\alpha)$ is the matrix representation of a rotation by $\alpha$, i.e.

$$M(\alpha) := \begin{pmatrix} \cos\alpha & -\sin\alpha \\ \sin\alpha & \cos\alpha \end{pmatrix}.$$

Let us suppose that $\theta$ is a suitable threshold for the identification of the atom dots described via the function $u$ and define the indicator function

$$\chi_{[u>\theta]}(x) := \begin{cases} 1; & u(x) > \theta \\ 0; & \text{else} \end{cases}.$$

Then, for a given lattice orientation $\alpha$ and a point $x$ with $\chi_{[u>\theta]}(x) = 1$, we expect $\chi_{[u>\theta]}(x + M(\alpha)q_i) = 1$ as well for $i = 1, \cdots, m$. Let us suppose that the average radius of a single atom dot is given by $r$ and define the maximal lattice spacing $d{:=}\max_{i=1\cdots,m}|q_i|$. Next, we consider an indicator function $f:\Omega\times\mathbb{R}\to\mathbb{R}$ depending on the position $x$ and on a lattice orientation $\alpha$ and given by

$$f(x,\alpha)=\frac{d^2}{r^2}\chi_{[u>\theta]}(x)\Lambda\left(\left(\chi_{[u>\theta]}(x+M(\alpha)q_i)\right)_{i=1,\cdots,m}\right). \tag{1}$$

Here, $(\chi_i)_{i=1,\cdots,m}$ with $\chi_i{:=}\chi_{[u>\theta]}(x+M(\alpha)q_i)$ is the vector of translated and rotated characteristic functions and $\Lambda:\{0,1\}^m\to\mathbb{R}$ a function attaining its global minimum at $(1,\cdots,1)$ and $\Lambda(1,\cdots,1)=0$. The scaling $\frac{d^2}{r^2}$ ensures a uniform upper bound of order 1 (in particular independent of $d$ and $r$) for the integral of $f$ over $\Omega$. One easily verifies that $f(x,\alpha)=0$ if $x$ is inside a grain with orientation $\alpha$ and its distance to the grain boundary is at least $d$. Possible choices for $\Lambda$ are for instance

$$\Lambda(\chi_1,\cdots,\chi_m):=\frac{1}{m}\sum_{i=1,\cdots,m}(1-\chi_i),\quad\text{or} \tag{2}$$

$$\Lambda(\chi_1,\cdots,\chi_m):=1-\prod_{i=1,\cdots,m}\chi_i\,. \tag{3}$$

In the numerical experiments we found (2) to be the most suitable choice. The achieved accuracy in the spatial position of the grain boundary is of the order of the lattice parameter $d$. In case of a smooth grain boundary, we expect a sub-lattice accuracy, due to the symmetric treatment of the indicator functions and the overlapping measurement of pattern consistency encoded in $\Lambda$.

For a fixed number of $n$ grains $\Omega_j$ with lattice orientation $\alpha_j$ for $j=1,\cdots,n$, we consider a piecewise constant function

$$\alpha=\sum_{j=1,\cdots,m}\alpha_j\chi_{\Omega_j}$$

reflecting the orientation within the grains. Here, the grain domains $\Omega_j$ and the grain orientations $\alpha_j$ form the set of unknowns. Now, we define a functional $E_{\text{grain}}$ acting on the set of lattice orientations $\alpha_j$ and open grain domains $\Omega_j$ in the spirit of the Mumford–Shah functional:

$$E_{\text{grain}}[(\alpha_j,\Omega_j)_{j=1,\cdots,m}]{:=}\sum_{j=1,\cdots,m}\left(\int_{\Omega_j}f(x,\alpha_j)\,\mathrm{d}x+\eta\mathrm{Per}(\Omega_j)\right),$$

where $\{\Omega_j\}_j$ is a partition of the domain $\Omega$, i.e. $\Omega_j\cap\Omega_i\neq\emptyset$ and $\bigcup_j\bar{\Omega}_j=\Omega$. $\mathrm{Per}(\Sigma)$ denotes the perimeter of a set $\Sigma$, i.e. the length of the boundary of the set. A minimizer of this energy is considered as a reliable identification of lattice orientations and corresponding grains. In case of only two different lattice orientations $\alpha_1$ and $\alpha_2$ and corresponding (possibly not connected) domains $\Omega_1$ and $\Omega_2$, we can formulate the variational problem as a problem on the binary function $\alpha$ and the interface $\Gamma_G$ between the two sets $\Omega_1$ and $\Omega_2$ and obtain (up to the constant term $\mathrm{Per}(\Omega)$) the energy

$$E_G[\alpha_1, \alpha_2, \Gamma_G] := \int_{\Omega_1} f(x, \alpha_1)\,\mathrm{d}x + \int_{\Omega_2} f(x, \alpha_2)\,\mathrm{d}x + 2\eta\mathcal{H}^1(\Gamma_G)$$

where $\mathcal{H}^1(\cdot)$ denotes the one-dimensional Hausdorff measure.

### 2.2 Simultaneously Detecting a Liquid–Solid Interface

Let us now incorporate the distinction between solid and liquid phase into our variational model. In particular in the simulation results, the solid state is characterized by prominent atom dots with large values of $u$. Indeed, taking into account threshold values $\theta_1$ and $\theta_2$, we suppose that $u(x) > \theta_2$ indicates an atom dot at position $x$ and vice versa inter–atom regions are characterized by low values of $u$, i.e. $u(x) < \theta_1$. In the liquid regime there are neither very high nor low values of $u$, i.e. $u(x) \in [\theta_1, \theta_2]$. Unfortunately, the converse is not true. In transition regions between atom and hole in a solid region $u$ will attain values between $\theta_1$ and $\theta_2$. But in these transition regions, the gradient of $u$ exceeds a certain threshold $\epsilon > 0$. Thus, we assume $x$ to be in the liquid phase $\Omega_L$ if $u \in [\theta_1, \theta_2]$ and $|\nabla u| \leq \epsilon$. A variational description of the domain splitting into a liquid phase $\Omega_L$ and a solid phase is encoded in the energy

$$E_{\mathrm{phase}}[\Omega_L] = \int_{\Omega_L} q(x)\,\mathrm{d}x + \int_{\Omega \setminus \Omega_L} 1 - q(x)\,\mathrm{d}x + \nu\mathrm{Per}(\Omega_L),$$

based on the indicator function

$$\begin{aligned} q(x) :&= 1 - \chi_{[u>\theta_1]}(x)\chi_{[u<\theta_2]}(x)\chi_{[|\nabla u|<\epsilon]}(x) \\ &= \begin{cases} 0 & ; \text{ for } \quad u \in [\theta_1, \theta_2] \wedge |\nabla u| < \epsilon \\ 1 & ; \text{ else}. \end{cases} \end{aligned}$$

Since $q = 1$ in the solid region the first term favors to remove solid phase points from $\Omega_L$. On the other hand $q = 0$, in the liquid region. Hence, it is preferable to remove liquid parts from $\Omega \setminus \Omega_L$. In the usual spirit of the Mumford–Shah approach, the third term regularizes the boundary of $\Omega_L$.

Now, we merge the two approaches above and simultaneously want to detect the liquid phase $\Omega_L$ and the solid phase $\Omega \setminus \Omega_L$ which itself is decomposed into grains $\Omega_j$, such that $\bigcup_j \Omega_j = \Omega \setminus \Omega_L$. We end up with the following energy

$$\begin{aligned} E[\Omega_L, (\alpha_j, \Omega_j)_{j=1,\cdots,m}] :=& \mu \int_{\Omega_L} q(x)\,\mathrm{d}x + \mu \int_{\Omega \setminus \Omega_L} (1 - q(x))\,\mathrm{d}x + \nu\mathrm{Per}(\Omega_L) \\ &+ \sum_{j=1,\cdots,m} \left( \int_{\Omega_j} f(x, \alpha_j)\,\mathrm{d}x + \eta\mathrm{Per}(\Omega_j) \right). \end{aligned} \tag{4}$$

The primal decomposition is the one into a liquid and a solid domain. To reflect this in the variational formulation above we consider a relatively large constant $\mu$. Indeed, we choose $\mu = 10$ and $\nu = \eta = 0.05$. Furthermore, in the application considered here we take into account the following values for the other constants involved: $\theta = 0.5$, $\theta_1 = 0.3$, $\theta_2 = 0.5$, $\epsilon = 1$ (artificial test data) or $\epsilon = 20$ (PFC simulation data).

In case of only two grain orientations $\alpha_1$ and $\alpha_2$ we can rewrite the energy and obtain

$$E[\Gamma_{LS}, \alpha_1, \alpha_2, \Gamma_G] := \mu \int_{\Omega_L} q(x)\,\mathrm{d}x + \mu \int_{\Omega \setminus \Omega_L} 1 - q(x)\,\mathrm{d}x \tag{5}$$

$$+ 2\nu \mathcal{H}^1(\Gamma_{LS}) + 2\eta \mathcal{H}^1(\Gamma_G)$$

$$+ \int_{\Omega_1} f(x, \alpha_1)\,\mathrm{d}x + \int_{\Omega_2} f(x, \alpha_2)\,\mathrm{d}x\,, \tag{6}$$

which now depends on a single grain interface $\Gamma_G$ and the liquid–solid interface $\Gamma_{LS}$.

## 3 Chan–Vese Type Level Set Approach

To solve the above free discontinuity problems, we consider a Chan–Vese type approach and rewrite the variational formulation based on an implicit description of the domains via level set functions. We focus here on two different unknown grain orientations $\alpha_1$ and $\alpha_2$ and consider two level set functions $\phi$ and $\psi$. Here, $\phi$ is supposed to define the decomposition into a liquid and a solid domain, i. e. $\Omega_L = [\phi \geq 0]$ and $\Omega_S = [\phi < 0]$. Furthermore, to simplify the algorithm, we suppose the second level set function $\psi$ to be defined on the whole domain $\Omega$ even though the grains described by this function are given as subsets of the solid domain. Indeed, we assume

$$\Omega_1 = [\phi < 0] \cap [\psi < 0]\,, \quad \Omega_2 = [\phi < 0] \cap [\psi > 0]\,.$$

Then, we end up with the following Chan–Vese type energy

$$\begin{aligned} E_{CV}[\phi, \alpha_1, \alpha_2, \psi] := \int_\Omega & \mu(H(\phi)\,q(x) + (1 - H(\phi))\,(1 - q(x))) + 2\nu|\nabla H(\phi)| \\ & +(1 - H(\phi))\,(1 - H(\psi))\,f(x, \alpha_1) \\ & +(1 - H(\phi))\,H(\psi)\,f(x, \alpha_2) \\ & +2\eta(1 - \beta H(\phi))|\nabla H(\psi)|\,\mathrm{d}x\,, \end{aligned} \tag{7}$$

depending on the two level set functions $\phi$, $\psi$ and the two grain orientations $\alpha_1$ and $\alpha_2$. Here $H(\cdot)$ denotes the Heavyside function with $H(s) = 1$ for $s > 0$ and $H(s) = 0$ else. In case $\beta = 1$, the interface $[\psi = 0] \cap [\phi \geq 0]$ is not at all controlled by the energy. For $0 \leq \beta < 1$ we observe a difference between the original energy and the new level set formulation in terms of the length of the extension of the grain interface $[\psi = 0]$ in the liquid domain $\Omega_L$, i.e. up to the constant length of the domain boundary $\partial\Omega$, $E_{CV} - E = 2\eta\beta\|H(\phi)\nabla H(\psi)\|_{var} = 2\eta\beta\mathcal{H}^1([\psi = 0]\cap[\phi \geq 0])$. Thus, the method will extend physical meaningful grain interfaces by shortest paths from the triple point on the liquid–solid interface to the boundary on the domain $\Omega$. In the applications considered, we observed no artifacts from this algorithmic simplification.

The variational modeling of more than two grain orientations can be based on the multiple domain segmentation method by Chan and Vese [25] in a straightforward way. In our current implementation, we confine to the case of only two orientations. The generalization is under development. As of now, we also only use $\beta = 0$.

## 4 Regularization and Numerical Minimization

Since $H$ is not continuous we replace it by a smeared out Heavyside function. Here we again follow [7] and consider $H_\delta(x) := \frac{1}{2} + \frac{1}{\pi}\arctan\left(\frac{x}{\delta}\right)$ where $\delta > 0$. Let us emphasize that a desired guidance of the initial zero contours to the actual interfaces to be detected requires a non-local support of the regularized Heavyside function. To numerically solve the problem, we apply a gradient descent in the two level set functions $\phi$ and $\psi$ and the two orientation values $\alpha_1$ and $\alpha_2$. Different from the grey value segmentation via the original approach by Chan and Vese, the energy is not quadratic in these two unknowns and thus minimization over these two angles is already a non-linear problem. Hence, we have to compute the variation of the energy with respect to the level set functions $\phi$, $\psi$ and the orientations $\alpha_1$ and $\alpha_2$. For the variation of the energy with respect to the level set function $\phi$ we obtain

$$
\begin{aligned}
\langle\delta_\phi E_{CV}, \theta\rangle =& \mu\int_\Omega H_\delta'(\phi)(2q(x)-1)\theta \,\mathrm{d}x - 2\nu\int_\Omega H_\delta'(\phi)\theta\; \nabla\cdot\left(\frac{\nabla\phi}{|\nabla\phi|}\right)\mathrm{d}x \\
& - \int_\Omega H_\delta'(\phi)\big((1-H_\delta(\psi))f(x,\alpha_1) + H_\delta(\psi)f(x,\alpha_2)\big)\theta \,\mathrm{d}x \\
& - 2\eta\beta\int_\Omega H_\delta'(\phi)\,|\nabla H(\psi)|\,\theta \,\mathrm{d}x
\end{aligned}
$$

which reflects the sensitivity with respect to modifications of the implicit description of the liquid–solid interface. Grain boundary sensitivity is encoded implicitly in the variation with respect to the level set function $\psi$ and we achieve

$$
\begin{aligned}
\langle\delta_\psi E_{CV}, \zeta\rangle =& \int_\Omega (1-H_\delta(\phi))H_\delta'(\psi)(f(x,\alpha_2) - f(x,\alpha_1))\zeta \,\mathrm{d}x \\
& - 2\eta\int_\Omega H_\delta'(\psi)(1-\beta H(\phi))\zeta\; \nabla\cdot\left(\frac{\nabla\psi}{|\nabla\psi|}\right)\mathrm{d}x\,.
\end{aligned}
$$

Finally, a variation of the energy with respect to one of the grain orientations – we exemplarily consider here $\alpha_1$ – leads to

$$
\partial_{\alpha_1}E_{CV} = \int_\Omega (1-H_\delta(\phi))(1-H_\delta(\psi))\partial_\alpha f(x,\alpha_1)\,\mathrm{d}x\,.
$$

Armijo's rule [13] is considered as a step size control in the descent algorithm. We consider bilinear finite elements on the regular grid for the spatial discretization of the two level set functions $\phi$ and $\psi$. Each pixel of an experimental image or each node of the regular simulation grid corresponds to a node of the finite element mesh.

The initialization of $\phi$ and $\psi$ has a significant impact on the results of the numerical descent method. If initialized improperly, the minimization might get stuck in local minima far away from a desired global minimum. In particular, if $\psi$ is initially set in such a way that the grain boundary interface is contained completely in the liquid phase, one frequently observes that $\psi$ does not move in the gradient descent at all.

## 5 Numerical Results

At first, we tested the algorithm on artificial test data, generated from homogeneous dots on a lattice with precisely the same lattice spacing as encoded in our algorithm. A simple blending between two such lattices with different orientation or with a constant grey image is used to artificially generate grain boundary type or liquid–solid type interfaces. Fig. 3 shows the identification of grain boundaries of different amplitude, whereas Fig. 4 demonstrates the extraction of a liquid–solid interface. Furthermore, in Fig. 4 a liquid–solid interface and a grain boundary have simultaneously been identified. Furthermore, we applied

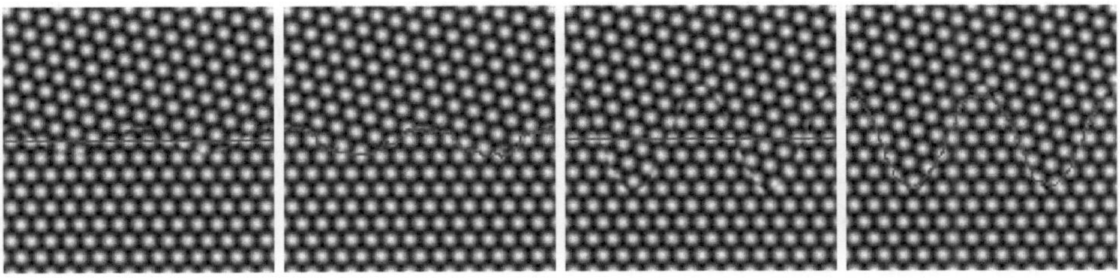

**Fig. 3.** Grain boundary detection on artificial test data: input images $u$ (first and third picture) with initial zero level set of $\psi$ and computed grain boundaries (second and fourth picture). Interfaces with different amplitude are considered in the first and second image pair.

our method to transmission electron microscopy images. These results, shown in Fig. 5, in particular demonstrate the robustness of the approach with respect to noise in the experimental data and natural fluctuations in the shape of the atom dots and the lattice spacing. In particular, let us emphasize that the variational method is capable to detect effects on an intermediate scale like the oscillating pattern of the interface in the second picture pair. Finally, we considered the identification of grain boundaries and liquid–solid interfaces in simulation data from a phase field crystal model. Fig. 6 shows the extraction of a grain boundary and the computation of both types of interfaces.

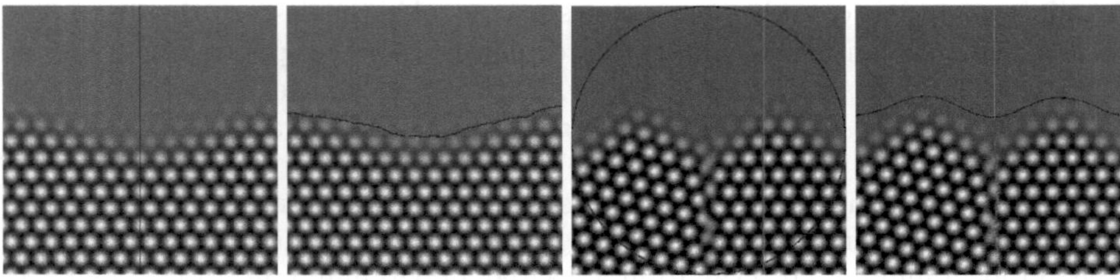

**Fig. 4.** Liquid–solid interface detection on artificial test data: input image $u$ (first picture) with initial zero level set of $\phi$ and computed interface (second picture). Combined grain boundary and liquid–solid interface detection on artificial test data: input image $u$ (third picture) with the initial grain boundary in red and the initial liquid–solid interface in blue, final grain boundary and liquid–solid interface location (fourth picture).

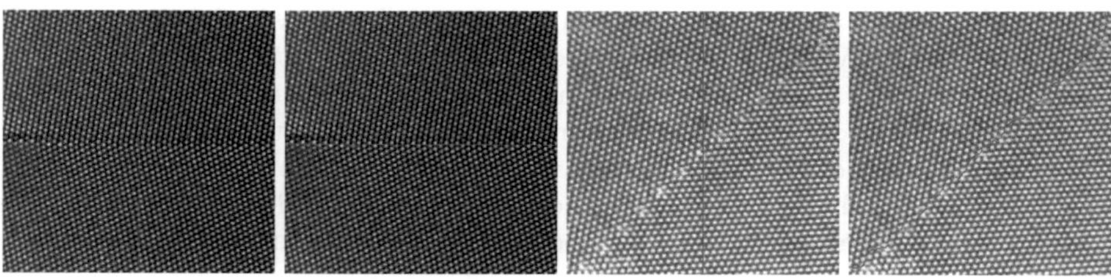

**Fig. 5.** Two results of grain boundary detection on TEM-images: input images $u$ (first and third picture) with initial position of the zero level set of $\psi$, finally detected grain boundaries (second and fourth picture). The TEM-image in the first picture pair is courtesy of Geoffrey H. Campbell, Lawrence Livermore National Laboratory (compare Fig. 1), the image used in the second picture pair is courtesy of David M. Tricker (Department of Materials Science and Metallurgy, University of Cambridge) showing a grain boundary in GaN.

**Fig. 6.** Grain boundary detection on PFC simulation data: crystal phase field function $u$ (first picture) with the initial zero level set of $\psi$, finally computed grain boundary (second picture). Combined grain boundary and liquid–solid interface detection on PFC simulation data: crystal phase field function $u$ (third picture) with the initial grain boundary in red and the initial liquid–solid interface in blue, final grain boundary and liquid–solid interface location (fourth picture).

## 6 Conclusion

We have presented a robust method for the reliable segmentation of grain boundaries in materials science on the atomic scale. The method is based on an explicit encoding of the lattice structure and its orientation in a Mumford-Shah type variational formulation. The numerical implementation is inspired by the segmentation approach by Chan and Vese. The algorithm works equally well on phase field crystal (PFC) simulations and on experimental transmission electron microscopy (TEM) images. The method has been extended to detect also liquid–solid interfaces. At first, we confine here to two different grain orientations. The straightforward extension to $2^n$ orientations is currently still work in progress. On still images, the demarcation of such interfaces might be done by hand as well. But for the validation of physical models with experimental data, it is the evolution of the grain boundaries which actually matters. Here, an accurate and robust extraction of interface velocities requires a reliable automatic tool. Thus, an extension of our model to temporal data is envisaged. So far, the lattice orientation is considered as the only local degree of freedom. The type of crystal structure and the atom spacing are preset. In a future generalization one might incorporate further lattice parameters in the variational approach or combine the lattice type classification directly with the variational parameter estimation.

## References


1. Antonin Aujol, Jean-François Chambolle. Dual norms and image decomposition models. *International Journal of Computer Vision*, 63(1):85–104, June 2005.
2. Jean-François Aujol, Gilles Aubert, and Laure Blanc-Feáaud. Wavelet-based level set evolution for classification of textured images. *IEEE Transactions on Image Processing*, 12(12):1634–1641, 2003.
3. Jean-François Aujol and Tony F. Chan. Combining geometrical and textured information to perform image classification. *Journal of Visual Communication and Image Representation*, 17(5):1004–1023, 2006.
4. Marc Berthod, Zoltań Kato, Shan Yu, and Josiane B. Zerubia. Bayesian image classification using markov random fields. *Image and Vision Computing*, 14(4):285–295, May 1996.
5. Charles Bouman and Michael Shapiro. Multiscale random field model for bayesian image segmentation. *IEEE Transactions on Image Processing*, 3(2):162–177, March 1994.
6. Vicent Caselles, Francine Catté, Tomeu Coll, and François Dibos. A geometric model for active contours in image processing. *Numer. Math.*, 66:1–31, 1993.
7. Tony F. Chan and Luminita A. Vese. Active contours without edges. *IEEE Transactions on Image Processing*, 10, no. 2:266 – 277, 2001.
8. Daniel Cremers and Christoph Schnörr. Statistical shape knowledge in variational motion segmentation. *Image and Vision Computing*, 21(1):77–86, January 2003.
9. G. Doretto, D. Cremers, P. Favaro, and S. Soatto. Dynamic texture segmentation. In B. Triggs and A. Zisserman, editors, *IEEE International Conference on Computer Vision (ICCV)*, volume 2, pages 1236–1242, Nice, Oct. 2003.

10. K.R. Elder and M. Grant. Modeling elastic and plastic deformations in nonequilibrium processing using phase field crystals. *Physical Review E - Statistical, Nonlinear, and Soft Matter Physics*, 70(5 1):051605–1–051605–18, November 2004.
11. Matthias Heiler and Christoph Schnörr. Natural image statistics for natural image segmentation. 63(1):5–19, 2005.
12. Wayne E. King, Geoffrey H. Campbell, Stephen M. Foiles, Dov Cohen, and Kenneth M. Hanson. Quantitative HREM observation of the $\Sigma 11(113)/[\bar{1}00]$ grain-boundary structure in aluminium and comparison with atomistic simulation. *Journal of Microscopy*, 190(1-2):131–143, 1998.
13. P. Kosmol. *Optimierung und Approximation.* de Gruyter Lehrbuch, 1991.
14. Sridhar Lakshmanan and Haluk Derin. Simultaneous parameter estimation and segmentation of gibbs random fields using simulated annealing. *IEEE Transactions on Pattern Analysis and Machine Intelligence*, 11(8):799–813, 1989.
15. B. S. Manjunath and Rama Chellappa. Unsupervised texture segmentation using markov random field models. *IEEE Trans. Pattern Anal. Mach. Intell.*, 13(5):478–482, 1991.
16. Yves Meyer. *Oscillating Patterns in Image Processing and Nonlinear Evolution Equations: The Fifteenth Dean Jacqueline B. Lewis Memorial Lectures.* American Mathematical Society, Boston, MA, USA, 2001.
17. D. Mumford and J. Shah. Optimal approximation by piecewise smooth functions and associated variational problems. *Comm. Pure Appl. Math.*, 42:577–685, 1989.
18. S. J. Osher and R. P. Fedkiw. *Level Set Methods and Dynamic Implicit Surfaces.* Springer-Verlag, 2002.
19. Nikos K. Paragios and Rachid Deriche. Geodesic active regions and level set methods for motion estimation and tracking. *Computer Vision and Image Understanding*, 97(3):259–282, March 2005.
20. Andreas Rätz and Axel Voigt. An unconditionally stable finite element discretization of the phase-field-crystal model. in preparation 2006.
21. Berta Sandberg, Tony Chan, and Luminita Vese. A level-set and gabor-based active contour algorithm for segmenting textured images. Technical Report 02-39, UCLA CAM Reports, 2002.
22. J.A. Sethian. *Level Set Methods and Fast Marching Methods.* Cambridge University Press, 1999.
23. Y. Singh. Density-functional theory of freezing and properties of the ordered phase. *Physics Reports*, 207(6):351–444, 1991.
24. Michael Unser. Texture classification and segmentation using wavelet frames. *IEEE Transactions on Image Processing*, 4(11):1549–1560, November 1995.
25. Luminita Vese and Tony F. Chan. A multiphase level set framework for image segmentation using the mumford and shah model. *International Journal of Computer Vision*, 50(3):271–293, December 2002.
26. Luminita Vese and Stanley Osher. Modeling textures with total variation minimization and oscillating patterns in image processing. *Journal of Scientific Computing*, 19(1-3):553–572, December 2003.

# Solving the Chan-Vese Model by a Multiphase Level Set Algorithm Based on the Topological Derivative

Lin He[1] and Stanley Osher[2]

[1] Johann Radon Institute for Computational and Applied Mathematics, 4040 Linz, Austria
lin.he@ricam.oeaw.ac.at
[2] UCLA Mathematics Department, Box 951555, Los Angeles, CA 90095-1555, U.S.A.
sjo@math.ucla.edu

**Abstract.** In this work, we specifically solve the Chan-Vese active contour model by multiphase level set methods. We first develop a fast algorithm based on calculating the variational energy of the Chan-Vese model without the length term. We check whether the energy decreases or not when we move a point to another segmented region. Then we draw a connection between this algorithm and the topological derivative, a concept emerged from the shape optimization field. Furthermore, to include the length term of the Chan-Vese model, we apply a preprocessing step on the image by using nonlinear diffusion. We show numerical experiments to demonstrate the efficiency and the robustness of our algorithm.

**Keywords:** Image Segmentation, Level Set Methods, Chan-Vese model, Topological Derivative.

## 1 Introduction

Image segmentation, an important problem in image analysis, is to partition a given image into disjoint regions, such that the regions correspond to the objects in the image. There are a wide variety of approaches to the segmentation problem. One of the popular approaches is active contour models or snakes, first introduced by Terzopoulos et al. [1][2]. The basic idea is to start with a curve around the object to be detected, the curve moves towards its interior normal and has to stop on the boundary of the object. The main drawbacks of snakes are their sensitivity to initial conditions and the difficulties associated with topological transformations. Caselles et al. [3] thus introduced the first level set formulation for the geometric active contour model in a non-variational setting (see also Malladi et al. [4]) and later in a variational form [5] (cf. Kachenassamy et al. [6]). A major advantage of the level set approach [7] is the ability to handle complex topological changes automatically. However, all above active contour models are dependent on the gradient of the given image to stop the evolution of the curve. Therefore these models can only detect objects with edges defined by a gradient.

F. Sgallari, A. Murli, and N. Paragios (Eds.): SSVM 2007, LNCS 4485, pp. 777–788, 2007.

Based on the Mumford-Shah functional [8] for segmentation, Chan and Vese [9][10] proposed a new level set model for active contours to detect objects whose boundaries are not necessarily defined by a gradient. In this paper, we will focus on solving this model. We denote the given image $f : D \rightarrow \mathbb{R}$, where $D \in \mathbb{R}^2$ is an open and bounded set. We let $\Omega$, an open subset of $D$, to be the detected objects and $\partial\Omega$ to be the boundary of the detected objects. The Chan-Vese model is to minimize the following variational functional,

$$F(\Omega, c_1, c_2) = \min_{\Omega, c_1, c_2} \int_{\Omega} (f - c_1)^2 \, dx + \int_{D \setminus \Omega} (f - c_2)^2 \, dx + \mu \int_{\partial\Omega} ds, \tag{1}$$

where $\mu > 0$ is the so called length parameter chosen by the user. To solve this minimization problem (1), the level set method [7] is used. A level set function $\phi(x)$ is represented by the region $\Omega$ as follows,

$$\phi(x) \begin{cases} > 0 & \text{if } x \in \Omega, \\ < 0 & \text{if } x \in D \setminus \bar{\Omega}. \end{cases}$$

Thus the minimization functional (1) can be reformulated in terms of the level set function $\phi(x)$ as

$$\begin{aligned} F(\phi, c_1, c_2) = \min_{\phi, c_1, c_2} & \textstyle\int_D (f - c_1)^2 H(\phi) \, dx + \int_D (f - c_2)^2 (1 - H(\phi)) \, dx \\ & + \mu \textstyle\int_D |\nabla H(\phi)| dx, \end{aligned} \tag{2}$$

where $H(x)$ is the heaviside function. This minimization problem is solved by taking the Euler-Lagrange equation and updating the level set function $\phi(x)$ by the gradient descent method,

$$\phi_t = -\delta(\phi)((f - c_1)^2 - (f - c_2)^2 - \mu \nabla \cdot \frac{\nabla \phi}{|\nabla \phi|}), \tag{3}$$

where $\delta(x)$ is the delta function and the constants $c_1$ and $c_2$ are updated at each iteration by

$$c_1 = \frac{\int_D f H(\phi(x)) \, dx}{\int_D H(\phi(x)) \, dx}, \; c_2 = \frac{\int_D f (1 - H(\phi(x))) \, dx}{\int_D (1 - H(\phi(x))) \, dx}. \tag{4}$$

The PDE (3) is a parabolic nonlinear PDE, due to the motion by mean curvature term $\nabla \cdot \frac{\nabla \phi}{|\nabla \phi|}$, which also imposes the CFL condition. Thus the computation is expensive.

To overcome the difficulty of solving the nonlinear PDE, Gibou and Fedkiw [11] made an observation of (3) that only the zero level set of the function $\phi(x)$ is important and the length constraint is not important for a clean image composed of distinct objects. So they neglect the term $\delta(\phi)$ and replace the length term by a nonlinear diffusion preprocessing step. Thus they end up having to solve the following ODE

$$\phi_t = -(f - c_1)^2 + (f - c_2)^2, \tag{5}$$

where $c_1$ and $c_2$ are updated according to (4). This approach converges fast since a large time step can be taken. They also draw a connection between (5) and the k-means procedure (cf. [12]). However, we will link (3) with the topological derivative, which has been studied in our previous work [13], to the application of structural optimal design. So this paper can also be considered a follow up application of the topological derivative.

Independently, Song and Chan [14] take a different approach to solve the Chan-Vese model. They focus on the variational energy (1) or (2) without the length term. They compute the difference of the energy functional when they move a pixel $x$ from inside the zero level set to the outside (or vice versa). If the energy does not decrease, the pixel is kept inside; otherwise, the pixel is moved outside by replacing $\phi(x)$ with $-\phi(x)$. This algorithm converges within a few iterations.

Based on [14], in Section 2 we introduce a fast algorithm by calculating a variational energy under the multiphase level set framework. Therefore we are able to deal with complex images with triple junctions, multiple layers, etc. Two simple examples are shown to demonstrate the robustness and the efficiency of our fast algorithm. In Section 3, we make a connection between the idea behind this fast algorithm and the concept of the topological derivative. Furthermore, by applying topological derivative to the level set method (cf. [15][?]), we make a link to the work in [11]. In addition, we take a preprocessing step (cf. [11]) to include the length term. In Section 4, we present more numerical examples to illustrate the effectiveness of our algorithm in more complicated applications, which include noisy images and medical images. They all converge in less than 3 iterations.

Last but not least, we want to mention a series of papers on image segmentation by different kinds of PDE based level set methods, such as the multilayer level set approach [16][17] and piecewise constant level set methods [18]. It would be interesting to compare our algorithm with theirs regarding the speed, the quality of segmented results and the sensitivity to initial conditions and noise. This could be our future work.

## 2 Our Fast Algorithm

### 2.1 The Two Level Set Framework

To segment images with multiple objects and/or junctions, we can easily generalize the above single level set method (2) to a multiphase level set method [10]. Particularly here we use a two level set method with the functions denoted as $\phi_1$ and $\phi_2$. Thus we can segment any image in one to four disjoint "color" regions, given by $\Omega_{11} := \{\phi_1 > 0, \phi_2 > 0\}$, $\Omega_{12} := \{\phi_1 > 0, \phi_2 < 0\}$, $\Omega_{21} := \{\phi_1 < 0, \phi_2 > 0\}$ and $\Omega_{22} := \{\phi_1 < 0, \phi_2 < 0\}$. And we define the associated energy functional without the length term as the following

$$\begin{aligned} F(\Omega_{11}, \Omega_{12}, \Omega_{21}, \Omega_{22}) &:= F(\phi_1, \phi_2, c) \\ &= \min_{\phi_1,\phi_2,c} \int_{\Omega_{11}} (f - c_{11})^2 \, dx + \int_{\Omega_{12}} (f - c_{12})^2 \, dx \\ &\quad + \textstyle\int_{\Omega_{21}} (f - c_{21})^2 \, dx + \int_{\Omega_{22}} (f - c_{22})^2 \, dx, \end{aligned} \tag{6}$$

where the four constants $c_{11}$, $c_{12}$, $c_{21}$ and $c_{22}$ are the average pixel values

$$c_{11} = \frac{\int_{\Omega_{11}} f\,dx}{\int_{\Omega_{11}} dx},\ c_{12} = \frac{\int_{\Omega_{12}} f\,dx}{\int_{\Omega_{12}} dx},\ c_{21} = \frac{\int_{\Omega_{21}} f\,dx}{\int_{\Omega_{21}} dx},\ c_{22} = \frac{\int_{\Omega_{22}} f\,dx}{\int_{\Omega_{22}} dx}. \tag{7}$$

## 2.2 The Algorithm

First we denote the number of pixels in the region of $\Omega_{11}$ as $m_{11}$, correspondingly the number of pixels in the region of $\Omega_{12}$, $\Omega_{21}$ and $\Omega_{22}$ as $m_{12}$, $m_{21}$ and $m_{22}$. And we further assume that the area a pixel occupies, i.e., a grid, is 1.

The idea of our fast algorithm is that for any pixel $x_0 \in D$, we consider the variation of the energy functional (6) corresponding to moving $x_0$ to any of the other three regions except the one $x_0$ originally belonged to. If the energy does not decrease, we keep $x_0$ at the current region; otherwise, we move $x_0$ to the region for which the energy decreases most.

We consider the case of moving the pixel $x_0 \in \Omega_{11}$ to the region $\Omega_{12}$. Denote the average of the new regions $\Omega_{11} - x$ and $\Omega_{12} + x$ correspondingly as $\bar{c}_{11}$ and $\bar{c}_{12}$, we have

$$\begin{aligned} \bar{c}_{11} &= \frac{\int_{\Omega_{11}-x_0} f\,dx}{m_{11}-1} = c_{11} - \frac{f(x_0)-c_{11}}{m_{11}-1}, \\ \bar{c}_{12} &= \frac{\int_{\Omega_{12}+x_0} f\,dx}{m_{12}+1} = c_{12} + \frac{f(x_0)-c_{12}}{m_{12}+1}. \end{aligned} \tag{8}$$

A simple computation gives us the variation of energy functional (6)

$$\begin{aligned} & F(\Omega_{11} - x_0, \Omega_{12} + x_0, \Omega_{21}, \Omega_{22}) - F(\Omega_{11}, \Omega_{12}, \Omega_{21}, \Omega_{22}) \\ &= -\frac{m_{11}}{m_{11}-1}(f(x_0) - c_{11})^2 + \frac{m_{12}}{m_{12}+1}(f(x_0) - c_{12})^2. \end{aligned} \tag{9}$$

We can obtain similar results for the variations of the energy functional (6) of moving $x_0$ to the other two regions, $\Omega_{21}$ or $\Omega_{22}$.

Thus for the pixel $x_0 \in \Omega_{11}$, we first find the smallest value among the following four values : $\frac{m_{11}}{m_{11}-1}(f(x_0) - c_{11})^2$, $\frac{m_{12}}{m_{12}+1}(f(x_0) - c_{12})^2$, $\frac{m_{21}}{m_{21}+1}(f(x_0) - c_{21})^2$ and $\frac{m_{22}}{m_{22}+1}(f(x_0) - c_{22})^2$. (Note: When there is more than one smallest value, the first one is chosen by the order of the regions: $\Omega_{11}$, $\Omega_{12}$, $\Omega_{21}$ and $\Omega_{22}$. Further, when there is an empty region and another region having an average pixel value equivalent to $f(x_0)$, the latter is chosen to prevent two regions having the same average pixel value.) Then we move $x_0$ to the corresponding region, denote it as $\Omega_{ij}$. If this corresponding region $\Omega_{ij}$ is not the original $\Omega_{11}$, then we need to change $\phi_1(x_0)$ and $\phi_2(x_0)$ based on the sign of both these level set functions represented by $\Omega_{ij}$. For example, if $i = 1$ and $j = 2$ then we just replace $\phi_2(x_0)$ by $-\phi_2(x_0)$. Furthermore, we need to update the constants $c_{11}$ and $c_{ij}$ following (8). Now we can advance to the next pixel in a prescribed order. For example, in our numerical experiments, we sweep the pixels row by row.

One iteration is finished when we have swept all the pixels in the image one time. Usually less than three iterations are needed before the energy $F$ remains unchanged.

The above algorithm finds the global minimizer and converges with two iterations when the image has less than or equal to four pixel values. For example, the given image $f$ has four pixel values $d_1 < d_2 < d_3 < d_4$, we denote it as $f = d_1\Omega_1 + d_2\Omega_2 + d_3\Omega_3 + d_4\Omega_4$. First because no two regions share the same average pixel value we derive that the minimizer $g$ has also four distinct values, denoted as $g = c_{11}\Omega_{11} + c_{12}\Omega_{12} + c_{21}\Omega_{21} + c_{22}\Omega_{22}$ with $c_{11} < c_{12} < c_{21} < c_{22}$. Next we prove any two pixels having the same value will belong to the same region. Otherwise, we assume that there exist two pixels $x_1 \in \Omega_1 \cap \Omega_{11}$ and $x_2 \in \Omega_2 \cap \Omega_{12}$. Since $g$ is a minimizer, we know that $\frac{m_{11}}{m_{11}-1}(d_1 - c_{11})^2 < \frac{m_{12}}{m_{12}+1}(d_1 - c_{12})^2$ and $\frac{m_{11}}{m_{11}+1}(d_1 - c_{11})^2 > \frac{m_{12}}{m_{12}-1}(d_1 - c_{12})^2$. That means $\frac{m_{11}}{m_{11}-1}\frac{m_{12}+1}{m_{12}} < \frac{(d_1-c_{12})^2}{(d_1-c_{11})^2} < \frac{m_{11}}{m_{11}+1}\frac{m_{12}-1}{m_{12}}$. However, the left term in the inequality is larger than one and the right term is smaller than one, therefore $x_1$ and $x_2$ can only belong to the same region. Based on this, we deduce that $f = g$, i.e., there exists only one minimizer, the true image. The proof of the fast convergence of the algorithm (within two iterations) for images having less than or equal to four pixel values will not greatly illuminate our understanding of the topic and because of space limitations will not be given here.

## 2.3 Examples

Figure 1 shows four different initial conditions used for this fast algorithm and the multiphase PDE based level set methods [10]. Our fast algorithm converges correctly within 2 iterations for all initial conditions. However, none of the four initial conditions would work for the PDE based level set method. They all converge to the same local minimum trapped at the outer circle. This is because the PDE based level set method evolves level sets mainly near the zero level set. When a pixel is far away from the zero level set, it is not likely to cross over it. Thus it demonstrates the efficiency and the robustness of our fast algorithm compared with the PDE based level set method. We will comment on this more in the next section after we introduce the concept of the topological derivative.

Then in Figure 2 we apply our fast algorithm on an image composed of five piecewise constant parts, where the gray value of the dark gray outer three-fourths circle is exactly the mean of the gray value of the light gray left square and the gray value of the dark center square. Therefore it is likely for some initial conditions that the optimal solution does not distinguish the center square from the three-fourths circle. See the bottom images. The top image shows a case where the algorithm works for the given initial condition.

***Remark:*** *Since the level set function values and the constants are updated along with the sweeping procedure, the sweeping order will matter. This is usually an issue of the Gauss Seidel iteration. However in our case, it is easier to just choose a different initial condition than a different sweeping order.*

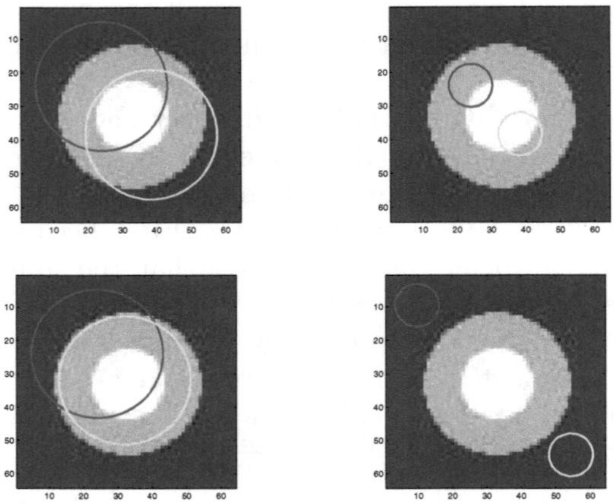

**Fig. 1.** Four different initial conditions used for comparisons between our fast algorithm and multiphase PDE based level set methods. Pink line: the zero level set of $\phi_1$; Yellow line: the zero level set of $\phi_2$.

## 3 The Connection to the Topological Derivative

Let us go back to (4), where both constants $c_1$ and $c_2$ are correspondingly defined as the average of the image $f$ inside or outside the subset $\Omega$. Therefore we can rewrite the minimization problem (1) or (2) as

$$F(\Omega) = \min_{\Omega} \int_{\Omega} (f - c_1(\Omega))^2\,dx + \int_{D\setminus\Omega} (f - c_2(\Omega))^2\,dx + \mu \int_{\partial\Omega} ds, \tag{10}$$

where again, $c_1(\Omega) = \frac{\int_\Omega f\,dx}{\int_\Omega dx}$ and $c_2(\Omega) = \frac{\int_{D\setminus\Omega} f\,dx}{\int_{D\setminus\Omega} dx}$.

Originally, the Chan-Vese model (1) or (2) is defined as a minimization problem determined by the shape of $\Omega$ and two constants $c_1$ and $c_2$. Now with this new form (10), the active contour model becomes a pure shape optimization problem, where a huge literature [19][20][21][22] can be accessed from the shape optimization field. Particularly in our work, the concept of the topological derivative [23][24][25] is employed to explain why our algorithm works efficiently and robustly.

### 3.1 The Topological Derivative

The idea of the topological derivative is to create a small ball $B_{\rho,x}$ with center $x$ and radius $\rho$ inside/outside the domain $\Omega$ and then consider the variation of the objective functional $\mathcal{F}$ with respect to the volume of this small ball. For $x \in \overline{\Omega}$, the topological derivative $d_{\mathcal{T}}\mathcal{F}(\Omega)(x)$ is defined as the limit (if it exists)

$$d_{\mathcal{T}}\mathcal{F}(\Omega)(x) := \lim_{\rho\to 0} \frac{\mathcal{F}(\Omega_{\rho,x}) - \mathcal{F}(\Omega)}{|B_{\rho,x} \bigcap \Omega|}, \tag{11}$$

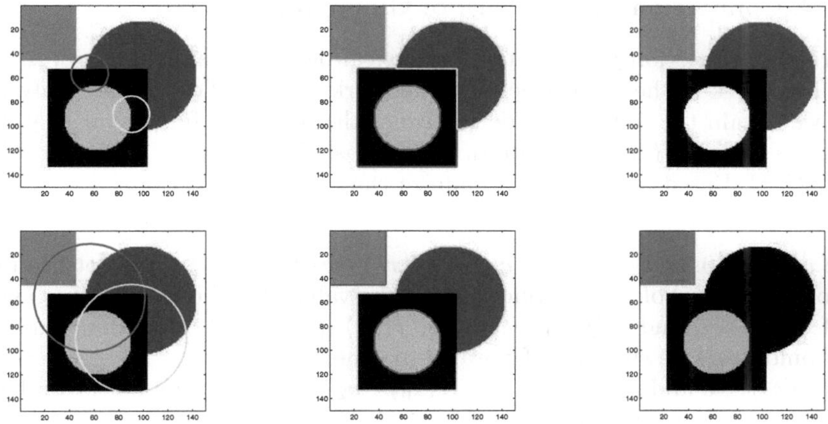

**Fig. 2.** Segmentation of a tricky synthetic image. Pink line: the zero level set of $\phi_1$; Yellow line: the zero level set of $\phi_2$. Top: an initial condition for which our algorithm works; Bottom: a different initial condition for which our algorithm does not distinguish the three-fourths circle from the dark center square. Left: the initial condition; Middle: zero level sets of $\phi_1$ and $\phi_2$ ; Right: the segmented result.

where $\Omega_{\rho,x} = \Omega - \overline{B(\rho,x)}$. (Remark: in order to define the topological derivative for $x \in D - \overline{\Omega}$, $\Omega_{\rho,x}$ is replaced by $\Omega + \overline{B(\rho,x)}$.) Thus to minimize the energy functional $\mathcal{F}$, a hole should be created at the point $x$ if the topological derivative is negative. This is the same idea as our fast algorithm, which replaces the hole having area $\pi\rho^2$ and $\rho \to 0$ with a grid having area 1.

The above definition of the topological derivative is based on two regions $\Omega$ and $D \setminus \Omega$. Nevertheless it can be generalized to multiple regions as the following. For $x \in \Omega_{11}$, if we create a small hole of $B_{\rho,x}$ and add it to the region $\Omega_{12}$, then with the notation of

$$\begin{aligned} c_{11}^{\rho} &= \frac{\int_{\Omega_{11}-B(\rho,x)} f dxdy}{\int_{\Omega_{11}-B(\rho,x)} dxdy} = c_{11} - \frac{\int_{B(\rho,x)} (f-c_{11})\, dx}{|\Omega_{11}|-\pi\rho^2} \\ c_{12}^{\rho} &= \frac{\int_{\Omega_{12}+B(\rho,x)} f dxdy}{\int_{\Omega_{12}+B(\rho,x)} dxdy} = c_{12} + \frac{\int_{B(\rho,x)} (f-c_{12})\, dx}{|\Omega_{12}|+\pi\rho^2}, \end{aligned} \tag{12}$$

we obtain the topological derivative of this perturbation as follows,

$$d_{\mathcal{T}}\mathcal{F}(\Omega_{11},\Omega_{12})(x) = \begin{cases} -(f(x)-c_{11})^2 + (f(x_0)-c_{12})^2, & \text{if } |\Omega_{12}| > 0; \\ -(f(x)-c_{11})^2, & \text{if } |\Omega_{12}| = 0. \end{cases} \tag{13}$$

Similarly, we can obtain the topological derivative of the perturbation, which is to create a hole centered at $x$ in $\Omega_{11}$ and to add it to $\Omega_{21}$ or $\Omega_{22}$. Comparing (13) with the expressions (9) in Section 2.2., we see the similarity to our fast algorithm.

### 3.2 A Connection to the Work of Gibou and Fedkiw

To draw a connection between the topological derivative and the work in [11], we will go back to the single level set framework again. Following the analysis of (13), we obtain the topological derivative of the objective functional $F$ at $\Omega$ is $-(f(x_0)-c_1)^2+(f(x_0)-c_2)^2$ for $x_0 \in \Omega$ and is $(f(x_0)-c_1)^2-(f(x_0)-c_2)^2$ for $x_0 \in D \setminus \Omega$. Based on these, to minimize the objective functional $F$, we know (cf. [15][13]):

- If $\phi(x_0) > 0$, and $-(f(x_0)-c_1)^2+(f(x_0)-c_2)^2 > 0$, then it is not favorable to generate a hole at $x_0$ which means the value of $\phi$ should not decrease.
- If $\phi(x_0) > 0$ and $-(f(x_0)-c_1)^2+(f(x_0)-c_2)^2 < 0$, then the value of $\phi$ should decrease since it is favorable to generate a hole.
- If $\phi(x_0) < 0$ and $(f(x_0)-c_1)^2-(f(x_0)-c_2)^2 > 0$, then it's not favorable to generate a hole and thus the value of $\phi(x,t)$ should not increase.
- If $\phi(x_0) < 0$ and $(f(x_0)-c_1)^2-(f(x_0)-c_2)^2 < 0$, then it's favorable to generate a hole which means the value of $\phi$ should increase.

Then to obtain the minimizer of the objective functional (10) (equivalent to (1) and (2)) without the length term, we choose to solve the ODE (5), which satisfies the above requirements. For more details, see [15][13]. This leads to Gibou and Fedkiw's work in [11]. To solve (5), we can take a large time step. Or, we can just update the level set function $\phi(x)$ by the sign of $-(f(x)-c_1)^2+(f(x)-c_2)^2$ as follows,

$$\phi^{n+1}(x) = \begin{cases} 1 & \text{if } -(f(x)-c_1(\phi^n(x)))^2+(f(x)-c_2(\phi^n(x)))^2 > 0 \\ -1 & \text{else.} \end{cases} \tag{14}$$

This method updates $\phi$ by the so called Jordan iteration compared with the the Gauss Seidel iteration used by our fast algorithm. The method of updating $\phi$ by the Jordan iteration converges within a few iterations since whether a pixel $x$ outside or inside the zero level set of $\phi$ is solely determined by the distance between the value of $f(x)$ and $c_1$ or $c_2$. For details of the proof, see also [14]. This analogy can be trivially extended to the multiphase level set framework we use here. Thus it also explains why our algorithm based on the Gauss Seidel iteration converges so fast.

Up to now, we have linked [11] and [14] together through the topological derivative. And we know that those PDE based level set methods solved in [9][10][16][17] can be considered as level set approaches based on shape derivatives [20]. As pointed out in [21][22], the level set approach based on the shape sensitivity may get stuck at shapes with fewer holes than the optimal geometry in some applications to structure design. This is also the case in image segmentation. That is why in the numerical experiments of those works mentioned above, the initial zero level set with many small circles are used. However, our algorithm together with [11][14] do not have this problem. The advantage of applying the topological derivative to the level set method is the ability to create holes far away from the zero level set.

### 3.3 The Length Term

Up until now, we have not mentioned how to deal with the length term. It is necessary particularly for noisy images or images with edges not defined by a gradient. To include the length term, special attention is paid in [14][11]. The authors of [14] apply several iterations of the PDE-based algorithm (3) to the obtained optimal solution from their fast algorithm. And the authors of [11] apply a preprocessing step on the given image $f$ and then apply their algorithm to this processed image. The idea of this preprocessing step is to use isotropic nonlinear diffusion [26] for denoising images while still keeping the image edges intact. The nonlinear equation they solve is

$$I_t(x,t) = \nabla \cdot (g(|\nabla I|)\nabla I), \tag{15}$$

where $I(x,t)$ defines the image and $g$ is an edge-stopping function such that $\lim_{s\to\infty} g(s) = 0$, i.e., diffusion stops at the location of large gradients.

The reason why they treat the length term different is not given in their work. Now based on the topological derivative, we know it is because the topological derivative of the objective functional $\mathcal{F}(\Omega) = |\partial\Omega|$ does not exist.

In our work, we also use the preprocessing step to re-introduce the notion of the scale term. Following [11], we choose $g(s) = \nu/(1+s^2/K^2)$, where $\nu$ is a parameter controlling the length scale and $K$ is fixed as 7. We use the Alternative Direction Explicit (ADE) technique to solve (15), see [27]. We only conduct a few iterations of (15) since our intention is not to denoise but to segment the image $f$.

## 4 More Numerical Examples

First in Figure 3, we apply our algorithm to a noisy image. The preprocessing step is taken with $\nu = 1$ and 5 ADE iterations. The convergence of the fast algorithm for this image is within two iterations. We note that our algorithm plus the preprocessing step is not sensitive to the noise. Even though the right triangle does not have a sharp contrast with the noisy background, the optimal solution still finds it correctly.

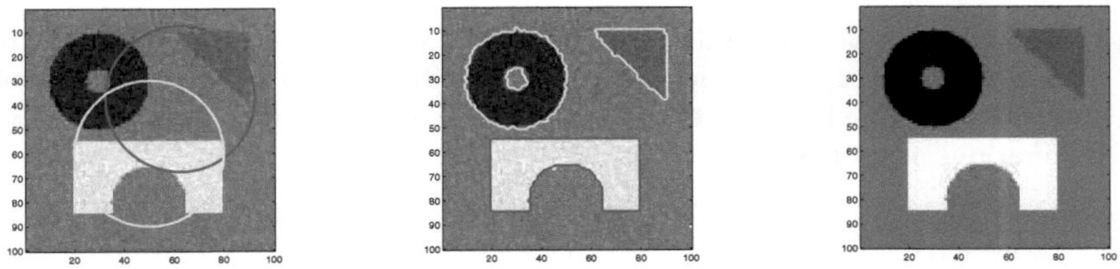

**Fig. 3.** Segmentation of a noisy image. Pink line: the zero level set of $\phi_1$; Yellow line: the zero level set of $\phi_2$. Left: the initial condition; Middle: zero level sets of $\phi_1$ and $\phi_2$ ; Right: the segmented result.

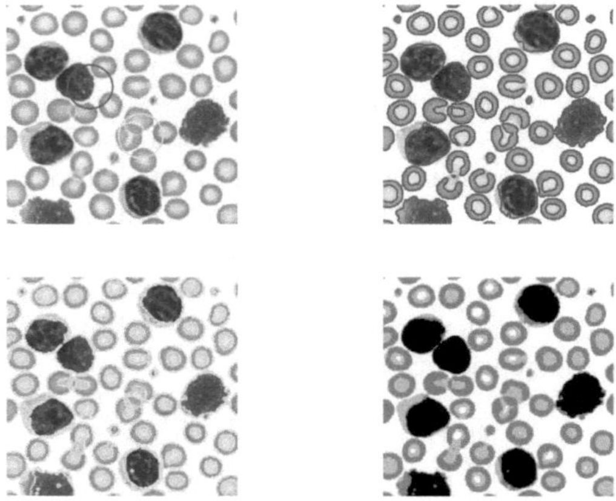

**Fig. 4.** Segmentation of blood cells. Pink line: the zero level set of $\phi_1$; Yellow line: the zero level set of $\phi_2$. Top left: the initial condition; Top right: the zero level set of $\phi_1$; Bottom left: the zero level set of $\phi_2$; Bottom right: the segmented result. $\nu = 1.0$, 5 ADE iterations are solved.

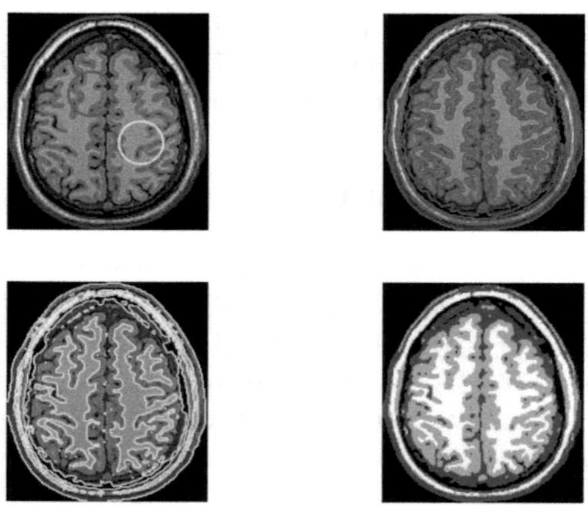

**Fig. 5.** Segmentation of a brain data image. Pink line: the zero level set of $\phi_1$; Yellow line: the zero level set of $\phi_2$. Top left: the initial condition; Top right: the zero level set of $\phi_1$; Bottom left: the zero level set of $\phi_2$; Bottom right: the segmented result. $\nu = 1.0$, 5 ADE iterations are solved.

We conclude the paper by showing the results of our algorithm on two medical images. For each image, a preprocessing step is taken and the fast algorithm converges within two iterations. The zero level sets of $\phi_1$ and $\phi_2$ are noticed overlapping sometimes since there is no restriction on the length of both interfaces. Nevertheless, the segmented results are still good. see Figure 4 and Figure 5.

## Acknowledgements

The work of and L.H. has been supported by the Austrian National Science Foundation FWF through project SFB F 013 / 08 and by the Johann Radon Institute for Computational and Applied Mathematics (Austrian Academy of Sciences ÖAW). The work of S.O. has been supported by the NSF through grants DMS-0312222, ACI-0321917 and DMI-0327077.

## References

1. Terzopoulos, D., Platt, J., Barr, A., Fleischer, K.: Elastically deformable models. In: Comp. Graphics Proc., ACM Press/ ACM SIGGRAPH (1987) 205–214
2. Terzopoulos, D., Fleischer, K.: Deformable models. The Visual Computer **4**(6) (1988) 306–331
3. Caselles, V., Catte, F., Coll, T., Dibos, F.: A geometric model for active contours in image processing. Numerische Mathematik **66** (1993) 1–31
4. Malladi, R., Sethian, J., Vemuri, B.: Shape modeling with front propagation: A level set approach. IEEE Transactions on Pattern Analysis and Machine Intelligence **17**(2) (1995) 158–175
5. Caselles, V., Kimmel, R., Sapiro, G.: Geodesic active contours. International Journal of Computer Vision **22**(1) (1997) 61–79
6. Kichenassamy, S., Kumar, A., Olver, P., Tannenbaum, A., Yezzi, Z.: Gradient flows and geometric active contour models. ICCV (1995)
7. Osher, S., Sethian, J.A.: Fronts propagating with curvature dependent speed; algorithms based on Hamilton-Jacobi formulations. J. Comput. Phys. **79** (1988) 12–49
8. Mumford, D., Shah, J.: Optimal approximation by piecewise smooth functions and associated variational problems. Comm. Pure Appl. Math **42** (1989) 577–685
9. Chan, T., Vese, L.: Active contours without edges. IEEE Transations on Image Processing **10**(2) (2001) 266–277
10. Vese, L., Chan, T.: A multiphase level set framework for image segmentation using the Mumford and Shah model. International Journal of Computer Vision **50**(3) (2002) 271–293
11. Gibou, F., Fedkiw, R.: Fast hybrid k-means level set algorithm for segmentation. In: The Procedings of the 4th Annual Hawaii International Conference on Statistics and Mathematics. (2002) Stanford Technical Report, Nov 2002.
12. MacQueen, J.: Some methods for classification and analysis of multivariate observation. In: Proceedings of the Fifth Berkeley Symposium on Mathematical Satistics and Probability. (1967) 281–297
13. He, L., Kao, C.Y., Osher, S.: Incorporating topological derivatives into shape derivatives based level set methods. Journal of Computational Physics (CAM Report 06-44) (2007)
14. Song, B., Chan, T.: A fast algorithm for level set based optimization. (CAM 02-68, UCLA, December 2002)
15. Burger, M., Hackl, B., Ring, W.: Incorporating topological derivatives into level set methods. J. Comp. Phys **194** (2004) 344–362
16. Chung, G., Vese, L.: Image segmentation using a multilayer level set approach. Technical Report 03-53, UCLA (2003)

17. Chung, G., Vese, L.: Energy minimization based segmentation and denoising using multilayer level set approach. Volume 3457/2005., EMMCVPR (2005) 439–455 LNCS Vol. 3757/2005.
18. Lie, J., Lysaker, M., Tai, X.C. In: Piecewise Constant Level Set Methods and Image Segmentation. Volume 3459 of Lecture Notes in Computer Science. Springer Berlin / Heidelberg (2005) 573–584
19. Murat, F., Simon, S.: Etudes de problems d'optimal design. Lectures Notes in Computer Science **41** (1976) 54–62
20. Sokolowski, J., Zolesio, J.P.: Introduction to Shape Optimization: Shape Sensitivity Analysis. Springer, Heidelberg (1992)
21. Allaire, G., Jouve, F., Toader, A.M.: A leve-set method for shape optimization. C.R. Acad. Sci. Paris, Ser. I **334** (2002) 1125–1130
22. Allaire, G., Jouve, F., Toader, A.M.: Structural optimization using sensitivity analysis and a level-set method. J. Comput. Phys. **194**(1) (2004) 363–393
23. Sokolowski, J., Zochowski, A.: On the topological derivative in shape optimization. SIAM J. Control Optim. **37** (1999) 1251–1272
24. Garreau, S., Guillaume, P., Masmoudi, M.: The topological asymptotic for PDE systems: The elasticity case. SIAM J. Control Optim. **39** (2001) 1756–1778
25. Amstutz, S., Andrae, H.: A new algorithm for topology optimization using a level-set method. J. Comp. Phys. **216** (2006) 573–588
26. Perona, P., Malik, J.: Scale-space and edge detection using anisotropic diffusion. IEEE Trans. PAMI **12(7)** (1990) 629–639
27. Leung, S., Osher, S.: Global minimization of the active contour model with TV-inpainting and two-phase denoising. Lecture Notes in Computer Science **3752** (2005) 149–160

# A Geometric Variational Framework for Simultaneous Registration and Parcellation of Homologous Surfaces*

Nicholas A. Lord, Jeffrey Ho, Baba C. Vemuri, and Stephan Eisenschenk

University of Florida
{nal,jho,vemuri}@cise.ufl.edu, stephan.eisenschenk@neurology.ufl.edu

**Abstract.** In clinical applications where structural asymmetries between homologous shapes have been correlated with pathology, the questions of definition and quantification of 'asymmetry' arise naturally. When not only the degree but the position of deformity is thought relevant, asymmetry localization must also be addressed. Asymmetries between paired shapes have already been formulated in terms of (non-rigid) diffeomorphisms between the shapes. For the infinity of such maps possible for a given pair, we define optimality as the minimization of deviation from isometry under the constraint of piecewise deformation homogeneity. We propose a novel variational formulation for segmenting asymmetric regions from surface pairs based on the minimization of a functional of both the deformation map and the segmentation boundary, which defines the regions within which the homogeneity constraint is to be enforced. The functional minimization is achieved via a quasi-simultaneous evolution of the map and the segmenting curve, conducted on and between two-dimensional surface parametric domains. We present examples using both synthetic data and pairs of left and right hippocampal structures, and demonstrate the relevance of the extracted features through a clinical epilepsy classification analysis.

## 1 Introduction

In the problem of quantifying differences between homologous shapes, we may be interested not only in the extent but the location of deformation. With regards to hippocampal shape analysis, for instance, it was shown as early as [18] that analysis of regional asymmetries could improve disease classification capability, relative to such commonly used global measures as volume, length, and surface area. Owing to the clinical relevance of the problem (hippocampal abnormality being suspected relevant in epilepsy, schizophrenia, and Alzheimer's disease), several methods for fine-grained regional hippocampal shape analysis have since

* This research was supported in part by the grant NIH R01-NS046812 to BCV and the UF Alumni Fellowship to NAL. Data sets were partially provided by Dr. Christiana Leonard of the UF McKnight Brain Institute. Thanks to S. Kodipaka for his invaluable assistance in the data analysis portion of this work.

F. Sgallari, A. Murli, and N. Paragios (Eds.): SSVM 2007, LNCS 4485, pp. 789–800, 2007.

been suggested. Medial and spherical harmonic representations form the basis of works by Gerig et al. [5], Shen et al. [13], Styner et al. [9], and Bouix et al. [2]. Davies et al. [10] developed a minimum description length framework in which modes of variation are identified across the surface dataset. Csernansky et al. [7] compute diffeomorphic maps between patient hippocampi and a reference using a viscous fluid flow model (producing inward/outward normal flows), and couple analysis with manual anatomical parcellation. All statistical analyses reported in the above works have indicated the benefit of incorporating regional information, which is unsurprising. However, said analyses are preliminary, and shape analysis results can actually appear contradictory across different studies (for instance, Styner notes in [14] the contrast between the primary abnormality localization in the hippocampal tail found in that work and the localization in the head reported in [6]). Furthermore, while statistically significant differences of hippocampal shape have been identified between diseased and normal sample populations, reliable classification of a sizeable number of patients with respect to those categories has not yet occurred. As such, the problem remains very much open.

The need to identify and isolate subregions of interest on the hippocampal surface suggests segmentation (which we henceforth use interchangeably with 'parcellation'), and the need to compare hippocampal surfaces between and within (i.e. detection of asymmetry between halves of the same structure, as with left and right hippocampal horns) patients represents an inherent requirement for registration. As such, the possibility of applying a unified segmentation and registration scheme is an intriguing one. The first instance of a fully unified (quasi-simultaneous) segmentation and registration scheme in the image processing literature known to us was given by Yezzi et al. [1], who unified mutual information (MI)-based flat 2D rigid image registration with piecewise constant (Chan-Vese) segmentation. Wyatt et al. [20] solved the same problem within a Markov random field (MRV) framework. Xiaohua et al. [21], Unal et al. [15], and F. Wang et al. [17] developed nonrigid extensions. Jin et al. [11] simultaneously evolved a surface segmentation and a radiance-discontinuity-detecting segmentation of the evolving surface itself. Pohl et al. [8] produced an expectation maximization (EM)-based algorithm for simultaneous multiscale affine registration and subcortical segmentation of MRIs using a labelled cortical atlas.

We now present a novel scheme for shape comparison that rests on the fundamental assumptions that: (1) the deformity of homologous anatomical structures can be quantified as the deviation from isometry of the deformation map between their surfaces, and (2) since deformities may be local, the evolution of the global correspondence must allow for a partial disconnection between the normal and abnormal regions. For the topologically spherical surfaces considered here, our intrinsic approach leads naturally to an elegant 2D parametric representation of both the segmentation and registration. Our use of Riemannian surface structure in the matching criterion is similar in spirit to work done by Y. Wang et al. (e.g. [19]), in which the authors presented a technique to nonrigidly register surfaces using a 2D (parametric) diffeomorphic map constructed from Riemannian surface structure information. The integrated manifold segmentation is enough to distinguish our approach from this work, but additionally, our definition of system energy in terms of piecewise isometry (i.e. using first fundamental forms in the matching criterion) inherently balances the consideration that the map should accurately match Riemannian surface characteristics with the idea that the map should incur as little deformation as possible in doing so. As such we eliminate the need for additional map-regularizing terms. In fact, the real novelty of the algorithm is more fundamental: this is the first proposal for simultaneous nonrigid surface (closed genus zero 2D manifold) registration and segmentation wherein the segmentation is driven by the evolving registration map itself.

The rest of the paper is organized as follows: Section 2 contains the mathematical formulation and algorithm conceptualization. In section 3, we delve into specifics of the implementation. Section 4 contains experimental results on both synthetic and real data sets. Finally in section 5, we draw conclusions.

## 2 Simultaneous Segmentation and Registration Algorithm

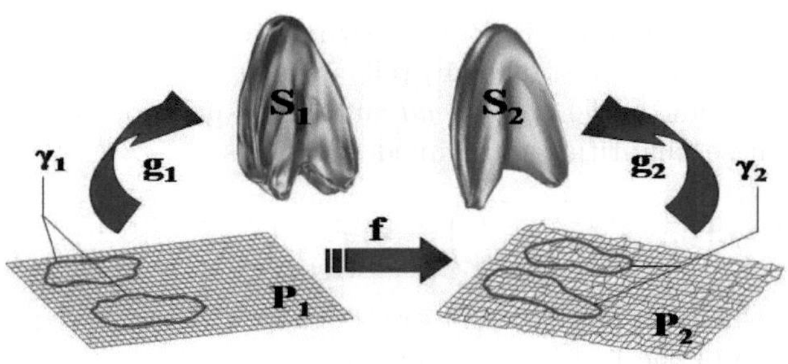


**Fig. 1. Illustration of framework.** Given homologous shapes $\mathbf{S}_1$ and $\mathbf{S}_2$, associate with flat rectangular patches (with appropriate BCs) $\mathbf{P}_1$ and $\mathbf{P}_2$ respectively. The patches store all Riemannian metric information $g_1$ and $g_2$ of their respective surfaces in terms of FFF matrices $G_1$ and $G_2$. Metric information is then matched between $\mathbf{P}_1$ and $\mathbf{P}_2$ through a homeomorphic map $f$ of $\mathbf{P}_1$ onto $\mathbf{P}_2$ (diffeomorphic except possibly on the curves $\gamma$) and a segmenting curve $\gamma_1$ in $\mathbf{P}_1$ which is carried by $f$ into $\mathbf{P}_2$ as $\gamma_2$. $\mathbf{S}_1$ and $\mathbf{S}_2$ are thus registered and segmented via the parametric correspondence. Note that $f$ can be visualized as the deformation of $\mathbf{P}_1$, and that the segmentation can be represented on $\mathbf{S}_1$ by shading interiors red and exteriors blue.

### 2.1 Differential Geometric Framework

Before elaborating on the matching criterion definition, we first show how it is possible to cast the shape registration and segmentation problem into a computational framework similar to that of flat 2D image registration/segmentation (save for the criterion itself). The main idea is to carry computations out on 2D rectangular domains through the use of appropriate parametrizations. To accomplish this, one need note only the following simple topological facts: (1)

hippocampi can be thought of as 2D Riemannian manifolds embedded in $\mathbb{R}^3$, (2) these manifolds are topologically equivalent to the sphere (closed and genus zero), (3) the exclusion of a pair of poles from the sphere produces a cylindrical topology, and (4) any surface topologically equivalent to the cylinder can be parametrized by a class of functions whose common domain is a single rectangular patch with periodic boundary conditions at a pair of opposite sides. Note that this last fact directly implies that any function defined on the surface in question can be fully represented as a function over a closed rectangular domain with the stated boundary conditions. In particular, a smooth map between two hippocampi can be represented as a smooth map (deformation field) between two parametric rectangular domains satisfying the boundary conditions. Through this simple and natural representation of the surfaces, our solution format has become quite similar to that of the more traditional 2D image segmentation and registration problem in that all of the computations are performed on and between rectangular domains. In contrast, when computation takes place in the embedding (3D) space before being restricted to an approximation to the surface within that space, one cannot draw nearly so direct a parallel.

Clearly, the matching criterion which we define in place of image intensity must communicate information about the shapes being registered and segmented. But a well-known theorem due to Bonnet (whose formal statement we omit for the sake of brevity: see for instance [4]) tells us that the first and second fundamental forms (FFF and SFF) contain *all* surface shape information. For the 2D manifold case, these quantities are defined as follows:

$$FFF = \begin{pmatrix} X_u \cdot X_u & X_u \cdot X_v \\ X_u \cdot X_v & X_v \cdot X_v \end{pmatrix} \quad SFF = \begin{pmatrix} N \cdot X_{uu} & N \cdot X_{uv} \\ N \cdot X_{uv} & N \cdot X_{vv} \end{pmatrix} \tag{1}$$

where $X$ is the 3D surface map, $u$ and $v$ are the map parameters (equivalently, the coordinate system of the parametric domain), $N$ is the surface normal (defined as $\frac{X_u \times X_v}{|X_u \times X_v|}$), and subscripts denote differentiation. There are different ways to conceptually distinguish the two quantities. For one, the geometric information contained in the FFF is of first order, while that of the SFF is second order in nature. Alternately, the FFF encodes the intrinsic geometry of the parameterized surface, while the SFF encodes the extrinsic geometry: the FFF encodes local length information, the SFF surface curvature (its trace, the mean curvature, is used as a matching criterion in [19]). Because of its lower differential order (and thus greater relative stability) and intrinsic nature, we choose to base our approach on comparison between the FFF tensors of the two surface parametric domains. This corresponds to the intuitive notion of shape identity, and indicates that the extent of failure to match this characteristic serves well as a measure of shape dissimilarity. We detail and formalize the algorithm in the following section on the system energy functional.

### 2.2 System Energy Functional

Let $\mathbf{S}_1$ and $\mathbf{S}_2$ be two surfaces in $\mathbb{R}^3$. The Euclidean metric in $\mathbb{R}^3$ induces Riemannian metrics $g_1$ and $g_2$ on $\mathbf{S}_1$ and $\mathbf{S}_2$, respectively. The goal is to segment regions in $\mathbf{S}_1$ and their corresponding regions in $\mathbf{S}_2$ while mapping those correspondents in such a manner as to optimally match their metric structures (optimality will be precisely defined immediately below). Specifically, the algorithm must compute a homeomorphism $f$ between $\mathbf{S}_1$ and $\mathbf{S}_2$ and a set of closed curves $\gamma_1$ on $\mathbf{S}_1$ and $\gamma_2$ in $\mathbf{S}_2$. The restriction of $f$ to the complement $\mathbf{S}_1 \backslash \gamma_1$ is a diffeomorphism between $\mathbf{S}_1 \backslash \gamma_1$ and $\mathbf{S}_2 \backslash \gamma_2$. If $U_1, \cdots, U_n$ and $V_1, \cdots, V_n$ denote the collections of open components of $\mathbf{S}_1 \backslash \gamma_1$ and $\mathbf{S}_2 \backslash \gamma_2$, respectively, then $f$ maps $U_i$ diffeomorphically onto $V_i$ for each $i$ and $f$ matches the Riemannian structures of $U_i$ and $V_i = f(U_i)$.

We solve the simultaneous segmentation and registration problem outlined above using a variational approach. The energy functional $\mathcal{E}$ is defined as a functional of a pair $(f, \gamma_1)$, where $\gamma_1$ is a set of closed curves on $\mathbf{S}_1$ and $f : \mathbf{S}_1 \rightarrow \mathbf{S}_2$ is a homeomorphism which is $C^\infty$ on $\mathbf{S}_1 \backslash \gamma_1$. Let $f^* g_2$ denote the pull-back metric on $S_1$. $\mathcal{E}(f, \gamma_1)$ is then defined as

$$\mathcal{E}(f, \gamma_1) = \int_{\mathbf{S}_1 \backslash \gamma_1} e \, dA + \alpha \int |d\gamma_1(t)/dt|_{g_1} dt$$
$$+\beta \left( \int_{\mathbf{S}_{1in}} |e - \bar{e}_{in}|^2_{g_1} dA + \int_{\mathbf{S}_{1out}} |e - \bar{e}_{out}|^2_{g_1} dA \right) \tag{2}$$

where

$$e = |f^* g_2 - g_1|^2_{g_1}, \tag{3}$$

$$\bar{e}_{in} = \frac{\int_{\mathbf{S}_{1in}} e \, dA}{\int_{\mathbf{S}_{1in}} dA}, \quad \bar{e}_{out} = \frac{\int_{\mathbf{S}_{1out}} e \, dA}{\int_{\mathbf{S}_{1out}} dA}, \tag{4}$$

$\mathbf{S}_{1in}$ and $\mathbf{S}_{1out}$ are the regions of $\mathbf{S}_1$ inside and outside of $\gamma_1$ respectively, and $\alpha, \beta$ are positive weighting constants.

The quantity $e$ defined in Eqn. 3 provides, at each point on $\mathbf{S}_1$, a measure of similarity between the Riemannian structures $g_1$ and $g_2$ as they correspond under the current estimate of $f$. As such it represents the local deviation from isometry, and its integral over the domain (we have adopted the convention of using $\mathbf{P}_1$, the parametric domain of $\mathbf{S}_1$) provides a global measure of how far the given deformation $f$ is from an isometry. This is the first term in the system energy represented in Eqn. 2. In local (parametric) coordinates, it is given by the distance between the two matrices $G_1$ and $J^t G_2 J$:

$$|J^t G_2 J - G_1|^2_{g_1}, \tag{5}$$

where $J$ is the Jacobian of $f$ when $f$ is expressed from the local coordinate system of $\mathbf{P}_1$ onto that of $\mathbf{P}_2$, and $G_1$ and $G_2$ are $2 \times 2$ positive definite matrices expressing the metrics $g_1$ and $g_2$ in the coordinate systems of $\mathbf{P}_1$ and $\mathbf{P}_2$ respectively, as in Eqn. 1. The second term in $\mathcal{E}$ restricts the length of the segmenting curve, so as to achieve the minimal necessary segmentation under the other energy constrains. Finally, the third and fourth terms place a homogeneity constraint on $e$

within each connected component defined by the segmenting curve. This corresponds to our notion that the curve should partition the surfaces in such a way as to isolate regions exhibiting one sort of deformation characteristic from those exhibiting another. Note that these terms are defined in terms of both $f$ and $\gamma_1$, and as such the minimization of their energy will drive both segmentation and registration.

## 3 Data Acquisition and Implementation of Energy Functional Minimization

The hippocampal surfaces were manually segmented from 3D MRI data by a trained neuroanatomist, and smooth surfaces were fit to the marked points using a deformable pedal surface as described by Vemuri and Guo in [16]. Each segmented hippocampal surface obtained from the application of this technique was represented by a $40 \times 21$ mesh of points on that surface, periodic in one direction. (This mesh is the instantiation of the surface parametric domain as discussed in Sec. 2.1.) Homologous hippocampi were brought into rigid alignment through application of the iterative closest points (ICP) algorithm following the segmentation. The rigid alignment step included extraction of volume data and normalization of the shapes with respect to this data.

The minimization of the system energy defined by Eqn. 2 is achieved by alternating between the deformation map and segmentation estimation processes. When optimization steps are sufficiently small, this alternation approximates minimization of $\mathcal{E}$ simultaneously with respect to both $f$ and $\gamma_1$. Because of the high differential order of the analytic Euler-Lagrange equation for $f$ (see [12]), we choose to minimize the functional $\mathcal{E}$ with respect to $f$ directly through a constrained optimization process (the 'LargeScale' version of Matlab's 'fmincon' function). The deformation is represented and evolved as a 2-vector field over the $40 \times 21$ grid representing $\mathbf{P}_1$ for each pair (periodic at one pair of ends, constant at the other). As the $\mathbf{P}_2$ grid is of equal size, this vector field implicitly carries $\mathbf{P}_1$ onto $\mathbf{P}_2$ (the constraint imposed upon the optimization is that this map is in fact onto). As the FFF is defined as in Eqn. 1, it is easily seen that at any given point it depends only on the derivatives of $X$ with respect to $u$ and $v$. The locality of the derivative operator leads naturally to a known Hessian sparsity structure. Exploitation of this sparsity structure makes this optimization computationally feasible. Minimization with respect to $\gamma_1$, on the other hand, is implemented via a level set segmentation as described by Chan and Vese in [3]. The only difference between this module of our method and that described therein is our generalization of the process to genus zero 2-manifolds, which entails respecting nonuniform surface length and area elements in accordance with the metric information.

## 4 Results

To illustrate the method's performance, we first present a visual display of its output and follow by demonstrating the classification power of the extracted features. The input dataset consisted of 60 L/R hippocampus pairs, collected and formatted as described in Sec. 3. Clinicians provided a trinary classification of this dataset: LATL (left atrophied temporal lobe) epileptic, RATL (right atrophied temporal lobe) epileptic, and Control. In this step, 6 samples were discarded due to ambiguity of their clinical class membership, leaving the final set at 54 patient samples (15 LATL, 16 RATL, 23 Control).

### 4.1 Output Visualization

Our result visualization format can best be understood through reference to Fig. 1, in which we present a holistic schematic. For each sample, we illustrate (1) the deformation map $f$ as the warp necessary to carry points in $\mathbf{P}_1$ to their correspondents in $\mathbf{P}_2$, (2) the segmentation $\gamma_1$ and the deformation energy function $e$ as they exist in $\mathbf{P}_1$, and (3) the segmentation $\gamma_1$ as it exists on $\mathbf{S}_1$ through the parametrizing function from $\mathbf{P}_1$ (there is a corresponding region on $\mathbf{S}_2$ as well, not depicted here but fully specified for given $f$ and $\gamma_1$ through the parametrizing functions: $\gamma_2$ is specified by $f$ and $\gamma_1$). As noted in Fig. 1, the segmentation is depicted on $\mathbf{S}_1$ through the shading of its interior and exterior with red and blue respectively: the curve(s) itself is then of course the boundary between the shaded regions.

We first consider the case of a pair of cylinders, where one member of the pair has had its surface distorted according to an outward normal vector field of magnitude dictated by a Gaussian function confined to a known support. This synthetic distortion produces visible protrusions on the affected surface. Since the support is known, we can use its segmentation as one quantification of accuracy. Fig. 2 illustrates the obtained results. Here, 86% of the distortion area is included in the segmentation, and 0% of the undistorted area is included in the segmentation. (Note that the 14% "misclassification" estimate is in a sense overly severe, as the two-mean framework is simply grouping the relatively mild distortion near the support boundary with the undistorted portion rather than with the severely distorted portion towards the support center.) We follow by demonstrating robustness of the segmentation in the face of reparametrization of the surface of comparison and random noise in the surface point cloud (see caption for parameter levels). When repeated under these conditions, the segmentation remains 100% consistent at the pixel resolution with that obtained prior.

For the real cases described previously, we choose to present the results obtained on one member of each class (RATL, LATL, and control), as Fig. 3. All figures depict a mirrored left hippocampus on the left side of each panel and the corresponding right hippocampus to which it is compared on the right of the panel. This aspect of the study (inherently) lacks ground truth, and visual inspection is the only way to confirm the "correctness" of results at this level. The provision of an input device to allow for expert segmentation (closed curves confined to the surface) is itself a research problem. This does not mean, however, that we cannot evaluate the quality of the data extracted from the real set. Immediately following, we present a classification analysis as evidence.

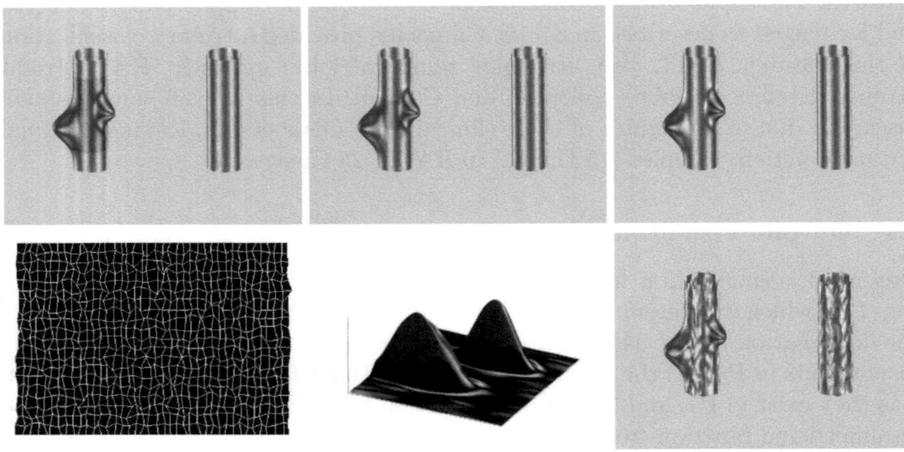

**Fig. 2. Top:** Segmentation of prominent distortions from cylindrical surface (left: initialization; middle: 5 iterations; right: 20 iterations). **Bottom:** Performance under surface noise ($\sigma = 0.01$ for unit cylinder) and reparametrization (left: warp field, $\sigma = 0.2$ in each parametric component; middle: segmented energy field in parametric domain; right: converged result, at 20 iterations)

### 4.2 Classification Analysis

If our algorithm does indeed produce more relevant shape information than is contained in the standard global measures, we should expect to see improved classification performance on incorporating this new information. Since we produce a segmented energy function over the entire surface (parametric) domain, we have options as to how to present features to the classifier algorithm. We can collect summary statistics that profit from both segmentation and deformation quantification, such as mean deformation energy ($e$ as defined in Eqn. 3) inside and outside of the segmenting curves. Alternately, once the shapes have been volume-normalized and mutually registered, we can compare the (quasi-)continuous $e$ functions across patients by forming feature vectors from the $e$ values at corresponding sample locations. There is also an array of supervised learning methods from which to select a preferred classification algorithm: we choose the ones observed to give best test performance on all feature sets examined, including volume alone (as further detailed below).

Fig. 4 demonstrates the success rates in classifying controls vs. epileptics (LATL and RATL groups combined into a single set) and Fig. 5 demonstrates analogous results for the problem of separating RATL and LATL members. The training and test statistics were collected through a standard leave-one-out cross validation procedure: test results should be regarded as the reliable performance indicators (and high discrepancy between training and test percentages as evidence of overfitting). The feature vectors being compared are (1) volume alone, (2) volume with a set of 6 summary statistics derived from the algorithm output

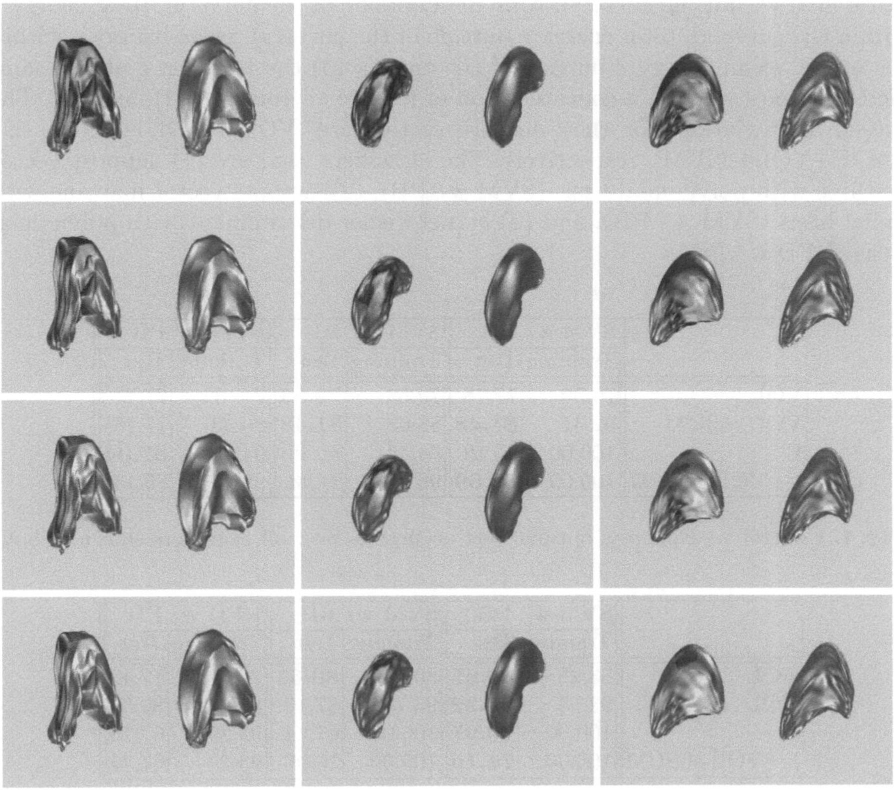

(a) Segmentation evolution on sample patients (top to bottom: initialization, 3, 10, 20 iterations.

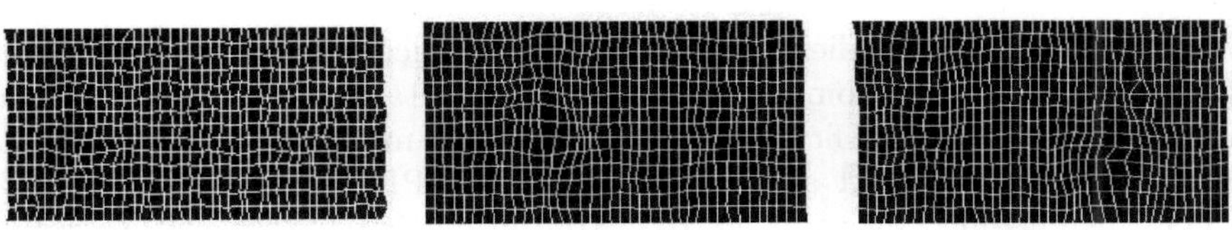

(b) Warp of parametric domain induced by evolution process.

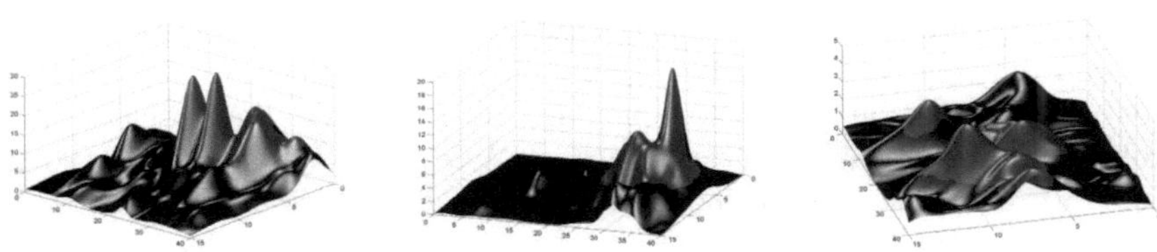

(c) Final segmented distortion energy $e$ graph as function over parametric domain

**Fig. 3.** Left column: LATL member; middle column: RATL member; right column: control member (total area within the curve(s), total area outside of the curve(s), total energy $e$ within the curve(s), total energy $e$ outside of the curve(s), mean energy $e$ within the curve, mean energy $e$ outside of the curve), (3) the function $e$ at 600 sampled locations, and (4) a concatenation of feature vectors (1), (2), and (3). The chosen abbreviations for these feature vectors are 'VOL', 'VOL+6SUM', 'E', and 'E+VOL+6SUM', respectively. The classifiers used are (1) support vector machine with polynomial basis (SVM w/ PB), (2) support vector machine with radial basis (SVM w/ RB), and (3) kernel Fisher discriminant with polynomial basis (KFD w/ PB).

| | SVM w/ PB | | SVM w/ RB | | KFD w/ PB | |
|---|---|---|---|---|---|---|
| | Training | Test | Training | Test | Training | Test |
| VOL | 79.59 | 77.78 | 79.63 | **79.56** | 77.81 | 77.78 |
| VOL+6SUM | 92.31 | **81.48** | 88.68 | **81.48** | 80.43 | 77.78 |
| E | 100.00 | 85.19 | 94.44 | 81.48 | 94.51 | **87.04** |
| E+VOL+6SUM | 100.00 | **88.89** | 96.26 | 81.48 | 99.97 | 85.19 |

**Fig. 4.** Control vs. Epilepsy, optimal test accuracies over all classifiers shown in bold

| | SVM w/ PB | | SVM w/ RB | | KFD w/ PB | |
|---|---|---|---|---|---|---|
| | Training | Test | Training | Test | Training | Test |
| VOL | 83.25 | **80.65** | 80.65 | **80.65** | 79.78 | 77.42 |
| VOL+6SUM | 93.55 | **90.32** | 93.76 | 87.10 | 93.87 | 80.65 |
| E | 100.00 | **70.97** | 100.00 | 67.74 | 99.89 | 67.74 |
| E+VOL+6SUM | 100.00 | **74.19** | 100.00 | 70.97 | 93.23 | 67.74 |

**Fig. 5.** LATL vs. RATL, optimal test accuracies over all classifiers shown in bold

The results of both studies make the method's success clear. In the case of distinguishing epileptics from controls, all experimental feature vectors demonstrated superior test classification accuracy, with a maximum performance of 88.89% on 'E+VOL+6SUM' as classified by SVM w/ PB, as compared to 79.56% for 'VOL' as classified by SVM w/ RB. Note in particular that feature vector 'E' outperforms 'VOL' as well, despite the fact that 'E' contains *no* volume information. This portion of the experiment confirms our notion that local distortion information is an important complement to volume data, and suggests that it can in some cases outweigh volume in independent relevance. While we do not observe this latter phenomenon in the LATL vs. RATL case, we do still observe substantial outperformance of 'VOL' by 'VOL+6SUM' (90.32% vs. 80.65%), again demonstrating the algorithm's extraction of clinically relevant shape information not represented by volume. The large difference between training and test accuracies for the other two feature vectors indicates overfitting, which is enabled by their higher dimensionalities. This indicates that an overly fine-grained approach can confound classification on this set, but that subregion identification and characterization remains nonetheless crucial. We have not come across clinical test classification accuracies of this level in any of the literature we have surveyed.

## 5 Conclusion

We have presented a novel scheme for simultaneous nonrigid registration and segmentation of homologous shapes, wherein the registration and segmentation processes are driven by intrinsic geometric characteristics of the shapes themselves. As such, we are able not only to identify surface pairs representing large deformations, but also to specify which subregions of the shapes appear most likely to be involved in the deformation, to produce an estimate of the implicit deformation field, and to quantify the deformation energy of the segmented subregions separately from that of the remaining surface patches. Given the number of considerations interplaying, the variational principle driving the evolution is notably compact. We have noted that the method's results conform to ground truth in cases where it can be meaningfully defined, and to expectation in cases where it cannot. The most important validation comes in the form of our substantial improvement of disease classification accuracy relative to standard volumetric analysis. In the case of epileptics versus controls, we have provided evidence that volume-independent local shape information is actually a more significant predictor than volume itself.

It should be noted that the two-region Chan-Vese-inspired framework (often used in segmentation applications for its strong balance of effectiveness and relative simplicity of implementation) is merely a good "first draft" of a segmentation scheme for this sort of problem. Extending the functional to the multiple mean case, or to that of generalized Mumford-Shah (of which the Chan-Vese approach is a special case), would potentially enable the identification of intermediate deformation levels currently being grouped with their nearest matches in the high/low system. We are currently investigating the application of norms based on geodesic distances between symmetric positive definite matrices (the class to which the FFF belongs) on their own theoretical manifold, as an improvement over the standard Frobenius norm. Finally, it goes without saying that all evolution/optimization-based methods (apart from the most trivial cases) suffer from some amount of initialization-dependence. Intelligent initialization schemes could be expected to further boost the algorithm's performance.

## References

1. A.Yezzi, L. Zollei, and T. Kapur. A variational framework for joint segmentation and registration. In *IEEE-MMBIA*, pages 44–51, 2001.
2. S. Bouix, J. C. Pruessner, D. L. Collins, and K. Siddiqi. Hippocampal shape analysis using medial surfaces. *Neuroimage*, 25(4):1077–1089, 2005.
3. T. F. Chan and L. A. Vese. Active contours without edges. *IEEE Transactions on Image Processing*, 10(2):266–277, 2001.

4. M. P. do Carmo. *Differential Geometry of Curves and Surfaces.* Prentice Hall, 1976.
5. G. Gerig et al. Age and treatment related local hippocampal changes in schizophrenia explained by a novel shape analysis method. In *MICCAI*, pages 653–660, 2003.
6. J. G. Csernansky et al. Hippocampal deformities in schizophrenia characterized by high dimensional brain mapping. *The American Journal of Psychiatry*, 159(12):2000–2006, 2002.
7. J.G. Csernansky et al. Preclinical detection of alzheimer's disease: hippocampal shape and volume predict dementia onset in the elderly. *Neuroimage*, 25(3):783–792, 2005.
8. K.M. Pohl et al. A bayesian model for joint segmentation and registration. *Neuroimage*, 31(1):228–239, 2006.
9. M. Styner et al. Statistical shape analysis of neuroanatomical structures based on medial models. *Medical Image Analysis*, 7(3):207–220, 2003.
10. R. H. Davies et al. Shape discrimination in the hippocampus using an mdl model. In *IPMI*, pages 38–50, 2003.
11. H. Jin, A. J. Yezzi, and S. Soatto. Region-based segmentation on evolving surfaces with application to 3d reconstruction of shape and piecewise constant radiance. In *ECCV*, number 2, pages 114–125, 2004.
12. F. Park and R. Brockett. Harmonic maps and the optimal design of mechanisms. In *Proc. 28th Conf. on Decision and Control*, pages 206–210,, 1989.
13. L. Shen, J. Ford, F. Makedon, and A. Saykin. Hippocampal shape analysis: Surface-based representation and classification. In *Medical Imaging 2003: Image Processing, SPIE Proceedings 5032*, pages 253–264, 2003.
14. M. Styner, J. A. Lieberman, and G. Gerig. Boundary and medial shape analysis of the hippocampus in schizophrenia. In *MICCAI*, pages 464–471, 2003.
15. G. B. Unal and G. G. Slabaugh. Coupled pdes for non-rigid registration and segmentation. In *CVPR*, pages 168–175, 2005.
16. B.C. Vemuri and Y. Guo. Snake pedals: Compact and versatile geometric models with physics-based control. *IEEE Trans. Pattern Anal. Mach. Intell.*, 22:445–459, 2000.
17. F. Wang and B. C. Vemuri. Simultaneous registration and segmentation of anatomical structures from brain mri. In *MICCAI*, pages 17–25, 2005.
18. L. Wang, S. C. Joshi, M. I. Miller, and J. G. Csernansky. Statistical analysis of hippocampal asymmetry in schizophrenia. *Neuroimage*, 14(3):531–545, 2001.
19. Y. Wang, M.-C. Chiang, and P. M. Thompson. Mutual information-based 3d surface matching with applications to face recognition and brain mapping. In *Proc. Int. Conf. on Computer Vision*, pages 527–534, 2005.
20. P. P. Wyatt and J. A. Noble. Map mrf joint segmentation and registration of medical images. *Medical Image Analysis*, 7(4):539–552, 2003.
21. C. Xiaohua, J.M. Brady, and D. Rueckert. Simultaneous segmentation and registration of medical images. In *MICCAI*, pages 663–670, 2004.

# Motion Compensated Video Super Resolution

Sune Høgild Keller[1], François Lauze[2], and Mads Nielsen[1,2]

[1] The Image Group, Department of Computer Science, University of Copenhagen, Universitetsparken 1, DK-2100 Copenhagen, Denmark
sunebio@diku.dk
http://www.image.diku.dk/sunebio
[2] Nordic Bioscience A/S, Herlev Hovedgade 207, 2730 Herlev, Denmark

**Abstract.** In this paper we present a variational, spatiotemporal video super resolution scheme that produces not just one but $n$ high resolution video frames from an $n$ frame low resolution video sequence. We use a generic prior and the output is artifact-free, sharp and superior in quality to state of the art home cinema video processors. Unlike many other super resolution schemes, ours does not limit itself to just translational or affine motion, or to certain subclasses of image content to optimize the output quality. We present a link between image reconstruction and super resolution and formulate our super resolution constraint with arbitrary up-scaling factors in space from that.

## 1 Introduction

In this paper we introduce a novel approach to motion compensated video super resolution. In Fig. 1(e) it is shown how our method is able to recreate the ground truth (Fig. 1(a)) of a skewing bar sequence while spatial methods (Fig. 1(c) and Fig. 1(d)) currently used in consumer electronics produce blurred outputs.

Super resolution (SR) is an ill-posed problem: one wants to reconstruct a hypothetical higher resolution image (sequence) from an image sub-sampling at a lower resolution. The inevitable loss of high frequency information follows immediately from the Nyquist-Shannon sampling theorem and SR algorithms hence normally use multiple low resolution views, perform registration on them and then extract a high resolution image. Super resolution is a thoroughly investigated subject in image processing (see for instance [6]) whereas video super resolution (VSR also known as multiframe super resolution) which by definition

(a) (b) (c) (d) (e)

**Fig. 1.** Frame no. 3 of the five frame skew sequence: (a) ground truth (arrows show skew motion), (b) down sampled by 0.5x0.5. 2x2 SR on (b): (c) bicubic, (d) bilinear and (e) our algorithm.

F. Sgallari, A. Murli, and N. Paragios (Eds.): SSVM 2007, LNCS 4485, pp. 801–812, 2007.

is the creation of $n$ high resolution (HR) video frames from $n$ low resolution (LR) video frames, is a lot less researched (see for instance [8]).

This paper is organized as follows. In Sec. 2 we present the low resolution image formation equation and the super resolution problem, and we discuss other work on super resolution and its relation to our work. In Sec. 3 we present our own motion compensated video super resolution scheme followed by a presentation of our experimental work in Sec. 4. Finally we conclude and look ahead in Sec. 5.

## 2 Overview of the Super Resolution Problem and Related Work

Following [12], a usual starting point for analysis of the SR problem is the continuous image formation equation that model the projection $\mathcal{R}$ of an HR image into an LR one:

$$u_0(y) = \mathcal{R}(u)(y) + e(y) := \int B(x, y)u(x)dx + e(y) \qquad (1)$$

where $u_0(y)$ is the low resolution irradiance field, $u(x)$ the high resolution one, $e(y)$ some noise and $B(x, y)$ is a blurring kernel. In general this kernel is assumed to be shift invariant and takes the form of a point spread function (PSF), $u_0 = B * u$. In the case of image sequences, $x$ can be replaced by $(x, t)$ where $t$ is the time variable.

The general numerical formulation is $L = RH + E$ where $H \in \mathbb{R}^N$, $L \in \mathbb{R}^n$ with $n < N$, $R$ a linear map $\mathbb{R}^N \to \mathbb{R}^n$ and $E \in \mathbb{R}^n$ a random vector. Inverting this equation is severely ill-posed and more information is needed to get a stable solution. First, more than one low resolution sample is used, and some prior on the distribution of HR images is introduced. A common type of prior imposes spatial or spatiotemporal regularity. Another one (that can be used simultaneously with the first one) relies on expected content of images. The general process can be described as follows: different LR images $L_1, \ldots, L_n$ are registered toward a common one, say $L_1$, via sub-pixel displacement fields $w_2, \ldots, w_n$ such that $L_1(y) = L_i(w_i(y))$, hoping to get as close as possible to the ground truth $H$. The earliest approach, modelled in the frequency domain, was proposed by Tsai and Huang in [16]. A different family of methods also emerged directly in the spatial domain where a maximum likelihood estimator (ML) will produce HR output $\bar{H}$ which minimizes the projection distances to all the sample LR images $L_i$. Adding priors will replace the ML estimator by a regularized ML or a maximum a posteriori (MAP) like estimator.

Borman and Stevenson have presented an extensive review of the different approaches in [2] and good references to recent work as well as another extensive bibliography can be found in [6]. In the work of Irani and Peleg [9] motion compensation (MC) is used to extract and register several frames from a given sequence and then create an HR image from them. Schultz and Stevenson [14] also integrate motion compensation as well as prior smoothness constraints more permissive for edges than the Gaussian one. Here too, authors aim at reconstructing one HR frame from a video sequence input. Generalizations of these methods have been used for spatiotemporal super resolution of video sequences by Shechtman *et al.* in [15], from a set of low resolution sequences (multi-camera approach). The recent video super resolution algorithm by Farsiu *et al.* [8] is a typical example of simplified modelling of the optical flow that prevents it from producing high quality results like the one given in Fig. 1(e).

Super resolution has its limits as shown by Baker and Kanade in [1] and recently by Lin and Shum in [12]. In order to overcome these limitations, Baker and Kanade have proposed to add a generative, trained model part to the reconstruction, a choice not possible for us, as we want a generic VSR algorithm applicable to any kind of video content.

The starting point of our work is Bayesian inference. Using (MAP) we derive a variational formulation via an energy minimization whose optimization takes the form of a non linear reaction-diffusion both spatially, and temporally along the motion trajectories, which are themselves computed using variational methods in an integrated framework. This is a technique known from image sequence inpainting [7] and [10]. The use of a highly precise variational optical flow algorithms enables us to capture flows even more complex and detailed then the one in the skew sequence example in Fig. 1. Even the skew motion would be a source of problems to the widely used block matching motion estimators.

Lillholm *et al.* [11] have presented a method for the reconstruction of images from a finite set of features measured on the image (linear filter responses) under smoothness constraints in a variational framework, a problem very close to ours. We follow their approach to find the correct energy minimization by orthogonally projecting the suggested energy back onto the solution hyperplane as dictated by the super resolution constraint that follows from the modelling of the image formation process and the definition of $R$ given above.

Our algorithm allows for arbitrary magnification factor but we will mainly look at two cases: a) The often presented 2x2 (magnification factor 2 in height and width) super resolution, and b) the upscaling from SD digital PAL video ($576 \times 720$) to 720p HD video ($720 \times 1280$). We can easily adapt our algorithm to do VSR to any other HD format (at least up to $1080 \times 1920$). In the two settings in focus we stay within the limits of SR and avoid the failure of to large magnification factors as shown in [1] and [12].

Our choice of focusing on video super resolution is motivated by real world needs. Today many high definition (HD) television formats exist along with the older standard definition (SD) formats like PAL, NTSC and SECAM. Screens for both broadcast and video viewing as well as computer screens have many different resolutions. People have large collections of SD DVDs and broadcasters have huge SD archives. The need for super resolution is there and as the display devices grow in size and viewing angle, the demand for high performance super resolution algorithms also grows. This also means that our main concern here is the output quality as judged by the human visual system (HVS). We focus on producing outputs sharper than the input while possibly removing errors (e.g. flickering) present in the input, and with no new artifacts (the typical ringing and other, see e.g. results given in [1] and [8]).

So far we have discussed scientific work on super resolution focussing on getting as close a possible to the ground truth. In actual products for video processing like the Faroudja and DVDO video processors, and in build-in processors in high-end DVD-players and displays (plasmas, projectors etc.) focus is on visual quality, but the majority of the resources are typically spent on deinterlacing, noise filtering, correction of MPEG-2 errors and color corrections. Unfortunately bilinear interpolation as in Fig. 1(d) is the standard method used for video super resolution as it is cheap, easy to implement and does not create artifacts. The smoothing it produces is not severely unpleasing to the HVS, but given the trend of larger and better displays it will not suffice over time.

## 3 Motion Compensated Video Super Resolution

We wish to model the image sequence content and optical flow using probability distributions and thus start formulating our problem using Bayesian inference. This has been done by Lauze and Nielsen in [10] for simultaneous image sequence inpainting and motion recovery, leaving out the locus of missing data, which we do not need, we are left with

$$p(u, \boldsymbol{v}|u_0) \propto \underbrace{p(u_0|u)}_{P_0} \underbrace{p(u_s)}_{P_1} \underbrace{p(u_t|u_s, \boldsymbol{v})}_{P_2} \underbrace{p(\boldsymbol{v})}_{P_3}. \tag{2}$$

where we $\boldsymbol{v}$ is the optical flow of the HR output sequence $u$, and $u_0$ is the input low resolution sequence. $u_s$ and $u_t$ are the spatial and temporal distribution of intensities respectively. On the left hand side we have the posterior distribution from which we wish to extract a MAP estimate. The right side terms are: $P_0$, the image sequence likelihood, $P_1$ the spatial prior on image sequences, $P_3$ the prior on motion fields and $P_2$ a term that acts both as likelihood term for the motion field and as spatiotemporal prior on the image sequence. In the sequel we assume that $P_0$ comes from a noiseless image formation equation (1) (i.e. $e(y) = 0$). This means that the optimal pair $(u, \boldsymbol{v})$ minimizes the constrained problem

$$\begin{cases} E(u, \boldsymbol{v}) = E_1(u_s) + E_2(u_s, u_t, \boldsymbol{v}) + E_3(\boldsymbol{v}) \\ Ru = u_0 \end{cases} \tag{3}$$

where $E_i = -\log P_i$ and $R$ the projection described in Sec. 2.

Applying Calculus of Variations, a minimizer for the above energy is characterized by the coupled system of equations, $\partial E(u, \boldsymbol{v})/\partial u = 0$ subject to $Ru = u_0$and $\partial E(u, \boldsymbol{v})/\partial \boldsymbol{v} = 0$, in our case solved first for the flow and secondly for the intensities. To get a reasonable tradeoff between tractability and modelling accuracy we generally chose to use total variation term for $E_i$ and for the optical flow related terms $E_2$ and $E_3$ we use the formulation of [3] as it is also done in [10] for motion compensated inpainting. We use [3] because it has been shown to yield some of the most precise flows (see [4]).

For the actual video super resolution part of our problem, $\partial E(u, \boldsymbol{v})/\partial u = 0$, we neglect the gradient matching term of [3], as it introduces a 4th order term in the Euler-Lagrange equation and we do not believe it will improve our results in the current settings. We are left with minimizing

$$E(u) = \lambda_s \int_\Omega \psi(|\nabla u|^2)dx + \lambda_t \int_\Omega \psi(|u(x+v, t+1) - u(x,t)|^2)dx \tag{4}$$

using both backwards and forwards flows (warps) in the second term as in [10]. $\Omega$ is the spatiotemporal domain of $u$, $u_t$ is the local temporal derivative of $u$ and $\nabla$ is the spatial gradient operator. Instead of using the $|\cdot|$ function which is non differentiable at the origin, we replace it by the approximation $\psi(s^2) = \sqrt{s^2 + \varepsilon^2}$, where $\varepsilon$ is a small positive constant. the $\lambda$'s are positive constants weighing the two terms with respect to each other.

### 3.1 The Super Resolution Constraint

Many things can be said about modelling image acquisition, but one thing is certain: it is a process with many variables and it changes with cameras and telecines (film scanners). As we wish to operate on all types of video, we can only do very little in this modelling step.

The point spread function (PSF) describes how the incoming light is sampled over the pixel area. Gaussian distributions are – as always – a popular choice along with the model we have chosen to use, the uniform distribution. In 2x2 SR the two models yield identical results, but in cases like SD to 720p HD their outcomes will be different, although the regularization part of our SR algorithm might minimize the difference. Replacing the uniform distribution with a gaussian would complicate modelling and whether it would improve results is an open question.

A second key element in super resolution is the magnification factor and our model allows for arbitrary, independent magnification factors in both frame height and width (integer to integer of course). Thus we do not require any preservation of aspect ratio in our super resolution constraint, which is fully on purpose as it enables us to change between different pixel aspect ratios in video, e.g. from PAL widescreen 1:1.422 pixels to square 1:1 HD pixels.

The super resolution constraint simply is: *The average pixel intensity over an area of interest is kept constant.* This essentially means that the projection operator $R$ is defined via an idealized point spread function $B$, whose support is an LR voxel (pixel + time spread), i.e. a moving average filter. This type of PSF is routinely used, often implicitly. We now describe the filters based on the super resolution constraint for both 1D and 2D signals.

**1D Super Resolution Filter.** The mapping $R_n^N : \mathbb{R}^n \to \mathbb{R}^N$ can be decomposed as a replication step $S : \mathbb{R}^n \to \mathbb{R}^{nN}$, where each source component is replicated $N$ times, followed by an average step $T : \mathbb{R}^{nN} \to \mathbb{R}^N$, where consecutive blocks of $n$ entries are replaced by their average. As an example shown in

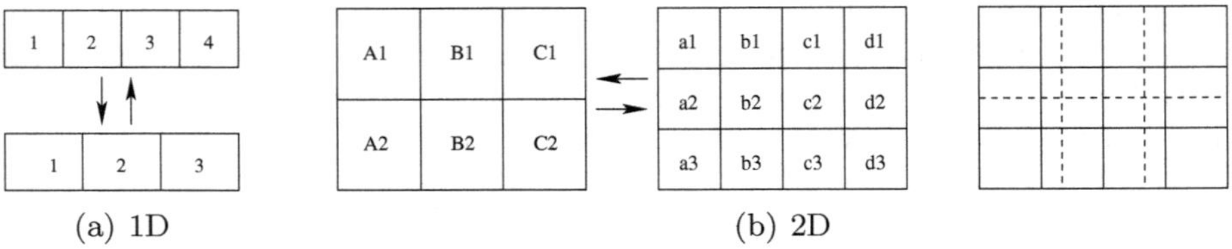


**Fig. 2.** Super resolution

Fig. 2(a), we present the $R_3^4$ mapping and its action on a vector $u \in \mathbb{R}^3$:

$$R_3^4 = \frac{1}{3}\begin{pmatrix} 3 & 1 & 0 & 0 \\ 0 & 2 & 2 & 0 \\ 0 & 0 & 1 & 3 \end{pmatrix}, \quad R_3^4\,(u_1, u_2, u_3) = \frac{1}{3}\,(3u_1, u_1 + 2u_2, 2u_2 + u_3, 3u_3)\,.$$

**2D Super Resolution Filter.** The 2D and higher dimensional filters of this form are separable, i.e. they can be obtained by cascading 1D filters along the appropriate dimensions. A 2D filter is simply obtained by defining the needed vertical and horizontal 1D filters, $R_v$ and $R_h$, and taking the Kronecker product of them (vectorizing the image matrix lexicographically column by column) as given here for the $m \times n = 2 \times 3$ to $M \times N = 3 \times 4$ example in Fig. 2(b)

$$R = \frac{mn}{MN} R_h \otimes R_v. \tag{5}$$

For larger image sizes, e.g. going from $576 \times 720$ SD to a $720 \times 1280$ HD, our projection will map pixel blocks of size $4 \times 9$ to blocks of size $5 \times 16$ as given by the greatest common divisors in height and width.

### 3.2 Minimization Algorithm

The full algorithm we use for video super resolution is described here. First we calculate the forward and backward flow on the low resolution sequence using the method from [3], that is finding the optimal solution of $\partial E(u, \boldsymbol{v})/\partial \boldsymbol{v} = 0$ by minimizing $E_2$ and $E_3$ in (3) using multiresolution. We then create the high resolution flow by applying $R$ (from (5)) on the LR flow, and in the same way we initialize the HR image sequence, from which we then compute the final HR sequence. We need to fix some notations first: $\nabla_3 u$ will denote the spatiotemporal gradient of $u$, $V = (\boldsymbol{v}^t, 1)^T)$ the spatiotemporal velocity field so that $u(x+v, t+1) - u(x, t) \approx \nabla u \cdot \boldsymbol{v} + u_t = \boldsymbol{V}^T \nabla_3 u$, with discretization suggested by the former expression (see [10] for details). We define $A(u) = 2\psi'(|\nabla u|^2)$ and $B(u) = 2\psi'(|V^T \nabla_3 u|^2)$. The gradient of the energy in (4) is

$$G(A(u), B(u)) := \frac{\partial E}{\partial u} = -\lambda_s \mathrm{div}_2\,(A(u)\nabla u) - \lambda_t \mathrm{div}_3\,\big(B(u)(V^T \nabla_3 u)V\big) \tag{6}$$

where $\mathrm{div}_2$ and $\mathrm{div}_3$ are the 2D and 3D divergence operators respectively.

Discretization is performed using schemes from [3] and [10]. In order to cope with nonlinearity in the gradient we use a fixed point approach. We only update the values $A(u)$ and $B(u)$ in each of a number of outer fixed point iterations in order to have a linear approximation of the discretized gradient coming from (6). For each outer iteration, we run a number of inner iterations on the now linearized system using a Gauss-Seidel relaxation scheme with the super resolution constraint incorporated by projection, a projection modulated by a positive weight, $\alpha$, in order to respect intensity bounds. Denoting the null space of $R$ by $T$ and by $P_T$ the orthogonal projection onto $T$, we sketch the algorithm here:

- Let $u^0$ be the initial guess for the HR sequence
- for $j = 0$ until $J - 1$
  1. Let $A^j := A(u^j)$, $B^j := B(u^j)$, $u^{j,0} := u^j$
  2. for $k = 0$ until $K - 1$
     (a) from $u^{j,k}$, compute $\bar{u}^{j,k+1}$ by one Gauss Seidel sweep on $G(A^j, B^j)$
     (b) form the update vector $\bar{\delta}^{j,k} := \bar{u}^{j,k+1} - u^{j,k}$
     (c) project it on $T$, $\delta^{j,k} := P_T(\bar{\delta}^{j,k})$
     (d) update current estimate $u^{j,k+1} := u^{j,k} + \alpha\delta^{j,k}$
  3. $u^{j+1} := u^{j,K}$
- output the HR sequence $u^J$.

Form classical linear algebra, the orthogonal projection is

$$P_T \equiv Id - R^T(RR^T)^{-1}R. \tag{7}$$

As mentioned in Sec. 3.1 the projection allows processing in blocks, so for any $M \times N$ block step 2(c-d) above is actually

$$u_l^{j,k+1} = u_l^{j,k} - \alpha \left( \bar{\delta}_l^{j,k} + \sum_{i=1}^{MN} P^T(l,i)\bar{\delta}_i^{j,k} \right), \qquad l = 1, ..., MN \tag{8}$$

The diffusion part of our algorithm follow the minimum-maximum principle ([13]) and keeps intensity values within bounds, but we need the weight $\alpha$ in (8) to ensure that the projection does not create values out of bounds. If a value in a block is detected out of bounds, $\alpha$ is recursively lowered from 1 to 0 in steps of 0.1, and all values in the block recalculated with the new $\alpha$, stopping when all values in the block are within bounds. This can potentially stop the evolution of the sequence, but it was found not to do so in testing. The problem was limited to very few regions around extremely high contrast edges and thus the computational overhead was also neglectable.

## 4 Experiments

The goal of our research is to improve the perceived quality by human viewers of super resolved video and we will therefore evaluate the visual quality of the output sequences. As mentioned in Sec. 2 we evaluate results at two different magnifications: The industry oriented 576p SD ($576 \times 720$) to 720p HD ($720 \times 1280$) and the classic 2x2 VSR. We work on the luminance channel (8 bit, [0-255]) of 5-20 frame sequences of standard DVD Video material telecined from 35mm film. First though, we wish to emphasize the importance of the super resolution constraint by a small experiment.

### 4.1 The Importance of the Super Resolution Constraint

There is no doubt that the super resolution constraint adds complexity to an already complex regularization scheme, so why not leave it out? An example of its importance as data attachment term is given here. In Fig. 3(a) we have one of four identical LR frames, a $4 \times 9$ matrix filled line by line with the values 1 to 36, and in Fig. 3(b) the corresponding HR frame, a $6 \times 12$ matrix initialized using the example matrix in (5).[1] In Fig. 3(c) we see how pure regularization without the SR constraint destroys the image content, whereas it is preserved in Fig. 3(d) with the SR constraint on. The destruction process is significantly slowed down if the temporal regularization is given high weight and the flow is accurate.

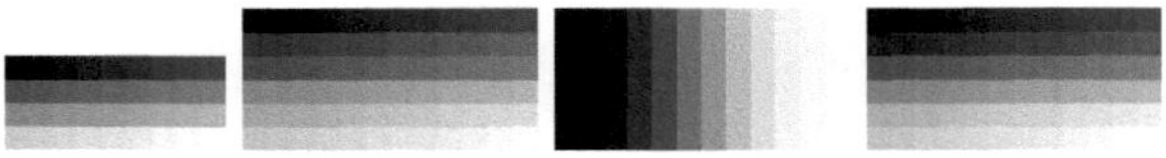

(a) (b) (c) (d)

**Fig. 3.** The importance of the SR constraint

### 4.2 Parameter Settings

We use the algorithm as given in Sec. 3.2 and give the used settings here.

After extensive testing we have found what we believe to be the best parameter setting of the optical flow calculations when used on real world SD resolution video data. We use 5 outer and 20 inner loops on a 100 level pyramid with a scale factor of 1.04 between the levels. The weight of the gradient constancy assumption is set to 200 and the weight on the smoothness term is 70 (see [3]). As there is no ground truth flow to compare to, we have no way of guaranteeing that this setting is optimal, but we believe it to be from extensive testing.

For the actual super resolution part, finding the optimal output sequence, we tested a wide range of settings for its four free parameters. First the number of inner and outer iterations as described in Sec. 3.2. We found that $5 \times 10$ (5 outer, 10 inner) iterations gives optimal results, but that even as little as $5 \times 5$ and $3 \times 5$ iterations yield practically the same quality. Even with the results given here using $5 \times 10$ iterations, this part is still much faster than the flow calculation part of our algorithm. For the spatial and temporal regularization weights $\lambda_s$ and $\lambda_t$ in (4) and (6) we have chosen four settings that we generally used in our experiments. They are: Temporal favoring $S_t$ ($\lambda_s$:$\lambda_t$ = 1:5), $S_{st}$ with equal weight (1:1), spatial favoring $S_s$ (5:1) and pure spatial $S_{ss}$ (1:0).

[1] The HR initialization looks so fine because there is no sharp edges in this example.

### 4.3 Results

The output quality for our two test cases, 2x2 video super resolution and SD to 720p HD VSR, are very similar on all tested sequences. Of the four test settings, spatial favoring $S_s$ gave the best results. First off one would think the $S_t$ setting favoring temporal regularization would be best, but the simple flow initialization we use causes some blockiness in the initialization to be preserved as (false) sharp details. The equal weighing $S_{st}$ does not have this problem and its output is often indistinguishable from the output of $S_s$. Switching off the temporal regularization completely in $S_{ss}$ gives a small loss of overall sharpness in all tests and the outputs are far sharper than the bilinear outputs.

Besides the problems with $S_t$, which will be discussed later in this section, we get sharp outputs for all 20 sequences in test. The outputs have no artifacts from the upscaling process and subsampling artifacts present in the LR inputs are reduced to some degree. All the $S_s$ (and most $S_{st}$) results are perceived quite a lot sharper than their LR inputs. Our evaluations was of course done on video

(a) LR input (b) Bilinear 2x2 SR

(c) Our 2x2 VSR (d) Bicubic 2x2 SR

(e) LR input (f) Our 720p VSR

**Fig. 4.** Inputs and results I. Best viewed electronically (zoomed to 125% or more).

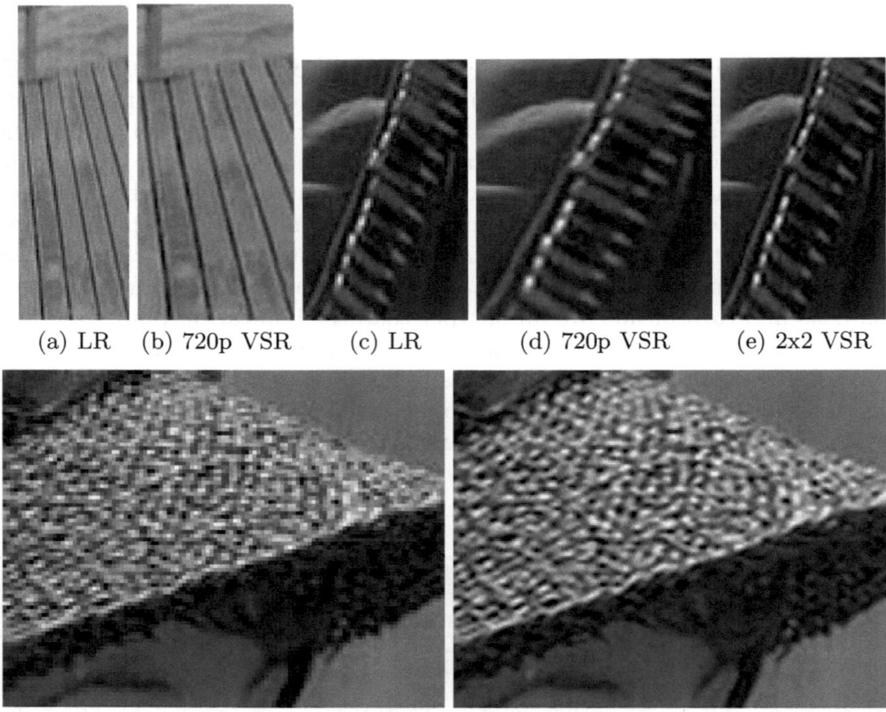

(a) LR (b) 720p VSR (c) LR (d) 720p VSR (e) 2x2 VSR

(f) LR (g) 2x2 VSR

**Fig. 5.** Inputs and results II. Best viewed electronically (zoomed to 125% or more).

but to give an impression of the quality obtained, Fig. 4 and Fig. 5 show some inputs and results produced with the $S_s$ setting. Due limited space $S_{st}$ and $S_{ss}$ will not be show – they are practically the same as $S_s$ – and $S_{st}$ neither – they are not unsharp but tend towards the LR inputs in blockiness. We show the HR outputs and the LR input at the (relative) same size to clearly show the enhancement resulting from the upscaling process. This also illustrate why high resolution is essential to large displays used with large viewing angles.

In 4(a) we see part of one frame of the LR input test sequence `Front Grill`. The initialized 2x2 HR input to our algorithm looks just like the LR. The result of bilinear 2x2 SR shown in 4(b) is clearly not as sharp as the 2x2 VSR output of our algorithm in 4(c). The bicubic 2x2 SR shown in 4(d) is not as sharp as our output and white details seem greyish, but it is a bit better than bilinear. The example given here is the best performance observed from bicubic, that generally lies closer to bilinear than our method in quality. Figure 4(f) is the 720p HD VSR output with correct aspect ratio as created by our algorithm from the anamorph LR input in 4(e). A lot of flickering is removed in the far away part of the wire fence (to see this video can be found at the authors home page). Note the good continuation of the wires in the lower left corner behind the policeman, and the contour of, and text on, the helmet. The cracks between the boards in the boardwalk in the 720p output in 5(b) look natural compared to how they look in the LR input in 5(a). In 5(d) and 5(e) we see how the bullets are depicted much better than in the input 5(c), and in 5(g) it is clearly seen how the straw hat suddenly get it 'true' structure represented when compared to the LR input in 5(f). In the `straw hat` and `wire fence` sequences, and seen very clearly in another test sequence (a helicopter flyby shot of Manhattan found as video on the authors home page), a major enhancement from the LR inputs to our HR outputs is flicker reduction, which cannot be seen in the figures.

We know the $S_t$ setting favoring temporal input has the potential to improve on the current results, especially it can increase the sharpness. This can be obtained by replacing the simple HR flow initialization we use now with a simultaneous upscaling of the flow and intensities. The $S_t$ setting already shows a tendency to push sharp details, although with the current simple flow it is easily overdone resulting in blockiness. The rather sharp $S_s$ and $S_{st}$ outputs show that we already gain from the temporal regularization as even our current flow will convey information of subsampled details when they are part of uniformly moving regions. The still relative sharp $S_{ss}$ outputs show that edges at motion boundaries – where flow accuracy suffers the most from the simple upscaling – are handled reasonably well by the spatial regularization.

## 5 Conclusion and Future Work

We have developed and tested a novel motion compensated video super resolution scheme with arbitrary upscale factors in space, and a super resolution constraint based on image reconstruction. In comparison to many other video super resolution algorithms we allow for any kind of flow, and our modelling is based on a generic prior as we aim at general applicability in video processing pipelines. From $n$ LR frames we produce $n$ sharp and artifact-free HR frames with reduction of flicker, clearly outperforming the (industrial) standard bilinear interpolation.

The major drawback of our algorithm in its current form is that it does not employ the full potential of our framework. We are currently working on an algorithm simultaneously updating the flow and intensities and expect to give even sharper outputs, possibly with an additional reduction of flicker. We also intend to reformulate our framework to enable temporal super resolution as well as adapting and applying it to 3D and 4D medical data. In terms of running time we are a factor of 500-1000 from realtime in the intensity part of our algorithm using unoptimized code. The flow part of the algorithm is much slower, but real time variational optical flow using a standard PC has been presented in [5]. Using modern, relatively cheap, HD capable field programmable gate arrays (FPGA) a real time implementation seems close at hand.

## References


1. S. Baker and T. Kanade. Limits on Super-Resolution and How to Break Them. *IEEE Trans. on Pattern Analysis and Machine Intelligence*, 24(9):1167–1183, 2002.
2. S. Borman and R. Stevenson. Spatial Resolution Enhancement of Low-Resolution Image Sequences: A Comprehensive Review with Directions for Future Research. Technical report, Laboratory for Image and Sequence Analysis (LISA), University of Notre Dame, July 1998.
3. T. Brox, A. Bruhn, N. Papenberg, and J. Weickert. High Accuracy Optical Flow Estimation Based on a Theory for Warping. In T. Pajdla and J. Matas, editors, *Proceedings of the 8th European Conference on Computer Vision*, volume 4, pages 25–36, Prague, Czech Republic, 2004. Springer–Verlag.
4. A. Bruhn, J.Weickert, and C.Schnörr. Lucas/Kanade Meets Horn/Schunck: Combining Local and Global Optic Flow Methods. *International Journal of Computer Vision*, 61(3):211–231, 2005.
5. A. Bruhn, J. Weickert, C. Feddern, T. Kohlberger, and C. Schnörr. Variational Optic Flow Computation in Real-Time. *IEEE Trans. on Image Processing*, 14(5):608–615, 2005.
6. S. Chaudhuri, editor. *Super-Resolution Imaging.* The International Series in Engineering and Computer Science. Springer, 2001.
7. J.P. Cocquerez, L. Chanas, and J. Blanc-Talon. Simultaneous Inpainting and Motion Estimation of Highly Degraded Video-Sequences. In *Scandinavian Conference on Image Analysis*, pages 523–530. Springer-Verlag, 2003. LNCS, 2749.
8. S. Farsiu, M. D. Robinson, M. Elad, and P. Milanfar. Fast and Robust Multiframe Super Resolution. *IEEE Trans. on Image Processing*, 13(10):1327–1344, 2004.
9. M. Irani and S. Peleg. Motion Analysis for Image Enhancement: Resolution, Occlusion, and Transparency. *Journal on Visual Communications and Image Representation*, 4(4):324–335, 1993.
10. F. Lauze and M. Nielsen. A Variational Algorithm for Motion Compensated Inpainting. In S. Barman A. Hoppe and T. Ellis, editors, *British Machine Vision Conference*, volume 2, pages 777–787. BMVA, 2004.
11. M. Lillholm, M. Nielsen, and L. Griffin. Feature-Based Image Analysis. *IJCV*, 52(2/3):73–95, 2003.
12. Z. Lin and H-Y Shum. Fundamental Limits of Reconstruction-Based Superresolution Algorithms under Local Translation. *IEEE Trans. on Pattern Analysis and Machine Intelligence*, 26(1):83–97, 2004.
13. P. Perona and J. Malik. Scale-Space and Edge Detection Using Anisotropic Diffusion. *IEEE Trans. on Pattern Analysis and Machine Intelligence*, 12(7):629–639, July 1990.
14. R.R. Schultz and R.L. Stevenson. Extraction of High Resolution Frames from Video Sequences. *IEEE Trans. on Image Processing*, 5(6):996–1011, 1996.
15. E. Shechtman, Y. Caspi, and M. Irani. Space-Time Super-Resolution. *IEEE Trans. on Pattern Analysis and Machine Intelligence*, 27(4):531–545, 2005.
16. R. Tsai and T. Huang. Multiframe Image Restoration and Registration. In *Advances in Computer Vision and Image Processing*, volume 1, 1984.

# Kullback Leibler Divergence Based Curve Matching Method

Pengwen Chen, Yunmei Chen*, and Murali Rao**

Department of Mathematics, University of Florida,
Gainesville, FL, USA 32611
pchen@math.ufl.edu,yun@math.ufl.edu, rao@math.ufl.edu

**Abstract.** In this paper, we propose a variational model for curve matching based on Kullback-Leibler(KL) divergence. This framework accomplishes the difficult task of finding correspondences for a group of curves simultaneously in a symmetric and transitive fashion. Moreover the distance in the energy functional has the metric property. We also introduce a location weighted model to handle noise, distortion and occlusion. Numerical results indicate the effective of this framework. The existence of this model is also provided.

## 1 Introduction

This paper presents a novel method of finding the optimal correspondences among curves. The curve matching problem is one of the main topics in computer vision and shape recognition. The matching task is a key issue in many applications, for instance, cursive script recognition [14], image tracking, video indexing, image segmentation and registration [6] [13].

One of the important approaches to finding optimal correspondences is based on minimizing a mixture of stretching and bending energies such that main features on curves can be matched. A variety of shape signature matching methods involving curvatures, bending angles, and orientations have been proposed depending on applications, e.g. [4] [12] [16]. Recently, Klassen et al. [9] [10](Chap. 12) developed an approach to shape matching by computing geodesics on a shape manifold. The numerical implementation later was improved by F. Schmidt et al. [11].

A study of these techniques compels us to posit that curve matching model having the following properties is preferable: First, the cost functional is differentiable, the correspondence is symmetric and robust to noise, distortion and occlusion (e.g. see [12]). Second, the distance between two corresponding curves has a metric property and leads to a nearest neighbor in a large database [15]. Third, the model is capable of finding the correspondences among a group of curves simultaneously, instead of two curves pairwise at a time.

* Chen is partially supported by NSF CCF-0527967 and NIH R01 NS052831-01 A1.
** Rao is partially supported by R01 NS42075.

F. Sgallari, A. Murli, and N. Paragios (Eds.): SSVM 2007, LNCS 4485, pp. 813–824, 2007.

Tagare [13] successfully solved the symmetry problem by introducing a bimorphism in the space of pairs of correspondences, whose two components determine the correspondence between the curves. This bimorphism optimizes a cost functional consisting of two parts. One of these forces the bimorphism close to uniform mapping, and the other minimizes the local curvature difference. Tagare et al. base their energy function on $L^2$ norm. However, an extension of this technique to matching a group of curves simultaneously requires searching of multi-morphisms in high dimension, which limits its scope. In [12] Sebastian et al. proposed a symmetric model based on minimizing a functional measuring the differential arc length and orientation in $L^1$ metric. They justified that this distance is in fact a metric. This metric property is helpful in indexing large databases efficiently [15]. However, their cost functional is not differentiable in its argument. This makes a mathematical treatment harder. Moreover, this model is limited to finding the correspondence between pairs of curves only. A more discussion about the variety of costs can be found in [4], but most of them are limited to pairs of curves only. Pairwise matching is less robust to noise, distortion, occlusion, and may lead to undesirable correspondences, since the information retrieved from a group is more reliable than from just any two. Moreover, it is clearly important to have the following transitivity property: if $f, g, h$ are the correspondence mappings from $a$ to $b$, $b$ to $c$, and $a$ to $c$, respectively, then $g \circ f = h$. It is difficult to construct symmetric pairwise mappings with this property, but as will be shown below this can be achieved by our method of matching a group of curves simultaneously.

In this paper, we propose a Kullback-Leibler (KL) divergence based model that aims to find the optimal correspondents among a group of curves with the features mentioned above. The strategy is to find an underlying curve called the "average curve" and the mappings from it to each curve in the group simultaneously by minimizing a cost functional consisting of two terms. The "stretching term" is measured by KL divergences [7] between the derivatives of correspondence mappings; the "bending term" is measured by Euclidean distance of local features.

Noise affects shapes of curves, and distorts the correspondence mapping. In this paper, we use KL divergence and Euclidean distance to quantify the distortions in stretching and bending, respectively. This approach has several features:

1. Assuming the distortion of the corresponding mapping follows a multinomial distribution, the optimal mapping whose stretching energy based on KL divergence preserves the order of sampling points ( i.e. the derivative of the mapping function is positive everywhere ) ensuring bijection among curves. While in models [12] [13] this constraint was enforced artificially.
2. Since the KL divergence and Euclidean distance both belong to the class of Bregman divergences [2] [5], alternate minimization algorithm can be used to perform our task [3].
3. The cost function is differentiable, making the mathematical analysis of its structure amenable, and we are able to construct models meeting specific

needs. For instance, it can be shown that the square root of the cost is also a metric in a slightly revised model.

Furthermore, since the selection of the parameter depends on the intensity of noise which may be location dependent, we also provide a location weighted model to deal with curves contaminated by location dependent noise, especially in cases of occlusion.

The result presented here mainly focuses on open curves. With an extra effort in searching for the optimal initial point, our discussion can be generalized to closed curves.(For instance, see [12].)

## 2 Model Description

KL-divergence is intimately connected with the maximal likelihood estimator for multi-nomial distributions. For instance, Stein's lemma [7]: Drawing $k$ times independently with distribution $Q$ from a finite set $X$, the probability of obtaining $x \in X^k := X \times \ldots \times X$ is $Q^k(x) = \prod_{a \in X} Q(a)^{N(a)} = e^{\sum_{a \in X} N(a) \log Q(a)}$, here $N(a)$ denotes the number of occurrences of $a \in X$. Now, given the empirical distribution $P$ defined by $P(a) = \frac{N(a)}{k}$, the probability of this empirical distribution is $e^{-kKL(P,Q)}$ to the first order in the exponent.

In the following, we model the curve matching as the dissimilarity of two distributions $P, Q$. Suppose given $n$ points $0 < x_1 < \ldots < x_n < 1$ in the interval $I_1 = [0, 1]$. Consider a distorted mapping function $f$ mapping $\{x_i, i = 1, \ldots, n\} \in I$ to $\{y_i, i = 1, \ldots, n\} \in I_1$. How should we arrange $n$ points $0 < y_1 < \ldots < y_n < 1$, such that the matching pairs $(x_i, y_i)$ is associated with " some likelihood function"?

We denote the lengths of $n$ subintervals $q_i := x_i - x_{i-1}, i = 1, \ldots, n$, then we have $\sum_{i=1}^n q_i = 1$. Let $X$ be a random variable, $X := \{1, \ldots, n\}$ with $\{Prob(X = i) = q_i, i = 1, \ldots, n\}$. Then $X$ follows a multi-nomial distribution. Denote $Q := \{q_1, .., q_n\}$. Denote $p_i := y_i - y_{i-1}, i = 1, \ldots, n$, and $P := \{p_i, i = 1, \ldots, n\}$. $KL(P, Q)$ measures the distortion in $f$. Clearly, let $y_i = x_i, i = 1, ..., n$, i.e. $f = id$. we have the minimal value $KL(P, Q) = 0$, which maximize $e^{-kKL(P,Q)}$.

In case of multiple curves, suppose $m$ multi-nomial distributions $P_1, \ldots, P_m$ are observed from $m$ independent random variables $X_1, \ldots, X_m$ based on the identical sampling number, which are supposed to follow the distribution $Q$, then $\sum_{i=1}^m KL(P_i, Q)$ quantify an overall distortion. To have nontrivial optimal matching, we consider a bending energy, measuring the likelihood of angle functions affected by Gaussian noises.

Give a group of curves $\{\mathcal{C}_i, i = 1, .., N\}$, our task is to determine an "average curve" $\mathcal{C}_0$, and the optimal correspondence functions $g_i : I_0 \to I_i$. We propose maximizing the conditional probability $P(\mathcal{C}_0, g_i, i = 1, \ldots, n | \mathcal{C}_i, i = 1, \ldots, n)$. Assume $P(\mathcal{C}_0, g_i)$,$P(\mathcal{C}_i | \mathcal{C}_0, g_i)$ follow Multinomial and Gaussian distributions respectively. By the Bayes formula, we have

$$P(\mathcal{C}_0, g_i | \mathcal{C}_i) = \frac{P(\mathcal{C}_0, g_i) P(\mathcal{C}_i | \mathcal{C}_0, g_i)}{P(\mathcal{C}_i)},$$

Our task is to find a optimal mapping with maximizing the conditional likelihood $P(\mathcal{C}_0, g_i|\mathcal{C}_i)$. Our approach is to maximize the log-likelihood function $\log P(\mathcal{C}_0, g_i|\mathcal{C}_i)$ to get the optimal correspondents among curves.

## 2.1 Definitions and Notations

To setup our model we begin by definitions.

**Definition 1.** *Given four intervals $I := [0, S]$, $I_1 := [0, S_1]$, $I_2 := [0, S_2]$ and $I_0 := [0, S_0]$. The feasible mapping set( from $I$ to $I_1$) is denoted by*

$$G_1 := \{g_1 | g_1 : I \to I_1, g_1'(s) > 0, g_1(0) = 0, g_1(s) = s_1, g_1(S) = S_1\}.$$

*Every mapping $g_1 \in G_1$ has the inverse mapping from $I_1$ to $I$, given by $g^{-1}(s)$, $s \in [0, S_1]$. Similarly we denote by $G_2, G_0$ the set of all mappings from $I$ to $I_2$, and $I_0$ respectively. We call $g(s) := \frac{S_1}{S} s$ as the uniform mapping in $G_1$. Let also $L^1_+ := \{f \in L^1, f \geq 0\}$.*

*Define a mapping divergence between two mappings $g_1 \in G_1, g_0 \in G_0$, by*

$$CKL_I(g_1', g_0') := \int_0^S g_1' \log \frac{g_1'}{g_0'} - g_1' + g_0' ds,$$

*(or just $CKL(g_1', g_0')$ if there is no possibility for confusion.) In case of equal arc length $S_1 = S_0 = 1$, this $CKL$ becomes KL divergences between two functions $g_1', g_0'$. And for any $g_1 \in G_1, g_2 \in G_2$, define a mapping distance:*

$$JS(g_1', g_2') := \inf_{S_0 \geq 0, g_0 \in G_0} (CKL(g_1', g_0') + CKL(g_2', g_0')).$$

Note that every bijection from $I$ to $I_i$ corresponds to a mapping in $G_i$, $i = 0, 1, 2$.

*Remark 1.* Note that any $g \in G_1$, $JS(g', \bar{g}')$ is always finite. However, $\|g' - \bar{g}'\|_{L^2}$, may be infinite. A simple example is, $g(s) := \sqrt{s}$ with $S_1 = S = 1$.

**Definition 2.** *Now, consider three planar curves $\mathcal{C}_i : \theta_i(s_i) \in C^\infty(I_i)$,( $\theta_i(s_i)$ is the absolute orientation at $s_i$ along the curve $\mathcal{C}_i$) with the associated mapping functions $g_i \in G_i$, $i = 0, 1, 2$. Given a tradeoff parameter $C$, define the elastic divergence $D(g_1, g_0)_{\mathcal{C}_1,\mathcal{C}_0} := G_1 \times G_0 \times C^\infty(I_1) \times C^\infty(I_0) \to \mathbb{R}$ by*

$$\int_0^S (g_1' \log \frac{g_1'}{g_0'} - g_1' + g_0')ds + C \int_0^S (\theta_1(g_1(s)) - \theta_0(g_0(s)))^2 ds,$$

*and similarly define $D(g_2, g_0)_{\mathcal{C}_2,\mathcal{C}_0}$. Define the elastic energy $E := C^\infty(I_1) \times C^\infty(I_2) \to \mathbb{R}$ by*

$$E(\mathcal{C}_1, \mathcal{C}_2) := \inf_{g_1, g_2, g_0, \mathcal{C}_0} (D(g_1, g_0)_{\mathcal{C}_1,\mathcal{C}_0} + D(g_2, g_0)_{\mathcal{C}_2,\mathcal{C}_0}).$$

*Then the $\mathcal{C}_0$ minimizing the above elastic energy will be called the "average curve".*

Now we give our theorem which states an important fact: given the correspondence mapping, the "optimal average curve" is uniquely determined. Due to the limit of pages, most of proofs are omitted.

**Theorem 1 (Optimal Condition on Average Curve).** *The above elastic energy $E(\mathcal{C}_1, \mathcal{C}_2)$ can be reformulated as*

$$\inf_{g_1,g_2} \Big( \int_0^S (g_1' \log \frac{2g_1'}{g_1'+g_2'} + g_2' \log \frac{2g_2'}{g_1'+g_2'}) ds + \frac{C}{2} \int_0^S (\theta_2(g_2(s)) - \theta_1(g_1(s)))^2 ds \Big), \tag{1}$$

*the minima taken over $G_1 \times G_2$. Then the optimal $g_0$ is given by $g_0(s) = (g_1(s) + g_2(s))/2$ for all $s \in I$, and the optimal curve $\mathcal{C}_0$ is described by $\theta_0(g_0(s)) = (\theta_1(g_1(s)) + \theta_2(g_2(s)))/2$.*

The following remark allows us to treat the "average curve" as the underlying template, which simplifies the matching task.

*Remark 2 (Parametrization Independence).* One of the remarkable properties of the mapping divergence $CKL$ or the mapping distance $JS$ is their independence of the underlying interval $I$. That is, choose another interval $I_\beta := [0, S_\beta]$, and any map $h : I_\beta \to I$, with $h(0) = 0, h(S_\beta) = S$, $h'(s) > 0$ for all $s \in [0, S_\beta]$. Then given two maps $g_i \in G_i$, $g_i \circ h : I_\beta \to I$, $i = 0, 1$, we have

$$CKL_{I_\beta}((g_1 \circ h)', (g_0 \circ h)') = CKL_I(g_1', g_0').$$

Therefore, **the choice of the underlying interval does not affect the mapping divergence or the mapping distance**.

Now we choose $I_\beta = I_0$. Then $I = I_0$, $S = S_0$, the map $g_0$ becomes the identity map $g_0(s) = s$, and $g_1'(s) + g_2'(s) = 2g_0'(s) = 2$ for all $s \in I$, $2S_0 = S_1 + S_2$, and

$$CKL(g_1', g_0') = \int_0^S (g_1' \log g_1' - g_1' + 1) ds, \text{ then we have}$$

$$D(g_1, g_0)_{\mathcal{C}_1,\mathcal{C}_0} = \int_0^S (g_1' \log g_1' - g_1' + 1) ds + C \int_0^S (\theta_1(g_1(s)) - \theta(s))^2 ds, \text{ and}$$

$$E(\mathcal{C}_1, \mathcal{C}_2) = \inf_{(g_1,g_2)\in\Delta} \int_0^S (g_1' \log g_1' + g_2' \log g_2') ds + \frac{C}{2} \int_0^S (\theta_1(g_1(s)) - \theta_2(g_2(s)))^2 ds. \tag{2}$$

Hereafter(in this paper), denote $\Delta := \{(g_1, g_2) : G_1 \times G_2, g_1' + g_2' = 2\}$ i.e $= \{(g_1, g_2) : g_1(s) + g_2(s) = 2s, g_1(0) = g_2(0) = 0, g_1(S) = S_1, g_2(S) = S_2\}$.

The above theorem can be easily generalized to the case of $n$ curves without any difficulty: The mapping $(g_1, \ldots, g_n)$ minimizing

$$E(\mathcal{C}_1, \mathcal{C}_2, \ldots, \mathcal{C}_n) := \inf_{g_1,g_2,\ldots,g_n,g,\mathcal{C}_0} \sum_{i=1}^n D(g_i, g_0)_{\mathcal{C}_i,\mathcal{C}_0}. \tag{3}$$

satisfies $\sum_{i=1}^{n} g_i(s) = ns$, and the optimal average curve $\mathcal{C}_0$ is given by the angle function $\theta_0(s)$ described by the equation:

$$\theta_0(s) = \frac{1}{n}\sum_{i=1}^{n} \theta_i(g_i(s)). \tag{4}$$

In other words, the "average curve" is constructed by taking the arithmetic mean of all the curves. This is a common property for all Bregman divergences. [3]. However, it can be shown that this mapping distance is the only Bregman divergence with the parametrization independence.

## 2.2 Existence and Uniqueness

The first question that needs to be addressed is the existence of the minimizer. Suppose the minimizers $(g_1, g_2)$ in Eq. ( 2) are $\overline{g}_1, \overline{g}_2, \mathcal{C}_0($ given by $\theta_0(s))$, and the curves are smooth, then we can show these $\overline{g}_1' > 0, \overline{g}_2' > 0$ a.e. in the following theorem. Then the optimal correspondence $g_{1,2} : [0, S_1] \to [0, S_2]$ is given by $g_{1,2} := \overline{g}_2 \circ \overline{g}_1^{-1} : \mathcal{C}_1 \to \mathcal{C}_2$, similarly $g_{2,1} := \overline{g}_1 \circ \overline{g}_2^{-1} : \mathcal{C}_2 \to \mathcal{C}_1$, i.e. those mappings are well-defined.( One point may not be mapped to an interval.)

**Theorem 2 (Existence).** *There exists a $\overline{g}_1' \in L^1_+(\Omega)$ which minimizes the above cost functional. Also $\overline{g}_1' > \epsilon > 0$, for some positive number $\epsilon$. (Proved in the appendix)*

**Theorem 3 (Euler-Lagrangian Equation).** *If the data functions $\theta_1, \theta_2$ are smooth, then the Euler-Lagrangian equation is given by*

$$(\log g_1')' - (\log g_2')' - C(\theta'(g_1) + \theta'(g_2))(\theta_1(g_1) - \theta_2(g_2)) = 0,$$

*subject to $(g_1, g_2) \in \Delta$.*

This is a second order nonlinear ordinary differential equation (ODE). By general theory the solution of this ODE is determined by $g_1'(0)$, and the assumed initial condition $g_1(0) = 0$. Therefore it is possible to have multiple solutions $g_1(s)$ with different initial conditions $g_1'(0)$. Although there is no uniqueness of the optimal solution in general, we still have the following result thanks to the strict convexity of the first term.

**Theorem 4 (Uniqueness).** *If the curves are smooth enough , then there exists a $c_0 > 0$, such that the optimal $(g_1, g_2)$ is unique for $C \leq c_0$.*

## 2.3 Variational Models with Desired Properties

Curve matching models can be designed to meet the users' desired needs. In this section, we provide several examples showing how to revise our model to fit those special needs.

First, suppose we want to have the metric property of the dissimilarity measurement, one way to achieve this is to modify the second term of our elastic energy:

$$E(\mathcal{C}_1, \mathcal{C}_2) := \min \int_0^S (g_1' \log \frac{2g_1'}{g_1' + g_2'} + g_2' \log \frac{2g_2'}{g_1' + g_2'})ds$$

$$+\frac{C}{2} \int_0^S (\theta_1(g_1(s))g_1'(s) - \theta_2(g_2(s))g_2'(s))^2 ds, \text{ subject to } g_1 \in G_1, g_2 \in G_2. \quad (5)$$

Using the fact that the square root of $JS$ divergence is a metric [8], we can prove the following result.

**Theorem 5.** *The square root of the cost functional $E(\mathcal{C}_1, \mathcal{C}_2)$ is a metric.*

Second , suppose a curvature based model with a square root metric property is desired. Such a model may be given as follows,

$$\int_0^S [(g_1' \log g_1' + g_2' \log g_2') + \frac{C}{2}(\frac{d\theta_1(g_1)}{ds} - \frac{d\theta_2(g_2)}{ds})^2] ds, \text{ subject to } g_1' + g_2' = 2. \quad (6)$$

Third, since the angle functions change as the reference coordinate system is rotated, sometimes a rotation invariance property is desired. To achieve this we introduce another parameter $\Theta$ to match the angle functions, i.e. we minimize this cost with respect to $(g_1, g_2, \Theta)$:

$$\int_0^S [(g_1' \log g_1' + g_2' \log g_2') + \frac{C}{2}(g_1'\theta_1(g_1) - g_2'\theta_2(g_2) - \Theta)^2] ds, \text{ subject to } g_1' + g_2' = 2. \quad (7)$$

In this setup we can prove the following result for the rotation angle.

**Theorem 6.** *Let $\Theta_1 := \frac{1}{S}\int_0^{S_1} \theta_1(s_1)ds_1$, $\Theta_2 := \frac{1}{S}\int_0^{S_2} \theta_2(s_2)ds_2$, then the optimal solution $\Theta$ of Eq. (7) is given by*

$$\Theta := \Theta_1 - \Theta_2 = \frac{1}{S}\int_0^S \theta_1(s_1)ds_1 - \frac{1}{S}\int_0^{S_2} \theta_2(s_2)ds_2.$$

The above result says in words: if rotation invariance is desired, one simply applies a rotation $\Theta^*$ to the curve $\mathcal{C}_1$.

## 2.4 Location Weighted Matching Model

Given a group of curves with equal arc lengths, it is easy to see that when $C$ is close to zero, the stretching term dominates, and matching takes place at points of equal arc lengths. On the other hand when $C$ is large, the model tends to match finer features, but it suffers from matching the noises. In general, the parameter $C$ needs to be chosen properly between these two extremes to get the desired matching. In reality, two issues need to be examined. Sometimes different scalings of noise occurs at different locations. For instance, the shape contours retrieved from medical images could have larger noise along some parts than others. The value $C$ on this portion should be chosen smaller. Second, the dissimilarity cost is exaggerated because the bending term is squared and a small portion has a large distortion. This is analogous to the well-known fact that Mean Squared Error paradigm performs poorly in data fitting when a large outlier exists. One way of fixing this drawback is to modify the cost given two curves: $\theta_1(s_1), \theta_2(s_2)$,

$$\min_{w\in L^1_+,(g_1,g_2)\in\Delta} \int_0^S \left( g_1' \log g_1' + g_2' \log g_2' + w(\theta_1 - \theta_2)^2 + \frac{1}{C_0}(w \log w - w + 1) \right) d\mu.$$

It is then possible to show that the minimizer is $w = \exp(-C_0(\theta_1 - \theta_2)^2)$, and the minimizer $(g_1, g_2)$ satisfy

$$\min_{(g_1,g_2)\in\Delta} \int_0^S \left( g_1' \log g_1' + g_2' \log g_2' + \frac{1}{C_0}(1 - \exp(-C_0(\theta_1 - \theta_2)^2)) \right) d\mu. \quad (8)$$

The underlying idea of the first equation is, use a variable weight $w$ instead of a fixed parameter $C$. And the cost in the last equation weights more the similarities, instead of weighting less the dissimilarities. In the presence of occlusion, results in these two approaches can be different(see the experiment). Clearly, a large term $(\theta_1 - \theta_2)^2$ has a greater effect in model (2) than in this model (8).

## 3 Numerical Algorithms

Since the cost functional is non-convex, the standard gradient descent optimization method may be trapped in local minima. Dynamical programming, which is applicable in matching situations involving sequentially ordered signal is known to be an efficient method of searching the global minima( [1]). This method has been frequently used in this curve matching problem, and its use is illustrated in [12] for instance.

For the sake of simplicity, we explain the numerical algorithm using the simplest model (2). Only a minor modification for the bending term is needed to implement other models. When matching a pair of curves, we can construct the dynamical programming table to find an optimal mapping (e.g. see [12]). To get the correspondences between a group of curves simultaneously we adopt the following approach.

Consider a group of curves $\mathcal{C}_1, \ldots, \mathcal{C}_N$ represented by the angle functions $\theta_k(s_k)$, $(k = 1, \ldots, N)$, parameterized by arc lengths $s_k$ $(0 \le s_k \le S_k$. Our task is to find their "average curve" $\mathcal{C}_0$ represented by $\theta_0(s)$ $(S = \frac{1}{N}\sum_{k=1}^N S_k)$ and the correspondence mapping from $\mathcal{C}_0$ to each $\mathcal{C}_k$ $(k = 1, \ldots, N)$. Denote $\{s_{0,i}, i = 0, 1, \ldots, P\}$ with $0 = s_{0,0} < \ldots < s_{0,P} = S$ are the sampling points on $\mathcal{C}_0$, and $s_{k,i_k}$ $(k = 1, \ldots, N)$ are the corresponding points of $s_{0,i}$ on curve $\mathcal{C}_k$. We represent the correspondence mappings by $\{\alpha_i := (s_{0,i}, s_{1,i_1}, s_{2,i_2}, \ldots, s_{N,i_N}),\ \ i = 1, ..., P\}$. By using these notations, our task can be described as follows:

1. Find the correspondents between each curve $\mathcal{C}_k$ and the "average curve" $\mathcal{C}_0$: sequences of pairs $\{(s_{k,i_k}, s_{0,i}), i = 0, 1, \ldots, P\}$, $k = 1, ..., N$. Then we have $\{\alpha_i,\ i = 0, 1, \ldots, P\}$.

2. Determine the "average curve" $\{(s_{0,i}, \theta_0(s_{0,i})), i = 0, \ldots, P\}$.

We do these two things alternately through a minimization algorithm.

*Remark 3 (Algorithm)*

1. Since the overall elastic energy is the sum of $N$ elastic divergences between each of the $N$ curves and the "average curve". Given an " average curve", one can improve the overall energy by minimizing each divergence via searching for a better correspondents $\{\alpha_i\}$.
2. On the other hand, given the correspondence sequence $\{\alpha_i,\ i = 0, ..., P\}$, one can improve the cost by "fine-tuning " the "average curve",

$$\theta_0(s_{0,i}) := \frac{1}{N}\sum_{k=1}^{N} \theta_k(s_{k,i_k}), \quad s_{0,i} := \frac{1}{N}\sum_{k=1}^{N} s_{k,i_k}.$$

Iterating these two steps will decrease the total cost until no changes occur. Then the algorithm stops. To determine the initial "average curve", one can choose one from the group to be the initial "average curve" and later pick the one with the minimal cost.

## 4 Experiments

In this section, we show our experimental results to validate our model. In the first experiment, given four normal ED-Endo, 4-chambered cardiac contours , we aim to find the "average curve" and the matching correspondence among them. We applied the curvature based model (6)(in the extended version to 4 curves) to the first 4 curves in Fig. 1. By using the algorithm described in the previous section we obtained the correspondence mappings $g_i$ $(i = 1, \ldots, 4)$ and then the "average curve" is computed using equation (4). The correspondences are presented through the identical symbols in Fig. 1. Each curve's cost divergence from the "average curve" is also shown in the top of each sub-figure. It is evident that the points with high curvature are well matched. This experiment also illustrates that our proposed method is useful for finding an "average shape" for a set of training shapes.

The aim of the second model is to show that the location weighted model improves the matching result greatly when partial occlusion occurs. Three sets of results on matching two fish curves are shown in Fig. 2. The matching results on the top row for two pairs of fish contours are obtained by using the Edit distance model [12]. Given two curves $\mathcal{C}$ and $\bar{\mathcal{C}}$ parameterized in arc lengths $s$ and $\bar{s}$, respectively. Find a mapping $g$ from $\bar{\mathcal{C}}$ to $\mathcal{C}$ that minimizes the dissimilarity in their curvatures $\bar{\kappa}(\bar{s})$ and $\kappa(s)$, the Edit distance model [12] model searches a optimal mapping $g$ to minimize the energy functional:

$$\int_0^S [|g'(s) - 1| + C|\bar{\kappa}(g(s))g'(s) - \kappa(s)|]ds.$$

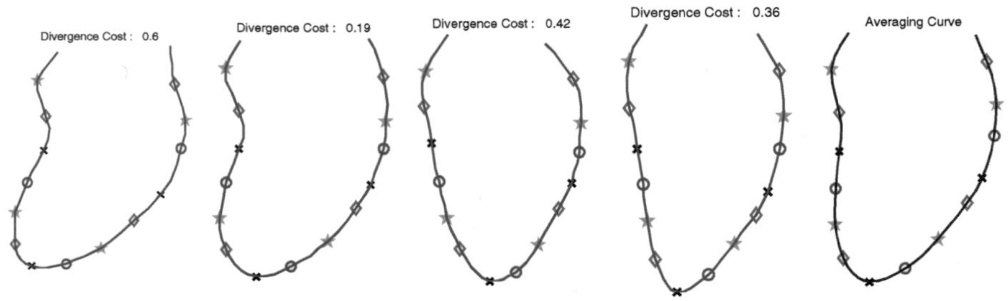


**Fig. 1.** (Experiment I) From left to right: the first 4 curves are given cardiac contours and the 5th curve is their "average curve" obtained using model (6) with $C = 10$. The same symbol indicates the correspondence among these 5 curves obtained by this model.

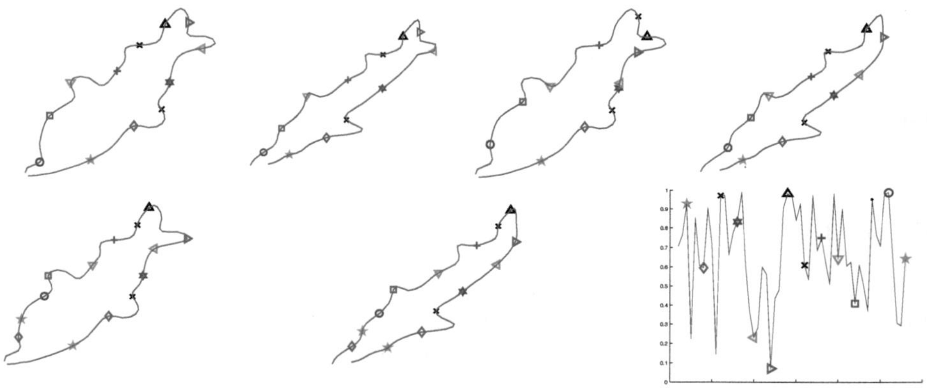

**Fig. 2.** (Experiment II) Top row: matching results for two pairs fish curves ( no occlusion and with occlusion, respectively) obtained by using Edit distance model. Bottom row: matching result for a pair of fish curves obtained by using model (9). The bottom right-most sub-figures shows the graph of weight function. The correspondences are indicated by the same symbols.

The left pair of fish contours does not have occlusion, while the right pair does. The parameter $C$ for both cases are chosen as $C = 0.3$.

To show the effectiveness of our location weighted idea we applied the following model to the same pair of the fish contours shown on the right of the top row:

$$\min_{w \in L^1_+, (g_1, g_2) \in \Delta} \int_0^S (g_1' \log g_1' + g_2' \log g_2') + w \left| \frac{d\theta_1}{ds} - \frac{d\theta_2}{ds} \right| + \frac{1}{C_0}(w \log w - w + 1) ds. \tag{9}$$

It can be reformulated as follows using the same computation in getting (8):

$$\min_{(g_1, g_2) \in \Delta} \int_0^S (g_1' \log g_1' + g_2' \log g_2') + \frac{1}{C_0} \left( 1 - \exp\left( -C_0 \left| \frac{d\theta_1}{ds} - \frac{d\theta_2}{ds} \right| \right) \right) ds, \tag{10}$$

with the weight function $w = \exp(-C_0|\frac{d\theta_1}{ds} - \frac{d\theta_2}{ds}|)$. The parameters used are $C = 0.3$, $C_0 = 0.1$ respectively. Note that the second term in the Edit distance model measures the change of angles, $\bar{\kappa}(g(s))g'(s) - \kappa(s) = \frac{d\bar{\theta}}{ds} - \frac{d\theta}{ds}$, it is comparable with the bending to term in this model if $w = 1$.

The bottom row in Fig. 2, shows the matching result obtained by our location weighted model with the presence of the occlusion in the fish tail. Comparing the matching results for these three pairs it clearly indicates that the edit distance model leads a poor matching in the presence of occlusion, while the location weighted model does much better job. The graph on the bottom row presents the the weight function whose value varies from 0.1 to 1.0. We can see that it is small on the location around the occlusion part (from the symbol $\triangleleft$ to the symbol $\triangle$). Then the correspondence map becomes more flexible to match those similar parts, i.e. the occlusion has less influence on the whole matching.

## 5 Conclusion

In this paper, we build a novel model based on KL divergences for curve matching. This model with slight modifications are capable of handling a group of curves simultaneously. Moreover the symmetry, transitivity and metric properties are inherent. We also introduce a location weighted model to handle noise, distortion and occlusion. Our experiments show the effectiveness of our model.

## References


1. A. A. Amini and T. E. Weymouth, R. C. Jain: Using Dynamic Programming for solving Variational Problems in Vision. *IEEE Trans. on Pattern Anal. and Mach. Intell.* 855–867, 12(9), 1990.
2. K. S. Azoury, M.K. Warmuth: Relative Loss Bounds for On-Line Density Estimation with the Exponential Family of Distributions,*Machine Learn.* , 43 , 211–246, 2001.
3. A. Banerjee, S. Merugu, I. S. Dhillon, and J. Ghosh: Clustering with Bregman Divergences, *JMLR*, Vol. 6, 1705–1749, October 2005.
4. R. Basri, L. Costa, D. Geiger, D. Jacobs: Determing the Similarity of Deformable Shapes, *IEEE Workshop on Phys. Based Model. Computer Vision.*, 135–143, 1995.
5. L. M. Bregman: The relaxation method of finding the common point of convex sets and its application to the solution of problems in convex programming. *USSR Computational Mathematics and Physics,* Vol. 7 , 200–217, 1967.
6. I. Cohen, N. Ayache, and P. Sulger: Tracking Points on Deformable Objects Using Curature Information, *Proc. European Conf. Computer Vision*, 458–466,1992.
7. T.M. Cover and J. A. Thomas: *Elements of Information Theory*, 1991.
8. D.M. Endres and J.E. Schindelin: A new metric for probability distributions, *IEEE Trans. Inform. Theory,* 49(7),1858–1860,2003.
9. E. Klassen, A. Srivastava, W. Mio, S. Joshi. Analysis of planar shapes using geodesic paths on shape spaces. *IEEE Trans. Pattern Anal. and Mach. Intell.*, 26(3): 372-383, 2004.
10. N. Paragios, Y. Chen, O. Faugeras: *Handbook of Mathematical Models in Computer Vision*, Springer, 2006.

11. F. Schmidt, M. Clausen, D. Cremers: Shape Matching by Variational Computation of Geodesic on a Manifold, *DAGM 2006, LNCS 4174,* 142–151, 2006.
12. T. B. Sebastian, P. N. Klein, and B. B. Kimia: On aligning Curves ,*IEEE Trans. Pattern Analysis and Machine intelligence*, 25(1), 116–124,Jan. 2003.
13. H. D. Tagare: Shape Based Nonrigid Correspondence with Application to Heart Motion Analysis, *IEEE Trans. Medical Imaging* , 18(7),570–578,1999.
14. C.C. Tappert: Cursive Script Recognition by Elastic Matching , *IBM J. Research Development,* 26(6), 765–771, 1982.
15. P. N. Yianilos: Data structures and algorithms for nearest neighbor search in general metric spaces, ACM–SIAM Symp. on Discrete algorithms, 311–321, 1993.
16. L. Younes: Computable Elastic Distance between Shapes, *SIAM J. Applied Math.*, Vol. 58, 565–586, 1998.

# A Appendix

*Proof (Proof of Theorem 2)* Denote the elastic energy

$$E(g) := \int_0^S (g' \log g' + (2-g') \log(2-g'))ds + \frac{C}{2} \int_0^S (\theta_1(g(s)) - \theta_2(2s - g(s)))^2 ds.$$

Let $M$ be the corresponding elastic energy $E(g)$ for $g(s) = 2sS_1/(S_1+S_2)$. Now consider a set $A := \{g_1' : g_1' \leq 2, E(g_1) \leq M\}$. Obviously the set $A$ is uniformly integrable, then taking a minimizing sequence $\{g_{1,k}', k = 1, \ldots\}$ by passing a subsequence if necessary , we have a function $\overline{g}_1'$ such that $g_{1,k}'$ converges weakly to $\overline{g}_1'$ in $L^2$, then we also know that $g_{1,k}$ converges to $\overline{g}_1$ a.e.

Also, we construct a sequence $\{\hat{g}_{1,k}' := \frac{1}{k}\sum_{i=1}^k g_{1,k}'\}$, which strongly converges in $L^2$, then $\{\hat{g}_{1,k}\}$ converges uniformly to $\overline{g}_1$.
Since both functions $\theta_1(s_1), \theta_2(s_2)$ are smooth, then

$$\lim_{k\to\infty} \|\theta_1(\hat{g}_1) - \theta_2(\hat{g}_2)\|_{L^2} = \|\theta_1(\overline{g}_1) - \theta_2(\overline{g}_2)\|_{L^2}, \quad \lim_{k\to\infty} E(\hat{g}_{1,k}) \geq E(\overline{g}_1).$$

Thus, we establish the existence of a minimizer.

Due to the limit of pages, the proof of the second part is omitted. Roughly, the proof is based on the fact, $(x \log x)' = \log x - 1 \to -\infty$ as $x \to 0^+$. □

# Beauty with Variational Methods: An Optic Flow Approach to Hairstyle Simulation

Oliver Demetz, Joachim Weickert, Andrés Bruhn, and Martin Welk

Mathematical Image Analysis Group
Faculty of Mathematics and Computer Science, Building E1.1
Saarland University, 66041 Saarbrücken, Germany
{demetz,weickert,bruhn,welk}@mia.uni-saarland.de

**Abstract.** Although variational models offer many advantages in image analysis, their successful application to real-world problems is documented only for some specific areas such as medical imaging. In this paper we show how well-adapted variational ideas can solve the problem of hairstyle simulation in a fully automatic way: A customer in a hairdresser's shop selects a new hairstyle from a database, and this hairstyle is automatically registered to a digital image of the customer's face. Interestingly already a carefully modified optic flow method of Horn and Schunck turns out to be ideal for this application. These modifications include an extension to colour sequences, an incorporation of warping ideas in order to allow large deformation rates, and the inclusion of shape information that is characteristic for human faces. Employing classical numerical ideas such as finite differences and SOR iterations offers sufficient performance for real-life applications. In a number of experiments we demonstrate that our variational approach is capable of solving the hairstyle simulation problem with high quality in a fully practical setting.

## 1 Introduction

By hairstyle simulation a customer in a hairdresser's shop can see herself wearing other hairstyles and decide whether a hairstyle suits herself or not. This process combines two images into one. The images being combined are on one hand the image of a face, and on the other hand an image of a hairstyle. The image containing the hairstyle is transparent/semi-transparent in areas where no hairs are shown.

The main problem of hairstyle simulation is not the combination itself. It is located in the fact that in general the proportions and the position of the hairstyle in its image will not fit the face. Furthermore, simple linear scalings do not solve the problem. Conventional software that is used in a hairdresser's shop requires a lot of manual interaction in order to register the desired hairstyle to a photo of the customer's face. This procedure is time-consuming and for hairdressers who lack experience with handling such a software, the results are not always optimal. Thus it would be desirable to exploit computer vision ideas in order to create a fully automised hairstyle simulation that does not require any interaction and creates a registration of high quality.

F. Sgallari, A. Murli, and N. Paragios (Eds.): SSVM 2007, LNCS 4485, pp. 825–836, 2007.

Addressing this problem is the goal of the current paper. Our work aims at computing a transformation, that, applied onto the hairstyle, makes a simple hairstyle combination possible. In order to avoid to recalculate this transformation again for each new hairstyle, all offered hairstyles have to be equally proportioned and positioned. To achieve this, they have been registered onto one reference face before. In this way a single mapping from the reference face to the customer face is sufficient to adapt any hairstyle to the customer's face. The corresponding displacement field is the transformation we are searching for. Figure 1 illustrates this procedure.

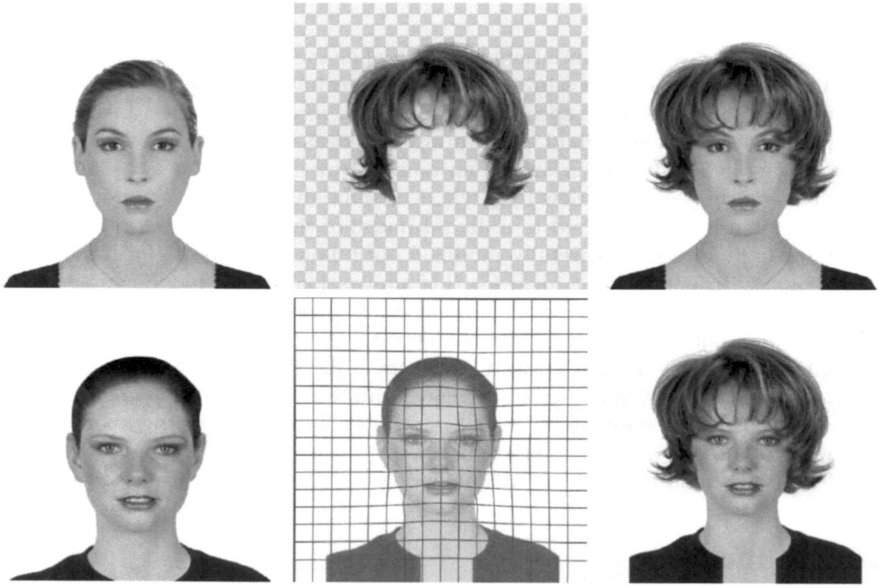

**Fig. 1.** Hairstyle simulation. **(a) Top left:** Reference face. **(b) Top middle:** Example of a hairstyle. Grey-chequered regions show transparency. **(c) Top right:** Hairstyle combined with the reference face. **(d) Bottom left:** Example of a customer face. **(e) Bottom middle:** Mesh representation of the resulting deformation flow field between the reference face (a) and the customer face (d). **(f) Bottom right:** Customer face combined with the new hairstyle transformed according to (e).

From a computer vision perspective, hairstyle simulation requires to find correspondences between two images. One searches for a displacement field that is used to deform one image onto the other. To this end it is highly desirable to compute a deformation field that is dense. Since it can happen that the local information is not sufficient to establish the desired correspondences, it is natural to apply variational methods that benefit from the filling-in effect of the smoothness term. Such approaches have been successfully used for computing optic flow fields in image sequences [10,17,8,18], for estimating the disparity in stereo reconstruction problems [13,12,1], and for calculating the deformation rates for medical image registration [2,7,15,19]. To our knowledge, however, there is no application of these techniques within the beauty and wellness industries. In our paper we shall see that the variational optic flow approach of Horn and Schunck [10] constitutes an ideal starting point to tackle the problem of hairstyle simulation if it is modified in order to meet some specific requirements: It has to incorporate colour information, it must be able to handle large displacements, and it must be tunable to the shape characteristics of human faces.

Our paper is organised as follows. In Section 2 we describe the basic variational optic flow model that is used for our approach. In Section 3 we discuss modifications that make the model more robust against non-optimal image acquisition. In order to cope with large deformation rates, a multiscale setting is used in which a warping strategy is applied when going from a coarser to a finer scale. This is described in Section 4. In Section 5 we sketch how we minimise the energy functional at each scale by computing the Euler-Lagrange equations, discretising them and solving the corresponding linear systems of equations in an iterative manner. Experimental results are presented in Section 6. We conclude the paper with a summary in Section 7.

## 2 Basic Variational Model

Our variational approach is based on an interpretation of the hairstyle simulation problem as an optic flow problem. We would like to find a displacement field that maps a reference face to a customer face. To this end we interpret these images as two subsequent frames in an image sequence. Then the displacement field between corresponding structures is nothing else than the optic flow field in this sequence. From a practical viewpoint it would be desirable to have a dense flow field within a rotationally invariant setting. This naturally suggests the use of continuous variational methods.

The oldest and simplest variational method for optic flow estimation goes back to Horn and Schunck [10]. Assume we are given some sequence $f(x, y, t)$ of greyscale images, where $(x, y)$ denote the location and $t$ is the time. The method of Horn and Schunck computes the optic flow field $(u, v)^\top = (u(x, y, t), v(x, y, t))^\top$ as the minimiser of the convex energy functional

$$E(u, v) \;=\; \int_{\Omega} \Big( (f_x u + f_y v + f_t)^2 + \alpha(|\nabla u|^2 + |\nabla v|^2) \Big)\, dx\, dy \qquad (1)$$

where $\Omega$ is the rectangular image domain, subscripts denote partial derivatives, and $\nabla$ is the spatial nabla operator. The first term of this energy functional is a data term that reflects the assumption that corresponding structures do not change their greyvalues over time (optic flow constraint), while the second term penalises spatial fluctuations of the flow field in a quadratic way. The positive regularisation parameter $\alpha$ determines the required amount of smoothness.

A colour image sequence may be regarded as a vector-valued function $f : \Omega \times [0, \infty) \to \mathbb{R}^3$ where the red, green and blue channels serve as components of

$f = (f_1, f_2, f_3)^\top$. If one searches for a joint optic flow field for all three channels and imposes colour constancy during the motion of image features, cf. e.g. [9], the Horn and Schunck method can be extended to

$$E(u,v) \;=\; \int_\Omega \Big( \sum_{i=1}^{3} (f_{i,x}u + f_{i,y}v + f_{i,t})^2 + \alpha(|\nabla u|^2 + |\nabla v|^2) \Big)\, dx\, dy. \qquad (2)$$

Since we work with colour images, we will make use of this adaptation.

## 3 Shape Weighting and Preregistration

So far we have assumed that there is a reasonable match between structures from both images throughout the image domain $\Omega$. While this condition appears fairly reasonable for the face region, it can be expected to fail in rather common situations: Objects in the background may lead to severe mismatches as well as clothes details and jewellery worn by the customer.

This problem is solved by incorporating a-priori knowledge. We modify our model (2) such that it takes into account which parts of the image are really important to match. The face must be matched accurately while background regions are of no interest and should not influence the algorithm. To this end, we introduce a space-variant weight function $w(x, y)$ that attains high values at the face region and low values in other image areas. This leads to

$$E(u,v) \;=\; \int_\Omega \Big( w \sum_{i=1}^{3} (f_{i,x}u + f_{i,y}v + f_{i,t})^2 + \alpha(|\nabla u|^2 + |\nabla v|^2) \Big)\, dx\, dy. \qquad (3)$$

This energy functional will be the main ingredient of our hairstyle simulation approach. Interestingly it turns out that typical features of more sophisticated optic flow models (such as discontinuity-preserving regularisers and other constancy assumptions; see e.g. [18]) are not required for our present task.

The downside of the weighting approach is that, in order to apply the mask appropriately, the customer and reference faces need to be adjusted to equal position and size beforehand, because otherwise some parts of the customer face might wrongly be excluded from the matching process. That is, the adjustment already requires to match the faces. To escape this vicious circle, we make use of a remarkable observation: There are two specific details which are already reliably matched without the weighting, namely the eyes. Knowing the eye coordinates in the reference image, we can locate the eyes in the customer face via the displacement field. From these, not only the position of the customer's face but even (via the eye distance) its approximate size can be inferred. The customer image is then shifted and linearly rescaled such that the face regions approximately coincide. Running the displacement computation again, with the weighting mask, yields the final flow field.

Another source of unmatched details are illumination differences between reference and customer image. While global illumination changes turn out to be rather uncritical, the positioning of light sources constitutes a radical distinction between lighting conditions in professional studios and in "normal" environments. In a studio setting, flashlights are installed also above and next to the model, thus widely eliminating shadows. In a typical non-studio situation there is just one flashlight located above the lens of the camera, causing shadows in the background as well as, in some cases, in the face itself. Moreover, overexposed areas tend to occur near the eyebrows. To rule out the influence of these differences, we choose a pragmatic solution: We ensure that the reference image is taken under similar lighting conditions as the customer images.

Focussing at the minimisation process, it should be noted that (3) incorporates a data term that involves temporal derivatives of the image sequence. If subsequent frames are fairly different – as can be the case if they are given by the images of two different faces, and even more in the preregistration step when their location and scale may differ substantially – local derivative approximations cannot capture these large displacements anymore. A remedy to this problem is the warping approach that is described in the next section.

## 4 Warping

Warping approaches are frequently used techniques to handle large displacements in the context of motion estimation [3,14,5] or medical image registration [11,15]. By incrementally computing the desired displacement field over multiple scales, they allow to decompose an originally difficult matching problem with large displacements into a series of simpler ones that only require the estimation of small displacements. In general, this decomposition is achieved via a coarse-to-fine strategy: Starting from a down-scaled version of the original problem, the image data and the solution are refined step by step. Thereby, previous solutions from coarser scales are used to compensate the image data for the already computed displacement field. As a consequence, only a small incremental deformation has to be estimated at each scale: the displacement field for the remaining difference problem. Once all these increments have been computed, they have to be summed up in order to form the final solution.

Let us now discuss how such an incremental coarse-to-fine strategy can actually be applied in the context of our hairstyle simulation problem. To this end, we have to define a pyramidal (hierarchical) representation of our image sequence $f$ that contains both the customer and the reference face:

$$f^0 \rightarrow f^1 \rightarrow \ldots \rightarrow f^{k_{\text{orig}}-1} \rightarrow f^{k_{\text{orig}}} . \tag{4}$$

Here, $k = 0$ denotes the coarsest (initial) scale of the coarse-to-fine process, while $k = k_{\text{orig}}$ stands for the finest (original) one. The ratio between two consecutive scales $k$ and $k-1$ is given by a factor $\eta$ in the interval $(0,1)$. One should note that this hierarchical representation is also required for our weighting mask $w$. Moreover, since we are interested in an incremental computation of the results, we also have to decompose the deformation field $u$ and $v$ at each scale into

$$u^k = u^{k-1 \rightarrow k} + \delta u^k , \tag{5}$$

$$v^k = v^{k-1\to k} + \delta v^k\,. \tag{6}$$

Here, $(u^{k-1\to k}, v^{k-1\to k})^\top$ is the already computed (and up-scaled) overall displacement field from the previous scale $k-1$, and $(\delta u^k, \delta v^k)^\top$ is the unknown deformation increment that we are looking for at the current scale $k$. At the coarsest scale the already known displacement field is initialised with zero, i.e. $(u^{-1\to 0}, v^{-1\to 0})^\top := (0,0)^\top$.

Due to the previous definitions we are now in the position to formulate our coarse-to-fine warping approach for the hairstyle simulation problem. It is given by the following three steps that have to be performed at each scale:

1) *Setting up the Difference Problem.* In order to derive the difference problem for the current scale, both the customer face $f(x,y,t)$ and the reference face $f(x,y,t+1)$ have to be scaled down. Moreover, the reference face has to be compensated by the already computed deformation field $(u^{k-1\to k}, v^{k-1\to k})^\top$ from the previous scale. Thus, at each scale, the following task remains to be solved: Find the deformation increment $(\delta u^k, \delta v^k)^\top$ that describes the mapping between $f^k(x,y,t)$ and $f^k(x+u^{k-1\to k}, y+v^{k-1\to k}, t+1)$.

2) *Solving the Difference Problem.* In order to compute this deformation increment, we propose the minimisation of the following energy functional:

$$\begin{aligned} E^k(\delta u^k, \delta v^k) \;=\; & \int_\Omega \Bigg( w^k \sum_{i=1}^{3} \left(f^k_{i,x}\,\delta u^k + f^k_{i,y}\,\delta v^k + f^k_{i,t}\right)^2 \\ & + \alpha \left( |\nabla(u^{k-1\to k} + \delta u^k)|^2 + |\nabla(v^{k-1\to k} + \delta v^k)|^2 \right) \Bigg)\, dx\, dy\,. \end{aligned} \tag{7}$$

As one can easily verify, this energy functional can be obtained from (3) by substituting (5)–(6). One should note that the overall displacement field $(u^{k-1\to k}, v^{k-1\to k})^\top$ from the previous scale must not occur explicitly in the data term: By using the deformation-compensated reference face for computing the temporal derivatives $f^k_{i,t}$, this displacement field is already considered implicitly in the formulation of the data term.

3) *Updating the Overall Displacement Field.* The final step at each scale is the update of the overall solution $(u^k, v^k)^\top$. To this end, the computed deformation increment has to be added to the overall displacement field from the previous scale (cf. (5)–(6)).

As soon as the update step on the finest (original) scale has been performed, the overall solution is obtained. However, due to the recursive definition of (5)–(6) it can also be computed by a simple summation of all deformation increments:

$$u = \sum_{i=0}^{k_{\text{orig}}} \delta u^{i\to k_{\text{orig}}}\,, \qquad v = \sum_{i=0}^{k_{\text{orig}}} \delta v^{i\to k_{\text{orig}}}\,. \tag{8}$$

After we have presented our warping approach for the hairstyle simulation problem in detail, let us now discuss its numerical realisation. This shall be done in the next section.

## 5 Numerical Realisation

In each level of our warping procedure we have to minimise an energy of type (7). At level $k$ this convex problem has a unique minimiser that satisfies the Euler-Lagrange equations

$$J_{1,1}^k\,\delta u^k + J_{1,2}^k\,\delta v^k + J_{1,3}^k \;-\; \alpha\,\Delta\left(u^{k-1\to k} + \delta u^k\right) = 0, \tag{9}$$

$$J_{1,2}^k\,\delta u^k + J_{2,2}^k\,\delta v^k + J_{2,3}^k \;-\; \alpha\,\Delta\left(v^{k-1\to k} + \delta v^k\right) = 0 \tag{10}$$

where $\Delta = \partial_{xx} + \partial_{yy}$ is the spatial Laplace operator, and

$$J^k \;:=\; (J_{n,m}^k) \;:=\; \sum_{i=1}^{3} w_k \begin{pmatrix} f_{i,x}^k f_{i,x}^k & f_{i,x}^k f_{i,y}^k & f_{i,x}^k f_{i,t}^k \\ f_{i,y}^k f_{i,x}^k & f_{i,y}^k f_{i,y}^k & f_{i,y}^k f_{i,t}^k \\ f_{i,t}^k f_{i,x}^k & f_{i,t}^k f_{i,y}^k & f_{i,t}^k f_{i,t}^k \end{pmatrix} \tag{11}$$

denotes a generalisation of the structure tensor from [4]. These continuous equations are discretised by finite difference approximations (see e.g. [16]). We use Sobel operators for discretising the spatial derivatives within the structure tensor, and we replace the temporal derivatives by differences between both frames.

This discretisation leads to a large linear system of equations, where the system matrix is sparse, symmetric and positive definite. In this case, an iterative SOR scheme is a simple and efficient convergent method [20].

In order to calculate the temporal derivatives, we have to compensate the reference frame for the deformation already computed at the coarser level. To this end we perform backward registration and evaluate values between the grid points by bilinear interpolation. Down- and upsampling within the warping scheme is done by area-based averaging and interpolation, respectively [6].

Typical computation times for images of 512 × 512 on a 3.2 GHz Pentium 540 with the previously described numerical scheme are in the order of 10 seconds.

## 6 Experiments

In this section, we present results of our approach in order to assess its usefulness. Beyond the fundamental correctness and applicability, emphasis will be put on aspects of robustness and suitability under practical conditions, i.e. whether the model can cope with the problems that occur upon the everyday usage in a hairdresser's shop. The section is split into two parts. First, we apply the basic method with just one displacement estimation run without weighting to images taken in a studio setting. This ideal situation serves to demonstrate the fundamental applicability of our approach, and discuss important parameter settings. After that, we show how the complete method copes with images taken under realistic conditions.

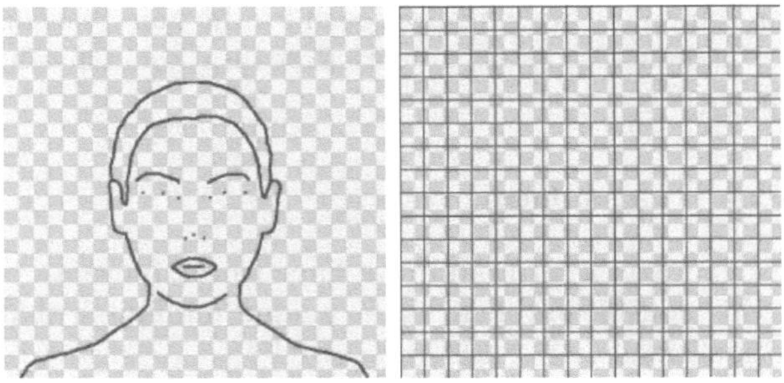

**Fig. 2.** Two synthetic hairstyles. Grey chequered regions show transparency. **(a) Left:** Contour representation. **(b) Right:** Mesh representation.

### 6.1 Test Scenario and Parameter Settings

In all tests, we use images of $512 \times 512$ pixels with 24 bit colour depth. The semi-transparent hairstyle images are given at the same resolution but with 32 bit colour depth due to the additional alpha channel. The combination of faces with hairstyles is performed by linear blending. For analytic purposes, we also use synthetic hairstyle images showing grids and contours as illustrated in Figure 2. With this image material, our scheme from Section 5 always converges to a steady state after 100 SOR iterations.

For the scaling parameter in our warping strategy, the most intuitive choice would be $\eta = \frac{1}{2}$. In practice, however, larger values such as $\eta = \frac{2}{3}$ yield better results: They slow down the refinement and thus reduce the gaps between scales.

While accuracy is a central requirement in estimating the displacement field, it has to be balanced with the conflicting goal of smoothness. The latter is equally important because the computed vector field is used to deform images. Discontinuities or steep changes would imply visible gaps or offsets in the hairstyle image, thus leading to substantially degraded results. The balance between accuracy and smoothness is controlled via the regularisation parameter $\alpha$. Its influence is studied in Figure 3. If $\alpha$ is chosen too small, the above-mentioned discontinuities and gaps are observed. Besides, the filling-in effect is less pronounced in this case. In practice, this is observed for $\alpha < 10\,000$. Too large values for $\alpha$, in contrast, lead to oversmoothed displacement fields that do not adapt well to image structures. Experimentally, one observes for $\alpha > 50\,000$ visible mismatches between face and hairstyle. A good compromise is achieved for $\alpha = 30\,000$.

Having discussed the choice of the model parameters, we demonstrate in Figure 4 that our model is capable of fitting hairstyles to different faces in good quality.

### 6.2 Experiments on Real-World Imagery

While the applicability of our approach is evident, we turn now to consider the robustness with respect to the perturbations that are common in realistic situations, such as illumination differences, noise, or objects in the background.

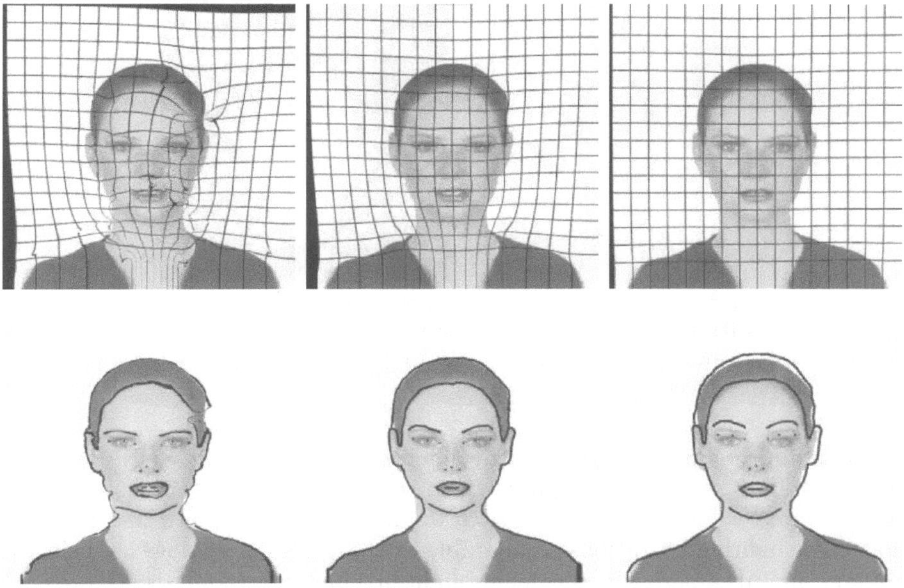

**Fig. 3.** Sensitivity of the solution w.r.t. the regularisation parameter $\alpha$. **From left to right:** $\alpha = 1\,000$, $\alpha = 30\,000$, $\alpha = 500\,000$. **Top row:** Mesh representation. **Bottom row:** Contour representation.

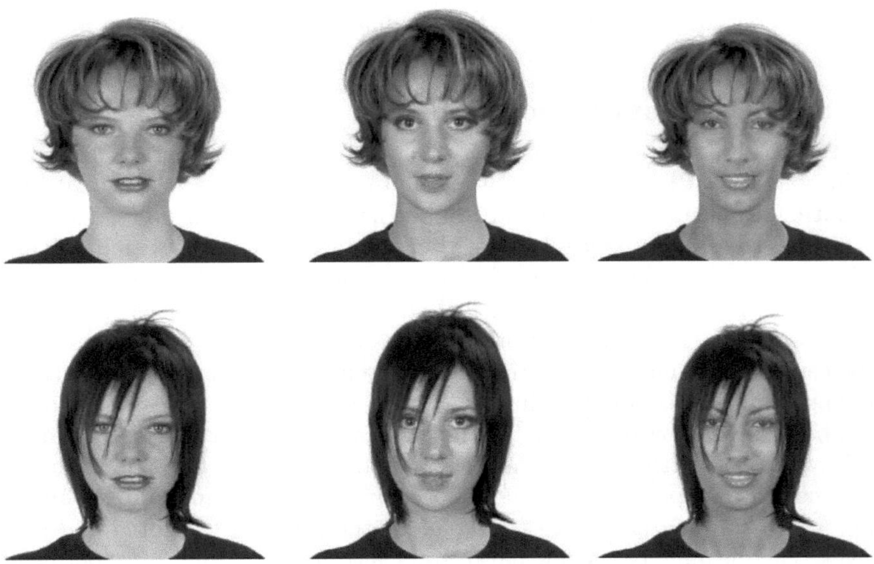

**Fig. 4.** Sample combinations ($\alpha = 30\,000$, 100 iterations, $\eta = \frac{2}{3}$). **Top row:** Hairstyle 1. **Bottom row:** Hairstyle 2. **From Left to Right:** Customer 1, 2, 3.

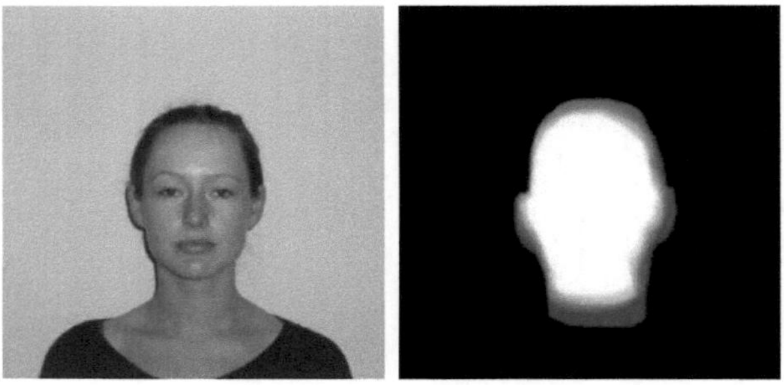

**Fig. 5.** **(a) Left:** Reference face used for real-world image material. **(b) Right:** Weighting mask obtained as the average of 15 different masks adopted on different example customer faces.

Let us address first the problem of different lighting conditions. As pointed out in Section 3, we aim at keeping lighting conditions similar for customer and reference images. To this end, we switch to another reference face that has been shot under lighting conditions comparable to the practical setting, i.e. using an ordinary digital camera with single flashlight in a badly lit room, see Figure 5(a).

To avoid misestimations caused by non-matched objects and structures in the customer image, we use the two-step procedure from Section 3. Our weighting mask, shown in Figure 5(b), is the average of numerous manually generated masks for individual customer faces. The results shown in Figure 6 were obtained by incorporating this preprocessing step. They demonstrate that the complete process is able to handle the kind of image material that occurs in practice.

## 7 Summary and Conclusions

In this paper, we have presented the complete solution of a practical visual computing problem by a variational approach, from a sound theoretical approach all the way to a fully applicable algorithm whose commercial use is imminent.

We have started out from the formulation of hairstyle simulation as a correspondence problem. The demand for a dense displacement field and the necessary balance between adaptation to image structures and smoothness of the displacement field made the variational approach with quadratic penalisation the method of choice for its solution. Within this setting, all partial problems could be addressed by specific well-founded measures. In particular, warping was employed to deal with large displacements, and a weighting function in the data term allowed to suppress the influence of spurious background and clothing details. A two-step matching procedure solved the problem of shifted and scaled images. The meaning of the few parameters is transparent, and by experimental guidance they were fixed to values that work for practical use. Careful design of the weighting function and the selection of an adequate reference face image rounded up the proceeding.

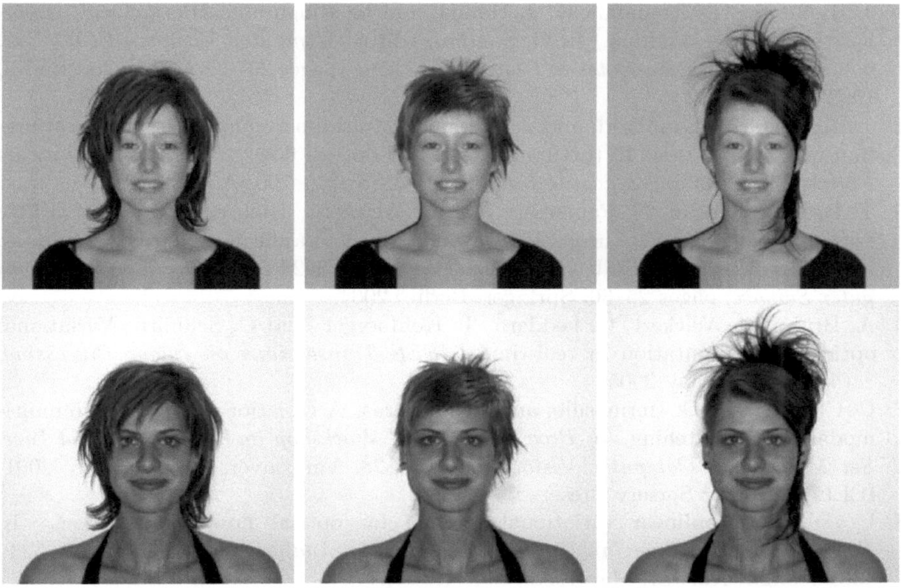

**Fig. 6.** Example combinations with three hairstyles. **Top:** Customer 1. **Bottom:** Customer 2. **Left to Right:** Hairstyle 1, 2, 3.

As a result, we have arrived at a method that combines results of high quality with robustness against the typical problems of the real-life setting. The algorithm processes input images taken by non-expert photographers with consumer cameras under simple lighting conditions, it does not require sophisticated user input like contours or marked features, and it achieves sufficient speed on a standard PC.

This contribution demonstrates that theoretical stringence and real-life applicability go well together. Moreover, by addressing an unprecedented field of application we want to emphasise that beyond a few well-established disciplines there are realms of visual computing applications awaiting future research efforts.

**Acknowledgements.** We are grateful to Jürgen Demetz (Style Concept SC KG, Beckingen) for supplying us with numerous images, and to Andrea Laick and Tina Scholl for volunteering for the real world experiments.

## References

1. L. Alvarez, R. Deriche, J. Sánchez, and J. Weickert. Dense disparity map estimation respecting image derivatives: a PDE and scale-space based approach. *Journal of Visual Communication and Image Representation*, 13(1/2):3–21, 2002.
2. R. Bajcsy and S. Kovacic. Multiresolution elastic matching. *Computer Vision, Graphics and Image Processing*, 46(1):1–21, 1989.

3. J. R. Bergen, P. Anandan, K. J. Hanna, and R. Hingorani. Hierarchical model-based motion estimation. In G. Sandini, editor, *Computer Vision – ECCV '92*, volume 588 of *Lecture Notes in Computer Science*, pages 237–252. Springer, Berlin, 1992.
4. J. Bigün, G. H. Granlund, and J. Wiklund. Multidimensional orientation estimation with applications to texture analysis and optical flow. *IEEE Transactions on Pattern Analysis and Machine Intelligence*, 13(8):775–790, August 1991.
5. T. Brox, A. Bruhn, N. Papenberg, and J. Weickert. High accuracy optical flow estimation based on a theory for warping. In T. Pajdla and J. Matas, editors, *Computer Vision – ECCV 2004, Part IV*, volume 3024 of *Lecture Notes in Computer Science*, pages 25–36. Springer, Berlin, 2004.
6. A. Bruhn, J. Weickert, C. Feddern, T. Kohlberger, and C. Schnörr. Variational optic flow computation in real-time. *IEEE Transactions on Image Processing*, 14(5):608–615, May 2005.
7. C. Chefd'Hotel, G. Hermosillo, and O. Faugeras. A variational approach to multi-modal image matching. In *Proc. First IEEE Workshop on Variational and Level Set Methods in Computer Vision*, pages 21–28, Vancouver, Canada, July 2001. IEEE Computer Society Press.
8. I. Cohen. Nonlinear variational method for optical flow computation. In *Proc. Eighth Scandinavian Conference on Image Analysis*, volume 1, pages 523–530, Tromsø, Norway, May 1993.
9. P. Golland and A. M. Bruckstein. Motion from color. *Computer Vision and Image Understanding*, 68(3):346–362, December 1997.
10. B. Horn and B. Schunck. Determining optical flow. *Artificial Intelligence*, 17:185–203, 1981.
11. M. Lefébure and L. D. Cohen. Image registration, optical flow and local rigidity. *Journal of Mathematical Imaging and Vision*, 14(2):131–147, March 2001.
12. A.-R. Mansouri, A. Mitiche, and J. Konrad. Selective image diffusion: application to disparity estimation. In *Proc. 1998 IEEE International Conference on Image Processing*, volume 3, pages 284–288, Chicago, IL, October 1998.
13. R. March. Computation of stereo disparity using regularization. *Pattern Recognition Letters*, 8:181–187, October 1988.
14. E. Mémin and P. Pérez. A multigrid approach for hierarchical motion estimation. In *Proc. Sixth International Conference on Computer Vision*, pages 933–938, Bombay, India, January 1998. Narosa Publishing House.
15. J. Modersitzki. *Numerical Methods for Image Registration.* Oxford University Press, Oxford, 2004.
16. K. W. Morton and L. M. Mayers. *Numerical Solution of Partial Differential Equations.* Cambridge University Press, Cambridge, UK, 1994.
17. H.-H. Nagel and W. Enkelmann. An investigation of smoothness constraints for the estimation of displacement vector fields from image sequences. *IEEE Transactions on Pattern Analysis and Machine Intelligence*, 8:565–593, 1986.
18. N. Papenberg, A. Bruhn, T. Brox, S. Didas, and J. Weickert. Highly accurate optic flow computation with theoretically justified warping. *International Journal of Computer Vision*, 67(2):141–158, April 2006.
19. O. Scherzer, editor. *Mathematical Models for Registration and Applications to Medical Imaging.* Springer, Berlin, 2006.
20. D. M. Young. *Iterative Solution of Large Linear Systems.* Dover, New York, 2003.

# A Variational Approach for 3D Motion Estimation of Incompressible PIV Flows

Luis Alvarez, Carlos Castaño, Miguel García, Karl Krissian, Luis Mazorra, Agustín Salgado, and Javier Sánchez

Departamento de Informatica y Sistemas
Universidad de Las Palmas de Gran Canaria, Spain
http://serdis.dis.ulpgc.es/~lalvarez/ami/index.html

**Abstract.** Estimation of motion has many applications in fluid analysis. Lots of work has been carried out using Particle Image Velocimetry to design experiments which capture and measure the flow motion using 2D images. Recent technological advances allow capturing 3D PIV image sequences of moving particles. In this context, we propose a new three-dimensional variational (energy-based) technique. Our technique is based on solenoidal projection to take into account the incompressibility of the real flow. It uses the result of standard flow motion estimation techniques like iterative cross-correlation or pyramidal optical flow as an initialization, and improves significantly their accuracies. The performance of the proposed technique is measured and illustrated using numerical simulations.

## 1 Introduction

"Particle Image Velocimetry (PIV) is a technique which allows one to record images of large parts of flow fields in a variety of applications in gaseous and liquid media and to extract the velocity information out of these images" [1]. The typical setting of a PIV experiment consists in the following components: the flow medium seeded with particles, droplets or bubbles, a double pulsed laser which illuminates the particles twice with a short time difference, a light sheet optics guiding a thin light plane within the flow medium, one or several CCD cameras which capture the two frames exposed by the laser pulses and a timing controller synchronizing the laser and the camera. Once the flow motion has been captured, software tools are needed to evaluate and display the flow motion. The standard techniques work in a planar domain (2D-PIV), permitting estimation of the 2 planar components of the fluid motion (2C-PIV). The third spatial component can also be extracted using stereo techniques, dual-plane PIV or holographic recording (3C-PIV) [2]. The extension of the observation to a volume (3D-PIV) is currently an active area of research. To this end, multicamera configuration or holographic techniques (see [3]) has been proposed.

In this paper, we propose a technique for 3D fluid motion estimation applied to 3D-PIV. The most widely used technique for motion estimation in 2D-PIV is based on local correlation between two rectangular regions of the two images

F. Sgallari, A. Murli, and N. Paragios (Eds.): SSVM 2007, LNCS 4485, pp. 837–847, 2007.

(see for instance [4]). This technique has a straightforward extension to 3D images. Another approach to motion estimation widely used in optical flow is a variational approach based on an energy minimization where on the one hand, we assume the conservation of the intensity of the displaced objects (in our case the particles) and on the other hand, we assume a certain regularity of the obtained flow. A variational approach was proposed in [5] in the context of 2D PIV. We propose to compare and combine both approaches in order to improve the accuracy of the flow estimation. The proposed method is very general and can be used in many applications of 3D flow estimation.

In the particular case of incompressible fluid motion, we have designed a method to include the incompressibility constraint in the flow estimation.

The paper is organized as follows: in section 2, we briefly describe the motion estimation using local cross-correlation; in section 3, we describe our variational approach and the solenoidal projection; in section 4, we present the numerical experiments followed by the conclusion.

## 2 Motion Estimation Using Local Cross-Correlation

Cross-correlation is the most common technique for fluid motion estimation in PIV and is described, for instance, in [1]. We will denote $I_1$ and $I_2$ the two images from which we compute the motion $\mathbf{u}$, $N$ the image dimension (in our case $N = 3$) and $\Omega$ the domain of definition of the images.

### 2.1 Basic Principle

Having the two volumes $I_1$ and $I_2$, for each voxel $\mathbf{v} = (v_x, v_y, v_z)$ of $I_1$, the method takes a rectangular subvolume $I_{1,\mathbf{v}}$ of $I_1$ centered on $\mathbf{v}$, and looks for a similar subvolume of $I_2$ centered on a neighbor $\mathbf{v} + \mathbf{d}$ of $\mathbf{v}$. The similarity measure between two rectangular subvolumes of the same dimensions is based on 2D cross-correlation and is defined as:

$$C_{\mathbf{v}}(I_1, I_2)(\mathbf{d}) = \sum_{\mathbf{y}=(-a,-b,-c)}^{(a,b,c)} I_1(\mathbf{v}+\mathbf{y})\, I_2(\mathbf{v}+\mathbf{d}+\mathbf{y}) \qquad (1)$$

The voxel $\mathbf{v}$ is assigned the displacement $\mathbf{d}$ which gives the maximal value of the cross-correlation. Doing this for every voxel in $I_1$ we obtain a complete vector field $\mathbf{u}$.

### 2.2 Implementation Using Fast Fourier Transform

Because the process of computing the cross-correlation for many subvolumes of $I_2$ and for each voxel is computationally heavy, the implementation takes advantage of the properties of the Fourier transform to improve the processing time. The Fourier transform has the property that a correlation in the spatial domain is equivalent to a multiplication in the Fourier domain.

$$C_{\mathbf{v}}(I_1, I_2) = \mathcal{F}^{-1}(\widehat{I_{1,\mathbf{v}}}\, \widehat{I_{2,\mathbf{v}}}^{*}), \qquad (2)$$

where $I_{1,\mathbf{v}}$ is a rectangular subvolume of $I_1$ centered on the voxel $\mathbf{v}$, $\widehat{I_{1,\mathbf{v}}}$ is the Fourier Transform of the subvolume $I_{1,\mathbf{v}}$, the operator $^*$ denotes the complex conjugate, and $\mathcal{F}^{-1}$ denotes the inverse Fourier transform. The image $C_\mathbf{v}(I_1, I_2)(\mathbf{d})$ gives the result of cross-correlation for all displacements $\mathbf{d}$ and the maximal value is a best estimate of the local displacement. Because of the hypothesis of periodicity introduced by the Fourier Transform, the window is usually chosen four times bigger than the expected displacement. The method is then extended to allow subvoxel accuracy by means of local interpolation of a Gaussian function close to the discrete maximum. When the correlation has been computed for every voxel, some kind of data validation procedure is needed to remove outliers.

Actually, we do not have to compute the correlation for each voxel, we can calculate the flow only for the voxels located on a given lattice. At the end of the process, we extrapolate the results and obtain a dense vector field. This improves not only the speed of the computation, but also in some cases the quality of the results because of the regularization induced by the extrapolation.

The whole process should be applied iteratively a few times using the current result as an initialization for the next iteration. The iterative process can be initialized with a null vector field $\mathbf{u}^0 = 0$, and $\mathbf{u}^{n+1}$ can be estimated at each voxel of the lattice using the displacement with maximal correlation for a window of $I_2$ displaced by $\mathbf{u}^n$:

$$C_\mathbf{v}(I_1, I_2, \mathbf{u}^n) = \mathcal{F}^{-1}(\widehat{I_{1,\mathbf{v}}}\ \widehat{I_{2,\mathbf{v}+\mathbf{u}^n(\mathbf{v})}}^{*}), \tag{3}$$

By doing this, we can improve the accuracy of the fluid motion estimation. It also permits the progressive reduction of the size of the correlation window.

## 3 Variational Approach

Variational approach to motion estimation are often used for optical flow computation [6,7,8]. It consists in minimizing an energy as a function of the displacement and that depends on a pair of images $I_1$ and $I_2$.

In this section, $E$ will denote the energy functional to minimize. For a given 3D vector field $\mathbf{u} = (u^x, u^y, u^z)^t$, the norm of its gradient $\| \nabla \mathbf{u} \|$ is defined as $\sqrt{\| \nabla u^x \|^2 + \| \nabla u^y \|^2 + \| \nabla u^z \|^2}$, and the Laplacian $\Delta \mathbf{u} = div(\nabla \mathbf{u})$ is defined as $(\Delta u^x, \Delta u^y, \Delta u^z)^t$.

The energy to minimize is expressed as :

$$E(\mathbf{u}) = \underbrace{\int_\Omega (I_1(\mathbf{x}) - I_2(\mathbf{x} + \mathbf{u}(\mathbf{x})))^2 \, d\mathbf{x}}_{data\ term} + \alpha \underbrace{\int_\Omega \|\nabla \mathbf{u}(\mathbf{x})\|^2 d\mathbf{x}}_{regularization\ term}, \tag{4}$$

where $\alpha$ is a scalar coefficient that weights the smoothing term. Under the assumption of intensity conservation for each voxel, the first term (*data term*) becomes zero when the first image matches the second one displaced by $\mathbf{u}$: $I_1(\mathbf{x}) = I_2(\mathbf{x} + \mathbf{u}(\mathbf{x}))$. This term tries to find the vector field that best fits the solution. The second term is a *regularization term* which smoothes the vector field. There are a lot of ways to define the regularization term, including, for instance, discontinuities preserving constraints, etc.. In this paper, since we deal with rather smooth flows we use the $L_2$ norm presented above. Euler-Lagrange equations yield:

$$(I_1(\mathbf{x}) - I_2(\mathbf{x}+\mathbf{u})).\nabla I_2(\mathbf{x}+\mathbf{u}) + \alpha div(\nabla \mathbf{u}) = 0 \tag{5}$$

The coefficient $\alpha$ is normalized to allow invariance under global intensity change. To this purpose, $\alpha$ is multiplied by

$$\alpha = \alpha_0 \left( \epsilon + \sqrt{\frac{1}{|\Omega|} \int_\Omega \|\nabla \mathbf{I_2}(\mathbf{x})\|^2 d\mathbf{x}} \right)^2 \tag{6}$$

with $\epsilon = 0.01$.

### 3.1 Numerical Scheme

We propose to look for the minimum of the energy by solving (5) directly using a fixed point approach. An alternative is to use a gradient descent with either explicit or semi-implicit scheme. We use an iterative method to find the vector field $\mathbf{u}$:

$$\begin{cases} \mathbf{u}^0 &= \mathbf{u_0} \\ \mathbf{u}^{n+1} &= \mathbf{u}^n + \mathbf{h}^{n+1} \end{cases} \tag{7}$$

where we update the vector field $\mathbf{u}$ at each iteration by adding another vector field $\mathbf{h}$ with small displacements. The displacement $\mathbf{h}$ being small, we can use first order Taylor expansions of $I_2$ and $\nabla I_2$ at $\mathbf{x} + \mathbf{u}^n$ to linearize (5), and we obtain:

$$d\mathbf{g} - \left[\mathbf{g}\mathbf{g}^t - dH'\right] \mathbf{h} + \alpha div(\nabla \mathbf{u}^n + \nabla \mathbf{h}) = 0 \tag{8}$$

denoting:

$$\mathbf{g}(\mathbf{x}) = \nabla I_2(\mathbf{x} + \mathbf{u}^n) \tag{9}$$

$$d(\mathbf{x}) = I_1(\mathbf{x}) - I_2(\mathbf{x} + \mathbf{u}^n) \tag{10}$$

$$H'(\mathbf{x}) = H(I_2)(\mathbf{x} + \mathbf{u}^n). \tag{11}$$

In the last equality, $H(I_2)(\mathbf{x})$ denotes the Hessian matrix of $I_2$ at the location $\mathbf{x}$. The term in second order spatial derivatives is usually neglected, supposing that the image varies slowly. Then, (8) becomes:

$$d\mathbf{g} + \alpha div(\nabla \mathbf{u}^n) - \mathbf{g}\mathbf{g}^t\mathbf{h} + \alpha div(\nabla \mathbf{h}) = 0 \tag{12}$$

After discretization using finite differences, the operator $div(\nabla \mathbf{h})$ can be divided in two terms $-2N\,I\,\mathbf{h}$ and $S(\mathbf{h})$, where the $N$ is the image dimension and $I$ is the identity matrix. The first term only depends on values of $\mathbf{h}$ at the current position $\mathbf{x}$ and the second term only depends on values of $\mathbf{h}$ at neighbor positions of $\mathbf{x}$: the vector $S(\mathbf{h})$ is written:

$$S(\mathbf{h}) = \begin{pmatrix} \sum_{\mathbf{y}\in N^*(\mathbf{x})} h^x(\mathbf{y}) \\ \sum_{\mathbf{y}\in N^*(\mathbf{x})} h^y(\mathbf{y}) \\ \sum_{\mathbf{y}\in N^*(\mathbf{x})} h^z(\mathbf{y}) \end{pmatrix}, \tag{13}$$

where $N^*(\mathbf{x})$ denotes the direct neighbors of $\mathbf{x}$ (4 in 2D and 6 in 3D), and $\mathbf{h} = (h^x, h^y, h^z)^t$.

Using $\mathbf{h}^{n+1}$ for the current location $\mathbf{x}$ and $\mathbf{h}^n$ for its neighbors, (12) becomes:

$$A\mathbf{h}^{n+1} = b \tag{14}$$

with $A = \mathbf{g}\mathbf{g}^t + \alpha 2N\ I$, and $b = d\mathbf{g} + \alpha div(\nabla \mathbf{u}^n) + S(\mathbf{h}^n)$. The matrix $A$ is real, symmetric and positive definite, so it can be inverted and we can compute for each position $\mathbf{x}$, $\mathbf{h}^{n+1} = A^{-1}b$. To improve the convergence rate, we use a Gauss-Seidel method which updates the displacement $\mathbf{h}^{n+1}$ at position $\mathbf{x}$ using the values of $\mathbf{h}^{n+1}$ already calculated. This scheme is recursive and to avoid privileging the direction of scanning the image, we apply two successive iterations of Gauss-Seidel in reverse directions. Furthermore, we use a pyramidal approach to compute the displacement flow at several scales, using the results from a given scale to initialize to the following higher scale.

## 4 Refined Variational Approach

We introduce two modifications to (4) to improve the solution. First, the regularization term is applied to the increment of the displacement vector $\mathbf{h}$ at each iteration instead of the whole vector $\mathbf{u}$. It allows the minimization to be invariant under the solution: if the data term is zero, no smoothing will be applied. This change implies removing the term $\alpha div(\nabla \mathbf{u}^n)$ from (12) while using the same numerical scheme. Second, we replace the solution by its solenoidal projection and re-iterate the minimization to take into account the incompressibility of the flow. This refinement step will use only one scale since it is initialized by the solution of one of the previous methods described in sections 2 or 3. The following paragraph describes the solenoidal projection.

### 4.1 Solenoidal Projection

In our experiments, the fluid flows are *incompressible*. As a consequence, the displacement vector field $\mathbf{u}$ should be divergence-free, i.e. $div(\mathbf{u}) = \frac{\partial u^x}{\partial x} + \frac{\partial u^y}{\partial y} + \frac{\partial u^z}{\partial z} = 0$. One way to fulfill this constraint is to project our estimated motion $\mathbf{u}$ into the space of divergence-free vector field. This new vector field $\mathbf{u}_s$ is called a solenoidal projection of $\mathbf{u}$. It can be expressed as:

$$\mathbf{u}_s = \mathbf{u} - \nabla v, \tag{15}$$

where $v$ is a scalar function of $\Omega \subset \mathbb{R}^3$, defined as a solution to the following *Poisson's equation*:

$$\begin{cases} div(\nabla v) = div(\mathbf{u}) & \text{in } \Omega \\ v = 0 & \text{in } \partial\Omega \end{cases} \tag{16}$$

This equation is solved using Gauss-Seidel technique.

# 5 Experiments and Results

In this section, we present experiments on synthetic data using both methods (correlation and variational). We used a 3D flow based on realistic flow models to check the performance of the proposed methods. In these experiments, we first apply the standard correlation or variational methods to obtain a good approximation of the flow and then we refine the results with the new variational approach.

## 5.1 Choice of the Parameters

The cross-correlation parameters are the window size in each dimension and the lattice spacing. The window size is approximately set to four times the expected maximal displacement and is the same in each dimension. In the following experiments, we use a lattice spacing of 2 voxels in each dimension, and the final result is interpolated to obtain a dense estimation. The variational approach uses the parameters $\alpha$ and the number of scales for the pyramidal approach. In the following experiments, we set $\alpha$ to 0.5 for both the standard and the refined variational approaches.

## 5.2 Description of the Models

In the first model (Fig. 1, left), we use an incompressible 3D flow model suggested to us by Professor F. Scarano that can be found in [9] (section 3-9.2). It corresponds to the Stokes's solution for an immersed sphere. The flow moves in the horizontal axis direction with a velocity $(U, 0, 0)$, and it avoids the sphere located at the center of the volume. The flow inside the sphere is null.

Having a sphere with radius $\alpha$ and center $(0, 0, 0)$, and a 3D point $(x, y, z)$ at a distance $r$ from the sphere center, the flow outside the sphere follows:

$$\begin{aligned} u &= U\left(1 - \frac{3\alpha}{4r^3}(2x^2 + y^2 + z^2) + \frac{\alpha^3}{4r^5}(2x^2 - y^2 - z^2)\right) \\ v &= U\left(-\frac{3\alpha}{4r^3}xy + \frac{3\alpha^3}{4r^5}xy\right) \\ w &= U\left(-\frac{3\alpha}{4r^3}xz + \frac{3\alpha^3}{4r^5}xz\right) \end{aligned} \tag{17}$$

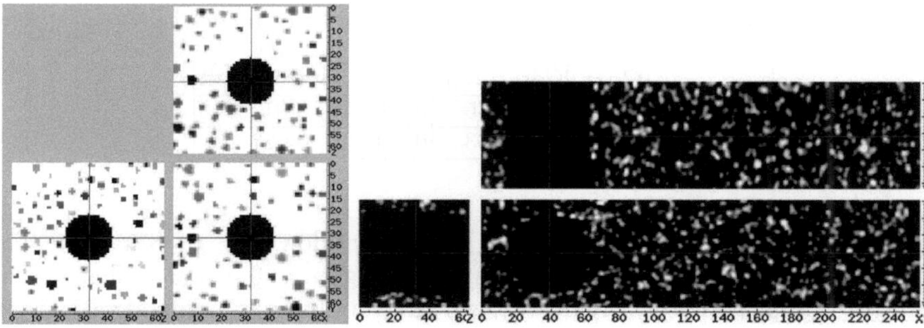

**Fig. 1.** Left, model 1 (sphere). Right, model 2 (cylinder).

The other model (Fig. 1, right) was provided to us by the Cemagref Rennes, France and it has been obtained using a Large Eddy Simulation of the incompressible Navier-Stokes equations which defines the turbulent motion after a cylinder [10]. It simulates a volume with synthetic particles following the horizontal axis and a cylinder situated on the z-axis obstructing the flow perpendicularly. We use two successive images from this sequence. The original model is a volume of 960 x 960 x 144 voxels but we limit our experiment to a window of 256 x 64 x 64 voxels to reduce the computation time. This window includes part of the cylinder and the turbulence behind it.

### 5.3 Experiments with Model 1 (Sphere)

Table 1 shows the average error and the standard deviation reached for the cross-correlation and the variational methods before and after the refinement. The error at each voxel is computed as the magnitude of the difference between the ideal displacement and the estimated one, and is measured in voxel unit.

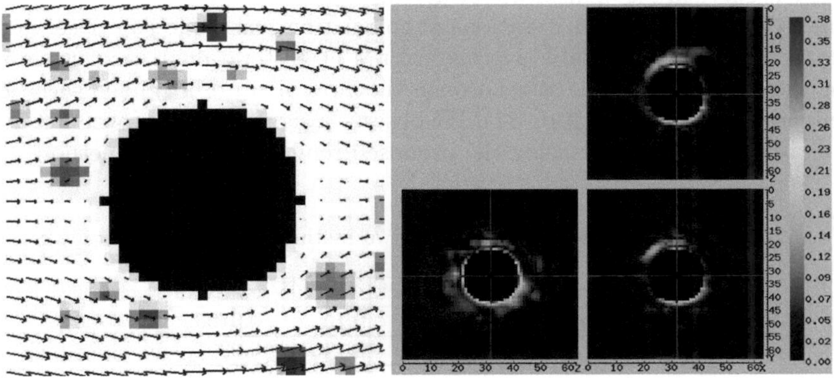

**Fig. 2.** Left, real flow (with zoom). Right, final error distribution (combined scheme).

**Table 1.** Comparison of the two methods for model 1

| | **Corr.** | Corr.+Ref. | **Var.** | **Var.+Ref.** |
|---|---|---|---|---|
| **av. error** | 0.029 | 0.0135 | 0.03164 | 0.01702 |
| **std. dev.** | 0.0296 | 0.0153 | 0.03033 | 0.02756 |

The correlation was applied 11 times with a window size of 8 voxels. The individual variational approach was applied using $\alpha = 0.5$ and 3 scales. The mean error is approximately divided by two after applying the refined variational approach, and the initialization with correlation gives a better result than the initialization with a variational approach. Figure 2 (right) shows the final average error distribution using the cross-correlation followed by the refined variational approach. We can observe that the highest error is located at the sphere boundaries.

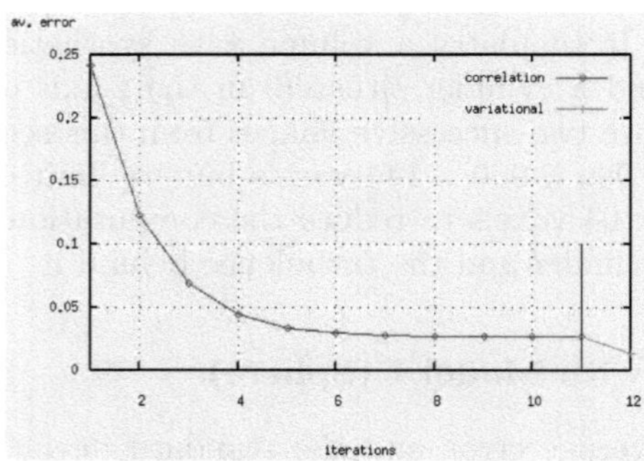


**Fig. 3.** Left, average error evolution using the combined scheme (11 times cross-correlation + variational)

The left curve in Fig. 3 displays the average error evolution using the combined scheme. First, we apply 11 iterations of correlation technique (we observe that the correlation reaches a stable average error after 11 iterations). Next, we use the output flow provided by the correlation as the input flow of the refined variational technique (curve after iteration 11). We observe a significant improvement in the flow estimation error after using the proposed refined variational method.

**Table 2.** Comparison of the two methods for model 2

| | **Corr.** | Corr.+Ref. | **Var.** | **Var.+Ref.** |
|---|---|---|---|---|
| **av. error** | 0.1670 | 0.0651 | 0.1579 | 0.0763 |
| **std. dev.** | 0.1375 | 0.0729 | 0.1461 | 0.0891 |

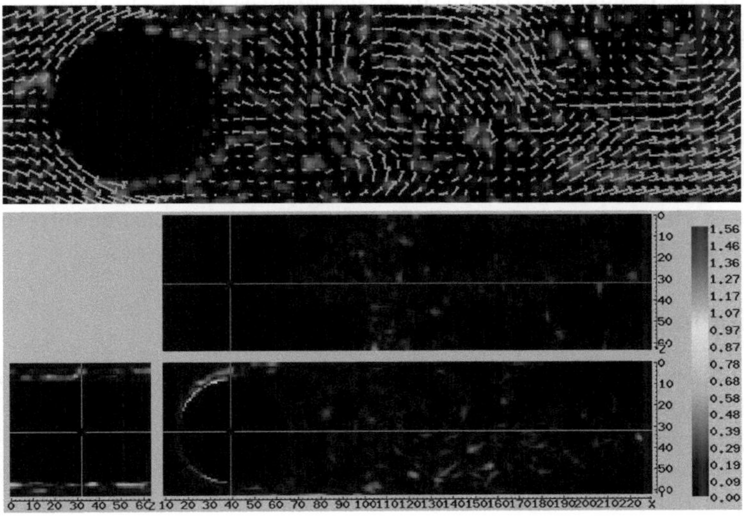


**Fig. 4.** Top, real flow. Bottom, final error distribution (combined scheme)

### 5.4 Experiments with Model 2 (Cylinder)

We ran the same experiments for this model. Table 2 shows the average error and standard deviation reached for the cross-correlation and the variational methods before and after the refinement. The correlation was applied 6 times with a sequence of different window sizes: 16, 16, 8, 8, 4, 4. The variational approach was applied using $\alpha = 0.5$ and 3 scales. Finally, the refined variational method was applied with $\alpha = 0.5$ and one scale. In this experiment, the variational and the correlation correlation methods alone reach similar accuracies, and after the refinement, the cross-correlation reaches a slightly better result. In both

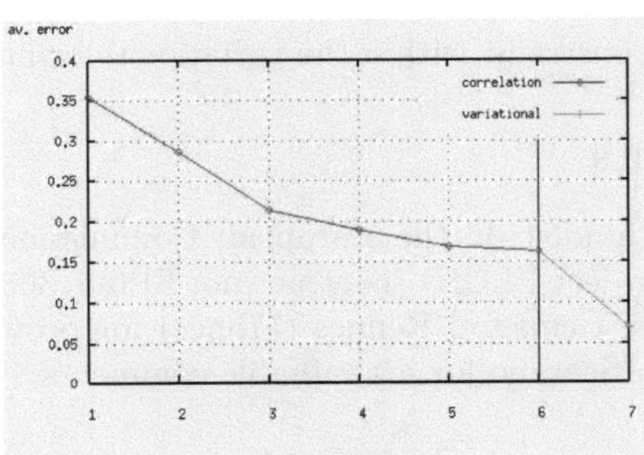


**Fig. 5.** Average error evolution using the combined scheme (6 times cross-correlation + variational)

cases, the new refined variational approach reduces the mean error by at least half. Figure 4 (bottom) shows the final average error distribution using the combination of the cross-correlation and the refined variational schemes. As in the previous model, the highest error is also located at the obstacle boundaries.

The curve in Fig. 5 displays the average error evolution using the combined scheme of correlation and refined variational approaches. It shows that the correlation reaches a stable average error after 6 iterations and that an additional iteration of the proposed variational approach reduces considerably the mean error.

## 6 Conclusion

In this paper, we presented an improvement to a standard variational 3D flow estimation technique based on soleinodal projections and a more flexible smoothing term. The proposed refined variational optical flow technique is initialized by standard techniques like cross-correlation or standard 3D optical flow. We have implemented these techniques and we have shown in the numerical experiments that the proposed technique improves the accuracy of the flow estimation and reduces the mean error by at least half. Slightly better results were obtained by the initialization from cross-correlation, which is probably due to the smoothing term of the standard variational approach that cannot deal with fast variations or discontinuities of the flow close to the obstacle.

Although we focused our attention to 3D fluid flow analysis, the proposed methodology is very general and can be applied to different fields. Correlation based techniques and energy minimization techniques have been developed in the research community in a completely independent way. Each one has its own advantages and limitations but we think that an adequate combination of both can improve the global estimation of the flow. On the other hand, we think that including physical 3D flow constraints, as for instance the incompressibility, to the 3D flow estimation, is a very important issue and allows combining the mathematical models of fluid motion with the experimental data.

In future work, we plan to investigate other regularization terms as proposed in [11,12], and to compare our current method with approaches which include an incompressibility constraint within the variational formulation [5,13].

## Acknowledgments

This work has been funded by the European Commission under the Specific Targeted Research Project FLUID (contract no. FP6-513663). We acknowledge the Research Institute Cemagref Rennes (France) for providing to us the PIV sequence and Prof. F. Scarano for his valuable comments.

## References

1. Raffel, M., Willert, C., Kompenhans, J.: Particle Image Velocimetry. A Practical Guide. Springer Verlag (1998)

2. Hinsch, K.D.: Three-dimensional particle velocimetry. Meas. Sci. Tech. **6** (1995) 742–753
3. Royer, H., Stanislas, M.: Stereoscopic and holographic approaches to get the third velocity component in piv. von Karman Institute for Fluid Dynamics, Lecture Series **03** (1996)
4. Scarano, F.: Iterative image deformation methods in piv. Measc. Sci. Technol. **13** (2002) R1–R19
5. Corpetti, T., Heitz, D., Arroyo, G., Mémin, E., Santa-Cruz, A.: Fluid experimental flow estimation based on an optical-flow scheme. Experiments in Fluids **40**(1) (2006) 80–97
6. Horn, B., Schunck, B.: Determining optical flow. MIT Aritificial Intelligence Laboratory (April 1980)
7. Beauchemin, S.S., Barron, J.L.: The computation of optical flow. ACM Computing Surveys **27**(3) (1995) 433–467
8. Barron, J., Fleet, D., Beauchemin, S.: Performance of optical flow techniques. International Journal of Computer Vision **12**(1) (1994) 43–77
9. White, F.: Viscous Fluid Flow. McGraw-Hill (2006)
10. Parnaudeau, P., Carlier, J., Heitz, D., Lamballais, E.: Experimental and numerical studies of the flow over a circular cylinder at reynolds number 3900. Physics of Fluids **(submitted)** (2006)
11. Weickert, J., Schnörr, C.: A theoretical framework for convex regularizers in pde-based computation of image motion. International Journal of Computer Vision **45**(3) (2001) 245–264
12. Alvarez, L., Weickert, J., Sánchez, J.: Reliable estimation of dense optical flow fields with large displacements. International Journal of Computer Vision **39**(1) (2000) 41–56
13. Yuan, J., Ruhnau, P., Memin, E., Schnörr, C.: Discrete orthogonal decomposition and variational fluid flow estimation. In Springer, ed.: Lecture Notes in Compute Science (Scale Space 2005). Volume 3459. (2005) 267–278

# Detecting Regions of Dynamic Texture

Tomer Amiaz[1], Sándor Fazekas[2], Dmitry Chetverikov[2], and Nahum Kiryati[1]

[1] School of Electrical Engineering
Tel Aviv University
Tel Aviv, Israel
[2] Computer and Automation Research Institute
Budapest, Hungary

**Abstract.** Motion estimation is usually based on the brightness constancy assumption. This assumption holds well for rigid objects with a Lambertian surface, but it is less appropriate for fluid and gaseous materials. For these materials a variant of this assumption, which we call the *brightness conservation* assumption should be employed. Under this assumption an object's brightness can diffuse to its neighborhood. We propose a method for detecting regions of dynamic texture in image sequences. Segmentation into regions of static and dynamic texture is achieved by using a level set scheme. The level set function separates the images into areas obeying brightness constancy and those which obey brightness conservation. Experimental results on challenging image sequences demonstrate the success of the segmentation scheme and validate the model.

**Keywords:** dynamic texture, level set, optical flow.

## 1 Introduction

A dynamic texture is a spatially repetitive, time-varying visual pattern possessing certain temporal stationarity [1,2,3]. Such a pattern is easily observed by the human eye but compared to a static texture it is much more difficult to discern by computer vision methodology [4]. Detecting, segmenting, recognizing, and classifying dynamic textures can rely on visual aspects such as geometry or motion, or both. The current methods of analysis are based mainly on optical flow estimation and geometric or model-based algorithms.

Dynamic textures, for example fire and smoke, flowing water, or foliage blown by the wind, are common in natural scenes. However, in many cases only parts of the scene form dynamic textures. In addition, their spatial extent can keep varying and they might be partially transparent, which makes it difficult to separate them from a textured background. Due to these problems the geometry (size and shape) can be misleading. The difference in dynamics, however, could be successfully employed to detect and segment them.

Segmentation is one of the classical problems of computer vision. Using motion-based features for segmentation – either alone or combined with other image cues – is a well-known practice [5]. Cremers and Soatto suggested a variational level set

F. Sgallari, A. Murli, and N. Paragios (Eds.): SSVM 2007, LNCS 4485, pp. 848–859, 2007.

method for motion based segmentation [6]. Doretto et al [7] used a similar level set approach for segmenting different types of dynamic textures based on statistical characteristics.

Motion estimators are usually built on the *brightness constancy assumption*. Under this assumption an object's brightness is constant from frame to frame. This assumption holds for rigid objects with a Lambertian surface, but fails for fluid and gaseous materials [8,9] which are typical for dynamic textures.

Dynamic textures are usually defined by extending the concept of self-similarity – well-established for static textures – to the spatiotemporal domain. *Weak dynamic textures* such as a simple moving texture are covered by this definition. In these dynamic textures, there exists a local moving coordinate system in which the texture becomes static. This local coordinate system can be computed using standard optical flow algorithms [10,11] relying on the brightness constancy assumption. However, a *strong dynamic texture*, possessing intrinsic dynamics, cannot be captured by this approach because of self-occlusion, material diffusion, and other physical processes not obeying the brightness constancy assumption.

In this paper, we present a variant of the brightness constancy assumption, which we call the *brightness conservation assumption*. Under this assumption the brightness of an image point (in one frame) can propagate to its neighborhood (in the next frame). While a static or weak dynamic texture obeys the brightness constancy assumption – as we are going to show – a strong dynamic texture is better modeled by the brightness conservation assumption. Using this, we suggest a level set segmentation scheme for detecting dynamic texture regions based on their specific motion characteristics.

We test our method on six videos of non-segmented dynamic textures (five of them taken from the DynTex database [12]) showing flowing water, steam, smoke, and fire; all in a natural context. Four of the sequences were recorded with a moving camera. The experimental results show the adequacy of our approach for detecting strong dynamic textures. This applies even in challenging sequences for which simple motion based segmentation could not be applied.

## 2 Background

Several examples can be taken from biological systems to emphasize the importance of motion in visual sensing. Studies of visual perception [13] revealed that humans use motion directly in recognizing aspects of their environment. Insects are essentially blind to anything that is standing still and the camouflage strategies of some animals are effective only as long as they are not moving. The observation that motion adds relevant information to visual patterns, explains the current effort in computer vision research to extend the already classical field of texture analysis from the spatial domain to the temporal domain.

### 2.1 Dynamic Textures

Dynamic textures studied so far include recordings of physical processes such as surface waves, fire and smoke, flag blown by the wind or the collective motion of different distinct elements such as a walking crowd, a flock of birds, or cars on a highway. All these exhibit spatiotemporal regularity with an indeterminate spatial and temporal extent. Dynamic texture analysis can be a basic building block of motion detection and recognition systems. Dynamic textures can be used to query multimedia databases.

Currently the most popular methods used to analyze dynamic textures (for a recent review see [4]) are based on optical flow calculation [1,14,15,16,17,18,19]. An alternate approach is to compute geometric properties in the spatiotemporal domain [20,21], while other methods are based on local or global spatiotemporal filtering [22] and spatiotemporal transforms [23,24]. There are also model-based methods [2,3,25,26,27,28] which use estimated model parameters as features.

The methods based on optical flow characterize the local dynamics of spatiotemporal textures by computing a velocity field describing the motion of small image regions. In this approach a dynamic texture can be viewed as a sequence of instantaneous motion patterns. When necessary, geometrical information and color can also be added to form a complete set of features for both motion and appearance based detection, segmentation, and recognition.

The optical flow concept arose from studies on the human visual system. Aside from algorithmic complexity and stability, one of the major difficulties in calculating optical flow is caused by the so-called *aperture problem* [29]. Independent of the imaging modality, a sensor "looking" at the world through a small hole cannot distinguish between different motion directions. This affects both human visual perception [30] and computer vision algorithms [31].

Due to the aperture problem, only the so called *normal flow* can be computed without ambiguity, unless the motion estimation is extended to a larger region using smoothness constraints [31]. By definition, the normal flow is orthogonal to contours and antiparallel to the local spatial image gradient. Its computation requires only the three partial derivatives of the spatiotemporal intensity function. Being purely local, the normal flow does not tend to extend motion over discontinuities, however, it is very noise-prone.

Most of the work done so far in studying dynamic textures used the normal flow, partly as an influence of the successful pioneering work of Nelson and Polana [1] and partly because the calculation of the normal flow is easy and fast. Even though it was recognized already at the early state of these studies [1] that the close relation of normal flow and spatial gradients (and hence contours and shapes) implies that the normal flow correlates with appearance features and thus it does not characterize the "pure dynamics", no solution was proposed. Later, to overcome this problem, Fablet and Bouthemy [16] used only the magnitude of the normal flow and recently Lu et al [17] as well as Fazekas and Chetverikov [19] stressed the necessity to apply a complete flow calculation in extracting characteristics of dynamic textures.

### 2.2 Motion Based Segmentation

Motion, when available, is a useful cue for segmentation and it has been used either alone [6,32] or to enhance other cues [33,34]. Motion based segmentation can proceed in either of two ways: Detect motion and apply a segmentation algorithm on the result [35] or simultaneously compute the motion and the segmentation [6,32]. Motion based segmentation schemes rely on optical flow algorithms to estimate the motion. It is interesting to note that recently segmentation has been used to enhance the optical flow estimation process itself [36,37].

Horn and Schunck's algorithm [10] is the base of all variational optical flow calculation methods. To define an optical flow field describing the motion, they use the *optical flow constraint*

$$I_t + uI_x + vI_y = 0, \tag{1}$$

where $I_t, I_x, I_y$ are the time and spatial derivatives of the image, respectively, and $(u, v)$ are the $x$ and $y$-dimension components of the flow. The above equation is the first-order Taylor approximation of the *brightness constancy assumption*

$$I(x + u, y + v, t + 1) = I(x, y, t). \tag{2}$$

In order to overcome the aperture problem, Horn and Schunck impose an additional smoothness constraint on $(u, v)$ obtaining the Lagrangian

$$L_{HS}(u, v) = (I_t + uI_x + vI_y)^2 + \alpha(u_x^2 + u_y^2 + v_x^2 + v_y^2), \tag{3}$$

where $\alpha$ is a parameter and $u_x$, $u_y$, $v_x$, and $v_y$ are the flow derivatives. The optical flow is calculated by minimizing the functional

$$F_{HS}(u, v) = \int_I L_{HS}(u, v)\, \mathrm{d}x\mathrm{d}y \tag{4}$$

based on the calculus of variations [31]. The accuracy of the method can be enhanced by using a coarse-to-fine scheme [38,39].

For simultaneous segmentation and motion estimation, Cremers and Soatto [6] use a variational scheme based on the level set method. Level set methods rely on introducing an evolving indicator function $\phi$, which classifies every point according to whether its value is positive or negative. This approach was pioneered by Osher and Sethian [40] (a similar technique was also suggested by Dervieux and Thomasset [41]). Chan and Vese [42,43] have formulated a variational segmentation problem using the Lagrangian

$$L_{CV}(I^+, I^-, \phi) = (I - I^+)^2 H(\phi) + (I - I^-)^2(1 - H(\phi)) + \nu|\nabla\phi|, \tag{5}$$

where $I^+$ and $I^-$ are the average intensity values of the two regions corresponding to the positive and negative value of $\phi$, $H(\cdot)$ is the Heaviside function, and $|\nabla\phi|$ is the norm of the gradient of $\phi$.

Cremers and Soatto [6] replace the image fidelity terms of (5) with terms based on fitting a parametric affine motion model to the images, obtaining

$$L_{CS}(p_1, p_2, \phi) = \frac{p_1^t T p_1}{|p_1|^2} H(\phi) + \frac{p_2^t T p_2}{|p_2|^2}(1 - H(\phi)) + \nu|\nabla H(\phi)|, \tag{6}$$

where $p_i^t T p_i / |p_i|^2$ represents the quality of the fit characterized by $p_i$.

## 3 Dynamic Texture Region Detector

We propose segmenting image sequences into regions obeying different motion models: One region should conform to the *brightness constancy assumption* (2), while the other to a variant which we call the *brightness conservation assumption.* In this section, we first describe this latter assumption and the rationale behind it, then we present our level set scheme simultaneously segmenting and detecting motion characteristics.

### 3.1 Brightness Conservation Assumption

The *brightness constancy assumption.* (see equation (2)) states that for two consecutive frames in a video sequence the brightness of a point $(x, y)$ in a frame is identical to the brightness of a point $(x + u, y + v)$ in the next frame. In other words, it states that by warping the image space according to a displacement field $(u, v)$ the images can be brought into point-by-point equality. There are situations in which this assumption does not hold: For example, in case of occlusions or when the captured scene includes glinting surfaces (e.g. water surface) or complex physical processes (e.g. smoke or fire). These situations are typical for dynamic textures.

The extent to which the brightness constancy assumption holds for an optical flow $(u, v)$ calculated with the Horn-Schunck method [10] or more accurate methods, e.g. [44] and [36], can be measured. This is done by computing the optical flow residual

$$\left\langle |I(x+u, y+v, t+1) - I(x, y, t)| \right\rangle_w, \tag{7}$$

where $\langle\cdot\rangle_w$ denotes a convolution with a Gaussian kernel $w$ and $I(x+u, y+v, t+1)$ is calculated with sub-pixel accuracy using bilinear interpolation.

Computing the above residual for the dynamic textures available in the DynTex database [12], we found that in certain image regions it is large and comparable to the null flow residual

$$\left\langle |I(x, y, t+1) - I(x, y, t)| \right\rangle_w, \tag{8}$$

even for the most accurate optical flow calculation methods [44,36] tested. This proves that the flow could not fully capture the dynamics.

It is important to note that the large residual is not a consequence of numerical errors or inaccuracy of the flow but is due to the fact that the classical brightness constancy assumption simply does not hold in certain conditions. One solution for this problem is to calculate a flow which can model not only space warps but also changes in brightness. A straightforward approach is to consider that the flow "carries" the brightness as a "physical quantity" and the changes in brightness are encoded in the divergence of the flow.

Considering an arbitrary region $\Omega$ of an image $I$, brightness conservation can be defined as the equilibrium of the total brightness change on $\Omega$ and the brightness carried in and out through its boundary. With notations usual in physics, this is

$$\int_{\Omega} \partial_t I \, \mathrm{d}A + \int_{\partial\Omega} I \, \boldsymbol{f} \cdot \boldsymbol{n} \, \mathrm{d}L = 0, \tag{9}$$

where $\partial_t I$ is the time derivative of $I$, $\boldsymbol{f} = (u, v)$ is the flow, $\partial\Omega$ denotes the boundary of $\Omega$, and $\boldsymbol{n}$ is the external normal of $\partial\Omega$.

The above equation is the integral form of the *continuity equation* known from physics and used for material quantities (e.g. mass or charge) carried by a flow. Through mathematical transformations and the divergence theorem one can derive its differential form

$$\partial_t I + \nabla I \cdot \boldsymbol{f} + I \, \mathrm{div}(\boldsymbol{f}) = 0. \tag{10}$$

Writing the above equation in the form of (1) we obtain

$$I_t + uI_x + vI_y + Iu_x + Iv_y = 0, \tag{11}$$

where $u_x$ and $v_y$ denote partial derivatives of the flow. It can be observed that this is the first order Taylor approximation of

$$I(x + u, y + v, t + 1) = I(x, y, t)(1 - u_x - v_y), \tag{12}$$

which we call the *brightness conservation assumption.* This should be compared to the brightness constancy assumption (2).

A flow satisfying equation (12) does not only encode a warp of the image space but also brightness diffusion through its divergence $u_x + v_y$. Consequently, such a flow can capture more dynamic information than a classical optical flow, and thus it is more suitable for the study of dynamic textures.

Similar to the derivation of the Horn-Schunck Lagrangian (3) from the optical flow constraint (1), we can formulate a Lagrangian based on the continuity equation (11). If $(u, v)$ satisfies equation (11), then, for any constant $\kappa$, the flow $(u + \kappa I_y, v - \kappa I_x)$ also satisfies it. In order to overcome this ambiguity we have to impose a certain level of smoothness, and thus we obtain

$$L_{BC}(u, v) = (I_t + uI_x + vI_y + Iu_x + Iv_y)^2 + \alpha(u_x^2 + u_y^2 + v_x^2 + v_y^2). \tag{13}$$

The above Lagrangian used in a variational scheme assures brightness conservation and it can be employed to detect actual particle density motion in gaseous materials (e.g. smoke or vapor). A similar technique was developed by Béréziat et al [8] and Cuzol et al [9].

### 3.2 Level Set Segmentation

We propose a new segmentation scheme which separates images into two regions: A regular region obeying the brightness constancy assumption (2) and a strong dynamic texture region obeying the brightness conservation assumption (12).

Embedding the Lagrangians (3) and (13) into a level set scheme, we obtain

$$\begin{aligned}&L_{DTS}(u,v,\tilde{u},\tilde{v},\phi) = \\ &(I_t + uI_x + vI_y)^2 H(\phi) + (I_t + \tilde{u}I_x + \tilde{v}I_y + I\tilde{u}_x + I\tilde{v}_y)^2 H(-\phi) + \\ &\alpha(u_x^2 + u_y^2 + v_x^2 + v_y^2) + \tilde{\alpha}(\tilde{u}_x^2 + \tilde{u}_y^2 + \tilde{v}_x^2 + \tilde{v}_y^2) + \tilde{\beta}(\tilde{u}^2 + \tilde{v}^2) + \\ &\nu|\nabla H(\phi)|. \qquad (14)\end{aligned}$$

Here, $(u, v)$ is a regular optical flow field, $(\tilde{u}, \tilde{v})$ is a field satisfying the brightness conservation assumption, $\phi$ is an indicator function (negative on strong dynamic texture regions and non-negative otherwise). Adding the term $\beta(\tilde{u}^2 + \tilde{v}^2)$ penalizes motion, and has been suggested to keep homogeneous areas inert [45]. We add this term to improve the numerical stability of the minimization process.

Simultaneous motion detection and segmentation is achieved by minimizing the objective functional

$$F_{DTS}(u,v,\tilde{u},\tilde{v},\phi) = \int_I L_{DTS}(u,v,\tilde{u},\tilde{v},\phi)\, \mathrm{d}x\mathrm{d}y. \qquad (15)$$

This is done by solving the Euler-Lagrange equations of $u$, $v$, $\tilde{u}$, $\tilde{v}$, and $\phi$.

Introducing the notations

$$R(u,v) = I_t + uI_x + vI_y, \qquad (16)$$

$$\tilde{R}(\tilde{u},\tilde{v}) = I_t + \tilde{u}I_x + \tilde{v}I_y + I\tilde{u}_x + I\tilde{v}_y, \qquad (17)$$

the equations for $u$ and $v$ are

$$I_x R(u,v)H(\phi) - \alpha(u_{xx} + u_{yy}) = 0, \qquad (18)$$

$$I_y R(u,v)H(\phi) - \alpha(v_{xx} + v_{yy}) = 0. \qquad (19)$$

For $\tilde{u}$ and $\tilde{v}$ we have

$$\tilde{\beta}\tilde{u} - I\tilde{R}_x(\tilde{u},\tilde{v})H(-\phi) - \tilde{\alpha}(\tilde{u}_{xx} + \tilde{u}_{yy}) = 0, \qquad (20)$$

$$\tilde{\beta}\tilde{v} - I\tilde{R}_y(\tilde{u},\tilde{v})H(-\phi) - \tilde{\alpha}(\tilde{v}_{xx} + \tilde{v}_{yy}) = 0. \qquad (21)$$

Here, $\tilde{R}_x(\tilde{u},\tilde{v})$ and $\tilde{R}_y(\tilde{u},\tilde{v})$ denote partial derivatives of $\tilde{R}(\tilde{u},\tilde{v})$ with respect to $x$ and $y$. The indicator function $\phi$ must satisfy

$$\delta(\phi)\left[R(u,v)^2 - \tilde{R}(\tilde{u},\tilde{v})^2 - \nu \operatorname{div}\left(\frac{\nabla\phi}{|\nabla\phi|}\right)\right] = 0, \qquad (22)$$

where $\delta(\cdot)$ is the derivative of the Heaviside function.

We discretize equations (18) and (19) as described in [10]. Equations (20) and (21) can be discretized in a similar way. Equation (22) is discretized according to the method presented in [43]. Below, we present in details the discretized form of equations (20) and (21).

Following the method described in [10], we assume central derivatives for the partial derivatives of $\tilde{u}$ and $\tilde{v}$ and solve iteratively the matrix equation

$$\begin{pmatrix} \tilde{A} & \tilde{C} \\ \tilde{C} & \tilde{B} \end{pmatrix} \begin{pmatrix} \tilde{u} \\ \tilde{v} \end{pmatrix} = \begin{pmatrix} \tilde{D} \\ \tilde{E} \end{pmatrix}, \tag{23}$$

where

$$\tilde{A} = I(2I - I_{xx})H(-\phi) + 4\tilde{\alpha} + \tilde{\beta}, \tag{24}$$
$$\tilde{B} = I(2I - I_{yy})H(-\phi) + 4\tilde{\alpha} + \tilde{\beta}, \tag{25}$$
$$\tilde{C} = -II_{xy}H(-\phi), \tag{26}$$
$$\tilde{D} = I(I_{tx} + I_x(2\tilde{u}_x + \tilde{v}_y) + I_y\tilde{v}_x + I(\tilde{v}_{xy} + \tilde{\tilde{u}}_x))H(-\phi) + \tilde{\alpha}(\tilde{\tilde{u}}_x + \tilde{\tilde{u}}_y), \tag{27}$$
$$\tilde{E} = I(I_{ty} + I_y(2\tilde{v}_y + \tilde{u}_x) + I_x\tilde{u}_y + I(\tilde{u}_{xy} + \tilde{\tilde{v}}_y))H(-\phi) + \tilde{\alpha}(\tilde{\tilde{v}}_x + \tilde{\tilde{v}}_y). \tag{28}$$

Here, we used the notations

$$\tilde{\tilde{u}}_x = \tilde{u}(x-1, y) + \tilde{u}(x+1, y), \tag{29}$$
$$\tilde{\tilde{u}}_y = \tilde{u}(x, y-1) + \tilde{u}(x, y+1). \tag{30}$$

The definitions of $\tilde{\tilde{v}}_x$ and $\tilde{\tilde{v}}_y$ are analogous to $\tilde{\tilde{u}}_x$ and $\tilde{\tilde{u}}_y$.

Iterations are carried out simultaneously on $u, v$, $\tilde{u}, \tilde{v}$ and $\phi$. Because, in the equations of the flow fields, a first order Taylor approximation was assumed, $(u, v)$ and $(\tilde{u}, \tilde{v})$ need to be small. This is achieved by using a coarse-to-fine scheme: The images are warped according to the flow calculated at a coarse scale and small corrections are added repeatedly at finer scales.

## 4 Experimental Results

We demonstrate our segmentation scheme on five video sequences from the DynTex database [12] and one additional sequence. The videos show non-segmented dynamic textures in a natural context: "`6482910`" shows a wide *stream* with water rippling over underwater pebbles; "`6483c10`" shows a different narrower *creek* winding between larger pebbles; "`6481i10`" shows a small *waterfall* with water falling in changing patterns over round pebbles and gathering into a small pool; and "`55fc310`" and "`648ea10`" show *steam* and *smoke*. The additional sequence shows a *camp fire*. The first three videos were recorded with a panning camera mounted on a tripod, the next two with a fixed camera, and the last one with a randomly moving handheld camera.

The images were processed in grayscale range of zero to one (i.e. no color information was used). They were pre-blurred with a standard $3 \times 3$ Gaussian kernel. Coarse-to-fine iterations were run over a Gaussian pyramid having in general 3 levels. The *Steam* and *Fire* sequences were processed in 4 levels. Each image in the pyramid was half the size of the image in the previous level. The number of iterations used to solve the Euler-Lagrange equations on each level was 100. Both flow fields $(u, v)$ and $(\tilde{u}, \tilde{v})$, and the indicator function $\phi$ were initialized

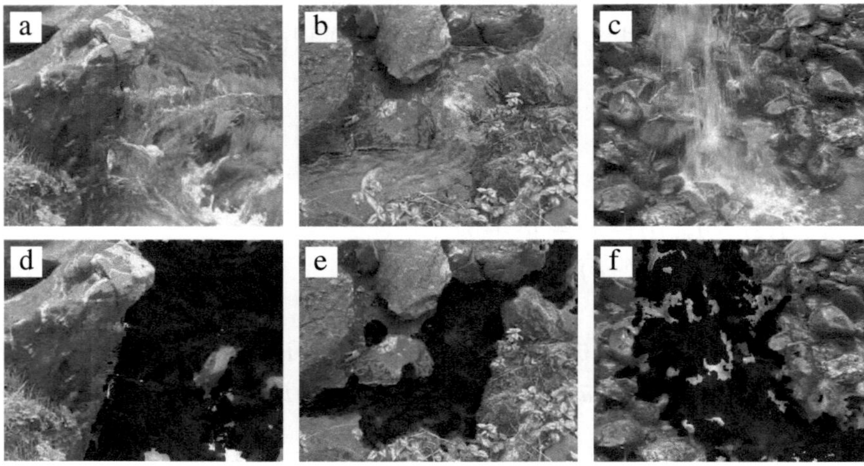


**Fig. 1.** (a) Frame 232 of the *Stream* sequence. (b) Frame 392 of the *Creek* sequence. (c) Frame 652 of the *Waterfall* sequence. All three sequences recorded by a panning camera. (d-f) Result of segmentation marked in blue.

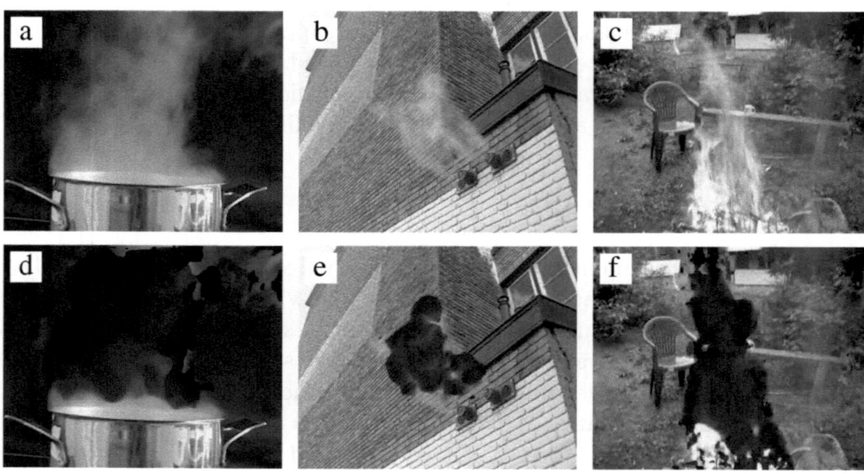


**Fig. 2.** (a) Frame 2 of the *Steam* sequence. (b) Frame 72 of the *Smoke* sequence. (c) Frame 99 of the *Fire* sequence. Recorded with a handheld moving camera. (d-f) Result of segmentation marked in blue.

to zero. The Heaviside function was implemented as $H(\phi) = 0.5 + \text{atan}(\phi/\phi_0)/\pi$ with $\phi_0 = 0.01$. We used $\alpha = 0.01$, $\tilde{\alpha} = 0.001$, $\tilde{\beta} = 0.01$, and $\nu = 0.001$ for all videos, with the exception of the *Fire* sequence, where $\nu = 0.003$ was used.

Figures 1 and 2 present the results of the segmentation experiments. The top panels show frames from each of the sequences tested. The bottom panels mark the region of strong dynamic texture in blue. Excellent agreement with the actual position of flowing water, steam, smoke, and fire can be observed. Only small regions were not detected as dynamic texture. These are typically regions where there is not too much activity and because all these materials are transparent they could not be distinguished from the background. There are also small non-dynamic areas (e.g. pebbles protruding from the water) being misdetected, probably due to over-smoothing of the level set function $\phi$.

## 5 Discussion

The motion based segmentation scheme presented in this paper is an effective tool to detect strong dynamic texture regions of an image sequence. It relies on the different dynamics of certain dynamic textures (demonstrated in this paper on turbulent water, smoke, steam, and fire) relative to the dynamics of moving objects. Strong dynamic textures possess intrinsic dynamics which cannot be diminished in any moving coordinate system. The segmentation achieved by our method can be refined with statistical data taking in consideration characteristics of specific dynamic textures. The information encoded in the calculated flows (regular and dynamic) should be further explored. Here we made use only of the calculated indicator function which defines the segmentation.

The slow convergence of the minimization process can be addressed by replacing the iteration scheme of the Horn-Schunck method with a faster solver of large sparse systems of linear equations. Modern optical flow methods (see for example [44]) have much faster convergence. We believe that adopting techniques employed in such methods will result in a much faster segmentation algorithm.

The functional (14) includes the brightness values of the image explicitly. Optical flow methods usually include only derivatives of the image brightness and are, therefore, invariant to the numerical value of black. The brightness conservation assumption as stated in (12) violates this principle. Future work should look into restating the assumption in an invariant formulation.

## Acknowledgment

This research was supported in part by MUSCLE: Multimedia Understanding through Semantics, Computation and Learning, a European Network of Excellence funded by the EC 6th Framework IST Programme. At Tel Aviv University this research was supported also by the A.M.N. Foundation.

## References

1. Nelson, R.C., Polana, R.: Qualitative recognition of motion using temporal texture. CVGIP: Image Understanding **56** (1992) 78–89
2. Szummer, M., Picard, R.: Temporal texture modeling. In: Proc. Int. Conf. Image Processing. Volume 3. (1996) 823–826

3. Doretto, G., Chiuso, A., Soatto, S., Wu, Y.N.: Dynamic textures. Int. J. Comp. Vision **51** (2003) 91–109
4. Chetverikov, D., Péteri, R.: A brief survey of dynamic texture description and recognition. In: 4th Int. Conf. on Computer Recognition Systems. (2005) 17–26
5. Murray, D.W., Buxton, B.F.: Scene segmentation from visual motion using global optimization. IEEE Trans. Pattern Analysis and Machine Intell. **9**(2) (1987) 220–228
6. Cremers, D., Soatto, S.: Motion competition: A variational approach to piecewise parametric motion segmentation. Int. J. Comp. Vision **62**(3) (2004) 249–265
7. Doretto, G., Cremers, D., Favaro, P., Soatto, S.: Dynamic texture segmentation. In: Ninth Int. Conf. on Computer Vision. (2003) 1236
8. Béréziat, D., Herlin, I., Younes, L.: A generalized optical flow constraint and its physical interpretation. In: Proc. Conf. Comp. Vision Pattern Rec. (2000) 487–492
9. Cuzol, A., Mémin, E.: Vortex and source particles for fluid motion estimation. In Kimmel, R., Sochen, N., Weickert, J., eds.: Lecture Notes in Computer Science. Volume 3459. (2005) 254–266
10. Horn, B.K.P., Schunck, B.G.: Determining optical flow. Artificial Intelligence **17**(1-3) (1981) 185–203
11. Lucas, B.D., Kanade, T.: An iterative image registration technique with an application to stereo vision. In: DARPA Image Understanding Workshop. (1981) 121–130
12. Péteri, R., Huskies, M., Fazekas, S.: DynTex: A comprehensive database of Dynamic Textures. `www.cwi.nl/projects/dyntex/` (2006)
13. Bruce, V., Green, P.R., Georgeson, M.: Visual Perception. Psychology Press (UK) (1996)
14. Bouthemy, P., Fablet, R.: Motion characterization from temporal co-occurrences of local motion-based measures for video indexing. In: Proc. of the Int. Conf. Pattern Recognition. Volume 1. (1998) 905–908
15. Peh, C.H., Cheong, L.F.: Synergizing spatial and temporal texture. IEEE Transactions on Image Processing **11** (2002) 1179–1191
16. Fablet, R., Bouthemy, P.: Motion recognition using nonparametric image motion models estimated from temporal and multiscale co-occurrence statistics. IEEE Trans. Pattern Analysis and Machine Intell. **25** (2003) 1619–1624
17. Lu, Z., Xie, W., Pei, J., Huang, J.: Dynamic texture recognition by spatiotemporal multiresolution histograms. In: Proc. of the IEEE Workshop on Motion and Video Computing (WACV/MOTION). (2005) 241–246
18. Péteri, R., Chetverikov, D.: Dynamic texture recognition using normal flow and texture regularity. Lecture Notes in Computer Science 3523. (2005) 223–230
19. Fazekas, S., Chetverikov, D.: Normal versus complete flow in dynamic texture recognition: A comparative study. In: Int. Workshop on Texture Analysis and Synthesis. (2005) 37–42
20. Otsuka, K., Horikoshi, T., Suzuki, S., Fujii, M.: Feature extraction of temporal texture based on spatiotemporal motion trajectory. In: ICPR. Volume 2. (1998) 1047–1051
21. Zhong, J., Scarlaroff, S.: Temporal texture recongnition model using 3D features. Technical report, MIT Media Lab Perceptual Computing (2002)
22. Wildes, R.P., Bergen, J.R.: Qualitative spatiotemporal analysis using an oriented energy representation. In: Proc. European Conf. on Computer Vision. (2000) 768–784
23. Smith, J., Lin, C.Y., Naphade, M.: Video texture indexing using spatiotemporal wavelets. In: Proc. Int. Conf. on Image Processing. Volume 2. (2002) 437–440

24. Wu, P., Ro, Y.M., Won, C.S., Choi, Y.: Texture descriptors in MPEG-7. Lecture Notes in Computer Science 2124 (2001) 21–28
25. Saisan, P., Doretto, G., Wu, Y.N., Soatto, S.: Dynamic texture recognition. In: Proc. Conf. Comp. Vision Pattern Rec. Volume 2., Kauai, Hawaii (2001) 58–63
26. Fujita, K., Nayar, S.: Recognition of dynamic textures using impulse responses of state variables. In: Int. Workshop on Texture Analysis and Synthesis. (2003) 31–36
27. Doretto, G., Jones, E., Soatto, S.: Spatially homogeneous dynamic textures. In: Proc. European Conf. on Computer Vision. Volume 2. (2004) 591–602
28. Yuan, L., Weng, F., Liu, C., Shum, H.Y.: Synthersizing dynamic texture with closed-loop linear dynamic system. In: Proc. European Conf. on Computer Vision. Volume 2. (2004) 603–616
29. Todorovic, D.: A gem from the past: Pleikart Stumpf's anticipation of the aperture problem, Reichardt detectors, and perceived motion loss at equiluminance. Perception **25** (1996) 1235–1242
30. Hildreth, E.C.: The analysis of visual motion: From computational theory to neural mechanisms. Annual Review of Neuroscience **10** (1987) 477–533
31. Horn, B.K.P.: Robot Vision. McGraw-Hill – New York (1986)
32. Paragios, N., Deriche, R.: Geodesic active regions and level set methods for motion estimation and tracking. Comp. Vision and Image Underst. **97**(3) (2005) 259–282
33. Zheng, H., Blostein, S.D.: Motion-based object segmentation and estimation using the MDL principle. IEEE Transactions on Image Processing **4**(9) (1995) 1223–1235
34. Galun, M., Apartsin, A., Basri, R.: Multiscale segmentation by combining motion and intensity cues. In: Proc. Conf. Comp. Vision Pattern Rec. Volume 1., Washington, DC, USA, IEEE Computer Society (2005) 256–263
35. Wang, J.Y.A., Adelson, E.H.: Representing moving images with layers. The IEEE Transactions on Image Processing **3**(5) (1994) 625–638
36. Amiaz, T., Kiryati, N.: Piecewise-smooth dense optical flow via level sets. Int. J. Comp. Vision **68**(2) (2006) 111–124
37. Brox, T., Bruhn, A., Weickert, J.: Variational motion segmentation with level sets. In: Proc. European Conf. on Computer Vision. Volume I. (2006) 471–483
38. Anandan, P.: A computational framework and an algorithm for the measurement of visual motion. Int. J. Comp. Vision **2**(3) (1989) 283–310
39. Black, M.J., Anandan, P.: The robust estimation of multiple motions: parametric and piecewise-smooth flow fields. Comp. Vision and Image Underst. **63**(1) (1996) 75–104
40. Osher, S., Sethian, J.A.: Fronts propagating with curvature-dependent speed: Algorithms based on Hamilton-Jacobi formulations. J. Comp. Phys. **79** (1988) 12–49
41. Dervieux, A., Thomasset, F.: A finite element method for the simulation of rayleigh-taylor instability. Lecture Notes in Mathematics **771** (1979) 145–158
42. Chan, T.F., Vese, L.A.: Active contours without edges. IEEE Transactions on Image Processing **10**(2) (2001) 266–277
43. Vese, L.A., Chan, T.F.: A multiphase level set framework for image segmentation using the Mumford and Shah model. Int. J. Comp. Vision **50**(3) (2002) 271–293
44. Papenberg, N., Bruhn, A., Brox, T., Didas, S., Weickert, J.: Highly accurate optic flow computation with theoretically justified warping. Int. J. Comp. Vision **67**(2) (2006) 141–158
45. Aubert, G., Kornprobst, P.: Mathematical Problems in Image Processing: Partial Differential Equations and the Calculus of Variations. Volume 147 of Applied Mathematical Sciences. Springer-Verlag (2002)

# A Method for the Transport and Registration of Images on Implicit Surfaces

Christophe Chefd'hotel

Siemens Corporate Research
Princeton, NJ, USA

**Abstract.** Image transport is at the heart of large deformation registration algorithms and level set methods. Here we propose to extend this concept to solve alignment and correspondence problems for images defined on general geometries. Our approach builds upon the framework proposed by Bertalmio et al. [1] for solving PDEs on implicit surfaces. The registration process is defined by a system of transport equations and is driven by the gradient of a similarity functional. The transformation is regularized using a nonlinear heat equation. Compared to recent developments for image registration on manifolds [4], the implicit representation of the image domain allows us to deal easily with arbitrary surfaces. We illustrate the potential of this technique with several synthetic experiments.

## 1 Introduction

Image registration is one of the fundamental problems in computer vision. It consists of finding a geometric transformation that aligns a pair of images. The recovered transformation can be used to compensate for motion, detect changes, or fuse complementary information provided by different sensors. For the estimation of highly nonrigid transformations, such as soft tissue deformations encountered in medical imaging, the so-called "fluid" or "large deformation" registration algorithms [3,10] have proven very effective. They describe the registration process as a flow on a set of smooth invertible transformations. The idea is to consider a family $t \mapsto \phi(t)$ of mappings (from $\mathbb{R}^n$ to $\mathbb{R}^n$) which is solution of a system of partial differential equations (PDEs)

$$\partial_t \phi + D\phi \mathbf{v} = 0, \ \phi(0) = \mathbf{id}, \tag{1}$$

where $\mathbf{v}$ is a smooth time-dependent vector field, $D\phi$ is the Jacobian matrix of $\phi$, and $\mathbf{id}$ is the identity transformation. In order to align an image $I_2$ to a reference image $I_1$, $\mathbf{v}$ is generally derived from the gradient of a similarity measure between $I_1$ and $I_2 \circ \phi$ (the warped version of $I_2$ under a transformation $\phi$).

In the following, we present an extension of this approach to images defined on general geometries. Such technique could be used in particular to align functional activation maps on the cortical surface. A generalization of large deformation registration techniques to the unit sphere was previously discussed in [4]. Here we

F. Sgallari, A. Murli, and N. Paragios (Eds.): SSVM 2007, LNCS 4485, pp. 860–870, 2007.

explore the case of images defined on an arbitrary implicit surface. We propose to map Eq. 1 to a curved geometry using the methodology presented by Bertalmio et al. [1] for partial differential equations on manifold.

The paper is organized as follows. We discuss the connection between large deformation registration and image transport in Section 2. In Section 3, we describe existing techniques for the implementation of the underlying transport equations. In Section 4, we develop our extension of the previous models and equations to the case of implicit surfaces. We present experimental results in Section 5. Concluding remarks are made in Section 6.

## 2 Image Registration and Image Transport

As a first step toward the construction of a registration method on implicit surfaces, we first show how image registration can been modeled as a problem of image transport. The idea is to reformulate Eq. 1 and translate a flow on a set of transformations into an equation acting directly on the space of images. Let $t \mapsto \phi(t)$ be solution of Eq. 1. We can track the registration process by considering the family of images $t \mapsto I(t)$ such that

$$\forall\, t,\ I(t) = (I_2 \circ \phi)(t).$$

If $I_2$ and $\phi$ are sufficiently regular, both in space and time, we can apply the chain rule of derivatives to obtain

$$\begin{aligned}\partial_t I &= (DI_2 \circ \phi)\partial_t \phi \\ &= -(DI_2 \circ \phi) D\phi \mathbf{v} \\ &= -D(I_2 \circ \phi)\mathbf{v} \\ &= -DI\mathbf{v}.\end{aligned}$$

Since $I$ is scalar-valued, we can replace the Jacobian $DI$ by $\nabla I^T$. Using the dot product notation $\mathbf{a} \cdot \mathbf{b} = \mathbf{a}^T\mathbf{b}$, the previous result becomes

$$\partial_t I + \nabla I \cdot \mathbf{v} = 0,\ I(0) = I_2. \tag{2}$$

Eq. 2 is a scalar transport equation which describes how $I_2$ is deformed during the alignment process. For a constant field $\mathbf{v}(t, \mathbf{p}) = \mathbf{a}$ (for all points $\mathbf{p}$ in the image domain), the solution of Eq. 2 reduces to

$$I(t, \mathbf{p}) = I_2(\mathbf{p} - t\mathbf{a}).$$

This corresponds to a translation of the original image by a vector $t\mathbf{a}$. In fact, one can observe that the evolution of $\phi$ is also characterized by a system of scalar transport equations. To see this, we simply rewrite Eq. 1 as

$$\forall\, i = 1, ..., n,\ \partial_t \phi_i + \nabla \phi_i \cdot \mathbf{v} = 0,\ \phi_i(0) = \phi_{i,0}. \tag{3}$$

This formulation shows that the components of $\phi$ and the evolving image $I$ are driven by the same evolution equation.

This type of equation plays a key role in various models of physical phenomena (see for instance [6] for transport equations in gas dynamics). It can also be found in computer vision where "level set methods" are used to model dynamic contours [8,7]. In the level set framework, $I_2$ would not be an image but a function whose zero level set is a closed submanifold of co-dimension 1 in $\mathbb{R}^n$ (curves in $\mathbb{R}^2$, surfaces in $\mathbb{R}^3$, etc.). It can be shown that a family $t \mapsto \mathcal{M}(t)$ of submanifolds evolving according to a velocity field $\mathbf{v}$ can be described as the zero-level set of the one-parameter family $t \mapsto I(t)$ solution of Eq. 2. In this framework, the design of velocity fields usually rely on intrinsic properties of $\mathcal{M}(t)$. The unit normal vector at point $\mathbf{p}$ on $\mathcal{M}(t)$ is given by $\mathbf{n}(\mathbf{p}) = \nabla I(t, \mathbf{p}) / \|\nabla I(t, \mathbf{p})\|$, and its mean curvature can be expressed as $\kappa(\mathbf{p}) = \mathrm{div}(\mathbf{n}(\mathbf{p}))$. For example, the mean curvature flow, defined by

$$\partial_t I - \nabla I \cdot \kappa \mathbf{n} = 0,$$

moves the evolving manifold in the direction of concavity. The velocity initially defined on $\mathcal{M}(t)$ is implicitly extended to all the level sets of $I(t)$. External forces can also be added to drive the evolving manifold according to a specific objective (for instance to detect object contours in an image [2]).

In the context of image registration, the velocity field $\mathbf{v}$ generally depends on $I_1$, $I_2$, and the deformation $\phi$ [3,10]. Hence, if we consider solving directly the image transport equation (Eq. 2), we also have to keep track of $\phi$ by simultaneously solving Eq. 1. However, we can also design a velocity field $\mathbf{v}$ that can be expressed in terms of $I_2 \circ \phi = I$. For instance, the gradient at identity of the sum of squared differences (SSD criterion) between $I_1$ and $I$, regularized by a smoothing operator $\mathcal{R}$, yields

$$\mathbf{v} = -\mathcal{R}((I_1 - I)\nabla I).$$

A simple choice for $\mathcal{R}$ is a convolution by a Gaussian kernel, as in Thirion's "demons" algorithm [9]. Eq. 2 then reduces to the autonomous equation

$$\partial_t I - \nabla I \cdot \mathcal{R}((I_1 - I)\nabla I) = 0, \; I(0) = I_2. \tag{4}$$

This flow can be used to deform directly the image without modeling explicitly the deformation $\phi$. This is particularly interesting for motion compensation problems where the goal is just to obtain a corrected (warped) image.

Eq. 4 is related to the work presented by Vemuri et al. in [12,11] on a level set approach to registration. They propose an image registration PDE defined by

$$\partial_t I = \|\nabla I\|(I_1 - I) \iff \partial_t I - \nabla I \cdot \left((I_1 - I)\frac{\nabla I}{\|\nabla I\|}\right) = 0. \tag{5}$$

This equation follows from a simple heuristic: image matching is performed by transporting an image in the direction $\mathbf{n} = \nabla I / \|\nabla I\|$ normal to its level sets with a speed proportional to the intensity difference with the reference image. It can be seen as a simplified version of Eq. 4 without regularization. Another difference is that in our case, the velocity along the normal directions not only depends on the image difference, but also on the norm of $\nabla I$ since $(I_1 - I)\nabla I = (I_1 - I)\|\nabla I\|\mathbf{n}$. In other words, strong edges in the image have a more active role in driving the registration process, whereas they all contribute equally in Eq. 5. As Eq. 5 lacks regularization, it is suggested in [12,11] to apply a Gaussian smoothing before computing the gradients of the transported image. Alternatively, in Eq. 4 the regularization operator $\mathcal{R}$ applies to the vector field itself. Note that Vemuri et al. also propose a way of tracking the deformation associated to Eq. 5. For this purpose, they introduce a vector equation which is analogous to Eq. 1, but where the flow is expressed in term of a global displacement $\mathbf{u} = \phi - \mathbf{id}$ instead of $\phi$.

## 3 Implementation of Transport Equations for Image Registration

Numerous schemes can be found in the computational physics and level set literature [6,7] to implement the previous equations. Some techniques take into account situations where the lack of smoothness of $\mathbf{v}$ generates singularities and shocks, that should be propagated properly during the evolution. In our case, the smoothness of $\mathbf{v}$ generally limits this type of problem. Unfortunately, another issue arises. The dissipative behavior of most numerical methods for

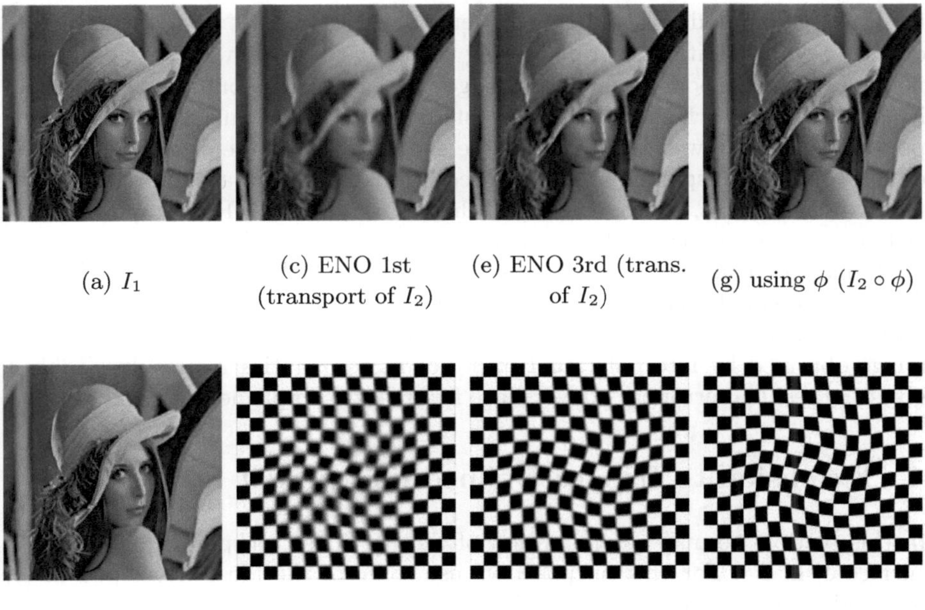

(a) $I_1$ (c) ENO 1st (transport of $I_2$) (e) ENO 3rd (trans. of $I_2$) (g) using $\phi$ $(I_2 \circ \phi)$

(b) $I_2$ (d) ENO 1st (transp. of $S$) (f) ENO 3rd (trans. of $S$) (h) using $\phi$ $(S \circ \phi)$

**Fig. 1.** Direct transport versus recovering $\phi$ + warping

**Table 1.** Residual registration errors (*lena* example)

| Method | Mean sq. err. |
|---|---|
| direct transport (ENO 1st order) | 193.5 |
| direct transport (ENO 2nd order) | 137.5 |
| direct transport (ENO 3rd order) | 112.1 |
| recovering $\phi$ + bilinear interpolation | **64.1** |

transport equations translates into a progressive blurring of the original data throughout the evolution. This phenomenon is due to the averaging of intensity values caused by the finite-difference approximation of derivatives. We refer to [6] for illustrations of this effect in the implementation of wave equations.

The deformation field is smooth, and the averaging artifact does not have a major impact when transporting the components of $\phi$. This effect, however, is highly visible when the equation is directly applied on images. They tend to lose their sharpness after a couple of iterations. One way to mitigate this problem is to use numerical methods of high-order accuracy. In fact, this point is instrumental in making the implementation of a direct image transport (Eq. 4) of any practical use.

To illustrate this problem, we show on Fig. 1 the results of the transport process of an artificially distorted "Lena" image. We implemented Eq. 4 using an "upwind" differencing scheme in combination with ENO (Essentially Non Oscillatory) derivative estimates. This approach has already proven its efficiency for level set methods [7] and can be generalized to achieve an arbitrary order of accuracy. In addition to transporting the "Lena" image, we applied identical flows to a checkerboard "shadow" image ($S$) that helps compare the diffusivity of the various schemes we implemented. We observe that all the algorithms performed equally well in capturing the geometry of the deformation, but the quality of the transported image is directly related to the accuracy of the numerical scheme (1st, 2nd or 3rd order schemes were tested). For comparison, the last column shows the result obtained by estimating $\phi$ and then computing the final image by computing $I_2 \circ \phi$ using a bilinear interpolation scheme.

Even if a small amount of blur might be acceptable, a high-order scheme is generally required to provide a transported image of quality comparable to the image obtained by interpolation $I_2 \circ \phi$. For more quantitative results, Table 1 gives the residual errors at convergence for different implementations.

## 4 Extension to Images Defined on Implicit Surfaces

In this section we map the definition of the previous transport equations to a surface $\mathcal{S}$ in $\mathbb{R}^3$. For this purpose, we use some of the ideas developed in [1] to build diffusion PDEs on implicit surfaces. We consider two images $I_1$ and $I_2$ now defined on a closed surface $\mathcal{S}$ in $\mathbb{R}^3$. The objective is now to find a smooth map $\phi : \mathcal{S} \rightarrow \mathcal{S}$ which maximizes the similarity between $I_1$ and $I_2 \circ \phi$. Following the previous discussion, we propose to transport $I_2$ on $\mathcal{S}$ according to a time-dependent vector field. By definition, the value of such field at point $\mathbf{p}$ on $\mathcal{S}$ lies into the tangent space $T_{\mathbf{p}}\mathcal{S}$. Here, we assume that $\mathcal{S}$ is implicitly defined as zero level set of a map $\psi : \mathbb{R}^3 \longrightarrow \mathbb{R}$,

$$\mathcal{S} = \{\mathbf{p} \in \mathbb{R}^3 \ / \ \psi(\mathbf{p}) = 0\},$$

and we identify its tangent space at $\mathbf{p}$ with

$$T_{\mathbf{p}}\mathcal{S} = \mathrm{Ker}(d\psi_{\mathbf{p}}) = \{\mathbf{h} \in \mathbb{R}^3 \ / \ d\psi_{\mathbf{p}}(\mathbf{h}) = D\psi(\mathbf{p})\mathbf{h} = \nabla\psi(\mathbf{p}) \cdot \mathbf{h} = 0\}. \quad (6)$$

Note that $\nabla\psi(\mathbf{p})$ is a vector perpendicular to $T_{\mathbf{p}}\mathcal{S}$, and we can express $\top(\mathbf{v})$, the projection of an arbitrary vector $\mathbf{v}$ (in $\mathbb{R}^3$) onto $T_{\mathbf{p}}\mathcal{S}$, as

$$\top(\mathbf{v}) = \mathbf{v} - \frac{(\mathbf{v} \cdot \nabla\psi(\mathbf{p}))\nabla\psi(\mathbf{p})}{\|\nabla\psi(\mathbf{p})\|}.$$

Following the methodology discussed in [1], we then extend the definition of $I_1$ and $I_2$ from $\mathcal{S}$ to $\mathbb{R}^3$. For this purpose, we introduce two maps $I_{1,e} : \mathbb{R}^3 \rightarrow \mathbb{R}$, $I_{2,e} : \mathbb{R}^3 \rightarrow \mathbb{R}$, such that $\forall\ \mathbf{p} \in \mathcal{S},\ I_{i,e}(\mathbf{p}) = I_i(\mathbf{p})$. Extended maps can be obtained, for instance, by giving to $I_{i,e}(\mathbf{p})$ the value of $I_i$ at the closest point of $\mathbf{p}$ on $\mathcal{S}$. Then, in order to transport $I_2$ according to a vector field $\mathbf{v}$, we simply introduce an extended vector field $\mathbf{v}_e$ on $\mathbb{R}^3$ such that $\forall\ \mathbf{p} \in \mathcal{S},\ \mathbf{v_e}(\mathbf{p}) = \mathbf{v}(\mathbf{p}) \in T_{\mathbf{p}}\mathcal{S}$, and consider the solution $I_e : [0, +\infty) \times \mathbb{R}^3 \rightarrow \mathbb{R}$ of

$$\partial_t I_e + \nabla I_e \cdot \mathbf{v}_e = 0,\ I_e(0) = I_{2,e}.$$

This equation leaves $\mathcal{S}$ invariant (as a set), and the restriction of $I_e$ to $\mathcal{S}$, noted $I$, is the transported version of $I_2$. It now remains to design the suitable extended field $\mathbf{v}_e$. In a simple SSD scenario, we would like extend to $\mathbb{R}^3$ the vector field

$$\mathbf{v} = -\mathcal{R}((I_1(\mathbf{p}) - I(\mathbf{p}))\nabla_{\mathcal{S}} I(\mathbf{p})),$$

where $\nabla_{\mathcal{S}} I$ denotes the intrinsic gradient of $I$ on $\mathcal{S}$ (note that $\nabla_{\mathcal{S}} I(\mathbf{p}) = \top(\nabla I_e(\mathbf{p})))$, and $\mathcal{R}$ is a suitable regularization operator for vector fields on $\mathcal{S}$. To design $\mathcal{R}$, we not only need to make sure the regularization is intrinsic to $\mathcal{S}$ but also that the resulting vectors stay inside their respective tangent spaces. We propose to circumvent this problem by delaying the projection onto the tangent space to the last step of the velocity field calculation. We set

$$\mathbf{v}_e(\mathbf{p}) = -\top(\mathcal{R}_\tau((I_{1,e}(\mathbf{p}) - I_e(\mathbf{p}))\nabla I_e(\mathbf{p}))).$$

In this formula, we define the regularization operator $\mathcal{R}_\tau$ as

$$\mathcal{R}_\tau(\mathbf{v}) = \mathbf{u}(\tau),$$

where $t \mapsto \mathbf{u}(t) = (u_1(t), u_2(t), u_3(t))$ is solution at time $\tau$ of

$$\partial_t u_i = \frac{1}{\|\nabla\psi\|}\mathrm{div}(\top(\nabla u_i)\|\nabla\psi\|),\ u_i(0) = v_i. \quad (7)$$

This PDE is the heat equation on implicit surfaces proposed in [1]. It corresponds to the gradient flow of the harmonic energy

$$e(u_i) = \int_{\mathcal{S}} \top(\nabla u_i) \cdot \top(\nabla u_i) \, dV_{\mathcal{S}},$$

where $dV_{\mathcal{S}}$ is the surface element on $\mathcal{S}$. We can see this regularization principle as an intrinsic formulation of the Gaussian smoothing discussed earlier. In this case, it is the parameter $\tau$ that controls the level of regularization.

In summary, the transport of $I_2$ on $\mathcal{S}$ is obtained by solving the equation

$$\partial_t I_e = \nabla I_e \cdot \top(\mathcal{R}_\tau((I_{1,e}(\mathbf{p}) - I_e(\mathbf{p}))\nabla I_e(\mathbf{p}))), \; I_e(0) = I_{2,e},$$

and defining $I(t)$ as the restriction of $I_e$ to $\mathcal{S}$. This method extends readily to the components of $\phi$ (using an extension $\phi_e : \mathbb{R}^3 \mapsto \mathbb{R}^3$). This allows us to estimate jointly the warped image and the corresponding deformation.

## 5 Numerical Experiments

To validate these results, we applied Eq. 4 to two pairs of images mapped onto a sphere and an arbitrary surface. Both surfaces were represented as zero level set of a function $\psi$ sampled on a volume of size $256 \times 256 \times 256$. The registration process is computationally intensive. It involves applying the regularizing PDE to each component of the velocity field at each iteration of the transport equation. Substantial gains in speed were obtained by limiting the integration of the evolution equations to a narrow band (4 voxel wide) around the zero level set of $\psi$. First-order accurate ENO schemes were used for the transport equations. In order to alleviate the lack of accuracy of this approximation, we tracked the evolution of both the image and the deformation $\phi$. At convergence, this allowed to compute the deformed image $I_2 \circ \phi$ by interpolation. For the visualization of our results, we extracted the zero-level of $\psi$ using the "marching-cube" algorithm, and colored the surface with the corresponding interpolated intensity values.

In Fig. 2, we show the alignment of two cartoon faces defined on a sphere. The domain of definition of the images was extended from the sphere to a narrow band in the volume by giving to each voxel the value of its nearest point on the surface. The superimposed images presented in Fig. 2e show the misalignment before registration. The final recovered image is shown in Fig. 2c. The accuracy of the alignment after registration can be observed on the superimposed view in Fig. 2f. Note how the final image obtained by direct transport (Fig. 2d) is blurred.

For the same example, we also visualized the deformation process by applying the same transport equation to a checkerboard shadow image and by mapping the transformation $\phi$ to a grid. We illustrate these results in Fig. 3 at three different time points during the registration process.

Finally, in Fig. 4, we show similar results for the registration of two images of "Lena" (one of them artificially deformed) mapped onto an arbitrary implicit surface. This surface was obtained by perturbing the distance map of a sphere

(a) $I_1$

(b) $I_2$

(c) $I_2 \circ \phi$

(d) $I_2$ reg. by direct transport

(e) $I_1 + I_2$

(f) $I_1 + I_2 \circ \phi$

**Fig. 2.** Image transport on an implicit surface (*face* example)

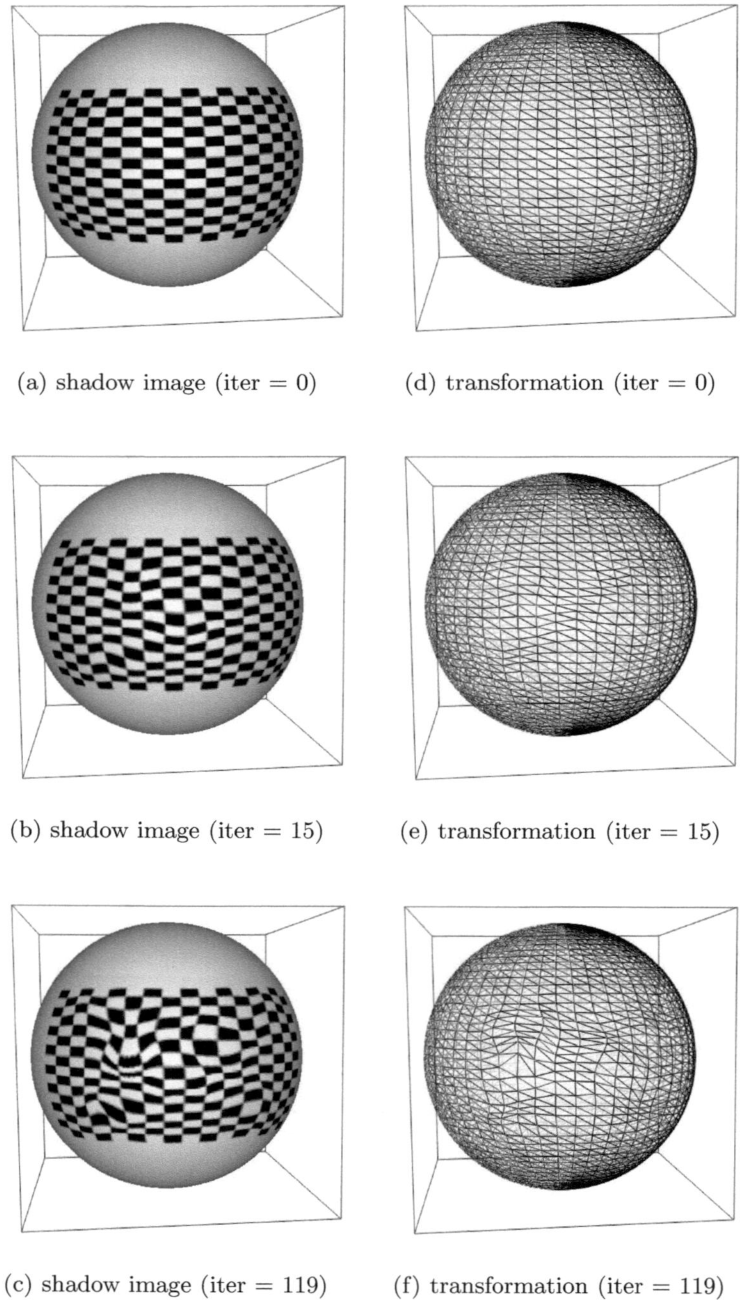


**Fig. 3.** Recovered transformation $\phi$ (*face* example). See Fig. 2

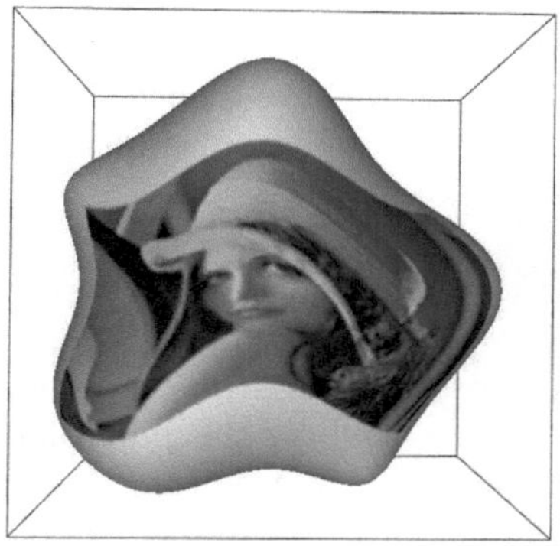

(a) $I_1$

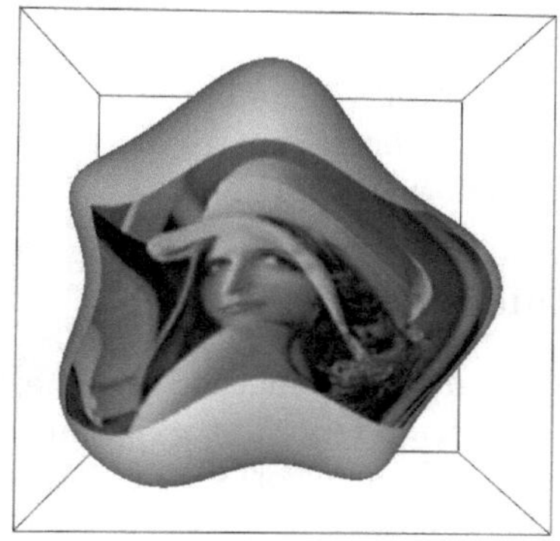

(b) $I_2$

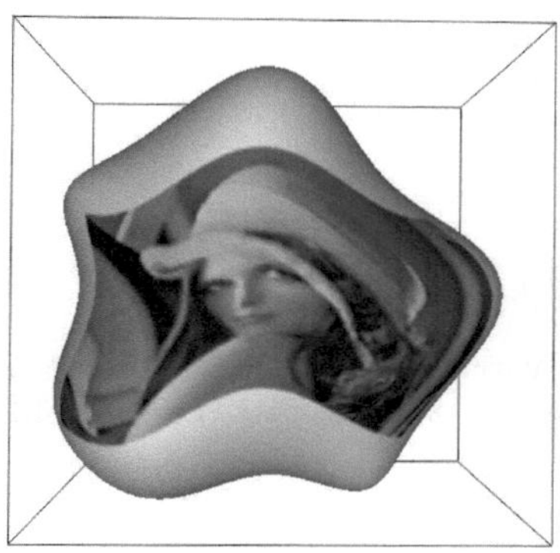

(c) $I_2 \circ \phi$

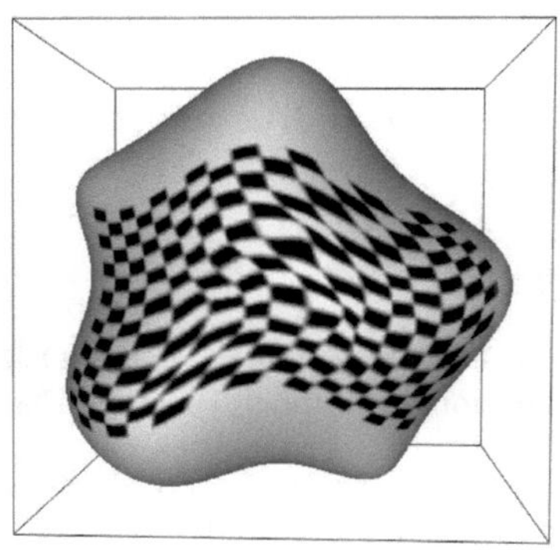

(d) $S \circ \phi$

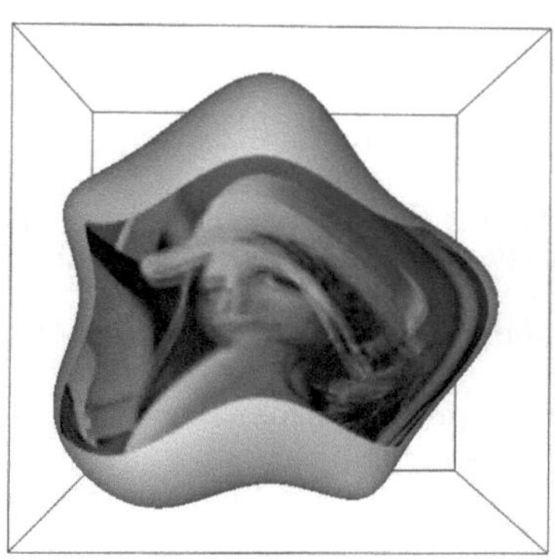

(e) $I_1 + I_2$

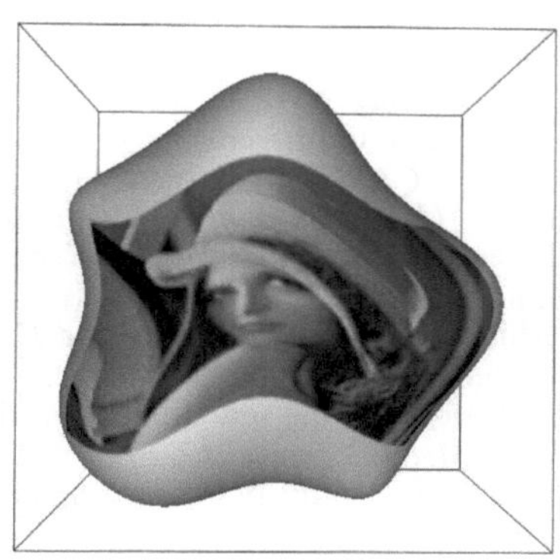

(f) $I_1 + I_2 \circ \phi$

**Fig. 4.** Image transport on an implicit surface (*Lena* example) with a geometric transformation generated by trigonometric functions. The effect of the registration can be seen by comparing the superimposed images presented in Fig. 4e and Fig. 4f. As in the previous example, we also applied the transport equation to a checkerboard shadow image to get a better assessment of the recovered deformation (Fig. 4d).

## 6 Conclusion

We presented a new method for the transport and registration of images on implicit surfaces. Our first experimental results confirm the potential of this technique. In future works, our approach could benefit from recent extensions [5] of the framework proposed by Bertalmio et al. [1]. One possible area of application is functional brain imaging. Level set segmentation methods are often used to extract the cortex from volumetric brain images. This model is then used to map functional activations and electric potentials that reflect the brain function. Using the output of the segmentation as function $\psi$, we could use our technique to capture, directly on the cortex, the spatial variability of the brain function over time and/or across a population.

## References


1. M. Bertalmio, L.-T. Cheng, and S. Osher. Variational problems and partial differential equations on implicit surfaces. *Journal of Computational Physics*, 174:759–780, 2001.
2. V. Caselles, R. Kimmel, and G. Sapiro. Geodesic active contours. *International Journal of Computer Vision*, 22(1):61–79, 1997.
3. G. Christensen, R. Rabbitt, and M. Miller. Deformable template using large deformation kinematics. *IEEE Transactions on Image Processing*, 5(10):1437–1447, 1996.
4. J. Glaunès, M. Vaillant, and M. Miller. Landmark matching via large deformation diffeomorphisms on the sphere. *Journal of Mathematical Imaging and Vision*, 20(1/2):179–200, 2004.
5. J. Greer. An improvement of a recent Eulerian method for PDEs on general geometries. CAM Report 05-41, UC Los Angeles, 2005.
6. C. Laney. *Computational Gasdynamics.* Cambridge Uniiversity Press, 2002.
7. S. Osher and R. Fedkiw. *Level Set Methods and Dynamic Implicit Surfaces.* Springer, 2002.
8. J. Sethian. *Level Sets Methods and Fast Marching Methods.* Cambridge University Press, 2nd edition, 1999.
9. J.-P. Thirion. Image matching as a diffusion process: An analogy with Maxwell's demons. *Medical Image Analysis*, 2(3):243–260, 1998.
10. A. Trouvé. Diffeomorphisms groups and pattern matching in image analysis. *International Journal of Computer Vision*, 28(2):213–221, 1998.
11. B. Vemuri and Y. Chen. *Joint Image Registration and Segmentation.* In S. Osher and N. Paragios, Geometric Level Set Methods, pages 251–269, Springer, 2003.
12. B. Vemuri, J. Ye, Y. Chen, and C. Leonard. A level-set based approach to image registration. In *Proceedings of the IEEE Workshop on Mathematical Methods in Biomedical Image Analysis*, pages 86–93, 2000.

# Direct Shape-from-Shading with Adaptive Higher Order Regularisation

Oliver Vogel, Andrés Bruhn, Joachim Weickert, and Stephan Didas

Mathematical Image Analysis Group
Faculty of Mathematics and Computer Science, Building E1.1
Saarland University, 66041 Saarbrücken, Germany
{vogel,bruhn,weickert,didas}@mia.uni-saarland.de

**Abstract.** Although variational methods are popular techniques in the context of shape-from-shading, they are in general restricted to indirect approaches that only estimate the gradient of the surface depth. Such methods suffer from two drawbacks: (i) They need additional constraints to enforce the integrability of the solution. (ii) They require the application of depth-from-gradient algorithms to obtain the actual surface. In this paper we present three novel approaches that avoid the aforementioned drawbacks by construction: (i) First, we present a method that is based on homogeneous higher order regularisation. Thus it becomes possible to estimate the surface depth directly by solving a single partial differential equation. (ii) Secondly, we develop a refined technique that adapts this higher order regularisation to semantically important structures in the original image. This addresses another drawback of existing variational methods: the blurring of the results due to the regularisation. (iii) Thirdly, we present an even further improved approach, in which the smoothness process is steered directly by the evolving depth map. This in turn allows to tackle the well-known problem of spontaneous concave-convex switches in the solution. In our experimental section both qualitative and quantitative experiments on standard shape-from-shading data sets are performed. A comparison to the popular variational method of Frankot and Chellappa shows the superiority of all three approaches.

**Keywords:** computer vision, shape-from-shading, variational methods, partial differential equations.

## 1 Introduction

The recovery of the 3-D shape of a surface from a single shaded image is one of the classical reconstruction problems in computer vision. Since the first prototypical approach of Horn three decades ago [5], a variety of algorithms have been developed; see e.g. [9,16]. In particular, two classes of shape-from-shading methods are frequently used in the literature: *propagation techniques* that recover the shape by propagating information from a set of known surface points (critical points) to the whole image [5,8,11,15,13], and *variational methods* that compute the solution as minimiser of a suitable energy functional [7,2,4]. In this paper we

F. Sgallari, A. Murli, and N. Paragios (Eds.): SSVM 2007, LNCS 4485, pp. 871–882, 2007.

focus on the latter class of techniques. Although variational methods have been very popular in the 80s, they have stopped being considered already one decade later, when propagation approaches started to dominate. In this context, they have been critisised to suffer from a number of drawbacks [11]:

- *Indirect Strategy.* In contrast to many other techniques that allow a direct estimation of the depth field [10,12], variational methods have the reputation to be only applicable for the recovery of the gradient field of the surface depth. Such indirect variational approaches are e.g. the methods of Horn and Brooks [2] and the algorithm of Frankot and Chellappa [4]. These methods compute first the gradient field of the depth and then have to rely on the subsequent application of a depth-from-gradient technique [4,1].

- *No Intrinsic Integrability.* Moreover, indirect techniques also require the use of integrability constraints to prevent impossible solutions [2,6,4]: If the two gradient functions in $x$- and $y$-direction – let them be given by $p(x, y)$ and $q(x, y)$, respectively – are computed independently from each other, there is no guarantee that there exists a common depth map $z(x, y)$ for which holds $p(x, y) = z_x(x, y)$ and $q(x, y) = z_y(x, y)$. Such integrability constraints have been first introduced by Horn and Brooks in [6]. However, since these constraints only encourage the integrability of the solution, but not enforce it, still intermediate steps are necessary that backproject the estimated gradient field into the range of admitted solutions [4].

- *Over-Regularisation.* Furthermore, since variational methods are based on a regularising smoothness assumption, it has often been remarked that this regularisation introduces a strong blurring in the solution that may deteriorate the quality of the reconstruction [11]. This issue has recently been researched by Agrawal *et al.* [1], however, only with respect to variational depth-from-gradient techniques. For direct variational shape-from-shading approaches such an investigation of adaptive regularisers is missing.

- *Spontaneous Concave-Convex Switches.* Finally, it is well-known that shape-from-shading in its original formulation (orthographic projection, Lambertian surface) is an ill-posed problem [5]. Due to the related concave-convex ambiguity, spontaneous switches in the solution may occur if the strength of the regularisation is too low. In this context one should note by incorporating knowledge on the surface at critical points [11] or assuming other conditions such as perspective instead of a orthographic projection [14], shape-from-shading can be turned into a partially or even fully well-posed problem. In this case the problem of spontaneous switches does not occur. However, since we restrict ourselves to the classical problem formulation without prior information, this issue remains relevant for us.

In our paper all four aspects are addressed. We show by the example of the methods of Horn and Brooks [2] and of Frankot and Chellappa [4] how the use of higher order smoothness terms allows a direct computation of the surface depth. Consequently, no shape-from-gradient algorithms are required and all integrability constraints become obsolete by construction. Moreover, we investigate how the smoothness term of our novel method can be adapted. This also includes a strategy to cope with spontaneous concave-convex switches in the solution.

Our paper is organised as follows. In Section 2, we give a short review on the shape-from-shading problem, while in Section 3 we discuss the indirect approaches of Horn and Brooks and Frankot and Chellappa. Based on the outcome of this discussion, we then develop a novel direct variational approach for shape-from-shading in Section 4. In Section 5 we extend this approach by adaptive smoothing strategies. We propose variants that are based on image- and depth-driven smoothness terms. In Section 6, experiments for all three novel algorithms are presented, while the summary in Section 7 concludes this paper.

## 2 The Shape-from-Shading Problem

Let us consider a shaded image as a function $f(x, y)$ where $(x, y)$ denotes the location within a rectangular image domain $\Omega$. Furthermore, let us assume that the surface $z(x, y)$ depicted in this image has been illuminated by a *single light source* only and that its reflectance properties can be expressed in terms of a reflectance map $\mathrm{R}(z_x(x, y), z_y(x, y))$. Such a reflectance map is a function that describes the amount of light reflected by the surface in viewing direction depending on its gradient $(z_x(x, y), z_y(x, y))^\top$. Then, solving the shape-from-shading problem means to find a suitable surface $z(x, y)$ such that the amount of light reflected by it equals the grey value that is observed at the corresponding pixel of the image:

$$f = \mathrm{R}(z_x, z_y). \tag{1}$$

In the literature this equation is known as the *image irradiance equation* [5,6]. Further assumptions in the classical shape-from-shading problem [5] are that the shaded image was obtained by an *orthographic projection* and that the surface is a *Lambertian* surface , i.e. the reflectance map reads

$$\mathrm{R}(z_x, z_y) = \rho \cdot \langle s, n \rangle \tag{2}$$

where $n$ denotes the unit surface normal

$$n = \frac{1}{\sqrt{1 + z_x^2 + z_y^2}} \begin{pmatrix} -z_x \\ -z_y \\ 1 \end{pmatrix}, \tag{3}$$

$s$ stands for the light source direction, and $\rho$ is the albedo of the surface – a constant that specifies the ratio of scattered to incident light. Evidently, this problem is ill-posed: For surfaces that are illuminated from above, i.e. $s = (0, 0, 1)^T$, both the surface $z$ and its negative counterpart $-z$ are solutions of the image irradiance equation. As we will see later, this ambiguity yields the well-known concave-convex switches in the solution.

## 3 Variational Shape-from-Shading

In order to solve the classical shape-from-shading problem, numerous variational approaches have been proposed in the literature. All these approaches, however, are based on an indirect two-step strategy: First the surface gradient is computed, then the actual surface is determined. Two typical representatives the method of Horn and Brooks [2] and its improved variant by Frankot and Chellappa [4]. In the following both approaches are discussed in detail.

### 3.1 Horn and Brooks

One of the first variational methods for shape-from-shading that still enjoys great popularity is the approach by Horn and Brooks [2]. This approach computes the surface gradient $\nabla z = (z_x, z_y) =: (p, q)$ as minimiser of the following energy functional:

$$\mathrm{E}(p,q) = \int_{\Omega} \left( (f - \mathrm{R}(p,q))^2 + \alpha \left( |\nabla p|^2 + |\nabla q|^2 \right) \right) dx\, dy \,. \tag{4}$$

While the first term (the data term) penalises deviations from the image irradiance equation, the second term (the smoothness term) assumes the recovered surface derivatives $p$ and $q$ to be smooth. The degree of smoothness of the solution is steered by a regularisation parameter $\alpha > 0$.

Obviously, this method suffers from two drawbacks: First of all, it is not considered explicitly in the formulation of the energy functional that $p$ and $q$ are derivatives of a common surface $z$. Thus it is not surprising that in most cases there might not even exist a surface $z$ with $\nabla z = (p, q)$ for the computed solution. Secondly, even if the integrability would be enforced by means of other constraints during the solution process of (4) as proposed in [6,4], it is not trivial to actually obtain the desired surface $z$. Additional depth-from-gradient techniques [4,1] are necessary anyway to compute the actual surface.

### 3.2 Frankot and Chellappa

Frankot and Chellappa [4] proposed a solution to both the integrability problem and the problem of recovering the depth from the gradient field. After each iteration step of the Horn and Brooks algorithm, they project the computed gradient $(p, q)$ onto the closest integrable pair of functions $(\widetilde{p}, \widetilde{q})$ by minimising

$$\int_{\Omega} \left( |p - \widetilde{p}|^2 + |q - \widetilde{q}|^2 \right) dx\, dy \,. \tag{5}$$

To this end, they compute the Discrete Fourier Transforms (DFT) of $p$ and $q$ and perform the projection step in the frequency domain. Moreover, they also propose a way to integrate $(p, q)$ in the frequency domain. However, from a modelling point of view such an alternating approach is neither desirable nor is its convergence mathematically understood. Hence, it would be much more natural to estimate the surface depth $z$ directly.

## 4 Who Dares Wins: Higher Order Regularisation

For solving both previously discussed problems more reliably, we propose the following strategy: By considering the actual approach of Horn and Brooks in (4) and replacing $p$ by $z_x$ and $q$ by $z_y$, we obtain a *direct* variational shape-from-shading method that overcomes all integrability problems by construction. The corresponding energy functional of this novel approach is given by

$$\mathrm{E}(z) = \int_{\Omega} \Big( (f - \mathrm{R}(z_x, z_y))^2 + \alpha \underbrace{(z_{xx}^2 + 2z_{xy}^2 + z_{yy}^2)}_{=\|\mathrm{Hess}(z)\|_F^2} \Big)\, dx\, dy\,. \tag{6}$$

Please note that all integrability problems only vanish if this energy functional is solved for $z$ – and not for $z_x$ and $z_y$ as proposed in [4]. Such a direct computation of $z$ also offers another advantage: Since it guarantees that $z_{xy} = z_{yx}$, we obtain a second order smoothness term in a natural way (the smoothness term in (4) is only a first order regulariser). This new smoothness term can be identified as the squared Frobenius norm of the Hessian.

Following the calculus of variations [3], we know that a minimiser of an energy functional must satisfy its Euler-Lagrange equation(s). For our novel approach this partial differential equation is given by

$$\begin{aligned} 0 =& \frac{\partial}{\partial x}\Big( (f - \mathrm{R}(z_x, z_y))\mathrm{R}_{z_x}(z_x, z_y) \Big) + \frac{\partial}{\partial y}\Big( (f - \mathrm{R}(z_x, z_y))\mathrm{R}_{z_y}(z_x, z_y) \Big) \\ &+ \alpha \Big( z_{xxxx} + 2z_{xxyy} + z_{yyyy} \Big). \end{aligned} \tag{7}$$

As one can see, our new regulariser results in a homogeneous fourth order diffusion term. Evidently, this makes its discretisation more complicated than the one of a standard second order diffusion expression. However, a common strategy proved to be very successful in this case: By discretising the continuous functional in (6) and computing its derivatives a suitable discretisation for (7) can be derived that even contains the correct boundary conditions.

In order to solve the resulting nonlinear system of equations, we apply a Jacobi-like method as proposed in [6]. Thereby the complete data term is taken from the old iteration step. In this context, one should note that the solution of this equation system may not be unique, since our energy functional is not strictly convex. This means that the solution we find might not be the global minimum of the functional, but only a local one. Although one can avoid this problem by manually providing the algorithm with the correct shape at the image boundary and occluding boundaries (and thus turn the classical shape-from-shading task into a well-posed problem), we want to tackle the original problem and thus intentionally refrain from providing this this prior knowledge.

## 5 Adaptive Higher Order Regularisation

It turns out that the shape of simple images can be reconstructed well using small weights $\alpha$ for the smoothness term. However, for complex images with many edges and occluding boundaries, it is necessary to use a stronger smoothness term to obtain a convergent iteration. Evidently this decreases the accuracy of the reconstruction. In the following we present two additional approaches that tackle this problem: By adapting the smoothness term either to the input image or to the depth map they reduce the amount of unnecessary regularisation and thus allow for a more precise reconstruction of the final surface.

### 5.1 Image-Driven Regularisation

So far, our variational approach from Section 4 only uses a homogeneous smoothness term, i.e. a regulariser that does not distinguish between different locations in the image. However, it is well known that at discontinuities the information provided by the input image is very poor. Consequently, the regularisation at such locations should be much stronger than in flat areas, where the shape of the surface is easy to reconstruct. In other words: One can improve the reconstruction quality in flat areas by regularising less.

In order to model this observation, we propose a modified approach that adapts its regularisation to the local image structure. It is given by the energy functional

$$\mathrm{E}(z) = \int_{\Omega} \left( (f - \mathrm{R}(z_x, z_y))^2 + \alpha\, \mathrm{g}(|\nabla f|^2)\, \|\mathrm{Hess}(z)\|_F^2 \right) dx\, dy \,. \tag{8}$$

that makes use of a weighting function $\mathrm{g}(|\nabla f|^2)$ in the smoothness term. Considering $|\nabla f|$ as an edge indicator, this weighting function should be large where $|\nabla f| \gg 0$, and small where $|\nabla f| \approx 0$. Thus, any positive, monotonically increasing function g can be chosen for this task. In the context of our image-driven approach, we use

$$\mathrm{g}(s^2) = \sqrt{s^2 + \epsilon^2} \tag{9}$$

where $\epsilon > 0$ is a small positive constant that ensures at least a small amount of regularisation. As one can easily verify, this function attains its minimum value $\epsilon$ at $s^2 = 0$ and approaches the identity function for large $s^2$.

As for our homogeneous method, the minimisation of (8) requires to solve its Euler-Lagrange equations. In the case of image-driven regularisation they are given by

$$\begin{aligned} 0 =& \frac{\partial}{\partial x}\Big( (f - \mathrm{R}(z_x, z_y)) \mathrm{R}_{z_x}(z_x, z_y) \Big) + \frac{\partial}{\partial y}\Big( (f - \mathrm{R}(z_x, z_y)) \mathrm{R}_{z_y}(z_x, z_y)) \Big) \\ &+ \alpha \left( \frac{\partial^2}{\partial x^2}\Big( \mathrm{g}(|\nabla f|^2) z_{xx} \Big) + 2 \frac{\partial^2}{\partial x \partial y}\Big( \mathrm{g}(|\nabla f|^2) z_{xy} \Big) + \frac{\partial^2}{\partial y^2}\Big( \mathrm{g}(|\nabla f|^2) z_{yy} \Big) \right). \end{aligned} \tag{10}$$

Compared to the Euler-Lagrange equation 7 of our shape-from-shading method with homogeneous higher order regularisation, the adaptive smoothness term induces a linear fourth order diffusion process, where $g$ plays the role of a diffusivity function. Again, we suggest to derive a suitable discretisation for this equation by discretising the continuous functional and computing its derivatives. The obtained equation system is then solved once more by a Jacobi-like iteration step.

### 5.2 Depth-Driven Regularisation

In our second approach, we proposed to use edge information of the input image $f$ to control the strength of the regularisation. However, in particular with respect to situations where frequent convex-concave switching artifacts occur, it may make much more sense to adapt the weight of the induced diffusion process to the edge information of the evolving depth map instead. This has the following reason: Since such artifacts manifest themselves in large gradients in the recovered surface, they can be tackled by increasing the smoothness at the corresponding locations.

Therefore, we propose to replace the image gradient $|\nabla f|$ in the Euler-Lagrange equations of the image-driven case by the depth gradient $|\nabla z|$. Thus, equation (10) turns into the following PDE:

$$\begin{aligned} 0 =& \frac{\partial}{\partial x}\Big((f - \mathrm{R}(z_x, z_y))\mathrm{R}_{z_x}(z_x, z_y)\Big) + \frac{\partial}{\partial y}\Big((f - \mathrm{R}(z_x, z_y))\mathrm{R}_{z_y}(z_x, z_y)\Big) \qquad (11) \\ &+ \alpha\left(\frac{\partial^2}{\partial x^2}\Big(\mathrm{g}(|\nabla z|^2) z_{xx}\Big) + 2\frac{\partial^2}{\partial x \partial y}\Big(\mathrm{g}(|\nabla z|^2) z_{xy}\Big) + \frac{\partial^2}{\partial y^2}\Big(\mathrm{g}(|\nabla z|^2) z_{yy}\Big)\right). \end{aligned}$$

One should note that in contrast to the image-driven approach, this approach is now based on a fourth order diffusion process which is nonlinear. However, essentially the same strategy as in the previous two cases can be applied to solve this equation numerically.

## 6 Experiments

Let us now investigate the reconstruction quality of all of our three novel techniques. To this end, we perform experiments using two popular shape-from-shading test images: the penny and the Mozart face [16]. Moreover, we compared the results of our techniques to the reconstruction of the algorithm of Frankot and Chellappa. This allows to analyse the advantages of the presented methods in a systematic way. With respect to the presented results for the method of Frankot and Chellappa, one should note that we also refrained in their case from providing information about singular points or occluding boundaries to the algorithm (this was also done in their original paper). In this way a fair comparison can be guaranteed. The parameter $\alpha$ was optimised manually to minimise the average $L_1$-error (12).

### 6.1 The Penny

In our first experiment, we used the penny image of size $128 \times 128$ pixels depicted in Figure 1. It shows the surface of a coin with Lambertian reflectance properties. This data set is challenging due to several reasons: One one hand the surface of the contain is a large-scale structure with a sharp edge at the top. Since we assume the surface to be smooth, this part should be difficult to reconstruct. On the other hand the coin is mainly concave, with convex engravings. The algorithm does not incorporate this shape information. So if the general shape of the coin was convex, this would be a good reconstruction as well. Moreover, the engravings on the coin are a problem themselves: While the head of Abraham Lincoln is pretty large, there are also writings on the coin, which are small-scale structures that are very difficult to recover.

In the left column of Figure 1, the computed reconstructions for the penny image are shown. Using the Frankot-Chellappa algorithm, we obtain a very curvy surface: Some parts of the surface are convex, some concave. Moreover, the edge of the coin is hardly recovered at all. Our homogeneously regularised algorithm improves this result: Lincoln is reconstructed relatively well, and the coin edges are also recovered much better. The surface at the edge, however, is partly estimated to be convex, partly to be concave. The image-driven algorithm has the same problem at the coin edge, however, the detailed structures are recovered much shaper. Nevertheless, we can observe that the surface tends to switch between concave and convex shape spontaneously – this can be observed very well at Lincoln's hair. Using depth-driven regularisation, these artifacts disappear almost completely. The reconstruction now looks somewhat like a coin (although its general shape is convex instead of concave). In the middle column, the backprojections of the reconstructions are compared. As one can see, the reconstructions using the Frankot-Chellappa algorithm and our homogeneously regularised algorithm are quite blurry, the writings are hardly readable and details of Lincoln are lost. With our algorithms based on adaptive higher order regularisation this blurring is reduced significantly: Much more details are preserved and in the case of the image-driven smoothness term even the engravings are readable very well. This is also confirmed by our error plots in the right column. There, the absolute differences between the ground truth and the backprojections of the reconstructions are shown (scaled by a factor 12 to improve visibility of the error).

### 6.2 The Mozart Face

In our second experiment we computed the reconstructions for the Mozart face of size $256 \times 256$. In this case, the task is even more difficult than in the case of the penny, since the original surface is very complex. As a consequence, the image contains many edges and singular points which may result in numerous concave-convex switchings in the reconstruction.

The computed surfaces presented in the first column of Figure 2 show similar tendencies as for the penny image. While the result for the Frankot-Chellappa algorithm is curvy and suffers significantly from a higher number of concave-convex switchings, the reconstruction using our homogeneously regularised algorithm is already slightly better. However, once again the approaches based on adaptive higher order regularisation yield the most detailed reconstructions: While the image-driven approach gives the sharpest results, it still suffers from several concave-convex switches. In the case of the depth-driven approach these switches are almost not existent. The backprojected images in the central column look once again very reasonable for all approaches. However, the error plots

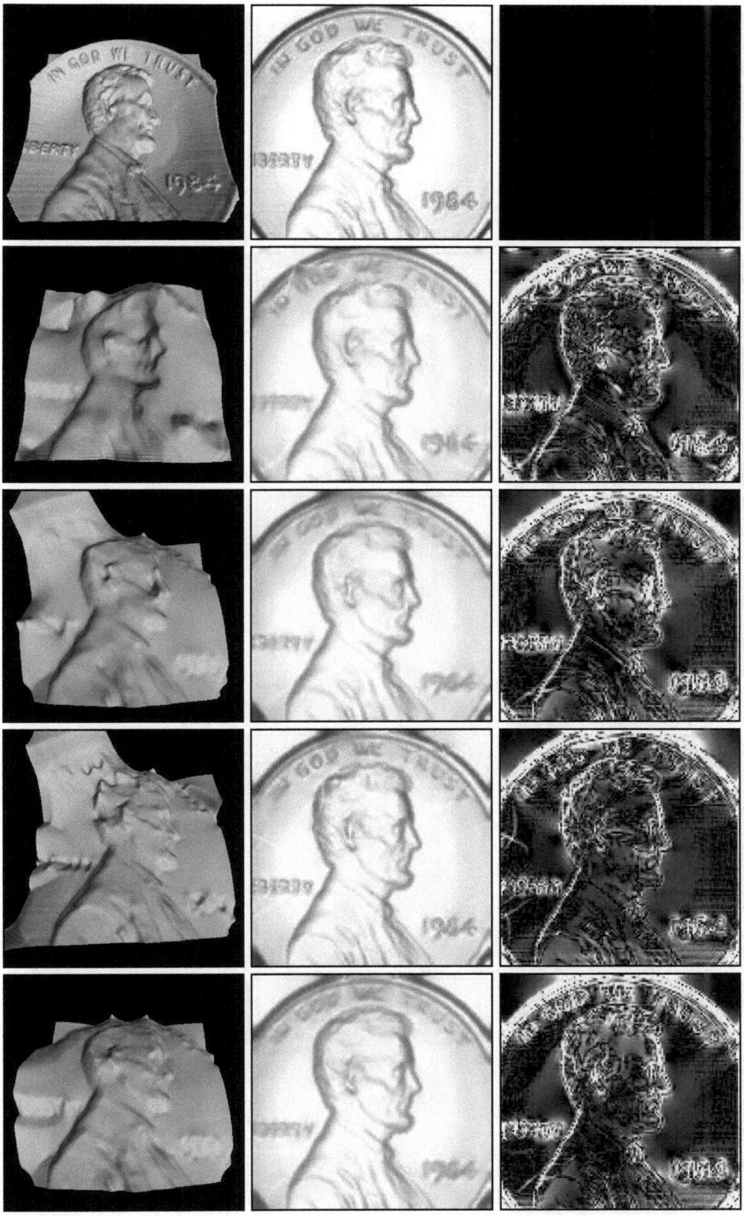

**Fig. 1.** The Penny. *Left to right:* Depth map, orthographic projection from above, difference image (scaled by factor 12). *Top to bottom:* Ground truth, Frankot-Chellappa, homogeneous regularisation, image-driven regularisation, depth-driven regularisation.

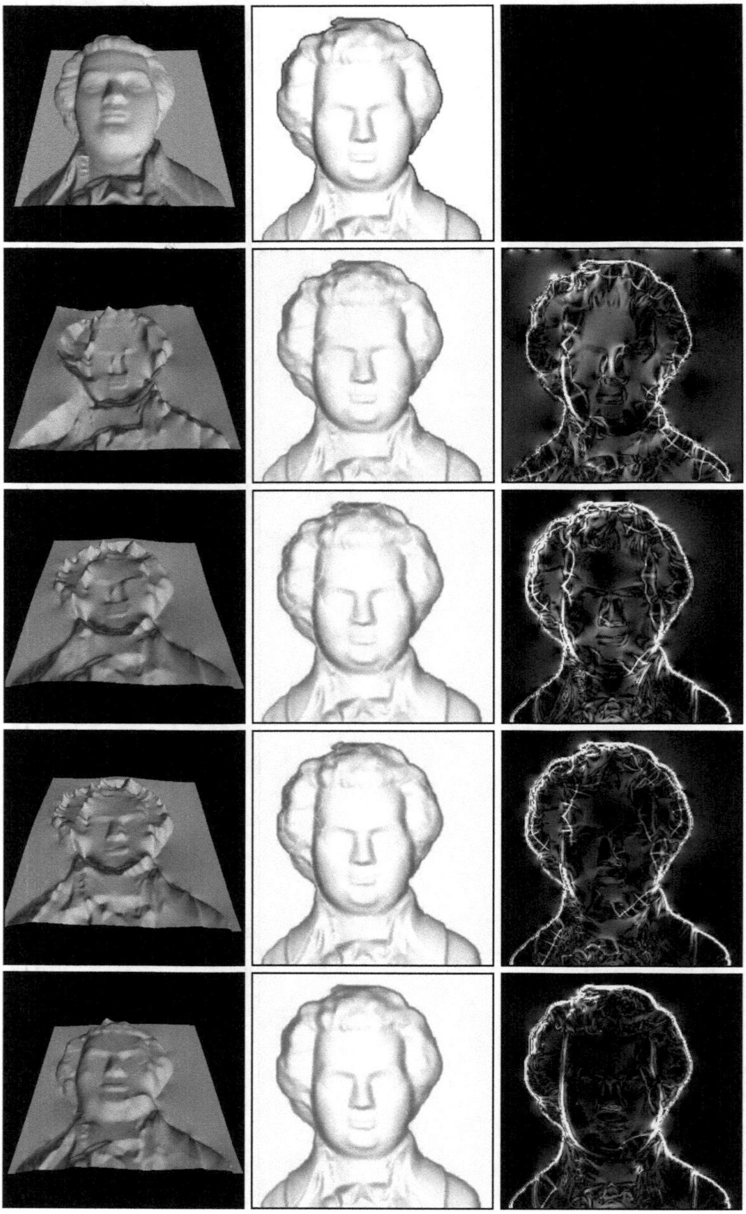

**Fig. 2.** Mozart's face. *Left to right:* Depth map, orthographic projection from above, difference image (scaled by factor 12). *Top to bottom:* Ground truth, Frankot-Chellappa, homogeneous regularisation, image-driven regularisation, depth-driven regularisation.

in the right column reveal the superiority of the proposed approaches: Again, for the Frankot-Chellappa algorithm, the error is spread all over the image, while it is mainly concentrated to edges in the adaptively regularised reconstructions. Thereby, the concave-convex switching artifacts lead to scar-like artifacts. These artifacts are greatly reduced in the case of our depth-driven algorithm.

For a quantitative inspection of our qualitative results, we used the *average* $L_1$*-error* of the image irradiance equation [16] (the error in the backprojection of the reconstructed surface), which is given by

$$E = \frac{1}{|\Omega|} \int_\Omega \big| f(x,y) - R(z_x(x,y), z_y(x,y)) \big| \, dx \, dy \, . \tag{12}$$

The error rates in 1 confirm our qualitative observations from Figure 2. As one can see, the reconstruction quality of the new algorithms is clearly better than the one of Frankot-Chellappa: Improvements up to 33% with the depth-driven approach are possible.

**Table 1.** Error rates for the Mozart image

| Approach | Error $E$ |
|---|---|
| Frankot-Chellappa | 0.0192 |
| Homogeneous Regularisation | 0.0176 |
| Image-Driven Regularisation | 0.0138 |
| Depth-Driven Regularisation | 0.0127 |

# 7 Summary and Conclusions

In this paper we introduced three novel variational approaches for solving the classical shape-from-shading problem. Unlike existing techniques that first compute the surface gradient and then recover the actual surface, all these approaches recover the desired surface directly within a single partial differential equation. Thus additional constraints to enforce the integrability of the solution and the subsequent application of depth-from-gradient algorithms became obsolete. Within this three methods, we developed one approach based on homogeneous regularisation, while the other two adapt their smoothing behaviour either to edges in the original image or edges the evolving depth map. Results show a clear advantage of all three concepts, whereby the image-driven approach yielded the sharpest results, while the depth-driven approach gave the best reconstructions itself. Ongoing work involves comparison to other methods.

We hope that by deriving such direct approaches, research on variational shape-from-shading methods will be pushed further. As other areas in computer vision show – such as stereo reconstruction or optical flow estimation – there is a large potential in variational methods. This potential has evidently not yet been exploited for the purpose of shape-from-shading.

## References

1. A. Agrawal, R. Raskar, and R. Chellappa. What is the range of surface reconstructions from a gradient field? In A. Leonardis, H. Bischof, and A. Pinz, editors, *Computer Vision – ECCV 2006, Part I*, volume 3951 of *Lecture Notes in Computer Science*, pages 578–591, Berlin, May 2006. Springer.
2. M. J. Brooks and B. K. P. Horn. Shape and source from shading. In *Proceedings of the International Joint Conference in Artificial Intelligence*, pages 932–936, Los Angeles, CA, August 1985. MIT Press.
3. L. E. Elsgolc. *Calculus of Variations.* Pergamon, Oxford, 1961.
4. R. T. Frankot and R. Chellappa. A method for enforcing integrability in shape from shading algorithms. *IEEE Transactions on Pattern Analysis and Machine Intelligence*, 10(4):439–451, July 1988.
5. B. K. P. Horn. *Shape from Shading: A Method for Obtaining the Shape of a Smooth Opaque Object from One View.* PhD thesis, Department of Electrical Engineering, MIT, Cambridge, MA, 1970.
6. B. K. P. Horn and M. J. Brooks. The variational approach to shape from shading. *Computer Vision Graphics and Image Processing*, 33:174–208, 1986.
7. K. Ikeuchi and B. K. P. Horn. Numerical shape from shading and occluding boundaries. *Artificial Intelligence*, 17:141–185, 1981.
8. R. Kimmel, K. Siddiqi, B. B. Kimia, and A. M. Bruckstein. Shape from shading: Level set propagation and viscosity solutions. *International Journal of Computer Vision*, 16:107–133, 1995.
9. R. Kozera. An overview of the shape from shading problem. *Machine Graphics and Vision*, 7(1):291–312, 1998.
10. Y. G. Leclerc and A. F. Bobick. The direct computation of height from shading. In *Proc. 1991 IEEE Computer Society Conference on Computer Vision and Pattern Recognition*, pages 552–558, Lahaina, HI, June 1991. IEEE Computer Society Press.
11. J. Oliensis. Shape from shading as a partially well-constrained problem. *Computer Vision, Graphics, and Image Processing: Image Understanding*, 54(2):163–183, 1991.
12. J. Oliensis and P. Dupuis. Direct method for reconstructing shape from shading. In *Proc. 1992 IEEE Computer Society Conference on Computer Vision and Pattern Recognition*, pages 453–458, Champaign, IL, June 1992. IEEE Computer Society Press.
13. E. Prados and O. D. Faugeras. Unifying approaches and removing unrealistic assumptions in shape from shading: Mathematics can help. In T. Pajdla and J. Matas, editors, *Computer Vision – ECCV 2004, Part IV*, volume 3024 of *Lecture Notes in Computer Science*, pages 141–154, Berlin, 2004. Springer.
14. E. Prados and O. D. Faugeras. Shape from shading: A well-posed problem? In *Proc. 2005 IEEE Computer Society Conference on Computer Vision and Pattern Recognition*, volume 2, pages 870–877, San Diego, CA, June 2005. IEEE Computer Society Press.
15. E. Rouy and A. Tourin. A viscosity solutions approach to shape-from-shading. *SIAM Journal of Numerical Analysis*, 29(3):867–884, 1992.
16. R. Zhang, P.-S. Tsai, J. E. Cryer, and M. Shah. Shape from shading: A survey. *IEEE Transactions on Pattern Analysis and Machine Intelligence*, 21(8):690–706, 1999.

# 3D Object Recognition by Eigen-Scale-Space of Contours

Tim K. Lee[1,2] and Mark S. Drew[2]

[1] Cancer Control Research Program, BC Cancer Research Centre, 675 West 10th Avenue, Vancouver, BC, Canada, V5Z 1L3
[2] School of Computing Science, Simon Fraser University, 8888 University Drive, Burnaby, BC, Canada, V5A 1S6
tlee@bccrc.ca, mark@cs.sfu.ca

**Abstract.** People often recognize 3D objects by their boundary shape. Designing an algorithm for such a task is interesting and useful for retrieving objects from a shape database. In this paper, we present a fast 2-stage algorithm for recognizing 3D objects using a new feature space, built from curvature scale space images, as a shape representation that is scale, translation, rotation and reflection invariant. As well, the new shape representation removes the inherent ambiguity of the zero position of arc length for a scale space image. The 2-stage matching algorithm, conducted in the eigenspaces of the feature space, is analogous to the way people recognize an object: first identifying the type of object, and then determining the actual object. We test the new algorithm on a 3D database comprising 209 colour objects in 2926 different view poses, and achieve a 97% recognition rate for the object type and 95% for the object pose.

**Keywords:** object recognition; shape; 3D object; scale-space filtering; eigen-analysis.

## 1 Introduction

Boundary contours of 3D objects contain rich information about the object. When such a contour is projected into our retinas as a planar 2D curve, we can often identify the object in spite of occlusions and accidental alignments. For example, Fig. 1 shows the 2D projections of the boundary contours of several objects from different viewing angles. Although no two of the closed curves are identical, we know all curves in the same column have similar shape and belong to the same type of object. In fact, we can name the objects as an axe, a baseball bat, a chair, a sailboat, a jet, and a bee. Designing a computer program to recognize 3D objects based on their projected boundary contours is interesting and useful for retrieving them from a database.

Building a shape recognition program requires two components: a shape representation and a matching algorithm. The curvature scale space (CSS) representation [1] has been shown to be a robust shape representation. Based on the scale space filtering technique applied to the curvature of a closed boundary curve, the representation behaves well under perspective transformations of the curve. Furthermore, a small local change applied to the curve corresponds to a small local

F. Sgallari, A. Murli, and N. Paragios (Eds.): SSVM 2007, LNCS 4485, pp. 883–894, 2007.

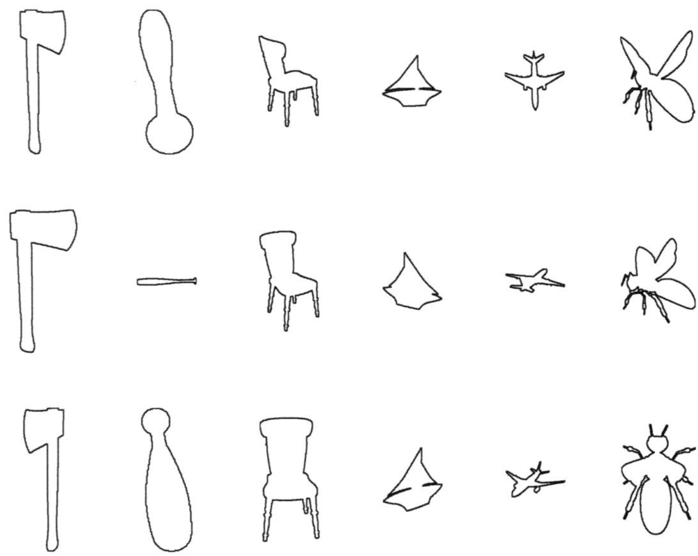

**Fig. 1.** Examples of the border contours for 3D objects from different viewing angles

change in the representation, and the amount of change in the representation corresponds to the amount of change applied to the curve.

More importantly, the representation supports retrieval of similar shape. Spurred partly by the success of the original CSS-derived shape retrieval algorithm [2], and because of the above properties, the CSS representation has been selected as the object contour-based shape descriptor for MPEG-7 [3]. In this paper, we propose an alternative fast matching algorithm for the CSS representation. The matching is carried out in the eigenspace of transformed CSS images. In spirit, then, this approach is inspired by Nayar et al.'s manifold projection scheme [4,5] for object and pose identification based upon appearance. In addition, matching images in eigenspace has been successfully applied to facial recognition [6], by finding an eigenface basis using a principal component analysis. Here, we are interested in developing eigenspaces that are expressive of CSS images, or rather eigenvectors of more compact expressions of these images. Objects can then be identified rapidly via a two-stage matching algorithm, which is an extension of our specialized shape matching algorithm [7]. The new algorithm is similar to the way in which many people recognize an object: first identifying the type of object, and then determining the actual object.

The paper is organized as follows. In §2 we briefly review the CSS representation. Section 3 presents a transformation applied to CSS images making them into a compact and expressive representation that is *scale, translation, rotation and reflection invariant* and is suitable for a fast matching algorithm. Section 4 describes our 2-stage matching algorithm, and §5 reports experiment results on a 3D shape database. A short discussion and conclusions follow in §6.

## 2 Curvature Scale Space

The CSS representation relies on a binary 2D image, called the curvature scale space image, to represent the shape of a closed curve $L_0$ parameterized by $t$

$$L_0(t) = L_0(x(t), y(t)) \tag{1}$$

over multiple scales (see Fig. 2). The $x$ dimension of the CSS image specifies the parameterized path length $t$ of the curve and the $y$ dimension specifies scales of the curve corresponding to the standard deviation $\sigma$ of a Gaussian function $g(t, \sigma) = 1/\sigma\sqrt{2\pi}e^{-t^2/2\sigma^2}$. The binary CSS image is constructed by convolution of the closed curve $L_0(t)$ by a series of Gaussians $g(t, \sigma)$ with increasing $\sigma$, case given by

$$L(t, \sigma) = L_0(x(t), y(t)) \otimes g(t, \sigma) = (X(t, \sigma), Y(t, \sigma)) \tag{2}$$

where $\otimes$denotes a convolution operation, $X(t, \sigma) = x(t) \otimes g(t, \sigma)$, and $Y(t, \sigma) = y(t) \otimes g(t, \sigma)$. The curvature functions $K(t, \sigma)$ of the smoothed curves $L(t, \sigma)$ are then calculated as

$$K(t,\sigma) = \frac{\dfrac{\partial X}{\partial t}\dfrac{\partial^2 Y}{\partial t^2} - \dfrac{\partial^2 X}{\partial t^2}\dfrac{\partial Y}{\partial t}}{\left[\left(\dfrac{\partial X}{\partial t}\right)^2 + \left(\dfrac{\partial Y}{\partial t}\right)^2\right]^{3/2}}. \tag{3}$$

For every zero-curvature point, i.e., $K(t, \sigma) = 0$ and $\partial K(t, \sigma)/\partial t \neq 0$, the corresponding location $(t, \sigma)$ in the binary CSS image is set to 1. The markings of the zero-curvature points form a set of contours, whose appearance captures the shape of the closed curve $L_0(t)$. Fig. 2 shows an example of the smoothing process of a closed boundary curve and its corresponding CSS image.

## 3 Transformation of Scale Space Images

CSS images are bedevilled by an inherent ambiguity: the zero position of arc length is not determined for a new curve, compared to the model one. As well, reflections form a problem. To handle these rotation and mirror transformations of the boundary curve, and to improve the execution speed and matching performance, we propose applying the Phase Correlation method [8] along the abscissa (arc length) dimension of the CSS images. This transform aligns every curve's zero-position — the resulting curve is a new kind of CSS image. We also need to compress the large amount of information in the CSS image in order to further speed up search. We form a new feature vector by summing separately over each of abscissa and ordinate (arc length and scale), and concatenating into a new feature vector, which we call the marginal-sum vector. This results in a feature vector not only invariant to rotations (starting position on the contour), but also invariant to reflections of the contour. Each of these steps is shown to result in a faster search and yet preserving the expressiveness of the original CSS formulation while simplifying the search procedure considerably. In addition, we propose conducting the matching in an eigenspace of reduced image vectors e.g. formed by the Singular Value Decomposition (SVD) method.

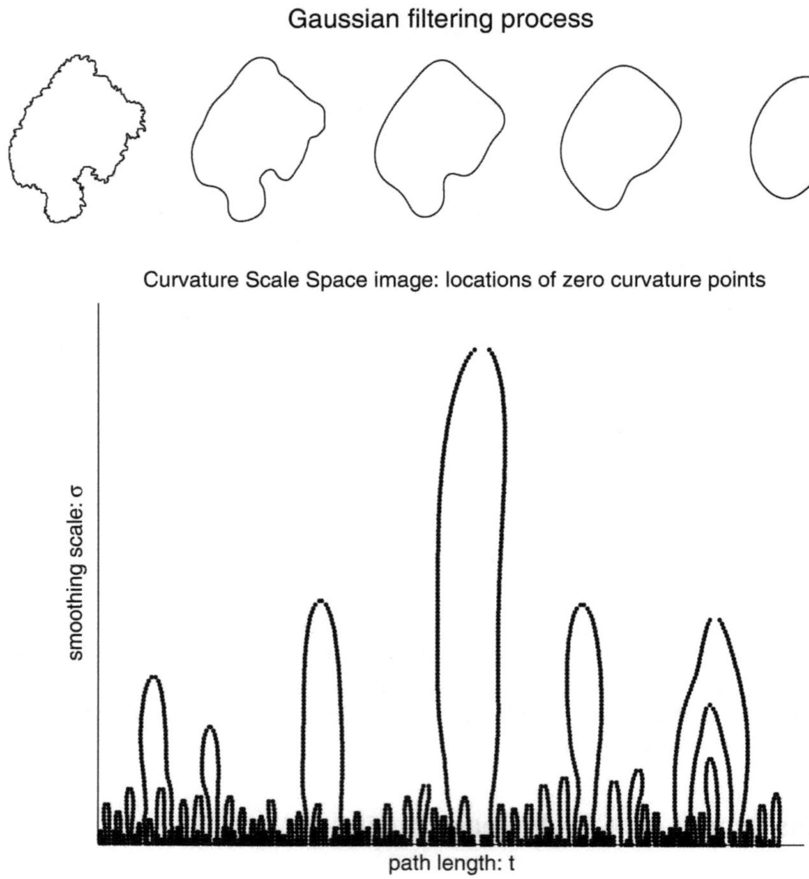


**Fig. 2.** (Top) Gaussian smoothing process of the leftmost closed curve. (Bottom) The corresponding curvature scale space image.

### 3.1 Eigenspace

Singular value decomposition is an efficient method for decorrelating vector information. For $n$ column vectors $\boldsymbol{x}_1, \boldsymbol{x}_2, \ldots, \boldsymbol{x}_n \in \boldsymbol{R}^m$, we form an $m \times n$ data matrix $\boldsymbol{X}$ of mean-subtracted vectors

$$\boldsymbol{X} = [\boldsymbol{x}_1 - \overline{x}\ , \boldsymbol{x}_2 - \overline{x}\ , \ldots, \boldsymbol{x}_n - \overline{x}\ ]\ . \tag{4}$$

The SVD operation produces factors

$$\boldsymbol{USV} = \boldsymbol{X}, \tag{5}$$

with orthogonal $m \times m$ matrix $\boldsymbol{U}$ of eigenfeatures, $\boldsymbol{S}$ the $m \times m$ diagonal matrix of singular values, and $\boldsymbol{V}$ an $m \times n$ matrix of loadings. The column vectors of $\boldsymbol{U}$ form the basis for the eigenspace. In the new representation, vector $\boldsymbol{x}$ goes into a new coefficient $m$-vector $\boldsymbol{u}$ via

$$\boldsymbol{u} = \boldsymbol{U}^{\mathrm{T}} \boldsymbol{x} . \tag{6}$$

Since the eigenvectors are ordered by variance-accounted-for, we may often be able to reduce the dimensionality by truncating $\boldsymbol{u}$: if the number of bases used is reduced to $k$ where $k < m$, the CSS eigenspace is truncated to $\boldsymbol{R}^k$. For any two vectors $\boldsymbol{u}_1$ and $\boldsymbol{u}_2$, here we use a Euclidean metric as our distance measure.

### 3.2 Marginal-Sum Feature Vectors

The entire CSS image can be vectorized to form a column vector $\boldsymbol{x}_i$ of the input matrix $\boldsymbol{X}$ for the above SVD operation. However, such a formulation will create two problems. First, the resultant input matrix $\boldsymbol{X}$, which is an ensemble of all the vectors $\boldsymbol{x}_i$, will be very large and lead to long execution time for the SVD operation. Second, raw CSS images are too *noisy* for matching. The parameterized curves for the type of object that we intend to match may have slight alternations at arbitrary locations, and the CSS contour points will be unlikely to line up in their corresponding CSS images. Matching the CSS images pixel-by-pixel does not achieve the optimum result. Therefore we derive special column vectors $\boldsymbol{x}_i$, namely, marginal-sum feature vectors, from the CSS images. (We also had experimented on using the entire CSS; results are reported in [9]. The present method performed better than raw CSS image matching.)

Let $\boldsymbol{C}(i, j)$ denotes the pixel at the $i^{th}$ row and $j^{th}$ column of an $r$ x $c$ CSS image. The marginal-sum feature vector $\boldsymbol{x}$ is defined as a column vector composed of row-sum and column-sum vectors: $\boldsymbol{r}$ sums down the rows of the CSS image, and $\boldsymbol{c}$ sums across the columns.

$$\boldsymbol{x} = [\boldsymbol{r} \;\; \boldsymbol{c}]^{\mathrm{T}} , \tag{7}$$

$$\boldsymbol{r} = [\textstyle\sum_i \mathbf{C}(i, 1), \sum_i \mathbf{C}(i, 2), \ldots \sum_i \mathbf{C}(i, c)]^{\mathrm{T}}, \text{ and} \tag{8}$$

$$\boldsymbol{c} = [\textstyle\sum_j \mathbf{C}(1, j), \sum_j \mathbf{C}(2, j), \ldots \sum_j \mathbf{C}(r, j)]^{\mathrm{T}} . \tag{9}$$

Vector $\boldsymbol{r}$ can be interpreted as the probabilities of zero curvature points, given a point $t$ along the parameterized closed curve, while vector $\boldsymbol{c}$ denotes the probabilities of zero curvature, given a smoothing level $\sigma$. Conceptually, we can view a feature vector $\boldsymbol{x}$ as the probability distribution function of the zero crossings of the CSS image.

### 3.3 Phase Correlation

Clearly, a rotation transformation on a closed boundary curve translates the initial point of the parameterization process, i.e., the CSS image is circularly shifted. On the other hand, a reflection transformation reverses the direction of the parameterization process, i.e., the CSS image is mirrored. These transformations pose a technical challenge to our algorithm; in particular, the vector $\boldsymbol{r}$ specified in eq. (8) will be affected, but the vector $\boldsymbol{c}$ specified in eq. (9) remains unchanged. Our solution to this problem is to carry out a phase correlation transform [8] on the vector $\boldsymbol{r}$, in the same way as Fourier phase normalization has been used to eliminate starting point dependency when matching contours using Fourier Descriptors [10,11]. This can be accomplished by converting the vector to the frequency domain, calculating the magnitude as a function of frequency, and transforming the results back to the spatial domain. The effect is translational alignment of the inverse-Fourier transformed functions. Note, however, that in carrying out this transform, we depart from a conventional CSS image, now going over to a phase-correlated version which is quite different from the original. But the phase correlation is carried out only in the abscissa, path-length direction, not in the ordinate, scale dimension. Mathematically, the phase-correlated vector $\tilde{r}$ can be expressed as

$$\tilde{r} = |F^{-1}(\,|F(\boldsymbol{r})|\,)|, \tag{10}$$

where $F$ denotes a 1D Discrete Fourier Transform.

Therefore, we replace the marginal-sum feature vector in eq. (7) by

$$\boldsymbol{x} = [\,\tilde{r}\ \ \boldsymbol{c}]^{\mathrm{T}}. \tag{11}$$

Notice that because of the nonlinearity of the absolute value operation, eq. (10) is *not* equivalent to forming a 2D CSS image which has been phase-correlated along the abscissa and then collapsed onto a row-sum. It is instead a much simpler operation. Our final representation is invariant to scale, translation, rotation and reflection transformation.

# 4 Two-Stage Matching Algorithm

Our matching program is designed for a shape database with categorization information. Many shape databases collect more than one image for an object or a class of objects. These related images can be grouped into a category. For example, the 3D database from Michael Tarr's laboratory at the Department of Cognitive and Linguistic Sciences, Brown University [12] collects a set of 209 objects with 14 viewpoints. All views of an object form a category and we form a separate eigenspace for each category. This turns out to be very effective with the following 2-stage recognition program: a high recognition rate is achieved even with a reduced subspace dimensionality.

## 4.1 Stage 1: Identify Image Category

Our recognition program is straightforward. In stage 1, we attempt to identify the category of a test object. This translates to the problem of finding the eigenspace that best describes the test object. To achieve this task, we project the feature vector of the test object (from Section 3) into each $\boldsymbol{R}^k$ category eigenspace in turn; the eigenspace giving the closest reconstruction of the test feature vector is defined as the best category. In other words, we are looking for the eigenspace that minimizes $\|\boldsymbol{x} - \hat{x}\|^2$, where $\boldsymbol{x}$ is the feature vector and $\hat{x} = \boldsymbol{U}\boldsymbol{U}^T\boldsymbol{x}$ is the reconstructed feature vector. By

Parseval's Theorem, the distance metric can be calculated simply in terms of basis vector coefficients, not whole feature vectors, and so is fast.

### 4.2 Stage 2: Determine the Image in a Category

Once the best matched category is identified, we can determine the best matched object in the category by finding the object in the category with the minimum distance $\| \boldsymbol{u} - \hat{\boldsymbol{u}} \|^2$, where $\boldsymbol{u}$ and $\hat{\boldsymbol{u}}$ are the coordinates in $\boldsymbol{R}^k$ eigenspace for the test object and the object in the category, respectively.

## 5 Experiment Results

The algorithms presented in §3 and §4 have been implemented in Matlab and tested using the Tarr 3D database [12].

### 5.1 Test Data Set

The Tarr database consists of 209 objects such as 'axe', 'baseball bat', 'chair, 'sailboat', 'jet', 'bee', with each object captured from 14 different viewpoints: {0°, 30°, 60°, 90°, ... 330°, bottom view, top view}, and are saved as 24-bit colour TIFF files. Fig. 3 shows the object 'axe' and 'bee'. Boundary contours are extracted using a simple contour-follower program [13], with segmentation results as in Fig. 4. Fig. 1 illustrates some other boundary contours for the database.

When we segment these objects from the background, we realize that some of images contain more than one objects and these objects are disjointed in one view but connected in other views (see Fig. 5.) In addition, highlights on some objects separate them into two objects after the segmentation (see Fig. 6.) In order to keep the segmentation step simple and avoid human intervention, we segmented the images as-is and retained the border contours of only the longest-perimeter objects, without any subsequent processing. This has a tendency to work against the accuracy of our algorithm, so provides an extra level of rigour in testing.

### 5.2 Results

We computed the CSS image of every boundary curve by smoothing it with a series of Gaussian functions with starting $\sigma$ value of 5, and incrementing by 1 until no zero curvatures remained on the smoothed curve. The small $\sigma$ values (from 1 to 4), associated with fine texture irregularities of the curve, were removed from the CSS image to reduce the computation error introduced by a discrete boundary curve. As a result, the $y$-axis (the $\sigma$ axis) of a CSS image had variable length, depending on the shape of the curve, but the $x$-axis had its length standardized to 200, the length of the parameterized curve. In order to standardize the size of all CSS images in a database, we padded the $y$-axis of all CSS images to 66, the maximum over all images.

The CSS images were used to construct the phase-correlated feature vector (eq. (11)). Grouping all phase-correlated marginal-sum vectors of the 14 views of an object into a matrix $\boldsymbol{X}$ (eq. (4)), we built 209 eigen-CSS eigenspaces, one for each

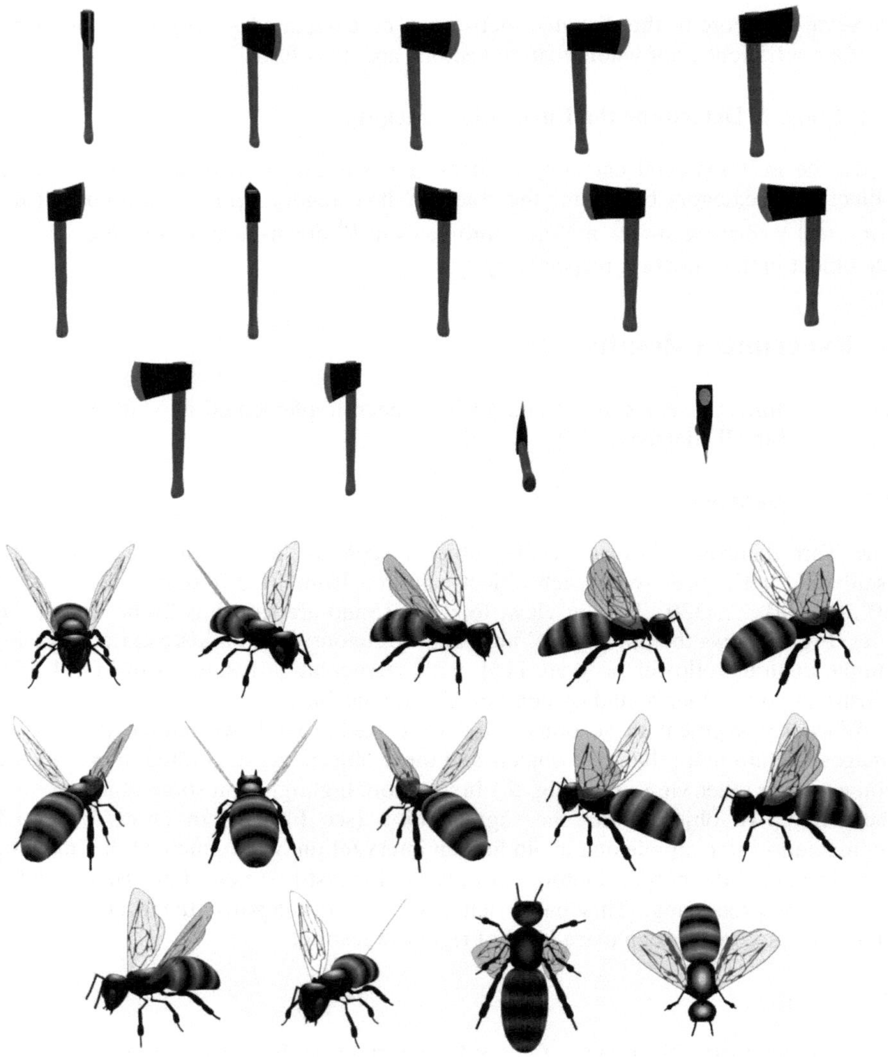

**Fig. 3.** Two sample objects, axe and bee, in the Tarr 3D database. Each object is captured from 14 viewpoints at 0°, 30°, 60°, …, 330°, bottom view, and top view.

object in the database, with the first $\boldsymbol{k}$ column vectors in SVD matrix $\boldsymbol{U}$, i.e., the eigenspace is in the subspace $\boldsymbol{R}^{\mathbf{k}}$.

We evaluated our recognition algorithm described in §4 by matching every boundary curve in a database against other curves in turn and determine whether our program can determine the test object's category (§4.1) and the specific object-pose (§4.2). Only the top best guess for the category and the pose itself were used to determine an average recognition rate, a stringent measure. For category identification

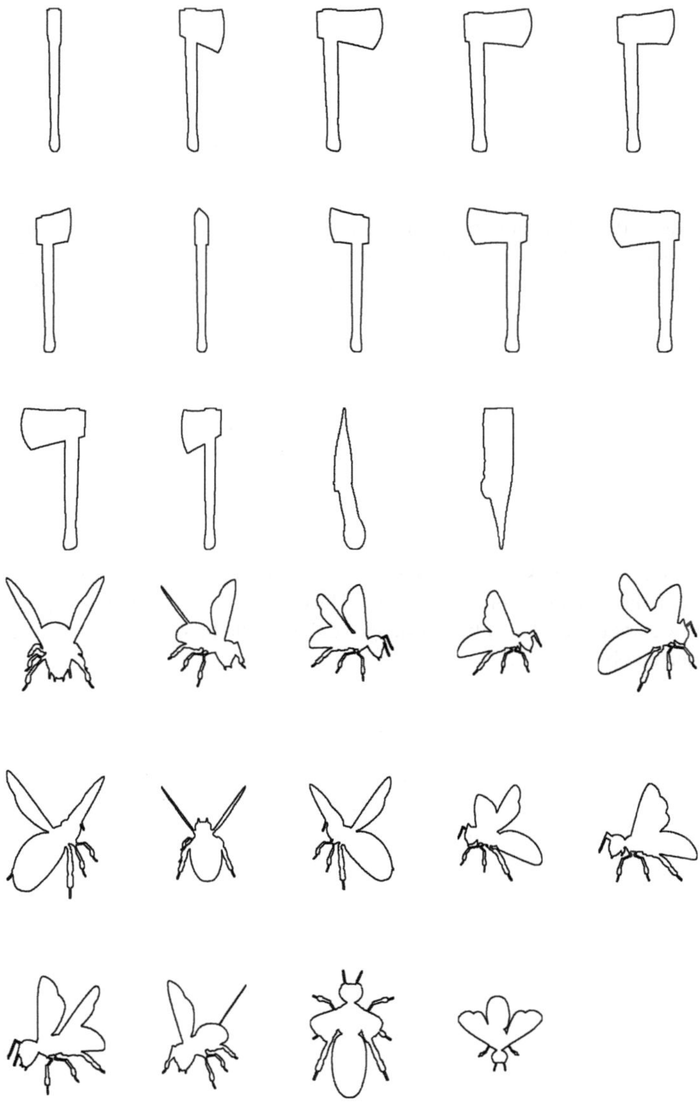

**Fig. 4.** The extracted boundary contours for Fig. 3. All contours are in fact closed curves, but slight gaps appearing in some of these figures are due to reproduction difficulties.

as described in §4.1, the average recognition rate was 92% for $\boldsymbol{k}$ = 5 (i.e. the eigenspace was reduced to $\boldsymbol{R}^5$) and the rate improved to 97% for $\boldsymbol{k}$ = 13. When we calculated the recognition rate for specific object-pose (§4.2), it dropped slightly to 90% for $\boldsymbol{k}$ = 5 and 95% for $\boldsymbol{k}$ = 13. Fig. 7 shows the category and pose recognition rate for $\boldsymbol{k}$ = 1 to 14. We consider these results excellent for such a simple recognition algorithm.

**Fig. 5.** A bottle and a glass are shown as separated objects in one view but as a connected object in another view. Our segmentation program retains only the border contour of the longest-perimeter connected object, e.g., just the bottle and not the glass for the leftmost image.

**Fig. 6.** The highlight on the second candle holder confused the simple segmentation program, which separated the holder from the candle. Thus, with our segmentation policy, only the candle holder was retained. The leftmost image of the candle holder and candle was segmented into one connected object.

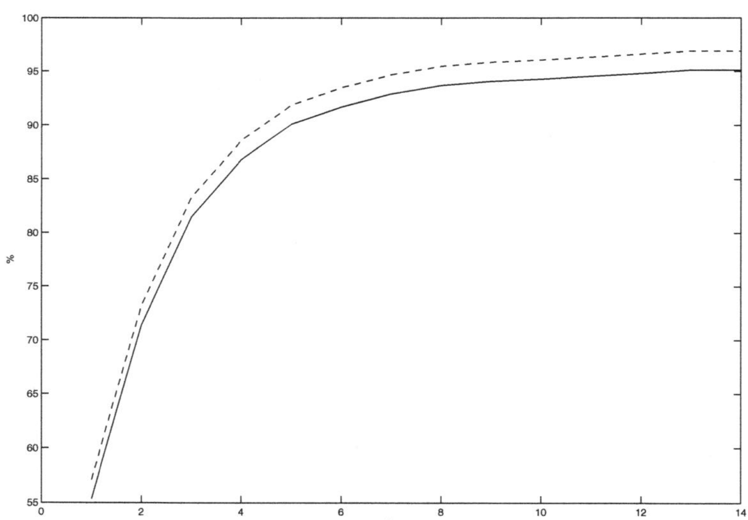


**Fig. 7.** Recognition rate for the Tarr 3D database: The dashed line shows the category recognition rate, while the solid line shows the pose recognition rate

## 6 Conclusions

The advantages of our matching program are simplicity and execution efficiency. The matching algorithm is straightforward and simple, and is scale, translation, rotation, and reflection invariant. In addition, our shape representation compresses the CSS images and allows them to be processed rapidly: the feature vector for the Tarr 3D database consisted of only 266 elements. In conjunction with the reduced dimensionality in the eigenspace, the matching is extremely efficient.

The experiment results for the Tarr database also demonstrate the effectiveness of our algorithm. This database is challenging to work with. Beside the hurdles mentioned in §5.1 regarding multiple objects and highlights, the top and bottom views of an object often look markedly different from other views and, moreover, the front (0°), back (180°), 90° and 270° views can deviate substantially from other side views. Furthermore, many objects in the database have similar shapes (for example axes, flags, gavels, and pipes; batteries, brushes, carrots, chisels, crayons, flashlights, knifes, pens pencils, rulers, and screwdrivers; different types of chairs, etc). Nevertheless, our matching program achieved a remarkable 97% recognition rate for categories and 95% for object-poses with only 13 eigen-bases, and not much less with only 5 bases. In sum, the method we present here is simple, fast and effective.

**Acknowledgments.** This work was supported in part by Discovery Grants from the Natural Sciences and Engineering Research Council of Canada (#288194 and # 611280).

## References

1. Mokhtarian F., Mackworth A.: Scale-based description and recognition of planar curves and two-dimensional shapes, PAMI 8 (1986) 34–43.
2. Mokhtarian F.: Silhouette-based isolated object recognition through curvature scale space, PAMI 17 (1995) 539–544.
3. Mokhtarian F., Bober M.: Curvature Scale Space Representation: Theory, Applications, and MPEG-7 Standardization, Kluwer, 2003.
4. Murase H., Nayar S.: Learning object models from appearance, in: AAAI93, 1993, 836–843.
5. Murase H.,Nayar S.: Illumination planning for object recognition in structured environments, in: CVPR94 (1994) 31–38.
6. Turk M., Pentland A.: Face recognition using eigenfaces, in: Computer Vision and Patt. Recog.: CVPR91 (1991) 586–591.
7. Drew M.S., Lee T.K., and Rova A.: Shape retrieval with eigen-css search, submitted for publication Feb. 24, 2005.
8. Kuglin C., Hines D: The phase correlation image alignment method, in: Int'l Conf. on Cybernetics and Society (1975) 163–165.
9. Drew M.S., Lee T.K., and Rova A.: Shape retrieval with eigen-css search, School of Computing Science, Simon Fraser University, Burnaby, BC, Canada Technical Report TR 2005-07, 2005.

10. Arbter K., Snyder W., Burkhardt H., Hirzinger G.: Application of affine invariant Fourier descriptors to recognition of 3-D objects, IEEE Trans. Patt. Anal. and Mach. Intell. 12 (1990) 640–647.
11. Kunttu I., Lepist L., Rauhamaa J., Visa A.: Recognition of shapes by editing shock graphs, in: Int. Conf. on Computer Vision: ICCV2001 (2001) 755–762.
12. Tarr M.J.: The object databank. Available from http://alpha.cog.brown.edu:8200/stimuli/objects/objectdatabank.zip.
13. Sonka M., Hlavac V., and Boyle R: Image Processing, Analysis and Machine Vision, Chapman & Hall Computing (1993), p. 129.

# Towards Segmentation Based on a Shape Prior Manifold

Patrick Etyngier, Renaud Keriven, and Jean-Philippe Pons

CERTIS Lab / Odyssée Team
Ecole des Ponts
Paris, France
{etyngier,keriven,pons}@certis.enpc.fr

**Abstract.** Incorporating shape priors in image segmentation has become a key problem in computer vision. Most existing work is limited to a linearized shape space with small deformation modes around a mean shape. These approaches are relevant only when the learning set is composed of very similar shapes. Also, there is no guarantee on the visual quality of the resulting shapes. In this paper, we introduce a new framework that can handle more general shape priors. We model a category of shapes as a finite dimensional manifold, the *shape prior manifold*, which we approximate from the shape samples using the Laplacian eigenmap technique. Our main contribution is to properly define a projection operator onto the manifold by interpolating between shape samples using local weighted means, thereby improving over the naive nearest neighbor approach. Our method is stated as a variational problem that is solved using an iterative numerical scheme. We obtain promising results with synthetic and real shapes which show the potential of our method for segmentation tasks.

## 1 Introduction

### 1.1 Motivation

Image segmentation is an ill-posed problem due to various perturbing factors such as noise, occlusions, missing parts, cluttered data, etc. When dealing with complex images, some prior shape knowledge may be necessary to disambiguate the segmentation process. The use of such prior information in the deformable models framework has long been limited to a smoothness assumption or to simple parametric families of shapes. But a recent and important trend in this domain is the development of deformable models integrating more elaborate prior shape information.

An important work in this direction is the *active shape model* of Cootes et al. [1]. This approach performs a principal component analysis (PCA) on the position of some landmark points placed in a coherent way on all the training contours. The number of degrees of freedom of the model is reduced by considering only the principal modes of variation. The active shape model is quite general and has been successfully applied to various types of shapes (hands, faces, organs). However, the reliance on a parameterized representation and the manual positioning of the landmarks, particularly tedious in 3D images, seriously limits it applicability.

Leventon, Grimson and Faugeras [2] circumvent these limitations by computing parameterization-independent shape statistics within the level set representation [3,4,5].

F. Sgallari, A. Murli, and N. Paragios (Eds.): SSVM 2007, LNCS 4485, pp. 895–906, 2007.

Basically, they perform a PCA on the signed distance functions of the training shapes, and the resulting statistical model is integrated into a geodesic active contours framework. The evolution equation contains a term which attracts the model toward an optimal prior shape. The latter is a combination of the mean shape and of the principal modes of variation. The coefficients of the different modes and the pose parameters are updated by a secondary optimization process. Several improvements to this approach have been proposed [6,7,8], and in particular an elegant integration of the statistical shape model into a unique MAP Bayesian optimization. Let us also mention another neat Bayesian prior shape formulation, based on a B-spline representation, proposed by Cremers, Kohlberger and Schnörr in [9].

Performing PCA on distance functions might be problematic since they do not define a vector space. To cope with this, Charpiat, Faugeras and Keriven [10] proposed shape statistics based on differentiable approximations of the Hausdorff distance. However, their work is limited to a linearized shape space with small deformation modes around a mean shape. Such an approach is relevant only when the learning set is composed of very similar shapes.

### 1.2 Contributions

In this paper, we introduce a new framework that can handle more general shape priors. We model a category of shapes as a smooth finite-dimensional submanifold of the infinite-dimensional shape space. In the sequel, we term this finite-dimensional manifold the *shape prior manifold*. This manifold cannot be represented explicitly. Let us mention related works of Duci et al. [11] and Zolésio [12]. The first one constructs shapes as elements of a linear space, as in *harmonic embedding* [11], the second assumes the Riemannian structure of the shape space.

We approximate *shape prior manifold* from a collection of shape samples using a recent manifold learning technique called Laplacian embedding [13]. Manifold learning is already an established tool in object recognition and image classification. Also, very recently, Charpiat *et al.* [14] have applied the Laplacian eigenmap to a set of fish shapes for the purpose of shape analysis, and obtained promising results. But to our knowledge such techniques have never been used in the context of image segmentation with shape priors.

A Laplacian embedding of the *shape prior manifold* is interesting in itself: it reveals the dimensionality of the shape category and a spatial organization of the associated shape samples. However, this embedding alone does not help to overcome noise, occlusion or other perturbations in a segmentation task. For the *shape prior manifold* to be really useful during a segmentation process, *we need the ability to compute the closest shape of the manifold to some current candidate shape* [6].

Unfortunately, the manifold learning literature does not give a solution to this problem. These approaches are mainly interested in recovering local properties of the manifold by analyzing graph adjacency of samples. They do not focus on recovering information in between samples.

A naive nearest neighbor approach is not an acceptable solution either. First, its answers are limited to the original finite and discrete set of shape samples, which does not account for the smoothness of the *shape prior manifold*. Second, in order to produce an acceptable guess, it would require a very dense sampling of the shape category of interest which is not affordable in practice. Third, it completely disregards the dimensionality and the spatial organization revealed during the manifold learning stage.

Our main contribution is to properly define this projection operator onto the *shape prior manifold*, by interpolating between some carefully selected shape samples using local weighted means. Our method is stated as a variational problem that is solved using an iterative numerical scheme.

The remainder of this paper is organized as follows. Section 2 is dedicated to learning the shape prior manifold from a finite set of shape samples using the Laplacian embedding technique. Section 3 presents a method for interpolation of the shape prior manifold and projection onto it. In Section 4, we report on some numerical experiments which yield promising results with synthetic and real shapes.

# 2 Learning the Shape Prior Manifold

## 2.1 Definitions

In the sequel, we define a *shape* as a simple (i.e. non-intersecting) closed curve, and we denote by $\mathcal{S}$ the space of such shapes. Please note that, although this paper only deals with 2-dimensional shapes, all ideas and results seamlessly extend to higher dimensions.

The space $\mathcal{S}$ is infinite-dimensional. We make the assumption that a category of shapes, i.e. the set of shapes that can be identified with a common concept or object, e.g. fish shapes, can be modeled as a finite-dimensional manifold.

In the context of estimating the shape of an object in a known category from noisy and/or incomplete data, we call this manifold the *shape prior manifold*. In practice, we only have access to a discrete and finite set of example shapes in this category. We will assume that this set constitutes a "good" sampling of the *shape prior manifold*, where "good" stands for "exhaustive" and "sufficiently dense" in a sense that will be clarified below.

## 2.2 Distances Between Shapes

The notion of regularity involved by the manifold viewpoint absolutely requires to define which shapes are close and which shapes are far apart. However, currently, there is no agreement in the computer vision literature on the right way of measuring shape similarity. Many different definitions of the distance between two shapes have been proposed.

One classical choice is the area of the symmetric difference between the regions bounded by the two shapes:

$$d_{SD}(S_1, S_2) = \frac{1}{2} \int |\chi_{\Omega_1} - \chi_{\Omega_2}| \ , \qquad (1)$$

where $\chi_{\Omega_i}$ is the characteristic function of the interior of shape $S_i$. This distance was recently advocated by Solem in [15] to build geodesic paths between shapes.

Another classical definition of distance between shapes is the Hausdorff distance, appearing in the context of shape analysis in image processing in the works of Serra [16] and Charpiat *et al.* [10]:

$$d_H(S_1, S_2) = \max\left\{ \sup_{x\in S_1} \inf_{y\in S_2} \|x-y\| \, , \, \sup_{y\in S_2} \inf_{x\in S_1} \|x-y\| \right\} . \tag{2}$$

Another definition has been proposed [2,6,10], based on the representation of a curve in the plane, of a surface in 3D space, or more generally of a codimension-1 geometric object in $\mathbb{R}^n$, by its signed distance function. In this context, the distance between two shapes can be defined as the $L^2$-norm or the Sobolev $W^{1,2}$-norm of the difference between their signed distance functions. Let us recall that $W^{1,2}(\Omega)$ is the space of square integrable functions over $\Omega$ with square integrable derivatives:

$$d_{L^2}(S_1, S_2)^2 = ||\bar{D}_{S_1} - \bar{D}_{S_2}||^2_{L^2(\Omega,\mathbb{R})} \, , \tag{3}$$

$$d_{W^{1,2}}(S_1, S_2)^2 = ||\bar{D}_{S_1} - \bar{D}_{S_2}||^2_{L^2(\Omega,\mathbb{R})} + ||\nabla\bar{D}_{S_1} - \nabla\bar{D}_{S_2}||^2_{L^2(\Omega,\mathbb{R}^n)} \, , \tag{4}$$

where $\bar{D}_{S_i}$ denotes the signed distance function of shape $S_i$ ($i = 1, 2$), and $\nabla\bar{D}_{S_i}$ its gradient.

### 2.3 Manifold Learning

Once some distance $d$ between shapes has been chosen, classical manifold learning techniques can be applied, by building an adjacency graph of the learning set of shape examples. Let $(S_i)_{i\in 1,\ldots,n}$ denote the $n$ shapes of the learning set. Two slightly different approaches can be considered to build the adjacency graph:

**$\varepsilon$-neighborhoods:** Two nodes $S_i$ and $S_j$ ($i \neq j$) are connected in the graph if $d(S_i, S_j) < \varepsilon$, for some well-chosen constant $\varepsilon > 0$.

**$k$ nearest neighbors:** Two nodes $S_i$ and $S_j$ are connected in the graph if node $S_i$ is among the $k$ nearest neighbors of $S_j$, or conversely, for some constant integer $k$.

The study of advantages and disadvantages of both approaches is beyond the scope of this paper. An adjacency matrix $(W_{i,j})_{i,j\in 1,\ldots,n}$ is then designed, the coefficients of which measure the strength of the different edges in the adjacency graph.

Once an adjacency graph is defined from a given set of samples, manifold learning consists in mapping data points into a lower dimensional space while preserving the local properties of the adjacency graph. This dimensionality reduction with minimal local distortion can advantageously be achieved using spectral methods, i.e. through an analysis of the eigen-structure of some matrices derived from the adjacency matrix.

Dimensionality reduction has enjoyed renewed interest over the past years. Among the most recent and popular techniques are the Locally Linear Embedding (LLE) [17], Laplacian eigenmaps [13] and the Locally Preserving Projections (LPP) [18].

Below, we present the mathematical formulation of Laplacian eigenmaps for data living in $\mathbb{R}^n$. An extension to shape manifolds is straightforward.

Let $\mathcal{M}$ be a manifold of dimension $m$ lying in $\mathbb{R}^n$ ($m << n$). For the moment, we take $m = 1$ since generalization to any dimension $m < n$ is immediate. The dimensionality reduction problem consists in finding a mapping $f : \mathcal{M} \longrightarrow \mathbb{R}$ such that if two points $x$ and $z$ are *close* in $\mathcal{M}$, so are $f(x)$ and $f(z)$. To characterize such an optimal mapping, Belkin and Niyogi [13] asserted the following inequality:

$$|f(z) - f(x)| \leq d_{\mathcal{M}}(x, z) \, ||\nabla f(x)|| + o\left(d_{\mathcal{M}}(x, z)\right) \tag{5}$$

where $d_{\mathcal{M}}$ is the geodesic distance on the manifold $\mathcal{M}$. The optimality condition then writes:

$$f^* = \arg \min_{||f||_{L^2(\mathcal{M})}=1} \int_{\mathcal{M}} ||\nabla f||^2 \tag{6}$$

$$= \arg \min_{||f||_{L^2(\mathcal{M})}=1} \int_{\mathcal{M}} \mathcal{L}(f) \, f \tag{7}$$

where $\mathcal{L} = -\mathrm{div}\,(\nabla f)$ is the Laplace operator. Note that the equivalence between (6) and (7) is due to the Stokes theorem.

Solving the minimization problem (7) is equivalent to solving the eigen problem $\mathcal{L}(f) = \lambda f$. The optimal mapping is then given by the eigen functions corresponding to the $m$ smallest non-zero eigenvalues, where $m$ is the target dimension. Note that the latter dimension can either be known *a priori* or be inferred from the profile of the eigen spectrum.

In practice, a discrete counterpart to this continuous formulation must be used. The discrete approximation of the Laplace operator is given by the matrix $L = D - W$ where $D$ is the diagonal matrix defined by $D_{i,i} = \sum_j W_{i,j}$. Optimal dimensionality reduction is then achieved by finding the eigenvectors of matrix $L$ corresponding to the $m$ smallest non-zero eigenvalues.

Although Laplacian eigenmaps are a powerful dimensionality reduction tool, they do not give access to an explicit projection operator on the low dimensional manifold. In contrast, the Locally Preserving Projection (LPP) [18] defines the mapping $f$ to be locally a linear operator on the high dimensional input data. This is a very interesting property since it allows to project new data points onto the low dimensional representation. Unfortunately, LPP cannot be considered in the context of shapes since it is limited to finite-dimensional problems.

Thus, existing manifold learning techniques are not able to describe the shape prior manifold in between the training shapes. However, in the context of shape estimation from noisy or incomplete images, the two following problems are fundamental:

**Problem 1.** How to interpolate the shape prior manifold from shape samples.

**Problem 2.** How to project a shape onto the shape prior manifold.

The next section deals with these two challenging problems.

## 3 Interpolation and Projection for the *Shape Prior Manifold*

From now on, we assume that applying the manifold learning techniques described in the previous section allowed us to determine the dimension $m$ of the shape prior manifold $\mathcal{M}$. Let $\mathcal{N} = (S_0, ..., S_m)$ be $m+1$ shapes of $\mathcal{M}$, close to each other according to the topology of $\mathcal{M}$. In practice, $S_0$ is given and $S_1, \cdots, S_m$ are its $m$ nearest neighbors in the adjacency graph. We will refer to $\mathcal{N}$ as a *neighborhood system*[1] of $\mathcal{M}$. Our first goal will be to interpolate $\mathcal{M}$ locally, given this neighborhood system.

When $m = 1$, a natural solution would be to choose as an interpolation, some geodesic path between $S_0$ and $S_1$, with respect to distance $d$. Yet, computing geodesic paths between shapes is still an open problem. See [16,19,20,21,15,11,12]. Moreover, extending this solution to the case $m > 1$ would require some minimal hyper-surface of $\mathcal{S}$ between the $S_i$. This seems out of reach with our current understanding of shape spaces.

### 3.1 Weighted Means as Interpolations

Charpiat *et al.* [10] define the empirical mean $\bar{C}$ of $l$ shapes $C_1, \cdots, C_l$ by:

$$\bar{C} = \arg\min_C \sum_{i=1}^{l} d(C_i, C)^2$$

Following the same path, we propose to use *weighted mean shapes* to locally interpolate $\mathcal{M}$ between our samples $S_0, \cdots, S_m$:

**Solution 1 (Problem 1: Local Interpolation of the Shape Manifold).**
*Let $\mathcal{M}$ be a finite $m$ dimensional shape manifold and $\mathcal{N} = (S_0, \cdots, S_m)$ be a* neighborhood system *as previously defined. Let $\Lambda = (\lambda_0, \cdots, \lambda_m)$ with $(\lambda_i \geq 0, \sum \lambda_i = 1)$ be some weights. We call a local interpolation of $\mathcal{M}$ according to $\mathcal{N}$, the following weighted mean:*

$$\bar{S}_{\mathcal{N}}(\Lambda) = \arg\min_S \sum_{i=0}^{m} \lambda_i d(S_i, S)^2$$

$\Lambda$ can be viewed as a local parametrization of $\mathcal{M}$ in the neighborhood system $\mathcal{N}$. The set covered by $\bar{S}_{\mathcal{N}}(\Lambda)$ for all the possible values of $\Lambda$ provides a continuous approximation of the manifold *between* the shapes of $\mathcal{N}$.

As in [10], the interpolation $\bar{S}_{\mathcal{N}}(\Lambda)$ is obtained by a gradient descent, a shape $S$ evolving according to a gradient flow:

$$-\sum_i \lambda_i d(S_i, S) \nabla d(S_i, S) \tag{8}$$

Figure 1 shows an example of such means for two given shapes. Although this involves two shapes only, please note that: (i) the number of shapes is not limited to $m = 1$, and (ii) even when $m = 1$, the path defined by the weighted means is neither a geodesic for some distance, nor a straight gradient descent from $S_0$ to $S_1$. Examples with more than two shapes are given in section 4.

Fitted with a way to locally complete the shape manifold, we can now proceed to the projection problem.

[1] More sophisticated choices are possible but beyond the scope of this paper.

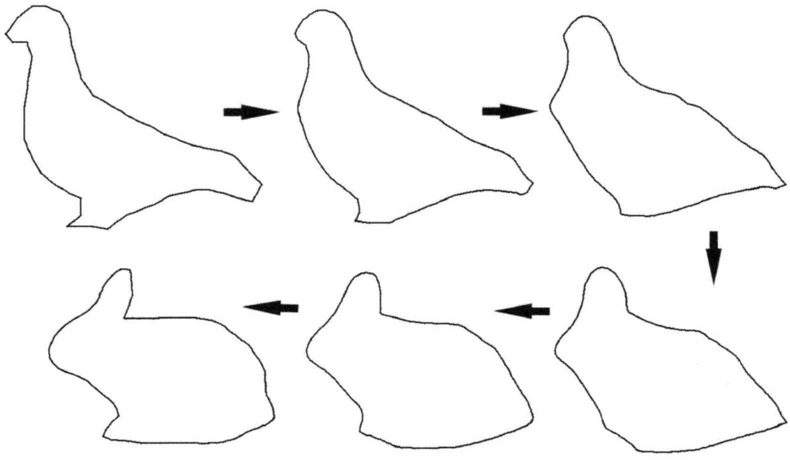

**Fig. 1.** Six weighted means.$\lambda = 0, 0.2, 0.4, 0.6, 0.8$ and 1 following the arrows.

### 3.2 Projection onto *the Shape Prior Manifold*

Image segmentation methods that take shape priors into account, generally require the projection (in some sense) of a shape candidate onto the set of shape samples. As previously mentioned, this projection is often just the mean of the samples, and sometimes a variation of this mean according to *deformation modes*. Here, we propose a projection based on our local interpolation:

**Solution 2 (Problem 2: projection onto the *shape prior manifold*)**
*Let $\mathcal{M}$ be a finite $m$ dimensional shape manifold. Let $M$ be a shape of $\mathcal{S}$. Let $\mathcal{N}(M) = (S_0, \cdots, S_m)$ be a neighborhood system of $\mathcal{M}$ close to $M$ (in practice $S_0$ is the nearest neighbor of $M$ and $S_1, \cdots, S_m$ are chosen as previously). We define the local projection $\Pi_{\mathcal{M}}(M)$ of $M$ onto $\mathcal{M}$ to be the interpolation according to $\mathcal{N}(M)$ that is the closest to $M$ :*

$$\begin{aligned} \Pi_{\mathcal{M}}(M) &= \bar{S}_{\mathcal{N}(M)}(\Lambda_\Pi) \\ \text{with } \Lambda_\Pi &= \arg\min_{\Lambda} d(M, \bar{S}_{\mathcal{N}(M)}(\Lambda)) \end{aligned} \tag{9}$$

While such a projection is clearly better than choosing the nearest neighbor, the energy involved in equation (9) cannot be minimized easily. The variations with respect to $\Lambda$ of the distance $d(M, \bar{S}_{\mathcal{N}(M)}(\Lambda))$ between the interpolation and shape $M$ are intricate. The gradient of this distance could be written, but, involving second order shape derivatives, it yields a complex minimization scheme that might not be useful for our purpose. Keeping shape priors in mind, it appears that an approximation of the projection $\Pi_{\mathcal{M}}(M)$ is sufficient.

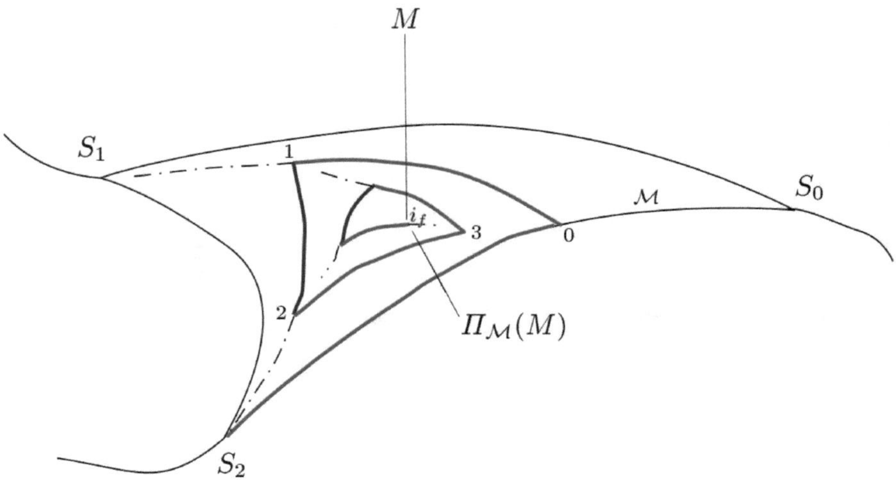


**Fig. 2.** The Snail algorithm : steps are indexed 1, 2, …, $i_f$

Many algorithms might be designed to get an approximate solution to (9). We suggest an iterative scheme illustrated figure 2 that we call the *snail algorithm*. Although it is not guaranteed to converge, it is fast and proved to give good approximations of the projection of a candidate shape onto the shape prior manifold, in only a few iterations. Actually, we investigated more extensive searches of the minimum of (9) without any significant improvement. The snail algorithm is defined by:

**Solution 3 (Approximation of minimization (9))**
*Let $\mathcal{M}$, $M$ and $\mathcal{N}(M)$ be defined as in solution 2. The snail algorithm proceeds as follows:*

1. Initialization: choose the shapes of the neighborhood system as initial guesses.
   *For $i = 0, \cdots, m$, let $\Lambda^i = (\lambda_0^i, \cdots, \lambda_m^i)$ be defined by $\lambda_j^i = \delta_{ij}$*

2. Iterations: look for a better projection between the latest estimate and the one computed $m+1$ steps before.
   *For $i = m, m+1, \cdots$ until convergence, estimate:*

$$\Lambda^{i+1} = \alpha_i \Lambda^i + (1-\alpha_i)\Lambda^{i-m}$$
$$\text{with } \alpha_i = \arg\min_{0 \le \alpha \le 1} d(M, \bar{S}_{\mathcal{N}(M)}(\alpha\Lambda^i + (1-\alpha)\Lambda^{i-m})) \quad (10)$$

3. Exit:
   *Let $i_f$ be the index of last iteration. Approximate the projection by:*

$$\Pi_{\mathcal{M}}(M) = \bar{S}_{\mathcal{N}(M)}(\Lambda^{i_f})$$

Note that we still need to design a minimization scheme to estimate the optimal $\alpha$ in (10). Again, a variational method is both too slow for our purpose and useless for an approximation. Computing a small number of interpolations and keeping the best one turns out to be satisfactory. Moreover, because these interpolations are obtained through a gradient descent (see [10]), estimating the interpolations for an increasing series of $\alpha$ is efficient, each interpolation being the initial guess for the next one.

## 4 Numerical Experiments

In this section, we present some results obtained with both synthetic and real shapes. Our *shape prior manifold* examples are based on a set of rectangles and a set of fishes built in [14]. We reproduce the graph Laplacian obtained in both cases in figure 3. The set of rectangle is randomly chosen such that the distribution of their corners is the uniform law in an authorized area (orientation between $-\frac{\pi}{6}$ and $\frac{\pi}{6}$, length between 2 and time the width). The set of fish is a subset of SQUID database.

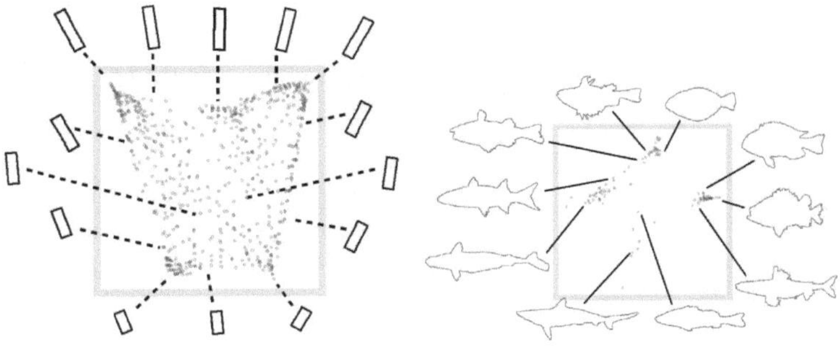

**Fig. 3.** Graph Laplacian (Courtesy [14])

In order to show the reliability of the method, we constructed corrupted shapes from which we extract the neighbor system defined above and computed the best projection onto the *shape prior manifold*. In the toy example, the dimension of the *shape prior manifold* is 2 and thus the interpolation is obviously between 3 shapes. A rectangle is chosen to lie between two angular positions and two differents sizes. this rectangle is corrupted in order to move it away from the *shape prior manifold*. We show in figure 4 the neigbor system chosen, the corrupted shape and its projection.

We provide also prominent results with the fish example. We have highly corrupted a fish shape $M$ : the head is deformed and the shape suffer from many occlusion. Of course, such a shape does not belong to the set used to build the graph Laplacian. Then, we determined the neighbor system $S_0, \dots S_3$ and the projection $\Pi_{\mathcal{M}}(M)$ onto the *shape prior manifold*. Such a projection is clearly better than the nearest neighbor as illustrated in figure 6. Our algorithm overcomes most of shape occlusions and deformations.

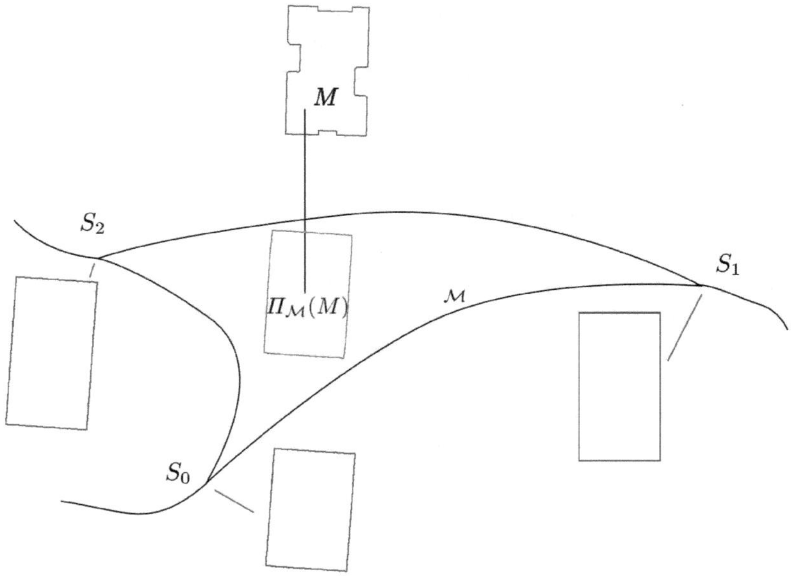


**Fig. 4.** Toy example : Projection onto *the shape manifold*

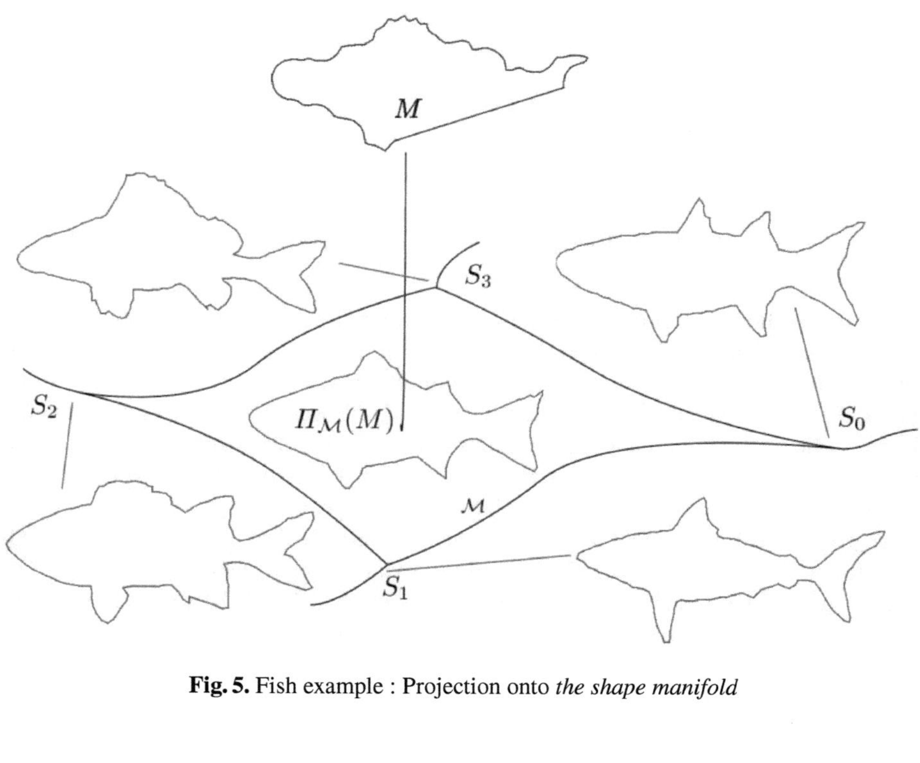


**Fig. 5.** Fish example : Projection onto *the shape manifold*

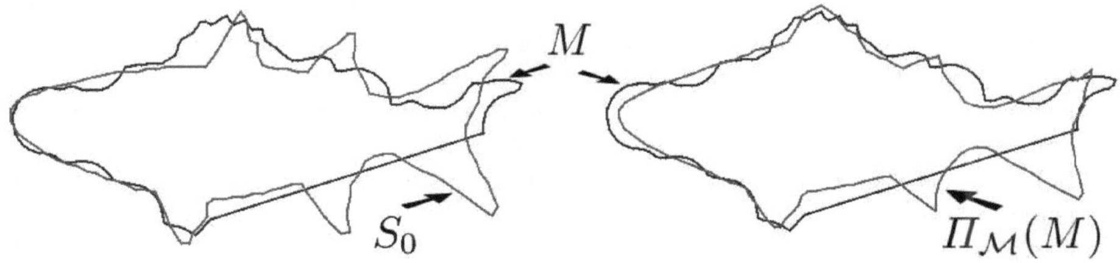


**Fig. 6.** Fish example : Comparaison between the nearest neighbor $S_0$ and the projection $\Pi_{\mathcal{M}}(M)$

## 5 Conclusion and Perspectives

We proposed a new framework for image segmentation that incorporates more general priors by learning a *shape prior manifold*. We provided a solution to interpolate between shape samples, defined a projection operator onto the *shape prior manifold* and suggested its fast estimation by means of an iterative process. Finally, numerical experiments on synthetic and real images showed promising results and the potential of the method for image segmentation. Incorporating it into a complete segmentation process is actually work in progress.

## References

1. Cootes, T., Taylor, C., Cooper, D., Graham, J.: Active shape models-their training and application. Computer Vision and Image Understanding **61**(1) (1995) 38–59
2. Leventon, M., Grimson, E., Faugeras, O.: Statistical shape influence in geodesic active contours. In: IEEE Conference on Computer Vision and Pattern Recognition. (2000) 316–323
3. Osher, S., Sethian, J.: Fronts propagating with curvature-dependent speed: Algorithms based on Hamilton–Jacobi formulations. Journal of Computational Physics **79**(1) (1988) 12–49
4. Sethian, J.: Level Set Methods and Fast Marching Methods: Evolving Interfaces in Computational Geometry, Fluid Mechanics, Computer Vision, and Materials Sciences. Cambridge Monograph on Applied and Computational Mathematics. Cambridge University Press (1999)
5. Osher, S., Fedkiw, R.: Level set methods: an overview and some recent results. Journal of Computational Physics **169**(2) (2001) 463–502
6. Rousson, M., Paragios, N.: Shape priors for level set representations. In: European Conference on Computer Vision. Volume 2. (2002) 78–92
7. Chen, Y., Tagare, H., Thiruvenkadam, S., Huang, F., Wilson, D., Gopinath, K., Briggs, R., Geiser, E.: Using prior shapes in geometric active contours in a variational framework. The International Journal of Computer Vision **50**(3) (2002) 315–328
8. Tsai, A., Yezzi, A., Wells, W., Tempany, C., Tucker, D., Fan, A., Grimson, W., Willsky, A.: A shape-based approach to the segmentation of medical imagery using level sets. IEEE Transactions on Medical Imaging **22**(2) (2003) 137–154
9. Cremers, D., Kohlberger, T., Schnörr, C.: Nonlinear shape statistics in mumford shah based segmentation. In: European Conference on Computer Vision. (2002) 93–108
10. Charpiat, G., Faugeras, O., Keriven, R.: Approximations of shape metrics and application to shape warping and empirical shape statistics. Foundations of Computational Mathematics **5**(1) (2005) 1–58

11. Duci, A., Yezzi, A.J., Mitter, S.K., Soatto, S.: Shape representation via harmonic embedding. In: ICCV '03: Proceedings of the Ninth IEEE International Conference on Computer Vision, Washington, DC, USA, IEEE Computer Society (2003) 656
12. Delfour, M.C., Zolésio, J.P.: Shapes and geometries: analysis, differential calculus, and optimization. Society for Industrial and Applied Mathematics, Philadelphia, PA, USA (2001)
13. Belkin, M., Niyogi, P.: Laplacian eigenmaps for dimensionality reduction and data representation. Neural Computation **15**(6) (2003) 1373–1396
14. Charpiat, G., Faugeras, O., Keriven, R., Maurel, P.: Distance-based shape statistics. In: IEEE International Conference on Acoustics, Speech and Signal Processing. Volume 5. (2006) 925–928
15. Solem, J.: Geodesic curves for analysis of continuous implicit shapes. In: International Conference on Pattern Recognition. Volume 1. (2006) 43–46
16. Serra, J.: Hausdorff distances and interpolations. In: International Symposium on Mathematical Morphology and its Applications to Image and Signal Processing. (1998) 107–114
17. Roweis, S., Saul, L.: Nonlinear dimensionality reduction by locally linear embedding. Science **290** (2000) 2323–2326
18. He, X., Niyogi, P.: Locality preserving projections. In: Advances in Neural Information Processing Systems 16. MIT Press (2004)
19. Beg, M.F., Miller, M.I., Trouvé;, A., Younes, L.: Computing large deformation metric mappings via geodesic flows of diffeomorphisms. Int. J. Comput. Vision **61**(2) (2005) 139–157
20. Michor, P.W., Mumford, D.: Riemannian geometries on spaces of plane curves. J. Eur. Math. Soc. **8** (2006) 1–48
21. Yezzi, A., Mennucci, A.: Conformal metrics and true "gradient flows" for curves. iccv **1** (2005) 913–919

# Geometric Sampling of Manifolds for Image Representation and Processing

Emil Saucan, Eli Appleboim, and Yehoshua Y. Zeevi

Electrical Engineering Department, Technion, Haifa, Israel
eliap,semil,zeevi@ee.technion.ac.il

**Abstract.** It is often advantageous in image processing and computer vision to consider images as surfaces imbedded in higher dimensional manifolds. It is therefore important to consider the theoretical and applied aspects of proper sampling of manifolds. We present a new sampling theorem for surfaces and higher dimensional manifolds. The core of the proof resides in triangulation results for manifolds with or without boundary, not necessarily compact. The proposed method adopts a geometric approach that is considered in the context of 2-dimensional manifolds (i.e surfaces), with direct applications in image processing. Implementations of these methods and theorems are illustrated and tested both on synthetic images and on real medical data.

## 1 Introduction and Related Works

In recent years it became common amongst the signal processing community, to consider images as Riemannian manifolds embedded in higher dimensional spaces (see, e.g. [10], [12], [22]). Usually, the embedding manifold is taken to be $\mathbb{R}^n$ yet, other possibilities are also considered ([5]). For example, a gray scale image is a surface in $\mathbb{R}^3$, whereas a color image is a surface embedded in $\mathbb{R}^5$, each color channel representing a coordinate. In both cases the intensity, either gray scale or color, is considered as a function of the two spatial coordinates $(x, y)$ and thus the surface may be equipped with a metric induced by this function. The question of smoothness of the function is in general omitted, if numerical schemes are used for the approximations of derivatives, whenever this is necessary. A major advantage of such a viewpoint of signals is the ability to apply mathematical tools traditionally used in the study of Riemannian manifolds, for image/signal processing as well. For example, in medical imaging it is often convenient to treat CT/MRI scans, as Riemannian surfaces in $\mathbb{R}^3$. One can then borrow techniques from differential topology and geometry and geometric analysis in the representation and analysis of the considered images.

Sampling is an essential preliminary step in processing of any continuous signal by a digital computer. This step lies at the heart of any digital processing of any (presumably continuous) data/signal. Undersampling causes distortions due to the aliasing of the post processed sampled data. Oversampling, on the other hand, results in time and memory consuming computational processes which, at the very least, slows down the analysis process. It is therefore important to have a

F. Sgallari, A. Murli, and N. Paragios (Eds.): SSVM 2007, LNCS 4485, pp. 907–918, 2007.

measure which is instrumental in determining what is the optimal sampling rate. For 1-dimensional signals such a measure exists, and, consequently, the optimal sampling rate is given by the fundamental sampling theorem of Shannon, that yielded the foundation of information theory and led technology into the digital era. Shannon's theorem indicates that a signal can be perfectly reconstructed from its samples, given that the signal is band limited within some bound on its highest frequency. Ever since the introduction of Shannon's theorem in the late 1940's, deducing a similar sampling theorem for higher dimensional signals has become a challenge and active area of research, especially recently, in view of methods based on representation of images as manifolds (mostly surfaces) embedded in higher dimensional manifolds. This is further emphasized by the broad interest in its applications in image processing, and by the growing need for fast yet accurate techniques for processing high dimensional data such as medical and satellite images.

Recently a surge in the study of *fat* triangulations (Section 2 below) and manifold sampling in computational geometry, computer graphics and their related fields has generated many publications (see [1], [4], [8], [9], [13], [14], [17], to name just a few). For example, in [1] Voronoi filtering is used for the construction of fat triangulations of compact, $\mathcal{C}^2$ surfaces embedded in $\mathbb{R}^3$. Note that Voronoi cell partitioning is also employed in "classical" sampling theory (see [23]). Cheng et al. [8] used these ideas for manifold reconstruction from point samples. In [14] an heuristic approach to the problem of the relation between curvature and sampling density is given. Again, in these studies the manifolds are assumed to be smooth, compact $n$-dimensional hyper-surfaces embedded in $\mathbb{R}^{n+1}$.

In this paper we present new sampling theorems for manifolds of dimension $\geq 2$. These theorems are derived from fundamental studies in three areas of mathematics: differential topology, differential geometry and quasi-regular maps. Both classical and recent results in these areas are combined to yield a rigorous and comprehensive sampling theory for such manifolds. Our approach lends itself also to a new, geometrical interpretation of classical results regarding proper interpretation of images.

In Section 3 we present geometrical sampling theorems for images/signals given as Riemannian manifolds, for both smooth and non-smooth images/signals. In preparation for that we provide, in Section 2, the necessary background regarding the main results on the existence of fat triangulations of manifolds, and the relation to sampling and reproducing of Riemannian manifolds. We also review the problem of smoothing of manifolds. Finally, in Section 4, we examine some delicate aspects of our study, and discuss extensions of this work, relating both to geometric aspects of sampling, as well as to its relationship to classical sampling theory.

## 2 Notations, Preliminaries and Background

### 2.1 Triangulation and Sampling

While for basic definitions and notation regarding triangulations and Piecewise Linear ($PL$) Topology we refer the reader to [15], we begin this section by recalling a few classical definitions:

**Definition 1.** *Let $f : K \to \mathbb{R}^n$ be a $C^r$ map, and let $\delta : K \to \mathbb{R}_+^*$ be a continuous function. Then $g : |K| \to \mathbb{R}^n$ is called a* $\delta$-approximation *to $f$ iff:*
*(i) There exists a* subdivision *$K'$ of $K$ such that $g \in C^r(K', \mathbb{R}^n)$ ;*
*(ii) $d_{eucl}\big(f(x), g(x)\big) < \delta(x)$ , for any $x \in |K|$ ;*
*(iii) $d_{eucl}\big(df_a(x), dg_a(x)\big) \le \delta(a) \cdot d_{eucl}(x, a)$ , for any $a \in |K|$ and for all $x \in \overline{St}(a, K')$.*
*(Here and below $|K|$ denotes* the underlying polyhedron *of $K$.)*

**Definition 2.** *Let $K'$ be a subdivision of $K$, $U = \overset{\circ}{U} \subset |K|$, and let $f \in C^r(K, \mathbb{R}^n)$, $g \in C^r(K', \mathbb{R}^n)$. $g$ is called a* $\delta$-approximation *of $f$ on $U$, iff conditions (ii) and (iii) of Definition 2.6. hold for any $a \in U$.*

The most natural and intuitive $\delta$-approximation to a given mapping $f$ is the *secant map induced by $f$*:

**Definition 3.** *Let $f \in C^r(K)$ and let $s$ be a simplex, $s < \sigma \in K$. Then the linear map: $L_s : s \to \mathbb{R}^n$, defined by $L_s(v) = f(v)$, where $v$ is a vertex of $s$, is called the* secant map induced by *$f$.*

**Fat Triangulation.** We now proceed to show that the apparent "naive" secant approximation of surfaces (and higher dimensional manifolds) represents a good approximation, insofar as distances and angles are concerned, provided that the secant approximation induced by a triangulations that satisfies a certain non-degeneracy condition called "fatness" (or "thickness").

We first provide the following informal, intuitive definition: A triangle in $\mathbb{R}^2$ is called *fat* (or *$\varphi$-fat*, to be more precise) iff all its angles are larger than some $\varphi$. In other words, fat triangles are those that do not "deviate" to much from being equiangular (regular), hence fat triangles are not too "slim". One can defined fat triangles more formally by requiring that the ratio of the radii of the inscribed and circumscribed circles of the triangle is bounded from bellow by $\varphi$, i.e. $r/R \ge \varphi$, for some $\varphi > 0$, where $r$ denotes the radius of the inscribed circle of $\tau$ (*inradius*) and $R$ denotes the radius of the circumscribed circle of $\tau$ (*circumradius*). This definition easily generalizes to triangulations in any dimension:

**Definition 4.** *A $k$-simplex $\tau \subset \mathbb{R}^n$, $2 \le k \le n$, is* $\varphi$-fat *if there exists $\varphi > 0$ such that the ratio $\frac{r}{R} \ge \varphi$; where $r$ denotes the radius of the inscribed sphere of $\tau$ and $R$ denotes the radius of the circumscribed sphere of $\tau$. A triangulation of a submanifold of $\mathbb{R}^n$, $\mathcal{T} = \{\sigma_i\}_{i \in \mathbf{I}}$ is $\varphi$-fat if all its simplices are* $\varphi$-fat. *A triangulation $\mathcal{T} = \{\sigma_i\}_{i \in \mathbf{I}}$ is* fat *if there exists $\varphi \ge 0$ such that all its simplices are $\varphi$-fat; for any $i \in \mathbf{I}$.*

One recuperates the "big" angle characterization of fatness through the following proposition:

**Proposition 1 ([7]).** *There exists a constant $c(k)$ that depends solely upon the dimension $k$ of $\tau$ such that*

$$\frac{1}{c(k)} \cdot \varphi(\tau) \le \min_{\sigma < \tau} \measuredangle(\tau, \sigma) \le c(k) \cdot \varphi(\tau) \,, \tag{1}$$

*and*

$$\varphi(\tau) \leq \frac{Vol_j(\sigma)}{diam^j \sigma} \leq c(k) \cdot \varphi(\tau)\,, \tag{2}$$

*where $\varphi$ denotes the fatness of the simplex $\tau$, $\measuredangle(\tau,\sigma)$ denotes the (internal) dihedral angle of the face $\sigma < \tau$ and $Vol_j(\sigma)$; and where $diam\,\sigma$ stand for the Euclidian $j$-volume and the diameter of $\sigma$ respectively. (If $dim\,\sigma = 0$, then $Vol_j(\sigma) = 1$, by convention.)*

Condition (1) is just the expression of fatness as a function of dihedral angles in all dimensions, while Condition (2) expresses fatness as given by "large area/diameter". Diameter is important since fatness is independent of scale.

The importance of fatness of triangulations, for attaining good approximations, is underlined in the following proposition that represents the desired approximation result:

**Proposition 2 ([15], Lemma 9.3).** *Let $f : \sigma \to \mathbb{R}^n$ be of class $\mathcal{C}^k$. Then, for $\delta, \varphi_0 > 0$, there exists $\varepsilon > 0$, such that, for any $\tau < \sigma$, such that $diam(\tau) < \varepsilon$ and such that $\varphi(\tau) > \varphi_0$, the secant map $L_\tau$ is a $\delta$-approximation of $f|\tau$.*

### 2.2 Fat Triangulation Results

In this section we review, in chronological order, existence theorems dealing with fat triangulations on manifolds. For detailed proofs see the original papers.

**Theorem 1 (Cairns, [6]).** *Every compact $\mathcal{C}^2$ Riemannian manifold admits a fat triangulation.*

**Theorem 2 (Peltonen, [18]).** *Every open (unbounded) $\mathcal{C}^\infty$ Riemannian manifold admits a fat triangulation.*

**Theorem 3 (Saucan, [19]).** *Let $M^n$ be an $n$-dimensional $\mathcal{C}^1$ Riemannian manifold with boundary, having a finite number of compact boundary components. Then, any uniformly fat triangulation of $\partial M^n$ can be extended to a fat triangulation of $M^n$.*

*Remark 1.* Theorem 3 above holds, in fact, even without the finiteness and compactness conditions imposed on the boundary components (see [20]).

**Corollary 1.** *Let $M^n$ be a manifold as in Theorem 3 above. Then $M^n$ admits a fat triangulation.*

In low dimensions one can also discard the smoothness condition:

**Corollary 2.** *Let $M^n$ be an $n$-dimensional, $n \leq 4$ (resp. $n \leq 3$), PL (resp. topological) connected manifold with boundary, having a finite number of compact boundary components. Then, any fat triangulation of $\partial M^n$ can be extended to a fat triangulation of $M^n$.*

## 2.3 Smoothing of Manifolds

In this section we focus our attention on the problem of smoothing of manifolds. That is, approximating a manifold of differentiability class $C^r$, $r \geq 0$, by manifolds of class $C^\infty$. Of special interest is the case where $r = 0$. Later, when stating our sampling theorem we will make a use of this in two respects. One of them will be as a post-processing step where, after reproducing a $PL$ manifold out of the samples, we can smoothen it to get a smooth reproduced manifold. Another aspect in which smoothing is useful is as a pre-processing step, when we wish to extend the sampling theorem to manifolds which are not necessarily smooth. Smoothing will take place followed by sampling of the smoothed manifold, yielding a sampling for the non-smooth one as well. As a major reference to this we use [15], Chap 4. Similar results can also be found in [11] and others.

**Lemma 1.** *For every $0 < \epsilon < 1$ there exists a $C^\infty$ function $\psi_1 : \mathbb{R} \to [0,\ 1]$ such that, $\psi_1 \equiv 0$ for $|x| \geq 1$ and $\psi_1 = 1$ for $|x| \leq (1-\epsilon)$. Such a function is called* partition of unity.

Let $c^n(\epsilon)$ be the $\epsilon$ cube around the origin in $\mathbb{R}^n$ (i.e. $X \in \mathbb{R}^n$ ; $-\epsilon \leq x_i \leq \epsilon\,,\ i = 1, ..., n$). We can use the above partition of unity in order to obtain a non-negative $C^\infty$ function, $\psi$, on $\mathbb{R}^n$, such that $\psi = 1$ on $c^n(\epsilon)$ and $\psi \equiv 0$ outside $c^n(1)$. Define $\psi(x_1, ..., x_n) = \psi_1(x_1) \cdot \psi_1(x_2) \cdots \psi_1(x_n)$.

We now introduce the main theorem regarding smoothing of $PL$-manifolds.

**Theorem 4 ([15]).** *Let $M$ be a $C^r$ manifold, $0 \leq r < \infty$, and $f_0 : M \to \mathbb{R}^k$ a $C^r$ embedding. Then, there exists a $C^\infty$ embedding $f_1 : M \to \mathbb{R}^k$ which is a $\delta$-approximation of $f_0$.*

The above theorem is a consequence of the following classical lemma concerning smoothing of maps:

**Lemma 2 ([15]).** *Let $U$ be an open subset of $\mathbb{R}^m$. Let $A$ be a compact subset of an open set $V$ such that $\overline{V} \subset U$, is compact. Let $f_0 : U \to \mathbb{R}^n$ be a $C^r$ map, $0 \leq r$. Let $\delta$ be a positive number. Then there exists a map $f_1 : U \to \mathbb{R}^n$ such that*

1. *$f_1$ is $C^\infty$ on $A$.*
2. *$f_1 = f_0$ outside $V$.*
3. *$f_1$ is a $\delta$-approximation of $f_0$*
4. *$f_1$ is $C^r$-homotopic to $f_0$ via a homotopy $f_t$ satisfying* (2) *and* (3) *above. i.e. $f_0$ can be continuously deformed to $f_1$.*

*Remark 2.* A modified version of the smoothing process presented herein was developed by Nash [16]. His idea was to define a radially symmetric convolution kernel $\varphi$, by taking its Fourier transform, $\hat{\varphi}$, to be a radially symmetric partition of unity. Nash's method renders an approximation that is faithful not only to the signal and its first derivative, as in the classical approach, but also to higher order derivatives (if they exist).

## 3 Sampling Theorems

We employ results regarding the existence of fat triangulations, to prove sampling theorems for Riemannian manifolds embedded in some Euclidean space.

**Theorem 5.** *Let $\Sigma^n, n \geq 2$ be a connected, not necessarily compact, smooth manifold, with finitely many compact boundary components. Then there exists a sampling scheme of $\Sigma^n$, with a proper density $\mathcal{D}$ with respect to the volume element on $\Sigma^n$, $\mathcal{D} = \mathcal{D}(p) = \mathcal{D}\Big(\frac{1}{k(p)}\Big)$, where $k(p) = \max\{|k_1|, ..., |k_{2n}|\} > 0$, and where $k_1, ..., k_{2n}$ are the principal (normal) curvatures of $\Sigma^n$, at the point $p \in \Sigma^n$.*

*Proof.* The existence of the sampling scheme follows immediately from Corollary 1, where the sampling points(points of the sampling) are the vertices of the triangulation. The fact that the density is a function solely of $k = max\{|k_1|, ..., |k_{2n}|\}$ follows from the proof of Theorem 2 (see [18], [19]) and from the fact that the osculatory radius $\omega_\gamma(p)$ at a point $p$ of a curve $\gamma$ equals $1/k_\gamma(p)$, where $k_\gamma(p)$ is the curvature of $\gamma$ at $p$; hence the maximal osculatory radius (of $\Sigma$) at $p$ is: $\omega(p) = \max\{|k_1|, ..., |k_{2n}|\} = \max\{\frac{1}{\omega_1}, ..., \frac{1}{\omega_{2n}}\}$. (Here $\omega_{2i}$, $\omega_{2i+1}$ , $i = 1, ..., n-1$ denote the minimal, respective maximal sectional osculatory radii at $p$.)

**Corollary 3.** *Let $\Sigma^n, \mathcal{D}$ be as above. If there exists $k_0 > 0$, such that $k(p) \leq k_0$, for all $p \in \Sigma^n$, then there exists a sampling of $\Sigma^n$ of finite density everywhere.*

*Proof.* Immediate from the theorem above.

In particular we have:

**Corollary 4.** *If $\Sigma^n$ is compact, then there exists a sampling of $\Sigma^n$ having uniformly bounded density.*

*Proof.* It follows immediately from a compactness argument and from the continuity of the principal curvature functions.

The implementation of Theorem 5 is illustrated in Figure 3. Note the fat triangulation and the good reconstruction (see below) obtained from it. Compare with the "flat" triangles in Figure 1, obtained by a "naive" sampling method.

*Remark 3.* Obviously, Theorem 5 above is of little relevance for the *space forms* $(\mathbb{R}^n, \mathbb{S}^n, \mathbb{H}^n)$. Indeed, as noted above, this method is relevant for manifolds considered (by the Nash embedding theorem [16]) as submanifolds of $\mathbb{R}^N$, for some $N$ large enough.

We approach the problem of sampling for non-smooth manifolds; a case that is of interest and practical importance in the context of image processing and computer vision (see, e.g. Figure 1). We begin by proposing the following definition:

**Definition 5.** *Let $\Sigma^n, n \geq 2$ be a (connected) manifold of class $\mathcal{C}^0$, and let $\Sigma^n_\delta$ be a $\delta$-approximation to $\Sigma^n$. A sampling of $\Sigma^n_\delta$ is called a* $\delta$-sampling of $\Sigma^n$.

**Theorem 6.** *Let $\Sigma^n$ be a connected, non-necessarily compact manifold of class $\mathcal{C}^0$. Then, for any $\delta > 0$, there exists a $\delta$-sampling $\Sigma^n_\delta$ of $\Sigma^n$, such that if $\Sigma^n_\delta \to \Sigma^n$ uniformly, then $\mathcal{D}_\delta \to \mathcal{D}$ in the sense of measures, where $\mathcal{D}_\delta$ and $\mathcal{D}$ denote the densities of $\Sigma^n_\delta$ and $\Sigma^n$, respectively.*

*Proof.* The proof is an immediate consequence of Theorem 2 and its proof, and the methods exposed in Section 2.3. We take the sampling of some smooth $\delta$-approximation of $\Sigma^n$.

**Corollary 5.** *Let $\Sigma^n$ be a $\mathcal{C}^0$ manifold with finitely many points at which $\Sigma^n$ fails to be smooth. Then every $\delta$-sampling of a smooth $\delta$-approximation of $\Sigma^n$ is in fact, a sampling of $\Sigma^n$ apart of finitely many small neighborhoods of the points where $\Sigma^n$ is not smooth.*

*Proof.* From Lemma 2 and Theorem 4 it folows that any such $\delta$-approximation, $\Sigma^n_\delta$, coincides with $\Sigma^n$ outside of finitely many such small neighborhoods.

*Remark 4.* In order to obtain a better approximation it is advantageous, in this case, to employ Nash's method for smoothing, cf. Remark 5 of Section 3 (see [16], [2] for details).

**Reconstruction.** We use the secant map as defined in Definition 3 in order to reproduce a $PL$-manifold as a $\delta$-approximation for the sampled manifold. As stated in the beginning of Section 2.3, we may now use smoothing in order to obtain a $\mathcal{C}^\infty$ approximation. This approach is illustrated in Figure 2, for the case of an analytic surface.

In the special case of surfaces (i.e. $n = 2$), more specific, geometric conditions can be obtained:

**Corollary 6.** *Let $\Sigma^2$ be a smooth surface. In the following cases there exist $k_0$ as in Corollary 3 above:*

1. *There exist $H_1, H_2, K_1, K_2$, such that $H_1 \leq H(p) \leq H_2$ and $K_1 \leq K(p) \leq K_2$, for any $p \in \Sigma^2$, where $H, K$ denote the mean, respective Gauss curvature. (That is both mean and Gauss curvatures are* pinched.*)*
2. *The Willmore integrand $W(p) = H^2(p) - K(p)$ and $K$ (or $H$) are pinched.*

*Proof.* 1. Since $K = k_1k_2$, $H = \frac{1}{2}(k_1 + k_2)$, the bounds for $K$ and $H$ imply the desired one for $k$.
2. Analogue reasoning to that of (*2.*), since $W = \frac{1}{4}(k_1 - k_2)^2$.

*Remark 5.* Condition (*ii*) on $W$ is not only compact, it has the additional advantage that the *Willmore energy* $\int_\Sigma W dA$ (where $dA$ represents the area element of $\Sigma$) is a *conformal invariant* of $\Sigma$.

Note that such geometric conditions, are hard to impose in higher dimensions, and the precise geometric constraints remain to be further investigated.

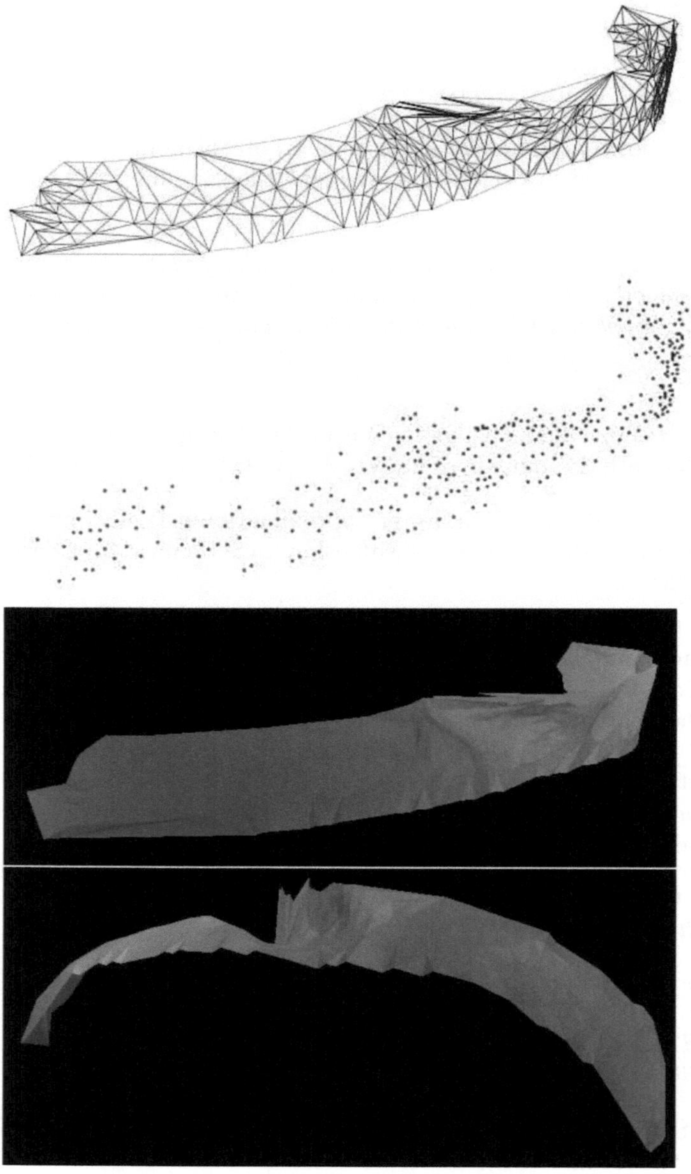

**Fig. 1.** The triangulation (upper image) obtained from a "naive" sampling (second image from above) resulting from a CT scan of the back-side of the human colon (second image from below). Note the "flat" triangles and the uneven mesh of the triangulation. This is a result of the high, concentrated curvature, as revealed in a view obtained after a rotation of the image (bottom). These and other images will be accessible through an interactive applet on the website [25].

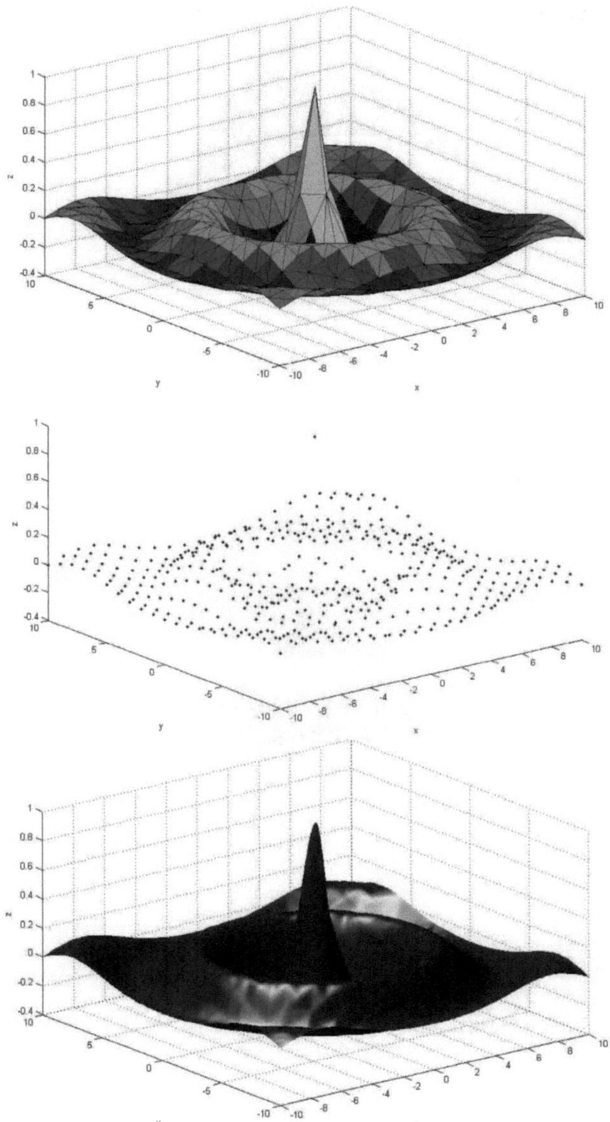

**Fig. 2.** The triangulation (upper image) obtained from the uniform sampling (second image from above) of the surface $S = \left(x, y, \cos\sqrt{x^2+y^2}/(1+\sqrt{x^2+y^2}\,)\right)$ (bottom image). Note the low density of sampling points in the region of high curvature.

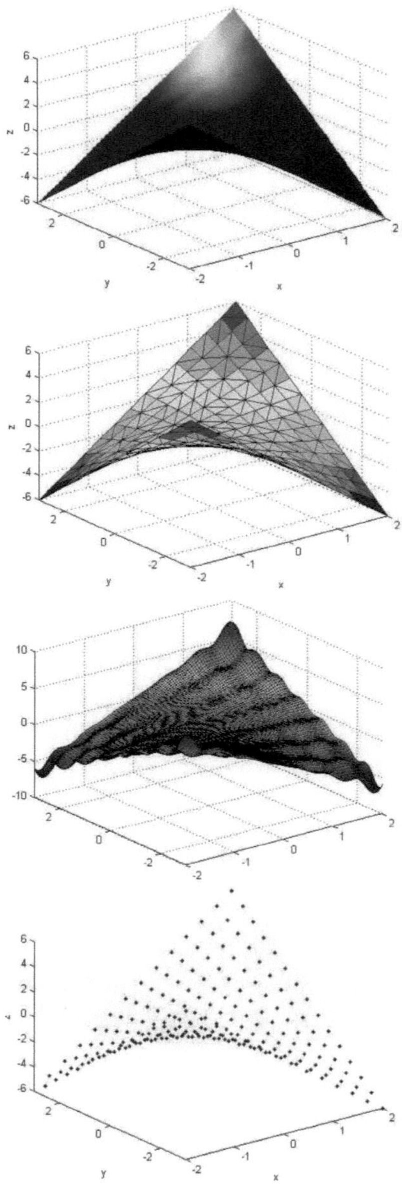


**Fig. 3.** Hyperbolic Paraboloid: Analytic representation, $z = xy$ – top image. Sampling according to curvature – second image from above. PL reconstruction – second image from bellow. Bottom – Nyquist reconstruction. To appreciate the triangulation results a full size display of color images [25] is required.

## 4 Conclusions and Discussion

The methods for sampling and reconstructing of images, introduced in this paper, extend previous studies based on the viewpoint that images and other types of data structures should be considered as surfaces and manifolds embedded in higher dimensional manifolds. In particular, the methods presented in this paper are based on the assertion that surfaces and manifolds should be properly sampled in Shannon's sense. This led to consideration of a sampling theorem for Riemannian manifolds. The sampling scheme presented in this paper, is based on the ability to triangulate such a manifold by a fat triangulation. This in turn, relies on geometric properties of the manifold and basically on its curvature. The sampling theorems are applicable to images/signals that can be represented as Riemannian manifolds, a well established viewpoint in image processing. Considering this viewpoint in rigorous manner still remains as a challenge for further study. It is common for instance, to consider a color image as a surface in $\mathbb{R}^5$ yet, it is more prone and probably more accurate to consider it as a three-dimensional manifold embedded in some higher dimensional Euclidian space. Another interesting issue currently under investigation, is whether the geometric framework for sampling of surfaces and manifolds present in this study can be degenerated to one-dimensional signals as an alternative to the classical sampling theorem of Shannon and how the two approaches are related. Some relevant results are already at hand [21]. Other theoretical and applied facets of this problem are currently under investigation.

## Acknowledgment

Emil Saucan is supported by the Viterbi Postdoctoral Fellowship. Research is partly supported by the Ollendorf Minerva Center and by the Fund for Promotion of Research at the Technion.

The authors would like to thank Efrat Barak, Ronen Lev, Leor Belenki and Uri Okun for their skillful programming that produced the images included herein.

## References

[1] Amenta, N. and Bern, M. *Surface reconstruction by Voronoi filtering*, Discrete and Computational Geometry, 22, pp. 481-504, 1999.

[2] Andrews, B. *Notes on the isometric embedding problem and the Nash-Moser implicit function theorem*, Proceedings of CMA, Vol. 40, 157-208, 2002.

[3] M. Berger, *A Panoramic View of Riemannian Geometry*, Springer-Verlag, Berlin, 2003.

[4] Bern, M., Chew, L.P., Eppstein, D. and Ruppert, J.: *Dihedral Bounds for Mesh Generation in High Dimensions*, 6th ACM-SIAM Symposium on Discrete Algorithms, 1995, 189-196.

[5] Bronstein, A. M., Bronstein, M. M. and Kimmel, R.: *On isometric embedding of facial surfaces into S3*, Proc. Intl. Conf. on Scale Space and PDE Methods in Computer Vision, pp. 622-631, 2005.

[6] Cairns, S.S.: *A simple triangulation method for smooth manifolds*, Bull. Amer. Math. Soc. **67**, 1961, 380-390.

[7] Cheeger, J., Müller, W., and Schrader, R.: *On the Curvature of Piecewise Flat Spaces*, Comm. Math. Phys., **92**, 1984, 405-454.

[8] Cheng, S. W., Dey, T. K. and Ramos, E. A.: *Manifold Reconstruction from Point Samples*, Proc. ACM-SIAM Sympos. Discrete Algorithms, 2005, 1018–1027.

[9] Edelsbrunner, H.: *Geometry and Topology for Mesh Generation*, Cambridge University Press, Cambridge, 2001.

[10] Hallinan, P. *A low-dimensional representation of human faces for arbitrary lighting conditions*, Proc. CVPR. pp. 995-999, 1994.

[11] Hirsh, M. and Masur, B.: *Smoothing of PL-Manifolds*, Ann. Math. Studies 80, Princeton University Press, Princeton, N.J., 1966.

[12] Kimmel, R. Malladi, R. and Sochen, N. *Images as Embedded Maps and Minimal Surfaces: Movies, Color, Texture, and Volumetric Medical Images*, International Journal of Computer Vision, **39**(2), pp. 111-129, 2000.

[13] Li, X.-Y. and Teng, S.-H.: *Generating Well-Shaped Delaunay Meshes in 3D*, SODA 2001, 28-37.

[14] Meenakshisundaram, G.: *Theory and Practice of Reconstruction of Manifolds With Bounderies*, PhD Thesis, Univ. North Carololina, 2001.

[15] Munkres, J. R.: *Elementary Differential Topology*, (rev. ed.) Princeton University Press, Princeton, N.J., 1966.

[16] Nash, J. *The embedding problem for Riemannian manifolds*, Ann. of Math. (2) **63**, 20-63, 1956.

[17] Pach, J. and Agarwal. P.K. *Combinatorial Geometry*, Wiley-Interscience, 1995.

[18] Peltonen, K. *On the existence of quasiregular mappings*, Ann. Acad. Sci. Fenn., Series I Math., Dissertationes, 1992.

[19] Saucan, E. *Note on a theorem of Munkres*, Mediterr. j. math., **2**(2), 2005, 215 - 229.

[20] Saucan, E. *Remarks on the The Existence of Quasimeromorphic Mappings*, to appear in Contemporary Mathematics.

[21] Saucan, E., Appleboim, E. and Zeevi, Y. Y. *Sampling and Reconstruction of Surfaces and Higher Dimensional Manifolds*, with Eli Appleboim and Yehoshua Y. Zeevi, Technion CCIT Report #591, June 2006 (EE PUB #1543 June 2006).

[22] Seung, H. S. and Lee, D. D. *The Manifold Ways of Perception*, Science 290, 2323-2326, 2000.

[23] Smale, S. and Zhou, D.X., *Shannon sampling and function reconstruction from point values*, Bull. Amer. Math. Soc., **41**(3), 2004, 279-305.

[24] Unser, M. and Zerubia, J.: *A Generalized Sampling Theory Without Band-Limiting Constrains*, IEEE Trans. Signal Processing, Vol. 45, No. 8, 1998, 956-969.

[25] http://visl.technion.ac.il/ImageSampling

# Modeling Foveal Vision*

Luc Florack

Eindhoven University of Technology, Department of Biomedical Engineering,
P.O. Box 513, NL-5600 MB Eindhoven, The Netherlands
L.M.J.Florack@tue.nl

**Abstract.** A geometric model is proposed for an artificial foveal vision system, and its plausibility in the context of biological vision is explored. The model is based on an isotropic, scale invariant two-form that describes the spatial layout of receptive fields in the the visual sensorium (in the biological context roughly corresponding to retina, LGN, and V1). It overcomes the limitation of the familiar log-polar model by handling its singularity in a graceful way. The log-polar singularity arises as a result of ignoring the physical resolution limitation inherent in any real (artificial or biological) visual system. The incorporation of such a limitation requires the introduction of a physical constant, measuring the radius of the geometric foveola (a central region characterized by maximal resolving power). The proposed model admits a description in singularity-free canonical coordinates that generalize the well-established log-polar coordinates, and that reduce to these in the asymptotic case of negligibly sized geometric foveola (or, equivalently, at peripheral locations in the visual field). Biological plausibility of the model is demonstrated by comparison with known facts on human vision.

**Keywords:** Generalized log-polar map, foveal vision, cortical magnification, scale invariance, receptive field scaling.

## 1 Introduction

The visual system of humans (and many other mammalian species) is triggered by a sensorium that is characterized by receptive fields of various sizes and profiles. Scale space theory potentially provides a feasible foundation for the description of receptive field profiles, their taxonomy and functional role as the basic operators for a "differential geometry engine", a rationale introduced by Koenderink in his seminal papers that boosted efforts in multiscale image analysis [1,2,3,4,5,6,7,8,9,10]. Established linear scale space theory, however, typically adopts spatial homogeneity as one of its axiomatic principles, and ignores the foveal properties and dynamic exploration degrees of freedom typical of many biological visual systems.

The visual system of humans shows a roughly linear decrease of visual acuity with eccentricity, is more or less rotationally invariant relative to the foveal point,

* The Netherlands Organisation for Scientific Research (NWO) is gratefully acknowledged for financial support.

F. Sgallari, A. Murli, and N. Paragios (Eds.): SSVM 2007, LNCS 4485, pp. 919–928, 2007.

and exhibits a large degree of invariance to object size. Taking into account the dynamic exploration degree of freedom, i.e. the ability to shift the foveal point arbitrarily relative to the visual world, this yields a homogeneous, isotropic, scale invariant visual sensorium. Numerous empirical results exist in the literature in support of these claims, including quantitative studies of retino-cortical magnification [11]. Among others, this phenomenon is responsible for the fact that, in the case of humans, about half of the striate cortex is devoted to a foveal region covering only one percent of the visual field, and is usually described with the help of so-called *log-polar coordinates*. The log-polar mapping arises naturally in simplified models of foveal systems endowed with aforementioned properties, cf. previous work [12]. Image processing algorithms as well as space-variant CMOS cameras [13] have been constructed to mimic this mapping. These have turned out useful in computer vision as they help to simplify certain tasks, such as time to impact calculation.

However, previously proposed theoretical models (to the best of my knowledge) fail to give a plausible and accurate account of the *entire retina*, including the *fovea centralis* (the immediate neighbourhood of the central fovea), where the log-polar model breaks down. This hiatus is reflected in the design of log-polar mapping algorithms and space-variant cameras, in which one typically employs some heuristics to implement the transient behaviour between the (peripheral) validity domain of the log-polar model and the central retina. This is especially regrettable, as the fovea happens to be the most crucial part of a foveal system!

In this paper an alternative model is presented that overcomes the difficulties of the log-polar paradigm. It is based on invariance principles, viz. rotational invariance (with respect to the foveal point) and spatial scale invariance, and explicitly incorporates the physical resolution limitation of any visual system. It should be regarded in the first place as a model for artificial foveal vision, but we will touch upon its biological merits by comparison with known results on human vision. As in previous work [12] a differential geometric rationale is adopted, but unlike previous work the physical resolution limitation is taken as a point of departure. For simplicity we ignore stereopsis, and model the retina as a flat disk. A rudimentary understanding of $n$-forms and exterior products may be helpful [14,15], but the theory should be more or less self-contained even without it.

## 2 Theory

### 2.1 Modeling the Sensorium

Consider the basic scale invariant non-exact one-forms

$$\boldsymbol{d\xi} = \frac{\boldsymbol{dx}}{\sqrt{x^2+y^2+a^2}} \quad \text{and} \quad \boldsymbol{d\eta} = \frac{\boldsymbol{dy}}{\sqrt{x^2+y^2+a^2}}, \tag{1}$$

in which $(x, y) \in \mathbb{R}^2$ are Cartesian coordinates. We will confine the region of interest to a disk of radius $R$, which represents the radius of the *geometric retina*.

This could model the radius of the human retina, but it could also delimit some circularly symmetric functional *attention window* within it (think of "tunnel vision"). The physical parameter $a > 0$ represents a transient radius separating two regions of a qualitatively different geometric nature, viz. the *geometric foveola* ($\sqrt{x^2+y^2} < a$) and its *periphery* ($\sqrt{x^2+y^2} > a$). (Terminology betrays a modest amount of foresight.)

The one-forms of Eq. (1) induce a scale invariant area two-form, which can be roughly viewed as the geometric counterpart of a (space variant) pixel element,

$$\boldsymbol{\Omega} = \sqrt{g}\, \boldsymbol{dx} \wedge \boldsymbol{dy} = \boldsymbol{d\xi} \wedge \boldsymbol{d\eta} \tag{2}$$

in which

$$\sqrt{g} = \frac{1}{x^2+y^2+a^2}, \tag{3}$$

is the squareroot of the metric determinant associated with the 2-dimensional spatial metric[1] in Cartesian coordinates,

$$\boldsymbol{g} = \frac{\boldsymbol{dx} \otimes \boldsymbol{dx} + \boldsymbol{dy} \otimes \boldsymbol{dy}}{x^2+y^2+a^2}. \tag{4}$$

### 2.2 Modeling Retino-Cortical Magnification

Let us now consider the infinitesimal area of the ring $\delta\Gamma : \rho < \sqrt{x^2+y^2} < \rho + d\rho$:

$$dV(\rho) = \oint_{\delta\Gamma} \sqrt{g}\, \boldsymbol{dx} \wedge \boldsymbol{dy} = d\rho \int_0^{2\pi} \frac{\rho}{\rho^2+a^2}\, d\phi = \frac{2\pi\rho\, d\rho}{\rho^2+a^2}. \tag{5}$$

Geometrically the quantity

$$V'(\rho) = \frac{2\pi\rho}{\rho^2+a^2} \stackrel{\text{def}}{=} P(\rho) \tag{6}$$

equals the perimeter of a circle of radius $\rho$ around the foveal point[2]. As an aside, observe that this perimeter is bounded by a maximal perimeter $P(\rho) \leq P_{\max}$; equality occurs precisely at the transition $\rho = a$, i.e.

$$P_{\max} = P(a). \tag{7}$$

[1] This metric defines a so-called Einstein manifold, i.e. the Ricci tensor is proportional to the metric tensor, viz. $\mathscr{R}_{\mu\nu} = \frac{1}{2}\mathscr{R}\, g_{\mu\nu}$, with curvature scalar $\mathscr{R} = \frac{4a^2}{x^2+y^2+a^2}$. The model proposed elsewhere [12] hinges on the assumption that $\mathscr{R}_{\mu\nu} = 0$, and thus fails to account for the metrical structure at the *fovea centralis*.

[2] Note that relative to the metrical units defined by Eq. (4) the length of a circular segment over a polar angle $d\phi$ equals $ds = \rho\, d\phi/\sqrt{\rho^2+a^2}$, while the proper length element in the radial direction equals $ds = d\rho/\sqrt{\rho^2+a^2}$. Asympotically we have $ds \to d\phi$, respectively $ds \to d\rho/\rho$ if $\rho \gg a$ ("log-polar coordinates"). Details on the connection with log-polar coordinates are discussed in Section 2.3.

This follows by solving the quadratic equation that results from fixing $P$ and $a$ in Eq. (6), which yields two distinct solutions for $\rho$ if $P\,a < \pi$, a unique solution if $P\,a = \pi$ (viz. $\rho = a$), and none if $P\,a > \pi$.

If we normalize $V(\rho)$ such that $V(0) = 0$, and introduce the dimensionless quantities

$$v(t,T) = \frac{V(a\,t)}{V(a\,T)}\,, \tag{8}$$

with $0 \leq t \leq T$, and

$$t = \frac{\rho}{a} \quad \text{and} \quad T = \frac{R}{a}\,, \tag{9}$$

then

$$v(t,T) = \frac{\ln\left(1+t^2\right)}{\ln\left(1+T^2\right)}\,. \tag{10}$$

This *integrated retino-cortical magnification* measures the relative capacity dedicated to the foveal region inside the disk of radius $\rho$ relative to that of the full retina. Obvious limiting cases are $v(0,T) = 0$ and $v(T,T) = 1$. Cf. Fig. 1 for an illustration.

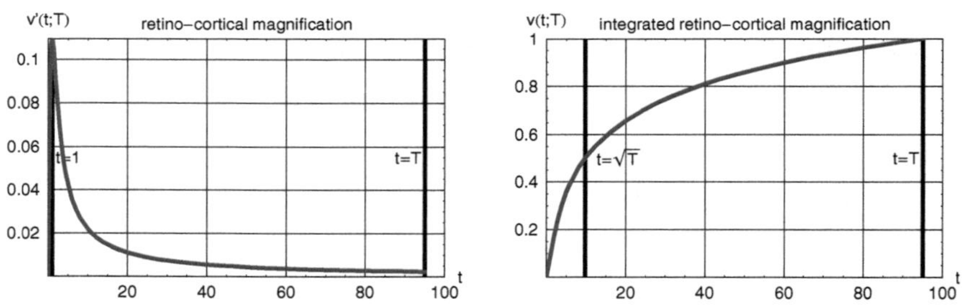


**Fig. 1.** Retinocortical magnification (left) and its integral (right) illustrated for the case $T = 95$, recall Eqs. (9–12). The peak on the left occurs at $t = 1$, marking the border of the geometric foveola. Half maximum on the right is reached at $t \approx \sqrt{T}$, corresponding to the geometric equipartitioning radius. In our model the geometric foveola has a relative processing capacity (not explicitly indicated in the right figure) of $v(t{=}1, T{=}95) \approx 8\%$, recall Eq. (10).

To test the biological plausibility of our model consider the equipartitioning case, i.e. Eq. (10) with equipartitioning radius $\rho_{\frac{1}{2}} = a\,t_{\frac{1}{2}}$, defined such that

$$v(t_{\frac{1}{2}}, T) \stackrel{\text{def}}{=} \frac{1}{2}\,. \tag{11}$$

A straightforward computation yields

$$t_{\frac{1}{2}} = \sqrt{\sqrt{1+T^2} - 1} \approx \sqrt{T}\,, \tag{12}$$

in which the approximation holds under the assumption that $T \gg 1$. In other words, assuming this assumption to hold, the theoretical equipartitioning radius equals the geometric mean of the radii of the geometric foveola and geometric retina[3]:

$$\rho_{\frac{1}{2}} = \sqrt{a\,R}\,. \tag{13}$$

In the case of humans it is known that about half of the striate cortex is devoted to the portion of the retina that lies within 7°–8° of the fovea[4] [11, Chapter 12]. The monocular visual field covers approximately 160° × 175° (width × height [16]), or roughly $(85 \pm 5)°$ in eccentricity when approximated by an isotropic figure for our model[5]. With these figures we have (for the biological counterparts of the quantities involved) $t_{\frac{1}{2}}/T = \rho_{\frac{1}{2}}/R \approx 0.1024$, from which one deduces with the help of Eq. (12) that $a = (t_{\frac{1}{2}}/T)^2\,R \approx 0.22\,\text{mm}$ for a retina with $R \approx 21\,\text{mm}$. The value $T = (R/\rho_{\frac{1}{2}})^2 \approx 95$ justifies our assumption $T \gg 1$, and the predicted size of the geometric foveola agrees well with that of the *human foveola*[6], which gives us a physiological interpretation of the geometric foveola $a$, and at the same time justifies its name. Cf. Figs. 2–4.

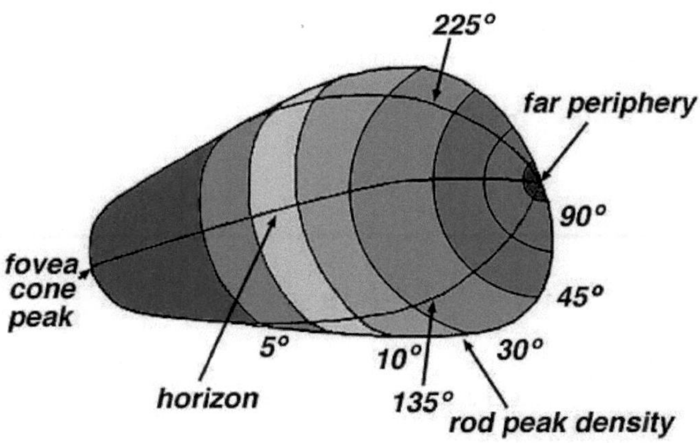


**Fig. 2.** Unfolded striate cortex. Source and further details: webvision.med.utah.edu.

Sofar results are consistent with those reported elsewhere [12] despite the different geometric approach. In the next section we consider the modification of the log-polar map by taking into account the physical resolution limitation of the fovea centralis, and show how it connects asymptotically to the log-polar map.

---

[3] Generalizing Eqs. (11–13) we may define $v(t_\alpha, T) \stackrel{\text{def}}{=} \alpha$ for $0 \le \alpha \le 1$, yielding $t_\alpha = \sqrt{(1+T^2)^\alpha - 1} \approx T^\alpha$, or, in terms of physical length scales, $\rho_\alpha = a^{1-\alpha}\,R^\alpha$.

[4] A typical retina measures $R \approx 21$ mm; 1° of visual angle corresponds to approximately 288$\mu$m. Source: webvision.med.utah.edu.

[5] Hartridge [17] reports a functional limit on visual field eccentricity of 104°.

[6] Rodieck [11, Chapter 9] reports a value of 0.20 mm.

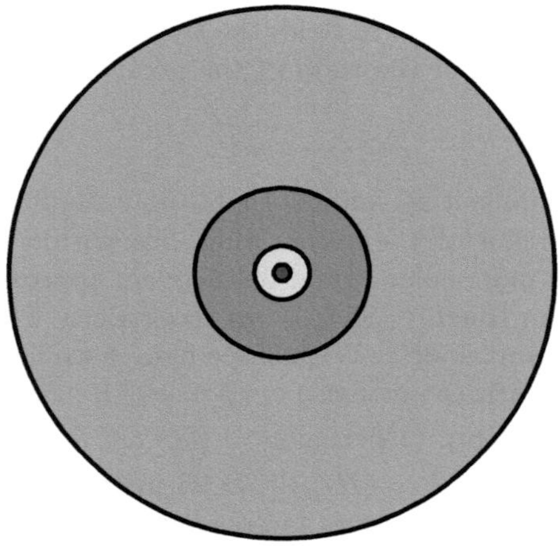

**Fig. 3.** The geometric retina can be partitioned into concentric disks with constant relative increments of spatial processing capacity based on Eqs. (9–10). The sketch shows some isocontours of the function $t(v) = \sqrt{(1+T^2)^v - 1}$, for fractional capacities $v_{\frac{1}{4}} = \frac{1}{4}$, $v_{\frac{1}{2}} = \frac{1}{2}$ (recall Eq. (11)), $v_{\frac{3}{4}} = \frac{3}{4}$, and $v_1 = 1$. Fig. 2 shows the actual realization of the retinotopic receptive field distribution in the visual cortex.

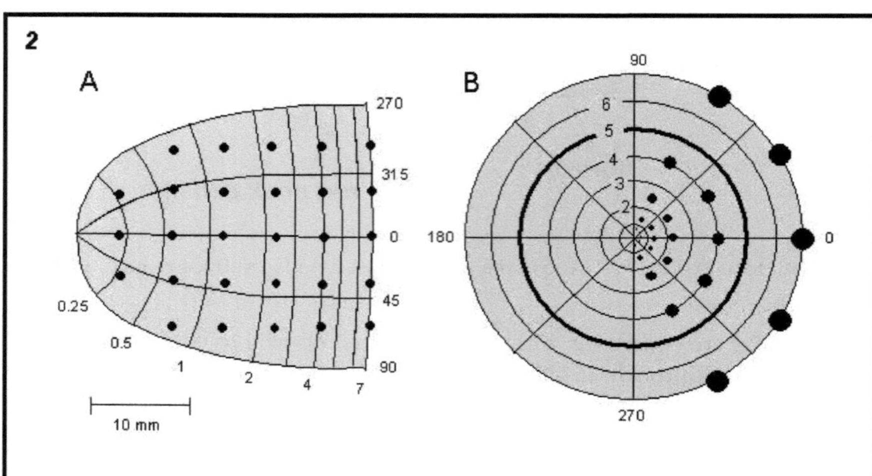


**Fig. 4.** Retino-cortical mapping. A: (hemifield) visual striate cortex, B: retina. Source: web.mit.edu/bcs/schillerlab (courtesy of Prof. Peter H. Schiller.)

### 2.3 Canonical Coordinates

Retinocortical magnification can be best described in terms of *canonical coordinates*. To this end, consider the following coordinate transformation:

$$\begin{cases} p = \operatorname{arcsinh} t = \ln\left(t + \sqrt{1+t^2}\right) \,, \\ q = \dfrac{t\,\phi}{\sqrt{1+t^2}} \qquad\qquad (0 \le t \le T \,,\ -\pi < \phi \le \pi) \,. \end{cases} \tag{14}$$

Since

$$\begin{cases} \boldsymbol{dp} = \dfrac{\boldsymbol{dt}}{\sqrt{1+t^2}} = \dfrac{\boldsymbol{d\rho}}{\sqrt{a^2+\rho^2}} \,, \\ \boldsymbol{dq} = \dfrac{t\,\boldsymbol{d\phi}}{\sqrt{1+t^2}} + \dfrac{\phi\,\boldsymbol{dt}}{(1+t^2)^{\frac{3}{2}}} = \dfrac{\rho\,\boldsymbol{d\phi}}{\sqrt{a^2+\rho^2}} + \dfrac{a^2\,\phi\,\boldsymbol{d\rho}}{(\rho^2+a^2)^{\frac{3}{2}}} \,, \end{cases} \tag{15}$$

it follows that, recall Eq. (2),

$$\boldsymbol{\Omega} = \boldsymbol{dp} \wedge \boldsymbol{dq} \,. \tag{16}$$

Note that Eq. (4) can be transformed into canonical coordinates, yielding

$$\boldsymbol{g} = \left(1 + \frac{4q^2}{\sinh^2(2p)}\right) \boldsymbol{dp} \otimes \boldsymbol{dp} - \frac{4q}{\sinh(2p)} \boldsymbol{dp} \otimes \boldsymbol{dq} + \boldsymbol{dq} \otimes \boldsymbol{dq} \,, \tag{17}$$

from which it becomes evident that the $(p,q)$-coordinate lines generally do not intersect perpendicularly, unlike the case of log-polar coordinates, cf. Fig. 5. For the angle of intersection $\alpha$ at a fiducial point $(p,q)$ we have

$$\cos\alpha = \frac{-2q}{\sqrt{4q^2 + \sinh^2(2p)}} \,, \tag{18}$$

which is, however, close to zero for typical (peripheral) points.

If $\Gamma_T$ denotes the full retinal domain, then in $(p,q)$-space it corresponds to the area in-between the graphs of

$$q_{\pm\pi}(p) = \pm\pi \frac{\sinh p}{\sqrt{1+\sinh^2 p}} = \pm\pi \tanh p \,, \tag{19}$$

and the lines $p{=}0$ and $p{=}\operatorname{arcsinh} T$. (To see this, express $q$ as a function of $p$ using Eq. (14) taking into account the domain boundaries.) By the same token, if $\Gamma_t$ denotes the fraction of the retinal domain on the left of the line $p{=}\operatorname{arcsinh} t$ instead of $p{=}\operatorname{arcsinh} T$, then a straightforward computation yields

$$\frac{\int_{\Gamma_t} \boldsymbol{dp} \wedge \boldsymbol{dq}}{\int_{\Gamma_T} \boldsymbol{dp} \wedge \boldsymbol{dq}} = \frac{\int_0^{\operatorname{arcsinh} t} q_\pi(p)\, dp}{\int_0^{\operatorname{arcsinh} T} q_\pi(p)\, dp} = v(t,T) \,, \tag{20}$$

as it should, recall Eq. (10). This once again confirms that the canonical coordinates $(p,q)$ given by Eq. (14) are indeed the natural ones to use in the context of an isotropic, scale invariant foveal system, as opposed to the commonly used log-polar coordinates. The latter arise only in the limit of vanishing $a$, i.e. the asymptotic case whereby $T \to \infty$. As such log-polar coordinates fail to describe both foveal as well as transient behaviour, and are suited only for the peripheral field. Although the periphery represents by far the largest part of the visual field, it is a serious shortcoming of the log-polar model that it fails to describe the structure of the immediate neighbourhood of the foveal point, quite an essential part of a foveal system!

The canonical coordinates, Eq. (14), on the other hand, can be used both for central, transient, and peripheral vision. Near the foveal point we have

$$\begin{cases} p \approx t \quad , \\ q \approx t\phi \; (0 \leq t \ll T \, , \; -\pi < \phi \leq \pi) \, . \end{cases} \tag{21}$$

These can be easily mapped to ordinary polar coordinates, $(\rho, \phi) = (ap, q/p)$, or to Cartesian coordinates, $(x, y) = (ap\cos(q/p), ap\sin(q/p))$.

In the far periphery we reobtain the familiar log-polar coordinates (up to an irrelevant offset):

$$\begin{cases} p \approx \ln 2t \; , \\ q \approx \phi \quad (1 \ll t \leq T \, , \; -\pi < \phi \leq \pi) \, . \end{cases} \tag{22}$$

In general, the physical part of the $(p, q)$-domain is shaped like a deformed isosceles triangle, with its tip representing the foveal point, cf. Fig. 5. This is

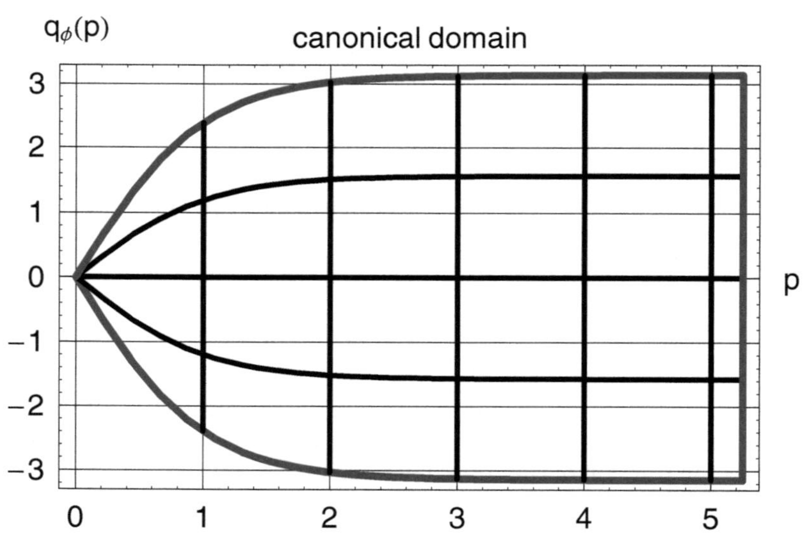


**Fig. 5.** The canonical $(p, q)$-domain is the region between the graphs of $q_{\pm\pi}(p) = \pm\pi \tanh p$ and the lines $p = 0$ and $p = \operatorname{arcsinh} T$ (with $T = 95.0$ for the sake of illustration), recall Eq. (19). The figure also shows the graphs of the functions $q_\phi(p) = \phi \tanh p$ for constant azimuth angles $\phi = 0, \pm\frac{\pi}{2}$, as well as the lines $p = \operatorname{arcsinh} t$ for $t = 1, 2, 3, 4, 5$. Cf. also Fig. 2.

somewhat reminiscent of the actually observed pear shape of the cortical surface of V1, recall Fig. 2.

## 3 Conclusion

We have established a biologically plausible geometric model for an isotropic, scale invariant foveal system that incorporates physical resolution limitations. The model is most naturally described in terms of a singularity-free canonical coordinate map that generalizes the familiar log-polar map typically used in the context of foveal systems. Unlike the latter, however, the generalized map has a globally valid domain of definition, and handles the transition from the peripheral field (where the log-polar paradigm is appropriate) into the fovea centralis (where a description in terms of standard polar coordinates is more appropriate) in a graceful manner.

The proposed model is consistent with certain known facts on biological vision, notably retino-cortical magnification. Further quantitative predictions could be inferred from it to evaluate its biological feasibility, and perhaps to even predict hitherto unexplored properties about biological foveal systems.

## References

1. Duits, R., Florack, L., de Graaf, J., ter Haar Romeny, B.: On the axioms of scale space theory. Journal of Mathematical Imaging and Vision **20**(3) (2004) 267–298
2. Florack, L.M.J., Haar Romeny, B.M.t., Koenderink, J.J., Viergever, M.A.: Linear scale-space. Journal of Mathematical Imaging and Vision **4**(4) (1994) 325–351
3. Iijima, T.: Basic theory on normalization of a pattern (in case of typical one-dimensional pattern). Bulletin of Electrical Laboratory **26** (1962) 368–388 (in Japanese).
4. Koenderink, J.J.: The structure of images. Biological Cybernetics **50** (1984) 363–370
5. Koenderink, J.J., Doorn, A.J.v.: Representation of local geometry in the visual system. Biological Cybernetics **55** (1987) 367–375
6. Koenderink, J.J., van Doorn, A.J.: Operational significance of receptive field assemblies. Biological Cybernetics **58** (1988) 163–171
7. Koenderink, J.J., van Doorn, A.J.: Receptive field families. Biological Cybernetics **63** (1990) 291–298
8. Koenderink, J.J.: The brain a geometry engine. Psychological Research **52** (1990) 122–127
9. Weickert, J.A., Ishikawa, S., Imiya, A.: Linear scale-space has first been proposed in Japan. Journal of Mathematical Imaging and Vision **10**(3) (1999) 237–252
10. Witkin, A.P.: Scale-space filtering. In: Proceedings of the International Joint Conference on Artificial Intelligence, Karlsruhe, Germany (1983) 1019–1022
11. Rodieck, R.W.: The First Steps in Seeing. Sinauer Associates, Inc., Sunderland, Massachusetts (1998)
12. Florack, L.M.J.: A geometric model for cortical magnification. In Lee, S.W., Bülthoff, H.H., Poggio, T., eds.: Biologically Motivated Computer Vision: Proceedings of the First IEEE International Workshop, BMCV 2000 (Seoul, Korea, May 2000). Volume 1811 of Lecture Notes in Computer Science., Berlin, Springer-Verlag (2000) 574–583

13. Tistarelli, M., Sandini, G.: On the advantages of polar and log-polar mapping for direct estimation of time-to-impact from optical flow. IEEE Transactions on Pattern Analysis and Machine Intelligence **15**(4) (1993) 401–416
14. Jost, J.: Riemannian Geometry and Geometric Analysis. Fourth edn. Springer-Verlag, Berlin (2005)
15. Koenderink, J.J.: Solid Shape. MIT Press, Cambridge (1990)
16. Wandell, B.A.: Foundations of Vision. Sinauer Associates, Inc., Sunderland, Massachusetts (1995)
17. Hartridge, H.: The limit to peripheral vision. Journal of Physiology **53**(17) (1919)

# Author Index

# Lecture Notes in Computer Science

For information about Vols. 1–4399

please contact your bookseller or Springer

Vol. 4534: I. Tomkos, F. Neri, J. Solé Pareta, X. Masip Bruin, S. Sánchez Lopez (Eds.), Optical Network Design and Modeling. XI, 460 pages. 2007.

Vol. 4526: M. Malek, M. Reitenspieß, A.P.A. van Moorsel (Eds.), Service Availability. X, 155 pages. 2007.

Vol. 4523: Y.-H. Lee, H.-N. Kim, J. Kim, Y. Park, L.T. Yang, S.W. Kim (Eds.), Embedded Software and Systems. XIX, 829 pages. 2007.

Vol. 4517: F. Boavida, E. Monteiro, S. Mascolo, Y. Koucheryavy (Eds.), Wired/Wireless Internet Communications. XIV, 382 pages. 2007.

Vol. 4515: M. Naor (Ed.), Advances in Cryptology - EUROCRYPT 2007. XIII, 591 pages. 2007.

Vol. 4514: S.N. Artemov, A. Nerode (Eds.), Logical Foundations of Computer Science. XI, 513 pages. 2007.

Vol. 4510: P. Van Hentenryck, L. Wolsey (Eds.), Integration of AI and OR Techniques in Constraint Programming for Combinatorial Optimization Problems. X, 391 pages. 2007.

Vol. 4509: Z. Kobti, D. Wu (Eds.), Advances in Artificial Intelligence. XII, 552 pages. 2007. (Sublibrary LNAI).

Vol. 4506: D. Zeng, I. Gotham, K. Komatsu, C. Lynch, M. Thurmond, D. Madigan, B. Lober, J. Kvach, H. Chen (Eds.), Intelligence and Security Informatics: Biosurveillance. XI, 234 pages. 2007.

Vol. 4504: J. Huang, R. Kowalczyk, Z. Maamar, D. Martin, I. Müller, S. Stoutenburg, K.P. Sycara (Eds.), Service-Oriented Computing: Agents, Semantics, and Engineering. X, 175 pages. 2007.

Vol. 4501: J. Marques-Silva, K.A. Sakallah (Eds.), Theory and Applications of Satisfiability Testing – SAT 2007. XI, 384 pages. 2007.

Vol. 4500: N. Streitz, A. Kameas, I. Mavrommati (Eds.), The Disappearing Computer. XVIII, 307 pages. 2007.

Vol. 4493: D. Liu, S. Fei, Z. Hou, H. Zhang, C. Sun (Eds.), Advances in Neural Networks – ISNN 2007, Part III. XXVI, 1215 pages. 2007.

Vol. 4492: D. Liu, S. Fei, Z. Hou, H. Zhang, C. Sun (Eds.), Advances in Neural Networks – ISNN 2007, Part II. XXVII, 1321 pages. 2007.

Vol. 4491: D. Liu, S. Fei, Z.-G. Hou, H. Zhang, C. Sun (Eds.), Advances in Neural Networks – ISNN 2007, Part I. LIV, 1365 pages. 2007.

Vol. 4490: Y. Shi, G.D. van Albada, J. Dongarra, P.M.A. Sloot (Eds.), Computational Science – ICCS 2007, Part IV. XXXVII, 1211 pages. 2007.

Vol. 4489: Y. Shi, G.D. van Albada, J. Dongarra, P.M.A. Sloot (Eds.), Computational Science – ICCS 2007, Part III. XXXVII, 1257 pages. 2007.

Vol. 4488: Y. Shi, G.D. van Albada, J. Dongarra, P.M.A. Sloot (Eds.), Computational Science – ICCS 2007, Part II. XXXV, 1251 pages. 2007.

Vol. 4487: Y. Shi, G.D. van Albada, J. Dongarra, P.M.A. Sloot (Eds.), Computational Science – ICCS 2007, Part I. LXXXI, 1275 pages. 2007.

Vol. 4486: M. Bernardo, J. Hillston (Eds.), Formal Methods for Performance Evaluation. VII, 469 pages. 2007.

Vol. 4485: F. Sgallari, A. Murli, N. Paragios (Eds.), Scale Space and Variational Methods in Computer Vision. XV, 931 pages. 2007.

Vol. 4484: J.-Y. Cai, S.B. Cooper, H. Zhu (Eds.), Theory and Applications of Models of Computation. XIII, 772 pages. 2007.

Vol. 4483: C. Baral, G. Brewka, J. Schlipf (Eds.), Logic Programming and Nonmonotonic Reasoning. IX, 327 pages. 2007. (Sublibrary LNAI).

Vol. 4482: A. An, J. Stefanowski, S. Ramanna, C.J. Butz, W. Pedrycz, G. Wang (Eds.), Rough Sets, Fuzzy Sets, Data Mining and Granular Computing. XIV, 585 pages. 2007. (Sublibrary LNAI).

Vol. 4481: J. Yao, P. Lingras, W.-Z. Wu, M. Szczuka, N.J. Cercone, D. Ślęzak (Eds.), Rough Sets and Knowledge Technology. XIV, 576 pages. 2007. (Sublibrary LNAI).

Vol. 4480: A. LaMarca, M. Langheinrich, K.N. Truong (Eds.), Pervasive Computing. XIII, 369 pages. 2007.

Vol. 4479: I.F. Akyildiz, R. Sivakumar, E. Ekici, J.C.d. Oliveira, J. McNair (Eds.), NETWORKING 2007. Ad Hoc and Sensor Networks, Wireless Networks, Next Generation Internet. XXVII, 1252 pages. 2007.

Vol. 4472: M. Haindl, J. Kittler, F. Roli (Eds.), Multiple Classifier Systems. XI, 524 pages. 2007.

Vol. 4471: P. Cesar, K. Chorianopoulos, J.F. Jensen (Eds.), Interactive TV: a Shared Experience. XIII, 236 pages. 2007.

Vol. 4470: Q. Wang, D. Pfahl, D.M. Raffo (Eds.), Software Process Dynamics and Agility. XI, 346 pages. 2007.

Vol. 4465: T. Chahed, B. Tuffin (Eds.), Network Control and Optimization. XIII, 305 pages. 2007.

Vol. 4464: E. Dawson, D.S. Wong (Eds.), Information Security Practice and Experience. XIII, 361 pages. 2007.

Vol. 4463: I. Măndoiu, A. Zelikovsky (Eds.), Bioinformatics Research and Applications. XV, 653 pages. 2007. (Sublibrary LNBI).

Vol. 4462: D. Sauveron, K. Markantonakis, A. Bilas, J.-J. Quisquater (Eds.), Information Security Theory and Practices. XII, 255 pages. 2007.

Vol. 4459: C. Cérin, K.-C. Li (Eds.), Advances in Grid and Pervasive Computing. XVI, 759 pages. 2007.

Vol. 4453: T. Speed, H. Huang (Eds.), Research in Computational Molecular Biology. XVI, 550 pages. 2007. (Sublibrary LNBI).

Vol. 4452: M. Fasli, O. Shehory (Eds.), Agent-Mediated Electronic Commerce. VIII, 249 pages. 2007. (Sublibrary LNAI).

Vol. 4451: T.S. Huang, A. Nijholt, M. Pantic, A. Pentland (Eds.), Artifical Intelligence for Human Computing. XVI, 359 pages. 2007. (Sublibrary LNAI).

Vol. 4450: T. Okamoto, X. Wang (Eds.), Public Key Cryptography – PKC 2007. XIII, 491 pages. 2007.

Vol. 4448: M. Giacobini et al. (Ed.), Applications of Evolutionary Computing. XXIII, 755 pages. 2007.

Vol. 4447: E. Marchiori, J.H. Moore, J.C. Rajapakse (Eds.), Evolutionary Computation,Machine Learning and Data Mining in Bioinformatics. XI, 302 pages. 2007.

Vol. 4446: C. Cotta, J. van Hemert (Eds.), Evolutionary Computation in Combinatorial Optimization. XII, 241 pages. 2007.

Vol. 4445: M. Ebner, M. O'Neill, A. Ekárt, L. Vanneschi, A.I. Esparcia-Alcázar (Eds.), Genetic Programming. XI, 382 pages. 2007.

Vol. 4444: T. Reps, M. Sagiv, J. Bauer (Eds.), Program Analysis and Compilation, Theory and Practice. X, 361 pages. 2007.

Vol. 4443: R. Kotagiri, P.R. Krishna, M. Mohania, E. Nantajeewarawat (Eds.), Advances in Databases: Concepts, Systems and Applications. XXI, 1126 pages. 2007.

Vol. 4440: B. Liblit, Cooperative Bug Isolation. XV, 101 pages. 2007.

Vol. 4439: W. Abramowicz (Ed.), Business Information Systems. XV, 654 pages. 2007.

Vol. 4438: L. Maicher, A. Sigel, L.M. Garshol (Eds.), Leveraging the Semantics of Topic Maps. X, 257 pages. 2007. (Sublibrary LNAI).

Vol. 4433: E. Şahin, W.M. Spears, A.F.T. Winfield (Eds.), Swarm Robotics. XII, 221 pages. 2007.

Vol. 4432: B. Beliczynski, A. Dzielinski, M. Iwanowski, B. Ribeiro (Eds.), Adaptive and Natural Computing Algorithms, Part II. XXVI, 761 pages. 2007.

Vol. 4431: B. Beliczynski, A. Dzielinski, M. Iwanowski, B. Ribeiro (Eds.), Adaptive and Natural Computing Algorithms, Part I. XXV, 851 pages. 2007.

Vol. 4430: C.C. Yang, D. Zeng, M. Chau, K. Chang, Q. Yang, X. Cheng, J. Wang, F.-Y. Wang, H. Chen (Eds.), Intelligence and Security Informatics. XII, 330 pages. 2007.

Vol. 4429: R. Lu, J.H. Siekmann, C. Ullrich (Eds.), Cognitive Systems. X, 161 pages. 2007. (Sublibrary LNAI).

Vol. 4427: S. Uhlig, K. Papagiannaki, O. Bonaventure (Eds.), Passive and Active Network Measurement. XI, 274 pages. 2007.

Vol. 4426: Z.-H. Zhou, H. Li, Q. Yang (Eds.), Advances in Knowledge Discovery and Data Mining. XXV, 1161 pages. 2007. (Sublibrary LNAI).

Vol. 4425: G. Amati, C. Carpineto, G. Romano (Eds.), Advances in Information Retrieval. XIX, 759 pages. 2007.

Vol. 4424: O. Grumberg, M. Huth (Eds.), Tools and Algorithms for the Construction and Analysis of Systems. XX, 738 pages. 2007.

Vol. 4423: H. Seidl (Ed.), Foundations of Software Science and Computational Structures. XVI, 379 pages. 2007.

Vol. 4422: M.B. Dwyer, A. Lopes (Eds.), Fundamental Approaches to Software Engineering. XV, 440 pages. 2007.

Vol. 4421: R. De Nicola (Ed.), Programming Languages and Systems. XVII, 538 pages. 2007.

Vol. 4420: S. Krishnamurthi, M. Odersky (Eds.), Compiler Construction. XIV, 233 pages. 2007.

Vol. 4419: P.C. Diniz, E. Marques, K. Bertels, M.M. Fernandes, J.M.P. Cardoso (Eds.), Reconfigurable Computing: Architectures, Tools and Applications. XIV, 391 pages. 2007.

Vol. 4418: A. Gagalowicz, W. Philips (Eds.), Computer Vision/Computer Graphics Collaboration Techniques. XV, 620 pages. 2007.

Vol. 4416: A. Bemporad, A. Bicchi, G. Buttazzo (Eds.), Hybrid Systems: Computation and Control. XVII, 797 pages. 2007.

Vol. 4415: P. Lukowicz, L. Thiele, G. Tröster (Eds.), Architecture of Computing Systems - ARCS 2007. X, 297 pages. 2007.

Vol. 4414: S. Hochreiter, R. Wagner (Eds.), Bioinformatics Research and Development. XVI, 482 pages. 2007. (Sublibrary LNBI).

Vol. 4412: F. Stajano, H.J. Kim, J.-S. Chae, S.-D. Kim (Eds.), Ubiquitous Convergence Technology. XI, 302 pages. 2007.

Vol. 4411: R.H. Bordini, M. Dastani, J. Dix, A.E.F. Seghrouchni (Eds.), Programming Multi-Agent Systems. XIV, 249 pages. 2007. (Sublibrary LNAI).

Vol. 4410: A. Branco (Ed.), Anaphora: Analysis, Algorithms and Applications. X, 191 pages. 2007. (Sublibrary LNAI).

Vol. 4409: J.L. Fiadeiro, P.-Y. Schobbens (Eds.), Recent Trends in Algebraic Development Techniques. VII, 171 pages. 2007.

Vol. 4407: G. Puebla (Ed.), Logic-Based Program Synthesis and Transformation. VIII, 237 pages. 2007.

Vol. 4406: W. De Meuter (Ed.), Advances in Smalltalk. VII, 157 pages. 2007.

Vol. 4405: L. Padgham, F. Zambonelli (Eds.), Agent-Oriented Software Engineering VII. XII, 225 pages. 2007.

Vol. 4403: S. Obayashi, K. Deb, C. Poloni, T. Hiroyasu, T. Murata (Eds.), Evolutionary Multi-Criterion Optimization. XIX, 954 pages. 2007.

Vol. 4401: N. Guelfi, D. Buchs (Eds.), Rapid Integration of Software Engineering Techniques. IX, 177 pages. 2007.

Vol. 4400: J.F. Peters, A. Skowron, V.W. Marek, E. Orłowska, R. Słowiński, W. Ziarko (Eds.), Transactions on Rough Sets VII, Part II. X, 381 pages. 2007.

MIX
Papier aus verantwortungsvollen Quellen
Paper from responsible sources
FSC® C105338

If you have any concerns about our products,
you can contact us on
**ProductSafety@springernature.com**

In case Publisher is established outside the EU,
the EU authorized representative is:
**Springer Nature Customer Service Center GmbH**
**Europaplatz 3, 69115 Heidelberg, Germany**

Printed by Libri Plureos GmbH
in Hamburg, Germany